SKY CATALOGUE 2000.0

Volume 2: Double Stars, Variable Stars and Nonstellar Objects

SKY CATALOGUE 2000.0

Volume 2: Double Stars, Variable Stars and Nonstellar Objects

Edited by

Alan Hirshfeld
Southeastern Massachusetts University

and

Roger W. Sinnott
Sky Publishing Corporation

1985

Cambridge University Press

Cambridge   London   New York   New Rochelle

Melbourne   Sydney

&

Sky Publishing Corporation

Cambridge, Massachusetts

Published by the Press Syndicate of the University of Cambridge
The Pitt Building, Trumpington Street, Cambridge CB2 1RP
32 East 57th Street, New York, New York 10022
296 Beaconsfield Parade, Middle Park, Melbourne 3206, Australia
and by Sky Publishing Corporation
49 Bay State Road, Cambridge, Massachusetts 02238-1290

First published 1985

Printed in the United States of America
by Murray Printing Company
Westford, Massachusetts

Library of Congress Cataloging in Publication Data

Main entry under title:

Sky catalogue 2000.0. Volume 2: Double stars, variable stars
and nonstellar objects

Includes bibliographical references and indexes.

1. Stars — Catalogs. I. Hirshfeld, Alan.

II. Sinnott, Roger W.

QB6.S54   523.8′908   81-17975
ISBN 0-521-25818-9 (v. 2)   AACR2
ISBN 0-521-27721-3 (pbk.: v. 2)

Preface

In 1964 appeared the final edition of the Skalnate Pleso Observatory's *Atlas Catalogue* by Antonin Becvar. With its wide availability and handy sections on stars of various types, star clusters, nebulae, and galaxies, Becvar's work remained uniquely important as a compendium of data for amateur and professional astronomers during the past two decades. However, the face of astronomy has changed remarkably since that time. Black holes and neutron stars have moved from the pages of science fiction stories into the everyday parlance of researchers. The cosmic microwave background was detected accidentally in 1965, and pulsars only two years later. The number of catalogued quasi-stellar objects has increased from a handful in 1964 to the thousands. A quark was not a fundamental unit of matter, but a fanciful term known only to readers of James Joyce. Optical astronomy reigned supreme two decades ago. Radio astronomy was still limited by poor resolution to broad surveys of the sky, and its X-ray relative was still in infancy. The profound influence of computers and solid state electronics on observational astronomy would be felt only in the next decade.

Sky Catalogue 2000.0 was conceived in 1978 to meet the growing needs of amateur astronomers and educators, as well as professional astronomers and other scientists, for a convenient source of modern data about astronomical objects. Volume 1 appeared in 1982 and contains information on 45,269 stars of visual magnitude 8.05 and brighter. Volume 2 has been in preparation since that time. The increased availability of astronomical data in computer-readable form has made it possible to expand this book far beyond the scope of Becvar's earlier work. In addition to the categories of astronomical objects covered in his catalogue, we have included new sections on suspected variables, dark nebulae, quasi-stellar objects, and X-ray sources. For every object class, there are more entries and a wider variety of information than in the *Atlas Catalogue*. Our Galaxies section, for instance, contains data for nearly twice as many systems; Becvar's Radio Sources chapter was limited to only 38 well-observed objects, but this volume has more than 700.

The pace of discovery did not slacken as Volume 2 was being compiled. We sometimes received computerized data catalogues merely a few years old, but which did not reflect the most recent advances in a particular field. Our colleagues at *Sky and Telescope* magazine handed us news releases and preprints as we scanned astronomical journals for the latest information. In an effort to make the book as up-to-date as possible, we manually entered a large amount of this new data into the computer, replacing many of the quantities already present. A scan of the References section starting on page xxxix shows that in some cases articles only a few months in advance of our press date were consulted.

We have attempted to create a book that is more than just a list of numerical data. It is a collection of observing challenges for both current amateurs and those of the next decade, when today's revolutionary technologies become tomorrow's backyard standard. After all, who could have suspected in 1964 that amateur astronomers would now be making telescopes nearly a meter in aperture, using microcomputer-controlled instruments, carrying out valuable photoelectric photometry, or building their own charge-coupled device (CCD) cameras? If advances proceed apace, what instruments and techniques will be available to amateurs at the turn of the century? We have therefore tried to include objects here that are certainly beyond the grasp of today's amateur instruments, but maybe not those of the near future. We also hope that this book will be a boon to "armchair astronomers," first by demonstrating the stunning variety of celestial objects and what researchers know (and do not know) about them, and second by providing an incentive to learn more about the field. To this end, there are many entries here that are not particularly inspiring telescopic sights, but that are nonetheless astrophysically significant.

We gratefully acknowledge the valuable assistance of the following people and organizations in the preparation of this volume: Wayne H. Warren, Jr., at the Astronomical Data Center, NASA Goddard Space Flight Center; Gene Campbell and the staff of the Smithsonian Astrophysical Observatory Computation Facility; Estelle Karlin and the Harvard College Observatory Library staff; Janet Mattei and others at the American Association of Variable Star Observers; the Centre de Données Stellaires, Observatoire de Strasbourg; and Betty Siegman of the Division of Radiophysics, Commonwealth Scientific and Industrial Research Organization. Paul Couteau, N. Kukarkina, Jean Meeus, Brian Skiff, R. B. Tully, and Charles E. Worley gave us much valuable data and advice in advance of publication. Richard Rodman translated an important article from Russian.

We also thank *Sky and Telescope* editor Leif J. Robinson, who originally conceived this volume and its predecessor, for his advice and critical comments, and the magazine's editorial staff, who consulted at all stages of this book and stepped in to cover our normal magazine duties during its lengthy preparation. In actually putting the book together in finished form, much time and expertise were contributed by project manager William E. Shawcross, Susan Bencuya, John Briggs, Jennifer Craddock, Alan MacRobert, and Steven Simpson. We especially thank Stephen De Luca for an incredible amount of painstaking data entry and proofreading, and Sasha Helper for her patience during the final stages of the project.

January, 1985

Alan Hirshfeld
Roger W. Sinnott

Contents

Introduction

THIS CATALOGUE consists of 15 main data sections, each devoted to a particular class of celestial object. The sections are arranged roughly in order of increasing scale and/or distance of the classes under study; hence, we progress from double and variable stars to clusters and nebulae. We then leave the confines of our Milky Way with tabulations of galaxies and quasi-stellar objects. In order to demonstrate the wide variety of astronomical systems that can be "seen" beyond the narrow optical band of the spectrum, there are concluding sections devoted to radio and X-ray sources. Several additional tables appear in the Introduction, including complete listings of Messier objects and the Local Group of galaxies. We have also prepared a special glossary to assist readers with the profusion of common astronomical names that have been coined over the years.

DOUBLE AND MULTIPLE STARS

More than half of all stars are believed to belong to true double or multiple star systems. Members of such a system travel through space together and remain gravitationally bound to each other. If there are only two components in a system, each star travels in an elliptical orbit around the common center of gravity of the pair. When there are three or more stars in a system, the motions are more complex. Multiple systems seem nearly always to consist of one or two close pairs that are well separated from each other or from additional isolated members.

Yet this modern picture is quite different from that held 200 years ago. Despite a few telescopic pairs discovered by accident, it was presumed that most stars, like the Sun, were single. No deliberate search for double stars was undertaken until 1776. In that year Christian Mayer at Mannheim, Germany, began a survey of the sky with an 8-foot Bird mural quadrant at 85x magnification. His 1779 catalogue contained over four dozen pairs in which the components were separated by less than a minute of arc. Mayer's belief that the fainter components were true satellites of their brighter neighbors, however, was quite radical for his day. Other astronomers believed that double stars were mere chance alignments, or so-called optical pairs, of a nearby star with a distant one in the same line of sight.

At first William Herschel shared the latter view. When he began his great series of measurements of double stars in 1779, his hope was to find the distances to nearby stars, using a parallax method that had been proposed by Galileo. The motion of the Earth around the Sun should cause a bright star to oscillate annually with respect to a fainter, and presumably much more remote, neighbor. But as Herschel's measurements accumulated, not only did they fail to reveal the long-sought parallax effect, but in many cases they showed a progressive relative motion of the components. In 1803 appeared the classic paper in which he concluded that Castor and five other doubles he had measured since the 1770's are "not merely double in appearance, but . . . real binary combinations of two stars, intimately held together by the bonds of mutual attraction."

William Herschel's measures did not show any definite *curving* of the path of any companion star with respect to the primary. The quarter century spanned by his work was simply not long enough for the pairs he knew. Moreover, his altazimuth reflectors had no clock drives, and his measures of position angle were far more reliable than those of separation.

The situation was changing rapidly, however. In 1823 a collaborative effort by James South and John Herschel brought about the publication of all of William's measures dating back to 1779, as well as many new measures and discoveries of their own. And in 1824, F. G. W. Struve began to observe doubles with the famous 9.6-inch Fraunhofer refractor at Dorpat, Russia. Pictured in Fig. 1, Struve's refractor was, both optically and mechanically, a thoroughly modern telescope fitted with a clock drive and filar micrometer. The reader has only to scan the present work to note how frequently the early linchpin in our knowledge of a double star dates from a measure in 1824-35, and how frequently the star's designation is Σ, to appreciate this astronomer's immortal contribution to double star astronomy. By R. G. Aitken's assessment, "Struve's measurements compare very favorably with the best modern ones," in their precision and in freedom from systematic errors.

Every careful measure·of a known pair has lasting value, and it consists of writing down the star's identity, the date to the nearest day, the position angle in degrees (°), and the

Fig. 1. With its tube of fir and mahogany veneer, this 9.6-inch f/18 refractor was used by Wilhelm Struve to discover some 3,000 double stars. The mounting is a very early example of the German equatorial style.

TABLE I. DOUBLE STAR DESIGNATION CODES

Designation	IDS Code	Discoverer or Work
β	BU	S. W. Burnham (1838-1921)
β pm	BUP	Burnham's measures of proper-motion stars, 1913 catalogue
Δ	DUN	J. Dunlop
δ	DAW	B. H. Dawson (d. 1960)
Σ	STF	F. G. Wilhelm Struve (1793-1864), Dorpat catalogue of 1827; *rej* means the pair was rejected from his later series of measures, usually because it was considered too wide
Σ I		W. Struve, first suppl.
Σ II		W. Struve, second supplement
φ	FIN	W. S. Finsen (1905-79)
A	A	R. G. Aitken (1864-1951)
Abetti	ABT	Giorgio Abetti (1882-1982)
AC	AC	Alvan Clark (1804-87)
A.G.	AG	*Astronomische Gesellschaft Katalog*, 1875
AGC	AGC	Alvan G. Clark (1832-97)
Alden	ALD	H. L. Alden (1890-1964)
Ali	ALI	A. Ali
Aller	ALL	R. M. Aller
Arg	ARG	F. W. A. Argelander (1799-1875)
B	B	Willem H. van den Bos (1896-1974)
Bail	BAL	R. Baillaud
Baize	BAZ	Paul Baize
Barnard	BAR	E. E. Barnard (1857-1923)
Bhas	BHA	T. P. Bhaskavan
Big	BIG	G. Bigourdan
Bird	BRD	F. Bird
Bloch	BLO	M. Bloch
BrsO	BSO	Brisbane Observatory, Australia
Brt	BRT	S. G. Barton
Btz	BTZ	E. Bernewitz
CapO	CPO	Cape Observatory, South Africa
Che	CHE	P. S. Chevalier
Com	COM	G. C. Comstock
CorO	COO	Cordoba Observatory, Argentina
Cou	COU	Paul Couteau
Cruls	CRU	L. Cruls
Danjon	DAN	Andre Danjon (1890-1967)
Dawes	DA	W. R. Dawes (1799-1868)
Dem	D	Ercole Dembowski
Dick	DIC	J. Dick
Dju	DJU	P. Djurkovic
Dob	DOB	W. A. Doberck
Dom	DOM	Jean Dommanget
Don	DON	H. F. Donner
Doo	DOO	Eric Doolittle (1869-1920)
Dorpat	DOR	Dorpat Observatory, Estonia
Edg	EDG	D. W. Edgecomb
Eggen	EGG	Olin J. Eggen
Es	ES	T. E. H. Espin (1858-1934)
Fle	FLE	J. O. Fleckenstein
For	FOR	L. Forgeron
Fox	FOX	Philip Fox
Franz	FRZ	J. Franz
Frh	FRH	R. Furuhjelm
Frk	FRK	W. S. Franks
Fur	FUR	H. Furner
GΣ	STG	G. Struve
Gale	GLE	W. F. Gale
Gallo	GAL	J. Gallo
GAn	GAN	G. Anderson
Giclas	GIC	Henry L. Giclas
Gli	GLI	J. M. Gilliss
Glp	GLP	S. de Glasenapp
Gol	GOL	H. Goldschmidt
Goyal	GYL	A. N. Goyal
Gsh	GSH	J. Glaisher
Gtb	GTB	K. Gottlieb
Gui	GUI	J. Guillaume
H I	H	William Herschel (1739-1822). Classes assigned in 1782-84 catalogues:
H II		I, difficult; II, close
H III		
H IV		but measurable; III,
H V		5"-15"; IV, 15"-30";
H VI		V, 30"-1'; VI, 1'-2'.
H N	H	W. Herschel's 1821 cat.
h	HJ	John Herschel (1792-1871)
HΣ	STH	Hermann Struve
Hall	HL	Asaph Hall (1829-1907)
HCWils	WHC	H. C. Wilson
HdO	HDO	Harvard Observatory, U. S. A. (and stations elsewhere)
HdZ	HDZ	Harvard zone catalogues
Hld	HLD	E. S. Holden
Hlm	HLM	E. Holmes
Hln	HLN	Frank Holden
Ho	HO	G. W. Hough (1836-1909)
Hooke		Robert Hooke (1635-1703)
Howe	HWE	H. A. Howe
Hrg	HRG	L. Hargrave
Hst	HTG	C. S. Hastings
Hu	HU	W. J. Hussey (1862-1926)
Huygens		Christiaan Huygens (1629-95)
Hynek		J. Allen Hynek
hz	HEI	Wulff D. Heintz
Hzg	HZG	E. Hertzsprung
I	I	R. T. A. Innes (1861-1933)
J	J	Robert Jonckheere
JAn	ANJ	John A. Anderson (d. 1959)
Jc	JC	W. S. Jacob
Jef	JEF	H. M. Jeffers
Joy	JOY	Alfred H. Joy (1882-1973)
Jsp	JSP	M. K. Jessup
Knott	KNT	G. Knott
Kr	KR	A. Krüger
Kru	KRU	E. C. Kruger
Ku	KU	F. Küstner
Kui	KUI	Gerard P. Kuiper (1905-73)
Lac	LCL	N. de Lacaille (1713-62)
Lal	LAL	F. de Lalande
Lam	LAM	J. von Lamont
Lau	LAU	H. E. Lau
Lbz	LBZ	P. Labitzke
LDS	LDS	W.J. Luyten's 1941 p.m. survey
Leon	LEO	F. C. Leonard (d. 1960)
Lewis	L	Thomas Lewis
LPO	LPO	La Plata Observatory, Argentina
Luy	LUY	W. J. Luyten
Lv	LV	F. P. Leavenworth
Ma	MA	J. H. Mädler (1794-1874)
Mh	MH	O. M. Mitchel
Mil	MIL	J. A. Miller
Milb	MLB	W. Milburn
MlbO	MLO	Melbourne Observatory, Australia
Mlf	MLF	Frank Muller
Mlr	MUL	Paul Muller
NZO	NZO	New Zealand Observatory
OΣ	STT	Otto Struve (1819-1905), Pulkovo catalogue, 1843; *rej* means rejected for various reasons in the 1850 revision, often because the pair was considered too wide to merit close attention
OΣΣ		O. Struve, Pulkovo catalogue supplement, 1843, containing very wide pairs
Ol	OL	C. P. Olivier (d. 1975)
Opik	OPI	Ernst O. Öpik (b. 1893)
Par	PAR	J. A. Parkhurst
Per	PER	J. Perrotin
Perry	PRY	J. J. M. Perry
Plq	PLQ	E. Paloque
Pol	POL	J. A. Pollock
Pou	POU	A. Pourteau
PrO	PRO	Perth Observatory, Australia
Prz	PRZ	E. Przbyllok
Ptt	PTT	Edison Pettit (1889-1962)
Pz	PZ	G. Piazzi
R	R	H. C. Russell
Richaud		Jean Richaud (1633-93)
Rmk	RMK	C. L. C. Rümker
Roe	ROE	E. D. Roe
Rst	RST	R. A. Rossiter (1886-1977)
S	S	James South (1785-1867)
S,h	SHJ	James South and J. Herschel (joint 1824 catalogue)
Schb	SHB	J. M. Schaeberle
Schj	SCJ	H. C. F. C. Schjellerup
Scott	SCT	J. L. Scott
Se	SE	A. Secchi
See	SEE	T. J. J. See (1866-1962)
Sei	SEI	J. Scheiner
Slr	SLR	R. P. Sellors
Smart	SMA	W. M. Smart (d. 1975)
Smyth	SMY	W. H. Smyth (1788-1865)
Sp	SP	G. V. Schiaparelli
St	ST	Carl L. Stearns (d. 1972)
Stein	STI	J. Stein
Stone	STN	Ormond Stone
SyO	SYO	Sydney Observatory, Australia
Tar	TAR	K. J. Tarrant
VBs	VBS	George Van Biesbroeck (1880-1974)
Vou	VOU	J. G. E. G. Voûte
Ward	WAR	I. W. Ward
Webb	WEB	T. W. Webb (1807-1885)
Weisse	WEI	M. Weisse
Wg	WG	R. W. Wrigley
Wils	WRH	R. H. Wilson, Jr.
Wirtz	WZ	Carl Wirtz
Wnc	WNC	F. A. Winnecke
WNO	WNO	U. S. Naval Observatory, Washington, D. C.
Worley	WOR	Charles E. Worley

separation in arc seconds ("). The standard system adopted by John Herschel, unchanged today, counts position angles counterclockwise from north in the field of view (Fig. 2). If other stars are seen in the eyepiece field but not measured, they might be noted to lie *preceding* (west of) or *following* (east of) the object of interest. Suggested by the drift of stars across the eyepiece field when a telescope's clock drive

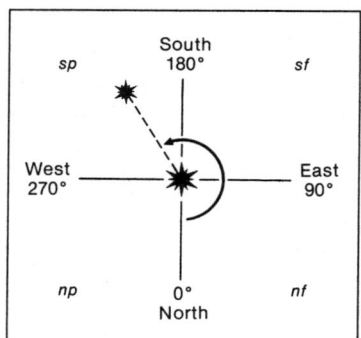

Fig. 2. Position angles are measured eastward from north. Hence, for the secondary star here, the value is about 210°. Both this angle and the separation (dashed line) are measured with a filar micrometer at the eye end of the telescope. South is up to match the inverted view with an astronomical telescope in the Northern Hemisphere.

is not running, these terms often occur in other visual descriptions as well. Many observers find them more natural than the geographical directions, which are often rotated or reversed by the optical system.

There is a remarkable parallel between double star astronomy and the rise of the achromatic refractor. Struve's success induced the Russian government to build a new Imperial Observatory at Pulkovo with a 15-inch refractor. As the Dorpat instrument had been in its day, this too was the world's largest refractor when placed in service in 1841. Its chief project, directed initially by Wilhelm Struve but a month later taken over by his son Otto, was to remeasure the Σ stars and search for new pairs. Otto Struve's numerous discoveries at Pulkovo have come to be known by the designation OΣ.

Meanwhile, in England such enthusiastic amateurs as W. H. Smyth, W. R. Dawes, and T. W. Webb were popularizing double star work in their writings and contributing a few discoveries of their own. Smyth drew seemingly fine distinctions among hues (pale topaz and violet, flushed white and sapphire blue, orange tint and cerulean blue, creamy white and lilac, etc.) that are now believed largely to have been subjective contrast effects, but his verbal flair has delighted generations of amateurs since.

The "eagle-eyed" Dawes, as he came to be called, performed a series of tests with refractors of various apertures to compare their separating, or resolving, powers. To make the test conditions uniform, he examined only pairs having equally bright components about three magnitudes more intense than the faintest stars visible with the aperture being tested. His famous *Dawes' limit,* enunciated in 1865, gives $4''.56/A$ as the resolving power of a telescope whose aperture in inches is A. Observers generally find this to be a rather stringent test of both atmospheric seeing and optical quality; achieving Dawes' limit requires high magnification, about 50x per inch of aperture. His practical experience agrees well with theoretical predictions of telescope performance, based on the wave nature of light (Fig. 3).

Dawes' success with refractors, including three Alvan Clark glasses of 7½- to 8¼-inch aperture, caught worldwide attention and brought considerable prestige to the Massachusetts optical firm. European astronomers must also have been excited by an 1873 paper in *Monthly Notices* of the Royal Astronomical Society written by a Chicago amateur, S. W. Burnham. Its title, "Catalogue of Eighty-one Double Stars, Discovered with a Six-Inch Alvan Clark Refractor," signaled a new age of double star discovery that profoundly increased the number of known pairs.

Closely involved in the resurgence of interest were successively larger Clark refractors, especially the 18½-inch Dearborn and 26-inch McCormick and Naval Observatory instru-

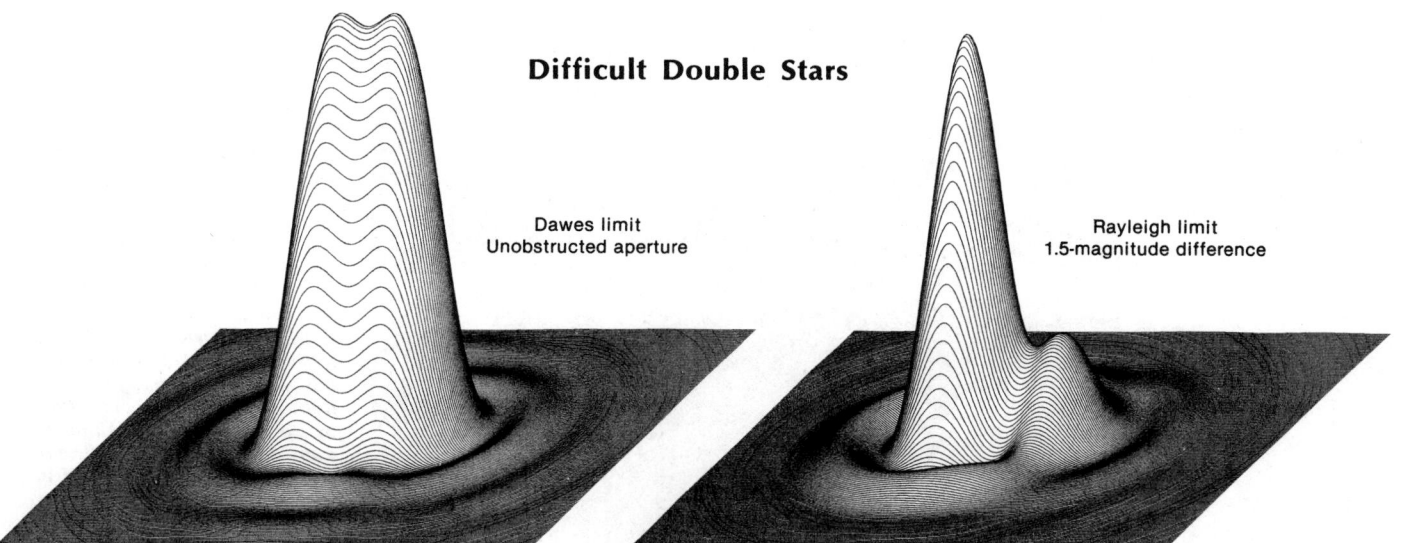

Difficult Double Stars

Dawes limit
Unobstructed aperture

Rayleigh limit
1.5-magnitude difference

Fig. 3. Even with an ideal telescope on an airless planet, stars would not be seen with perfect clarity. Because of the wave nature of light, a telescope with a circular aperture produces star images that consist of a central, spurious disk and surrounding diffraction rings. When two stars are very close together, their individual diffraction patterns overlap, as illustrated in these three-dimensional computer plots by David E. Stoltzmann, where intensity is represented by height. At left, we see two equally bright stars that are as close as they can be and yet be resolved, according to W. R. Dawes. At right, the stars differ considerably in brightness, but the companion can still be glimpsed because it lies in the first dark ring of the primary star's diffraction pattern. This slightly greater separation, called the Rayleigh limit, occurs (for yellow-green light and a telescope unobstructed by a secondary mirror) when the pair are separated by $5''.53/A$, where A is the telescope aperture in inches.

ments, followed by the 36-inch Lick and 40-inch Yerkes instruments that so dominated the public attention toward the end of the 19th century (Fig. 4). Apparently it was not true, as James South had lamented back in 1842, that "Struve has reaped the golden harvest among the double stars, and there is little now for me to hope or expect."

By long-standing tradition, a double star is named for the observer who either discovers or first measures it. The name is usually abbreviated in print and followed by a serial number; thus the 20th star in S. W. Burnham's second list of discoveries is called "Burnham 101" and written $\beta\,101$. (It happens to be a pair in rapid orbital motion, and it is described on pages 55 and 176 of the present work.) A modified system of designation codes was introduced in the Lick *Index Catalogue of Visual Double Stars* (IDS), 1963, where the use of 80-column punch cards forced the adoption of one, two, or three roman capital letters. In the IDS, the traditional Σ was replaced by STF, OΣ by STT, β by BU, and so on. But in *Sky Catalogue 2000.0* we have restored the former designations. For some of the less prolific discoverers, we spell out the full last name when there is room. Table I lists, for each designation code used here, the corresponding IDS code and the name of the discoverer.

After Burnham turned professional, he reobserved many of his own discoveries with larger instruments and added many new ones, for a lifetime total exceeding 1,300. Today's amateurs can gain a feel for Burnham's skill by trying their own instruments on $\beta\,396$ (page 8), $\beta\,15$ (page 37), or $\beta\,341$ (page 83). These are a few of the discoveries he made with the 6-inch Clark, and they happen to have extremely slow relative motions. They look today almost exactly as they did in the 1870's.

Our Double and Multiple Stars section is based on the 1976 magnetic-tape version of the Lick IDS, as maintained and updated at the U. S. Naval Observatory. This compre-

hensive work, with over 70,000 entries, is a direct successor to the earlier general catalogues of Burnham (the BDS, 1906) and Aitken (ADS, 1932). We have extracted all doubles whose components have a combined magnitude of 8.0 or brighter, and all multiple systems in which any two components meet this selection rule. All told, 11,788 individual pairings are described here, and they form 8,315 separate double or multiple systems.

After the selection was made, we cross-checked all positions against the stellar data base of *Sky Catalogue 2000.0*, Vol. 1. This allowed the substitution of improved coordinates for a great many pairs and the incorporation of Bayer and Flamsteed designations, which were missing from the IDS. At the same time, we copied the visual magnitudes from Vol. 1 for comparison. Extreme caution was needed because, while the magnitudes in Vol. 1 are often combined magnitudes of close components (especially when photoelectrically determined), those in the IDS are old visual values of the individual stars. And these were many times extremely crude estimates by observers whose prime concern was use of the filar micrometer.

We experimented with a number of ways to adjust the magnitudes from the IDS and make them more in agreement with the modern values of Vol. 1. Angular separation proved to be a poor clue as to whether a Vol. 1 magnitude refers to the primary star or a combined pair. In the end we decided to adopt whichever interpretation would lead to a *smaller* adjustment of the IDS values.

Consider, for example, the star $\Sigma\,738 = \lambda$ Orionis, at $5^h\,35^m.1$, $+9°\,56'$, whose **V** magnitude is listed in Vol. 1 as 3.39. The IDS gives the magnitudes of the A and B components as 3.7 and 5.6, respectively, which implies a combined magnitude of 3.53. If the Vol. 1 magnitude refers to the primary star alone, those of the A and B components must in fact be 3.4 and 5.3; but if the Vol. 1 magnitude is com-

Fig. 4. During an observing career that began in 1903, the Belgian-born astronomer George Van Biesbroeck made 35,915 double star measures, a lifetime total exceeded only by W. H. van den Bos and W. Rabe. Many of these measures were made with the Yerkes 40-inch refractor in Williams Bay, Wisconsin. The telescope's 60-foot tube was aimed downward for this Yerkes Observatory portrait, taken in about 1960 by Joseph Tapscott. Although Van Biesbroeck had officially retired from the Yerkes staff in 1945, he maintained a vigorous observing schedule with this and other large telescopes until his death in 1974.

bined the values are 3.6 and 5.5. Since the latter choice leads to a smaller adjustment of the IDS, it is the one we adopted.

Such a procedure is not infallible, of course, but it does seem better than leaving the IDS magnitude values alone. Note that for any star needing this correction we applied the same increment to all components, preserving the magnitude differences among them. (It is just these differences that would have been estimated by the observer, rather than actual values.) No systematic correction has been made to the magnitudes of stars not matched in Vol. 1. But many pairs have been looked up in other sources, particularly A. Wallenquist's 1981 work, and corrected when necessary.

Since angular separation did not enter into our selection at all, a great variety of systems is represented. Some pairs are so close together that they are only resolvable in the world's most powerful refractors; others are extremely wide pairs that have attracted attention because of one star's proper motion or the pair's common proper motion. Many measures of this kind, published by Burnham in 1913, are identified with the symbol "β pm" here. Sometimes, an observer has made numerous measures in an open cluster like NGC 6231, or in a mysterious nebulous region such as the η Carinae complex, not knowing whether the measures may prove useful to future generations of astronomers.

Readers may be dismayed at the number of stars for which our notes say "single," "duplicity doubted," and so on. Such remarks have been taken from the printed version of the IDS. Once a star has been entered in a double star catalogue it is quite hard to remove it, as W. J. Hussey pointed out many years ago. There are countless examples of stars that have appeared single in some years, then barely double in others through orbital motion; astronomers have understandably been reluctant to wipe the slate clean on a star without being certain. But many mistakes are sure to remain among these questioned cases. It is important to note that the letters *rej* (for "rejected") after a Σ or OΣ designation do not necessarily mean that the Struves mistrusted their earlier measures. Often they thought very wide pairs would not prove interesting enough to study further. But Hussey and other later workers have occasionally found just the opposite to be true.

As mentioned earlier, the IDS was compiled on 80-column punch cards. This caused formatting problems with large angular separations; if a value exceeded 99″, a code letter at the end of the line directed users to the notes section where the hundreds digit could be found. Needless to say, infrequent users of the work could easily miss this subtlety. We have looked up all the IDS notes and corrected our list accordingly.

The vast majority of double stars show no more than gradual relative motion, amounting to less than 15° in position angle per century, with a similar small change in separation. For these systems, current knowledge is rather well summarized by the two-date ephemeris of the IDS repeated here: the earliest reliable measure and a fairly recent one. In many cases the stars show no definite motion and only the early measure is given.

Consider the pair in the constellation Phoenix known as h 5437; this is one of John Herschel's discoveries, listed near the top of page 3. The first measure was made in 1852 and the second in 1959, a span of just over a century. But the motion is so gradual that we may extrapolate with some confidence to 1985 and expect a current position angle of about 317° and a separation 1″.7. Thus, thanks to the early benchmarks furnished by the Struves, Herschels, and other pioneers, it often matters little whether the second date is in the 1950's or 1970's. We have nevertheless scanned the U. S. Naval Observatory's published photographic measures (by O. G. Franz, V. V. Kallarakal, F. J. Josties, and their colleagues, 1963-78) and updated the IDS values for some 300 pairs.

Double star observers make their contributions not only through discovery of new pairs, but also in patiently monitoring those long known. It is one of the few areas of astronomy (or any other science) where instrumental techniques have remained largely unchanged for 160 years, where little can be learned except in the passage of time, and where the opinions of experienced observers are given equal weight, whether expressed in 1830, 1910, or today. Satellite-borne telescopes promise dramatic gains in astrometric accuracy, but nothing can speed up the leisurely apparent motions of stars in their orbits.

Visual Binaries. After measures of a double star have accumulated over many years, study may reveal a definite curving motion of the secondary star. At this stage the system commands much greater interest, for the opportunity arises to compute an orbit. Strictly speaking each star moves in an ellipse around the common center of gravity of the pair, but it is customary to regard the primary star as fixed and determine the somewhat larger (but in every respect similar) orbit of the secondary star relative to the primary. Moreover, this true orbit in space is haphazardly inclined to our view, so what we see is called the apparent orbit. The apparent orbit is an exact ellipse, as is the true one, but it may have a very different shape. Here the primary star is not generally located at one geometric focus, as must be the case for the true orbit.

Less than a thousand visual binary stars have known orbits, and for most of these the elements are but roughly determined. Yet their importance in astronomy is very great. The only method by which the masses of other stars may be measured directly is to determine the trigonometric parallax of a binary star and apply Kepler's third law to the observed period of revolution and semimajor axis of the orbit. Conversely, as knowledge of stellar masses has increased, a reverse procedure gives the distances to more remote binaries using the so-called dynamical parallax method. Provided the components' masses can be estimated, the dynamical parallax is more accurate than the trigonometric parallax for any star beyond about 15 parsecs. Some visual binaries are members of open clusters that contain Cepheid variable stars. Dynamical parallaxes have played key roles in refining the distances to these yardsticks of the universe.

The Visual Binary Stars section of this catalogue gives the orbital elements of 518 distinct stellar pairs, several of which belong to triple or quadruple systems. Our chief source was the 1982 magnetic tape received from P. Couteau and M. Fulconis, which listed the elements of pairs with a primary star of magnitude 8.7 or brighter. These were then checked against a preprint of the 1983 catalogue of C. E. Worley and W. D. Heintz; elements from the latter were substituted if they appeared to be significantly better. To resolve discrepancies, we computed position angles and separations corresponding to early published measures, for example those in the classic 1879 handbook by E. Crossley, J. Gledhill, and J. M. Wilson.

To illustrate each pair's current motion, we include short

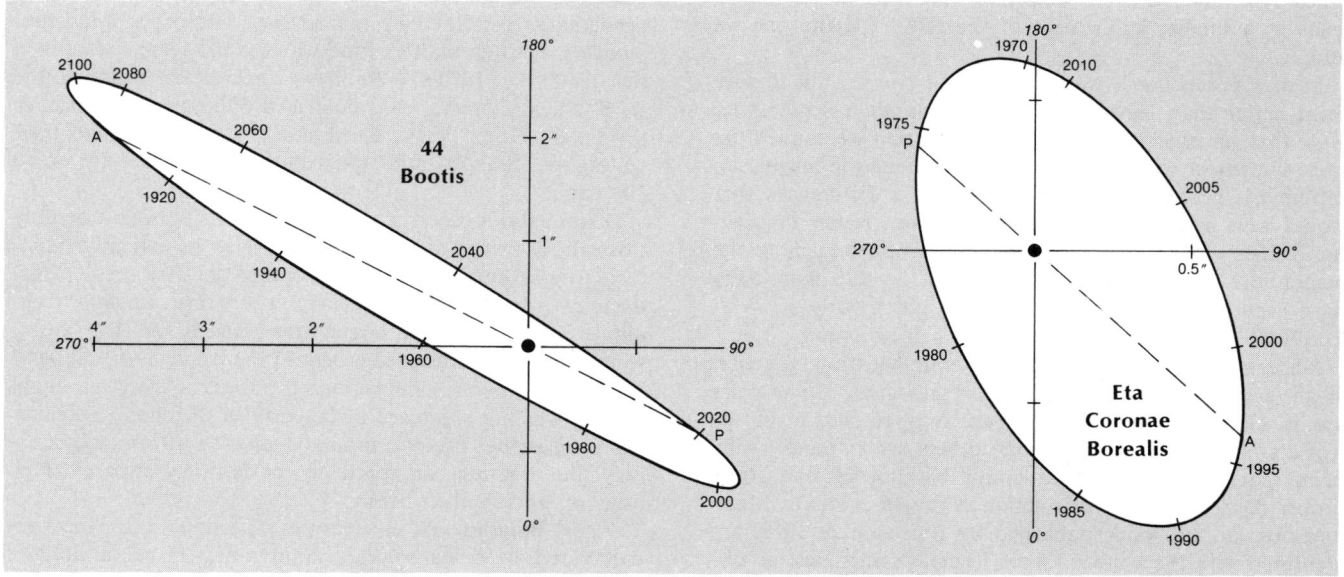

Fig. 5. *These diagrams of the apparent orbits of 44 Bootis and η Coronae Borealis have been prepared from the orbital elements on pages 185 and 186. We see both orbits considerably inclined from face on, so that the line of apsides, joining the secondary star's positions at apastron (A) and periastron (P), is not the same as the major axis of the apparent ellipse.*

ephemerides centered on the 1980's. The ephemeris interval has been chosen to aid mental interpolation. Readers can easily extend an ephemeris, or find a pair's separation and position angle for a specific date, by direct calculation from the elements. A concise method for doing so is given in Jean Meeus' 1982 *Astronomical Formulae for Calculators*. When enough points have been computed, they can be con-

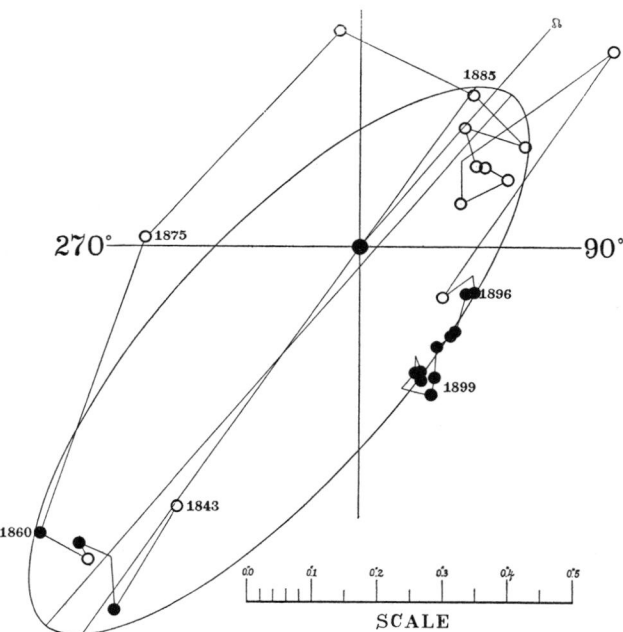

Fig. 6. *In the binary system OΣ 400, the separation of the two 8th-magnitude components never exceeds 1". The difficulty of measuring such a pair is well illustrated in this diagram by W. J. Hussey, which appeared in Lick Observatory* **Publications,** *Vol. V, 1901. Fine lines join the dots to show the chronological sequence of observation, while the ellipse is Hussey's orbit. More recent elements appear on page 194. This system has now been observed through nearly two complete revolutions.*

nected with a smooth curve to make a complete picture of the orbit (Fig. 5).

The foreshortening of a binary star orbit produces some interesting illusions. The ellipse traced in space by the companion of 42 α Comae is fully ⅞ as wide as it is long, according to H. Haffner's orbital elements on page 182. Yet this system appears so nearly edge on that eclipses may even be possible (see note 63 on page 160). At the opposite extreme, however, is OΣ 2 on page 163; here the true orbit is fairly elongated (eccentricity 0.84), but it is tipped toward us in just such a manner that the companion appears to move almost in a circle.

Computing the orbital elements from a set of observations is a rather advanced art. The key idea is that Kepler's law of areas must be obeyed in both the true and apparent orbits, because the latter is simply a projection of the former. One might be tempted simply to fit an ellipse to the positions of the secondary, ignoring the dates on which they were observed. But this is a bad idea, as R. G. Aitken's book points out, because the dates themselves are known more accurately than anything else. Fig. 6 gives some idea of the task being faced. Early workers often used pins, string, and planimeters; today a least-squares solution on a computer is more appropriate, though the results for a close pair may not be any more reliable.

Spectroscopic Binaries. In 1889, E. C. Pickering discovered that the A component of the famous double Mizar in Ursa Major was a spectroscopic binary. (Over two centuries earlier, Mizar AB had been the first visual double star discovered telescopically, by G. Riccioli in about 1650.) Algol, β Aurigae, and Spica soon followed. The radial velocity of such a star, as measured by the locations of lines in its spectrum, varies rhythmically over the course of hours or days. Usually the lines of only one component can be measured. But sometimes two overlapping spectra are seen, the two patterns weaving back and forth across each other as the components move around one another.

The data in this section come from the 1978 catalogue of A. H. Batten, J. M. Fletcher, and P. J. Mann. We've selected the 485 systems brighter than magnitude 6.3 and several dozen fainter ones of special interest. They include Vela X-1 and Castor C (YY Geminorum), as well as the binary pulsar 1913 + 16, which has been used in recent tests of general relativity. The extensive notes in Batten's work have enabled us to flag many tabular quantities and indicate variants in meaning of some elements.

Most spectroscopic orbits refer to extremely close components that cannot be resolved visually. The periods of these systems are often so short that, just as for variable stars, the epochs are best expressed by Julian dates and fractions of a day. (Simple computer programs for converting between Julian dates and calendar dates appeared in *Sky and Telescope*, **67**, May, 1984, page 454.) The spectroscopic elements have the same meanings as those of visual binaries, except that the orbit's angular size is replaced by linear size. This linear size cannot be known absolutely, but only as a function of assumed mass and orbital inclination, unless other than spectroscopic evidence can be brought to bear. Aitken's 1918 work explains the differences further.

VARIABLE STARS

The study of variable stars is a branch of astronomy where amateurs can, and do, make valuable observations. Although a backyard research station might have a permanently mounted telescope equipped for sophisticated photoelectric measurement, a great deal can also be accomplished with much more modest equipment. A telescope used for visual observation of variable stars does not need a clock drive, and no great demands are placed on optical resolution or light grasp. Some observations can even be made in the presence of streetlights or strong moonlight. In fact, hundreds of variables are within reach of hand-held binoculars, and over a dozen (like Algol and Betelgeuse) can be followed with the naked eye.

Professional astronomers have time to study only a small fraction of the tens of thousands of known variables. Since early in this century, amateurs have eagerly taken up the challenge and monitored hundreds of stars, often through organizations like the American Association of Variable Star Observers (AAVSO). Such an affiliation helps ensure that consistent observing practices are followed and that an individual's results will be collected for efficient use by professional astronomers. In the 1982-83 season alone, the AAVSO processed 194,580 magnitude estimates made by over 500 of its members worldwide.

Variable stars play a crucial role in astronomical research because they are abundant and encompass a wide range of stellar spectral types, sizes, masses, densities, and ages. Spectroscopic studies of variables have revealed much about the internal constitution of stars at different stages in their evolution. A landmark in establishing the extragalactic distance scale was Henrietta Leavitt's discovery, in 1912, that the luminosities of Cepheid variables in the Small Magellanic Cloud are directly related to their periods. RR Lyrae stars, novae, and supernovae have also been used as "standard candles" by which to gauge the distance to star clusters and galaxies. Astronomers study eclipsing binaries for a wealth of data about stellar interactions. Since the mid-1970's, observations of RS Canum Venaticorum variables have even confirmed the existence of spots on stars other than the Sun.

Variable stars are named according to a rather awkward convention that grew bit by bit over the years. The first variable to be designated in a constellation is assigned the letter R and the possessive form of the constellation name, as in R Andromedae. (The letters A through Q are skipped, to avoid confusion with some old designations of nonvariable stars.) Those subsequently confirmed in the same constellation are labeled S, T, . . . , Z, and then RR, RS, . . . , RZ, SS, ST, . . . , SZ, and so on up to ZZ. Once these 54 designations have been used up, the naming continues with AA, AB, . . . , AZ, BB, BC, . . . , BZ and so on up to QZ, except that combinations with the letter J are omitted. In this way, 334 variables in each constellation can be named. Beyond this point they are labeled V335, V336, and so on. Variables that already have Greek-letter (Bayer) names, such as β Persei or δ Cephei, generally do not receive new designations.

In making the acquaintance of a variable star, there is no substitute for starting to compile a nightly series of magnitude estimates. An observer normally compares the variable with two nearby constant stars, one a little brighter and the other a little fainter. The comparison stars should be of nearly the same color as the variable and at the same altitude in the sky (preferably in the same telescope field). The observer jots down the date, time, and brightness estimate, along with the identities or locations of the comparison stars.

Sometimes the visual magnitudes of the comparison stars are unknown, especially if the variable is a nova that has appeared suddenly in an unfamiliar part of the sky. But an estimate of lasting value can still be made if the observer knows how to establish, on the spot, an arbitrary brightness scale. For example, suppose that four comparison stars have been selected and labeled A, B, C, and D in order of decreasing brightness on a sketch of the eyepiece field. If the variable appears a little fainter than C but much brighter than D, the observer might write down "C 1 *v* 3 D," to indicate that the variable's brightness is one-fourth of the way from that of C toward that of D. Later, when comparison-star magnitudes become known, that of the variable can easily be calculated. (The validity of this procedure is guaranteed by Fechner's law of 1858, which states that sensation in the eye, to a very close approximation, varies directly as the logarithm of the stimulus. The magnitude scale is likewise logarithmic.)

Skilled observers, working in this way, can pin down the brightness of a variable star to within 0.1 magnitude or better using no special equipment, if good comparison stars are available. Visual photometers have been devised to assist such estimates, usually by creating artificial comparison-star images, and there are excellent photographic techniques as well. But in recent years photoelectric photometry has displaced all other methods for precision work. Light intensity is gauged using a photomultiplier tube or even a photodiode of known wavelength sensitivity, in conjunction with standard filters. A series of measurements is carried out on the variable, a nearby comparison star, and the vacant sky background. With certain other precautions, including allowance for differential atmospheric extinction, it is not too difficult to achieve 0.01-magnitude accuracy with an 8-inch telescope on stars brighter than 11th magnitude, and to go several magnitudes fainter if expectations are relaxed.

After amassing brightness estimates of a variable star, the ultimate goal is to construct a *light curve* that shows how the brightness changes over time. The star's maximum and minimum light (brightest and faintest magnitudes, respec-

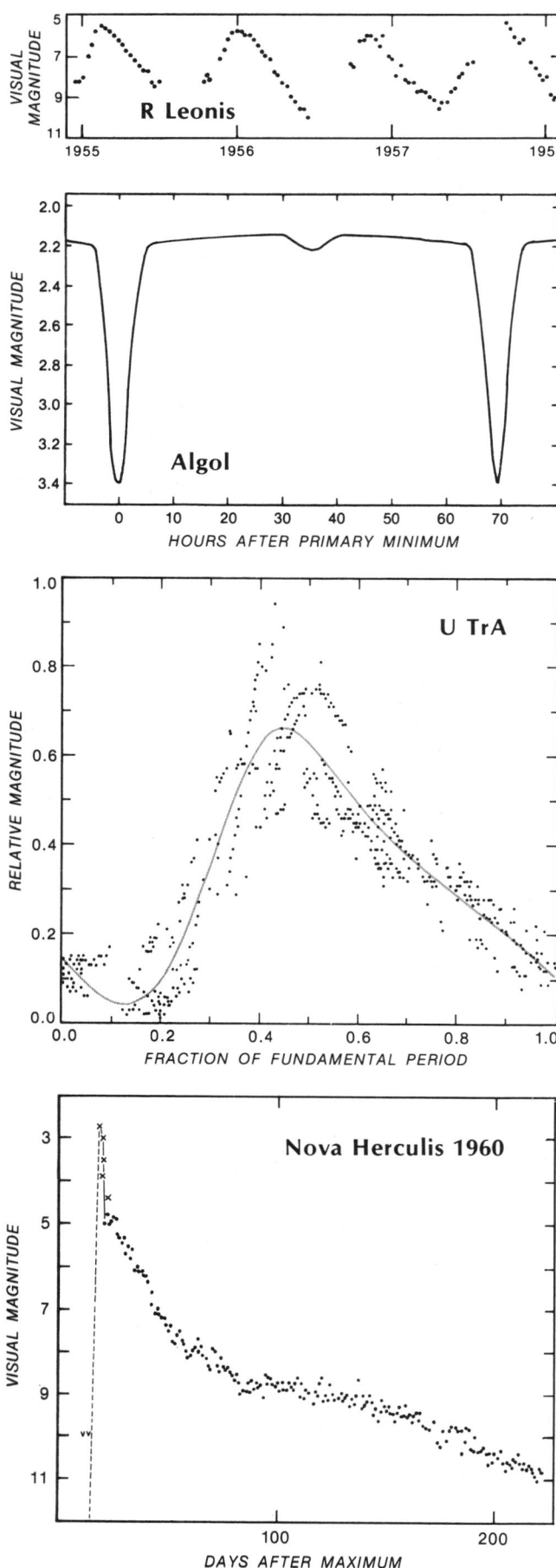

tively), period, and variable type can be determined from the curve's shape. And if the variations are regular enough, times of future maxima and minima can be predicted. Fig. 7 shows some recent light curves compiled by the AAVSO.

Among the most dramatic variables to watch are the eclipsing binaries. Often, as is true of the famous naked-eye star Algol, nearly the entire fading and subsequent recovery from an eclipse can be witnessed in a single night. The changes may be so rapid that a careful observer, making perhaps 10 estimates each hour, can later fix the instant of least light to within two or three minutes. Such timings, often linked to a precise count of cycles spanning many decades, explain the large number of significant digits to which many periods are listed. Algol and many other eclipsing binaries undergo gradual as well as sudden (but slight) changes in period over the years, underscoring the importance of continued monitoring.

The epochs and periods given in this catalogue can be used to make accurate predictions of eclipsing-binary minima. The listed epochs are heliocentric; that is, they give the Julian dates when an observer *located at the Sun* would see a minimum. When the Earth is between the variable star and the Sun, we observe the event as much as eight minutes earlier, because the star's light reaches the Earth before the Sun. When the Sun lies between us and the star, the event is delayed by a similar amount.

By repeatedly adding the star's listed period (or multiples of the period) to the heliocentric epoch, we can generate as many other dates of heliocentric minima as we desire. When one is found near a date of interest, the prediction can be refined by applying a light-time correction to give the *geocentric* or Earth-based time of the event. This correction

Fig. 7a. *At the top is the light curve for R Leonis, a Mira (M) variable located about 5° west of Regulus and described on page 235. This well-observed star is bright enough to be followed in binoculars for most of its 312-day period. Among the largest stars known, if placed at the center of our solar system it would extend out to the planet Jupiter.*

Fig. 7b. *Algol, or β Persei, is described on page 226. It is the prototype of a major class of eclipsing binaries, designated in the catalogue by "EA." These systems consist of two well-separated stars whose eclipses give rise to a light curve similar to the one shown here. Algol's brightness remains almost constant for about 90 percent of the time. Deep minima occur every 2.867 days as the brighter member is partially eclipsed by its fainter companion. Shallower dips result when the dim component passes behind the brighter one. For the β Lyrae (EB) and W Ursae Majoris (EW) binary classes, where the stars are much closer together or even in contact, the light curves are more rounded.*

Fig. 7c. *U Trianguli Australis is a classical Cepheid (Cδ) that displays two simultaneous and different pulsation periods. They beat against each other so that the amplitude varies between 0.6 and 1.0 magnitude. The solid curve shows the basic, or fundamental, period of 2.56842 days. The light curve given here was compiled by A. N. Cox at Los Alamos Scientific Laboratory and is the result of six years of study. This star attains a maximum visual magnitude of about 7.5 and is listed on page 243.*

Fig. 7d. *The nova outburst that occurred in the constellation Hercules in 1960 was one of the brightest in recent decades. Note the abrupt rise and the gradual decline that is characteristic of novae; the star is currently fainter than 18th magnitude. See the listing on page 247 under V446 Herculis for more information.*

depends on where a line from the Sun falls perpendicularly on the line of sight between the Earth and the star (in fact, it is just the time it takes for light to travel between this intersection and the Earth). The following method, easily programmed on a calculator or computer, gives the light-time correction without the need of an almanac; the accuracy is better than one second of time, and the formulae are valid for at least several centuries around the present.

Assume we have already determined a heliocentric Julian date, H, and that the star lies at right ascension α and declination δ referred to the 2000.0 equinox. Then calculate, successively:

$$T = (H - 2451545.0)/36525,$$
$$L = 280°.4659 + 35999°.371946\,T,$$
$$M = L - 282°.9405 - 0°.322204\,T,$$
$$\lambda = L + 1°.9148 \sin M + 0°.0200 \sin 2M,$$
$$\nu = \lambda - L + M,$$
$$R = 0.99972/(1 + 0.01671 \cos \nu),$$
$$l = \cos \delta \cos \alpha,$$
$$m = 0.91748 \cos \delta \sin \alpha + 0.39778 \sin \delta,$$
$$\tau = 0.005775\,R\,(l \cos \lambda + m \sin \lambda).$$

The last quantity, τ, is the correction (expressed as a fraction of a day) to be applied to a heliocentric date to convert it to geocentric; that is, $G = H + \tau$. Alternatively, the light time can be used with reversed sign to convert a geocentric timing to a heliocentric one; this operation is necessary if observations of different minima are being analyzed to determine the star's period, for example. The motion of the Earth during the light time can be disregarded in these calculations. But it is very important to use the star's 2000.0 position in these formulae, and not that referred to another equinox.

As an example of the calculation, consider the eclipsing binary λ Tauri, whose abrupt half-magnitude dimmings at intervals of just under 3 days 23 hours were first noticed by the English amateur J. Baxendell in 1848. On page 227, we find that a heliocentric minimum of this star occurred on Julian date 243 5089.204, and that the period is 3.952955 days; also, $\alpha = 4^h\ 00^m.7$ and $\delta = +12° 29'$ (2000.0). Treating these numbers as exact for the sake of illustration, suppose we want to predict a minimum in the fall of 1986. Some experimenting shows that 2,943 cycles, multiplied by the period and added to the epoch above, brings us up to Julian date $H = 244\,6722.750565$. Working through the equations we find that the light-time correction, τ, is -0.004673 day at this instant. Adding τ to H gives 244 6722.745892 as the date of *geocentric* minimum; in other words, the minimum will be seen on Earth 404 seconds early. Conversion of the Julian date tells us that the minimum can be expected at about 5:54 Universal time on October 19, 1986. (This is uncertain by several minutes, not because of light time, but because the variable's elements are not known with perfect accuracy.)

The primary data source for our Variable Stars section is the Moscow *General Catalogue of Variable Stars* (GCVS) by B. V. Kukarkin *et al*. After selecting stars from the magnetic tape of the third edition (1969-70), we introduced all changes and revisions that have appeared in the three printed supplements (1971, 1974, and 1976). In the final stages of our work the first volume of the fourth edition became available on tape. Its improved data, covering the constellations alphabetically from Andromeda through Crux, were substituted.

We compiled our list by selecting every variable with a maximum brighter than or equal to magnitude 9.0, regardless of the photometric system on which the value is based. As a result, occasional red variables that are *visually* somewhat brighter than 9.0 may not appear here if their quoted magnitudes in the GCVS are either photographic or photoelectric **B** (the blue band of **UBV** photometry). To this list we have added stars in the *AAVSO Variable Star Atlas* with visual maxima between 9.0 and 9.5 having a light amplitude of a magnitude or more, plus all variables in the 1982 edition of the Yale *Bright Star Catalogue*. There are altogether 2,427 entries in this section. Whenever possible, values were updated according to the most recent source cited above or in the References. A notes section beginning on page 255 gives additional information on nearly half of the stars in the main listing.

Immediately following the Variable Stars section, on page 263, is a list of 706 stars whose variability remains unconfirmed. They represent all objects in the Moscow *New Catalogue of Suspected Variable Stars* (NSV), B. V. Kukarkin *et al.*, 1982, that have quoted maxima brighter than magnitude 10.3 and amplitudes reported to be 0.5 magnitude or more, regardless of the wavelength band in which the brightness was measured. Several dozen other stars for which there are tantalizing reports have also been included, even if the information about these sightings is quite scanty. A few notes follow this section on page 272.

The NSV itself lists all stars for which variability has been reported but not confirmed as of 1980. Many of these will eventually prove to be *bona fide* variables. Many other entries probably stem from erroneous observations, defects on photographic plates, or even positions that have been reported incorrectly. We offer this section as a working list of unexplained mysteries, stars that may not exist in some cases, other stars that could be quite ordinary after all. Perhaps an enterprising amateur will delve further into the observational histories of some of them and make a habit of checking up on them from time to time. Something interesting may turn up!

VARIABLE STAR TYPES

Some variables change in brightness because of intrinsic phenomena, such as pulsations or eruptions. Others vary due to external circumstances, of which binary-star eclipses are the main example. These two broad categories have been further subdivided by astronomers into the many individual types defined below. Each type occupies a region of the color-magnitude diagram based on its absolute magnitude, color index (or spectral type), and age, as shown in Fig. 8, adapted from P. N. Kholopov's 1981 article on variable star classification. Not all of these types are represented by stars in the Variable Stars section.

BL Her. BL Herculis stars (see "CW" below).

BY. BY Draconis variables are young, rapidly rotating, type K or M main-sequence stars that have emission-line spectra. They are quasi-periodic in the range of a fraction of a day to 120 days, with amplitudes normally less than about 0.3 magnitude. Their variability is due to nonuniform surface brightness ("starspots").

Cep. Cepheids (after δ Cephei, whose changing light was first noted by the English amateur John Goodricke in 1784) are supergiants of spectral type F, G, or K with pulsation periods from 1 to 70 days and amplitudes in the range 0.1 to 2 magnitudes. Their characteristic, asymmetric light curve shows a steeper rise than decline. The Cepheids are highly evolved stars that have exhausted the hydrogen and helium fuel in their cores. They are subdivided into classical Cepheids (Cδ) and Population II Cepheids (CW).

cst. Constant stars incorrectly designated as variables.

CW, CWa, CWb. Cepheids belonging to the spherical, or old, component of the galaxy. They are called Population II or Type II Cepheids and are often found in globular clusters. Almost all fall within one of two period ranges: from about 10 to 30 days (subclass "a," the W Virginis stars) and 1 to 4 days (subclass "b," the BL Herculis stars).

Cδ, Cδs. Long-period classical Cepheids belonging to the disk, or younger, component of the galaxy; also called Population I or Type I Cepheids. They are typically one to two magnitudes more luminous than Population II Cepheids (class CW) of the same period. The variable δ Cephei is the prototype; Polaris is another well-known example. Classical Cepheids are sometimes found in open clusters. Those with small amplitudes and nearly symmetric light curves are labeled with the suffix "s."

E, EA, EB, EW. Eclipsing binaries are classified by the shape of their light curves. In Algol-type (EA) systems, the component stars are well separated and nearly spherical, so the light curves are almost level between eclipses. Periods range from a fraction of a day to many years. The β Lyrae (EB) binaries have gravitationally distorted, ellipsoidal components that are usually of early spectral type (O, B, or A) and unequal brightness. Their light curves vary continuously with amplitudes less than 2 magnitudes. Periods are typi-

cally greater than a day. W Ursae Majoris (EW) systems also display continuously changing light curves, but most have amplitudes less than about a magnitude. Their highly distorted dwarf components are in contact, or nearly so, and generally orbit each other in less than a day. The stars most often are cooler than those in β Lyrae systems; that is, they have later spectral types, such as *F, G,* or *K.*

A new classification scheme for eclipsing binaries, used in the fourth edition of the GCVS and therefore partially included in our Variable Stars section, is based on the intrinsic properties of the binary as deduced from its light curve, spectral type, and other observations. The codes, described fully in the article by P. N. Kholopov mentioned earlier, are often appended to the former designations with slashes (/):

D. Detached components; that is, each member is well inside its Roche lobe (the surface around each star within which the star's gravity dominates over that of its companion; if matter from one member passes beyond the Roche lobe, it may be captured permanently by the other star).

DM. Detached main-sequence components.

DS. Detached subgiant and main-sequence components.

DW. Detached main-sequence and subdwarf components.

AR. Detached subgiant components, for example as in the AR Lacertae system.

SD. Semidetached binary, in which only one member has filled its Roche lobe.

KE. Contact binary whose components have early spectral types (in the range *O* to *A4*).

KW. Contact binary with main-sequence or subdwarf components of intermediate to late spectral type (*A5* to *K*). W Ursae Majoris is the prototype.

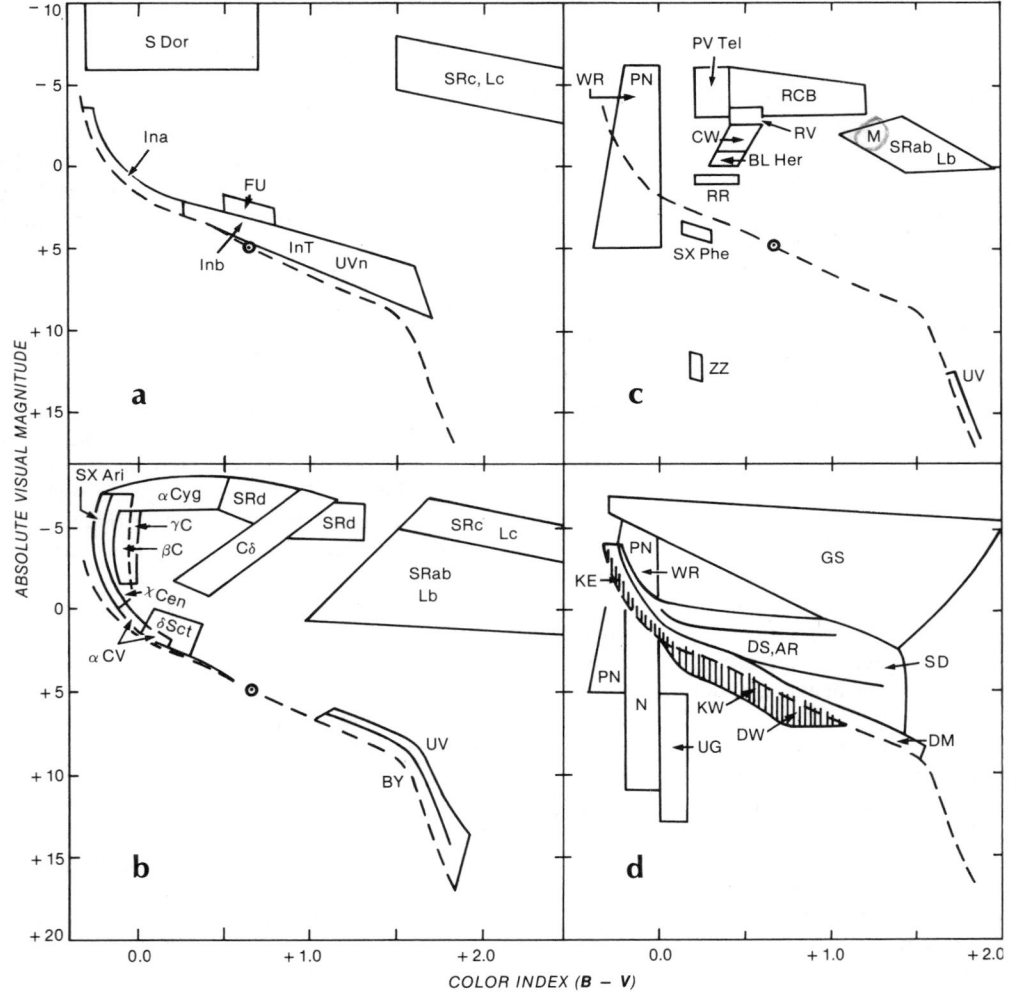

Fig. 8. The regions in the color-magnitude diagram occupied by the types of variable stars defined in the text. Variability is a transient phenomenon whose occurrence depends on a star's evolutionary stage, mass, chemical composition, and in some cases proximity to a companion. Therefore, the figure is divided into three sections by age and a fourth that is reserved for binary systems regardless of age: **a,** *variables younger than 10 million years, including newly formed T Tauri stars (InT) just entering the main-sequence phase of development;* **b,** *age between 10 million and one billion years (the classical Cepheids, Cδ, fall in this category);* **c,** *stars older than one billion years, including the common RR Lyraes (RR), Population II Cepheids (CW), and Mira-type variables (M);* **d,** *variables that belong to binary systems, displayed here according to the separation and evolutionary stage of the members. Novae (N), Wolf-Rayet (WR), and U Geminorum (UG) stars, and planetary-nebula nuclei (PN) are also shown. In the first three frames, the Sun's position is indicated by the symbol ⊙. This diagram is based on that of P. N. Kholopov, which appeared in his 1981 article cited in the References section.*

GS. At least one component is a giant or supergiant.

PN. The binary is the nucleus of a planetary nebula.

RS. An eclipsing RS Canum Venaticorum system. (See the definition later under "RS.")

WD. One member of the binary is a white dwarf.

WR. One member is a Wolf-Rayet star.

Ell. Binary systems with gravitationally distorted, ellipsoidal components that do not eclipse each other. The binary varies in brightness because the orbital motion changes the amount of stellar surface visible to the observer (for example, viewing the ellipsoidal stars more nearly end on than broadside).

FK Com. These are stars similar in nature to FK Comae Berenices, which is a type *G* giant with 2.4-day brightness and spectrum variations, extreme chromospheric activity, and an anomalously fast rotation rate. Its rapid spin is probably due to a close, low-mass companion in a decaying orbit near the surface of the giant. The components may eventually coalesce, if they have not done so already.

FU Ori. FU Orionis stars are eruptive, novalike variables of spectral type *A* to *F* that are associated with diffuse nebulae. They typically brighten by up to 6 magnitudes over several months and then stay at maximum light for decades. Emission lines in their spectra grow more prominent during the process.

I, Ia. Irregular variables are pulsating stars whose observed light fluctuations are not periodic. Those of early spectral type (*O, B,* or *A*) are labeled with the suffix "a."

In, Ina, Inas, Inb, Inbs, Ins, InT. Orion-type irregular variables associated with diffuse nebulae. Most may be low-mass stars just prior to the main-sequence stage. The suffix "a," "b," or "s" is added to denote variables of early spectral type (*O* to *A*), those of intermediate or late spectral type (*F* to *M*), or those with rapid light fluctuations, respectively. T Tauri variables, very young stars of intermediate mass that are just entering the main-sequence stage, are indicated by the suffix "T." They were named after the prototype T Tauri by A. H. Joy in 1945.

Is, Isa, Isb, IsT. Formerly known as RW Aurigae stars, these are rapid irregular variables apparently not connected with diffuse nebulae. They display 0.5- to 1-magnitude fluctuations over several hours or days. Sometimes long-term variations of up to 4 magnitudes are also observed. The suffixes "a" and "b" have the same meanings as for the type In variables. Those whose spectra display T Tauri characteristics are indicated by "T."

L, Lb, Lc. Slow irregular variable stars, some of which merely have been observed insufficiently to assign them to a more appropriate category. The suffix "b" indicates either late-type (*K, M, C, S*) giants or red irregulars whose spectral class and luminosity are unknown. Late-type supergiants are designated by the suffix "c."

M. Mira variables, formerly called long-period variables, are named after the prototype star Mira (*o* Ceti), whose changing light was first noted by Fabricius in 1596. They are red giants with periods from 80 to 1,000 days, amplitudes greater than 2.5 magnitudes (the average is 5 or 6), and *M-, C-,* or *S*-type spectra showing emission lines.

N, Na, Nb, Nc, Nr. Novae, which are thought to be close binary systems that include a hot white dwarf accreting material from its companion. The light increases suddenly by 7 to 16 magnitudes over several days or more and then fades to the original brightness in the course of years or decades. Small variations are observed at minimum. The suffix "a," "b," or "c" denotes a nova whose light decay is fast (taking about 100 days or less to decline by three magnitudes), slow (on the order of 150 days), or extremely slow (years), respectively. Recurrent novae, those for which more than one outburst has been observed, are indicated by the suffix "r"; outbursts are typically separated by 10 to 80 years.

Nl. The novalike variables are a poorly defined category of stars whose light variations or spectra resemble those of novae. Many, however, are not related to the novae.

PV Tel. The PV Telescopii variables are type *B* supergiants that undergo small-amplitude pulsations on a time scale of 0.1 to 1 day. Their spectra display enhanced helium and carbon lines.

RCB. R Coronae Borealis variables. These highly luminous stars of spectral type *F, G, K,* or *R* spend most of their time near maximum light. They are characterized by sudden, unpredictable drops in brightness of 1 to 9 magnitudes. Full brightness returns slowly in several dozen to several hundred days. The dimming may be caused by carbon particles — soot — condensing around the star.

RR, RRab, RRc. RR Lyrae stars, sometimes referred to as short-period Cepheids, or as cluster variables because they are often found in globular clusters. The RR Lyraes are old (Population II) type *A* or *F* giants whose light varies by about a magnitude in a very regular period ranging from 0.2 to 1.2 days. The "ab" subclass (formerly divided into RRa and RRb) encompasses those with highly asymmetric, Cepheidlike light curves, while the designation "c" refers to those with nearly symmetric, almost sinusoidal curves of somewhat smaller amplitude. The mean periods of the two groups are approximately 0.6 and 0.3 day, respectively. Dwarf Cepheids (type RRs) were formerly a distinct category, but are now considered to be part of the δ Scuti class.

RS. The RS Canum Venaticorum stars are binary systems containing a rapidly rotating type *F* to *K* subgiant or giant. They have periods ranging from a fraction of a day to several months. Their variability is due to starspot activity and, in some instances, eclipses; the latter are indicated by the codes EA/RS, EB/RS, and so on. Many type-RS variables are radio and X-ray emitters because of chromospheric and coronal disturbances.

RV, RVa, RVb. RV Tauri stars are pulsating supergiants of type *F* to *K*; their light curves exhibit alternate deep and shallow minima (sometimes causing them to be mistaken for eclipsing binaries) with amplitudes up to 3 or 4 magnitudes. The time between successive primary minima often varies for a given star but is usually in the range of 30 to 150 days. The suffix "a" or "b" is added to indicate constant or periodically changing mean brightness, respectively.

S Dor. Among the most luminous blue stars known, the S Doradus variables are very young and massive. They have absolute visual magnitudes as bright as -10, with variations of 1 to 3 magnitudes that can be irregular or cyclic. Their spectra fall in the range of type *B* to *F* and often display P Cygni line profiles; these are broad emission lines flanked at shorter wavelengths by sharp absorption features, indicating high-speed ejection of the star's atmosphere. The prototype, S Doradus, is located in the open cluster NGC 1910, which is part of the Large Magellanic Cloud.

SN, SNI, SNII. Supernovae are exploding stars that brighten abruptly by up to 20 magnitudes and then fade over many months. Type I supernovae, which can achieve an absolute visual magnitude of -20, are found among the old, low-mass Population II stars. Type II supernovae are generally less luminous at maximum. They are observed in the spiral arms of galaxies and are therefore thought to originate among the relatively young, massive Population I stars.

SR, SRa, SRb, SRc, SRd. Semiregular variables. This is a diverse group of giants and supergiants whose pulsations are generally periodic (in the range of 30 to more than 1,000 days) but have occasional irregularities. The amplitudes are generally much less than in the Miras, and the light curves have a variety of forms. Semiregular giants of late spectral type (*M, C,* or *S*) that have relatively stable cycles are flagged with the suffix "a," while "b" refers to those with poorly expressed periodicity. The abbreviation SRc designates young semiregular supergiants of late spectral type. The letter "d" is added to indicate semiregular stars of spectral type *F, G,* or *K.*

SX Ari. Sometimes called helium variables, the SX Arietis stars are the high-temperature analogue of the α^2 Canum Venaticorum class. Their spectral types fall in the range *B*0 to *B*7, with anomalously intense lines of helium, silicon and other elements. The SX Arietis stars have strong magnetic fields which vary with their brightness and spectra.

SX Phe. SX Phoenicis stars (see the δSct type).

UG. U Geminorum stars, also called dwarf novae or SS Cygni variables, are close binaries that consist of a type *K* to *M* dwarf or subgiant, which has filled its Roche lobe, plus a white dwarf surrounded by an accretion disk of infalling matter. They are usually quiescent but from time to time undergo outbursts of 2 to 6 magnitudes lasting several days. The interval between eruptions is 10 days to several years. The SU Ursae Majoris subclass displays both normal maxima and "supermaxima," which are about 2 magnitudes brighter and last five times as long. The Z Camelopardalis variables described below are also related to U Geminorum stars.

unq. Unique variable stars that do not conform to any recognized category.

UV, UVn. UV Ceti variables, or flare stars, are main-sequence, type *K*e or *M*e dwarfs that exhibit abrupt flares of 1 to 6 magnitudes lasting several minutes. They were discovered by W. J. Luyten in 1948. Flash variables, which display similar behavior but are associated with nebulae, are designated UVn. They are characterized by slightly earlier spectral types and somewhat greater luminosity than the UV Ceti stars.

X. X-ray sources.

Z And. Z Andromedae variables, or symbiotic stars, are close binaries whose composite spectra indicate the pairing of a cool giant with a very hot companion. The irregular variability at visual and ultraviolet wavelengths appears to be caused by a combination of the cool star's pulsations plus some interaction (probably involving mass exchange) between the two components. The stars are often embedded in nebulosity. The RR Telescopii subcategory represents systems whose variability is halted by a 4- to 6-magnitude brightening with no observed return to the original luminosity. Although their high-excitation emission spectra suggest that these may be planetary nebulae in formation, most planetaries have single stars at their centers.

Z Cam. Z Camelopardalis variables are binary systems similar to U Geminorum stars. They undergo outbursts of 2 to 5 magnitudes every 10 to 40 days. However, this behavior is sometimes interrupted by quiescent periods when the star remains between maximum and minimum.

ZZ. ZZ Ceti variables are nonradially pulsating white dwarfs with periods as short as 30 seconds, or as long as about 25 minutes, and amplitudes generally less than 0.2 magnitude. One-magnitude flares, perhaps due to interactions with a close companion, are occasionally seen.

αCV. Also known as spectrum or magnetic variables, α^2 Canum Venaticorum stars have strong magnetic fields that vary periodically along with their brightness and spectra. Most are of spectral type *A*p, with anomalously intense lines of silicon, strontium, chromium, and rare-earth elements. Periods range from about 0.5 to 160 days. Amplitudes are normally less than 0.1 magnitude.

αCyg. The α Cygni (Deneb) class comprises nonradially pulsating supergiants of spectral type *B* to *A.* Their typical 0.1-magnitude variations appear irregular but may be due to the superposition of many oscillations with nearly the same period. There is speculation that the α Cygni stars may be older S Doradus variables.

βC, βCs. The β Cephei stars, sometimes called β Canis Majoris variables, are short-period (mostly between 0.1 and 0.3 day) giants or subgiants of spectral type *B*0 to *B*3 with amplitudes of 0.01 to 0.3 magnitude. They are thought to be relatively massive stars that have nearly exhausted the hydrogen fuel in their cores. While many are both radial and nonradial pulsators, some (such as 53 Persei) exhibit only nonradial oscillations. There is an additional group of ultra-short-period (several hundredths of a day) β Cephei stars designated by the suffix "s"; χ Centauri is an example.

γC. With γ Cassiopeiae as the prototype of the class, these rapidly rotating, emission-line *B* stars undergo equatorial mass loss as they evolve off the main sequence. They sometimes vary irregularly with amplitudes up to 1.5 magnitudes.

δSct. The δ Scuti variables are stars of spectral type *A*2 to *F*5 with periods in the range 0.02 to 0.4 day. Most have amplitudes of several thousandths or hundredths of a magnitude, but a small percentage vary by as much as 0.8 magnitude. (Those with amplitudes greater than 0.3 magnitude are sometimes called dwarf Cepheids, type-RRs variables, or AI Velorum stars.) Their light curves are highly variable. Multiple radial and nonradial pulsations are often present. Most δ Scuti variables are Population I stars either on or evolving from the main sequence. However, several have been found that are deficient in heavy elements, a characteristic of older, Population II objects. The subdwarf examples of this old component are referred to as SX Phoenicis variables.

χCen. Variables similar to χ Centauri (see the βC type).

OPEN AND GLOBULAR CLUSTERS

Astronomers have used star clusters and their relatives the OB associations to trace the size and structure of the Milky Way, study stellar evolution and dynamics, reveal the presence of interstellar dust, and even measure the distance to other galaxies. Clusters fall into two categories — open (formerly called galactic) and globular.

Open Clusters. These irregular, gravitationally bound congregations of stars, usually a few dozen to several hundred in number, lie in the disk or spiral arms of the Milky Way. Most are less than 10 parsecs (32.6 light-years) in diameter. Although open clusters range in age from about a million to somewhat more than five billion years, the majority are relative newcomers on the galactic scene. In fact, some of the youngest are still accompanied by the diffuse nebulae in which they formed. Notable examples are the Pleiades (M45) in Taurus and the Lagoon nebula (M8) in Sagittarius. Almost 1,200 open clusters have been discovered in our galaxy. Most are relatively nearby since cosmic dust in the galactic plane obscures those farther away. There may be as many as 100,000 more open clusters hidden by this interstellar veil.

The main information source for the Open Clusters section is Gösta Lynga's *Catalogue of Open Cluster Data*, which was compiled in 1981 and updated in 1983. All objects with total magnitude of 11.5 or brighter, with a diameter greater than 10 minutes of arc, or whose most prominent star has a photographic magnitude brighter than 12, are included here. Many of the sparse clusters are probably not true systems, but what we've called *asterisms* (for want of a better word): stars that appear clumped in the sky but that are really at quite different individual distances from our solar system. Several clusters in the Large and Small Magel-

TABLE II. OB ASSOCIATIONS

Name	α (2000) δ h m ° '	Diam. '	Dist. (pc)	O Stars	B Stars	Earl. Spec.	RV (km/sec)	Clusters	Stars
Cas OB4	0 28.4 +62 42		2,880	5	12	O	−43	103	
Cas OB14	0 28.8 +63 22		1,110	0	3	B0.5	−8		χ Cas
Cas OB1	1 00.8 +61 30	120	2,510	0	5	B0.5	−38	381?	
Cas OB8	1 46.2 +61 19		2,880	1	10		−30	581, 663; 654?	
Per OB1	2 14.5 +57 19	360	2,290	9	56	O5	−41	h, χ Per	
Cas OB6	2 43.2 +61 23	480	2,190	17	8	O4	−47	IC 1805	
Cam OB1*	3 31.6 +58 38		1,000	3	9	O5	−6	1444? 1502?	
Per OB3	3 27.8 +49 54		170						α, δ Per
Per OB2*	3 42.2 +33 26	480 x 300	400	1	3	O7.5			ζ, o, χ Per
Aur OB2	5 28.3 +34 54		3,160	5	3	O4	−13	1893, IC 410	
Aur OB1	5 21.7 +33 52	360 x 300	1,320	5	5	O7	−3	1912,60; 1931?	
Gem OB1	6 09.8 +21 35	300	1,510	4	13	O6.5	+13	2175?	χ² Ori
Ori OB1*	5 31.4 −2 41	960	460	9	6	O7		Trapezium	θ, β, γ, δ, ε Ori
Mon OB1	6 33.1 +8 50	840 x 300	550	1	0	O7	+22	2264	S Mon
Mon OB2	6 37.2 +4 50	360 x 250	1,510	10	7	O4	+28	2244	Plaskett's star
CMa OB1	7 07.0 −10 28	240	1,320	4	3	O6.5	+27	2335,53; 2343?	
Pup OB1	7 54.8 −27 05	240 x 180	2,510	7	0	O5	+43	2467?	
Vel OB2	8 11.8 −47 50		460						Vela pulsar?
Vel OB1	8 49.9 −45 00	360 x 240	1,400	5	11	O5.5		2659?	
Car OB1	10 46.7 −59 05	120 x 66	2,510	6	15	O5		3293; IC 2581?	
Car OB2*	11 06.0 −59 51	330 x 150	2,000	8	6	O4		3572, Tr 18	
Car OB4	11 08.3 −60 31	114 x 54						3590	
Cen OB2	11 35.3 −62 36	84 x 48	2,500					IC 2944	λ Cen
Cen OB1	13 04.8 −62 04	360	2,510	2	19	O9		4755	χ Cru
Nor OB1	15 58.7 −54 30		2,500	0	6	B0		6031?	
Sco-Cen*	16: −25:		160			O		IC 2602?	α CMa, α Car, α Eri
Ara OB1	16 39.5 −46 46	270 x 180	1,380			O5		6169,93	μ Nor
Sco OB1	16 53.5 −41 57	96 x 66	1,910	18	10	O6	−18	6231	ζ¹ Sco
Sco OB2*	16 14.9 −25 55		160	0	3	B0			α, β¹, δ Sco
Sgr OB1	18 07.9 −21 28	570 x 240	1,580	8	9	O4	−4	6514,30-1	μ Sgr
Sgr OB4	18 14.4 −19 03		2,400	1	6	O7	+3	6603	
Ser OB1	18 20.8 −14 35	300 x 180	2,190	9	9	O4	−23	6611	
Ser OB2	18 18.6 −11 58	500:	2,000:	9	6	O5	+6	6604?	
Vul OB1	19 44.0 +24 13		2,000	5	7	O6	+8	6823	
Cyg OB3	20 04.7 +35 50		2,290	9	15	O6.5	0	6871?	Cyg X-1
Cyg OB1	20 17.8 +37 38	420 x 240	1,820	12	28	O5	−7	6913, IC 4996	
Cyg OB9	20 23.3 +39 56		1,200	7	7	O6.5	−20	6910	
Cyg OB2*	20 32.4 +41 17	30	1,820	13	2	O3			
Cyg OB4	21 13.1 +37 52		1,000						σ Cyg
Cyg OB7	21 02.7 +49 43		830	3	6	O6	−10		α Cyg
Lac OB1	22 41.2 +39 05	900 x 540	600	1	0	O9			10 Lac
Cep OB2	21 47.9 +61 04	460	830	8	9	O6	−20	7160, IC 1396	μ, ν, λ Cep
Cep OB1	22 24.6 +55 14	210	3,470	7	26	O5	−51	7380?	β Cep
Cep OB3	23 00.4 +64 03		870	3	3	O7	−21		
Cas OB5	23 58.7 +60 22	150	2,510	5	10	O6	−46	7788; 7790?	ϱ Cas
Cep OB4	23 59.5 +67 35		840	2	0	O7			

Supplementary notes:

Cam OB1. Source of the runaway star α Cam.

Car OB2. Also contains the clusters Hogg 10 and 11.

Cyg OB2. The 12th member star in R. M. Humphrey's 1978 list is one of the visually most luminous stars in the galaxy with an absolute magnitude of −9.9. Its apparent magnitude is only 11.5 because of interstellar absorption, without which the star would appear in our skies as bright as Regulus.

Ori OB1. Also contains the clusters NGC 1981 and Collinder 70, the stars ζ, η, ι, χ, λ, σ, and υ Ori, and many other bright stars. Source of the runaway stars AE Aur, μ Col, and 53 Ari.

Per OB2. Source of the runaway star ξ Per.

Sco OB2. Part of the Sco-Cen association. Source of the runaway star ζ Oph.

Sco-Cen. The nearest OB association, extending from Scorpius to Crux (galactic longitude 292° to 2°, galactic latitude −10° to +30°). It includes α and β Cru, β and δ Cen, and many other bright stars.

TABLE III. THE MESSIER CATALOGUE

M	NGC	h	m	°	′	Const.	Dim.	Mag.	Type
1	1952	5	34.5	+22	01	Tau	6 x 4	8.4:	Di
2	7089	21	33.5	−0	49	Aqr	13	6.5	Gb
3	5272	13	42.2	+28	23	CVn	16	6.4	Gb
4	6121	16	23.6	−26	32	Sco	26	5.9	Gb
5	5904	15	18.6	+2	05	Ser	17	5.8	Gb
6	6405	17	40.1	−32	13	Sco	15	4.2	OC
7	6475	17	53.9	−34	49	Sco	80	3.3	OC
8	6523	18	03.8	−24	23	Sgr	90 x 40	5.8:	Di
9	6333	17	19.2	−18	31	Oph	9	7.9:	Gb
10	6254	16	57.1	−4	06	Oph	15	6.6	Gb
11	6705	18	51.1	−6	16	Sct	14	5.8	OC
12	6218	16	47.2	−1	57	Oph	14	6.6	Gb
13	6205	16	41.7	+36	28	Her	17	5.9	Gb
14	6402	17	37.6	−3	15	Oph	12	7.6	Gb
15	7078	21	30.0	+12	10	Peg	12	6.4	Gb
16	6611	18	18.8	−13	47	Ser	7	6.0	OC
17	6618	18	20.8	−16	11	Sgr	46 x 37	7	Di
18	6613	18	19.9	−17	08	Sgr	9	6.9	OC
19	6273	17	02.6	−26	16	Oph	14	7.2	Gb
20	6514	18	02.6	−23	02	Sgr	29 x 27	8.5:	Di
21	6531	18	04.6	−22	30	Sgr	13	5.9	OC
22	6656	18	36.4	−23	54	Sgr	24	5.1	Gb
23	6494	17	56.8	−19	01	Sgr	27	5.5	OC
24		18	16.9	−18	29	Sgr	90	4.5:	
25	IC 4725	18	31.6	−19	15	Sgr	32	4.6	OC
26	6694	18	45.2	−9	24	Sct	15	8.0	OC
27	6853	19	59.6	+22	43	Vul	8 x 4	8.1:	Pl
28	6626	18	24.5	−24	52	Sgr	11	6.9:	Gb
29	6913	20	23.9	+38	32	Cyg	7	6.6	OC
30	7099	21	40.4	−23	11	Cap	11	7.5	Gb
31	224	0	42.7	+41	16	And	178 x 63	3.4	S
32	221	0	42.7	+40	52	And	8 x 6	8.2	E
33	598	1	33.9	+30	39	Tri	62 x 39	5.7	S
34	1039	2	42.0	+42	47	Per	35	5.2	OC
35	2168	6	08.9	+24	20	Gem	28	5.1	OC
36	1960	5	36.1	+34	08	Aur	12	6.0	OC
37	2099	5	52.4	+32	33	Aur	24	5.6	OC
38	1912	5	28.7	+35	50	Aur	21	6.4	OC
39	7092	21	32.2	+48	26	Cyg	32	4.6	OC
40		12	22.4	+58	05	UMa		8	
41	2287	6	47.0	−20	44	CMa	38	4.5	OC
42	1976	5	35.4	−5	27	Ori	66 x 60	4	Di
43	1982	5	35.6	−5	16	Ori	20 x 15	9	Di
44	2632	8	40.1	+19	59	Cnc	95	3.1	OC
45		3	47.0	+24	07	Tau	110	1.2	OC
46	2437	7	41.8	−14	49	Pup	27	6.1	OC
47	2422	7	36.6	−14	30	Pup	30	4.4	OC
48	2548	8	13.8	−5	48	Hya	54	5.8	OC
49	4472	12	29.8	+8	00	Vir	9 x 7	8.4	E
50	2323	7	03.2	−8	20	Mon	16	5.9	OC
51	5194-5	13	29.9	+47	12	CVn	11 x 8	8.1	S
52	7654	23	24.2	+61	35	Cas	13	6.9	OC
53	5024	13	12.9	+18	10	Com	13	7.7	Gb
54	6715	18	55.1	−30	29	Sgr	9	7.7	Gb
55	6809	19	40.0	−30	58	Sgr	19	7.0	Gb
56	6779	19	16.6	+30	11	Lyr	7	8.2	Gb
57	6720	18	53.6	+33	02	Lyr	1	9.0:	Pl
58	4579	12	37.7	+11	49	Vir	5 x 4	9.8	S
59	4621	12	42.0	+11	39	Vir	5 x 3	9.8	E
60	4649	12	43.7	+11	33	Vir	7 x 6	8.8	E
61	4303	12	21.9	+4	28	Vir	6 x 5	9.7	S
62	6266	17	01.2	−30	07	Oph	14	6.6	Gb
63	5055	13	15.8	+42	02	CVn	12 x 8	8.6	S
64	4826	12	56.7	+21	41	Com	9 x 5	8.5	S
65	3623	11	18.9	+13	05	Leo	10 x 3	9.3	S
66	3627	11	20.2	+12	59	Leo	9 x 4	9.0	S
67	2682	8	50.4	+11	49	Cnc	30	6.9	OC
68	4590	12	39.5	−26	45	Hya	12	8.2	Gb
69	6637	18	31.4	−32	21	Sgr	7	7.7	Gb
70	6681	18	43.2	−32	18	Sgr	8	8.1	Gb
71	6838	19	53.8	+18	47	Sge	7	8.3	Gb
72	6981	20	53.5	−12	32	Aqr	6	9.4	Gb
73	6994	20	58.9	−12	38	Aqr			
74	628	1	36.7	+15	47	Psc	10 x 9	9.2	S
75	6864	20	06.1	−21	55	Sgr	6	8.6	Gb
76	650-1	1	42.4	+51	34	Per	2 x 1	11.5:	Pl
77	1068	2	42.7	−0	01	Cet	7 x 6	8.8	S
78	2068	5	46.7	+0	03	Ori	8 x 6	8	Di
79	1904	5	24.5	−24	33	Lep	9	8.0	Gb
80	6093	16	17.0	−22	59	Sco	9	7.2	Gb
81	3031	9	55.6	+69	04	UMa	26 x 14	6.8	S
82	3034	9	55.8	+69	41	UMa	11 x 5	8.4	Ir
83	5236	13	37.0	−29	52	Hya	11 x 10	7.6:	S
84	4374	12	25.1	+12	53	Vir	5 x 4	9.3	E
85	4382	12	25.4	+18	11	Com	7 x 5	9.2	E
86	4406	12	26.2	+12	57	Vir	7 x 6	9.2	E
87	4486	12	30.8	+12	24	Vir	7	8.6	E
88	4501	12	32.0	+14	25	Com	7 x 4	9.5	S
89	4552	12	35.7	+12	33	Vir	4	9.8	E
90	4569	12	36.8	+13	10	Vir	10 x 5	9.5	S
91	4548	12	35.4	+14	30	Com	5 x 4	10.2	S
92	6341	17	17.1	+43	08	Her	11	6.5	Gb
93	2447	7	44.6	−23	52	Pup	22	6.2:	OC
94	4736	12	50.9	+41	07	CVn	11 x 9	8.1	S
95	3351	10	44.0	+11	42	Leo	7 x 5	9.7	S
96	3368	10	46.8	+11	49	Leo	7 x 5	9.2	S
97	3587	11	14.8	+55	01	UMa	3	11.2:	Pl
98	4192	12	13.8	+14	54	Com	10 x 3	10.1	S
99	4254	12	18.8	+14	25	Com	5	9.8	S
100	4321	12	22.9	+15	49	Com	7 x 6	9.4	S
101	5457	14	03.2	+54	21	UMa	27 x 26	7.7	S
102									
103	581	1	33.2	+60	42	Cas	6	7.4:	OC
104	4594	12	40.0	−11	37	Vir	9 x 4	8.3	S
105	3379	10	47.8	+12	35	Leo	4 x 4	9.3	E
106	4258	12	19.0	+47	18	CVn	18 x 8	8.3	S
107	6171	16	32.5	−13	03	Oph	10	8.1	Gb
108	3556	11	11.5	+55	40	UMa	8 x 2	10.0	S
109	3992	11	57.6	+53	23	UMa	8 x 5	9.8	S
110	205	0	40.4	+41	41	And	17 x 10	8.0	E?

Column headings: Messier number (*M*), number in J. L. E. Dreyer's *New General Catalogue* (NGC) or *Index Catalogue* (IC), right ascension (α) and declination (δ) for equinox 2000.0, constellation, dimensions in minutes of arc, magnitude, and type. Most magnitudes are photoelectric **V** (visual) values rounded to tenths. When these were not available, we have taken approximate visual magnitudes from M. V. Zombeck's 1982 compilation and other sources; such values are followed by a colon or rounded to the nearest whole magnitude. Types: diffuse nebula (Di), globular cluster (Gb), open cluster (OC), planetary nebula (Pl), or galaxy (E for elliptical, Ir for irregular, S for spiral).

Common names and notes:

M1. Crab nebula. Remnant of the supernova of A.D. 1054. Often incorrectly called a planetary nebula.

M8. Lagoon nebula.

M11. Described by W. H. Smyth in 1844 as resembling a "flight of wild ducks."

M13. Hercules cluster.

M17. Omega nebula.

M20. Trifid nebula.

M24. Messier described the brightening of the Milky Way here; he did not mention the small open cluster NGC 6603.

M27. Dumbbell nebula.

M31. Andromeda galaxy.

M40. The wide double star Winnecke 4; magnitudes 9.0 and 9.6, separation 50″.

M42. Great Nebula in Orion.

M44. Praesepe, or Beehive cluster.

M45. Pleiades, or Seven Sisters.

M47. Messier's original position was erroneous. It is now generally agreed that Messier must have seen the bright cluster NGC 2422.

M48. Again, no cluster fitting Messier's description lies at the position he gave. In *Sky and Telescope*, **20**, October, 1960, page 196, Owen Gingerich concluded that NGC 2548 is the cluster most likely observed.

M51. Whirlpool galaxy.

M57. Ring nebula in Lyra.

M64. Black-eye galaxy.

M73. An asterism consisting of only four stars.

M76. Little Dumbbell nebula.

M91. Another of the "missing" Messier objects. W. C. Williams explains M91 as NGC 4548, on the assumption that Messier made certain errors in recording the position (see *Sky and Telescope*, **38**, December, 1969, page 376).

M97. Owl nebula.

M102. An accidental reobservation of M101, according to Messier's colleague P. Méchain. Many sources incorrectly equate M102 and the galaxy NGC 5866.

M104. Sombrero galaxy.

lanic Clouds have been added from the 1968 list of S. van den Bergh and G. Hagen along with a few other interesting clusters that for one reason or another would have been excluded by a strict cutoff procedure, bringing the total number to 750.

Brian Skiff's 1983 work on cluster brightnesses has been incorporated into our tabulation. He has computed total, or integrated, **V** magnitudes for most of the open clusters, using previously published **UBV** and RGU photometry. (The RGU photographic system is described in W. Becker's 1963 paper and the references therein.) Total magnitude is calculated by adding together the brightnesses of all the measured member stars. However it can misrepresent the cluster's telescopic visibility, which also depends on the size, concentration, and range in brightness of the stars. If the magnitude of the brightest star is nearly equal to the total magnitude of all the members, an observer may not even see a cluster. The fainter stars might be invisible or merely contribute to a faint background glow. Only for very compact systems will the total magnitude convey an accurate idea of observability.

Cluster stars move through space together. Therefore, if the group as a whole is receding (or approaching) along the observer's line of sight, over the years the members will seem to converge toward (or radiate from) the center. If a component of the cluster's motion is in the plane of the sky, the convergent point (or radiant for approaching systems) will be located somewhere away from the center. Only the nearest clusters have measurable proper motions that can be used to find convergent points. They are called moving clusters, of which the Hyades group is the best known example. Studies of moving clusters are important because their distances can be calculated geometrically beyond the range to which trigonometric parallaxes are reliable.

The most luminous young stars, those of spectral types *O* and *B*, are almost always situated in loose groupings called OB associations that are scattered among the galaxy's spiral arms. These systems typically contain between 10 and 100 members. They are larger than open clusters, spanning anywhere from 30 to more than 200 parsecs (about 100 to over 650 light-years). Because of their sparseness and consequent weak internal gravitation, associations tend to disperse over time. Table II lists some of the more prominent OB associations, along with position, apparent dimensions, distance in parsecs, number of *O* and *B* stars with absolute visual magnitudes brighter than −6 according to the 1978 work of R. M. Humphreys, the earliest (hottest) observed spectral type, and the average radial velocity of a sample of member stars. Open clusters and bright or interesting stars that belong to the association are also given. Most of the clusters are identified by their number in J. L. E. Dreyer's *New General Catalogue of Nebulae and Clusters of Stars* (NGC); otherwise, the *Index Catalogue* (IC) or another designation is listed. An asterisk after the association name means that there is a supplementary note for that object at the end of the list. The actual size of each system in parsecs can be computed by dividing the angular dimensions by 3,440 and multiplying the result by the distance.

Globular Clusters. These spherical swarms of up to 100,000 stars are found both in the disk and halo of the Milky Way. They are among the oldest objects in the galaxy, having ages of 10 billion years or more. Because globulars formed early in galactic history, the interstellar gas from which they coalesced had not yet been significantly en-

Fig. 9. The well-known open cluster M45, the Pleiades, in a Harvard Observatory photograph.

riched with heavy elements thrown into space by generations of supernovae. Therefore, globular cluster stars are generally poor in heavy elements compared to their counterparts, the open clusters. They are also larger than the latter, ranging in size from about 10 to more than 100 parsecs (about 30 to more than 300 light-years).

The 150 entries in our Globular Clusters section include the 138 galactic systems that were known as of 1980, along with 12 others in the Large and Small Magellanic Clouds. Many are splendid telescopic sights because of their brightness and high contrast with the surrounding star fields. Fully 29 of them appear in the famous catalogue of deep-sky objects compiled by Charles Messier in the second half of the 18th century (Table III). To sweep up one of these

Fig. 10. The sparse system NGC 4372 in the southern constellation Musca resembles an open cluster. However, its lack of bright, blue stars and its low heavy-element abundance show that it is actually a globular cluster passing through the galaxy's disk. The bright star, HD 107947, is of visual magnitude 6.8.

Fig. 11. This 202-minute exposure of the globular cluster M13 in Hercules was made with the 61-inch reflector at Flagstaff, Arizona. Official U. S. Navy photograph.

round glows must have brought sudden excitement to the Parisian comet hunter. With his small telescopes he could not resolve the individual stars of any globular except M4 in Scorpius, which he described as an ''amas d'étoiles trés-petites'' (cluster of very small stars).

Dozens of the brighter globulars are partially resolved with the 6-inch and 8-inch telescopes of many amateurs today. Stars are detected most readily at the fringes of a cluster and often extend across its face, but there usually remains a bright unresolved core. Sometimes quite a large aperture is required to glimpse individual members, even though the glow itself is quite easy. Writing in 1975 about NGC 2419 in Lynx, R. Racine and W. E. Harris noted, ''Visual inspection under good seeing at the prime focus of the 200-inch telescope shows the brightest cluster stars barely resolved'' (*Astrophysical Journal,* **196,** 413, 1975).

The remote outlying globulars possessing only Palomar designations are too dispersed and faint to be seen visually at all. Most of the systems compiled by A. Terzan in 1971 are detectable only in the infrared because their visible light is so highly attenuated by interstellar dust. Probably fewer than 100 globulars in our galaxy remain undiscovered, either blocked from view by intervening dust or too sparse and faint to have been noticed.

The primary references for our Globular Clusters section are as follows: positions from S. J. Shawl and R. E. White (1984); apparent diameters, spectral types, and concentration classes from G. Alcaino (1977); apparent and absolute magnitudes, observed and intrinsic color indexes, and distances from W. E. Harris and R. Racine (1979); and radial velocities and metallicity indexes from R. F. Webbink (1981), S. J. Shawl *et al.* (1981), and R. Zinn and M. J. West (1984). The published radial velocities and metallicity indexes were noted to be discordant for several clusters. Recently discovered objects and updated information were added from other sources cited in the References section. Because its nature is still uncertain, the peculiar globularlike cluster E3 at $\alpha = 9^h\ 21^m.0,\ \delta = -77°\ 17'$ (equinox 2000.0) has not been included in the data base. Further information about it can be found in the 1984 paper by J. E. Hesser *et al.* and references therein.

Globular clusters are important in many branches of astronomical research. As early as 1914, Harlow Shapley studied their distribution in order to locate the galactic center geometrically. More recently, they have been used as proving grounds for theories of stellar evolution since each is a homogeneous sample of old stars. Variable star researchers prize them for their rich harvest of RR Lyraes (M3 has about 200!) and Population II Cepheids. In addition, distant globulars provide clues to past and present conditions in the Milky Way's halo. And because of their high luminosity, clusters are visible around other galaxies. Over 300 have been detected near the Andromeda galaxy M31, while the giant elliptical system M87 in Virgo is surrounded by a swarm of at least 6,000. Because of their fairly uniform absolute magnitude, rich globulars are used as ''standard candles'' by which to estimate the distance to nearby galaxies.

The space density of stars in a globular cluster is clearly far greater than that near the solar system. In fact, the bright central regions of many globulars often appear completely filled in on photographs. This is merely the result of blurring by the Earth's atmosphere, the limited resolution of the telescope, and light scattering in the photographic emulsion. There is actually a great deal of space between the stars. However, the illusion has still led to speculation about the appearance of the night sky to an inhabitant near a globular's center. Anatole Boyko evaluated some of these notions in *Sky and Telescope,* **28,** November, 1964, page 269. George Abell arrived at a similar picture in his book *Exploration of the Universe:* '' . . . the nearest neighboring stars, light-months away, would appear as points of light. Thousands of stars, however, would be scattered uniformly over the sky. The Milky Way would be hard, if not impossible, to see, and even on the darkest of nights the brightness of the sky would be comparable to what it is on earth in bright moonlight.''

NEBULAE

To the amateur astronomer, nebulae are challenging subjects for photography or for testing the limits of visual acuity. To the professional, they are low-density interstellar laboratories, spiral-arm tracers, or stellar birthplaces. However, both amateur and professional marvel at their ghostly beauty. Nebulae can be bright or dark, depending on whether they shine under the influence of a nearby star or are seen in silhouette against a glowing background. The term ''planetary nebula'' has evolved from one merely descriptive of compact, roughly circular clouds to an astrophysical definition related to a late stage in a star's life. The categories outlined below represent just the visible manifestations of the interstellar gas and dust which permeate our galaxy.

Bright Nebulae. M42 in Orion and M8 in Sagittarius are well-known examples of emission nebulae. They are so described because of the bright lines in their spectra, first seen by the British astronomer William Huggins in 1864. Intense ultraviolet radiation from embedded *O* and *B* stars keeps the tenuous gas highly ionized; that is, constituent atoms have lost at least some of their electrons. Every time one of the free electrons recombines with an atom, a photon is generated that contributes to the intensity of a bright spectral line.

A considerable fraction of nebular light is emitted in the hydrogen line known as Hα, one of the Balmer series, at a wavelength of 6563 angstroms. Since this wavelength falls

Fig. 12. A section of NGC 6960, the Filamentary nebula, which forms the western part of the Veil supernova remnant. The bright star is 4th-magnitude 52 Cygni; south is left. Lick Observatory photograph, made in 1960 with the newly completed 120-inch reflector.

in the red portion of the visible spectrum, the characteristic reddish glow of emission nebulae on color photographs is not surprising. For the same reason, such objects are much more striking on red- rather than blue-sensitive plates. Many of their details could not be recorded at all in the early years of astrophotography because the early emulsions did not respond to wavelengths much longer than 5000 angstroms. Commercial "nebula filters" are effective visual aids because they can be made to transmit light in a narrow wavelength band centered on the Hα line.

Spectroscopic studies have shown the bright nebulae to be incredibly tenuous, containing only 1,000 or so particles in every cubic centimeter. Temperatures can exceed 10,000° Kelvin. Astronomers often refer to emission nebulae as H II regions, a notation that stands for the ionized hydrogen that is so abundant in these clouds. An H II region encompasses the entire realm of gas ionized by a hot star or cluster. As a result, it may contain several emission clouds that have historically been recognized as individual entities. The complex of nebulosity surrounding γ Cygni is an example.

Bengt Strömgren first outlined the theory for predicting the actual size of an H II region in 1939. He reasoned that the largest bright nebulae should be associated with the hottest stars because the latter radiate the most ultraviolet light. Strömgren concluded that an extreme O star would be capable of ionizing a cloud several hundred parsecs across, while one of type $B0$ might produce only a 60-parsec H II region. Already at $A0$ the predicted size is less than a parsec, dramatically illustrating that only the hottest stars cause visible bright nebulosities. Adopting angular diameters from our catalogue and distances from the astronomical literature, the actual sizes of the Orion and Omega nebulae are found to be about 7 and 25 parsecs, respectively, while the Lagoon,

Rosette, and North America nebulae are all nearly 40 parsecs in diameter.

Reflection nebulae are different in nature from their emission counterparts. In 1912 Vesto M. Slipher obtained a spectrum of the Pleiades nebulosity. The following year he noted in a *Popular Astronomy* article the curious fact that "... the whole spectrum is a true copy of that of the brighter stars in the Pleiades." In other words, he saw the same sort of continuous spectrum, crossed by absorption

Fig. 13. The diffuse nebula NGC 7635, known also as the Bubble, in a 1954 photograph made with the 200-inch Hale reflector on Palomar Mountain.

Fig. 14. *IC 405, the Flaming Star nebula discovered by J. M. Schaeberle in 1892, is illuminated by the variable AE Aurigae, a 6th-magnitude O star. This photograph was taken in 1922 with the 100-inch telescope on Mount Wilson.*

lines, that is characteristic of stars. Here the light of the cluster is being reflected by adjacent clouds of dust particles.

Since these minute grains are more effective scatterers at short wavelengths, a reflection nebula always appears bluer than its illuminating stars, which are most often blue themselves. Red reflection nebulae are rare; the outer part of IC 4592 has this tint due to the glow of Antares. Such cases also occur for very distant reflection nebulae because of interstellar reddening, a process described more fully on the facing page.

Our sample of nebulae includes the well-known clouds easily visible to the naked eye or with slight optical aid, along with many fainter objects that will challenge the seasoned observer. The presence of an NGC number and brightness code (such as eF, vB, and so on) under "Description" means that the nebulosity has been detected visually. There are other entries in our list that can be recorded only on photographs. They are included here if noteworthy in some respect other than brightness; supernova remnants and nebulosities associated with variable stars are examples.

The data for each object were synthesized from a variety of sources, including the NGC, IC, and S. Cederblad's informative compilation from 1946. Some of the major nebula catalogues, such as those by S. Sharpless (1959), B. Lynds (1965), and S. van den Bergh (1966), are based on the National Geographic Society-Palomar Observatory Sky Survey and consequently extend in declination only down to −33°. We have therefore obtained additional objects in the southern sky from Colin Gum's 1955 study, that of A. W. Rodgers, C. T. Campbell, and J. B. Whiteoak from 1960, and the 1975 work by van den Bergh and Herbst on southern reflection nebulosities. Several other papers cited in the References section were consulted for a sampling of the brightest nebulae in the Magellanic Clouds.

The coordinates given in our tabulation are generally those of the most prominent region of the nebula. Howev-

er, the location of the geometric center is sometimes listed for objects whose light emanates from individual filaments in an extended or annular configuration. A nebula's angular extent is quite understandably a poorly defined measure. The apparent dimensions of a diffuse object depend on whether it is viewed optically or photographically, and if the latter the exposure time and color sensitivity of the film are critical. The values given here are generally those quoted in one of the references and are measured on red-sensitive photographic plates. As such, they will almost always be larger than what is seen visually. Since the dimensions are from different sources, they do not form a homogeneous set; treat them as a guide and not as the "final word."

Dark Nebulae. A glance at the Milky Way on a clear, moonless evening in July or August reveals an obvious "dark rift" interrupting the diffuse glow of the multitude of stars. The breach extends from Cygnus to Sagittarius and beyond. Smaller areas of sky with few or no stars were first noted by William Herschel in the 18th century. Some astronomers argued that the dark regions are, in the skeptical words of H. D. Curtis, " . . . 'holes' torn in the star fabric of the Milky Way by some rapidly rushing star cluster . . . ." His photographs of dark lanes in neighboring galaxies spoke to the contrary. Those of E. E. Barnard (see Fig. 15) and Max Wolf also implied the existence of intervening clouds of light-absorbing material. Subsequent work by Edwin Hubble and Henry Norris Russell led to the inevitable conclusion that the dark regions are in fact concentrations of interstellar dust.

Although quite tenuous, the dark nebulae are of sufficient extent to block completely our view of the luminous galactic center. Nearly all of those that we see, including the clouds that make up the dark rift in the Milky Way, are within 1,000 parsecs of the Sun. For instance, the well-known Coalsack, shown in Fig. 16, is only 170 parsecs away. This means that clouds severely hamper optical study of our galaxy's overall spiral-arm structure. They may also have rendered invisible a nearby supernova that occurred around

Fig. 15. *This 4½-hour exposure of Messier 24, a star cloud in Sagittarius, was taken in 1905 by E. E. Barnard. He aptly named the dark, oval nebula near the center of the photograph the "Black Hole," never suspecting that the same term would now have a vastly different meaning. Note the 12th-magnitude foreground star and its faint companion superimposed on the nebula, also known as Barnard 92.*

Fig. 16. This view of the southern Milky Way has the Coalsack dark nebula prominent in the center and, from left, the bright stars α and β Centauri and β and α Crucis. Immediately to the left of the Coalsack is the large, bright nebula Cederblad 122, while the η Carinae complex (NGC 3372) appears at the extreme right. The four stars of the Southern Cross, situated to the upper right of the Coalsack, have very unequal brightnesses here because a blue-sensitive emulsion was used for this early Harvard photograph.

1600 in the constellation Cassiopeia. We "see" it now as the powerful radio source Cassiopeia A.

Individual dust grains in the dark clouds are only several ten-thousandths of a millimeter across. They are probably similar in composition to sand or graphite and likely originate in the cool, tenuous atmospheres of red-giant stars. Astronomers use the term "extinction" to express the fact that the particles absorb as well as scatter light. Because their average size is slightly smaller than the wavelength of visible light, dust grains scatter the blue portion of starlight more effectively than the red. As a result, starlight passing through the interstellar medium is not only attenuated, but also reddened.

This phenomenon explains the paradox of red stars of spectral type *O*, which were first noted in the 1920's. The light of these blue objects is merely reddened in its long passage through interstellar space. The same effect also renders distant star clusters nearly invisible on blue-sensitive photographic plates, as dramatically illustrated in Bart and Priscilla Bok's book *The Milky Way* (pages 255 and 256 of the 5th edition).

In addition to isolated dark nebulae, which are typically several parsecs across, and the broad obscuring complexes that may span more than 100 parsecs, there are the compact *globules* that range in diameter from several hundredths of a parsec to about a parsec. The smallest are often seen as tiny, dark specks projected against emission nebulosities, such as the Lagoon and Rosette nebulae. They have high opacities, typically attenuating background starlight by at least five magnitudes. The importance of globules as possible sites of star formation was first pointed out by Bart Bok in the late 1940's.

Our Dark Nebulae section contains a sample of 150 objects, a small subset of the thousands that have been catalogued to date. Most were selected on the basis of high opacity, contrast with stars in the adjacent field, interesting shape, or association with a bright nebula. The primary sources are the compilation by Beverly Lynds from 1962 and E. E. Barnard's "Catalogue of 349 Dark Objects in the Sky," incorporated in his beautiful 1927 photographic atlas of the Milky Way. Instead of listing individual Barnard objects on separate lines, we often combine them based on their proximity in the sky. For example, the region in Taurus around $4^h\ 33^m$, $+26°.1$, has nearly two dozen clouds assigned numbers by Barnard. In our tabulation the dimensions, area, and description of the overall complex are given, along with a list of all of Barnard's designations. Similarly, each line may incorporate more than one object with a Lynds number; here, only the number of the most prominent cloud is recorded.

Additional southern-hemisphere nebulae were selected from articles by Aa. Sandqvist and K. P. Lindroos (1976) and Sandqvist (1977). The "Catalogue of Bright Nebulosities in Opaque Dust Clouds" by C. Bernes was consulted to determine physical associations between bright and dark nebulae. Several globules are also included; their identification as such was verified from the 1971 tabulations by Bart Bok and his collaborators.

We have adopted the opacity-class system of Lynds, where the "darkest" nebulae are assigned the number 6 and those barely noticeable on the National Geographic Society-Palomar Observatory Sky Survey are class 1. Although this gives an indication of the apparent darkness of a cloud, the actual observability depends on the contrast with surrounding stars and bright nebulae. That is, a class 3 object silhouetted against the Milky Way's glow will be easier to see than one of class 6 in front of a sparse gathering of stars. Since small dark clouds are particularly elusive, we often give their locations relative to a nearby star or bright nebula.

Visual observers should be aware that a moderate-aperture telescope is necessary to view most of the objects listed here. Some can only be observed photographically. Pictures of some of these dark nebulae have appeared in recent issues of *Sky and Telescope,* as follows: Barnard 86, 91, 303, and objects in M8 and M20 are in **63**, March, 1982, page 254; for the Horsehead nebula (Barnard 33), see **62**, November, 1981, page 416; Barnard 108 and 112 are shown in **62**, August, 1981, page 171; Barnard 64 is in **61**, June, 1981, page 557; Barnard 68 and 72 are pictured in **61**, April, 1981, page 284.

Planetary Nebulae. William Herschel invented the term "planetary nebula" to describe the class of faint, circular objects whose greenish glow reminded him of the disk of the planet Uranus. These systems, which have nothing to do with planets, consist of an extremely hot star surrounded by a tenuous shell of ionized gas. The greenish tint is due to light emitted in the spectral lines of doubly ionized oxygen. (Before their laboratory identification in 1927, the lines had been attributed to a hypothetical element "nebulium.")

Often, planetaries appear ringlike because the observer sees more glowing material at the periphery (where the line of sight passes obliquely through the shell) than at the center. Upon closer inspection, they display an incredible variety of forms. Many have faint, complex halos that become visible only in long time-exposure photographs. The appearance also depends on the wavelength in which they are viewed. Although the gas is tenuous, there may be several tenths of a solar mass in a planetary's shell.

Fig. 17. NGC 3132, also known as the Eight-burst planetary, lies in Vela near the Antlia border. The nebula arises from the hot, 16th-magnitude companion of the conspicuous central star. South is up.

Planetaries can easily be confused with other objects of similar appearance. Thus H. D. Curtis misclassified the Crab nebula as a planetary; it is actually a supernova remnant. Herbig-Haro objects, which are extremely young stars embedded in gas, have sometimes been called planetaries too. The mistake has also been made in the case of nova shells, compact H II regions, and other associations of stars and nebulae. A correct identification can often be made with a detailed photograph and spectroscopic evidence of the distinctive oxygen emission lines described above. Nearly 1,500 confirmed and suspected galactic planetaries have been catalogued to date, but 50 times as many may still await discovery. They have also been observed in the Andromeda galaxy M31, the Large and Small Magellanic Clouds, and other nearby systems.

The central stars of planetary nebulae have extremely high surface temperatures, usually from 30,000 to several hundred thousand degrees Kelvin. Most are stars of spectral type *O,* Wolf-Rayet stars, or their relatives. (The abbreviation OVI in the data base refers to stars whose spectra display emission lines of five-times-ionized oxygen, not to the spectral type *O.* These objects form a high-temperature extension of the Wolf-Rayet class.) Sometimes a central star cannot be classified because it exhibits a continuous spectrum — that is, one without lines. There are other cases in which the spectral type implies a surface temperature too cool to excite a nebula. These are probably binaries that contain unresolved, hot components.

The progenitors of planetary nebulae ("protoplanetaries") are bloated red-giant Mira variables. Although each starts with one to six times the Sun's mass, it continuously loses material in a slow stellar wind of typically 20 kilometers per second. When the bulk of the atmosphere is gone and the hotter, underlying layers are revealed, the surrounding gas is ionized and begins to glow. The wind increases sharply to perhaps 1,000 kilometers per second and plows the matter into a shell that is typically 0.15 parsec across. (This is about 30,000 times the distance from the Sun to the Earth.) The star then quickly evolves into a white dwarf, and the glowing nebula remains visible for less than 100,000 years. The entire process is described in more detail by Sun Kwok in *Sky and Telescope,* **62**, May, 1982, page 449.

The distances to planetaries must be known in order to deduce their galactic distribution, luminosity, birthrate, mass, and central-star diameter. With few exceptions, they are too far away to have measurable trigonometric parallaxes. Although the central stars are usually so abnormal that they cannot be directly compared to stars of known absolute magnitude, several are binaries with a normal component. The luminosity class of the latter is then determined from its spectrum. Also, the expansion velocity of the nebula, as measured from its Doppler-shifted spectral lines, can be combined with the observed *angular* expansion rate to arrive at a distance. Sometimes a planetary belongs to a stellar group at a known distance, for example a star cluster or a nearby galaxy. In 1928 F. G. Pease discovered one in the globular cluster M15. Other methods involve statistical procedures, spectral-line intensities, stellar-atmosphere models, interstellar-extinction measurements toward the planetary, or radio observations. No matter which approach is taken, the resulting distance is always very uncertain. Those listed here should therefore be used with caution.

The objects in this section were selected from the *Strasbourg Catalogue of Galactic Planetary Nebulae,* which was compiled in 1981. Data were added from the *Catalogue of Central Stars of True and Possible Planetary Nebulae*, from its 1982 and 1983 supplements, and from the *Catalogue of Galactic Planetary Nebulae* by L. Perek and L. Kohoutek. Radial velocities have been updated according to the 1983 paper of S. E. Schneider *et al.*

GALAXIES

Galaxies are gravitationally bound aggregations of stars, gas, and dust that are the next step up in scale from star clusters. This is our familiar modern view, and we often forget the long and tortuous path of observationally supported inferences by which it was reached. The concept quite belies appearances, for a small telescope shows a galaxy as scarcely more than a dim round or oval glow, or perhaps a streak with occasional hints of a dark lane or other feature. By 1845, when Lord Rosse's 72-inch reflector was first turned on the sky, some were seen for the first time to have an intriguing spiral structure. Yet for many years they continued to be lumped into the broad category of *nebulae* because nearly all of them gave no clear sign of being resolvable into stars.

As the 20th century dawned, astronomers still disputed whether these spiral nebulae were within our own Milky Way or were "island universes" in their own right scattered beyond its borders. Heber D. Curtis provided the first direct evidence of their extragalactic nature in 1918. He found that novae in the Andromeda nebula M31 attained a peak brightness far less than those observed elsewhere in the sky. This could be explained if M31 were situated well outside our galaxy. Five years later Edwin Hubble discovered a Cepheid variable in M31. Its faintness compared to "local" Cepheids was incontrovertible proof that the Andromeda nebula lay far beyond the Milky Way. Shortly thereafter Hubble confirmed a linear relationship between galaxies' radial velocities and distances from which was deduced that the universe as a whole is expanding. These seminal discoveries paved the way for our current understanding of the overall structure and evolution of the cosmos.

Galaxies display a wide array of sizes and luminosities. The smallest dwarf-irregular systems are only 300 parsecs (1,000 light-years) in diameter. At the opposite extreme, within the galaxy cluster Abell 1413 is a gargantuan elliptical that spans almost two million parsecs (six million light-years). If substituted for the Milky Way, it would extend beyond our neighbor M31. While the faintest dwarf ellipticals emit no more light than a large globular star cluster, the most luminous "normal" galaxies have absolute visual magnitudes as bright as −23 (over 10 times brighter than the Milky Way). The majority of galaxies are constant in visible light, but radio, infrared, and X-ray observations have revealed dramatic activity in the nuclei of systems as apparently benign as our own. Other objects have been wracked by explosions and are spewing matter into intergalactic space. Still others, such as the supergiant elliptical mentioned above, have grown to outsize proportions by gravitationally swallowing their neighbors.

Most of the objects in our Galaxies section have photoelectric blue magnitudes (defined below) brighter than 15 and have come from the *Second Reference Catalogue of Bright Galaxies* by Gerard de Vaucouleurs *et al.,* or else from R. B. Tully's *Atlas and Catalog of Nearby Galaxies.* Yet the list is not a complete survey to magnitude 15, nor does it contain all galaxies with NGC designations. At the

Fig. 18. IC 5152, a Local Group galaxy of type Ir+, is less than one-tenth the size of our Milky Way. The bright foreground star is HD 209142 with a visual magnitude of 7.6. From the Cape Photographic Atlas of Southern Galaxies.

time the sources were compiled, the brightnesses of many galaxies were still unknown or only crudely estimated. Systems brighter than the chosen limit have subsequently appeared in the astronomical literature, and only some of those with NGC designations have been added to our data base. However, many magnitudes and radial velocities have been updated according to more recent sources (cited in the References). We have also inserted the number of each object in the *Uppsala General Catalogue of Galaxies* (UGC) or its supplement (UGCA). A cross-check with the 1,246 objects in the *Revised Shapley-Ames Catalog of Bright Galaxies* has insured that they are all included here.

Several fainter systems of special interest have also been added, bringing the total number to 3,116. To place the entire sample into perspective, consider that in the 1930's Hubble counted 44,000 galaxies on photographs of selected areas of the sky. By 1957, approximately one million had been tallied on Harvard Observatory plates. And according to P. Nilson's history of galaxy catalogues in the UGC, Fritz Zwicky surveyed nearly 15 million objects in compiling the six-volume *Catalogue of Galaxies and Clusters of Galaxies.* Probably well over a billion are within reach of the 200-inch (5-meter) Hale telescope on Palomar Mountain.

Galaxy classification. In his 1954 *Encyclopaedia Britannica* article on classification, A. Wolf wrote: "Classification is . . . probably the simplest method of discovering order in the world. By noting similarities between numerous distinct individuals as forming one class or kind, the many are in a sense reduced to one, and to that extent simplicity and order are introduced into the bewildering multiplicity of nature."

Indeed, the appearance of galaxies in time-exposure photographs varies widely. They range from seemingly amor-

Fig. 19. NGC 1300, a classic barred spiral (type SBb) in Eridanus, has a visual magnitude of about 10. Note the faint, offset dust lanes along the bar. This photograph was taken with the 200-inch Hale reflector and appeared in the **Hubble Atlas of Galaxies.**

phous clouds to highly organized spirals resolvable into stars. These shapes are sometimes complemented by bar structures, rings, jets, or filaments. In 1926, Edwin Hubble divided the galaxies into three broad categories based on their form. With some subdivisions that have been introduced more recently by other astronomers, the Hubble types are as follows:

E. Elliptical galaxies, disklike or spheroidal assemblages of stars with practically no interstellar gas or dust. The lack of gas is explained if the initial rate of star formation was very fast. The material that would otherwise have gone to newer generations of stars was rapidly exhausted. The light of ellipticals is therefore dominated by old giants of Population II, the stars characteristic of the halo and nucleus of our galaxy. Elliptical galaxies are subdivided into categories according to their *apparent* degree of flattening, from the roundest (E0) to the flattest (E7). The appended number comes from multiplying by 10 the difference between the object's major and minor diameters, dividing the result by the major diameter, and then rounding to the nearest integer. Dwarf systems are indicated by the prefix "d." The largest known galaxies are those of the cD class, where "c" denotes a supergiant system and "D" an E-type whose appearance is dominated by a large, diffuse halo. Since they are usually found near the centers of rich clusters, astronomers suspect that their huge size comes from disrupting and incorporating smaller cluster members.

S and *SB.* Galaxies whose stars, nebulae, and dust display a spiral pattern. Those with tightly-wound arms and prominent central bulges are designated type Sa, while loosely coiled systems with relatively inconspicuous nuclei are la-

beled Sc. In this scheme, the Milky Way is an Sb type with intermediate characteristics. Galaxies between Sa and Sb are called Sb − systems, while those between Sb and Sc are designated Sb + . Sometimes the stars and gas in the inner region of a spiral are organized into a straight "bar" that extends diametrically across the nucleus for thousands of light-years. These barred spirals are classified SBa, SBb or SBc. If the presence of a bar is uncertain, the code S(B) is used.

A spiral galaxy's nuclear bulge and tenuous halo comprise mostly old Population II stars. The younger Population I is found in the outer disk and spiral arms where gas is still coalescing into stars. Apparently, star formation took place in an initial burst in the bulge and halo, but at a more leisurely pace in the disk and arms.

Ir. Irregular galaxies, which reveal no symmetry in their structure. Some irregulars (variously designated Ir + , type I, Im, or Magellanic irregulars) are resolvable into individual stars and nebulae, while others (Ir − , type II, I0) appear completely amorphous, sometimes with prominent dark lanes of dust. The abundance of interstellar gas and luminous young stars in Ir + systems implies that star formation is still very active in these galaxies. Some Ir − objects show evidence of explosions in their nuclei.

The basic galaxy types were illustrated in Hubble's well-known "tuning fork" diagram of 1936 (shown in *Sky & Telescope,* **52,** December, 1976, page 410). He added the hypothetical lenticular category (S0), whose members are composite in nature. Discovered several years later, the lenticulars are shaped like spirals and may even show a thin dust lane; however, like ellipticals, they contain only old stars. The tuning-fork diagram suggested to some astronomers an

Fig. 20. Two nearly edge-on spirals. At left is Messier 104, the Sombrero galaxy, in Virgo; its Hubble type is Sb −. Note the prominent dust lanes and the unusual extent of the Sombrero's nuclear bulge. Official U. S. Navy photograph. At right, NGC 4565 is an Sb galaxy in Coma and resembles how the Milky Way would look from a similar distance and aspect.

evolutionary progression from the roundest to the flattest ellipticals and subsequently into spiral galaxies. This speculation is now believed to be false.

Although his basic guidelines are still used, Hubble's classification system has been expanded over the years to keep pace with the wealth of information revealed by improved photographs. In 1960, Sidney van den Bergh of the David Dunlap Observatory (DDO) recognized that the standard Hubble spiral types were defined by the appearance of only high-luminosity galaxies. But the brightest and faintest spirals differ in luminosity by a factor of more than 100. Van den Bergh therefore introduced the DDO classification scheme, which assigns luminosity categories analogous to those used for stars; that is, supergiant galaxies are designated by the Roman numeral I, bright giants II, giants III, subgiants IV and dwarfs V.

The average absolute photographic magnitude of each category depends on the adopted value of the Hubble constant (to be discussed later), but is roughly −21, −20, −19, −18, and −15 for classes I through V, respectively. Van den Bergh found that only I, II, and III galaxies display prominent arms and can be given a Hubble class. Less luminous systems usually have only traces of arm structure and do not correlate well with Hubble's guidelines. Therefore, intrinsically faint spirals are labeled S+ if they are resolvable into stars and S− if they are not. (There is a rough correspondence between luminosity classes III, IV, and V and G. de Vaucouleurs' extension to the Hubble sequence Sd, Sm, and Im, respectively. The latter are adopted here if a DDO type is not available.)

One should be cautious when scanning the Hubble-type column in search of trends. Although most of the galaxies listed here are spirals, dwarf elliptical systems are actually the most common type in the universe. Because of their low luminosity they are underrepresented in any sample chosen by apparent brightness (such as the one in this book). In addition, most rich clusters, which account for a substantial fraction of all known galaxies, are dominated by ellipticals. These clusters are generally far away and therefore too faint for their members to be included here.

Some of the galaxies are classified as Seyfert objects. These systems, first described by Carl Seyfert in 1943, appear relatively normal except for their brilliant, almost starlike nuclei. The broad emission lines in their spectra imply that gas is being expelled at great speed from their cores. Seyfert galaxies may represent less energetic versions of quasars. In fact, the most extreme Seyferts are comparable in luminosity to the mildest quasars. The photographic images of N galaxies, a related high-energy category, are dominated even more by their bright, compact centers.

Magnitudes and dimensions. A galaxy's total magnitude and size are difficult to measure because its light merges gradually into the surrounding sky. For example, most textbook photographs of the Andromeda galaxy M31 show just its inner half. The outer spiral arms are so tenuous that they reveal themselves only on longer exposures. As the exposure time increases, so does the apparent size and the total brightness. How can the magnitude and dimensions be measured if the galaxy's edge is not well defined?

Since 1958, de Vaucouleurs has studied how the **B** (blue) magnitude in **UBV** photometry declines in the outer regions of a diverse sample of galaxies. He obtained a set of standard brightness profiles that can be applied to other less well observed systems. For each object, the total magnitude B_T (simply called **B** in our tabulation) is derived by inferring where the surface brightness would reach zero and then summing all the light within this boundary. The total color index $(B − V)_T$, a measure of the average surface temperature of the galaxy's stars, is computed from the composite light in the **B** and **V** photoelectric bands. It ranges from about 0.3 to more than 1.0. In general, the smaller the value, the greater the number of young, hot, blue stars. Therefore, elliptical galaxies usually have color indexes at the high end of the range and irregulars at the low end, with spirals in between.

The dimensions of a galaxy are measured in a manner similar to that for the total magnitude. Starting at the center of the image, the size is determined by moving outward until the surface brightness (the brightness per unit area) has dropped to a specified level. This catalogue lists the maximum and minimum angular diameters out to a standard brightness level equivalent to one star of **B** magnitude 25 per square arc second. These dimensions will nearly match those measured from the blue plates of the National Geographic Society-Palomar Observatory Sky Survey. They are usually larger than what is seen visually in a telescope. (A galaxy's light is always superimposed, of course, on that of the sky itself, which even under the very darkest conditions contributes the light of one star of blue magnitude 22 per square arc second.)

Once the total magnitude and size are known, the galaxy's average surface brightness m can be derived. Consider an ellipse whose major and minor axes are D and d, respectively. According to the *Second Reference Catalogue*, this m is given by the equation:

$$m = \mathbf{B_T} + \Delta m + 5\log D - 2.5\log(D/d) - 0.26,$$

where $\mathbf{B_T}$ is the total blue magnitude extrapolated to zero surface brightness, as defined earlier. The magnitude increment Δm must be added to correct for the light of the galaxy's outermost regions, where the brightness level is fainter than the cutoff mentioned above. The increment is approximately 0.25 magnitude for elliptical systems, 0.13 for type S0, and 0.11 for spirals and irregulars. The equation gives m in terms of the equivalent stellar **B** magnitude per square *arc minute*. Thus M31 has about the same average surface brightness as a field containing one 14th-magnitude star in each 1'-by-1' area of sky. Normal galaxies usually fall within the range 12 to 16.

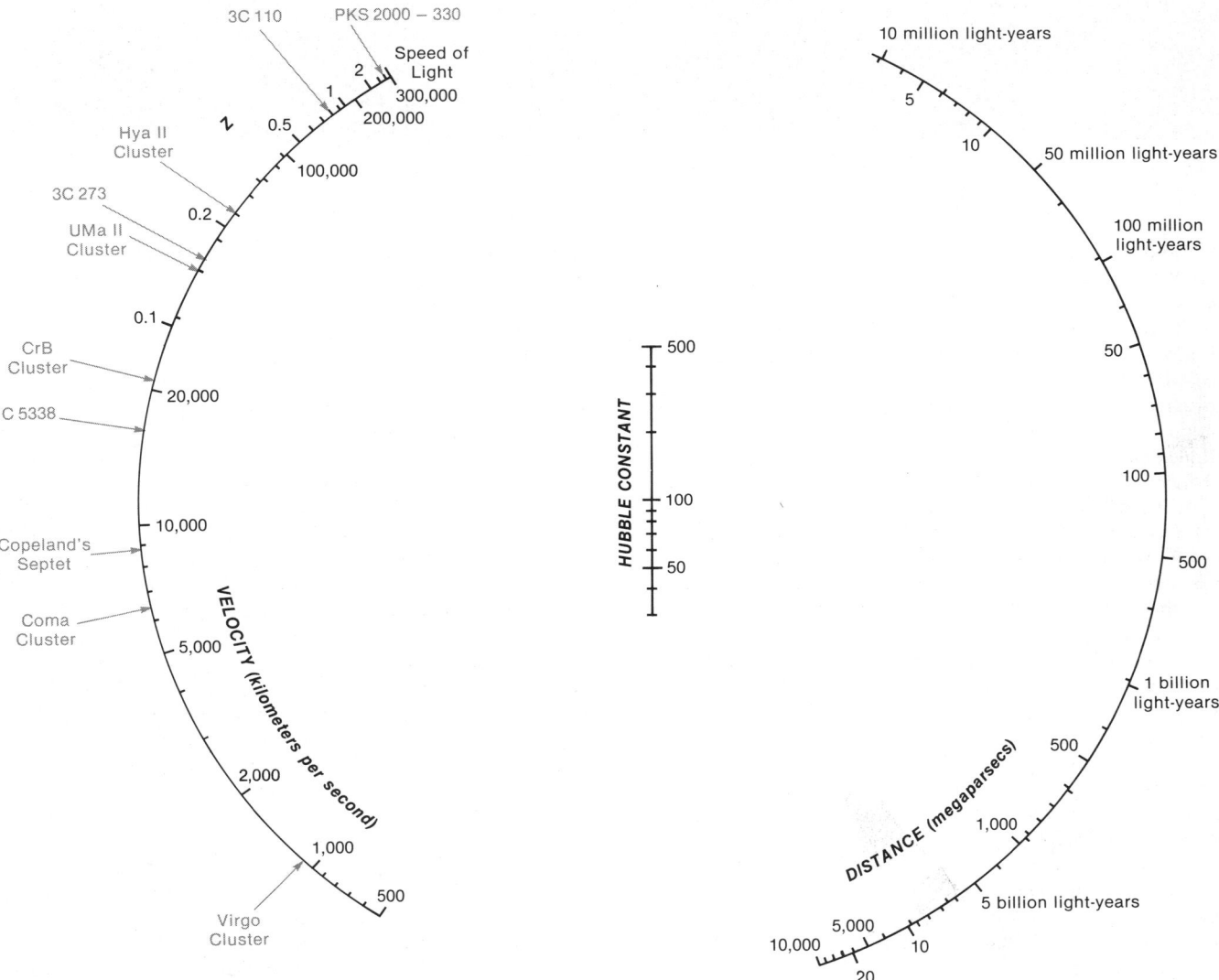

Fig. 21. This nomogram expresses Hubble's law in graphical form. The distance to a galaxy, cluster of galaxies, or quasi-stellar object is found by laying a straightedge between the object's recessional velocity (or alternatively, redshift z) on the left-hand scale and the value of the Hubble constant H, in kilometers per second per megaparsec, plotted near the middle. Read the distance in either megaparsecs or light-years at the point where the straightedge crosses the right-hand scale. In 1929, Hubble found H to be about 500, whereas today's estimates are anywhere from 55 to 100. Notice how the early value leads to significantly smaller distances than when modern ones are used. In other words, the established scale of the universe has ballooned by a factor of perhaps 10 since Hubble's day! As discussed in the text, caution should be exercised when interpreting the nomogram at large redshifts.

Fig. 22. The interacting pair of 12th-magnitude galaxies VV 304 in Pavo consists of NGC 6769 (right) and 6770, both classified as S(B)bp. The third galaxy in this picture from the Cape Photographic Atlas of Southern Galaxies is NGC 6771, type SB0. North is up.

The ease with which a galaxy can be photographed is often indicated better by this calculated surface brightness than by its magnitude. For example, the Seyfert galaxy M77 has a B of 9.5, while the value for the Fornax dwarf galaxy is 9.0 (both objects may be found on page 321). Yet their calculated surface brightnesses are 13.4 and 15.1, respectively! No doubt this explains why the Fornax dwarf eluded discovery until as recently as 1938; it consequently bears no NGC designation.

Similar comparisons can help an observer predict the ease of detecting a galaxy visually. After doing the calculation for blue light, subtract the listed color index to obtain a rough visual surface brightness. There is another factor to consider as well: A galaxy's light can be spread out in any number of ways. One spiral system might have a brilliant nucleus surrounded by a faint disk, while another might be of almost uniform brightness over its entire image. Even if their total magnitudes are identical, the first will probably be more visible in the field of a telescope because its contrast with the background sky is greater. For the same reason, edge-on systems are often easier to detect than those seen face on. Further details can be found in Vol. 4 of the *Webb Society Deep-Sky Observer's Handbook.*

Radial velocities. Just as for stars, radial velocities of galaxies are determined by measuring the Doppler shifts of absorption lines in their spectra. This was first accomplished by V. M. Slipher of Lowell Observatory in 1912. Over 10 years of painstaking work revealed that, except for a few of the closest systems, all galaxies display redshifted lines. In other words, they are receding from us. If the perturbing effects of galaxy interactions are subtracted out, the recessional velocities are directly proportional to the distance. This relationship, named the Hubble law in honor of its discoverer, indicates that the universe is expanding. (Actually, the geometry of our universe may cause the linear dependence of distance on velocity to break down at very large distances.)

According to extensive observations made in the 1970's by Allan Sandage and Gustav Tammann, the constant of proportionality between velocity and distance — the Hubble constant *H* — is close to 55 kilometers per second per megaparsec. This means that, for every additional million parsecs in distance, galaxies recede from us an extra 55 kilometers per second due to the expansion of the universe. Some astronomers differ with the galaxy luminosities adopted by Sandage and Tammann and therefore favor a larger number, perhaps even as high as 100. Since Hubble's initial determination in 1929, the accepted value for *H* has dropped by a factor of 10. This is because of an increased understanding of the "standard candles" — such as luminous blue stars, Cepheids, novae, H II regions, and supernovae — used to establish the scale of the universe.

The Hubble law can be used to estimate a galaxy's distance, which simply equals the observed recessional velocity divided by the agreed-on value of *H*. The velocity for this purpose should have the solar system's local motion removed, as is already the case for velocities in our tabulation. However, the method cannot be applied to objects in the nearest congregations such as the Local Group or the Virgo cluster (discussed below), where individual velocities are greater than those due to the feeble effect of the cosmic expansion at such close range. Also, the distances of the remotest systems are subject to question, as described in the Quasi-stellar Objects section. These caveats should be kept in mind when using the nomogram in Fig. 21.

Groups, Clusters, and Superclusters. Only within the past few years have astronomers concluded that most, if not all, galaxies belong to some type of aggregation. These can range from simple pairs to systems of increasingly large scale: groups, clusters, and superclusters. For example, the

Fig. 23. A small portion of the Coma galaxy cluster with the 12th-magnitude elliptical system NGC 4889 at center. This photograph was taken with the 200-inch Hale reflector.

Milky Way and M31 are part of the Local Group, a loose collection of at least two dozen galaxies about a megaparsec (over three million light-years) across. Table IV lists the Local Group's five spirals, 14 ellipticals, and 10 irregulars in order of decreasing absolute magnitude, according to van den Bergh's 1981 compilation. It gives each object's name, position, total visual magnitude, maximum and minimum apparent diameter in the system described earlier, Hubble type, absolute visual magnitude, approximate distance and computed size in kiloparsecs, plus other names. These include the Messier or NGC number, or a designation from H. Arp's *Atlas of Peculiar Galaxies* or the David Dunlap Observatory list of dwarf galaxies (D). Although 29 objects are described here, other faint members may be discovered in the future, especially those partially hidden by intervening dust in our galaxy's plane. Also, several systems whose distances are highly uncertain may yet prove to be members.

The Local Group is approximately 15 megaparsecs (50 million light-years) from the center of a cluster in the constellation Virgo (but which spills over into Coma Berenices) that has hundreds of known members. Many of the sky's brighter galaxies, including the Messier objects M49, M58 to M60, M84 to M90, and M98 to M100, belong to this system. The Virgo cluster in turn forms the hub of an even larger assemblage called the Local Supercluster. (For further details, see R. B. Tully's article in *Sky and Telescope*, **63**, June, 1982, page 550.)

Moving farther out into space, one finds such swarms in every direction. The well-known cluster in Coma Berenices,

for example, is about 120 megaparsecs (390 million light-years) away and encompasses thousands of members. The first major catalogue of rich clusters was completed by George Abell in 1958 and contains 2,712 entries. Some of the more prominent systems from his and other catalogues are listed in Table V along with Abell designation (if one), position, angular diameter, radial velocity, richness class (from the relatively sparse clusters of class 0 up to the populous class 5 congregations), Rood-Sastry type (see next paragraph), brightest member galaxies, and radio source designation, plus notes indicating if they belong to a supercluster or have been detected in X-rays.

The classification scheme introduced by H. J. Rood and G. N. Sastry in 1971 is based primarily on the apparent distribution of the 10 brightest cluster members. For instance, a system such as Abell 2199 is called cD because its appearance is dominated by one of these highly luminous galaxies. The Coma cluster, on the other hand, is centered on a pair of supergiant galaxies; it is placed in the B (for binary) category. If three or more of the brightest members are centrally located along a line, as in the Perseus cluster, type L is assigned. The Corona Borealis system contains a core of at least four bright galaxies surrounded by a halo of fainter members; such a case is designated by the letter C. Type F indicates that the cluster appears flattened, whereas a congregation whose brightest galaxies are distributed irregularly is labeled with the letter I. In their study of 111 clusters, Rood and Sastry found that I and cD types predominate.

Much as for individual galaxies, the approximate distance

TABLE IV. THE LOCAL GROUP

Name	α (2000) δ h m ° ′	Mag.	Dim. ″, ′	Type	M_v	Dist. (kpc)	Diam. (kpc)	Notes
Andromeda galaxy	0 42.7 +41 16	3.4	178 x 63	Sb	−21.1	730	38	M31, NGC 224
Milky Way	(17 45.6 −28 56)			Sb +:	−20.5	(8.5)	30:	
Triangulum galaxy	1 33.9 +30 39	5.7	62 x 39	Sc	−18.9	900	16	M33, NGC 598
Large Magellanic Cloud	5 23.6 −69 45	0.1	650 x 550	SBm	−18.5	50	9.5	
IC 10	0 20.4 +59 18	10.3	5 x 4	Ir +:	−17.6	1,300	1.9	
Small Magellanic Cloud	0 52.7 −72 50	2.3	280 x 160	SBmp	−16.8	60	4.9	
NGC 205	0 40.4 +41 41	8.0	17 x 10	E6:	−16.4	730	3.6	M110
NGC 221	0 42.7 +40 52	8.2	8 x 6	E2	−16.4	730	1.7	M32, Arp 168
NGC 6822	19 44.9 −14 48	9:	10 x 10	Ir +	−15.7	520	1.5	Barnard's galaxy
NGC 185	0 39.0 +48 20	9.2	12 x 10	dE0	−15.2	730	2.5	
NGC 147	0 33.2 +48 30	9.3	13 x 8	dE4	−14.9	730	2.8	D3
IC 1613	1 04.8 +2 07	9.3	12 x 11	Ir +	−14.8	740	2.6	D8
WLM system	0 02.0 −15 28	10.9	12 x 4	Ir +	−14.7	1,600	5.6	D221
Leo A	9 59.4 +30 45	12.6	5 x 3	Ir +	−14.1	2,300:	3.3	Leo III, D69
Fornax dwarf galaxy	2 39.9 −34 32	8:	20: x 14:	dE3	−13.6	130	2.2:	
IC 5152	22 02.9 −51 17	11:	5 x 3	Ir +	−13.5	1,500	2.2	
Pegasus dwarf galaxy	23 28.6 +14 45	12.0	5 x 3	Ir +	−13.4	1,300	1.9	D216
Sculptor dwarf galaxy	0 59.9 −33 42	10:		dE3	−11.7	85	1.5:	
Leo I	10 08.4 +12 18	9.8	11 x 8	dE3	−11.0	230	0.7	Regulus dwarf, D74
Andromeda I	0 45.7 +38 00	13.2		dE0	−11:	730		
Andromeda II	1 16.4 +33 27	13:		dE0	−11:	730		
Andromeda III	0 35.4 +36 31	13:		dE2	−11:	730		
Aquarius dwarf galaxy	20 46.9 −12 51			Ir	−11:	1,500		D210
Sagittarius dwarf galaxy	19 30.0 −17 41	15:		Ir−	−10:	1,100		
Leo II	11 13.5 +22 10	11.5	15 x 13	dE0	−9.4	230	1.0	Leo B, D93
Ursa Minor dwarf galaxy	15 08.8 +67 12	12:	27 x 16	dE6	−8.8	75	0.6	D199
Draco dwarf galaxy	17 20.2 +57 55	11:	34 x 19	dE3	−8.6:	80	0.8	D208
LGS 3	1 03.8 +21 53	15:	2	Ir	−8.5:	900	0.5	
Carina dwarf galaxy	6 41.6 −50 58			dE		170		

Notes: Tabulated by absolute magnitude. The linear sizes of the Fornax and Sculptor dwarf galaxies are computed from their actual extent of more than a degree. Because of the sparseness of these galaxies, the angular diameter according to the standard system defined in the text is underestimated. For the Milky Way, the coordinates and distance are those of the galactic center in Sagittarius.

TABLE V. CLUSTERS OF GALAXIES

Name	Abell No.	α (2000) δ h m ° ′	Diam. °	RV (km/sec)	D	R	RS Type	NGC	Radio Source	Notes
Haufen A	151	1 08.9 −15 25		15,800	3	1	cD			
	194	1 25.6 −1 30	0.3	5,320	1	0	L	541,5,7	3C 40	In Perseus supercluster
	400	2 57.6 +6 02		7,200	1	1	I		3C 75	
Perseus	426	3 18.6 +41 32	4	5,460	0	2	L	1275	3C 84	In Perseus supercluster; XRS
Fornax II		3 28 −20 45	7	1,560				1232		
Fornax I		3 32 −35 20	7	1,500				1316	For A	
Gemini	568	7 07.6 +35 03	0.5	23,400	3	0	C			
Cancer		8 21 +20 56	3	4,800				2563		
Hydra II		8 58 +3 09		60,900						
Leo	1020	10 27.8 +10 25	0.6	19,500	4	1				Also Abell 1016?
Hydra I	1060	10 36.9 −27 32		3,000	0	1	C	3309,11		XRS
Ursa Major II		10 58 +56 46	0.2	41,000						
Leo A	1185	11 10.9 +28 41		10,500	2	1	C	3550		
	1367	11 44.5 +19 50		6,150	1	2	F	3842,62	3C 264	In Coma supercluster; XRS
Ursa Major I	1377	11 47.1 +55 44	0.7	15,300	3	1	B			
Virgo		12 30 +12 23	12	1,200				4472,86	Vir A	In Local supercluster; XRS
Centaurus		12 50 −41 18	2	3,200				4696	PKS	XRS
Coma	1656	12 59.8 +27 59	4	6,650	1	2	B	4889		In Coma supercluster; XRS
Bootes	1930	14 33 +31 33	0.3	39,300						
Corona Borealis	2065	15 22.7 +27 43	0.5	21,600	3	2	C			
Hercules	2151	16 05.2 +17 45	1.7	11,200	1	2	F	6040,47	4C + 17.66	In Hercules supercluster; XRS
	2152	16 05.4 +16 27		11,500	1	1	I:			In Hercules supercluster
	2197	16 28.2 +40 54		9,100	1	1	L	6173		In Hercules supercluster
	2199	16 28.6 +39 31	0.2	9,200	1	2	cD	6166	3C 338	In Hercules supercluster; XRS
Pegasus II		23 10 +7 36	2	12,700				7720	4C + 07.61	
Pegasus I		23 22 +9 02	1	4,000				7619	PKS	

to each cluster (except the nearby one in Virgo) can be deduced from the nomogram, using the listed radial velocity. Taken together, these "galaxian archipelagos," as Abell called them, exhibit a filamentary, almost weblike distribution surrounding huge volumes of empty space. The curious texture of the universe's galactic population is strikingly illustrated in a map conceived by Princeton University scientists that appeared in *Sky and Telescope*, **59**, May, 1980, page 365. Experienced amateurs should refer to Vol. 5 of the *Webb Society Deep-Sky Observer's Handbook*, which is devoted to the visual study of galaxy clusters.

QUASI-STELLAR OBJECTS

Quasi-stellar objects, also called QSO's, quasi-stellar sources (QSS's), or quasars, appear starlike and usually blue on photographic plates, sometimes with a luminous jet extending out from the pinpoint image. Their spectra display broad, highly redshifted emission lines. The large redshifts are generally thought to be "cosmological" in origin; that is, they arise from the expansion of the universe as with galaxies. The Hubble law, discussed in the previous section, implies that QSO's are very distant and must therefore be extraordinarily luminous. Their absolute visual magnitudes fall within the range of −24 to −30, while that of the most luminous galaxy is about −23 (assuming the Hubble constant is 55 kilometers per second per megaparsec).

Many also vary in brightness by up to a magnitude over the course of months or years. However, 3C 279 (the 279th object in the third Cambridge radio survey) sometimes changes by 0.25 magnitude in a day, while three-magnitude variations have been observed in 3C 446. For such a distant object to vary in so short a time, the energy source must be relatively small compared to the size of a galaxy. The nature of the "powerhouse" at the heart of a QSO remains an enigma. The most promising theory at this time presumes the accretion of matter onto a supermassive black hole, with the concurrent release of tremendous quantities of energy at many wavelengths.

The first optical identification of a QSO was made in 1960 by Thomas Matthews and Allan Sandage. In the constellation Triangulum, they found a 16th-magnitude blue "star" at the position of the radio source 3C 48. Three years later, 3C 273 in Virgo was detected visually as a 13th-magnitude object, and it is still the brightest QSO known. At the time, the emission lines in their spectra defied explanation since none matched any spectral features normally observed in stars and galaxies.

Maarten Schmidt solved the mystery by recognizing the familiar pattern of hydrogen's Balmer series; it was shifted so far toward the red that it had previously escaped notice. However, other QSO's had redshifts so great that lines usually found in the ultraviolet portion of the spectrum were displaced into the visible. Since our atmosphere is opaque below 2900 angstroms, certain identification of these features had to await rocket- and satellite-based studies of stellar ultraviolet lines. The largest redshift in our list is 3.78, for the object in Sagittarius known as PKS 2000−330. This means that a line near the middle of the visible spectrum (about 5500 angstroms) is shifted far into the infrared at 26,290 angstroms and is replaced by one that normally appears at 1150 angstroms!

The earliest known QSO's had attracted attention because of their strong radio emission. However, a growing number are now being detected initially through optical surveys. This is a formidable task since a QSO's appearance on a photograph is virtually indistinguishable from those of the numerous foreground stars. Candidates are singled out for further study either on the basis of their blue color or their so-called "ultraviolet excess," which is measured photoelectrically. (That is, QSO's emit more energy than do most stars at ultraviolet and violet wavelengths relative to blue.) Actual QSO's are then isolated by obtaining spectra and looking for the telltale broad, redshifted emission lines.

Such surveys have revealed that most QSO's in a randomly chosen sample are radio-quiet or weak emitters at best. Radio and optical searches have uncovered over 2,200 QSO's to date. In a 1979 review, K. Davidson and H. Netzer estimate that the actual number of QSO's brighter than 21st magnitude may approach a million. A small number of known QSO's, such as 3C 108, 418, and 422, are uncharacteristically red. Since they are viewed through our galaxy's disk, the blue component of their light has been attenuated by interstellar dust. These three objects were discovered only because they are radio sources. How many *radio-quiet*, red QSO's have thus far eluded astronomers?

The 297 QSO's in our list were selected primarily from the 1981 catalogue of A. Hewitt and G. Burbidge. Objects were included if brighter than 17th magnitude, or if they have a redshift greater than 3.3, a radio strength exceeding 10 janskys, or a designation in the revised third Cambridge (3C) radio survey. Other bright systems were selected from the 1983 article by Schmidt and R. F. Green, and from the recent work of M.-P. Véron-Cetty and P. Véron. Our list also contains several newly discovered multiple QSO's, which are discussed below.

Over two dozen entries in our compilation are designated as BL Lacertae objects. When the highly variable prototype of this class was discovered by Cuno Hoffmeister in 1929, it was thought to be a star in the Milky Way; hence, its variable star designation. But while stellar in appearance BL Lacertae and its counterparts do not display the broad emission lines characteristic of QSO's. In fact, their spectra are practically featureless. Some are radio sources or have faint envelopes whose spectra exhibit weak, highly redshifted absorption lines, confirming their common nature with the QSO's.

During the past few years, high-resolution images of nearby QSO's have been obtained with both conventional and electronic cameras. Susan Wyckoff and Peter Wehinger presented some of these pictures in *Sky and Telescope,* **61,** March, 1981, page 200. Here the QSO's no longer appear starlike, but each is embedded in a faint "haze." Spectroscopic study of the haze surrounding 3C 48, first accomplished in 1981 by T. A. Boroson and J. B. Oke, proves that it contains a mixture of hot stars and gas similar to that found in the spiral arms of normal galaxies. Therefore, QSO's are probably distant, young spirals with superluminous nuclei. In light of this evidence, the distinction between QSO's and Seyfert galaxies is perhaps an artificial one; the former are merely a more energetic extension of the latter. There may be no difference between a low-energy QSO and a high-energy Seyfert.

Another advance was made in 1979 with the discovery of the first double QSO. The two components, designated 0957+561 A and B, are separated by six seconds of arc but have nearly identical spectra and redshifts. This object and the four other multiples discovered to date are each the result of gravitational lensing by an intervening massive galaxy. The light rays from a single QSO are bent and focused by the galaxy's gravitational field in a manner prescribed by Einstein's general theory of relativity. As a result, several images of the same QSO can be seen from Earth. The lens-

ing effect is discussed and illustrated by M. Gorenstein in *Sky and Telescope,* **66,** November, 1983, page 390.

The redshift z is defined as $(\lambda - \lambda_0)/\lambda_0$, where λ is the measured wavelength of a spectral line and λ_0 is its rest, or laboratory, wavelength. As long as z is not too large, the apparent recessional velocity of the QSO due to the expansion of the universe can be computed from the classical Doppler relation $v = cz$, where c is the speed of light, 300,000 kilometers per second. For larger z, relativistic effects modify the equation so that it becomes:

$$v = c(z^2 + 2z)/(z^2 + 2z + 2).$$

The results from the classical and relativistic forms already differ by more than 10 percent at a redshift of 0.2.

For example, PKS 2000−330, with its redshift of 3.78, is receding from us at 275,000 kilometers per second, or nearly 92 percent of the speed of light. For the sake of illustration, let us assume that even a remote QSO obeys Hubble's law. Its distance can then be deduced once its redshift is measured, and the nomogram of Fig. 21 used in the same manner as for galaxies. Start by finding either the redshift z or the recessional velocity at upper left. Using the adopted value of the Hubble constant H, locate the approximate distance to the object, in megaparsecs or light-years, on the right-hand scale. However, the reader should be aware that the interpretation of redshift as a Doppler effect is a useful, but artificial, construct.

Strictly speaking, the cosmological redshift arises from the expansion of space itself, not from the motion of galaxies *through* space. Coupling a derived recessional velocity with Hubble's law gives only a crude estimate of the distance to a remote object. The actual distance depends on the curvature of space in the universe, and there is still no agreement as to which of various models is the correct one. A more complete discussion of this subject is given on page 373 of Frank Shu's textbook *The Physical Universe* (Mill Valley, Calif., 1982: University Science Books).

RADIO AND X-RAY SOURCES

The universe as it became known at optical wavelengths was largely a quiescent one, punctuated by an occasional supernova or the suggestion of intense activity in a galactic nucleus. However, when astronomers began to peek beyond the visible range, they saw an unexpected vista of cataclysmic and sometimes bizarre phenomena, such as quasars, pulsars, and black holes. The concluding sections of this catalogue are devoted not to particular types of objects, but rather to two wavelength domains that have contributed immensely to astronomical knowledge. These are radio waves, in the long-wavelength portion of the electromagnetic spectrum, and X-rays, in the short-wavelength region.

Radio Sources. Extraterrestrial radio waves were first detected in 1932 by Karl Jansky, an engineer for Bell Telephone Laboratories. At Holmdel, New Jersey, he had built a rotatable wood-and-wire antenna to identify sources of interference in radio-telephone transmissions. In addition to the expected disruptions from thunderstorms, Jansky's 1932 article in the *Proceedings* of the Institute of Radio Engineers mentioned a "very steady hiss type static the origin of which is not yet known." Within two years he had identified the Milky Way as the source, with the strongest emission coming from the galactic center, known now as Sagittarius A.

The subsequent development of radio astronomy was slow. In its infancy, the study of cosmic radio waves was quite literally a backyard science. Grote Reber built a parabolic antenna of 9.5-meter diameter beside his Wheaton, Illinois, home in 1937. Over the course of the next decade, he used it to map the sky at frequencies of 160 and 480 megahertz, corresponding to wavelengths of 188 and 63 centimeters, respectively.

In addition to the source at the galactic center, Reber detected Cassiopeia A, the strongest of all radio sources, which has subsequently been found to be a supernova remnant. He also found Cygnus A, the active galaxy that is the "brightest" extragalactic source, and our own Sun. Solar radio waves were actually received slightly earlier by J. S. Hey in England during World War II, but the discovery remained classified until 1946.

H. C. van de Hulst suggested in 1944 that the Milky Way's interstellar neutral hydrogen clouds might emit detectable radio waves at 1420 megahertz (a wavelength of about 21 centimeters). The hypothesis was confirmed in 1951, after which began the detailed mapping of the galaxy's spiral arms. These surveys demonstrated the potential of radio astronomy as a research tool and provided the impetus for the accelerated development of the field. Today, radio telescopes rival optical instruments in both sensitivity and resolution.

As with visible light, radio energy from celestial sources can be continuous over a range of frequencies, or it can appear in discrete spectral "lines." The continuous emission is in turn broadly referred to as *thermal* in origin if the energy stems from the random vibrations of a source's constituent electrons, or as *nonthermal* if it is produced by a process unrelated to the source's temperature.

The Sun, for instance, glows brightest in visible light, but it is a strong thermal source at centimeter wavelengths as well. H II regions also emit detectable continuous energy of this sort. However, a thermal origin for the radio emission from pulsars, supernova remnants, active galaxies, QSO's, and the like would imply absurdly high temperatures in these systems. Instead, this energy is due largely to the synchrotron process, in which electrons radiate as they spiral through the object's magnetic field. Thermal and nonthermal sources can also be differentiated by studying how their energy is distributed over frequency. The emission of the former typically rises with frequency and then levels off, while that of the latter often drops with increasing frequency.

Astronomers represent the average trend of an object's radio spectrum by the index α, defined as the logarithm (to the base 10) of S_1/S_2 divided by the logarithm of f_1/f_2, where S_1 and S_2 are the radio strengths, or flux densities, at the frequencies f_1 and f_2. Depending on the frequency range, thermal sources usually have indexes that are near zero or slightly positive. Negative indexes are common among nonthermal sources, but exceptions are seen. Some systems display both types of spectra. For example, above about 1400 megahertz the Milky Way's emission is dominated by the thermal radiation of bright nebulae; below this frequency the nonthermal (synchrotron) background from electrons traversing the galactic magnetic field is greater.

The most important radio spectral line is the one cited earlier at 21 centimeters. It arises from the so-called spin-flip transition of neutral atomic hydrogen. This occurs when an electron whose spin direction is identical to that of its central proton (the high-energy state) spontaneously flips over to spin in the opposite direction (the low-energy state). With each transition, a photon of 21-centimeter wavelength is released. Since galaxies such as our Milky Way contain widely distributed neutral-hydrogen clouds, radio astronomers use the 21-centimeter line to map the contours of spiral arms and study galactic rotation.

High-resolution studies of radio galaxies like Cygnus A have shown that their radio waves often emanate not from the galaxy itself, but rather from two lobes separated by a megaparsec or more on either side. The lobes mark clouds of material presumably ejected from the nuclei of these systems. A radio galaxy's total energy, as "stored" in the

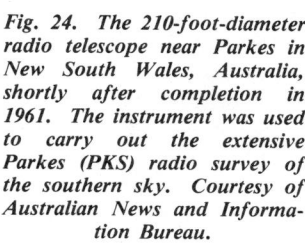

Fig. 24. The 210-foot-diameter radio telescope near Parkes in New South Wales, Australia, shortly after completion in 1961. The instrument was used to carry out the extensive Parkes (PKS) radio survey of the southern sky. Courtesy of Australian News and Information Bureau.

form of high-speed particles and magnetic fields, is truly colossal. In his textbook *The Physical Universe,* Frank Shu equates it with the energy that would be released by 10 billion simultaneous supernovae! How radio lobes attain their energy is still a matter of speculation, but processes involving supermassive black holes are often proposed.

The entries in our Radio Sources tabulation are primarily from two sources. One is the Parkes (PKS) radio survey, covering all declinations south of $+27°$. The other is the revised Third Cambridge survey, here abbreviated 3C (although it is sometimes designated 3CR in the astronomical literature), which was designed to detect the strongest sources north of $-5°$. PKS sources are listed here if the flux density exceeds 5 janskys at 408 megahertz, and the entire 3C catalogue has been included. The journal articles cited in the References have been consulted to add strong emitters in the Milky Way's disk (such as the galactic center), supernova remnants, and H II regions, plus several pulsars.

In keeping with the scope of our catalogue, the Sun and planets have not been listed. Wherever possible, flux densities are given at frequencies of 178, 408, 1400, 2700, and 5000 megahertz (168.5-, 73.5-, 21.4-, 11.1-, and 6.0-centimeter wavelengths) to convey an idea of each object's radio spectrum. Since these values sometimes come from more than one reference, there is an unavoidable heterogeneity in our list. However, the variation among flux densities quoted by several different authors was rarely found to exceed 10 percent.

X-ray Sources. Unlike visible light and radio waves, X-rays cannot penetrate the Earth's atmosphere to reach ground-based telescopes. As a result, the development of X-ray astronomy had to await the means of transporting detectors into space. A rocket-borne device that was launched in 1948 confirmed the suspicion that the Sun emits X-rays. In 1962 the first source beyond the solar system was detected. Dubbed Scorpius X-1, it was identified four years later as a 12th-magnitude variable star, and it is now known to be binary. Despite an unimpressive visual appearance, Scorpius X-1 emits 8,000 times more energy in X-rays than the Sun does at all wavelengths. The Crab nebula was also found to generate X-rays.

A long-term observatory was established in space with the launch in late 1970 of the Uhuru satellite, resulting in a catalogue of over 150 X-ray emitters. One of these was the galaxy M87 in Virgo, the first extragalactic source to be discovered. The sophisticated Einstein observatory became operational in 1978. Sensitive enough to detect objects one ten-millionth the strength of Scorpius X-1, it could also form X-ray images of extended sources. Its high-resolution imager was capable of 4-arc-second resolution over a field 20 arc minutes across.

X-rays are generated by a wide variety of celestial objects ranging from individual stars to clusters of galaxies. Coronal disturbances and surface flares on normal stars exemplify the former. There are also binaries in which one member, a white dwarf or neutron star, reacts explosively as matter from the companion accretes onto its surface. The cataclysmic variables U Geminorum and AM Herculis are cases of this sort. The simultaneous X-ray and optical nova V616 Monocerotis also falls in this category. In 1976 its X-ray output intensified a millionfold, while its light jumped by seven magnitudes. In the well-known Cygnus X-1 system, gas from a blue supergiant is thought to be spiraling in toward a companion black hole, heating up and emitting

X-rays in the process. The X-ray intensity fluctuates irregularly over the course of milliseconds.

Other candidates for being close binaries are *bursters,* which probably consist of a neutron star and a low-mass companion. But duplicity has been proved in only a few cases. These powerful sources occasionally flare up for several seconds and then subside. William Liller discovered an unusual one, now known as the Rapid Burster, in the core of the faint globular cluster named Liller 1. It flares every few seconds or minutes, the recycling time increasing with the energy of the preceding burst. This behavior is maintained for weeks at a time, after which the bursts mysteriously disappear for about six months. Although it was initially suspected that bursters are associated with massive black holes in the centers of globular clusters, the idea was abandoned when similar noncluster sources were found.

The term *transient* was originally coined to describe newly discovered sources that intensified over the course of days, gradually declined in the months that followed, and did not recur. "Fast" transients last only minutes or hours. More recently, recurrent cases have been observed. These puzzling transients may be similar in nature to bursters. However, some are possibly RS Canum Venaticorum binaries, described in the Variable Stars section of this Introduction.

Over 100 galactic and extragalactic supernova remnants have also proved to be X-ray sources. Five of the most prominent are listed in our tabulation. High-resolution studies, tracing out the shock-wave fronts where the remnant plows into the interstellar medium, yield information about the shape and intensity distribution of the remnant as well as the density and composition of both regions. Central pulsars, typified by those in the Crab and Vela supernova remnants, are also investigated in X-rays.

The nuclei of normal galaxies, such as the Milky Way and M31 in Andromeda, are weak X-ray sources. However, those of radio galaxies and QSO's can be extremely luminous in this regard, as our tabulation indicates. The most promising hypothesis to account for this energy is the accretion of matter onto supermassive black holes at the centers of these systems.

Finally, the detection of extended X-ray emission from rich galaxy clusters implies that the space between their members is filled with tenuous gas whose temperature may be as high as $100,000,000°$ Kelvin. Since the gas is enriched with heavy elements, it is not virgin primordial material, but reprocessed gas that has in some manner been stripped from the galaxies after they formed. How this occurred is uncertain, but the cause may have been supernova explosions or galaxy collisions and interactions. Some of the clusters may also have a central galaxy that is X-ray dominant, such as M87 in the nearby Virgo cluster.

The 74 sources in our X-ray tabulation are but a small fraction of the total number now known. The sample was chosen to illustrate the variety of types discussed above, and it consists primarily of sources for which optical identifications have been confirmed. Most of the information was obtained from the 1983 catalogue of galactic and Magellanic Cloud objects by H. V. D. Bradt and J. E. McClintock. Data on extragalactic sources and supernova remnants have been selected from the compilations of P. R. Amnuel *et al.* and M. V. Zombeck's *Handbook of Space Astronomy and Astrophysics.* The optical magnitudes and spectral types were sometimes obtained from other sections of the present volume.

References

General References

Becvar, A., *Atlas of the Heavens — II: Catalogue 1950.0,* 4th edition, Prague and Cambridge, Mass., 1964: Czechoslovak Academy of Sciences.

Blanco, V. M., S. Demers, G. G. Douglass, and M. P. FitzGerald, *Publications* of the U. S. Naval Observatory, 2nd Series, **21,** Washington, D. C., 1968.

Burnham, Jr., R., *Burnham's Celestial Handbook,* New York, 1978: Dover Publications, Inc.

Cannon, A. J., and E. C. Pickering, *The Henry Draper Catalogue* and *The Henry Draper Extension,* Harvard College Observatory Annals, **91-99, 100,** and **112,** Cambridge, Mass., 1918-24, 1936, 1949. [HD]

Dreyer, J. L. E., *New General Catalogue of Nebulae and Clusters of Stars (1888), Index Catalogue (1895), Second Index Catalogue (1908),* London, 1962: Royal Astronomical Society. [NGC, IC]

Hoffleit, D., *The Bright Star Catalogue,* 4th revised edition, New Haven, Conn., 1982: Yale University Observatory. [YBS; star designations HR]

Houk, N., and A. P. Cowley, *University of Michigan Catalogue of Two-Dimensional Spectral Types for HD Stars,* **1** and **2,** Ann Arbor, Mich., 1975-78.

Schaifers, K., and H. H. Voigt, editors, *Landolt-Börnstein Numerical Data and Functional Relationships in Science and Technology,* Group VI, Vol. 2, New York, 1981: Springer-Verlag.

Smithsonian Institution, *Smithsonian Astrophysical Observatory Star Catalog,* Washington, D. C., 1966, 1971: Smithsonian Institution. [SAO]

Sulentic, J. W., and W. G. Tifft, *The Revised New General Catalogue of Nonstellar Astronomical Objects,* Tucson, 1980: University of Arizona Press.

Vehrenberg, H., *Atlas of Deep-Sky Splendors,* 4th English edition, Cambridge, Mass., 1983: Sky Publishing Corp.

Zombeck, M. V., *Handbook of Space Astronomy and Astrophysics,* Cambridge, 1982: Cambridge University Press.

Glossary of Selected Astronomical Names

All short abbreviations of references used in the Glossary are grouped here for convenience. For many of them, more complete bibliographical information appears elsewhere in the References section.

AA. Astronomy and Astrophysics.

AAS. Astronomy and Astrophysics Supplement Series.

AfA. Arkiv for Astronomi, Stockholm.

AJ. Astronomical Journal.

Allen. R. H. Allen, *Star Names: Their Lore and Meaning,* New York, 1963: Dover.

ApJ. Astrophysical Journal.

ApJL. Astrophysical Journal Letters.

ApJS. Astrophysical Journal Supplement Series.

ApSS. Astrophysics and Space Science.

BBR. H. J. Bernhard, D. A. Bennett, and H. S. Rice, *New Handbook of the Heavens,* New York, 1941: Whittlesey House.

BC. B. J. Bok and C. S. Cordwell, *A Study of Dark Nebulae,* 1971.

Becvar. A. Becvar, *Atlas of the Heavens — II: Catalogue 1950.0,* 4th edition, 1964.

BIS. F. Spite and R. Lahmek, "Stars Named After Astronomers' Names," *Bulletin d'Information du Centre de Données Stellaires,* **22,** 105, 1982.

Burnham. R. Burnham, Jr., *Burnham's Celestial Handbook,* 1978.

Cederblad. S. Cederblad, "Catalogue of Bright Diffuse Galactic Nebulae," 1946.

CSCA. J. Ruprecht *et al., Catalogue of Star Clusters and Associations,* Supplement I, 1981.

CU. P. Murdin and D. Allen, *Catalogue of the Universe,* New York, 1979: Crown Publishers, Inc.

Denning. W. E. Denning, *Telescopic Work for Starlight Evenings,* London, 1891: Taylor and Francis.

FDN. A. Fernandez, M.-C. Lortet, and F. Spite, "The First Dictionary of the Nomenclature of Celestial Objects," *Astronomy and Astrophysics Supplement Series,* **52,** 1983.

GAA. J. Hopkins, *Glossary of Astronomy and Astrophysics,* 2nd edition, Chicago, 1980: University of Chicago Press.

HA. A. Sandage, *Hubble Atlas of Galaxies,* 1961.

J. Herschel. J. F. W. Herschel, *Results of Astronomical Observations at the Cape of Good Hope,* London, 1847: Smith, Elder, and Co.

Holmberg. E. Holmberg, "Photographic Photometry of Extragalactic Nebulae," *Meddelanden* fran Lunds Astronomiska Observatorium, Ser. 2, **14,** No. 136, 1958.

IAU. International Astronomical Union.

IBVS. Information Bulletin on Variable Stars.

LB. K. Schaifers and H. H. Voigt, editors, *Landolt-Börnstein Numerical Data and Functional Relationships in Science and Technology,* 1981.

MK. J. H. Mallas and E. Kreimer, *The Messier Album,* Cambridge, Mass., 1978: Sky Publishing Corp.

MNRAS. Monthly Notices of the Royal Astronomical Society.

MWLC. Mount Wilson-Las Campanas Observatory *Report.*

Noonan. T. W. Noonan, "List of Clusters of Galaxies with Published Redshifts," 1981.

PA. Popular Astronomy.

PASJ. Publications of the Astronomical Society of Japan.

RASC. Observer's Handbook 1984, R. L. Bishop, editor, Toronto, 1983: Royal Astronomical Society of Canada.

RC2. G. de Vaucouleurs *et al., Second Reference Catalogue of Bright Galaxies,* 1976.

RVG. G. G. C. Palumbo *et al., Catalogue of Radial Velocities of Galaxies,* 1983.

S&T. Sky and Telescope.

SA. A. Sandage and G. A. Tammann, *A Revised Shapley-Ames Catalog of Bright Galaxies,* 1981.

Serviss. G. P. Serviss, *Astronomy with an Opera-Glass,* New York, 1888: D. Appleton and Co.

Sh2. S. Sharpless, "A Catalogue of H II Regions," 1959.

Shapley. H. Shapley, *Galaxies,* 3rd edition, Cambridge, Mass., 1972: Harvard University Press.

Smyth. W. H. Smyth, *A Cycle of Celestial Objects,* London, 1844: J. W. Parker.

SovA. Soviet Astronomy.

SW. Sterne und Weltraum.

Vehrenberg. H. Vehrenberg, *Atlas of Deep-Sky Splendors,* 4th English edition, Cambridge, Mass., 1983: Sky Publishing Corp.

Webb. T. W. Webb, *Celestial Objects for Common Telescopes,* 1st edition, London, 1859: Longman, Green, Longman, and Roberts.

WS2. K. Glyn Jones, editor, *Webb Society Deep-Sky Observer's Handbook,* Vol. 2, Planetary and Gaseous Nebulae, 1979.

WS4. K. Glyn Jones, editor, *Webb Society Deep-Sky Observer's Handbook,* Vol. 4, Galaxies, 1981.

WS5. K. Glyn Jones, editor, *Webb Society Deep-Sky Observer's Handbook,* Vol. 5, Clusters of Galaxies, 1982.

YBS. D. Hoffleit, *The Bright Star Catalogue,* 4th revised edition, 1982.

ZAp. Zeitschrift für Astrophysik.

Double and Multiple Stars

Aitken, R. G., *New General Catalogue of Double Stars Within 120° of the North Pole,* Washington, D. C., 1932: Carnegie Institution. [ADS]

Burnham, S. W., *A General Catalogue of Double Stars Within 121° of the North Pole,* Washington, D. C., 1906: Carnegie Institution. [BDS]

Burnham, S. W., *Measures of Proper Motion Stars Made With the 40-inch Refractor,* Washington, D. C., 1913: Carnegie Institution. [β pm]

Crossley, E., J. Gledhill, and J. M. Wilson, *A Handbook of Double Stars,* London, 1879: Macmillan.

Franz, O. G., *et al.,* "Photographic Measures of Double Stars," *Publications* of the U. S. Naval Observatory, 2nd Series, **18,** Part 1, Washington, D. C., 1963. Also, the continuations by V. V. Kallarakal *et al.,* **18,** Part 7 (1969); F. J. Josties *et al.,* **22,** Part 6 (1974); and F. J. Josties *et al.,* **24,** Part 5 (1978).

Jeffers, H. M., W. H. van den Bos, and F. M. Greeby, *Index Catalogue of Visual Double Stars, 1961.0,* Lick Observatory, 1963: University of California. The magnetic tape of this catalogue (U. S. Naval Observatory, 1976) includes some more recent measures and corrections. [IDS]

Jones, K. Glyn, editor, *Webb Society Deep-Sky Observer's Handbook,* Vol. 1, Double Stars; Short Hills, N. J., 1979: Enslow Publishers.

Russell, H. C., *Results of Double Star Measures,* Sydney, 1882: Thomas Richards.

Wallenquist, A., *A Catalogue of Photoelectric Magnitudes and Colours of Visual Double and Multiple Systems,* Uppsala, 1981: Almqvist & Wiksell.

Visual Binary Stars

Aitken, R. G., *The Binary Stars,* New York, 1918: McMurtrie.

Couteau, P., and M. Fulconis, magnetic tape, 1982.

Couteau, P., *Observing Visual Double Stars,* Cambridge, Mass., 1981: MIT Press.

Meeus, J., *Astronomical Formulae for Calculators,* Richmond, Va., 1982: Willmann-Bell.

See, T. J. J., *Researches on the Evolution of the Stellar Systems,* Vol. 1, Orbits and General Characteristics of Binary Stars; Lynn, Mass., 1896: Nichols Press.

Worley, C. E., and W. D. Heintz, "Fourth Catalog of Orbits of Visual Binary Stars," *Publications* of the U. S. Naval Observatory, 2nd Series, **24,** Part 7, Washington, D. C., 1983.

Spectroscopic Binaries

Aitken, R. G., *The Binary Stars,* New York, 1918: McMurtrie.

Batten, A. H., J. M. Fletcher, and P. J. Mann, *Seventh Catalogue of the Orbital Elements of Spectroscopic Binary Systems,* Publications of the Dominion Astrophysical Observatory, **25,** No. 5, 1978.

Variable Stars and Suspected Variable Stars

Bidelman, W. P., "Addenda and Corrections to the New Catalogue of Suspected Variable Stars," *Information Bulletin on Variable Stars,* No. 2515, 1984.

Clark, D. H., and F. R. Stephenson, *The Historical Supernovae,* New York, 1977: Pergamon Press.

Hall, D. S., "The RS Canum Venaticorum Binaries," in *Solar Phenomena in Stars and Stellar Systems,* R. M. Bonnet and A. K. Dupree, editors, Dordrecht, 1981: D. Reidel.

Kholopov, P. N., "On the Classification of Variable Stars," *Variable Stars,* **21,** 465, 1981.

Kukarkin, B. V., *et al., General Catalogue of Variable Stars,* 3rd edition, Moscow, 1969-70; 1st, 2nd and 3rd Supplements, 1971, 1974, 1976; 4th edition (Vol. 1 only), 1984: Soviet Academy of Science. [GCVS]

Kukarkin, B. V., *et al., New Catalogue of Suspected Variable Stars,* Moscow, 1982: Publishing Office Nauka. [NSV]

Rudnicki, K., editor, *Rocznik Astronomiczny Obserwatorium Krakowskiego,* International Supplement, No. 54, Cracow, 1983.

Scovil, C. E., *The AAVSO Variable Star Atlas,* Cambridge, Mass., 1980: Sky Publishing Corp.

Open Clusters

Alter, G., J. Ruprecht, and V. Vanysek, *Catalogue of Star Clusters and Associations,* 2nd edition, Budapest, 1970: Akademiai Kiado.

Becker, W., "Applications of Multicolor Photometry," Chapter 13 in *Basic Astronomical Data,* K. Aa. Strand, editor, Chicago, 1963: University of Chicago Press.

Blaauw, A., "The O Associations in the Solar Neighborhood," *Annual Review of Astronomy and Astrophysics,* **2,** 213, 1964.

Humphreys, R. M., "Studies of Luminous Stars in Nearby Galaxies. I. Supergiants and O stars in the Milky Way," *Astrophysical Journal Supplement Series,* **38,** 309, 1978.

Janes, K., and D. Adler, "Open Clusters and Galactic Structure," *Astrophysical Journal Supplement Series,* **49,** 425, 1982.

Jones, K. Glyn, editor, *Webb Society Deep-Sky Observer's Handbook,* Vol. 3, Open and Globular Clusters; Hillside, N. J., 1980: Enslow Publishers.

Kron, G. E., "Star Clusters in the Small Magellanic Cloud: I. Identification of 69 Clusters," *Publications* of the Astronomical Society of the Pacific, **68,** 125, 1956.

Lynga, G., *Catalogue of Open Cluster Data,* Lund, 1981, revised 1983: Lund Observatory.

Lynga, G., "Open Clusters in Our Galaxy," *Astronomy and Astrophysics,* **109,** 213, 1982.

Ruprecht, J., B. Balazs, and R. E. White, *Catalogue of Star Clusters and Associations,* Supplement I, Budapest, 1981: Akademiai Kiado.

Shapley, H., and E. M. Lindsay, "A Catalogue of Clusters in the Large Magellanic Cloud," *Irish Astronomical Journal,* **6,** 74, 1963.

Skiff, B., private communication, 1983.

van den Bergh, S., and G. L. Hagen, "*UBV* Photometry of Star Clusters in the Magellanic Clouds," *Astronomical Journal,* **73,** 569, 1968. [vdB-Ha]

Globular Clusters

Aaronson, M., R. A. Schommer, and E. W. Olszewski, "AM-1: A Very Distant Globular Cluster," *Astrophysical Journal,* **276,** 221, 1984.

Alcaino, G., "Basic Data for Galactic Globular Clusters," *Publications* of the Astronomical Society of the Pacific, **89,** 491, 1977.

Alcaino, G., "Basic Morphological Data for Galactic Globular Clusters," *Vistas in Astronomy,* **23,** 1, 1979.

Arp, H., "Globular Clusters in the Galaxy," Chapter 19 in *Galactic Structure,* A. Blaauw and M. Schmidt, editors, Chicago, 1965: University of Chicago Press.

Gascoigne, S. C. B., "Colour-Magnitude Diagrams for Nine Globular-like Clusters in the Magellanic Clouds," *Monthly Notices* of the Royal Astronomical Society, **134,** 59, 1966.

Grindlay, J. E., and P. Hertz, "Discovery of an Obscured Globular Cluster Associated with GX 354+0 (=4U/MXB 1728-34)," *Astrophysical Journal Letters,* **247,** L17, 1981.

Hanes, D., and B. Madore, editors, *Globular Clusters,* Cambridge, 1980: Cambridge University Press.

Harris, W. E., "Spatial Structure of the Globular Cluster System and the Distance to the Galactic Center," *Astronomical Journal,* **81,** 1095, 1976.

Harris, W. E., "The Galactic Distance Scale: Globular Clusters," in *Star Clusters,* J. E. Hesser, editor, Boston, 1980: D. Reidel.

Harris, W. E., and R. Racine, "Globular Clusters in Galaxies," *Annual Review of Astronomy and Astrophysics,* **17,** 241, 1979.

Harris, W. E., and S. van den Bergh, "The Distant Globular Clusters Palomar 14 and Palomar 15," preprint, 1984.

Hesser, J. E., *et al.,* "A New Color-Magnitude Diagram for the Peculiar Star Cluster E3 = C0921-770," *Publications* of the Astronomical Society of the Pacific, **96,** 406, 1984.

Hodge, P. W., "Studies of the Large Magellanic Cloud. I. The Red Globular Clusters," *Astrophysical Journal,* **131,** 351, 1960.

Kron, G. E., "Star Clusters in the Small Magellanic Cloud: I. Identification of 69 Clusters," *Publications* of the Astronomical Society of the Pacific, **68,** 125, 1956.

Kron, G. E., and N. U. Mayall, "Photoelectric Photometry of Galactic and Extragalactic Star Clusters," *Astronomical Journal,* **65,** 581, 1960.

Kukarkin, B. V., *The General Catalogue of Globular Star Clusters of Our Galaxy,* Moscow, 1974: Publishing Office Nauka.

Liller, W., "Searches for the Optical Counterparts of the X-ray Burst Sources MXB 1728-34 and MXB 1730-33," *Astrophysical Journal Letters,* **213,** L21, 1977.

Lindsay, E. M., "The Cluster System of the Small Magellanic Cloud," *Monthly Notices* of the Royal Astronomical Society, **118,** 172, 1958.

Madore, B. F., and H. C. Arp, "Three New Faint Star Clusters," *Astrophysical Journal Letters,* **227,** L103, 1979. [AM]

Malken, M., D. Kleinmann, and J. Apt, "Infrared Studies of Globular Clusters Near the Galactic Center," *Astrophysical Journal,* **237,** 432, 1980.

Schuster, H.-E., and R. M. West, "Two New Stellar Systems Detected on ESO Schmidt Plates," *The Messenger,* No. 10, 13, 1977.

Shapley, H., and E. M. Lindsay, "A Catalogue of Clusters in the Large Magellanic Cloud," *Irish Astronomical Journal,* **6,** 74, 1963.

Shawl, S. J., J. E. Hesser, and J. E. Meyer, "Image-Tube Radial Velocities of Selected Globular Clusters," in *Astrophysical Parameters for Globular Clusters,* A. G. D. Philip and D. S. Hayes, editors, Schenectady, N. Y., 1981: L. Davis Press.

Shawl, S. J., and R. E. White, "Accurate Optical Positions for the Centers of Galactic Globular Clusters," preprint, submitted to *Astronomical Journal,* 1984.

Terzan, A., "Quatre nouveaux amas stellaires dans la direction de le région centrale de la Galaxie," *Astronomy and Astrophysics,* **12,** 477, 1971. [Ter]

van den Bergh, S., and G. L. Hagen, "*UBV* Photometry of Star Clusters in the Magellanic Clouds," *Astronomical Journal,* **73,** 569, 1968. [vdB-Ha]

Webbink, R. F., "Catalogue of Radial Velocities in Galactic Globular Clusters," *Astrophysical Journal Supplement Series,* **45,** 259, 1981.

Zinn, R., and M. J. West, "The Globular Cluster System of the Galaxy. III. Measurements of Radial Velocity and Metallicity for 60 Clusters and a Compilation of Metallicities for 121 Clusters," *Astrophysical Journal Supplement Series,* **55,** 45, 1984.

Bright Nebulae

Bok, B. J., M. J. Bester, and C. M. Wade, "Catalogue of H II Regions in the Milky Way," *Daedelus,* **86,** 9, 1955.

Cederblad, S., "Catalogue of Bright Diffuse Galactic Nebulae," *Meddelanden* fran Lunds Astronomiska Observatorium, Ser. 2, **12,** No. 119, 1946. [Ced]

Davies, R. D., K. H. Elliot, and J. Meaburn, "The Nebula Complexes of the Large and Small Magellanic Clouds," *Memoirs of the Royal Astronomical Society,* **81,** 89, 1976.

Dorschner, V. J., and J. Gürtler, "Untersuchungen über Reflexionsnebel am Palomar Sky Survey I. Verzeichnis von Reflexionsnebel," *Astronomische Nachrichten,* **287,** 257, 1964.

Gum, C. S., "A Survey of Southern H II Regions," *Memoirs of the Royal Astronomical Society,"* **67,** 21, 1955. [Gum]

Henize, K. G., "Catalogue of Hα-Emission Stars and Nebulae in the Magellanic Clouds," *Astrophysical Journal Supplement Series,* **2,** 315, 1956.

Innes, R., "Catalogue of Clusters and Nebulae near the Large Magellanic Cloud," Union Observatory of South Africa *Circular,* No. 61, 243, 1924.

Jones, K. Glyn, editor, *Webb Society Deep-Sky Observer's Handbook,* Vol. 2, Planetary and Gaseous Nebulae; Hillside, N. J., 1979: Enslow Publishers.

Lynds, B. T., "Catalogue of Bright Nebulae," *Astrophysical Journal Supplement Series,* **12,** 163, 1965. [LBN]

Maran, S. P., J. C. Brandt, and T. P. Stecker, editors, *The Gum Nebula and Related Problems,* Washington, D. C., 1973: NASA (SP-332).

Marsalkova, P., "A Comparison Catalogue of H II Regions," *Astrophysics and Space Science,* **27,** 3, 1974.

Minkowski, R., "New Emission Nebulae," *Publications* of the Astronomical Society of the Pacific, **58,** 305, 1946. [M1]

Rodgers, A. W., C. T. Campbell, and J. B. Whiteoak, "A Catalogue of Hα-Emission Regions in the Southern Milky Way," *Monthly Notices* of the Royal Astronomical Society, **121,** 103, 1960. [RCW]

Sharpless, S., "A Catalogue of H II Regions," *Astrophysical Journal Supplement Series,* **4,** 257, 1959. [Sh2]

van den Bergh, S., "A Study of Reflection Nebulae," *Astronomical Journal,* **71,** 990, 1966. [vdB]

van den Bergh, S., "A Catalogue of Galactic Supernova Remnants," in *Supernova Remnants and their X-ray Emission,* J. Danziger and P. Gorenstein, editors, Boston, 1983: D. Reidel.

van den Bergh, S., and W. Herbst, "Catalogue of Southern Stars Embedded in Nebulosity," *Astronomical Journal,* **80,** 208, 1975. [vdBH]

van den Bergh, S., and W. Herbst, "R Associations. I. **UBV** Photometry and MK Spectroscopy of Stars in Southern Reflection Nebulae," *Astronomical Journal,* **80,** 212, 1975.

van den Bergh, S., A. P. Marscher, and Y. Terzian, "An Optical Atlas of Galactic Supernova Remnants," *Astrophysical Journal Supplement Series,* **26,** 19, 1973.

Dark Nebulae

Barnard, E. E., "Catalogue of 349 Dark Objects in the Sky," in *A Photographic Atlas of Selected Regions of the Milky Way,* Washington, D. C., 1927: Carnegie Institution. [B]

Bernes, C., "A Catalogue of Bright Nebulosities in Opaque Dust Clouds," *Astronomy and Astrophysics Supplement Series,* **29,** 65, 1977. [Be]

Bok, B. J., *The Distribution of Stars in Space,* Chicago, 1937: University of Chicago Press; pages 72-87.

Bok, B. J., and C. S. Cordwell, *A Study of Dark Nebulae,* Tucson, 1971: Steward Observatory, University of Arizona.

Bok, B. J., C. S. Cordwell, and R. H. Cromwell, "Globules," Chapter 4 in *Dark Nebulae, Globules, and Protostars,* B. T. Lynds, editor, Tucson, 1971: University of Arizona.

Feitzinger, J. V., and J. A. Strüwe, "Catalogue of Dark Nebulae and Globules for Galactic Longitudes 240 to 360 Degrees," preprint, submitted to *Astronomy and Astrophysics Supplement Series,* 1984.

Khavtassi, D. Sh., "The Statistical Study of Dark Nebulae," *Byulletin* Abastumanskoy Astrofizicheskoy Observatorii, No. 18, 29, 1955 (NASA Technical Translation TT F-1741b).

Lynds, B. T., "Catalogue of Dark Nebulae," *Astrophysical Journal Supplement Series,* **7,** 1, 1962. [LDN]

Lynds, B. T., "Dark Nebulae," Chapter 3 in *Nebulae and Interstellar Matter,* B. M. Middlehurst and L. H. Aller, editors, Chicago, 1968: University of Chicago Press.

Malin, D. F., "Dust Clouds of Sagittarius," *Sky and Telescope,* **63,** 254, March 1982.

Sandqvist, Aa., "More Southern Dark Dust Clouds," *Astronomy and Astrophysics,* **57,** 467, 1977. [Sa]

Sandqvist, Aa., and K. P. Lindroos, "Interstellar Formaldehyde in Southern Dark Dust Clouds," *Astronomy and Astrophysics,* **53,** 179, 1976. [SL]

Schoenberg, E., "Katalog von 1456 Dunkelwolken der nördlichen Milchstrasse bis zur südlichen Deklination = −36°," *Veröffentlichungen* der Sternwarte München, **5,** No. 21, 1964.

Sim, M. E., "Results of a Search for Globules in *OB* Clusters and Associations," *Publications* of the Royal Observatory, Edinburgh, **6,** 181, 1968.

Planetary Nebulae

Acker, A., *et al.,* Catalogue of the Central Stars of True and Possible Planetary Nebulae, Publication Speciale du Centre de Données Stellaires, Strasbourg, No. 3, 1982.

Acker, A., J. Marcout, and F. Ochsenbein, "Catalogue and Bibliographical Index of Planetary Nebulae" (Strasbourg Catalogue of Galactic Planetary Nebulae), *Astronomy and Astrophysics Supplement Series,* **43,** 265, 1981.

Acker, A., J. Marcout, and F. Ochsenbein, *Complement I to the Catalogue of the Central Stars of True and Possible Planetary Nebulae,* Strasbourg, 1982: Observatoire de Strasbourg.

Acker, A., J. Marcout, and F. Ochsenbein, *Index of Discovery Lists of True, Probable, and Possible Planetary Nebulae,* Strasbourg, 1982: Observatoire de Strasbourg.

Acker, A., J. Marcout, and C. Schohn, *Complement II to the Catalogue of Central Stars of True and Possible Planetary Nebulae,* Strasbourg, 1983: Observatoire de Strasbourg.

Jones, K. Glyn, editor, *Webb Society Deep-Sky Observer's Handbook,* Vol. 2, Planetary and Gaseous Nebulae; Hillside, N. J., 1979: Enslow Publishers.

Kohoutek, L., "New and Misclassified Planetary Nebulae," in *Planetary Nebulae,* D. R. Flower, editor, Dordrecht, Holland, 1983: D. Reidel.

Perek, L., and L. Kohoutek, *Catalogue of Galactic Planetary Nebulae,* Prague, 1967: Academia Publishing House of the Czechoslovak Academy of Sciences. [PK]

Pottasch, S. B., *Planetary Nebulae,* Dordrecht, Holland, 1984: D. Reidel.

Schneider, S. E., Y. Terzian, A. Purgathofer, and M. Perinotto, "Radial Velocities of Planetary Nebulae," *Astrophysical Journal Supplement Series,* **52,** 399, 1983.

Galaxies

Abell, G. O., "The Distribution of Rich Clusters of Galaxies," *Astrophysical Journal Supplement Series,* **3,** 211, 1958. [A]

Arp, H., "Atlas of Peculiar Galaxies," *Astrophysical Journal Supplement Series,* **14,** 1, 1966.

Bahcall, N. A., and R. M. Soneira, "A Supercluster Catalog," *Astrophysical Journal,* **277,** 27, 1984.

Binggelli, B., and A. Sandage, "Studies of the Virgo Cluster. I. Photometry of 109 Galaxies near the Cluster Center To Serve as Standards," *Astronomical Journal,* **89,** 64, 1984.

Christian, C. A., and R. B. Tully, "The Local Group Irregular Galaxies LGS 3 and Pegasus," *Astronomical Journal,* **88,** 934, 1983.

Davidson, K., and H. Netzer, "The Emission Lines of Quasars and Similar Objects," *Reviews of Modern Physics,* **51,** 715, 1979.

Demers, S., *et al.,* "Leo A: A Color-Magnitude Diagram of its Brightest Stars," *Astronomical Journal,* **89,** 1160, 1984.

de Vaucouleurs, G., and G. Bollinger, "Contributions to Galaxy Photometry. VI. Revised Standard Total Magnitudes and Colors of 228 Multiply Observed Galaxies," *Astrophysical Journal Supplement Series,* **34,** 469, 1977.

de Vaucouleurs, G., A. de Vaucouleurs, and H. G. Corwin, Jr., *Second Reference Catalogue of Bright Galaxies,* Austin, Texas, 1976: University of Texas Press. [RC2]

de Vaucouleurs, G., and C. Head, "Contributions to Galaxy Photometry. VII. Standard Total Magnitudes of 139 Bright Galaxies in the Virgo Cluster Area," *Astrophysical Journal Supplement Series,* **36,** 439, 1978.

de Vaucouleurs, G., and W. D. Pence, "Contributions to Galaxy Photometry. IX. Reduction to the Standard B_T System of the Magnitudes of 1180 Galaxies in the Ames, Reiz, and Zwicky Catalogs in the Virgo Area," *Astrophysical Journal Supplement Series,* **40,** 425, 1979.

Dickel, J. R., and H. J. Rood, "Integrated Masses of Galaxies," *Astrophysical Journal,* **223,** 391, 1978.

Fisher, J. R., and R. B. Tully, "Neutral Hydrogen Observations of a Large Sample of Galaxies," *Astrophysical Journal Supplement Series,* **47,** 139, 1981.

Hoessel, J. G., and J. R. Mould, "Photometry of Resolved Galaxies. I. The Pegasus Dwarf Irregular," *Astrophysical Journal,* **254,** 38, 1982.

Humason, M. L., N. U. Mayall, and A. R. Sandage, "Redshifts and Magnitudes of Extragalactic Nebulae," *Astronomical Journal,* **61,** 97, 1956.

Jones, K. Glyn, editor, *Webb Society Deep-Sky Observer's Handbook,* Vol. 4, Galaxies; Hillside, N. J., 1981: Enslow Pub.

Jones, K. Glyn, editor, *Webb Society Deep-Sky Observer's Handbook,* Vol. 5, Clusters of Galaxies; Hillside, N. J., 1982: Enslow Publishers.

Kraan-Korteweg, R. C., and G. A. Tammann, "A Catalogue of Galaxies within 10 MPC," *Astronomische Nachrichten,* **300,** 181, 1979.

Nilson, P. N., *Uppsala General Catalogue of Galaxies,* Uppsala, 1973: Uppsala Astronomical Observatory. [UGC]

Nilson, P. N., *Catalogue of Selected Non-UGC Galaxies,* Uppsala, 1974: Uppsala Astronomical Observatory. [UGCA]

Noonan, T. W., "List of Clusters of Galaxies with Published Redshifts," *Astrophysical Journal Supplement Series,* **45,** 613, 1981.

Palumbo, G. G. C., G. Tanzella-Nitti, and G. Vettolani, *Catalogue of Radial Velocities of Galaxies,* New York, 1983: Gordon and Breach Science Publishers.

Rood, H. J., and G. N. Sastry, " 'Tuning Fork' Classification of Rich Clusters of Galaxies," *Publications* of the Astronomical Society of the Pacific, **83,** 313, 1971.

Sandage, A., *Hubble Atlas of Galaxies,* Washington, D. C., 1961: Carnegie Institution.

Sandage, A., "Classification and Stellar Content of Galaxies Obtained from Direct Photography," Chapter 1 in *Galaxies and the Universe,* A. Sandage, M. Sandage, and J. Kristian, editors, Chicago, 1975: University of Chicago Press.

Sandage, A., and B. Binggelli, "Studies of the Virgo Cluster. III. A Classification System and an Illustrated Atlas of Virgo Cluster Dwarf Galaxies," *Astronomical Journal,* **89,** 919, 1984.

Sandage, A., and G. A. Tammann, *A Revised Shapley-Ames Catalog of Bright Galaxies,* Washington D. C., 1981: Carnegie Institution.

Sastry, G. N., and H. J. Rood, "Rectangular Coordinates of Rich Clusters of Galaxies in the Palomar Sky Survey Charts," *Astrophysical Journal Supplement Series,* **23,** 371, 1971.

Tully, R. B., *Atlas and Catalog of Nearby Galaxies,* magnetic tape, 1981.

van den Bergh, S., "Luminosity Classifications of Dwarf Galaxies," *Astronomical Journal,* **71,** 922, 1966. [D or DDO]

van den Bergh, S., "The Classification of Normal Galaxies," *Journal* of the Royal Astronomical Society of Canada, **69,** 57, 1975.

van den Bergh, S., "The Classification of Active Galaxies," *Journal* of the Royal Astronomical Society of Canada, **69,** 105, 1975.

van den Bergh, S., "A New Classification System for Galaxies," *Astrophysical Journal,* **206,** 883, 1976.

van den Bergh, S., "The Galaxies of the Local Group," in *The Structure and Evolution of Normal Galaxies,* S. M. Fall and D. Lynden-Bell, editors, Cambridge, 1981: Cambridge University Press.

Weedman, D. W., "Seyfert Galaxies," *Annual Review of Astronomy and Astrophysics,* **15,** 69, 1977.

Weedman, D. W., "More Seyfert Galaxies," *Monthly Notices* of the Royal Astronomical Society, **184,** 11P, 1978.

Quasi-stellar Objects

Burbidge, G., and M. Burbidge, *Quasi-stellar Objects,* San Francisco, 1967: W. H. Freeman.

Craine, E. R., *A Handbook of Quasi-stellar and BL Lacertae Objects,* Tucson, Ariz., 1977: Pachart.

Djorgovski, S., and H. Spinrad, "Discovery of a New Gravitational Lens," *Astrophysical Journal Letters,* **282,** L1, 1984.

Hewitt, A., and G. Burbidge, "A Revised Optical Catalogue of Quasi-stellar Objects," *Astrophysical Journal Supplement Series,* **43,** 57, 1980; **46,** 113, 1981.

Lawrence, C. R., *et al.,* "Discovery of a New Gravitational Lens System," *Science,* **223,** 46, 1984.

Schmidt, M., "Quasars," Chapter 8 in *Galaxies and the Universe,* A. Sandage, M. Sandage, and J. Kristian, editors, Chicago, 1975: University of Chicago Press.

Schmidt, M., and R. F. Green, "Quasar Evolution Derived from the Palomar Bright Quasar Survey and other Complete Quasar Surveys," *Astrophysical Journal,* **269,** 352, 1983.

Sramek, R. A., and D. W. Weedman, "An Optical and Radio Study of Quasars," *Astrophysical Journal,* **221,** 468, 1978.

Véron-Cetty, M.-P., and P. Véron, "A Catalogue of Quasars and Active Nuclei," European Southern Observatory *Report,* No. 1, 1984.

Walsh, D., R. F. Carswell, and R. J. Weymann, "0957 + 561A,B: Twin Quasi-stellar Objects or Gravitational Lens?" *Nature,* **279,** 381, 1979.

Weedman, D. W., *et al.,* "Discovery of a Third Gravitational Lens," *Astrophysical Journal Letters,* **255,** L5, 1982.

Weymann, R., *et al.,* "The Triple QSO PG 1115 + 08: Another Probable Gravitational Lens," *Nature,* **285,** 641, 1980.

Radio Sources

Bennett, A. S., "The Preparation of the Revised 3C Catalogue of Radio Sources," *Monthly Notices* of the Royal Astronomical Society, **125,** 75, 1962.

Bennett, A. S., "The Revised 3C Catalogue of Radio Sources," *Memoirs* of the Royal Astronomical Society, **68,** 163, 1962. [3C or 3CR]

Bolton, J. G., and A. J. Shimmins, "The Parkes 2700 MHz Survey (Fifth Part): Catalogue for the Declination Zone −35° to −45°," *Australian Journal of Physics,* Astrophysical Supplement, No. 30, 1, 1973. [PKS]

Bolton, J. G., A. J. Shimmins, J. V. Wall, and P. W. Butler, "The Parkes 2700 MHz Survey (Seventh, Eighth, Ninth and Tenth Parts)," *Australian Journal of Physics,* Astrophysical Supplement, No. 34, 1, 1975. [PKS]

Burbidge, G., and A. H. Crowne, "An Optical Catalogue of Radio Galaxies," *Astrophysical Journal Supplement Series,* **40,** 583, 1979.

Dixon, R. S., "A Master List of Radio Sources," *Astrophysical Journal Supplement Series,* **20,** 1, 1970.

Dixon, R. S., and J. D. Kraus, "A High-Sensitivity 1415 MHz Survey Between Declinations of 0° and 20° North," *Astronomical Journal,* **73,** 381, 1968.

Edge, D. O., *et al.,* "A Survey of Radio Sources at a Frequency of 159 Mc/s," *Memoirs* of the Royal Astr. Soc., **68,** 7, 1959.

Ekers, J. A., "The Parkes Catalogue of Radio Sources," *Australian Journal of Physics,* Astrophysical Suppl., No. 7, 1, 1969. [PKS]

Ekers, R. D., "Interferometric Observations of the Brightness Distribution of Southern Radio Sources," *Australian Journal of Physics,* Astrophysical Supplement, No. 6, 1969.

Fitch, L. T., R. S. Dixon, and J. D. Kraus, "A High-Sensitivity 1415 MHz Survey between Declinations of 0° and 20° North," *Astronomical Journal,* **74,** 612, 1969.

Fomalont, E. B., "A General Discussion of the Distribution of Brightness of Extragalactic Radio Sources," *Astrophysical Journal,* **157,** 1027, 1969.

Fomalont, E. B., "Two-Dimensional Structure of 76 Extragalactic Radio Sources at 1425 MHz," *Astronomical Journal,* **76,** 513, 1971.

Galt, J. A., and J. E. D. Kennedy, "Survey of Radio Sources Observed in the Continuum near 1420 MHz, Declinations −5° to +70°," *Astronomical Journal,* **73,** 135, 1968.

Gower, J. F. R., P. F. Scott, and D. Wills, "A Survey of Radio Sources in the Declination Ranges −07° to 20° and 40° to 80°," *Memoirs* of the Royal Astr. Soc., **71,** 49, 1967. [4C, part 2]

Gunn, J. E., *et al.,* "Investigations of the Optical Fields of 3CR Radio Sources to Faint Limiting Magnitudes — IV," *Monthly Notices* of the Royal Astronomical Society, **194,** 111, 1981.

Jenkins, C. J., G. G. Pooley, and J. M. Riley, "Observations of 104 Extragalactic Radio Sources with the Cambridge 5-km Telescope at 5 GHz," *Memoirs* of the Royal Astr. Soc., **84,** 61, 1977.

Kellerman, K. I., I. I. K. Pauliny-Toth, and W. C. Tyler, "Measurements of the Flux Density of Discrete Radio Sources at Centimeter Wavelengths. I. Observations at 2695 MHz (11.3 cm)," *Astronomical Journal,* **73,** 298, 1968.

Komesaroff, M. M., "A Southern Milky Way Survey at 408 Mc/s," *Australian Journal of Physics,* **19,** 75, 1966.

Kristian, J., "Optical Identification of 3CR Sources," Appendix to Chapter 6 in *Galaxies and the Universe,* A. Sandage, M. Sandage, and J. Kristian, editors, Chicago, 1975: Univ. of Chicago.

Laing, R. A., *et al.,* "Investigations of the Optical Fields of 3CR Radio Sources to Faint Limiting Magnitudes — II," *Monthly Notices* of the Royal Astronomical Society, **183,** 547, 1978.

Large, M. I., D. S. Mathewson, and C. G. T. Haslam, "A Radio Survey of the Galactic Plane at a Frequency of 408 Mc/s," *Monthly Notices* of the Royal Astronomical Society, **123,** 113, 1961.

Long, R. J., J. B. Haseler, and B. Elsmore, "A Survey of Radio Sources at 408 Mc/s," *Monthly Notices* of the Royal Astronomical Society, **125,** 313, 1963. [LHE]

Longair, M. S., and J. E. Gunn, "An Investigation of the Optical Fields of 35 3CR Radio Sources to Faint Limiting Optical Magnitudes," *Monthly Notices* of the Royal Astr. Soc., **170,** 121, 1975.

Macdonald, G. H., S. Kenderline, and A. C. Neville, "Observations of the Structure of Radio Sources in the 3C Catalogue — I," *Monthly Notices* of the Royal Astr. Soc., **138,** 259, 1968.

Mackay, C. D., "Observations of the Structure of Radio Sources in the 3C Catalogue — II," *Monthly Notices* of the Royal Astronomical Society, **145,** 31, 1969.

Maltby, P., and A. T. Moffet, "Brightness Distribution in Discrete Radio Sources. III. The Structure of the Sources," *Astrophysical Journal Supplement Series,* **7,** 141, 1962.

Manchester, R. N., and J. H. Taylor, "Observed and Derived Parameters for 330 Pulsars," *Astronomical Journal,* **86,** 1953, 1981.

Matthews, T. A., W. W. Morgan, and M. Schmidt, "A Discussion of Galaxies Identified with Radio Sources," *Astrophysical Journal,* **140,** 35, 1964.

Pacholczyk, A. G., *A Handbook of Radio Sources,* Tucson, Arizona, 1978: Pachart.

Pauliny-Toth, I. I. K., and K. I. Kellerman, "Measurements of the Flux Density and Spectra of Discrete Radio Sources at Centimeter Wavelengths. II. The Observations at 5 GHz (6 cm)," *Astronomical Journal,* **73,** 953, 1968.

Pauliny-Toth, I. I. K., C. M. Wade, and D. S. Heeschen, "Positions and Flux Densities of Radio Sources," *Astrophysical Journal Supplement Series,* **13,** 65, 1966.

Pilkington, J. D. H., and P. F. Scott, "A Survey of Radio Sources Between Declinations 20° and 40°," *Memoirs* of the Royal Astronomical Society, **69,** 183, 1965. [4C, part 1]

Riley, J. M., M. S. Longair, and J. E. Gunn, "Investigations of the Optical Fields of 3CR Radio Sources to Faint Limiting Magnitudes — III," *Monthly Notices* of the Royal Astronomical Society, **192,** 233, 1980.

Scheer, D. J., and J. D. Kraus, "A High-Sensitivity Survey of the North Galactic Polar Region at 1415 MHz," *Astronomical Journal,* **72,** 536, 1967.

Shimmins, A. J., "The Parkes 2700 MHz Survey: Catalogue for 03ʰ, 11ʰ, 19ʰ, and 23ʰ Zone, Declinations −33° to −75°," *Australian Journal of Physics,* Astrophysical Supplement, No. 21, 1, 1971. [PKS]

Shimmins, A. J., "Accurate Flux Densities at 5009 MHz of 1007 Radio Sources," *Australian Journal of Physics,* Astrophysical Supplement, No. 23, 1, 1972.

Shimmins, A. J., and J. G. Bolton, "The Parkes 2700 MHz Survey (Fourth Part): Catalogue for the South Polar Cap Zone, Declinations −75° to −90°," *Australian Journal of Physics,* Astrophysical Supplement, No. 26, 1, 1972. [PKS]

Shimmins A. J., and J. G. Bolton, "The Parkes 2700 MHz Survey (Sixth Part): Catalogue for the Declination Zone −30° to −35°," *Australian Journal of Physics,* Astrophysical Supplement, No. 32, 1, 1974. [PKS]

Smith, H. E., H. Spinrad, and E. O. Smith, "The Revised 3C Catalogue of Radio Sources: A Review of Optical Identifications and Spectroscopy," *Publications* of the Astronomical Society of the Pacific, **88,** 621, 1976.

Wall, J. V., A. J. Shimmins, and J. K. Merkelijn, "The Parkes 2700 MHz Survey: Catalogues for the ±4° Declination Zone and for the Selected Regions," *Australian Journal of Physics,* Astrophysical Supplement, No. 19, 1, 1971. [PKS]

Westerhout, G., "A Survey of the Continuous Radiation from the Galactic System at a Frequency of 1390 Mc/s" *Bulletins* of the Astronomical Institute of the Netherlands, **14,** 215, 1958. [W]

Wilson, R. W., "A Catalogue of Radio Sources in the Galactic Plane," *Observations* of the Owens Valley Radio Obs., No. 2, 1963.

Wyndham, J. D., "Optical Identification of Radio Sources in the 3C Revised Catalogue," *Astrophysical Journal,* **144,** 459, 1966.

X-ray Sources

Amnuel, P. R., O. H. Guseinov, and Sh. Yu. Rakhamimov, "A Catalog of X-ray Sources," *Astrophysical Journal Supplement Series,* **41,** 327, 1979.

Amnuel, P. R., O. H. Guseinov, and Sh. Yu. Rakhamimov, "Second Catalogue of X-ray Sources," *Astrophysics and Space Science,* **82,** 3, 1982.

Bradt, H. V., R. E. Doxsey, and J. G. Jernigan, "Positions and Identifications of Galactic X-ray Sources," *Advances in Space Exploration,* **3,** 3, 1979.

Bradt, H. V. D., and J. E. McClintock, "The Optical Counterparts of Compact Galactic X-ray Sources," *Annual Review of Astronomy and Astrophysics,* **21,** 13, 1983.

Giacconi, R., editor, *X-ray Astronomy with the Einstein Satellite,* Boston, 1981: D. Reidel.

Manchester, R. N., and J. H. Taylor, "Observed and Derived Parameters for 330 Pulsars," *Astronomical Journal,* **86,** 1953, 1981.

Glossary of Selected Astronomical Names

THE FARAWAY SIGHTS observed nightly by astronomers are exotic, beautiful, and often poorly understood. Perhaps the urge to cope with the unknown explains why so many homespun names have been coined for them, in conscious rebellion against the cold alphanumerics of catalogue designations.

Some of the names collected here are widely known, others are obscure, but they have all appeared in the astronomical literature. We have excluded most classical names of stars, constellations, and asterisms, since entire books are available on the subject (notably R. H. Allen's *Star Names: Their Lore and Meaning*). Sometimes an object has several common names, and these are cross-referenced to the one that seems to be in widest use.

Immediately after the name is the type of object, as follows: *CG*, cluster or group of galaxies; *Di*, diffuse (bright) nebula; *DN*, dark nebula; *Gb*, globular cluster; *Gx*, galaxy; *IR*, infrared object; *OC*, open cluster; *Pl*, planetary nebula; *QSO*, quasi-stellar object; *RS*, radio source; *SN*, supernova; *SNR*, supernova remnant; *St*, star or double star; *XR*, X-ray source.

After the type is the object's identity and position (always equinox 2000.0), along with a reference. Abbreviations adopted here for books and periodicals are explained on page xxxix. The reference is not intended to single out the original appearance of the name, but merely to illustrate usage or indicate where further information may be found.

30 Doradus. *Di.* (See Tarantula nebula.)

47 Tucanae. *Gb.* NGC 104 at 0^h $24^m.1$, $-72° 05'$.

γ Cassiopeiae nebula. *Di.* Sh2-185 (IC 59 and 63) at 0^h $58^m.1$, $+60° 56'$; Vehrenberg, page 29.

γ Cygni nebula. *Di.* IC 1318? Sh2-108? At 20^h $28^m.5$, $+39° 57'$; Vehrenberg, page 217.

η Carinae nebula. *Di.* NGC 3372, at 10^h $43^m.8$, $-59° 52'$; Burnham, page 466.

κ Crucis cluster. *OC.* NGC 4755, also called the Jewel Box, at 12^h $53^m.6$, $-60° 20'$; Burnham, page 730.

λ Centauri nebula. *Di.* RCW 62 (IC 2944 and 48) at 11^h $38^m.3$, $-63° 22'$; *MNRAS*, **121**, 103, 1960.

λ Orionis nebula. *Di.* Sh2-264, at 5^h 35^m, $+10°$; *CU*, page 113.

ρ Ophiuchi complex. *Di.* IC 4603-4 and dark nebulae, at 16^h $25^m.6$, $-24°$; *GAA*.

ρ Ophiuchi dark cloud. *DN.* (See ρ Oph complex.) *CU*, page 133.

ρ Ophiuchi nebula. *Di.* IC 4604, at 16^h $25^m.6$, $-23° 26'$; Vehrenberg, page 161.

ω Centauri. *Gb.* NGC 5139, at 13^h $26^m.8$, $-47° 29'$; Burnham, page 558.

Abt's star. *St.* HD 98088 = ADS 8115 at 11^h $17^m.0$, $-7° 08'$; *FDN*.

Ambartsumian's knot. *Gx.* NGC 3561C, Arp 105(C), at 11^h $11^m.6$, $+28° 43'$; *RC2*.

America nebula. *Di.* (See North America nebula.) M. Wolf's name, according to Cederblad.

Andrew's star. *St.* Suspected variable at 5^h $50^m.6$, $+31° 22'$; *S&T*, **29**, January, 1965, page 62.

Andromeda I. *Gx.* Satellite of M31, at 0^h $45^m.7$, $+38° 00'$; *LB*.

Andromeda II. *Gx.* Satellite of M31, at 1^h $16^m.4$, $+33° 27'$; *LB*.

Andromeda III. *Gx.* Satellite of M31, at 0^h $35^m.4$, $+36° 31'$; *LB*.

Andromeda IV. *Gx.* Satellite of M31, at 0^h $42^m.5$, $+40° 34'$; *LB*.

Andromeda nebula (or galaxy). *Gx.* (See Great Nebula in Andromeda.)

Antares nebula. *Di.* Ced 132, near Antares at 16^h $29^m.2$, $-26° 27'$; *LB*.

Antennae. *Gx.* NGC 4038-9 = VV 245, at 12^h $01^m.9$, $-18° 52'$; *CU*.

Ant nebula. *Di.* Bipolar nebula PK 331−1.1 at 16^h $17^m.2$, $-51° 59'$; *ApJ*, **221**, 151, 1978.

Aquarius dwarf galaxy. *Gx.* Local Group member? Position 20^h $46^m.9$, $-12° 51'$; *Die Sterne*, **58**, 303, 1982.

Arp's galaxy. *Gx.* Compact dwarf galaxy at 11^h $19^m.6$, $+51° 30'$; *ApJ*, **142**, 402, 1965.

Baade A, B. *Gx.* A pair with discordant redshifts at 0^h $49^m.9$, $+42° 35'$; *RVG*.

Baade's Windows (Sgr I, II). Low-absorption regions near globular cluster NGC 6522. Positions 17^h $59^m.4$, $-29° 03'$, and 18^h $15^m.3$, $-27° 53'$; *MNRAS*, **174**, 169, 1976.

Babcock's magnetic star. *St.* HD 215441 at 22^h $44^m.2$, $+55° 35'$; *FDN*.

Barbon's galaxy. *Gx.* Mrk 328 at 23^h $37^m.7$, $+30° 08'$; *RVG*.

Barnard's Loop. *Di.* Sh2-276, 12° ring centered at 5^h 35^m, $-3°$; *ApJS*, **4**, 257, 1959.

Barnard's galaxy. *Gx.* NGC 6822, in Local Group, at 19^h $44^m.9$, $-14° 48'$; Shapley, page 128.

Barnard's star. *St.* The star with the largest known proper motion. Position for epoch and equinox 2000.0: 17^h $57^m.8$, $+4° 42'$; *S&T*, **60**, August, 1980, page 111.

Baxendell's unphotographable nebula. *Di.* NGC 7088 = Ced 193, at 21^h $33^m.4$, $-0° 23'$. Seen by J. Baxendell in 1880 as a very large diffuse nebulosity, 52' by 75', lying north of the globular cluster M2. Confirmed visually by J. L. E. Dreyer (10-inch refractor) and listed by Cederblad, even though it seems not to exist.

B Cas. *SN.* (See Tycho's star.)

Bear Claw galaxy (or nebula). *Gx.* (See Bear Paw galaxy.) *FDN*.

Bear Paw galaxy. *Gx.* NGC 2537 at 8^h $13^m.2$, $+46° 00'$; WS4, page 94.

Becklin's star. *St.* IRC −10093 in M42, at 5^h $35^m.3$, $-5° 23'$; *BIS*.

Becklin-Neugebauer object. *IR.* In M42 near Becklin's star at 5^h $35^m.2$, $-5° 22'$; *GAA*; *CU*, page 104.

Beehive cluster. *OC.* (See Praesepe.)

Bidelman's helium variable star. *St.* V761 Cen = HD 125823, at 14^h $23^m.0$, $-39° 31'$; *FDN*.

Bidelman's peculiar star. *St.* KS Per = HD 30353, at 4^h $48^m.9$, $+43° 17'$; *FDN*.

Bidelman's star. *St.* HD 127617, at 14^h $31^m.9$, $+18° 45'$; *FDN*.

Binary pulsar. *St.* Discovered by R. Hulse and J. Taylor at 19^h $15^m.5$, $+16° 06'$; see IAU *Circular* 2704, 1974.

Black-eye galaxy (or nebula). *Gx.* M64 = NGC 4826, at 12^h $56^m.7$, $+21° 41'$; *S&T*, **53**, June, 1977, page 492.

"Black Hole" in Sagittarius. *DN.* Barnard 92, at 18^h $15^m.5$, $-18° 11'$; *PA*, **37**, 13, 1929.

Blaze star. *St.* The recurrent nova T CrB at 15^h $59^m.5$, $+25° 55'$; Burnham, page 708.

Blinking planetary. *Pl.* NGC 6826, at 19^h $44^m.8$, $+50° 31'$; Burnham, page 1178.

Blue planetary. *Pl.* NGC 3918 at 11^h $50^m.3$, $-57° 11'$; Allen, page 152.

Blue Snowball. *Pl.* NGC 7662 = PK 106−17.1 at 23^h $25^m.9$, $+42° 32'$; *S&T*, **19**, February, 1960, page 214.

Bode's nebula. *Gx.* M81, at 9^h $55^m.6$, $+69° 04'$; *CU*, page 34.

Bode's nebulae. *Gx.* M81 and M82, 9^h $55^m.7$, $+69° 23'$; BBR, page 201.

Bok's Valentine. *DN.* ESO 210−6A, at 8^h $25^m.5$, $-51° 01'$; *S&T*, **67**, March, 1984, page 227.

Boomerang nebula. *Di.* Bipolar nebula at 12^h $44^m.8$, $-54° 31'$; *MNRAS*, **193**, 321, 1980.

Bond's flare star. *St.* V3885 Sgr, 23^h $31^m.7$, $-2° 45'$; *FDN*.

Bootes cluster. *CG.* Abell 1930, 14^h 33^m, $+31° 33'$; Noonan.

Box. *CG.* NGC 4169, 4173-5, at 12^h $12^m.4$, $+29° 10'$; MWLC, 1981-2, page 638.

Box nebula. *Pl.* NGC 6309 at 17^h $14^m.1$, $-12° 54'$; *S&T*, **68**, August, 1984, page 185.

Branchett's object. *St.* Possible nova Sct 1981, at 18^h $46^m.9$, $-4° 57'$; IAU *Circular* 3571, 1981.

Brewer's star. *St.* HD 50169; magnetic star at 6^h $52^m.0$, $-1° 39'$; *FDN*.

Bridal Veil nebula. *Di.* (See Veil nebula.) Burnham, page 801.

Brocchi's cluster. *OC.* Collinder 399 at 19^h $25^m.4$, $+20° 11'$; *S&T*, **60**, September, 1980, page 257.

Bubble nebula. *Di.* NGC 7635, 23^h 20^m.7, +61° 12 ′; Burnham, page 522.

Bug nebula. *Di.* NGC 6302, 17^h 13^m.7, −37° 06 ′; Cederblad.

Burbidge chain. *CG.* Galaxy chain northeast of NGC 247, at 0^h 47^m.5, −20° 26 ′; *ApJ,* **185,** 797, 1973.

Burnham's nebula. *Di.* T Tau nebula (not NGC 1555), at 4^h 22^m.0, +19° 32 ′; *SW,* **19,** 205, 1980.

Butler's star. *St.* Flare star in the Small Magellanic Cloud. *FDN.*

Butterfly cluster. *OC.* M6 = NGC 6405, at 17^h 40^m.1, −32° 13 ′; Burnham, page 1706.

Butterfly nebula. *Pl.* (See Little Dumbbell.)

Butterfly nebula. *Di.* IC 2220, near HD 65750, at 7^h 56^m.9, −59° 08 ′; *AA,* **63,** 353, 1978.

BW Tauri. *Gx.* Seyfert galaxy 3C 120 at 4^h 33^m.2, +5° 21 ′; *RC2.*

California nebula. *Di.* NGC 1499, at 4^h 00^m.7, +36° 37 ′; Vehrenberg, page 53.

Campbell's star. *St.* HD 184738; nucleus of the planetary nebula PK 64 + 5.1 at 19^h 34^m.8, +30° 31 ′; *S&T,* **68,** August, 1984, page 185.

Cancer cluster. *CG.* Contains NGC 2563 at 8^h 21^m, +20° 56 ′; Noonan.

Capricorn dwarf system. *Gb.* Pal 12, at 21^h 46^m.5, −21° 14 ′.

Carafe galaxy. *Gx.* Seyfert galaxy with ring, 4^h 28^m.0, −47° 54 ′; *MNRAS,* **186,** 495, 1976.

Carafe group. *CG.* NGC 1595, 1598, and Carafe, 4^h 28^m.3, −47° 51 ′; *CU,* page 33.

Carina arm. Milky Way spiral arm at galactic longitude 280° to 300°; *PASP,* **88,** 647, 1976.

Carina dwarf galaxy. *Gx.* Type-dE galaxy in the Local Group, at 6^h 41^m.6, −50° 58 ′; *CU,* page 57.

Carina-Sagittarius arm. The Carina and Sagittarius arms of the Milky Way at galactic longitude 280° to 30°; *PASP,* **88,** 647, 1976.

Cartwheel galaxy. *Gx.* Ring galaxy at 0^h 37^m.4, −33° 45 ′; *CU,* page 32.

Cassiopeia A. *RS.* 3C 461, SNR. This is the strongest radio source in the sky. Located at 23^h 23^m.4, +58° 50 ′; Burnham, page 516.

Cave nebula. *Di.* Near HD 216629 and 217061, at 22^h 56^m.8, +62° 37 ′. Cederblad attributes the name to M. Wolf.

Centaurus A. *RS.* Galaxy NGC 5128 at 13^h 25^m.5, −43° 01 ′; *CU,* page 41.

Centaurus chain. *CG.* Chain of 10 galaxies at 12^h 44^m.3, −40° 43 ′; *ApSS,* **19,** 387, 1972.

Centaurus cluster. *CG.* Contains NGC 4696 at 12^h 50^m, −41° 18 ′; Noonan.

Centaurus Spur. Part of the Carina-Sagittarius spiral arm, at galactic longitude 310°. *PASP,* **88,** 647, 1976.

Centaurus X-3. *XR.* (See Krzeminski's star.)

Chanal's variable. *St.* Suspected variable NSV 2229, at 5^h 34^m.8, −5° 34 ′; *S&T,* **68,** November, 1984, page 443.

Chavira's supernova. *SN.* SN 1965h in NGC 4666 at 12^h 45^m.2, −0° 27 ′; *FDN.*

Checkmark nebula. *Di.* (See Omega nebula.) *FDN.*

Chevremont's star. *St.* Variable in globular cluster M2; the latter's coordinates are 21^h 33^m.5, −0° 49 ′. *S&T,* **38,** October, 1969, page 235.

Christmas Tree cluster. *OC.* NGC 2264 at 6^h 41^m.1, +9° 53 ′; Burnham, page 1206.

Chuadze's supernova. *SN.* SN 1967c in NGC 3389, 10^h 48^m.5, +12° 32 ′; *SovA,* **13,** 423, 1969.

Chu's object. *Di.* *B* star and nebula at 3^h 56^m.8, +51° 26 ′; *PASJ,* **31,** 417, 1979.

Circinus galaxy. *Gx.* Highly obscured spiral galaxy at 14^h 13^m.2, −65° 20 ′; *RC2.*

Cirrus nebula. *Di.* (See Veil nebula.) J. Herschel: like cirrostratus clouds.

Cloud nebula. *Di.* Recorded on an 1895 photograph by E. E. Barnard. (See Antares nebula.) Allen, page 367.

Clown Face nebula. *Pl.* (See Eskimo nebula.) Burnham, page 940.

Coalsack. *DN.* The dark 7°-by-5° region of the Milky Way near α Cru, 12^h 53^m, −63°; Serviss, page 2.

Coat Hanger. *OC.* (See Brocchi's cluster.)

Cocoon nebula. *Di.* IC 5146 at 21^h 53^m.5, +47° 16 ′; Vehrenberg, page 229.

Coddington's nebula. *Gx.* IC 2574 at 10^h 28^m.4, +68° 25 ′; *HA,* page 39.

Cohen-Schwartz star. *St.* T Tau star and IR source at 5^h 36^m.4, −6° 46 ′; *ApJ,* **268,** 766, 1983.

Coma cluster. *CG.* Abell 1656 at 12^h 59^m.8, +27° 59 ′; Burnham, page 691.

Coma star cluster. *OC.* Melotte 111 at 12^h 25^m, +26°; Burnham, page 668.

Cone nebula. *Di.* NGC 2264, at 6^h 40^m.9, +9° 54 ′; Burnham, page 1210.

Conus nebula. *Di.* (See Cone nebula.) Becvar.

Copeland's Septet. *CG.* NGC 3745-6, 48, 50-1, 53-4, at 11^h 37^m.8, +22° 00 ′; *LB.*

Cor Caroli. *St.* α CVn, at 12^h 56^m.0, +38° 19 ′; Burnham, page 359.

Cork nebula. *Pl.* (See Little Dumbbell.)

Corona Borealis cluster. *CG.* Abell 2065, at 15^h 22^m.7, +27° 43 ′; Burnham, page 714.

Crab nebula. *SNR.* M1 = NGC 1952, 5^h 34^m.5, +22° 01 ′; Webb, page 237, attributes the name to Earl of Rosse.

Crab pulsar. *St.* CM Tau; pulsar in M1. Position 5^h 34^m.5, +22° 01 ′; Burnham, page 1858.

Crescent nebula. *Di.* NGC 6888 at 20^h 12^m.0, +38° 21 ′; Burnham, page 796.

Crimson star. *St.* (See Hind's Crimson star.) Burnham, page 1093.

Cygnus A. *RS.* Galaxy; 3C 405. Located at 19^h 59^m.5, +40° 44 ′; *CU,* page 40.

Cygnus Egg. *Pl.* (See Egg nebula.) *AJ,* **87,** 1207, 1982.

Cygnus loop. *Di.* (See Veil nebula.) WS2, page 40.

Cygnus nebula. *Di.* (See γ Cyg nebula.) Sh2.

Cygnus superbubble. *Di.* 13° SNR around Cyg X, centered at 20^h 32^m, +42°; *ApJL,* **238,** L71, 1980.

Cygnus X. *RS.* Cyg OB2 and H II complex at 20^h 32^m, +42°; *ApJL,* **238,** L71, 1980.

Cygnus X-1. *XR.* Binary, possible black hole, at 19^h 58^m.4, +35° 12 ′; *CU,* page 87.

Cynosura. *St.* (See North Star.) *FDN.*

Dark Rift. *DN.* (See Great Rift.) *FDN*

Dark Bay nebula. *DN.* (See Horsehead nebula.) BBR, page 193.

Dark S nebula. *DN.* (See Snake nebula.) *FDN.*

Delle Caustiche. The star cloud M24, at 18^h 16^m.9, −18° 29 ′; Allen, page 359.

Demon star. *St.* Algol = β Per, at 3^h 08^m.2, +40° 57 ′; *GAA.*

Diabolo nebula. *Pl.* (See Dumbbell nebula.) *FDN.*

Dog star. *St.* Sirius = α CMa. Brightest star in the night sky, at 6^h 45^m.1, −16° 43 ′; Burnham, page 387.

Double cluster. *OC.* NGC 869 and 884 in Perseus at 2^h 20^m.7, +57° 08 ′; *S&T,* **64,** December, 1982, page 614.

Double-double star. *St.* ε Lyr, at 18^h 44^m.3, +39° 38 ′; Burnham, page 1151.

Double-headed Shot. *Pl.* (See Dumbbell nebula.) Allen, page 474.

Double quasar (or QSO). *QSO.* First double QSO discovered. At 10^h 01^m.3, +55° 54 ′; *Nature,* **279,** 381, 1979.

Draco dwarf galaxy. *Gx.* DDO 208, of type dE in the Local Group. Position 17^h 10^m.2, +57° 55 ′; *GAA.*

Dragon. *DN.* In the Lagoon nebula (M8), 18^h 04^m.8, −24° 30 ′; *AA,* **63,** 345, 1978.

Dumbbell nebula. *Pl.* M27 = NGC 6853 at 19^h 59^m.6, +22° 43 ′; Webb, page 246.

Eagle nebula. *Di.* M16 = NGC 6611, located at 18^h 18^m.8, −13° 47 ′; Burnham, page 1783. But the name has also been applied to IC 2177, 7^h 05^m.1, −10° 42 ′; Vehrenberg (3rd ed., pages 83 and 245).

Egg nebula. *Pl.* CRL 2688 = AFGL 2688 at 21^h 02^m.3, +36° 42 ′; *S&T,* **49,** January, 1975, page 21.

Eggen's nearby star. *St.* CoD −31° 622, briefly thought to be near the solar system but later found not to be. At 1^h 32^m.3, −30° 41 ′; *S&T,* **53,** February, 1977, page 107.

Eight-burst planetary. *Pl.* NGC 3132, 10^h 07^m.7, −40° 26 ′; Burnham, page 2041.

Electra nebula. *Di.* Part of the Pleiades. At 3^h 44^m.9, +24° 07 ′; *LB.*

Emu. *DN.* Bok's "dark constellation" (Crux to Sagittarius); WS2, page 33.

Eskimo nebula. *Pl.* NGC 2392, at 7^h 29^m.2, +20° 55 ′; Burnham, page 940.

Exclamation Mark galaxy. *Gx.* West of galaxy at this position: 0^h 39^m.3, −43° 06 ′; *Messenger,* No. 21, 25, 1980.

The Eyes. *Gx.* NGC 4435,38 at 12^h 27^m.7, +13° 03 ′; WS5, page 58.

Fath 703. *Gx.* A 12th-magnitude spiral, 15^h 13^m.8, −15° 28 ′; *LB.*

Fehrenbach's star. *St.* HD 116745 in ω Cen, 13^h 26^m.5, −47° 16 '; *BIS.*

Field of the nebulae. *CG.* (See Virgo cluster.) Serviss, page 51.

Filamentary nebula. *Di.* NGC 6960, at 20^h 45^m.7, +30° 43 '; *BBR*, page 204. Part of the Veil nebula.

Fish on the Platter. *DN.* Barnard 144, 19^h 59^m, +35°; *BC.*

Fish's Mouth. *DN.* In M42, northeast of the Trapezium, at 5^h 35^m.4, −5° 23 '; *WS2*, page 34.

Flaming Star nebula. *Di.* IC 405 = Ced 42. At 5^h 16^m.2, +34° 16 '; Burnham, page 285.

Flying star. *St.* The star 61 Cygni, with large proper motion. Position 21^h 06^m.9, +38° 45 '; Burnham, page 768.

Fly's Wing galaxy. *Gx. CU*, page 29.

Footprint nebula. *Di.* M1-92, 19^h 36^m.3, +29° 33 '; *WS2*, page 36 and 116.

Fornax A. *Gx.* NGC 1316 and radio source at 3^h 22^m.7, −37° 12 '; *CU*, page 40.

Fornax I cluster. *CG.* Contains Fornax A? Located at 3^h 32^m, −35° 20 '; Burnham, page 899.

Fornax II cluster. *CG.* Contains NGC 1232, at 3^h 28^m, −20° 45 '; Noonan.

Fornax dwarf galaxy. *Gx.* Galaxy of type dE in the Local Group, at 2^h 39^m.9, −34° 32 '; *GAA.*

Fourcade-Figueroa object. *Gx.* Situated 3° from NGC 5128 at 13^h 34^m.8, −45° 33 '; *PASP*, **90**, 237, 1978.

Garnet star. *St.* (See Herschel's Garnet star.)

Gem of the Milky Way. E. E. Barnard's name for the Scutum star cloud. Burnham, page 1749.

Geminga. Gamma-ray source, at 6^h 33^m.6, +17° 47 '; *S&T*, **66**, September, 1983, page 213.

Gemini cluster. *CG.* Abell 568, at 7^h 07^m.6, +35° 03 '.

Ghost of Jupiter. *Pl.* NGC 3242 at 10^h 24^m.8, −18° 38 '; Burnham, page 1029.

Gould's Belt. Orion spiral arm extension, after B. A. Gould. *CU*, page 79.

GR 8 (Gibson Reaves 8). *Gx.* DDO 155, at 12^h 58^m.7, +14° 13 '; *LB.*

Graham's object. *Gx.* Ring galaxy at 6^h 41^m.4, −74° 19 '; *ApJ*, **208**, 650, 1976.

Great Cluster in Hercules. *Gb.* M13 = NGC 6205 at 16^h 41^m.7, +36° 28 '; Burnham, page 978.

Great Looped nebula. *Di.* (See Tarantula nebula.) *GAA*; *BBR*, page 205.

Great Nebula in Andromeda. *Gx.* M31 = NGC 224 at 0^h 42^m.7, +41° 16 '; *BBR*, page 201.

Great Nebula in Orion. *Di.* M42, at 5^h 35^m.9, −5° 27 '; Burnham, page 1317.

Great Rift. *DN.* Dark clouds in the Aquila-Cygnus region of the Milky Way; *WS2*, page 34.

Great Square of Pegasus. The stars α, β, γ Peg, and α And; Serviss, page 69.

Grus Quartet. *CG.* NGC 7552, 82, 90, and 99 at 23^h 18^m, −42° 24 '; *FDN.*

Gulf of Mexico. *DN.* In the North America nebula; *S&T*, **59**, April, 1980, page 280.

Gum nebula. *Di.* Largest nebula in sky, centered at 8^h 30^m, −45°; *CU*, page 115.

Gyulbudaghian's nebula. *Di.* The variable cometary nebula at 20^h 45^m.9, +67° 58 '; *S&T*, **58**, August, 1979, page 115.

h and χ Persei. *OC.* (See Double Cluster.)

Hammerhead. *Di.* Southwest of the Trapezium in M42, at 5^h 33^m.7, −5° 40 '; *MK*, page 99.

Hardcastle nebula. *Gx.* Barred spiral at 13^h 13^m.0, −32° 42 '; *RC2.*

Harp star. *St.* Vega = α Lyr, at 18^h 36^m.9, +38° 47 '; Burnham, page 2134.

Haufen A. *CG.* Abell 151, 1^h 08^m.9, −15° 25 '; *ApJS*, **45**, 613, 1981.

Heiles clouds (1,2,4). *DN.* High-opacity dust clouds; *ApJ*, **151**, 919, 1968.

Helical nebula. *Pl.* (See Helix nebula.)

Helix galaxy. *Gx.* NGC 2685, at 8^h 55^m.6, +58° 44 '; *S&T*, **61**, May, 1981, page 380.

Helix nebula. *Pl.* NGC 7293 at 22^h 29^m.6, −20° 48 '; Burnham, page 192; *RASC.*

Hercules A. *Gx.* Radio source 3C 348 at 16^h 51^m.1, +5° 00 '; *LB.*

Hercules cluster. *CG.* Abell 2151, at 16^h 05^m.2, +17° 45 '; Noonan.

Hercules cluster. *Gb.* (See Great Cluster in Hercules.)

Herschel's Garnet star. *St.* μ Cep, at 21^h 43^m.5, +58° 47 '; Allen, page 158.

Hertzsprung's object. Plate defect? Located at 4^h 16^m.0, +60° 01 '; *PASP*, **95**, 1019, 1983.

Hind's Crimson star. *St.* R Lep; discovered by J. R. Hind, 1845, at 4^h 59^m.6, −14° 48 '; Burnham, page 1093.

Hind's new star. *St.* Nova Oph 1848 at 16^h 59^m.5, −12° 54 '; Webb, page 217.

Hind's variable nebula. *Di.* NGC 1555, 1 ' west of T Tau, at 4^h 21^m.8, +19° 32 '; Burnham, page 1831.

Hoffmeister's cloud. *DN.* About 20 square degrees, centered at 20^h 47^m, −42°; *ZAp*, **55**, 46, 1962.

Hoffmeister's star. *St.* V442 Cas = Sonneberg 9484, at 23^h 40^m.3, +53° 58 '; *S&T*, **38**, December, 1969, page 388.

Holmberg I. *Gx.* DDO 63, at 9^h 40^m.5, +71° 11 '; Holmberg.

Holmberg II. *Gx.* DDO 50 = Arp 268, at 8^h 18^m.9, +70° 43 '; Holmberg.

Holmberg III. *Gx.* A spiral of magnitude 13 at 9^h 14^m.8, +74° 14 '; Holmberg.

Holmberg IV. *Gx.* DDO 185 at 13^h 54^m.8, +53° 54 '; Holmberg.

Holmberg V. *Gx.* Barred spiral of magnitude 13 at 13^h 40^m.7, +54° 20 '; Holmberg.

Holmberg VI. *Gx.* NGC 1325A at 3^h 24^m.8, −21° 20 '; Holmberg.

Holmberg VII. *Gx.* DDO 137 at 12^h 34^m.7 , +6° 18 '; Holmberg.

Holmberg VIII. *Gx.* DDO 166 at 13^h 13^m.3, +36° 13 '; Holmberg.

Holmberg IX. *Gx.* DDO 66, type Ir V. Of magnitude 16.5 and in the M81 group at 9^h 57^m.6, +69° 03 '; *AfA*, **5**, No. 20, 1969.

Homunculus nebula. *Di.* Core of η Car nebula at 10^h 43^m.8, −59° 52 '; *S&T*, **44**, July, 1972, page 4.

Honda's variable. *St.* Long-period variable (not nova) at 21^h 43^m.9, +31° 28 '; *IAU Circular* 3565, 1981.

Horsehead nebula. *DN.* Barnard 33, southeast of Orion's Belt at 5^h 40^m.9, −2° 28 '; Burnham, page 1339.

Horseshoe nebula. *Di.* (See Omega nebula.) *BBR*, page 204.

Hourglass nebula. *Di.* J. Herschel's name for the brightest part of M8, at 18^h 03^m.7, −24° 23 '; *ApJ*, **263**, 130, 1982.

Hubble's variable nebula. *Di.* NGC 2261, at 6^h 39^m.2, +8° 44 '; *CU*, page 125.

Huruhata's object. *St.* Eclipsing binary discovered in 1974, at 7^h 30^m.3, +4° 32 '; *IBVS*, No. 1574, 1979.

Hyades. *OC.* Melotte 25, the naked-eye V-shaped cluster forming the Head of Taurus, at 4^h 26^m.9, +15° 52 '; Burnham, page 1817.

Hydra A. *Gx.* Radio source 3C 218, at 9^h 18^m.1, −12° 06 '; *LB.*

Hydra I. *CG.* Abell 1060 at 10^h 36^m.9, −27° 32 '; Noonan.

Hydra II. *CG.* 8^h 58^m, +3° 09 '; Noonan.

Innes's star. *St.* Proxima Centauri, at 14^h 29^m.7, −62° 41 '; Burnham, page 550.

Integral Sign galaxy. *Gx.* UGC 3697 at 7^h 11^m.4, +71° 50 '; *FDN.*

Intergalactic Wanderer. *Gb.* NGC 2419 at 7^h 38^m.1, +38° 53 '; Burnham, page 1128.

Iron star. *St.* XX Oph, 17^h 43^m.9, −6° 16 '; Burnham, page 1246.

Jewel Box. *OC.* The name properly refers to NGC 4755, at 12^h 53^m.6, −60° 20 '. This is the name given by J. Herschel; Burnham, page 730. Sometimes also applied to M6 = NGC 6405, at 17^h 40^m.1, −32° 13 '; *S&T*, **48**, October, 1974, page 227 (but see Butterfly cluster).

Jupiter nebula. *Pl.* (See Ghost of Jupiter.) Smyth.

Kapteyn's star. *St.* HD 33793, at 5^h 10^m.6, −44° 52 '; Burnham, page 1462.

Keenan's system. *Gx.* NGC 5216,18 = Arp 104, at 13^h 32^m.2, +62° 44 '; *RASC.*

Kemble's Cascade. Star chain northeast of NGC 1502, at 3^h 57^m.0, +63° 00 '; *S&T*, **60**, December, 1980, page 547.

Kepler's star (or supernova). *SN.* 3C 358 = SN Oph 1604, at 17^h 30^m.6, −21° 29 '; Burnham, page 1249.

Keyhole nebula. *DN.* In η Car nebula at 10^h 43^m.8, −59° 52 '; *BBR*, page 205.

Kleinmann-Low nebula. *IR.* Infrared source in M42, at 5^h 35^m.3, −5° 22 '; *CU*, page 104.

Klemola's star. *St.* BD +10° 2179 = SAO 99230 at 10^h 38^m.9, +10° 04 '; *BIS.*

Krzeminski's star. *St.* Cen X-3 optical component, at 11^h 21^m.3, −60° 37 '; *BIS.*

Kurtz's light variable. *St.* HD 188136, Ap star? Its place is 20^h 03^m.6, −78° 50 '; *ApJ*, **247**, 969, 1981.

Kutner's cloud. *DN.* LDN 1524, 29, 31, 33, and 35 at about 4^h 33^m, +24° 36 '; *AA*, **96**, 202, 1981.

Kuwano's object. *St.* The novalike object PU Vul, at 20^h 21^m.2, +21° 36 '; *IAU Circular* 3344, 1979.

La Superba. *St.* Y CVn, at 12^h 45^m.1, +45° 26 '; Burnham, page 361.

Lace-work nebula. *Di.* NGC 6960, at 20^h $45^m.6$, $+30°$ $42'$; Allen, page 195.

Lagoon nebula. *Di.* M8 = NGC 6523, at 18^h $03^m.8$, $-24°$ $23'$; Burnham, page 1574.

Lamont's star. *St.* Near the nucleus of M31, at 0^h $45^m.4$, $+41°$ $32'$; *S&T,* **35,** February, 1968, page 97.

Large Magellanic Cloud (LMC). *Gx.* Milky Way satellite, at 5^h $23^m.6$, $-69°$ $45'$; Burnham, page 837.

Leo I. *Gx.* DDO 74, of type dE in the Local Group, at 10^h $08^m.4$, $+12°$ $18'$; RC2.

Leo II, Leo B. *Gx.* DDO 93, of type dE in the Local Group, at 11^h $15^m.5$, $+22°$ $10'$; RC2.

Leo III, Leo A. *Gx.* DDO 69, Ir in Local Group, at 9^h $59^m.4$, $+30°$ $45'$; RC2.

Leo A cluster. *CG.* Abell 1185, at 11^h $10^m.9$, $+28°$ $41'$; Noonan.

Leo cluster. *CG.* Abell 1020, at 10^h $27^m.8$, $+10°$ $25'$; Noonan.

Leo Triplet. *CG.* Leo Group = M65, M66, and NGC 3628, at 11^h $18^m.9$, $+13°$ $06'$; RVG.

Liller's star. *St.* Near Cen X-3, at 11^h $21^m.2$, $-60°$ $36'$; *BIS.*

Lindsay-Shapley ring. *Gx.* Ring galaxy, at 6^h $43^m.1$, $-74°$ $14'$; *LB.*

Little Dog star. *St.* Procyon = α CMi, at 7^h $39^m.3$, $+5°$ $14'$; Burnham, page 448.

Little Dumbbell. *Pl.* M76 = NGC 650-1, at 1^h $42^m.4$, $+51°$ $34'$; Burnham, page 1435.

Little Gem. *Pl.* NGC 6445, at 17^h $49^m.3$, $-20°$ $00'$; RASC.

LMC. *Gx.* (See Large Magellanic Cloud.)

Local Arm. (See Orion arm.)

Local Group. *CG.* Galaxies near Milky Way; *GAA.*

Local Supercluster. *CG.* Includes the Virgo cluster and the Milky Way; *ApJ,* **257,** 389, 1982.

Longmore's group. *CG.* Sparse; *CU,* page 25.

Loop nebula. *Di.* (See Tarantula nebula.) *S&T,* **47,** January, 1974, page 10. Also a name for the Veil nebula; *S&T,* **53,** February, 1977, page 112.

Lord Rosse's nebula. *Gx.* (See Whirlpool nebula.)

Lovas's supernova. *SN.* SN 1964e in MCG 9-20-51, at 11^h $59^m.7$, $+52°$ $43'$; *FDN;* IAU *Circular* 1858, 1964.

Lower's nebula. *Di.* Sh2-261 = part of Ced 64, at 6^h $08^m.9$, $+15°$ $49'$; Cederblad.

Lupus loop. *RS.* Radio SNR, at 15^h 12^m, $-39°$; *GAA.*

Luyten's flare star. *St.* Nearby UV Ceti variable, at 1^h $38^m.8$, $-17°$ $58'$; Burnham, page 641.

Luyten's star. *St.* BD $+5°$ 1668, at 19^h $42^m.1$, $+6°$ $00'$; *BIS.*

M81 dwarf B. *Gx.* UGC 5423, at 10^h $05^m.5$, $+70°$ $22'$.

M81 group. *CG.* Nearby group, at 10^h $02^m.1$, $+68°$ $45'$; *ApJ,* **257,** 423, 1982.

M101 group. *CG.* Nearby group, at 14^h $09^m.4$, $+54°$ $55'$; *ApJ,* **257,** 423, 1982.

Maffei I. *Gx.* UGCA 34, at 2^h $36^m.3$, $+59°$ $39'$; *GAA.*

Maffei II. *Gx.* UGCA 39, at 2^h $41^m.9$, $+59°$ $36'$; *GAA.*

Magellanic Clouds. *Gx.* (See under Large and Small.)

Magellanic Stream. *RS.* Gas stream from LMC; *CU; GAA.*

Magellanic System. *RS.* LMC, SMC hydrogen envelope; *GAA.*

Maia nebula. *Di.* NGC 1432 in the Pleiades, at 3^h $45^m.8$, $+24°$ $22'$; *LB.*

Markarian's chain. *CG.* M86 to M88 in Virgo cluster, at 12^h 30^m, $+13°$ $30'$; WS5, page 58.

Martial star. *St.* Betelgeuse, at 5^h $55^m.2$, $+7°$ $24'$; Burnham, page 1281.

Mayall's object. *Gx.* Arp 148 = VV 32, a cigar-shaped galaxy with ring located at 11^h $03^m.9$, $+40°$ $51'$; *GAA; LB.*

McLeish's object. *Gx.* Located at 20^h $09^m.7$, $-66°$ $13'$; RASC.

Medusa nebula. *Pl.* Abell 21 = YM 29, at 7^h $29^m.0$, $+13°$ $15'$; *FDN.*

Medusa star. *St.* (See Demon star.) Burnham, page 1409.

Merope nebula. *Di.* NGC 1435, in the Pleiades at 3^h $46^m.1$, $+23°$ $47'$; Burnham, page 1882.

Merrill's star. *St.* High-velocity Wolf-Rayet star, nucleus of *Pl* PK 50+3.1, at 19^h $11^m.5$, $+16°$ $52'$; *AA,* **114,** 135, 1982.

Mice. *Gx.* Arp 242, also called NGC 4676A and B. Located at 12^h $46^m.2$, $+30°$ $44'$; WS4, page 133; *LB.*

Milky Way. *Gx.* The band of light produced by innumerable faint stars in the plane of our galaxy, and which circles the sky. The glow is brightest in Sagittarius, the direction to the galactic center being 17^h $45^m.6$, $-28°$ $56'$. The name Milky Way is quite ancient; Allen, page 475.

Millisecond pulsar. *St.* Period 0.001557806449022 second. Located at 19^h $39^m.6$, $+21°$ $35'$; *S&T,* **65,** February, 1983, page 131.

Miniature spiral. *Gx.* NGC 3928 = Mrk 190, at 11^h $51^m.8$, $+48°$ $41'$; *PASP,* **92,** 409, 1980.

Minkowski's Footprint. *Di.* (See Footprint nebula.)

Minkowski's object. *Gx.* Peculiar galaxy near NGC 541, at 1^h $25^m.8$, $-1°$ $22'$; WS5, page 103.

Monoceros loop. *Di.* SNR? Located at 6^h 38^m, $+6°$ $27'$; *AAS,* **31,** 261, 1978.

Monogem ring. *XR.* SNR? Diameter $20°$, centered on 7^h 13^m, $+14°$; *ApJ,* **248,** 152, 1981.

Moving nebula. *Di.* Ejected from Mrk 335 nucleus? Position 0^h $06^m.4$, $+20°$ $12'$; *ApJ,* **247,** 32, 1981.

Network nebula. *Di.* NGC 6992-5 (part of the Veil nebula) at 20^h $56^m.4$, $+31°$ $43'$; BBR, page 204.

New 1. *Gx.* A spiral of 12th magnitude at 1^h $05^m.1$, $-6°$ $13'$; *SA.*

New 2. *Gx.* NGC 4507, a spiral at 12^h $35^m.6$, $-39°$ $55'$; *SA.*

New 3. *Gx.* A 12th-magnitude spiral at 12^h $49^m.4$, $-10°$ $07'$; *SA.*

New 4. *Gx.* UGC 8041, a barred spiral at 12^h $55^m.2$, $+0°$ $07'$; *SA.*

New 5. *Gx.* A 12th-magnitude spiral at 20^h $24^m.0$, $-44°$ $00'$; *SA.*

New 6. *Gx.* A 13th-magnitude spiral at 21^h $23^m.33$, $-45°$ $47'$; *SA.*

Ney-Allen nebula. *IR.* Infrared source around θ^1 Ori, at 5^h $35^m.3$, $-5°$ $18'$; *GAA.*

Niebelungen ring. *Gx.* (See Graham's object.)

Nile star. *St.* Sirius. (See Dog star.) Burnham, page 387.

North America nebula. *Di.* NGC 7000, at 20^h $58^m.8$, $+44°$ $20'$; Burnham, page 811.

Northern Coalsack. *DN.* LDN 906, at 20^h 40^m, $+42°$; Allen, page 197.

North Polar Sequence. *St.* Photometric standard stars, near the north celestial pole. *GAA.*

North Polar Spur. *RS.* SNR, from galactic plane toward north galactic pole; *GAA.*

North Star. *St.* Polaris = α UMi, at 2^h $31^m.8$, $+89°$ $16'$; Burnham, page 2009.

Nubecula Major. *Gx.* (See Large Magellanic Cloud.) Burnham, page 837.

Nubecula Minor. *Gx.* (See Small Magellanic Cloud.) Burnham, page 1915.

Omega nebula. *Di.* M17 = NGC 6618, at 18^h $20^m.8$, $-16°$ $11'$; BBR, page 204.

Origem loop. *RS.* SNR? $6°$ diameter, at 6^h 18^m, $+16°$ $23'$; *AA,* **35,** 429, 1974.

Orion A. *RS.* M42 region = 3C 145, at 5^h $35^m.3$, $-5°$ $23'$; *CU,* page 103.

Orion arm. Milky Way spiral arm containing the solar system. *GAA.*

Orion B. *RS.* The IC 434 region, 3C 147.1, at 5^h $41^m.7$, $-1°$ $54'$.

Orion loop. *Di.* (See Barnard's Loop.) Burnham, page 1329.

Orion molecular clouds (OMC). *IR.* In the Trapezium region. *GAA.*

Orion nebula. *Di.* (See Great Nebula in Orion.)

Orion's Cloak. *Di.* Around Ori OB1 association, at 5^h $31^m.4$, $-2°$ $41'$; *ApJ,* **230,** 469, 1979.

Orion Spur. The Sun's location in the Orion arm; *GAA.*

Osawa's star. *St.* V436 Cas = HD 221568, at 23^h $32^m.8$, $+57°$ $54'$; *FDN.*

Ostrich. *DN.* (See Emu.) WS2, page 33.

Owl nebula. *Pl.* M97 = NGC 3587, at 11^h $14^m.8$, $+55°$ $01'$; Burnham, page 1996.

Papillon. *Gx.* IC 708, in Abell 1314, at 11^h $34^m.0$, $+49°$ $05'$; *AA,* **77,** 183, 1979.

Parrot's Head. *DN.* Barnard 87 (and named by Barnard), at 18^h $04^m.3$, $-32°$ $30'$; *CU,* page 132.

Pazmino's cluster. *OC.* Stock 23, at 3^h $16^m.3$, $+60°$ $02'$; *S&T,* **55,** March, 1978, page 275.

Peacock star. *St.* α Pav, at 20^h $25^m.65$, $-56°$ $44'.1$; Burnham, page 1355.

Pearce's star. *St.* AO Cas, at 0^h $17^m.7$, $+51°$ $26'$; Burnham, page 503.

Pegasus I cluster. *CG.* Contains NGC 7619, at 23^h 22^m, $+9°$ $02'$; Noonan.

Pegasus II cluster. *CG.* Contains NGC 7720, at 23^h 10^m, $+7°$ $36'$; Noonan.

Pegasus dwarf galaxy. *Gx.* DDO 216, at 23^h $28^m.6$, $+14°$ $45'$; *LB.*

Pelican nebula. *Di.* IC 5067 and 70 at 20^h 50^m.8, +44° 21'; Vehrenberg, page 223.

Perseus A. *Gx.* Seyfert galaxy NGC 1275, at 3^h 19^m.8, +41° 31'; Burnham, page 1448.

Perseus arm. Milky Way spiral arm at galactic longitude 100° to 150°. *GAA.*

Perseus cluster. *CG.* Abell 426, at 3^h 18^m.6, +41° 32'; *GAA.*

Piazzi's flying star. *St.* 61 Cyg, at 21^h 06^m.9, +38° 45'; Burnham, page 768.

Pickering's triangular wisp. *Di.* Part of the Veil nebula, at 20^h 48^m.5, +31° 09'; *ApJ,* **23,** 261, 1906.

Pin-wheel nebula. *Gx.* NGC 4254 = M99, at 12^h 18^m.8, +14° 25'; Allen, page 172.

Pinwheel galaxy. *Gx.* (See Triangulum galaxy.) Burnham, page 1897. Also, M101 = NGC 5457 at 14^h 03^m.2, +54° 21'; RASC.

Pipe nebula. *DN.* Barnard 59, 65-7, and 78. The "bowl" of the pipe lies at 17^h 33^m, −26°. Burnham, page 1267.

Plaskett's star. *St.* Massive star HD 47129, at 6^h 37^m.5, +6° 08'; Burnham, page 1193.

Pleiades. *OC.* The naked-eye cluster M45 in Taurus, at 3^h 47^m.0, +24° 07'; *Job, 9:9, 38:31.*

Polaris Australis. *St.* The name given to σ Oct by W. H. Wollaston in 1789; it is the nearest bright star to the south celestial pole, at 21^h 08^m.7, −88° 57'. YBS.

Polarissima Australis. *Gx.* NGC 2573, at 1^h 37^m, −89° 21'; J. Herschel.

Polarissima Borealis. *Gx.* NGC 3172, at 11^h 50^m.2, +89° 07'; J. Herschel; *S&T,* **64,** September, 1982, page 291.

Pole star. *St.* (See North Star.) Burnham, page 2009.

Popovic's object. Defect on a 1911 photographic plate? Position 0^h 57^m.5, −1° 21'; *PASP,* **95,** 1020, 1983.

Popper's star. *St.* HD 124448, a hydrogen-deficient star at 14^h 15^m.0, −46° 17'.

Praesepe. *OC.* M44 = NGC 2632 in Cancer at 8^h 40^m.1, +19° 59'; Webb says this is the ancient name. (See Beehive cluster.)

Proxima Centauri. *St.* The nearest star to our Sun, at 14^h 29^m.7, −62° 41'. Discovered by R. Innes, 1915.

Przybylski's star. *St.* V816 Cen = HD 101065, at 11^h 37^m.6, −46° 43'; *FDN.*

Pulcherrima. *St.* ε Boo, at 14^h 45^m.0, +27° 05'; Allen, page 104.

Pup. *St.* The white-dwarf companion to Sirius, often called Sirius B, at 6^h 45^m.2, −16° 43'. The name derives from the fact that Sirius itself is the Dog star.

Puppis A. *RS.* SNR, about 10,000 years old, at 8^h 24^m.1, −43° 00'; *GAA.*

Puppis Window. Low-absorption region in galactic disk at longitude 245°. *AAS,* **50,** 377, 1982.

Question Mark galaxy. *Gx.* (See Whirlpool galaxy.)

Rapid Burster. *XR.* MXB 1730−33 in globular cluster Liller 1, at 17^h 33^m.4, −33° 23'; *FDN.*

Rat-Tail galaxies. *Gx.* Burnham.

Red Rectangle. *Pl.* AFGL 915; HD 44179 and nebula at 6^h 20^m.0, −10° 39'; *AJ,* **87,** 1207, 1982.

Regulus dwarf galaxy. *Gx.* (See Leo I.)

Reinmuth 80. *Gx.* NGC 4517A, at 12^h 32^m.5, +0° 23'; RC2.

Reticulum dwarf system. *Gb?* Intergalactic? LMC member? At 4^h 36^m.2, −58° 50'; *GAA; LB.*

Rheingold, Das. *Gx.* (See Graham's object.)

Ring nebula in Lyra. *Pl.* M57 = NGC 6720, at 18^h 53^m.6, +33° 02'; Burnham, page 1163.

Ring-tail galaxy. *Gx.* (See Antennae.) Burnham, page 719.

Ring-tail Snorter. *Gx.* (See Antennae.) *S&T,* **23,** January, 1962, page 17.

Roberts-Altizer variable. *St.* Galactic U Gem star near NGC 3147, at 10^h 15^m.6, +73° 26'; *FDN.*

Rosette nebula. *Di.* NGC 2237-9 and 46, at 6^h 32^m.3, +5° 03'; WS2, page 109.

Rosino's supernova. *SN.* Galactic U Gem variable? *FDN.*

Rosino-Zwicky object. *St.* Variable star near M88, at 12^h 32^m.4, +14° 19'; *FDN.*

Runaway star. *St.* (See Barnard's star.) Burnham, page 125.

Running Chicken nebula. *Di.* IC 2944, at 11^h 35^m.8, −63° 01'; Vehrenberg, page 111.

S nebula. *DN.* (See Snake nebula.) Burnham, page 1268.

SS 433. *St.* Neutron star in radio source W50 at 19^h 11^m.8, +4° 59'; *S&T,* **58,** December, 1979, page 510. ("SS" stands for Sanduleak-Stephenson object.)

Sagittarius A. *RS.* Galactic center, at 17^h 46^m.1, −28° 51'; Burnham, page 1267.

Sagittarius arm. Milky Way spiral arm at galactic longitude 340° to 30°. *GAA.*

Sanduleak's star. *St.* Possible symbiotic star in the Large Magellanic Cloud at 5^h 45^m.3, −71° 16'; *BIS.*

Sanduleak-Pesch binary. *St.* White-dwarf binary, at 17^h 05^m.5, +48° 03'; *AA,* **127,** 25, 1983.

Saturn nebula. *Pl.* NGC 7009, at 21^h 04^m.2, −11° 22'; BBR, page 205.

Schaeberle's Flaming Star. *Di.* (See Flaming Star nebula.) Cederblad.

Scheiner's star. *St.* BD +15° 2083 = HD 83225, at 9^h 37^m.2, +15° 14'; *S&T,* **42,** December, 1971, page 344.

Schweizer-Middleditch star. *St.* Near center of SNR 1006, at 15^h 02^m.8, −41° 59'; *ApJ,* **241,** 1039, 1980.

Sco-Cen association. Nearest OB association; extends from Scorpius to Crux. Burnham, page 1668.

Sculptor dwarf galaxy. *Gx.* Galaxy of type dE in the Local Group, at 0^h 59^m.9, −33° 42'; *GAA.*

Sculptor dwarf irregular galaxy. *Gx.* (See SDIG.)

Sculptor group. *CG.* Nearby group, at 0^h 38^m.0, −30° 58'; *ApJ,* **257,** 423, 1982.

Sculptor galaxy. *Gx.* NGC 253, in Sculptor group, at 0^h 47^m.6, −25° 17'; *CU,* page 22.

Scutum star cloud. Part of the southern Milky Way, at 18^h 40^m, −7°; Burnham, page 1749.

SDIG. *Gx.* Sculptor dwarf irregular galaxy, at 0^h 08^m.1, −34° 34'; *CU,* page 21; *FDN.*

Seashell galaxy. *Gx.* Interacting with NGC 5291, at 13^h 47^m.4, −30° 25'; *CU,* page 30; *LB.*

Serpens dwarf galaxy. *Gx.* A galaxy of type dE near the Local Group, at 15^h 16^m.1, +0° 08'; *LB.*

Seven Sisters. *OC.* (See Pleiades.)

Sextans A. *Gx.* DDO 75, a dwarf irregular galaxy at 10^h 11^m.1, −4° 43'; *LB.*

Sextans B. *Gx.* DDO 70, a dwarf irregular galaxy at 10^h 00^m.0, +5° 20'; *LB.*

Sextans C. *Gx.* A galaxy of type dE near the Local Group, at 10^h 05^m.6, +0° 04'; *LB.*

Seyfert's Sextet. *CG.* NGC 6027A-E = VV 115, at 15^h 59^m.2, +20° 45'; *GAA; CU,* page 44.

Shakhbazian 1. *CG.* Group of compact galaxies, at 10^h 54^m.8, +40° 28'; *CU,* page 13.

Shapley 1. *Pl.* Ring-shaped, at 15^h 51^m.7, −51° 31'; *CU,* page 140.

Siamese Twins. *Gx.* NGC 4567-8 = VV219, at 12^h 36^m.6, +11° 14'; Burnham, page 2080.

Sidus Ludoviciana (-um). *St.* A field star near Mizar and Alcor, mistaken for a new planet in 1722. Position 13^h 25^m.0, +54° 53'; *S&T,* **16,** April, 1957, page 265; June, 1957, page 378.

Small Magellanic Cloud (SMC). *Gx.* Milky Way satellite, at 0^h 52^m.7, −72° 50'; Burnham, page 1915.

Small Sagittarius cloud. Star cloud M24, at 18^h 16^m.9, −18° 29'; Burnham, page 1603.

SMC. *Gx.* (See Small Magellanic Cloud.)

Snake nebula. *DN.* Barnard 72, at 17^h 23^m.5, −23° 38'; BC.

Snickers. *RS.* Possible nearby dwarf galaxy, discovered in a radio survey at 6^h 28^m, +15°; *ApJL,* **201,** L103, 1975.

Sombrero galaxy. *Gx.* M104 = NGC 4594, at 12^h 40^m.0, −11° 37'; Burnham, page 2097.

Southern Pleiades. *OC.* IC 2602 (θ Car), at 10^h 43^m.2, −64° 24'; *S&T,* **67,** April, 1984, page 344.

Spectral Gem of the Southern Skies. *St.* γ^2 Vel, a Wolf-Rayet star at 8^h 09^m.5, −47° 20'; YBS, page 399.

Spider. *Gx.* Peculiar galaxy VV 794, at 10^h 42^m.7, +34° 26'; *FDN.*

Spindle galaxy. *Gx.* NGC 3115, at 10^h 05^m.2, −7° 43'; Burnham, page 1797.

Star Queen nebula. *Di.* M16 = IC 4703, at 18^h 18^m.8, −13° 47'; Burnham, page 1783.

Stepanian's star. *St.* LX Ser; 14th-magnitude flare star? Its place is 15^h 38^m.0, +18° 52'; *BIS.*

Stephan's Quartet. *CG.* NGC 7317-19, at 22^h 36^m.0, +33° 58'; *CU,* page 27.

Stephan's Quintet. *CG.* NGC 7317-20 = Arp 319, at 22^h 36^m.0, +33° 58'; Burnham, page 1389.

Struve's Lost nebula. *Di.* NGC 1554, at 4^h 22^m.9, +19° 32'; Cederblad.

Sunflower galaxy. *Gx.* M63 = NGC 5055, at 13^h 15^m.8, +42° 02'; Burnham, page 373.

Swan nebula. *Di.* (See Omega nebula.) Cederblad.

Sword of Orion. *Di.* M42 and vicinity, at 5^h 35^m, −5° 30'; Vehrenberg, page 64.

Sword-hand of Perseus. *OC.* (See Double Cluster.) Webb, page 226.

Tarantula nebula. *Di.* NGC 2070 = 30 Dor, at 5^h 38^m.7, −69° 06'; Becvar.

Taurus A. *RS.* Crab nebula radio source, at 5^h 34^m.5, +22° 01'; *CU,* page 152.

Taurus dark cloud. *DN.* Large complex, 150 parsecs away, at 4^h 30^m, +27°; *GAA.*

Tempel's nebula. *Di.* Near Merope, at 3^h 46^m.1, +23° 47'; Denning, page 330.

Thumbprint nebula. *DN.* "Bright" dark nebula, at 12^h 42^m.9, +78° 16'; *ApJ,* **208,** 709, 1976.

Toby Jug nebula. *Di.* IC 2220, bipolar. HD 65750, at 7^h 56^m.9, −59° 07'; *MNRAS,* **196,** 403, 1981.

Trapezium. *St.* The four bright components of θ^1 Ori in M42, at 5^h 35^m.3, −5° 23'; MK, page 97.

Triangulum galaxy. *Gx.* M33 = NGC 598, at 1^h 33^m.9, +30° 39'; *LB.*

Trifid nebula. *Di.* M20 = NGC 6514, at 18^h 02^m.6, −23° 02'; Burnham, page 1591.

Triple quasar (or QSO). The three components of this QSO have magnitudes of 16, 18, and 18, with a redshift of 1.722. Located at 11^h 18^m.3, +7° 46'; *Nature,* **285,** 641, 1980.

Tweedledee and Tweedledum. *St.* A tiny and difficult double-double star, φ 332.

Name suggested by W. S. Finsen, who was struck by the nearly identical position angles and separations at the time of his 1953 discovery. Located in Serpens Cauda at 18^h 45^m.5, +5° 30'; *S&T,* **22,** November, 1961, page 263.

Tycho's star (or supernova). *SN.* SN 1572 = B Cas = 3C 10, at 0^h 25^m.3, +64° 09'; Burnham, page 503.

Ursa Major moving cluster. *OC.* Collinder 285, centered at about 12^h 03^m, +58°; *CSCA.*

Ursa Major I cluster. *CG.* Abell 1377, at 11^h 47^m, +55° 44'; Noonan.

Ursa Major II cluster. *CG.,* at 10^h 58^m, +56° 46'; Noonan.

Ursa Minor dwarf galaxy. *Gx.* DDO 199, a galaxy of type dE in the Local Group, at 15^h 08^m.8, +67° 12'; *FDN.*

Van Biesbroeck's star. *St.* Among the least luminous stars known, with an absolute visual magnitude of 19. Position 19^h 16^m.97, +5° 08'.9; Burnham, page 226.

Van Maanen's star. *St.* Nearby star of large proper motion at 0^h 49^m.17, +5° 23'.3; Burnham, page 1474.

Veil nebula. *Di.* NGC 6960, 92, and 95 at 20^h 50^m, +31°; Burnham, page 801.

Vela pulsar. *St.* In Vela SNR, at 8^h 35^m.4, −45° 11'; *CU,* page 149.

Vela supernova remnant. *SNR.* Radio source Vela X at center. Position 8^h 34^m, −45°; *CU,* page 149.

Velorum. *OC.* More properly called the *o* Velorum cluster (IC 2391), at 8^h 40^m.2, −53° 04'; *AA,* **90,** 269, 1980.

Virgo A. *Gx.* M87 = NGC 4486 = Arp 152, at 12^h 30^m.8, +12° 24'; Burnham, page 2092.

Virgo XR-1. *XR.* (See Virgo A.) *CU,* page 37.

Virgo cluster. *CG.* Near the Local

Group; centered at 12^h 30^m, +12° 23'. Burnham, page 2074.

Virgo supercluster. *CG.* (See Local Supercluster.)

Wachmann's flare star. *St.* V371 Ori, at 5^h 33^m.8, +1° 57'; *FDN.*

Walborn's star. *St.* Wolf-Rayet star in the Large Magellanic Cloud at 4^h 55^m.3, −67° 11'; *FDN.*

Warren and Penfold's (WP) star. *St.* Optical counterpart of LMC X-3, at 5^h 39^m.0, −64° 05'; *FDN.*

Whirlpool galaxy. *Gx.* M51 = NGC 5194-5, at 13^h 29^m.9, +47° 12'; Burnham, page 369.

Wild Duck cluster. *OC.* M11 = NGC 6705, at 18^h 51^m.1, −6° 16'. From Smyth's description, "flight of wild ducks."

Wild's supernova. *SN.* SN 1966j in galaxy NGC 3198, at 10^h 19^m.9, +45° 33'; *SovA,* **13,** 423, 1969.

Wild's Triplet. *CG.* Arp 248 = VV 35, at 11^h 46^m.8, −3° 50'; *LB.*

Wischnjewsky's supernova. *SN.* In Fornax A, 1980, at 3^h 23^m.0, −37° 13'; *MNRAS,* **196,** 65P, 1981.

Witch Head nebula. *Di.* IC 2118, at 5^h 06^m.9, −7° 13'; Cederblad.

Wolf-Lundmark-Melotte (WLM) system. *Gx.* DDO 221, in Local Group, at 0^h 02^m.0, −15° 28'; *LB.*

Zealey-Lee supernova. *SN.* In the nucleus of an anonymous galaxy at 23^h 06^m.6, −44° 27'; IAU *Circular* 3282, 1978.

Zwicky 2. *Gx.* DDO 105, at 11^h 58^m.4, +38° 04'; *LB.*

Zwicky's Triplet. *CG.* Arp 103, at 16^h 49^m.5, +45° 30'; *LB,* RC2, and RASC. The name has also been applied to Arp 175 (IC 3481, 81A, and 83), at 12^h 33^m.0, +11° 23'; *FDN.*

Double and Multiple Stars

Double and Multiple Stars

Column Headings

α 2000 and **δ 2000** — Right ascension and declination, referred to the 2000.0 equinox. In the case of wide pairs, the position refers to the *brighter* component, not necessarily the A component. If a star is multiple, two or more lines are used to describe its components, and the position is given only on the first line (or the top of a new page).

Star Name — The standard designation, usually consisting of the discoverer's name and serial number. (A key to abbreviations appears in Table I of the Introduction.)

In the case of multiple systems, traditional Roman letters also identify the components. These letters form another system of nomenclature that is wholly independent of that by discoverer. Often, the brightest star is called A and its closest companion B, with outlying members denoted C, D, and so on. But numerous exceptions exist, partly because of the historical order in which discoveries have been made. (Thus, a newly discovered companion to A still closer than B might be called "a" to avoid relettering the whole system and introducing confusion with previously published measures.) The lettering scheme of this Catalogue is that of the Lick *Index Catalogue of Visual Double Stars, 1961.0* (IDS).

Const — Constellation. The standard three-letter abbreviations are used.

Years — The year of the first satisfactory measure of a pair, followed by the year of a more recent measure if available. These dates refer to the position angles and separations tabulated immediately to the right.

Pos Ang — The position angle, as measured in each of the two years listed to the left. Position angles give direction on the sky and are counted in degrees from north (0°) through east (90°), south (180°), and west (270°). For simple doubles the position angle is that of the fainter star referred to the brighter one. In the case of multiple stars, the particular components measured are identified by the letters under Star Name. For example, "AC" there implies that the position angles are for C with respect to A; "A × CD" means the observer has measured the position angle, at A, to the close pair CD (perhaps bisecting an elongated image). Occasionally, an observer has not actually measured a position angle, but has noted the rough direction: north (n); following or east of (f), south preceding (sp), and so on.

Separation — The angular distance of a pair, in seconds of arc, for each of the two years given at left. Sometimes the separation, if very large, is expressed in minutes of arc and enclosed in parentheses.

Both the position angle and separation are omitted for the second date if there has been no definite relative motion during the pair's observed history. Such a pair is often called "fixed."

m_1 and **m_2** — Visual magnitudes of the two components. The left-right order matches the letter designations under Star Name. Thus, for a line describing components AC, m_1 is the magnitude of A and m_2 is the magnitude of C. However, m_1 will be blank if the magnitude of A has already been listed in a previous line of the same multiple system.

The magnitudes of components are often difficult to measure separately, and for faint components they must be regarded as very rough visual estimates. The values given here are based mainly on the IDS, but with the adjustments described in the Introduction.

Spec — The Henry Draper spectral type, according to the IDS. As with magnitudes, the spectral types have a left-right correspondence to the letter designations of components on the same line, and are listed only once (if known) for each component of a multiple system.

ADS — The star's number in R. G. Aitken's *New General Catalogue of Double Stars* (1932).

Orb — An "o" appears in this column if the orbit of a pair has been calculated. (If orbital motion is rapid, the measures are omitted.) The letter "o" usually, but not always, means that the orbital elements are given in the section Visual Binary Stars, beginning on page 161.

Notes — Under this heading are given common names of stars, and occasionally an alternate discoverer's designation. Many other notes about questionable duplicity, optical components, composite spectra, and difficulty of measurement with large telescopes are inserted here from the IDS (often without attribution). We've added some descriptive comments from early visual observers, especially John Herschel (h), W. H. Smyth, T. W. Webb, and H. C. Russell (R). Visual impressions of colors are strongly influenced by magnitude differences and contrast effects, as the spectral types reveal. Numbers in brackets guide the reader to more extended notes that start on page 159.

α 2000	δ 2000	Star Name		Const	Years	Pos Ang	Separation	m₁	m₂	Spec	ADS	Orb	Notes
0ʰ00.ᵐ0	+38°52′	β 860		And	1881 1933	107°	6.″7	6.6	11.4	B9	17157		
0 00.3	−44 17	I 1477		Phe	1926 1959	250 292	0.3 0.5	6.8	7.3	K0			
0 00.4	+26 55	A.G. 299		Peg				6.5		F8			Single (IDS)
0 00.6	−53 06	h 5437		Phe	1852 1959	292 312	2.6 1.9	6.8	10.3	G0			
0 00.6	−66 41	Gli 289		Tuc	1889 1938	267	3.7	7.6	9.4	F5			
0 01.1	+69 35	OΣΣ 253		Cep	1875 1920	353	100.5	7.8	8.0	B9			
0 01.2	+60 21	OΣΣ 254	AB	Cas	1874 1953	90 89	58.9 58.1	7.6	8.7	N A			
—			Aa		1879 1918	324	155.7		11.1				
—			Bb		1879 1918	129	134.4		11.0				
0 01.5	+30 44	Ho 208		Peg	1884 1956	236 208	0.7 0.9	8.2	10.2	G0	17167		
0 01.9	−41 37	I 1060		Phe	1912 1944	340	0.7	7.9	10.9	F2			
0 02.2	+27 05	β 733	AB	Peg	1878 1960			5.8	8.9	G0	17175	o	85 Peg
—			AC		1852 1932	114 330	33.0 75.5		8.6		17175		
—			AD		1878 1921	277 296	61.7 109.9		13.0		17175		
0 02.2	−27 09	h 5440		Scl	1834 1941	288	3.3	8.6	8.9	F8	17177		
0 02.3	−29 43	B 631		Scl		320	3.0	5.0	13.0	B7			ζ Scl
0 02.4	+10 46	A 1249	AB	Peg	1905 1958	240 245	0.3 0.3	9.2	9.9	G0	17180		Too close 1940,47
—		Bgh	ABxC			120	63.2		8.5	F8 G0	17180		
0 02.6	+66 06	Σ 3053	AB	Cas	1832 1958	70	15.2	5.9	7.3	G0 A2	1		
—			AC		1911 1912	290	98.5		10.8		1		
0 02.8	+2 08	β 281	AB	Psc	1877 1959	217 179	1.1 1.5	7.8	11.0	G0	9		
—		h 998	AC		1877 1922	336 332	30.4 35.0		11.7		9		
0 02.8	+35 49	Ho 622		And	1896 1902	87	23.6	7.3	12.3	A0	8		
0 02.8	+80 17	Σ 3051		Cep	1832 1925	23	16.9	7.9	9.8	F2	5		
0 02.9	+47 15	A 800		And	1904 1959	290	1.5	8.8	8.8	A0	10		
0 02.9	+71 21	Σ 3052		Cep	1831 1925	8	34.0	8.5	9.1	K0			
0 03.0	−0 56	h 999		Psc	1905 1916	78	27.9	8.0	13.4	K0			
0 03.3	−4 09	Rst 4738		Psc	1939	76	1.5	8.0	12.3	F2			
0 03.9	+57 23	h 1926	AB	Cas	1906 1908	312	26.5	8.0	11.8	A2			
—			AC		1908	82	28.3		12.1				
—			AD		1908	160			14.0				
0 04.0	+12 09	Σ 3055	AB	Peg	1831 1943	0	5.6	7.2	11.4	F0	26		
—			AC		1912	32	122.6		12.6		26		
0 04.0	−48 08	CorO 264		Phe	1887 1943	64	3.2	7.8	9.6	F0			
0 04.2	+49 59	h 1931		Cas	1907	113	21.9	8.0	10.7	K2			
0 04.2	+62 17	S 838		Cas	1824 1904	196	246.0	5.9	9.9	A0			9 Cas
0 04.6	+42 06	OΣ 514		And	1847 1958	168	5.2	6.1	8.7	A2	30		
0 04.7	+34 16	Σ 3056	AB	And	1831 1959	158 148	0.6 0.6	8.0	8.0	K0	32		
—			ABxC		1831 1926	355 1	20.5 23.1		9.0		32		
—			ABxD		1912	238	95.7		9.7		32		
0 04.9	+26 39	Fox		Peg	1907	122	23.1	6.2	10.7	K2	42		
0 04.9	+58 32	Σ 3057		Cas	1832 1953	299	3.7	6.6	8.7	B3	36		
0 05.0	+45 40	β 997		And	1879 1959	340 338	4.0 4.0	8.1	9.1	F8	41		
0 05.2	+45 14	Es 1293	AB	And	1914 1925	120	12.9	6.5	13.5	A0	46		
—		β 997	AC		1906 1934	235	21.6		9.5		46		
0 05.4	+34 06	Hu 1201	AB	And	1905 1960	307	0.2	8.1	9.5	F8	51		
—		Ho 490	AC		1892 1948	168 163	20.8 23.3		13.4		51		
0 05.7	+17 50	Σ 3061		Peg	1829 1972	148	7.7	8.8	8.8	F2	56		
0 06.0	+49 37	Hu 502		Cas	1902 1935	113	2.4	7.6	10.9	A3	59		
0 06.2	−64 14	I 433		Tuc	1911 1956	135	5.7	7.9	10.1	F2			
0 06.3	+58 26	Σ 3062		Cas	1831 1960			6.4	7.2	G5	61	o	
0 06.3	−49 05	HdO 180		Phe	1900 1928	176	5.4	5.8	11.5	G0			
0 06.6	+29 01	β pm	AB	And	1853 1909	244 250	143.2 158.6	6.3	8.9	K0	69		
—			AxCD		1886 1909	182 185	156.4 153.2				69		
—		β	CD		1907 1950	210	3.3	10.0	10.5		69		J 866 same star
0 06.6	+56 30	Stein 1275		Cas	1917	166	8.4	7.4	12.8				
0 06.8	+54 30	Es 611		Cas	1908 1937	292	10.4	8.2	9.9	K0	72		
0 07.6	+31 40	β 1014		And	1891 1955	335	1.5	7.0	12.5	G5	83		
0 07.6	+40 09	Σ 3064 rej		And	1879 1955	354 3	23.8 24.8	7.0	10.0	G5	82·		
0 07.6	+52 46	A 1253		Cas	1906 1943	86 87	2.3 3.0	7.6	12.7	G5	81		
0 07.7	+37 11	A 1501		And	1907 1959	227 240	1.0 1.1	7.6	12.6	F5	86		
0 07.7	−60 39	I 434		Tuc	1911 1926	31	5.4	8.1	10.8	G5			
0 07.8	−22 31	See 2		Cet	1897 1959	174 153	2.1 1.8	5.9	11.0	A0	89		
0 08.0	+31 24	OΣΣ 256	AB	And	1876 1923	117 115	103.1 109.5	7.9	7.9	A5			
—			Aa		1912	260	20.1		13.6				
0 08.4	+29 05	H V 32		And	1836 1954	266 280	64.9 81.5	2.1	11.3	A0	94		21 α And, Alpheratz
0 08.8	+8 01	HdO 2		Psc	1877 1958	120 127	12.8 16.9	7.3	12.3	F0	101		
0 09.0	−21 12	B 1903		Cet	1930 1959	125	1.6	7.2	11.7	K0			
0 09.0	−54 00	HdO 181		Phe	1901 1959	292 206	0.7 0.4	6.6	7.9	G0			
0 09.1	+40 51	β 483	AB	And	1878 1932	43	2.8	7.2	11.5	G5	105		
—			AC		1879 1909	261	144.0		7.9	G5	105		
—			CD		1909	280	74.1		10.9		105		
0 09.1	−50 10	h 3347		Phe	1914 1916	81	25.0	7.1	13.4	B8			
0 09.2	+59 09	AGC 15		Cas	1889 1935	189 243	22.6 31.3	2.3	13.6	F5	107		11 β Cas, Caph
0 09.3	+79 43	Σ 2		Cep	1830 1959			6.6	6.9	A3	102	o	

3

α 2000	δ 2000	Star Name		Const	Years	Pos Ang	Separation	m₁	m₂	Spec	ADS	Orb	Notes
0ʰ09.3	-27°59'	β 391	AB	Scl	1876 1954	277° 265°	0".9 1".4	6.1	6.2	F2 F2	111		κ¹ Scl
--		LDS 2095	AC		1954	243	70.0		18		111		
0 09.5	+19 07	Cou 247		Peg	1967	31	0.5	8.0	9.2	G5			
0 09.7	+52 02	β 484	AB	Cas	1878 1941	156	1.9	8.0	12.2	K0	116		
--			AC		1914	52	82.6		9.2		116		
0 10.0	+11 09	Σ 5		Psc	1830 1958	160	7.7	5.5	9.4	B8	122		34 Psc
0 10.0	+46 23	Σ 3		And	1831 1938	84	5.0	8.1	9.1	A3	119		[1]
0 10.4	+49 53	Es		Cas	1909 1917	294	77.4	8.8	8.8				
0 10.6	-73 13	I 43	AB	Tuc	1900 1931	329 320	1.0 0.5	6.9	8.0	G			Spectrum composite G+A5
--			ABxC				20.0		13.8				
0 10.8	-34 52	Rst 2236		Scl	1933 1945	99	1.7	6.9	13.2	K0			
0 11.3	-40 22	Rst 1180		Phe	1931 1945	53	1.1	7.6	12.0	G5			
0 11.6	-3 05	Σ 8		Psc	1831 1953	292	7.8	7.8	9.4	F8	144		
0 11.6	+55 58	Σ 7		Cas	1831 1955	217 212	1.3 1.3	7.9	9.4	B8	143		
0 11.6	+59 46	β 254	AB	Cas	1875 1945	237	7.4	8.0	12.0	G5	140		
--			AC		1893 1915	241	38.2		12.7		140		
0 11.6	-27 48	LDS 2099		Scl	1954	291	46.0	7.2	21				κ² Scl
0 11.8	+66 08	OΣ 1		Cas	1850 1947	204 210	1.4 1.6	7.3	10.0	A0	145		
0 11.9	+28 25	β 255		And	1875 1960	99 86	0.4 0.6	8.5	8.8	F5	147		
0 11.9	-75 29	Jsp 2		Hyi	1928	295	4.9	7.4	11.9	K0			
0 12.1	+53 37	β 1026		Cas	1888 1960			7.3	8.1	F0	148	o	
0 12.1	-58 32	Rst 4739		Tuc	1942	283	0.3	8.2	8.7	G5			
0 13.2	+15 11	β pm	AB	Peg	1879 1907	286 285	161.8 163.4	2.8	11.7	B2			88 γ Peg
--			BC		1879 1907	199 196	21.0 20.4		12.3				
0 13.3	+51 50	Es 748		Cas	1909	359	10.5	8.0	12.1	K5	158		
0 13.3	-20 37	B 1905		Cet	1930 1959	110	3.2	7.3	12.0	F0			
0 13.4	+26 59	OΣ 2	AB	And	1851 1957			6.7	7.5	F5	161	o	
--			AC		1851 1944	225	17.8		10.4		161		
0 13.7	+36 11	β	AB	And	1903	318	20.8	11.1	11.8	F8 G0			
--			AC		1903	227	177.7		8.0	A0			
0 14.0	+76 02	OΣΣ 1		Cep	1875 1923	103	76.3	7.6	7.9	M			Optical pair
0 14.4	+43 50	A 1255		And	1893 1935	348	4.0	8.1	10.9	F8	179		
0 14.5	-7 47	β 486		Cet	1878 1958	4	3.0	5.1	11.1	M	180		AD Cet
0 14.7	-12 45	HdO 7		Cet	1867 1933	7	7.1	8.0	12.0	K0	185		
0 14.8	-0 19	Bail 642		Psc	1898	62	6.2	8.1	11.3	A5			
0 14.8	+62 50	Σ 10		Cas	1832 1950	176	17.6	8.0	8.7	A0	183		
0 14.9	+21 33	β 1027		Peg	1888 1949	187	1.6	7.4	10.5	K2	190		
0 15.0	+8 49	Σ 12		Psc	1832 1958	148	11.6	6.0	7.6	F0	191		35 Psc = UU Psc
0 15.2	+27 17	Kui 2		And	1958	130	29.2	6.1	12.9	A0			
0 15.2	+44 06	A 1256	AB	And	1906 1957	4 42	0.1 0.2	7.3	7.5	B9	197		
--			ABxC		1906 1945	344	18.8		13.0		197		
--			ABxD		1906	134	27.6		14.5		197		
0 15.2	-40 06	Wg 2		Phe	1904	3	13.8	7.4	8.9	G5			
0 15.9	-5 36	Σ 15		Psc	1831 1944	198	4.7	7.8	10.3	G5	205		
0 16.1	+8 07	β		Psc	1904 1908	142	6.0	8.0	12.5	A5	211		
0 16.1	-31 26	LDS 2106		Scl	1954	192	64.0	7.1	17				
0 16.2	+76 57	Σ 13		Cep	1831 1959			7.0	7.3	B9	207	o	
0 16.4	+43 36	h 1947		And	1879 1958	76	9.2	6.2	10.2	A0	215		
0 16.7	+36 30	OΣ 4		And	1854 1960			8.2	8.9	G0	221	o	
0 16.7	+36 38	Σ 19		And	1836 1954	133 138	2.3 2.4	7.1	9.6	A0	220		
0 16.7	+54 40	Σ 16		Cas	1832 1934	38	5.7	7.8	9.1	A3	218		
0 17.0	+61 32	β 392		Cas	1879 1956	69	19.4	5.7	12.5	G5	222		
0 17.4	+8 53	A 1803	AB	Psc	1908 1958	310	0.1	7.9	8.0	F5	238		[2]
--		Σ 22	ABxC		1836 1962	236	4.3		7.8		238		
--			AD		1910	142	63.7		11.9		238		
0 17.5	+0 19	Σ 23	AB	Psc	1836 1954	0 256	12.7 3.6	7.9	10.2	F8	241		Optical
--			AC		1911 1925	281	103.0		11.8		241		
0 17.9	+13 03	h 3		Psc	1909 1974	262 263	11.5 12.0	8.6	8.9				
0 18.2	+72 56	A 803		Cep	1904 1958	175 218	0.3 0.1	8.1	8.4	A3	243		
0 18.3	-21 08	β 393		Cet	1879 1954	12 22	0.7 0.7	7.1	7.9	A0	251		
0 18.4	+49 30	Es 41		Cas	1901 1924	218	6.0	7.8	11.0	F8	248		
0 18.5	+26 08	Σ 24		And	1831 1943	248	5.2	7.6	8.4	A2	252		A pretty pair (Webb)
0 18.5	-27 57	h 1949		Scl	1892	322	74.6	8.6	8.9	K2 K5			
0 18.7	+43 47	OΣ 5		And	1847 1958	240	6.2	6.0	9.7	B9	254		26 And
0 19.2	+59 42	Kr 4		Cas	1890 1954	190 185	1.9 2.2	8.3	8.8	F0	263		
0 19.4	-8 49	h 1953	AB	Cet	1880 1910	16 17	62.0 63.4	3.5	12.9	K0			8 ι Cet
--			AC		1897 1959	198 194	110.1 108.8		8.4				
0 20.0	+38 14	S 384	AB	And	1824 1926	13 20	45.7 76.4	7.2	9.7	G0	271		
--			Aa		1902 1910	258 267	20.2 20.6		13.6		271		
0 20.1	+6 18	h 1955	AB	Psc	1902 1910	29	50.5	7.2	11.3	K0			
--			AC		1891 1920	10	61.0		12.5				
0 20.2	-35 54	I 701		Scl	1912 1959	58	0.6	8.3	8.5	K0			
0 20.6	+45 31	A 647		And	1904 1958	216 171	0.7 0.5	7.2	9.2	F5	277		
0 20.7	+12 18	β 1015		Psc	1888 1960			8.7	8.9	F5	281	o	
0 20.9	+10 59	β 1093		Psc	1889 1958	54 103	0.4 0.7	7.0	7.9	A0	287		

α 2000	δ 2000	Star Name		Const	Years	Pos Ang	Separation	m_1	m_2	Spec	ADS	Orb	Notes
0h 20.9	+32° 59'	AC 1		And	1861 1959	276° 288°	0".6 1".6	7.5	8.0	F5	285		
0 21.4	+67 00	OΣ 6	AB	Cep	1849 1957			7.7	8.3	A0	293	o	
--		Σ 26 rej	AC		1849 1950	114	13.5		10.2		293		
0 21.9	-23 00	h 1957	AxBC	Cet	1846 1952	19 24	6.1 6.1	7.6		G0	302		
--		Rst 5493	BC		1949 1951	91 85	0.2 0.2	9.9	9.9		302		
0 22.4	+13 29	Σ 27		Psc	1829 1954	344 324	31.7 28.5	6.2	10.1	K0	303		42 Psc; optical (Cou)
0 22.9	-12 13	β pm		Cet	1909	300	185.1	6.4	11.8	G0			9 Cet
0 23.2	-76 44	I 436		Hyi	1911 1931	305	0.4	8.7	8.9	F5			
0 24.0	-3 29	β 488	AB	Psc	1878 1939	348 343	3.3 3.4	7.8	10.8	A0	326		
--			AC		1914	58	40.5		12.7		326		
0 24.3	+52 01	Hu 506		Cas	1902 1960	217 16	0.2 0.2	5.6	8.1	B3	328		Single in 1953 (M1r)
0 24.7	-53 59	h 3364		Phe	1895 1955	222 239	31.5 39.8	7.4	10.5	G0			
0 25.4	+1 56	OΣ 8 rej		Psc				5.8		G5			44 Psc; single?
0 26.2	+56 47	OΣ 9	AB	Cas	1847 1950	61 52	1.5 1.8	6.8	10.0	G0	350		
--			AC		1897 1910	4	22.8		9.8		350		
0 26.3	+3 50	HdO 18		Psc	1908	175	60.0	6.9	10.4	B8			
0 27.2	+21 15	β 1225		Psc	1891 1959	189 196	1.2 1.4	8.0	11.7	A3	362		
0 27.2	+49 59	Σ 30	AB	Cas	1831 1968	296 308	21.2 15.2	6.9	8.8	B9	361		
--			AC		1910	98	64.0		12.6		361		
--		F1e	AD		1956	250	148.5		10.0		361		
0 27.2	+60 40	Stein 64		Cas	1903	259	11.3	7.9	12.1	K0			
0 27.5	+16 02	OΣ 10 rej	AB	Psc	1866 1919	237 238	96.3 102.8	6.4	10.2	A5 F8			
--			AC		1907	154	276.2		9.6				
--			BD		1907	157	153.4		14.1				
0 27.7	-16 25	h 1968	AB	Cet	1877 1954	73 230	8.5 17.3	7.3	9.8	F8	366		Probably an optical pair
--		β	AC		1891 1921	121 123	93.4 88.2		11.0		366		Also likely optical
0 27.9	-24 51	I 702		Scl	1914 1954	82	1.3	8.1	9.9	F8	370		
0 28.4	-20 20	B 1909		Cet	1929 1959			7.2	7.2	G0		o	
0 28.6	-11 14	Rst 4146		Cet	1939 1943	154	1.9	6.9	12.6	F5			
0 28.8	-24 38	LDS 2126		Cet	1954	193	5.0	7.1	19				
0 29.1	+11 19	A 805		Psc	1904 1953	304	4.0	8.0	14.2	F0	392		
0 29.1	-27 01	Rst 1189		Scl	1930 1950	54	0.4	8.1	9.9	G5			
0 29.3	+64 15	β 1157		Cas	1890 1954	88	1.6	8.1	11.0	A0	388		
0 30.0	-3 57	h 322		Cet	1866 1939	185 194	8.6 10.3	5.7	10.5	K5	410		12 Cet
0 30.1	+29 45	β 1095	AB	And	1889 1948	0 6	2.4 2.4	5.2	13.0	F0	409		28 And
--			AC		1912 1924	130 130	142.4 141.4		11.3		409		
0 30.3	+59 59	β 1094	AB	Cas	1889 1948	245 254	0.7 0.7	6.0	9.8	B9	412		
--			AC		1912 1959	338 338	120.3 120.8		11.8		412		
0 30.7	+32 08	OΣ 11 rej		And	1914 1920	317	197.8	8.0	7.6	F2 F5			
0 30.8	+16 02	Σ 32		Psc	1829 1921	108 103	13.7 20.9	6.9	10.7	A0	420		
0 30.8	+47 32	β 394		Cas	1876 1959	284	1.1	8.5	8.7	G0	416		
0 31.0	+34 06	Σ 33		And	1831 1960	206 210	2.5 2.7	8.7	8.8	F5	421		
0 31.0	-10 05	h 1981	AxBC	Cet	1890 1929	87	79.2	6.9		A3 G0	426		
--		β 1158	BC		1890 1958	138 186	0.3 0.5	9.2	9.2	G0	426		
0 31.4	+33 35	h 5451		And	1875 1932	85	56.4	6.0	8.4	K0			
0 31.4	-48 48	Rst 11		Phe	1929 1946	299 295	28.6 30.3	4.8	13.6	A2			λ¹ Phe
0 31.5	-62 58	B 7	AB	Tuc	1926 1932	151	2.4	4.4	13.5				β¹ Tuc
--		Lac 119	AC		1835 1952	172 169	27.1 27.1		4.8	B9 A2			C is β² Tuc
--		I 260	CD		1900 1959	5			6.0			o	
0 31.8	+54 31	OΣ 12		Cas	1845 1959	123 176	0.5 0.5	5.3	5.6	B8	434	o	14 λ Cas
0 32.1	-18 13	B 1910		Cet	1933 1959	214 235	0.2 0.2	8.0	8.2	K0			
0 32.4	+6 57	Σ 36		Psc	1833 1933	83	27.5	5.7	9.5	A0	449		51 Psc
0 32.4	+58 20	β 1227	AB	Cas	1891 1936	202	2.9	7.0	11.3	F2	443		
--			AC		1897 1936	86	22.5		11.2		443		
--			AD		1897 1936	115	32.3		11.7		443		
0 32.6	+20 18	h 1982		Psc	1879 1924	306 302	38.3 44.5	5.4	11.6	G5	452		52 Psc
0 32.7	+23 12	β 1310	AB	And	1903 1958	212	3.7	7.0	11.8	G0	451		
--			AC		1903 1925	301 294	15.2 17.7		12.5		451		
--			AD		1902 1926	145 147	96.5 95.2		9.2		451		
0 32.7	-25 22	h 3442		Scl	1902 1948	203 196	25.6 21.9	6.9	11.0	K0	456		
0 32.7	-63 02	B 8		Tuc	1925 1954	143	0.1	5.8	6.0	A2			
0 32.8	+28 17	OΣ 14	AB	And	1847 1955	160	8.7	6.3	11.3	G5	455		
--			AC		1912	84	60.0		12.3		455		
0 32.9	+16 10	J 183		Psc	1910 1952	101	21.6	6.9	12.9	F5	458		
0 32.9	+35 51	Ho 211		And	1888 1924	18	1.4	7.9	12.2	A0	454		
0 33.4	+47 39	A 911		Cas	1905 1956	318	0.6	8.4	9.1	A2	461		
0 33.5	+40 06	Ho 3		And	1885 1954	121 134	0.4 0.4	8.0	10.3	F8	463		Single 1900-06 (Doo)
0 33.5	-55 20	I 45	AB	Phe	1900 1959	274 231	0.5 0.5	7.9	8.3	A2			
--		h 3376	ABxC		1837 1930	247	7.0		11.0				
0 33.5	-61 09	h 3378		Tuc	1916	354	7.6	8.0	12.8	A0			
0 33.6	-26 06	h 3377		Scl	1879 1954	57 60	18.4 20.1	7.5	9.7	K0 M5	466		
0 33.7	+62 47	Stein 97		Cas	1904 1908	5	13.0	8.3	9.7	G5			
0 33.7	-35 00	h 3375		Scl	1835 1954	164 168	6.4 5.3	6.6	8.4	G0			
0 33.9	+3 19	HdO 22		Psc	1868			7.8		G5			Single in 1903
0 34.5	-4 33	Dem 2	AB	Cet	1870 1959	239 252	0.8 0.6	7.6	8.1	G0	475		

α 2000	δ 2000	Star Name	Const	Years	Pos Ang	Separation	m_1	m_2	Spec	ADS	Orb	Notes
0h34.5m	−4°33′	Σ 39 ABxC	Cet	1830 1953	45° 45°	20″.1 19″.7	9.3			475		
--		Aller ABxD		1932	337	205.4				475		Smyth wondered if D is var.
0 34.6	+62 54	β 108 AB	Cas	1875 1916	358	4.2	7.7	10.8	F5	470		
--		AC		1898	338	20.9		15		470		
--		AD		1898	214	24.1		14.0		470		
--		AE		1898	151	27.2		13.0		470		
--		AF		1898 1917	219	41.8		10.1		470		
0 35.2	−3 36	Ho 212 AB	Cet	1887 1959			5.6	6.3	G0	490	o	13 Cet
--		β 490 ABxC		1877 1899	65 43	37.1 24.5		12.5		490		
0 35.2	+36 50	Σ 40 AB	And	1831 1961	313	11.7	6.6	9.2	K0	486		
--		AC		1913	123	24.9		12.8		486		
0 35.6	+34 29	Hu 1011	And	1904 1959	132 114	0.4 0.4	8.0	9.5	A2	492		
0 35.8	+49 01	OΣ 15	Cas	1890 1960	302 314	0.2 0.2	7.3	8.3	A2	493		Single in 1954 (M1r)
0 36.1	−20 46	B 2066	Cet	1929	230	7.0	7.9	13.0	F0			
0 36.2	+24 02	Ho 623	And	1899 1922	152	8.3	7.3	12.3	G5	499		
0 36.6	−49 08	LDS 21	Phe	1954	273	329.6	6.8	8.6	G0 G5			
0 36.7	−27 25	h 3379	Scl	1877 1919	231	14.6	7.9	11.5	F2	514		
0 36.9	+33 43	H V 17 AB	And	1821 1937	173	35.9	4.4	8.6	B3	513		29 π And = h 1030
--		AC		1879 1924	357	55.2		13.0		513		
0 37.0	+2 06	HdO 29	Cet	1903	342	23.5	7.8	9.2				No br. star here 1909 (β)
0 37.2	+58 01	β 1097	Cas	1889 1957	72 72	0.8 0.5	8.2	8.2	B9	515		Spectrum composite A2+K
0 37.3	−24 46	β 395	Cet	1886 1959			6.3	6.4	K0	520	o	
0 37.4	−37 17	I 705	Scl	1925 1944		0.1	7.7	7.9	G0			
0 37.8	+72 54	h 1989 AB	Cas	1908	51	32.3	7.0	10.8	B3	516		
--		BC		1908	343	10.8		12.1		516		
0 38.1	−0 30	HdO 31	Cet	1901 1908	307	30.5	6.7	13.5	K0	534		
0 38.7	+46 57	Σ 45 AB	Cas	1829 1930	83 88	8.8 14.5	6.9	9.9	G5	538		
--		β AC		1911	261	109.8		12.8		538		
0 38.8	−25 06	h 1991	Scl	1931 1953	95	46.9	6.6	9.7	K0			
0 38.8	−85 25	LDS 25	Oct		nf	22.0	7.7	12.9	G5			
0 38.9	−25 36	h 1992	Scl	1895	246	45.4	7.8	8.4	A3 K			
0 39.2	+49 21	OΣ 16	Cas	1845 1955	26 23	14.8 13.3	5.4	9.9	K2	546		
0 39.3	+30 52	β 491 AB	And	1878 1934	299 298	27.9 28.7	3.3	12.4	K2 M2	548		31 δ And
--		AC		1910 1959	111 108	47.5 41.4		15		548		
0 39.4	+21 15	β pm	Psc	1852 1907	104 92	98.5 122.0	5.9	10.8	K0			54 Psc
0 39.6	+36 47	OΣ 17	And	1846 1932	161	8.6	7.9	11.1	F0	553		
0 39.9	+21 26	Σ 46	Psc	1830 1939	194	6.5	5.4	8.7	K0	558		55 Psc; yellow, blue
0 40.0	+76 52	OΣΣ 5	Cas	1875 1923	143	116.3	7.0	8.7	K0 G5			
0 40.2	+47 15	β 257	Cas	1876 1954	237 240	0.5 0.6	8.0	9.1	G0	559		
0 40.3	+24 03	Σ 47 AB	And	1832 1962	205	16.5	7.4	9.2	A3	562		
--		β AC		1881 1922	230	42.4		11.1		562		
--		AD		1912	244	28.9				562		
--		BC		1881	243	27.9				562		
0 40.5	+36 27	Cou 1051	And	1973	72	0.4	8.4	9.4	A0			
0 40.5	+56 32	β AB	Cas	1889 1934	272 275	17.6 19.8	2.2	13.7	K0	561		18 α Cas, Schedar
--		AC		1878 1908	109 105	40.1 38.3		12.7		561		
--		H V 18 AD		1781 1913	275 280	56.2 64.4		8.9		561		Also h 1993; optical
0 40.5	−16 31	β 109 AB	Cet	1898 1917	355 354	103.1 105.0	6.6	9.1	G5			
--		AC		1893	357	91.5		11.0				
--		BC		1876 1916	162	11.0						
0 40.7	−4 21	h 323	Cet	1879 1922	290 289	65.2 64.3	6.0	8.5	G5			Optical
0 40.8	−7 14	Σ 49	Cet	1830 1958	322 320	4.5 7.8	7.0	10.0	G0	566		Optical
0 41.6	+24 38	Wils	And	1949	90		6.0		A5			Hynek 1; probably single
0 41.8	−56 30	h 3387	Phe	1886 1933	253	13.2	5.8	10.2	F0			ξ Phe
0 42.0	−55 47	MlbO 1	Phe	1877 1953	164 162	6.3 6.3	7.8	8.3	G0			
0 42.4	+4 10	OΣ 18 AB	Psc	1845 1959			7.8	9.4	F8	588	o	
--		A AC		1896 1913	270	42.8		12.1		588		
0 42.6	+57 36	Es 1806	Cas	1920	236	5.5	7.8	12.1	B9	587		
0 42.6	−54 07	h 3388	Phe	1873 1951	239 240	16.7 16.7	8.1	8.5	G5			
0 42.7	+71 22	Σ 48 AB	Cas	1836 1955	333	5.5	8.0	8.2	A0	582		
--		Baize BC		1931	323	55.7				582		
0 42.7	−38 28	HdO 182	Scl	1901 1954	322 3	0.5 0.7	6.7	7.0	A0			λ¹ Scl
0 42.7	−65 37	I 440	Tuc	1928 1944	354	0.1	7.9	7.9	A3			
0 42.8	−9 55	h 1995	Cet	1904 1921	136	39.3	6.6	9.6	F8			
0 43.3	−45 11	h 3390	Phe	1900 1936	313	14.2	7.1	9.7	G5			Orange, blue (h)
0 43.4	−57 28	h 3391	Phe	1887 1933	217	19.8	4.4	11.4	A0			η Phe
0 44.2	+46 14	Σ 52	And	1831 1956	26 9	1.4 1.5	7.9	8.9	F5	616		
0 44.4	+33 37	Σ 55	And	1831 1950	323 328	2.1 2.1	8.1	8.9	A3	618		
0 44.5	−62 30	CorO 3	Tuc	1899 1942	68	2.4	6.3	8.3	F5			
0 44.7	+48 17	β 231	Cas	1876 1960	304 302	32.8 33.6	4.5	11.0	B2	622		22 o Cas
0 45.1	−0 16	LDS 836	Cet	1935	43	59.0	7.5	14.8	F8			
0 45.3	+55 13	β 492 AB	Cas	1878 1933	152	2.2	5.5	11.5	A0	625		
--		AC		1909	24	88.3		10.4		625		
0 45.5	+43 24	β 865 AB	And	1880 1941	197 190	1.2 1.5	8.5	9.0	A5	627		
--		Fox AC		1920	147	37.9		12.0		627		

α 2000	δ 2000	Star Name		Const	Years		Pos Ang		Separation		m₁	m₂	Spec	ADS	Orb	Notes
0ʰ 45ᵐ.5	−12° 53′	β pm		Cet	1909	1959	232°	238°	77″.0	69″.5	6.2	12.5	G0			18 Cet
0 45.7	+54 59	Arg 2	AB	Cas	1901	1933	118	115	13.8	11.0	8.4	9.4	F8	630		Also named Es 43
			AC		1907	1928	94	90	26.2	24.8		12.8		630		
0 45.7	+74 59	H N 122		Cas	1903	1918	160	160	35.6	36.1	5.7	9.4	A2	624		YZ Cas; A is a sp. binary
0 45.7	−16 25	M1f 1		Cet	1887	1955	195		2.5		6.5	9.5	F0	636		
0 45.8	−41 55	h 3395		Phe	1836	1957	63	91	8.9	5.7	8.4	8.9	K0			Both alike, almost red (h)
0 45.8	−47 33	HdO 183		Phe	1901		307		14.3		5.8	13.5	K0			
0 46.0	−16 39	Rst 2251		Cet	1934	1943	20		6.1		8.0	14.5	F0			
0 46.4	+30 57	Σ I 1		And	1834	1933	55	50	46.4	46.8	7.4	7.6	G5 G5	639		
0 46.6	−54 06	h 3397		Phe	1916	1927	175		12.2		7.9	14.2	F5			
0 47.3	+24 16	β pm	AB	And	1910		349		28.4		4.1	15	K0			34 ζ And
			AC		1910		230		96.4			13.4				
			AD		1910		258		167.9			10.6				
0 47.5	+51 06	H V 82	AB	Cas	1823	1920	78	76	47.1	52.2	8.0	8.3	K2 K0			
			AC		1909		281		104.6			13.5				
0 47.9	−29 21	I 261		Scl	1901	1954	53		0.3		8.0	9.0	A5	664		
0 48.0	+51 27	Σ 59		Cas	1832	1953	147		2.2		7.2	8.1	A0	659		
0 48.2	+65 07	M1r 81		Cas	1970		95		4.5		7.1	10.5	G5			
0 48.4	+5 17	β pm		Psc	1909	1929	200	208	194.7	181.2	5.8	10.4	G5			96 G. Psc; optical
0 48.7	+7 35	β pm		Psc	1911		16		132.2		4.4	13.0	K5			63 δ Psc
0 48.8	+18 42	β 495		Psc	1878	1959					8.4	8.4	G0	673	o	
0 49.0	+16 56	β pm	AB	Psc	1911	1927	322	325	69.0	71.7	5.1	12.5	F5			64 Psc
			AC		1911	1927	165	165	78.9	75.9		12.9				
0 49.1	+57 49	Σ 60	AB	Cas	1832	1959					3.4	7.5	F8	671	o	24 η Cas
			AC		1912	1922	234	238	154.5	158.9		11.3		671		
			AD		1912	1921	34	30	159.8	159.6		11.5		671		
			AE		1913	1921	120	120	203.2	190.6		8.9		671		
			AF		1913		268		281.7					671		
			AG		1913	1915	243	243	338.5	339.2		8.8		671		
		Pav	BC		1913		118		209.1					671		
		Σ 60	BH		1928		2		(10′.8)			8.5		671		
0 49.2	−24 08	B 13		Cet	1926		167		8.8		6.1	13.8	G5	679		
0 49.3	−21 24	β 301	AB	Cet	1891	1943	319		1.0		8.4	13.0	F5	681		
			AC		1891	1917	300		11.0			9.4		681		
0 49.4	−13 34	β 1160		Cet	1890	1933	114		1.3		5.8	12.0	K2	680		
0 49.6	−54 10	h 3402		Phe	1894	1954	60	58	12.1	7.8	8.1	10.8	F8			
0 49.8	+70 27	β pm		Cas	1893	1911	301	296	62.7	66.5	7.9	10.4	K0			
0 49.9	+27 43	Σ 61		Psc	1832	1959	297		4.4		6.3	6.3	F0	683		65 Psc
0 49.9	+30 27	OΣΣ 9	AB	Psc	1875	1924	234	238	91.8	100.2	7.6	9.2	G5 G5			
			AC		1892	1924	324	322	105.7	109.3		10.3				
			Cc		1908	1924	220		19.7			11.1				
0 50.4	+50 38	β 232	AB	Cas	1876	1960					8.4	8.8	F5	684	o	
			ABxC		1875	1933	293	295	28.7	27.1		10.2		684		
0 51.6	+12 47	β 496		Psc	1878	1914	2		5.3		6.7	12.7	F0	702		
0 51.6	+22 38	A 1808		And	1908	1954	29	41	0.2	0.1	8.1	8.1	K0	701		
0 51.8	+48 03	A 812		Cas	1904	1945	323		1.8		7.6	11.5	K0	699		
0 51.9	−43 43	I 47		Phe	1900	1914	1	11	1.1	1.1	7.2	7.6	F2			
0 52.0	+33 53	h 628		And	1901	1911	69		42.5		7.0	10.8	K2			
0 52.2	−22 37	Stone 3	AB	Cet	1877	1959	273	256	2.4	1.8	7.6	8.3	G0	716		
			AC		1897	1908	197	196	35.7	32.6		11.9		716		
0 52.4	+25 47	Weisse 2		Psc							7.7		K0			Probably single (IDS)
0 52.4	−69 30	Δ 2		Tuc	1836	1952	79	81	21.2	20.7	6.5	7.9	F8 F8			
0 52.7	−24 00	β 734		Cet	1879	1934	346		10.8		5.5	9.7	K0	726		
0 52.8	+56 38	β 1	AB	Cas	1875	1936	82		1.4		7.8	9.8	B2	719		
			AC		1875	1936	133		3.8			8.8		719		
			AD		1875	1936	194		8.9			9.3		719		
			AE		1889	1914	332		15.7			12.0		719		
			CD		1936		220		7.6					719		
0 52.8	+68 52	Σ 65		Cas	1832	1967	35	39	3.0	3.1	8.0	8.0	A2	710		
0 53.0	−61 04	Gli 4		Tuc	1877	1943	70		6.0		8.4	9.0	G0			
0 53.1	+57 13	Es 552		Cas	1908	1913	86		13.0		7.1	13.1	B9	723		
0 53.1	+61 07	β 497	AB	Cas	1878	1916	172	171	121.1	129.7	4.8	9.3	F8	721		
			AD		1922		143		93.2					721		
			AE		1922		39		130.3					721		
			BC		1877	1946	151	162	0.9	0.9		12.1		721		
0 53.2	−24 47	WNO 1		Cet	1876	1952	12	9	5.4	5.4	6.5	8.4	F2	733		
0 53.3	+4 05	A 2307		Psc	1911	1957	32	230	0.2	0.2	7.4	8.4	F0	732	o	Close after 1945
0 53.3	−45 30	B 644		Phe	1926	1954	260	300	0.2	0.2	8.6	8.8	F0			
0 53.8	+52 41	Σ 70	AB	Cas	1832	1962	245		8.0		6.3	9.3	A0	735		
			AC		1909	1913	148	148	69.7	78.5		9.8		735		
			Cc		1909		88		1.7			10.2		735		
0 53.9	−54 36	CapO		Phe	1938		323		12.9		7.8	10.8	G5			
0 54.6	+9 26	Σ 74		Psc	1830	1950	300		3.1		8.2	9.2	F0	754		
0 54.6	+19 11	OΣ 20	AB	Psc	1847	1959					6.2	6.9	A0	746	o	66 Psc
		β	ABxC		1912		10		150.9			12.8		746		

α 2000	δ 2000	Star Name	Const	Years	Pos Ang	Separation	m₁	m₂	Spec	ADS	Orb	Notes
0ʰ54.9ᵐ	+49°24'	Hu 802		Cas	1903 1955	214° 209°	0″.3 0″.4	8.1	8.7	A0	749		
0 55.0	+23 38	Σ 73	AB	And	1836 1959			6.0	6.4	K0	755	o	36 And
--		Fox	ABxC		1921	221	160.2		10.9		755		
0 55.0	+24 06	β pm		And	1893 1918	217 215	65.8 60.2	7.5	9.8	G5			
0 55.0	+58 58	β 1098	AB	Cas	1889 1934	75 70	12.8 14.3	4.8	12.3	K0	748		26 υ¹ Cas
--			AC		1912 1959	128 127	93.6 94.1		11.6		748		
0 55.4	+30 39	β 500		Psc	1878 1958	289 298	1.0 0.5	8.7	8.7	A2	768		
0 55.4	+40 23	A 1511		And	1907 1929	36	1.3	6.9	11.4	A2	766		
0 56.7	+60 43	β 1028	AB	Cas	1888 1956	256 248	2.2 2.1	2.5	11.2	B0	782		27 γ Cas; variable
--		β 499	AC		1879 1959	348	52.4		13.2		782		
0 56.8	+38 30	h 1057	AB	And	1878 1924	314 307	37.3 42.2	3.9	12.9	A2	788		37 μ And
--			AC		1878 1924	117 126	38.4 34.2		11.4		788		
--			AD		1903	41	272.7		9.8		788		
0 56.8	+60 22	β 1099	AB	Cas	1889 1959			6.1	6.8	B9	784	o	
--		β	ABxC		1910 1959	156	41.4		13.6		784		
0 57.0	+57 29	Es 44		Cas	1907 1910	252	8.2	8.1	10.1	K0	787		
0 57.2	+23 25	Fox		And	1913 1921	232	133.4	4.4	11.3	G5			38 η And
0 57.3	+52 14	Es 940	AB	Cas	1910	356	62.6	7.5	9.5	K2	792		
--			BC		1910	33	7.0		13.7		792		
0 57.6	-19 00	h 2004		Cet	1850 1943	240	3.4	7.2	10.2	A0	799		
0 57.9	-66 34	I 48		Tuc	1900 1933	333 339	0.6 0.5	7.6	8.6	G0			
0 58.1	-57 42	B 1027		Phe	1927 1954	65	0.8	7.8	11.0	F8			
0 58.2	-15 41	S 390		Cet	1824 1975	213 216	7.8 6.4	7.9	7.9	F5	806		
0 58.3	+21 24	β 302		Psc	1876 1958	92 136	0.8 0.6	6.7	8.1	A0	805		
0 59.3	-0 40	A 1902		Cet	1908 1958	199 293	0.3 0.2	8.7	8.8	F5	819		
0 59.4	+0 47	Σ 80	AB	Cet	1833 1975	300 333	18.3 26.4	8.3	8.7	K0	818		
--		β	BC		1907	258	178.0		10.5		818		
0 59.5	-75 49	Gli 8		Hyi		83	27.0	8.2	9.2	F5 F8			
0 59.8	+6 29	Kui 4		Psc	1958	312		6.1	13.2	M2			
1 00.1	-2 01	Σ 81 rej		Cet	1891 1911	69 68	17.8 17.4	7.3	10.1	F5	825		
1 00.1	+44 43	Σ 79		And	1832 1967	192	7.8	6.0	6.8	B9 B9	824		
1 00.3	+18 12	A 2210		Psc	1910 1958	211 249	5.1 4.2	7.4	13.5	G5	827		
1 00.4	+18 03	Brt 1927		Psc	1934	178	1.9	8.7	8.7	F8			
1 00.6	+47 19	Ma 1		And	1845 1959	45 17	1.0 0.9	8.1	8.6	A2	829		
1 00.7	-52 35	I 49		Phe	1900 1942	43 38	0.7 0.7	7.8	8.2	F0			
1 01.1	+60 21	A 926		Cas	1905 1958	248 299	0.3 0.3	8.6	9.0	F5	832		
1 01.4	+35 35	Cou 854		And	1972	60	0.2	8.8	8.8	F8			
1 01.4	+49 33	Es 45		Cas	1901 1938	243	7.7	6.9	10.8	G5	842		
1 01.5	+69 21	A 2901		Cas	1917 1958	10 49	0.2 0.3	7.5	7.5	B9	836		
1 02.3	+81 53	Knott 1	AB	Cep	1881 1899	62	13.8	6.6	11.2	A0	830		U Cep; A is variable
--			AC		1881 1899	321	21.2		12.2		830		
1 02.4	-21 37	See 10		Cet	1897 1959	324 318	4.8 5.0	8.2	10.2	G5	860		
1 02.6	+27 45	Ho 493		Psc	1893 1931	21 16	33.4 32.9	6.7	12.7	F5	857		
1 02.8	+47 42	h 2010		And	1879 1903	270	10.0	8.0	9.5	A0	856		
1 02.9	+41 21	h 1064		And	1902 1933	4	20.1	6.0	12.4	A2	863		39 And
1 02.9	+51 48	β 1161		Cas	1890 1959	324 353	0.5 0.4	7.2	8.0	B5	859		
1 03.0	-4 50	β pm		Cet	1911	241	108.9	5.4	11.8	G5			25 Cet
1 03.0	+47 23	OΣ 21		And	1847 1959			6.8	8.1	F0	862	o	
1 03.3	-60 06	h 3416	AB	Tuc	1836 1951	128	5.0	7.4	7.5	F5			
--			AC		1916	41	99.5		11.8				
1 03.5	+50 19	Σ 83 rej		Cas	1900 1908	313	21.9	8.1	11.1	G0			
1 03.6	+61 05	β 396		Cas	1877 1945	66	1.2	6.0	9.1	F0	868		
1 03.6	+63 41	Mlr 87	AB	Cas	1970	130	0.2	8.4	8.4	G0			
--		hz	CD		1973	38	3.7	11.0	11.4				
1 03.7	+50 26	Hu 517		Cas	1902 1955	14 20	0.5 0.6	8.6	9.0	A0	871		
1 03.8	+1 22	Σ 84	AB	Cet	1832 1926	253	16.0	6.2	8.6	F0	875		26 Cet; pale y., bl. (Webb)
--		β	AC		1909	290	107.2		12.7		875		[3]
1 03.8	+6 46	A 2004		Psc	1909 1957	245	1.4	7.0	9.9	A0	874		
1 03.9	+35 28	Ho 213		And	1887 1958	196 258	0.3 0.3	8.0	8.0	A3	873		
1 03.9	-40 39	h 3415		Phe	1857 1947	154 144	1.1 1.0	7.6	8.5	A2			
1 04.0	-40 32	HdO 184		Phe	1900 1938	259	3.7	7.5	11.5	F5			
1 04.5	-33 32	Stone 60		Scl	1877 1931	219	8.6	6.6	10.6	G5			
1 05.1	+14 57	h 1068		Psc	1877 1911	262	55.2	5.7	12.7	F2	889		72 Psc
1 05.6	+21 28	Σ 88	AB	Psc	1832 1959	159	30.0	5.6	5.8	A2 A0	899		74 ψ Psc
--			AC		1907 1959	122 123	94.2 92.6		11.2		899		
--			BC		1910 1914	108	73.4				899		
1 05.8	+4 55	Σ 90	AB	Psc	1833 1954	83	33.0	6.8	7.6	F2 F2	903		77 Psc
--			AC		1921	286	152.3				903		
--			Aa		1910	313	31.8		14.6		903		
--			Ab		1910	352	77.2		13.9		903		
1 05.9	+32 59	Ho 5		Psc	1885	310	0.4	8.7	8.7	F0			Probably single (Doo)
1 06.1	-46 43	Slr 1	AB	Phe	1891 1954	25 346	1.0 1.4	4.0	4.2	K0			β Phe
--		h 3417	AC		1913 1927	52 52	56.6 57.5		11.5				
1 06.2	+32 11	S 393	AB	Psc	1851 1921	293 294	55.7 56.9	6.7	9.6	K0			
--			AC		1879 1921	234 234	138.4 139.2		10.6				

α 2000	δ 2000	Star Name		Const	Years	Pos Ang		Separation		m₁	m₂	Spec	ADS	Orb	Notes
1ʰ06.ᵐ4	+62°40′	h 1066		Cas	1828 1935	295°		7".4		8.1	10.9	K0	905		
1 06.6	+64 24	M1r 90		Cas	1970	310		1.8		7.5	13.0	G0			
1 06.9	−81 39	h 3420		Hyi	1918	32		22.6		7.8	11.9	K0			
1 07.0	+47 44	A 931		And	1905 1959	275		0.3		8.8	8.8	A2	916		
1 07.1	+11 33	OΣ 22		Psc	1847 1940	197		8.6		7.2	10.2	F0	920		
1 07.2	−1 44	Σ 91		Cet	1831 1972	318		4.2		7.4	8.2	F5	923		
1 07.2	+38 39	A 1516	AB	And	1907 1960					8.3	8.3	F8	918	o	
—		OΣΣ 11	AC		1875 1932	158 161		63.0	61.5		8.8		918		
1 07.2	+53 30	H IV 66	AB	Cas	1783 1935	77 75		24.0	21.7	6.5	10.1	K0	915		
—			AC		1880 1909	122		199.1			10.9		915		
1 07.8	−41 29	Rst 3352		Phe	1936 1959	307 289		0.2	0.2	5.8	6.2	A3			υ Phe
1 07.9	+46 51	β 397	AB	And	1876 1968	142		8.7		7.6	9.8	K2	926		
—			AC		1891	64		16.6			12.9		926		
1 08.3	+54 55	β pm	AB	Cas	1855 1910	164 221		238.9	190.5	5.2	11.0	G5			30 μ Cas
—			AC		1907	145		205.8			11.0				
—			Aa		1907	145		87.7			12.2				
—			CD		1907	114		4.2			12.7				
1 08.3	+60 08	Smart		Cas	1921	316		8.6		8.0	13.0				
1 08.4	+5 39	β pm	AB	Psc	1907	286		86.1		5.5	12.0	F0			80 Psc
—			AC		1853 1907	133 129		152.2	159.3		9.3				
1 08.4	−55 15	Rst 1205	AB	Phe	1931 1949	22 40		0.6	0.8	4.1	6.9	B8			ζ Phe; A is variable
—		Rmk 2	ABxC		1835 1953	243		6.4			6.9				Very beautiful (h)
1 08.6	−10 11	β pm		Cet	1879 1907	304 305		225.2	233.5	3.5	10.2	K0			31 η Cet
1 08.6	−46 40	S1r 2	AB	Phe	1892 1954	200 191		1.1	1.2	7.3	8.3	A5			
—			ABxC		1913	81		47.5			13.7				
1 08.8	+45 12	AC 13	AB	And	1876 1957	257		0.5		8.4	8.5	A0	936		
—		h 2018	ABxC		1876 1937	359 2		15.3	15.7		11.0		936		
1 08.8	−29 37	I 262		Scl	1900 1954	169 144		0.7	0.5	8.3	9.3	F5			
1 09.4	−56 36	Hu 1342		Phe	1913 1945	348 314		0.3	0.3	7.7	8.1	F8			
1 09.5	+47 15	OΣ 515		And	1851 1960					4.6	5.5	B8	940	o	42 φ And
1 09.7	+23 48	β 303		Psc	1876 1954	284 290		0.6	0.7	7.3	7.5	F0	955		
1 09.7	+35 37	Barnard 1	AB	And	1898 1934	186 202		28.4	27.0	2.1	14.1	M	949		43 β And, Mirach
—			AC		1879 1937	268 273		84.9	94.8		12.6		949		
—			AD		1879 1938	141 144		90.8	80.4		11.8		949		
—		Kui 5	Bb		1934	126		24.6			13.4		949		
1 09.8	+38 07	Ho 214		And	1887 1944	246		2.9		7.6	11.6	F0	948		
1 10.0	+52 02	β 868	AB	Cas	1880 1938	233		9.2		7.9	9.7	F5	950		Magnitude of A is for Aa
—			AC		1913	10		49.8			11.2		950		
—		Eggen 1	Aa		1964			0.1		8.7	8.7		950		
1 10.0	−72 57	Gli 9		Tuc	1887 1928	355		3.9		7.2	10.2	A0			
1 10.1	+51 45	OΣ 23	AB	Cas	1847 1950	192		14.6		8.2	8.7	F8	956		
—			AC		1909	94		50.0			12.2		956		
1 10.6	+51 01	β 235	AB	Cas	1868 1953	287 285		43.8	43.4	6.9	10.4	F5	963		Magnitude of A is for Aa
—		OΣΣ 12	AC		1847 1953	66 68		60.6	59.4		9.0		963		
—		β 235	Aa		1875 1959	74 114		0.5	1.0	7.5	7.9		963		
—			Bb		1878 1953	77 80		8.5	9.0		11.9		963		
—		OΣ 24	Cc		1847 1953	45 48		7.8	8.0		11.3		963		
1 10.7	+42 56	β pm	AB	And	1893 1911	297 299		155.3	154.4	7.8	9.9	K0			
—			BC		1907	224		34.5			12.0				
1 10.7	+68 47	H V 16		Cas					25.0		5.3		A0			31 Cas [4]
1 10.9	+9 34	h 634		Psc	1879 1925	271 255		35.6	41.1	7.2	11.7	G0	972		
1 11.0	+49 15	h		Cas	1831	128		8.7		8.0	11.0				[5]
1 11.1	+55 09	β pm		Cas	1853 1907	132 135		154.7	145.6	4.3	10.2	A5			33 θ Cas
1 11.2	+37 43	Ho 215		And	1886	280		0.4		6.6	6.6	B8			45 And [6]
1 11.3	+41 13	A 655		And	1904 1960					8.5	8.9	G5	974	o	
1 11.7	+22 43	Cou 147		Psc	1967	283		3.4		7.1	11.8	A3			
1 11.9	+47 48	β 398		And	1877 1957	48		1.8		8.7	8.8	A2	978		
1 12.5	+1 28	OΣ 27 rej		Cet					1.0		6.9	9.2	F8			[7]
1 12.5	−9 13	Rst 4167		Cet	1938 1943	306		1.7		7.8	13.2	G0			
1 12.7	+65 01	Σ 96		Cas	1831 1956	283		1.1		7.9	8.9	F0	983		
1 12.9	+32 05	Σ 98		Psc	1832 1933	248		19.5		7.0	8.0	A0	988		
1 13.0	+30 04	OΣ 26	AB	Psc	1849 1933	258		10.8		6.2	10.0	K0	990		
—			AC		1910	342		113.9			12.0		990		
—		Ma	AD			105		54.0					990		
1 13.2	+61 42	β 258	AB	Cas	1875 1934	260 263		0.8	1.1	6.4	9.2	B9	987		
—		Lv	AC		1916	203		41.2			12.2		987		
1 13.6	+41 09	Es 157	AB	And	1916	92		15.8		8.0	13.8	G5	994		
—			AC		1916	48		19.7			14.2		994		
1 13.7	+7 35	Σ 100	AB	Psc	1832 1974	64 63		23.6	23.0	5.6	6.5	A5 F8	996		86 ζ Psc
—		β 1029	BC		1888 1958	248 80		0.9	1.0		12.2		996		Single 1914-22,36,51
1 13.7	+24 35	Σ 99	AB	Psc	1832 1936	227		7.8		4.7	10.1	K0	995		85 φ Psc
—			AC		1912	173		144.0			13.0		995		
1 13.7	−31 07	PrO 4		Scl	1903			9.6		7.7	11.1	G5			
1 13.7	−50 39	h 3421		Phe	1879 1954	62 74		45.9	55.0	8.6	8.6	F0 F0			
1 14.3	+30 33	h 636		Psc	1902 1911	288		20.4		7.5	10.0	A0	1000		

α 2000	δ 2000	Star Name		Const	Years		Pos Ang		Separation		m₁	m₂	Spec	ADS	Orb	Notes
1ʰ14.4ᵐ	−7°55′	Σ I 3		Cet	1836	1931	331°		49″.7		5.2	8.7	F0 G	1003		37 Cet
1 14.7	+60 57	β 1100		Cas	1889	1960					8.0	8.0	F5	999	o	
1 14.8	−0 58	β pm	AB	Cet	1909		276		114.1		5.8	12.1	F5			38 Cet
			AC		1909		227		217.8							C is the galaxy NGC 442
1 15.0	−55 38	h 3422		Phe	1892	1954	56		14.0		7.6	11.6	K5			
1 15.4	+38 29	Ho 6		And	1881	1925	111	118	1.3	1.3	7.9	10.9	G5	1009		
1 15.6	+59 18	A 935		Cas	1905	1959	40	21	0.3	0.3	8.2	9.4	B9	1007		
1 15.8	+9 47	A 2102		Psc	1909	1958	329	149	0.5	0.4	7.2	10.2	F2	1016		Single 1948−54
1 15.8	−68 53	h 3423	AB	Tuc	1836	1954	16	336	4.7	5.4	5.1	7.3	F8			κ Tuc
−−			AxCD		1874	1898	309		319.3							
−−		I 27	CD		1897	1954					7.8	8.2	G5		o	39 Cet
1 16.6	−2 30	β		Cet	1913		84		177.6		5.4	11.1	G0			
1 16.6	+74 02	h 2028	AB	Cas	1875	1917	204		61.4		7.1	8.6	B9 A0			
−−			BC		1908		210		46.8			12.2				
1 17.0	+38 28	Σ 104		And	1830	1929	323		13.5		8.1	10.1	G5	1028		
1 17.0	+53 46	h 2030	AB	Cas	1900	1908	49		23.7		9.4	13.2	A			
−−			AC		1900	1908	194		38.5			8.5	F2			
1 17.1	+56 31	β 3		Cas	1875	1926	28		4.5		8.1	10.5	K0	1024		
1 17.1	−66 24	h 3426		Tuc	1837	1928	337		2.5		6.3	8.7	A0			
1 17.2	+2 01	HdO 45		Cet	1904	1913	100		36.9		8.0	10.6	G5			
1 17.2	+10 36	β 503		Psc	1878	1932	137		5.5		8.0	12.0	F5	1035		
1 17.3	+1 50	β 504	AB	Cet	1878	1929	277		1.6		7.9	12.4	K0	1042		
−−		HdO 46	AC		1904	1913	176		49.6			11.3		1042		
1 17.4	−37 16	I 50	AB	Scl	1926		247		70.9		7.7	9.7	K0 F8			
−−			BC		1900	1929	142		0.9			10.7				
1 17.7	+44 38	h 1077		And	1907	1910	295		38.5		7.4	9.4	K5			
1 17.8	+49 01	Σ 102	AB	And	1833	1960	309	288	0.6	0.5	7.1	8.3	A0	1040		
−−			AC		1833	1934	224		10.0			8.5		1040		
−−			AD		1832	1934	67	62	29.9	26.8		10.6		1040		
−−			CD		1904		58		36.8					1040		
−−			DE		1904	1908	355		97.8			11.2		1040		
			Ee		1908		54		10.2			13.1		1040		
1 17.8	+49 46	Hu 520		And	1902	1958	162		0.2		8.7	8.9	A2	1039		M1r reverses quadrant
1 18.0	+53 55	Es 119	AB	Cas	1901		113		4.7		8.2	9.8	A2	1044		
−−			AC		1910		274		40.7			10.9		1044		
1 18.7	+32 05	Cou 663		Psc	1971		169		0.3		8.0	9.0	A5			
1 18.8	+37 23	Σ 108		And	1830	1958	62		6.2		6.4	9.2	A3	1055		
1 18.9	+39 58	OΣ 29 rej		And	1866	1925	265	264	19.8	21.2	7.5	11.7	G5	1053		
1 18.9	+66 10	Σ 105	AB	Cas	1832	1957	184		2.9		8.3	9.5	A0	1050		
−−			AC		1904		209		180.0					1050		
1 19.1	+80 52	OΣ 28	AB	Cep	1847	1958	324	299	0.5	0.8	7.2	8.7	F0	1030		
−−		OΣΣ 14	AC		1875	1930	26		130.3			7.8		1030		
−−		β	CD		1908		156		66.5			11.8		1030		
1 19.3	−34 30	β 1229		Scl	1891	1948	293	286	1.0	1.0	8.6	8.8	F0			
1 19.8	−0 31	Σ 113	AxBC	Cet	1836	1959					6.5		F5	1081		42 Cet
−−		φ 337	BC		1958	1960	47	91	0.1	0.1	7.5	7.7	F5	1081	o	
1 19.9	+35 20	Pop 53		And	1970		70		2.2			4.0				
1 20.0	−15 49	h 2036		Cet	1835	1959	45	353	1.2	1.8	7.5	7.7	G0	1087		
1 20.1	+43 58	Dawes 8		And	1859	1953	139		2.5		8.3	9.6	G5	1079		
1 20.1	+52 40	Hu 522		Cas	1902	1921	88		3.9		8.0	14.5	A0	1076		
1 20.1	+58 14	H III 23	AB	Cas	1783	1925	208		48.6		5.1	12.2	F5	1073		34 φ Cas
−−			AC		1903	1956	231		133.8			7.8	B5	1073		
−−			AD		1880	1909	287		179.0			10.5		1073		
−−			AE		1912		239		170.2					1073		
−−			CD		1907	1956	265		41.5					1073		
			CE		1912		264		42.6					1073		
1 20.3	−48 41	h 3428		Phe	1882	1914	157		20.6		7.8	9.6	F5			
1 20.5	−57 20	h 3430		Phe	1857	1942	246	228	2.6	2.6	7.2	9.6	F8			
1 20.7	+51 36	Hu 523		Cas	1902	1948	94		0.4		7.2	10.7	B9	1083		
1 20.8	+56 13	β 782		Cas	1881	1934	79		3.0		8.1	9.7	F2	1089		
1 21.1	+64 40	S 397		Cas	1824	1967	353	344	50.4	55.5	6.3	8.7	A0	1088		35 Cas
1 21.3	+11 32	β 4	AB	Psc	1877	1958					7.4	8.0	F0	1097	o	
−−			AC		1899	1917	248	250	22.2	23.3		13.4		1097		
1 21.9	+14 22	Σ 116		Psc	1904		102		21.7		8.0			1099		
1 21.9	+47 17	A 938		And	1905	1932	293		3.6		7.7	11.8	A2	1099		
1 22.0	−69 43	I 263		Tuc	1900	1959	60	233	0.7	0.6	7.7	8.2	F2		o	Direct motion
1 22.5	−19 05	h 2043		Cet	1846	1952	74		5.0		6.5	8.8	F5	1106		
1 22.6	−3 48	h 637		Cet	1877	1899	168		31.5		7.9	12.6	K0			
1 23.4	+58 09	Σ 115	AB	Cas	1836	1959	150	142	0.8	0.7	7.1	7.3	F5	1105		
−−			ABxC		1909	1925	280		45.5			12.8		1105		
−−			ABxD		1880	1909	182		249.1			10.5		1105		
1 23.6	−24 21	Se 1		Cet	1855	1945	82	85	2.7	2.7	6.9	8.9	A5	1113		
1 23.7	−40 57	HdO 185		Phe	1900	1930	240		4.6		7.4	11.5	K0			
1 24.0	−8 00	h 1079		Cet	1877	1924	300	298	76.4	84.6	6.2	10.9	A5			44 Cet
1 24.0	−8 11	β 505		Cet	1877	1924	60	57	58.8	65.4	3.6	14.6	K0	1118		45 θ Cet

α 2000	δ 2000	Star Name		Const	Years		Pos Ang		Separation		m₁	m₂	Spec	ADS	Orb	Notes
1ʰ24.ᵐ1	+72°51′	Σ 114		Cas	1832	1926	356°		3″.7		7.3	10.5	A0	1107		
1 24.3	−6 55	β 1163		Cet	1890	1959					6.6	6.8	F0	1123	o	
1 24.4	+58 57	Es 1712	AB	Cas	1918		297		5.7		10.2	14.8		1112		
--			AC		1917	1918	185		49.4			8.0	K0	1112		
1 25.0	−5 57	Σ 120		Cet	1831	1933	278		7.3		6.8	10.6	A0	1131		
1 25.1	−39 23	I 1137		Phe	1925	1945	237		3.7		7.8	11.6	G5			
1 25.3	−59 30	h 3435		Hyi	1916				25.5		7.1	9.1	F2			
1 25.6	+31 33	OΣ 30	AB	Psc	1855	1923	238		4.4		8.2	11.8	F8	1134		
--			AC		1852	1931	105		56.8			8.1	G0	1134		
--			Aa		1907	1923	162	165	26.1	24.6		14.0		1134		
1 25.7	+2 58	J 1899	AB	Psc	1941	1951	42		130.9		7.0	10.2	F0	1138		
--		HdO 51	BC		1908	1951	56		5.2			11.1		1138		
1 25.7	+36 21	A 1907		And	1908	1932	220		1.6		7.8	13.4	F8	1136		
1 25.8	+60 14	β pm		Cas	1909	1925	67	66	136.0	131.7	2.7	11.4	A5			37 δ Cas
1 25.9	+68 08	H V 83	AC	Cas	1783	1937	100	113	33.4	25.0	4.7	9.6	K0	1129		36 ψ Cas = S,h 18
--		β 1101	AD		1902	1937	113	118	25.0	22.8		9.7		1129		
--		Σ 117	CD		1831	1936	254		2.9					1129		
1 25.9	−47 54	Rst 33		Phe	1928	1959					8.4	9.1	G5		o	
1 26.0	+10 24	A 2213		Psc	1910	1932	124		4.6		7.2	14.2	A0	1139		
1 26.9	+3 32	Σ 122		Psc	1833	1968	328		6.0		6.6	8.6	B8	1148		
1 27.0	−0 09	Σ 125		Cet	1833	1923	33	339	17.0	49.2	8.0	10.4	G5			
1 27.0	−32 33	LDS 2199		Scl	1955		234		10.0		8.0	19				
1 27.1	−30 14	h 3436		Scl	1846	1918	126		9.8		6.9	9.7	K0			
1 27.4	+41 06	Ho 7		And	1885	1959	159	165	13.5	15.3	6.4	13.4	A3	1151		
1 27.4	−22 20	Rst 2261		Cet	1935	1949	321	331	0.6	0.6	6.5	9.0	K0			
1 27.7	+5 21	β 1164	AB	Psc	1890	1959					7.6	7.9	G0	1158	o	95 Psc
--			ABxC		1910	1925	221	222	148.3	147.4		11.4		1158		
1 27.7	+45 24	β 999	AB	And	1881	1943	92	117	2.3	2.0	4.8	11.5	F5	1152		48 ω And
--			AC		1881	1925	110	111	134.3	119.0		10.2		1152		
--		β 82	CD		1881	1943	138		4.9			10.2		1152		
1 27.7	+48 08	Es 450	AB	And	1907		147		9.5		8.0	13.5		1153		
--			AC		1907		269		20.1			13.0		1153		
1 27.8	−10 54	β 399		Cet	1876	1942	302		1.6		6.1	9.8	K0	1162		
1 28.1	−17 16	h 3437		Cet	1836	1976	247	247	12.3	12.0	7.3	8.9	F0	1171		
1 28.3	+42 47	AC 14		And	1859	1960	95		0.8		8.1	9.1	G5	1161		
1 28.4	+7 58	S 398		Psc	1825	1924	99		69.2		6.6	8.7	K0			
1 28.7	+74 12	h 2045		Cas	1830		86		20.0		7.3	13.3	G5			
1 29.1	+21 43	Ho 9	AB	Psc	1893	1924	80	73	94.3	79.8	7.7	10.3	G5	1176		
--			AD		1907	1918	185	189	82.1	80.4		9.9		1176		
--			BC		1883	1943	93		2.7			11.7		1176		
1 29.6	−21 38	See 14		Cet	1897	1909	250		22.0		5.1	12.6	A0	1184		48 Cet
1 29.7	+22 50	A 1910	AB	Psc	1908	1960					7.4	7.7	A0	1183	o	
--			ABxC		1917		350		18.2			12.7		1183		
1 30.0	−22 31	I 445	AB	Cet			nf		67.0		7.8	10.1	G0	1186		
--			BC		1913	1942	356		2.2			10.3		1186		
1 30.2	+6 09	β pm	AB	Psc	1883	1907	303	302	176.7	182.5	4.8	10.4	K2			98 μ Psc
--			BC		1907		307		27.4			11.4				
1 30.4	−26 12	β 1230		Scl	1891	1959	226		2.9		5.9	10.9	K0	1193		
1 30.6	+43 39	Es 4		And	1892	1945	102		3.1		8.2	10.2	F2	1192		
1 31.5	+15 21	β 506		Psc	1878	1958	12	36	1.0	1.0	3.6	10.6	G5	1199		99 η Psc
1 31.6	−19 01	h 2052		Cet	1904	1917	117		78.4		7.1	7.4	A3 K0			BDS 796, h 3373 are same
1 31.6	−53 22	I 264	AB	Eri	1900	1959	121	63	0.5	0.8	8.4	8.8	F5			
--		h 3444	ABxC		1900	1930	7		38.8			10.8				
1 31.7	+61 03	Σ 128 rej		Cas	1901	1918	305		13.0		7.8	9.5	B8	1196		Stein 227
1 32.1	+16 57	Σ 132	AB	Psc	1783	1921	28	348	15.8	43.4	6.9	9.9	G5	1202		
--			AC		1888	1921	229		68.7			10.9		1202		
--			AD		1888	1921	114	113	140.2	133.0		10.0		1202		
--			Aa		1908		154		14.2			14.0		1202		
--			Dd		1908	1924	288		5.5			10.6		1202		
1 32.3	−18 34	Rst 2262	AB	Cet	1935	1941	163		2.0		8.0	13.8	G5			
--		Bhas 1	AC		1917	1929	97		9.7			10.6				
1 32.3	−26 33	Arg 4		Scl	1877	1952	72		17.9		8.2	8.9	A3	1212		
1 32.8	+35 51	Σ 133	AB	And	1833	1945	179	186	3.0	3.1	6.8	10.3	K0	1211		
--			AC		1833	1925	199	195	29.1	24.0		10.6		1211		
--			AD		1863	1935	198	195	27.3	22.9		10.6		1211		
--			CD		1833	1946	346	350	4.8	4.9				1211		
1 33.2	+60 41	Σ 131	AB	Cas	1830	1956	142		13.8		7.3	10.5	B3	1209		
--			AC		1910	1956	145		28.2			10.8		1209		
--		F1e	AD		1956		117		56.2					1209		
--			AE		1956		138		82.8					1209		
--		Σ 131	BC		1901		150		14.4					1209		
1 33.2	−24 11	LDS 2209		Cet	1951		228		336.0		8.0	13.0				
1 33.3	+8 13	OΣ 31		Psc	1850	1959	85	78	4.0	4.1	6.6	10.7	K0	1214		
1 33.4	+60 36	β pm	AB	Cas	1878	1909	186		60.0		9.2	8.6				
--			AC		1878	1909	125		44.3			8.2				

α 2000	δ 2000	Star Name		Const	Years	Pos Ang	Separation	m₁	m₂	Spec	ADS	Orb	Notes
1ʰ 33.7ᵐ	−78°30′	h 3453		Hyi	1835 1918	285° 304°	50″.2 50″.1	6.1 13.5		G5			
1 34.2	−57 00	Hu 1345		Eri	1913 1922	201	5.6	6.9 14.0		B9			
1 34.3	−45 41	I 51		Phe	1900 1942	11	1.5	7.0 9.5		G5			
1 34.8	+12 34	Σ 136	AB	Psc	1831 1976	79 77	16.0 15.5	7.3 8.4		A3	1238		
			AC		1911	311	75.0	13.0			1238		
--			AD		1911	8	124.1	11.0			1238		
1 34.9	−23 42	Rst 2264		Cet	1933 1944	132	11.3	6.5 13.3		K0			
1 35.0	−29 55	δ 31	AB	Scl	1920 1959			7.8 7.9		G5		o	
--		β 1000	ABxC		1881 1959			11.5				o	
--			ABxD		1891 1909	20 20	142.1 140.4	8.8		G0			
1 35.4	+63 05	Mlb 280	AB	Cas	1923	79	58.4	6.8 11.3		K0	1233		
--			BC		1923	45	6.4	11.8			1233		
1 35.7	+72 26	A 816		Cas	1904 1957	313 310	0.4 0.6	8.5 8.6		G5	1226		
1 35.7	−53 12	h 3449		Eri	1894 1933	168 179	24.6 26.1	7.8 12.8		K0			
1 35.9	−17 32	h 2061		Cet	1890 1916	322 319	63.0 70.1	7.2 9.4		G5			
1 36.0	+7 39	Σ 138	AB	Psc	1830 1959			7.7 7.7		F8	1254		
--			AC		1875 1910	63 67	22.2 24.3	15			1254		
--			AD		1914 1931	71 72	74.9 75.6				1254		
1 36.1	−29 54	h 3447		Scl	1837 1959	76 122	3.0 1.1	6.0 7.1		F0		o	τ Scl
1 36.3	+4 19	β 869		Psc	1880 1924	197	5.1	7.9 11.6		K0	1257		
1 36.8	+41 24	β pm	AB	And	1909	128	114.0	4.1 12.5		G0			50 υ And
--			AC		1903	282	280.2						
1 37.1	+48 43	A 817		And	1904 1956	46 38	0.4 0.5	8.5 9.0		A0	1263		
1 37.4	+58 38	OΣ 33		Cas	1846 1972	75 77	24.3 26.2	7.4 8.5		B8	1262		
1 37.4	−82 17	Gli 14		Oct	1887 1939	55	5.5	7.6 8.4		G5			
1 37.6	−9 24	Kui 7		Cet	1934 1958	157 344	0.2 0.1	6.8 7.2		F5		o	[8]
1 37.6	+22 34	OΣΣ 20	AB	Psc	1875 1923	314	94.3	8.1 9.1		F5 F5			
			Aa		1903 1910	257	27.2	13.6					
1 38.3	+72 35	h 2053		Cas	1906 1911	28	31.7	8.1 10.8		G5			
1 38.5	+73 02	h 2054		Cas	1879 1925	237	53.3	5.3 11.3		K0	1268		40 Cas
1 38.7	−44 36	Pol 1		Phe	1887 1947	39	1.7	8.5 8.9		F8			
1 38.8	+38 39	β 1166	AB	And	1890 1928	347	2.7	7.8 10.9		G5	1284		
--			AC		1898	9	24.8	12.8			1284		
1 38.8	−25 01	I 448		Scl	1913 1930	350	20.3	6.3 12.5		A0	1298		
1 38.8	−53 26	Δ 4		Eri	1836 1934	104	10.4	7.2 8.2		F0			
1 39.3	+16 38	β 5	AB	Psc	1875 1957	289 285	1.3 1.0	7.1 9.1		G5	1300		
--			ABxC		1910	186	90.5	11.7			1300		B not seen 1956 (M1r 19-in)
1 39.3	+54 36	A 1266		Per	1906 1933	237 246	0.2 0.1	7.8 8.8		A0	1286		Also h 3380
1 39.4	−17 48	h 2067		Cet	1878 1919	92	33.9	7.6 11.6		A5	1303		
1 39.7	−37 28	h 3452		Scl	1895 1930	276	20.2	7.2 9.0		K0 K0			
1 39.8	−56 12	Δ 5		Eri	1835 1957			5.8 5.8		G5 G5		o	p Eri; superb double (h)
1 40.0	+38 59	Σ 141		And	1833 1954	301	1.7	8.2 8.7		F5	1305		
1 40.4	+34 21	Σ 143	AB	Tri	1831 1916	320 319	30.3 37.8	8.1 9.4		G0			
--		Fox	AC		1911 1921	137	70.8	9.9					
1 40.6	+40 35	β		And	1880 1925	328	52.3	4.9 10.1		B8	1311		53 τ And
1 40.6	+54 57	A 1267		Cas	1906 1958	341 357	0.3 0.3	8.0 8.6		A0	1309		
1 40.9	+77 58	h 2056	AB	Cep	1911	213	33.6	6.6 12.9		B9	1288		
--		VBs	Aa		1911	11	31.7	13.7			1288		
1 41.3	+25 45	Σ 145	AB	Psc	1832 1974	32 31	11.3 10.5	6.2 10.8		F5	1326		
--			AC		1905 1959	339 338	80.0 82.4	11.0			1326		
1 41.4	−79 09	h 3467		Hyi	1918	349	16.0	6.3 12.3		K0			τ¹ Hyi
1 41.6	+62 41	Kr 12		Cas	1890 1953	300	0.6	8.6 8.6		K0	1318		
1 41.7	−11 19	Σ 147		Cet	1831 1959	88 90	4.0 2.1	6.1 7.4		F5	1339		
1 42.2	−17 54	β pm		Cet	1911	292	85.4	7.4 12.0		G0			
1 42.3	+58 38	h 1088		Cas	1901 1933	168	19.6	6.3 8.8		B9	1334		
1 42.4	−6 46	A 1		Cet	1899 1959	165 226	0.3 0.6	8.5 9.0		F2	1345		
1 42.5	+20 16	h 2071	AB	Psc	1879 1910	222 248	38.5 19.0	5.2 11.6		G5			107 Psc
--			AC		1852 1924	328 353	65.2 104.4	12.0					An optical pair
1 42.6	+60 33	β	AB	Cas	1912	101	12.7	7.4 14.2		K2	1337		
--			AC		1912	12	26.7	13.9			1337		
1 43.2	−21 37	h 3456		Cet	1897 1930	349	14.9	8.0 9.5		G5	1358		
1 43.3	+60 33	β 1103	AB	Cas	1889 1934	4 358	1.7 1.6	5.8 12.1		B9	1344		44 Cas
--			AC		1912	310	66.0	9.6			1344		
--			AD		1912	229	140.2	9.2			1344		
1 43.6	+52 53	β 1104		Per	1889 1923	196	2.9	6.8 11.4		A3	1354		
1 43.8	+55 53	OΣ 35		Cas	1847 1958	115 99	9.8 12.8	7.2 10.4		A2	1353		
1 44.1	−15 56	β pm		Cet	1911 1939	243 215	110.8 89.8	3.5 13.0		K0			52 τ Cet; optical
1 44.3	+57 32	β 870		Cas	1880 1957	68 19	1.0 1.0	6.4 7.8		A2	1359		
1 44.7	−6 46	β 6	AB	Cet	1877 1943	167	2.6	6.6 9.4		G0	1376		
			AC		1913 1921	84	92.8	11.0			1376		
1 44.8	+85 13	OΣ 32		Cep	1847 1936	134 143	9.5 7.3	8.0 12.5		A5	1366		
1 45.1	+43 42	Σ 154		And	1833 1969	126	5.2	8.4 8.6		F0	1374		
1 45.4	−77 21	Jsp 26		Hyi	1929	265	1.2	7.3 10.5		F5			
1 45.6	−25 03	h 3461	AB	Scl	1836 1959	70 34	4.7 4.7	5.4 8.6		F0	1394		ε Scl
--			AC		1926	41	15.3	14.9			1394		

α 2000	δ 2000	Star Name		Const	Years		Pos Ang		Separation		m₁	m₂	Spec	ADS	Orb	Notes

α 2000	δ 2000	Star Name		Const	Years	Pos Ang	Separation	m1	m2	Spec	ADS	Orb	Notes
1ʰ45.6ᵐ	−25°03′	h 3461	AD	Scl	1913	197°	141″.8		11.4		1394		
1 47.3	+58 01	Stein 1684		Cas	1910 1911	310	13.6	7.0	13.2	K0			
1 47.6	−43 58	I 52		Phe	1901 1956	180 140	0.6 0.2	8.5	8.8	F5			
1 47.7	+63 51	β pm		Cas	1892 1911	135 138	102.7 91.1	5.7	8.8	K0			
1 47.8	−80 11	h 3474		Hyi	1918	28	39.7	6.1	13.5	F0			τ² Hyi
1 48.0	+53 53	β 1312		Per	1902 1922	288	5.0	8.0	12.2	A0	1416		
1 48.7	+7 41	h 644	AB	Psc	1879 1921	278	17.5	7.3	11.7	K0	1435		
—			AC		1913 1921	226	126.8		9.3		1435		
1 48.7	−38 39	I 1610		For	1927	115	0.3	7.8	9.8	F2			
1 49.3	+47 54	Σ 162	AB	Per	1836 1960	224 207	1.9 1.9	6.5	7.0	A2	1438		
—			ABxC		1929	176	19.2		8.4		1438		
—			AC		1836 1956	179	20.4				1438		
—			AD		1881 1956	95 95	134.7 137.2		10.2		1438		
—			BC		1908	176	18.8				1438		
1 49.6	−10 41	β pm		Cet	1891 1916	250	183.8	4.9	6.9	F0			53 χ Cet
1 49.8	−38 24	HdO 186		For	1901	352	28.0	6.4	13.4	F5			
1 49.9	+80 53	OΣ 34		Cep	1843 1954			7.8	8.1	A0	1411	o	
1 50.1	+22 17	Σ 174		Ari	1830 1967	170 166	2.6 2.8	6.2	7.4	F5	1457		1 Ari; sp. compos. F5+A2
1 51.0	+25 51	Cou 452		Tri	1970	183	0.3	8.0	8.5	A0			
1 51.2	+24 39	Ho 311		Ari	1890 1959	174 197	0.4 0.4	7.6	7.8	A5	1473		
1 51.3	+60 21	A 951		Cas	1905 1956	190 209	0.4 0.4	8.6	8.9	B9	1461		
1 51.3	+64 51	Σ 163	AB	Cas	1831 1936	35	34.8	6.8	8.8	K5	1459		
—			AC		1879 1908	254	114.8		10.1		1459		
1 51.4	+43 29	h 2089		And	1908	305	29.3	8.6	9.0	A5			
1 51.5	−10 20	β pm		Cet	1879 1907	41 41	185.7 187.0	3.7	9.9	K0 K0			55 ζ Cet
1 51.6	+47 05	A 952		And	1905 1927	70	2.2	7.5	13.0	G5	1469		
1 51.9	−23 09	I 450		Cet	1912 1959	202 213	0.5 0.5	8.6	8.9	A5	1490		
1 52.0	+10 49	Σ 178		Ari	1828 1968	193 201	3.1 3.1	8.5	8.5	F0	1487		Both lucid white (Smyth)
1 52.3	+4 40	OΣ 36 rej		Psc			15.0	7.0	10.0				Hu found nothing here
1 52.7	+57 17	Arg 6		Per	1904 1928	137	14.4	8.1	10.4	B9	1480		
1 53.1	+29 35	β pm	AB	Tri	1879 1909	304 309	80.6 85.0	3.4	12.7	F5			2 α Tri
—			AC		1879 1909	182 182	228.3 222.3		12.0				
1 53.2	+37 19	Σ 179		And	1831 1955	160	3.5	7.4	8.4	F5	1500		
1 53.3	+40 44	h 1094		And	1879 1913	356	59.8	5.6	10.9	K0			55 And
1 53.5	+19 18	Σ 180	AB	Ari	1830 1969	0 0	8.6 7.8	4.8	4.8	A0	1507		5 γ Ari [9]
—			AC		1823 1924	85 84	228.8 221.3		9.6		1507		
1 54.1	−77 29	Jsp 31		Hyi	1929	354	0.2	7.9	8.0	F0			
1 54.7	+59 55	A 953		Cas	1905 1956	266 253	0.4 0.6	8.8	8.8	F5	1509		
1 55.1	+28 48	Σ 183	AB	Tri	1833 1958	22 278	0.6 0.2	7.7	8.4	F2	1522	o	Too close in 1955
—			ABxC		1832 1969	164	5.6		8.7		1522		
1 55.1	+59 58	A 955		Cas	1905 1948	119	0.9	7.8	11.4	A0	1513		
1 55.3	−60 19	h 3475		Hyi	1837 1957	36 64	2.4 2.4	7.1	7.3	F0			
1 55.4	+76 13	Σ 170		Cas	1830 1955	246	3.3	7.4	8.4	A5	1504		
1 55.9	+1 51	Σ 186		Cet	1831 1959			6.8	6.8	G0	1538	o	A white pair
1 56.0	−51 37	h 3473		Eri	1899 1947	196 202	6.2 5.0	3.7	10.7	G5			χ Eri
1 56.2	+37 15	Σ I 4	AB	And	1836 1928	302 300	177.5 190.4	5.7	6.0	K0 K2	1534		56 And
—		β	Aa		1903 1934	79	18.4		11.1		1534		
—			Bb		1880 1918	258	203.8		9.3		1534		
1 56.4	+61 17	Σ 182	AB	Cas	1836 1953	123	3.6	8.2	8.2	A0	1531		
—			AC		1913 1953	65 71	28.4 29.9		13.4		1531		
1 57.0	+31 01	A 819	AB	Tri	1904 1958	131 170	0.5 0.6	8.2	8.7	F5	1548		
—			ABxC		1906 1917	270	66.5		9.0		1548		
1 57.2	−10 15	LDS 60		Cet		sf	30.0	6.4	10.4	G5			
1 57.5	−52 12	CorO 10	AxBC	Eri	1887 1942	40	3.3	7.9		F5			
—		Rst 43	BC		1928 1948	55 45	0.6 0.6	10.5	10.6				
1 57.6	+44 33	A 1526		And	1907 1958	291 283	0.3 0.2	8.7	8.8	A0	1554		
1 57.9	+23 36	H V 12	AB	Ari	1781 1933	46	37.4	4.9	7.7	A5	1563		
—			AC		1923	75	179.2		9.6		1563		
—			AD		1923	84	261.2				1563		
1 58.0	−2 04	β 7	AB	Cet	1875 1933	12 16	2.8 2.8	6.6	11.6	A0	1567		
—			AC		1913	197	150.1		9.6	A0	1567		
1 58.1	+41 23	S 404	AB	And	1783 1925	67 78	15.8 25.0	7.6	9.6	G5	1560		
—			BC		1912	100	81.6		12.5		1560		
1 58.4	−33 04	B 24		For	1927	122	6.1	6.4	13.3	G5			
1 58.5	−41 10	I 1139		Phe	1925	75	7.3	7.3	14.0	F0			
1 58.8	−56 53	LDS 63		Eri		sf	21.0	8.5	9.0	G5			
1 59.0	−22 55	H II 58		Cet	1822 1959	304	8.4	7.4	7.7	F2	1581		
1 59.1	+33 13	β pm	AB	Tri	1893 1911	139 139	136.6 129.5	7.2	9.4	G5			
—			Aa		1908	31	86.6		12.2				
1 59.3	+24 51	Σ 194		Ari	1831 1956	264 274	1.2 1.3	8.4	8.7	A3	1579		
1 59.5	+21 01	Σ 196	AB	Ari	1832 1958	56 48	2.4 2.2	9.5	12.0	G	1582		
—			AC		1832 1918	167 163	39.5 30.9		10.6		1582		
—			AD		1862 1914	1 4	183.7 188.1		6.1	K0	1582		
1 59.6	+64 37	h 1100		Cas	1901 1959	310 310	38.8 41.5	5.3	9.6	A0	1571		
2 00.4	−8 31	h 3476		Cet	1900 1919	193	61.7	5.5	9.6	M			

α 2000	δ 2000	Star Name		Const	Years	Pos Ang	Separation	m₁	m₂	Spec	ADS	Orb	Notes
2ʰ00ᵐ.5	−62°46′	h 3479		Hyi	1916	275°	35″.3	7.8	9.8	F2 F5			
2 00.7	+58 32	Σ 192	AB	Cas	1832 1920	182	4.5	8.1	10.7	F0	1587		
—			AC		1881 1921	146	76.2		9.2		1587		
2 00.9	−43 50	I 265		Phe	1901 1956	25 49	0.8 0.5	8.4	9.4	F8			
2 01.2	+16 34	β 515		Ari	1878 1934	243 267	1.4 1.4	7.7	12.5	F2	1605		Increasing position angle?
2 01.3	+60 31	Σ 193		Cas	1832 1905	194	2.9	8.0	10.4	F0	1599		
2 02.0	+2 46	Σ 202		Psc	1831 1972			4.2	5.1	A2	1615	o	113 α Psc
2 02.0	+70 54	β 513	AB	Cas	1878 1959			4.7	6.4	A3	1598	o	48 Cas
—			AC		1891 1923	51	23.7		13.1		1598		
—			AD		1898 1959	83 85	47.1 51.1		12.5		1598		
2 02.2	+36 43	A 1813	AB	And	1908 1960			8.7	8.9	G5	1613	o	
—			ABxC		1908 1958	339 261	0.7 0.8		11.5		1613		
2 02.2	+75 30	Σ 185	AB	Cas	1831 1955	40 16	1.4 1.3	6.7	8.2	A0	1588		
—			AC		1913 1921	174	116.5		8.8		1588		
2 02.7	−30 20	h 3478		For	1846 1954	139 146	40.2 42.0	8.0	8.8	G5 G5			
2 03.0	+33 17	Σ 201		Tri	1833 1935	118	3.9	5.4	11.4	A2	1621		3 ε Tri
2 03.2	+48 37	Hu 806	AB	Per	1902 1920	154	1.7	7.9	13.0	G0	1620		
—			AC		1904 1920	128 140	11.9 8.5		14.4		1620		
2 03.2	+73 51	Σ 191		Cas	1832 1928	192	5.5	6.3	8.6	A3	1606		
2 03.7	+25 56	Σ 208	AB	Ari	1833 1960			5.9	7.3	F5	1631	o	10 Ari
—			ABxC		1912	144	98.9		13.4		1631		
2 03.8	−0 20	H V 102	AB	Cet	1877 1975	193 194	42.7 43.0	5.9	10.4	G5	1634		61 Cet
—			AC		1877 1909	326 326	80.4 83.0		11.8		1634		
2 03.9	+42 20	Σ 205	AxBC	And	1830 1967	62 63	10.3 9.8	2.3		K0	1630		γ And [10]
—		OΣ 38	BC		1842 1959			5.5	6.3	A0	1630	o	
—		Barnard	ABxD		1898	245	27.9		15		1630		
2 03.9	−45 25	Rst 2272		Phe	1932 1946	178	1.0	7.3	11.5	G0			Last p.a. 335°?
2 04.6	−65 08	h 3482		Hyi	1916	216	50.0	7.5	12.8	F0			
2 05.0	+64 23	β 873	AB	Cas	1880 1938	24	2.2	7.2	10.8	B3	1635		
—			AC		1913	144	87.3		10.6		1635		
2 05.1	+77 17	S,h 22		Cas	1821 1922	192 199	93.6 95.6	5.4	11.4	F0			47 Cas
2 05.1	−49 41	B 1431		Eri	1928	24	7.3	7.8	11.7	A5			
2 05.5	+76 07	β 785	AB	Cas	1881 1911	246	5.4	5.3	12.3	G5	1625		49 Cas
—			AC		1911	128	28.0		13.5		1625		
2 05.6	−24 22	I 454	AB	For	1913 1959	135 142	1.4 1.5	8.9	11.0	G5	1652		
—			ABxC		1933	258	56.3		8.7	G5 G5			
2 05.8	+13 21	A.G. 30		Ari	1892			7.8		F5			
2 06.0	+63 10	Stein 344		Cas	1904	36	11.8	7.5	11.1	A0			
2 06.6	−37 07	B 665		For	1927 1941	350	1.4	7.5	11.5	K0			
2 06.7	+10 29	h 21		Cet	1891 1924	348	25.8	7.7	12.5	F8			
2 06.8	+25 42	Ho 312		Ari	1890 1943	330 342	1.1 1.6	6.0	11.5	B8	1658		11 Ari
2 07.4	−59 41	h 3484		Hyi	1919	61	52.6	7.6	9.6	F5			
2 08.2	−55 32	I 455		Eri	1911 1945	202	5.7	7.7	11.0	K0			
2 08.7	−0 26	Σ 218		Cet	1832 1951	248	4.9	8.3	9.3	F0	1673		
2 09.0	+35 41	Cou 1067		Tri	1973	34	0.2	8.7	8.7	F5			
2 09.4	+25 56	H VI 69	AB	Ari	1879 1915	36	93.2	5.1	8.7	F0 F2			14 Ari
—			AC		1875 1923	278	105.9		7.7				
2 09.6	+42 52	Es 48		And	1901 1957	183 241	10.9 10.2	7.2	11.0	F5	1675		
2 09.7	+20 21	Σ 221	AB	Ari	1836 1935	146	8.4	8.3	9.5	A3	1678		
—			AC		1856 1904	226 229	61.0 63.4		12.3		1678		
2 09.9	+32 33	Cou 861		Tri	1972	74	2.2	7.5	12.3	G5			
2 10.1	−64 21	h 3486		Hyi	1916	243	58.9	7.5	12.5	G0			
2 10.3	+33 22	Σ 219		Tri	1831 1935	183	11.6	8.2	9.0	A0	1681		
2 10.4	+56 18	Stein 1797		Per	1915	134	14.8	7.6	10.9	K0			
2 10.5	+81 29	OΣ 37		Cep	1843 1937	224 209	1.4 1.2	6.9	9.1	A3	1659		
2 10.9	+13 41	Σ 224		Ari	1830 1942	243	5.5	8.4	8.9	G5	1689		H N 105, BDS 1146
2 10.9	+39 02	Σ 222		And	1831 1949	35	16.6	6.1	6.8	A0 A2	1683		59 And
2 11.4	+62 22	Σ 216		Cas	1831 1953	270 246	0.6 0.3	8.1	9.0	F0	1682		
2 11.5	+57 39	β 874		Per	1880 1935	272	5.6	6.4	12.4	B3	1685		5 Per
2 11.9	−70 57	h 3489	AB	Hyi	1916 1931	244 244	22.4 21.9	7.1	11.5	G5			
—		δ	BC		1916 1931	270 270	8.0 8.4		14.0				
2 12.1	−32 17	Jc 7	AB	For	1847 1933	280	6.7	8.0	10.2	G0			
—			AC		1847 1931	182 180	150.5 149.9		8.7	F0			
2 12.3	+23 58	Σ 226		Ari	1832 1953	250 239	2.4 1.8	7.8	9.7	G5	1696		
2 12.4	+30 18	Σ 227		Tri	1836 1973	80 71	3.8 3.9	5.3	6.9	G0	1697		6 Tri; exquisite (Smyth)
2 12.7	−66 09	I 266		Hyi	1900 1933	160	3.0	7.3	11.5	K0			
2 12.8	−2 24	Σ 231	AB	Cet	1832 1975	232 234	16.2 16.5	5.7	7.5	G0	1703		66 Cet
—			AC		1908	61	172.7		11.4		1703		
2 12.8	+79 42	S 405	AB	Cep	1823 1923	275	55.8	6.5	7.1	A3 A2			
—			BC		1912	225	51.0		13.9				
2 12.9	+57 12	OΣΣ 24	AB	Per	1875 1924	332 332	55.7 68.2	7.0	8.9	G0 G0			
—			AC		1907 1925	217	102.3		10.9				
—			BC		1907	192	142.5						
2 13.4	+51 06	β pm		Per	1852 1921	76 69	46.6 27.3	5.4	9.8	K0			
2 13.7	−3 02	H VI 110		Cet	1895 1921	136	81.8	7.3	10.1	A2			Spectrum composite A2+G0

α 2000	δ 2000	Star Name		Const	Years	Pos Ang		Separation		m₁	m₂	Spec	ADS	Orb	Notes
2ʰ14ᵐ.0	+47°29′	Σ 228		And	1831 1958	°	°	″	″	6.6	7.1	F0	1709	o	
2 14.7	+30 24	Σ 232		Tri	1832 1955	246		6.6		8.0	8.0	B8	1723		
2 15.0	+58 29	Σ 230		Per	1831 1925	258		24.1		7.9	9.1	B9	1717		
2 15.7	+25 03	Cou 79		Ari	1965	50		0.2		6.2	6.5	F5			21 Ari
2 15.7	+67 40	β pm		Cas	1875 1918	109 101		62.4	36.2	7.4	9.3	K0			
2 15.8	−18 14	Hst 1		Cet	1879 1959					8.3	9.3	K0	1733	o	
2 16.3	−9 49	Σ 242 rej		Cet	1877 1924	228 238		50.5	59.2	7.3	9.8	G0			
2 16.5	−51 31	Δ 6		Eri	1916 1936	219 220		86.3	86.6	3.6	8.8	B8			φ Eri
2 16.9	+57 03	OΣΣ 25		Per	1875 1924	204		103.0		6.5	7.3	B1 B1			
2 17.1	+34 13	Dorpat 66		Tri	1902 1907	342 337		62.6	65.4	4.9	13.4	G0	1739		8 δ Tri
2 17.4	+23 53	Σ 240		Ari	1832 1944	51		4.8		8.4	8.9	F0	1749		
2 17.4	+28 45	Σ 239		Tri	1832 1968	210 211		14.1	13.8	7.0	8.0	F5	1752		Both silvery white (Smyth)
2 17.5	−30 43	Stone 5		For	1877 1942	201		2.7		8.1	8.8	F5			
2 18.0	+57 31	S 409	AD	Per	1824 1918	136 136		124.5	121.9	6.3	8.9	K0	1753		
--		β 1170	AxBC		1879 1911	354		70.2					1753		
--			BC		1890 1958	313		0.3		11.5	11.7		1753		
2 18.1	+14 28	A.G. 35		Ari						8.0		F0			Single in 1902
2 18.6	+34 30	Σ 246		Tri	1932 1955	122 123		10.5	9.2	8.2	9.4	F8	1764		H III 46, BDS 1165
2 18.6	+40 17	Σ 245	AB	And	1832 1943	292		11.0		7.2	8.2	F2	1763		Magnitude of A is for Aa
--		Eggen 2	Aa		1964			0.1		8.0	8.0		1763		
2 18.7	+4 12	β 437		Cet	1877 1947	32 30		7.2	7.7	8.0	12.0	G5	1772		
2 19.0	+28 39	h 1115		Tri	1900 1924	205		57.1		5.0	11.0	A2	1770		10 Tri
2 19.1	+57 08	h 1114		Per	1902	338		27.8		6.6	11.7	B0			
2 19.2	+50 09	β pm		And	1879 1908	354		136.6		7.3	10.8	K5			
2 19.2	+57 10	β		Per	1902	336		27.4		6.7	12.0	B0	1766		
2 19.3	−2 59	β	AB	Cet	1878 1911	90 85		74.7	73.1	2.0	12.0	M	1778		68 o Cet; Mira (var.)
--		H VI 1	AC		1782 1925	92 78		114.6	118.7		9.3		1778		
--		Joy 1	Aa		1923 1958						9.5		1778	o	
--		H VI 1	BC		1911	69		45.5					1778		Both B and C are optical
2 19.4	+60 29	A 821		Cas	1904 1929	60		0.5		8.0	12.3	F0	1765		
2 19.7	+60 02	OΣΣ 26	AB	Cas	1875 1925	200		63.3		6.9	7.4	A2 G5			
--			AC		1911	133		72.2			11.0				
--			BC		1909 1911	82		75.0							
2 19.8	−23 52	h 2130		For	1912	108		42.9		8.5	8.5	K2 F0			
2 19.9	−55 57	h 3497		Hor	1899 1913	82		33.8		5.8	10.7	K5			
2 21.1	+80 44	Σ 223		Cep	1831 1943	46		0.7		7.6	10.0	A0	1754		
2 21.3	+8 52	β 8		Cet	1875 1960	200 218		1.0	1.3	8.2	9.1	F0	1801		
2 21.5	+46 26	Es 619		And	1894 1908	241		5.3		7.9	10.6	K5	1793		
2 21.6	+44 36	Σ 249		And	1831 1933	194		2.3		7.2	9.2	A2	1795		
2 21.6	−48 19	I 147		Eri	1900 1943	338		1.3		7.5	10.3	G5			
2 21.7	+38 30	OΣ 40		And	1850 1960	56 48		0.6	0.6	8.4	9.2	A3	1799		
2 22.0	−29 21	CorO 12		For	1919 1929	79		9.9		7.9	10.3	G0	1816		
2 22.4	+55 51	β 875		Per	1880 1918	162		11.6		5.2	12.0	A2	1802		9 Per
2 22.8	+41 24	β pm	AB	And	1880 1917	358 2		53.3	56.3	5.8	10.8	F0			
--			AC		1909	7		295.5			7.6	B0			
2 23.1	+70 21	Mlr 377		Cas	1972	162		0.5		8.2	8.2	F5			
2 23.2	−29 52	β 738		For	1879 1960	177 297		0.6	0.1	7.6	7.9	F8		o	Retrograde motion
2 23.9	+55 22	β		Per	1906	122		8.3		6.3	14.4	G0	1820		
2 24.3	+79 46	OΣ 39 rej		Cep						7.8		F0			Probably single (IDS)
2 24.8	−29 52	β 739		For	1879 1959	265 262		1.6	1.6	8.6	8.9	F5			
2 24.9	−3 54	β 517	AB	Cet	1877 1916	248		11.2		7.1	12.1	K0	1840		
--			AC		1878 1914	287 290		55.0	56.1		11.1		1840		
2 25.0	+25 29	Cou 357		Ari	1968	152		0.3		8.2	9.2	G0			
2 25.2	+55 15	Es 7		Per	1899 1928	256		10.9		7.6	14.2	B9	1829		
2 25.4	+58 12	Hld 7		Per	1881 1943	185		1.9		7.9	10.2	A2	1832		
2 25.6	+50 17	β pm	AB	And	1908	303		47.9		4.7	12.7	K5			65 And
--			AC		1879 1908	307		191.7			10.4				
2 25.7	+61 33	Σ 257		Cas	1830 1960					7.5	8.0	B8	1833	o	
2 26.0	−15 20	H III 80	AB	Cet	1783 1923	293		12.2		5.9	8.9	A2	1849		
--			AC		1913 1922	30		105.8			10.8		1849		
2 26.8	+10 34	OΣΣ 27	AB	Ari	1875 1974	31		73.8		6.8	8.3	A3 F0			
--			AC		1898 1924	155		62.3			11.3				
2 28.0	+1 58	Kui 8		Cet	1939 1960	20 29		0.3	0.5	7.1	7.5	K0			
2 28.0	−58 08	δ 1	AB	Hor	1916 1948	226 219		0.9	0.9	8.2	8.6	F8			
--		h 3503	ABxC		1916 1932	300		17.6			10.2				
2 28.2	+29 52	Σ 269		Tri	1832 1945	340 343		1.9	1.7	7.9	10.2	G0	1868		
2 28.2	+38 50	A 1815		And	1908 1931	135		1.9		6.9	10.8	F0	1864		
2 29.1	+37 20	A 1816		And	1908 1929	248		1.6		6.6	11.3	A5	1881		
2 29.1	+67 24	Σ 262	AB	Cas	1829 1971					4.6	6.9	A5	1860	o	z Cas; variable
--			AC		1829 1968	107 114		7.6	7.2		8.4		1860		
--			BC		1900 1937	102		9.4					1860		
--			CD		1880 1956	57 58		205.7	207.2				1860		
2 29.3	−21 02	B 2072		Cet		310		10.0		7.2	15	F5			
2 29.4	+55 32	Σ 268		Per	1831 1970	129		2.7		6.8	8.1	A2	1878		
2 29.6	+9 34	β 518		Cet	1878 1953	138 142		1.6	1.6	6.1	10.6	K0	1896		

α 2000	δ 2000	Star Name		Const	Years	Pos Ang		Separation		m₁	m₂	Spec	ADS	Orb	Notes
2ʰ29.ᵐ6	+60°40′	Σ 263	AB	Cas	1832 1902	103°		14.″6	″	8.3	11.5	B	1877		
--		Σ 264	Ab		1961	252		53.2					1877		
--			A′B′		1832 1904	226		16.6		8.5	9.5	B	1877		
--			AA′		1832 1904	263		38.8				B B	1877		
2 29.9	−22 41	h 3502		Cet	1906 1919	85		28.4		6.8	12.5	A2	1906		
2 30.3	−30 21	h 3504		For	1836 1934	269		6.7		8.2	9.0	F8			
2 30.4	+45 26	A 967		And	1905 1927	220		3.8		7.3	12.8	G5	1899		
2 30.5	+25 14	Σ 271	AB	Ari	1831 1935	182		12.4		5.9	10.4	F5	1904		In a poor field (Smyth)
--			AC		1912 1925	34		112.5					1904		
2 30.8	+55 33	Σ 270	AB	Per	1829 1967	303		21.2		7.4	9.2	F5	1901		
--			AC		1902 1926	339 338		35.4	38.3		10.6		1901		
--			AD		1902 1927	270		42.4			12.5		1901		
2 30.9	+53 11	h 2139	AB	Per	1902 1937	120		3.3		8.8	8.8	F5	1903		
--			AC		1902 1923	18		19.3			10.4		1903		
--			AD		1931	202		56.2					1903		
--			BC		1926	15		18.6					1903		
2 31.5	+1 06	Σ 274		Cet	1833 1932	219		13.5		7.3	7.8	A2	1924		
2 31.5	+37 27	β 304		And	1879 1925	283 283		17.9	19.9	7.7	11.7	F0	1914		
2 31.5	+42 33	A 660		And	1904 1960	303 315		0.2	0.4	8.4	8.5	A0	1913		
2 31.6	−27 30	B 38		For	1925 1948	252		1.1		7.6	10.0	A2	1931		
2 31.8	+38 07	Es		And	1923	330		19.9		7.2	11.7	K0	1919		
2 31.8	+89 16	Σ 93	AB	UMi	1834 1955	210 218		18.3	18.4	2.0	9.0	F8	1477		1 α UMi, Polaris [11]
--			AC		1884 1890	88 83		43.3	44.7		13.0		1477		
--			AD		1884	172		82.7			12.0		1477		
2 31.9	+57 42	β 1314	AB	Per	1902 1904	120		3.6		7.1	12.8	B3	1911		
--			AC		1904	121		6.8			14.3		1911		
--			AD		1902 1921	334		13.4			11.4		1911		
--			AE		1902 1904	162		14.3			13.6		1911		
--			AF		1902 1921	268		25.2			11.4		1911		
2 32.3	+28 33	A 2020		Tri	1909 1930	39		2.9		8.0	13.9	A0	1930		
2 32.6	+3 44	A 2334		Cet	1911 1957	5 330		0.2	0.2	8.7	8.8	F0	1941		Round in 1956
2 32.6	+35 42	A 1927		Tri	1908 1958	171 188		0.8	0.9	8.1	10.6	G5	1927		
2 32.7	+61 27	Stein 368		Cas	1898 1902	95		10.2		7.8	11.5	B	1920		
2 33.2	+58 26	Σ 272		Per	1830 1955	42 35		1.7	1.8	8.3	8.3	A2	1933		
2 33.2	+60 00	A 823		Cas	1904 1929	250		0.6		7.6	11.6	A0	1932		
2 33.3	+52 19	OΣ 42	AB	Per	1847 1960	110 273		0.4	0.2	7.0	7.5	A2	1938		Rapid; measures discordant
--			ABxC		1903 1909	348		90.9					1938		
--			AD		1875 1909	83		125.1			9.1		1938		
2 33.3	−75 54	h 3522	AB	Hyi	1918	290		34.3		7.0	10.7	F2			
--		δ	BC		1918	354		11.4			12.6				
2 33.5	+57 32	h 2143		Per	1904 1921	19		23.4		8.0	8.6	B3	1937		
2 33.6	+31 24	h 653		Tri	1904 1925	42		23.0		7.4	11.1	K2	1947		
2 33.7	−48 23	Rst 53		Eri	1929 1947	201 197		9.4	9.2	8.0	13.9	F8			
2 33.8	−28 14	h 3506		For	1836 1952	242 244		10.8	10.8	5.0	7.7	B9	1954		ω For
2 34.1	−5 38	Σ 280		Cet	1831 1959	346		3.6		7.9	8.1	K2	1953		
2 34.2	−31 31	h 3509		For	1918	59		23.5		7.5	11.5	F0			
2 34.7	−7 52	β pm		Cet	1911	47		95.1		5.8	12.5	K0			77 Cet
2 35.1	+1 08	HdZ		Cet						8.0					[12]
2 35.5	+39 40	h 1120	AB	And	1894 1959	91 83		17.3	16.3	6.3	10.8	B9	1961		
--			AC		1906 1959	321 321		39.6	41.5		10.6		1961		
2 35.6	+37 19	Σ 279	AB	And	1831 1933	71		16.9		5.7	10.7	K0	1964		
--			AC		1909 1933	207 207		22.9	45.6		11.5		1964		
2 35.8	+34 41	A.G.		Tri	1906 1959	17 17		138.9	141.3	5.7	6.9	M A5			15 Tri
2 35.9	+5 36	Σ 281		Cet	1831 1972	83 81		8.0	8.1	4.9	9.5	G5	1971		78 v Cet
2 36.0	−7 50	β pm	AB	Cet	1911	123		117.3		5.5	13.1	K5			80 Cet
--			AC		1900 1911	189		146.7			9.1				
2 36.0	−21 24	h 3511		Cet	1877 1953	98		14.8		7.4	9.2	G0	1978		Yellow, pale blue (h)
2 36.1	+6 53	Plq 32		Cet	1912 1929	109		164.8		5.8	11.9	K0		o	Bright star is 268 G. Cet
2 36.9	+59 53	Σ 277	AB	Cas	1831 1915	137		3.1		7.6	10.9	A0	1970		
--			AC		1881 1908	22		161.5			10.0		1970		
2 37.0	+24 39	Σ I 5		Ari	1835 1937	274		38.6		6.6	7.4	F5 F5	1982		30 Ari; yel., pale lilac
2 37.2	+30 24	Cou 671		Ari	1971	121		0.5		7.8	9.7	A0			
2 37.6	+50 14	Arg 8		And						7.3		B8			No close pair found near
2 38.3	+37 44	Fox	AB	And	1922 1934	260 262		9.2	8.5	6.2	11.2	F5	1996		
--		β 305	AC		1875 1925	205		20.8			10.4		1996		
2 38.4	−1 25	A 450		Cet	1903 1957	220 213		0.4	0.4	8.5	9.0	F2	2005		[13]
2 38.6	+3 27	Kui 9		Cet	1937 1955	308 315		0.9	0.9	6.3	9.5	G5			
2 38.6	+19 44	h 2152		Ari	1903 1913	59		42.6		7.5	13.2	K0			
2 38.7	−52 57	CorO 14		Hor	1917 1928	129		9.0		7.4	8.3	F8 G5			
2 38.8	+33 25	Σ 285		Tri	1832 1959	178 167		1.8	1.7	7.5	8.2	K0	2004		
2 38.9	+69 18	Σ 278		Cas	1830 1958	82 44		0.4	0.5	8.6	8.9	A2	1985		
2 39.0	+8 55	OΣΣ 30		Cet	1875 1924	214		68.7		8.2	9.9	A0			
2 39.0	+14 52	Σ 287		Ari	1830 1967	73 74		6.5	6.8	7.4	9.7	K0	2008		
2 39.0	+62 36	OΣΣ 28		Cas	1875 1924	147		67.8		6.7	7.4	B9 A0			
2 39.0	−54 50	h 3520		Hor	1879 1931	204		20.8		7.5	8.6	F5			

α 2000	δ 2000	Star Name		Const	Years	Pos Ang		Separation		m₁	m₂	Spec	ADS	Orb	Notes
2ʰ39ᵐ.6	−11°52′	ø 312		Cet	1951 1960	°	°	″	″	5.8	5.8	F5			o 83 ε Cet
2 39.7	−59 34	Δ 7	AB	Hor	1836 1953	96		36.6		7.2	7.5	K0 A5			Magnitude of B is for BC
--		I 386	BC		1901 1959	320	1	0.3	0.4	8.0	8.5				
2 39.8	+33 57	Σ 286		Tri	1830 1938	254		2.9		7.9	10.2	G5	2020		
2 39.8	−42 54	B 2075		Eri	1932	5		23.8		4.8	14.8	A2			
2 40.5	+61 29	Σ 283	AB	Cas	1831 1959	207		1.8		8.2	9.0	G5	2014		
--			AC		1904	15		18.3			13.3		2014		
2 40.5	−24 08	See 19		For	1897 1956	324	308	0.5	0.5	8.0	8.5	F5	2044		
2 40.7	+27 04	Σ 289		Ari	1831 1921	0		28.6		5.5	8.4	A2	2033		33 Ari
2 40.8	+61 17	Σ 284	AB	Cas	1830 1911	195		5.5		7.9	9.9	F0	2018		
--			AC		1911	9		53.1					2018		
2 40.9	+4 52	OΣ 45		Cet	1847 1953	296	281	1.6	1.2	7.9	10.1	F0	2043		
2 41.1	+18 48	Σ 291	AB	Ari	1832 1955	118		3.4		7.7	8.0	B9	2042		
--			AC		1892 1923	241		65.7			9.3		2042		
--			BC		1907	244		67.4					2042		
2 41.2	−0 42	Σ 295		Cet	1831 1970	335	310	4.8	4.0	5.8	9.0	F5	2046		84 Cet
2 41.4	+4 26	A 2337	AB	Cet	1911 1955	244	252	2.2	2.2	6.9	12.7	A0	2050		
--			ABxC		1920 1922	131	132	26.0	28.3		13.0		2050		
--			ABxD		1920 1922	204		25.8			11.7		2050		Plq calls ABxD optical
--			AC		1911 1918	130	130	26.7	26.5				2050		
--			AD		1911 1918	204	204	26.8	25.8				2050		
2 41.7	+55 29	A 1280		Per	1906 1958	339	4	0.3	0.3	7.9	9.1	A2	2040		
2 42.0	+42 42	h 1124		Per	1904	151		16.6		8.0	11.6	B8	2049		
2 42.0	+42 47	h 1123		Per	1879 1916	248		20.0		8.0	8.0	A0 A0	2048		
2 42.2	+42 42	OΣ 44	AB	Per	1850 1959	58	55	1.4	1.4	8.4	9.1	B9	2052		
--			AC		1908	290		86.3					2052		
2 42.4	+20 01	β 522		Ari	1878 1934	264		19.1		5.7	12.2	A0	2062		34 μ Ari
2 42.7	+42 48	h 2155		Per	1869 1904	321		17.1		8.0	10.4	B9	2060		
2 43.0	+20 14	A 2220		Ari	1910 1953	225		1.6		8.0	12.0	G5	2072		
2 43.0	+48 16	β 521	AB	Per	1878 1928	153		5.8		6.6	11.6	G5	2064		
--			AC		1913	233		72.8			10.7		2064		
2 43.0	+53 32	Es 8		Per	1899 1935	43	29	12.8	12.4	5.8	14.8	K0	2059		
2 43.0	−20 17	h 3524		Cet	1892 1934	144	152	19.4	19.4	7.5	9.3	G0	2079		
2 43.1	+58 53	A 970		Cas	1905 1929	100		5.2		7.6	13.9	A0	2058		
2 43.3	+3 14	Σ 299		Cet	1836 1955	289	294	2.7	2.8	3.5	7.3	A2	2080		86 γ Cet
2 43.3	−40 32	h 3527		Eri	1836 1954	45	41	1.5	2.0	7.1	7.3	A0 A0			
2 43.4	−66 43	ø 333		Hor	1955 1960	37	35	0.5	0.1	7.1	7.1	F8			
2 43.8	−27 54	β 261	AB	For	1870 1943	100		3.0		7.9	9.4	G5	2092		
--			AC		1916	131		70.2			9.8		2092		
2 43.9	+25 38	β 306	AB	Ari	1870 1938	18		3.0		6.4	11.0	A2	2082		
--			AC		1910 1925	90		50.3			12.1		2082		
2 44.2	+35 07	h 654		Per	1903 1924	45		32.2		7.3	9.9	B5	2084		
2 44.2	+49 14	Σ 296	AB	Per	1783 1972					4.1	9.9	F8	2081	o	13 θ Per
--			AC		1853 1924	211	229	66.1	77.2				2081		
--			BC		1912	213		72.1					2081		
2 44.2	−25 30	ø 379	Aa	For	1963	143		0.1		7.8	7.8		2098		
--		BrsO 1	AaxB		1836 1954	185	191	11.5	12.1	7.0	9.0	G0	2098		
2 44.6	+29 28	Σ 300		Ari	1832 1975	299	312	2.9	3.1	7.8	8.0	F0	2091		
2 45.0	+14 14	Hu 1046		Ari	1904 1958	93	112	1.1	0.7	7.8	11.5	F5	2101		
2 45.4	+56 34	Σ 297	AB	Per	1831 1926	277		15.8		8.5	8.8	A0	2094		
--			AC		1830 1904	106		28.4			10.6		2094		
2 45.5	−63 42	HdO		Hor		70		20.0		5.7	13.0	K0			
2 45.6	−41 14	CapO 30		Eri	1900 1955	29		2.1		8.2	9.4	F5			
2 45.7	−39 43	B 2077		Eri		37		7.0		8.0	15	G5			
2 46.0	−4 57	β 83		Eri	1876 1959	121	59	1.4	0.6	8.0	10.6	F2	2111		
2 46.4	+53 10	Es 9	AB	Per	1899 1929	30		2.7		8.0	11.5	K5	2102		
--			AC		1908	304		10.3			14.0		2102		
2 46.7	+75 24	h 2151		Cas	1885 1912	118	113	15.8	15.6	7.1	13.3	A3	2087		
2 47.1	+35 33	β 9	AB	Per	1875 1959	161	188	1.5	1.7	6.4	8.5	F2	2117		
--			AC		1913	138		31.6			13.2		2117		
2 47.1	+50 07	Arg 9		Per	1900 1969	145	148	3.1	2.8	8.8	8.8	F8	2112		
2 47.5	+19 22	Σ 305	AB	Ari	1830 1973					7.4	8.2	G0	2122	o	
--			AC		1910	34		88.6			12.6		2122		
2 47.6	+29 41	β 307		Ari	1876 1914	316		15.4		7.2	11.6	B5	2126		
2 47.6	+53 57	Σ 301		Per	1830 1938	16		8.1		7.8	8.8	A0	2115		
2 48.0	+1 42	A 2411		Cet	1912 1943	268		0.4		8.4	9.4	G0	2133		
2 48.1	−36 59	δ 35		For	1919 1934	172		1.8		8.0	11.8	G0			
2 48.6	−37 24	h 3532		For	1877 1952	145		5.4		7.0	8.2	F2			
2 48.8	+49 11	Σ 304 rej		Per	1901 1925	289		25.2		7.6	9.1	B9	2139		
2 49.1	−32 24	B 1031		For	1928	67		4.7		4.5	14.0	K0			β For
2 49.3	+17 28	Σ 311	AB	Ari	1832 1953	120		3.2		5.2	8.7	B5	2151		42 π Ari
--			AC		1832 1938	110		25.2			10.8		2151		
--		Hall	BC		1879	108		21.8					2151		
2 49.4	+33 56	Σ 310		Tri	1832 1938	91		2.6		7.3	10.5	A2	2150		
2 49.8	−20 15	h 3533		Eri	1892 1935	273	271	39.7	39.1	7.8	8.8	K5 G5			

α 2000	δ 2000	Star Name		Const	Years	Pos Ang		Separation		m₁	m₂	Spec	ADS	Orb	Notes
2ʰ49.8ᵐ	−24°34′	β 877	AB	For	1880 1954	145°	°	12.0″	″	6.1	12.5	G5	2167		γ¹ For
--		h 2161	AC		1880 1954	157 143		48.8	40.9		10.5		2167		
2 49.9	+8 57	Σ 313		Cet	1831 1926	192		5.5		8.5	8.8	F8	2163		
2 50.0	+27 16	OΣ 47 rej	AB	Ari	1852 1922	255 277		18.8	24.6	3.6	10.7	B8	2159		41 Ari
--		H V 116	AC		1783 1922	189 213		39.3	31.3		10.5		2159		
--		H VI 5	AD		1821 1922	227 232		127.6	124.9		9.0		2159		
--		S1v	BC		1967	195		30.4					2159		
2 50.0	+30 32	OΣ 46	AB	Ari	1852 1935	74		4.9		6.8	10.0	F0	2158		
--		h 656	AC		1877	168		20.1			12.7		2158		
2 50.0	+48 48	Hu 206		Per	1900 1955	338		1.8		7.8	12.0	A2	2154		Spectrum composite A2+G
2 50.0	+64 38	Σ 302		Cas	1832 1913	167		5.1		7.5	10.2	B9	2142		
2 50.2	+72 55	h 2157	AB	Cas	1903 1911	288		11.2		7.9	11.4	B8	2135		
--			AC		1903 1911	21		24.5			10.8		2135		
--			AD		1903 1911	65		33.6			11.2		2135		
2 50.2	−35 51	h 3536		For	1897 1933	14		5.0		5.9	10.1	K0			η² For
2 50.4	−4 59	β 10		Eri	1874 1939	99		2.7		7.1	11.0	A0	2170		
2 50.6	+38 19	β pm	AB	Per	1911	121		97.1		4.2	12.7	F0			16 Per
--			AC		1857 1924	144 146		263.0	249.3		8.9				
2 50.7	+55 54	Σ 307	AB	Per	1836 1932	300		28.3		3.8	8.5	K0	2157		15 η Per [14]
--			AC		1878 1925	268		66.6			9.8		2157		
--		S,h 34	AE		1821 1925	205		238.1					2157		
--		Ward	CD		1904 1925	114		5.2			10.3		2157		
2 51.0	−21 00	See 20		Eri	1897 1904	129		46.7		4.8	15	K0	2179		2 τ² Eri
2 51.1	+60 25	Σ 306	AB	Cas	1831 1938	93		2.1		7.3	9.2	B0	2161		
--			AC		1867 1901	157		27.4			11.5		2161		See BDS 1430
--			AD		1892	74		17.0			13.8		2161		
--			AE		1892	112		19.2			13.5		2161		
--			AF		1892	106		27.4			13.0		2161		
--			AG		1868 1918	156		123.7					2161		
2 51.3	+1 42	Vou 36		Cet	1936 1959	355 1		0.2	0.3	8.4	8.9	A3			
2 51.4	+3 03	Ho 218		Cet	1889 1953	210 215		0.4	0.1	8.8	8.8	F2	2178		Rapid changes
2 51.4	−0 41	Rst 4752		Cet	1942 1949	46		1.2		8.0	12.8	F2			
2 51.8	+58 19	β pm		Cas	1879 1910	147		192.7		6.5	9.3	A0			
2 52.7	+6 28	Σ 323		Cet	1830 1948	283 278		2.6	2.7	8.7	8.7	B9	2193		
2 52.7	−41 24	I 412		Eri		290		16.0		6.7	14.5	A2			
2 52.9	+53 00	A 2906	AB	Per	1917 1955	116		0.2		6.5	8.8	B9	2185		Round in 1954
--		Σ 314	ABxC		1830 1955	295 308		1.5	1.6		7.1		2185		
2 53.0	+60 28	β	AB	Cas	1903 1912	194		15.6		7.4	11.3	A3	2184		
--			AC		1912	203		75.5			11.4		2184		
--		M1r 107	Bb		1970	320		1.2			14.0		2184		
2 53.4	+48 34	OΣ 48		Per	1854 1936	317		6.7		6.5	10.6	K0	2192		
2 53.5	−8 51	Rst 4216		Eri	1938 1942	314		7.2		7.8	13.0	G0			
2 53.5	−11 52	Gallo 79		Eri	1903 1936	175		5.6		8.5	8.9		2206		
2 53.7	+38 20	β 524	AB	Per	1880 1960					5.6	6.7	F0	2200	o	20 Per
--		Σ 318	ABxC		1829 1970	237		14.1			10.1		2200		
--			ABxD		1913	49		56.2					2200		
--			AC		1928 1929	237		15.0					2200		
2 53.8	−44 37	I 1480		Eri	1927 1946	126 145		0.4	0.3	8.3	9.0	G0			
2 54.0	+47 10	β 1293		Per	1900 1940	352 346		1.7	1.8	6.7	10.3	B9	2199		
2 54.3	+52 46	Edg	AB	Per	1878 1923	106 106		50.7	51.7	4.0	10.6	G0	2202		18 τ Per [15]
--		β	BC		1900 1925	82 87		4.6	3.5		11.7		2202		
2 54.6	+37 45	A.G. 58		Per						7.9		G5			Single in 1902
2 55.4	+27 43	Ho 316		Ari	1891 1907	284		19.6		6.7	12.7	B9	2215		
2 55.5	+76 32	h 2158		Cas	1909	161		15.6		7.7	13.4	G5	2187		
2 55.6	+26 52	Σ 326	AB	Ari	1831 1976	216 220		9.0	5.9	7.6	9.8	K0 K5	2218		
--		LDS 883	AC		1936	264		44.0			15		2218		
2 55.8	−38 39	B 679		For	1927 1938	56		2.2		7.1	12.0	K0			
2 55.9	+31 35	Es 325		Per	1906			12.9		8.0	12.6	F5	2219		
2 55.9	+61 31	β pm	AB	Cas	1911 1925	256 256		50.8	52.6	5.6	13.1	F5			
--			AC		1911 1925	43 43		91.5	89.6		11.1				
2 56.2	+72 53	Σ 312	AB	Cas	1832 1972	14 34		3.6	2.3	8.0	8.9	G0	2204		
--			AC		1831 1957	127 129		42.3	42.9		9.9		2204		
--			AD		1911	279		54.5			12.7		2204		
--			BC		1909	133		43.4					2204		
2 56.4	+64 20	β pm		Cas	1879 1910	129		115.6		6.2	9.6	K5			
2 56.4	−28 57	δ 77		For	1920 1942	73		0.3		8.5	9.0	F5	2231		
2 56.5	−39 27	CorO 17		For	1900 1954	184 188		3.4	3.4	8.5	8.9	G0			
2 56.6	+47 10	Σ 324 rej	AB	Per	1830	342		8.0		6.0	15	K0			No companion in 40-inch (β)
--			AC		1900 1934	203		24.8			11.5				
2 57.0	−63 02	Rst 56		Hor	1929 1945	143		5.4		7.5	13.0	F8			
2 57.2	+1 53	A 2413		Cet	1912 1960					8.2	8.4	G0	2236	o	
2 57.2	−0 34	Σ 330		Cet	1832 1930	192		8.8		7.3	9.3	G5	2237		Very yellow, bluish (Webb)
2 57.2	−24 59	β 741	AB	For	1879 1959					8.1	8.2	G5	2242	o	
--		S 423	ABxC		1824 1954	219 224		27.7	28.6		7.8	G5 G5	2242		
2 58.0	−15 50	Hu 811		Eri	1902 1946	220 226		1.8	1.8	7.8	11.3	G0	2247		

α 2000	δ 2000	Star Name		Const	Years	Pos Ang	Separation	m_1	m_2	Spec	ADS	Orb	Notes
$2^h58.1^m$	$+69°12'$	Σ 317		Cas	1831 1958	88° 83°	3″.2 3″.8	7.9	9.6	F2	2226		[16]
2 58.3	−40 18	Pz 2		Eri	1835 1952	82 88	8.2 8.2	3.4	4.5	A2 A2			θ Eri
2 58.6	+24 08	β 1173	AB	Ari	1890 1959	325 61	0.1 0.2	8.3	8.4	K0	2246	o	
——			ABxC		1890 1946	283	4.6		13.0		2246		
2 58.7	−2 47	Kui 10		Eri	1934 1955	239 237	3.4 2.3	5.2	12.5	A2			
2 58.8	+43 22	LDS 2816		Per	1957	280	245.0	8.1	10.8	K0			
2 58.9	+21 37	β 525		Ari	1877 1959			7.5	7.5	A3	2253	o	
2 59.2	+21 20	Σ 333	AB	Ari	1830 1959			5.2	5.5	A2	2257		48 ε Ari
——			ABxC		1912 1922	192 191	145.4 146.3		12.7		2257		
2 59.4	+6 39	Σ 334		Cet	1830 1958	323 312	1.6 1.2	7.9	8.4	F0	2261		
2 59.6	+32 33	h 5455		Per	1823	195	20.0	8.0	12.0				Not found in 1906
3 00.5	+18 00	OΣ 49	AB	Ari	1843 1958	71 54	1.7 2.1	7.0	10.0	A0	2279		
——			AC		1913	206	146.1		11.9		2279		
3 00.6	+47 53	A 1529		Per	1907 1946	166 142	0.3 0.1	8.0	9.0	A2	2271		Round 1951−54
3 00.9	+52 21	Σ 331		Per	1828 1954	85	12.1	5.3	6.7	B5	2270		
3 00.9	+59 40	OΣΣ 31		Cas	1875 1916	230	73.6	7.4	8.8	A0 A0			
3 01.5	+32 25	Σ 336		Per	1831 1971	8	8.4	6.9	8.4	G5	2286		Spectrum composite G5+A5
3 02.3	−71 54	φ 360		Hyi	1961	179	0.1	6.3	6.3	B8			θ Hyi
3 02.4	+72 36	A 827		Cas	1904 1948	259	0.3	8.5	8.6	A0	2276		
3 02.7	−7 41	β 11		Eri	1875 1959	87 75	2.7 1.8	5.3	9.5	G5	2312		9 ρ² Eri
3 03.0	−2 05	Σ 341		Eri	1831 1936	226	8.6	7.7	9.7	F5	2316		
3 03.5	−10 58	β 1174		Eri	1890 1948	306 293	1.2 0.8	7.2	10.8	F8	2323		
3 03.6	+36 27	β pm	AB	Per	1889 1918	265 266	81.2 85.9	7.8	9.8	F8			
——		Cou 869	Aa		1972	105	4.4		13.0				
3 03.8	−21 22	h 3548		Eri	1877 1936	123	12.2	7.7	11.5	G0	2326		
3 04.2	+61 42	Kui 11		Cas	1931	132	12.7	6.6	12.4	G0			
3 04.6	−51 19	Δ 10		Hor	1837 1916	70	38.2	7.5	8.8	G0 K0			
3 04.8	+53 30	h 2170	AB	Per	1879 1938	324 326	57.7 57.0	2.9	10.6	F5	2324		γ Per; sp. compos. F5+A3
——		Wils	Aa		1939	49	0.1				2324		
3 05.0	−13 26	β 527		Eri	1877 1959	60 90	0.8 1.1	8.2	8.7	F5	2347		
3 05.4	+25 15	Σ 346	AB	Ari	1832 1958	264 165	0.7 0.1	6.9	6.9	B8	2336		52 Ari [17]
——			ABxC		1832 1938	356	5.1		10.8		2336		
——			ABxD		1879 1925	83	102.1		10.8		2336		
——			ABxE		1911	191	133.6		12.3		2336		
3 05.7	+32 02	Es 326	AB	Per	1906	36	102.3	8.2	10.0	G5	2338		
——			BC		1899 1927	34	4.4		11.0		2338		
3 05.8	+43 42	β 1175		Per	1890 1959	281 276	0.3 0.4	7.6	9.0	F5	2334		Spectrum composite F5+A3
3 06.1	+79 25	Σ 320		Cep	1831 1934	229	4.6	5.6	8.8	M	2294		
3 06.6	−6 05	Rst 4223		Eri	1938 1942	231	15.1	5.3	12.5	M			
3 06.7	+45 45	Es 558		Per	1908 1917	358	8.4	7.8	9.7	B9	2341		
3 06.7	−13 19	Σ 356		Eri	1831 1919	12	15.7	8.1	11.2	F5	2356		
3 07.5	−78 59	h 3568		Hyi	1870 1939	224	15.2	5.6	9.3	F0			
3 07.6	+75 48	h 2166	AB	Cas	1903 1957	249	59.4	7.6	9.4	M	2339		
——			AC		1903 1957	192	60.2		9.5		2339		
——			AD		1903 1957	140	64.2		9.7		2339		
——			Cc		1903	339	7.4		13.3		2339		
3 07.7	+36 37	β pm		Per	1893 1918	243 247	116.2 119.3	7.9	9.3	G0			
3 08.2	+40 57	β 526	AB	Per	1878 1920	155 156	59.1 57.5	2.1	12.7	B8	2362		26 β Per, Algol (var.)
——			AC		1878 1924	145 145	68.1 67.2		12.5		2362		
——			AD		1879 1924	192	81.9		10.5		2362		
——			DE		1878 1922	116 118	10.8 11.4		12.5		2362		
3 08.5	+56 39	A 975		Per	1905 1925	201	1.5	8.1	10.6	K0	2360		
3 08.8	+35 28	Σ 352		Per	1831 1940	2	3.5	7.7	9.8	A0	2364		
3 09.1	+7 28	Σ I 6		Cet	1835 1935	163	81.2	8.3	8.5	G0 G0			
3 09.1	+49 37	β pm		Per	1911	132	146.2	4.1	12.3	G0			ζ Per
3 09.5	+44 51	Es		Per	1914 1933	334 329	21.8 27.7	3.8	13.3	K0	2368		27 κ Per
3 09.9	−83 32	h 3582		Oct	1919	297	19.5	7.6	10.5	F0			
3 10.1	−63 55	h 3559		Hor	1879 1916	40	43.1	6.7	10.7	A2			
3 10.1	−64 54	φ 361		Hor	1961	72	0.1	7.4	7.4	A2			
3 10.2	+21 45	β 1030		Ari	1888 1959	165 124	0.6 0.7	8.6	8.6	F0	2375		
3 10.8	+63 47	Σ 349		Cas	1832 1937	320	6.1	7.9	8.6	F8	2371		
3 10.9	−11 08	β pm	AB	Eri	1894 1907	57 56	144.2 143.7	7.2	8.8	G5 A3			
——			BC		1907	155	86.3		12.5				
3 11.3	−3 49	β 400		Eri	1879 1956	53 55	22.2 24.4	6.1	11.7	M	2389		
3 11.3	+39 37	β pm		Per	1879 1908	111	177.4	4.6	10.7	K0			28 ω Per
3 11.7	+81 28	Σ 327 rej		Cep	1904 1935	283	24.2	6.0	10.7	A2	2348		
3 12.1	+37 12	Σ 360		Per	1831 1970	146 128	1.3 2.5	8.1	8.3	G0	2390	o	
3 12.1	−28 59	h 3555		For	1836 1959			4.0	7.0	F8	2402	o	α For; B is variable
3 12.4	+47 44	β pm	AB	Per	1908	7	7.4	6.4	14.3	K0			
——			AC		1908	176	28.0		12.7				
——			AD		1880 1908	294 295	198.6 202.1		10.3				
——			DE		1908	94	19.9		13.5				
3 12.4	−44 25	Jc 8	AB	Eri	1897 1959			6.6	6.9	F2		o	
——		h 3556	ABxC		1851 1959	230 200	2.7 3.5		8.9				A most beautiful object (h)
3 12.5	+71 33	OΣ 50	AB	Cas	1847 1959	232 179	0.9 1.3	8.4	8.4	F8	2377	o	

α 2000	δ 2000	Star Name		Const	Years	Pos Ang	Separation	m₁	m₂	Spec	ADS	Orb	Notes
3ʰ12.ᵐ5	+71°33′	h 2172	AC	Cas	1900 1931	296° 299°	27″.0 27.2		13.2		2377		
3 12.8	−1 12	h 663		Cet	1871 1958	253 231	5.1 3.3	5.1	11.5	F8	2406		94 Cet
3 14.3	+18 21	Cou 359		Ari	1969	170	0.1	8.7	8.7	G0			
3 14.3	+22 57	Σ 366 rej	AB	Ari	1879 1915	42 40	48.9 47.2	6.9	9.6	K0	2414		
—		β 530	BC		1879 1943	196 192	1.8 1.7		10.3		2414		
3 14.8	−14 27	h 3557		Eri	1880 1919	5	27.1 27.0	8.0	11.1	K0			
3 14.9	−59 31	h 3564		Ret	1916	279	29.7	7.0	12.0	K0			
3 15.1	+73 52	h 2173		Cas	1904	163	47.0	6.8	10.7	K0			
3 15.2	−64 27	Δ 12	AxBC	Ret	1836 1946	103	19.1	6.9		F5			
—		Rst 67	BC		1928 1945	48 38	0.3 0.4	9.5	9.8				
3 15.6	+52 45	Es 768	AB	Per	1909	315	73.5	8.1	10.2	A0	2423		
—			BC		1909	47	2.6		11.5		2423		
3 15.8	+32 51	h 332	AB	Per	1903	109	32.0	6.3	12.8	F0	2431		
—			AC		1903	10	45.2		12.5		2431		
3 15.8	+50 57	Hu 544		Per	1902 1954	98 99	0.6 1.2	6.8	9.1	A0	2425		
3 15.8	+57 08	Es 11		Per	1899 1935	66	10.8	5.8	14.0	A0	2424		
3 16.0	−5 55	β 84		Eri	1875 1959	23 14	0.7 1.0	6.9	7.1	B9	2440		
3 16.0	+34 41	Ho 503		Per	1896	99	30.8	6.3	11.8	A2	2433		
3 16.3	+19 20	A 2224		Ari	1910 1960	328 334	0.8 0.9	8.2	9.5	K0	2439		
3 16.3	+60 02	Σ 362	AB	Cam	1831 1955	142	7.1	8.5	8.8	A0	2426		In beaut. wide group (Webb)
—			AC		1893 1915	43	26.1		10.5		2426		
—			AD		1915	286	30.9		11.1		2426		
—			AE		1866 1915	242	35.3		9.9		2426		
3 16.5	+78 31	Σ 345		Cep	1831 1968	82	6.7	8.0	9.8	K0	2403		
3 17.2	+40 29	Σ 369		Per	1829 1958	28	3.5	6.7	8.0	A0	2443		
3 17.4	+7 39	β 1039	AB	Cet	1889 1898	209	2.2	7.4	13.4	F8	2451		
—		Bgh	AC			38	156.0		8.6	G0	2451		
3 17.5	+65 40	OΣ 52	AB	Cam	1846 1959	153 84	0.5 0.5	6.8	7.3	A2	2436	o	
—			ABxC		1912	223	80.5		12.7		2436		
3 17.6	+42 41	A 1704	AB	Per	1907 1917	231	74.0	7.7	12.2	F0	2445		
—			AxBC		1931	232	74.4				2445		
—			BC		1907 1931	248	0.6		13.5		2445		
3 17.7	+38 38	OΣ 53		Per	1845 1959			7.8	8.3	G0		2446	o	
3 18.2	−62 30			Ret	1952	216	310.0	5.2	5.5	G3 G4			ζ Ret; large p.m.
3 18.4	−0 56	AC 2		Cet	1854 1959			5.6	7.5	G5	2459	o	95 Cet
3 18.4	−22 31	See 23		Eri	1897 1959	288 295	0.3 0.2	5.1	6.6	K0	2463		15 Eri [18]
3 18.7	−18 34	h 3565		Eri	1835 1954	110 118	5.8 7.2	5.7	9.1	F0	2465		
3 19.4	+3 22	β pm	AB	Cet	1853 1909	153 157	271.6 268.7	4.8	9.3	G5			96 κ Cet
—			BC		1909	272	215.1		11.7				
3 19.5	−21 45	Jc 1	AB	Eri	1857 1937	288	5.7	3.7	9.2	M2	2472		16 τ⁴ Eri
—			AC		1877 1955	99 112	40.0 39.2		10.5		2472		
—			AD		1879	293	123.1		10.4		2472		
—			AE		1879	276	130.0		10.4		2472		
—			AF		1836 1880	241 236	150.0 160.2		9.7		2472		
3 19.6	+67 14	Hu 1056		Cam	1904 1958	282 272	0.6 1.0	8.6	8.6	F8	2452		
3 20.0	+65 39	β pm		Cam	1911	58	121.5	4.8	12.5	B3			
3 20.1	+69 44	h 1133		Cas	1896 1908	203	30.8	6.7	12.7	A0	2455		
3 20.1	−28 51	LDS 93		For		n	255.0	7.4	8.4	G0 G0			
3 20.3	+19 44	Σ 376		Ari	1830 1949	252	7.0	8.4	8.5	A2	2475		
3 20.3	+77 44	β 1176	AB	Cep	1890 1899	276	1.2	5.5	12.3	F0	2450		
—			AC		1890 1934	228 246	11.0 12.0		13.1		2450		
3 20.4	+23 41	Σ 375	AB	Ari	1832 1943	315	2.3	7.5	9.6	A5	2473		
—			AC		1911	290	64.9		12.6		2473		
3 20.6	+19 17	Σ 377	AB	Ari	1831 1956	115	1.1	8.5	8.9	A0	2478		
—			ABxC		1870 1879	221	25.0		11.5		2478		
—			AC		1829 1912	222	25.0				2478		
3 20.9	+1 10	Ho 320		Cet	1890 1943	172	1.7	8.0	10.5	G0	2484		
3 21.0	+33 13	A.G. 65		Per				7.9		K0			No companion 1903 (40-in)
3 21.2	+21 09	Cou 259		Ari	1968	246	0.7	5.3	8.7	B3			61 τ Ari
3 21.5	+45 23	Ho 319	AB	Per	1892 1914	46	12.0	7.5	11.8	B8	2483		
—			AC		1914	308	18.1		14.5		2483		
3 21.7	−20 19	h 3570		Eri	1900 1920	256	34.2	6.6	12.6	A0			
3 21.9	+49 04	Webb		Per	1913	328	205.8	6.2	9.7	F5			
3 22.1	+62 44	Σ 373 rej	AB	Cam	1875 1923	117	20.1	7.7	10.0	F8			
—		OΣΣ 33	AC		1875 1918	110	117.5		7.8	A0			
3 22.9	+29 49	Σ 379		Ari	1830 1919	102	10.3	8.7	8.7	A2	2499		H III 91, BDS 1688
3 23.3	−7 48	β 531		Eri	1877 1945	60 48	2.4 3.9	6.3	12.7	G0	2507		
3 23.3	+20 58	Σ 381		Ari	1830 1955	98	1.0	7.6	8.3	G5	2504		
3 23.3	−11 21	h 2187		Eri	1901 1902	240	56.5	7.7	10.4	F5			
3 23.5	−14 13	Rst 3381		Eri	1936 1951	73 80	0.5 0.5	8.0	10.0	G5			
3 23.6	−11 13	Σ 387 rej		Eri				8.0	10.0				Probably this is h 2187 Error in position?
3 23.6	−40 05	I 468		Eri	1927	221	2.1	6.8	11.0	G0			
3 23.7	+4 53	β 1178		Tau	1890 1954	349	1.0	6.4	12.1	G0	2509		
3 24.0	−26 13	h 3572		For	1919 1932	95	20.7	8.6	8.8	F5 F5			
3 24.2	+67 27	Σ 374		Cam	1831 1923	295	10.9	7.8	9.3	F8	2494		

α 2000	δ 2000	Star Name		Const	Years	Pos Ang		Separation		m₁	m₂	Spec	ADS	Orb	Notes
3ʰ24.3	+49°52′	β pm		Per	1879 1909	196° 196°		166″.0 167″.0		1.8	11.8	F5			33 α Per
3 24.4	−14 00	β 12		Eri	1875 1953	274		2.3		6.9	9.8	A0	2523		
3 24.4	−15 39	A 2909	AB	Eri	1918 1953					8.3	8.3	G0	2524	o	
--			ABxC		1918 1921	252		16.8			13.4		2524		
3 24.5	+33 32	Σ 382	AB	Per	1831 1969	154 153		3.6	4.4	5.6	9.1	A0	2514		
--		β	AC		1903	173		27.8			12.6		2514		
3 24.6	−45 40	h 3576		Hor	1836 1953	341		2.9		7.4	9.0	A2			
3 24.6	−51 04	h 3575		Hor	1835 1916	44		34.7		6.7	10.2	A0			
3 24.7	+40 46	A 1287		Per	1906 1931	94		2.8		7.8	13.8	B9	2513		
3 25.0	+40 13	Hu 1058		Per	1904 1957	114		0.8		8.4	9.1	K0	2518		
3 25.3	+60 29	A 978		Cam	1905 1925	243		2.1		7.9	13.7	B8	2510		
3 25.6	+45 31	Ho 321		Per	1893 1939	35		1.5		7.7	10.2	B8	2525		
3 25.7	−41 20	I 1484		Eri	1926	315		1.9		8.0	13.0	F0			
3 26.0	−35 57	B 1449		For	1928 1953	216		0.2		8.1	8.3	A5			
3 26.1	+20 15	A 2344		Ari	1911 1953	188		1.1		8.5	9.2	A5	2529		
3 26.2	−30 37	B 2082		For		sf		10.0		7.9	14.9	G0			
3 26.7	−28 34	See 25		For	1897 1936	18		10.2		7.7	11.7	K0	2543		
3 27.1	−37 53	CorO 19		For	1919 1929	310		12.9		8.0	9.7	F5			
3 27.7	−14 22	A 2911		Eri	1918 1941	114		1.9		7.2	14.1	A0	2551		
3 28.0	+20 28	Σ 394	AB	Ari	1828 1967	162		6.8		7.1	8.1	A3	2546		Magnitude of A is for Aa
--		Cou 260	Aa		1968	16		0.2		7.6	8.3				
3 28.0	−77 37	Rst 2310		Men	1935	140		2.1		7.7	13.5	K0			
3 28.2	−35 51	I 58		For	1898 1939	248		6.3		6.5	10.5	A0			χ³ For
3 28.4	+22 48	β 878	AB	Ari	1881 1958	78 69		1.1	0.8	6.0	12.2	G5	2552		66 Ari
--			AC		1911	226		151.3			12.4		2552		
3 28.4	+60 15	A 980	AB	Cam	1905 1958	176 95		0.3	0.2	6.8	7.8	B8	2538	o	
--		B	ABxC		1958	194		28.0			12.5		2538		
3 28.5	+59 54	Σ 384	AB	Cam	1830 1933	270		2.0		7.9	9.1	F8	2540		Gold, blue (Webb)
--			AC		1879 1908	340		116.9			10.5		2540		
3 28.6	+11 23	β 879		Tau	1878 1925	71		24.6		6.8	12.8	G5	2561		
3 28.7	+50 26	Σ 388		Per	1831 1938	210		2.8		8.1	9.1	F0	2548		
3 29.0	+40 11	A.G. 67	AB	Per	1902 1925	349		23.6		7.6	10.0	G5	2553		
--			AC		1911 1915	354		52.7			11.1		2553		
3 29.1	+59 56	Σ 385		Cam	1829 1937	162		2.4		4.2	8.5	B9	2544		
3 29.2	+45 03	Σ 391		Per	1831 1974	95		3.8		7.7	8.4	B3	2559		
3 29.4	+46 56	OΣ 55	AB	Per	1867 1904	293		27.3		6.2	10.8	B5	2560		
--		A 982	BC		1904 1932	234		3.5			13.8		2560		
--			BD		1916	301		22.2			14.5		2560		
--			BE		1916	273		44.0			14.0		2560		
--			EF		1916	308		2.8			14.7		2560		
3 29.4	+49 31	β 1179		Per	1890 1959	163 155		0.7	0.9	4.7	10.4	B5	2558		34 Per
3 29.4	−62 56	h 3580		Ret	1916	125		54.1		4.7	10.3	F5			κ Ret
3 30.0	+55 27	Σ 390	AB	Cam	1832 1920	159		14.8		5.1	9.5	A2	2565		
--			AC		1910	172		110.0			10.3		2565		
3 30.2	+59 22	Σ 389		Cam	1831 1953	67		2.8		6.5	7.5	A0	2563		
3 30.3	+20 07	Σ 399 rej	AB	Tau	1879 1914	147		20.2		8.1	9.6	G5			H IV 89
--			AC		1909	204		50.6			12.1				
3 30.3	+52 54	Σ 392		Per	1831 1921	347		25.8		7.4	9.6	K0	2566		
3 30.7	−4 17	Σ 408		Eri	1831 1956	348 331		1.4	1.4	8.3	8.5	A3	2581		
3 30.7	−27 55	B 2083		For		nf		8.0		7.8	14.0	F5			
3 30.8	+5 09	Σ 406		Tau	1836 1940	124		9.3		8.1	10.1	F0	2580		
3 31.1	+27 44	Σ I 7		Tau	1836 1933	233		44.1		7.4	8.4	B9 A0			
3 31.3	+19 46	Σ 403		Tau	1829 1948	182 175		2.9	2.6	8.2	9.6	F8	2584		
3 31.3	+27 34	Σ 401		Tau	1830 1949	270		11.3		6.4	6.9	A0	2582		
3 31.6	+47 52	OΣ 56 rej		Per	1843 1947	351 352		22.8	28.7	6.8	10.3	B9	2579		
3 32.0	+67 35	OΣ 54		Cam	1850 1947	354 358		25.8	23.6	7.5	9.7	F0	2576		
3 32.1	−49 10	Rst 77		Hor	1928 1954	349 344		2.1	2.1	7.6	13.4	K0			
3 32.2	+11 33	A.G. 68		Tau	1895 1925	248		17.6		6.9	10.1	A0	2591		H IV 44
3 32.3	−7 05	Σ 411 rej	AB	Eri	1880 1936	88		19.1		7.8	8.8	F8	2596		
--			AC		1880 1890	28		38.2			11.2		2596		
3 32.7	−60 34	Jsp 48		Ret	1929 1954	138 114		0.5	0.5	8.2	8.6	F5			
3 33.4	+23 22	OΣ 57	AB	Tau	1854 1918	320		9.5		8.4	11.4	A3	2605		
--			AC		1854 1923	35		70.8			8.5	G5	2605		
--			Cc		1907	170		35.0			13.5		2605		
3 33.5	+58 46	Σ 396	AB	Cam	1829 1972	243		20.4		6.3	8.2	A2	2592		
--			AC		1879 1908	102		165.3			10.8		2592		
3 33.6	+58 16	Es 121		Cam	1902	326		6.9		8.0	13.5	A0	2593		
3 33.9	−31 05	B 52		For	1926 1960					6.8	7.1	F5		o	
3 34.2	+48 37	β 787	AB	Per	1881 1958	228 282		2.0	3.8	7.3	11.3	A0	2609		
--			AC		1898	176		12.7					2609		C does not exist (B)
--			AD		1898 1958	176 178		36.8	35.6		10.4		2609		
3 34.4	+24 28	Σ 412	AB	Tau	1830 1959					6.6	6.7	A2	2616	o	7 Tau
--			ABxC		1830 1958	63 56		22.4	22.4		10.0		2616		
3 34.6	−31 52	B 53		For	1926 1942	225		1.4		6.4	10.2	K0			
3 35.0	+32 01	Σ 410		Per	1831 1927	208		5.4		6.6	10.6	F0	2622		IX Per; variable

α 2000	δ 2000	Star Name		Const	Years		Pos Ang		Separation		m₁	m₂	Spec	ADS	Orb	Notes
3ʰ35ᵐ.0	+60°02'	Σ 400	AB	Cam	1829	1960	°	°	″	″	6.8	7.6	F5	2612	o	
--			ABxC		1879	1908	237	238	92.7	92.2		10.3		2612		
3 35.3	+43 51	A 1534		Per	1907	1932	273	275	4.2	3.5	8.0	14.5	K2	2621		
3 35.4	+33 41	Σ 413		Per	1831	1951	128		2.5		8.6	8.6	F0	2625		
3 35.4	+35 29	Pop 83		Per	1973		268		0.6		8.6	8.9	G5			
3 35.6	+31 41	β 533		Per	1878	1957	66	43	0.4	1.1	7.6	7.6	F0	2628		
3 35.6	+42 53	β pm	AB	Per	1909		191		53.9		7.4	12.7	F8			
--			AC		1907	1909	94		94.0			8.4	K0			
3 36.1	+63 17	Σ 402 rej		Cam	1902	1905	161		12.5		7.7	10.3	K0	2617		
3 36.2	+29 59	β 1040		Tau	1888	1928	337		3.6		7.8	11.5	A0	2633		
3 36.8	+0 35	Σ 422		Tau	1832	1975	232	265	6.1	6.6	5.9	8.8	G0	2644	o	
3 36.9	-49 57	See 29		Hor	1897	1955	257	259	10.2	14.6	7.5	11.8	G0			
3 37.2	+1 21	A 2419		Tau	1912	1951	98		0.7		8.6	8.7	A2	2647		
3 38.2	-59 47	Δ 14		Ret	1836	1916	271		57.5		7.1	8.9	F2 F5			
3 38.3	+44 48	S 430	AB	Per	1823	1925	95		41.1		8.0	8.0	A0 A0			
--		β	AC		1904		0		55.3							
--			Aa		1904		38		29.4							
3 38.8	-8 30	β 534		Eri	1879	1943	194		2.5		7.5	11.1	G5	2660		
3 39.1	+24 42	OΣ 60		Tau	1841						7.1		A0			Probably not double [19]
3 39.4	+16 32	h 3250		Tau	1904	1959	150	152	39.4	36.9	6.2	11.7	G5	2661		
3 39.8	-40 21	Δ 15		Eri	1836	1952	328		7.8		7.3	8.5	A2			
3 40.0	+63 52	OΣΣ 36	AB	Cam	1875	1923	69		46.1		6.8	8.6	F5	2650		Magnitude of B is for BC
--		Hu 1062	BC		1905	1953	234	170	0.2	0.1	9.1	9.6		2650		P.a. uncertain; round 1924
3 40.1	+34 07	Σ 425		Per	1830	1973	105	76	2.9	1.8	7.6	7.6	F5	2668		
3 40.4	+21 24	Hu 813		Tau	1902	1920	290		3.4		7.2	14.7	A5	2673		
3 40.5	+5 08	Σ 430	AB	Tau	1831	1935	56		26.3		6.8	9.8	G5	2681		
--			AC		1831	1935	301		37.1			10.5		2681		
3 40.6	+28 46	Σ 427		Tau	1831	1954	208		6.8		7.3	8.1	A0	2679		
3 40.7	+46 02	OΣ 59		Per	1850	1959	352		2.7		8.1	8.4	G5	2669		
3 40.8	+39 07	Σ 426	AB	Per	1829	1930	341		19.8		7.8	9.3	A3	2677		
--			BC		1904		37		12.6			14.3		2677		
3 40.9	-12 37	Σ 436		Eri	1832	1935	232	236	30.2	40.1	7.5	9.1	F2	2690		Optical
3 41.2	-11 48	Rst 3385		Eri	1937	1940	8		14.7		6.5	14.2	F2			
3 41.8	-48 14	Slr 5		Hor	1892	1943	188		1.7		8.0	9.8	F0			
3 41.9	+43 31	A 1707	AB	Per	1907	1917	342		8.3		7.9	14.4	G5	2688		
--			AC		1907	1917	145		68.3			10.1		2688		
--			CD		1907	1917	105		4.0			14.3		2688		
3 42.1	-17 09	Hu 436		Eri	1901	1959	288	298	1.3	1.4	7.9	9.4	F5	2706		
3 42.2	-15 58	A 2912		Eri	1918	1921	27		1.4		8.0	12.2	K2	2707		
3 42.4	+33 58	Σ 431		Per	1830	1925	238		20.0		5.0	9.5	B2	2699		40 Per
3 42.4	-80 01	h 3612		Men	1879	1918	162		19.4		7.9	9.2	F8			
3 42.5	+32 56	h 336		Per	1902	1913	317	318	36.3	38.4	6.7	10.1	F8	2701		
3 42.5	-85 16	R 38		Oct	1871	1945	246		2.2		6.7	8.2	B9			
3 42.7	+59 58	Webb	AB	Cam	1892	1913	96	95	18.7	21.4	5.9	13.7	K0	2691		
--			AC		1903	1915	300		34.8			12.9		2691		
--			AD		1863	1925	35		54.7			8.5		2691		
--			AE		1879	1909	160		167.7			10.7		2691		
3 42.8	+69 51	Σ 419	AB	Cam	1828	1955	74		3.0		7.9	7.9	A3	2678		Magnitude of B is for BC
--		A 984	BC		1894	1929	167	154	0.4	0.5	8.0	10.6		2678		
3 42.8	-37 19	B 1034		Eri	1927	1942	15		5.4		4.6	12.2	K2			
3 42.9	+47 47	β pm		Per	1879	1907	313	312	99.4	99.1	3.0	10.3	B5			39 δ Per
3 43.1	+25 41	Σ 435		Tau	1832	1934	2		13.0		7.3	8.8	F5	2708		
3 43.8	+42 36	β pm	AB	Per	1893	1911	278	282	41.7	48.4	7.5	10.5	G0			
--			BC		1907		184		55.4			12.7				
3 44.0	+38 22	Σ 434	AB	Per	1830	1925	88	85	28.3	31.0	7.8	8.6	K5	2717		
--			AC		1921		350		90.6					2717		
3 44.1	+48 31	β 1182	AB	Per	1890	1934	260		4.4		6.1	13.9	K0	2712		
--			AC		1890	1910	243	244	19.3	18.6		13.2		2712		
3 44.1	-40 40	h 3589		Eri	1837	1951	349		5.2		6.6	9.2	K0			
3 44.2	-64 48	LDS 104		Ret			f		(24'.7)		3.9	8.1	K0 G0			β Ret
3 44.3	+32 17	β 535		Per	1877	1958	60	37	1.0	1.0	3.8	8.3	B1	2726		38 o Per; variable
3 44.6	+27 54	OΣΣ 38	AB	Tau	1875	1920	38	43	122.6	126.7	6.7	7.0	F0 G0	2735		
--		β 1041	BC		1888	1923	348	331	7.9	15.2		13.5		2735		
3 44.6	-54 16	h 3592		Ret	1836	1951	15		5.2		6.4	9.0	K0			Straw-yellow, sky blue (R)
3 45.1	+40 05	Frk		Per	1918		133		49.9		7.6	9.9	G5			
3 45.2	+24 28	h 3251		Tau	1880	1925	330	329	66.4	68.9	4.4	8.1	B5			19 Tau
3 45.2	+42 35	Es		Per	1916		47		31.4		3.8	11.8	F5	2738		41 ν Per
3 45.3	-27 52	β 1003		For	1881	1959	20	78	2.6	1.8	8.2	11.8	K0	2756	o	
3 45.4	+49 51	Hu 103	AB	Per	1900	1954	205		1.0		8.6	8.9	A0	2736		
--			AC		1916		296		28.6			11.4		2736		
3 45.6	+24 20	Cou 560		Tau	1970		4		0.3		7.4	9.1	B9			
3 45.7	+6 03	h 2204		Tau	1904	1927	65		64.7		5.4	11.6	B3	2750		29 Tau
3 46.0	+45 41	β 1183		Per	1890	1934	138		6.5		5.6	14.0	B9	2746		
3 46.2	+71 37	Σ 421		Cam	1829	1911	234		12.4		7.1	11.1	G5	2718		
3 46.3	+23 57	HdO		Tau							4.2		B5			23 Tau; probably single

α 2000	δ 2000	Star Name		Const	Years		Pos Ang		Separation		m₁	m₂	Spec	ADS	Orb	Notes
3ʰ47ᵐ.4	+23°55′	Σ 450		Tau	1832	1952	265°		6″.1		7.3	9.3	B9	2767		
3 47.5	+24 06	Σ I 8	AB	Tau	1836	1920	289		117.2		2.9	8.0	B5 A0			25 η Tau
--			AC		1868	1903	312		180.8			8.0	A0			
--			AD		1868	1903	295		190.5			8.6	G0			
			BC		1824		344		85.6							
--			BD		1824		304		74.7							
3 47.9	+33 36	Σ 448		Per	1831	1934	16		3.2		6.7	9.2	B3	2772		BDS 1870, h 5457
3 48.0	+38 21	Σ 447		Per	1830	1925	178	165	26.5	28.9	8.2	9.4	F2			
3 48.3	+11 09	Σ 452		Tau	1830	1933	59		9.0		5.1	10.2	B3	2778		30 Tau
3 48.3	+50 44	OΣ 63		Per	1848	1940	270		7.0		6.1	11.3	B8	2769		
3 48.5	-31 47	h 3596		For	1836	1953	137		9.2		8.2	8.5	A3			
3 48.6	-37 37	Δ 16		Eri	1836	1957	200	212	7.3	7.9	4.8	5.3	B8 A0			
3 49.2	+24 03	Σ 453		Tau	1830	1929			0.4		3.7	6.7	B8	2786		27 Tau, Atlas; single? [20]
3 49.3	-1 27	Rst 4760		Eri	1942	1952	288		0.7		6.9	9.4	G5			
3 49.3	+57 07	S 436		Cam	1823	1975	75		58.3		6.5	7.3	B9 A0			
3 49.4	+24 23	OΣΣ 40		Tau	1875	1923	308		87.1		6.6	8.1	B9 A0			
3 49.5	+52 39	Σ 446	AB	Per	1830	1934	253		8.6		6.9	9.1	B0	2783		
--		Es 12	AC		1892	1910	39		12.1			12.0		2783		
--			AD		1909		336		66.5			10.3		2783		
--			DE		1909		232		2.6			10.7		2783		
3 49.8	+9 24	h 666		Tau	1900	1920	17		30.7		6.8	13.0	B9	2796		
3 50.0	+23 51	OΣ 64	AB	Tau	1847	1943	238		3.2		6.9	9.8	B9	2795		
--			AC		1847	1937	236		10.2			9.0		2795		
--		Hu	BC		1898		55		7.0					2795		
3 50.2	+29 39	Fox		Tau	1923		297		28.2		7.9	11.6	A0			
3 50.2	+34 49	Es 277	AB	Per	1906		142		30.2		6.8	9.8	F0	2794		
--			BC		1906		290		7.4			13.8		2794		
3 50.3	-1 31	β 401	AB	Eri	1877	1953	254		4.5		6.5	10.5	F2	2803		
--			AC		1913		289		40.6			11.2		2803		
3 50.3	+25 35	OΣ 65		Tau	1846	1960					5.8	6.2	A3	2799	o	
3 50.4	+71 20	h 2200	AB	Cam	1904	1909	240		56.2		4.6	12.3	A0			γ Cam
--			AC		1909		84		106.3			8.5				
3 51.7	+70 30	h 1139		Cam	1904	1918	176		47.3		7.4	9.7	A3			
3 51.7	-22 56	h 3601		Eri	1897	1932	300		10.6		7.7	9.7	G5	2825		
3 51.9	+34 22	Kui 14		Per	1958		31		15.1		5.8	13.3	B3			
3 52.0	+6 32	Kui 15		Tau	1937	1960	215	207	0.3	0.4	6.4	6.5	B9			31 Tau = Kui III
3 52.1	-1 09	β pm		Eri	1886	1919	154		95.5		6.8	9.7	B9			
3 52.1	+40 48	OΣ 66		Per	1846	1954	136	141	0.5	0.9	8.0	8.5	A2	2815		
3 52.7	-5 22	h 338		Eri	1863	1952	135		8.2		5.5	10.6	B8	2832		30 Eri
3 52.8	+1 45	A 2347		Tau	1911	1929	259		4.0		7.9	12.1	K0	2831		
3 52.9	+14 23	Wirtz		Tau	1910	1934	103		8.6		8.0	11.5	G5	2829		
3 53.7	+53 17	A 1293		Cam	1906	1959					8.5	8.9	G0	2828	o	Too close after 1954
3 54.1	+31 53	Σ 464	AB	Per	1830	1968	208	208	12.6	12.9	2.9	9.5	B1	2843		44 ζ Per
--			AC		1880	1923	286		32.8			11.3		2843		
--			AD		1880	1957	198	195	89.1	94.2		9.5		2843		
--			AE		1881	1925	185		120.3			10.2		2843		
--		S1v	BD		1967		194		82.5					2843		
3 54.3	-2 57	Σ 470	AB	Eri	1833	1955	347		6.8		4.8	6.1	G5 A2	2850		32 Eri
--			AC		1913	1921	5		165.8			11.4		2850		
3 54.4	-40 21	φ 344	AB	Eri	1960		119		0.1		6.5	6.5	F5			Spectrum composite F5+A3
--		HdO 188	ABxC		1901	1928	171		23.2			12.9				
3 54.5	+5 10	OΣ 41	AxBC	Tau	1875	1927	357		59.0		7.7		F0	2849		
--		A 1831	BC		1908	1958	35	142	0.2	0.2	9.8	9.8		2849		Sep. <0″.1 in 1954 (82-in)
3 54.6	-52 41	HdO 189		Dor	1897		268		22.8		6.5	12.6	A2			
3 55.4	+17 38	β 85		Tau	1875	1930	216		4.1		8.0	10.2	G5	2864		
3 55.4	+31 03	Ho 325		Per	1891	1934	12		22.5		6.1	12.0	B0	2859		X Per; A is variable
3 56.1	-47 46	Hu 1361		Hor	1912	1945	83	77	4.0	4.0	7.6	11.6	F8			
3 56.2	+59 39	β pm	AB	Cam	1893	1911	355	357	153.4	150.2	6.7	9.9	G0			
--			AC		1893	1911	39	40	187.2	188.1		10.0				
--			Aa		1907		216		6.8			12.9				
--			Ab		1907		67		33.4			12.4				
3 56.6	+50 42	S 440	AB	Per	1825	1908	30		75.3		5.3	10.1	F5			43 Per
--			AC		1908	1925	133		101.0			11.2				
--			Aa		1901		279					12.3				
3 56.6	-39 55	h 3611		Hor	1879	1945	140		4.1		8.2	8.8	A3			
3 56.7	-1 34	See		Eri	1900		193		0.3		7.9	11.9	A2	2886		
3 57.1	+61 07	OΣ 67		Cam	1847	1959	44		1.9		5.3	8.5	K0	2867		Spectrum composite K0+A0
3 57.2	+69 30	Σ 455	AB	Cam	1827	1922	166		11.9		8.5	9.0	G5	2851		
--			AC		1910		332		58.3			14.0		2851		
--			AD		1909		211		96.3			12.4		2851		
3 57.3	+41 53	Σ 469		Per	1828	1958	148		9.1		6.8	10.3	A2	2884		
3 57.8	+22 55	Cou 364		Tau	1968		150		2.0		8.0	9.7	K0			
3 57.9	+40 01	Σ 471	AB	Per	1832	1938	10		8.8		2.9	8.1	B1	2888		45 ε Per
--			AC		1912		11		78.3			13.8		2888		
3 58.0	-13 31	h 3608		Eri	1877	1909	242		52.8		3.0	12.5	K5	2904		34 γ Eri

α 2000	δ 2000	Star Name		Const	Years	Pos Ang		Separation		m₁	m₂	Spec	ADS	Orb	Notes
$3^h 58^m.6$	$-2° 39'$	β 1042	AD	Eri	1899 1910	250°	°	39″.2	″	7.1	10.8	G5	2909		
--			AxBC		1888 1913	93		55.6					2909		
--			BC		1888 1943	34		1.2		8.7	9.5		2909		
3 58.9	+51 30	Es 878		Per	1910	224		12.3		7.7	9.9	K2	2896		
3 59.1	+9 48	Hu 27		Tau	1899 1959	211 261		0.6	0.3	8.6	8.8	G0	2911		
3 59.7	+38 49	OΣ 69		Per	1849 1950	328 320		1.6	1.5	6.5	9.2	A0	2910		
3 59.7	+48 10	OΣ 68 rej	AB	Per	1867 1925	175		39.3		8.1	9.4	B9			
--			BC		1912	184		31.6			15				
4 00.8	+18 12	β pm		Tau	1909 1921	277 278		161.3 164.2		5.9	9.2	F0			
4 00.9	+23 12	Σ 479	AB	Tau	1831 1954	128		7.3		7.0	7.9	B9	2926		H N 93
--			AC		1831 1936	241		58.0			9.1		2926		
4 01.0	-54 24	Δ 17	AB	Ret	1916	142		62.9		7.7	8.2	A2 A2			
--		I 269	BC		1900 1933	79		3.4			10.8				
--			BD		1900	196		27.7			12.8				
4 01.6	+38 40	Σ 476	AB	Per	1831 1950	284 287		17.6	21.9	7.9	9.1	K2	2932		
--			BC		1910	213		69.5			12.1		2932		
4 01.8	+10 00	OΣ 70		Tau	1848 1914	227		12.0		5.7	11.7	B8	2938		
4 01.9	+62 31	Bird 1	AB	Cam	1897 1954	224		2.0		8.3	9.3	B8	2924		
--			AC		1897 1954	173		8.8			9.6		2924		
4 02.1	+41 51	Σ 477		Per	1830 1927	213		3.0		8.4	9.4	A2	2934		
4 02.1	-34 29	β 1004	AB	Eri	1881 1959	154 106		1.8	1.5	7.2	7.8	G0			
--			AC		1881 1955	131 149		63.0	47.9		11.5				
4 02.2	+28 08	Σ 481	AB	Tau	1832 1958	106		2.4		7.7	11.3	G5	2944		
--			AC		1832 1958	329 326		18.8	15.5		9.8		2944		
--			AD		1906						9.0		2944		Dec. difference 11″.18
4 02.4	-28 31	δ 79		Eri	1920 1959	132 88		0.5	0.4	8.1	8.5	F8	2952		
4 02.5	-61 21	LDS 109		Ret		sp		29.0		6.7	11.9	G5			
4 02.9	-59 35	Pol 2		Ret	1896 1942	107		1.8		8.6	8.9	F2			
4 03.5	+42 11	A 1709	AB	Per	1907 1960	223		0.9		8.0	9.9	A2	2950		
--			AC		1907 1917	339		53.2			10.1		2950		
--			CD		1907 1934	206		3.5			13.8		2950		
4 03.8	+37 58	Es 2085		Per	1924 1951	266		4.4		7.8	9.9	A0	2956		
4 04.1	+39 31	Σ 483		Per	1830 1956					7.5	8.6	G5	2959	o	
4 04.4	+24 06	β 544		Tau	1877 1927	257		25.5		5.5	12.0	F5	2965		36 Tau [21]
4 04.6	+55 04	Ho 221	AB	Cam	1888 1921	94		4.8		6.6	11.3	F5	2957		
--			AC		1913 1921	277		136.2			10.4		2957		
4 04.7	+22 05	β pm		Tau	1853 1921	187 190		138.7 137.2		4.4	9.4	K0			37 Tau
4 04.9	-35 27	I 152		Eri	1900 1954	67		0.7		8.3	8.6	G0			
4 05.0	+37 05	Ku		Per	1902	278		49.2		8.3	9.7	A0			
4 05.3	+22 01	β pm	AB	Tau	1887 1918	4 2		168.5 170.1		6.1	8.8	G5			39 Tau
--			BC		1887 1907	128 128		61.6 61.1			11.5				
4 06.5	+14 21	S 443	AB	Tau	1904 1908	110		42.2		8.5	9.4	G			
--			AC		1904 1908	305		184.7			7.8	F2			
4 06.5	+43 27	A 1710		Per	1907 1959					8.2	8.2	G5	2980	o	Quadrant change in 1950?
4 06.7	-80 44	Rst 2341		Men	1935	346		0.3		8.7	8.7	F8			
4 06.9	+33 27	OΣ 71	AB	Per	1846 1957	206 220		1.0	0.8	6.9	8.9	B3	2990		AG Per; variable
--			AC		1878 1948	119 118		34.7	36.0		12.7		2990		
4 07.0	-22 00	Hu 1363		Eri	1919 1959	133 121		0.3	0.3	7.1	7.5	A3	3000		
4 07.4	+38 01	β 545	AB	Per	1878 1953	307		1.0		8.0	11.5	K0	2992		
--			AC		1851 1918	25 30		239.8 233.1			7.1	G5	2992		
4 07.4	+62 23	Σ 484	AB	Cam	1830 1925	132		5.4		10.0	10.5		2984		
--			AC		1830 1908	334		22.6			10.0		2984		
4 07.6	+38 04	OΣ 531	AB	Per	1852 1960					7.4	8.9	G5	2995	o	
--			AC		1904	209		234.2					2995		
4 07.7	+15 10	Σ 495		Tau	1830 1942	221		3.8		6.0	8.8	F0	2999		
4 07.9	+62 20	Σ 485	AB	Cam	1830 1967	304		17.9		7.0	7.1	B0	2984		B is SZ Cam
--		Hzg	AC		1907	278		69.7			9.8		2984		In open cluster NGC 1502
--		Σ 485	AD		1881 1910	76		139.1			10.4		2984		
--		H1m	AF		1902	129		14.0			12.4		2984		
--		Es	Aa		1902 1925	257		6.0			12.9		2984		
--		Σ 485	Ab		1902 1910	359		11.6			13.6		2984		
--			Ac		1910	131		14.4			14.1		2984		
--		Hzg	Ad		1907 1957	259		60.2					2984		
--		Σ 485	BE		1867 1957	333		19.6			11.9		2984		
4 08.0	+17 20	OΣ 72		Tau	1854 1938	326		4.4		6.0	9.1	K0	3006		
4 08.0	+43 11	A.G.		Per	1907	74		133.8		7.1	6.7	K0 B8			
4 08.0	+62 20	H1m 3		Cam	1902 1925	215		5.6		10.4	11.4		2989		
4 08.1	+34 07	Cou 1082		Per	1973	64		0.3		8.0	8.7	A0			
4 08.3	+19 44	β 309		Tau	1875 1927	278		5.7		7.9	11.2	A2	3010		
4 08.3	+41 29	Σ 492 rej	AB	Per	1900 1921	202 200		94.4	97.0	7.6	11.0	A2	3001		
--			BC		1896 1921	137		5.3			11.0		3001		
4 08.3	-32 51	I 153		Eri	1897 1955	338 343		0.9	0.9	8.1	8.2	A3			
4 08.6	-45 52	Rst 98	AB	Hor	1930 1949	312		0.9		6.6	10.6	F0			
--			AC		1929	305		17.6			14.7				
4 08.9	+23 06	Σ 494		Tau	1830 1955	187		5.2		7.6	7.6	A3	3019		

α 2000	δ 2000	Star Name		Const	Years	Pos Ang	Separation	m₁	m₂	Spec	ADS	Orb	Notes
4ʰ08.9	+46°14'	A 998		Per	1905 1955	333° 300°	0".3 0".3	7.7	8.0	A0	3007		
4 09.2	+40 10	Σ 3114		Per	1832 1948	190 161	1.9 2.8	7.6	10.1	F8	3015		
4 09.4	-7 56	A 469		Eri	1903 1958	343 37	0.2 0.2	8.1	8.1	A2	3032		
4 09.5	-17 29	β pm	AB	Eri	1894 1908	139 139	126.3 125.2	8.2	8.7	K0 F0			
--			Aa		1908	39	115.6		13.8				
4 09.6	+31 39	Ho 327	AB	Per	1892 1927	322 310	16.3 15.0	6.9	12.6	F5	3029		
--			AC		1900 1927	174 160	12.8 14.8		14.1		3029		
4 10.0	+80 42	Σ 460		Cep	1828 1959			5.5	6.3	F8	2963	o	Spectrum composite F8+A2
4 10.0	-25 03	Rst 2334		Eri	1933 1943	65	4.2	7.5	14.6	K0			
4 10.7	-4 52	A 2801		Eri	1914 1960			8.1	8.1	G0	3041	o	
4 11.3	+5 31	β pm		Tau	1888 1921	43 42	128.5 124.1	5.7	9.8	F0			45 Tau
4 11.4	+41 52	β 546		Per	1878 1960	24 41	0.9 0.9	8.8	8.8	F8	3038		
4 11.7	+31 33	Cou 880		Per	1972	46	0.6	7.9	9.5	A0			
4 11.8	-59 10	Jsp 54		Ret	1930	291	3.9	8.0	13.5	F5			
4 11.9	-8 50	h 2224		Eri	1906 1910	311	51.5	6.8	8.8	G5			
4 12.2	+0 44	Σ 510		Tau	1831 1925	300	11.0	6.9	9.9	G5	3054		
4 12.3	+9 39	OΣ 74		Tau	1849 1955	270 280	0.5 0.5	8.3	8.8	F0	3053		
4 12.4	+23 34	Cou 703		Tau	1971	89	0.9	7.4	9.1	G5			
4 12.5	-36 09	h 3628		Eri	1838 1933	50	50.3	7.1	8.0	F5 F5			
4 13.0	-28 32	Howe 10		Eri	1876 1955	34 44	2.4 2.0	8.2	8.3	F8			
4 13.1	-23 08	Hu 30		Eri	1900 1928	175	5.1	6.8	13.5	F0	3067		
4 13.5	+8 53	β 1278	AB	Tau	1898 1938	303 302	7.4 8.4	6.5	13.7	A3	3063		
--			AC		1898 1959	252 255	55.3 54.3		12.5		3063		
4 13.6	+7 43	A 1938		Tau	1908 1960			6.0	6.1	F0	3064	o	46 Tau
4 13.6	-25 32	See 34		Eri	1897 1945	54 60	19.3 18.1	8.0	13.0	K0	3075		
4 13.9	+9 16	β 547	AB	Tau	1877 1958	359 351	0.9 1.1	4.9	7.4	G5	3072		47 Tau
--			AC		1877 1959	223 226	32.2 29.8		11.8		3072		
4 13.9	+42 35	A 1711		Per	1908 1960	135 106	0.5 0.6	8.2	10.0	F0	3062		
4 14.2	-46 08	Rst 2338		Hor	1934 1960			7.3	7.7			o	
4 14.4	-10 15	Σ 516	AB	Eri	1836 1967	154 146	6.3 6.4	5.0	8.0	K0	3079		39 Eri
--			AC		1911	159	148.9		9.5		3079		
4 14.4	-62 28	h 3638		Ret	1917	354	48.5	3.4	12.0	G5			α Ret
4 14.8	-62 12	h 3641		Ret	1836 1955	288 230	6.9 10.2	5.5	11.1	K0			
4 14.9	+48 25	OΣ 73	AB	Per	1851 1934	349	14.8	4.1	11.6	G0	3071		51 μ Per
--		H VI 20	AC		1822 1925	232 232	91.6 83.8				3071		S,h 364
--		OΣ 73	Ab		1901	124	50.4		12.4		3071		
4 15.1	-30 04	h 3632		Eri	1836 1937	163	11.0	7.7	9.0	A2			
4 15.2	-7 39	Σ 518	AB	Eri	1836 1970	107 104	83.5 83.4	4.4	9.5	G5 A	3093		40 o² Eri
--			Aa		1850 1907	197 51	128.3 114.2		12.2		3093		
--			Ab		1850 1907	279 8	99.4 211.0		12.5		3093		
--			BC		1783 1975				11.2		3093	o	
--			Ba		1922	196	147.0			A	3093		
--			Bb		1922	356	279.5				3093		
4 15.5	+6 11	H VI 98	AB	Tau	1783 1937	315	65.5	6.3	7.0	G0 G0	3085		
--			AC		1885 1907	47	214.5		10.0	M2	3085		
--			CD		1907	139	52.7		10.0		3085		
4 15.8	+15 24	β pm	AB	Tau	1909 1959	32 29	138.9 137.5	6.3	11.8	F5			48 Tau
--			AC		1909 1959	309 308	146.8 151.4		11.0				
4 15.8	+45 24	Σ 512		Per	1830 1950	224	5.2	8.8	8.8	G5	3080		
4 15.9	+31 42	OΣ 77	AB	Per	1846 1958			8.0	8.1	F8	3082	o	
--		OΣΣ 43	ABxC		1847 1970	42	56.3		8.5	F8	3082		
--		OΣ 77	ABxD		1873 1924	313 314	127.1 129.2		8.0		3082		
4 16.0	+0 27	Σ 517		Tau	1830 1934	12	3.4	7.4	9.1	A0	3095		
4 16.0	+30 02	OΣ 78		Tau	1847 1942	247	2.5	7.7	9.7	G0	3089		
4 16.2	+34 52	OΣ 76		Per	1849 1903	210	3.9	7.7	12.4	A3	3088		
4 16.4	+70 31	Σ 496 rej	AB	Cam	1906	48	31.7	8.8	8.7				
--			AC		1830	42	35.0		11.				
--			BC		1830	330	18.0						
4 16.4	-60 57	Gale 1		Ret	1897 1959	342 42	1.1 0.4	6.9	7.4	A0			
4 16.5	-59 18	Jsp 56		Ret	1930	34	13.7	4.4	12.5	K2			ε Ret
4 16.6	+2 48	A 2618		Tau	1913 1929	29	1.7	8.1	11.4	A5	3101		
4 16.6	-10 05	β 548		Eri	1877 1939	345	6.2	7.5	12.0	A2	3103		
4 17.0	+19 41	Ho 328		Tau	1890 1960			7.7	8.1	F5	3102	o	
4 17.1	+64 10	Σ 503		Cam	1830 1974	226	4.4	8.8	8.8		3081		
4 17.2	+53 22	h 2225		Cam	1903 1926	231	30.2	8.1	10.3	K0			
4 17.3	+46 13	OΣΣ 44		Per	1875 1924	322	58.4	7.2	8.6	A2 B			
4 17.5	+62 35	Σ 505		Cam	1830 1906	117	9.3	8.1	10.8	G0	3087		
4 17.7	-63 15	Rmk 3		Ret	1835 1943	6 4	5.8 4.1	6.2	8.2	B9			
4 17.9	+58 47	Σ 511		Cam	1829 1960			7.5	7.9	A0	3098	o	
4 17.9	+67 05	β 1233		Cam	1891 1926	38	5.1	6.9	12.1	B8	3086		
4 17.9	-33 48	I 270	AB	Eri	1900 1954		0.1	3.6		B9			41 Eri [22]
--		h 3636	ABxC		1919	13	49.2		11.8				
4 18.2	+61 55	Σ 509	AB	Cam	1830 1904	20	11.8	7.7	11.2	B9	3097		
--			AC		1830 1919	248	38.0		10.0		3097		
4 18.3	+22 49	Σ 520		Tau	1837 1960	99 154	1.0 0.3	8.2	8.3	F5	3114	o	

α 2000	δ 2000	Star Name		Const	Years		Pos Ang		Separation		m₁	m₂	Spec	ADS	Orb	Notes
4ʰ18.ᵐ4	+21°35′	β pm		Tau	1909	1918	62°	°	166″.4	″	5.7	10.8	A5			51 Tau [23]
4 18.6	+60 30	OΣ 75		Cam	1843	1960	166	182	0.4	0.4	8.0	8.4	B9	3105		
4 18.7	+16 32	OΣ 79		Tau	1846	1959					7.2	8.2	G0	3135	o	
4 18.7	−52 52	φ 87		Dor	1929	1952	154	133	0.6	0.6	6.1	9.1	F5			
4 18.8	−34 07	I 724		Eri	1927	1935	206		0.5		8.3	9.3	G0			
4 18.9	+1 45	OΣΣ 49		Tau	1875	1934	145		103.1		8.2	8.2	G5 A2			
4 19.0	−33 54	h 3642		Eri	1837	1948	159		6.0		6.4	8.4	A2			
4 19.2	+61 35	Σ 513	AB	Cam	1830	1958	57		5.5		8.0	9.7	A3	3109		Magnitude of B is for BC
--		β 1333	BC		1904	1958	93	78	0.2	0.3	9.9	11.4		3109		
4 19.3	−44 16	h 3643		Hor	1879	1913	114		70.4		6.7	9.1	K0 G0			
4 19.7	+39 55	A 1833		Per	1908	1917	94		3.8		8.0	13.8	F0	3122		
4 19.8	+23 44	Σ 523	AB	Tau	1829	1914	163		10.3		7.7	9.7	A0	3131		
--			AC		1893	1915	48		108.8					3131		
4 20.0	+14 02	β pm		Tau	1909		359		34.6		5.6	13.5	F0			57 Tau
4 20.2	−16 26	Hu 438		Eri	1901	1923	162		4.3		6.6	14.3	B9	3140		
4 20.4	+27 21	S,h 40		Tau	1821	1925	240	250	56.8	52.1	5.0	8.4	K0	3137		52 φ Tau
4 20.4	+34 34	β		Per	1913		313		110.0		4.9	12.8	G5			54 Per
4 20.5	−1 19	Rst 4769		Eri	1942	1952	18		0.4		7.5	7.5	A3			
4 20.8	−19 20	See 36		Eri	1897	1933	346		8.0		7.1	12.3	G0	3145		
4 20.9	+50 23	Σ 519 rej		Per	1892	1926	346		18.3		7.9	9.4	K2	3136		
4 20.9	−51 03	Rst 1271		Dor	1932	1937	185		0.2		7.9	8.9	G5			
4 21.0	+50 15	S 445	AB	Per	1823	1919	327	327	75.2	73.6	7.3	8.2	K2 F8			
--			AC		1824	1918	261		149.6			10.8				
--		Es	Aa		1913		326		26.1			14.8				
--			Bb		1892	1913	257	258	21.7	22.8		11.3				
4 21.1	+55 32	OΣΣ 46	AB	Cam	1875	1924	160		99.0		8.0	8.0	A3 A0			
--			Aa		1901	1907	189		33.5			13.0				
4 21.5	−0 06	Ho 329	AB	Eri	1891	1909	66	64	33.0	35.8	5.9	12.9	K2	3152		
--			AC		1909		279		197.8			10.9		3152		
4 21.5	−25 44	β 744	AB	Eri	1891	1959					6.6	6.9	F0	3159	o	
--		h 3644	ABxC		1891	1951	7	3	35.4	38.6		12.3		3159		
--			ABxD		1891	1951	41		44.5			8.9	G0	3159		
4 21.8	+50 02	Σ 521		Per	1830	1933	257		2.1		7.5	9.6	G0	3141		
4 21.8	−42 47	I 271		Cae	1900	1948	145		2.3		7.7	10.0	G5			
4 21.9	−41 13	h 3646	AB	Cae	1900	1926	137		38.0		7.9	10.1	F5			
--		I 272	BC		1900	1926	191	203	1.3	1.3		12.0				
4 22.0	+39 56	β 310		Per	1878	1927	172		19.3		7.3	11.3	F8	3155		
4 22.1	+14 05	β pm	AB	Tau	1909		233		79.3		5.7	13.1	A3			60 Tau
--			AC		1909		333		107.0			12.6				
4 22.2	−4 41	Σ 536		Eri	1832	1959	152	177	1.8	1.6	8.0	8.6	A5	3163		
4 22.4	+20 49	β 87		Tau	1875	1946	170		1.9		6.0	9.1	K5	3158		Spectrum composite K5+A0
4 22.4	+51 37	Σ 522		Per	1831	1958	38	32	1.5	1.5	8.8	8.8	A0	3147		
4 22.6	+25 38	Σ 528		Tau	1830	1931	24		19.4		5.5	7.6	B9	3161		59 χ Tau
4 22.7	+15 03	OΣ 82	AB	Tau	1848	1959					7.3	8.5	G0	3169	o	
--		LDS 1166	AC		1962		285		62.0			17		3169		
4 22.9	+17 33	β pm		Tau	1909		341		106.6		3.8	12.5	K0			61 δ¹ Tau
4 22.9	−82 54	h 3692		Men	1919		182		47.7		6.8	12.5	A2			
4 23.0	+59 37	Kui 16	AB	Cam	1937	1949	212		1.1		6.2	11.2	A0	3146		
--		Arg	ABxC		1904	1925	59		32.3			8.8		3146		
4 23.1	−35 33	I 384		Eri	1900		187		18.7		6.4	13.4	G5			
4 23.2	−5 00	h 342	AB	Eri	1783	1916	235		17.4		8.1	9.1	K2	3176		
--			AC		1877	1904	83		28.6			11.9		3176		
4 23.3	−7 42	Ho 331		Eri	1890	1911	350		15.8		7.5	13.2	K0	3178		
4 23.3	+11 23	Σ 535		Tau	1831	1959	354	304	2.0	1.2	7.1	8.6	A2	3174		
4 23.6	+42 26	OΣ 80		Per	1848	1958	188	166	0.5	0.5	6.5	7.0	B9	3172		
4 23.9	+9 28	Hu 304		Tau	1901	1960					5.8	5.9	A2	3182	o	66 Tau
4 23.9	−66 44	h 3654		Ret	1917		113		18.3		7.1	12.7	A3			
4 24.0	+24 18	Σ 534	AB	Tau	1831	1968	290		28.9		6.2	8.6	B8	3179		62 Tau
--			BC		1911		336		110.5			12.0		3179		
4 24.1	+17 27	β pm		Tau	1909		246		137.3		4.8	13.3	A5			64 δ² Tau
4 24.2	−40 03	I 1146		Cae	1926	1939	42		6.3		7.3	11.7	B9			
4 24.2	−57 04	Rmk 4		Dor	1836	1955	230	242	6.6	5.9	6.9	7.3	G0			
4 24.4	+34 19	Σ 533	AB	Per	1831	1932	61		19.6		7.2	8.7	B9	3185		
--			AC		1880	1908	190		110.3					3185		
4 24.5	+22 45	β 1235		Tau	1891	1956	61	55	0.4	0.3	8.7	8.8	F2	3191		
4 24.6	+33 58	OΣ 81		Per	1847	1976	53	22	4.5	4.2	5.9	8.7	F5	3188		56 Per
4 24.8	−50 37	Rst 1273		Dor	1932	1937	254		4.3		7.9	13.8	G5			
4 24.9	−34 45	I 59	AB	Eri	1897	1928	198		42.4		6.6	10.0	F5			
--			BC		1897	1928	281		3.7			11.0				
4 24.9	−77 41	h 3673		Men	1879	1939	64	67	10.2	10.2	8.3	8.5	F8			
4 25.2	−64 05	h 3655		Ret	1917		162		48.7		7.9	12.1	G0			
4 25.4	+22 18	Σ I 9	AB	Tau	1836	1923	173		339.5		4.4	5.4	A3 F0			
--			Aa		1909		266		135.5			11.9				
--			Bb		1909		209		106.9			12.2				
4 25.4	+56 23	A 834	AB	Cam	1904	1953	220		0.4		8.5	9.1	F0	3184		

α 2000	δ 2000	Star Name		Const	Years	Pos Ang	Separation	m₁	m₂	Spec	ADS	Orb	Notes
4ʰ 25ᵐ.4	+56°23′	A 834	AC	Cam	1914 1953	83°	25″.3		9.4		3184		
4 25.5	+17 56	Kui 17	AB	Tau	1938 1958	323 333	1.3 1.4	4.2	7.5	A2	3206		68 δ³ Tau
--		H VI 101	AC		1783 1925	235 233	63.6 77.2		8.7		3206		[24]
4 25.5	-53 07	Slr 6		Dor	1892 1942	99	1.0	7.1	8.9	F0			
4 25.6	+15 56	β pm	AB	Tau	1910	343	124.4	6.3	12.4	F8			70 Tau
--		φ 342	Aa		1959 1960	148 136	0.1 0.1	7.1	7.1			o	
4 25.6	+18 52	β 1185		Tau	1890 1960			8.2	8.8	G0	3210	o	
4 25.7	-2 14	β 403		Eri	1877 1959	101 92	2.0 1.5	8.0	9.4	F8	3214		
4 25.8	-1 23	Σ 547	AB	Eri	1831 1958	344 105	4.2 4.1	9.6	12.6	K0	3217		AB is optical
--			AC		1906	258	211.2		7.0	G5	3217		
--			CD		1906	233	26.1		12.4		3217		
4 26.1	+42 39	Es 1521		Per	1916	293	6.3	8.0	13.0	A0	3205		
4 26.2	+34 43	Hu 609		Per	1902 1960			8.6	9.0	F5	3211	o	
4 26.3	+22 49	β pm		Tau	1909	289	104.3	4.3	12.4	A5			69 υ Tau
4 26.6	-40 32	h 3650		Cae	1836 1937	183	3.0	7.1	8.3	A0			
4 26.8	+55 39	Σ 531		Cam	1830 1959	292 312	0.8 1.0	7.6	8.8	F0	3207		
4 26.9	-24 05	β 311		Eri	1877 1959			6.8	7.0	A2	3230	o	
4 27.0	+19 08	Σ 546		Tau	1836 1954	190 184	6.6 6.8	7.9	9.7	G0	3224		
4 27.0	+57 35	Opik		Cam	1926 1958	171 176	22.1 20.6	6.2	13.2	A0	3203		
4 27.0	-72 38	B 2091		Hyi	1933	140	1.3	7.9	10.2	G5			
4 27.1	+18 12	Σ 545		Tau	1830 1924	58	18.9	6.9	8.7	A0	3226		
4 27.5	+11 13	β 1186		Tau	1890 1952	182 166	0.6 0.5	5.9	8.8	B8	3228		
4 27.7	+43 11	Es 568		Per	1908	304	5.1	7.5	11.8	G5	3225		
4 27.8	-62 31	B 688		Ret	1927	125	10.2	5.8	12.5	K0			
4 27.9	-21 30	β 184		Eri	1877 1957	263 252	1.1 1.5	7.3	7.7	F5	3247		
4 28.0	+10 04	A 2033		Tau	1909 1919	251	0.7	7.5	10.3	A0	3238		
4 28.0	+21 37	β pm		Tau	1909	233	161.4	5.7	10.5	A5			
4 28.6	+19 11	β pm		Tau	1901	268	181.6	3.5	10.5	K0			74 ε Tau
4 28.6	-29 45	B 68		Eri	1925 1960		0.1	8.6	8.7	F0	3253		
4 28.7	+15 52	Σ I 10		Tau	1836 1921	346	337.4	3.4	3.8	F0 K0			θ²,θ¹ Tau [25]
4 28.7	+20 41	Cou 264		Tau	1968	69	3.3	7.8	14.0	B8			
4 28.7	-11 58	β 549		Eri	1877 1914	189	8.4	7.9	12.4	G5	3252		
4 28.9	+30 22	Σ 548	AB	Tau	1831 1932	35	14.4	6.5	8.5	F5	3243		
--			AC		1880 1908	194	121.4		10.6		3243		
--			Cc		1908	52	16.6		13.3		3243		
4 28.9	-25 12	B 69	AB	Eri	1926 1932	354	2.4	7.8	12.8	G5	3257		
--		Stone 8	AC		1875 1932	351	7.0		9.3		3257		
4 28.9	-49 36	h 3658		Dor	1895 1935	121	5.7	8.3	8.9	G0			
4 29.0	+16 10	Hu 1080		Tau	1904 1960			7.0	7.7	F8	3248	o	
4 29.0	+71 40	A 1004		Cam	1905 1929	203	4.6	8.0	13.7	F0	3215		
4 29.5	+17 52	Bgh		Tau		9	109.1	7.1	8.7	F8 G8			
4 29.8	+17 41	Cou 567		Tau	1971	27	0.2	8.5	8.5	B9			
4 30.1	+15 38	Σ 554		Tau	1831 1959			5.7	8.0	F0	3264	o	80 Tau
4 30.6	+13 43	β pm		Tau	1910	105	111.9	5.4	11.2	F0			83 Tau
4 30.6	+15 42	β pm		Tau	1910 1919	339	161.8	5.5	9.4	A5 K0			81 Tau
4 30.6	+16 12	LDS 2246		Tau	1955	131	250.0	4.8	6.7				
4 30.6	+32 27	OΣ 83		Per	1842			6.2		B9			Probably not double (IDS)
4 30.7	-35 39	h 3659		Eri	1919	36	36.6	6.0	14.3	K0			
4 31.1	+6 47	OΣ 84		Tau	1847 1932	254	9.5	7.3	8.2	G5	3279		
4 31.3	-35 46	I 154		Eri	1900 1951	133	0.4	8.1	8.2	K0			
4 31.4	+40 01	Σ 552		Per	1831 1949	114	9.0	7.0	7.2	B8	3273		
4 31.4	-13 39	Σ 560 rej		Eri	1901 1919	44	29.8	6.2	9.1	A2	3284		
4 31.5	+37 39	β 789		Per	1881 1953	321	1.3	8.5	9.2	F8	3275		
4 31.5	-35 54	β 746		Eri	1895 1947	12 5	1.2 1.2	8.3	9.1	F0			
4 31.8	-24 06	Rst 2347		Eri	1935 1960	8 356	0.2 0.2	8.6	8.8	A2			
4 32.0	+53 55	Σ 550	AB	Cam	1830 1967	308	10.3	5.7	6.8	B1	3274		1 Cam
--			BC		1879 1908	215	150.0		11.1		3274		
4 32.4	+38 50	A 1839		Per	1908 1944	258 268	0.2 0.2	8.8	8.8	F5	3283		
4 32.6	-3 13	H VI 64		Eri	1900 1918	117	124.5	5.8	9.2	B9			
4 32.6	-25 12	Rst 2348		Eri	1935 1950	16 7	0.3 0.3	8.3	9.6	A5			
4 33.0	-38 17	β 747		Cae	1894 1942	220	2.9	8.0	9.6	A2			
4 33.3	+52 48	Es		Cam	1909 1918	129	47.6	7.4	9.3	K5			
4 33.4	+43 04	S,h 44	AB	Per	1875 1913	199 198	113.7 116.2	6.1	6.8	F0 F0			57 Per
--			AC		1908	225	103.7		13.4				
--			AD		1908	354	76.4		12.1				
--			BC		1908	314	51.4						
4 33.5	+18 01	Σ 559		Tau	1830 1968	277	3.1	6.9	7.0	B8	3297		
4 33.5	+72 32	h 2228		Cam	1904	240	39.6	6.0	13.0	A5	3267		
4 33.6	-62 49	h 3670		Ret	1872 1917	99	31.9	5.9	9.2	K0 G5			
4 33.9	-6 44	β 881		Eri	1879 1936	56	1.5	5.7	10.5	B9	3305		46 Eri
4 34.0	-55 03	B 2092	AB	Dor	1956 1960	154 182	0.3 0.2	3.8	4.3	A0			α Dor
--		h 3668	ABxC		1836 1938	108 101	82.3 77.7		9.8				
4 34.6	+28 58	h 5461	AB	Tau	1899 1933	102	25.5	5.7	10.6	B9	3304		
--			AC		1904 1924	133	50.0		11.7		3304		
4 34.6	+51 00	Hu 551	AB	Per	1901 1933	308	1.5	7.8	11.7	F0	3298		

α 2000	δ 2000	Star Name		Const	Years	Pos Ang	Separation	m_1	m_2	Spec	ADS	Orb	Notes
4h34.6	+51 00'	Hu 551	AC	Per	1915	64° °	80".2 "		11.5		3298		
4 34.8	+22 42	Σ 562		Tau	1830 1942	274	2.0	6.8	10.5	F2	3311		
4 34.9	-25 02	h 3664		Eri	1837 1896	192	21.6	7.4	10.5	F5			
4 35.2	-9 44	Σ 570		Eri	1830 1955	259	12.8	6.7	7.7	A0 A0	3318		
4 35.7	+10 10	S,h 45		Tau	1822 1920	299	69.7	4.3	8.4	A3	3317		88 Tau
4 35.7	+19 53	Lewis 4		Tau	1902	192	0.3	7.1	7.6	F8	3316		Doubtful [26]
4 35.7	+19 58	LDS 2256		Tau	1955	326	43.0	8.0	20				
4 35.7	+39 44	Hu 1084		Per	1904 1958	44 26	0.7 0.2	8.6	9.0	F5	3315		
4 35.9	+16 31	β 550	AB	Tau	1877 1934	110	30.4	0.9	13.4	K5	3321		87 α Tau, Aldebaran
---		Σ II 2	AC		1832 1923	36 34	109.0 121.7		11.1		3321		
---		β 1031	AE		1899	331	31.0				3321		
---			AF		1908	124	271.4		13.4		3321		
---			CD		1888 1923	281 274	2.3 1.7		13.5		3321		
4 36.0	-3 37	Σ 571	AB	Eri	1830 1904	258	17.5	6.3	11.0	B9	3328		
---			AC		1909	200	45.1		12.6		3328		
---			AD		1909	274	59.6		12.6		3328		
4 36.1	+8 13	A 1840	AB	Tau	1908 1958	41 158	0.3 0.2	8.8	8.8	A0	3326		
---		OΣ 87	ABxC		1846 1967	235	6.0		9.2		3326		
4 36.3	-3 21	β pm		Eri	1909	170	50.9	3.9	13.1	B2			48 ν Eri
4 36.3	+47 22	S 451		Per	1825 1924	199	58.4	7.5	7.6	F5 F			
4 36.6	+19 45	OΣ 86		Tau	1845 1958	79 32	0.6 0.5	8.2	8.2	A2	3329		
4 36.6	-27 03	B 70		Eri	1927 1938	66	1.3	7.2	11.0	F5	3342		
4 36.7	+41 05	Σ 563		Per	1828 1918	31	11.8	8.0	9.7	A2	3324		
4 37.1	+48 24	OΣ 85		Per	1846 1943	29	1.4	7.9	10.4	B5	3325		
4 37.4	+0 34	OΣΣ 53		Tau	1876 1922	173	78.1	8.3	8.4	G5 G5			
4 37.6	-2 28	β 88		Eri	1891 1916	90 88	32.4 31.7	5.2	11.7	A5	3350		51 Eri
4 37.6	-29 54	B 72		Eri	1926	92	5.0	8.0	14.0	F5			
4 37.8	+44 42	A 1010		Per	1905 1955	336	0.5	8.4	9.2	B8	3332		
4 38.0	-13 02	Σ 576		Eri	1830 1952	172	12.4	6.8	8.1	A0	3355		
4 38.1	+42 07	Σ 565	AB	Per	1831 1943	180 172	1.6 1.3	7.5	8.8	K0	3338		
---			AC		1906 1914	108 109	32.1 29.5		11.5		3338		
---			AD		1906	52	71.6		11.0		3338		
4 38.1	-17 49	B 1939		Eri	1955	155	0.1	8.5	8.5	K0			
4 38.2	+12 31	β pm	AB	Tau	1910	322	44.1	4.3	13.3	A3			90 Tau
---			AC		1910	312	115.0		10.4				
4 38.2	+16 02	β pm		Tau	1910	50	142.7	5.8	11.7	F0			89 Tau
4 38.2	+71 28	h 1146		Cam	1904 1926	40	26.1	8.2	10.2	A2			
4 38.2	-14 18	Kui 18		Eri	1934 1959	1 29	1.1 0.7	4.0	7.0	K0			53 Eri
4 38.2	-44 38	h 3675		Cae	1913	53	37.3	7.8	11.8	A5			
4 38.5	+26 56	Σ 572		Tau	1830 1959	210 194	3.2 4.0	7.3	7.3	F0	3353		
4 39.1	+7 52	β pm	AB	Tau	1884 1921	247 249	69.2 71.6	5.5	10.4	F0			
---			AC		1884 1921	317 316	299.4 297.8		9.0				
4 39.3	+15 55	Σ I 11		Tau	1836 1922	193	431.2	4.7	5.1	A3 A2			92 σ² Tau
4 39.6	-21 15	β 1236	AB	Eri	1891 1951	118 99	1.4 1.4	7.2	10.8	G5	3375		
---			AC		1891 1938	314	40.4		8.8	G0	3375		
4 39.9	+53 05	β 1043		Cam	1888 1919	297	3.8	5.1	12.1	K0	3359		3 Cam
4 40.0	+26 56	Lewis 5		Tau	1899	213	0.8	8.0	9.0				No double star here [27]
4 40.0	+53 28	β 1295	AB	Cam	1901 1958	140 118	0.2 0.1	5.8	7.4	F0	3358	o	2 Cam
---		Σ 566	ABxC		1829 1959	311 258	1.6 1.3		7.3		3358		
---			ABxD		1888 1911	210 215	23.7 22.6		13.3		3358		
4 40.3	-58 57	h 3683		Dor	1836 1959			7.2	7.3	G0		o	
4 40.4	-19 40	Stone 9		Eri	1877 1923	161	0.3	4.9	5.2	M	3380		54 Eri; round after 1926
4 40.6	-41 52	HdO 190		Cae	1898 1933	112 121	5.6 6.6	4.5	12.5	F2			α Cae
4 40.8	-56 02	φ 88	AB	Dor	1929 1935	30	1.4	7.8	12.0	F0			
---			AC		1929	211	6.5		14.0				
4 40.9	+0 58	Σ 583	AB	Tau	1831 1927	326	5.7	8.0	9.6	A0	3383		
---			AC		1904 1920	265	99.7		9.8		3383		
4 41.1	-39 35	Rst 5208		Cae	1947	352	0.2	8.7	8.7	F5			
4 41.3	+28 37	h 346		Tau	1901 1923	55	43.5	5.7	11.0	A0	3379		
4 41.7	-61 13	h 3686		Dor	1916 1936	220	7.3	8.4	8.5	A0			
4 41.8	-47 16	h 3681		Pic	1912	254	41.0	6.8	10.8	G0			
4 41.9	-47 50	CorO 23		Pic	1896 1943	234 229	3.6 3.6	7.4	9.9	F0			
4 42.0	-54 48	Hu 1375		Dor	1912 1938	183	3.7	8.0	12.6	K0			
4 42.2	+2 59	A 2424		Tau	1911 1953	63 37	0.2 0.1	8.6	8.6	A0	3395		P.a. uncertain [28]
4 42.2	+22 57	Ho 642	AB	Tau	1899 1909	np	0.1	4.3	8.9	B5			94 τ Tau; AB is single
---		S 455	AC		1824 1926	213	62.8		8.6	A0			
4 42.3	+9 37	Bgh		Tau		84	60.0	7.8	9.0	F2 F8			
4 42.3	+37 30	Σ 577		Per	1829 1959	99 36	1.6 1.2	8.6	8.6	F8	3390	o	
4 42.5	-20 58	Don 75		Eri	1929 1954	43 59	0.2 0.2	8.4	8.6	F5			Spectrum composite F5+A3
4 42.9	+18 43	LDS 2266	AB	Tau	1955	102		7.5	10.8				
---			BC		1955	132	168.0		18				
4 42.9	+53 08	Σ 574		Cam	1830 1934	312	4.0	8.2	10.0	F5	3392		
4 42.9	+57 12	A 1014		Cam	1905 1958			8.8	8.8	F8	3389	o	
4 43.1	+33 56	h 348		Aur	1903 1921	285	30.2	7.6	10.9	K0	3397		
4 43.3	+59 31	A 1013		Cam	1905 1960	311 356	0.5 0.3	7.3	7.3	A3	3391		

α 2000	δ 2000	Star Name		Const	Years	Pos Ang		Separation		m₁	m₂	Spec	ADS	Orb	Notes
4ʰ43ᵐ.4	+49°58′	Kui 19		Per	1958	41°		20″.8		5.8	13.1	B8			
4 43.6	−8 48	Σ 590		Eri	1831 1968	317		9.2		6.7	6.8	F5	3409		55 Eri
4 44.1	+42 25	Σ 582	AB	Per	1831 1934	23		5.5		7.5	10.2	B9	3400		
--			AC		1904	141		97.2			10.7		3400		
--		Σ 581 rej	CD		1904 1913	340		7.3			10.8		3400		
4 44.4	+11 09	β pm	AB	Ori	1910 1922	97		79.2		5.4	11.2	A3			
--			BC		1910	359		12.4			12.6				
4 45.4	+43 47	Es 13	AB	Per	1900 1916	218 216		17.8	15.4	7.9	14.9	K5	3414		
--		β	Aa		1911 1916	347		16.4			15		3414		
4 45.4	−80 39	h 3721		Men	1879 1936	221 228		3.1	3.1	8.3	9.1	F2			
4 45.8	+28 40	Cou 706		Tau	1971	123		5.5		7.1	12.0	G5			
4 45.8	−11 57	Hu 104	AB	Eri	1900 1953	263		1.1		7.7	11.5	F0	3428		
--		Σ 596	ABxC		1831 1923	281 292		11.1	10.4		10.2		3428		
4 46.0	+11 42	β pm	AB	Ori	1911	204		38.9		5.4	13.2	A0			
--			AC		1911	232		90.6			12.9				
4 46.0	+19 10	A.G.		Tau	1870 1917	112		33.1		8.5	8.9	A0			
4 47.0	−21 13	Don 79		Eri	1929 1943	327		9.0		7.2	13.4	G5			
4 47.1	+10 03	A 2037		Ori	1909 1919	79		2.9		7.2	12.2	A2	3448		
4 47.4	−61 29	CapO 4		Dor	1891 1942	40		2.8		7.5	10.0	F0			
4 47.8	+53 18	Hu 612		Cam	1902 1958					7.0	8.8	F2	3434	o	
4 48.0	+53 07	Σ 587	AB	Cam	1830 1925	185		21.0		7.4	8.9	A3	3442		
--		Σ 586 rej			1913	129		28.4		9.4	9.8				[29]
4 48.0	+56 45	β	AB	Cam	1913 1958	197 201		13.4	13.5	5.3	13.1	A2	3432		4 Cam
--			AC		1913 1958	219 223		102.4	99.0		11.9		3432		
4 48.0	+59 16	Stein 537		Cam	1912	38		11.1		8.0	12.8	A2			
4 48.6	+17 49	Σ 598		Tau	1828 1937	318		9.4		8.1	9.6	F2	3460		
4 48.6	−20 25	Ara 840		Eri	1917	117		1.4		8.3	8.7				
4 48.8	+75 56	β		Cam	1913	60		96.9		6.1	12.1	F0			
4 49.2	+5 13	OΣΣ 55		Ori	1875 1913	15		37.7		8.2	9.2	K F			
4 49.6	+2 13	A 2621		Ori	1913 1957	339 39		0.2	0.2	8.4	8.4	A0	3465	o	
4 49.6	−53 53	I 342		Dor	1901 1951	176 148		1.0	2.1	8.2	8.6	G5			
4 49.7	+15 54	h 3261	AB	Tau	1878 1925	57 57		30.8	29.3	6.1	11.1	G5	3464		96 Tau
--		β 551	BC		1878 1924	205		6.3			12.9		3464		
4 49.8	+6 58	β pm		Ori	1852 1907	128 138		112.5	94.6	3.2	8.7	F8			1 π³ Ori
4 49.8	−54 27	I 343		Dor	1901 1929	46		2.5		7.8	12.2	B8			
4 50.1	+66 32	Σ 584		Cam	1831 1907	122		11.8		7.8	10.5	K0	3452		
4 50.3	+6 57	H V 83		Ori	1900 1917	4		95.8		7.2	8.7	A2			
4 50.3	−41 19	h 3697		Cae	1881 1955	282 268		15.1	13.6	6.1	10.1	F0			
4 50.5	−38 34	h 3695		Cae	1919	46		40.3		7.3	13.0	F5			
4 50.9	−53 28	Δ 18		Pic	1826 1952	58		12.3		5.6	6.4	F0 F0			ι Pic; both straw-yel.(R)
4 51.0	+44 59	Σ 599	AB	Aur	1831 1925	336		10.2		8.3	9.6	B9	3468		
--			AC		1907	30		17.8			13.5		3468		
--			BC		1908 1914	66		14.6					3468		
4 51.2	+11 04	β 883	AB	Ori	1879 1960					7.5	7.6	F5	3475	o	
--			ABxC		1879 1924	148 162		18.4	16.6		13.7		3475		
4 51.2	−43 03	Don 83		Cae	1931 1947	292		1.1		7.8	12.0	F8			
4 51.3	−68 04	φ 362		Dor	1961	7		0.1		7.6	7.6	G0			
4 51.4	+18 50	β pm		Tau	1909 1922	303 303		174.5	175.8	5.1	10.4	F0			97 Tau
4 51.5	−34 54	φ 320		Cae	1952 1960	45 34		0.2	0.2	6.5	6.9	A0			
4 51.6	−14 35	Gallo 153		Eri	1902	64		13.0		7.8	12.8	F5	3485		
4 51.7	+45 50	β pm		Aur	1909	87		101.6		7.0	12.3	G0			
4 51.8	+13 39	β 552	AB	Ori	1874 1960					6.5	8.1	F5	3483	o	
--			AC		1898 1906	213		44.9			12.3		3483		
4 52.1	+63 30	β		Cam	1913	350		115.4		5.4	9.9	M			
4 52.6	−30 40	See 41		Cae	1897 1927	122		9.4		7.8	12.1	A2			
4 53.1	−1 16	S 457		Ori	1824 1918	354		41.2		8.4	8.3	A A			
4 53.2	+8 33	A 2040		Ori	1909 1919	260		2.5		7.8	12.6	A2	3502		
4 53.2	+49 35	Σ 603		Aur	1830 1938	239		8.4		8.6	8.8	G0	3498		
4 53.2	−20 46	h 3700		Eri	1877 1904	347		26.5		6.9	11.5	A3	3511		
4 53.6	+25 22	A 1843		Tau	1908 1955	334 280		0.2	0.4	7.3	10.0	A0	3501		
4 53.7	+23 33	β 1237		Tau	1891 1941	59		4.5		7.5	10.1	K0	3504		
4 53.9	+67 29	Hu 1087		Cam	1905 1923	110		1.3		7.5	12.0	F2	3479		
4 54.2	+7 23	Σ 612	AB	Ori	1831 1951	198		16.4		8.5	8.8	K0	3514		
--			AC		1898 1927	242 236		31.6	46.7		13.8		3514		
--			BC		1906	245		32.2					3514		
4 54.5	−3 14	Rst 5501		Ori	1950 1960	95 70		0.2	0.3	7.4	7.4	F5			
4 54.9	+8 36	OΣ 90	AB	Ori	1845 1943	344		2.0		6.9	8.9	A0	3517		
--			AC		1909	95		38.1			11.9		3517		
--			CD		1909	320		11.8			12.7		3517		
4 54.9	+10 09	β pm	AB	Ori	1887 1919	248		171.6		4.7	8.9	A0 A			7 π¹ Ori
--			Aa		1907	4		33.1			12.7				
4 55.1	−0 32	A 1019	AB	Ori	1905 1958	120 124		0.2	0.1	8.6	8.6	A5	3522		Too close in 1948 (B)
--		Σ 614	AC		1832 1932	68		4.3					3522		
4 55.1	+55 16	β 1187		Cam	1890 1933	245		12.9		5.6	12.9	A0	3508		5 Cam
4 56.2	−17 44	Rst 2360		Lep	1934 1943	340		1.0		7.5	13.1	G5			

α 2000	δ 2000	Star Name		Const	Years	Pos Ang	Separation	m₁	m₂	Spec	ADS	Orb	Notes
4ʰ56ᵐ.3	+3°11'	OΣ 91		Ori	1851 1956	242° 230°	0".8 0".6	8.5	9.0	B9	3542		
4 56.4	−5 10	S,h 48		Eri	1821 1913	75 75	65.9 67.3	5.5	9.1	B9			62 Eri
4 56.4	+13 31	β 553	AB	Ori	1877 1916	48 51	28.6 32.4	4.1	11.1	K0	3540		9 o² Ori. Optical
--			AC		1918	69	100.4		11.4		3540		
4 56.8	−16 08	h 3705		Lep	1880 1920	141	22.3	7.8	9.9	A0	3556		
4 57.3	+53 45	Dem 5	AB	Cam	1865 1959	309 258	1.2 0.3	4.5	7.8	A2	3536		7 Cam [30]
--		Σ 610	ABxC		1831 1933	240	25.8		11.3		3536		
4 57.3	+61 45	OΣ 88		Cam	1854 1943	302 311	0.7 0.9	7.1	8.8	G0	3526		
4 57.8	+23 57	β 1045	AB	Tau	1889 1938	6	6.2	5.8	12.1	K0	3557		99 Tau
--			AC		1912 1938	353	103.5		12.8		3557		
4 58.0	−50 35	B 1476		Pic	1929 1960	292	0.5	7.6	9.0	G5			
4 58.2	−2 13	h 689		Ori	1880 1909	276	20.9	6.4	10.9	A0	3570		
4 58.2	+25 03	β	AB	Tau	1912	358	41.8	5.6	13.3	B9	3547		98 Tau
--			AC		1912	289	94.6		9.9		3547		
4 59.0	+14 33	S,h 49	AB	Ori	1822 1972	305	39.2	6.0	8.0	B8 B9	3579		Same as BDS 2461, H V 57?
--			AC		1874 1957	88	54.6		9.8		3579		
--			BC		1907	114	89.3				3579		
4 59.0	−16 23	β 314	AB	Lep	1876 1960			5.9	7.3	F2	3588	o	
--			ABxC		1889 1914	29 34	54.4 52.9		8.2		3588		
4 59.3	+37 53	Σ 616		Aur	1828 1950	352 359	6.5 5.4	5.0	8.0	A0	3572		4 Aur
4 59.5	−49 27	h 3715		Pic	1880 1953	112	9.9	7.3	9.0	F0			
4 59.7	+15 55	β pm		Tau	1910	146	106.4	6.8	12.0	F5			
4 59.7	+31 35	Ho 222		Aur	1887 1943	222	1.9	8.1	10.9	F8	3581		
4 59.9	+27 20	Σ 623		Tau	1829 1934	205	20.5	6.8	8.3	B9	3587		
4 59.9	+53 28	A 1303		Cam	1906 1958	37 336	0.2 0.2	8.8	8.8	F5	3573		
5 00.1	+69 58	Mlr 399	AB	Cam	1972	3	0.2	8.0	8.3	A0			
--			ABxC		1972	336	37.9		9.8	A0			
--			ABxD		1972	37	27.6		13.0				
5 00.2	−78 18	h 3741		Men	1835 1919	122	46.4	6.3	10.3	K0			
5 00.3	+39 24	OΣ 92		Aur	1849 1975	230 265	2.8 3.7	6.0	9.7	F5	3589		5 Aur
5 00.5	+5 06	OΣ 93		Ori	1847 1959	66 255	1.4 0.6	8.5	9.0	G5	3596		
5 00.6	+3 16	Σ 628 rej		Ori	1902	47	23.0	8.2	10.2	B9			
5 00.6	+3 37	Σ 627		Ori	1831 1932	260	21.3	6.6	7.0	A0 A0	3597		
5 00.7	−13 30	Σ 631		Eri	1831 1937	106	5.5	7.5	9.0	A0	3606		
5 01.1	−11 13	A 2629		Lep	1913 1960	147 110	0.2 0.1	8.1	8.1	A0	3610		
5 01.2	−74 20	Hrg 2		Men	1883 1947	170	1.0	7.5	7.9	A0			
5 01.3	−23 42	See 44		Lep	1897 1960	335 334	1.6 0.7	7.6	9.8	F2	3618		
5 01.6	+61 05	Hu 1093		Cam	1904 1920	5	5.4	6.0	11.5	F5	3590		
5 01.7	+20 50	Hu 445		Tau	1901 1960			8.6	8.9	G0	3614	o	
5 01.7	+26 40	A 1844	AB	Tau	1908 1954	354 294	0.3 0.3	7.1	8.4	F5	3608	o	Two rev. in interval?
--		S 461	ABxC		1824 1967	159	78.7		8.6		3608		
5 01.8	+11 22	S 463		Ori	1825 1919	30 29	33.6 32.0	7.2	9.5	B8			
5 01.9	−76 38	Rst 2368		Men	1934	95	0.9	7.5	11.2	F5			
5 02.0	+1 37	Σ 630	AB	Ori	1832 1932	50	14.2	6.5	7.7	B8	3623		
--		A 2630	AD		1909	99	130.3		9.5		3623		
--			BC		1913 1930	25	0.5		10.8		3623		
5 02.0	+28 17	A 480		Tau	1903 1917	317 327	0.5 0.5	8.0	11.8	A0	3613		
5 02.0	+43 49	β 554	AB	Aur	1878 1925	224	21.2	3.0	14.0	F5	3605		7 ε Aur; A is variable
--			AC		1878 1925	275	43.0		11.7		3605		
--			AD		1879 1924	317	46.2		12.0		3605		
--			AE		1913 1921	48	207.6		9.2		3605		
5 02.1	+69 10	β 313	AB	Cam	1874	250	10.0	7.1	12.1	F5			No companion seen 1900-04
--			AC		1898	150	40.3						
--			AD		1898	179	43.0						
5 02.3	−56 05	LDS 135		Dor		sf	78.0	7.0	11.3	G0			
5 02.8	−49 09	Rst 120		Pic	1928 1946	203 198	10.3 10.6	5.4	13.0	F5			η¹ Pic
5 03.0	−8 40	Σ 636		Eri	1830 1949	102	3.7	7.1	8.2	A0	3640		
5 03.4	+60 27	S 459	AB	Cam	1825 1923	208	80.8	4.0	8.6	G0 A5	3615		10 β Cam
--		Es 58	BC		1892 1919	168	14.8		11.2		3615		
5 03.5	+63 05	Σ 618	AB	Cam	1831 1925	211	32.6	8.6	8.6	G0 G0			
--			BC		1909 1925	184 185	87.1 91.6		12.1				
5 03.6	+63 05	LDS 889		Cam	1936	211	33.0	8.6	8.8	G0 G0			
5 03.7	−2 32	J 307	AB	Ori	1916	322	52.2	6.8	9.2	K0	3650		Mag. of B possibly for BC
--			BC		1910	196	0.3	10.0	10.0		3650		Reality of BC doubted [31]
5 04.3	−6 02	A 481		Eri	1903 1958	357 320	0.2 0.3	7.1	8.1	B9	3662		
5 04.4	+21 39	Cou 154	AB	Tau	1967	310	0.2	8.0	8.1	B5			
--			AC		1967	191	2.4		13.0				
5 04.4	−35 29	Jc 9		Cae	1847 1942	316 308	2.9 2.9	4.6	8.1	K0			γ Cae
5 04.5	+27 42	Weisse 6		Tau				8.0					No double found here 1904
5 04.5	+29 37	A 1024		Aur	1905 1953	355 342	0.5 0.6	8.4	9.2	A0	3651		
5 04.7	+74 04	OΣ 89		Cam	1848 1957	306 296	0.4 0.6	6.4	7.4	A2			Uncertain 1869-1935 [32]
5 05.0	+29 59	A 1026		Aur	1905 1940	42	1.1	8.0	11.2	K0	3660		
5 05.2	+23 48	Pou 535		Tau	1897	103	12.7	8.0	11.9	K0			
5 05.5	+19 48	OΣ 95		Tau	1845 1960	344 311	0.6 1.1	6.9	7.5	A2	3672		
5 05.5	+46 55	A 1023		Aur	1905 1951	67	0.4	6.8	8.3	F8	3659		

α 2000	δ 2000	Star Name		Const	Years	Pos Ang	Separation	m₁	m₂	Spec	ADS	Orb	Notes
5ʰ05ᵐ.6	+23°04′	OΣ 97		Tau	1852	158°	0″.5	6.7		B5			Single 1889 (see BDS note)
5 05.8	+35 51	Ho 223		Aur	1890 1946	43	1.4	7.8	11.8	F5	3670		
5 05.9	−13 55	A 3009		Lep	1926 1959	110 192	0.9 0.8	8.1	10.4	G5	3686	o	
5 06.1	+58 58	Σ I 13	AB	Cam	1836 1924	8	180.5	5.4	6.5	B3 K0			11 Cam
--			BC		1879 1908	18	173.4		10.9				
--			CD		1908	9	16.2		13.4				
5 06.4	+40 02	Hu 1095		Aur	1904 1958	358 15	0.3 0.3	8.1	9.3	A0	3678		
5 06.4	+54 24	Es 888	AB	Cam	1910 1925	178 190	7.7 6.7	7.3	11.0	G5	3667		
--			AC		1910 1925	246	26.5		12.0		3667		
5 06.7	+51 36	β 1046	AB	Aur	1888 1958	94 82	6.3 5.2	5.0	12.2	F0	3675		9 Aur
--		H VI 35	AC		1888 1923	61	90.1		9.4		3675		
--		β 1046	AE		1922	348	131.6		12.3		3675		
--			CD		1909 1925	123	47.0		12.6		3675		
5 06.8	−4 39	Σ 642 rej		Eri	1879 1922	9	52.8	5.2	8.4	B9	3698		66 Eri
5 07.1	−19 24	ø 376		Lep	1962	46	0.1	7.5	7.5	G0			
5 07.2	+22 24	Cou 155		Tau	1967	318	0.2	8.6	8.7	F5			
5 07.4	+50 18	OΣ 94 rej	AB	Aur	1843 1921	304 304	15.6 17.9	7.4	11.1	B9	3684		
--			AC		1899 1921	64	25.4		11.0		3684		
--			Aa		1900	341	26.1		13.6		3684		
5 07.5	+18 39	A 3010		Tau	1912 1955			5.6	5.6	G0	3701	o	104 Tau
5 07.5	+55 32	β 749		Cam	1879 1956	226 234	0.9 1.2	7.6	9.7	F8	3683		
5 07.6	+9 28	β pm	AB	Ori	1852 1909	255 264	128.2 124.0	6.2	10.0	G0			13 Ori
--			AC		1904	268	402.7						
5 07.8	−5 05	β pm		Eri	1879 1907	143	116.7	2.8	10.8	A3			67 β Eri
5 07.8	+54 59	Σ 635		Cam	1830 1951	281 300	0.4 0.9	8.7	8.7	B9	3689		
5 07.9	+8 30	OΣ 98		Ori	1844 1959			5.8	6.5	F0	3711	o	14 Ori
5 07.9	+21 42	S 466		Tau	1825 1925	251	111.5	6.0	9.0	B3			105 Tau
5 08.0	+73 35	Σ 615		Cam	1831 1934	337 356	1.3 1.4	8.2	10.0	A0	3658		
5 08.1	+24 16	Edg	AB	Tau	1878 1925	150	13.3	5.5	12.0	B3	3709		103 Tau
--		H V 114	AC		1878 1924	197	35.3		8.6		3709		
5 08.3	−8 40	Σ 649	AB	Eri	1831 1932	81	21.6	5.8	9.8	B8	3722		White, very ruddy (h)
--			AC		1913	3	85.7				3722		
--			BC		1913	347	82.3				3722		
5 08.5	−18 07	Rst 2365		Lep	1934 1942	53	7.2	7.6	14.9	K0			
5 08.5	−41 13	h 3728		Col	1853 1951	260	10.0	6.7	10.1	G0			
5 08.9	+3 13	A 2636		Ori	1913 1957			7.1	7.6	A0	3728	o	
5 08.9	+39 29	Es 169		Aur	1905	176	4.5	8.0	11.7	B5	3718		
5 08.9	−12 55	Rst 3413		Lep	1938 1943	150 144	1.3 1.2	8.1	10.5	G0			
5 09.1	+49 07	OΣ 96 rej		Aur	1897 1912	104	21.0	6.6	11.1	A3	3715		
5 09.3	+9 50	β pm	AB	Ori	1909	122	88.6	5.4	12.2	A2			16 Ori
--			AC		1909	228	168.0		9.9				
5 09.7	+15 36	OΣ 99 rej		Ori	1923	37	0.3	4.8		F0			15 Ori; round (Hu)
5 09.7	+29 48	OΣΣ 61		Aur	1874 1913	244	69.1	6.8	8.3	F8			
5 09.7	−22 30	See 47		Lep	1897 1933	56 60	3.5 3.5	7.1	12.5	F0	3742		
5 09.8	+28 02	Σ 645	AxBC	Tau	1829 1951	27	11.8	6.1		A3	3730		
--		β 1047	BC		1889 1960	75 72	0.4 0.3	9.0	9.5		3730		
5 09.8	+42 41	β 751	AB	Aur	1891 1958	258 249	3.1 4.0	8.2	9.8	G5	3727		
--			AC		1899 1958	204 176	24.4 12.5		11.4		3727		
5 09.9	+10 54	J 323	AB	Ori	1911 1957	164	3.4	8.1	10.1	G5	3736		
--			AC		1940	186	84.6		10.4		3736		
5 10.0	+8 10	OΣ 100		Ori	1848 1968	252 257	3.9 3.8	7.2	10.0	F8	3737		
5 10.0	+75 41	A 841	AxBC	Cam	1904 1919	340 340	48.8 47.7	7.3		A0	3681		
--			BC		1904 1929	216	0.5	9.0	9.8		3681		
5 10.1	+27 33	Bgh		Tau		352	14.0	7.3	8.9	F5 G5			
5 10.1	+40 09	Brt 112		Aur	1934	184	5.1	8.2	10.2	B9			
5 10.3	+37 18	Σ 644	AB	Aur	1828 1959	221	1.6	6.7	7.0	B2	3734		
--		VBs	AC		1903	15	72.6		9.3		3734		
5 10.3	+57 04	A 1029		Cam	1905 1945	11	1.5	8.0	12.0	B9	3725		
5 10.7	+63 36	Σ 633		Cam	1831 1922	342	12.3	6.7	10.3	F0	3723		
5 10.7	−20 45	Bhas 3		Lep	1917	240	17.2	7.3	10.0	B9			
5 10.7	−48 27	Rst 125		Pic	1928 1947	194	2.3	7.7	11.0	A2			
5 10.9	−1 46	β 885		Ori	1880 1950	193	0.6	8.5	8.6	B9	3755		
5 11.0	+32 03	Σ 648	AB	Aur	1831 1955	74 66	4.7 4.8	8.4	9.1	G5	3744		
--			AC		1906 1927	116 114	36.6 40.2		12.8		3744		
--			AD		1906 1927	63 66	36.7 38.4		13.2		3744		
5 11.7	+0 31	Hu 33		Ori	1899 1936	324 332	0.2 0.1	7.2	7.7	B9	3767		Round 1946,51,56
5 11.8	+1 02	Σ 652		Ori	1830 1953	183	1.7	6.1	7.6	F5	3764		
5 11.9	−9 07	β pm		Ori	1890 1910	227 231	131.5 118.1	8.0	10.4	K0			
5 12.0	+6 50	OΣΣ 62		Ori	1875 1922	49	124.2	7.7	7.9	K0 K0			
5 12.3	−11 52	Σ 655		Lep	1832 1933	337	12.7	4.5	10.8	B8	3778		3 ι Lep
5 12.5	−17 27	S 470		Lep	1825 1919	278	47.3	8.7	8.6	A0			
5 13.2	−12 56	Σ 661		Lep	1832 1959	358	2.6	4.5	7.4	B8	3800		4 κ Lep
5 13.3	+2 52	Σ 654	AB	Ori	1832 1972	64	7.0	4.5	8.3	K0	3797		17 ρ Ori
--			AC		1913	155	182.4		11.7		3797		
5 13.3	+37 20	Sei 105		Aur	1895	354	16.7	6.5	11.0	F8			

α 2000	δ 2000	Star Name		Const	Years	Pos Ang	Separation	m₁	m₂	Spec	ADS	Orb	Notes
5ʰ13ᵐ.5	+1°58′	OΣ 517	AB	Ori	1854 1958			6.9	7.1	A2	3799	o	Spectrum composite A2+G
--			ABxC		1878 1958	137	6.8		12.9		3799		
5 13.5	-4 39	h 2257		Ori	1902	268	34.5	8.0	11.7	A0			
5 13.5	-31 54	h 3735		Col	1836 1952	153	7.2	8.4	8.6	F2			
5 13.5	-55 34	h 3742		Pic	1916 1955	279 288	22.8 25.6	7.1	12.7	G0			
5 13.7	+46 59	OΣ 101		Aur	1848 1934	184	5.8	7.7	10.2	B8	3781		
5 13.8	+0 34	OΣ 102		Ori				6.3		K2			Almost surely single (IDS)
5 13.9	-22 59	β 317	AB	Lep	1876 1938	12	8.6	7.4	10.7	G5	3819		
--			AC		1898 1933	47	18.1		12.6		3819		
5 14.0	+51 26	Hu 821		Aur	1904 1944	190 246	0.9 0.5	8.2	10.2	F8	3780		
5 14.3	+40 01	Es 280		Aur	1906 1947	305	3.3	8.2	9.7		3754		
5 14.3	+69 49	Σ 638		Cam	1831 1934	222	5.2	7.6	8.6	K0	3759		
5 14.5	-8 12	Σ 668	AB	Ori	1831 1954	202	9.5	0.1	6.8	B8	3823		19 β Ori, Rigel [33]
--		β 555	AD		1878 1921	2 1	44.5 43.9		15		3823		
--			BC		1878 1959	173 170	0.4 0.1				3823		Always difficult to measure
5 14.5	-18 23	B 1943		Lep	1934 1938	284	1.4	7.7	11.1	K0			
5 14.6	+76 28	Hu 1097		Cam	1904 1922	116	1.5	6.3	10.8	B9	3738		
5 14.7	-7 04	Σ 667		Ori	1830 1946	313	4.2	7.1	8.6	K2	3825		
5 14.8	+12 32	Hu 1224		Ori	1905 1956	132 123	0.9 1.0	7.9	8.7	A0	3822		
5 15.0	+42 29	A 1555		Aur	1907 1944	186	1.0	8.0	10.7	G5	3813		
5 15.1	-36 39	h 3740		Col	1880 1933	287	23.9	6.8	8.5	G5			
5 15.2	+8 26	Σ 664		Ori	1829 1949	170	4.9	7.7	8.2	F0	3827		
5 15.3	+47 10	A 1031		Aur	1905 1951	349 30	0.4 0.3	7.1	10.4	G5	3812		
5 15.4	+32 41	Σ 653	AB	Aur	1830 1909	342 352	12.6 11.1	5.1	11.1	A2	3824		14 Aur
--			AC		1830 1933	226	14.6		7.4		3824		
--			AD		1880 1908	320 321	184.5 184.0		10.4		3824		
5 15.5	+22 17	Cou 158		Tau	1967	349	1.9	6.3	12.5	A0			108 Tau
5 15.9	+34 25	Es 170		Aur	1895 1917	21	13.4	7.9	10.0	F8	3834		
5 16.0	+25 57	Σ 662		Tau	1831 1927	103	5.4	8.0	11.1	A0	3839		
5 16.1	+51 57	h 2253	AB	Aur	1900 1908	24	26.8	8.0	13.0	B8			
--			AC		1908	11	52.0		9.7				
5 16.2	-41 04	B 2096		Col		n	12.0	8.0	13.0	G5			
5 16.3	+34 19	Sei 136		Aur	1895	355	8.4	6.0		B0	3843		Is B a nebula? [34]
5 16.5	+42 40	Es 573		Aur	1908 1944	123 126	5.0 5.7	8.0	10.5	F0	3842		
5 16.5	-21 06	Don 97		Lep	1930 1950	3 330	0.3 0.3	8.1	9.6	A2			
5 16.7	+18 26	Σ 670		Tau	1830 1953	171 164	2.3 2.5	7.7	8.2	B3	3854		Both bluish (Smyth)
5 16.7	+46 00	Barnard	AB	Aur	1898	23	46.6	0.1	17	G0	3841		13 α Aur, Capella [35]
--		β	AC		1878	318	78.2		15		3841		
--			AD		1878	183	126.2		13.6		3841		
--			AE		1878	316	143.2		12.1		3841		
--		h 2256	AF		1878 1922	146 143	158.0 141.5		11.1		3841		
--		S,h 51	AG		1821 1895	348 348	454.2 484.6		10.1		3841		
--		Frh	AH		1895	141	723.3		11.7		3841		
--		JAn	Aa		1919 1935						3841	o	
--		St 3	HL		1935 1965	117 115	1.8 3.2		13.0		3841		
5 17.2	+16 46	Σ 672 rej		Tau				8.0	10.0				[36]
5 17.2	+33 20	Σ 666		Aur	1830 1953	74	3.0	8.4	8.4	A3	3853		
5 17.2	+39 28	Hu 1101		Aur	1904 1956	287	0.4	7.5	9.5	K0	3851		
5 17.3	+53 35	Es		Aur	1909 1918	339	47.5	6.4	8.6	M	3845		R Aur; A is variable
5 17.4	-71 32	Alden 14		Men	1928 1931	242	4.7	7.8	11.4	K0			
5 17.5	+20 08	Σ 674		Tau	1828 1940	148	10.2	6.9	9.9	F5	3866		CD Tau
5 17.6	-6 51	H V 25	AB	Ori	1868 1919	250	35.2	3.6	13.6	B5	3877		20 τ Ori
--		h 2259	AD		1876 1919	60	36.0		10.8		3877		
--		β 188	BC		1876 1922	51	3.7		11.9		3877		
5 17.6	-15 13	S 473		Lep	1825 1953	305	20.6	6.7	8.7	B8	3883		
5 18.2	+33 22	OΣ 103		Aur	1848 1936	56	4.2	4.5	10.3	K0	3872		16 Aur
5 18.2	+37 39	β 885½		Aur	1880 1935	69	2.5	8.1	10.1	B8	3871		
5 18.4	+24 02	Pou 650		Tau	1891	221	9.3	8.5	8.8				
5 18.4	-41 03	I 61		Col	1897 1950	100 88	0.8 0.8	8.2	8.8	F5			
5 18.7	+33 32	Es 59		Aur	1882 1968	10	14.0	8.5	9.0	A3	3879		
5 18.8	-18 08	See 50	AB	Lep	1897 1951	199 234	28.9 46.0	6.0	13.8	G0	3899		
--			BC		1907 1951	99 101	14.0 15.8		15		3899		
5 18.9	+36 12	Sei 178		Aur	1895	197	26.5	8.0	10.7	B5			
5 18.9	+45 15	Σ 669		Aur	1831 1968	276	9.9	8.4	8.9	F0	3878		
5 19.1	+40 06	Σ II 3	AB	Aur	1900	274	29.1	4.7	13.3	G0	3886		15 λ Aur
--			AC		1879 1934	198 268	40.5 41.7		12.0		3886		
--			AD		1836 1921	29 2	103.6 146.6		8.5		3886		
--		Dob	AE		1911	60	168.8				3886		
--		Kui 20	CB		1934	351	27.2				3886		
--		Dob	DE		1912	113	145.6				3886		
--		GΣ			1921	211	92.5				3886		[37]
5 19.2	+20 08	Σ 680		Tau	1827 1938	204	9.0	6.1	10.0	K0	3894		Deep yellow, bluish (Smyth)
5 19.3	+34 53	Sei 180		Aur	1895	358	7.2	7.0	10.5	B8	3888		
5 19.3	-10 45	Σ 688		Ori	1832 1950	274	10.5	8.6	8.7	F0	3909		BDS 2665, H III 94 same
5 19.3	-18 31	S 476		Lep	1824 1952	18	39.4	6.2	6.4	B8 B8	3910		Superb star (h)

α 2000	δ 2000	Star Name	Const	Years		Pos Ang		Separation		m₁	m₂	Spec	ADS	Orb	Notes
5ʰ19.4ᵐ	+33°59′	Ho 18		Aur	1883	1922	165°	°	4".1	"	6.5	11.8	A5	3893		18 Aur
5 19.7	+68 00	Mlr 402		Cam	1972		179		0.9		7.4	10.0	A2			
5 20.1	+39 21	h 3272	AB	Aur	1879	1909	342		19.3		7.8	11.0	K5	3898		
--			AC		1879	1906	295		29.2			11.5		3898		
--			AD		1879	1906	43		33.4			10.6		3898		
--		β pm	AE		1878	1909	318		190.0			9.4	A	3898		
5 20.4	-5 22	β 189		Ori	1875	1933	285		4.4		6.4	11.1	B9	3926		
5 20.4	-8 02	β 190	AB	Ori	1876	1958	358	343	0.7	0.5	8.4	9.2	F2	3927		
--		Σ 692	AC		1831	1976	4		34.8			9.3		3927		
5 20.4	-21 14	h 3750		Lep	1876	1959	282		4.2		4.7	8.5	A0	3930		Most beautiful double (h)
5 20.6	+63 11	Σ 656		Cam	1831	1955	215		2.7		8.2	9.9	A0	3889		
5 20.7	+37 26	Sei 201		Aur	1895		51		25.6		6.8	12.8	O			
5 20.7	+46 58	Σ 681		Aur	1831	1928	181		23.2		6.7	8.7	F0	3903		Spectrum composite F0+A
5 21.0	+78 23	Σ 632		Cam	1831	1934	44		2.2		8.2	10.2	F0	3840		
5 21.3	-34 21	δ 117		Col	1922	1933	7		2.2		6.1	10.9	B5			
5 21.5	-0 25	h 697	AB	Ori	1878	1901	60		32.7		5.7	11.4	B3	3941		
--			AC		1878	1901	110		37.6			10.9		3941		
5 21.6	-2 03	Σ 693		Ori	1831	1937	9		3.6		8.6	8.9	A2	3944		
5 21.6	+12 40	OΣ 105		Ori	1848	1959	110	94	0.7	0.2	8.6	8.6	A3	3936		
5 21.7	+18 54	Ku		Tau	1901		104		36.3		7.5	9.9	G0			
5 21.8	-24 46	h 3752	AB	Lep	1837	1953	110	97	3.2	3.2	5.4	6.6	G0 A3	3954		
--			AC		1837	1898	106	105	58.8	61.2		9.1	K0	3954		
5 21.9	+42 03	A.G.		Aur	1907		83		5.9		8.4	8.7				
5 22.1	+5 24	OΣ 106		Ori	1844	1917	42		9.3		7.2	10.7	F5	3949		
5 22.1	-81 02	Rst 2387		Men	1935	1936	124		0.5		7.9	8.0	A2			
5 22.2	+45 05	Σ 684		Aur	1830	1958	139		1.5		7.7	9.5	A0	3932		
5 22.6	+79 14	Σ 634	AB	Cam	1834	1943	349	91	34.0	10.4	5.1	9.1	F8	3864		An optical pair
--			AC		1900		150		192.6					3864		
5 22.8	+3 33	Σ 696		Ori	1831	1934	28		32.1		5.0	7.1	B3 A	3962		
5 22.8	+66 12	Σ 663		Cam	1831	1934	77		2.8		8.1	11.3	A0	3917		
5 22.9	+14 22	Hu 1225		Ori	1905	1932	320	318	3.4	3.0	7.4	11.9	K0	3961		
5 23.1	+1 03	Σ 700		Ori	1831	1941	5		4.7		7.7	7.9	A0	3968		
5 23.1	+1 17	Webb		Ori	1919		12		64.3		7.7	9.9	K2			
5 23.2	+47 01	OΣ 104		Aur	1847	1958	191	190	15.8	19.7	7.1	11.1	K5	3948		
5 23.2	-31 45	h 3757		Col	1919	1933	307	315	15.7	15.1	7.6	11.6	F0			
5 23.3	-8 25	Σ 701		Ori	1830	1933	142		5.9		6.0	7.8	A0	3978		
5 23.5	+16 02	Σ 697		Ori	1829	1938	285		26.0		6.9	8.4	B8 A	3969		
5 23.5	+57 33	β		Cam	1913		219		106.2		5.2	12.9	A0			16 Cam
5 23.6	-10 26	Stone 10		Ori	1877	1942	121		1.2		8.8	8.8	A0	3987		
5 23.7	-22 17	CorO 31		Lep	1913	1933	283		18.0		7.4	9.7	A3	3993		
5 23.8	-62 35	B 1945		Dor	1930	1951	331	337	1.1	1.1	8.1	10.1	F5			
5 23.9	-0 52	Wnc 2	AxBC	Ori	1866	1970	171	161	1.4	2.7	6.1		F5	3991		
--		A 847	BC		1904	1956					7.8	7.9		3991	o	
5 24.4	+17 23	S 478		Tau	1825	1923	271	271	61.8	85.7	5.0	8.8	G0 K0			111 Tau
5 24.4	+42 37	Es 576	AB	Aur	1908		343		8.4		8.0	13.7	A2	3975		
--			AC		1908		236		42.4			8.6	A0	3975		
--		A 1719	CD		1907	1932	95		0.7			10.0		3975		
5 24.5	-2 24	Dawes 5	AB	Ori	1849	1959	87	80	1.0	1.5	3.8	4.8	B1	4002		28 η Ori; variable
--		H VI 67	AC		1904		51		115.1			9.4		4002		
5 24.6	+1 48	S 479	AB	Ori	1825	1919	220		46.4		8.1	8.3	A A0			
--			AC		1825	1903	35		158.4			4.7	B3			
--		β	Aa		1903		279		14.7							
5 24.6	+9 10	A 2703		Ori	1914	1937	303	175	0.2	0.2	8.4	9.0	F5	3997		Too close 1937,51,54
5 24.6	-2 30	β 556		Ori	1878	1922	239		1.0		8.0	13.0	B8	4007		
5 24.7	+37 23	β 888	AB	Aur	1880	1922	171	167	7.9	8.7	5.0	11.0	K5	3984		21 σ Aur
--			AC		1898	1914	330		27.3			13.2		3984		
--			CD		1898	1914	348	342	4.4	6.8		15		3984		
5 24.7	+63 23	Σ 677		Cam	1831	1960					7.9	8.2	G0	3956	o	
5 24.8	+64 44	Σ 676		Cam	1831	1958	282	270	0.8	1.3	8.0	9.0	F8	3955		
5 24.8	-52 19	I 345	AB	Pic	1901	1960	197	152	0.5	0.2	6.9	7.2	A0			θ Pic
--		Δ 20	ABxC		1826	1938	285	287	38.2	38.2		6.8	A0 A0			
5 25.0	-2 53	H VI 68		Ori	1900	1917	282		136.0		8.4	9.2	B8			
5 25.0	-19 22	B 1944		Lep	1930	1940	75		2.6		6.9	10.9	A0			
5 25.1	+6 21	β pm		Ori	1879	1909	144		179.0		1.6	12.1	B2			24 γ Ori
5 25.2	+34 51	Σ 698		Aur	1831	1951	346		31.2		6.6	8.7	K0 K	4000		Beautiful (Webb)
5 25.5	-0 33	A 848		Ori	1904	1960	36	138	0.2	0.2	6.7	7.3	B9	4020		
5 25.6	+38 03	Σ 699		Aur	1830	1952	343		8.8		8.4	9.1	A0	4004		
5 25.9	-35 21	h 3760	AB	Col	1835	1935	222		7.4		7.7	8.3	F5			
--			AC		1898	1935	280	282	25.2	26.0		10.6				
5 26.0	-19 42	h 3759		Lep	1837	1918	315	318	28.7	27.1	5.8	8.6	F5 F5	4034		
5 26.0	-32 13	Jsp 76		Col	1929	1944	50		13.6		7.0	13.0	K0			
5 26.3	+28 36	Barnard		Tau	1898		239		33.4		1.7		B8			112 β Tau
5 26.4	-20 43	β 319		Lep	1876	1939	231		3.9		7.4	10.5	A3	4042		
5 26.4	-43 24	h 3763		Pic	1879	1935	253		12.0		8.5	9.2	F2			
5 26.5	+2 58	Σ 712		Ori	1831	1975	45	63	3.1	3.1	7.5	9.5	B9	4033		

α 2000	δ 2000	Star Name		Const	Years	Pos Ang	Separation	m₁	m₂	Spec	ADS	Orb	Notes
5ʰ26ᵐ.8	+3°06′	Knott 3	AB	Ori	1864 1934	323°	2.7″	4.6	10.2	B2	4039		30 ψ Ori
--			AC		1912	196	83.4				4039		
5 27.0	+27 37	Ho 226	AB	Tau	1887 1959	230 255	0.5 0.7	8.7	8.7	F8	4032		
--			ABxC		1906 1934	274 274	23.6 25.4		10.6		4032		
5 27.0	-68 37	I 276		Dor	1900 1959	213 172	0.5 1.2	6.7	6.9	F0			
5 27.1	-40 57	I 346		Col	1928	172	20.3	5.9	14.6	A2			
5 27.2	+17 58	OΣ 107	AB	Tau	1849 1914	306	10.1	5.4	10.2	B3	4038		115 Tau
--			AC		1899 1908	341	9.8		11.9		4038		
--			BC		1899	58	5.8				4038		
5 27.6	+1 06	A 2643		Ori	1913 1954	13	0.6	7.7	11.8	B9	4056		
5 27.6	-0 38	Bail 670		Ori	1892	314	15.4	8.0	10.2				
5 27.6	+21 56	h 365	AB	Tau	1877 1925	348	37.8	4.9	11.0	B3	4048		114 Tau
--			AC		1877 1925	194	58.8		10.5		4048		
--			AD		1877 1925	280	74.2		11.7		4048		
5 27.6	+34 29	β pm	AB	Aur	1908	351	20.6	5.2	14.5	K0			24 φ Aur
--			AC		1908	71	61.3		10.6				
--			AD		1879 1908	14	206.8		8.5	B0			
5 27.6	-20 55	See 53		Lep	1897 1959	32 160	0.3 0.2	8.7	8.9	F0			
5 27.7	+74 33	A 846		Cam	1904 1937	347	1.1	7.2	10.7	A0	3982		
5 27.8	-32 25	h 3762		Col	1898 1933	212 212	26.2 24.9	6.9	13.1	G5			
5 28.0	+33 46	S 483	AB	Aur	1825 1918	59 49	87.6 97.4	6.8	8.8	F5	4050		
--			AC		1891 1907	340 341	100.3 103.4		10.1		4050		
--			AD		1891 1907	317 319	106.6 108.6		10.1		4050		
--		Es 334	Bd		1906 1907	348 350	15.1 15.7		13.8		4050		
5 28.0	-38 54	B 1482		Col	1929 1952	278	1.1	7.7	10.7	G5			
5 28.2	-20 46	β 320	AB	Lep	1875 1957	282 330	2.8 2.5	2.8	7.3	G0	4066		9 β Lep
--		h 3761	AC		1898 1921	146 145	65.6 64.3		11.8		4066		
--			AD		1879	75	206.4		10.3		4066		
--			AE		1879	58	241.5		10.3		4066		
5 28.7	-8 23	h 2268		Ori	1879 1918	300	26.0	7.0	9.6	G5	4071		
5 28.7	-65 27	Rst 137		Dor	1929	339	10.0	6.8	13.0	G0			
5 28.8	+1 39	h 2267		Ori	1904	111	25.0	6.9	10.6	B3			
5 28.9	-3 18	Dawes 6		Ori	1854 1960	80 128	0.8 0.2	7.0	7.3	B9	4078		
5 29.0	+67 55	Σ 689		Cam	1831 1910	324	6.1	8.2	10.2	B8	4026		
5 29.3	+18 22	OΣ 108		Tau	1849 1937	133	3.4	6.7	10.2	A2	4073		
5 29.3	+25 09	Σ 716	AB	Tau	1829 1957	197 204	4.9 4.8	5.8	6.6	A0	4068		118 Tau
--			AC		1912	99	141.3		11.6		4068		
5 29.4	-7 16	H V 101		Ori	1899 1901	116	48.6	6.7	9.7	B5			
5 29.6	+3 09	Σ 721	AxBC	Ori	1783 1921	149	24.6	7.2		B3	4088		
--		β 557	BC		1878 1951	142 174	0.5 0.2	9.2	9.2		4088		
5 29.7	-0 48	A 850		Ori	1904 1917	155	2.1	7.9	12.3	A2	4096		
5 29.7	-1 06	Σ 725		Ori	1829 1933	87	12.7	4.7	9.9	K5	4097		31 Ori
5 29.7	+35 23	Hu 217		Aur	1900 1960	257 251	0.6 0.6	6.8	8.3	B5	4072		
5 29.8	+18 25	β 891	AB	Tau	1879 1917	125	10.6	7.6	13.6	A0	4087		
--		h 3275	AC		1875 1932	21 21	52.9 54.2		7.6	F	4087		
--		A 2433	CD		1912 1931	248	1.1		12.9		4087		
5 30.1	-1 45	Rst 4781		Ori	1942 1950	171	0.3	8.3	8.5	B9			
5 30.1	+29 33	Σ 719	AB	Aur	1833 1955	330	1.0	7.3	9.8	G5	4086		
--			AC		1833 1934	351	15.0		9.3		4086		
5 30.2	-47 05	h 3767	AB	Pic	1835 1933	223 257	25.9 25.9	5.5	11.7	G5			
--		Δ 21	AD		1837 1913	264 268	197.0 197.0		6.5	F0			
--		Rst 136	BC		1930 1949	96 90	0.6 0.6		12.7				
5 30.3	+41 17	Σ 715	AB	Aur	1831 1943	201	0.9	8.1	8.8	A0	4083		
--			AC		1904	51	19.5		11.0		4083		
5 30.3	-63 56	HdO 192		Dor	1900 1930	68	9.2	6.2	11.3	F0			
5 30.6	+31 31	h 701		Aur	1901 1912	137	35.2	7.5	11.2	K5			
5 30.8	+5 57	Σ 728		Ori	1830 1960			4.5	5.8	B3	4115	o	32 Ori
5 30.8	+8 24	A 2706		Ori	1914 1921	99	3.8	7.7	13.7	F8	4114		
5 30.8	+39 50	OΣΣ 63		Aur	1874 1925	274	75.4	6.5	7.8	K0 K0			
5 30.9	-4 15	h 2270		Ori	1896 1911	338	26.9	7.6	10.6	A2			
5 30.9	+10 15	Σ 726		Ori	1831 1953	259	1.2	8.0	8.5	B9	4113		
5 31.0	-6 12	Σ 732 rej		Ori	1903	252	44.7	8.6	8.9	B9			
5 31.2	+3 18	Σ 729	AB	Ori	1831 1957	27	1.8	5.8	7.1	B3	4123		33 Ori; third body? (Dom)
--			AC		1912 1923	52	95.3		13.4		4123		
5 31.2	-42 18	Δ 22		Col	1826 1952	171 169	7.5 7.5	7.5	8.0	A5			
5 31.6	+54 40	Σ 711	AB	Aur	1830 1968	234 228	9.0 8.0	7.8	9.5	G0	4099		
--			BC		1881 1908	243 245	197.1 190.6		10.3		4099		
--			CD		1908	308	15.4		13.4		4099		
--			Cc		1908	174	0.6		11.0		4099		
5 31.7	+2 50	S,h 61		Ori	1822 1918	353	67.2	8.1	8.6	A2			
5 31.9	-76 20	h 3795		Men	1918	107	38.2	5.2	11.2	K0			γ Men
5 32.0	-0 18	β 558	AB	Ori	1878 1922	227	32.8	2.2	13.7	B0	4134		34 δ Ori
--		Σ I 14	AC		1781 1932	359	52.6		6.3	B0	4134		Spectroscopic binary
5 32.2	+17 03	Σ 730		Tau	1831 1972	141	9.6	6.0	6.5	B9	4131		
5 32.3	+42 24	A 1722		Aur	1907 1940	31	1.5	8.1	10.1	F0	4124		

α 2000	δ 2000	Star Name		Const	Years	Pos Ang	Separation	m₁	m₂	Spec	ADS	Orb	Notes
5ʰ32ᵐ4	+49°24′	Σ 718	AB	Aur	1829 1955	74° °	7″7 ″	7.5	7.5	F5	4119		
--			AC		1881 1910	185	119.4		9.2		4119		
5 32.5	+70 49	A 1034		Cam	1905 1959	271 156	0.3 0.4	8.5	8.9	G5	4076		
5 32.7	-1 36	β 1048		Ori	1889 1937	357	2.1	5.3	9.8	B2	4141		
5 32.7	-17 49	h 3766	AB	Lep	1877 1920	156	35.8	2.6	11.1	F0	4146		11 α Lep
--			AC		1913	186	91.4		11.8		4146		
5 33.0	-1 45	HCWils 3		Ori	1884	153	2.8	7.0	9.0				No such pair near [38]
5 33.1	-1 43	Σ 734	AB	Ori	1832 1943	354	1.8	6.5	8.2	B3	4150		
--			AC		1783 1934	243	29.4		8.2		4150		H V 119
--		β 1049	CD		1888 1943	295	0.7		9.3		4150		
5 33.5	-24 20	h 3770		Lep	1881 1959	19	4.0	7.7	11.2	F8	4157		
5 33.6	+36 27	Sei 323		Aur	1895	265	8.8	7.8	11.0	B1	4143		
5 33.7	-54 54	h 3777	AB	Pic	1834 1938	349 345	50.8 54.7	6.4	10.0	F5			
--			BC		1897 1926	104	11.3		12.2				
5 33.8	+44 47	Σ 727		Aur	1830 1935	59	2.2	8.0	9.5	F5	4137		
5 34.1	-1 02	H V 118		Ori	1901	264	27.5	6.2	9.8	B3	4159		
5 34.1	+69 40	Σ 704		Cam	1831 1915	8 14	26.5 20.5	8.0	10.3	F5	4118		
5 34.3	+3 46	OΣ 110 rej		Ori	1923	210	0.4	5.4		A2			38 Ori; single (Hu)
5 34.5	-4 29	β 13		Ori	1876 1943	138	1.1	8.2	10.2	B9	4172		
5 34.6	+51 55	Σ 723		Aur	1830 1912	105	4.4	8.1	10.2	A0	4147		
5 34.7	-4 24	Σ 743		Ori	1830 1954	280	1.9	8.3	9.4	B8	4176		
5 34.9	-0 07	Σ 741		Ori	1831 1894	285	10.0	8.0	11.0	B5	4180		
5 35.0	-6 00	Σ 747		Ori	1833 1924	223	35.7	4.8	5.7	B1 B1	4182		
5 35.1	+9 56	Σ 738	AB	Ori	1830 1957	43	4.4	3.6	5.5	O5 O5	4179		39 λ Ori
--			AC		1880 1957	184	28.6		11.1		4179		
--			AD		1905 1957	271	78.3		11.1		4179		
5 35.1	+30 56	β 1267		Aur	1892 1959	218 203	0.8 0.6	8.8	8.8	F5	4166		
5 35.2	+10 14	OΣ 111		Ori	1857 1943	351	2.9	5.6	9.8	B8	4181		
5 35.3	-5 23	Σ 748	AB	Ori	1836 1975	32 31	8.7 8.8	6.7	7.9	B5 B2	4186		41 θ¹ Ori, the famous
--			AC		1836 1975	132 132	13.0 12.8		5.1	B3	4186		Trapezium [39]
--			AD		1836 1975	95 96	21.5 21.5		6.7		4186		
--			AE		1832 1934	351	4.1		11.1		4186		
--			AH		1889	178	7.9		15		4186		
--			BC		1836 1957	162	16.8				4186		
--			CF		1858 1957	122	4.0		11.5		4186		
--			CG		1888	34	7.4		16		4186		
--			CH		1889	276	8.6				4186		
--			DB		1836 1932	300	19.3				4186		
--			DC		1836 1957	241	13.4				4186		
--			DG		1925 1936	272	7.0				4186		
--			HH′		1889	274	1.3		16		4186		
5 35.3	-71 08	I 277		Men	1900	190	3.9	7.8	11.0	K2			
5 35.4	-4 26	φ 345		Ori	1960	101 120	0.2 0.2	6.7	7.4	B0			
5 35.4	-4 50	Dawes 4		Ori	1876 1936	213	1.6	4.7	7.9	B3	4187		42 Ori
5 35.4	-5 25	Σ I 16	AB	Ori	1836 1937	92	52.5	5.2	6.5	B1 B1	4188		43 θ² Ori
--			AC		1869 1926	97	128.7		9.1	B8	4188		
--		Σ I 17			1836 1926	314	135.3	4.9	5.0		4188		Meas. of θ² A and θ¹ C
5 35.4	-5 55	Σ 752	AB	Ori	1831 1932	141	11.3	2.8	6.9	O5 B9	4193		44 ι Ori
--			AC		1824 1904	103	49.5				4193		
5 35.4	-8 39	A 489		Ori	1903 1933	78	2.8	8.1	11.1	A0	4170		
5 35.4	-33 16	Hu 1393		Col	1922 1946	341 335	0.6 0.6	7.3	8.1	F5			
5 35.5	-1 51	A 2917		Ori	1915 1922	265	0.6	8.1	10.1	F0	4191		
5 35.5	-4 22	Σ 750		Ori	1831 1940	60	4.2	6.5	8.5	B5	4192		
5 35.5	+37 54	Hu 1229		Aur	1905 1948	197	1.8	7.5	13.0	A5	4168		
5 35.6	-3 15	Rst 4281		Ori	1939 1947	352	1.1	6.3	12.5	B5			
5 35.6	+38 01	A.G. 98		Aur				7.8		A2			Single in 1902
5 35.7	-4 51	H1d	AB	Ori	1877 1904	169	19.3	5.3	14.3	F0	4196		45 Ori
--			AC		1904	34	19.1		14.3		4196		
5 35.8	-0 59	Σ 751		Ori	1831 1931	124	15.6	8.5	8.9	B9 A	4195		
5 36.2	-1 12	β pm		Ori	1879 1907	57	179.9	1.7	10.4	B0			46 ε Ori
5 36.4	+22 00	Σ 742		Tau	1837 1975	251 270	3.3 3.9	7.2	7.8	F8	4200	o	
5 36.6	-6 04	Σ 754		Ori	1830 1933	287	5.2	5.7	8.9	B3	4212		
5 36.6	-34 43	δ 118		Col	1921 1960	209 204	0.3 0.3	8.2	8.9	K0			
5 37.1	+26 55	Σ 749	AB	Tau	1829 1959	23 333	0.7 1.0	6.4	6.5	B8	4208		
--			CD		1897 1924	290	4.4	10.4	10.9		4208		
5 37.1	+41 50	Σ 736		Aur	1830 1974	342 357	2.0 2.5	7.4	8.7	F8	4204		
5 37.2	+4 46	A 2651	AB	Ori	1913 1952	158 172	0.4 0.4	7.8	9.5	B9	4217		
--			ABxC		1913 1921	326	8.2		12.6		4217		
5 37.3	+64 09	Hu 1107		Cam	1905 1941	47	1.3	6.0	10.0	B9	4177		19 Cam
5 37.3	+66 42	Mlr 314		Cam	1972	307	0.3	6.5	7.3	A5			
5 37.8	-54 34	h 3787		Pic	1882 1938	248	24.5	7.7	10.2	G5 G5			
5 37.9	+0 58	OΣΣ 65	AB	Ori	1875 1923	31	80.1	7.4	8.1	B9 B9			
--		β	BC		1903	62	21.0		14.4				
5 38.1	-0 10	Σ 757	AB	Ori	1831 1943	239	1.6	8.6	8.8	B9	4234		
--		Σ 758	AC		1831 1932	86 86	50.9 51.8		8.9		4234		

α 2000	δ 2000	Star Name		Const	Years	Pos Ang	Separation	m₁	m₂	Spec	ADS	Orb	Notes
5ʰ38ᵐ.1	-0°10′	Σ 758	AD	Ori	1842 1934	79°	41″.5		8.8		4234		
--		S 493	AE		1825 1904	262	138.4		10.1		4234		
--		Σ 758	CD		1831 1934	298	11.1				4234		
5 38.2	-46 06	h 3784		Pic	1879 1941	59 67	5.7 5.0	7.4	9.5	G0			
5 38.6	+30 30	β 1240	AB	Aur	1892 1958			6.0	6.3	A2	4229	o	26 Aur
--		Σ 753	ABxC		1828 1967	267	12.4		8.0		4229		
--		β 90	ABxD		1877 1915	113 114	31.5 33.1		11.5		4229		
5 38.6	-41 18	h 3781		Col	1883 1919	135	15.9	7.9	9.4	F5			
5 38.7	-2 36	β 1032	AB	Ori	1888 1960			4.0	6.0	B0	4241	o	48 σ Ori
--		Σ 762	ABxC		1831 1973	237 238	11.1 11.4		10.3		4241		
--			ABxD		1831 1969	84 84	12.8 12.9		7.5		4241		
--			ABxE		1869 1970	61 61	41.6 42.6		6.5	B3	4241		
--			ED		1831 1932	231	30.1				4241		
5 38.7	+32 29	Ara		Aur	1933	260	51.7	7.8	10.1	G0			
5 39.1	-50 08	h 3789		Pic	1847 1935	1	9.1	8.5	9.0	G5			
5 39.2	+23 17	Σ 755	AB	Tau	1830 1922	316	6.0	9.1	9.8	F8	4239		
--			AC		1879 1907	34	146.2		7.8	K5	4239		
--			CD		1879 1907	176	79.1		10.4		4239		
5 39.3	+43 43	A 1564		Aur	1907 1953	341 35	0.3 0.1	8.7	8.7	F5	4236		
5 39.3	-17 51	β 321	AB	Lep	1877 1947	146	0.8	6.4	7.9	B9	4254		
--		h 3780	ABxI		1914	102	89.2				4254		
--			AC		1876 1916	136	89.2		8.5		4254		
--			AE		1876 1915	7	76.1		8.4		4254		
--			AF		1876 1915	299	128.8		8.1		4254		
--			AG		1878 1916	49	59.8		9.5		4254		
--			AH		1878	310	41.8		12.4		4254		
--			CD		1877 1946	357	1.5		9.2		4254		
5 39.6	-34 04	HdO 193		Col	1898 1950	359 359	11.4 13.5	2.6	12.3	B8			α Col
5 39.7	-20 26	La1	AB	Lep	1870 1950	123	11.0	6.9	7.9	B8	4260		
--		HdO 77	AC		1870 1945	83	32.2		11.3		4260		
5 39.8	+35 39	Hu 824		Aur	1904 1926	160	2.8	7.8	12.6	K0	4242		
5 39.9	+13 01	OΣ 113		Ori	1847 1935	28	10.2	7.2	10.9	A0	4253		
5 39.9	+37 57	OΣ 112		Aur	1848 1960	85 52	0.6 0.6	7.8	8.5	B9	4243		
5 40.0	+36 01	Hu 825		Aur	1902 1952	343	0.4	8.4	8.6	A0	4249		
5 40.3	+15 21	Σ 766		Ori	1829 1938	275	10.0	7.0	8.2	F0	4256		
5 40.8	-1 57	Σ 774	AB	Ori	1836 1970	151 162	2.6 2.4	1.9	4.0	B0 B0	4263	o	50 ζ Ori
--			AC		1880 1930	10	57.6		9.9		4263		
--			BC		1909	8	59.7				4263		
5 41.3	+16 32	β 1007		Tau	1881 1960	266 238	2.6 0.3	5.3	5.9	B3	4265	o	126 Tau [40]
5 41.3	+29 29	Σ 764		Aur	1831 1932	14	26.0	6.9	7.5	A A	4262		
5 41.3	+53 29	β pm	AB	Aur	1891 1925	71	97.5	6.3	9.8	K0			
--			AD		1874 1909	277 279	686.4 688.0		9.1	F5			
--			Aa		1907 1925	302 307	82.2 87.0		12.3				
--			Bb		1907 1910	160 159	131.2 129.7		12.7				
5 41.3	-26 21	h 3788		Lep	1898 1929	154	25.9	7.8	9.6	F5	4281		
5 41.5	-48 15	h 3793		Pic	1879 1956	110 127	11.7 12.1	7.2	11.0	A0			
5 41.7	-2 54	β 1052		Ori	1889 1943	189 172	0.7 0.7	6.8	7.8	F0	4279		
5 42.0	+71 38	OΣ 109		Cam	1847 1931	129	10.8	8.3	9.6	A2	4238		
5 42.5	-0 01	Σ 782	AB	Ori	1831 1925	309 307	36.2 42.3	8.3	8.6	K0 K2			
--			AC		1910	164	131.7		11.7				
5 42.5	+29 51	β 14		Aur	1874 1933	194	5.7	7.4	10.5	G0	4280		
5 42.7	-67 08	I 745		Dor	1913 1942	251	0.8	7.8	10.5	F8			
5 42.9	-6 48	A 494	AB	Ori	1903 1960			6.5	6.9	F5	4299	o	
--			ABxC		1903 1971	232 231	99.0 101.6		10.0		4299		
--			CD		1903 1915	199	1.0		13.5		4299		
5 42.9	+18 59	Ho 649		Tau	1906	p	0.4	6.8	10.1	B9			Single (IDS)
5 42.9	-33 58	h 3794		Col	1898 1919	277	23.5	7.3	12.0	G5			
5 43.0	+33 19	Ho 509		Aur	1897 1923	204	11.5	6.8	11.8	F5	4287		
5 43.0	+73 59	A 1037	AB	Cam	1904 1930	357	0.8	6.8	11.5	F0	4246		
--			AC		1922	345	27.1		14.0		4246		
5 43.3	+41 07	Σ 768 rej		Aur	1905	221	18.6	7.5	10.1	B8	4288		
5 43.4	-49 51	Rst 148		Pic	1928 1949	96	3.1	7.6	13.2	B9			
5 43.6	+12 59	A 117	AB	Tau	1901 1942	255	0.5	8.5	8.7	A0	4304		
--		A	ABxC		1901	191	26.0		14.5		4304		
5 44.4	+27 44	Σ 779	AB	Tau	1831 1909	253	8.3	8.0	10.0	A0	4308		
--			AC		1911	36	52.1		11.5		4308		
5 44.5	-7 45	A 497		Ori	1903 1939	182	2.2	8.2	10.2	A0	4330		
5 44.5	+15 04	OΣ 115	AB	Tau	1847 1958	123 119	0.8 0.6	7.5	8.3	G0	4323		
--			AC		1850 1911	256	92.8		11.0		4323		
5 44.5	-22 27	H VI 40	AB	Lep	1825 1957	350 350	94.0 96.3	3.7	6.3	F8 G5	4334		13 γ Lep
--		H V 50	BC		1832	345	45.0		10.9		4334		
5 44.7	+3 50	Σ 788	AB	Ori	1831 1927	89	7.4	7.8	9.5	B9	4329		
--			AC		1831 1905	148	35.9		10.1		4329		
5 44.8	-21 40	Ho 336		Lep	1890 1945	238	19.4	6.7	11.7	B3	4339		
5 44.9	+26 21	A 496		Tau	1903 1956	11 18	0.2 0.3	7.7	8.3	B9	4324		

α 2000	δ 2000	Star Name		Const	Years		Pos Ang		Separation		m₁	m₂	Spec	ADS	Orb	Notes
h m 5 45.0	+4° 00'	Σ 789 rej	AB	Ori	1905	1958	150°	150°	17″.5	15″.7	6.2	8.7	F0	4333		
---		A 2655	BC		1913	1958	108	112	1.2	1.1		12.2		4333		
5 45.0	+28 12	Cou 762		Tau	1972		75		0.2		8.0	8.0	A0			
5 45.9	+25 55	Σ 785	AB	Tau	1830	1943	348		14.2		8.1	9.1	B9	4343		
---		OΣ 116	AC		1846	1908	66		18.1			13.2		4343		
---			AD		1911		10		201.4			11.1		4343		
---			Ac		1908		72		31.9			13.5		4343		
---			DE		1890	1911	252		6.4			12.8		4343		
5 46.0	-4 16	Σ 790		Ori	1830	1933	89		6.9		6.4	8.7	K0	4361		
5 46.0	+21 19	Σ 787	AB	Tau	1832	1953	78	65	1.4	1.0	8.4	8.8	F2	4349		
---			AC		1911		40		12.7			12.9		4349		
5 46.0	+37 17	β pm		Aur	1880	1909	114	111	48.3	29.3	7.4	11.1	K0			
5 46.4	+47 54	β 752		Aur	1879						8.0		B9			Single (IDS)
5 46.5	-46 36	h 3801		Pic	1913		194		37.0		5.3	12.7	K0			
5 46.9	+9 31	J 251		Ori	1910	1957	274	288	16.3	17.2	5.8	11.8	G5	4369		
5 47.0	-44 48	h 3803		Pic	1836	1932	114		20.2		8.0	10.5	A2			
5 47.2	+28 37	Cou 763		Aur	1972		196		0.8		8.0	10.3	B9			
5 47.4	+29 39	β 560		Aur	1877	1959	208	136	0.9	1.5	8.0	8.0	F8	4371		
5 47.4	-36 29	B 1489		Col	1929	1934	208		3.0		7.1	12.0	K2			
5 47.5	+20 56	β 91		Tau	1875	1943	85		1.6		8.1	10.6	F	4378		Composite spectrum F+A
5 47.7	+13 54	h 3279	AB	Tau	1877	1927	298		17.8		5.3	12.3	B5	4381		133 Tau
---			AC		1877	1929	181		24.9			11.6		4381		
5 47.8	-2 18	β 15		Ori	1875	1906	178		2.1		8.0	12.2	K0	4391		
5 47.9	+7 58	OΣ 119		Ori	1848	1954	304	334	0.6	0.8	7.9	8.7	F8	4388		
5 47.9	+24 41	OΣΣ 66		Tau	1874	1923	166		94.2		7.2	8.0	K2 K2			
5 47.9	+35 10	Ho 19		Aur	1886	1958	346		7.0		6.6	12.6	G0	4377		
5 47.9	-13 32	β 405		Lep	1877	1901	126		14.3		8.1	10.6	K2	4397		
5 48.0	+6 27	Σ 795		Ori	1831	1959	200	210	1.8	1.6	6.1	6.1	A3	4390		52 Ori
5 48.0	+12 25	β 561	AB	Ori	1878	1919	4	3	19.7	18.9	6.7	12.7	B5	4386		
---			AC		1917		34		77.0			9.3		4386		
5 48.2	+3 54	J 36		Ori	1910	1952	113	109	1.7	1.6	7.8	10.6	F8	4395		
5 48.2	-8 23	Σ 798		Ori	1830	1917	181		20.7		7.3	9.3	B9	4402		
5 48.2	+30 32	OΣ 117		Aur	1847	1914	30		11.9		7.3	10.0	K5	4383		
5 48.2	-48 55	I 63	AB	Pic	1897	1959	12	15	1.4	1.1	7.5	9.0	A0			
---			AC		1901		211		31.9			13.0				
5 48.4	+20 52	OΣ 118	AB	Tau	1854	1953	318		0.6		6.1	7.6	B9	4392		
---		OΣΣ 67	ABxC		1849	1933	161		75.5			8.6	A	4392		
5 48.5	+4 42	Σ 797		Ori	1832	1931	16		7.1		7.3	10.1	A0	4401		
5 48.5	-13 22	Σ 801 rej	AxBC	Lep	1902	1905	326		27.1		7.4		K0	4410		
---		Rst 5502	BC		1951		96		0.2		10.8	10.8		4410		
5 48.8	+12 00	h 5465		Ori	1823		45		12.0		6.9		B8			Single (IDS)
5 49.1	+62 49	Σ 3115		Cam	1831	1958	36	5	1.7	1.0	6.5	7.6	A2	4376		
5 49.1	-32 30	Jsp 84		Col	1929	1944	81		3.9		6.8	14.4	K0			
5 49.2	+39 11	β 192	AB	Aur	1877	1916	352		39.4		4.5	11.5	K0	4398		29 τ Aur
---		H V 21	AC		1877	1926	33	35	47.8	49.6		11.5		4398		
5 49.4	+36 11	Hu 1233		Aur	1905	1954	24		0.7		8.4	9.4	F5	4406		
5 49.5	+12 39	β pm		Tau	1909		129		18.9		4.9	10.3	B9			134 Tau
5 49.5	-12 13	Fox	AB	Lep	1910		304		9.0		8.0	11.7	K5	4430		
---			AC		1910		290		46.0			12.7		4430		
5 49.6	-14 29	β 94		Lep	1876	1943	179	173	2.7	2.3	5.5	8.9	G5	4432		BDS 3207 prob. same star
5 49.9	+31 47	Σ 796	AB	Aur	1830	1943	62		3.8		7.0	8.1	A3	4421		
---			AC		1880	1909	324		207.5			10.5		4421		
5 50.1	-5 54	Rst 4285		Ori	1939	1941	283		0.7		8.0	12.5	G5			
5 50.1	+39 35	Σ 791	AB	Aur	1830	1938	91		4.8		8.5	9.1	A0	4420		
---			AC		1880	1909	213		60.6			9.7		4420		
5 50.1	-34 33	B 1490		Col	1929	1936	155		6.1		7.3	13.5	K5			
5 50.2	-55 37	Rst 159		Pic	1929	1935	296		0.4		8.7	8.8	G0			
5 50.5	-37 37	B 2102		Col	1960		59		9.5		7.0	13.6	K5			
5 50.5	-52 46	B 1493		Pic	1929	1959	270	250	0.3	0.2	6.8	7.6	F5			Spectrum composite F5+A
5 50.6	-1 26	β 1188	AB	Ori	1890	1937	105		1.1		7.9	10.3	G5	4442		
---		Σ 809	AC		1831	1923	98		25.2			8.8	F0	4442		
5 50.6	+56 55	H IV 125		Cam	1900	1924	131		25.1		6.5	9.5	A2	4412		29 Cam
5 50.8	+14 27	Ku 23		Tau	1902	1953	106		0.9		6.8	8.8	B9	4441		
5 51.0	+27 58	β 1054		Tau	1889	1958	232		15.0		5.6	11.6	K0	4474		
5 51.0	+65 45	Σ 780	AB	Cam	1831	1954	104		3.8		6.9	8.1	F8	4405		
---			AC		1831	1954	154	150	10.9	12.3		10.0		4405		
---			AD		1904	1954	53	56	18.0	19.3		13.4		4405		
5 51.0	-69 54	h 3820		Dor	1872	1917	91		26.5		7.8	10.8	F2			
5 51.4	+32 08	Es 415		Aur	1907	1934	17		15.2		6.3	11.9	M	4443		
5 51.4	-11 39	A 2512		Lep	1912	1959	278		1.0		7.3	8.9	B9	4458		
5 51.5	+39 09	H V 90		Aur	1878	1911	206		54.6		4.0	9.3	K0	4440		32 ν Aur
5 51.6	-41 40	h 3807		Col	1894	1934	272		5.0		8.1	11.0	G0			
5 52.0	-7 19	β 95	AB	Ori	1878	1915	297		13.9		8.0	12.0	B9	4462		
---			AC		1914		92		87.7			10.8		4462		
5 52.0	+37 50	Sei 397		Aur	1895		150		24.9		8.1	10.6	G5			

α 2000	δ 2000	Star Name		Const	Years	Pos Ang	Separation	m₁	m₂	Spec	ADS	Orb	Notes
5ʰ52.ᵐ2	+38°34′	Σ 799		Aur	1829 1957	192° 171°	1″.1 0″.8	7.1	8.2	B8	4452		
5 52.4	+1 51	β		Ori	1901	121 2	43.4	4.8	13.3	K0	4467		56 Ori
5 52.4	+46 48	Es 1321	AB	Aur	1914	351	5.6	7.9	9.7	K0	4449		
--			AC		1914	179	23.8		14.7		4449		
--			AD		1914	317	33.2		14.7		4449		
5 52.6	-37 38	I 64	AB	Col	1900 1960	233 250	18.7 16.5	5.6	11.7	K0			
--			AC		1899	16	44.7		11.9				
5 52.8	-38 32	I 16		Col	1896 1937	126	1.3	6.8	10.3	K0			
5 52.9	+29 07	Cou 898		Aur	1972	143	0.2	8.7	8.8	G5			
5 52.9	+34 27	Σ 807		Aur	1829 1934	140 147	2.2 2.2	7.9	9.9	F8	4463		
5 53.2	-47 57	h 3816		Pic	1913	180	22.9	6.6	11.6	K2			
5 53.2	-61 50	Slr 15		Dor	1895 1959	325 286	0.8 0.2	7.5	7.9	A0			
5 53.3	-32 49	B 1495		Col	1929 1960	21 54	0.2 0.3	8.4	8.6	A2			
5 53.5	+37 20	β 1053		Aur	1889 1959	283 350	0.4 1.4	6.9	8.9	F5	4472		
5 53.6	-56 40	ø 93		Pic	1929	129	0.3	8.6	8.8	G5			
5 54.0	+12 25	LDS 891		Ori	1936	230	86.0	7.8	15	F8			
5 54.0	+28 58	Cou 900		Aur	1972	84	0.1	8.8	8.8	B9			
5 54.2	+10 15	OΣ 123		Ori	1846 1959	176 185	2.4 1.9	7.2	8.9	G5	4491		
5 54.2	-29 09	ø 382		Col	1962 1965	17 324	0.1 0.1	7.2	7.2	F2			
5 54.3	+30 30	Σ 811		Aur	1829 1934	231	5.0	7.7	9.2	B5	4483		
5 54.4	+18 53	Σ 813		Ori	1831 1954	148	3.1	8.5	8.5	A0	4490		
5 54.4	+38 35	Weisse 11		Aur	1904 1919	164	24.7	8.6	8.7	A0			
5 54.4	-15 44	A 3019	AB	Lep	1926 1933	42 41	45.6 44.8	7.7	11.0	B9	4500		
--			BC		1926 1933	4	3.2		13.2		4500		
5 54.5	+11 46	Kui 21		Ori	1958	196	22.7	6.5	12.0	B9			
5 54.5	-19 42	GAn 2		Lep	1876 1941	19 20	9.1 9.8	7.4	9.7	G5	4503		
5 54.6	+5 21	Σ 815	AB	Ori	1832 1905	137	12.8	8.0	10.2	G5	4496		
--			AC		1864 1908	306	88.2		8.8		4496		
5 54.7	+13 51	S 502		Ori	1874 1913	130	45.7	8.7	8.7	A0 A0			
5 54.8	-53 35	Rst 5503		Pic	1948	63	0.2	8.6	8.6	A3			
5 54.9	+5 52	Σ 816		Ori	1830 1943	289	4.3	6.8	9.3	B9	4499		
5 54.9	+14 13	Ho 20	AB	Ori	1886 1937	276	7.8	6.8	11.8	K0	4497		
--			AC		1886	287	50.2		11.3		4497		
5 55.2	+7 24	H VI 39	AB	Ori	1891	110	39.8	0.5	14.1	M	4506		58 α Ori, Betelgeuse [41]
--			AC		1891	290	62.0		13.8		4506		
--			AD		1877 1912	348	76.8		13.1		4506		
--			AE		1786 1917	152 153	161.8 174.4		10.6		4506		
--			Aa		1940	92	0.1				4506		
5 55.4	+29 58	Mil		Aur	1904	231	4.7	7.3	9.5				
5 55.6	+53 28	OΣ 120 rej		Aur	1843 1914	133 136	43.7 45.8	8.0	9.2	M F5			
5 55.6	-18 58	B		Lep	1933	198	10.0	7.6	14.6	K0			
5 55.8	+36 56	OΣ 122		Aur	1847 1958		7.6 8.3	A5			4505	o	
5 55.9	-21 41	See 57		Lep	1897 1899	109	24.6	6.8	14.2	G0	4527		
5 56.0	+11 31	Ho 227		Ori	1890	241	2.1	7.9	12.4	B9	4520		
5 56.1	+13 56	S 503	AB	Ori	1825 1935	134 280	39.9 14.2	6.7	8.7	G5	4519		
--			AC		1878 1934	157 323	28.1 31.5		10.7		4519		
--			AD		1825 1923	337 334	201.8 257.7		8.5	G5	4519		
--			BD		1914	335	233.5				4519		
5 56.1	-22 50	Ara 1632		Lep	1921	12	12.2	7.2	12.1				
5 57.2	-53 26	h 3822	AB	Pic	1882 1938	304	55.9	6.5	7.5	K2 K0			
--			BC		1917	125	20.1		13.1				
5 57.4	+0 02	β 1190	AB	Ori	1890 1945	340 340	1.4 1.2	6.8	10.2	A0	4542		
--			AC		1890 1945	96	6.5		11.9		4542		
5 57.5	-35 17	h 3819		Col	1919	110	33.8	4.4	12.7	B3			γ Col
5 57.7	-7 39	Σ 823		Mon	1831 1936	339	7.7	8.5	9.2	K0	4547		
5 57.8	-14 13	Gallo		Lep	1902	73	7.0	8.0	12.5	K2	4549		
5 58.3	-20 09	Lal 43		Lep	1825 1957	74 68	3.6 3.5	8.8	8.8	F0	4567		
5 58.4	+1 50	H V 100		Ori	1878 1925	204	36.7	6.1	9.7	A5	4555		59 Ori
5 58.4	-4 39	A 322	AB	Mon	1902 1932	359	4.1	6.8	13.6	G0	4557		
--			AC		1892 1911	296 297	176.8 179.4		9.0		4557		
5 58.4	+29 38	Σ 821		Aur	1830 1939	12 4	2.2 2.1	8.0	9.8	B9	4544		
5 58.8	+0 33	Kui 22		Ori	1958	30	19.1	5.2	11.8	A0			60 Ori
5 58.9	+12 49	OΣ 124		Ori	1845 1960	309 302	0.5 0.3	6.0	7.8	G5	4562		Spectrum composite G5+A5
5 59.2	-33 49	I 155	AB	Col	1898 1927	23	2.0	8.1	11.1	A5			
--			AC		1900 1927	98	17.2		13.1				
5 59.4	+17 49	OΣ 126		Ori	1846 1938	60	10.5	7.9	10.4	F5	4571		
5 59.4	+45 37	A 1726		Aur	1908 1919	274	5.0	6.6	13.6	A0	4551		
5 59.5	+44 57	H VI 88	AB	Aur	1841 1924	39	184.6	1.9	10.4	A0	4556		34 β Aur; variable
--		Barnard	Aa		1901 1934	181 174	12.6 12.8		13.9		4556		
5 59.5	+54 17	β pm	AB	Aur	1888 1908	270 271	113.4 115.4	3.7	9.5	K0			33 δ Aur
--			AC		1877 1907	68 67	199.6 197.1		9.5				
--			Ca		1908	111	93.5		10.8				
5 59.7	+22 28	OΣ 125		Ori	1847 1953	0	1.5	7.5	9.0	A0	4577		
5 59.7	+37 13	OΣ 545	AB	Aur	1871 1976	6 313	2.2 3.6	2.6	7.1	A0	4566		37 θ Aur
--			AC		1852 1939	291 297	43.2 50.0		10.6		4566		

α 2000	δ 2000	Star Name		Const	Years	Pos Ang	Separation	m₁	m₂	Spec	ADS	Orb	Notes
5ʰ 59.7ᵐ	+37° 13'	OΣ 545	AD	Aur	1840 1924	351° 350°	123".3 130".7		9.1		4566		
5 59.8	+43 59	A 1571		Aur	1907 1919	90	4.7	7.9	13.2	K0	4563		
6 00.1	−0 30	Σ 827 rej		Ori	1906 1920	222	23.6	7.8	10.1	B9			
6 00.2	+43 11	Σ 822 rej		Aur	1904	51	15.4	7.9	10.4	B9	4574		
6 00.3	+44 36	β 1055	AB	Aur	1888 1953	333 324	1.6 1.7	6.4	11.2	G5	4576		
---		H VI 91	AC		1783 1939	315 331	30.0 34.2		9.0		4576		
6 00.4	−31 02	Hu 1399	AB	Col	1915 1960			8.9	9.7	K5		o	
---		h 3823	AC		1836 1960	131 42	4.4 1.7		8.7				
---		B	AD		1927 1960	272 191	17.6 11.9		12.8				
6 00.9	−21 00	h 3821	AB	Lep	1914 1920	214	17.9	8.2	9.4	A2	4602		
---		I	AC		1920	87	19.8		12.5		4602		
6 01.0	+27 34	Ho 21		Gem	1884 1940	240	9.3	6.1	12.4	B8	4589		
6 01.0	−37 04	B 1045		Col	1928 1933	192	1.0	7.2	10.0	G5			
6 01.3	+31 32	Sei 453		Aur	1899	212	18.8	8.1	9.1	G5			
6 01.3	+65 32	Σ 812 rej		Cam	1905	66	25.4	6.7	11.0	A3			
6 01.5	+36 32	Σ 825	AB	Aur	1829 1959	146	8.3	8.1	9.3	A0	4592		
---			AC		1906	139	37.8		11.5		4592		
6 01.7	−35 03	B 1954		Col	1931 1945	229 224	0.2 0.2	7.8	8.1	A2			
6 01.7	−56 59	φ 95		Pic	1929 1944	112	0.4	8.6	8.8	F0			
6 01.8	−10 36	β 16		Mon	1872 1935	354	1.8	5.0	8.5	B8	4615		3 Mon
6 02.1	+38 43	OΣ 127		Aur	1848 1958	333	1.5	7.2	11.0	G5	4597		
6 02.1	−27 26	h 3825		Col	1920	338	32.6	7.1	10.9	F5			
6 02.4	+9 39	A 2715	AB	Ori	1914 1960			4.4	6.0	A2	4617	o	61 μ Ori
---		β 1056	ABxC		1899 1955	272 279	16.8 18.2		14.1		4617		
6 02.9	+25 53	A 120		Gem	1901 1959	163 173	0.6 0.7	8.3	9.5	F8	4619		
6 03.0	+82 45	Hu 1112		Cam	1905 1946	316	0.2	8.0	8.7	F8	4492		
6 03.2	+11 08	A 2807		Ori	1914 1943	96 108	0.5 0.5	8.1	9.6	B9	4634		
6 03.3	−26 17	B 96	AB	Lep	1927 1959	131	2.2	5.0	12.3	K2	4645		
---		See 59	AC		1897	203	21.3		14.8		4645		
6 04.1	+23 16	Kui 23	AB	Gem	1948 1960	187 342	0.2 0.3	4.7	5.1	G5		o	1 Gem [42]
---		β	ABxC		1911	26	96.6		12.9	G5			
6 04.3	−32 10	Vou		Col	1951 1953	40	0.1	6.5	6.5	B4			
6 04.3	−41 09	h 3831	AB	Col	1854 1942	136 129	2.7 2.5	8.4	8.5	F5			
---			AC		1938	186	15.1		14.0				
6 04.5	+51 35	OΣ 128 rej	AxBC	Aur	1843 1944	13 13	40.0 39.0	6.5		A5	4633		
---		Hu 559	BC		1902 1951	340 321	0.5 0.6	9.2	10.2		4633		
6 04.6	+45 35	A 1729		Aur	1908 1960	9 51	0.3 0.4	7.4	9.4	A2	4639		
6 04.7	−45 05	h 3834	AB	Pup	1854 1951	237 220	2.6 4.8	5.9	9.4	F5			
---			AC		1854	321	196.7		6.2	F8			
6 04.8	−48 28	Δ 23		Pup	1836 1959			7.2	7.4	G5		o	Superb double star (h)
6 04.9	−5 21	H V 14		Mon	1900	259	48.0	7.9	10.1	B9			
6 05.0	+37 58	β 893	AB	Aur	1878 1925	130	17.9	6.4	12.7	F8	4649		
---			AC		1916	140	83.9		11.6		4649		
6 05.1	+0 52	Σ 838		Ori	1830 1925	327	40.0	7.1	9.2	K0	4662		
6 05.2	+7 08	A 1951		Ori	1909 1946	50	0.4	8.3	9.0	A0	4660		
6 05.3	+74 00	OΣ 121		Cam	1843 1959	191 138	0.4 0.2	7.6	8.8	F8	4603	o	
6 05.3	−25 01	Arg 12		Lep	1876 1954	296	4.6	8.4	8.7	A0	4676		
6 05.4	+14 35	A.G.		Ori	1894 1903	187	37.2	8.1	9.2	F0			
6 05.4	−28 40	h 3830		Col	1854 1959	3	6.3	8.4	8.6	F2	4682		
6 05.9	+48 15	Es 1234		Aur	1913	267	9.8	6.8	11.2	B0	4655		
6 05.9	−13 15	Ho 512		Lep	1898 1904	354	14.9	7.7	13.7	F0	4685		
6 06.0	+12 29	Ho 228		Ori	1887 1935	268	2.1	8.0	11.0	B9	4675		
6 06.0	−55 58	h 3837	AB	Pic	1894 1913	291	12.0	8.0	12.0	K0			
---			AC		1913	26	20.1		12.3				
6 06.1	+35 23	β pm		Aur	1892 1911	15 16	116.0 122.0	6.1	9.5	G0			
6 06.4	−0 58	A 1048		Ori	1905 1916	284	2.8	7.7	13.2	F2	4690		
6 06.4	+29 31	OΣ 129		Aur	1848 1958	209	10.0	6.0	10.7	M0	4673		
6 06.5	+10 45	Σ 840	AxBC	Ori	1830 1936	247	21.2	7.8		A0	4687		
---			BC		1830 1958	184 156	0.9 0.5	8.8	9.0		4687		
6 06.5	−23 07	h 3833		Lep	1905 1920	72	44.3	5.5	10.6	A2	4704		
6 06.6	−4 12	J 187		Mon	1910 1957	144	28.9	5.4	11.6	B3	4698		
6 06.7	+20 07	A.G. 105		Ori	1902 1943	193	1.4	8.3	9.6	A0	4686		
6 06.9	−23 06	h 3835		Lep	1905	83	30.2	7.5	11.0	A0			
6 07.3	+18 48	Cou 471		Ori	1970	167	0.3	8.0	9.7	F0			
6 07.4	+36 16	OΣ 131		Aur	1847 1934	280	1.5	7.0	10.2	B9	4691		
6 07.6	+21 52	Cou 271		Gem	1968	148	0.5	7.9	9.7	B2			
6 07.8	+42 40	OΣ 130		Aur	1843 1960	187 204	0.4 0.4	7.2	8.6	B5	4696		
6 07.9	−42 09	φ 70		Col	1928 1951	150	0.1	6.2	6.3	A0			π² Col
6 08.0	+21 18	OΣ 133		Ori	1853 1938	34	3.2	7.9	11.1	F0	4716		
6 08.2	+37 59	OΣ 132		Aur	1847 1953	314 324	1.6 1.6	7.1	10.3	A2	4709		
6 08.2	−25 25	B 99	AB	Lep	1934 1959	316 76	0.1 0.1	8.2	8.3	A2	4742		Last p.a. is uncertain
---			ABxC		1926 1959	25	1.2		10.8		4742		
6 08.3	−49 45	Rst 176		Pup	1929 1946	315	0.5	8.5	9.2	A0			
6 08.4	+31 13	h 5468		Aur	1827	74	17.0	8.4	9.4	F8			
6 08.4	−11 09	β 17	AB	Lep	1872 1935	179 184	3.3 3.1	6.8	10.8	A2	4741		

α 2000	δ 2000	Star Name		Const	Years		Pos Ang		Separation		m₁	m₂	Spec	ADS	Orb	Notes
6ʰ08ᵐ.4	−11°09′	β 17	AC	Lep	1876	1933	248°		9″.0			11.8		4741		
6 08.5	+13 58	Σ 848	AB	Ori	1831	1951	108		2.5		8.3	9.0	B2	4728		In open cluster NGC 2169
—			AC		1844	1912	295		15.1			12.5		4728		
—			AD		1830	1932	121		28.4			9.1		4728		
—			AE		1830	1932	183		43.4			9.7		4728		
—			Aa		1932		204		14.5			12.0		4728		
—		S1v	BD		1970		123		25.8					4728		
—		Σ 848	CE		1932		111		56.5					4728		
—		S1v	DE		1970		227		39.1					4728		
6 08.6	−44 21	See 61		Pup	1897	1913	120		33.1		6.3	13.1	B9			
6 08.8	−52 08	Hu 1573		Car	1914	1951	183	152	0.2	0.5	8.4	9.2	F8			
6 09.0	+2 30	Σ 855	AB	Ori	1831	1929	114		29.3		6.0	7.0	A0	4749		
—			AC		1915	1923	106		118.1			8.9		4749		
6 09.0	+11 40	Σ 853		Ori	1830	1957	340	1	24.1	33.2	8.3	8.8	G5	4747		An optical pair
6 09.1	+31 16	h 379		Aur	1877	1926	116		9.4		7.5	11.5	A2	4734		
6 09.3	−18 08	B		Lep	1933	1958	358		23.0		6.4	11.4	A0			
6 09.6	+5 40	Σ 859		Ori	1829	1921	249	245	31.4	38.3	8.3	8.3	G0 F8			
6 09.7	+23 07	β 1241	AB	Gem	1891	1958	339		0.5		5.8	9.9	B1	4751		3 Gem
—			AC		1891	1958	63		18.4			14.4		4751		
6 09.7	+43 08	Webb		Aur	1913		216		43.6		7.2	8.7	A0			
6 09.8	+29 14	A 54	AB	Aur	1900	1957	344	338	0.5	0.5	8.4	8.7	F5	4750		
—			AC		1901		312		51.9					4750		
6 09.8	−22 46	Rst 3442		Lep	1935	1960					6.4	6.5	F6		o	
6 10.0	+8 06	Che		Ori	1911		170		6.3		8.5	9.0				
6 10.3	+15 54	H VI 114		Ori	1903	1917	108		54.7		7.4	8.9	G5			
6 10.5	+23 00	β 1058		Gem	1889	1960	284	252	0.4	0.3	7.3	7.6	B9	4768		
6 11.3	+30 40	Σ 861	AxBC	Aur	1831	1935	15	18	67.1	63.7	8.0		A	4779		
—			BC		1830	1942	318		1.6		9.3	9.3		4779		
6 11.5	+43 10	Es		Aur	1913		215		15.4		7.1	10.8	G5	4774		
6 11.6	+48 43	Σ 845		Aur	1830	1957	353	356	8.0	7.7	6.3	7.0	A0 A0	4773		41 Aur
6 11.6	−27 56	I 750		Col	1927	1951	7	349	0.5	0.7	7.6	9.9	F2	4806		
6 11.7	−4 40	AC 3		Mon	1854	1958	174	193	1.1	0.8	6.2	8.7	B9	4799		
6 11.7	+17 23	Σ 867		Ori	1831	1953	155		2.1		7.1	8.6	A0	4789		
6 11.7	−27 14	B 2106		Lep			n		10.0		7.9	12.0	K0			
6 11.9	+14 13	J 2016	AB	Ori	1942		185		40.0		4.5	13.1	B3			70 ξ Ori
—			BC		1942		150		4.0			12.4				
6 12.0	+19 47	H VI 72		Ori	1899	1925	214	212	85.0	86.3	5.8	9.3	B9			68 Ori
6 12.1	+29 30	Σ 862		Aur	1831	1906	336		6.5		7.3	11.1	G5	4790		
6 12.2	−65 32	Δ 26		Dor	1826	1917	112	117	20.9	20.9	7.1	8.5	F5 G			
6 12.3	−25 15	B 104		CMa	1926	1959	193	187	0.5	0.7	8.2	8.6	F0	4817		
6 12.5	−61 28	I 3		Pic	1896	1951	5		0.9		7.3	7.8	B9			
6 13.4	+10 14	Ho 22		Ori	1886	1952	198		0.8		8.6	8.6	A2	4823		
6 14.1	+23 59	OΣΣ 70		Gem	1875	1924	178		116.2		8.0	8.1	F0			
6 14.3	+14 30	S 509	AB	Ori	1825	1924	198		169.8		7.7	8.4	A2 F0	4835		
—		Ho 23	BC		1884	1926	249		2.7			12.9		4835		
6 14.5	+11 48	OΣΣ 71		Ori	1875	1921	310		90.3		7.3	7.7	K0 A0			
6 14.5	+17 54	Kui 24		Ori	1934	1960	137	139	0.3	0.5	6.5	6.5	A5			
6 14.5	−29 36	I 349		CMa	1901	1933	43		6.3		7.1	10.8	A0	4858		
6 14.6	+2 17	OΣ 135		Ori	1847		154		0.6		7.5	9.5	B9			Probably not a double star
6 14.6	−4 34	β 566		Mon	1878	1938	220	209	1.4	1.7	5.8	11.8	A0	4846		
6 14.7	−4 27	A 505		Mon	1903	1917	254		0.7		7.3	13.3	K0	4851		
6 14.7	+14 35	Σ 877		Ori	1829	1952	263		5.6		7.5	8.0	B9	4840		
6 14.8	+19 09	h 2302	AB	Ori	1886	1922	202	202	32.0	25.4	5.2	11.2	F5	4842		71 Ori
—			AC		1886	1923	265	269	80.7	76.1		10.5		4842		
—			AD		1886	1922	252	255	91.2	85.7		10.0		4842		
6 14.9	−6 16	h 384		Mon	1905		27		51.4		4.0	13.0	K0	4853		5 γ Mon
6 14.9	+22 30	β 1008		Gem	1882	1958	301	266	1.0	1.4	3.3	8.8	M	4841	o	7 η Gem; A is var.
6 15.0	+76 30	Σ 824		Cam	1831	1958	215	200	1.7	1.1	8.0	10.0	F8	4771		
6 15.1	+13 51	Kui 25		Ori	1958		115		21.3		5.9	12.5	B2			
6 15.4	−9 02	A 668		Mon	1904	1960	338	29	0.2	0.2	6.8	6.8	B9	4866		
6 15.5	+3 57	β 193	AB	Ori	1892	1926	90		18.3		7.5	10.5	B5	4863		
—			AC		1898	1906	230		58.8			9.8		4863		
6 15.5	−4 55	β 567		Mon	1879	1958	250	242	3.8	4.1	6.0	10.2	A2	4865		
6 15.6	+36 09	Σ 872	AB	Aur	1828	1949	217		11.3		6.9	7.9	F0	4849		Pale y., pale lilac (Webb)
—			AC		1879	1925	285		201.8			11.5		4849		
—			CD		1908		164		120.3			11.6		4849		
6 15.6	+66 10	OΣΣ 69		Cam	1874	1924	126		69.9		8.0	9.4	F5			
6 15.7	+60 00	β pm		Cam	1908		356		102.7		5.4	10.5	K0			40 Cam
6 15.9	+1 10	Rst 5225		Ori	1946	1960	160	284	0.2	0.2	7.1	7.1	F5		o	
6 16.1	−2 53	β 1019		Ori	1892	1944	274		0.8		8.2	9.8	A0	4882		
6 16.3	−59 13	Δ 27	AB	Pic	1826	1950	223	229	58.0	40.1	6.4	8.0	G0 F8			
—		LDS 157	AC				np		39.0			14.2				
6 16.4	+12 16	Kui 26	AB	Ori	1958		164		24.0		5.0	12.4	F5			74 Ori
—		Lau	AC		1909		88		204.0			9.2				
6 16.5	+19 01	β 894		Ori	1881	1928	138		5.2		7.3	11.6	F5	4881		

α 2000	δ 2000	Star Name		Const	Years	Pos Ang		Separation		m₁	m₂	Spec	ADS	Orb	Notes
$6^h16^m.7$	$-12°03'$	β 18		CMa	1876 1942	277°	°	1".7	"	7.0	8.7	F2	4891		
6 17.1	+9 57	β 96	AB	Ori	1892 1925	255 258		62.9	62.7	5.4	9.5	A2	4890		
--			AC		1892 1925	159 159		119.9	117.3		8.5		4890		75 Ori; mag. of A is for Aa
--		ø 331	Aa		1954 1959	128 186		0.1	0.1	6.2	6.2		4890		Too close in 1956 (0".1)
--		J 408	CD		1877 1933	226		4.9			11.0		4890		
6 17.1	+15 51	H V 23		Ori	1903	229		44.6		7.3	9.3	B9			
6 17.1	-22 43	h 3845		CMa	1898 1959	22	6	34.3	47.6	6.1	11.1	G0			Large star ruddy (h)
6 17.3	+5 06	β pm	AB	Ori	1891 1911	246		189.7		5.7	9.6	G0			
--			Aa		1911	265		58.5			11.6				
--			Ab		1911	231		69.3			11.6				
6 17.6	-24 27	B 110		CMa	1926 1932	49		6.7		7.3	11.2	F5	4908		
6 18.1	+14 23	Ho 229	AB	Ori	1914 1958	326 336		2.3	2.1	6.2	13.2	A0	4901		
--			AC		1890 1958	186		25.4			12.6		4901		
6 18.2	+68 19	Mlr 317		Cam	1972	283		0.2		8.0	8.4				
6 18.3	-5 39	A 323		Mon	1902 1943	218		1.1		6.7	10.0	A0	4910		
6 18.8	+11 10	A 2809		Ori	1914 1921	196		0.3		7.2	9.3	A0	4918		
6 19.3	-12 31	Ho 231		CMa	1887 1920	48		6.5		8.0	11.0	B9	4941		
6 19.3	-24 58	S 516	AB	CMa	1825 1959	3	7	64.7	61.5	7.2	8.3	A0			
--			AC		1825 1950	242 243					7.1	B9			
6 19.4	+13 26	OΣΣ 73	AB	Ori	1875 1923	44		73.1		6.9	7.7	F5 G5	4924		
--			Aa		1904	30		15.3			14.4		4924		
6 19.4	+23 16	H V 55	AB	Gem	1899	87		59.6		7.3	8.8	B9			
--			AC		1903	59		62.8			9.7				
6 19.7	+12 18	Σ 891		Ori	1830 1904	293		21.9		7.8	10.8	B8	4939		
6 19.7	+38 26	A 2116		Aur	1910 1952	23	33	1.6	1.5	7.1	13.6	F0	4922		
6 19.8	-39 29	h 3849		Col	1837 1950	53		39.6		6.7	8.3	K0 G5			
6 19.9	+25 01	Σ 889		Gem	1830 1957	222 237		22.0	21.3	7.6	9.9	K2	4930		
6 20.0	+28 26	β 895	AB	Aur	1879 1960					7.9	7.9	A3	4929	o	
--		Σ 888	ABxC		1831 1976	246 263		2.7	2.8		9.4		4929		
6 20.0	-17 41	Hu 1240		CMa	1905 1952	241 232		0.9	1.2	7.9	11.4	A2	4960		
6 20.3	+7 43	A 2719		Ori	1914 1951	60		0.4		7.4	7.6	B9	4951		
6 20.3	-30 04	Smyth		CMa	1833 1950	338		175.5		3.0	7.6	B8 K0			1 ζ CMa
6 20.6	+18 03	OΣΣ 75		Gem	1876 1921	128		47.3		7.6	8.3	K0			
6 20.6	+25 11	OΣΣ 74	AB	Gem	1874 1924	264		58.0		7.2	9.4	B9 A0			
--			Aa		1911	290		41.0			12.4				
6 20.7	+72 09	Mlr 404		Cam	1972	243		1.8		6.8	11.8	G5			
6 21.0	+2 34	A 2666		Ori	1913 1920	304		0.9		7.5	11.4	M	4966		
6 21.1	+36 19	A 1954		Aur	1908 1922	113		0.4		7.9	10.3	K0	4955		
6 21.2	+21 08	S 513	AB	Gem	1825 1908	258		59.4		7.7	9.5	B8 A0	4962		
--			AC		1908	247		42.0			10.8		4962		
--			AD		1914	23		265.2			7.7	A0	4962		
--			AE		1962	255		84.2					4962		
--			Ad		1908	295		38.4					4962		
--			BC		1843 1914	67	86	16.3	15.1				4962		
--			Bd		1908 1914	326		41.8					4962		
--			Be		1908	296		38.4					4962		
--			CE		1914	325		41.5					4962		
6 21.4	+2 16	A 2667		Mon	1913 1960					6.8	7.4	A5	4971	o	
6 21.4	-11 46	Σ 3116	AB	CMa	1831 1935	23		4.2		5.6	9.8	B2	4978		
--		Fox	AC		1911	274		56.5					4978		
6 22.1	+59 22	Σ 881	AB	Lyn	1830 1959	90 124		0.8	0.8	6.2	7.7	A2	4950		4 Lyn
--			ABxC		1908 1935	96		26.2			12.9		4950		
--			ABxD		1879 1908	356		100.4			11.0		4950		
6 22.4	+46 12	A 1319		Aur	1906 1952	139		0.7		7.4	9.8	A2	4970		
6 22.5	-33 52	See 65		Col	1897 1933	109		17.5		7.9	11.9	G0			
6 22.5	-60 13	Jsp 101		Pic	1930	51		16.0		6.6	10.6	G0			
6 22.6	+32 46	Hu 829		Aur	1904 1933	325		2.9		8.0	11.5	F5	4982		
6 22.7	-17 57	β pm		CMa	1879 1908	339 340		185.0	185.9	2.0	9.8	B1			2 β CMa
6 22.8	+17 34	Σ 899		Gem	1831 1953	20		2.3		7.3	8.3	A0	4991		
6 22.9	+22 31	β 1059	AD	Gem	1880	77		72.7		3.2		M	4990		
--			AxBC		1889 1899	141 141		122.5	121.7				4990		
--			BC		1889 1939	267 260		0.8	0.8	9.8	10.7		4990		
6 22.9	-45 38	h 3856		Pup	1880 1913	4		34.4		6.7	9.7	K0			
6 23.2	+16 31	Ho 233		Ori	1887 1936	36		2.1		7.8	10.6	B9	4997		
6 23.5	+27 07	OΣ 138 rej		Gem				1.0		7.9	10.9	A0			Probably single (IDS)
6 23.5	-29 52	B 1046		CMa	1928	26		1.5		7.0	12.0	K0			
6 23.8	+2 40	J 53	AB	Mon	1910 1957	127		1.8		7.1	10.7	K0	5013		
--			AC		1913 1943	103		33.9			9.8		5013		
6 23.8	+4 36	Σ 900	AB	Mon	1831 1934	27		13.4		4.5	6.5	A5 A5	5012		8 Mon
--			AC		1911	254		93.7			12.7		5012		
6 23.8	-19 47	β 568		CMa	1878 1956	155		0.8		7.2	7.5	B8	5023		
6 24.0	-36 42	h 3857	AB	Col	1882 1960	256		12.9		5.7	10.8	G5			
--		Δ 28	AC		1826 1960	64	72	68.9	64.8		6.9	G5			
6 24.0	-45 57	LDS 161		Pup		nf		70.0		7.9	13.1	G0			
6 24.1	-50 32	I 282		Pup	1901 1959	314 306		1.6	1.3	7.9	9.9	F0			

α 2000	δ 2000	Star Name		Const	Years		Pos Ang		Separation		m_1	m_2	Spec		ADS	Orb	Notes
$6^h24^m.2$	+7°54′	J 259		Mon	1910	1918	322°	°	5″.0	″	7.9	11.8	A3		5018		
6 24.3	-12 58	Σ 903		CMa	1829	1909	295		23.2		6.1	10.8	B8		5030		
6 24.3	-16 13	S 518		CMa	1825	1917	88		16.6		6.7	9.1	A3		5034		
6 24.4	-28 01	I 753	AB	CMa	1913	1960	40	6	0.4	0.3	9.3	9.3			5037		
--		CorO 36	ABxC		1879	1960	242		9.2			9.1			5037		
6 24.6	-1 25	β 97		Mon	1876	1947	258	265	1.2	1.0	6.9	8.9	B8		5029		
6 24.6	+59 42	OΣΣ 72	AB	Lyn	1874		300		43.5		8.1	12.1	K0				
--			AC		1874	1914	322		34.4			7.5		A3			
6 24.9	+42 33	A 2356		Aur	1911	1956	80		0.7		8.8	8.8	A3		5016		
6 25.0	+10 31	Σ 901		Ori	1829	1913	248		20.0		7.9	9.7	B9		5035		
6 25.2	+1 30	φ 343		Mon	1959		92		0.1		8.1	8.1	A0				
6 25.2	-10 56	β 569		Mon	1877	1943	121	119	1.8	1.4	7.8	10.1	K0		5048		
6 25.5	-35 04	h 3858	AB	Col	1846	1932	48	48	127.2	132.3	6.3	7.4	K0	A3			
--			BC		1846	1959	316	311	3.7	3.8		8.5					
6 25.6	+22 27	OΣ 139		Gem	1847	1959	309	238	0.8	0.3	7.9	9.5	A3		5042	o	
6 25.7	-48 11	I 156		Pup	1894	1946	133	129	1.2	1.2	5.9	8.4	B9				
6 25.8	-40 59	h 3860		Col	1847	1947	227		8.6		6.9	8.7	A3				
6 26.2	+18 45	β 1191		Gem	1890		162		1.3		6.8	13.6	K0		5054		
6 26.6	+15 31	OΣ 140	AB	Gem	1847	1943	118		2.8		6.9	9.4	B9		5062		
--			AC		1933		34		49.7						5062		
6 26.7	+0 27	Σ 910	AxBC	Mon	1831	1932	151		66.6		7.7		G5	G5	5069		
--			BC		1829	1943	171	158	0.7	0.6	9.4	10.1	G5		5069		
6 26.7	-7 31	Σ 914		Mon	1831	1938	298		21.1		6.4	8.7	A0		5070		BDS 3401 is probably same
6 26.7	-58 14	φ 98		Pic	1929	1942	335	328	0.2	0.2	8.6	8.9	F5				
6 26.8	+58 25	S 514	AB	Lyn	1879	1924	139	139	30.3	31.4	5.3	9.8	K2		5036		5 Lyn
--			AC		1879	1924	272		96.0			7.9			5036		
6 27.1	-37 54	HdO 194		Col	1928	1944	56		32.8		6.5	13.5	F0				
6 27.2	+0 08	Rst 5231		Mon	1946		337		6.0		7.9	11.0	B8				
6 27.2	-25 51	LDS 162		CMa			nf		43.0		6.1	12.6	F8				
6 27.4	-25 45	B 114		CMa	1926	1960	294	28	0.4	0.5	8.7	8.9	K0		5092		
6 27.6	-15 18	A 3029		CMa	1926		197		4.9		7.4	12.4	B8		5087		
6 27.8	+20 47	S,h 70	AB	Gem	1822	1957	205	204	32.7	27.1	6.6	8.0	K0		5080		
--			AC		1911		34		77.7			12.4			5080		
6 28.0	-4 46	β pm	AB	Mon	1879	1907	256		77.2		5.1	9.3	B3				10 Mon
--			AC		1879	1907	231		80.7			9.3					
6 28.0	+36 12	Ali 325		Aur	1928	1930	53		15.9		7.7	12.6	G0				
6 28.2	+5 16	Σ 915	AB	Mon	1833	1933	40		5.9		7.4	8.4	B9		5097		
--			AC		1905	1912	126		38.8			11.0			5097		
6 28.2	+70 32	OΣ 136		Cam	1847	1937	80		5.7		6.0	9.8	A2		5039		
6 28.3	+24 41	Cou 914		Gem	1972		125		0.2		8.6	8.6	A0				
6 28.6	-78 54	h 3888		Men	1872	1919	116		35.7		7.4	9.8	F0				
6 28.7	-32 22	β 753		CMa	1892	1942	45		1.3		5.9	7.9	B3				
6 28.8	-7 02	Σ 919	AB	Mon	1831	1955	132		7.3		4.7	5.2	B2	B2	5107		11 β Mon
--			AC		1831	1955	124		10.0			6.1		B2	5107		
--		β 570	AD		1878	1932	56		25.9			12.2			5107		
--		Σ 919	BC		1831	1973	106		2.8						5107		
6 28.8	+52 27	A 1732		Aur	1908	1918	14		3.5		7.2	12.4	G5		5078		
6 28.9	+40 07	Σ 905	AB	Aur	1833	1935	117	125	1.8	1.8	8.0	10.0	A0		5088		
--			AC		1880	1908	148		151.4			10.4			5088		
6 28.9	-45 47	I 350		Pup	1902	1932	10		15.9		7.7	12.7	A0				
6 29.0	+20 13	OΣΣ 77	AB	Gem	1876	1924	329		112.5		4.2	8.7	B5		5103		18 ν Gem; 2nd mag. BC?
--		β 1192	Aa		1890	1958	358	0	22.6	23.7		15			5103		
--			Ab		1890		13		53.9			14.0			5103		
--			Ac		1890	1924	255	254	56.8	55.6		12.7			5103		
--		β	Ad		1890		12		92.1			13.1			5103		
--		β 1192	BC		1890	1944	346	282	0.2	0.2					5103		[43]
--		Btz	AA′		1920	1944	121	129	0.2	0.2			B5		5103		
6 29.1	-40 22	Δ 29		Col	1826	1950	109	113	71.3	65.9	7.4	7.8	K0	K0			
6 29.2	+2 39	Bail 1695		Mon	1910	1958	294	299	19.4	19.3	6.2	10.2	M				Also Kui
6 29.4	-17 49	h 3865		CMa	1898	1920	66		23.5		8.0	10.8	A0				
6 29.4	-22 35	h 3863		CMa	1877	1959	119		2.6		6.9	8.7	A2		5128		
6 29.6	+16 59	A 2450		Gem	1912	1933	54		2.5		8.0	11.3	G5		5112		
6 29.8	+36 36	Σ 912		Aur	1830	1936	27		3.4		8.2	10.2	A0		5106		
6 29.8	-50 14	R 65	AB	Pup	1879	1960					6.0	6.1				o	
--		Δ 30	AC		1835	1938	317	314	13.4	12.4		9.0	F2				
--		HdO 195	CD		1903	1960						9.7				o	
6 29.9	+7 07	OΣ 142		Mon	1848	1901	353		8.6		7.7	11.2	B3		5124		
6 30.0	+17 54	OΣ 141		Gem	1848	1953	142		2.2		7.6	9.7	A0		5121		
6 30.4	-3 01	Bail 63		Mon	1896		222		13.3		7.1	11.0	G5				
6 30.4	+17 57	Ho 340		Gem	1891	1927	22		6.5		7.6	13.4	G5		5132		
6 30.4	+22 33	Ho 514	AB	Gem	1895	1927	128	123	19.3	19.7	6.8	12.5	A0		5129		
--			AC		1895		35		39.0			12.8			5129		
6 30.5	+4 20	Σ 920		Mon	1829	1928	210		9.3		8.0	11.2	B3		5137		
6 30.5	-14 57	h 3864		CMa	1903	1920	44		21.6		6.9	9.8	B8		5144		
6 30.6	-13 09	Gallo 242		CMa	1902		66		36.6		6.2	11.2	B3		5148		

α 2000	δ 2000	Star Name		Const	Years		Pos Ang		Separation		m₁	m₂	Spec	ADS	Orb	Notes
6ʰ 30ᵐ.7	+20° 25′	H N 111		Gem	1905		167°		42″.5		8.0		A0			
6 30.7	−40 27	I 4		Col	1896	1960	303		0.9		7.3	7.6	B8			
6 30.8	+58 10	β pm		Lyn	1851	1908	122	116	188.2	179.6	6.0	9.0	G5			6 Lyn
6 31.2	+9 56	A 2816	AB	Mon	1914	1937	318		0.9		7.7	9.9	B8	5154		
—			AC		1920		41		23.0			10.9		5154		
6 31.2	+11 15	Σ 921		Mon	1831	1967	4		16.3		6.1	8.3	B0 A5	5153		
6 31.2	+16 56	OΣ 143	AB	Gem	1852	1958	103		7.9		6.3	9.4	K0	5146		
—		Fox	AC		1911	1956	343	344	42.6	44.8		10.9		5146		
6 31.3	+4 50	Bail 2666		Mon	1910		337		10.2		8.3	9.7	B0			
6 31.3	+15 44	OΣ 519		Gem	1847	1934	78		8.2		7.9	10.2	G5	5152		
6 31.4	+2 55	OΣ 144 rej		Mon	1898	1928	156	145	20.7	22.5	7.7	12.4	G0	5161		
6 31.4	−12 23	Rst 3460		CMa	1938	1943	293		14.8		5.3	13.6	K0			
6 31.4	−20 37	h 2321		CMa	1877	1946	307		5.6		8.2	9.2	F5	5164		
6 31.4	−57 37	h 3873		Pic	1912		294		18.8		8.0	10.5	K0			
6 31.6	+32 10	β 896	AB	Aur	1879	1955	199	190	0.9	1.0	7.0	10.0	B9	5151		
—			AC		1879	1920	211	210	18.4	18.1		12.9		5151		
6 31.7	+5 46	Σ 926		Mon	1829	1974	287	288	10.7	10.9	7.1	8.5	A0	5162		
6 31.8	+28 22	β 1021		Aur	1892	1943	84		0.8		8.3	9.6	A0	5157		
6 31.9	+4 57	GAn 3	AB	Mon	1865	1916	282	284	2.7	3.2	6.8	12.7	B2	5165		
—			AC		1876	1917	319		6.9			11.8		5165		
—			AD		1876	1916	288		12.5			12.3		5165		
—			AE		1876	1916	197		13.2			9.6		5165		
6 31.9	−23 25	See 68	AB	CMa	1897	1900	144		24.6		4.3	13.9	B1	5176		4 ξ¹ CMa. Variable
—			AC		1897	1900	303		28.2			14.1		5176		
6 32.0	−58 45	h 3874		Pic	1894	1937	231		2.4		5.8	9.0	B9			μ Pic
6 32.3	+15 42	OΣ 145		Gem	1846	1957	338		1.5		7.2	10.0	F5	5168		
6 32.3	+17 47	Σ 924	AB	Gem	1830	1956	210		20.0		6.3	6.9	F8	5166		BDS 3424 prob. same star
—			BC		1913		136		47.4			13.0		5166		
6 32.4	−5 52	Rst 4308		Mon	1939	1948	95		4.2		5.6	13.3	A5			
6 32.4	+11 40	OΣ 146 rej		Mon	1867	1957	142	140	33.3	31.1	6.0	9.6	K0	5170		
6 32.5	−21 01	Rst 4309		CMa	1940		266		5.1		7.1	14.5	B5			
6 32.6	+10 08	Abetti		Mon	1921		255		103.6		8.1	10.9	K0			
6 32.6	−32 02	h 3869		CMa	1897	1930	258		24.9		5.7	7.7	B3 A0			
6 32.6	−48 08	Rst 199		Pup	1928	1949	28	18	1.3	1.4	7.6	11.1	K2			
6 32.7	−5 21	β 98		Mon	1876	1951	140	151	1.0	0.8	8.2	8.2	B8	5183		
6 32.8	−27 24	I 755		CMa			150		30.0		7.0	14.0	F5			
6 33.0	+39 47	A 2519		Aur	1913	1921	70		4.1		7.2	12.2	A2	5167		
6 33.1	+25 18	h 392		Gem	1906	1918	37		24.0		8.4	9.0	A0 A0			
6 34.0	+52 28	Σ 918	AB	Aur	1829	1972	322	333	4.4	4.7	7.2	8.2	A3	5178		
—			BC		1880	1919	27	27	133.8	136.6		10.0		5178		
6 34.1	+22 07	S 524	AB	Gem	1824	1924	243		53.6		8.0	8.0	A3 A3			
—			AC		1824	1917	149		106.1			10.0				
6 34.1	−29 38	h 3871		CMa	1848	1955	354		7.6		7.1	8.2	A2	5214		
6 34.2	−42 06	See 69		Col	1897	1935	8		11.5		7.3	12.7	G5			
6 34.3	+14 44	Σ 932		Gem	1830	1955	342	317	2.4	1.8	8.1	8.3	F5	5197		
6 34.3	+38 05	OΣ 147	AB	Aur	1849	1926	73		43.2		6.6	10.0	K0	5188		
—			AxCD		1849	1926	117		46.3					5188		
—			CD		1849	1957	115	109	0.6	0.5	10.6	11.0		5188		
6 34.4	+3 18	A 2673	AB	Mon	1913	1943	304		1.3		7.5	10.0	K0	5202		
—			AC		1913	1920	107		28.7			11.7		5202		
6 34.4	−3 04	h 394		Mon	1897	1910	332		57.2		7.3	9.3	K2			
6 34.5	−11 14	Ho 234		CMa	1888	1959					8.1	8.2	F0	5212	o	
6 34.6	−61 30	Jsp		Pic	1930		293		13.3		7.3	14.0	K0			
6 34.7	+38 32	Σ 928	AB	Aur	1829	1952	133		3.5		7.6	8.2	F5	5191		
—			BC		1880	1910	123		128.6			11.0		5191		
6 34.7	−34 01	β 754		Col	1892	1947	34	43	0.9	0.9	7.4	7.7	F0			
6 34.8	+7 34	Σ 938		Mon	1831	1958	209		10.5		6.5	10.7	A0	5211		14 Mon
6 34.9	−32 18	I 178		CMa	1898	1940	199		0.8		8.1	10.2	B8			
6 35.0	+11 07	Smart		Mon	1911		16		6.1		8.0	11.8	A3			
6 35.1	+37 04	OΣ 148		Aur	1849	1958	74		2.6		7.4	11.1	G5	5201		
6 35.4	+37 43	Σ 929		Aur	1830	1949	25		6.0		7.2	8.3	G5	5208		
6 35.4	−36 47	β 755	AB	Col	1887	1959	255	258	0.9	1.3	6.0	6.8	B9			
—		h 3875	AC		1887	1932	301		21.4			11.5				
6 35.4	−38 49	CapO 6		Col	1893	1959	240		1.1		8.2	8.3	B8			
6 35.7	+28 16	A 506		Aur	1903	1959	25	21	0.3	0.3	8.5	9.0	B9	5218		
6 35.8	−16 06	Howe 13		CMa	1876		300		11.2		7.8	8.8	B9	5242		
6 35.9	+5 18	Σ 939	AB	Mon	1832	1919	106		30.1		8.3	9.6	B5 B8			These stars form roughly an
—			AC		1832	1919	50		39.7			9.7				equilateral triplet
—			BC		1832	1902	3		33.7							
6 35.9	−36 05	ø 19		Col	1925	1959					6.8	7.3	F5		o	
6 35.9	−48 17	CorO 39		Pup	1903	1959	201	204	10.7	10.2	7.5	9.2	K0			
6 36.0	−35 10	I 1118		Col	1916	1943	100	93	1.2	1.2	7.6	9.8	F5			
6 36.1	+13 42	Ho 341	AB	Gem	1891		134		1.4		7.0	12.0	A0	5231		
—			AC		1891		155		65.0			9.2		5231		
6 36.1	−24 07	Arg 15		CMa	1907	1933	234	234	37.9	39.1	7.9	9.0	B9 F8			

α 2000	δ 2000	Star Name		Const	Years	Pos Ang	Separation	m₁	m₂	Spec	ADS	Orb	Notes
6ʰ36ᵐ2	−36°08′	Rst 4816		Col	1942 1951	312° 34°	0″.2 0″.2	7.9	7.9	G0			
6 36.4	+27 17	OΣ 149		Gem	1848 1960			7.1	8.7	G0	5234	o	
6 36.4	−18 40	S,h 73		CMa	1821 1926	262	17.5	5.8	8.5	G5 G0	5253		6 ν¹ CMa, H IV 81
6 36.7	−22 37	H II 60		CMa	1878 1933	336	9.1	6.4	10.0	B8	5260		
6 36.8	+41 08	Σ 933		Aur	1829 1915	75	25.6	8.4	8.9	A2 A2	5233		
6 36.9	−22 16	Don 136		CMa	1930 1943	150	1.1	7.9	11.5	M3			
6 37.0	−38 09	HdO 196	AB	Col		270	18.0	6.0	12.0	G5			
--			AC			180	25.0		10.0				
--			AD			100	35.0		13.0				
6 37.1	+5 31	A 2821		Mon	1914 1922	213	3.3	7.2	13.2	A2	5256		
6 37.1	−50 29	B 1514		Pup	1931 1944	195	0.6	8.0	10.2	M			
6 37.2	−36 59	Rst 4819		Col	1942 1951	4 356	0.6 0.6	6.1	6.9	B9			
6 37.5	+9 09	Ho 515	AB	Mon	1895 1909	253	10.1	7.5	11.7	G5	5262		
--			AC		1909	104	13.1		13.1		5262		
--			AD		1895	55	30.0		12.5		5262		
6 37.6	+12 11	S 529	AB	Gem	1825 1918	163 151	92.0 70.3	7.6	9.0	G5			
--			AC		1825 1918	171 168	187.9 163.0		8.3	F8			
6 37.6	−55 21	h 3884		Car	1913 1918	282	25.6	6.9	10.8	K0			Large star red (h)
6 37.7	+16 24	β pm	AB	Gem	1880 1907	295 295	133.0 135.5	1.9	11.2	A0			24 γ Gem, H V 71
--			AC		1880 1907	336 335	141.7 143.5		10.9				
6 37.7	+61 29	β pm	AB	Lyn	1854 1909	85 80	144.5 157.0	6.0	9.0	G0			8 Lyn
--			AC		1904	95	374.4						
--			Bb		1907	253	71.0		12.7				
6 37.9	−8 47	A 509	AB	Mon	1903 1943	141	1.4	7.4	9.9	K0	5277		
--			AC		1903 1916	72	9.6		14.3		5277		
6 37.9	+23 36	Pou 1700		Gem	1906	48	19.5	6.8	14.3	K0			
6 38.0	−61 32	I 5		Pic	1896 1954	270 270	2.9 2.4	6.4	8.4	G0			
6 38.1	+59 27	Σ 923 rej		Lyn	1903 1926	145	30.4	7.6	9.8	G5			
6 38.3	+24 27	Ho 625	AB	Gem	1900 1906	194	13.4	6.7	11.9	A5	5270		
--			AC		1906	251	47.4		9.3		5270		
6 38.3	+64 44	Σ 922	AB	Cam	1831 1909	135	10.5	7.4	11.2	F5	5236		
--			AC		1831 1908	3	25.6		10.8		5236		
6 38.6	+40 20	Frk	AB	Aur	1918	253	48.9	7.7	9.4	K0			
--			AC		1918	90	61.5		9.2				
6 38.6	−48 13	Δ 31		Pup	1826 1937	321	13.0	5.0	8.3	G7 A0			
6 38.7	+41 35	Σ 941	AB	Aur	1830 1959	81	2.0	7.2	8.2	B9	5269		
--			AC		1894 1909	134	82.8		10.2		5269		
--			CD		1905	316	6.4		12.8		5269		
6 38.7	−45 04	Hu 1415	AB	Pup	1913 1933	25	1.9	7.7	12.0	B8			
--		h 3882	AC		1847 1933	331	18.1		9.9				
6 38.9	+27 48	OΣ 151 rej	AB	Gem	1843 1914	135	29.3	7.3	10.2	A2			
--			AC		1898	235	29.9		13.4				
6 38.9	−40 35	h 3881		Col	1904 1920	160	28.3	7.9	11.5	F8			
6 39.1	−29 09	I 1156		CMa	1926 1928	204	0.7	8.2	10.2	K0	5311		
6 39.2	+9 39	HΣ		Mon	1894 1943	278	0.8	8.6	8.6	B5	5290		
6 39.3	+42 29	β pm	AB	Aur	1894 1907	110 109	53.2 52.5	4.8	10.3	G5			50 ψ² Aur
--			AC		1907	75	99.4		11.0				
6 39.3	−31 50	Vou 21	AB	CMa	1929	28	23.8	7.8	10.7	B3			
--			BC		1928 1929	161	0.6		10.9				
6 39.5	+66 10	Milb 343		Cam	1924	158	2.6	7.6	12.1	B8	5255		
6 39.6	+28 16	OΣ 152		Aur	1850 1960	36	0.9	6.0	7.8	B8	5289		54 Aur
6 39.6	−23 42	φ 321		CMa	1952 1959	196	0.1	6.7	7.1	A0			
6 39.8	+12 59	β 571		Gem	1877 1915	318	2.8	5.9	11.9	A2	5302		
6 40.2	+67 19	Σ 925		Cam	1831 1924	92	3.4	8.0	10.5	B9	5267		
6 40.3	+3 32	h 2331		Mon	1905	293	27.4	7.8	12.3	K2			
--			AC		1905	50	24.8		11.8				
6 40.3	−4 28	Rst 4820		Mon	1941	298	6.0	7.5	11.2	K5			
6 40.4	+40 58	Σ 945		Aur	1830 1960	249 299	1.1 0.5	7.3	8.2	F2	5296		
6 40.5	+9 49	Σ 951	AB	Mon	1830 1924	310	21.4	7.9	10.1	B8	5316		
--		Ward	BC		1878 1924	229	11.5		11.4		5316		
6 40.5	−21 24	B 1961		CMa	1931 1944	10	0.5	8.1	8.4	A3			
6 40.6	+17 39	Brt 1208		Gem	1902 1946	175	5.3	8.1	9.7	F8			
6 40.8	+48 16	Σ 944	AB	Aur	1829 1902	54	6.7	8.4	10.4	F8	5300		
--			AC		1881 1910	181	195.6		9.2		5300		
6 41.0	+9 54	Σ 950	AB	Mon	1831 1957	209 213	2.8 2.8	4.7	7.5	O5	5322		15 Mon = S Mon
--			AC		1831 1957	13	16.6		9.8		5322		
--			AD		1841 1938	308	41.3		9.6		5322		
--			AE		1874 1957	139	73.9		9.9		5322		In open cluster NGC 2264
--			AF		1874 1924	222	156.0		7.7		5322		
--			AK		1923	56	105.6		8.1		5322		
--			Ad		1914	261	136.6		9.6		5322		
--			FG		1873 1924	262	39.8		8.2		5322		
--			Fd		1914	342	98.5				5322		
--			Gd		1914	5	98.3				5322		
6 41.0	−71 47	I 351		Vol	1900	334	10.7	6.5	10.0	K0			

α 2000 δ 2000		Star Name		Const	Years	Pos Ang		Separation		m₁	m₂	Spec	ADS	Orb	Notes
6ʰ 41ᵐ.1	+20° 39′	Ho 236		Gem	1890 1906	203°		17″.4		7.6	13.4	G5	5320		
6 41.2	+8 59	Σ 953		Mon	1832 1932	330		7.1		7.2	7.7	F5	5328		
6 41.2	+9 28	Σ 954		Mon	1829 1924	153		12.8		7.1	9.6	B3	5327		
6 41.2	−8 00	Σ 955	AB	Mon	1830 1942	270		1.0		8.7	9.0	B9	5336		
---		S 532	AC		1831 1935	190		11.5			8.6		5336		
6 41.2	−40 21	h 5443		Pup	1835 1933	107		15.4		6.1	10.5	B5			
6 41.3	+28 12	h 397	AB	Gem	1904	46		30.7		6.4	11.6	K0			25 Gem
---			AC		1878 1904	57		55.9			10.3				
6 41.4	+0 57	Bail 1325		Mon	1903	268		12.7		7.5	12.5	A0			
6 41.4	−7 10	A 1055		Mon	1905 1927	281		2.7		6.8	14.1	B8	5340		
6 41.4	+15 26	A 2728		Gem	1914 1953	232		5.0		7.5	13.2	A5	5330		
6 41.8	+30 41	A 218		Gem	1901 1960	246 155		0.2	0.2	8.8	8.8	F5	5332		
6 42.0	−16 00	β 19		CMa	1874 1956	166		3.6		7.0	9.3	B9	5358		
6 42.1	+3 15	Ho 237	AB	Mon	1887	150		0.3		7.8	7.8	A2			Probably single (IDS)
---			AC		1887	53		120.0			9.6				
6 42.1	+25 28	OΣ 153 rej		Gem	1843 1917	71		20.2		7.8	11.8	M			
6 42.3	−38 24	Δ 32		Pup	1835 1951	277		8.0		6.5	8.0	A3			
6 42.4	+2 52	A 2677	AB	Mon	1913 1923	16		0.8		7.7	12.2	B9	5359		
---			AC		1923	312		21.4			14.2		5359		
6 42.5	+66 12	Mlr 318		Cam	1972	309		1.3		7.1	8.7	F8			
6 42.5	−23 14	β 195	AB	CMa	1877 1933	215		5.8		7.4	10.4	B5	5371		
---			AC		1892 1917	178		34.9			11.8		5371		
6 42.5	−42 34	I 283		Pup	1900 1951	182 186		2.4	2.4	8.0	10.5	G0			
6 42.5	−61 45	I 6		Pic	1896 1946	251		0.9		7.8	8.0	F5			
6 42.7	+1 43	Σ 956	AB	Mon	1830 1934	189 189		4.6	6.2	7.9	10.9	B2	5364		
---			AC		1830 1934	155 157		35.0	35.8		9.0		5364		
---			AD		1928	180		56.0			10.9		5364		
---			CD		1905	215		27.9					5364		
6 42.8	−22 27	S 534		CMa	1825 1951	143		18.2		6.2	8.7	F0	5377		
6 42.8	−50 27	h 3889		Pup	1836 1917	266		42.3		6.9	9.4	B5 B9			
6 42.9	−55 07	I 480		Car	1911 1921	1		5.6		8.1	11.2	B8			
6 43.2	−0 01	Bail 1018		Mon	1897					7.7		G5			
6 43.9	+25 08	S 533		Gem	1825 1925	94		110.3		3.0	9.0	G5	5381		27 ε Gem
6 44.0	+13 14	Lam 3		Gem	1836 1904	185 184		32.0	27.2	4.5	11.0	K0	5387		30 Gem
6 44.2	−54 42	Δ 34		Car	1826 1938	191		130.1		6.6	6.8	A0 A0			
6 44.3	+40 37	OΣ 154		Aur	1846 1941	137 110		30.4	23.8	6.9	9.4	M	5379		
6 44.5	−19 40	B 2122		CMa		85		6.0		7.6	13.5	A3			
6 44.5	−31 04	HdO 197		CMa	1901	97		36.3		5.2	10.6	B3			10 CMa
6 44.9	+10 45	A 2825		Mon	1914 1957	324 58		0.2	0.1	8.6	8.8	G0	5407		
6 45.0	−30 35	CorO 44		CMa	1897 1933	223		4.5		6.5	10.5	B5			
6 45.1	−16 43	AGC 1	AB	CMa	1862 1960					−1.5	8.5	A0	5423	o	9 α CMa, Sirius [44]
---			AD		1915	183		31.6			14.0		5423		
---			BC		1920 1929	69 129		1.8	1.4				5423		Only suspected
6 45.2	+30 50	Σ 957	AB	Gem	1831 1942	93		3.5		7.3	8.8	A0	5403		
---			AC		1907	101		18.9			13.1		5403		
6 45.4	+24 40	OΣ 155		Gem	1843 1925	261		15.5		7.3	10.2	K2	5409		
6 45.5	−30 57	h 3891		CMa	1838 1959	223		4.9		5.7	8.0	B3			
6 45.6	+4 36	Bail 2701		Mon	1910	342		14.1		7.7	12.5	A0			
6 45.7	−20 41	h 2341		CMa	1836	86		45.0		8.3	9.1	A0 A0			
6 46.1	+0 45	A 3036		Mon	1922 1947	44		0.4		8.1	10.3	B9	5432		
6 46.1	−20 45	I 760		CMa		315		1.2		8.0	12.0	A0	5437		
6 46.2	+2 35	Bail 1720		Mon	1910					7.5					
6 46.2	+59 27	Σ 948	AB	Lyn	1831 1959					5.4	6.0	A2	5400	o	12 Lyn
---			AC		1831 1959	308		8.7			7.3		5400		
---			AD		1879 1910	256		170.0			10.6		5400		
---			BC		1903 1957	304		10.3					5400		
6 46.7	+43 35	S,h 75		Aur	1823 1958	17 31		55.4	36.2	5.3	8.3	G0	5425		56 ψ⁵ Aur. Optical
6 46.7	−47 48	h 3895		Pup	1912 1920	64		26.2		6.9	11.0	G5			Orange, blue (h)
6 46.7	−54 42	I 157		Car	1901 1943	347		1.6		6.5	9.9	G5			
6 46.8	−6 09	Σ 967 rej		Mon	1832 1958	192 191		11.5	13.4	7.9	11.9	A0	5445		
6 47.2	−30 57	B 1519		CMa	1929 1932	72		3.3		7.9	13.5	F5			
6 47.4	+18 12	OΣ 156		Gem	1844 1959					6.8	7.0	A0	5447	o	
6 47.4	−37 56	h 3893		Pup	1920	299		65.4		5.3	11.1	B9			
6 47.4	−48 34	I 158		Pup	1902 1943	186		1.8		6.8	10.3	G0			
6 47.5	−76 51	h 3911		Men	1836 1918	43 47		20.7	21.8	6.9	10.4	G5			
6 47.7	−81 04	Rst 2448		Men	1935	163		3.8		7.4	12.5	K5			
6 47.8	+0 20	OΣ 157		Mon	1847 1958	8 273		0.7	0.4	7.5	7.8	A2	5455		
6 48.0	−11 06	Σ 969		Mon	1830 1923	317		7.2		7.8	10.8	B9	5463		
6 48.2	+0 18	Rst 5248		Mon	1946	143		7.0		7.8	11.5	G0			
6 48.2	+55 42	Σ 958	AB	Lyn	1830 1956	257		4.8		6.3	6.3	F5 F5	5436		
---			BC		1880 1910	264		164.3			11.2		5436		
6 48.3	+23 13	Pou 2028		Gem	1925	311		15.4		8.0	11.8	G5			
6 48.5	−4 03	A 58		Mon	1900 1950	147 154		4.1	4.5	8.2	8.9	F8	5471		
6 48.8	−16 13	β 20		CMa	1876 1943	32		3.1		7.1	11.1	K0	5484		
6 49.0	−15 09	AC 4		CMa	1877 1956	286 307		1.1	0.9	5.4	8.1	B5	5487		

α 2000	δ 2000	Star Name		Const	Years		Pos Ang		Separation		m_1	m_2	Spec	ADS	Orb	Notes
6h 49m.1	−26° 04′	B 120		CMa	1926	1943	278°	278°	2″.3	2″.0	7.4	10.4	K2	5492		
6 49.3	−2 16	ø 322		Mon	1952	1959	81	91	0.1	0.1	6.5	6.5	A0			
6 49.4	−28 39	B 1520		CMa	1929		198		3.3		7.4	13.5	K0			
6 49.5	−45 34	Don 147		Pup	1929	1947	233		1.4		7.1	11.4	B8			
6 49.6	+53 02	Σ 960		Lyn	1829	1924	67		21.9		8.0	9.9	F0	5462		
6 49.6	−24 09	S 538		CMa	1825	1926	3		27.2		7.2	8.5	A2			
6 49.7	−24 05	β 324	AB	CMa	1877	1942	206		1.8		6.3	7.6	A0	5498		
—		S 537	AC		1825	1933	281		30.5			8.6	A2	5498		
—			AD		1898	1933	2		29.3			12.8		5498		
6 49.8	+16 12	J 2028		Gem	1942	1959	45		27.5		5.7	13.0	B8			33 Gem
6 49.8	+39 29	β 756		Aur							7.6		B9			Not a double star (IDS)
6 50.0	+46 11	Es 15		Aur	1899	1916	275		28.8		7.3	10.7	K0	5480		
6 50.0	−45 27	I 159		Pup	1900	1939	324		6.6		6.6	11.2	K0			
6 50.4	−16 02	β 898	AB	CMa	1879	1916	355		3.2		8.0	11.5	B9	5503		
—			AC		1879	1917	282		96.9			10.2		5503		
—			CD		1879	1917	269		1.9			11.0		5503		
6 50.4	−31 42	H V 108	AB	CMa	1783	1919	66		42.9		5.6	8.2	B8 A3			Magnitude of B is for BC
—		Rst 2435	BC		1932	1944	106		0.6		8.3	10.8				
6 50.8	−0 32	β 897		Mon	1879	1933	30		5.9		5.8	11.3	F2	5505		
6 50.8	+41 47	β pm	AB	Aur	1888	1925	108	100	41.0	40.9	5.0	10.0	K0			58 ψ⁷ Aur
—			AC		1907	1925	76	75	117.7	118.0		11.6				
6 51.0	−56 14	h 3898		Car	1873	1912	310		16.8		8.6	8.7	G0 F8			
6 51.1	−28 44	I 431	AB	CMa	1903	1939	319		0.4		9.4	9.6	F0	5527		
—		h 3896	ABxC		1898	1935	163		11.3			8.9	F0	5527		
6 51.2	−10 05	h 741		Mon	1904	1905	226		16.1		7.4	10.9	K5	5516		
6 51.5	+21 46	β 1193		Gem	1890	1935	355		10.6	11.3	5.3	14.1	A0	5511		36 Gem
6 51.7	−44 18	LDS 173		Pup	1944		337		45.8		8.0	9.8	K0 G0			
6 51.8	−26 35	β 325		CMa	1877	1935	36		1.7		8.0	9.0	B5	5539		
6 52.3	−5 10	WNO		Mon	1971		179		58.3		6.6	10.4	K6 M2			
6 52.7	−26 07	B 2127		CMa			255		8.0		7.5	14.0	K5			
6 52.8	+33 58	Fox	AB	Gem	1913	1922	322		78.7		3.6	12.6	A2	5532		34 θ Gem
—			AC		1913	1922	185		102.4			12.7		5532		
6 52.8	+47 12	A 1736	AB	Aur	1908	1919	229		1.1		8.0	12.0	F2	5525		
—			AC		1908	1919	152		12.3			13.5		5525		
6 52.8	−43 59	I 1498		Pup	1927	1933	288		10.0		6.5	12.5	B9			
6 53.0	+38 52	Σ 974	AB	Aur	1831	1934	223		22.3		6.2	9.5	F2	5534		59 Aur
—			BC		1908		219		25.8			12.8		5534		
6 53.0	−26 57	See 71	AB	CMa	1897	1899	99		10.4		6.5	14.0	M4	5548		
—			AC		1897	1899	247		20.2			14.0		5548		
6 53.1	+49 32	h 2342	AB	Aur	1908		49		21.6		7.9	13.4	K2			
—			AC		1908		75		27.9			13.8				
6 53.1	+59 27	Σ 963	AB	Lyn	1830	1960					5.7	6.9	F5	5514	o	14 Lyn; sp. compos. F5+A2
—			ABxC		1879	1910	122		181.0			11.0		5514		
6 53.4	+51 31	OΣ 158 rej	AB	Lyn	1868	1925	304		16.8		7.0	11.3	F5	5531		
—			AC		1910	1925	63		56.0			10.7		5531		
6 53.9	−4 25	Σ 985 rej		Mon	1904	1918	322		32.6		8.3	8.9	K5			
6 54.1	+6 41	OΣΣ 79		Mon	1875	1922	89		116.1		7.2	7.3	G5 A0			[45]
6 54.1	−5 51	Σ 987		Mon	1831	1953	163	170	1.1	1.2	7.1	7.2	A3	5557		
6 54.1	−77 47	h 3932		Men	1879	1918	284		8.3		7.7	9.9	A2			
6 54.2	−34 13	h 3900		Pup	1879	1959	284	279	2.2	2.1	7.7	9.0	A0			
6 54.4	+21 10	OΣ 160		Gem	1848	1956	167	181	1.3	1.5	6.7	9.7	G5	5553		
6 54.6	+13 11	Σ 982	AB	Gem	1829	1976					4.7	7.7	F0	5559	o	38 Gem
—			AC		1912	1932	328	328	109.3	111.5		10.3		5559		
6 54.9	+21 34	OΣ 161 rej		Gem	1868	1912	171		20.1		6.8	11.1	B9	5564		
6 55.0	+11 58	A 2833		Gem	1914	1942	169		0.6		8.2	8.7	F5	5571		
6 55.0	−20 24	H V 65	AB	CMa	1783	1915	147		44.4		5.8	9.3	A2 K5	5585		17 CMa
—			AC		1825	1915	184		50.5			9.0		5585		
—			AD		1825	1915	186		129.9			9.5		5585		
6 55.1	+32 07	β pm	AB	Gem	1912		16		11.8		3.8	13.5				Nova Gem 1912
—			AC		1912		99		79.6			12.0				
6 55.1	−38 33	ø 20		Pup	1927	1936	290		1.5		7.4	12.0	K0			
6 55.2	−29 02	Rst 1329		CMa	1932	1944	4	9	0.5	0.5	7.1	8.9	F5			
6 55.3	+48 33	Σ 977	AB	Aur	1831	1957	129		1.9		8.2	9.7	F2	5555		
—			AC		1881	1910	223		152.5			10.8		5555		
6 55.5	+37 55	Σ 978		Aur	1831	1918	99	88	14.8	17.0	6.8	9.6	K0	5574		
6 55.5	−22 02	See 72		CMa	1897	1933	39		13.0		6.8	12.8	B5	5601		
6 55.6	−20 08	H N 123		CMa	1876	1933	18		11.6		4.7	9.7	F2	5602		19 π CMa
6 55.9	−25 31	Stone 16		CMa	1876	1933	97		3.7		7.4	10.4	F0	5606		
6 55.9	−31 47	Jsp 125		CMa	1930	1944	10		18.8		6.4	14.0	B8			
6 56.1	−14 03	Σ 997	AB	CMa	1831	1944	340		3.0		5.3	8.6	G5	5605		μ CMa; most beautiful (h)
—			AC		1865	1912	288		88.4			10.5		5605		
—			AD		1865	1912	61		101.3			10.7		5605		
6 56.2	+2 18	β 326		Mon	1876	1943	63	51	1.2	1.2	8.1	9.6	G5	5603		
6 56.6	+46 32	Σ 979		Aur	1830	1958	209		7.4		8.4	9.2	A0	5582		
6 56.6	−22 39	S 541		CMa	1825	1929	44		23.2		8.0	9.0	K0 F			

α 2000	δ 2000	Star Name		Const	Years	Pos Ang	Separation	m$_1$	m$_2$	Spec	ADS	Orb	Notes
6^{h}57^m.0	+24°57′	Σ 991	AB	Gem	1830 1937	167°	3″.8	8.0	9.0	A0	5608		
--			AC		1907	202	134.4		10.6		5608		
--			CD		1907	359	13.2		11.1		5608		
6 57.0	-19 26	Ho 517		CMa	1906 1959	336 319	3.3 2.9	6.8	12.8	F5	5623		
6 57.2	-25 09	CorO		CMa		sp	10.0	8.2	10.0	K5			
6 57.3	+58 25	OΣ 159	AB	Lyn	1844 1959	323 33	0.5 0.9	4.8	5.9	G0	5586		15 Lyn
--			AC		1878 1924	342 346	23.6 29.0		12.4		5586		
--			AD		1850 1924	167 167	206.6 197.4		8.9		5586		
6 57.3	-35 30	I 65		Pup	1898 1960			6.9	7.1	F5		o	
6 57.4	+2 53	A 2681		Mon	1913 1958	294 340	0.2 0.1	8.5	8.7	A5	5625	o	
6 57.6	-24 38	B 122		CMa	1926 1959	271 264	1.0 1.0	5.8	7.1	F5	5629		
6 57.8	-42 02	I 1161		Pup	1926 1942	24	1.9	8.0	12.5	A5			
6 58.1	+14 14	OΣΣ 80	AB	Gem	1876 1914	53	124.0	7.6	7.3	B9 A5			
--			AC		1914	112	78.8		8.2	G5			
--			BC		1876	192							
6 58.1	-27 10	B 707		CMa	1927 1959	20	0.2	7.0	7.4	B3			
6 58.3	-35 25	I 66	AB	Pup	1898 1942	254	1.9	8.1	9.8	B9			
--		h 3905	AC		1898 1920	269	14.8		9.5				
6 58.5	-3 01	β 327	AB	Mon	1876 1956	98	0.7	7.8	8.3	B3	5640		
--			AC		1876 1930	100	13.4		11.3		5640		
6 58.6	-25 25	B 124		CMa	1927 1933	63	10.2	5.6	12.6	B3	5651		
6 58.6	-28 58	CapO 7		CMa	1850 1951	161	7.5	1.5	7.4	B1	5654		21 ε CMa
6 58.7	-31 00	HdO 198	AB	CMa		315	35.0	6.4	9.0	B6			
--			AC			320	70.0		10.0				
6 58.8	+26 05	β pm		Gem	1910 1925	28 33	28.2 28.8	6.1	12.1	F5			39 Gem
6 58.9	+3 36	β 1060		Mon	1889 1953	58 59	3.0 3.6	6.0	11.0	K0	5648		
6 59.0	-4 01	Rst 4833		Mon	1941 1951	300	0.7	7.9	13.2	A0			
6 59.0	-26 28	B 708		CMa	1927 1944	34	3.3	7.2	11.4	B8			
6 59.5	+37 06	Σ 994	AB	Aur	1831 1949	57 56	25.6 26.5	7.9	8.2	B	5642		
--		Ali 574	AC		1930	121	9.2		12.8		5642		
--		Gui	AD		1913	353	73.2				5642		The D and E stars could not
--		Bloch	AE		1924	346	196.5				5642		be found in photographic
--		Bot	AF		1958	353	7.2		8.5		5642		catalogues (IDS)
--			AG		1958	123	23.4		11.0		5642		
--		Es			1892	221	9.1		12.0		5642		
6 59.6	-27 54	See 73		CMa	1897 1944	3	0.3	8.3	8.4	A2	5673		
6 59.8	+15 56	A 2461	AB	Gem	1912 1947	286 320	0.3 0.3	7.2	8.7	G0	5660		
		OΣ 162 rej	ABxC		1843 1924	155	21.4		10.0		5660		
7 00.0	-22 43	S 543		CMa	1825 1905	272	91.0	8.5	8.6	K0 A0			
7 00.3	-22 07	ø 334	Aa	CMa	1955 1959	126	0.1	7.3	7.3	B5	5687		
--		See 74	AaxB		1897 1899	231	13.2		14.4		5687		
--			AaxC			n	128.4		10.7		5687		
--			CD		1897 1943	347	2.4		11.7		5686		
7 00.3	-60 52	h 3922		Car	1917	237	17.0	7.9	12.1	B8			
7 00.4	-8 24	Rst 4329		Mon	1939 1941	359	1.5	6.0	12.7	A0			
7 00.6	+12 43	Σ 1007 rej	AB	Gem	1905 1916	300	15.2	7.8	11.8	A3	5676		β 99
--			AC		1905 1916	247	22.2		10.5		5676		
--			AD		1905 1916	28	68.1		8.8		5676		
7 00.6	-20 38	β 572		CMa	1877 1933	143	5.2	6.6	10.6	A3	5692		
7 00.8	-25 39	I 183		CMa	1897 1933	144	3.4	7.4	10.2	B3	5699		
7 00.9	+12 24	β 100		Gem	1875 1944	258	3.3	7.7	11.5	K0	5684		
7 01.2	+11 46	OΣ 163	AB	Mon	1848 1958	321 4	0.6 0.2	7.1	8.4	A2	5689		
--			ABxC		1879	158	14.2		12.0		5689		
7 01.2	-30 40	Jsp		CMa		324	21.0	7.4	13.0	F0			
7 01.3	+32 25	Es 339	AB	Gem	1907 1925	187	16.1	6.6	13.1	F0	5680		
--			AC		1925	293	32.1				5680		
7 01.4	-3 07	A 518	AB	Mon	1903 1929	187	2.5	7.9	15	B3	5705		
--		Σ 1010	AC		1833 1931	5	23.2		9.2		5705		
7 01.4	+70 48	β pm		Cam	1879 1910	356	117.0	5.7	11.0	K0			
7 01.5	-9 42	A 3042	AB	Mon	1922 1952	86 169	0.2 0.2	8.4	8.9	F2	5707		
--			AC		1922	77	11.0		14.6		5707		
--			AD		1922	330	12.6		13.6		5707		
7 01.7	-27 56	B 126		CMa	1926	153	10.9	3.5	14.0	M0	5719		22 σ CMa
7 01.8	-10 53	β 573		Mon	1878 1960	250 285	0.8 1.0	7.1	7.6	F8	5712		
7 01.8	-11 18	Hu 112		CMa	1900 1946	189	0.6	7.1	7.8	B3	5713		
7 01.9	-1 21	Kui 28		Mon	1958	247	24.5	6.2	10.9	K0			
7 02.0	-57 36	Rst 227		Car	1930	143	0.6	8.1	10.3	G5			
7 02.1	-23 30	h 3914		CMa	1897 1929	313	11.1	8.0	11.5	G5	5728		
7 02.1	-50 26	B 1053		Pup	1931	213	2.4	7.8	12.8	A0			
7 02.3	+12 34	h 3288		Gem	1831	254	50.0	8.1	9.1	A2			
7 02.8	+13 05	Ho 342		Gem	1891 1958	71 86	0.8 1.1	8.1	8.9	F5	5725		
7 02.8	-16 42	Wirtz		CMa	1908	87	8.7	6.9	12.3	K0	5736		
7 03.0	+54 03	A 1575		Lyn	1907 1957	280	0.6	8.2	9.7	F5	5704		
7 03.1	+54 10	Σ 1001	AB	Lyn	1831 1956	65	9.0	7.7	9.3	G0	5706		
--			AC		1903 1956	56 66	9.9 9.7		9.6		5706		

α 2000	δ 2000	Star Name		Const	Years	Pos Ang		Separation		m₁	m₂	Spec	ADS	Orb	Notes
7ʰ 03.1ᵐ	+54° 10′	Σ 1001	BC	Lyn	1831 1939	355°	3°	1″.6	1″.7				5706		
7 03.3	−57 30	I 483		Car	1911 1912	34		11.1		7.7	11.0	K2			
7 03.3	−59 11	Δ 39		Car	1826 1952	76	82	2.5	1.7	6.0	7.1	B9			
7 04.0	−43 36	Δ 38	AB	Pup	1826 1932	122		20.5		5.6	7.2	G0 G0			
--			AC		1900	334		184.8			8.1	K2			
7 04.1	+20 34	S,h 77	AB	Gem	1880 1924	84		87.0		3.8	10.5	G0	5742		43 ζ Gem; A is variable
--			AC		1821 1925	355 350		91.0	96.5		8.0		5742		
--			Aa		1905	83		80.0			12.0		5742		
--			BC		1913	310		133.0					5742		
--			Cc		1907	166		27.2			12.9	G	5742		
7 04.1	+75 14	Σ 973	AB	Cam	1831 1971	27 31		11.9	12.6	7.1	8.1	G0	5669		
--		LDS 1642	AC		1965	17		60.0			16		5669		
7 04.3	+1 29	OΣΣ 82		Mon	1876 1925	318		90.4		6.5	7.4	B9 B9			
7 04.4	+4 34	Ho 241		Mon	1887	183		8.8		7.4	12.4	B9	5747		
7 04.4	−10 27	Rst 3489		Mon	1938 1943	297		0.6		7.3	8.6	B0			
7 04.6	−11 31	Σ 1016		CMa	1831 1938	150		5.3		7.4	9.4	B5	5761		
7 04.7	+5 56	h 2360		Mon	1898	154		19.9		7.9	10.0	K0	5755		
7 05.1	+0 20	A 2841		Mon	1914 1954	240 227		0.3	0.4	8.3	9.3	K0	5768		
7 05.2	−0 52	A 1741		Mon	1908 1958	9 18		0.7	1.0	8.2	8.5	F5	5769		
7 05.3	−34 09	Jsp 132	AB	Pup	1929 1945	226		0.4		7.9	8.6	F5			
--			ABxC		1929	19		8.8			13.0				
7 05.4	−52 37	CorO 47		Car	1919 1923	89		22.0		7.7	10.3	K5			
7 05.5	−34 47	h 3928	AB	Pup	1836 1959	158 149		4.9	3.1	6.5	7.6	F0			
--			AC		1836 1920	287		37.2			10.0				
--			AD		1879 1920	125		38.7			10.8				
--			AE			301		18.0			13.5				
7 05.7	+52 45	Σ 1009	AB	Lyn	1830 1976	159 150		2.9	4.1	6.9	7.0	A2	5746		
--			BC		1879 1910	133		179.1			11.0		5746		
7 05.8	−10 40	Dem 12	AB	Mon	1870 1933	281		6.2		6.5	10.4	B3	5782		
--		Σ 1019	AC		1830 1920	294		37.8			10.1		5782		
--		Rst 3491	CD		1938 1943	306		0.9			11.3		5782		
7 06.0	−19 13	HdO 97		CMa	1869	np		5.0		0.9					
7 06.0	−42 20	h 3931	AB	Pup	1920	213		57.3		7.1	10.0	B9			
--			AC		1835 1920	40		72.5			9.0	F5			
7 06.1	−45 58	Rst 4838		Pup	1942 1947	278		2.3		7.6	13.0	K5			
7 06.2	+24 52	OΣ 164 rej		Gem	1843 1905	50		13.7		7.3	10.3	K2	5775		
7 06.7	−11 18	β 328	AB	CMa	1875 1950	128 116		0.3	0.6	5.7	6.9	B3	5795		
--		Σ 1026 rej	ABxC		1879 1958	350		17.8			9.1		5795		
7 06.9	+62 33	Σ 1006		Cam	1831 1922	72 73		30.6	29.6	8.3	9.3	G5	5758		
7 07.3	+24 10	h 412		Gem	1820	20		35.0		6.9	16	M0			
7 07.5	−1 13	Che		Mon	1911	336		9.5		8.0	11.0		5808		Star place uncertain
7 07.8	−67 56	NZO 6		Vol	1911	21		2.0		8.0	10.5	B8			
7 07.9	−4 40	Σ 1029		Mon	1833 1953	26		1.8		8.2	8.9	F0	5813		
7 07.9	−15 42	A 3043		CMa	1926 1952	338 13		0.2	0.1	7.6	8.6	G0	5814		
7 08.0	+15 32	β pm	AB	Gem	1891 1911	99		172.0		7.5	7.4	F8 F8			
--			BC		1908	87		88.9			10.7				
7 08.0	−20 05	HdO 98		CMa	1869	np		5.0		0.9			5816		
7 08.3	+16 55	Σ 1027		Gem	1830 1938	355		6.9		8.3	8.4	K2	5816		
7 08.4	+15 56	OΣ 165	AB	Gem	1847 1953	130 15		3.8	10.0	5.4	11.1	K0	5812		45 Gem
--			AC		1913	328		52.7			13.3		5812		
7 08.4	−20 52	I 772		CMa	1913 1933	326		16.0		6.9	12.0	K2	5823		
7 08.8	−24 03	Ara 2035		CMa	1922	292		15.0		6.4	11.5				
7 08.8	−70 30	Δ 42		Vol	1826 1941	300		13.6		4.0	5.9	K0 G0			γ² Vol; superb (h)
7 09.0	−28 10	B 2135		CMa		sf		6.0		7.8	12.8	B8			
7 09.1	+71 58	h 2355		Cam	1909	248		64.2		7.1	11.3	K0			
7 09.1	−60 35	I 184	AB	Car	1901 1951	179 165		0.5	0.4	8.6	9.0	F0			
--			AC		1900	340		24.5			12.0				
7 09.2	−56 22	Δ 40		Car	1826 1912	141		37.5		8.2	8.5	F F0			
7 09.4	+36 34	Σ 1022		Aur	1831 1936	129		5.6		6.8	10.0	A0	5820		
7 09.4	−60 23	h 3941		Car	1836 1951	311 291		0.8	0.6	7.4	7.9	K0			
7 09.6	+25 44	Ho 519	AB	Gem	1891 1927	124 102		19.7	22.2	7.0	13.0	G0	5827		
--		OΣΣ 83	AC		1874 1921	87 84		105.4	111.5		7.7	F0	5827		
7 09.6	−16 14	β 329		CMa	1880 1915	98 100		29.5	32.5	6.0	11.3	B3	5837		
7 09.7	−25 14	B 2136		CMa	1945	133		10.6		5.7	12.9	B3			
7 10.2	−4 14	β pm	AB	Mon	1909	238		54.7		4.9	11.9	K0			20 Mon
--			AC		1886 1909	340 339		190.6	186.2		9.8				
--			CD		1910	126		85.6			11.3				
7 10.2	−18 41	Rst 2456	AB	CMa	1936 1950	338		0.8		6.2	11.2	F0			
--			AC		1935 1936	90		7.9			13.4				
7 10.4	−30 04	B 1531		CMa	1928 1959	9		0.9		8.6	8.7	A5			
7 10.4	−55 35	Rmk 5		Car	1835 1954	225		7.0		7.5	7.6	K0			
7 10.5	−29 21	B 1532		CMa	1928 1949	321 330		0.7	0.7	8.2	8.7	F5			
7 10.7	−41 16	Rst 2458		Pup	1933 1945	55		0.4		7.7	9.8	A2			
7 11.1	+30 15	β 1009	AB	Gem	1882 1925	178		1.9		4.4	10.9	K0	5846		46 τ Gem
--			AC		1913 1923	342		58.4			12.3		5846		

α 2000	δ 2000	Star Name		Const	Years	Pos Ang		Separation		m₁	m₂	Spec	ADS	Orb	Notes
7ʰ11ᵐ.1	+49°55′	Es		Lyn	1910 1918	127°	°	46″.0	″	8.2	9.6	G5			
7 11.2	−10 33	A 2122		Mon	1910 1959	134 105		0.3	0.2	8.8	8.8	B8	5857		
7 11.3	−21 48	h 3934	AB	CMa	1876 1937	236		13.6		7.0	9.4	B5 B9	5863		
−−		Rst 4840	BC		1942 1949	273		0.6			11.5		5863		
7 11.4	−39 06	ø 377		Pup	1961	28		0.1		7.6	7.6	A2			
7 11.7	−84 28	h 3996		Men	1900 1919	255		16.4		7.6	11.6	B9			
7 11.8	+19 53	Cou 925		Gem	1973	89		0.4		8.0	9.6	A2			
7 11.9	−0 30	J 58		Mon	1910	170		32.0		4.2	13.1	A0	5864		22 δ Mon
7 12.0	+22 17	Σ 1035		Gem	1829 1949	41		8.7		8.2	8.2	F5	5858		Yel. pair, pretty (Webb)
7 12.0	+57 31	Σ 1020		Lyn	1830 1910	284		13.3		8.1	10.3	F5	5840		
7 12.0	−63 11	HdO 199		Car	1894 1951	157 132		0.5	0.3	6.6	7.0	A0			
7 12.1	−27 20	Rst 1341		CMa	1932 1944	32		12.5		6.9	12.8	B3			
7 12.4	−31 59	B 2137		CMa		217		7.0		7.1	12.5	M			
7 12.4	−36 33	β 757		Pup	1881 1943	66		2.5		6.0	8.5	B5			
7 12.5	−8 13	A 522		Mon	1903 1917	352		1.3		8.0	12.4	A0	5877		
7 12.8	−7 09	β 197	AB	Mon	1876 1951	147		2.2		7.9	10.4	F2	5888		
−−		Lv	AC		1917	20		20.1			11.5		5888		
−−			AD		1917	185		39.3			10.8		5888		
7 12.8	+15 11	Weisse 14		Gem	1901 1953	157		2.2		7.9	8.4	A0	5875		
7 12.8	+21 21	OΣ 168 rej	AB	Gem	1868 1900	67 67		22.7	24.0	7.7	11.8	G5	5873		
−−			AC		1899 1900	114		51.9			11.3		5873		
7 12.8	+27 13	Σ 1037	AB	Gem	1830 1960					7.2	7.2	F5	5871	o	
−−		OΣ 166 rej	AC		1899 1925	111 102		16.3	15.3		12.3		5871		
−−			BC		1953	94		15.6					5871		
7 12.8	−57 29	HdO 304		Car		350		15.0		6.8	10.0	A0			
7 12.9	+55 49	Σ 1025	AB	Lyn	1830 1955	141 133		22.7	25.6	8.3	8.6	K0	5854		
−−			AC		1880 1908	253		236.8			11.2		5854		
−−			Cb		1908	202		22.7			13.1		5854		
−−			Cc		1908	250		3.1			11.8		5854		
7 13.2	−15 05	Hu 455		CMa	1902 1943	203		4.0		8.1	10.6	A0	5901		
7 13.2	−30 58	h 3940		CMa	1898 1934	99		6.8		7.9	9.9	B9			
7 13.4	+16 10	H VI 74	AB	Gem	1901 1903	20		149.2		5.3	10.5	M			51 Gem
−−			AC		1901 1903	37 37		222.9	221.2		10.5				
7 13.5	+32 09	OΣ 167		Gem	1850 1933	158		5.3		7.3	10.4	A5	5884		
7 13.5	−38 29	I 1505		Pup	1927 1941	195		2.9		7.6	10.4	A2			
7 13.6	−8 55	Σ 1049		Mon	1830 1946	40		3.6		7.9	9.7	A0	5904		
7 13.8	+28 30	OΣ 520		Gem	1850 1959	344 12		0.6	0.6	7.9	9.9	A0	5893		
7 13.8	−22 54	h 3938		CMa	1879 1939	250		19.7		6.5	8.3	B3 A5	5912		
7 13.9	−3 43	A 523	AxBC	Mon	1903 1918	229		98.2		7.9		K0	5906		
−−			BC		1903 1918	323		1.2		12.0	13.3		5906		
7 13.9	+48 30	Σ 1032	AB	Lyn	1831 1935	100 110		2.6	2.6	7.7	11.0	A0	5879		
−−			AC		1880 1908	319		32.7			9.8		5879		
7 14.0	−23 40	B 131		CMa	1926 1928	36		2.6		7.5	14.0	K0	5914		
7 14.2	−3 54	A 524		Mon	1903 1958	148 157		2.9	2.7	5.8	10.8	K5	5911		
7 14.3	−26 21	ø 323		CMa	1952 1960	112 140		0.1	0.1	5.5	5.5	B5			27 CMa
7 14.4	+24 43	β pm	AB	Gem	1886 1909	108		27.5		6.7	10.2	B9			
−−			Aa		1907	17		49.0			11.0				
−−			Bb		1907	193		93.4			10.4				
7 14.5	−19 59	B 1967		CMa	1931	179		2.2		7.4	11.9	B3			
7 14.6	−25 18	B 2138		CMa		sp		8.0		7.4	11.0	B8			
7 14.6	−31 30	CorO 48		CMa	1911 1940	6 10		7.6	7.8	8.0	9.8	B9			
7 14.6	−48 16	Rst 241		Pup	1929 1946	205		18.5		4.8	13.4	B8			
7 14.7	+24 53	Ho 343		Gem	1890 1910	257 265		22.4	23.9	5.8	12.5	K0	5909		52 Gem
7 14.8	+52 33	Σ 1033	AB	Lyn	1829 1959	279		1.5		7.7	8.3	F0	5896		
−−			AC		1783 1921	272		79.5					5896		
7 14.8	−15 29	β 575	AB	CMa	1878 1959	199 75		0.7	0.7	8.0	8.0	F8	5925	o	
−−		Σ 1057	AC		1831 1919	2		15.6			9.7		5925		
7 15.0	−2 49	A 525	AB	Mon	1903 1930	262		2.0		7.9	12.6	A2	5924		
−−			AC		1892 1930	36		12.0			11.7		5924		
7 15.0	+11 01	h 753		CMi	1908	4		22.6		8.2	9.0	A2			
7 15.3	−0 10	β 1268		Mon	1892 1946	311		3.8		6.4	12.2	G5	5933		24 Mon
7 15.4	+19 04	A 2527		Gem	1913 1921	289		1.9		7.7	13.0	A5	5922		
7 15.4	−42 05	Rst 2463		Pup	1933 1945	185		1.6		7.0	12.5	A0			
7 15.5	−1 52	Σ 1056		Mon	1830 1933	299		3.8		8.1	9.1	G0	5935		
7 15.5	−75 52	I 312		Men	1901 1944	168 161		1.1	1.1	8.1	8.1	F5			
7 15.6	+21 58	β pm		Gem	1879 1919	264		95.1		7.5	10.3	K5			
7 16.2	−54 03	h 3952		Car	1892 1917	277		16.2		7.2	11.9	K0			
7 16.5	+47 39	Σ 1044		Lyn	1828 1930	168		12.4		8.1	8.3	A2	5932		
7 16.6	−23 19	h 3945		CMa	1837 1959	65 55		27.2	26.6	4.8	6.8	K5 F0	5951		High y., contrasted bl. (h)
7 16.8	+24 32	Σ 1053		Gem	1831 1894	310		13.8		7.1	9.8	A0	5945		
7 17.0	−30 54	BrsO 2		CMa	1838 1928	182		37.9		6.3	8.1	A5 A5			
7 17.1	−12 02	A 2123	AB	CMa	1910 1958	336 30		0.3	0.3	7.6	7.6	F5	5956		
−−		Σ 1064	AC		1831 1930	240		15.6			9.1		5956		
7 17.1	−13 59	Hu 113	AB	CMa	1900 1923	54		1.8		7.8	12.3	B8	5957		
−−			AC		1923	125		8.5					5957		

α 2000	δ 2000	Star Name		Const	Years	Pos Ang	Separation	m₁	m₂	Spec	ADS	Orb	Notes
7ʰ17ᵐ.1	−37°06′	Δ 43		Pup	1826 1936	213°	69″.2	2.7	8.0	K5 B9			π Pup
7 17.2	−46 42	Rst 242		Pup	1930 1946	47	3.3	7.6	12.6	K0			
7 17.5	−46 59	I 7		Pup	1896 1960			7.1	7.8	K0		o	
7 17.6	+9 18	OΣ 170		CMi	1844 1959	133 95	1.0 1.5	7.6	7.9	G0	5958		
7 17.8	−25 59	See 75		CMa	1897 1899	7	12.4	8.0	13.5	F0	5966		
7 18.0	−13 42	Σ 1069		CMa	1831 1924	194	25.3	8.3	8.3	A A			
7 18.1	+16 32	Σ 1061		Gem	1829 1953	33	9.6	3.6	10.7	A2	5961		54 λ Gem
7 18.1	+24 05	Cou 585		Gem	1971	153	0.4	8.5	9.0	F5			
7 18.1	+34 58	Σ 1054	AB	Gem	1830 1930	291	18.6	8.3	9.5	F2	5954		
—			AC		1880 1917	268	79.4		10.1		5954		
7 18.3	−36 44	Jc 10	AB	Pup	1846 1936	97	239.9	4.7	5.1	B3 B3			
—			BC		1846 1936	216	117.8		8.7	A0			
—			CD		1846 1937	211	2.9		9.1				
7 18.4	−57 21	Rst 244	AxBC	Car	1930	240	2.4	7.9		G0			
—			BC		1930	205	0.3	13.6	14.0				
7 18.6	+58 55	β pm		Lyn	1878 1910	6	168.9	7.4	11.0	F0			
7 18.6	−30 48	h 3949		CMa	1837 1955	77	3.1	7.7	8.0	B3			
7 18.7	−24 57	h 3948	AB	CMa	1876 1957	90	8.2	4.4	10.5		5977		30 τ CMa
—			AC		1876 1898	79	14.5		11.2		5977		
—			AD		1834 1957	74	85.0		8.8		5977		
—		φ 313	Aa		1951 1959	117	0.2			O9			
7 18.8	+13 32	J 1993		Gem	1941	150	10.0	7.4	13.4	K2			
7 19.0	−52 32	I 485		Car	1911 1926	356	3.1	8.0	10.5	F5			
7 19.1	+66 44	Hzg		Cam	1932	33	8.8	7.6	12.8	F8			
7 19.2	−24 57	B 133		CMa	1926	99	4.6	7.3	14.2	B3	5985		
7 19.3	−22 03	Lal 53		CMa	1848 1949	346	3.9	7.5	7.6	A2	5986		
7 19.6	−15 53	Hu 1243		CMa	1906 1937	70	0.8	8.7	8.7	A0	5988		
7 19.9	+54 55	Σ 1050	AB	Lyn	1829 1924	20	19.3	8.1	8.8	A0	5968		
—			AC		1880 1908	226	198.4		11.8		5968		
7 20.1	−4 59	Ho 242		Mon	1887 1924	64	4.4	7.1	12.1	A5	5992		
7 20.1	+21 59	Σ 1066		Gem	1829 1954			3.5	8.2	F0	5983	o	55 δ Gem
7 20.1	−24 25	β 331		CMa	1877 1943	116	2.0	8.5	9.2	B8	5998		
7 20.4	−52 19	Rmk 6		Car	1834 1955	16 23	10.1 9.5	6.0	6.6	F2			
7 20.5	+0 24	Σ 1074	AB	CMi	1831 1958	115 161	0.5 0.6	7.4	7.8	B9	5996		
—		β 577	AC		1892 1924	101	12.8		12.5		5996		
—			AD		1878 1922	11	15.3		12.0		5996		
—			AE		1892 1919	278	53.7		9.9		5996		
7 20.5	−1 36	A 1963		Mon	1909 1953	266 64	0.2 0.1	8.7	8.7	B9	6001		Rapid; too close 1957
7 20.5	+68 32	Σ 1038		Cam	1831 1925	96	11.3	7.9	10.3	A5	5962		
7 20.6	−52 12	h 3958		Car	1836 1955	281 281	31.8 29.7	7.2	9.0	A0 F0			
7 20.8	+9 39	A 2939		CMi	1914 1921	244	3.7	7.4	13.6	F5	6000		
7 20.9	−20 39	Don 170		CMa	1933 1941	283	2.3	7.5	13.5	A3			
7 20.9	−26 58	See 76		CMa	1896 1933	217	8.8	6.0	13.2	B3	6015		
7 20.9	−56 47	h 3962	AB	Car	1874 1915	105	8.6	8.0	9.0	B9			
—		Rst 247	BC		1930	257	1.1		12.2				
7 21.0	+10 12	Σ 1073		CMi	1830 1927	67	8.8	7.8	9.8	A0	6002		
7 21.3	+34 02	Σ 1070	AB	Gem	1830 1939	319 321	1.9 1.7	8.4	9.4	A0	5999		
—			AC		1880 1908	122	87.5		10.6		5999		
7 21.4	−48 32	Δ 45		Pup	1826 1912	157	22.7	7.1	8.3	A A			
7 21.6	−79 10	h 3987		Men	1919	236	47.2	7.7	12.7	K0			
7 21.7	−61 57	BrsO		Car	1836	210	179.0	7.1	8.4	B8 A3			
7 21.9	+20 27	β	AB	Gem	1901 1935	205 202	17.8 16.5	5.2	12.2	K2	6016		56 Gem
—			BC		1900 1935	245 246	7.7 8.2				6016		
7 21.9	+46 14	Es		Lyn	1913 1918	266	80.8	8.0	8.0	K2 F8			
7 22.0	+36 46	β 901	AB	Aur	1879 1934	8 10	10.6 11.4	5.1	11.6	K0	6009		65 Aur
—			AC		1879 1931	27 36	36.0 39.7		12.0		6009		
7 22.3	+50 09	Σ 1065		Lyn	1830 1950	254	15.0	7.3	7.4	F0 F0	6004		
7 22.3	+59 54	Σ 1055		Cam	1830 1955	344 317	2.4 1.9	6.3	10.8	A5	5995		47 Cam
7 22.3	−35 55	h 3957		Pup	1877 1935	194	7.7	7.5	8.8	F8			
7 22.9	+55 17	Σ 1062	AB	Lyn	1829 1956	315	14.8	5.6	6.5	B8 A	6012		19 Lyn
—			AD		1905 1956	3	214.9		8.9		6012		
—			BC		1879 1908	287	74.2		10.9		6012		
7 23.0	−3 54	J 1489		Mon	1941 1958	120	11.6	7.9	13.1	F5			
7 23.0	−25 46	B 719	AB	CMa	1927 1959	172 200	0.6 0.6	7.8	9.1	M0	6033		Involved in nebulosity [46]
—		See 78	ABxC		1897 1936	288 291	2.0 2.8				6033		
—			ABxD		1897 1936	13 348	2.9 3.5				6033		
—			ABxE		1897 1900	30	6.9				6033		
—			ABxF		1936 1947	107 103	1.4 1.1				6033		
7 23.0	−55 35	h 3967		Car	1913	139	46.5	6.9	13.0	K0			
7 23.5	+0 12	Bail 1092		CMi	1897	211	7.0	8.1	11.1				
7 23.5	+67 14	Milb 402		Cam	1925	168	5.6	8.1	10.1	G0	6003		
7 23.8	+10 42	Σ 1082	AB	CMi	1830 1927	326	19.9	8.5	9.2	A3	6035		
—			AC		1904	22	15.6		13.5		6035		
—			BC		1880	100					6035		
7 24.0	−3 59	Σ 1084		Mon	1830 1919	286 286	13.3 14.1	7.1	9.6	K0	6047		

α 2000	δ 2000	Star Name		Const	Years	Pos Ang		Separation		m₁	m₂	Spec	ADS	Orb	Notes
h m 7 24.0	-35° 50′	h 3965		Pup	1915 1933	307°	°	25.0	″	6.3	12.5	B8			
7 24.1	+21 28	Σ 1081	AB	Gem	1828 1956	216	232	1.3	1.7	8.5	9.2	B9	6038		
--			AC		1865	64		107.2			9.2		6038		
7 24.1	-29 18	Smyth		CMa	1833 1909	285		178.7		2.4	6.9	B7 A0			31 η CMa
7 24.2	-58 30	LDS 181		Car	sp			128.0		6.6	12.3	G5			
7 24.7	-31 49	δ 129	AB	CMa	1922 1949	309		1.9		5.5	11.0	K2			
--		Δ 47	AC		1922	342		99.2			7.6	A0			
--		B 1540	CD		1929 1949	204		1.0			10.8				
7 24.8	-19 01	Kui 29		CMa	1935 1937	310		2.7		6.3	11.2	A3			
7 24.8	-37 17	h 3966		Pup	1835 1951	322		7.0		6.9	7.0	A3			
7 24.9	+44 51	Es 585		Lyn	1908 1944	237	244	2.7	3.4	7.8	11.8	K0	6044		
7 25.1	-21 10	β 199	AB	CMa	1870 1957	22		1.8		7.1	8.1	B2	6065		
--		Ho 522	AC		1898 1959	117		6.7			12.8		6065		
7 25.1	-43 23	Hu 1423		Pup	1912 1931	299	303	6.0	6.2	8.0	12.0	A0			
7 25.3	+56 34	OΣΣ 84	AB	Lyn	1875 1924	325		114.2		7.6	7.7	K2 F8			
--			BC		1911	194		37.5			12.7				
7 25.3	-21 59	Rst 4360		CMa	1940 1949	58		2.4		6.1	13.5	A5			
7 25.6	+20 30	Σ 1083		Gem	1828 1950	44		6.4		7.2	8.3	A5	6060		
7 25.9	+18 09	Ho 346		Gem	1891 1930	58		12.8		6.9	11.7	G5	6064		
7 26.0	+14 06	Σ 1088	AB	Gem	1829 1914	195		11.1		7.6	9.6	B9	6068		
--		Σ 1087	A′B′		1829 1913	42	40	19.8	21.3	8.8	12.1	A2	6068		
--		Σ 1088	AA′		1829 1913	238		112.4				B9 A2	6068		
7 26.0	-18 21	Stone 17		CMa	1877 1944	76		4.9		7.9	9.9	A0	6078		
7 26.0	-42 07	h 3968		Pup	1920	141		25.9		7.3	12.0	K5			Large star very red (h)
7 26.1	+37 49	A1i 834		Aur	1930	48		5.7		8.3	9.4	K0			
7 26.1	-47 46	Rst 5263		Pup	1944 1949	12		0.5		8.0	10.0	K0			
7 26.2	-20 24	Don 181		CMa	1933 1959	120		0.5		8.3	8.5	F0			
7 26.3	+69 29	Σ 1059 rej		Cam	1906	305		25.7		8.1	10.0	G0			
7 26.3	-28 10	See 79		CMa	1897 1959	296	303	0.5	0.7	8.5	8.7	F0	6084		
7 26.5	+18 31	Σ 1090	AB	Gem	1830 1921	97		60.8		7.3	8.2	F0	6073		
--			BC		1830 1915	320		19.9			9.9		6073		
7 26.6	+73 05	Σ 1051	AB	Cam	1831 1953	268	284	1.2	1.1	7.1	9.2	F0	6028		
--			AC		1831 1935	82		31.5			7.8	F0	6028		
--			CD		1912	126		46.3			14.1		6028		
7 26.7	+31 37	OΣ 171		Gem	1851 1953	132		1.1		7.3	10.1	G5	6072		
7 26.8	+64 57	Hu 840		Cam	1905 1921	90		0.9		8.0	11.7	G0	6053		
7 27.0	-34 19	h 3969		Pup	1879 1933	226		17.4		7.0	7.7	F8 F8			
7 27.1	-17 52	β 578		CMa	1878 1942	54	46	2.4	1.8	5.6	10.9	F0	6093		
7 27.2	+8 17	β pm	AB	CMi						2.9		B8			3 β CMi
--			AC		1879 1907	23	23	97.0	98.5		11.0				
--			AD		1879 1907	77	77	123.2	124.9		10.9				
--			AE		1879 1907	310	311	139.0	138.7		10.7				
7 27.4	+8 45	Σ 1095		CMi	1831 1932	78		10.1		8.4	8.9	A2	6088		
7 27.4	+15 19	Σ 1094		Gem	1829 1953	96		2.5		7.4	8.4	B8	6086		
7 27.6	-18 30	S 550		Pup	1825 1918	116		39.5		7.4	8.5	A0 B9			
7 27.7	+21 27	S,h 368	AB	Gem	1863 1917	324		42.9		5.2	9.4	F5	6089		63 Gem
--			AC		1909	219		145.9			10.4		6089		
7 27.7	+22 08	Es	AB	Gem	1892 1901	22		11.3		6.9	12.1	K5	6087		
--		S 548	AC		1825 1921	276		35.6			10.0		6087		
7 27.8	-11 33	β 332	AB	CMa	1875 1958	169	175	0.8	0.7	6.1	8.1	F5	6104		Spectrum composite F5+A0
--		Σ 1097	AC		1832 1958	313		20.0			8.5		6104		
--		β 332	AD		1878 1958	157		23.2			9.5		6104		
--			AE		1878 1958	43		32.2			12.2		6104		
7 27.8	-28 22	B 1061		Pup	1928	273		1.7		6.8	11.3	B5			
7 28.0	+6 57	β 21		CMi	1875 1934	25		4.0		5.3	11.1	A5	6101		5 η CMi
7 28.0	-26 50	h 2391		Pup	1930 1933	290		16.7		6.5	11.0	K5			
7 28.2	+8 56	Lam 4	AB	CMi	1836 1911	247	240	34.6	30.3	4.3	13.6	K0	6100		4 γ CMi
--			AC		1911	262		119.2			12.0		6100		
7 28.4	-12 57	Rst 3518		Pup	1938 1943	92		1.8		8.0	13.2	A0			
7 28.5	+42 45	Σ 1086		Aur	1830 1914	102		12.2		7.9	9.4	K0	6094		
7 28.7	+24 39	OΣΣ 85	AB	Gem	1875 1923	27	24	56.3	58.9	7.4	9.8	F8			
--			Aa		1911	132		34.4			11.5				
7 28.9	+48 11	Kui 30	AB	Lyn	1934 1951	93		1.0		5.6	10.5	B9	6095		
--		β 758	ABxC		1883 1934	94		16.6			10.2		6095		
7 28.9	-31 51	Δ 49		Pup	1836 1950	53		8.9		6.5	7.2	B3			
7 29.1	+31 47	A 2124	AB	Gem	1910 1935	11	8	2.8	3.4	4.2	12.5	F0	6109		62 ρ Gem
--			AC		1886 1909	293	291	211.6	213.7		10.6		6109		
--			CD		1909	267		104.1			12.2		6109		
7 29.1	-18 53	Rst 2481		Pup	1935 1949	184		0.8		7.2	12.9	K0			
7 29.2	+14 22	OΣΣ 86		Gem	1875 1923	349		55.9		7.0	8.8	A3			
7 29.2	-43 18	Δ 51		Pup	1836 1952	74	74	22.5	22.3	3.3	9.4	M0 G5			σ Pup
7 29.3	+3 23	A 2739		CMi	1914 1921	225		3.8		7.7	11.7	A5	6121		
7 29.3	-13 08	Rst 3519		Pup	1938 1943	233		10.7		7.3	13.6	K0			
7 29.4	-15 00	Σ 1104	AB	Pup	1831 1954	292	358	2.4	2.1	6.4	7.5	F8	6126	o	
--			AC		1882 1910	188		20.4			10.7		6126		

α 2000	δ 2000	Star Name		Const	Years	Pos Ang	Separation	m₁	m₂	Spec	ADS	Orb	Notes
7ʰ29ᵐ.4	−15°00′	Σ 1104	AD	Pup	1882 1916	359° 8°	33″.6 42″.4		11.2		6126		
--			AE		1909	53	81.6				6126		
7 29.5	+34 48	OΣ 172 rej		Gem	1878 1929	247	14.9	7.9	11.9	A0	6111		BDS 4152, h 3295
7 29.5	+47 18	A 2046		Lyn	1909 1959	234 244	1.1 1.4	8.1	9.4	G5	6107		
7 29.8	+27 55	β 1194		Gem	1890 1935	289 288	13.9 12.8	5.0	13.5	K0	6119		65 Gem
7 29.9	+49 40	β pm	AB	Lyn	1888 1909	304 304	165.9 169.3	5.4	10.2	F5			22 Lyn
--			BC		1909	150	110.8		12.0				
7 29.9	−23 01	Rst 2482		Pup	1935 1943	179	2.8	4.9	10.8	A3			
7 30.3	+41 36	Es 586		Aur	1908	17	13.4	7.8	11.2	K0	6122		
7 30.3	+49 59	Σ 1093		Lyn	1831 1959			8.8	8.8	F5	6117	o	
7 30.5	+7 43	A 2869		CMi	1914 1958	84 59	0.4 0.3	8.3	8.5	A5	6138		
7 30.6	+5 15	Σ 1103		CMi	1832 1949	243	4.4	7.7	9.2	B9	6140		
7 30.8	−28 06	See 80		Pup	1897 1949	86 67	0.3 0.4	8.4	8.8	A0	6151		
7 31.0	−2 10	A 1967		Mon	1909 1942	356	1.3	7.3	9.4	G5	6147		
7 31.3	−1 08	h 760		Mon	1878 1904	355	30.3	6.6	13.0	G5			
7 31.6	+62 30	Es 1895		Cam	1906 1921	286	9.9	6.9	9.2	A0	6125		
7 31.7	−35 53	Jsp 167	AB	Pup	1929	345	91.5	6.6	11.0	A0			
--			BC		1929 1945	35 26	1.4 1.4		13.2				
7 31.7	−43 17	B 1549	AB	Pup	1933	195	4.3	8.0	14.5	K0			
--		CorO 50	AC		1919	208	16.6		10.6				
7 31.9	+50 10	Σ 1096		Lyn	1902 1918	324	28.8	8.4	9.4	A0			
7 31.9	−20 56	h 3973		Pup	1897 1947	38	9.0	8.3	9.3	B8	6159		Pure white, brick red (h)
7 32.0	+75 20	A 1069		Cam	1905 1956	350 343	0.5 0.6	8.4	9.4	A2	6103		
7 32.1	−8 53	Σ 1112 rej		Mon	1905 1938	112	23.4	6.1	8.8	F5	6158		
7 32.4	+18 22	A 2871		Gem	1914 1921	114	4.0	7.8	11.9	F5	6156		
7 32.4	+21 06	Cou 378		Gem	1969	309	4.8	8.0	12.3	A2			
7 32.4	−35 58	Rst 4855		Pup	1942	35	0.3	6.7	7.9	B9			
7 32.8	+22 53	Σ 1108		Gem	1827 1934	178	11.5	6.5	8.3	G5	6160		
7 33.0	−28 20	B		Pup	1940	178	9.7	8.0	12.5	B3			
7 33.0	−55 10	h 3984	AB	Car	1913	298	14.3	8.2	13.0	F8			
--			AC		1837 1938	257 250	64.5 65.2		9.3	A5			
7 33.2	+26 07	Cou 1109		Gem	1974	74	3.5	7.9	13.9	A2			
7 33.2	+32 52	β 22		Gem	1875 1930	149	6.4	7.9	10.9	G5	6162		
7 34.0	−23 42	Howe 18		Pup	1877 1959	204	1.8	8.2	9.0	B9	6184		
7 34.1	−50 50	h 3986		Pup	1883 1917	220	43.8	7.7	9.2	M G			
7 34.1	−60 15	Jsp 173		Car	1930	64	0.8	8.1	10.6	F5			
7 34.2	+46 09	A 2047		Lyn	1909 1921	257	2.0	7.3	13.5	K5	6167		
7 34.3	+3 22	J 2835	AB	CMi	1945	110	90.0	5.8	11.0	A0			9 δ³ CMi
--			BC		1945	195	5.0		13.0				
7 34.3	−23 28	H N 19		Pup	1825 1952	105 114	9.3 9.6	5.8	5.9	F4 F5	6190		
7 34.4	+33 07	β 579	AB	Gem	1878 1945	216	1.0	7.7	12.0	A2	6173		
--		OΣ 173 rej	AC		1843 1916	233	18.7		12.5		6173		
--			AD		1867 1916	348	43.7		9.7		6173		
7 34.4	−42 04	I 1167		Pup	1926 1943	310	1.5	7.4	9.9	A0			
7 34.5	+12 18	Σ 1116		Gem	1828 1957	111 99	1.8 1.8	7.9	8.6	B8	6180		
7 34.5	+28 41	Cou 1110		Gem	1974	306	2.8	6.7	11.2	G5			
7 34.5	−13 52	Doo	AB	Pup	1906 1958	91 114	8.5 12.7	6.7	11.6	F5	6189		
--			AC		1906 1958	314 304	22.3 17.4		13.0		6189		
--			AD		1906 1958	17 26	28.7 24.4		11.9		6189		
--			AE		1906 1958	125 129	35.6 40.7		12.4		6189		
--			AF		1906 1958	128 131	61.5 65.7		9.7		6189		
--			FG		1905 1958	108	6.6		10.5		6189		
7 34.6	+31 53	Σ 1110	AB	Gem	1826 1960			1.9	2.9	A0 A0	6175	o	66 α Gem, Castor [47]
--			AC		1835 1955	164	72.5		8.8	M1	6175		
--			AD		1902 1924	222	204.4				6175		
--			BC		1895 1957	160 163	70.4 69.5				6175		
7 35.1	+24 16	Σ 1113 rej	AB	Gem	1892	179	91.3	8.1	10.1	M0			
--		h 425	BC		1892 1914	46	8.4		10.7				
7 35.1	+30 58	OΣ 175	AB	Gem	1847 1960	334 330	0.5 0.4	5.8	6.4	K0	6185		
--			AC		1893 1908	195	81.3		9.3		6185		
7 35.1	−26 07	See 83		Pup	1897 1899	202	9.0	7.9	12.9	K0	6195		
7 35.4	−28 22	h 3982	AB	Pup	1920	156	38.4	4.6	9.3		6205		
--			BC		1920	130	42.2		10.0		6205		
7 35.4	−74 17	h 3997		Vol	1836 1947	100 120	2.0 2.0	7.0	7.1	B9 B9			
7 35.9	+26 54	β pm		Gem	1911	46	46.1	4.1	13.1	K5			69 υ Gem
7 35.9	+43 02	OΣ 174		Lyn	1851 1957	85	2.0	6.5	8.1	F0	6191		
7 36.0	−22 18	Alden		Pup	1932 1935	145	5.1	7.9	11.0	A0			
7 36.1	−14 30	A 3092	AB	Pup	1928	176	5.2	5.7	12.3	B5	6208		
--		Σ 1120	ABxC		1830 1926	36	19.6		9.6		6208		
7 36.2	−14 25	S 555		Pup	1825 1923	228	95.8	7.9	9.1	B9			
7 36.2	−23 50	HdO 107	AB	Pup	1898 1918	156	20.5	8.2	10.2	K0			
--			AC		1916	336	50.1		10.0				
7 36.2	−53 49	I 777		Car	1911 1942	47	1.3	7.7	9.9	K5			
7 35.4	−54 11	Rst 2493		Car	1936 1948	302 340	0.2 0.2	7.5	8.0	A0			
7 36.5	−25 20	B 729		Pup	1927 1959	188	0.1	7.6	7.6	B5			

α 2000	δ 2000	Star Name		Const	Years		Pos Ang		Separation		m₁	m₂	Spec	ADS	Orb	Notes
7ʰ36.ᵐ6	−14°29'	Σ 1121	AB	Pup	1831	1952	305°	°	7″.4	″	7.9	7.9	B9 B9	6216		In open cluster NGC 2422
--			AC		1893	1909	133		11.8			13.0		6216		
--			AD		1893	1909	98		64.7			10.7		6216		
--			AE		1893	1909	239		72.2			11.8		6216		
--			AF		1893	1909	315		84.0			12.0		6216		
--			AG		1893	1913	1		84.3			9.1	A	6216		
--			AH		1893	1909	268		149.9			10.8		6216		
--			AI		1893	1909	26		165.1			7.9	B9	6216		
--			AJ		1893	1909	354		196.5			9.2	A	6216		
--		S1v	DE		1970		260		128.7					6216		
--			DH		1970		271		213.8					6216		
--			FC		1970		135		101.5					6216		
--			GJ		1970		349		113.1					6216		
--			IG		1970		228		95.8					6216		
7 37.1	−27 25	B 730		Pup	1927	1951	67		0.3		7.8	7.9	B8			
7 37.3	−4 07	β pm	AB	Mon	1909	1958	267	266	28.2	25.8	5.1	13.1	F5			25 Mon
--			AC		1887	1909	349	350	122.7	121.7		10.4				
7 37.4	+38 52	Hu 842	AB	Lyn	1904	1944	16	36	0.4	0.5	7.9	10.2	A0	6211		
--		Σ 1118	ABxD		1925		92		97.0					6211		
--		Σ 1118 rej	AC		1904	1930	16		23.6			10.2		6211		
7 37.4	−34 58	ø 324	AB	Pup	1953	1959	118		0.2		5.6	6.1	B8			
--			ABxC		1953	1959	87	72	0.4	0.5		6.2				
7 37.5	−2 02	Ho 244	AB	Mon	1887		199		11.8		7.7	13.7	K5	6226		
--			AC		1887		195		34.0			13.7		6226		
--			AD		1887		109		69.0			9.9		6226		
--		Bail 485			1893		346		18.0		7.7	11.1	K5			
7 37.6	−14 26	S 557		Pup	1825	1918	337	338	66.4	65.0	6.5	9.3	B9			
7 37.8	−2 35	B 2525	AB	Mon	1936	1937	132	148	0.1	0.1	8.4	8.6	A5	6232		
--		A 534	ABxC		1903	1943	289	303	0.8	1.1		10.2		6232		
7 38.0	−32 08	B 1554		Pup	1930	1936	331	335	0.4	0.4	6.9	8.4	F5			
7 38.3	−25 22	B 731		Pup	1926	1959			0.1		5.4	5.6	B8	6246		
7 38.4	−47 43	h 3990		Pup	1883	1913	342		37.2		8.3	9.3	G A			
7 38.5	+0 30	OΣ 176	AB	CMi	1855	1953	214		1.5		7.1	9.1	B9	6240		
--		A 2529	AC		1913	1920	335		3.1			13.7		6240		
7 38.5	+33 43	Σ 1119		Gem	1829	1934	349		3.0		8.2	9.5	F0	6230		
7 38.5	+35 03	β 200	AB	Gem	1876	1926	190	191	98.4	100.4	5.6	11.6	G5	6229		70 Gem
--			AC		1876	1926	99	100	162.0	159.9		10.6		6229		
--			CD		1876	1916	242		1.6			11.6		6229		AC is H VI 70
--			CE		1880	1900	207		17.5			14.1		6229		
7 38.7	−1 27	Ho 245	AB	Mon	1887	1954	181		0.5		8.7	8.7	A0	6244		
--			AC		1887		98		50.0			9.3		6244		
--			AD		1887		155		60.0			9.3		6244		
7 38.7	−4 59	A 535		Mon	1903	1959	148	167	0.3	0.3	8.6	8.7	F0	6245		
7 38.8	−26 48	H III 27	AB	Pup	1826	1951	318		9.9		4.5	4.7	B5 B8	6255		
--		β 1061	BC		1889	1927	229	222	6.5	7.2		13.8		6255		
7 38.8	−35 48	Jsp 177		Pup	1929	1945	148		8.6		8.0	11.8	K0			
7 38.9	+42 29	OΣΣ 87		Lyn	1875	1924	178		64.4		7.9	8.3	F5 F5			
7 39.0	+0 58	A 2531		CMi	1913	1939			1.0		7.9	9.9	G5	6249		
7 39.0	−20 17	Arg 47a		Pup	1876	1955	331		2.9		7.9	8.3	A3	6254		
7 39.0	−53 16	B 2148		Car			230		15.0		6.1	13.9	A0			
7 39.3	+5 14	Schb	AB	CMi	1896	1932					0.4	12.9	F5	6251	o	10 α CMi, Procyon [48]
--		Lam	AC		1836	1958	262	13	56.6	119.0		11.6		6251		
--		Dick	AD		1925	1958	272	291	129.2	112.6				6251		Vicinity is rich in small
--		Dem	AE		1874		81		342.8					6251		pairs and triplets (Webb)
7 39.3	−49 03	h 3994	AB	Pup	1836	1936	17		14.8		7.2	10.0	B8			
--			AC		1835	1936	216		22.9			10.4				
7 39.5	−26 55	δ 90		Pup	1920	1933	129		6.7		6.7	13.0	B3	6264		
7 39.7	−38 08	I 160		Pup	1902	1959	149		1.2		5.8	8.5	B9			
7 39.7	−43 17	I 353		Pup	1902	1942	36		0.9		7.7	8.2	A0			
7 39.8	−6 15	Σ 1128 rej		Mon	1905	1938	169		15.8		7.9	9.9	G5	6262		
7 40.0	−3 36	Bgh		Mon			112		58.1		7.7	9.1	K2 K2			
7 40.1	+5 14	Σ 1126	AB	CMi	1829	1960	132	158	1.5	1.1	6.6	6.9	A0	6263		
--			AC		1914		249		44.4			10.8		6263		
7 40.2	−19 40	See 84		Pup	1897	1959	287	279	9.3	8.3	5.9	11.1	K0	6273		
7 41.3	−60 00	Hu 1583		Car	1914	1933	239		1.1		8.5	8.8	A0			
7 41.7	+37 26	OΣ 177		Lyn	1845	1941	150	76	0.6	0.3	8.0	9.0	A3	6276	o	Nearly round 1952,53,56,59
7 41.8	−72 36	Δ 57		Vol	1873	1917	116		16.7		4.0	9.8	K0			ζ Vol
7 42.2	−3 31	Σ 1132		Mon	1829	1922	238	235	19.2	19.8	8.5	9.1	K5	6302		
7 42.3	−58 40	h 4000		Car	1895	1942	236	241	1.4	1.6	7.6	10.2	B9			Beautiful double star (h)
7 42.7	−42 33	I 354		Pup	1902	1939	126		0.9		7.8	9.0	B8			
7 43.0	+58 43	h 2405		Lyn	1901	1911	320		54.7		5.0	9.5	A2	6285		24 Lyn
7 43.0	−17 04	Hu 710		Pup	1902	1960					7.1	7.6	G5	6315	o	
7 43.1	+0 11	B 2526	AB	CMi	1936	1939	332		0.1		7.0	7.2	G5	6313		Nearly round 1953,54,56,59
--		A 2534	ABxC		1913	1956	208	217	0.6	0.6		8.2		6313		
7 43.3	+28 53	β pm		Gem	1877	1909	315	316	178.6	182.2	4.3	10.8	K0			75 σ Gem

α 2000	δ 2000	Star Name		Const	Years	Pos Ang	Separation	m₁	m₂	Spec	ADS	Orb	Notes
7ʰ43ᵐ.5	+3°29′	Σ 1134	AB	CMi	1832 1928	147° °	10″.2 ″	7.5	10.7	F8	6317		
--			AC		1906 1927	347 347	83.6 85.5		10.1		6317		
7 43.5	-28 25	See 85		Pup	1897 1899	33	26.2	4.6	13.5	K5	6324		1 Pup
7 43.5	-54 15	Rst 2503		Car	1936 1948	355 349	1.7 1.9	6.9	12.5	G5			
7 43.7	-25 30	B 738		Pup	1927	304	5.3	6.6	14.1	A3			
7 43.8	-39 52	Gli 71		Pup	1904 1954	9	7.5	8.5	9.0	B9			
7 44.2	-50 27	Δ 55		Pup	1837 1917	133	51.8	6.9	8.0	F5 G0			
7 44.3	+45 22	Es		Lyn	1913	278	11.1	7.6	11.1	K0	6316		
7 44.3	-30 15	Rst 272		Pup	1930 1944	235	0.7	7.9	10.0	A0			
7 44.4	+24 24	OΣ 179		Gem	1853 1971	233 240	6.2 7.1	3.6	8.1	G5	6321		77 κ Gem
7 44.6	+31 07	A 674		Gem	1904 1934	128	1.0	7.8	10.6	A5	6323		
7 44.6	-17 17	A 3095	AB	Pup	1929	278	18.7	8.2	9.8	F5			
--			BC		1929	37	1.5		14.1				
7 45.1	-50 17	h 4002		Pup	1913 1933	91	19.5	7.5	11.5	F2			
7 45.3	-0 26	h 767		Mon	1893 1897	164	21.6	7.8	9.5	A0			
7 45.3	+28 02	β 580	AB	Gem	1880 1900	275 280	41.4 29.6	1.1	13.6	K0	6335		78 β Gem, Pollux
--		H VI 42	AC		1781 1922	66 73	116.7 201.1		8.8		6335		
--		β 580	AE		1879 1924	90 90	206.3 234.0		9.6		6335		
--		Σ II 5	AF		1836 1924	74 77	203.8 258.2		10.3		6335		
--		β 580	AG		1904 1924	328 332	170.1 164.5				6335		
--			CD		1878 1922	128 135	1.4 1.4		11.6		6335		
--			CE		1898 1922	146	71.1				6335		
--			CF		1851 1922	90	57.8				6335		
7 45.5	-14 41	Σ 1138	AB	Pup	1829 1933	339	16.8	6.1	6.8	A0 A0	6348		2 Pup
--			AC		1893 1932	228	100.5		10.4		6348		
7 45.6	-56 43	h 4005		Car	1836 1915	218	36.2	6.1	11.1	F0			
7 45.7	+34 33	OΣ 181		Gem	1848 1898	260	6.3	8.0	12.3	A3	6337		
7 45.9	+65 09	Σ 1122		Cam	1830 1928	5	15.4	7.8	7.8	F2	6319		BDS 4203, H V 135
7 46.1	+21 07	Ho 247		Gem	1887 1958	101 200	0.4 0.4	8.0	8.4	F2	6347		
7 46.3	+17 01	Ho 347		Gem	1892 1906	280 277	13.7 14.6	7.8	12.0	F5	6349		
7 46.3	-59 49	CorO 58		Car	1917	43	22.9	8.5	8.6	G5 G5			
7 46.6	+1 51	A 2879		CMi	1914 1935	158	1.2	7.4	11.4	A0	6361		
7 46.6	+4 08	Σ 1137		CMi	1828 1940	132	2.9	8.3	9.3	F5	6360		
7 46.8	-20 50	I 781		Pup	1912 1944	339	3.8	7.2	11.2	B9	6368		
7 46.8	-46 48	Hu 1428		Pup	1913 1951	354 16	0.4 0.3	7.4	8.4	B8			
7 46.9	+26 43	β pm	AB	Gem	1910	295	45.1	10.9	14.0				
--			AC		1910	190	55.4		13.1				
--			AD		1900 1910	273 273	803.0 800.4		8.1	B8			
7 47.0	+64 03	Σ 1127	AB	Cam	1830 1956	340	5.3	7.0	8.8	A2	6336		BDS 4378 (also H II 101?)
--			AC		1830 1956	175	11.3		9.9		6336		
--			BC		1956	170	17.4				6336		
7 47.1	+18 47	Cou 772		Gem	1972	77	0.2	8.6	8.6	A0			
7 47.1	-41 30	Δ 56		Pup	1836 1920	177	49.6	7.0	7.8	B2			
7 47.2	-43 24	Hu 1429		Pup	1928 1943	284	0.5	8.1	8.6	B8			
7 47.4	-38 31	I 161		Pup	1897 1933	83	10.9	5.1	10.9	B3			
7 47.4	-53 20	h 4008		Car	1917	230	20.8	7.6	13.0	B9			
7 47.5	+33 25	Σ 1135	AB	Gem	1831 1923	212 214	22.6 21.0	5.1	11.2	K2	6364		80 π Gem
--			AC		1903 1922	341	91.9		10.2		6364		
7 47.6	-15 59	β pm	AB	Pup	1864 1914	311 311	128.8 130.5	6.8	6.7	K2 K0			
--			Aa		1909	54	10.5		14.3				
7 47.6	-47 04	Rst 276		Pup	1930 1949	180 190	1.6 1.6	7.9	12.2	G0			
7 47.8	-3 32	Rst 4375		Mon	1939 1951	84	0.2	8.7	8.7	A0			
7 47.9	-12 12	Σ 1146		Pup	1831 1960	18 5	3.3 2.2	5.6	7.7	F5	6381		5 Pup
7 48.0	+60 18	Hu 1247		Cam	1900 1959			7.8	7.8	F5	6354	o	
7 48.0	-19 24	B 1077	AB	Pup	1928 1959	279 297	0.3 0.4	8.2	8.9	A2			
--			ABxC		1933	64	41.3		9.8				
7 48.1	+5 25	Σ 1143		CMi	1825	152	9.3	7.0	11.0	K0			Pos. error? (see BDS II)
7 48.1	+32 51	β pm		Gem	1879 1908	47	44.3	7.2	11.3	G5			
7 48.1	-25 56	See 86		Pup	1897 1900	198	27.0	4.6	12.6	B2	6384		o Pup
7 48.2	+28 41	AGC 2		Gem	1879 1937	116	0.8	7.7	10.7	G5	6371		
7 48.3	-56 28	φ 109		Car	1930 1943	215	1.1	6.3	8.5	K0			Spectrum composite K0+A2
7 48.4	+18 20	Σ 1140		Gem	1829 1942	273	6.3	7.9	9.6	G5	6376		
7 48.5	+28 46	S 560		Gem	1825 1917	359	89.9	7.1	8.8	G5			
7 48.5	+59 05	OΣ 180		Lyn	1848 1912	204 205	14.9 16.3	8.0	11.9	F0	6363		
7 48.6	+23 08	Wils 15	AB	Gem	1937 1958	99 71	0.2 0.2	6.9	7.0	F2	6378		82 Gem; sp. compos. F2+A0
--		β 1062	ABxC		1899 1958	34	4.0		13.5		6378		
--			ABxD		1958	25	67.0		12.0		6378		
7 48.6	-63 41	h 4014		Car	1879 1917	154	11.1	8.1	9.3	A0			
7 49.0	+0 40	OΣΣ 88	AB	CMi	1875 1935	5	57.1	7.6	8.6	F2			
--			AC		1935	52	94.6		9.8				
7 49.1	+28 34	Σ 1144		Gem	1829 1921	357	8.0	8.2	10.2	F0	6382		
7 49.1	-56 25	See 88		Car	1897 1933	183	6.9	5.6	13.6	K0			
7 49.2	-46 22	Jc 11		Pup	1846 1913	104	59.2	4.1	9.0	B0 A0			
7 49.2	-60 17	h 4012		Car	1917 1933	133 137	21.5 23.3	5.8	12.7	F2			
7 49.3	-24 52	β 1063		Pup	1889 1942	190	4.8	3.3	12.8	G6	6393		7 ξ Pup

α 2000	δ 2000	Star Name		Const	Years	Pos Ang		Separation		m₁	m₂	Spec	ADS	Orb	Notes
7ʰ 49ᵐ.3	-33°20′	Jsp		Pup	1929	259°		10″.5		7.7	13.6	B			
7 49.4	-30 33	I 186		Pup	1901 1954	199 225		1.3	1.3	8.2	8.4	F5			
7 49.5	+3 13	Σ 1149		CMi	1830 1927	41		21.7		7.9	9.6	G0	6391		Pretty (Webb)
7 49.6	-55 05	CorO 60		Car	1897 1933	55		4.0		7.6	9.1	A0			
7 49.8	-20 12	See 87		Pup	1897 1928	147 143		4.6	5.6	6.6	13.6	G0	6398		
7 49.8	-36 06	Jsp 192		Pup	1929 1945	129		2.8		7.0	15	K0			
7 50.8	+3 17	A 2880		CMi	1914 1959	182 237		0.2	0.2	7.0	7.0	G5	6405		
7 50.8	-46 52	Rst 4866		Pup	1942 1949	340		0.2		8.4	8.4	F0			
7 50.8	-70 47	h 4023	AB	Vol	1901 1917	219		2.0		8.5	8.9	F0			
—			AC			18		12.0			13.5				
7 51.0	+33 14	β pm		Gem	1879 1909	146		77.8		6.0	10.2	A0			
7 51.0	-24 32	B 146		Pup	1927 1959	52 67		1.1	0.9	6.5	10.0	A0	6414		
7 51.1	+37 13	h 3301		Lyn	1905 1906	62		23.5		7.7	13.5	F0	6399		
7 51.3	-9 24	β 1195		Mon	1891 1950	81 91		0.5	0.3	7.3	7.6	A0	6412		
7 51.8	-13 54	β 101		Pup	1875 1960					5.6	6.2	G0	6420	o	9 Pup
7 52.0	-31 38	B 1566		Pup	1929 1951	100		1.2		8.1	9.0	B9			
7 52.1	-3 03	Σ 1154		Mon	1827 1941	354		2.5		7.0	9.2	A5	6421		
7 52.2	-59 37	h 4018	AB	Car	1837 1933	327		5.1		7.5	10.0	B9			Miniature of α Cru (R)
—			AC		1838	259		60.4			11				
7 52.3	-34 42	Howe 65		Pup	1877 1959	292 274		3.0	3.0	5.1	8.1	F3			
7 52.5	-13 52	H III 28		Pup	1901	195		24.0		6.6	10.2	K0	6426		
7 52.7	+3 23	OΣ 182		CMi	1853 1958	47 21		1.1	1.0	7.5	8.0	A2	6425		
7 52.8	-5 26	φ 325		Mon	1952 1959	182 198		0.3	0.1	6.4	6.7	F2		o	
7 52.9	+64 54	Σ 1136		Cam	1830 1935	248 227		11.6	6.2	7.6	11.3	K5	6406		
7 53.0	-33 18	Rst 2508		Pup	1934 1960	84 72		0.2	0.3	8.5	8.5	A0			
7 53.3	+15 44	A 2881		Gem	1914 1922	324		4.2		7.5	13.2	K0	6430		
7 54.0	-2 22	Bail 190		Mon	1892	224		12.5		8.5	9.0				
7 54.0	+16 02	OΣ 183 rej		Gem	1878 1901	20		15.9		7.1	11.1	A2	6440		
7 54.1	-18 20	h 4013	AB	Pup	1902 1915	197		19.0		7.1	10.7	K5	6452		
—			AC		1902	271		67.0			9.9		6452		
—			CD		1902			3.0			13.5		6452		
7 54.2	-58 35	h 4021		Car	1917 1933	296 307		7.5	8.3	7.8	13.0	G0			
7 54.3	-43 05	Don 209		Pup	1929 1947	37		5.4		7.9	14.0	K0			
7 54.5	-2 48	Σ 1157		Mon	1831 1960	267 219		1.6	1.0	8.0	8.0	F0	6454		
7 54.9	+19 14	β pm	AB	Gem	1893 1919	251 258		88.1	87.0	8.1	10.3	K2			
—			AC		1893 1919	89 83		124.0	122.1		10.6				
—			Aa		1907	37		36.8			11.7				
7 55.0	-13 17	A 2539	AxBC	Pup	1912 1930	330 322		11.6	10.5	8.0		F5	6461		
—			BC		1913 1930	6		1.8		11.5	11.8		6461		
7 55.2	+48 09	Hu 711		Lyn	1903 1930	200		4.1		7.8	12.1	A0	6447		
7 55.3	-41 50	h 4019		Pup	1895 1932	155		5.4		7.8	10.0	G0			
7 55.4	-53 37	Gli 77		Car	1880 1928	341		34.8		7.8	9.2	K0			
7 55.6	-43 47	CorO 64		Pup	1897 1931	146		5.6		7.5	11.0	K0			
7 55.8	-43 51	See 91		Pup	1897 1959	302 332		0.4	0.6	6.7	6.9	B5			
7 55.9	-19 38	HdO 111		Pup	1870	170		8.0		8.0	8				
7 56.0	+23 42	Cou 929		Gem	1973	33		0.2		7.4	7.7	G0			
7 56.0	+24 40	Σ 1156		Gem	1827 1892	159		18.6		7.8	10.1	K0			BDS 4352, Σ 1163 same
7 56.0	-35 55	LPO 12	AB	Pup	1919	211		5.0		8.2	12.3	B8			
—			AC		1919	148		15.7			9.8				
7 56.3	+10 52	A 2882		Cnc	1914 1923	215		1.9		7.6	14.3	A0	6473		
7 56.4	+4 15	Bail 2798		CMi	1910	194		19.1		7.9	12.3	A3			
7 56.6	+19 54	Cou 930		Gem	1972	322		0.5		8.1	10.1	K2			
7 56.9	-3 27	Schj		Mon	1905	357		77.1		7.8	10.3	K0			
7 57.0	-41 12	I 1173		Pup	1926 1934	29		2.8		8.1	11.1	B8			
7 57.1	+20 58	Ho 250	AB	Gem	1907 1944	158		0.7		7.2	9.2	G5	6478		
—			AC		1887	154		9.4			13.0		6478		
7 57.3	+1 08	OΣ 185		CMi	1847 1960					7.0	7.1	F8	6483	o	
7 57.3	-44 07	See 92	AB	Pup	1897 1933	37		10.0		5.1	14.0	B3			
—			AC		1897 1933	87		22.7			14.5				
7 57.3	-47 53	I 26		Pup	1895 1951	28 45		0.8	0.5	6.7	7.3	B5			
7 57.4	+13 12	Σ 1162		Cnc	1829 1907	328		9.1		7.9	9.8	G5	6482		
7 57.6	-46 36	Rst 296		Pup	1929 1944	237		2.3		7.2	11.7	B5			
7 57.8	-60 18	LDS 198	AB	Car		f		60.0		5.6	9.9	F8			
—		Jsp 208	BC		1930 1940	270		2.3			13.5				
7 58.1	-10 53	β 902		Mon	1879 1916	244		1.2		8.0	11.0	K0	6490		
7 58.3	+2 13	S,h 87	AB	CMi	1822 1925	66 77		76.0	88.6	5.4	8.4	K0			14 CMi
—			AC		1822 1925	153 150		112.2	120.0		9.3				
—			AD		1906 1925	285 284		20.1	16.3						
—			BC		1912	192		127.8							
7 58.4	-60 51	h 4031		Car	1835 1921	357		5.5		7.4	8.3	B8			
7 58.8	+5 37	Σ 1168		CMi	1831 1924	215 220		5.9	6.2	6.8	10.6	B9	6492		
7 58.8	-47 41	I 30		Pup	1896 1947	349		1.7		7.3	8.8	B8			
7 59.0	-0 27	Rst 4872		Mon	1943 1950	288		0.4		8.1	10.2	A2			
7 59.0	-48 23	Rst 297		Pup	1929 1944	258		3.8		6.8	11.2	B3			
7 59.2	-49 59	Δ 59		Pup	1836 1954	47		16.4		6.5	6.5	B3 B3			

α 2000	δ 2000	Star Name		Const	Years	Pos Ang	Separation	m₁	m₂	Spec	ADS	Orb	Notes
7ʰ59ᵐ.3	+16°49'	A 2884		Cnc	1914 1922	113°	2".5	7.4	13.4	F8	6493		
7 59.7	+13 41	Σ 1170		Cnc	1830 1953	101	2.2	8.5	8.5	F5	6499		
7 59.7	+46 37	Σ 1161		Lyn	1830 1954	194	2.7	8.0	9.9	G0	6491		
7 59.7	−47 18	I 1070	AB	Pup	1912 1959	336 353	0.5 0.3	8.0	8.3	B9			
−−		h 4032	ABxC		1836 1913	351	29.2		9.3				
8 00.0	−30 23	B 1572		Pup	1931	331	6.1	7.7	14.3	G0			
8 00.2	+4 52	Kui 33		CMi	1958	73	30.1	5.7	12.4	A0			Error in position?
8 01.0	+23 35	Σ 1171		Cnc	1828 1936	339 326	2.8 2.3	6.4	10.9	K0	6513		
8 01.0	−40 35	I 163		Pup	1902 1943	120	0.8	8.2	10.2	G0			
8 01.3	−22 20	β 333	AB	Pup	1877 1942	45	1.6	7.8	10.0	A2	6524		
−−			AC		1885 1898	73	42.3		8.6	G0	6524		
8 01.4	+16 57	Σ 1173		Cnc	1830 1932	51	10.0	7.9	9.6	G0	6519		
8 01.4	−54 31	Δ 60		Car	1837 1938	161	40.4	6.0	8.2	B5 B8			
8 01.5	+23 55	A 2540		Cnc	1913 1923	164	1.4	7.4	12.6	K0	6518		
8 01.7	−8 36	A 1580		Mon	1907 1951	133 188	0.3 0.1	7.4	8.9	F5	6526		Spectrum composite F5+A2
8 01.7	+25 05	H VI 75	AB	Cnc	1904 1924	23	45.5	6.2	11.0	A0			4 Cnc
−−			AC		1904	295	109.3		10.0				
8 01.9	−27 13	β 202	AB	Pup	1876 1933	161	7.6	6.6	9.3	B9	6535		
−−			AC		1897	77	19.4		13.6		6535		
−−			AD		1897 1916	240	29.6		11.8		6535		
8 02.3	+2 20	Gsh		CMi	1842	316	241.4	4.4	9.1	K0			
8 02.5	+3 05	β 23		CMi	1875 1936	181	2.7	7.8	11.6	K0	6533		
8 02.5	+12 27	Ho 349	AB	Cnc	1891 1911	226	9.8	7.5	12.0	K0	6531		
−−			AC		1891	290	63.2		11.8		6531		
8 02.5	+63 05	S,h 86		Cam	1823 1924	82	48.6	6.0	8.4	F8 F8			
8 02.5	−20 19	See 95		Pup	1897 1899	191	13.0	7.5	14.7	A0	6539		
8 02.5	−44 40	I 8		Pup	1896 1947	312 305	2.3 2.3	6.7	9.2	A0			
8 02.6	−27 33	h 4037	AB	Pup	1876 1954	244	7.1	7.1	8.8	G5	6544		
−−			AC		1902 1954	74	63.5		9.0	B8	6544		
8 02.7	−41 19	h 4038		Pup	1837 1920	346	27.0	5.5	8.5	A0			
8 02.8	−31 13	I 785		Pup	1911 1927	280	6.1	7.7	11.4	B3			
8 03.1	−32 28	h 4035		Pup	1902 1954	134 134	34.4 35.0	5.8	9.5	K2			
8 03.2	+9 57	A 2956		Cnc	1915 1921	171	4.5	7.7	13.5	K2	6542		
8 03.3	+26 16	OΣ 186		Cnc	1847 1957	73	0.8	7.5	8.2	A2	6538		
8 03.4	+13 40	Hu 848		Cnc	1905 1921	154	1.9	7.6	12.8	G0	6543		
8 03.5	−8 10	h 2426		Mon	1830	145	25.0	7.4	11.4	F8			
8 03.5	+12 11	Ho 350		Cnc	1891 1929	189 198	4.2 4.8	6.7	10.8	F8	6546		
8 03.5	+27 48	Es	AB	Gem	1907	188	60.1	4.9	11.9	K0			χ Gem
−−			AC		1907	81	78.6		10.9				
8 04.1	+33 02	OΣ 187		Gem	1844 1960			7.1	7.7	A0	6549	o	
8 04.4	+12 17	β 581	AB	Cnc	1878 1960			8.7	8.7	G5	6554	o	
−−			ABxC		1878 1974	185 213	4.8 5.0		10.5		6554		
−−			ABxD		1921	166	227.6		10.8		6554		
8 04.5	−25 42	I 487		Pup	1911 1959	22	1.8	6.5	10.0	K2	6566		
8 04.5	−42 46	Don 222		Pup	1930 1947	298	4.2	7.4	11.6	B3			
8 04.6	+54 45	Σ 1172		Lyn	1829 1954	243	1.6	7.7	9.5	A0	6545		
8 04.7	+47 17	A 2050	AB	Lyn	1909 1954	291 318	0.2 0.2	8.5	8.5	F5	6552		
−−		Σ 1174	ABxC		1830 1943	214	5.7		8.5		6552		
8 04.7	−62 50	Δ 62		Car	1836 1917	262	87.1	6.5	7.8	B5 K0			
8 04.8	+79 29	Kui 31		Cam	1935 1958	161	6.7	5.3	13.6	A0			
8 04.9	−54 59	I 190	AB	Car	1901 1943	118	2.3	7.4	10.0	A2			
−−			AC		1901	144	24.0		14.0				
8 05.0	−22 26	h 4041		Pup	1920	127	36.4	7.8	13.8	G0			
8 05.0	−34 19	B 1579		Pup	1929 1959	293 312	0.5 0.7	8.4	8.5	F5			
8 05.0	−60 23	MlbO 2		Car	1877 1944	351	1.6	7.7	8.7	A0			
8 05.2	−45 25	Jc 12		Vel	1847 1913	16	27.1	8.6	8.6	A A			
8 05.4	+5 50	Σ 1182		CMi	1831 1944	73	4.5	8.0	10.0	B9	6571		
8 05.4	+8 12	Σ 1181		Cnc	1830 1927	140	5.2	8.0	9.5	G5	6570		
8 05.5	−13 34	Hu 623		Pup	1900 1959	64	5.3	7.4	12.9	A0	6579		
8 05.5	−69 51	h 4055		Vol	1837 1918	190	6.0	8.5	8.6	K0			
8 05.6	+27 32	Σ 1177		Cnc	1828 1960	351	3.5	6.6	7.5	B9	6569		
8 05.7	−33 34	h 4046	AB	Pup	1838 1954	88	22.1	6.0	8.4	G5 A			
−−		I 189	BC		1903 1919	6	14.1		11.0				
8 05.7	−45 10	Don 223		Vel	1932 1947	243	1.5	7.1	11.7	B5			
8 05.9	+68 25	Σ 1164 rej		Cam	1904 1926	345 357	26.4 29.9	7.7	10.1	F5			
8 06.0	+59 15	h 2424		Lyn	1908	150	42.4	6.6	10.9	A0			
8 06.0	+71 47	Σ 1159 rej		Cam	1910 1918	95	35.3	7.4	9.7	K0			
8 06.3	+41 59	Σ 1176		Lyn	1830 1918	28	22.3	8.1	9.7	F5	6572		
8 06.3	−40 06	I 164		Pup	1902 1943	79	0.6	7.5	9.0	G5			
8 06.5	−9 15	Σ 1183	AB	Mon	1831 1935	327	30.9	6.0	8.7	A0 A0	6588		
−−		A 543	BC		1903 1936	324	1.2		13.1		6588		
−−			BD		1903 1917	20	14.0		14.9		6588		
8 06.8	−31 10	I 488		Pup	1911 1937	21	0.8	8.6	9.0	B8			
8 06.9	−27 07	Δ 61		Pup	1920	35	70.7	7.0	8.5	B9 K0			
8 07.0	+20 53	Ho 351	AB	Cnc	1892 1945	234	2.0	8.0	12.7	G5	6590		

α 2000	δ 2000	Star Name		Const	Years	Pos Ang	Separation	m₁	m₂	Spec	ADS	Orb	Notes
8ʰ07.0	+20°53′	Ho 351	AC	Cnc	1892 1941	56°	43″.1		13.0		6590		
8 07.0	+54 07	A 1333		Lyn	1906 1959	248 213	0.2 0.3	8.8	8.8	A0	6578		
8 07.1	−48 11	Rst 5285		Vel	1944 1949	278	0.9	7.8	11.3	A0			
8 07.5	−24 18	Rst 5284		Pup	1947	15	29.1	2.8	13.7	F5			15 ρ Pup
8 07.9	−1 21	Σ 1189 rej		Mon	1896 1931	335	8.9	8.0	10.3	A2	6611		
8 07.9	−68 37	Rmk 7		Vol	1835 1922	24	6.1	4.4	8.0	B8			ε Vol
8 08.2	−61 05	h 4053	AB	Car	1917	98	11.6	6.6	9.4	K0			
−−			AC		1917	318	19.4		10.8				
8 08.3	−22 08	β 334		Pup	1877 1955	354 349	2.8 2.8	7.9	8.3	G5	6614		
8 08.3	−23 37	B 149		Pup	1926	310	4.0	6.6	14.0	B5	6618		
8 08.4	−19 51	S 563	AB	Pup	1825 1918	56	134.3	7.5	8.0	A0 K0	6622		
−−		β	BC		1903	300	7.8		13.5		6622		
8 08.5	+51 30	Es 70	AB	Lyn	1901 1926	266	45.6	4.8	12.8	A2	6600		27 Lyn
−−			BC		1916 1926	349	7.3		13.3		6600		
8 08.5	−52 37	B 1586		Car	1929 1934	245	0.3	7.4	9.0	F2			
8 08.6	−2 59	Σ 1190	AB	Mon	1827 1936	105	32.0	4.3	10.0	G0	6617		29 ζ Mon
−−			AC		1831 1936	245	66.5		7.8		6617		
−−			Aa		1905	282	40.3				6617		
8 08.7	−20 22	B		Pup		221	20.2	6.4	13.0	A3			
8 08.8	+27 29	Σ 1186		Cnc	1828 1934	218	3.2	6.9	10.2	K0	6612		
8 08.9	−68 59	I 192		Vol	1900 1918	173	2.0	7.5	10.3	A0			
8 09.0	−39 31	Gli 80		Pup		246	40.3	7.6	8.5	K0 K2			
8 09.1	−26 14	B 756		Pup	1927 1955	175	4.3	8.0	10.4	K2			
8 09.1	−30 19	B 1583		Pup	1929 1944	90	1.1	6.7	10.9	K5			
8 09.5	+32 13	Σ 1187		Cnc	1829 1975	71 27	1.6 2.8	7.1	8.0	F2	6623	o	
8 09.5	−16 15	Ho 352		Pup	1890 1906	185	5.3	5.7	12.4	B3	6632		
8 09.5	−47 20	Δ 65	AB	Vel	1826 1951	220	41.2	1.9	4.2	O9 B3			γ Vel
−−			AC		1835 1907	151	62.3		8.2	A0			
−−			AD		1846 1902	141	93.5		9.1	A0			
−−		I 1175	DE		1928	146	1.8		12.5				
8 09.7	+35 42	Fox		Lyn	1922	182	68.6	7.3	11.3	K0			
8 09.8	−42 38	Δ 63		Pup	1836 1951	81	5.5	6.6	7.7	A0			
8 10.2	+25 51	β pm		Cnc	1910	51	87.6	6.8	9.8	K0			
8 10.2	+35 27	h 3308	AB	Lyn	1901 1926	263 270	45.7 51.4	6.6	10.4	F8			
−−			AC		1926	133	65.6						
8 10.2	+55 48	A 1335		Lyn	1906 1960	198 224	1.4 1.4	8.0	11.3	A2	6620		
8 10.5	+25 30	β pm	AB	Cnc	1910 1925	293 297	73.7 73.8	5.7	12.1	G5			14 ψ Cnc
−−			AC		1910 1925	212 213	117.3 112.1		10.6				
8 10.7	−13 48	LDS 204		Pup		sp	93.0	5.5	12.2	G0			18 Pup
8 11.0	−37 18	δ 43	AB	Pup	1920	206	16.8	6.4	13.6	B0			
−−		h 4051	AC		1920	265	18.0		13.4				
8 11.0	−39 05	B 2169		Pup		160	6.0	7.4	13.9	B9			
8 11.3	+1 17	Σ 1198		CMi	1829 1924	157	33.3	7.8	9.1	G0 F8			
8 11.3	−7 02	A 545		Hya	1903 1936	251	3.6	8.0	12.0	F5	6645		
8 11.3	−12 56	β 1064	AB	Pup	1889 1899	247	2.1	4.7	11.2	K0	6647		19 Pup. Single 1936,38 (B)
−−			AC		1898	299	33.3		13.2		6647		
−−			AD		1899	276	60.5		8.9		6647		
−−		S,h 291	AE		1826 1898	256	71.0		7.8		6647		
8 11.4	−42 59	h 4057		Pup	1836 1933	298	25.2	4.8	9.4	A3			
8 11.6	+32 27	β pm	AB	Cnc	1857 1907	280 321	58.0 53.4	6.9	10.9	G0			
−−			AC		1850 1907	84 76	193.0 225.4		10.3				
8 11.8	+13 27	Hu 851		Cnc	1905	230	2.3	7.5	14.1		6649		
8 12.1	−15 40	h 4050		Pup	1898 1903	303	23.0	8.7	8.7	A0 A0			
8 12.2	+17 39	Σ 1196	AB	Cnc	1826 1960			5.6	6.0	F7	6650	o	16 ζ Cnc
−−			ABxC		1841 1969	149 88	5.1 5.7		6.2	G2	6650	o	AB/2 x C
−−			ABxD		1913 1921	108 108	289.6 287.9		9.7		6650		
−−			AC		1826 1960	155 83	5.3 6.1				6650		
−−			BC		1840 1957	155 98	5.6 5.6				6650		
−−			CD		1913 1922	108 108	283.6 281.0				6650		
8 12.5	−46 16	See 96		Vel	1897 1942	290	0.4	6.4	7.2	B4			
8 12.6	+28 49	Cou 1114		Cnc	1974	229	0.2	8.7	8.8	A0			
8 12.9	+9 35	Σ 1201		Cnc	1831 1927	183	6.6	7.8	9.5	A5	6659		
8 13.3	−24 17	See		Pup	1898 1954	260	1.4	7.4	9.0	F0	6670		
8 13.6	+10 23	β 204		Cnc	1875 1935	300	1.0	7.9	10.9	G0	6665		
8 13.6	+10 51	Σ 1202		Cnc	1829 1954	336 311	2.4 2.4	7.4	9.5	F8	6663		
8 13.6	−31 44	HdO 201	AB	Pup		355	12.0	6.8	12.1	G0			
−−			AC			280	18.0		11.1				
8 13.6	−45 57	CorO 69	AB	Vel	1920 1930	46	8.7	8.1	8.5	F2			
−−			AC			96	20.0		13				
8 13.6	−47 00	Gli 87		Vel		341	35.0	5.1	9.0	B5			
8 13.7	−33 34	B 2173		Pup	1938	280	6.3	6.4	13.1	K2			
8 13.8	+1 59	β 1244		Hya	1891 1957	50 26	0.7 0.9	8.3	8.5	A0	6671		
8 13.9	+27 47	Ho 38	AB	Cnc	1886 1905	82	7.4	7.6	12.6	F2	6662		
−−			AC		1896 1904	140	30.0		11.8		6662		
−−			CD		1904	219	7.4		13.3		6662		

α 2000	δ 2000	Star Name		Const	Years	Pos Ang	Separation	m_1	m_2	Spec	ADS	Orb	Notes
8h 14m.0	−36° 19′	Δ 67		Pup	1837	175°	67.4″	5.1	6.0	B3 B8			
8 14.0	−40 21	h 4062		Pup	1920	341	51.1	4.4	9.5	K2			
8 14.2	+17 41	β 1243	AB	Cnc	1891 1957	345 332	1.4 1.5	6.4	12.3	F0	6673		
—		H VI 78	AC		1898 1924	302 301	64.6 63.0		9.2		6673		
8 14.2	−32 08	h 4059		Pup	1898 1919	330	29.6	6.1	13.1	B5			
8 14.3	+13 03	Kui 34		Cnc	1958	41	22.8	6.5	12.1	K0			
8 14.4	−45 50	φ 113	AB	Vel	1932 1936	293	0.4	6.3	7.8	B3			
—		h 4069	ABxC		1836 1959	255 250	33.8 32.4		9.0	K0			
8 14.7	−73 48	I 9		Vol	1897 1941	310 300	1.0 1.0	7.4	7.6	A5			
8 14.8	+43 02	OΣ 189		Lyn	1846 1934	293	4.2	6.8	9.9	A0	6675		
8 14.9	−35 41	See 98		Pup	1897 1920	68	5.5	6.7	12.7	K2			
8 15.0	−21 40	B 1980		Pup	1931	306	4.4	8.0	13.4	K0			
8 15.3	−62 55	Rmk 8		Car	1877 1943	63 66	3.9 3.9	5.3	8.0	A2			
8 15.5	−37 22	h 4063		Pup	1879 1920	350	17.7	7.5	9.6	B8			White, very red (h)
8 15.8	+2 48	Σ 1210		Hya	1829 1929	113	15.5	7.2	9.5	A0	6698		
8 15.8	+60 23	Σ 1192	AB	UMa	1832 1935	258	2.8	6.5	10.2	F0	6680		
—			AC		1832 1921	224	48.6		9.9		6680		
—			AD		1921	195	99.4		12.0		6680		
8 15.9	−30 56	β 454	AB	Pup	1892 1942	16 11	2.4 2.0	6.4	8.3	G5			
—			AC		1898	288	19.1		14.1				
8 16.0	+18 42	Ho 524		Cnc	1894 1933	340	4.0	7.5	10.5	K0	6696		
8 16.1	+57 06	β pm	AB	Lyn	1893 1919	164 160	134.2 131.4	8.0	9.6	G5			
—			BC		1908 1919	153 153	100.4 101.3		11.5				
8 16.2	−35 53	Jsp 250		Pup	1929 1945	195	11.8	7.0	11.0	G5			
8 16.4	−3 14	Howe 21		Hya	1879 1945	241	1.4	7.1	10.6	F8	6707		
8 16.5	+9 11	β 1065		Cnc	1889 1910	294	29.2	3.5	14.0	K2	6704		17 β Cnc; see ADS note
8 16.5	+79 30	Σ 1169		Cam	1832 1974	10 14	20.7 20.7	8.4	8.7	G0	6646		
8 16.5	−16 19	β 905	AB	Pup	1879 1933	12	3.8	7.7	10.3	K0	6711		
—		Fox	AC		1914	29	72.2		10.8		6711		
8 16.6	−49 54	Rst 5291		Vel	1944 1949	88	7.2	8.0	15	B5			
8 16.8	−9 01	β 102		Hya	1875 1944	120	3.3	7.0	10.5	A0	6714		
8 16.9	−15 09	h 4070		Pup	1916 1920	105	30.2	7.2	11.8	K2			
8 17.0	+59 11	β pm	AB	Lyn	1851 1909	8 9	95.0 96.1	6.7	9.5	K0			
—			AC		1851 1909	308 308	222.4 223.8		8.7	G5			
—			AD		1851 1908	63 63	235.0 236.6		9.5				
—			DE		1909	108	155.8		10.5				
—			DF		1909	166	173.0		11.2				
8 17.1	−34 30	Rst 5290		Pup	1944	55	3.7	7.6	13.5	A0			
8 17.3	−5 22	A 337		Hya	1902 1960			8.4	8.7	F2	6719	o	
8 17.3	−42 31	Don 237		Pup	1929 1947	230	4.9	7.2	11.2	B2			
8 17.7	−55 47	See		Car	1897 1933	188	7.0	7.1	12.7	K0			
8 17.7	−55 54	h 4079		Car	1913	171	30.4	7.3	12.0	K0			
8 17.9	+2 28	A 2889		Hya	1914 1947	45 22	0.2 0.2	8.3	8.8	A0	6726		
8 17.9	−59 10	h 4084	AB	Car	1917	155	43.8	6.4	9.7	F8 A			
—			BC		1880 1917	87	3.1		9.9				
8 18.2	−37 22	h 4073		Pup	1878 1951	179 176	2.0 1.8	7.2	7.9	A0			
8 18.3	−47 08	h 4080		Vel	1879 1913	218	5.9	8.5	8.8	F2			
8 18.9	−45 02	BrsO 3	AB	Vel	1836 1940	327	5.2	8.7	8.9	A0			
—			AC		1851	143	77.6		9.2	A2			
8 19.1	−48 12	h 4081		Vel	1920	185	42.5	6.5	10.9	B3			
8 19.2	−46 12	Rst 4884		Vel	1942 1947	136	7.8	7.2	13.7	K0			
8 19.5	+35 03	OΣΣ 91	AB	Lyn	1875 1925	226 220	92.5 92.6	7.1	8.8	A5 G0			
—			Aa		1911	213	34.0		14.0				
8 19.8	+3 57	φ 346		Hya	1959	91	0.2	6.9	6.9	G5			
8 19.8	−71 31	BrsO	AB	Vol	1835 1917	57	65.0	5.4	5.7	B9 A0			κ Vol
—			BC		1835 1917	30	37.7		8.5				
8 20.5	−22 55	Don 238		Pup	1930 1941	8	5.6	6.1	12.9	K0			
8 20.6	−77 29	I 799		Cha	1911	250	31.0	4.4	12.1	K5			θ Cha
8 20.7	+72 24	Σ 1193	AB	UMa	1831 1925	85 87	44.4 43.1	6.1	9.1	K5	6724		
—			BC		1911	2	54.7				6724		
8 20.9	−22 27	Don 241		Pup	1930 1941	225	1.7	8.0	13.0	A3			
8 21.2	+4 23	Bail 2815		Hya	1910	85	10.5	8.1	10.2	A2			
8 21.2	+47 25	Hu 224	AB	Lyn	1898 1920	316	4.2	8.2	12.2	F8	6746		
—		OΣ 190 rej	AC		1867 1975	168	38.6		9.7		6746		Mag. of C here is for Cc
—			AD		1867 1970	280 285	78.3 78.1		8.0	G5	6746		
—		A 1745	Cc		1907 1959			10.2	10.9		6746	o	
8 21.3	−1 36	Σ 1216		Hya	1831 1959			7.1	7.4	A0	6762	o	
8 21.4	−36 29	h 4085		Pup	1895 1933	273	6.8	5.2	12.0	B3			
8 21.4	−44 59	Rst 4886		Vel	1942 1947	229	1.8	7.9	13.8	A0			
8 21.6	+33 56	β 576		Lyn	1878 1941	142	1.5	7.2	11.9	K0	6757		
8 21.7	+17 17	Par		Cnc	1917	270	9.6	6.8		M	6763		V Cnc; A is variable
8 22.0	−13 46	Rst 3589		Pup	1938 1943	334	2.1	8.0	13.0	F5			
8 22.1	−41 00	h 4087	AB	Pup	1837 1955	324 278	1.5 1.5	7.7	8.0	F5			
—			AC		1897 1934	341	13.2		12.5				
—			AD		1920	305	30.5		13.9				

α 2000	δ 2000	Star Name		Const	Years	Pos Ang	Separation	m₁	m₂	Spec	ADS	Orb	Notes
8ʰ22ᵐ.1	−73°24′	h 4103	AB	Vol	1915 1917	287°	30″.8	5.3	11.8	A6			η Vol
--			AC		1917	162	42.4		11.6				
8 22.2	+74 49	OΣ 188		Cam	1847 1914	193	10.5	6.5	9.8	K0	6736		
8 22.5	−48 29	I 67		Vel	1901 1946	139	0.8	5.2	6.2	B2			
8 22.6	−29 42	B 1600		Pup	1929 1960	228 251	0.3 0.3	7.3	8.7	F0			
8 22.8	−26 21	B 767	AB	Pup	1926 1954		0.1	6.6	6.7	F0	6782		
--		Ho 353	ABxC		1890 1915	223	32.6		13.0		6782		
8 22.8	−76 26	h 4109		Cha	1879 1940	128	26.1	7.1	8.9	A2 A			
8 23.5	−30 48	B 2177		Pup	1938	284	9.7	7.0	14.0	K2			
8 23.8	+57 26	OΣΣ 92		Lyn	1875 1914	180	57.9	8.2	9.7	F2			
8 23.8	−28 58	h 4088	AB	Pup	1898 1920	285	26.9	6.7	12.0	A0	6794		
--			AC		1898	242	9.9				6794		
8 23.9	+10 38	A 2961		Cnc	1915 1934	259	1.0	6.1	11.1	K5	6787		21 Cnc
8 24.1	−35 31	B 1985		Pup	1930 1942	241	2.4	7.4	13.4	A2			
8 24.3	+44 57	Σ 1217		Lyn	1830 1920	241 241	29.8 29.0	7.7	9.2	G0 K0	6783		
8 24.4	−41 32	I 1180		Pup	1926 1955	44	0.5	8.4	9.2	B9			
8 24.6	−1 09	B 2527	AB	Hya	1938 1958	144 318	0.2 0.1	7.9	7.9	G0			Less than 0″.1, 1951–59
--		β pm	AC		1891 1919	46 44	153.2 155.5		9.8				
--			AD		1908	13	66.2		13.4				
8 24.7	+42 00	S 565		Lyn	1824 1924	165 171	73.0 79.2	6.1	8.6	K5			
8 24.7	−32 55	Jsp 271	AB	Pup	1929	208	130.0	8.0	11.5	K5			
--			BC		1929 1945	240	1.8		11.6				
8 24.8	+20 09	OΣ 191 rej		Cnc	1867 1923	191	37.5	7.2	9.2	A5 A5			
8 25.0	+9 26	β 1066		Cnc	1889 1944	188	2.4	6.9	13.3	K0	6798		
8 25.0	−42 46	Rst 4888		Pup	1942 1947	135	0.4	6.6	7.0	B5			
8 25.1	−24 03	S 568		Pup	1825 1917	86	41.0	5.3	8.8	K5 K1	6800		
8 25.1	−49 10	Rst 321		Vel	1928 1960			8.4	8.7	G5		o	
8 25.2	−55 28	Rst 3593		Car	1937	73	0.3	7.1	8.1	B8			
8 25.4	+37 24	Hu 856		Lyn	1904 1960	265 202	0.2 0.2	8.4	8.9	F5	6796	o	Round or single 1944–56
8 25.4	−55 45	Hu 1438		Car	1928 1933	201	8.4	7.0	11.7	A2			
8 25.5	−51 44	Δ 69		Vel	1920 1931	220	25.5	5.2	10.2	B3			
8 25.8	−0 25	Schj 10	AxBC	Hya	1902 1946	90 84	42.8 39.0	7.3		K0			
--		Rst 5297	BC		1946 1950	75	0.3	10.8	10.8				
8 25.9	+7 34	H VI 109		Cnc	1783 1911	342	31.5	5.1	9.2	K0	6805		h 785
8 26.0	−8 46	J 1520		Hya	1952	303	7.1	7.8	12.5	K0			
8 26.0	−32 10	Ho 528		Pup	1906 1930	226 228	12.4 13.0	7.2	12.2	F0			
8 26.3	−39 04	h 4093	AxBC	Pup	1847 1953	122 124	8.1 8.1	6.7		A0 A0			
--		B 1605	BC		1929 1956	79 351	0.2 0.1	7.9	8.1				
8 26.5	−3 59	H VI 118		Hya	1901 1924	3	72.8	5.6	9.4	A5			2 Hya, BDS 4606
8 26.5	+27 54	S 566		Cnc	1825 1923	22 21	120.9 130.6	5.6	9.9	K2			22 φ¹ Cnc
8 26.7	+24 32	Σ 1224	AB	Cnc	1830 1957	37 47	5.8 5.8	7.0	7.8	A3 G	6811		24 Cnc. Mag. of B is for BC
--		A 1746	BC		1908 1960			8.6	8.6		6811	o	
8 26.8	+26 56	Σ 1223		Cnc	1829 1958	212 218	4.6 5.1	6.3	6.3	A2 A2	6815		23 φ² Cnc
8 27.0	−52 42	B 1606		Vel	1929 1960			7.2	7.4	F2		o	
8 27.4	+23 09	Σ 1227		Cnc	1828 1921	163	24.6	8.2	9.5	A0	6816		
8 27.5	−55 01	φ 116		Car	1929 1951	170 137	0.2 0.2	7.3	7.3	A0			
8 27.6	−20 51	B 2179		Pyx	1959	203	0.3	6.9	8.0	A0			
8 27.7	−4 25	A 550		Hya	1936 1958			7.5	7.6	F0	6825	o	
8 28.0	−35 07	φ 314	Aa	Pyx	1951 1959	65 96	0.1 0.1	6.6	6.6	B5			
--		Gli 96	AaxB		1881 1919	146 144	24.0 25.4		10.0				
8 28.0	−41 15	CorO 71		Vel	1902 1928	107 105	6.2 7.0	7.8	10.8	A0			
8 28.1	+33 32	OΣ 193		Lyn	1851 1916	295	14.1	7.6	11.6	K0	6821		
8 28.5	−2 31	A 551	AB	Hya	1903 1960			7.1	7.2	F0	6828	o	
--		Σ 1233	ABxC		1828 1930	330	18.1		11.6		6828		
8 28.8	−17 32	Arg 20		Pyx	1877 1918	172	15.3	8.6	8.9	F5 F5	6832		
8 28.9	−20 57	Bhas 9		Pyx	1917	30	18.9	6.7	12.5	B8			
8 28.9	−42 35	h 4102		Vel	1913	281	68.5	6.4	9.3	B3 G0			
8 28.9	−48 11	Rst 329		Vel	1929 1949	111	0.3	8.4	8.7	A0			
8 29.1	−44 10	B 1101		Vel	1928 1951	215 230	0.3 0.3	6.5	6.6	B3			
8 29.1	−47 56	φ 315	Aa	Vel	1951 1959	113 120	0.1 0.1	6.4	6.4	B5			
--		h 4104	AaxB		1836 1951	241 245	3.8 3.6		7.7				Beautifully triple (h)
--			AaxC		1835 1934	40 39	18.8 18.8		9.3				
8 29.5	−18 45	B		Pyx		265	8.0	7.7	13.2	B8			
8 29.5	−44 44	Δ 70		Vel	1826 1951	350	4.5	5.2	6.8	B5			
8 29.5	−54 13	Rst 2562		Car	1936 1948	8	2.3	6.6	13.6	B9			
8 29.6	−55 11	φ 117		Car	1932	20	2.4	6.4	12.3	G0			
8 29.7	−67 08	h 4110		Vol	1918	215	24.9	7.5	12.0	A0			
8 29.8	+26 12	Weisse 19		Cnc			3.0	7.0	8.2	A0			Single in 1902
8 30.2	−18 22	h 4100		Pyx	1915 1917	178	22.5	8.1	10.3	K0			
8 30.3	+60 43	β 1067	AB	UMa	1889 1899	192	7.1	3.4	15	G0	6830		1 o UMa
--			AC		1888 1923	152	143.0		10.9		6830		
--			AD		1888 1924	208 207	177.2 173.2		10.4		6830		
8 30.4	−26 02	B 158	AB	Pyx	1926 1959	43 39	0.4 0.6	8.6	9.0	G0	6847a		
--		S 569	ABxC		1825 1959	342 336	39.3 24.1		9.4				
8 30.6	−40 31	Δ 71		Vel	1826 1920	50	63.8	7.2	7.6	B9 K5			

α 2000	δ 2000	Star Name		Const	Years		Pos Ang		Separation		m₁	m₂	Spec	ADS	Orb	Notes
8ʰ30.ᵐ7	−44°44′	I 168		Vel	1902	1934	79°	74°	3″.4	3″.4	6.5	11.0	B5			
8 30.8	−41 31	I 313	AB	Vel	1901	1937	216		3.8		6.8	9.2	K2			
--		Rst 2563	BC		1933	1945	320		2.4			13.2				
8 31.4	−36 42	h 4106	AB	Pyx	1881	1955	143	202	8.0	1.0	7.7	9.7	K0			
--		LDS 223	AC		1943		218		47.9			10.7				
8 31.4	−39 04	h 4107	AB	Vel	1847	1920	329		4.4		6.4	8.3	B3			
--			AC		1847	1920	100		30.8			9.6				A white, C plum-colored (h)
8 31.5	−19 35	I 489		Pyx	1911	1960	108	66	0.5	0.3	5.9	6.5	A0	6862		
8 31.5	−25 41	I 805		Pyx	1913	1956	44		1.9		7.7	10.2	A0	6863		
8 31.6	+18 06	h 2452		Cnc	1862	1902	59	61	58.4	63.1	5.4	9.8	M			31 θ Cnc
8 31.6	+34 58	Hu 716		Lyn	1902	1960					7.5	9.1	G5	6851	o	
8 31.7	−26 01	I 807		Pyx	1913	1959	105	44	0.4	0.2	8.7	8.9	A0	6865		
8 32.1	−53 13	S1r 8		Vel	1892	1947	302	292	0.8	0.8	6.1	7.1	G5			Spectrum composite G5+A0
8 32.3	−43 56	B 1612		Vel	1929		146		1.7		7.6	13.8	B3			
8 32.6	+5 43	Σ 1241 rej		Hya							7.4		K0			
8 32.6	−15 02	B 2528		Hya	1936	1959	201	212	0.3	0.3	6.5	8.6	A5			
8 32.7	−55 28	Rst 332		Car	1929				11.4		7.5	14.5	A0			
8 32.9	−38 11	Rst 2565		Vel	1935		264		0.9		7.9	11.9	K0			
8 33.0	−45 45	HdO 202		Vel			225		15.0		6.5	10.3	B5			
8 33.1	+55 21	Σ 1234		Lyn	1831	1957	71	67	20.8	23.5	8.0	9.3	K0	6858		
8 33.1	−24 36	β 205	AB	Pyx	1877	1960					6.9	7.0	A5	6871	o	
--			ABxC		1898		354		26.4			14.0		6871		
8 33.2	+33 26	Σ 1240	AB	Lyn	1830	1958	70	78	22.2	26.8	7.7	10.7	A0	6866		
--			AC		1898	1909	246		51.1			10.5		6866		
8 33.9	+1 35	Σ 1243		Hya	1830	1943	228		1.8		7.7	10.0	A0	6874		
8 34.0	+10 35	A 2896		Cnc	1914	1921	343		1.9		7.6	12.8	F0	6873		
8 34.2	+8 27	Kui 35		Cnc	1958		202		31.2		6.0	13.6	F0			
8 34.2	+56 55	Σ 1235		Lyn	1831	1958	84		1.2		8.0	10.0	A0	6867		
8 34.5	−32 36	ø 335		Pyx	1955	1959	134	185	0.2	0.1	7.1	7.2	G5		o	
8 34.5	−37 37	I 195		Vel	1900	1954	42		1.9		6.4	8.7	K5			
8 35.4	−25 07	β 206		Pyx	1877	1954	279		1.8		8.2	8.4	G0	6887		
8 35.8	+6 37	Σ 1245	AB	Cnc	1832	1975	25		10.3		6.0	7.2	F6 G5	6886		
--			AC		1893	1910	121	120	92.2	93.2		10.7		6886		
--			AD		1893	1910	280	282	120.2	117.0		12.0		6886		
--			AE		1893	1926	211	210	131.9	122.4		8.8		6886		
8 36.2	+13 47	OΣΣ 94		Cnc	1875	1925	133		43.5		7.8	8.9	A0 A0			
8 36.7	+22 10	Brt 2390		Cnc	1939		75		4.6		8.0	10.0	G0			
8 36.7	−47 30	h 4116		Vel	1847	1938	2		7.6		7.6	9.1	A5			
8 36.8	+74 43	OΣ 192 rej		Cam	1871	1937	234		1.8		6.3	9.8	A5	6872		
8 37.1	−43 49	B 2183		Vel	1938		143		6.3		7.6	12.7	G5			
8 37.2	−49 26	h 4119		Vel	1879	1913	226		10.1		7.4	10.4	A0			
8 37.3	−62 51	h 4125		Car	1872	1918	237		7.6		5.5	11.1	G5			
8 37.5	−33 45	h 4115	AB	Pyx	1919		158		22.4		6.5	11.8	A5			
--			AC		1919		20		45.0			12.5				
--			AD		1919		197		29.9			13.1				
8 37.7	+5 42	β pm		Hya	1879	1909	312		243.7		4.2	10.4	A0			4 δ Hya
8 37.8	−6 48	h 99	AB	Hya	1880	1918	211	202	64.6	61.0	6.8	9.1	G0 A0	6900		
--			BC		1880	1903	210	211	9.2	9.8		13.4		6900		
8 37.9	−26 15	Rst 2571		Pyx	1933	1943	97		16.0		5.3	13.1	A0			η Pyx
8 38.7	−19 44	β 207		Pyx	1876	1938	103		4.3		6.4	9.4	K5	6903		
8 38.7	−53 05	HdO 203		Vel	1900		294		22.2		6.5	14.0	B9			
8 38.8	−33 52	B 1621		Pyx	1929	1951	5	2	0.6	0.5	8.1	8.6	G5			
8 38.9	−6 51	Rst 4414		Hya	1939	1941	90		7.6		7.1	13.7	A0			
8 39.1	−22 40	β 208	AB	Pyx	1877	1960					5.3	6.7	G5	6914	o	
--			AC		1918		202		85.0			10.7		6914		
8 39.1	−55 57	Hu 1443		Car	1913	1943	80	56	0.6	0.7	7.9	8.6	F5			
8 39.1	−70 23	Don 271	AB	Vol	1930		66		21.4		5.3	15	A0			θ Vol
--		h 4134	AC		1915	1917	108		45.0			10.3				
8 39.2	−40 25	B 1623		Vel	1929	1934	250		0.6		7.6	8.4	O5			
8 39.2	−60 19	h 4128		Car	1836	1946	222	210	2.0	1.4	6.9	7.5	A0			
8 39.3	−16 02	A 3063	AB	Hya	1926	1952	296		0.3		8.3	8.8	A0	6912		
--			ABxC		1926	1936	234		12.7			13.0		6912		
8 39.4	−9 35	Rst 4415		Hya	1939	1951	176		1.1		8.2	10.2	A0			
8 39.4	−28 23	See 101		Pyx	1897	1899	89		7.3		8.0	11.8	F0	6918		
8 39.4	−36 36	I 314		Pyx	1900	1959					6.5	7.6	F0		o	
8 39.7	+5 46	Σ 1255	AB	Hya	1831	1925	31		26.5		7.5	8.6	G5 G5	6913		
--			BC		1904	1928	32	14	39.9	44.5		13.6		6913		
--			BD		1904	1928	54	46	56.4	56.9		13.1		6913		
8 39.7	+20 05	Cou 47		Cnc	1965		152		0.4		8.1	9.0	A3			
8 39.7	−29 34	h 4120		Pyx	1905		61		52.4		4.9	9.1	G4 G0	6923		ζ Pyx
8 39.8	+11 31	β pm	AB	Cnc	1891	1911	321	324	130.2	137.0	7.7	10.3	G5 G			
--			BC		1907		13		30.4			12.1				
8 39.9	+19 33	β 584	AB	Cnc	1878	1934	292		1.4		6.9	11.9	A0	6915		
--		S 571	AC		1825	1952	156		45.2			7.2	A0	6915		C is a two-line sp. binary
--			AD		1825	1952	241		92.9			6.7	K0	6915		

α 2000	δ 2000	Star Name		Const	Years	Pos Ang	Separation	m_1	m_2	Spec	ADS	Orb	Notes
8^{h}39^m.9	+19°33'	β 584	DC	Cnc	1875 1952	88°	99".7				6915		
8 40.0	−53 03	h 4126		Vel	1835 1933	30	16.6	5.4	9.8	B5			
8 40.1	+20 00	β pm	AB	Cnc	1887 1921	151	149.8	6.5	6.5	K0 A0			
——			Aa		1887 1921	309	134.1		8.9				39 Cnc
——			Ab		1887 1921	110	135.0		9.2				
——			Bc		1907 1921	147	140.0		10.4				
8 40.1	−35 19	Rst 4894		Pyx	1943	119	12.8	4.0	12.5	G5			β Pyx
8 40.1	−53 51	ø		Vel			10.0	7.4		A3			
8 40.3	−40 16	CorO 74		Vel	1894 1935	61 64	3.8 4.0	5.2	8.5	B9			
8 40.4	+19 33	S 574		Cnc	1825 1924	249 249	132.8 134.9	6.3	7.4	A2 A0			41 ε Cnc
8 40.4	+19 40	Σ 1254	AB	Cnc	1831 1956	54	20.5	6.4	8.9	G5	6921		
——			AC		1863 1956	342	63.2		8.6	A0	6921		
——			AD		1863 1956	43	82.6		8.9		6921		
——		S 572	CD		1825 1956	90	75.9				6921		
8 40.4	+51 45	Es 909	AB	UMa	1910	317	49.1	7.8	11.3	K0	6906		
——			BC		1910	276	6.4		12.5		6906		
8 40.4	−42 23	Δ 72		Vel		358	131.4	7.2	8.0	A2 G0			
8 40.5	−41 04	Wg 105		Vel	1904	322	17.0	8.0	11.7	B9			
8 40.6	−46 39	h 4127		Vel	1913 1915	58	37.5	3.8	10.2	F8			
8 40.6	−59 46	Jsp 295		Car	1930	336	16.5	4.3	13.2	B1			
8 40.7	−11 56	Σ 1261		Hya	1831 1912	302	29.6	7.9	10.6	G5			
8 40.7	−12 11	Σ 1260		Hya	1830 1947	301	5.1	8.3	8.8	A2	6929		
8 40.7	−57 33	h 4130		Car	1850 1955	222 238	3.8 3.8	6.5	8.4	A2			
8 40.9	−17 21	A 3065		Hya	1926 1960	251 273	0.2 0.2	8.6	8.9	A0	6932		
8 41.3	+20 29	β 585		Cnc	1878 1958	106 98	0.4 0.5	6.9	8.4	A2	6930		
8 41.6	−45 00	Don 270		Vel	1929 1947	107	7.8	6.7	13.9	B9			
8 41.7	−15 57	h 4124		Hya	1898 1920	116 112	33.5 31.0	4.9	11.7	K0	6937		9 Hya
8 42.1	−48 15	Rst 5303		Vel	1944 1949	246	0.7	8.0	11.0	B8			
8 42.1	−52 45	B 1624		Vel	1929 1959	118 278	0.2 0.6	8.2	9.2	G5		o	Direct motion
8 42.3	−17 26	Ho 529		Hya	1894 1959	163 180	0.4 0.4	8.3	8.3	F5	6944		
8 42.3	−48 06	Rst 5304		Vel	1944 1949	303 313	0.3 0.3	5.8	7.1	B3			
8 42.4	−53 07	BrsO	AB	Vel	1836 1938	310	76.6	5.0	5.8	B5 B9			
——		B 1625	BC		1929 1959	112 130	0.5 0.5		8.0				
——		BrsO	BD		1846 1938	266	60.8		9.6				
8 42.5	−8 30	H VI 107		Hya	1899 1905	152	91.4	7.0	10.7	B9			
8 42.6	−48 05	Rst 344		Vel	1929	342	9.5	8.0	15	B8			
8 42.6	−48 10	See 103		Vel	1897 1933	208	13.9	7.3	11.8	B9			
8 43.1	−12 25	Rst 3603		Hya	1938 1958	286 180	0.3 0.3	8.5	8.6	F5		o	
8 43.2	−28 41	B 163		Pyx	1927	140	5.6	8.0	13.8	K0	6953		
8 43.3	+21 28	β pm	AB	Cnc	1888 1919	66 66	103.0 106.3	4.7	8.7	A0			43 γ Cnc
——			Aa		1908	258	103.3		12.4				
——			Bb		1908	102	94.6		12.9				
8 43.3	+38 48	β 209		Lyn	1875 1959	355 3	1.6 1.3	8.4	8.7	F5	6946		
8 43.4	+48 52	Σ 1258		UMa	1830 1968	331	9.9	7.5	7.8	F0	6945		
8 43.5	+2 43	Bail 2346		Hya	1910			7.7					
8 43.5	−14 03	Rst 3605		Hya	1938 1943	102	8.2	7.6	13.5	A0			
8 43.7	−7 14	S 579	AB	Hya	1824 1925	309 310	77.9 78.9	4.6	8.7	G0 A0			
——		β	Aa		1903	340	57.1		12.6				
8 43.8	−17 01	LDS 231		Hya		nf	263.0	8.1	10.3	F8			
8 44.0	−38 47	B 1628		Vel	1929	272	5.1	7.0	13.8	K0			
8 44.3	−20 16	I 813		Pyx	1911 1960	300	3.3	7.5	13.0	K0	6969		
8 44.4	−4 12	A 552		Hya	1903 1951	49 16	0.2 0.1	8.2	9.2	A2	6964		Less than 0".1 in 1953,54
8 44.4	−42 39	h 4133		Vel	1835 1959	61 63	47.2 45.3	4.1	10.3	G5			
8 44.6	−37 58	Jsp 298		Vel	1929 1945	286	6.6	8.0	15	K0			
8 44.7	+18 09	h 457		Cnc	1852 1958	121 90	45.8 38.4	3.9	11.9	K0	6967		47 δ Cnc; optical
8 44.7	−23 47	See 106	AB	Pyx	1897 1955	224 238	16.7 17.9	6.8	11.5	K0	6974		
——			BC		1897 1955	333	3.5		12.0		6974		
8 44.7	−41 17	I 815	AB	Vel	1911 1937	2	4.4	7.2	11.3	B5			
——			AC		1933 1938	130	8.3		13.3				
——			AD			225	35.0		11.0				
——		Wg 109	DE		1904	325	10.0		12.5				
——			EF			7	10.0		13.5				
8 44.7	−54 43	I 10	AB	Vel	1894 1952	175 153	2.6 2.6	2.1	5.1	A0			δ Vel
——		h 4136	AC		1879 1913	61	69.2		11.0				
——			CD		1930 1935	102	6.2		13.5				
8 45.1	−32 15	HdO 204	AB	Pyx		310	10.0	6.9	12.0	F2			
——			AC			275	13.0		12				
8 45.1	−58 44	Rmk 9	AB	Car	1836 1951	292	4.1	6.9	7.0	B8			
——			AC		1872 1913	359	50.9		11.0				
——			AD		1872 1913	222	61.4		10.8				
8 45.3	−2 36	Σ 1270		Hya	1830 1955	262	4.7	6.4	7.4	F5	6977		
8 46.1	+25 21	A 2130	AB	Cnc	1910 1956	14 31	0.7 1.0	7.9	11.4	A2	6981		
——			AC		1910 1919	309	15.0		13.8		6981		
8 46.1	−11 00	β 1069		Hya	1889 1958	61 68	2.1 2.4	6.3	9.8	K5	7026		
8 46.3	+0 39	h 3313		Hya	1912	61	35.2	7.0	10.0	A5			

α 2000	δ 2000	Star Name		Const	Years	Pos Ang		Separation		m₁	m₂	Spec	ADS	Orb	Notes
h m	° ′					°	°	″	″						
8 46.4	−13 33	Rst 5507		Hya	1951	266		22.2		4.3	13.6	G5			12 Hya
8 46.4	−52 51	Hu 1590		Vel	1914 1933	338		0.3		8.3	8.8	A3			
8 46.5	−42 34	Jc 13		Vel	1856 1941	313		2.1		7.3	8.8	B9			
8 46.7	+28 46	Σ 1268		Cnc	1828 1968	307		30.5		4.2	6.6	G5 A5	6988		z Cnc; nice contr.(Webb)
8 46.8	+6 25	Sp	AB	Hya	1888 1959					3.8	4.7	F8	6993	o	11 ε Hya
—		Σ 1273	ABxC		1830 1976	195 281		3.2	2.8		6.8		6993		
—			ABxD		1878 1938	192 196		20.0	19.2		12.4		6993		
—			ABxE		1921	359		336.2			10.0		6993		
—			ABxF		1921	265		424.5			10.1		6993		
8 46.8	−39 30	h 4138		Vel	1904 1920	324		8.4		7.6	11.6	A0			
8 47.2	+11 10	Σ 1276		Cnc	1831 1938	354		12.5		8.3	8.5	A0	6995		
8 47.2	−51 55	Rst 352		Vel	1930 1948	59		5.7		7.8	13.6	G0			
8 47.2	−63 49	Slr 9		Car	1893 1943			1.4		8.3	9.4	A0			
8 47.3	−38 07	Jsp 302		Vel	1929	245		0.2		8.7	8.7	K0			
8 47.4	−17 03	β 586		Hya	1878 1960	53 81		0.8	0.3	6.7	9.2	G0	6999		Spectrum composite G0+A2
8 47.4	−34 36	Gli 101	AB	Pyx		131		51.8		6.7	10.6	K5 A2			
—		Rst 4898	BC		1943	271		1.6			11.7				
8 47.6	+0 05	Rst 5306		Hya	1946 1958	23		0.2		8.6	8.6	B9			
8 47.6	−28 22	WNO 3		Pyx	1877 1898	286		25.6		8.2	8.6	F5 F5			
8 47.6	−40 14	B 1630		Vel	1928 1953	126		4.0		7.8	12.0	K5			
8 47.7	−38 57	I 70		Vel	1901 1953	113		1.3		7.2	9.2	A0			
8 47.9	−43 21	Don 282	AxBC	Vel						7.2		K0			
—			BC		1929 1947	183		2.0		11.8	11.9				
8 48.1	+17 24	h 4135		Cnc	1904	53		42.2		7.5	12.6	K2			
8 48.2	+2 35	β 335		Hya	1875 1936	269		2.7		7.5	10.8	F5	7003		
8 48.3	+0 33	OΣ 194		Hya	1849 1927	56		12.6		7.3	10.8	K2	7004		
8 48.3	+34 36	Σ 1272		Lyn	1831 1924	343		20.4		8.1	9.6	F8	7000		
8 48.4	+5 50	AGC 3		Hya	1878 1935	145		12.4		4.4	11.9	A0	7006		13 ρ Hya
8 48.4	−3 57	h 106		Hya	1820	340				6.0					Some error in position
8 48.4	−65 26	Hrg 19		Vol	1882 1930	149 165		3.2	4.5	7.5	11.0	K0			
8 48.7	+0 57	A 2552		Hya	1913 1956	259 151		0.3	0.2	8.5	8.5	F5	7012	o	May have a short period
8 48.9	+72 01	Σ 1253		UMa	1831 1912	244		25.9		8.0	10.0	K0			
8 49.0	+38 21	Σ 1274		Lyn	1830 1938	41		8.9		7.4	9.1	A2	7005		
8 49.1	−5 14	h 2468		Hya	1894 1916	345		23.8		7.9	10.3	K0			
8 49.5	+8 52	β 1068	AB	Cnc	1889 1954	190 122		0.4	0.2	7.9	10.0	A5	7021		
—			ABxC		1889 1903	312		17.7			12.8		7021		
8 50.0	+14 50	Σ 1283		Cnc	1829 1974	123		16.4		7.6	8.6	F0	7031		
8 50.2	−49 55	I 1181		Vel	1928 1933	338		3.8		7.6	12.5	F0			
8 50.3	−44 43	Hu 1447		Vel	1913 1934	220 217		3.3	3.3	7.5	12.5	G5			
8 50.4	−4 12	h 107		Hya	1894 1911	58 52		26.5	27.3	7.0	10.6	K0	7040		
8 50.4	−35 56	h 4144		Pyx	1879 1935	315		2.4		7.0	9.2	B9			
8 50.6	−46 32	HdO 205		Vel	1894 1959	84		3.2		4.9	9.0	B9			
8 50.6	−66 48	HdO 206		Vol		230		20.0		5.4	12.0	F0			
8 50.7	+18 00	A 2473		Cnc	1912 1960	325 15		0.3	0.3	7.5	7.5	G5	7039		
8 50.7	+35 04	Σ 1282		Lyn	1830 1956	279		3.6		7.5	7.5	F8	7034		
8 50.8	−54 07	h 4148		Vel	1874 1917	111		5.8		8.1	11.3	A2			
8 51.3	+8 20	Per		Cnc	1884 1956	350		0.8		8.3	9.5	K0	7046		
8 51.5	+12 08	Σ 1287	AB	Cnc	1830 1942	99 91		1.4	1.9	7.9	10.2	F0	7049		
—			AC		1883 1926	109 104		15.6	14.9		11.8		7049		
8 51.5	+57 32	Σ 1275	AB	UMa	1832 1958	196		2.0		8.4	8.4	F0	7033		
—		h 2466	AC		1905 1919	81 79		38.4	39.8		12.6		7033		
8 51.6	−7 11	β 587	AB	Hya	1878 1958	160 125		0.4	0.9	5.6	8.6	A2	7050		15 Hya
—		H V 20	AC		1878 1924	358		45.7			9.6		7050		
—		β 587	AD		1878 1924	53 54		50.0	51.9		10.7		7050		
8 51.6	−26 12	B 2190		Pyx		n		10.0		7.6	15	K5			
8 52.0	+25 43	OΣΣ 96	AB	Cnc	1875 1925	314 313		41.9	45.0	7.5	8.7	K0 G0			
—			AC		1903 1921	261		34.3			11.5				
—			BC		1874	185									
8 52.1	+4 28	Σ 1290		Hya	1834 1930	320		3.0		7.5	9.4	A2	7055		
8 52.1	+47 34	β pm	AB	UMa	1873 1919	321 322		155.9	166.8	8.0	9.8	G5			
—			AC		1893 1919	280 281		220.4	222.9		10.0				
—			BC		1908	233		146.4							
8 52.4	+5 20	A 2900		Hya	1914 1949	295 301		0.8	0.6	6.3	9.5	A3	7061		
8 52.4	−55 21	ø 37		Car	1927 1943	46		0.9		6.9	9.9	A0			
8 52.5	+28 16	h 460		Cnc	1903 1924	333		43.1		6.2	9.7	M0			53 Cnc
8 52.5	−13 14	h 4146		Hya	1901 1920	104		33.6		6.1	11.4	K0	7062		
8 52.6	+32 28	Hu 1125	AB	Cnc	1905 1934	274 272		3.9	4.4	5.7	13.2	A3	7057		51 σ¹ Cnc
—		S 583	AC		1825 1924	23 23		82.1	79.0		9.3		7057		
8 52.6	−36 33	ø 296		Pyx	1930 1959	119 74		0.1	0.1	7.2	7.2	F5			More than one revolution?
8 52.7	−52 08	CapO 9		Vel	1877 1937	80		2.9		6.8	7.8	A0			
8 53.1	+54 58	A 1584		UMa	1907 1959					8.1	8.1	G0	7054	o	
8 53.8	−47 31	ø 316		Vel	1951 1959					6.1	6.1	A5		o	
8 53.9	+1 49	A 2554		Hya	1913 1941	143 255		0.3	0.3	7.9	9.5	F0	7074		
8 53.9	+19 58	Cou 773		Cnc	1972	30		0.2		7.3	7.7	F8			
8 53.9	−41 50	h 4150		Vel	1915 1920	265		17.5		7.4	10.0	B9			

α 2000	δ 2000	Star Name		Const	Years	Pos Ang		Separation		m₁	m₂	Spec	ADS	Orb	Notes
8ʰ54.ᵐ0	+8°25′	OΣ 195		Cnc	1848 1955	139°		9″.6		8.0	8.5	F8	7073		
8 54.2	-8 46	β 24		Hya	1875 1960	175		1.1		7.8	8.9	F0	7080		Spectrum composite F0+A2
8 54.2	+30 35	Σ 1291	AB	Cnc	1829 1960	333	316	1.5	1.4	6.0	6.5	K0	7071		57 Cnc
--			ABxC		1921 1953	199		55.6			9.1		7071		
8 54.7	-39 29	B 1636		Vel	1929 1953	238		0.7		8.1	10.3	K0			
8 54.8	+43 35	Σ 1289		Lyn	1830 1955	5		3.9		8.2	9.0	F8	7075		
8 54.9	-7 49	β 103		Hya	1875 1940	73		2.9		7.9	11.1	A5	7086		
8 54.9	+26 12	A 2131	AB	Cnc	1910 1958					7.0	8.1	G0	7082	o	
--		Ho 357	ABxC		1892 1925	8	3	31.1	44.0		13.0		7082		
8 54.9	-36 03	B 2194		Pyx		165		10.0		7.4	14.5	K0			
8 55.0	-60 39	h 4156		Car	1897 1933	319		21.1		3.8	12.6	B8			
8 55.1	-43 28	I 317		Vel	1901 1942	303		2.0		8.4	9.4	B3			
8 55.1	-46 22	Don 296	AB	Vel						7.7	10.5	A0			
--			BC		1929 1947	75		2.4			12.5				
8 55.2	-18 14	S 585		Hya	1901 1918	147		66.7		5.8	7.0	K0 K2			
8 55.3	-11 22	S 584		Hya	1825 1919	211	215	71.2	66.0	6.8	9.2	A0 K0			
8 55.3	-45 02	h 4153		Vel	1913	131		35.0		6.3	12.8	B3			
8 55.5	-7 58	Σ 1295		Hya	1831 1968	0	2	4.4	4.1	7.0	8.0	A3	7093		17 Hya
8 55.5	-27 41	See 107	AB	Pyx	1897 1899	268		23.8		4.9	14.0	A2	7095		δ Pyx
--		Alden	CD		1927	17		2.5		11.0	11.0		7095		
8 55.7	+41 41	A 2132		Lyn	1910 1958	202	242	0.2	0.1	8.2	8.3	F0	7084		[49]
8 56.2	-55 32	Δ 73	AB	Vel	1836 1938	356	359	66.5	65.2	7.7	8.2	K0 K0			
--			AC		1913	240		27.5			10.5				
8 56.3	-37 07	Rst 2593		Pyx	1935 1947	83	65	0.7	0.7	7.0	9.7	G0			
8 56.3	-52 43	R 87		Vel	1881 1938	341	339	2.7	2.7	4.8	7.4	B5			
8 56.4	-53 48	CorO 78		Vel	1874 1919	144		10.9		8.6	9.0	A0			
8 56.8	-17 26	Arg 72		Hya	1875 1953	183		3.0		7.3	7.5	F2	7103		
8 57.0	-59 14	Δ 74		Car	1836 1917	75		40.4		5.1	6.8	B5 B8			
8 57.1	+37 15	Hu 859		Lyn	1904 1944	199		0.3		8.2	9.8	A3	7100		
8 57.1	-29 51	CorO 77		Pyx	1920 1930	200		10.2		8.4	9.3	F5	7108		
8 57.1	-43 15	See 108	AB	Vel	1897 1935	46		3.1		7.6	9.6	B3			
--		Gli 102	AC			2		43.1			10.1				
--			AD			240		48.1			9.2				
8 57.2	-51 04	Rst 5511		Vel	1947	274		0.2		8.6	8.6	F5			
8 57.3	-66 12	h 4164		Vol	1901 1918	145		10.7		7.3	10.1	K0			Very neat; y., pale bl. (h)
8 57.8	-18 53	Ho 358		Hya	1892 1933	298		1.9		7.1	12.2	A0	7112		
8 58.0	+30 14	Ho 252		Cnc	1925 1945	143	225	0.3	0.1	7.1	7.1	F5	7107		61 Cnc [50]
8 58.1	+1 32	Kui 36		Hya	1933 1958	197		2.5		6.6	12.1	A0			
8 58.4	+32 21	Dju		Cnc	1955	246		0.3		8.5	8.7	K5			
8 58.5	+11 51	h 110		Cnc	1836 1934	325		11.3		4.3	11.8	A3	7115		65 α Cnc
8 58.9	-47 14	h 4161		Vel	1913 1929	337	338	25.1	24.2	5.2	11.2	F0			
8 59.0	+63 26	β 408		UMa	1877 1946	343		2.9		7.2	9.7	G5	7106		
8 59.2	+48 02	h 2477	AB	UMa	1841 1958	350	16	10.7	4.5	3.1	10.8	A5	7114	o	9 ι UMa
--		Hu 628	BC		1903 1960						11.0		7114	o	
8 59.4	-59 05	I 318		Car	1900 1914	20	27	29.0	26.2	5.2	12.0	F0			
8 59.5	+32 25	S,h 100		Cnc	1823 1914	295		89.6		5.6	9.4	G5			64 σ³ Cnc
8 59.5	-23 10	Don		Pyx	1930	84		13.3		8.0	13.0	K2			
9 00.0	+26 26	A 1975		Cnc	1909 1920	82		3.5		6.6	14.0	A2	7123		
9 00.0	-49 33	Gli 104		Vel	1900 1934	307		9.1		7.0	9.7	K0			
9 00.1	-12 28	Hu 225	AB	Hya	1900 1959					8.7	8.9	F5	7131	o	
--		Worley 36	ABxC		1972	132		5.6			14.0	F5	7131		
9 00.6	+41 47	Kui 37	AB	Lyn	1936 1960					4.1	6.2	F5		o	Formerly called 10 UMa
--		β pm	AC		1893 1908	122	120	140.0	141.9		10.7				
--			AD		1851 1916	206	199	150.3	126.9		9.7				
--			AE		1851 1916	114	108	204.6	225.2		10.4				
9 00.8	-9 11	β 409		Hya	1878 1901	185		9.7		7.3	9.8	A0	7136		
9 01.4	+32 15	Σ 1298	AB	Cnc	1831 1955	137		4.6		5.9	8.0	A2	7137		66 σ⁴ Cnc
--			AC		1879 1908	319		187.4			10.8		7137		
9 01.7	-52 11	h 4165		Vel	1837 1954	87	118	1.4	0.9	5.5	6.7	B9			
9 01.8	+27 54	S,h 101		Cnc	1888 1924	324	325	101.5	103.9	6.1	8.9	A5			67 Cnc
9 02.0	+2 40	β 211		Hya	1875 1943	265		1.1		7.1	9.6	A2	7152		
9 02.8	+0 25	Ho 361		Hya	1892 1945	92		4.4		7.8	11.8	B9	7159		
9 03.1	-8 59	Brt 545		Hya	1931 1971	319	321	5.7	6.1	8.0	9.5	A2			
9 03.2	+47 40	Hu 720		UMa	1904 1958	144		0.7		8.5	8.9	F5	7153		
9 03.2	-62 21	h 4175		Car	1915 1917	132		19.9		7.7	9.8	F0			
9 03.3	-33 36	h 4166	AxBC	Pyx	1903 1952	153		13.7		6.7		A0 A			
--		Rst 3619	BC		1936 1947	77		0.8		8.6	11.8				
9 03.6	+47 09	A 1585		UMa	1907 1956					4.2	4.4	A0	7158	o	12 κ UMa
9 03.8	-64 41	Slr 16		Vol	1895 1947	310		1.0		7.4	9.5	B9			
9 04.4	-33 06	Rst 2599		Pyx	1935 1947	235		0.5		8.4	9.1	F0			
9 04.5	-56 21	Rst 3620	AB	Vel	1938	126		0.6		7.6	9.8	B9			
--		h 4177	ABxC		1897 1959	263	241	14.1	12.4		9.3				
--			ABxD		1897 1913	296		35.5			9.7				
9 04.8	-57 51	h 4178		Car	1900 1934	162		3.3		6.4	10.0	A3			
9 05.7	+50 18	Es		UMa	1910 1918	258		79.2		8.1	8.6	G5 G5			

α 2000	δ 2000	Star Name		Const	Years		Pos Ang		Separation		m₁	m₂	Spec	ADS	Orb	Notes
9ʰ05ᵐ.7	−50°09′	I 196		Vel	1900	1956	171°		0″.5		8.4	8.7	A0			
9 06.5	−24 09	Ho 530		Pyx	1905	1931	81		17.2		8.0	12.5	F2	7185		
9 06.5	−74 45	I 288		Car	1900	1947	254	267	0.6	0.6	8.6	8.8	A0			
9 06.6	+2 49	Σ 1309		Hya	1834	1937	273		11.5		8.6	8.9	F5	7182		
9 06.9	−18 12	Rst 3621		Hya	1936	1943	37		1.5		7.1	14.5	K0			
9 07.4	+22 59	Σ 1311	AB	Cnc	1831	1956	200		7.5		6.9	7.3	F5	7187		
--		Ho 644	AC		1892	1906	118		27.8			12.6		7187		
9 07.5	−39 50	B 1645		Vel	1929	1941	122		3.5		7.9	10.8	A2			
9 07.7	−10 29	h 804		Hya	1903	1912	329		12.8		7.8	10.7	A0	7197		
9 07.7	−44 38	I 492		Vel	1913	1945	147		0.3		7.3	7.8	B5			
9 07.8	−32 02	B 1644		Pyx	1929		44		3.3		8.0	14.5	K5			
9 08.0	+27 34	OΣΣ 97		Cnc	1875	1914	57		51.5		8.5	8.5	G0 G0			
9 08.0	+81 02	Σ 1284		Cam	1833	1936	168		2.4		8.0	9.7	F2	7163		
9 08.0	−25 52	I 491		Pyx	1911		263		2.1		4.6	9.8	M0	7202		κ Pyx
9 08.0	−43 26	See 109		Vel	1897		137		18.2		2.2	14.8	K4			λ Vel
9 08.0	−64 02	h 4185		Car	1880	1918	244		10.5		8.7	8.9	G5 G5			
9 08.2	+11 57	Hu 866		Cnc	1905	1944	7		0.6		7.4	13.4	K0	7200		
9 08.6	−25 50	Rst 2610		Pyx	1935	1943	345		1.6		6.8	11.8	F8			
9 08.7	−8 35	Kui 38		Hya	1936	1958	293	296	1.2	1.1	5.5	10.5	B8			19 Hya
9 08.8	+26 38	β pm	AB	Cnc	1851	1911	43	40	92.6	112.0	6.0	9.1	G5			75 Cnc
--			BC		1907		338		128.8			10.3				
9 09.0	−14 11	Hu 227	AB	Hya	1900	1960	215		2.2		7.9	11.5	K0	7210		
--			AC		1909		318		33.2			11.3		7210		
9 09.1	−1 35	h 119		Hya	1896	1915	328		53.2		7.8	9.5	K5			
9 09.4	+22 03	Wils 16		Cnc	1937	1944	62	108	0.2	0.1	5.6	6.1	G5			77 ξ Cnc [51]
9 09.5	+2 56	OΣ 197		Hya	1847	1955	62		1.4		7.9	9.5	F0	7215		
9 09.8	−25 48	β 410		Pyx	1877	1954	162	158	1.7	1.7	7.4	8.9	A0	7220		
9 09.9	−30 22	H N 96	AD	Pyx	1898		340		35.4		5.6	13.5				
--			AxBC		1883	1933	147		17.8				A3 A3			ε Pyx
--		B 1113	BC		1928	1951	106	88	0.3	0.3	10.5	10.8				
9 10.0	−44 10	CorO 80		Vel	1901	1936	46		2.5		8.5	9.1	A0			
9 10.2	−16 52	h 4182		Hya	1906	1915	107		25.2		7.1	10.7	F0	7222		
9 10.3	+22 00	Wils 17		Cnc	1940		107		0.1		6.0		G5			79 Cnc [52]
9 10.4	+67 08	Σ 1306	AB	UMa	1832	1960					4.8	8.2	F8	7203	o	13 σ² UMa
--			AC		1879	1919	148	148	206.3	204.6		9.3		7203		
9 10.9	+63 31	H V 73		UMa	1899	1906	49		57.2		4.7	10.3	F5	7211		14 τ UMa
9 11.1	−60 58	ø 129		Car	1929	1954	292		0.2	0.1	8.1	8.1	A2			Too close 1951,54
9 11.5	+0 17	β 104		Hya	1875	1932	108	103	3.3	2.5	6.8	11.6	K0	7227		
9 11.5	−57 58	h 4190		Car	1879	1932	22		8.2		6.6	10.1	B3			
9 11.7	−46 54	See 110		Vel	1897	1952	52		1.0		8.4	9.0	A2			
9 11.9	+18 03	β pm	AB	Cnc	1882	1919	252	252	134.6	133.4	6.9	9.5	A0			
--			BC		1909	1919	317		148.2			10.3				
9 12.1	−53 58	h 4189		Vel	1898	1917	107		20.0		7.8	10.8	B9			
9 12.3	+15 00	β pm	AB	Cnc	1907		118		35.4		6.4	13.2	G5			81 Cnc
--			AC		1855	1907	236	229	237.1	222.4		10.6				
--		ø 347	Aa		1959	1960	156	127	0.1	0.2	7.2	7.2			o	
9 12.4	−6 56	A 2973		Hya	1916	1929	7		2.6		7.6	12.4	K0	7240		
9 12.5	−43 37	ø 317	AB	Vel	1951	1959	130		0.1		6.7	7.2	B8			Too close 1952-59
--		h 4188	ABxC		1835	1951	286	282	2.8	2.7		6.9				
9 12.7	+16 31	Σ 1322		Cnc	1830	1957	54		1.8		8.1	8.6	A0	7236		Above two wide pairs (Bird)
9 12.8	+61 41	Σ 1315		UMa	1831	1925	26		24.9		7.7	7.7	A0 A0	7226		
9 12.8	−60 55	HdO 207		Car	1926	1959	70	219	0.2	0.2	7.2	7.3	A0			Direct motion
9 12.9	+10 40	A 2974		Cnc	1915	1931	43		1.2		7.9	11.3	F5	7242		
9 13.6	+46 59	Σ 1318		UMa	1830	1958	245	237	3.5	2.9	7.8	9.0	F8	7243		
9 14.3	+61 25	H V 15		UMa	1782		190		49.0		5.1		F8			16 UMa; single
9 14.4	+2 19	h 2489	AB	Hya	1853	1953	172	197	62.1	29.4	3.9	9.9	A0	7253		22 θ Hya
--			AC		1924		286		80.9					7253		
--		B	AD		1962		147		96.4			12.5		7253		
9 14.4	+52 41	Σ 1321	AB	UMa	1832	1977					7.6	7.7	K2 K2	7251	o	
--			AC		1907		283		28.4			14.7		7251		
--			AD		1879	1957	208	172	188.9	121.4		10.8		7251		
9 14.4	−32 33	Rst 2619		Pyx	1934	1945	98		1.4		6.5	11.5	K0			
9 14.4	−43 14	h 4191		Vel	1897	1935	14		5.8		5.3	9.4	B5			Beautiful and delicate (h)
9 14.8	+4 13	β 455		Hya	1877	1954	65	72	1.9	1.8	8.4	9.4	K0	7257		
9 14.9	−8 45	Weisse 21		Hya	1880	1918	14		25.8		7.4	8.6	A0	7258		
9 15.1	−20 07	Ho 363		Hya	1890	1957	176	181	1.6	2.4	7.5	9.5	A0	7263		
9 15.1	−30 22	δ 131		Pyx	1923	1939	124		0.9		7.9	9.5	A0			
9 15.2	+23 23	β pm	AB	Cnc	1908		170		59.8		8.0	11.2	G5			
--			AC		1841	1908	87	83	51.2	60.4		9.6				
9 15.2	+34 38	β pm		Lyn	1912	1925	262	262	152.6	150.6	6.0	9.5	G0			
9 15.2	−45 33	I 11		Vel	1896	1960	271	284	0.9	0.8	6.7	7.6	A0			
9 15.4	+22 48	A 2136		Cnc	1910	1933	119		1.9		7.9	14.0	F2	7261		
9 15.5	+27 55	Σ 1327	AB	Cnc	1831	1958	81	61	16.1	8.0	8.2	9.4	F8	7260		
--			AC		1831	1925	28	22	25.1	26.3		9.3		7260		
--			CB		1831	1912	167	179	20.2	21.4				7260		

α 2000	δ 2000	Star Name		Const	Years	Pos Ang	Separation	m₁	m₂	Spec	ADS	Orb	Notes
9ʰ15ᵐ.5	+27° 55′	Σ 1327	CD	Cnc	1913	351°	2″.2		14.3		7260		
9 15.7	−1 14	Σ 1329	AB	Hya	1834 1954	246 253	27.2 14.5	8.7	8.7	G5 G5	7266		An optical pair
−−			AC		1911	102	157.3		12.9		7266		
9 15.8	−37 25	See 111		Vel	1897	126	11.3	4.6	14.4	F5			
9 16.0	−27 12	B 2205		Pyx		np	10.0	7.8	12.0	K0			
9 16.1	−8 21	β 212		Hya	1875 1960	230 209	1.5 1.3	7.6	8.3	A5	7270		
9 16.1	−44 54	Don 329		Vel	1929 1947	142 148	6.8 6.8	6.0	14.0	B5			
9 16.2	+23 24	OΣ 198		Cnc	1879 1958	167 136	16.1 13.9	7.5	12.0	K0	7268		
9 16.3	−23 08	h 4193		Hya	1877 1960	118	3.1	8.4	9.2	A5	7277		
9 16.7	+0 44	β 588		Hya	1878 1953	123 142	2.4 2.1	6.5	11.0	F5	7276		
9 16.7	−6 21	Kui 40		Hya	1936 1959	282	1.6	5.2	10.8	K0			23 Hya
9 16.8	−21 57	I 493		Hya	1934 1937	145	0.1	8.7	8.8	A0			Too close; p.a. uncertain
9 17.2	+7 16	OΣΣ 98	AB	Cnc	1873 1915	169 170	113.1 111.1	8.6	8.9	F8 F8			
−−			BC		1912	89	145.2		13.9				
9 17.3	+23 39	Σ 1332		Cnc	1829 1975	16 26	5.6 5.9	7.8	8.1	F5	7281		
9 17.3	−68 41	φ 363	AB	Car	1960	16	0.1	6.1	6.1	F2		o	
−−		I 358	ABxC		1901	132	18.1		12.2				
9 17.4	−74 54	I 12	AB	Car	1902 1935	264 254	0.3 0.3	5.9	6.4	A0			
−−		h 4206	ABxC		1900 1935	341 343	7.1 7.1		9.9				
−−			ABxD		1919	353	48.2		10.2				
9 17.5	+0 33	Σ 1336 rej		Hya	1905 1911	180	40.1	7.0	10.2	A0			
9 17.5	−70 23	CapO 43		Car	1903 1940	338	4.7	8.5	9.0	G5			
9 17.9	+11 30	h 128		Cnc	1879 1958	278 278	23.7 21.8	6.3	12.8	A0	7285		
9 17.9	+28 34	Σ 3121		Cnc	1832 1960			7.9	8.0	K0	7284	o	
9 17.9	−69 47	Rmk 10		Car	1838 1926	18	10.4	7.8	8.1	A0			
9 18.3	+18 47	Cou 384		Cnc	1969	66	0.2	8.2	8.3	F0			
9 18.4	+35 22	Σ 1333		Lyn	1828 1959			6.4	6.7	A5	7286	o	
9 18.4	−20 22	S 595		Hya	1825 1916	280 284	61.2 67.5	8.3	9.0	A3 F0			
9 18.8	+36 48	Σ 1334	AB	Lyn	1829 1968	240 229	2.7 2.7	3.9	6.6	A2	7292		38 Lyn
−−			BC		1909	212	87.7		10.8		7292		
−−			BD		1879 1909	256	177.9		10.7		7292		
9 18.9	−39 26	B 2208		Vel	1940	277	9.3	7.8	12.6	K0			
9 19.0	+10 57	A 2975		Cnc	1915 1923	358	0.7	7.9	11.3	A2	7296		
9 19.6	−15 50	A 3077		Hya	1925 1943	190	4.1	5.8	10.8	K0	7302		
9 19.8	−11 58	B 2529		Hya	1936 1958	139 12	3.0 3.2	4.8	12.3	G5			26 Hya [53]
9 20.0	−27 46	h 4199	AB	Pyx	1898 1920	111	11.7	8.2	9.5	A0 A0			
−−			AC		1898	271	8.4						
9 20.2	−23 28	HCWils 6		Hya	1886 1954	37 45	1.5 1.5	7.8	9.4	F0	7312		
9 20.3	−45 10	Rst 4907		Vel	1944	319	3.5	6.6	14.5	K0			
9 20.5	−9 33	S,h 105	AB	Hya	1823 1923	211	229.4	5.0	6.9	G5 F2	7311		27 Hya
−−		Doo	BC		1898 1969	197 198	9.6 9.3		9.1		7311		
9 20.7	+51 16	OΣ 199	AB	UMa	1847 1973	117 133	5.7 5.7	6.1	10.2	F2	7303		
−−			AC		1890 1924	4 5	147.3 142.1		10.0		7303		
9 20.7	−29 13	I 198		Pyx	1898 1952	180 200	0.4 0.2	8.6	9.0	A2	7318		
9 20.7	−31 46	h 4200		Pyx	1891 1954	73	3.1	7.3	7.9	A0			
9 20.9	+61 21	Σ 1331	AB	UMa	1833 1954	153	1.0	8.4	8.4	A5	7300		
−−			AC		1833 1954	201 186	11.4 13.2		11.1		7300		
−−			AD		1904 1954	116 118	19.5 20.4		13.6		7300		
−−			AE		1863 1910	223	139.7		9.7		7300		
9 21.0	+38 11	Σ 1338	AB	Lyn	1829 1959			6.5	6.7	F2	7307	o	
−−			ABxC		1879 1909	167	143.4		11.4		7307		
9 21.2	−35 21	I 170		Pyx	1898 1943	35	1.4	7.5	10.0	F2			
9 21.6	−60 50	I 494		Car	1911 1934	238	1.9	7.5	10.1	A0			
9 21.7	−47 19	Rst 393		Vel	1929 1944	255	0.5	8.0	8.7	A2			
9 21.8	+88 34	Cou 10		UMi	1959	65	1.7	7.1	10.4	A0			
9 22.0	−46 46	Rst 4909		Vel	1943 1949	21	1.6	7.7	13.1	G5			
9 22.1	−73 32	B 2213		Car	1908 1940	77	7.0	8.0	12.0	G5			
9 22.5	−1 05	A 1760		Hya	1908 1940	144	1.7	8.0	11.3	A0	7329		
9 22.5	+49 33	Σ 1340	AB	UMa	1830 1925	319	6.2	7.1	8.9	B9	7324		
−−			AC		1879 1909	83	135.0				7324		
9 22.5	−17 54	β 337		Hya	1876 1932	321 334	7.7 8.3	6.8	10.8	F2	7331		
9 22.5	−53 44	Rst 2624		Vel	1936 1948	204	3.4	8.0	14.7	F5			
9 22.8	−9 50	A 1342	AB	Hya	1906 1959	29 27	0.2 0.1	7.3	7.3	A2	7334	o	Close; quadrant uncertain
−−		Rst	ABxC		1938 1958	204	1.8		11.6		7334		
9 23.1	−53 49	φ		· Vel			11.0	6.8		G5			
9 23.3	+3 30	Σ 1347		Hya	1832 1937	311	21.2	7.3	8.6	F0	7342		
9 23.4	−77 53	h 4217		Cha	1837 1919	279 s	20.0	7.1	13.0	G0			Change due to p.m. of A
9 23.8	−23 39	HdO 123	AB	Hya	1868 1957	5	4.4	7.8	10.4	F2	7350		
−−			AC		1910 1957	194 191	55.1 54.5		10.7		7350		
9 24.3	−39 26	φ 348		Vel	1959	126	0.1	6.9	6.9	A2			
9 24.5	+6 21	Σ 1348		Hya	1831 1976	334 317	1.1 1.9	7.5	7.6	F5	7352		
9 24.5	+18 08	A 2477		Leo	1912 1958	236 301	0.3 0.3	7.3	8.8	G0	7341	o	
9 24.6	−70 25	I 199		Car	1901 1915	138	2.5	8.0	10.5	B9			
9 24.7	+26 11	β 105	AB	Leo	1876 1959	204 208	3.0 2.1	4.5	10.3	K0	7351		1 κ Leo
−−			AC		1912 1918	211	52.6		10.5		7351		

α 2000	δ 2000	Star Name		Const	Years	Pos Ang	Separation	m_1	m_2	Spec	ADS	Orb	Notes
9^{h}24.9	+51°34′	OΣ 200		UMa	1847 1959	335°	1.4″	6.5	8.1	G0	7348		
9 25.5	−61 57	h 4213		Car	1874 1932	327	8.7	5.8	10.3	A2			
9 25.6	+54 01	Σ 1346	AB	UMa	1830 1937	311	5.7	7.8	8.8	A2	7354		
--			BC		1879 1908	189	97.3		12.0		7354		
9 25.6	−25 20	Rst 2627		Pyx	1934 1945	314	0.6	7.0	9.3	B9			
9 25.6	−53 15	CorO 83		Vel	1894 1920	152	19.4	7.1	10.1	F8			
9 26.0	−56 05	φ 71		Vel	1928	249	3.3	7.3	13.9	B8			
9 26.4	−42 15	B 1122		Vel	1928 1956	150 177	0.2 0.2	7.5	7.6	A0			
9 26.5	+78 26	Σ 1326		Dra	1832 1926	172	29.2	8.6	9.0	G			
9 26.6	+6 32	β 589		Leo	1878 1928	216	3.0	7.9	12.9	F5	7371		
9 26.7	−28 47	Jc 5		Pyx	1858 1947	234 264	0.6 0.6	6.5	7.2	B8	7379		
9 26.8	+14 30	Ho 365		Leo	1890 1906	153	12.8	7.8	13.8	A3	7373		
9 27.2	−9 13	A 1588	AB	Hya	1907 1956	184	0.2	7.3	7.3	A0	7382		29 Hya
--		β 590	ABxC		1878 1934	175	10.8		11.8		7382		
9 27.3	+6 14	Σ 1355		Hya	1832 1957	328 345	2.8 2.5	7.5	7.5	F5	7380		
9 27.6	−8 40	H VI 111	AB	Hya	1841 1899	153	283.1	2.0	9.5	K2			30 α Hya
--			AC		1783	f	210.0						
9 27.6	−35 00	B 2215		Ant	1938 1956	116 200	0.1 0.1	7.4	7.5	A3			Last p.a. is uncertain
9 27.8	−6 04	B 2530		Hya	1938 1960	319 153	0.2 0.4	5.8	6.6	G0			Quadrant reversal 1951?
9 27.8	−42 35	h 4212		Vel	1898 1915	62	21.0	7.2	13.0	A			
9 27.9	−57 20	I 830		Car	1893 1911	58	17.9	7.5	9.5	M			
9 28.3	−8 05	β 213		Hya	1875 1936	177	1.6	8.1	10.6	A0	7392		
9 28.3	−9 59	Σ 1357		Hya	1831 1944	54	7.5	6.9	10.4	K0	7393		
9 28.5	+8 11	H IV 47		Leo	1852 1958	82 80	25.7 25.2	5.7	10.4	K0	7391		3 Leo
9 28.5	+9 03	Σ 1356		Leo	1825 1960			5.9	6.5	G0	7390	o	2 ω Leo
9 28.6	−45 30	See 112	AB	Vel	1897 1933	270	9.3	7.8	12.2	B5			
--		Δ 76	AC		1826 1913	98	60.7		7.8	B8			
9 28.7	+45 36	S 598	AB	UMa	1824 1924	162 162	86.6 77.3	5.5	8.0	G5 F8			
--			AC		1879 1923	80	83.5		9.7				
9 28.7	−20 48	I 495		Hya	1911 1959	135 139	0.9 0.8	8.3	9.5	B9	7395		
9 28.8	−43 48	B 2216		Vel		sf	10.0	8.0	13.5	F8			
9 29.0	−1 15	A 1763		Hya	1908 1949	106 115	1.6 1.3	6.3	11.3	A5	7396		
9 29.0	+19 17	Cou 936		Leo	1973	239	0.5	8.2	9.1	F5			
9 29.0	−20 26	B 1124		Hya	1928 1948	254 254	2.9 2.1	8.0	13.3	F0			
9 29.1	−1 32	Bail 518		Hya	1897	40	7.5	8.3	9.2				
				Hya	1821 1935	3	65.7	4.9	7.9	F5			31 τ¹ Hya
				Ant	1934 1944	73	6.7	7.5	13.8	M			
				Vel	1837 1913	77	108.4	7.2	7.3	F5 F5			
				Hya	1878 1942	32	0.7	7.9	8.7	F5	7400		Spectrum composite F5+A3
				Ant	1897 1928	178 135	4.1 11.6	5.5	14.1	K3	7405		
				UMa	1909 1957	35 27	0.8 1.3	8.6	8.6	F5	7398		
				Car	1928 1934	244	1.2	8.0	11.3	A2			
				Car	1930	98	2.3	5.9	13.6	M			
				Leo	1830 1955	242	14.2	8.3	8.6	G5	7406		
					1910 1928	70 71	82.0 85.0		12.3		7406		
					1910	57	157.9		11.2		7406		
9 30.7	+33 39	H 1100		LMi	1876 1914	130	62.8	6.0	9.3	K0			7 LMi
--			AC		1912	213	97.8		9.7				
9 30.7	−25 37	h 2498		Ant	1898	33	16.4	7.7	13.0	A3			
9 30.7	−40 28	Copeland		Vel	1897 1956			4.1	4.6	A7		o	ψ Vel
9 30.8	−31 53	Δ 78		Ant	1836 1952	210 212	8.0 8.0	6.2	7.1	A0			ζ¹ Ant
9 30.9	+44 41	Σ 1358		UMa	1831 1920	153 164	24.4 23.5	7.9	9.4	M	7403		
9 31.2	+67 32	Σ 1349		UMa	1831 1924	165	19.2	7.5	8.7	A2	7399		
9 31.4	−73 45	I 201		Car	1901 1936	152 144	1.9 1.9	8.0	10.2	F0			
9 31.5	+1 28	Σ 1365		Hya	1830 1968	163 158	3.1 3.4	7.4	8.4	F8	7412		
9 31.5	+63 04	Σ 1351	AB	UMa	1830 1975	270	22.7	3.7	8.9	F0	7402		23 UMa
--			AC		1886 1957	227 231	90.6 99.6		10.4		7402		
9 31.5	−35 43	See 114	AB	Ant	1897	283	23.6	5.9	14.1	K0			
--			BC		1897	233	4.1		15				
9 31.8	−72 29	B 1994		Car	1934	217	4.6	7.7	14.2	B3			
9 32.0	+9 43	S,h 107		Leo	1781 1921	77 75	35.1 37.4	5.2	8.2	K0	7416		6 Leo
9 32.1	−61 14	I 200		Car	1901 1938	355	1.1	7.2	9.6	A0			
9 32.2	+21 51	β 909		Leo	1879 1924	91	6.0	7.5	12.3	K0	7419		
9 32.2	−10 51	Σ 1367		Hya	1829 1937	183	5.4	7.9	9.4	G0	7423		
9 32.3	−40 39	Rst 1435		Vel	1935 1945	83	1.0	5.4	9.0	G5			
9 32.5	−57 57	Rst 411		Car	1930	8	0.9	8.1	11.3	B9			
9 32.7	+1 52	φ 349		Hya	1959	73	0.1	6.9	6.9	F5			
9 32.8	−57 06	SyO		Car	1895 1937	19	10.8	7.1	10.6	B5			
9 32.9	+51 41	β 1071		UMa	1889 1958	75 101	5.1 4.1	3.2	13.9	F8	7420		25 θ UMa
9 32.9	−14 00	β 910	AB	Hya	1879 1942	305	6.8	7.2	9.7	G5	7427		
--			AC		1913	279	166.5		8.7		7427		
9 33.1	−39 08	Jc		Ant	1847	204	55.2	6.4	9.9	F2 G0			
9 33.1	−86 01	Δ 82		Oct	1837 1940	267 271	15.7 15.7	7.4	8.0	F2			
9 33.2	−36 24	h 4218		Ant	1879 1942	27 30	5.9 5.9	7.6	10.5	A0			
9 33.3	+28 22	H N 29		Leo	1840 1958	257 260	35.0 30.5	6.5	10.5	A2	7426		

α 2000	δ 2000	Star Name		Const	Years	Pos Ang	Separation	m₁	m₂	Spec	ADS	Orb	Notes
9ʰ33.3	−57°58′	R 123		Car	1873 1947	34° 0°	1″.9 ″	7.8	7.9	B9			The f * of sm. triangle (R)
9 33.6	−49 45	Δ 79		Vel	1826 1913	32	135.4	7.1	7.5	G5 G5			
9 33.7	−49 00	h 4220		Vel	1836 1955	200 214	2.6 2.0	5.6	6.2	B3			
9 34.3	+66 48	Σ 1350	AB	UMa	1831 1958	247	10.4	8.3	8.4	F5	7425		
—			AC		1900 1921	214	130.4		9.2	G	7425		
—			BC		1833 1957	210 212	121.4 120.9				7425		
9 35.2	+14 05	OΣΣ 102		Leo	1875 1920	41 43	50.1 48.2	8.3	9.7	B9			
9 35.2	+60 54	Σ 1363		UMa	1832 1904	354	10.7	7.2	10.9	F0	7432		
9 35.3	+9 08	Che		Leo	1911	177	9.2	8.0	10.0				Not in BD
9 35.4	+3 54	Σ 1371	AB	Hya	1831 1928	278	7.5	8.0	10.5	G0	7444		
—			AC		1910 1922	270	122.9		11.7		7444		
9 35.4	+39 57	Σ 1369	AB	Lyn	1831 1956	148	24.7	7.0	8.0	F2	7438		
—			AC		1824 1956	325	117.8		8.7	G0	7438		
9 35.4	−38 54	B 2222		Ant		250	6.0	7.9	13.0	K2			
9 35.6	−19 35	S 604		Hya	1825 1915	91	51.5	6.3	9.6	A0			
9 35.7	+35 49	Hu 1128		LMi	1905 1937	35 31	5.8 2.0	5.4	13.9	K0	7441		11 LMi [54]
9 35.7	−68 13	I 359		Car	1901 1945	177 169	1.6 1.6	7.4	10.9	A0			
9 35.9	+14 23	H V 58		Leo	1783 1946	80	41.2	6.2	10.0	A0	7448		7 Leo
9 36.0	−27 31	B 185		Ant	1926 1955	205	3.6	7.6	10.6	A0	7452		
9 36.1	−31 14	h 4224		Ant	1848 1938	117	7.5	8.3	8.8	G A3			
9 36.4	−48 45	R 125		Vel	1880 1948	165 179	3.2 3.2	6.3	8.7	F0			
9 36.5	−11 16	Rst 3648		Hya	1937 1943	110	2.1	7.5	13.7	A0			
9 36.6	−24 42	ø 383		Ant	1964	166	0.1	7.3	7.3	F2			
9 36.8	−49 21	Rst 5335		Vel	1944 1949	323 323	26.6 25.7	4.4	13.1	A5			
9 37.1	+16 14	Σ 1372		Leo	1829 1959	53 61	0.5 0.5	8.5	8.6	F8	7456		
9 37.2	−53 40	See 115		Vel	1902 1949	168 180	0.5 0.5	6.3	6.3	A2			
9 37.5	−49 59	Hu 1465	AB	Vel	1926 1933	194	4.5	7.0	13.5	G0			
—			AC			s	13.2		11.0				
9 37.7	+1 42	A 2557		Hya	1913 1957	60	1.3	7.0	12.0	G5	7462		
9 37.9	−0 19	A 2558		Hya	1913 1957	322	1.5	8.0	11.8	A2	7464		
9 37.9	+45 54	A 1765		UMa	1908 1958	304 182	0.2 0.1	8.7	8.7	F5	7457		Round or single 1934−51
9 37.9	+73 05	Σ 1362		Dra	1836 1955	129	4.9	7.2	7.2	F0	7446		
9 38.1	−48 46	I 1187		Vel	1928	181	3.5	7.3	12.7	F0			
9 38.4	+7 43	A 2759		Leo	1914 1957	152	1.8	7.6	12.9	A2	7467		
9 38.4	−57 32	h 4232		Car	1871 1951	302	10.9	7.8	8.2	A0			
9 38.5	−62 33	I 203		Car	1902 1947	311	0.3	7.9	9.0	B9			
9 38.7	−39 37	I 202		Ant	1902 1960	176 148	1.3 0.6	6.8	8.8	F5			
9 38.8	+10 47	OΣ 204		Leo	1846 1940	102	8.4	7.6	11.6	A3	7471		
9 38.8	−50 30	Hu 1467		Vel	1926 1932	310	6.3	8.0	12.5	A2			
9 38.9	+69 17	A 1084		UMa	1905 1921	247	3.3	8.0	15	F0	7459		
9 39.7	−41 39	h 4231	AxBC	Vel	1920 1930	211	7.7	8.5		F8			
—		B 1133	BC		1928 1930	285	0.5	9.6	10.0				
9 39.8	−50 08	Rst 4917		Vel	1943 1949	8 27	0.2 0.2	7.9	8.4	F0			
9 39.8	−59 17	Gli 112	AxBC	Car		352	53.1	7.5		A5			
—		I 321	BC		1901 1933	29	2.2	10.8	10.8	A			
9 40.7	−37 10	Rst 2646		Ant	1935 1945	21	4.7	7.8	13.0	F8			
9 40.7	−57 59	B 780		Car	1927 1959			6.0	6.1	A2		o	
9 41.2	+9 54	H VI 76		Leo	1783 1919	40 44	63.5 85.4	3.5	9.5	F5	7480		14 o Leo
9 41.2	−36 02	B 2223		Ant		88	11.0	7.3	13.0	F8			
9 41.3	−23 36	H N 20		Hya	1904	292	54.7	4.8	10.0	B3			
9 41.4	+38 57	Σ 1374		LMi	1838 1976	275 301	3.3 2.9	7.3	8.6	G5	7477		Yellowish, v. bl. (Webb)
9 41.5	−18 29	β 214		Hya	1875 1959	261 250	3.1 3.2	7.6	11.4	F8	7488		
9 41.6	+31 24	Kui 42		Leo	1958	335	28.4	6.1	13.6	K5			Error in position?
9 42.1	−23 43	Rst 2648		Hya	1935 1946	136	1.2	8.0	12.9	F0			
9 42.2	−66 55	h 4241		Car	1918	304	34.2	6.6	10.6	A0			
9 42.3	+2 19	A 2559		Sex	1910 1929	256	3.9	7.9	12.8	K2	7493		
9 42.3	+78 50	h 1168		Dra	1908	52	23.0	8.0	12.6	A5			
9 42.7	−55 50	R 129		Vel	1873 1959	290 294	4.1 3.2	7.4	7.6	G5			
9 42.8	−22 58	Don		Hya	1930	284	15.7	8.0	12.0	F2			
9 43.2	+67 08	Milb 405	AB	UMa	1925	107	6.3	10.2	12.7		7487		Magnitude of A incorrect?
—		Mlr 323	Aa		1972	59	0.2	8.4	8.9	A5	7487		
9 43.3	−60 02	h 4240		Car	1882 1917	57	12.4	7.7	9.7	B8			
9 43.4	−16 20	h 821		Hya	1909 1941	353	12.7	8.1	11.1	K0	7501		
9 43.5	+2 38	Σ 1377		Sex	1830 1944	142 136	3.3 3.9	7.4	10.6	F8	7500		
9 43.5	−51 14	B 1658		Vel	1929 1944	200	2.1	6.2	10.9	B8			
9 43.7	+14 01	β pm		Leo	1880 1909	140 138	280.3 281.8	5.4	10.7	M			16 ψ Leo
9 43.8	+55 57	Es 1825		UMa	1920	50	7.0	7.9	13.1	G0	7494		W UMa; A is variable
9 44.2	−27 46	ø 326		Ant	1952 1959	71 5	0.1 0.1	5.4	5.6	F7		o	θ Ant
9 44.5	+18 52	h 469		Leo	1898 1904	252	30.5	6.6	12.6	K0			
9 44.8	−27 38	B		Ant		p	15.0	7.0	12.0	F0			
9 44.9	−50 01	B 2226		Vel		np	6.0	7.5	12.6	G5			
9 44.9	−57 34	ø 145		Car	1929	290	3.3	7.0	13.2	G5			
9 45.1	−49 30	Δ 80		Vel	1836 1919	70	18.6	8.4	8.4	F8 F8			
9 45.1	−69 46	Don 356	AB	Car	1932	140	3.1	7.2	10.2	A0			
—		HdO 208	AC		1932	46	13.3		15				

α 2000	δ 2000	Star Name		Const	Years		Pos Ang		Separation		m₁	m₂	Spec	ADS	Orb	Notes
9ʰ45ᵐ.2	−59°29′	Hrg 28		Car	1902	1934	285°		2″.1		7.5	10.0	G5			
9 45.3	+8 53	Σ 1379		Leo	1830	1927	173		9.6		7.9	11.6	F5	7508		
9 45.7	−41 40	h 4242		Vel	1882	1920	358		7.7		7.8	9.8	B9			
9 45.7	−70 07	I 204	AB	Car	1902	1911	126		1.7		7.2	9.7	B8			
—			AC		1901		320		3.7			11.7				
9 45.9	+63 39	Kui 43		UMa	1935	1958	27		6.3		6.3	11.5	F2			28 UMa
9 46.0	−71 47	B 1997		Car	1932		332		5.0		8.0	12.8	K2			
9 46.1	−45 55	h 4245		Vel	1897	1939	216		9.4		6.8	9.4	G5			
9 46.8	−39 19	I 1519	AB	Ant	1927	1956	354	6	0.3	0.2	8.1	8.4	F0			
—			ABxC				f		8.0			13				
9 47.1	−65 04	Rmk 11		Car	1836	1943	127		5.0		3.1	6.1	F0			υ Car
9 47.5	−37 43	I 172		Ant	1901	1959	312	321	1.0	1.0	7.7	10.2	F8			
9 47.6	−38 11	h 4246	AB	Ant	1920		142		32.4		6.7	13.8	K2			
—			AC		1920		24		35.0			13.9				
9 47.7	+20 36	Cou 284		Leo	1968		72		0.2		7.7	7.7	F0			
9 47.8	+10 04	Ho 253		Leo	1887	1951	292		1.1		7.6	12.6	G5	7517		
9 48.2	−54 24	ø 33		Vel	1928		191		8.6		7.3	13.2	K0			
9 48.3	−25 12	B		Ant			np		8.0		8.0	15	A2			
9 48.5	−37 38	Jsp 351		Ant	1929	1945	71		2.3		7.0	12.5	K2			
9 48.7	−26 25	I 205		Ant	1898	1959	22	337	2.3	1.7	6.8	10.8	F0	7524		
9 48.8	−35 01	h 4249		Ant	1836	1952	126	123	4.3	4.3	8.0	8.1	A3			
9 48.8	−52 37	B 1663		Vel	1929	1943	277		1.4		7.9	12.0	K0			
9 49.0	+34 05	Σ 1382 rej	AB	LMi	1903	1926	106		27.8		7.1	11.1	A3	7520		
—			AC		1903	1926	255		32.7			11.8		7520		
9 49.1	−43 29	I 1520		Vel	1927	1942	250		3.1		6.5	10.9	K0			
9 49.2	−77 01	Rst 1459		Cha	1932		38		1.7		8.0	14.0	K0			
9 49.8	+21 11	Kui 44		Leo	1935	1960	214		0.3		6.7	7.0	F0			20 Leo
9 49.9	+16 50	OΣ 207 rej		Leo	1867	1910	322	322	19.0	20.4	8.0	11.1	K2			
9 49.9	−68 25	Don 367		Car	1932		92		5.7		6.8	14.7	G5			
9 50.0	−45 44	h 4254		Vel	1913		46		66.3		5.1	9.6	B8 B9			
9 50.2	−49 37	CorO 92		Vel	1920	1955	22		5.8		8.1	9.1	F8			
9 50.8	+11 51	J 389		Leo	1911	1957	142	145	20.3	24.5	6.7	13.2	A0	7539		
9 50.9	+19 19	OΣΣ 103	AB	Leo	1875	1914	125		78.2		8.5	9.0	A3			
—			AC		1908		178		95.7			9.9				
—			BC		1908		230		79.4							
9 51.0	+59 02	OΣ 521		UMa	1855	1934	295		11.3		3.8	11.4	F0	7534		29 υ UMa
9 51.2	+36 29	Ho 369	AB	LMi	1891	1960	98	281	0.3	0.3	8.5	8.6	F2	7541		P.a.'s are scattered
—			AC		1891		101		61.7			12.6		7541		
9 52.1	+54 04	OΣ 208		UMa	1843	1960					5.3	5.4	A2	7545	o	30 ø UMa
9 52.3	+40 58	A 2139		LMi	1910	1956	282	291	1.6	1.5	8.0	10.5	A3	7546		
9 52.5	−8 06	AC 5	AB	Sex	1854	1960					5.6	6.1	A2	7555	o	8 γ Sex
—		h 4256	AC		1880	1940	325		35.8			12.0		7555		
9 53.0	−27 20	Rst 5341		Ant	1944	1959	5		1.2		6.3	10.3	F8			
9 53.2	−23 41	B		Hya			sf		10.0		7.7	14.0	K0			
9 53.3	+50 37	OΣ 209		UMa	1846	1934	307		4.8		7.5	10.5	G5	7554		
9 53.5	−19 29	h 4261		Hya	1916	1941	84		8.2		7.8	9.6	G5	7567		
9 53.9	+64 47	OΣ 522		UMa	1851	1915	122	124	15.0	14.2	7.5	11.2	K0	7556		
9 54.1	+4 57	S 605		Sex	1825	1924	290		52.5		7.0	9.2	K5 K0			
9 54.1	−28 00	β 215		Ant	1877	1943	343		1.7		7.0	9.0	B9	7570		
9 54.3	−45 17	Δ 81		Vel	1836	1952	240		5.4		5.8	7.9	B5			
9 54.5	−12 55	h 4262	AB	Hya	1904	1941	103		7.7		8.7	10.3	A0	7571		
—			AC		1904		129		129.3			7.0	F8	7571		
9 54.5	−34 54	I 842		Ant	1911	1943	31		3.9		7.5	11.3	A2			
9 55.0	−16 12	β 592		Hya	1879	1947	192		9.9		6.7	12.1	G0	7576		
9 55.1	+61 07	Kui 45		UMa	1958		273		12.2		6.3	13.1	K0			
9 55.1	−69 11	Rmk 12		Car	1879	1918	213		9.2		7.0	8.8	B9			
9 56.3	−45 15	Don 372		Vel	1929	1947	217		5.0		7.7	13.9	K0			
9 56.3	−46 23	B 1137		Vel	1929		214		1.5		8.0	12.8	K0			
9 56.7	−55 05	ø 150		Vel	1929		258		0.2		8.6	8.7	A3			
9 56.8	−26 33	β 216		Hya	1877	1957	163	156	3.4	3.4	6.3	10.7	A2	7591		
9 56.9	−54 34	I 396		Vel	1901		8		37.2		3.5	11.8	B7			ø Vel
9 56.9	−63 23	ø 151		Car	1931	1938	345		0.9		8.2	10.2	B9			
9 57.0	+19 46	Σ 1399		Leo	1828	1958	175		30.3		7.7	9.6	G0	7589		
9 57.0	−34 50	Jsp 360		Ant	1929	1945	193		10.3		7.5	13.4	F5			
9 57.2	−49 52	Hu 1472		Vel	1926	1932	8		1.6		7.2	12.2	B8			
9 57.7	−1 57	A 1767		Sex	1908	1933	13	19	1.6	1.8	6.7	10.3	G5	7596		
9 57.7	−48 25	h 4269		Vel	1897	1933	321		14.1		6.1	10.7	B5			
9 57.9	−60 45	ø 152		Car	1931	1943	298		0.4		8.1	9.4	G5			
9 58.1	+20 59	Cou 286		Leo	1968		183		1.0		7.7	10.9	F5			
9 58.2	−67 01	Don 379		Car	1929		268		1.6		8.0	13.2	A0			
9 58.2	−72 04	B		Car			195		8.0		7.5	12.5	B8			
9 58.7	+10 58	H V 63		Leo	1899	1921	356		50.7		7.4	9.5	F2			
9 58.9	−35 53	h 4271		Ant	1898	1920	318		31.0		5.2	11.2	F0			η Ant
9 59.0	−62 19	Jsp 362		Car	1930		233		1.2		7.8	12.5	A5			
9 59.4	−44 57	h 4273		Vel	1898	1929	135		15.5		6.9	10.8	A0			

α 2000	δ 2000	Star Name		Const	Years	Pos Ang		Separation		m₁	m₂	Spec	ADS	Orb	Notes
10ʰ00ᵐ.2	+6°15'	Σ 1401		Sex	1829 1958	21°	0	23″.9	″	7.8	10.8	F5	7604		
10 00.6	−56 57	ø 153		Vel	1931	256		7.6		6.5	13.0	G5			
10 01.0	+31 55	β pm		LMi						5.5	8.7	G5			20 LMi
10 01.2	−56 06	HdO 209		Vel		55		15.0		6.4	9.9	B8			
10 01.3	−22 46	HdO 124	AB	Hya	1898 1927	8		13.2		7.6	10.3	A0	7610		
−−			AC		1898 1928	50		42.0			11.9		7610		
10 01.3	−70 44	I 291		Car	1901 1938	322 302		1.1	1.2	7.3	10.0	F2			
10 01.5	+68 43	Σ 1398		UMa	1832 1960	229 125		3.7	2.2	7.9	11.1	F0	7603		
10 01.9	−28 41	h 4277		Ant	1886 1915	32		21.9		8.5	8.9	A0 A0	7614		
10 02.1	−15 02	Hu 1253		Hya	1906 1945	96 84		0.4	0.5	7.6	9.2	A0	7615		
10 02.1	−54 59	Δ 83		Vel	1837 1938	223 225		109.5 111.2		7.7	7.9	K0 B8			
10 02.8	+49 53	h 2515		UMa	1906	10		51.7		7.5	12.0	K0			
10 02.9	+68 47	Σ 1400		UMa	1832 1958	228 228		1.8	3.0	8.0	11.2	F5	7611		Easy to confuse w. Σ 1398
10 03.2	+50 07	β		UMa	1906	231		28.2		7.0	12.6	K2			
10 03.2	−52 03	h 4282		Vel	1836 1917	199		47.7		7.7	8.8	A2 A0			
10 03.6	−61 53	Hrg 47		Car	1883 1947	350		1.2		6.3	7.8	B8			
10 03.7	−25 19	Ho 370		Hya	1892 1959	330 322		15.5	16.1	6.7	12.2	A0	7625		
10 03.9	+38 01	h 2517		LMi	1902 1926	153		44.3		6.8	11.6	F5	7621		
10 04.0	+32 39	Hu 631		LMi	1903 1960	272 263		0.8	0.8	7.8	9.4	F8	7624		
10 04.0	−18 06	β 1072	AB	Hya	1889 1959	43 48		10.9	11.9	5.9	12.2	A0	7627		
−−		S,h 110	AC		1822 1935	273		21.2			8.1	A0	7627		
10 04.3	−28 23	I 292		Ant	1883 1960	217 142		1.2	0.3	8.0	8.2	F8	7629		
10 04.5	−51 48	h 4283		Vel	1836 1941	181		8.0		7.5	9.2	B9			A very fine double star (h)
10 04.9	+55 29	Σ 1402		UMa	1831 1924	96 102		21.1	27.2	8.1	9.6	K5 G0			
10 05.0	−51 19	Hu 1594		Vel	1915 1960	260 278		0.2	0.4	7.2	7.5	B9			
10 05.1	−45 54	h 4284		Vel	1897 1935	66		6.6		7.3	9.3	K0			
10 05.1	−61 41	Gtb		Car	1946	60		1.0		7.6		K5			
10 05.2	−28 12	I 293		Ant	1899 1960	324 324		0.5	0.3	7.2	8.5	A0	7635		
10 05.6	−84 05	h 4310		Oct	1871 1941	272 269		3.9	3.9	7.8	8.5	G5			
10 05.7	+31 05	Σ 1406		LMi	1830 1956	227		1.0		8.4	9.1	A3	7632		
10 05.7	+41 03	A 2142		LMi	1910 1960	308 304		0.6	0.9	8.0	8.9	F0	7631		
10 05.8	−30 53	Ho 371		Ant	1891 1938	43		6.2		6.6	12.1	K0			
10 05.9	+39 35	Σ 1405 rej		LMi	1904 1926	251		21.9		7.3	10.5		7633		
10 06.2	−47 22	I 173		Vel	1901 1959					5.3	6.9	G6		o	
10 06.7	−23 08	h 4285		Hya	1898 1959	358		8.7		7.7	9.9	F5	7642		
10 06.8	−24 43	β 217		Hya	1868 1959	93 121		1.9	1.9	7.9	7.9	F8	7644		
10 07.1	−52 32	Rst 467		Vel	1929 1948	61 69		7.9	9.4	7.9	12.8	F2			
10 07.1	−54 34	I 499		Vel	1911 1959	305 301		1.7	1.7	7.7	9.8	B9			
10 07.3	+16 46	Wils 18		Leo	1937 1940	88		0.1		3.5		A0			30 η Leo; <0″.1 in 1954
10 07.4	−19 43	β 218		Hya	1875 1959	123 132		1.0	0.6	7.8	8.3	A0	7647		
10 07.9	+10 00	GAn 5		Leo	1878 1934	44		7.9		4.4	13.4	K2	7649		31 Leo
10 07.9	−71 16	B 2231		Car		335		8.0		8.0	13.0	K2			
10 08.3	+31 36	Kui 48	AB	LMi	1958	174		0.2		6.8	7.1	F5	7651		
−−		h 475	ABxC		1903 1958	172		27.6			13.0		7651		
−−		Cou	Aa		1973								7651		
10 08.4	+11 58	Σ II 6	AB	Leo	1836 1924	307		176.9		1.4	7.7	B8 G	7654		32 α Leo, Regulus
−−		HdO 127	AD		1924	274		217.0					7654		
−−			BC		1867 1943	93 86		3.9	2.5		13.2		7654		
10 08.4	−19 45	β 911	AB	Hya	1880 1959	312		4.6		7.3	11.0	G0	7655		
−−			AC		1880 1959	83 62		47.3	66.4		9.3		7655		
10 08.7	−65 49	h 4292		Car	1894 1919	123		60.2		5.3	9.5	G7			
10 09.1	−15 30	Rst 3682		Hya	1937 1943	63		2.4		7.5	12.7	G5			
10 09.2	+21 47	h 2520		Leo	1892 1925	343		34.0		8.0	11.9	A0			
10 09.2	−23 34	B 784		Hya	1927 1959	191		0.5		8.2	9.8	G5			
10 09.3	+20 20	A 2145		Leo	1910 1960					7.4	7.4	F5 A2	7662	o	
10 09.5	−35 51	Rst 2678		Ant	1935 1945	272		2.0		6.1	10.8	G0			
10 09.5	−68 41	I 13	AB	Car	1900 1947	148 131		1.0	0.7	6.5	6.6	A0			
−−		h 4295	ABxC		1901 1933	40		26.0			11.2				
10 09.7	+3 10	Σ 1412 rej	AB	Sex	1905 1926	295		30.6		8.7	10.7	K0			
−−			AC		1875 1926	48 50		98.7	95.2		8.7	G5			
10 10.1	−45 47	h 4290		Vel	1913 1915	312		15.2		8.0	11.0	F2			
10 10.6	+23 53	A 1989		Leo	1909 1935	286		3.3		7.9	14.4	G5	7669		
10 10.6	−12 21	β 593	AB	Hya	1878 1922	118 115		50.8	56.4	3.6	13.4	K0	7671		41 λ Hya
−−			AC		1921	167		112.2			11.3		7671		
10 10.6	−36 36	Howe 67		Ant	1885 1952	129		3.9		8.6	9.0	A2			
10 11.4	+49 27	β pm	AB	UMa	1894 1918	196 187		183.3 165.4		6.8	9.0	K5			
−−			BC		1877 1908	254		303.5			10.1				
10 11.6	+13 21	Hu 874		Leo	1905 1960	289 286		0.2	0.2	6.8	7.6	F5	7674		34 Leo; too close 1949−56
10 12.0	+20 07	h 476		Leo	1843 1925	49		24.2		8.0	10.3	G5			
10 12.0	−28 36	B 194		Ant	1926 1959	18 45		0.3	0.2	7.0	7.1	A0	7681		
10 12.1	−6 13	Ho 44		Sex	1884 1959	191 212		0.4	0.3	8.5	8.7	F2	7675		
10 12.6	+60 32	Hu 1129		UMa	1905 1920	308		0.6		7.9	12.9	F8	7676		
10 13.0	−47 29	I 361		Vel	1901 1959	135 127		4.6	5.3	7.5	10.1	K0			
10 13.0	−65 10	Slr 17	AB	Car	1895 1944	341		3.8		7.1	11.2	B9			
−−			BC		1931	287		4.5			12.8				

α 2000	δ 2000	Star Name		Const	Years	Pos Ang	Separation	m₁	m₂	Spec	ADS	Orb	Notes
10ʰ 13.3	−50° 37′	CapO		Vel		p	8.0	7.9	12.9	K2			
10 13.4	−50 55	h 4299		Vel	1917	327	31.2	8.4	9.0	K A2			
10 13.5	−51 45	Rst 5517		Vel	1949	241	0.3	6.0	7.5	A2			
10 13.8	−40 21	Rst 1488		Ant	1935 1945	104	4.7	6.0	13.0	K0			
10 13.8	−65 43	h 4301		Car	1879 1918	25	6.8	8.6	8.9	A0			
10 13.9	−40 19	Rst 1489		Ant	1935 1945	215	12.7	6.3	13.3	K0			
10 15.1	−55 33	See 118	AB	Vel	1897 1934	148 142	13.0 15.2	6.7	12.5	K0			
--		I 850	CD		1934	8	5.8	12.0	12.5				
--			EF		1934	142	7.2	11.5	11.6				
10 15.1	−67 17	CapO 48		Car	1880 1926	333	2.1	7.9	9.6	G5			
10 15.2	+19 07	Σ 1417		Leo	1830 1942	259	2.4	8.8	8.8	F2	7695		
10 15.6	−35 29	Jsp 384		Ant	1929 1941	319	9.2	7.8	14.8	K2			
10 16.1	−59 54	Hu 1597		Car	1914 1960	116 260	0.3 0.4	6.9	7.0	A2		o	Direct motion
10 16.3	+17 44	OΣ 215		Leo	1844 1960			7.2	7.5	F0	7704	o	[55]
10 16.3	−28 59	I 851		Ant	1911 1960	233 286	0.5 0.2	8.4	8.5	F5	7706		Retrograde motion
10 16.7	+23 25	Σ I 18		Leo	1836 1925	343 340	314.4 325.9	3.5	5.8	F0 G0			36 ζ Leo
10 16.8	−20 40	H1d 101		Hya	1888 1959	116 110	1.4 1.4	6.6	10.1	F5	7711		
10 16.9	+25 52	β pm		Leo	1910	261	852.4	7.6	9.3	G0			
10 17.1	−61 20	B 1673	AB	Car	1931	164	16.4	3.4	12.6	K5			
--			AC		1932	6	26.0		12.7				
10 17.2	+23 06	OΣ 523		Leo	1851 1972	299	7.4	5.8	11.4	F5	7712		39 Leo
10 17.2	−66 35	Gli 137		Car	1930 1944	277	3.1	8.5	9.1	B9			
10 17.9	+71 03	Σ 1415	AB	UMa	1832 1968	167	16.7	6.7	7.3	A3 A3	7705		
--			AC		1879 1956	10 14	148.6 150.1		10.6		7705		
10 18.1	+27 31	Σ 1421		Leo	1830 1953	330	4.4	8.3	9.3	F2	7715		
10 18.1	−40 14	CorO 100		Ant	1903 1934	323	6.0	8.3	9.4	A0			
10 18.2	−50 49	Hu 1598		Vel	1915 1959	244 237	0.5 0.4	8.5	9.1	A3			
10 18.4	+37 31	Hu 875		LMi	1905 1954	73	1.0	8.0	10.8	A2	7717		
10 18.6	−49 55	B 2235		Vel		f	10.0	8.0	14.0	K5			
10 18.9	+44 03	β pm		UMa	1893 1918	105 103	152.4 150.6	6.7	9.0	G5			
10 19.0	+46 46	Kui 49		UMa	1958	57	27.8	6.5	12.2	K0			
10 19.0	−56 01	R 140		Vel	1881 1954	279 281	3.2 3.2	7.8	8.3	A2			
10 19.1	−64 41	h 4306		Car	1836 1941	139 134	2.1 2.1	7.0	7.0	A0			
10 19.3	−12 32	Rst 3688		Hya	1937 1959	307 314	0.7 0.5	6.4	7.7	F0			
10 19.3	−61 06	B 2237	AB	Car		n	10.0	8.0	12.0	B8			
--			BC			340	2.0		12.5				
10 19.9	−51 34	h 4307	AB	Vel	1903 1940	263 262	14.3 13.5	7.0	10.5				
--		Rst 5518	Aa		1948	40	0.8		10.0	F2			
--		I 852	BC		1911 1940	127	2.2		11.5				
10 20.0	+19 51	Σ 1424	AB	Leo	1831 1976			2.2	3.5	K0	7724	o	41 γ Leo, Algieba
--			AC		1904 1915	291 291	251.7 259.9		9.2		7724		
--			AD		1904	302	333.0		9.6		7724		
--			BC		1913	291	264.4				7724		
--			CD		1856 1911	321 334	111.5 98.3				7724		
10 20.1	−67 10	R 141		Car	1880 1947	44	1.9	7.5	8.4	B9			Fine object (R)
10 20.2	−33 08	h 4304		Ant	1898 1934	286	9.5	7.6	10.1	A2			
10 20.4	+11 21	h 159		Leo	1906 1909	10	42.1	8.1	10.6	K0			
10 20.4	−63 20	ø		Car		258	10.0	8.0	11.2	A0			
10 20.5	+6 26	Σ 1426	AB	Leo	1832 1957	257 300	0.6 1.0	8.4	8.9	F5	7730		
--			ABxC		1832 1939	9	7.6		9.3		7730		
--			ABxD		1876	45	34.4		15		7730		
10 20.6	−23 39	h 4305		Hya	1879 1921	216	17.5	7.5	9.8	K0	7735		
10 20.7	−53 40	ø		Vel			10.0	7.7		F5			
10 20.8	−42 52	Don 408		Vel	1930 1946	188	8.5	7.3	13.5	B9			
10 20.8	−67 31	h 4314		Car	1879 1918	12	19.0	8.6	8.8	A A			
10 20.9	−56 03	Rmk 13	AB	Vel	1836 1952	104 102	7.2 7.2	4.7	8.4	B5			
--			AC		1835 1931	190	36.8		9.5				
10 21.6	−22 32	β 219		Hya	1876 1960	187 187	2.1 1.8	6.7	8.2	A0	7739		
10 21.8	−9 46	β 25		Sex	1875 1959	180 153	1.8 1.8	8.2	8.8	G5	7738		
10 22.0	+43 54	Σ 1427		UMa	1829 1973	214	9.4	7.6	8.1	F5	7737		
10 22.4	−47 58	h 4312		Vel	1879 1913	265	25.2	7.4	10.5	A0			
10 22.7	+15 21	OΣ 216		Leo	1845 1960			7.5	9.6	G5	7744	o	
10 23.2	+5 42	S,h 115		Sex	1823 1924	330 353	60.4 58.6	6.6	9.1	F2			
10 23.3	−13 23	h 4311		Hya	1867 1944	122	4.2	6.6	9.7	F8	7749		
10 23.4	−50 21	B 2240		Vel		160	6.0	8.0	13.0	F5			
10 23.7	+2 38	β pm	AB	Sex	1842 1911	24 25	82.1 102.1	8.5	8.7	G5			
--			AC		1908	145	167.5		10.7				
10 23.8	−44 15	I 208	AB	Vel	1902 1960	28 23	0.8 0.7	7.7	8.9	A0			
--			·ABxC		1902 1960	22 24	24.2 25.4		14.0				
10 24.2	+2 22	β 1322	AB	Sex	1904 1954	326 312	7.8 10.2	6.3	12.6	K0	7755		
--		h 2530	AC		1800 1922	64	212.2		6.6	K0	7755		
10 24.4	+34 11	OΣΣ 104		LMi	1875 1928	286	207.8	7.8	8.3	M K0			
10 24.4	−38 35	I 209		Ant	1901 1954	137 131	0.9 1.2	8.4	8.6	K0			
10 24.4	−61 33	Hu 1599		Car	1914 1943	1	1.3	7.8	9.4	A0			
10 25.6	+8 47	Σ 1431		Leo	1832 1934	68	3.2	7.9	9.6	F0	7764		

α 2000	δ 2000	Star Name		Const	Years		Pos Ang		Separation		m₁	m₂	Spec	ADS	Orb	Notes
10ʰ26ᵐ.0	+2°56′	A 2570		Sex	1913	1958	330°	313°	0″.4	0″.4	7.6	7.6	A0	7769		
10 26.0	+52 37	Σ 1428		UMa	1831	1954	84	87	3.8	3.1	7.9	8.4	F5	7762		BDS 5386, H I 71 same?
10 26.2	+3 56	β 1280	AB	Sex	1899		192		116.3		6.8	8.7	A2	7773		
--			BC		1899	1924	19		0.9			11.3		7773		
10 26.3	-53 54	h 4319		Vel	1893	1934	123		12.1		7.3	11.8	K5			
10 26.7	-47 40	Rst 4928		Vel	1943	1949	314		0.5		6.7	8.7	A5			
10 26.9	+17 13	OΣ 217		Leo	1844	1956	145	167	0.4	0.4	7.9	8.4	F8	7775	o	
10 26.9	+19 31	Cou 292		Leo	1968		40		0.2		8.0	8.4	F8			
10 27.0	+29 40	Σ 1432		LMi	1829	1924	123		29.3		8.0	10.0	F2			
10 27.1	-68 31	M1bO		Car	1929		131		10.9		7.9	13.5	B9			
10 27.5	+3 34	OΣ 218		Sex	1844	1955	61	98	1.0	0.8	7.4	8.9	K0	7779		
10 27.6	+9 46	h 832		Leo	1904	1924	132	132	38.6	37.2	6.0	11.0	A0	7781		45 Leo
10 27.8	-38 42	I 210		Ant	1898	1936	237		0.9		7.6	10.1	F8			
10 27.9	+36 42	Hu 879		LMi	1904	1957					4.4	6.1	K0	7780	o	31 β LMi
10 28.1	+48 47	Kui 50		UMa	1935	1958	15		4.7		6.4	12.4	G0			
10 28.3	-52 34	CorO 103		Vel	1920	1938	286		5.5		8.4	8.5	G5			
10 29.0	-48 59	I 73	AB	Vel	1897	1938	221		5.1		8.2	10.2	B5			
--			AC		1932		31		24.4			12.5				
10 29.2	+10 09	OΣ 220		Leo	1853	1960	62	85	1.3	0.8	7.8	9.7	F8	7792		
10 29.4	+6 15	A 2766		Sex	1914	1929	319		1.1		7.8	12.9	K2	7794		
10 29.6	-30 36	H N 50		Ant	1848	1932	226		11.0		5.6	9.6	A0			δ Ant
10 29.9	+28 35	OΣΣ 105		LMi	1875	1918	225		131.3		7.3	8.3	K0			
10 30.2	+51 00	OΣ 219		UMa	1847	1925	297		13.4		7.7	11.0	F0	7796		BDS 5471, h 2535 same star?
10 30.3	+63 21	Es 1905	AB	UMa	1921		66		47.3		7.4	11.6	K0	7793		
--			BC		1921		200		4.0			13.4		7793		
10 30.6	+55 59	LDS 2863		UMa	1955		304		120.0		4.8	8.7	F5			36 UMa
10 30.7	-54 29	h 4327		Vel	1837	1938	172		114.0		7.8	8.6	K5 K0			
10 30.7	-61 21	Δ 87		Car	1835	1919	331		82.6		6.4	7.6	M B9			
10 31.0	-7 38	Σ 1441	AB	Sex	1830	1958	168		2.6		6.4	9.9	K5	7808		
--			AC		1909	1925	312		64.7					7808		
10 31.0	-44 22	Don 415		Vel	1931	1946	17		2.6		6.9	14.0	G5			
10 31.0	-61 37	I 174		Car	1902	1939	45		1.2		8.0	10.0	B9			
10 31.0	-73 13	h 4333		Car	1916	1917	101		32.0		4.9	12.5	M1			
10 31.1	-68 54	R 151		Car	1880	1913	192		3.4		7.9	9.7	A0			
10 31.2	-42 14	Δ 86	AB	Vel	1838	1920	291		83.3		7.5	8.1	B9 A2			
--			BC		1920		75		42.0			10.2				
10 31.4	-53 43	h 4329		Vel	1837	1938	17	94	17.5	46.4	4.9	8.4	F6 K5			
10 31.7	-44 29	B 1161		Vel	1929	1959	27	146	0.1	0.1	7.8	8.1	B3			Measures uncertain
10 31.9	+32 23	h 482		LMi	1903	1924	245	245	40.2	43.3	5.8	11.8	B9	7813		33 LMi
10 31.9	-45 04	Pz		Vel	1826	1951	218		13.5		6.2	6.5	B8			
10 31.9	-52 14	CapO 10		Vel	1881	1940	345		2.2		7.5	9.0	B8			
10 31.9	-72 07	Δ 91		Car	1826	1940	60		10.0		8.5	8.8	A0			
10 31.9	-81 55	h 5444		Cha	1919		235		41.9		7.1	9.6	B8			
10 32.0	+22 02	Σ 1442		Leo	1831	1953	155		13.4		8.0	8.6	F0	7817		
10 32.5	-6 04	β 1073		Sex	1889	1940	47		3.4		6.9	11.4	K0	7822		
10 32.8	+9 18	Wils 19		Leo	1937	1940	68	86	0.1	0.1	3.9		B0			47 ρ Leo
10 32.9	-39 56	φ 26		Ant	1926	1959	180		0.2		8.7	8.8	B9			
10 32.9	-47 00	h 4330		Vel	1836	1960	163		40.2		5.0	8.9	K4 A			
10 33.2	+40 26	h 2534		UMa	1879	1958	316	337	24.6	19.3	4.8	11.6	F5	7826		
10 33.3	-55 23	Δ 89	AB	Vel	1837	1938	30		26.0		6.6	8.4	G5 A0			
--		H1d 106	BC		1892	1954	250		1.5			8.8				
10 33.5	-46 59	h 4332		Vel	1836	1933	162		28.4		7.2	9.6	B9			
10 33.8	+23 21	Σ 1447		Leo	1830	1943	124		4.2		7.3	9.1	A2	7833		
10 34.0	-23 45	β 1269		Hya	1892	1946	61		19.1		5.1	13.8	K4	7834		44 Hya
10 34.1	+12 22	A.G.		Leo							7.7		A3			[56]
10 34.2	+73 50	Σ 1437		Dra	1832	1955	290		23.5		7.6	10.1	A3	7824		
10 34.3	-37 23	I 1202		Ant	1926	1934	131		3.7		7.2	11.0	B9			
10 34.3	-46 29	B 1164		Vel	1928	1942	81		0.6		8.4	9.4	K0			
10 34.4	+21 36	Hu 1338	AB	Leo	1910	1920	182		3.6		7.3	13.4	K0	7836		
--		Σ 1448	AC		1827	1931	259		11.0			9.0		7836		
10 34.7	-22 12	Rst 2698		Hya	1931	1959	26		1.9		7.7	10.5	K0			
10 34.9	-64 08	Δ 93	AB	Car	1826	1939	39		25.0		7.9	9.0	A0 A0			
--		I 74	BC		1902	1939	232		3.0			10.7				
10 35.0	+8 39	Σ 1450		Leo	1830	1971	157		2.4		5.8	8.5	A0	7837		49 Leo = TX Leo
10 35.1	-57 41	CorO 107		Car	1902	1934	239		5.0		7.1	9.5	K0			
10 35.2	-43 40	Don 421		Vel	1930	1948	176	167	13.8	14.7	6.1	14.5	G5			
10 35.5	+45 39	β 1074		UMa	1889	1926	206		2.5		7.2	11.2	K0	7838		
10 36.1	+11 37	Big		Leo	1880	1917	338		56.3		7.9	9.3	K2			
10 36.1	-26 40	β 411		Hya	1877	1960					6.7	7.5	F3	7846	o	
10 36.1	-16 21	β 1075		Hya	1889	1959	280		3.4		6.0	13.0	K5	7847		
10 36.3	-47 51	I 175		Vel	1902	1960	156		1.9		7.3	10.8	G0			
10 36.5	-12 14	Kui 51		Hya	1958		75		14.4		5.4	10.5	F8			
10 36.6	-28 46	I 857		Hya	1913	1954	303	284	0.4	0.3	7.5	7.9	A0	7852		
10 37.0	-8 50	A 556		Sex	1903	1960	54	102	1.3	1.3	7.1	10.3	G0	7854		
10 37.3	-48 14	See 119		Vel	1897	1960					4.2	5.1	F2		o	Spectrum composite F2+A3

α 2000	δ 2000	Star Name		Const	Years		Pos Ang		Separation		m₁	m₂	Spec	ADS	Orb	Notes
10ʰ38ᵐ3	+1°15′	Σ 1456		Sex	1833	1954	46°		13″.5		8.2	9.9	F5	7862		
10 38.3	+60 07	OΣ 222		UMa	1847	1957	340		4.4		7.0	11.0	F8	7855		
10 38.7	+5 44	Σ 1457		Sex	1829	1959	288	326	0.7	1.8	8.0	9.0	F2	7864		
10 38.8	−42 45	φ 338		Vel	1956	1959	45		0.1		6.8	6.9	F5			
10 38.8	−59 11	Δ 94		Car	1826	1932	21		14.5		4.7	8.1	M1 A			
10 38.9	−64 30	R 152		Car	1874	1947	7	16	2.8	2.1	7.2	9.0	B9			
10 39.0	−58 49	Gli 152		Car	1871	1959	70	78	18.7	23.5	6.0	8.8	M2 B			
10 39.3	−55 36	Δ 95	AB	Vel	1826	1938	105		51.9		4.4	6.6	G0 B8			
—		h 4341	BC		1913	1934	175		20.2			11.2				
10 39.4	−67 10	NZO 19		Car	1913		351		4.6		8.0	10.2	K0			
10 39.7	+8 51	OΣ 224		Leo	1843	1960					7.8	9.0	F5	7871	o	
10 40.2	+38 24	Σ 1459		LMi	1829	1969	153		5.2		8.5	9.0	K0	7873		
10 40.6	+12 05	h 167		Leo	1924		330		55.2		7.9	11.9	K2			
10 40.7	+42 09	Σ 1460		UMa	1830	1944	166		3.5		8.6	8.6	F2	7878		
10 40.8	+41 32	OΣ 226 rej		UMa	1878	1904	60		17.9		7.9	12.7	K0	7880		
10 40.9	−35 44	B 2001		Ant	1931	1960	63		0.7		6.4	8.9	G5			
10 41.2	−59 58	Jsp 416		Car	1929		216		5.3		7.3	13.3	B3			
10 41.6	−0 16	Σ 1464	AB	Sex	1831	1928	302		5.4		8.2	10.9	F0	7885		
—			AC		1903		222		59.2			10.0		7885		
10 41.7	+10 44	OΣ 227		Leo	1845	1958	326	353	0.5	0.8	8.0	9.0	A2	7888		
10 42.0	−63 30	I 860		Car	1911	1959	75		0.6		8.0	8.2	A5			
10 42.2	+31 42	β pm		LMi	1879	1909	173	173	119.1	116.8	6.0	10.0	M			
10 42.4	−47 13	Rst 2705		Vel	1935	1944	70		1.3		7.1	11.8	K0			
10 42.6	+3 35	A 2768		Sex	1914	1957	271	207	0.4	0.5	6.7	8.8	F5	7896	o	
10 42.7	−32 43	See 121		Ant	1897	1945	219	217	17.8	17.4	5.6	13.1	A0			
10 42.9	+50 48	Σ 1462	AB	UMa	1831	1934	173		8.3		7.4	9.3	A3	7894		
—			AC		1880	1909	58		194.5			9.5		7894		
10 43.0	+26 20	β 913	AB	LMi	1880	1958	123	112	10.9	18.4	5.5	12.5	A2	7899		40 LMi
—			AC		1898	1923	77	78	29.8	32.5		13.5		7899		
—			AD		1899	1919	276	278	55.8	54.1		13.0		7899		
10 43.2	−61 10	Δ 97		Car	1837	1950	174		12.4		6.8	8.7	B5			
10 43.3	+4 45	Σ 1466	AB	Sex	1832	1958	240		6.8		6.3	7.4	K0	7902		
—			AC		1907		210		336.3					7902		
10 43.5	+46 12	Smart	AB	UMa	1874	1925	88	88	288.1	286.9	5.3	8.1	F0 F8			
—			AC		1907		5		139.3							
—			BD		1907		170		273.1							
—			Ba		1907		270		120.0			12.9				
10 44.0	−59 33	h 4356	AB	Car	1836	1934	149		2.8		7.3	8.8	B			
—			AC				270		5.0			13.0				
—			BD				187		3.0			13.5				
—		See 123			1897	1934	306	311	3.7	4.5	9.6	12.2				
10 44.1	−48 54	h 4352	AB	Vel	1920		207		20.0		7.3	11.6	F5			
—			AC		1920		230		47.4			13.4				
10 44.1	−59 35	h 4360	AB	Car	1835	1939	115		2.0		9.0	9.1	A3			
—			AC		1836	1934	288		12.6			8.4	B			
—			AE		1934		97		7.7							
—	φ		CD		1934	1939	298		3.3			12.5				
10 44.3	−70 52	Δ 99	AB	Car	1826	1917	75		63.0		6.3	6.5	A3 A3			
—			AC		1917		42		39.9			10.5				
10 44.4	−60 00	δ 8	AB	Car	1913	1934	278		7.3		6.2	12.5	B0			
—			AC		1934		92		8.8							
10 44.7	−38 09	B 794		Ant	1927	1935	177	190	0.2	0.2	8.3	8.5	F8			
10 45.1	−59 41	B 2256	Aa	Car	1933	1945	195		0.2		8.5	8.6	P			η Car [57]
10 45.1	−59 40	I 1092	AaxB	Car	1915	1947	317	317	1.0	1.6		11.1				
—			AaxC		1914	1947	74	63	1.0	1.3		10.9				
—			AaxD		1928	1934	304		4.1			13.6				
—	h 4366		AaxE		1913	1945	60		13.6			12.0				
—			AaxF		1897	1945	41		13.9			11.7				
—			AaxG		1859	1913	67		38.5			10.2				
—	Δ 98		AaxH		1826	1913	17		61.1			8.6	O5			
—	B 2256		Aaxb				315		0.8			12				
—	I 1092		Aaxc		1926		270		2.0			13.0				
—			Aaxd		1933		194		1.9			13.5				
—			Aaxe				140		2.0			14.5				
—			Aaxf		1933		141		4.0			14.3				
—			Aaxg		1934		103		4.5			14.4				
10 45.3	−80 28	I 294		Cha	1901	1946	61	76	0.6	0.6	6.1	6.4	K1			δ¹ Cha
10 45.5	−25 02	I 502	AB	Hya	1911	1959	35	21	0.5	0.3	7.8	9.0	K0	7918		
—			AC		1950	1959	137		1.9			13.3		7918		
10 45.8	−10 43	β 914		Sex	1880	1960	339		1.3		7.3	10.9	A0	7917		
10 45.8	−59 29	h 4369		Car	1837	1934	41		14.2		7.9	11.1	M			
10 45.9	+30 41	S 612		LMi	1825	1924	173	173	200.3	197.1	5.3	8.1	B9 K2			42 LMi
10 46.2	−59 19	R 156		Car	1876	1946	30		1.8		8.6	8.7	A5			
10 46.3	−60 36	Rst 4463		Car	1939		294		4.5		6.3	14.1	A0			
10 46.5	−64 16	φ 364		Car	1960		124		0.1		5.9	6.0	B8			

α 2000	δ 2000	Star Name		Const	Years	Pos Ang	Separation	m₁	m₂	Spec	ADS	Orb	Notes
10^h 46^m.8	−49°25′	R 155		Vel	1880 1942	55° 90°	2″.8 0″.7	2.7	6.4	G5			μ Vel [58]
10 47.0	+13 03	Σ 1472		Leo	1828 1921	40 38	33.7 38.2	8.1	8.8	K0			
10 47.6	−15 16	Σ 1474	AB	Hya	1831 1933	24	69.6	6.7	7.8	A0	7930		
−−			AC		1889 1924	23	75.8		6.8	F5	7930		
−−			BC		1831 1949	17	6.7				7930		
10 47.6	−15 37	Σ 1473	AB	Hya	1832 1916	10	30.7	7.7	8.6	F8			
−−			AC		1908	326	98.7		10.1				
10 47.7	+65 28	Σ 1469		UMa	1831 1924	322	11.0	7.7	10.7	F8	7925		
10 48.0	+41 07	OΣ 229		UMa	1846 1959	347 293	0.7 0.8	7.4	7.8	A3	7929		
10 48.9	−59 27	h 4374		Car	1836 1913	115 119	13.8 13.8	7.9	10.4	B0			
10 49.3	−4 01	Σ 1476		Sex	1832 1958	354 10	1.9 2.2	7.0	7.8	A2	7936		40 Sex
10 49.3	−26 49	I 503		Hya	1913 1959	109 115	1.4 1.2	7.9	8.7	F2	7939		
10 49.4	+17 09	β 596	AB	Leo	1878 1925	278	2.5	7.2	13.7	K0	7935		
−−		Fox	AC		1914	273	110.0		12.0		7935		
10 49.4	+41 23	Σ 1475 rej		UMa	1904 1916	201	23.6	7.8	11.3	G0			
10 49.4	−59 19	R 161		Car	1874 1960	258 284	0.8 1.0	6.2	7.4	A0			
10 50.0	−40 57	I 1205	AB	Vel	1926	286	29.2	8.0	11.5	G0			
−−			BC		1926	201	2.6		13.5				
10 50.3	−8 54	h 838	AB	Sex	1878 1934	303	27.3	5.8	11.5	A2	7942		41 Sex
−−			AC		1879	72	233.4				7942		
10 51.0	+16 03	A 2372		Leo	1911 1958	72 90	0.2 0.3	7.8	9.3	F8	7944		
10 51.0	−59 57	h 4378		Car	1836 1918	345	31.0	7.1	10.4	A0			
10 51.2	−21 31	Ho 375		Hya	1890 1959	174 176	12.3 12.4	8.0	12.1	K5	7947		
10 52.0	+9 04	A 2772		Leo	1914 1931	97	2.0	8.0	12.2	F5	7953		
10 52.0	+16 06	A 2373		Leo	1911 1960	223 178	0.2 0.2	8.8	8.8	F5	7952		
10 52.2	+1 02	β pm		Leo	1912	271	39.2	6.4	12.2	A2			
10 52.2	+7 28	Σ 1482		Leo	1831 1954	305	11.7	8.3	9.2	G5	7955		
10 53.0	−63 05	I 418		Car	1902 1943	202	2.3	8.4	8.6	A3			
10 53.4	−2 15	S 617		Leo	1824 1925	178	35.2	6.1	10.1	K0	7967		Probably same as BDS 5610
10 53.5	+69 51	β pm	AB	UMa	1894 1908	51 52	70.1 75.1	6.0	9.0	G5			
−−		Kui 52	Aa		1958	115	25.4		13.4				
−−		β pm	Bb		1908	223	59.6		12.8				
10 53.5	−20 08	β	AB	Hya	1886 1924	208 212	135.3 130.0	5.2	9.2	F5			
−−		Ho 533	BC		1890 1906	131 131	13.4 14.2		10.3				
−−			BD		1890	294	45.3		13.2				
10 53.5	−58 51	Δ 102	AB	Car	1917	202	153.9	3.9	6.6	K0 B3			
−−		Δ 103	AC		1837 1917	13 11	65.3 61.5		8.5				
10 53.7	−2 08	Ma 5		Leo	1843	15	0.4	5.5		K0			β could not see companion
10 53.7	−70 43	h 4383		Car	1837 1946	280 285	1.5 1.5	6.6	7.2	B8			
10 54.0	−26 45	B		Hya		s	15.0	7.3	13.0	B9			
10 54.5	−61 50	Rst 4468		Car	1939	152	7.2	5.9	10.9	K5			
10 54.6	−38 45	h 4381		Ant	1879 1933	42	25.8	7.0	8.4	B9 A			
10 55.0	+52 07	Σ 1486		UMa	1831 1915	102 102	28.3 30.0	8.2	9.5	K5			
10 55.3	−35 30	I 1206	AB	Ant	1926 1947	163 174	0.2 0.2	8.6	9.1	A0			
−−		HCWils	ABxC		1885	169	34.0		9.5				
10 55.3	−52 39	Rst 516		Vel	1929 1949	246 249	7.4 8.0	7.7	12.9	F0			
10 55.6	+24 45	Σ 1487		Leo	1830 1958	103 110	6.2 6.5	4.5	6.3	A0	7979		54 Leo
10 55.7	+0 44	β 1076		Leo	1889 1959			6.1	8.0	F2	7982	o	55 Leo; single in 1933,35
10 56.4	−57 33	h 4387		Car	1913	152	23.6	7.3	10.6	B0			
10 56.8	+52 11	Σ 1488 rej		UMa	1907 1910	214	20.1	8.1	11.2	A3	7983		
10 57.1	−71 21	h 4392		Car	1879 1940	160 157	24.9 24.9	8.3	8.4				
10 57.3	−69 02	h 4393		Car	1835 1918	131	8.4	7.0	9.2	B8			
10 57.6	+30 39	Σ 1492 rej		LMi	1906 1918	165	21.5	8.1	10.0	A2	7988		
10 57.6	−45 53	h 4388		Vel	1913 1930	208 206	35.4 34.7	6.8	11.5	K0			
10 58.2	−31 34	h 4389		Hya	1896 1933	337	10.4	8.3	9.5	F0			
10 58.2	−35 40	B 1175		Hya	1928 1959	215 238	0.4 0.4	8.5	9.0	G0			
10 58.7	+69 49	Prz		UMa	1922 1934	87	13.7	8.0	11.4	F0	7993		
10 59.0	+75 43	LDS 1724		Dra	1965	332	160.0	8.0	16				
10 59.1	−34 53	h 4391		Hya	1920	59	26.7	7.8	11.4	B9			
10 59.2	−33 44	I 211		Hya	1898 1960	180 207	1.9 1.9	5.8	8.8	F0			
10 59.2	−61 19	R 164		Car	1873 1939	83 78	3.7 4.0	6.2	9.7	B9			
10 59.2	−81 33	I 212		Cha	1901 1947	150 183	0.7 0.7	7.4	7.5	F5			
10 59.4	+79 41	Hu 883	AB	Cam	1905 1920	74	3.6	7.3	11.8	F8	7992		
−−		VBs	AC		1920	176	12.6		13.6		7992		
10 59.4	−43 08	h 4394		Vel	1913	261	28.6	8.3	9.5	G5 G			
10 59.5	+9 56	A 2774		Leo	1914 1939	95	1.5	7.0	12.0	A2	8003		
10 59.8	+58 54	Σ 1495		UMa	1833 1924	38	34.4	7.2	9.5	K2	8001		
11 00.0	−3 28	Σ 1500		Leo	1825 1958	331 305	1.1 1.6	7.5	9.1	F8	8007		
11 00.1	−51 49	Rst 1550		Vel	1934 1949	259 260	10.7 10.3	6.2	13.3	A3			
11 00.2	−43 23	See 126		Vel	1897 1953	168	0.9	8.3	8.6	F0			
11 00.3	+42 55	Kui 53		UMa	1958	151	37.1	6.0	11.6	G0			
11 00.3	−44 55	Rst 4933		Vel	1944	312	0.7	7.3	10.0	K5			
11 00.4	−40 15	I 864	AB	Ant	1911 1947	356	1.0	8.8	8.8	G0			
−−			ABxC		1911	39	20.9		12.8				
11 00.7	+6 06	β 598		Leo	1878 1919	221 220	46.8 44.8	5.0	12.5	A5	8019		59 Leo

α 2000	δ 2000	Star Name		Const	Years	Pos Ang	Separation	m₁	m₂	Spec	ADS	Orb	Notes
11ʰ 00.ᵐ7	+42° 44′	Es		UMa	1913 1934	201°	13.″4	6.7	10.7	K0	8015		
11 00.8	+29 13	Cou 960		LMi	1973	100	0.1	8.7	8.7	F0			
11 00.9	−40 30	ø 365		Vel	1960	110	0.1	7.6	7.6	F8			
11 01.1	+0 03	h 1182		Leo	1828	130	22.0	7.9	12.9	A2			
11 01.6	+66 27	β	AB	UMa	1905 1926	12	7.6	7.9	10.6	K0	8020		
—		Σ 1498 rej	AC		1905 1926	291	29.0		10.7		8020		
11 01.8	+29 52	Cou 961		LMi	1973	316	1.0	7.4	9.2	K0			
11 02.1	−15 41	H I 177		Crt	1876 1959	18	2.9	8.5	8.9	F0	8025		
11 02.4	−26 50	B 208		Hya	1926 1959		0.1	6.9	7.0	F0	8028		Too close
11 02.6	−3 31	h 173		Leo	1907	177	36.6	7.1	13.6	G5			
11 02.8	−59 44	Jsp 455		Car	1930	31	7.0	7.0	13.4	B8			
11 03.2	−11 18	A 1774		Crt	1908 1952	271	3.7	5.6	10.6	G5	8037		
11 03.2	−51 00	B 2263		Vel	np		7.0	7.8	13.0	G5			
11 03.3	−27 31	Howe 25		Hya	1877 1959	331 339	2.6 2.5	7.8	9.3	A0	8038		
11 03.7	+44 20	h 2554		UMa	1894 1926	278	49.0	7.5	8.7	G5			
11 03.7	+61 45	β 1077		UMa	1889 1945			1.9	4.8	K0	8035	o	50 α UMa, Dubhe [59]
11 04.0	+3 38	Σ 1504		Leo	1829 1959	276 296	1.1 1.2	7.8	7.9	F0	8043		
11 04.5	+38 14	Ho 377	AB	UMa	1891 1958	250	8.4	6.1	12.6	A2	8046		
—		Hzg	AC		1907	83	150.3				8046		
11 04.6	+12 40	h 174		Leo	1905	5	61.2	6.7	10.5	A0			
11 04.7	−4 13	Σ 1506	AxBC	Leo	1829 1970	212 220	10.4 11.4	7.6		G5	8048		
—		A 676	BC		1904 1960			10.1	10.1		8048	o	
11 04.7	−46 06	Don 457		Vel	1931 1946	68	6.9	7.7	13.4	A2			
11 04.9	−61 03	Δ 105		Car	1826 1935	221	24.0	7.8	9.6	B3			
11 05.0	+7 20	Kui 54	AB	Leo	1934 1958	257 262	2.6 3.3	4.6	10.9	F0			63 χ Leo
—		OΣ	AC		1882 1924	303 305	287.7 276.4		8.9				
—		Dick	AD		1924	294	314.1						
11 05.0	+38 25	Ho 378		UMa	1891 1960	219 234	0.4 0.8	8.5	8.7	F2	8047		
11 05.0	−48 12	Rst 4934		Vel	1943 1949	280	1.4	7.8	13.2	A0			
11 05.1	−59 43	Gli 159		Car	1920	274	17.6	8.1	9.3	B0			
11 05.3	−27 18	ø 47		Hya	1927 1958			5.6	5.7	F4		o	χ¹ Hya
11 06.2	−85 45	Rst 2733		Oct	1935	81	3.8	7.2	13.0	A0			
11 06.5	−53 16	h 4405		Cen	1917	48	19.2	8.1	10.2	B8			
11 06.5	−58 41	HdO 210	AB	Car		230	12.0	6.0	11.9	K0			
—			AC			180	15.0		12.9				
11 06.8	−70 50	B 2006		Car	1931 1954	12 144	0.2 0.1	7.9	8.0	F2			
11 06.8	−70 53	B 2268		Car		140	15.0	5.6	13.8				
11 06.9	+1 57	β 599		Leo	1878 1958	82 102	1.8 2.4	5.5	9.3	A0	8060		65 Leo
11 07.0	−24 09	B 2267		Crt		sf	10.0	7.8	14.0	K0			
11 07.3	−41 27	I 869	AB	Cen	1911 1941	332	1.5	8.1	10.3	A2			
—		h 4408	AC		1920 1932	345	20.4		11.5				
—			AD		1920 1932	110	22.1		13.3				
11 07.3	−42 38	h 4409		Cen	1837 1960	278 260	2.1 1.4	5.3	7.8	A5			
11 07.5	+21 09	h 2558		Leo	1911	272	22.3	7.2	15	A2			
11 08.0	+52 49	Σ 1510		UMa	1832 1976	341 330	3.9 5.2	7.7	9.0	F5	8065		
11 08.1	−60 50	B 1180		Car	1928	256	2.1	7.3	13.7	B8			
11 08.3	−15 58	h 4410		Crt	1898 1959	220 235	20.8 19.8	7.9	13.4	F5	8069		
11 08.5	+3 14	H V 68	AB	Leo	1899 1904	228	150.0	8.1	9.5	K0			
—			BC		1899 1904	207	42.2		10.8				
11 08.8	+24 39	A 677		Leo	1904 1958	244	4.8	5.7	14.2	A2	8071		67 Leo
11 09.7	+63 20	H IV 106		UMa	1783 1939	132	20.2	7.9	11.4	F5	8073		Σ 1513 rej
11 10.0	−29 36	h 4412		Hya	1879 1954	266	12.6	8.3	9.1	A0	8080		
11 10.6	−32 34	I 213		Hya	1901 1959	126 122	0.9 0.6	7.3	9.3	F5			
11 11.3	−74 28	I 874		Car	1911 1941	80	0.5	8.1	8.8	F8			
11 11.7	−60 27	BrsO	AxBC	Car		nf	53.0	8.3		B B			
—		I 1065	BC		1911 1926	19	0.4	9.0	9.8				
11 11.8	+42 50	β pm	AB	UMa	1893 1911	247	135.2	7.3 8.8		F8 G5			
—			Bb		1908	336	132.5		12.8				
11 12.1	+35 00	A 2156		UMa	1910 1958	257 235	0.3 0.4	8.3	9.1	A2	8085		
11 12.1	−71 13	I 231		Car	1901 1941	2	2.6	7.0	10.8	B9			
11 12.5	+35 50	OΣΣ 108	AB	UMa	1876 1924	72 70	128.4 138.7	6.3	7.9	G0 K0			
—			AC		1910 1925	160 157	88.2 86.7		12.3				
—			BD		1825 1910	88 88	(35′)		8.2	G5			
11 12.5	−18 30	β 220		Crt	1875 1960	144 13	0.6 0.1	6.7	7.1	A0	8086	o	
11 12.6	−60 19	h 4414		Car	1889 1932	277	21.7	4.7	11.5	F5			
11 12.8	−64 10	h 4415		Car	1916 1918	125	19.1	5.2	11.9	B8			
11 13.1	−47 03	R 165		Cen	1880 1952	60 67	3.2 3.4	7.6	7.6	F8			
11 13.3	+33 27	Ho 254		UMa	1887 1923	163	2.5	6.8	12.8	K0	8091		
11 13.3	−50 07	Rst 4940		Cen	1944 1949	114	0.4	7.5	8.5	A2			
11 13.6	+55 25	A 1353		UMa	1906 1958	200 60	0.5 0.2	7.9	8.7	F2	8092		
11 13.7	+20 08	Σ 1517		Leo	1829 1958	288 349	1.0 0.3	7.7	7.7	G0	8094	o	Changed quadrant 1925?
11 13.7	+41 05	Ho 50		UMa	1882 1958	32	3.1	6.4	9.4	K0	8093		
11 13.8	−51 48	CapO		Cen	1928	288	12.0	7.8	12.0	K0			
11 14.0	+8 04	Kui 56		Leo	1958	238	21.5	5.8	11.3	K0			
11 14.1	+20 31	β 1282	AB	Leo	1899 1924	344 344	187.3 191.4	2.6	8.6	A3			68 δ Leo

α 2000	δ 2000	Star Name		Const	Years	Pos Ang	Separation	m₁	m₂	Spec	ADS	Orb	Notes
11ʰ14ᵐ1	+20°31′	β 1282	Aa	Leo	1879 1911	43° °	95″4 ″		12.1				
11 15.1	−39 29	See 128		Cen	1897 1959	208 85	0.5 0.1	7.8	7.9	A0			Direct motion
11 15.3	+27 34	Σ 1521		Leo	1829 1957	96	3.7	7.7	8.0	A5	8105		
11 15.4	+47 28	Hu 639		UMa	1902 1958	274 101	0.3 0.1	8.2	8.2	A5	8104		
11 15.4	+73 28	Σ 1516	AB	Dra	1831 1940	299 102	9.9 36.2	7.6	8.1	K5	8100		
—		OΣ 539	AC		1858 1976	294 317	8.2 6.7		11.1		8100		
11 15.4	−66 31	φ 179		Car	1932 1942	107	0.9	8.1	10.3	K0			Spectrum composite K0+A2
11 15.7	+45 51	OΣΣ 109		UMa	1877 1925	257	80.1	8.6	8.8	K2 K0			
11 15.7	−16 21	S,h 372		Crt	1879 1917	300	19.2	8.2	10.0	A3			
11 15.9	−47 55	h 4421		Cen	1902 1933	67	23.3	6.9	10.5	A0			
11 16.1	+52 46	Σ 1520		UMa	1831 1958	344	12.7	6.6	7.9	F2	8108		
11 16.5	−45 53	h 4423		Cen	1836 1959	276 276	1.7 2.4	6.9	7.2	F2			
11 16.6	−55 33	I 875		Cen	1911 1959	238	1.9	7.8	9.5	G5			
11 16.7	−3 39	S,h 121		Leo	1821 1923	287 289	106.2 96.9	4.5	9.2	A5 K5			74 φ Leo
11 17.0	−7 08	β 600	AB	Crt	1878 1959	226 210	1.2 1.0	6.1	11.6	F0	8115		
—			AC		1823 1959	98 98	67.1 57.2		9.9		8115		
11 17.0	−55 37	φ 181		Cen	1929 1936	252 243	0.3 0.3	8.0	8.5	A0			
11 17.3	−67 49	Don 471		Car	1931	202	7.0	6.1	15	M			
11 17.5	−59 06	R 163	AxBC	Car	1879 1959	58	1.6	7.0		A0			
—		Rst 4472	BC		1939 1959	24	0.4	8.1	8.6				
11 17.5	−63 29	φ		Car		332	7.0	7.7	11.5	A0			
11 18.2	+31 32	Σ 1523		UMa	1826 1977			4.3	4.8	G0	8119	o	53 ξ UMa
11 18.5	+33 06	Σ 1524		UMa	1830 1958	147	7.2	3.5	9.9	K0	8123		54 ν UMa
11 18.8	+66 41	OΣ 233	AB	UMa	1849 1934	335	4.9	7.2	10.1	F0	8122		
—			AC		1912	303	41.7		12.2		8122		
11 18.9	−11 46	Hu 130		Crt	1900 1959	134 123	1.2 1.0	8.7	8.9	F5	8129		
11 19.0	+14 16	Σ 1527		Leo	1829 1970	10 30	3.9 1.7	7.0	8.1	G0	8128	o	
11 19.1	−31 01	I 506		Hya	1911 1960	98 127	1.0 0.8	8.2	9.5	G0			
11 19.4	−1 39	Σ 1529		Leo	1833 1955	252	9.6	7.0	8.0	F8	8131		Yellow-white, ashy (Webb)
11 19.8	−53 29	See 129		Cen	1897 1933	7	6.7	7.9	11.5	A2			
11 20.1	−67 03	Don 474		Car	1929	334	0.8	8.0	11.0	B9			
11 20.3	−28 20	B 2275	AB	Hya		p	15.0	6.7	15	K0			
—		LDS 347	AC			sf	49.0		13.5				
11 21.0	−54 29	I 879		Cen	1912 1960			4.3	5.0	B5		o	π Cen
11 21.1	−35 31	I 1539		Hya	1927 1937		0.1	8.1	8.3	A3			P.a. uncertain
11 21.4	−24 00	Rst 2743		Crt	1935 1940	144	2.5	7.8	14.4	K0			
11 21.6	−43 33	h 4426		Cen	1898 1947	174	13.3	7.2	10.5	K0			
11 21.8	+45 20	A 1847		UMa	1908 1934	326 322	1.5 1.5	7.9	14.2	A2	8139		
11 21.8	−24 11	B 2277		Crt		nf	10.0	8.0	13.0	G5			
11 22.5	−53 22	Rst 1578		Cen	1934 1947	343	7.8	7.0	13.9	B9			
11 23.1	+4 08	A 2776	AB	Leo	1914 1957	274 78	0.2 0.1	8.8	8.8	F5	8145		
—			ABxC		1914 1945	61	6.2		13.5		8145		
11 23.2	−36 10	See 131		Cen	1897 1899	127	31.3	5.1	14.6	K6			
11 23.4	−64 57	h 4432		Mus	1836 1947	288 303	2.3 2.3	5.4	6.6	B5 B5			
11 23.9	+10 32	Σ 1536		Leo	1832 1959			4.0	6.7	F5	8148	o	78 ι Leo
11 24.7	−61 39	BrsO 5		Cen	1838 1959			7.6	8.6	K5		o	
11 24.9	−17 41	h 840		Crt	1877 1955	96	5.2	4.1	9.6	A5	8153		15 γ Crt
11 25.0	−59 39	Hrg 56		Cen	1883 1930	150 146	5.3 5.9	8.0	10.5	G5			
11 25.3	−84 57	Rst 2752		Oct	1935	202	3.3	7.8	13.0	F5			
11 25.6	+16 27	h 4433		Leo	1840 1924	339 351	58.5 55.7	5.6	9.2	F2			81 Leo
11 25.6	−37 45	B 796		Cen	1927 1940	59 60	5.0 5.2	5.9	11.4	M			
11 25.9	+52 08	A 1592		UMa	1907 1928	63	4.4	7.2	13.9	F0	8158		
11 26.6	−61 07	Rst 4477		Cen	1939	190	13.0	5.5	14.0	B5			
11 26.8	+3 01	Σ 1540	AB	Leo	1832 1970	150 150	29.6 28.5	6.2	7.9	K0 K0	8162		
—			AC		1936 1937	188	90.3		9.9		8162		
11 26.8	−53 10	I 883		Cen	1911 1960	278 315	0.3 0.2	6.4	6.8	G0			Spectrum composite G0+A2
11 27.0	−69 03	HdO 211		Mus		190	12.0	7.8	11.0	A2			
11 27.2	−12 21	Kui 57		Crt	1958	336	27.9	5.9	12.9	F4			16 κ Crt
11 27.2	−15 39	Hu 462		Crt	1902 1960			8.2	8.6	K0	8166	o	
11 27.3	−78 31	h 4440		Cha	1919	162	22.4	7.8	12.5	K0			
11 27.5	−39 53	h 4438		Cen	1898 1920	196	23.0	7.4	11.4	A0			
11 27.7	+46 18	Σ 1541		UMa	1831 1968	29	7.7	7.9	10.3	F8	8168		
11 27.9	+2 51	Σ I 19	AB	Leo	1834 1932	170 176	94.8 91.1	5.1	8.0	K0 G5			84 τ Leo
—			AD		1875 1910	92	764.8		9.9				
—			BC		1910	234	106.5		14.3				
—			BD		1875 1910	85	755.3						
11 27.9	−1 42	Rst 4944		Leo	1943 1958	327	0.3	6.6	8.0	K0			
11 27.9	+44 34	Σ 1542		UMa	1831 1939	265	3.0	6.9	10.4	F0	8171		
11 28.2	+55 40	A 1355		UMa	1906 1928	0	1.4	8.0	12.4	A0	8173		
11 28.3	−72 28	B 1699		Mus	1930 1959	356	0.2 0.3	6.8	6.9	B3			
11 28.6	−42 40	BrsO 6		Cen	1835 1947	167	13.1	5.2	7.9	B9			
11 28.6	−45 08	I 885		Cen	1913 1944	185 176	1.0 1.0	8.0	10.0	G0			
11 28.6	−54 53	φ		Cen		230	0.4	7.1	8.9	A3			
11 29.1	+39 20	Σ 1543	AB	UMa	1831 1958	11 359	5.4 5.4	5.3	8.3	A2	8175		57 UMa
—			AC		1879 1910	9	216.5		11.5		8175		

α 2000	δ 2000	Star Name		Const	Years		Pos Ang		Separation		m_1	m_2	Spec		ADS	Orb	Notes
11^{h}29^m.1	+39°20'	Σ 1543	AD	UMa	1927		252°	°	346".3	"		7.7	K0		8175		
--			DE		1895	1910	184		98.2			10.1			8175		
--			DF		1895	1910	168		128.5			10.2			8175		
11 29.1	−58 48	φ 186	AB	Cen	1928		82		6.6		7.6	12.7	A0				
--		CorO 125	AC		1920		123		20.0			10.2					
11 29.2	−4 27	A 70		Leo	1900	1916	353		5.0		7.9	14.6	F8		8178		
11 29.3	+30 25	Lewis 11		UMa	1900	1960	7	331	0.9	0.9	7.1	11.1	F2		8177		
11 29.8	−24 29	Jc 16	AB	Crt	1847	1954	77	80	8.2	8.2	5.8	8.8	F1		8183		
--			AC		1910	1933	115		169.3			8.9		F2	8183		
11 29.9	−55 19	φ 43		Cen	1928	1940	285		1.6		7.7	12.0	K2				
11 30.1	+29 58	OΣΣ 111		UMa	1875	1914	33		67.0		6.9	8.9	F0				
11 30.2	−49 39	Rst 543		Cen	1929	1953	301		1.8		8.3	9.0	F5				
11 30.3	+81 02	Σ 1539		Cam	1832	1919	313		19.0		8.1	10.5	F5		8181		
11 30.8	−6 43	Σ 3072		Crt	1831	1909	330		9.4		7.7	10.7	F8		8190		
11 30.8	+41 17	OΣ 234		UMa	1844	1959					7.6	7.9	F5		8189	o	
11 30.8	+41 21	S,h 126		UMa	1823		90		13.0		7.0	8.0					No such pair in this place
11 30.8	−58 49	φ 187		Cen	1928	1960	221	233	0.4	0.3	8.2	9.1	G0				
11 30.9	−50 57	B 1192		Cen	1928	1940	120		3.1		8.0	10.4	G5				
11 31.1	−43 33	CapO		Cen			np		8.0		8.0	13.0	K2				
11 31.3	+59 42	Σ 1544		UMa	1831	1927	90		12.4		7.2	8.2	A5		8191		
11 31.4	+3 04	J 85		Leo	1900	1953	164		4.7		8.0	12.0	F0		8193		
11 31.7	+7 52	Ho 51		Leo	1882	1941	174		2.7		7.9	12.9	K0		8195		
11 31.7	+14 22	Σ 1547		Leo	1782	1958	318	328	14.6	15.4	6.4	8.4	G0		8196		88 Leo
11 31.8	−59 27	h 4445		Cen	1917	1930	125		13.5		5.1	11.5	G4				o¹ Cen
11 31.8	−73 54	h 4450		Mus	1915	1917	40		21.0		6.9	10.8	B9				
11 32.2	+36 15	Hu 1134		UMa	1905	1960	122	117	0.1	0.2	7.4	7.4	K0		8198		Round in 1925 (VBs)
11 32.3	+56 06	Σ 1546		UMa	1832	1920	345		11.5		8.0	10.3	F8		8199		
11 32.3	+61 05	OΣ 235		UMa	1844	1960					5.8	7.1	F5		8197	o	
11 32.3	−29 16	H III 96		Hya	1783	1952	210		9.2		5.8	5.9	F6 F7		8202		
11 32.4	+12 13	Bgh		Leo			106		96.0		8.3	9.1	F5 G5				
11 32.7	−65 52	NZO 23		Mus	1914	1939	238		1.1		8.6	8.9	B9				
11 32.9	−36 13	I 1544		Cen	1927	1936	86		2.2		6.7	11.8	G0				
11 33.0	−23 27	HdO 130		Crt	1903	1932	344		10.0		7.3	12.8	A2		8212		
11 33.0	−31 51	h 4449		Hya	1919		150		67.8		3.5	10.5	G7				ξ Hya
11 33.3	−50 55	B 1193		Cen	1928	1941	78	82	2.0	2.0	8.1	10.4	K0				
11 33.6	−40 35	I 78		Cen	1894	1959	88	97	1.0	1.0	6.2	6.2	A2				
11 34.3	−41 09	Rst 1597		Cen	1934	1945	88		0.5		8.0	10.0	A0				
11 34.7	+16 48	Σ 1552	AB	Leo	1829	1958	209		3.3		6.0	7.3	B3		8220		90 Leo
--			AC		1822	1938	234		63.1			8.7			8220		
--			BC		1901	1938	236	235	59.9	64.6					8220		
11 35.2	−47 22	Rst 548		Cen	1929	1949	178	163	6.6	6.7	5.7	13.3	M				
11 35.7	−61 17	B 797		Cen	1927	1952	23		4.0		6.7	12.9	A0				
11 35.8	−63 01	Rst 3746		Cen	1937	1944	316		16.3		3.1	11.3	B9				λ Cen
11 36.2	+3 18	A 2777		Leo	1914	1928	90		3.6		7.2	12.9	G5		8233		
11 36.3	+27 47	Σ 1555	AB	Leo	1829	1959	339	315	1.2	0.3	6.4	6.8	A3		8231		
--		h 503	AC		1844	1958	148		21.2			10.2			8231		
11 36.6	+56 08	Σ 1553		UMa	1832	1968	171	167	5.3	6.0	7.9	8.4	G5		8236		
11 36.6	−33 34	h 4455	AB	Hya	1836	1954	243		3.3		5.8	7.9	K0				
--			AC		1897		84		47.6			13.3					
11 36.6	−60 54	Gli 165		Cen	1883	1944	5		1.9		8.3	9.3	B9				
11 36.8	−24 26	h 4456	AB	Crt	1868	1959	123	122	20.9	15.2	7.2	10.9	K2		8240		
--			AC		1910		328		39.5			13.2			8240		
11 37.0	−38 58	Δ 113		Cen	1920	1935	151		145.5		6.9	7.3	A0 K2				
11 37.5	−61 29	I 421		Cen	1902	1944	117		1.5		7.2	9.9	B5				
11 37.8	+49 49	Hu 728		UMa	1904	1955	106		0.3		8.4	9.4	F5		8244		
11 38.0	+12 57	h 183		Leo	1907		11		95.1		7.6	13.1	K5				
11 38.2	−63 12	CorO 126	AB	Cen	1920		271		11.0		7.3	12.4	B3				
--			AC		1920		203		13.5			11.1					
11 38.3	−63 22	I 422	AB	Cen	1902	1946	95	103	0.4	0.4	7.6	7.8	B2				
--			ABxC		1902	1943	3		1.7			11.0					
--			ABxD		1902	1943	323		9.6			12.0					
11 38.4	−2 26	Σ 1560		Vir	1831	1958	279		5.2		6.2	10.4	K0		8247		
11 38.7	+45 07	Σ 1561	AB	UMa	1831	1970	266	252	10.5	9.4	6.3	8.4	G0		8250		
--			AC		1783	1931	86	90	32.3	120.7		9.4			8250		
--			AE		1923	1930	307	309	94.0	91.1		12.3			8250		
--			BC		1909		88		117.8						8250		
--			BD		1931		314		84.7			8.5		F2	8250		
--			DC		1895	1910	251		546.4						8250		
11 38.7	−13 12	Kui 58		Crt	1934	1958	226		1.4		5.5	10.9	G0				24 ι Crt
11 38.8	+39 10	h 506		UMa	1902	1906	143		29.4		7.3	13.3	K5				
11 38.8	+64 21	Σ 1559		UMa	1836	1958	322		2.1		6.8	7.8	A2		8249		
11 38.9	−70 53	B 1703		Mus	1930	1960	330	334	0.4	0.4	8.5	8.7	A0				
11 39.2	−57 44	h 4460		Cen	1874	1932	176		8.6		7.7	8.9	A0				
11 39.2	−58 36	Jsp 493	AxBC	Cen	1929		108		49.1		7.6		G5				
--			BC		1929	1947	61		1.6		12.1	12.5					

α 2000	δ 2000	Star Name		Const	Years	Pos Ang	Separation		m_1	m_2	Spec	ADS	Orb	Notes
11^{h}39^m.5	+25° 18′	A 678		Leo	1904 1960	156° 197°	1″.2	1″.1	7.7	11.2	G0	8255		
11 39.5	+52 11	Σ 1563		UMa	1902 1910	158	13.7		8.1	10.3	G0	8253		
11 39.5	−37 26	Howe 70		Cen	1879 1954	104	3.3		8.1	8.4	F8			
11 39.5	−65 24	B 1705	AB	Mus	1930 1954	119 178	0.2	0.2	5.6	6.4	G0			Composite spectrum G0+A0
--		I 34	ABxC		1900 1901	234	38.7			12.1				
--			ABxD		1900 1901	32	41.8			12.6				
11 39.6	+19 00	Σ 1565		Leo	1829 1955	304	21.6		7.0	8.9	F5 F5	8257		
11 39.8	−63 29	BrsO 7		Cen	1838 1918	277	27.8		7.7	8.4	B3 B3			
11 39.9	−14 28	β 1078		Crt	1889 1934	50 52	8.2	7.8	6.2	12.2	A0	8259		
11 39.9	−50 29	Rst 4949		Cen	1942 1949	14 25	1.1	1.1	6.8	11.0	A0			
11 40.0	−30 29	I 1545		Hya	1927 1957	298 302	1.4	1.4	7.9	11.0	K0			
11 40.0	−33 27	I 232		Hya	1900 1960	162	2.1		6.9	9.4	K0			
11 40.0	−38 07	Δ 114		Cen	1903 1934	95	17.0		6.7	9.5	G5			
11 40.4	+0 57	A 2578		Vir	1913 1933	140	0.9		6.8	11.5	A3			
11 40.6	+21 02	Σ 1566		Leo	1829 1947	349	2.7		8.2	9.7	G5	8263		
11 40.6	−62 34	CapO 11		Cen	1888 1941	220	2.6		7.4	7.8	B2			
11 41.0	+34 12	β pm		UMa	1850 1907	108 101	163.3 159.3		5.3	10.2	G5			61 UMa
11 41.0	−83 06	h 4468		Cha	1871 1919	153 147	20.1 22.2		6.3	11.5	K0			
11 41.2	−61 08	HdO 212		Cen	1913 1946	325	1.0		7.3	7.7	K0			
11 41.6	+31 45	β pm		UMa	1850 1909	297 300	105.8 84.3		5.7	10.0	F5			62 UMa
11 41.7	−32 30	h 4465	AB	Hya	1897 1934	346 346	26.6 28.2		5.3	12.8	M1			
--			AC		1919	44	67.0			8.4	G0			
--		Rst 2760	CD		1934 1944	113	4.6			13.6				
11 43.1	−52 27	See 134		Cen	1897	323	12.2		8.0	13.0	K0			
11 43.2	−39 26	LDS 358	AxBC	Cen	1959	332	24.7		8.4		G0			
--		I 1547	BC		1927 1959	230 236	3.1	2.6	9.5	9.8				
11 43.8	+24 01	A 679		Leo	1904 1933	93	5.0		7.4	15	K0	8282		
11 44.2	+25 13	OΣ 239 rej		Leo	1867 1924	20 24	38.1 37.3		6.0	10.3	K5	8285		
11 44.8	+60 03	h 3334		UMa	1904	156	44.5		7.9	11.6	K2			
11 44.8	−55 28	I 890		Cen	1912 1959	308 304	2.0	2.0	7.8	10.3	A0			
11 45.5	+45 36	Σ 1570		UMa	1831 1973	49	10.7		8.5	9.0	F0	8292		
11 45.6	−66 44	h 4471		Mus	1918	275	40.6		3.6	12.6	A5			λ Mus
11 46.5	−58 02	Hu 1485	AB	Cen	1913 1934	320	2.9		8.0	12.5	B8			
--			AC		1913 1934	276	7.9			11.5				
11 46.5	−61 11	ø		Cen		320	25.0		4.1	12.9	G0			
11 46.8	+15 00	β 602		Leo	1878 1960	73 109	0.6	0.5	8.0	10.5	A5	8302		
11 47.5	+20 02	S,h 130		Leo	1823 1918	25 28	76.9 73.8		8.3	9.7	F5			
11 47.8	+49 49	Hu 729		UMa	1902 1955	0 351	1.5	1.3	7.0	11.5	A0	8307		
11 47.9	+8 15	S,h 131	AB	Vir	1899 1960	356 357	151.7 150.2		5.3	12.1	A0			4 Vir
--			AC		1823	323								
11 47.9	−28 19	Rst 2765		Hya	1935 1944	100	1.9		7.7	14.3	A2			
11 48.0	+20 13	Σ II 7		Leo	1836 1925	355	74.3		4.5	9.6	F8			93 Leo
11 48.4	−10 19	H VI 115	AB	Crt	1899 1960	70 67	83.9 88.4		6.3	9.6	G0			
--			BC		1910 1960	349 351	100.3 101.5			12.8				
11 48.6	+14 17	β 603		Leo	1879 1933				6.0	8.3	A5	8311	o	
11 49.1	+14 34	β 604	AB	Leo	1898	346	39.7		2.1	15	A2	8314		94 β Leo, Denebola
--			AC		1878 1914	344 358	77.1 80.3			13.1		8314		
--			AD		1914	203	264.0					8314		
11 49.2	+67 20	Σ 1573		Dra	1832 1953	178	11.2		7.6	8.6	F8	8313		
11 49.2	−57 42	Rst 2768		Cen	1936 1947	291	1.0		7.1	11.1	B9			
11 49.5	−46 04	ø 366		Cen	1960	147	0.2		7.9	8.1	F5			
11 49.7	−32 26	I 1550		Hya	1927 1960	162 143	0.7	0.9	8.1	10.1	F8			
11 50.4	−2 25	A 1357		Vir	1906 1938	215	1.8		7.9	12.6	G0	8318		
11 50.7	+1 46	β pm	AB	Vir	1850 1908	283 284	200.7 245.6		3.6	10.4	F8			5 β Vir
--			AC		1852 1884	86 85	539.1 512.3			8.6				
11 50.9	+12 17	h 1201		Leo	1877 1934	190	14.9		6.4	11.2	A3	8320		
11 51.2	+33 22	Arg		UMa	1905 1960	274 274	45.2 46.6		6.3	8.8	F2			Single
11 51.2	−36 56	B 2287		Cen	1930 1941	226	9.0		7.9	10.5	A0			
11 51.4	−11 38	Kui 59		Crt	1958	73	30.7		6.2	13.4	F0			Error in position?
11 51.8	−64 36	Gli 169		Cen	1883 1954	228	4.3		7.5	8.8	B8			
11 51.9	−65 12	CorO 130		Mus	1894 1940	159	1.8		5.2	7.4	B7			
11 52.0	+8 50	Σ 1575		Vir	1832 1940	210	30.6		8.1	8.7	G0 F5	8327		
11 52.2	−56 59	I 892	AB	Cen	1911 1960	124 122	49.1 54.6		5.6	10.9	A2			
--			BC		1911 1960	128 122	4.1	5.1		11.9				
11 52.8	+15 26	S,h 132	AB	Leo	1823 1925	14	38.9		6.9	9.9	A2			
--			AC		1914 1922	264	90.6			11.6				
11 52.9	−33 54	h 4478		Hya	1834 1959	340 8	2.2	0.9	4.7	5.5	B9			β Hya
11 53.6	+42 55	OΣ 240		UMa	1847 1912	318	8.6		8.0	10.8	F5	8338		
11 53.6	−16 07	A 2381	AB	Crt	1911 1960	356	4.4		7.5	12.0	K0	8339		
--			AC		1923	321	68.7			11.1		8339		
11 53.7	+73 45	β 794	AB	Dra	1881 1960				7.2	8.4	F8	8337	o	
--			AC		1886 1890	72	5.6			14.0		8337		
--			AD		1890 1898	78	26.9			13.3		8337		
11 54.1	+71 55	A 75		Dra	1900 1958				8.0	8.8	F5	8344	o	
11 54.4	−37 45	Howe 71		Cen	1897 1956	278 275	1.8	1.3	6.8	8.1	F8			

α 2000	δ 2000	Star Name		Const	Years	Pos Ang		Separation		m₁	m₂	Spec	ADS	Orb	Notes
11ʰ 54.6ᵐ	+19° 25′	OΣΣ 112		Leo	1875 1920	35°		73.2″	″	8.4	8.4	G0 G5			
11 54.9	−42 23	I 79		Cen	1900 1947	103 99		0.8	0.6	8.5	9.0	F0			
11 55.0	−56 06	H1d 114		Cen	1887 1956	207 180		1.8	3.2	7.3	7.7	G0			
11 55.0	−62 35	BrsO		Cen	1838 1956	266 267		17.7	18.8	7.8	9.3	M0 M0			
11 55.1	+46 29	A 1777	AB	UMa	1908 1958	3 59		0.3	0.2	6.7	8.5	A0	8347		65 UMa
--		Σ 1579	ABxC		1832 1969	38		3.7			8.3		8347		
--			ABxD		1833 1969	114		63.1			6.5		8347		
11 55.4	−41 54	I 80		Cen	1900 1954	107 99		1.3	1.3	7.9	8.1	A2			
11 55.7	+15 39	H VI 13		Leo	1782	nf		90.0		5.5		A2			95 Leo [60]
11 56.0	−22 10	Hu 1489		Crt	1920 1952	212		1.5		7.8	11.5	K2	8353		
11 56.1	+21 59	Σ 1582		Leo	1827 1970	77		12.0		8.2	9.7	F8	8352		
11 56.3	+35 27	OΣ 241	AB	UMa	1849 1959	119 139		1.4	1.7	6.8	8.7	F2	8355		
--		Dom	Aa		1952	48		30.4			7.4		8355		[61]
11 56.7	−32 16	Δ 116	AB	Hya	1826 1959	263 262		19.8	19.0	8.2	8.3	G0 G0			
--			AC		1919 1959	320 333		29.0	24.3		11.4				
11 57.3	−22 32	h 4481		Crv	1876 1954	194		3.6		8.0	8.1	F5	8361		
11 57.3	−49 22	Rst 564		Cen	1929 1942	3		2.6		6.7	11.0	A3			
11 57.3	−80 30	I 893		Cha	1912 1941	118		1.1		7.6	10.3	A0			
11 57.8	−43 43	B 1203	AB	Cen	1928 1952	310		0.2	0.1	8.0	8.2	A0			P.a. decreasing 1928-40
--		h 4482	ABxC		1913 1929	290 289		22.8	23.3		12.1				
11 58.1	+32 16	β 918	AB	UMa	1880 1934	231		7.4		6.4	12.6	F0	8368		
--			AC		1913	66		122.5			10.6		8368		
11 58.3	−40 57	h 4484		Cen	1897 1943	310		3.1		6.9	9.5	K0			
11 58.9	−25 55	I 510		Hya	1909 1959					7.1	7.3	A0	8371		Duplicity doubtful
11 59.1	+0 32	J 92		Vir	1910 1958	175		14.2		6.2	13.9	K0	8372		
11 59.3	+33 10	β 919		UMa	1880 1958	16		4.5		6.0	12.0	K0	8374		
11 59.6	−78 13	h 4486		Cha	1835 1941	178 188		1.7	0.9	5.4	6.0	B9			ε Cha
12 00.1	+70 40	β 795	AB	Dra	1881 1942	329		14.2		7.9	13.2	A	8379		
--		OΣ 242	AC		1868 1942	150 154		33.7	32.2		8.0	K	8379		
--		β 795	CD		1881 1942	114		6.0			12.5		8379		
12 00.2	+36 44	OΣΣ 114		UMa	1875 1925	81		86.7		8.0	8.6	A0			
12 00.2	+87 00	Σ 1583		UMi	1833 1925	283		11.1		8.2	9.2	A2	8382		Same as OΣ 238 rej
12 00.5	−24 27	h 4489		Crv	1879 1935	153		9.9		8.5	9.0	A2	8385		
12 00.6	+69 11	A 1088		Dra	1905 1960	223 246		0.3	0.3	7.8	8.5	F0	8387		
12 00.7	−21 50	β 1079		Crv	1889 1930	148		11.7		6.3	13.4	K0	8389		
12 01.5	−0 23	Σ 1591	AB	Vir	1831 1932	354 352		53.8	51.7	8.5	8.5	K A			
--			AC		1905 1915	53		24.1			12.6				
12 01.7	+70 41	Σ 1590		Dra	1832 1944	235		5.1		7.5	10.5	K0	8395		
12 01.8	−34 39	I 215		Hya	1897 1960	238 121		0.7	0.6	7.4	8.1	G0			
12 01.9	+0 06	OΣΣ 116		Vir	1875 1932	181		74.4		8.3	8.8	A2			
12 02.1	+43 03	For	AB	UMa	1909	52		310.5		5.1	6.8	A3 K0			67 UMa
--			AD		1909	269		345.7							
--			BC		1909	323		198.2							
12 02.2	−52 32	Rst 570		Cen	1929 1959	237 228		0.5	0.5	8.2	8.4	F2			Spectrum composite F2+A5
12 02.3	−85 38	h 4490		Oct	1871 1940	145		25.0		6.1	10.4	K2			
12 03.0	−63 19	CapO 55		Cru	1921	325		4.5		4.3	13.6	A5			θ¹ Cru
12 03.5	−2 26	Σ 1593	AB	Vir	1829 1956	17		1.3		8.7	8.7	F0	8403		
--			AC		1893 1941	2 3		52.7	49.7		11.4		8403		
12 03.6	−39 01	See 143		Cen	1897 1959					7.2	7.4	F5		o	
12 03.7	−44 07	h 4491		Cen	1913 1933	41		23.4		8.2	8.5	F8 F8			
12 03.8	−54 43	h 4492		Cen	1900 1917	273		15.8		7.6	11.6	F2			
12 04.3	+21 28	Σ 1596		Com	1829 1958	237		3.7		5.9	7.4	F0	8406		2 Com
12 04.3	−21 02	β 458		Crv	1879 1901	233		30.5		7.9	10.0	A2			
12 04.8	−62 00	Δ 117	AB	Cru	1826 1918	149		23.0		7.7	8.1	B8 F5			
--			AC		1837 1918	18		25.0			10.2				
12 05.1	−61 11	ø		Cru		26		7.0		6.6	12.1	K0			
12 05.4	+38 39	h 2595		UMa	1830	315		15.0		8.0	18	F5			Single in 1901 (40-inch)
12 05.6	−5 51	h 198		Vir	1918	271		80.2		6.9	10.2	M			
12 05.6	+51 56	Σ 1600		UMa	1832 1941	93		7.7		7.4	8.4	F5	8414		
12 05.6	+52 53	OΣ 244		UMa	1820 1934	321		3.2		8.1	10.1	F5	8416		
12 05.6	+68 48	Σ 1599	AB	Dra	1831 1923	167		10.2		7.4	10.4	K5	8413		
--			AC		1911	332		105.8			12.3		8413		
--			AD		1875 1911	88		127.0			9.2		8413		
--			AE		1875 1911	180 180		129.7 126.8			7.8	M	8413		
12 05.7	+62 56	Hu 1136		UMa	1905 1958	223		1.8		6.1	11.5	K0	8417		
12 05.9	−35 42	See 144		Cen	1897 1900	176		25.4		6.2	13.9	B9			
12 06.0	+68 42	Σ 3123	AB	Dra	1832 1960					7.9	7.9	F5	8419	o	
--			ABxD		1953	181		26.0					8419		
--			AC		1895 1924	309		3.0			15		8419		
--			Dd		1953	69		0.2		8.0	8.0		8419		
12 06.1	−32 58	h 4495	AB	Hya	1836 1940	317		6.6		6.7	8.9	G0			
--			AC		1897	64		26.2			13.2				
12 06.4	−65 43	ø 367	AB	Mus	1960	119		0.1		7.0	7.0	F5			
--		h 4498	ABxC		1871 1940	60		8.7			8.0	A3			
12 06.6	−37 52	h 4500		Cen	1920	31		50.3		6.7	9.0	K0			

α 2000	δ 2000	Star Name		Const	Years	Pos Ang	Separation	m₁	m₂	Spec	ADS	Orb	Notes
12ʰ06ᵐ.9	-64°37'	h 4501		Cru	1918	299°	44.0	4.2	11.7	F0			η Cru
12 07.0	+69 05	Σ 1602	AB	Dra	1831 1954	180 179	13.0 17.1	8.1	9.6	G5	8428		
--			AC		1911	317	155.2				8428		
12 07.2	-6 05	A 2982	AxBC	Vir	1917	195	141.8	7.6		K2	8429		
			BC		1917 1957	76	0.4	11.0	11.5		8429		
12 08.0	-20 22	AC 6		Crv				6.0					No bright double star here
12 08.1	+55 28	Σ 1603		UMa	1838 1955	82	22.3	7.8	8.2	F8 F8	8434		
12 08.3	-18 34	β 412		Crv	1877 1952	160	2.0	8.4	8.9	F2	8436		
12 08.4	+43 54	Es		CVn	1912 1918	278	19.1	8.3	9.5	F5			
12 08.4	-50 43	Jc 2	AB	Cen	1847 1913	325	268.6	2.8	4.7	B3 B5			δ Cen
--			AC		1847 1913	227	216.6		6.4	B9			
12 09.3	+11 18	Σ 3078		Vir	1830 1927	305	9.7	8.0	10.8	F5	8438		
12 09.5	-11 51	Σ 1604	AB	Crv	1831 1970	93 89	12.0 9.9	6.8	9.3	G0	8440		
--			AC		1831 1959	97 75	58.0 19.1		9.2		8440		
--			BC		1900 1913	90 89	25.4 21.0				8440		
12 09.7	+1 54	β pm		Vir	1909	101	39.5	6.0	13.3	K0			10 Vir
12 09.8	-51 47	I 216		Cen	1900 1951	39 27	1.1 1.1	7.0	9.5	F8			
12 10.0	-34 42	Jc 17		Hya	1848 1951	20	3.4	6.4	8.2	A0			
12 10.6	-36 50	See 146		Cen	1897 1959	58	0.9	8.1	10.1	A2			
12 10.8	+39 53	Σ 1606		CVn	1831 1960	349 300	1.4 0.5	7.3	8.0	A3	8446	o	
12 11.0	+81 43	S,h 136		Cam	1876 1924	76	66.7	6.5	8.5	K0			BDS 6074 probably same
12 11.0	-45 25	I 423		Cen	1902 1935	166	2.6	6.6	10.4	K0			
12 11.2	-52 13	R 192		Cen	1895 1942	99	3.2	8.0	10.0	A2			
12 11.4	+53 25	Σ 1608		UMa	1832 1968	224 221	10.6 12.9	8.0	8.2	K0	8450		
12 11.4	-16 47	S 634		Crv	1877 1960	280 291	6.8 5.5	7.2	8.4	G5	8444		
12 11.7	-30 36	h 4505	AB	Hya	1900 1947	271	10.2	7.6	10.5	K0			
--		Knp	AC		1967	119	36.1		11.2				
12 12.0	+28 32	LDS 1285		Com	1963	164	150.0	7.0	17				
12 13.6	-33 48	Howe 72		Hya	1897 1959	173 164	1.4 1.4	6.5	8.3	B9			
12 13.7	-27 19	B 221		Hya	1926 1956	122 112	0.3 0.3	8.7	8.7	F5	8463		
12 14.0	-45 43	Rmk 14		Cen	1872 1954	244	2.9	5.6	6.8	M0			Both stars yellow (h)
12 14.1	+12 09	h 2603		Vir	1904 1925	17	23.7	7.9	12.3	K0			
12 14.1	+32 47	Σ 1615	AB	Com	1831 1931	88	26.7	6.9	9.7	K0	8470		
--		A 2058	BC		1909 1935	276	2.7		14.0		8470		
12 14.1	-36 33	See 148		Cen	1897 1900	63	19.5	7.9	13.4	B8			
12 14.1	-36 44	See 147	AB	Cen	1927 1960	307 334	0.3 0.3	7.6	8.6	A2			
--			ABxC		1897 1928	303	29.5		12.0				
12 14.3	-5 43	h 203	AB	Vir	1878 1909	352 352	30.2 25.1	6.6	12.5	A5	8471		
--			AC		1909	195	74.4		11.7		8471		
12 14.5	+8 47	Σ 1616	AB	Vir	1828 1953	296	23.3	7.6	9.8	G0 K2	8473		
--			AC		1864 1920	292 293	149.5 156.8		9.7		8473		
12 14.7	-24 47	B 222		Crv	1926 1935	95	4.3	7.5	14.2	K0	8474		
12 15.0	-1 20	h 204		Vir	1896 1911	56	35.7	7.7	10.9	K0			
12 15.0	-36 13	Slr 10		Cen	1895 1953	245	1.9	8.0	9.5	A2			
12 15.1	-7 15	Σ 1619	AB	Vir	1829 1974	288 271	7.8 7.1	8.1	8.4	G5	8477		
--			AC		1910 1925	176	98.7		10.6		8477		
12 15.2	-10 19	β pm		Vir	1909 1925	149 144	89.4 73.3	6.1	12.9	F8			
12 15.4	+57 02	β pm	AB	UMa	1878 1909	73 73	192.4 189.6	3.3	9.9	A2			69 δ UMa
--			AC		1879 1907	124 125	188.6 186.4		11.5				
12 15.7	-80 47	φ 195		Cha	1928 1941	317	0.4	8.5	9.1	A0			
12 15.8	-23 21	β 920		Crv	1879 1959	232 293	0.8 1.2	6.8	7.9	F5	8481		
12 16.1	+40 40	Σ 1622		CVn	1832 1958	260	11.4	5.8	8.1	K5	8489		2 CVn; v. gold, blue (Webb)
12 16.2	+80 08	Σ 1625		Cam	1832 1953	219	14.4	7.3	7.8	F0 F0	8494		
12 16.7	+39 36	Σ 1624		CVn	1831 1916	150	6.1	7.2	10.1	A2	8495		
12 17.5	+28 56	OΣ 245		Com	1848 1958	275 280	8.2 8.6	5.7	9.8	A0	8501		
12 17.8	-36 06	R 193		Cen	1881 1954	168 164	1.0 0.6	6.8	7.1	A0			
12 17.9	-24 01	β 921	AxBC	Crv	1879 1943	219	3.4	7.0		B9	8503		
--		Don 521	BC		1932 1951	35 47	0.3 0.3	11.2	11.2		8503		
12 18.0	+68 48	OΣ 246 rej		Dra				7.8		G5			Probably single
12 18.1	+18 26	Cou 181		Com	1967	148	8.9	7.5	13.8	F5			
12 18.2	-3 57	Σ 1627		Vir	1830 1958	196	20.1	6.6	6.9	F0 F0	8505		
12 18.2	-52 18	I 1220		Cen	1926 1959	331	0.2 0.1	7.7	8.0	B9			Too close since 1928
12 18.4	-64 00	h 4512		Cru	1918	339	33.8	4.0	13.0	B3			ζ Cru
12 19.0	-33 18	h 4513		Hya	1919	99	47.0	7.7	9.8	K0			
12 19.0	-55 09	HdO 214		Cen	1900	87	35.7	5.0	12.0	M			
12 19.3	-8 55	A 145		Vir	1901 1939	165	3.3	7.0	14.7	K0	8509		
12 19.4	+17 44	A 2059		Com	1909 1958	314 344	0.6 0.6	8.1	9.7	F5	8508		
12 19.9	-26 44	B 2292		Hya	1945	318	12.5	6.8	13.8	K5			
12 20.1	+13 51	β 27		Com	1875 1955	106	3.5	6.9	10.8	K0	8514		
12 20.2	+37 54	Σ 1632		CVn	1831 1971	193	10.2	6.8	10.0	K0	8516		
12 20.2	-22 11	β 605		Crv	1878 1959	139 187	1.1 0.4	6.2	8.0	G5	8515		
12 20.3	+3 19	β pm		Vir	1909 1925	3 4	130.3 132.2	5.0	11.5	K0			16 Vir
12 20.6	-22 13	β 1245		Crv	1891 1946	42 66	4.8 11.2	5.2	13.6	B8	8517		5 ζ Crv
12 20.7	+17 48	Ho 52		Com	1883 1958	44	9.1	4.7	12.7	K0	8521		11 Com
12 20.7	+27 03	Σ 1633		Com	1831 1958	245	9.0	7.0	7.1	F2	8519		V. pretty, solitary (Webb)

α 2000	δ 2000	Star Name		Const	Years	Pos Ang	Separation	m₁	m₂	Spec	ADS	Orb	Notes
12ʰ20.ᵐ9	−13°34′	β pm		Crv	1909	313°	48.″6	5.1	13.2	K0			
12 21.1	−42 34	I 1221		Cen	1926 1959	83	1.7	6.8	10.4	A0			
12 22.1	−67 31	HdO 215		Mus	1900 1930	130	32.4	5.2	10.6	A5			ζ² Mus
12 22.3	−73 30	φ 198		Mus	1933	108	2.1	6.8	10.5	A2			
12 22.4	−24 14	B 2294		Crv	1933	141	10.5	7.1	13.0	K0			
12 22.5	+5 18	Σ 1636		Vir	1829 1958	337	20.0	6.6	9.4	F8	8531		17 Vir
12 22.5	+25 51	S,h 143	AB	Com	1904 1935	54	35.0	4.8	11.8	F5	8530		12 Com
--			AC		1821 1972	167	65.2		8.3	A3	8530		
12 23.2	−37 29	Rst 2793		Cen	1935 1960	198 194	0.3 0.3	7.5	8.4	F0			
12 23.8	+54 10	OΣ 249	AB	UMa	1853 1959	315 280	0.5 0.4	8.1	8.9	G5	8535		
--			ABxC		1855 1925	149	13.2		11.2		8535		
12 23.8	−30 20	See 150		Hya	1897 1949	100	18.0	6.6	13.5	M0			
12 24.2	+5 58	OΣ 248 rej		Vir				8.0		M			Single
12 24.2	−38 55	HdO 216		Cen	1927 1960	54	4.6	7.0	12.8	A5			
12 24.4	+25 35	Σ 1639	AB	Com	1836 1959			6.8	7.8	A5	8539	o	
--			AC		1952	160	90.6				8539		
12 24.4	+43 05	OΣ 250		CVn	1845 1959	331 341	0.4 0.4	8.4	8.7	F0	8540		
12 24.7	−41 23	h 4518		Cen	1881 1959	208	10.0	6.3	9.6	K0			
12 24.8	−58 07	BrsO 8		Cru	1913 1943	335	5.3	7.6	7.9	G0			
12 25.0	+44 44	Σ 1642		CVn	1832 1954	180	2.6	8.4	9.2	F5	8546		Beautiful field (Bird)
12 25.1	+63 48	β		Dra	1905	245	52.3	6.3	10.0	G5			
12 25.5	−69 29	h 4522		Mus	1837 1918	67	12.8	7.9	8.9	B8			
12 26.0	−14 57	β 606		Crv	1878 1959	98 78	1.4 0.3	7.3	9.3	F5	8547		
12 26.0	−49 35	See 153		Cen	1897	203	13.6	7.0	12.8	A0			
12 26.5	−51 27	See 154		Cen	1897 1933	285	21.7	4.8	13.6	B3			
12 26.6	−63 06		AB	Cru	1826 1955	121 115	5.6 4.4	1.4	1.9	B1 B3			α Cru
--			AC		1826 1913	202	90.1		4.9	B5			
12 26.8	−0 11	AGC 4		Vir	1876 1956	234 200	0.8 0.7	7.7	11.2	A5	8550		
12 26.8	−5 36	A 78		Vir	1900 1959	86 114	0.3 0.3	8.0	8.5	F2	8551		
12 27.2	−19 58	S 637		Crv	1825 1919	203	60.6	8.6	9.0	F5 G0			
12 27.3	+7 14	Fox		Vir	1913	323	52.3	8.3	9.6	K0			
12 27.3	−55 44	Jsp 533		Cru	1930 1947	303	8.0	7.1	13.1	B9			
12 27.4	−28 43	B 228		Hya	1926 1959	133 280	0.3 0.1	8.2	8.4	F0	8555		Direct motion
12 28.0	−39 29	Rst 1669		Cen	1934 1945	333	0.9	7.8	11.5	K5			
12 28.1	+44 48	Σ 1645		CVn	1832 1976	161 158	10.4 9.9	7.4	8.0	F5	8561		Composite spectrum F5+K
12 28.2	−70 33	Don 528		Mus	1932 1948		9.6	7.7	12.5	K2			
12 28.3	+4 24	β 923	AB	Vir	1879 1925	60	2.6	6.8	13.5	A0	8563		
--		Fox	AC		1913	77	60.1				8563		
12 28.3	−61 46	Rst 4499	AD	Cru	1960	330	9.2	7.0	13.4				
--		CapO 12	AxBC		1880 1960	270 221	2.0 1.9			G0			
--		Rst 4499	BC		1939 1960	357 311	0.3 0.3	8.5	8.5				
12 28.6	−56 24	HdO 217	AB	Cru	1903	275	27.2	6.2	13.0	K0			
--			AC		1902	290	49.0		12.0				
12 28.9	+25 55	Σ I 21	AB	Com	1836 1928	251	145.4	5.3	6.6	A0 A3	8568		17 Com
--		β 1080	BC		1889	156	1.8		14.6		8568		
12 28.9	−37 58	I 218		Cen	1902 1954	234	2.3	7.8	10.4	K0			
12 28.9	−61 52	I 36		Cru	1902	325	21.8	6.9	10.8	B5			
12 29.8	−29 19	Rst 5367		Hya	1944 1951	298	0.5	8.1	10.1	K0			
12 29.9	−16 31	S,h 145		Crv	1823 1958	214	24.2	3.0	9.2	A0	8572		7 δ Crv
12 30.0	−6 01	Rst 4500		Vir	1938 1943	276	3.2	7.1	13.0	F2			
12 30.0	+51 32	β pm	AB	CVn	1908	172	109.2	6.2	10.4	F8			7 CVn
--			AC		1908 1919	326 327	231.8 229.0		9.0	A5			
12 30.1	−13 24	β 28	AB	Crv	1875 1960			6.5	8.6	G0	8573	o	
--		Lv	AC		1918	294	91.0		11.1		8573		
--			AD		1918	183	79.2		12.1		8573		
12 30.6	+3 30	Σ 1648		Vir	1829 1968	40	8.0	7.6	9.6	K0	8576		
12 30.6	+9 43	Σ 1647		Vir	1830 1959	202 237	1.2 1.3	8.5	8.8	F2	8575		
12 30.9	−30 41	I 514		Hya	1926 1959	135	0.1	8.2	8.3	A3			
12 31.2	+1 20	S,h 146	AB	Vir	1823 1921	290	49.9	7.8	9.5	A5	8582		
--		A 2583	BC		1912 1952	336 336	4.6 5.0		12.9		8582		
12 31.2	−57 07	Δ 124	AB	Cru	1826 1919	42 31	93.1 110.6	1.6	6.7	M4 A2			γ Cru
--			AC		1879	82	155.2		9.5				
12 31.3	−41 30	I 82		Cen	1898 1959	352 5	0.8 0.7	8.2	9.2	F0			
12 31.6	−11 04	Σ 1649		Vir	1830 1955	194	15.3	8.0	8.4	A5	8585		
12 32.1	+74 49	Σ 1654		Dra	1832 1973	24	3.7	7.6	9.1	K0	8591		
12 32.5	−59 54	Jsp 539		Cru	1929 1942	294	0.2	8.4	8.7	A0			
12 33.1	−47 03	Rst 3796		Cen	1936 1941	26	7.2	7.6	13.2	K0			
12 34.0	−49 55	HdO 218		Cen	1897 1932	121	7.8	6.4	12.9	F2			
12 34.2	−18 12	Lv 5		Crv	1888 1959	33 356	1.4 1.0	7.3	9.9	G0	8597		
12 34.2	−32 06	h 4528		Hya	1897 1960	154 145	23.6 26.8	7.1	11.7	A5			
12 34.7	−44 40	B 1718		Cen	1929		1.0	5.8	9.8	G5			
12 34.9	−5 09	Rst 4502		Vir	1938 1958	69 89	0.2 0.2	8.6	8.6	A5			
12 34.9	+22 38	Wils		Com	1937	90	0.2	4.8		A0			23 Com; single in 1940
12 35.1	+7 27	Σ 1658	AB	Vir	1830 1958	342 9	2.0 2.5	7.9	9.7	F8	8601		
--			AC		1865 1923	257 261	101.9 112.1		9.7		8601		

α 2000	δ 2000	Star Name		Const	Years		Pos Ang		Separation		m₁	m₂	Spec	ADS	Orb	Notes
12ʰ35ᵐ.1	+18°23′	Σ 1657		Com	1830	1958	271°		20.3		5.2	6.7	K0 A3	8600		24 Com
12 35.6	−34 53	LDS 413		Cen			sf		97.0		7.8	10.8	K0			
12 35.7	−16 50	h 1218	AB	Crv	1875	1933	259		11.8		6.6	11.0	F2	8603		Here mag. of A is for Aa
—		ø 368	Aa		1960		118		0.1		7.4	7.4		8603		
12 35.8	+56 35	Σ 1662		UMa	1831	1958	230	238	20.2	19.4	8.1	10.4	K2	8605		
12 36.0	−39 52	ø 1	AB	Cen	1931		227		6.9		5.9	13.3	A0			
—		h 4533	AC		1920	1931	74		40.0			12.0				
12 36.2	+56 50	A 1601		UMa	1906	1916	67		2.0		7.2	14.5	K2	8607		
12 36.8	+20 14	A.G. 180		Com							7.9		F5			Single
12 36.9	−32 11	I 1222	AB	Hya	1926	1951	216		0.4		8.6	8.8	K2			
—			ABxC		1926	1951	313		12.8			13.7				
12 37.2	+21 12	Σ 1663		Com	1830	1957	117	89	0.8	0.6	8.1	9.0	F2	8611		
12 37.2	−69 08	Don 541		Mus	1932	1948	316		29.6		2.7	12.8	B5			α Mus
12 37.7	−27 08	B 230		Hya	1926	1959	152	170	1.7	1.3	5.5	11.4	F0	8612		
12 38.1	−55 56	H1d 116		Cru	1880	1956	193		1.9		7.2	9.0	A2			
12 38.3	−11 31	Σ 1664	AB	Vir	1830	1923	272	237	17.1	26.3	8.1	9.3	K0 G5			
—			AC		1907		306		62.5			11.5				
—			AE		1907		110		118.8							
—			BC		1907		329		56.7							
—			CD		1898	1907	265	266	33.6	32.5		11.6				
—			EF		1907		111		93.7							
12 38.7	−4 22	S 639		Vir	1840	1919	106	108	50.2	53.5	7.0	9.7	K5			
12 38.9	−67 12	h 4535		Mus	1915	1918	339		17.0		6.3	12.0	B2			
12 39.0	+22 40	Kui 60		Com	1958		224		32.8		6.4	12.4	K0			
12 39.2	−8 00	β pm	AB	Vir	1886	1919	138		173.1		4.7	9.0	K0 F0			26 χ Vir
—			AC		1886	1919	111		221.2			10.3				
—			AD		1886	1909	330		321.2			9.0	K2			
12 39.2	+14 20	Σ 1666		Com	1830	1908	190		7.2		8.1	10.2	A3	8616		Sp. of A composite A3+G
12 39.2	−75 22	I 296		Mus	1901	1931	273		2.0		6.7	8.7	B9			
12 40.3	−71 39	B 1720		Mus	1930		58		0.8		8.1	10.0	A0			
12 40.9	+8 50	Σ 1668		Vir	1830	1958	197	190	1.7	1.4	7.6	8.1	F2	8625		
12 40.9	+27 08	Cou 596		Com	1971		209		0.2		7.8	7.9	F5			
12 40.9	−27 43	B 231		Hya	1926	1932	82		4.3		7.3	13.7	B9	8624		
12 41.3	−13 01	Σ 1669	AB	Crv	1828	1973	299	311	5.4	5.4	6.0	6.1	F5	8627		
—			AC		1900	1930	235		59.0			10.5		8627		
—			BC		1909	1916	228		58.0					8627		
12 41.5	−48 58	h 4539	AB	Cen	1835	1959					2.9	2.9	A0		o	γ Cen
—		See 159	ABxC		1897	1902	118		39.5			14.3				
12 41.6	+10 26	H VI 81		Vir	1899	1918	282		85.5		6.2	10.0	A5			27 Vir
12 41.7	−1 27	Σ 1670	AB	Vir	1825	1976					3.5	3.5	F0 F0	8630	o	29 γ Vir, Porrima
—			AC		1889		159		53.1			15		8630		
—			AD		1912	1923	88		123.6			12.1		8630		
—			BD		1880		88		102.8					8630		
12 41.8	+9 53	Ho 54	AB	Vir	1882	1904	102		119.5		7.7	10.7	F0	8631		
—			BC		1882	1959	151	141	1.5	2.2		10.7		8631		
—			BD		1904	1906	62		10.4			14.3		8631		
12 42.0	+6 48	β 924		Vir	1880	1958	29	37	3.7	4.0	5.6	11.4	A0	8633		31 Vir
12 42.1	−54 46	ø 200		Cen	1930	1959	297	273	0.2	0.3	7.2	7.5	K5			
12 43.5	−58 54	Rst 606	AB	Cru	1939		163		2.5		6.5	11.0	K0			
—		h 4543	AC		1913		96		31.1			9.8				
12 43.8	−12 01	Hu 738		Crv	1900	1959	244	259	6.5	9.1	6.8	11.8	K0	8645		
12 44.0	+21 10	OΣ 253		Com	1847	1934	237		6.6		8.0	11.2	F0	8649		
12 44.1	+35 46	Ho 256		CVn	1887	1959	102	113	0.5	0.8	7.1	9.1	A3	8651		
12 44.6	−57 17	ø 65		Cru	1928	1959	238		0.2		7.3	7.5	A0			
12 44.9	−52 45	h 4546		Cen	1882	1917	223		15.0		7.8	9.8	F2			
12 45.0	−8 32	Rst 4965		Vir	1943		97		7.6		7.8	11.2	K0			
12 45.1	+27 24	h 521		Com	1875		3		32.8		7.5	13.8	K0			
12 45.2	−62 13	CorO 140		Cru	1901	1934	97		4.9		7.9	9.9	F8			
12 45.3	−3 53	Σ 1677		Vir	1830	1938	349		15.9		6.8	8.3	A3	8657		
12 45.4	+14 22	Σ 1678		Com	1832	1958	212	181	32.6	34.1	6.8	8.5	A0	8659		
12 45.5	−37 27	I 1559		Cen	1927	1942	309		6.6		7.6	13.2	G0			
12 45.6	−60 59	h 4547		Cru	1882	1942	40	22	27.1	26.9	4.7	9.5	G8			ι Cru
12 46.2	+2 28	OΣ 255 rej		Vir	1878	1929	338	343	20.2	17.9	7.7	12.7	A5	8662		
12 46.3	+8 48	Bgh		Vir			271		20.0		8.4	8.6	F8 K0			
12 46.3	−68 06	R 207		Mus	1880	1955	317	14	0.8	1.4	3.7	4.0	B3			β Mus
12 46.4	+9 32	β pm		Vir	1852	1909	183	191	192.6	171.5	5.8	8.9	K0			33 Vir
12 46.4	−56 29	h 4548		Cru	1913		169		52.6		4.7	9.0	B5			
12 46.5	+24 09	β pm		Com	1894	1911	13	12	156.0	160.3	5.9	9.8	F5			
12 46.8	−33 19	HdO 219		Cen			226		65.0		5.9	11.9	K0			
12 47.2	+11 57	β pm		Vir	1909	1960	4	2	138.9	139.4	6.1	9.3	A3			34 Vir
12 47.6	−6 18	Rst 4966		Vir	1943	1958	243		16.3		6.3	12.4	F8			
12 47.7	−59 41	I 362	AB	Cru	1901		322		44.3		1.3	11.2	B1			β Cru
—		Δ 125	AC		1853		23		369.9			7.3	B8			
12 48.3	−67 08	h 4550		Mus	1873	1918	98		13.6		8.1	9.1	A2			
12 48.9	+12 06	β pm	AB	Vir	1893	1919	358	356	155.2	158.7	7.1	10.1	G5			

α 2000	δ 2000	Star Name		Const	Years		Pos Ang		Separation		m₁	m₂	Spec	ADS	Orb	Notes
12ʰ48ᵐ9	+12°06′	β pm	BC	Vir	1893	1918	328°	328°	136″7	135″6		9.5				
12 49.0	−65 36	Gli 185		Mus	1883	1920	9		8.6		7.6	9.5	A0			
12 49.1	+42 13	Es		CVn	1908	1918	50		47.4		7.7	8.1	G5			
12 49.2	+83 25	Σ 1694		Cam	1832	1958	326		21.6		5.3	5.8	A2 A0	8682		B is a sp. binary
12 49.3	+27 33	h 522		Com	1878	1912	10	13	43.2	42.5	5.8	11.5	A0	8674		30 Com
12 50.7	−34 00	See 163		Cen	1897		60		27.6		4.9	14.9	A0			
12 51.4	−10 20	Σ 1682	AB	Vir	1831	1959	309	301	33.6	30.2	6.5	9.3	K0	8684		
—			AC		1911		201		143.9			10.9		8684		
12 51.7	−31 05	h 4554		Cen	1898	1920	25		31.6		7.3	11.8	M0			
12 51.9	+19 10	Σ 1685	AB	Com	1829	1949	202		16.0		7.3	7.9	F2	8690		Spectrum composite F2+A2
—		S,h 123	AC		1823	1918	327		247.4			8.3		8690		
—			BC		1910	1918	328		56.4					8690		
12 51.9	+25 40	Σ 1684 rej		Com	1905		268		29.2		7.7	10.7	K5			
12 52.2	+17 04	Σ I 23	AB	Com	1836	1922	49		95.2		6.3	6.7	K5 F8			32 Com
—			AC		1910		262		897.3			8.7				
12 52.4	−53 50	HdO 220		Cen	1903	1947	210		6.4		6.3	12.3	K0			
12 52.8	−39 19	B 1216		Cen	1928	1960	301		0.3		8.3	9.3	K2			
12 53.2	−39 23	See 165	AB	Cen	1900	1936	167		3.5		6.8	12.5	K2			
—			AC		1897	1920	233		20.6			14.0				
12 53.3	+21 14	Σ 1687	AB	Com	1829	1959					5.1	7.2	K0	8695	o	35 Com
—			AC		1830	1958	126		28.7			9.1		8695		
12 53.6	−43 05	h 4555		Cen	1898	1919	304		22.8		7.2	12.2	A5			
12 53.9	−29 19	Stone 26		Hya	1867	1948	33		2.9		7.7	9.9	K5	8698		
12 54.0	−18 02	S 643		Crv	1825	1915	294		23.4		6.8	9.3	A0 A2	8699		
12 54.0	−60 20	B 805		Cru	1927		264		4.3		6.8	13.2	B3			
12 54.2	+82 31	OΣ 258		Cam	1848	1911	70		10.4		7.3	10.5	K0	8375		
12 54.2	−27 58	h 4556		Hya	1836	1954	82		6.0		7.4	8.9	G0	8700		
12 54.6	−57 11	Δ 126		Cru	1826	1952	17		34.9		4.3	5.3	B3 B3			
12 55.0	+58 09	Σ 1691	AB	UMa	1831	1937	276		18.7		8.6	9.4	F0	8702		
—			AC		1910		86		131.0			9.7		8702		
12 55.0	−85 07	Rst 2819		Oct	1935	1941	230		0.6		6.0	6.5	K0			ι Oct
12 55.5	+11 30	Σ 1689		Vir	1827	1927	198	211	28.7	29.0	7.1	9.4	M	8704		
12 55.6	+3 24	β pm		Vir	1879	1925	142	138	152.0	164.5	3.4	10.4	M			43 δ Vir
12 55.9	−56 50	HdO 221		Cru	1900	1913	315		29.1		5.3	10.3	O9			
12 56.0	+38 19	Σ 1692		CVn	1830	1970	228	229	19.6	19.4	2.9	5.5	A0	8706		12 α CVn, Cor Caroli
12 56.2	−4 52	Σ 1690		Vir	1832	1943	148		5.7		7.0	8.5	A0	8707		
12 56.3	+54 06	Σ 1695	AB	UMa	1832	1958	289	283	3.3	3.7	6.0	7.9	A2	8710		
—			AC		1880	1909	142		124.1			10.4		8710		
12 56.4	−0 57	OΣ 256		Vir	1848	1958	57	91	0.7	0.9	7.2	7.6	F5	8708		
12 56.6	+43 33	β 925		CVn	1879	1926	211		6.9		7.0	12.5	A0	8713		
12 56.7	−11 57	Rst 3816		Vir	1937	1943	83		7.0		7.4	14.0	K5			
12 56.7	−40 16	Wg 159		Cen	1897	1903	163		23.4		8.0	11.6	K0			
12 56.7	−47 41	I 83		Cen	1897	1959					7.4	7.6	F5		o	
12 56.8	+45 36	OΣ 257		CVn	1846	1926	353		13.0		8.5	9.2	F2	8714		
12 57.2	+8 18	ø		Vir	1963		162		0.1		7.9	8.1	F5			
12 57.3	−48 38	CorO 142		Cen	1913	1933	134		5.0		8.2	10.1	F5			
12 57.5	−9 46	A 146		Vir	1901	1960	308	296	1.8	1.8	7.1	10.1	F5	8715		
12 57.5	+24 57	Cou 397		Com	1969		64		0.4		8.5	9.2	G5			
12 58.0	−38 55	h 4560		Cen	1898	1920	252		29.7		6.9	12.3	A0			
12 58.2	−54 11	CorO 143		Cen	1919	1938	112		16.5		7.4	8.9	A0 F			
12 58.4	−51 31	Rst 1697	AB	Cen	1934	1947	100		1.0		7.8	9.9	G5			
—			AC				p		10.0			12.0				
12 58.7	+27 28	Σ 1699		Com	1830	1957	1	6	1.5	1.5	8.8	8.8	G5	8721		
12 59.0	−9 50	B 2541		Vir	1945	1960	326	300	1.5	1.6	7.6	12.5	G5			
12 59.3	+6 30	Σ 1701		Vir	1829	1927	306		21.6		7.5	9.5	G5	8724		
12 59.7	−3 49	Σ 1704		Vir	1830	1958	55		20.9		5.8	11.0	A0	8727		44 Vir
12 59.8	−55 55	Δ 127		Cen	1826	1938	126		16.8		8.4	8.9	B9 B9			
13 00.0	−41 23	I 1224		Cen	1925	1956	340	353	0.3	0.3	8.1	8.3	F2			
13 00.1	+0 18	Σ 1706 rej		Vir	1905	1911	172		29.0		8.0	10.2	M			
13 00.2	−23 55	I 1225		Hya	1926	1959			0.1		7.9	8.1	A2	8728		Measures uncertain
13 00.3	+30 47	β 1081		Com	1889	1958	351		5.2		5.1	14.4	K0	8731		37 Com
13 00.3	−48 36	CapO 13		Cen	1889	1947	67		5.1		7.1	9.2	K0			
13 00.6	−3 22	AGC 5	AB	Vir	1876	1958	150	165	1.2	1.1	6.0	11.0	K0	8732		46 Vir
—			AC		1878	1912	117	122	34.0	36.0		13.0		8732		
13 00.6	+18 22	β 112	AB	Com	1875	1926	350		149.9		6.1	9.5	F5	8735		
—			AxBC		1892	1898	349		151.0					8735		
—			BC		1875	1953	293		2.0			9.9		8735		
13 00.7	+56 22	β 1082		UMa	1889	1959					5.0	7.4	F0	8739	o	78 UMa
13 01.1	−33 37	h 4563		Cen	1837	1952	237		6.4		7.0	8.4	G5			
13 01.6	−24 40	B 2306		Hya	1933		88		16.2		7.0	13.5	F0			
13 01.8	+63 37	LDS 2662		Dra	1966		321		119.0		6.3	15	F5			
13 01.8	−30 50	δ 164		Cen	1924	1934	292		5.9		7.4	12.0	K0			
13 02.2	+10 58	β pm		Vir	1879	1909	120	120	240.9	248.7	2.8	11.7	K0			47 ε Vir
13 02.4	−29 16	B 239		Hya	1926	1932	256		4.0		7.1	13.9	A0	8745		
13 02.5	+22 58	A.G. 183		Com							8.0		K0			Single; possibly BDS 6354

α 2000	δ 2000	Star Name		Const	Years	Pos Ang	Separation	m_1	m_2	Spec	ADS	Orb	Notes
13^{h}02^m.5	+23°30′	Σ 1709		Com	1831 1955	250°	2″.4	7.7	10.5	F2	8749		
13 02.8	−55 40	h 4564		Cen	1894 1913	219	22.0	8.2	9.5	K2			
13 03.0	+22 38	A.G. 184		Com				7.0		A0			Double, A.G. Berlin. Single
13 03.8	−20 35	β 341		Vir	1876 1947	312	0.8	6.3	6.4	G0	8757		
13 03.8	−34 15	See 169		Cen	1897 1942	235	11.2	7.1	11.5	G5			
13 03.8	−40 51	B 2308		Cen		s	10.0	7.0	13.0	K0			
13 03.9	−3 40	β 929		Vir	1879 1960	229 205	0.5 0.8	7.2	7.5	F0	8759		48 Vir
13 04.5	+8 39	Σ 1716		Vir	1831 1944	148	2.7	8.1	10.9	F0	8763		
13 04.8	+73 02	β 799	AB	UMi	1881 1959	239 260	0.6 1.2	6.5	8.5	A5	8772		
—			ABxC		1955	195	91.1				8772		
13 05.5	+14 44	h 220		Com	1896 1926	42	17.6	7.8	12.1	F8	8766		
13 05.5	+41 41	Es 125		CVn	1902 1936	121	2.4	7.9	10.5	F5	8769		
13 05.6	−22 04	I 915		Vir	1912 1960	67 65	0.6 0.3	8.4	8.5	F5	8765		
13 05.9	+45 16	β 930		CVn	1879 1958	109 119	2.7 2.7	5.6	11.9	K0	8775		
13 06.2	+29 02	h 2638	AB	Com	1878 1958	219	6.3	6.5	12.0	A2	8777		
—			AD		1878 1901	8	39.6				8777		
—		β 1083	BC		1889 1901	237 221	0.5 0.4		12.2		8777		Single in 1958 (B)
—		β			1909	36	149.2			A2	8777		[62]
13 06.2	+40 55	h 2639	AB	CVn	1906 1916	160	31.4	7.5	10.5	K5			
—			AC		1906 1916	137	57.2		9.5				
—			BC		1916	112	31.3						
13 06.3	−48 28	h 4567		Cen	1858 1933	79	11.4	4.7	11.2	B3			
13 06.4	+21 09	Cou 11		Com	1959 1960	323	1.0	6.1	9.1	F5			39 Com
13 06.6	−46 02	I 917		Cen	1911 1959	303 289	1.2 1.2	8.1	8.3	F2			
13 06.9	−49 54	Δ 128		Cen	1836 1933	100	25.1	4.3	9.4	B3			ξ² Cen
13 07.3	+0 35	Σ 1719		Vir	1830 1950	1	7.2	7.6	8.1	F5	8786		
13 07.4	−59 52	R 213		Cen	1874 1956	24	0.8	6.7	6.9	B9			Both alike and orange (R)
13 07.6	+26 39	β pm		Com	1911	324	185.9	10.5	7.6	K0			
13 07.6	−73 35	B 1728		Mus	1930	286	4.4	7.6	14.1	A0			
13 07.8	−18 00	Com		Vir	1888 1960	181	3.3	7.9	11.8	G5	8787		
13 07.9	−19 02	B 2014		Vir	1933 1959	102	2.2	7.8	11.0	K2			
13 08.0	−56 41	h 4569		Cen	1873 1934	242	4.8	7.5	9.2	A0			
13 08.1	−65 18	Rmk 16		Mus	1826 1952	187	5.3	5.7	7.3	B0			θ Mus
13 08.2	+38 44	Σ 1723		CVn	1832 1934	8	6.6	8.3	8.6	K0	8795		
13 08.3	−78 27	h 4566		Cha	1900 1918	228 229	29.8 30.5	6.6	13.2	A0			
13 08.4	+15 29	Σ 1722		Com	1829 1946	344 339	3.6 2.9	8.3	9.3	K0	8796		
13 08.5	−2 41	S 647		Vir	1825 1919	213	41.9	8.0	9.5	F2			
13 09.3	−51 41	CorO 149		Cen	1920	256	12.7	8.2	10.2	A0			
13 09.8	+62 14	OΣΣ 121		UMa	1877 1914	9	107.8	6.6	9.8	A0			
13 09.9	−5 32	Σ 1724	AB	Vir	1830 1958	343	7.1	4.4	9.4	A0	8801		51 θ Vir
—		H III 50	AC		1782 1934	298	69.6		10.4		8801		
—			BC		1912 1934	295	66.2				8801		
13 09.9	−47 33	Rst 4974		Cen	1943 1949	36	1.8	7.8	13.8	K2			
13 10.0	+17 32	Σ 1728	AB	Com	1827 1960			5.1	5.1	F5	8804	o	42 α Com [63]
—			AC		1851 1923	322 330	135.4 110.3		10.1		8804		
13 10.1	+38 30	Σ I 24	AB	CVn	1835 1922	298 297	90.0 84.4	6.0	6.2	F0 B9	8805		17 CVn
—		β 608	BC		1878 1958	285 276	1.2 1.2		11.2		8805		
13 10.7	−4 56	β 609		Vir	1878 1959	356 351	0.9 0.4	7.6	11.6	F2	8807		
13 10.8	+13 18	β 931		Vir	1879 1928	204	5.0	7.3	12.4	K0	8810		
13 10.8	−24 11	Hu 1500		Hya	1913 1959	29 34	3.6 2.9	7.7	11.5	G5	8809		
13 10.9	+21 14	Cou 96		Com	1893 1966	316 312	10.9 10.7	7.0	10.8	F2			
13 11.5	+21 55	Cou 54		Com	1893 1965	288 291	10.6 10.3	6.8	10.4	G5			
13 11.5	−35 08	h 4571		Cen	1879 1930	267	23.5	6.6	9.0	K0			
13 11.7	−26 33	φ 305		Hya	1933 1960	f	0.2	7.2	7.3	A3		o	Maximum separation 0″.2
13 11.9	+27 53	β pm		Com	1851 1907	279 251	131.1 90.8	4.3	10.1	G0			43 β Com
13 11.9	−69 57	HdO 222		Mus	1900	280	30.7	5.9	11.8	F2			
13 12.0	+32 05	OΣ 261		CVn	1843 1959	359 342	0.6 2.2	7.2	7.7	F8	8814		
13 12.1	+36 54	Σ 1730		CVn	1832 1957	338	1.8	8.2	9.9	F0	8815		
13 12.1	−16 12	h 2645	AB	Vir	1878 1908	10 6	70.9 79.3	5.0	12.4	F2			53 Vir
—			AC		1897	354	184.2						Not found in 1908 (β)
—			AD		1908	328	235.0		10.0				
13 12.2	+16 08	A 2225		Com	1910 1930	69	3.0	7.4	13.1	F2	8816		
13 12.3	−59 55	See 170	AB	Cen	1897 1960	sf	0.3	5.0	5.7	B8			Obs'd arc short; oft single
—		I 424	ABxC		1902 1960	346 2	1.7 1.7		8.4				
—		HdO 223	ABxD		1901 1931	258	49.0		12.6				
13 12.4	+80 28	Kui 61		Cam	1935 1958	175 178	0.8 1.0	6.3	10.3	G5			
13 12.5	−34 45	LDS 436		Cen	1897 1943	221	31.5	7.8	9.8	F5			
13 12.6	+58 27	Σ 1732	AB	UMa	1832 1915	128	26.3	8.2	9.7	G0	8821		
—		β	BC		1905	291	2.6		13.7		8821		
13 12.6	−39 42	I 1566		Cen	1927 1942	12 13	4.9 5.8	7.7	12.7	K0			
13 12.7	−65 18	I 919		Mus	1915 1960	42	0.5	8.2	9.9	B9			
13 12.9	−59 49	CorO 152		Cen	1920	147	25.8	6.3	9.5	F5			
13 13.0	−29 06	Howe 27	AB	Hya	1877 1934	293	2.7	8.0	9.5	A0	8818		
—			AC		1897 1916	281	38.0		13.2		8818		
13 13.2	−2 33	Σ 1731		Vir	1831 1930	301	9.0	8.0	10.1	F8	8823		

α 2000	δ 2000	Star Name		Const	Years	Pos Ang	Separation	m_1	m_2	Spec	ADS	Orb	Notes
13^{h}13^m.3	−15°28′	β 221		Vir	1875 1959	47° °	1.6″ ″	8.1	9.6	G5	8822		
13 13.4	−18 50	S,h 151		Vir	1823 1958	34	5.4	6.8	7.3	A0	8824		54 Vir
13 13.4	−50 42	I 1227		Cen	1926 1960	122 69	0.4 0.4	6.7	7.0	A0			
13 13.5	+67 17	Σ I 25	AB	Dra	1835 1924	296	178.8	6.5	6.7	K0 K0			
--			AC		1835 1924	233 229	124.9 114.5		9.2				
--			BC		1912	335	171.1						
13 13.6	+56 42	OΣΣ 122	AB	UMa	1876 1926	210 213	115.1 117.5	6.8	8.7	G0 G5			
--			BC		1924 1926	245	62.2						
13 13.7	+29 49	Ho 55	AB	Com	1905	166	0.7	7.4	11.4	K0	8826		Single in 1958 (B, Cou)
--			AC		1893	154	73.4		10.5		8826		
13 14.0	−62 39	I 399		Cen	1902 1931	138	8.2	6.7	11.2	F5			
13 14.2	−59 06	φ 205		Cen	1928	341	2.7	4.9	10.2	F4			
13 14.5	+11 20	h 2647		Vir	1905 1960	211 213	47.5 49.1	5.7	10.7	K5	8832		
13 14.5	−24 17	φ 297	AB	Hya	1931 1959	333 23	0.3 0.1	7.3	7.5	A2	8831		
--		Stone 28	ABxC		1879 1931	335 333	12.4 12.4		11.4		8831		
13 14.7	−63 35	MlbO 3		Cen	1883 1959	40	1.7	7.0	9.5	B3			
13 14.9	−11 22	S,h 162	AB	Vir	1823 1922	62 49	44.8 79.3	7.0	7.9	G0 K0			
--		Rst 3829	Aa		1937 1958	264 291	0.7 0.8		9.1				
--		S,h 162	BC		1909	135	68.8		12.7				
13 15.2	−67 54	Δ 131		Mus	1826 1836	332	60.0	5.0	8.2	B8 A0			η Mus
13 15.3	−34 09	CorO 153	AB	Cen	1897 1942	347 350	4.4 4.6	7.6	11.0	A0			
--		See 172	AC		1897	227	20.0		14				
--			AD		1897	109	49.3		15				
13 16.1	−57 04	h 4576		Cen	1872 1933	128	5.6	7.7	9.7	F0			
13 16.5	+19 47	Bgh		Com		58	3.0	6.5	8.2	A3 A2			
13 16.8	+9 25	Kui 62		Vir	1958	89	34.3	5.2	14.3	F8			59 Vir
13 16.8	−41 17	I 233	AB	Cen	1897 1953	114 110	3.4 3.4	7.3	10.3	G5			
--		See 173	AC		1897	4	44.0		10.3				
13 16.9	+17 01	β 800	AB	Com	1881 1977	122 106	1.3 6.8	6.6	9.7	K0	8841		
--			AC		1909 1925	9 3	89.5 92.5		10.4		8841		
13 16.9	−34 36	I 1567		Cen	1897 1960			8.6	9.0	G5		o	
13 17.0	−13 10	h 2648		Vir	1906	98	42.6	7.8	10.6	K2			
13 17.4	−21 33	β 222		Vir	1867 1953	12 18	1.7 1.7	8.4	9.1	F0	8844		
13 17.5	−0 41	φ 350		Vir	1959	27	0.1	7.2	7.2	F0			
13 17.6	−37 01	h 4578		Cen	1889 1940	153 150	8.0 8.6	7.4	10.6	A2			
13 17.9	−68 30	HdO 224		Mus	1902 1936	213	0.5	7.2	8.0	B9			
13 18.4	−18 19	H VI 90		Vir	1862 1907	22 29	169.2 231.5	4.7	10.2	G5			61 Vir
13 18.6	+39 55	Es 1544		CVn	1916 1936	268	7.1	8.2	10.1	K0	8852		
13 18.9	+0 30	A 2585	AB	Vir	1913 1960	237 229	0.6 0.9	8.6	8.9	G0	8855		
--			AC		1913 1921	233	13.6		13.0		8855		
13 18.9	−23 10	Smyth		Hya	1879 1891	95	138.4	3.0	9.7	G6			46 γ Hya
13 19.1	−52 39	I 516		Cen	1912 1956	190	0.6	8.1	9.6	F5			
13 19.5	+35 08	h 529	AB	CVn	1896 1935	120 124	17.3 17.3	10.0	11.0	M2	8861		
--			AC		1909	75	152.6		12.3		8861		
--			AD		1879 1902	264 268	311.5 319.0		6.0	A5	8861		
13 19.6	+9 42	A 1787		Vir	1908 1944	358	1.6	7.9	11.3	F0	8860		
13 19.6	−63 45	Rst 632		Cen	1930 1944	305	5.0	8.0	10.7	F5			
13 19.7	−11 40	Hu 740		Vir	1901 1944	271	3.9	7.1	12.6	A2	8858		
13 20.2	+15 34	h 223		Com	1900 1912	348	36.4	7.4	9.9	G5			
13 20.3	+17 46	A 2166		Com	1910 1960	7 213	0.2 0.1	7.8	7.8	F5	8863		Quadrant change in 1940?
13 20.6	−59 46	See 176		Cen	1897	209	13.1	6.2	12.6	F2			
13 20.7	+2 57	Σ 1734		Vir	1830 1959	198 183	0.7 1.0	6.8	7.5	A0	8864		
13 21.4	−22 48	Hu 1503		Hya	1913 1959	192	1.2	7.2	11.2	F5	8867		
13 21.4	−64 03	h 4579		Cen	1871 1943	98	4.6	8.4	9.2	G0			
13 21.8	−55 25	I 924		Cen	1912 1942	87	0.6	8.4	8.9	A0			
13 22.3	+30 22	Stone 29		CVn	1879	175	0.4	7.5	7.5				Single (IDS)
13 22.5	−22 57	Arg 26		Hya	1903 1925	80	27.8	8.5	8.6	G0 G0			
13 22.6	+26 07	Ho 259		Com	1887 1908	242	9.7	7.5	13.5	F8	8879		
13 22.6	−60 59	φ 208	AB	Cen	1930 1959	168	0.1	5.4	5.4	B5			Measures uncertain
--		Δ 133	ABxC		1826 1879	343	60.0		6.5	B3			
13 22.8	−13 11	H IV 119		Vir	1783 1924	310	19.4	7.8	10.8	A2	8878		
13 22.9	−47 57	Slr 18		Cen	1895 1960	224 239	0.6 0.6	6.9	7.2	A2			
13 22.9	−72 09	B 1736		Mus	1930	316	0.3	6.4	7.8	B5			
13 23.2	−14 55	Σ 1738		Vir	1830 1955	281	3.9	8.5	8.6	F8	8881		
13 23.7	+2 43	Σ 1740		Vir	1833 1968	76 75	27.3 26.5	7.4	7.6	G5 G5	8883		
13 23.9	+54 56	Σ 1744	AB	UMa	1830 1977	148 152	14.4 14.4	2.3	4.0	A2 A2	8891	o	79 ζ UMa, Mizar [64]
--			AC		1893 1966	72 71	708.6 708.7		4.0		8891		C is 80 UMa = Alcor
13 24.0	−20 55	β 610		Vir	1878 1933	18	3.9	6.6	10.6	K0	8885		
13 24.3	+1 24	Σ 1742		Vir	1831 1958	353	1.2	7.6	8.1	A2	8890		
13 24.6	−51 30	I 1231		Cen	1926 1953	290	0.2 0.1	7.9	8.0	A2			Too close after 1927
13 24.8	−34 38	I 220		Cen	1902 1943	7 10	0.5 0.5	8.1	9.6	G5			
13 25.1	−15 38	β 460		Vir	1877 1937	35	2.2	7.6	9.9	A2	8893		
13 25.2	−11 10	β pm		Vir	1879 1909	33 32	144.4 147.9	1.0	11.8	B2			67 α Vir, Spica
13 25.2	−64 29	h 4583		Cen	1918 1935	209 203	27.1 25.7	5.3	11.0	F2			
13 25.3	+0 51	OΣ 265 rej	AB	Vir	1851 1958	275 285	17.9 23.6	8.0	11.0	K0	8895		

α 2000	δ 2000	Star Name		Const	Years	Pos Ang		Separation		m_1	m_2	Spec	ADS	Orb	Notes
13^h25^m.3	+0° 51'	OΣ 265 rej	AC	Vir	1909	80°		73".7			13.0		8895		
13 25.3	+40 28	h 1231	AB	CVn	1903	10		21.7		8.8	12.5	F8			
--			AC		1903	233		94.6			8.8	K2			
13 26.1	+72 23	Kui 63		UMi	1958	58		25.7		5.8	13.2	M1			
13 26.1	-32 33	B 249		Cen	1926 1959	13 315		0.3	0.4	7.1	8.6	F0			
13 26.6	-24 17	Bgh		Hya		88		19.0		8.1	8.5	K0 K0			
13 27.1	+64 44	OΣΣ 123	AB	Dra	1876 1924	147		68.9		6.7	7.0	F0 F0			Striking object (Webb)
--			BC		1910 1960	92 94		32.9	36.4		12.2				
13 27.1	-22 22	β 1107		Vir	1889 1937	133		1.2		8.5	8.6	F8	8906		
13 27.1	-49 09	φ 351		Cen	1959 1960	166 156		0.1	0.1	7.2	7.2	A0			
13 27.2	-40 10	h 4588		Cen	1903 1920	139 138		44.4	45.4	6.3	12.0	K0			
13 27.4	-62 39	I 517		Cen	1912 1939	275		2.0		7.9	11.1	B			
13 28.1	-43 46	R 218		Cen	1881 1943	170		2.4		7.3	9.8	K0			
13 28.2	+9 28	Σ 1746		Vir	1829 1958	251 247		29.6	25.0	7.7	10.3	K0	8912		
13 28.3	+2 14	A 2490		Vir	1912 1937	97		1.1		7.6	11.1	K0	8913		
13 28.4	+13 47	β pm		Vir	1852 1909	144 137		304.5	286.4	5.0	8.6	G0			70 Vir
13 28.4	-67 52	h 4586		Mus	1837 1929	150 144		3.1	3.1	7.3	9.1	A3			
13 28.5	+15 43	OΣ 266		Com	1844 1959	329 351		1.0	2.0	8.4	8.9	F5	8914		
13 28.9	+59 56	Σ 1752	AB	UMa	1832 1958	149 128		1.6	1.1	9.0	11.0	F8	8919		
--		S 649	CA		1824 1924	111		182.3		5.5		A0	8919		
13 29.1	+56 14	A 1362		UMa	1906 1935	115		2.1		7.7	10.8	G5	8921		
13 29.2	+22 11	Σ 1748	AB	Com	1832 1935	182		5.7		8.1	11.1	F8	8918		
--			AC		1911	353		150.1			10.8		8918		
13 29.2	-54 55	h 4589		Cen	1913	99		14.7		7.8	10.4	G5			
13 29.2	-66 04	LDS 444		Mus		sf		29.0		8.6	8.9	F8 G			
13 29.7	-23 17	Ho 381		Hya	1891 1931	324		21.2		5.0	12.0	M8	8920		R Hya; A is variable
13 29.8	+1 06	J 2091		Vir	1952	336		22.8		6.7	13.1	A2			
13 30.0	+60 21	Σ 1754 rej		UMa	1906	32		23.5		8.0	10.2	G5			
13 30.4	-6 28	Σ 1750		Vir	1831 1938	16		29.8		6.1	11.4	A5	8924		72 Vir
13 30.5	-23 39	Ho 540		Hya	1895 1903	196		13.3		7.9	13.0	A0	8923		SS Hya; A is variable
13 30.7	+9 19	Σ 1751		Vir	1831 1935	60		5.7		7.1	10.3	K0	8928		
13 30.9	+24 14	OΣ 268 rej	AB	Com	1878 1930	77		19.3		7.5	13.0	F2	8929		
--		H V 70	AC		1783 1930	263 258		56.9	71.4		8.0	G5	8929		
--			AD		1908	30		84.6			12.7		8929		
13 31.0	-39 24	See 179		Cen	1897 1960					4.5	4.7	K0		o	
13 31.4	-42 28	See 180		Cen	1897 1942	231		3.7		6.7	9.2	K0			
13 32.0	-18 44	B 2542		Vir	1940 1959	199 183		0.1	0.1	6.7	6.9	A3			73 Vir; too close in 1951
13 32.0	-65 19	φ 369		Mus	1960	56		0.1		7.4	7.4	A3			
13 32.1	-63 02	Δ 137		Cen	1826 1918	358		16.1		7.7	9.0	B0 B			
13 32.4	+36 49	Σ 1755		CVn	1832 1950	131		4.4		7.2	8.1	G5	8934		
13 32.4	-12 40	S, h 165		Vir	1823 1922	79		47.9		7.9	9.3	F0 F0			
13 32.5	-62 21	Hrg 86		Cen	1883 1955	239		1.6		7.2	7.6	B9			
13 32.5	-69 14	I 298		Mus	1901 1947	203 187		0.8	0.8	7.2	8.8	F2			
13 32.8	+24 21	A 567		Com	1903 1958	262		1.5		6.1	12.3	G5	8937		
13 32.8	+34 54	OΣ 269	AB	CVn	1844 1960					7.3	7.8	A5	8939	o	
--			AC		1879 1910	333		116.3			9.3		8939		
13 32.9	-15 22	h 2658	AB	Vir	1879 1902	110 110		78.3	79.6	5.6	11.2	K0			75 Vir
--			Aa		1902	320		18.9			13.5				
13 33.0	-2 25	Hld 15	AB	Vir	1881 1909	296		16.2		7.9	12.1	K0	8938		
--			AC		1909	249		140.4			11.9		8938		
13 33.4	+39 56	h 2659		CVn	1830	315		10.0		7.8	17	F0			[65]
13 33.7	-41 19	Rst 1727		Cen	1934 1945	179		2.2		7.5	13.4	A2			
13 34.3	-0 19	Σ 1757	AB	Vir	1831 1974					7.8	8.7	K0	8949	o	
--			AC		1907 1921	153 153		42.2	45.0		11.7		8949		
--			AD		1921	72		128.0					8949		
13 34.3	-8 37	β 114		Vir	1842 1960	127 156		1.4	1.3	7.9	8.4	F8	8950		
13 34.3	-11 32	Hu 470		Vir	1901 1945	254		3.6		7.9	12.9	F2	8947		
13 34.4	+26 16	Σ 1760		Com	1831 1950	65		8.6		8.8	8.8	F2	8953		
13 34.4	+38 47	h 1234		CVn	1896 1960	22 14		32.4	30.6	6.4	11.2	A3	8956		
13 34.5	-48 16	Rst 4985		Cen	1943 1949	160 166		0.5	0.5	6.5	8.5	A0			
13 34.6	-13 26	S 650		Vir	1825 1919	149 136		45.5	50.0	8.5	8.8	G0			
13 34.7	-13 13	β 932	AB	Vir	1879 1960					6.5	6.9	A0	8954	o	
--			ABxC		1879 1959	155 149		23.8	25.7		12.5		8954		
13 34.8	+65 15	Mlr 162		Dra	1971	306		2.0		7.7	14.0	F2			
13 35.1	-58 22	Jsp 588		Cen	1929 1947	278		0.3		8.5	9.2	G0			
13 35.6	+10 12	h 228		Vir	1828 1922	16		70.2		6.7	8.7	K0			
13 36.8	+6 50	A 1611		Vir	1907 1957	138 132		0.6	0.7	8.6	8.7	A5	8968		
13 36.8	+69 47	Σ 1771		UMi	1831 1958	71 79		1.7	1.8	8.5	9.2	F5	8976		
13 36.8	-26 30	H N 69	AB	Hya	1825 1953	191		10.1		5.9	6.8	A2 A	8966		
--			AC		1922	232		198.3			11.3		8966		
--			AD		1922	18		218.2			10.0		8966		
13 36.8	-32 24	I 221		Cen	1898 1959	170 132		0.5	0.5	8.1	8.9	F2			
13 36.9	-56 09	I 518		Cen	1912 1944	178 173		2.6	2.6	8.1	11.0	K0			
13 37.1	-16 27	Rst 3845		Vir	1937 1943	306		1.0		8.0	11.0	F2			
13 37.2	-61 42	I 365	AB	Cen	1900 1960					6.2	6.5	F5		o	

α 2000	δ 2000	Star Name		Const	Years		Pos Ang		Separation		m₁	m₂	Spec	ADS	Orb	Notes
13ʰ37ᵐ.2	−61°42′	I 365	ABxC	Cen	1926		229°		45″.0			12.0				
13 37.4	−64 56	h 4596		Mus	1879	1947	282		1.4		8.3	8.5	B9			
13 37.5	+36 18	Σ 1768	AB	CVn	1831	1959					5.0	6.9	F0	8974	o	25 CVn
—			AC		1918	1959	140	141	220.2	217.7		8.6		8974		
13 37.6	−7 52	Σ 1763	AB	Vir	1830	1957	41		2.8		7.9	7.9	K2	8972		
—			AC		1912		328		14.1			11.0		8972		
13 37.7	+2 23	Σ 1764	AB	Vir	1832	1933	31		15.8		7.0	8.7	K0	8975		
—			AC		1909		139		172.2			9.4		8975		C is A of Σ 1765 rej
13 37.7	+50 43	Σ 1770		UMa	1831	1956	121		1.8		6.8	8.3	K5	8979		
13 37.8	−35 04	See 184		Cen	1897	1959	312	303	1.1	2.4	7.4	9.4	G0			
13 38.0	+39 10	Σ 1769	AB	CVn	1832	1958	24	35	2.8	2.1	7.8	10.2	G5	8981		
—			AC		1832	1956	259		56.1			9.6		8981		
—			AD		1909	1922	194		167.6			11.7		8981		
—			AE		1909	1922	175		151.5					8981		
13 38.1	−58 25	R 223		Cen	1913	1960	23		2.5		6.4	11.0	K0			
13 38.8	−57 37	φ 339		Cen	1957	1959	30		0.1		6.8	6.8	K0			
13 39.2	−49 00	h 4600		Cen	1882	1933	119		16.7		7.9	9.2	K0 G5			
13 39.6	+10 45	β 612	AB	Boo	1878	1960					6.3	6.3	F2	8987	o	
—			ABxC		1912	1960	275	274	130.3	125.2		11.0	F2	8987		
13 39.8	−52 41	B 1230		Cen	1929	1939	107		3.5		8.0	12.0	K2			
13 39.9	−53 28	Rst 5376		Cen	1948		158		36.0		2.3	12.7	B2			ε Cen
13 40.4	+50 31	Σ 1774 rej		UMa	1879	1958	134		17.6		6.4	9.7	F8	8992		
13 40.7	+19 57	Σ 1772	AB	Boo	1831	1976	149	136	4.8	4.6	5.8	8.7	A2	8991		1 Boo
—			AC		1913		22		88.6					8991		
13 40.7	+28 04	β pm		CVn	1879	1960	228	226	91.4	91.2	6.3	9.6	K0			
13 40.7	+76 51	h 2682	AB	UMi	1905	1956	279		26.3		6.7	9.7	A5	8997		
—			AC		1905	1956	315	316	47.8	45.9		9.0		8997		
13 40.8	−28 14	Ho 382	AB	Hya	1891	1905	334		14.0		8.2	12.5	F5	8990		
—		h 4604	AC		1880	1905	281		15.9			10.0		8990		
13 41.5	−23 27	φ 352	AB	Hya	1959	1960	350	327	0.1	0.1	7.4	7.4	A0	8994		
—		h 4606	ABxC		1904	1931	353		31.1			9.9		8994		
13 41.7	−54 34	Δ 141		Cen	1835	1956	163		5.3		5.3	6.7	B9			
13 41.7	−75 07	h 4598		Mus	1871	1918	46		13.1		6.7	11.7	A0			
13 42.3	−33 59	h 4608		Cen	1848	1952	174	185	4.4	4.2	7.4	7.5	F5			
13 42.4	−57 19	Rst 2861		Cen	1934	1947	234		0.6		8.0	10.3	A2			
13 42.7	−61 54	Rst 4515		Cen	1939		69		6.4		6.8	12.8	K0			
13 43.0	+50 02	h 2676		UMa	1901	1925	125		29.7		7.6	9.5	K0			
13 43.0	−30 11	δ 166		Cen	1924	1929	157		0.9		7.0		A2			B variable?
13 43.1	+3 32	Σ 1777		Vir	1828	1958	235	229	3.4	2.9	5.5	7.9	K0	9000		84 Vir
13 43.5	−4 16	Σ 1775	AB	Vir	1829	1938	336		27.7		7.1	9.8	K2	9002		
—			AC		1898		178		38.5			13.5		9002		
13 43.6	−57 35	Rst 1739		Cen	1932	1947	60		1.5		7.3	12.4	K2			
13 43.7	−42 04	φ 353	AB	Cen	1959	1960	39	48	0.1	0.1	6.8	6.8	B8			
—		Rst 1741	ABxC		1934	1960	65	71	0.8	0.8		9.1				
13 43.8	−40 11	Howe 95		Cen	1897	1947	187		1.6		7.6	7.9	A5			
13 44.0	−59 14	Δ 142		Cen	1826	1879	90		32.8		6.7	8.4	B9 B9			
13 44.2	+59 22	Hu 1261		UMa	1905	1922	110		1.2		7.9	13.9	K0	9009		
13 45.1	+18 22	Σ 1782		Boo	1828	1915	186		29.8		7.8	9.3	F5			
13 45.3	−66 45	Don 616		Mus	1929		198		3.6		7.6	12.1	B8			
13 45.5	+3 30	A 1612		Vir	1907	1959	317	334	1.0	1.2	8.0	11.0	F8	9014		
13 45.6	−15 46	h 2677		Vir	1879	1960	312	313	43.2	43.9	6.2	11.7	A0			85 Vir
13 45.9	−12 26	β 935	AB	Vir	1879	1958	298	303	1.6	1.2	5.8	10.8	K0	9018		86 Vir
—		Σ 1780 rej	AC		1879	1958	164		26.9			11.9		9018		
—			CD		1879	1958	274	272	1.7	2.4		13.1		9018		
13 46.1	+5 07	Σ 1781		Vir	1830	1960					7.8	8.2	G0	9019	o	
13 46.1	+41 02	Σ 1783		CVn	1832	1952	49		2.2		8.1	10.3	K0	9020		
13 46.6	+54 26	LDS 2914		UMa	1955		89		70.0		5.5	18	A0			84 UMa
13 46.7	−51 26	HdO 225		Cen			310		40.0		4.7	11.0	G5			
13 46.9	−36 15	h 4612		Cen	1897	1931	341		26.3		5.2	12.5	A0			
13 47.0	+38 33	S 654		CVn	1825	1925	238		71.3		5.6	8.8	K0			
13 47.2	−9 43	Kui 65		Vir	1935	1960	264	253	0.4	0.4	6.5	7.5	K0			
13 47.2	−62 35	CorO 157		Cen	1900	1933	318		9.4		6.5	10.5	G5			
13 47.3	+17 27	OΣ 270		Boo	1849	1960	348	11	10.3	4.8	4.5	11.1	F5	9025		4 τ Boo
13 48.6	+48 21	β 802		UMa	1881	1950	223		3.6		7.5	10.7	F0	9030		
13 48.8	−52 50	Rst 2871		Cen	1935	1948	121		1.7		8.0	13.4	F8			
13 48.9	−35 42	Howe 94		Cen	1890	1938	355		11.6		6.6	9.6	F8			
13 49.1	+26 59	Σ 1785		Boo	1830	1976					7.6	8.0	N2	9031	o	
13 49.3	−40 31	Δ 146		Cen	1826	1920	86	86	51.4	60.0	7.1	7.2	F0 K5			
13 49.3	−62 06	Δ 143		Cen	1826	1933	36		12.5		7.6	8.1	G5 B8			
13 49.4	−44 13	CorO 161		Cen	1920	1930	150		14.1		8.2	9.2	A0			
13 49.6	−42 28	HdO 226		Cen	1897	1925	128		48.0		3.0	13.7	B3			μ Cen
13 49.6	−47 22	Δ 144		Cen	1836	1935	256		9.2		8.0	8.8	F5			
13 49.7	−12 33	Rst 3851		Vir	1937	1943	217		2.4		7.7	14.0	M0			
13 50.4	+21 17	S 656		Boo	1825	1923	208		85.8		6.8	7.3	G0 G0			BDS 6648, H VI 15 same?
13 50.7	−29 52	h 4617		Hya	1880	1954	261		4.9		7.7	9.6	G5	9033		

α 2000	δ 2000	Star Name		Const	Years		Pos Ang		Separation		m₁	m₂	Spec	ADS	Orb	Notes
13ʰ50ᵐ.8	+68°18′	LDS 2329		Dra	1955		66°		74″.0		8.0	9.0				
13 51.0	+68 19	OΣΣ 127	AB	Dra	1844	1924	68	65	74.1	79.1	6.4	8.2	K0 G5			
--			BC		1912		65		44.9			13.0				
13 51.4	+64 43	h 3342	AB	Dra	1879	1924	26	26	56.3	54.8	4.7	12.9	M	9039		10 Dra
--			AC		1879	1911	65	63	90.9	90.2		11.7		9039		
13 51.5	−48 18	CapO 61		Cen	1928	1932	131		30.5		7.4	7.7	G0 F8			
13 51.8	−33 00	H III 101		Cen	1783	1954	113	108	8.5	7.9	4.5	6.0	B5 B8			3 Cen
13 51.9	+10 08	B 2543		Boo	1932	1949	210	141	0.3	0.6	7.0	8.4	K5			
13 52.0	−1 29	Doo		Vir	1899		73		9.7		6.0	6.5				Some error; no such star
13 52.0	−31 37	β 343		Cen	1877	1959	130	66	1.4	0.7	6.5	7.5	F8			
13 52.0	−47 52	h 4619		Cen	1881	1913	199		23.6		7.1	9.5	G0			
13 52.0	−52 49	Rmk 18		Cen	1837	1954	289		18.0		5.4	7.6	B8 A3			
13 52.3	−30 47	See 189		Cen	1897	1900	255		13.0		7.8	12.8	A3			
13 52.6	−50 55	B 2323		Cen	1936		70		6.0		7.4	12.9	G5			
13 53.1	−73 16	B 1743		Aps	1930		288		3.1		7.0	11.6	A0			
13 53.2	−3 33	Rst 5530		Vir	1950		357		0.4		7.6	10.0	F8			
13 53.2	−31 56	H N 51		Cen	1837	1951	185		14.9		4.7	8.4	B7			4 Cen
13 53.3	+42 11	h 1244		CVn	1876	1958	131		6.8		6.9	11.4	A0	9044		
13 53.5	−35 40	Howe 28	AB	Cen	1889	1960	84	264	1.2	0.2	6.3	6.5	F2		o	Motion direct
--		β 1108	AC		1889	1959	168	158	27.5	27.7		12.5				
--		I	AD				149		30.0			14.5				
--		H V 124	AE		1889	1959	359	4	65.2	66.7		8.5				
13 53.5	−41 38	B 1233		Cen	1928	1959	306		0.2		8.6	8.8	A5			
13 53.5	−80 16	h 4610		Aps	1919		310		16.5		7.8	13.2	F0			
13 53.9	−14 40	Rst 3852		Vir	1937	1959	320	245	0.3	0.2	7.9	8.2	F0			
13 53.9	−47 08	h 4624		Cen	1897	1959	350		21.4		6.1	11.2	B3			
13 54.4	+29 55	OΣ 272		CVn	1849	1954	24	7	1.9	1.6	7.5	10.4	F2	9051		
13 54.5	−22 15	Hu 1262		Vir	1905	1959	282	295	0.8	1.0	6.7	10.8	K0	9050		
13 54.5	−51 31	B 1234		Cen	1929	1960	317	310	0.4	0.3	8.0	9.6	K0			
13 54.6	−34 36	See 191		Cen	1897	1900	156		19.5		6.9	13.0	F2			
13 54.6	−66 54	Δ 145		Cir	1837	1918	50		24.1		8.2	9.2	B9 A0			
13 54.7	+18 24	S,h 169		Boo	1822	1925	120	103	126.2	112.6	2.7	8.7	G0			8 η Boo
13 54.7	−50 41	BrsO 9		Cen	1879	1941	77		17.6		8.5	8.8	F5 F5			
13 55.0	−8 04	Σ 1788	AB	Vir	1831	1977	54	86	2.4	3.4	6.5	7.7	F8	9053		
--			AC		1886	1922	293		127.7			10.3		9053		
--			AD		1886	1924	215		156.8			10.9		9053		
13 55.0	+78 24	Σ 1798		UMi	1832	1968	13	12	7.2	7.5	8.1	9.9	F2	9069		
13 55.3	−32 06	Howe 74		Cen	1877	1908	117		6.2		7.1	9.6	G5			
13 55.5	−12 13	Rst 3854		Vir	1937	1943	273		4.0		7.9	13.0	G5			
13 55.6	−9 32	Ho 261		Vir	1887	1924	180		7.1		7.6	12.1	F0	9056		
13 56.2	−61 12	B 1744		Cen	1929	1949	326		3.1		8.1	11.0	F8			
13 56.3	+5 18	OΣ 273	AB	Vir	1844	1960	101	112	0.7	1.1	8.4	8.9	F5	9060		
--			AC		1875	1910	236		237.4					9060		C is the galaxy NGC 5363
13 56.3	−38 40	I 224		Cen	1897	1944	184		2.8		7.6	11.0	F2			
13 56.3	−54 08	R 227	AB		1880	1953	348	4	1.5	1.8	6.5	7.5	A2			
--			ABxC		1897		286		27.9			12.6				
13 56.5	−54 42	LDS 463		Cen			nf		33.0		6.1	13.0	G0			
13 56.7	+2 59	β 461	AB	Vir	1879	1899	235		33.3		6.9	11.9	A5			
--			AC		1879	1899	216		40.6			11.2				
13 56.8	−49 06	I 937		Cen	1912	1944	207		0.6		8.3	9.6	F8			
13 56.9	−27 40	See 193		Hya	1897	1926	163		6.5		7.8	12.9	F5	9062		
13 57.3	−56 02	Δ 151		Cen	1835	1943	310	45	11.5	23.2	7.6	9.4	G5 A2			
13 57.4	−62 29	ø 370		Cen	1960		65		0.1		7.5	7.5	G0			
13 58.5	−65 48	h 4632		Cir	1881	1933	14		6.4		6.2	10.2	K0			
13 58.6	−34 24	I 938		Cen	1914	1931	203		4.0		7.8	11.6	K5			
13 58.9	+53 06	Σ 1795		UMa	1832	1916	3		7.6		6.8	10.0	A2	9077		
13 59.1	+25 49	Σ 1793		Boo	1831	1950	242		4.6		7.5	8.5	A5	9076		
13 59.4	+24 18	Cou 402		Boo	1969		99		4.8		8.0	11.3	F0			
14 00.9	−66 16	HdO 227		Cir	1901		197		46.2		6.0	11.9	A5			
14 01.3	−40 13	I 1573		Cen			350		10.0		6.1	12.3	K0			
14 01.5	−62 57	I 225		Cen	1901	1929	301		2.4		7.6	11.4	K0			
14 01.6	+1 33	S,h 171	AB	Vir	1823	1927	290		80.0		4.3	9.6	A2	9085		93 τ Vir
--			BC		1912		21		158.5			13.2		9085		
14 02.0	+57 13	A 1097	AB	UMa	1905	1960					8.4	8.7	F5	9089	o	
--		Σ 1800 rej	ABxC		1906	1976	20	20	27.8	28.2		10.6		9089		
14 02.8	+62 16	M1r 136		UMa	1971		28		5.5		7.7	12.5	F0			
14 02.9	−35 11	I 1574		Cen	1927	1933	306		0.1		8.6	8.8	F8			Measures uncertain
14 03.0	−2 42	Bail 231		Vir	1896		224		5.9		8.1	11.0				
14 03.0	−31 41	β 1197		Cen	1890	1947	179	211	0.9	1.9	6.5	7.6	F5			
14 03.4	−3 21	Bail 232		Vir	1896		30		7.1		8.2	8.5				
14 03.5	−10 58	Rst 3858		Vir	1937	1943	59		1.9		7.5	13.5	A2			
14 03.7	+8 29	β 1270		Boo	1892	1960					8.3	8.4	F5	9094	o	
14 03.8	−60 22	Vou 31		Cen	1935	1960	259	251	1.3	1.3	0.7	3.9	B3			β Cen
14 04.0	−44 37	I 939	AB	Cen	1913	1960	142	166	0.8	0.8	8.7	8.8	G0			
--		Hu	AC		1913	1960	74	79	8.1	8.1		12.7				

α 2000	δ 2000	Star Name		Const	Years	Pos Ang		Separation		m₁	m₂	Spec	ADS	Orb	Notes
14ʰ 04.ᵐ1	−37° 16′	h 4643		Cen	1897 1920	135° 134°		21.″4	22.″2	7.2	12.2	F8			
14 04.6	+34 25	Hu 646		CVn	1903 1956	25 20		1.9	1.9	7.9	14.5	A0	9102		
14 04.6	−35 39	I 941		Cen	1910 1960			0.1		7.9	8.2	F0			Measures uncertain
14 04.8	−6 33	Σ 1799		Vir	1830 1944	294		4.2		8.0	9.2	F0	9100		
14 04.8	+25 49	Bgh		Boo		33		96.5		7.0	8.9	F5 K0			
14 05.5	−36 33	Δ 154		Cen	1836 1934	130		20.6		8.3	9.7	A5			
14 05.5	−50 24	CorO 165		Cen	1891 1953	68		2.5		8.2	8.6	K0			
14 06.0	−13 04	Howe 30		Vir	1879 1918	7		13.4		8.4	9.4	F2	9104		Sp. of A composite F2+A3
14 06.3	−26 35	β 938		Hya	1879 1952	118 91		0.8	0.3	8.4	8.5	F0	9106		
14 06.4	+38 25	Σ 1803		CVn	1831 1925	43		17.8		8.1	9.9	K0	9111		
14 06.7	+34 47	OΣ 274		CVn	1845 1958	71 59		14.8	13.1	7.2	10.2	K0	9112		
14 06.7	−36 22	See 196		Cen	1897	129		69.9		2.1	14.1	K0			5 θ Cen
14 07.1	−63 27	h 4642	AB	Cen	1918	12		9.1		6.7	12.4	G0			Spectrum composite G0+A3
—		δ	AC		1918	104		20.8			13.3				
—		h 4642	AD		1918	335		26.4			10.2				
—		Rst 1765	DE		1930 1944	8		0.6			12.0				
14 07.7	−49 52	Slr 19		Cen	1895 1959	230 295		1.0	1.4	7.2	7.4	G0			
14 07.8	−53 41	Δ 155		Cen	1837 1959	28 12		24.5	19.5	7.6	8.6	F2			
14 07.9	−61 31	LPO 45		Cen	1919 1934	112		7.7		8.1	11.3	B8			
14 08.1	−12 56	Σ 1802		Vir	1830 1959	286 279		4.2	5.6	7.6	8.9	G0	9115		
14 08.3	+49 27	H VI 112		Boo	1879 1917	275 274		83.5	79.7	5.3	9.8	M			13 Boo
14 08.9	−43 28	h 4653		Cen	1913	35		28.6		6.2	12.8	B9			
14 09.2	+7 23	OΣ 275		Vir	1844 1944	353		5.0		7.2	10.7	K0	9127		
14 09.6	−51 30	h 4651		Cen	1837 1959	134 132		62.7	64.1	6.2	9.2	B9 K2			
14 09.9	−44 17	See 198		Cen	1897 1929	137		9.2		7.7	12.0	B9			
14 09.9	−53 26	Rst 5532		Cen	1947	295		30.4		4.8	14.0	G5			
14 10.4	−30 05	See 199		Hya	1897 1926	226		8.3		7.0	12.5	A2	9135		
14 10.5	−70 18	Don 646	AB	Cir	1931 1948	162		5.2		6.1	14.6	K0			
—			AC		1931 1948	146		8.8			15				
14 11.0	−2 40	β 803		Vir	1881 1946	226		5.4		7.8	12.0	K0	9140		
14 11.0	+15 13	LDS 1406		Boo	1963	158		24.0		8.0	16				
14 11.0	−46 55	Pol 3		Cen	1887 1938	55		3.9		8.2	9.0	F2			
14 11.8	−19 01	Don 649		Vir	1932 1939	311 312		1.6	2.0	7.9	13.7	K0			
14 12.0	+44 11	OΣ 278		Boo	1843 1959					8.4	8.6	F2	9159	o	
14 12.3	+2 25	h 3343		Vir	1905 1959	214 212		62.6	59.9	5.0	12.0	A0	9152		CU Vir; variable
14 12.4	+28 43	OΣ 277	AB	Boo	1845 1957	334 19		0.4	0.4	8.3	8.5	F2	9158		
—		Σ 1812	AC		1832 1957	108		14.2			9.3		9158		Some measures are for ABxC
—			AD		1908 1953	153		72.6			11.8		9158		Some measures are for ABxD
14 12.6	−55 29	h 4659		Cen	1913	106		18.6		8.0	10.5	K0			
14 12.9	−25 30	Rst 5382		Hya	1945 1950	245		0.4		7.4	9.0	G5			
14 13.2	−63 59	Rst 688		Cen	1929 1944	238 246		0.4	0.5	8.8	8.8	F0			
14 13.4	+5 24	Σ 1813		Vir	1829 1955	193		4.8		8.4	8.5	A3	9163		
14 13.5	+18 10	LDS 1409		Boo	1963	249		129.0		7.0	19				
14 13.5	+51 47	Σ 1821		Boo	1832 1968	238 236		12.6	13.4	4.6	6.6	A5	9173		17 κ Boo
14 13.8	+9 01	A 1100		Boo	1905 1959	173		0.3		8.4	9.3	A5	9169		
14 13.8	+12 00	OΣ 279		Boo	1845 1952	253		2.2		6.7	8.9	K0	9168		
14 13.8	+41 11	Ho 58		Boo	1884 1931	226 229		3.8	3.9	8.0	12.2	K2	9175		
14 13.9	+29 06	Σ 1816		Boo	1831 1960	80 88		1.9	1.1	7.5	7.6	F0	9174		Spectrum composite F0+A2
14 14.5	−75 46	h 4652		Aps	1918	68		21.1		8.0	12.1	A2			
14 14.6	−66 09	Rst 3867		Cir	1936 1944	8		5.1		6.9	13.7	B5			
14 14.7	−65 42	Rst 690		Cir	1929 1944	282		1.8		6.8	10.1	B5			
14 14.8	+10 06	Kui 66		Boo	1936 1960	121		0.8		5.4	8.0	G5			15 Boo
14 14.9	−57 05	HdO 228	AB	Cen	1901 1913	170		33.9		5.1	10.9	B3			
—			AC			220		36.0			12.4				
14 15.0	−61 42	CorO 167		Cen	1887 1956	159		2.8		6.6	8.4	O5			
14 15.2	−67 40	Don 652		Cir	1929 1948	263		1.0		7.6	11.6	G0			
14 15.3	+3 08	Σ 1819		Vir	1830 1959					7.8	7.9	F8	9182	o	
14 15.5	+50 26	Σ 1829		Boo	1831 1938	150		5.4		8.0	8.5	F5	9187		
14 15.8	+56 40	Σ 1830	AB	UMa	1830 1956	264 303		4.8	8.1	9.7	11.0	G5	9191		
—			AC		1886 1923	81		35.5			13.7		9191		
—			AD		1872 1956	61 63		143.2	138.6		6.6	F8	9191		
14 16.2	+51 22	Σ I 26	AB	Boo	1836 1942	33		38.5		4.9	7.5	A5	9198		21 ι Boo
—			AC		1911 1925	197		85.9			12.6		9198		
14 16.2	+56 43	Σ 1831	AB	UMa	1830 1956	143 139		6.0	6.2	7.1	9.8	F0	9197		
—			AC		1872 1956	227 222		105.0	107.7		6.6	F8	9197		C is D of ADS 9191
—			CD		1924	119		114.4					9197		
14 16.4	+6 05	Σ 1824		Vir	1829 1920	281		5.3		8.1	10.1	A0	9188		
14 16.5	+20 07	Σ 1825		Boo	1830 1958	186 163		3.4	4.4	6.5	8.2	F5	9192		
14 16.5	−57 18	BrsO 10		Cen	1913	116		30.2		7.2	9.8	A0			
14 16.6	−66 35	HdO 229		Cir	1901	307		23.8		5.8	12.9	B2			
14 17.1	+54 33	β 1271		Boo	1892 1946	355 350		2.8	2.4	7.1	12.3	G5	9208		
14 18.0	−7 33	β pm		Vir	1907 1960	8 350		44.0	57.1	6.5	11.0	G0			
14 18.8	−58 41	I 523		Cen	1913 1956	17 41		0.3	0.3	8.0	8.1	B9			
14 19.0	−25 49	β 1246	AB	Hya	1891 1898	188		3.1		5.9	13.3	F4	9212		
—			AC		1891 1959	89 112		36.3	64.0		11.0		9212		

α 2000	δ 2000	Star Name		Const	Years	Pos Ang	Separation	m₁	m₂	Spec	ADS	Orb	Notes
14ʰ19ᵐ3	+13°00′	β pm		Boo	1887 1909	217° 217°	154″4 156″4	5.4	10.3	F0			18 Boo
14 19.4	−18 31	H1d 18	AB	Vir	1881 1960	355	3.7	7.9	11.3	A2	9218		
−−			AC		1909	336	134.5		12.1		9218		
14 19.6	−5 09	Kui 67		Vir	1935	107	15.3	7.6	12.9	K0			
14 19.7	−36 52	β 1110		Cen	1889 1933	132	3.9	6.9	11.5	M			
14 19.9	+67 47	Σ 1840		UMi	1831 1924	222	27.4	6.8	9.5	A0	9231		
14 20.2	−43 04	h 4672		Lup	1881 1935	302	3.7	5.7	8.1	G5			
14 20.3	+8 35	OΣ 281		Boo	1843 1944	151 163	1.1 1.5	7.9	11.4	G5	9227		
14 20.3	+48 30	Σ 1834		Boo	1831 1959			8.0	8.3	F8	9229	o	Close white pair
14 20.5	+26 34	Danjon		Boo	1938	46	0.2	8.2	8.2	A2			
14 20.6	−37 53	See 204		Cen	1897 1931	87	36.0	4.1	13.6	A0			ψ Cen
14 20.6	−58 31	Jsp 622		Cen	1929 1947	143	8.3	7.9	13.7	B8			
14 20.8	−42 25	I 1241	AB	Cen	1927 1959	131	0.3	7.6	9.1	A3			
−−			ABxCD		1890	211	80.9	7.3	8.0				
−−		CorO 168	CD		1889 1959	211 205	1.7 1.7	8.7	8.8	F0			
14 21.1	+67 48	Σ 1841 rej		UMi	1905 1921	265 265	34.8 36.4	7.1	10.7	A2			
14 22.0	−8 05	Ho 384		Vir	1891 1937	50 51	25.6 29.4	7.3	12.8	K2	9233		
14 22.6	−7 46	Σ 1833	AB	Vir	1832 1954	167 172	4.9 5.7	7.6	7.6	G0	9237		
−−			BC		1910	197	103.8		12.9		9237		
14 22.6	+32 30	Ho 262		Boo	1886 1925	272	5.3	7.1	13.1	K0	9243		
14 22.6	−48 19	R 244	AB	Lup	1881 1936	122	4.4	6.1	9.8	B2			
−−		Knp	Aa		1963	136	0.2						
14 22.6	−58 28	Δ 159	AB	Cen	1826 1931	160	9.3	5.0	7.1	G0			
−−			AC		1903	255	20.7		13.5				
−−			AD		1871 1902	4	45.6		10.5				
14 22.6	−73 33	h 4667		Aps	1879 1946	141	2.4	8.4	8.9	A0			
14 23.3	−50 46	B 1245			1929 1940	145	1.3	6.0	9.6	K0			
14 23.4	+8 27	Σ 1835	AxBC	Boo	1832 1958	186 192	6.1 6.2	5.1		A0	9247		
−−		β 1111	BC		1889 1960			7.6	7.7		9247	o	
14 24.0	−19 48	h 2714		Lib	1879 1959	276 283	20.0 30.7	7.7	11.9	K0			
14 24.1	+11 15	Σ 1838		Boo	1832 1946	334	9.1	7.4	7.5	F5	9251		BDS 6833, H III 20 same?
14 24.6	+47 50	Σ 1843		Boo	1830 1922	188	20.1	7.6	9.1	F5	9259		
14 24.7	−11 40	Σ 1837		Lib	1829 1958	327 292	1.4 1.2	6.7	8.3	F2	9254		
14 25.2	+51 51	OΣ		Boo	1854 1918	182	69.2	4.1	11.1	F8			23 θ Boo
14 25.2	−28 08	Stone 31		Hya	1897 1956	284 271	0.6 0.6	7.7	8.5	K2	9256		
14 25.3	−13 21	h 546		Lib	1902 1911	42	39.7	6.6	10.0	K0			BDS 6823 is same star
14 25.5	−19 58	S,h 179	AB	Lib	1822 1955	296	35.1	6.4	7.6	A0 A0	9258		
−−		β 225	BC		1875 1955	102 93	1.4 1.2		8.8		9258		
14 25.5	−43 39	B 2335		Lup		f	6.0	8.0	14.0	G0			
14 25.9	−22 08	h 4679		Lib	1879 1933	306	16.6	8.2	9.3	F2	9260		
14 26.1	−45 13	Δ 160		Lup	1913	204	158.2	4.6	9.3	B3 M0			τ¹ Lup
14 26.2	−45 23	I 402		Lup	1902 1960	183	0.3	5.1	5.2	F1			τ² Lup; often too close
14 26.3	−24 13	CorO 170		Lib	1902 1944	132	2.4	8.4	9.2	F8	9261		
14 26.7	+16 25	A 2069		Boo	1909 1960			8.9	9.1	F8	9264	o	
14 26.7	+78 30	Mlr 335		UMi	1972	94	1.0	7.2	12.5	F5			
14 27.2	−46 08	HdO 230		Lup	1900	143	26.9	5.8	11.4	A3			
14 27.5	+75 42	h 2733	AB	UMi	1905 1958	124	21.7	4.3	13.3	K2	9286		5 UMi
−−			AC		1879 1959	129 131	56.4 58.8		9.8		9286		
14 28.2	−2 14	Σ 1846	AB	Vir	1829 1958	110	4.8	4.8	9.3	K0	9273		105 φ Vir
−−			AC		1911 1960	209 205	95.1 92.8		12.4		9273		
14 28.2	−29 30	φ 306	AB	Hya	1933	130	0.1	5.8	5.8	B8	9270		52 Hya
−−		β 940	ABxC		1879 1935	279	4.2		10.0		9270		
−−			ABxD		1914	282	140.8		12.0		9270		
14 28.4	−23 01	Ho 386	AB	Lib	1893 1945	326	4.4	8.0	12.0	F8	9274		Mag. of A is for Aa
−−		Mlr	Aa		1966	178	0.2				9274		
14 28.6	+28 17	Σ 1850		Boo	1832 1958	262	25.6	7.0	7.4	A0 A0	9277		
14 29.1	+69 17	Mlr 170		UMi	1971	45	0.2	8.8	8.8	K5			
14 29.4	−80 06	h 4671		Aps	1836 1940	127	5.0	8.1	9.0	F8			
14 29.5	−37 02	See 205		Cen	1897 1900	66	11.0	7.9	14.0	G0			
14 29.8	+31 47	Σ 1854 rej		Boo	1905 1958	256	25.8	6.1	9.6	B9	9288		
14 29.8	−25 33	CorO 172		Hya	1897 1932	58	12.6	7.7	10.0	K0	9280		
14 30.0	−4 15	Σ 1852 rej		Vir	1879 1937	267	25.0	7.1	10.2	F2	9284		
14 30.1	−45 19	I 426		Lup	1902 1932	310	10.5	5.5	11.8	B9			
14 30.3	−49 31	HdO 232		Lup	1900 1913	18	22.1	5.4	11.9	A2			
14 30.8	+4 46	β		Vir	1903 1960	196	55.5	6.0	9.5	K2			See BDS 6932
14 31.0	−5 48	Rst 4529		Vir	1938 1960			8.4	8.4	G5		o	
14 31.3	−15 38	β 117	AB	Lib	1876 1960	96 84	2.4 2.1	8.3	9.2	G5	9291		
−−			AC		1911	335	107.3		12.0		9291		
14 31.3	−67 43	HdO 231		Cir	1901	194	35.4	5.8	14.0	K0			
14 31.8	+30 22	h 2728		Boo	1879 1953	334 339	53.2 42.2	3.6	11.3	K0	9296		25 ρ Boo
14 31.8	−76 16	I 326	AB	Aps	1898 1933	122 118	2.3 2.3	7.0	10.3	K0			
−−			AC		1933	10	14.8		13.5				
14 32.0	−32 11	See 206		Cen	1897 1932	20	23.1	7.5	13.9	K2			
14 32.1	+38 19	β 616		Boo	1878 1925	99 111	26.2 33.4	3.0	12.7	F0	9300		27 γ Boo
14 32.3	+26 41	A 570		Boo	1903 1960			6.5	6.7	A2	9301	o	

α 2000	δ 2000	Star Name		Const	Years		Pos Ang		Separation		m₁	m₂	Spec	ADS	Orb	Notes
14ʰ32ᵐ3	+80°20'	OΣΣ 130		UMi	1876	1914	300°	°	48".5	"	8.5	8.5	K5 F0			
14 32.5	+49 11	Hu	AC	Boo	1898		131		6.5		7.8			9306		
--		OΣ 283	AxBC		1848	1913	130		5.2				F5	9306		
--		Hu 57	BC		1887	1921	140		1.3		11.7	12.4		9306		
14 33.2	-30 43	β 1112		Cen	1889	1960	6	12	2.5	2.5	6.2	9.5	K0			
14 33.3	+85 56	Σ 1915		UMi	1832	1935	326	321	2.5	2.5	7.1	10.1	K0	9358		
14 33.5	-33 19	HdO 234		Cen	1902	1941	158		14.2		7.0	10.8	K0			
14 33.5	-52 41	HdO 233		Lup			200		20.0		5.9	14.0	K0			
14 33.6	+35 35	Σ 1858	AB	Boo	1831	1955	36		2.7		8.0	8.8	G5	9312		
--			AC		1910	1925	311	315	37.1	33.2		13.4		9312		
14 33.9	-46 28	Δ 162		Lup	1913		241		72.2		7.5	10.5	G5			
14 34.2	+32 32	Kui 68		Boo	1958		328		24.7		6.3	11.3	F0			
14 34.4	+24 24	OΣΣ 129		Boo	1874	1924	68		78.6		7.9	8.2	A0 G0			
14 34.7	+29 45	β pm	AB	Boo	1852	1907	82	82	248.0	237.1	4.5	9.8	F0			28 σ Boo
--			AC		1907		97		237.0			11.3				
--			BC		1853	1907	181	180	57.6	60.2						
14 35.5	-41 31	HdO 235		Cen	1898		130		26.9		5.8	11.0	B8			
14 35.6	-36 33	h 4687		Cen	1837	1959	82	106	2.0	1.1	8.6	8.8	F8			
14 35.8	+0 15	β 941	AB	Vir	1879	1960	218	184	0.8	0.5	9.6	9.6	F8	9318		
--			AC		1914		116		195.5			7.8	G5	9318		
14 35.8	+6 20	Ma 6		Vir	1843		196		19.3		7.5	10.0				Not found by β
14 35.9	-69 58	Don 664		Cir	1931	1948	194		9.8		7.8	14.8	M			
14 36.3	-46 15	HdO 236		Lup			100		40.0		5.6	12.7	K2			
14 36.8	-36 40	Rst 3887		Cen	1936	1945	270		0.7		7.8	10.3	B8			
14 36.9	+48 13	A 347		Boo	1902	1960					8.4	8.6	F2	9324	o	
14 37.1	-62 30	R 249		Cen	1881	1944	34		3.0		8.2	9.8	K0			
14 37.2	-37 32	Howe 75		Cen	1889	1960	215		4.1		7.9	8.4	A0			
14 37.3	-46 08	φ 318	AB	Lup	1951	1959	41	35	0.1	0.1	6.2	6.2	K0			
--		h 4690	ABxC		1846	1933	25		19.3			9.2	A			Y., bl.; very beautiful (R)
14 37.5	+2 17	A 2227		Vir	1910	1956	139		1.9		6.6	10.6	F8	9323		
14 38.0	+51 35	Σ 1863		Boo	1830	1960	110	74	0.6	0.6	7.4	7.7	F2	9329		
14 38.0	-54 31	Δ 163		Lup	1826	1938	114	106	57.0	60.7	8.1	8.5	F0 B8			
14 38.3	-49 54	φ 371		Lup	1960		36		0.1		7.4	7.4	A0			
14 38.4	-35 17	See 208		Cen	1897	1899	20		23.2		6.8	11.8	A2			
14 39.5	-68 12	M1bO		Cir	1929		254		16.4		6.8	14.3	B3			
14 39.6	-60 50	Richaud	AB	Cen	1689	1960					0.0	1.2	G0 K5		o	α Cen [66]
--		I	AC		1915		sp		(131')			11.0	M5			C is Proxima Centauri
14 39.6	-64 43	I 367		Cir	1901	1929	357		4.5		8.0	11.4	F2			
14 39.7	-26 43	β 805	AB	Hya	1881	1899	134		23.6		7.0	12.2	F5	9330		
--			AC		1881	1899	42		124.1			9.5	M	9330		
--			CD		1881	1959	240	285	2.0	1.1		11.7		9330		
14 40.1	+5 00	A 1107		Vir	1905	1956	75	87	0.3	0.4	8.3	9.3	F0	9334		
14 40.3	-40 51	HdO 237		Cen	1932		22		8.7		6.7	12.4	B9			
14 40.3	-45 48	See 209		Lup	1897	1959	241	244	12.8	10.4	6.6	12.4	A0			
14 40.4	-26 15	β 806	AB	Hya	1890	1938	97		0.7		7.2	9.2	A2	9333		
--			AC		1881	1931	67		71.7			9.1		9333		
--			Aa		1890	1899	329		17.8			13.5		9333		
--			CD		1881	1942	344		1.2			10.4	A5	9333		
14 40.5	-57 02	LDS 495		Cen			np		490.0		7.4	11.2	G5			
14 40.7	+16 25	Σ 1864	AB	Boo	1830	1957	99	108	5.8	5.6	4.9	5.8	A0	9338		29 π Boo
--			AC		1905	1960	161	162	126.2	127.7		10.0		9338		
14 40.7	+31 17	Σ 1867		Boo	1831	1960	22	5	1.6	1.0	8.4	8.9	F5	9340		
14 40.8	-66 57	NZO 52		Cir	1911	1927	56		2.1		8.2	9.2	F5			
14 40.9	-32 20	h 2736		Cen	1898	1959	81	77	20.1	17.4	8.0	9.5	K0			
14 41.0	+57 57	Σ 1872	AB	Dra	1830	1972	38	47	7.5	7.6	7.4	8.4	K0	9346		
--			AC		1912	1925	58		76.0			11.2		9346		
14 41.1	+13 44	Σ 1865	AB	Boo	1830	1960					4.5	4.6	A2	9343	o	30 ζ Boo
--		H IV 104	AC		1901	1911	259		99.3			10.9		9343		
14 41.1	-22 37	Rst 2917		Lib	1935	1959	243	210	0.3	0.3	7.9	8.5	A3			
14 41.5	-2 02	Rst 5004		Vir	1943		210		1.6		7.6	13.0	A2			
14 41.6	+51 24	Σ 1871		Boo	1829	1971	283	305	1.8	1.8	8.0	8.0	F0	9350		
14 41.8	+9 32	Σ 1866		Boo	1829	1956	21		0.9		8.6	8.6	G5	9345		
14 41.8	-29 42	β 345		Hya	1877	1960	306	292	0.9	0.9	7.8	8.1	F0	9344		
14 41.9	-30 56	β 414		Cen	1889	1960	347		1.0		6.9	7.8	B9			
14 41.9	-47 23	HdO 238		Lup	1901		232		27.6		2.3	13.4	B2			α Lup
14 42.1	+61 16	Σ 1878		Dra	1832	1974	336	319	3.1	4.1	6.3	8.5	F2	9357		
14 42.5	-64 59	Δ 166		Cir	1837	1951	244	232	15.7	15.7	3.2	8.6	A0			α Cir
14 42.8	+6 35	A 1109		Vir	1905	1960	32	74	0.6	1.3	7.3	9.8	F8	9353		
14 42.8	-55 11	Δ 168		Lup	1837	1951	202		5.7		8.3	8.6	F5			
14 42.9	+8 05	Σ 1870		Boo	1829	1937	230		4.4		8.0	10.9	F2	9355		
14 43.0	+1 46	h 5486		Vir	1823		150				8.0	9.0				Not able to identify (IDS)
14 43.5	-62 58	B 1759	AB	Cir	1930	1960	50	59	0.3	0.2	7.0	7.4	B3			
--			ABxC		1930	1960	77		1.9							C is variable
14 43.6	-39 39	I 1250		Cen	1925	1932	149		4.7		7.9	12.5	K0			
14 44.0	+2 43	h 5488		Vir	1823		50				8.0	8.5				No star in or near (IDS)

α 2000	δ 2000	Star Name		Const	Years		Pos Ang		Separation		m₁	m₂	Spec	ADS	Orb	Notes
14ʰ44ᵐ.1	+61°06'	Σ 1882	AB	Dra	1831	1931	2°	°	11".5	12".2	6.9	8.4	F2	9371		
--			Aa		1902	1916	88	83	7.8	9.1		10.5		9371		
14 44.4	-6 23	Rst 4530		Vir	1938	1943	57		6.6		7.6	12.5	K0			
14 44.8	-70 05	LDS 499		Cir			np		23.0		8.0	11.0	G0			
14 45.0	+27 04	Σ 1877	AB	Boo	1829	1971	321	339	2.6	2.8	2.5	4.9	K0 A0	9372		36 ε Boo, Izar
			AC		1912	1960	257	256	178.7	177.0		11.8		9372		
14 45.2	-55 36	Δ 169		Cir	1872	1938	106		68.0		6.2	7.6	B3 K0			
14 45.3	-36 09	I 528		Cen	1927				0.1		8.0	8.2	A2			Measures uncertain
14 45.3	-62 53	HdO 239		Cir	1900		64		36.3		5.4	10.9	A5			
14 45.5	+42 23	OΣ 285		Boo	1845	1960					7.9	8.2	F5	9378	o	
14 45.5	-26 56	B 278		Hya	1926	1933	25		0.3		8.2	9.6	A3	9369		
14 45.7	-22 25	B 1762		Lib	1929	1939	166		4.6		8.0	12.1	A0			
14 46.0	-15 28	H1d 20		Lib	1881	1955	246		2.7		6.6	11.3	K0	9376		5 Lib
14 46.0	-25 27	H III 97		Hya	1783	1954	139	126	9.6	8.6	5.1	7.1	F1 F9	9375		54 Hya
14 46.0	-44 52	h 4696		Lup	1913		206		35.3		6.9	13.0	B9			
14 46.2	-21 11	φ 309		Lib	1935	1960					7.1	7.2	F9		o	
14 46.3	+9 39	Σ 1879	AB	Boo	1829	1960					7.8	8.4	F8	9380	o	
--			ABxC		1910	1925	206	208	56.7	53.0		12.2		9380		
--			ABxD		1910	1925	218	219	133.7	130.4		10.9		9380		
14 46.4	-7 23	Σ 1876	AB	Vir	1832	1947	52	92	1.2	1.4	8.4	8.9	G0	9379		
--			AC		1891	1908	302	300	70.7	65.6		10.9		9379		
14 47.0	-52 23	h 4698		Lup	1897	1933	260		8.9		5.2	13.0	K0			
14 47.1	+0 58	Σ 1881		Vir	1830	1948	359		3.5		6.7	9.0	A0	9383		
14 47.2	-52 12	HdO 240		Lup	1900	1929	290		39.1		6.3	11.6	A0			
14 47.4	+2 02	β 1113		Vir	1889	1924	137		4.3		7.6	13.2	A2	9384		
14 47.7	+24 06	Ho 263		Boo	1887	1955					7.6	10.6	K0	9386		Suspected double by Ho
14 47.9	-68 31	I 235		Cir	1900	1927	111		0.5		7.6	8.4	A0			
14 48.4	+24 22	Σ 1884		Boo	1829	1957	55		1.7		6.1	7.7	F5	9389		
14 48.5	-17 20	β 346		Lib	1877	1960	236	269	1.3	2.0	7.4	8.2	G0	9387		
14 48.5	-35 50	B 2024	AB	Cen	1932	1960	139	35	0.1	0.1	7.2	7.9	K0			
--		h 4702	ABxC		1897	1932	215		9.8			9.3				
14 48.7	-66 36	I 369	AB	Cir			80		60.0		6.0	9.0	B3 A0			
--			BC		1900	1929	241		2.7			11.0				
14 48.9	+5 57	Σ 1883		Vir	1830	1958					7.6	7.6	F8	9392	o	
14 49.2	-10 50	Hu 141		Lib	1900	1958	323	258	0.4	0.1	7.8	9.0	F8	9395		
14 49.2	-59 24	φ 298	AB	Cir	1931	1953	145	302	0.1	0.1	7.3	7.3	G5			Motion direct
--		h 4699	ABxC		1871	1917	125		37.0			10.0	A			
14 49.3	-14 09	β 106	AB	Lib	1875	1958	335	355	1.4	1.8	5.8	6.7	A2	9396		7 μ Lib
--			AC		1889	1958	283	289	18.2	15.0		14.5		9396		
--			AD		1889	1958	185	174	26.1	25.0		13.9		9396		
--			AE		1878	1958	232		27.3			12.5		9396		
14 49.3	-24 15	H VI 117	AB	Lib	1825	1917	219	220	56.7	61.0	5.8	8.7	K0 F8	9394		
--		β 617	BC		1878	1944	338		2.7			11.1		9394		
14 49.4	-67 14	Don 680		Cir	1929	1948	64	49	1.0	1.0	7.5	11.2	K0			
14 49.5	+51 22	Σ 1889 rej		Boo	1900	1958	88		15.7		6.5	9.8	F2	9405		
14 49.6	+46 34	OΣ 286 rej		Boo							7.6		F8			Likely single (IDS)
14 49.7	+7 59	A 1110	AB	Boo	1905	1959	274	252	0.2	0.6	7.7	7.9	F5	9400		
--			ABxC		1905	1916	203		19.6			12.0		9400		
--			ABxD		1905	1916	327		23.0			12.5		9400		
14 49.7	+48 43	Σ 1890		Boo	1830	1970	44	45	3.7	2.9	6.2	6.9	F5	9406		39 Boo
14 50.0	+28 37	h 5489		Boo	1878	1959	23	22	112.3	111.2	5.7	10.6	A2			
14 50.0	-29 02	Rst 2927		Hya	1935	1940	329		1.5		7.3	13.5	K0			
14 50.0	-62 20	Rst 5008		Cir	1942		117		2.8		7.4	12.3	B8			
14 50.1	-37 24	LDS 507		Cen			nf		69.0		7.8	12.0	F8			
14 50.1	-41 51	I 1256		Cen	1926	1931	134		0.2		8.7	8.9	A3			
14 50.7	+74 09	β pm		UMi	1879	1908	342	342	207.8	209.1	2.1	11.3	K5			7 β UMi, Kochab
14 50.9	-16 02	S,h 186		Lib	1823	1913	314		231.0		2.8	5.2	A2 F5			α², α¹ Lib
14 51.0	+9 43	Σ 1886		Boo	1827	1937	227		7.7		7.6	9.6	K0	9410		BDS 7052 is same star
14 51.3	-47 24	h 4706		Lup	1836	1933	220		6.7		8.0	8.9	K0			
14 51.4	+19 06	Σ 1888	AB	Boo	1836	1977					4.7	7.0	G5	9413	o	37 ξ Boo
--			AC		1900	1953	353	348	59.2	66.7		12.6		9413		
--			AD		1907	1953	284	286	141.3	148.9		9.6		9413		
--			BC		1921	1932	101		282.2					9413		
14 51.4	+44 55	OΣ 287		Boo	1843	1959	94	161	0.4	1.0	8.4	8.6	G0	9418	o	
14 51.5	+2 44	h 5490	AB	Vir	1823		253				12.0	13.0				No such star here (IDS)
--			CD		1823		310				7.0	10.0				Not in BD
14 51.5	+44 36	h 2752		Boo	1843	1958	124		5.9		8.2	10.2	F5	9419		
14 51.5	-74 56	h 4695		Aps	1901	1918	290		17.7		6.8	12.0	B9			
14 51.6	-43 35	φ 319		Lup	1951	1959	115		0.1		5.1	5.1	B5			o Lup
14 51.7	-30 53	H N 48		Cen	1904	1919	180		24.3		8.2	9.2	F5 F5			
14 52.1	+20 17	Ho 389		Boo	1892	1957	95		1.5		7.0	9.3	A0	9420		
14 52.5	+18 44	β 31	AB	Boo	1874	1957	182	210	1.1	1.8	8.2	9.9	K0	9423		
--			AC		1878	1954	161	167	9.0	9.1		12.0		9423		
--			AD		1910	1924	277	274	94.9	93.5		11.2		9423		
14 52.5	+23 48	Cou 305		Boo	1968		66		3.2		6.8	12.5	K0			

DOUBLE AND MULTIPLE STARS

α 2000	δ 2000	Star Name		Const	Years	Pos Ang	Separation	m₁	m₂	Spec	ADS	Orb	Notes
14ʰ53ᵐ.1	+78°11'	Hu 908	AB	UMi	1904 1935	266° 258°	1".2 1".2	6.5	10.0	K0	9445		
			AC		1914	142	113.1		9.1		9445		
14 53.2	−73 11	I 236		Aps	1900 1947	102 115	1.7 2.0	5.8	8.0	G5			
14 53.4	+15 42	OΣ 288		Boo	1845 1959			6.8	7.5	G0	9425	o	
14 53.4	−45 51	Δ 171	AB	Lup	1837 1932	226	17.5	7.1	9.5	B8			
—			AC			332	17.0		12.6				
14 53.6	+69 46	Σ 1897 rej		UMi	1905	324	24.7	7.7	10.2	K5			
14 53.8	−0 24	OΣΣ 131	AB	Vir	1873 1932	210 212	89.7 86.9	8.0	8.0	K0 K0			
—			AC		1912	219	273.4		10.8				
—			BC		1912	222	186.3						
14 54.0	−5 34	Rst 4531		Lib	1938 1950	309	0.5	7.8	11.0	K0			
14 54.2	−66 25	h 4707		Cir	1837 1959			7.6	7.9	G0		o	
14 54.4	−34 09	I 226		Cen	1902 1935	218	2.8	7.0	10.5	A0			
14 54.5	−39 21	I 1578		Cen	1927 1959	86 49	0.1 0.1	8.1	8.4	G0			
14 54.6	−33 18	β 347	AB	Cen	1889 1933	319	13.3	6.0	10.8	K0			
—			AC		1889 1903	243	58.1		10.0				
14 54.9	−36 26	I 84	AB	Cen	1897 1943	259	4.6	7.2	11.0	A0			
—		See 214	AC		1897 1900	57	39.0		11.5				
14 55.4	+6 47	h 1259		Vir	1897 1911	83	35.2	6.9	10.8	K5			
14 55.4	−55 26	h 4712		Lup	1835 1946	228	7.1	8.3	9.1	A2			
14 55.7	−33 51	Ho 390		Cen	1892 1922	170	24.3	5.3	12.5	A0			
14 55.8	+39 39	A 1627		Boo	1907 1960			8.8	8.8	F0	9441	o	
14 56.0	+32 18	OΣ 289		Boo	1846 1976	120 112	4.6 4.8	6.1	9.6	A0	9442		
14 56.0	−60 54	B 2342		Cir		90	6.0	6.8	12.0	G5			
14 56.5	+59 22	Σ 1898		Dra	1832 1958	206 215	2.6 2.8	8.2	10.2	F8	9450		
14 56.5	−34 38	I 227	AB	Cen	1897 1960			8.0	8.2	F8		o	
—		B	ABxC		1927 1960	114	6.5		14.0				
—		See 215	ABxD		1897 1899	24	48.6		10.3				
14 56.5	−47 53	h 4715		Lup	1835 1952	278	2.4	6.0	6.8	B9			
14 56.6	−14 06	h 561	AB	Lib	1890 1910	162	126.3	8.7	8.5	K2 A5			
—			AC		1890 1910	261	156.3		10.1				
14 56.6	−62 22	B 2026		Cir	1931 1959	285	0.2 0.2	7.0	7.4	B3			
14 56.7	−62 47	ø 372		Cir	1960	96	0.1	6.2	6.2	B3			θ Cir
14 56.8	+70 50	Σ 1905		UMi	1832 1935	160	3.7	8.8	8.8	F8	9460		
14 56.8	+74 54	S 666		UMi	1879 1918	35	171.7	7.0	8.5	K0			
14 57.1	+35 29	h 243		Boo	1843 1899	23	18.0	7.3	13.0	F2	9449		
14 57.1	−66 53	I 327		Cir	1900 1934	78	2.5	8.0	11.0	K0			
14 57.2	−42 08	See 216		Cen	1897 1959	78 78	15.9 20.7	7.2	14.0	K0			
14 57.5	+40 09	Σ 1895		Boo	1831 1934	43	12.5	8.1	8.6	A2	9451		
14 57.5	−21 25	H N 28	AB	Lib	1823 1976	270 303	10.8 23.0	5.7	8.0	K5 M0	9446		33 G. Lib; A yellow, B more
—			AC		1878 1911	171 285	52.5 26.8				9446		yellow, according to h,
—			AD		1878 1911	52 12	69.4 101.4				9446		who called them not a bad
—			AE		1878 1911	166 191	105.5 45.4		12.8		9446		miniature of α Cen
—			AF		1878 1911	322 325	121.6 188.7		11.8		9446		
14 57.6	−0 10	H V 51		Vir	1899 1924	224	85.8	5.6	8.5	K0			
14 57.6	−35 23	h 4718		Cen	1856 1960	63	1.9	7.1	8.4	K0			
14 58.1	−63 33	h 4714		Cir	1871 1918	145	22.6	8.2	8.4	A A			
14 58.5	−47 26	CapO 62		Lup	1904 1959	158 162	25.7 25.1	7.3	8.9	G A			
14 58.7	−22 24	h 2757		Lib	1879 1932	95	12.0	7.7	9.9	F5			
14 58.7	−27 39	β 239		Hya	1878 1953	310 335	0.8 0.8	6.3	6.6	A5	9453	o	59 Hya
14 58.9	−4 59	β 1085		Lib	1889 1934	21	9.3	6.1	13.3	F5	9457		
14 58.9	−11 09	Σ 1894	AB	Lib	1831 1955	39	19.7	5.8	10.0	K0	9456		18 Lib
—			AC		1894 1923	41	162.3		11.3		9456		
14 59.2	−42 06	I 1260		Cen	1926 1960	82	3.9	3.1	11.2	B2			κ Cen
14 59.3	−43 28	I 1261		Lup	1926 1960	78	3.9	7.0	12.9	F0			
14 59.5	−30 43	h 4722	AB	Cen	1885 1956	341 337	8.6 8.6	7.2	9.3	F0			
—			AC		1897	295	32.3		13				
—			AD		1897	226	39.2		14				
14 59.6	+53 52	S,h 191		Boo	1823 1940	342	40.5	6.8	7.4	F0 F0	9474		
14 59.7	+35 06	h 562		Boo	1905 1926	307	24.4	7.9	10.1	G5			
14 59.9	−75 02	HdO 241		Aps		35	30.0	6.3	13.0	B9			
15 00.5	+3 06	h 5491		Vir	1834	60		8.0	9.9				Some error in place
15 00.6	+47 17	OΣ 291 rej		Boo	1867 1932	156	35.6	6.4	8.9	A0	9477		
15 01.3	−67 59	Gli 213		Cir	1872 1943	334	5.2	7.1	9.2	B9			
15 01.6	−3 10	Σ 1899		Lib	1825 1937	67	28.2	6.8	9.3	K2	9479		
15 01.6	−41 50	I 1264		Cen	1926 1960	257	1.2	8.1	9.9	G0			
15 01.8	−0 08	β 348	AB	Vir	1875 1958	115 111	0.5 0.4	5.8	8.1	K0	9480		
—			AC		1899 1960	216 217	32.9 31.7		14.1		9480		
15 02.0	−51 55	h 4723		Lup	1872 1937	169	5.5	7.1	10.5	K0			
15 03.0	−32 39	See 217		Lup	1897 1899	119	36.1	5.4	13.7	B5			
15 03.6	−27 51	h 4727		Lib	1876 1960	35 39	7.6 7.6	8.6	8.7	G0	9488		
15 03.8	+47 39	Σ 1909		Boo	1832 1960			5.3	6.2	G0	9494	o	44 i Boo; B is ecl. binary
15 04.1	+5 30	Σ 1904		Vir	1829 1958	347	9.9	7.2	7.2	F0	9493		
15 04.1	−6 53	Ho 391	AB	Lib	1891 1957	142 117	1.8 1.9	7.8	10.8	G5	9492		
—			AC		1891 1957	297 307	44.3 33.8		12.2		9492		

α 2000	δ 2000	Star Name		Const	Years		Pos Ang		Separation		m₁	m₂	Spec	ADS	Orb	Notes
15ʰ04ᵐ.1	+29°22′	h 564		Boo	1820		20°	°	15″.0	″	6.0	20				Not found by β
15 04.3	+29 05	β		Boo	1901		33		41.3		7.8	11.2	A2			This may be h 564 (β)
15 04.5	+19 50	Hu 745		Boo	1903	1955	22		0.4		7.8	9.7	K0	9495		
15 04.6	−17 54	S 665		Lib	1825	1919	91		25.0		8.3	9.2	K0 K0			
15 05.1	−47 03	h 4728		Lup	1835	1956	111	73	0.8	1.4	4.6	4.7	B5 B5			π Lup
15 05.3	−29 59	Rst 1817		Lib	1933	1940	126		7.5		7.2	13.5	K0			
15 05.3	−41 04	HdO 242	AB	Lup			75		25.0		5.3	12.5	G5			
--			AC				180		30.0			9				
15 05.4	+48 09	β 1086		Boo	1889	1958	254		6.2		5.6	13.3	A0	9500		47 Boo
15 05.5	−7 01	β 119	AB	Lib	1875	1960	313	287	1.5	1.9	8.0	8.5	G0	9497		
--			AC		1906		102		53.6					9497		
15 06.4	−72 10	CapO 15		Aps	1880	1947	43		1.6		7.5	9.0	A0			
15 06.8	−16 29	β pm	AB	Lib	1909		275		11.1		6.4	13.0	A0			
--			AC		1909		234		35.7			12.9				
15 07.3	+18 26	A 2385		Boo	1910	1958					6.7	6.8	A0	9505	o	Less than 0″.1 in 1952,59
15 07.3	+24 52	β pm	AB	Boo	1888	1909	55	53	108.2	107.4	4.9	10.7	F0			45 Boo
--			AC		1887	1918	50	41	252.3	244.1		9.8				
--			BC		1907		37		141.7							
15 07.5	+9 14	Σ 1910		Boo	1832	1957	211		4.3		7.5	7.5	G5	9507		
15 07.8	+63 07	Σ 1918 rej		Dra	1905	1919	21		17.9		6.8	10.8	F2	9520		
15 08.6	+25 07	h 2766		Boo	1903	1960	331	331	56.3	57.5	5.8	9.9	K0			
15 08.8	−45 17	See 219		Lup	1897	1960	178	178	0.3	0.2	4.6	4.9	B3		o	λ Lup; one rev., direct
15 08.9	−6 10	Rst 4534	AB	Lib	1938	1959	347	9	0.2	0.3	8.6	8.6	F5	9515		
--		Ho 392	ABxC		1891	1958	173		7.1			12.0		9515		
15 09.0	+1 41	β 349		Vir	1876	1960	39		3.8		7.8	12.1	F0	9517		
15 09.1	−50 52	B 1264		Lup	1931		53		1.0		8.0	10.8	G5			
15 09.6	−68 43	Don 714	AB	TrA	1932		116		6.6		7.0	15	A0			T TrA
--			AC		1932		179		12.3			14.0				
--		HdO 243	AD		1900		41		41.3			10.5				
--			AE				311		55.0			11.0				
--		Don 714	Ac				34		20.0			15				
--		I	EF													
15 09.9	+38 59	Σ 1916		Boo	1829	1925	330		9.9		8.0	10.5	F8	9527		
15 09.9	−23 59	B 2350		Lib	1953		200		10.3		6.8	12.5	K5			
15 10.1	+37 41	h 567		Boo	1925		153		10.7		8.0	12.0	A2			
15 10.3	−41 01	CorO 178		Lup	1897	1933	75		4.9		8.2	8.8	A0			
15 10.7	−43 44	CapO		Lup	1902		22		50.6		7.3	7.8	G0 G0			
15 11.3	−55 21	h 4734		Lup	1887	1933	246		11.2		5.4	12.1	G5			
15 11.6	+10 08	A 1116		SerCp	1905	1960	21	41	0.4	0.6	8.5	8.5	A5	9530		
15 11.6	−45 17	B 1267	AB	Lup	1929	1959	312		1.1		6.7	10.5	K0			
--		Δ 178	AC		1826	1959	275	264	40.3	32.2		7.1	K0			
15 11.8	−5 28	Σ 1914		Lib	1827	1915	336		31.1		8.5	9.2	A0			
15 11.8	+61 52	Σ 1927		Dra	1832	1970	354		16.1		7.8	8.7	G0	9537		
15 11.9	+38 39	Σ 1921		Boo	1830	1925	283		30.4		8.5	8.7	A2 A2			
15 11.9	−32 50	h 4743		Lup	1890	1919	197		11.0		8.5	8.8	A A			
15 11.9	−48 44	Δ 177		Lup	1826	1951	144		26.8		3.9	5.8	B9 A0			
15 12.1	+18 59	Cou 189		SerCp	1967		141		0.4		5.9	7.8	M4			
15 12.1	−45 01	I 238		Lup	1901	1939	139		3.2		8.0	11.0	B9			
15 12.2	−19 47	H IV 44	AB	Lib	1782	1919	111		57.8		5.1	9.4	A0	9532		24 ι Lib
--		B 2351	Aa		1940	1959	342	16	0.2	0.1		5.6		9532	o	Motion retrograde
--		β 618	BC		1878	1943	20	17	1.9	1.9		11.1		9532		
15 12.3	−52 06	Δ 176		Lup	1826	1938	249		71.9		3.4	7.0	G5 F8			ζ Lup
15 12.7	+19 18	Σ 1919		SerCp	1832	1958	10	10	24.8	23.9	6.7	7.6	G5	9535		
15 12.7	+48 35	Es		Boo	1908		343		25.8		7.4	10.6	K0	9539		
15 12.7	−14 42	Weisse 28		Lib							8.0		F2			Single 1910 with 40-inch
15 12.8	−60 24	h 4735		Cir	1883	1933	31		7.1		7.5	11.5	F8			
15 12.9	+27 55	H V 125		Boo	1823	1925	228		32.5		8.0	9.5	G5 G0			
15 13.0	+56 03	OΣ 294		Dra	1848	1958	248		3.2		7.4	11.9	A2	9546		
15 13.1	−37 15	CorO 179		Lup	1885	1953	227		6.5		8.0	8.2	A3			
15 13.8	−1 21	A 691		SerCp	1904	1958	226	231	0.1	0.1	7.5	7.5	K0	9544		Rapidly moving
15 13.9	−26 12	B 288		Lib	1926	1959	82		1.8		6.1	10.6	K0	9538		
15 13.9	−43 21	δ		Lup	1913	1932	231		19.0		8.0	12.8	G5			
15 14.0	−43 48	I 228		Lup	1913	1956	38	25	1.3	1.3	8.0	8.2	A2			
15 14.0	−61 21	I 329		Cir	1900	1947	330	336	0.7	0.8	6.9	7.6	B3			
15 14.1	+31 47	OΣ 292 rej		Boo	1899	1959	155	157	125.9	122.1	6.2	7.7	K5			
15 14.2	−38 30	B 1274		Lup	1928	1956	40	52	0.4	0.6	7.7	8.3	A2			
15 14.5	−18 26	S,h 195		Lib	1823	1916	140		47.4		7.1	8.1	F5			
15 14.5	−43 23	Δ 179		Lup	1836	1959	47	46	10.5	10.5	7.1	8.6	A2			
15 14.6	−59 50	CorO 180		Cir	1912	1920	288		12.2		8.2	9.8	G5			
15 14.9	+38 18	Σ 1926		Boo	1830	1960	261	252	1.6	0.6	7.2	9.5	F0	9553		
15 15.0	+36 49	OΣ 295		Boo	1846	1959	128	137	0.7	0.8	8.1	9.7	F5	9554		
15 15.3	−44 09	HdO 244		Lup	1901	1934	39		13.9		6.8	9.5	B9			
15 15.5	+33 19	Σ I 27		Boo	1835	1976	79		104.9		3.5	8.7	K0	9559		
15 15.7	−27 36	β 350		Lib	1876	1959	163	141	1.3	0.7	7.1	8.1	F2	9552		
15 15.8	+50 56	OΣΣ 137		Boo	1876	1920	107	104	75.8	72.9	6.7	8.5	G5			

α 2000	δ 2000	Star Name		Const	Years	Pos Ang		Separation		m₁	m₂	Spec	ADS	Orb	Notes
15ʰ15.8ᵐ	−37°09′	CorO 271		Lup	1897 1935	197°		21″.4		7.9	8.7	A2			
15 15.9	−37 00	See 221		Lup	1897 1935	42		5.3		7.9	10.9	F2			
15 15.9	−48 04	h 4750		Lup	1897 1933	21 19		13.3 13.3		6.0	10.2	A2			
15 15.9	−56 17	Hrg 107		Cir	1883 1930	50		5.4		8.5	9.1	F0			
15 16.0	−4 54	Σ 3091	AB	Lib	1832 1958					7.9	8.0	F8	9557	o	
—			ABxC		1875 1905	247		12.6			14.0		9557		
—			ABxD		1905	280		28.0			12.5		9557		
—			ABxE		1908	52		41.5			12.0		9557		
15 16.4	+16 48	Ho 547		SerCp	1895 1939	298		5.2		7.9	12.0	G0	9562		
15 16.6	−60 54	B 1777	AB	Cir	1930	163		1.2		5.8	8.7	B1			
—		I 428	AC		1901 1934	315		10.8			11.0				
15 16.9	−8 17	Σ 1925	AB	Lib	1831 1939	7 15		4.2	5.7	8.5	9.0	G5	9564		
—			AC		1911	270		66.1			13.4		9564		
—			AD		1911	276		125.2			13.0		9564		
—			AE		1911	271		217.2			12.2		9564		
15 16.9	−60 57	HdO 245		Cir		270		50.0		5.1	13.4	O8			δ Cir
15 17.2	−34 35	h 4752		Lup	1919	6		18.2		8.0	10.8	A2			
15 17.3	+71 13	H V 86	AB	UMi	1905 1922	132		55.6		7.3	11.0	F8			
—			AC		1905 1922	114		89.5			10.6				
15 17.9	−27 00	β 352		Lib	1879 1931	66		14.0		7.8	9.8	K2	9569		
15 18.3	+26 50	Σ 1932		CrB	1830 1960					7.3	7.4	F8	9578	o	
15 18.4	+0 56	β 943		SerCp	1879 1940	94		2.6		6.7	12.3	K0	9574		
15 18.5	−47 53	h 4753	AB	Lup	1836 1955	174 142		2.0	1.2	5.1	5.2	B8			μ Lup; v. fine double (R)
—		Δ 180	AC		1836 1955	131 130		24.0	23.7		7.2	A			
15 18.6	−46 12	I 1267		Lup	1925 1960	335 25		0.3	0.3	8.5	9.2	F0			
15 18.6	−78 28	Rst 2943		Aps	1933	156		0.5		7.6	7.9	F0			
15 18.7	+10 25	Σ 1931	AB	SerCp	1832 1973	172 168		13.1	13.4	7.1	8.5	F8	9580		
—			AC		1908	93		162.1			12.0		9580		
15 18.7	−31 13	See 226		Lup	1897 1900	70		21.1		6.1	14.0	K0			
15 18.8	−60 30	I 370	AB	Cir	1897 1934	116		5.4		5.5	12.0	O9			
—		HdO 246	AC		1900	244		44.5			11.0	A2	9579		
15 19.2	−24 16	β 227		Lib	1876 1959	182 168		1.9	1.9	7.5	8.8	A2	9579		5 Ser
15 19.3	+1 46	Σ 1930	AB	SerCp	1831 1958	41 36		10.1	11.2	5.1	10.1	G0	9584		
—			AC		1887 1924	51 40		124.6	127.2		9.1		9584		
—			AD		1852 1911								9584		See ADS, BDS for measures
15 19.4	−36 42	h 4755		Lup	1848 1945	202		4.4		8.4	9.4	F0			
15 19.7	−24 16	h 4756		Lib	1876 1959	330 290		1.0	0.7	7.9	8.1	F2	9586		
15 20.1	+29 37	β pm		CrB	1887 1918	337		147.3		5.5	9.4	K0			1 o CrB
15 20.1	+60 23	OΣ 138	AB	Dra	1876 1919	199		150.5		7.5	7.7	F0 F2			
—			AC		1905 1919	165		82.0			9.3				
—			BC		1919	47		94.1							
15 20.2	−75 34	h 4742		Aps	1900 1918	200		31.9		6.9	12.7	K0			
15 20.6	+55 20	h 2779		Dra	1864 1914	349		11.1		8.1	10.7	F5	9606		
15 20.7	−67 29	I 332		TrA	1901 1929	107		1.1		6.5	8.5	B5			
15 21.0	+0 43	β 32		SerCp	1875 1958	13 20		2.3	3.1	5.4	10.0	K0	9596		6 Ser
15 21.0	−15 33	β		Lib	1903 1960	351 350		47.1	44.4	6.2	8.4	F5			29 o Lib
15 21.1	−5 49	Rst 4538		Lib	1938 1958	5		12.2		5.5	12.6	K2			
15 21.5	−38 13	Howe 76		Lup	1885 1933	123		5.6		6.6	9.1	A0			
15 21.8	−36 16	See 229	AB	Lup	1897	240		16.7		3.6	15	K5			φ¹ Lup
—			AC		1897	119		17.3			14.5				
15 21.9	+16 30	Ho 264	AB	SerCp	1887 1955	318 311		0.9	1.1	7.6	11.6	F5	9609		
—			AC		1893	47		51.9			10.5		9609		
15 22.4	+25 37	h 2777		CrB	1906	345		40.8		7.3	10.2	K0			
15 22.5	−34 09	Howe 77		Lup	1897 1939	251		8.9		7.7	9.7	A0			
15 22.6	−47 55	Slr 20		Lup	1895 1960	217 77		1.4	0.7	8.3	8.6	G0			Motion retrograde
15 22.6	−59 10	CorO 186		Cir	1897 1937	98 80		6.0	6.3	7.6	10.3	F0			
15 22.7	−44 41	Copeland	AB	Lup	1896 1960	282 247		1.3	0.6	3.7	5.2	B3			ε Lup
—		Δ 182	AC		1826 1955	174 171		26.5	26.5		8.8				
15 22.8	−60 22	Rst 5014	AB	Cir	1942	282		1.0		8.5	9.5	A0			
—		CorO 187	ABxC		1919	233		11.5			8.7				
15 23.2	+30 17	Σ 1937	AB	CrB	1826 1960					5.6	5.9	G0	9617	o	2 η CrB
—			ABxD		1879 1921	49 47		212.8	215.0		10.9		9617		
—			AC		1856 1920	26 12		48.6	57.7		12.5		9617		
15 23.4	−59 19	h 4757		Cir	1836 1949	108 49		1.2	0.9	5.1	5.5	B5 F8			γ Cir; split at 320x (h)
15 24.0	−27 18	See 233		Lib	1897 1900	224		14.0		7.5	14.0	G5	9616		
15 24.5	+37 23	Σ I 28	AxBC	Boo	1834 1956	171		108.3		4.3		F0	9626		51 μ Boo
—		Σ 1938	BC		1826 1974					7.0	7.6	K0	9626	o	
15 24.6	+54 13	Hu 149		Dra	1900 1958	296 279		0.2	0.5	7.5	7.6	K0	9628		
15 24.6	−48 35	B 1288	AB	Lup	1930 1959	303 309		0.3	0.3	8.4	8.5	B9			
—		CapO	ABxC		1930	343		8.7			12.0				
15 24.7	−39 43	Rst 1839		Lup	1934 1945	38		1.4		5.4	10.9	A0			υ Lup
15 24.9	+58 58	β pm		Dra	1879 1909	50		254.6		3.3	9.0	K0			12 ι Dra
15 25.1	−38 10	Rst 2955		Lup	1934 1959	224		0.2		7.8	7.8	A2			
15 25.2	−26 45	h 4767		Lib	1880 1930	143		32.6		8.0	10.5	M1			
15 25.3	−38 44	Δ 183	AD	Lup	1836 1935	136 134		144.5	148.5	4.7	9.3	F0			

α 2000	δ 2000	Star Name	Const	Years	Pos Ang	Separation	m₁	m₂	Spec	ADS	Orb	Notes
15ʰ25ᵐ.3	−38°44′	I 87	AE		30.0			11				
−−		Δ 183	AxBC	1836 1935	209 205	89.5 92.4			A0			
−−		I 87	BC	1897 1953	247 230	1.3 1.3	9.3	9.6	G0			
15 25.4	−21 55	h 4769	Lib	1868 1940	192	9.6	7.8	9.1	K0	9625		
15 26.2	+18 10	Σ 1940	SerCp	1830 1955	324	1.0	8.3	8.8	F8	9634		
15 26.2	−68 19	HdO 247	TrA	f	50.0		5.9	11.9	K0			
15 26.3	−42 52	Δ 184	Lup	1837 1959	118 101	13.1 18.7	8.2	9.8	G0			
15 26.4	+44 00	OΣ 296	AB Boo	1845 1959	328 286	1.5 1.8	7.6	9.2	G5	9639		
−−			AC	1911	316	67.3		12.5		9639		
15 27.3	+9 42	A 1120	SerCp	1905 1960			8.5	9.1	G0	9643	o	
15 27.3	−36 46	HdO 248	AB Lup			20.0	5.5	14.0	B5			
−−			AC		210	30.0		12.5				
15 27.4	+17 38	A 2074	SerCp	1909 1960			8.2	8.9	F8	9645	o	
15 27.5	−10 58	Σ 1939	Lib	1830 1937	135 130	9.3 9.4	8.1	9.1	G0	9640		
15 27.7	+6 05	Σ 1944	SerCp	1832 1957	342 316	1.3 1.0	8.4	9.0	F8	9647		
15 27.7	+42 53	Ku	Boo	1901 1919	318	41.4	7.6	9.7	G5			
15 27.8	+29 06	Jef	CrB	1945	79	0.1	3.7		F0			3 β CrB
15 28.0	+46 12	Lewis 13	Boo	1900 1959	340 332	2.9 2.9	7.5	9.0		9662		
15 28.2	−9 21	S,h 202	AB Lib	1823 1970	133	51.9	6.8	8.1	K0 K0			
−−			BC	1910	110	166.4		11.2				BDS 7220, H V 27 same star
15 28.2	−37 22	Rst 3920	Lup	1936 1945	277 277	8.1 8.5	7.0	12.9	M			
15 28.5	−51 36	See 234	Lup	1897 1934	32	13.3	6.2	12.5	K0			
15 28.8	−31 29	I 239	Lup	1901 1960	358 325	0.4 0.3	7.7	8.4	A2			
15 28.9	+57 27	β 945	AB Dra	1879 1913	13 49	16.4 16.2	6.9	12.8	F8	9672		
−−		Fox	AC	1913	359	77.1		12.0		9672		
−−		Giclas	AD	1967	131	195.0		13.1		9672		
15 29.0	−28 52	β 1114	AB Lib	1889 1959	325 317	0.6 1.1	8.1	8.6	F8	9659		
−−		h 4774	ABxC	1877 1931	6	9.2		10.0		9659		
15 29.2	+80 27	Σ 1972	AB UMi	1832 1959	83 80	30.2 31.1	6.6	7.3	G5	9696		
−−			AC	1911 1925	104 104	131.4 135.4		11.0		9696		
15 29.4	+47 43	β 944	AB Boo	1879 1912	127	10.8	6.8	12.8	A0	9673		
−−			AC	1911 1913	68	56.2		12.8		9673		
15 29.5	−40 31	Rst 1842	Lup	1935 1945	253	2.4	7.4	12.9	G5			
15 29.5	−51 23	h 4772	Nor	1836 1921	250 280	12.6 7.7	7.8	12.2	K2			
15 29.5	−58 21	CapO 16	Cir	1891 1936	23 27	2.3 2.3	7.2	8.2	A3			
15 29.6	+10 27	h 253	SerCp	1820	10	15.0	8.0	9.0				Not seen; error by h?
15 30.0	+25 30	Σ 1950	SerCp	1830 1954	93	3.2	8.1	9.6	K2	9675		
15 30.4	−41 55	h 4776	Lup	1835 1951	225 229	5.8 5.8	6.8	8.5	A0			
15 30.6	−58 07	h 4771	Nor	1837 1933	187	5.3	8.6	8.9	A2			
15 30.7	+38 10	Hu 1163	CrB	1905 1958	266 308	0.4 0.2	8.5	8.8	G5	9682		
15 30.8	+72 56	Mlr 191	UMi	1971	270	1.5	7.8	10.0				
15 31.2	−18 51	Ho 393	AB Lib	1891 1912	275 285	3.7 7.3	10.7	14.6		9678		
−−			AC	1909	314	168.6		8.0		9678		
15 31.3	−13 00	β 33	AB Lib	1875 1959	47 40	2.8 3.0	7.7	10.0	F8	9680		
−−			AC	1898 1922	138	247.6		10.8		9680		
−−		Doo	AE	1898 1959	132 132	31.5 23.3		11.9		9680		Fox reverses quadrant 1926
−−		β 34	CD	1898 1959	56	6.5		10.5		9680		
15 31.3	−33 49	B 2036	AB Lup	1931 1960	185 185	0.1 0.3	7.7	7.9	A2			
−−		Howe 78	ABxC	1885 1960	161 138	1.6 1.6		9.0				
15 31.5	−73 23	h 4764	Aps	1918	255	27.0	5.5	12.4	B5			κ¹ Aps
15 31.7	+14 05	Ho 549	AB SerCp	1895	134	118.8	8.2	10.2	G5	9685		
−−			BC	1895 1959	70 74	0.4 0.5		10.2		9685		
15 31.7	−20 10	S 672	Lib	1825 1958	281	11.1	6.2	8.5	A5 F0	9681		
15 31.8	+40 54	A 1634	AB Boo	1907 1960	237	0.1 0.1	5.8	5.8	A2	9688		53 ν² Boo; rapid changes?
−−		β	ABxC	1913 1960	89 90	91.6 93.6		13.0		9688		
15 31.9	+76 46	Hu 911	UMi	1905 1958	260	1.2	8.0	12.0	G0	9712		
15 32.5	+8 35	OΣΣ 140	SerCp	1874 1914	180	112.6	8.7	8.9	F8			
15 32.7	−57 24	h 4777	Nor	1837 1933	298	5.7	7.7	9.3	F0			
15 32.8	+19 45	Hu 577	SerCp	1902 1960			8.8	8.8	F5	9692	o	
15 32.9	+31 22	Cou 610	CrB	1971	203	0.5	4.1		B5			4 θ CrB
15 33.2	−24 29	Lal 123	AxBC Lib	1825 1951	298 300	9.2 9.2	7.5		A3 A3	9689		
−−		See 238	BC	1897 1959			7.8	8.0		9689	o	
15 33.3	−13 01	Rst 3922	Lib	1937 1943	343 348	0.8 0.8	7.6	12.8	G5			
15 33.6	−47 32	Δ 187	Lup	1836 1922	238 229	34.1 28.3	7.1	10.0	F0 G0			
15 34.4	−57 04	Rst 2962	Nor	1934 1947	333	0.9	7.3	10.8	K0			
15 34.5	−47 14	See 239	Lup	1897 1913	7	13.5	7.8	11.8	K0			
15 34.6	+44 54	h 2788	Boo	1903 1923	304	57.2	8.2	8.4	K0			
15 34.6	−43 27	I 1580	Lup	1927	261	0.8	8.1	11.1	B9			
15 34.7	+26 55	Cou 798	CrB	1972	31	0.2	8.6	8.6	G0			
15 34.8	+10 32	Σ 1954	AB SerCp	1833 1974			4.2	5.2	F0	9701	o	13 δ Ser
−−			AC	1911 1959	12 14	65.8 65.2		14.7		9701		
−−			CD	1911	339	4.4		15		9701		
15 35.1	−41 10	h 4786	Lup	1835 1959			3.5	3.6	B3		o	γ Lup
15 35.5	−14 47	Go1	AB Lib	1878 1960	153	41.7	3.9	11.1	K0	9704		38 γ Lib; mag. of A is Aa
−−		Wils 20	Aa	1937 1954	61	0.1 0.1	4.6	4.8		9704		P.a. uncertain (Wils, ∅)

α 2000	δ 2000	Star Name		Const	Years	Pos Ang	Separation	m₁	m₂	Spec	ADS	Orb	Notes
15ʰ 35.9ᵐ	+12° 55′	Σ 1957		SerCp	1831 1960	163° 150°	1″.4 0″.6	8.0	9.7	F8	9708		
15 35.9	−44 58	h 4788		Lup	1836 1955	349 4	2.7 2.2	4.7	6.7	B3			
15 36.0	+39 48	OΣ 298	AB	Boo	1843 1960			7.4	7.6	K0	9716	o	
—			ABxC		1867 1937	328	121.9		8.2	K0 K0	9716		
—			ABxD		1888 1934	236 232	203.3 189.3				9716		
—			ABxE		1921 1934	333 332	356.4 474.0				9716		
—			CE		1934	333	352.7				9716		
15 36.2	−33 06	φ 231		Lup	1929 1933	304	0.1	7.0	7.0	B9			
15 36.2	−65 07	I 240		TrA	1902 1927	190	2.4	7.2	10.7	B9			
15 36.3	−65 37	Rst 778		TrA	1930	324	4.4	6.5	13.5	F0			
15 36.7	−66 19	Δ 188		TrA	1918	218	83.2	4.1	9.5	K0 A5			ε TrA
15 36.9	−6 02	Rst 4546		Lib	1938 1943	294	2.1	7.2	12.8	K2			
15 36.9	−53 43	φ 230		Nor	1930	164	4.0	8.0	13.1	A5			
15 37.0	−28 08	I 1271		Lib	1925 1944	161	3.3	3.6	10.6	K5	9705		39 υ Lib
15 37.4	−54 22	Rst 2967		Nor	1936 1947	45	0.7	7.9	10.9	G5			
15 37.9	−31 15	I 243		Lup	1897 1959	6 340	0.6 0.7	8.0	8.5	F5			
15 38.1	−42 34	HdO 250		Lup	1897 1933	28	11.8	4.3	11.0	M0			ω Lup
15 38.2	+36 15	Hu 1167	AB	CrB	1905 1934	87	0.9	7.0	12.1	F5	9731		
—		Σ 1964	AC		1830 1933	86	15.2		7.6		9731		Here the C mag. is for CE
—			AD		1900	80	15.6		8.7		9731		
—			CD		1830 1941	8 14	1.3 1.6				9731		
—		Wak 1	CE		1970	82	0.1	8.3	8.4		9731		
15 38.4	+22 40	Weisse 29		SerCp				7.6		K0			Single; nothing near (IDS)
15 38.7	−8 47	Σ 1962		Lib	1830 1958	188	11.9	6.5	6.6	F8 F8	9728		
15 38.7	−39 08	HdO 251		Lup	1923	51	10.9	6.0	14.0	A3			
15 38.8	−52 22	Δ 189	AB	Nor	1897 1914	280	53.4	5.4	9.9	A0			
—		I 88	BC		1897 1914	355	2.8		10.6				
15 38.9	+57 28	OΣΣ 141	AB	Dra	1876 1924	205	91.2	7.6	8.8	M			
—			AC		1911	335	234.6		8.1	K0			
—			Aa			158	37.9		14.1				
15 39.0	+25 45	Cou 612		CrB	1971	19	0.2	8.4	8.4	G5			
15 39.1	−72 18	I 969		Aps	1911 1930	320 319	1.7 2.2	7.6	10.6	A0			
15 39.4	+36 38	Σ 1965		CrB	1829 1973	301 305	6.0 6.3	5.1	6.0	B8	9737		7 ζ CrB
15 39.6	+79 59	Σ 1989		UMi	1832 1960			7.4	8.2	F2	9769	o	
15 39.6	−27 39	β 121		Lib	1876 1947	280	1.6	8.4	8.4	A0	9733		
15 39.9	−19 46	β 122		Lib	1875 1959	204 218	1.8 1.7	7.6	7.8	F2	9735		
15 40.2	+12 03	OΣ 300		SerCp	1848 1968	261	15.3	6.4	9.5	G5	9740		
15 40.4	−73 27	HdO 249	AB	Aps		120	15.0	5.7	12.4	B8			κ² Aps
—			AC				15.0		13.4				
15 40.5	+18 41	A 2076		SerCp	1909 1960	146 176	0.2 0.6	8.4	8.4	A2	9742		
15 40.5	−63 50	B 842		TrA	1927 1960	263	0.6	7.7	8.5	A0			
15 41.0	−47 44	I 971		Lup	1913 1934	293	18.1	6.3	12.2	K5			
15 41.1	−39 59	I 89		Lup	1897 1953	150 158	1.2 1.2	7.0	7.9	F5			
15 41.4	−2 38	Rst 4549		SerCp	1938 1943	75	3.1	8.0	13.5	G5			
15 41.6	+19 40	Hu 580	AB	SerCp	1902 1960			5.2	5.3	A2	9744	o	21 ι Ser
—		β	ABxC		1912 1960	351 352	141.7 143.2		13.4		9744		
—			ABxD		1912 1960	111 110	149.2 151.4		12.6		9744		
15 41.9	−19 41	β pm		Lib	1886 1910	279	172.0	4.7	9.7	M0			43 κ Lib
15 41.9	−30 09	Arg 28	AB	Lup	1880 1960	24	35.0	7.7	9.2	K0 F5			
—			AC		1904 1960	327 331	35.0 35.6		11.4				
—			AD		1880 1960	322 323	88.8 88.8		9.9				
15 42.0	+0 27	A 2176		SerCp	1910 1960			8.0	8.0	A0	9747	o	
15 42.7	+26 18	Σ 1967		CrB	1826 1960			4.1	5.5	A0	9757	o	8 γ CrB
15 42.8	−16 01	β 35	AB	Lib	1875 1946	99 104	2.4 2.3	7.3	8.4	G4	9751		
—			AC		1910 1923	36	112.9		10.0		9751		
15 43.0	−58 07	Δ 190	AB	Nor	1871 1934	93	5.6	7.4	9.7	M A3			
—		I 372	AC		1900	49	33.6		11.7				
15 43.2	+13 40	β 619		SerCp	1878 1960	1 358	0.5 0.5	6.9	7.4	G5	9758		
15 43.2	−25 25	β 354		Lib	1876 1938	289	5.6	7.3	9.4	F0	9754		
15 43.6	−51 38	Rst 790		Nor	1930 1947	38	2.6	6.7	12.9	K2			
15 43.8	−39 28	See 247		Lup	1897 1900	278	13.0	7.7	13.7	A0			
15 44.0	+2 31	A 2230	AB	SerCp	1910 1972	103 48	3.6 4.4	5.9	12.1	G5	9763		23 ψ Ser
—			AC		1908 1918	208 208	208.0 207.3		8.9		9763		
—			AD		1887 1918	280 281	172.1 171.5		10.5		9763		
—			CE		1908 1918	236	171.9		7.2		9763		
15 44.0	+22 20	Cou 106		SerCp	1966	278	0.3	8.3	8.8	F8			
15 44.2	−50 13	h 4797		Nor	1913	255	22.4	6.8	10.7	F8			
15 44.3	+6 26	h 1277	AB	SerCp	1836 1911	0 350	61.5 58.2	2.7	11.7	K0	9765		24 α Ser
—			AC		1836 1960	271 268	112.7 136.1				9765		
15 44.3	−54 19	I 1099		Nor	1914 1949	338 359	0.4 0.3	8.5	8.6	A3			
15 44.4	−41 49	Howe 79		Lup	1890 1952	350 343	3.7 3.7	6.1	7.7	A0			
15 45.0	−50 47	Hld 124		Nor	1890 1953	209 201	2.5 2.5	6.8	8.4	A2			
15 45.0	−59 07	h 4795	AB	Nor	1881 1933	222	7.5	7.6	11.0	A2			
—			AC		1917	136	20.9		12.0				
—			AD		1917	177	45.3		10.2				

α 2000	δ 2000	Star Name		Const	Years		Pos Ang		Separation		m₁	m₂	Spec		ADS	Orb	Notes
15ʰ45ᵐ.0	−59°07′	h 4795	DE	Nor	1917 1933	234°		8″.0				11.4					
15 45.0	−63 12	φ 299		TrA	1929	46		0.2			8.3	8.3	A2				
15 45.1	−35 06	See 248		Lup	1926	136		0.2			8.4	8.7	A0				
15 45.2	−28 32	Rst 1862		Lib	1934 1940	352		4.2			8.0	13.6	K0				
15 45.2	−58 41	φ 234	AB	Nor	1928	173		0.3			8.1	8.8	F8				Spectrum composite F8+A5
−−		Δ 191	ABxC		1836 1917	297		32.5				8.6		A2			
15 45.6	−53 26	Rst 2978		Nor	1935 1947	239		1.7			8.0	12.0	K0				
15 45.7	−22 45	B 2367		Lib	1937 1959	312	31	0.1	0.1		7.6	7.8	K0				
15 46.2	+15 25	Σ 1970	AB	SerCp	1832 1940	265		30.6			3.7	9.9	A2		9778		28 β Ser
−−			AC		1912 1960	209	210	200.1	201.1			10.7			9778		
15 46.2	+42 28	OΣ 301		Boo	1849 1935	31		3.9			7.4	11.0	K0		9782		
15 46.2	−28 04	β 620	AB	Lib	1878 1959	164		0.5			7.2	7.2	A5		9775		
−−		h 4803	ABxC		1878 1914	214		50.7				9.1		G0			
15 46.5	+36 26	Σ 1973		CrB	1829 1925	322		30.6			8.0	9.2	F5				
15 46.7	−34 41	B 847		Lup	1927 1937			0.1			6.2	6.5	B9				Measures uncertain
15 46.8	−58 08	I 974		Nor	1911 1959	9	46	0.6	0.1		7.6	7.8	A2				
15 46.9	−44 15	I 245		Lup	1900 1941	328	331	0.9	0.7		8.0	9.0	K0				
15 47.1	−35 31	B 2038	AB	Lup	1931 1960	15	42	0.3	0.3		7.2	8.2	B9				
−−		Δ 192	ABxC		1890 1938	143		34.7				7.5		B9			
15 47.4	+18 51	Cou 192		SerCp	1967	185		1.5			8.0	14.0					
15 47.4	+59 29	A 1127		Dra	1905 1958	263	284	0.2	0.3		8.5	8.9	F5		9794		
15 47.5	−21 14	h 4807		Sco	1880 1954	3		10.9			8.0	12.5	G5		9784		
15 47.5	−37 55	See 249	AB	Lup	1897 1953	130	133	15.2	14.9		6.0	13.4	G6	DA			
−−			AC		1897 1953	226	135	24.8	8.1			14.4					
15 47.6	+55 23	β 946		Dra	1879 1958	152	136	1.3	1.8		5.8	11.5	A2		9793		
15 47.9	−65 27	Rmk 20		TrA	1835 1947	156	149	2.4	1.9		6.3	6.3	A5	A5			
15 48.7	+83 37	Σ 2034		UMi	1831 1956	115		1.4			7.6	8.1	A3		9853		
15 48.8	−52 26	B 1792		Nor	1931	292		1.2			6.1	10.6	K0				
15 49.2	+60 32	Hu 912		Dra	1905 1960	138	227	0.3	0.3		8.5	8.7	F5		9806	o	
15 49.8	+44 31	β 621		Her	1891 1960	62	38	0.6	0.6		7.9	9.1	A0		9802		
15 50.0	−3 55	Rst 4553	AB	Lib	1938 1943	311		7.4			6.9	12.5	B9				
−−			BC		1938 1943	303		1.5				13.0					
15 50.1	−53 13	h 4805		Nor	1836 1917	132	127	32.9	28.4		5.8	11.6	A0				
15 50.3	−45 24	I 548	AxBC	Nor	1911 1934	178		2.9			6.1		A5				
−−		Don 764	BC		1928 1948	284	298	0.7	0.6		11.5	11.8					
15 50.8	+19 11	A 2078		SerCp	1909 1954	155		0.9			8.5	9.0	F2		9809		
15 51.1	−55 03	Δ 193		Nor	1836 1960	19	15	22.8	18.0		5.8	9.0	B5				Spectrum of B may be K
15 51.2	+35 39	β pm		CrB	1909 1925	202	202	139.7	134.6		4.8	11.5	K0				11 κ CrB
15 51.2	+52 54	Σ 1984	AB	Dra	1830 1944	273		6.5			6.6	8.9	A0		9816		
−−			AC		1910	98		17.1				12.8			9816		
15 51.9	−12 32	Hu 153		Lib	1900 1959	78		0.4			8.2	8.4	F2		9812		
15 52.2	−29 53	Ho 397	AB	Lup	1892 1904	88	87	29.0	29.9		6.4	12.8	K0		9813		
−−			AC		1892 1904	125	125	41.6	42.9			13.3			9813		
15 53.2	+13 12	β pm		SerCp	1862 1907	130	116	106.4	98.3		6.2	11.8	G0				39 Ser
15 53.6	+16 05	A 2079		SerCp	1909 1958	59		3.6			6.1	12.5	F2		9828		
15 53.6	−25 20	β 36		Sco	1855 1946	278	274	2.5	2.5		4.7	7.4	B3		9823		2 Sco
15 54.1	+16 59	A 2080		SerCp	1909 1960						8.5	8.5	F2		9831	o	
15 54.8	−50 20	Δ 195	AB	Nor	1836 1954	10		11.9			7.2	8.0	A2				
−−			AC		1900 1942	290	293	28.8	27.8			11.3					
15 54.9	+34 22	OΣ 302		CrB	1846 1967	52		28.6			7.2	9.2	A2		9838		
15 54.9	−60 45	Slr 11	AB	TrA	1891 1943	97		1.2			6.5	8.8	B8				
−−		Δ 194	AC		1836 1917	51	48	43.5	45.0			9.1					
−−			AD		1836 1917	257	257	49.3	48.1			8.7		B8			
15 55.0	−19 23	Hu 1274		Lib	1904 1958	153	129	0.4	0.5		6.1	8.1	B5		9834		47 Lib; single 1924 (VBs)
15 55.1	−63 26	LDS 542		TrA		sp		155.0			2.9	13.9	F0				β TrA
15 55.4	+29 32	Ho 399		CrB	1891 1954	119		3.2			7.8	10.3	A0		9844		
15 55.5	−60 11	h 4813		Nor	1871 1933	100		3.9			5.9	8.9	A3				Pale yellow and green (R)
15 55.7	−26 45	I 977		Sco	1914 1959	33	104	0.4	0.5		7.8	8.3	F8		9836	o	
15 55.8	+37 57	H VI 94		CrB	1891 1960	64	66	98.3	94.3		5.5	10.0	F2				12 λ CrB
15 55.9	−2 10	Σ 1985		SerCp	1831 1976	327	348	5.4	5.9		7.0	8.1	G0		9842	o	
15 56.0	−60 04	Rst 5034		Nor	1942	238		0.7			8.2	10.0	B9				
15 56.1	−39 52	HdO 252		Lup		300		18.0			6.0	11.9	B9				
15 56.5	+15 40	β pm	AB	SerCp	1850 1908	299	315	139.2	201.5		3.9	10.5	F5				41 γ Ser
−−			BC		1908	165		177.1				10.9					
15 56.8	+12 29	Σ 1988		SerCp	1830 1959	266	259	2.9	2.1		7.4	8.1	F2		9850		
15 56.9	−29 13	See 251		Sco	1897 1899	97		38.3			3.9	12.7	B4		9846		5 ρ Sco
15 56.9	−33 58	Pz		Lup	1826 1951	49		10.4			5.3	5.8	A0	A0			ξ Lup; splendid (h)
15 57.0	−0 45	Bail 883		SerCp	1892						7.9						
15 57.1	−16 02	h 1281		Lib	1890 1930	232		34.8			6.6	12.1	M		9848		
15 57.2	+3 24	Σ 1987		SerCp	1831 1968	322		10.4			7.2	8.7	A0		9855		
15 57.6	+26 53	AGC 7	AB	CrB	1877 1958	353	3	4.9	1.8		4.2	12.6	K0		9859		13 ε CrB
−−			AC		1912 1959	176	174	105.2	101.4			11.5			9859		
15 58.4	+70 54	A 1134		UMi	1905 1947	47	38	1.9	2.3		7.3	12.3	G5		9878		
15 58.8	+21 48	Σ 1990	AB	SerCp	1832 1940	60		56.3			8.7	8.7	K	A2	9865		
−−			AC		1866 1940	58		60.0				9.7			9865		

α 2000	δ 2000	Star Name		Const	Years	Pos Ang		Separation		m₁	m₂	Spec	ADS	Orb	Notes
15ʰ 58.ᵐ8	+21° 48′	Σ 1990	CB	SerCp	1831 1935	207°		3″.9					9865		
15 58.9	−3 04	Σ 3101		SerCp	1831 1959	60 69		2.0	2.2	8.6	8.9	G0	9864		
15 58.9	−26 07	β 622		Sco	1878 1914	132		50.4		2.9	12.1	B3	9862		6 π Sco
15 59.0	−65 02	B 854		TrA	1927	284		9.7		5.8	12.4	B8			
15 59.5	−24 12	B 2374		Sco		nf		10.0		7.5	14.0	G0			
15 59.5	−71 07	B 2372	AB	Aps		240		8.0		7.8	14.0	G5			
--		BrsO	AC			314		35.0			9.0	A2			
15 59.9	−58 03	Jsp 677		Nor	1930	283		0.8		7.8	9.3	B9			
15 59.9	−78 02	I 333		Aps	1900 1932	134 106		0.7	0.3	7.1	7.4	F5			
16 00.1	−38 24	Rmk 21	AB	Lup	1834 1934	20		15.0		3.6	7.8	B3			η Lup
--			AC		1847 1935	248		115.0			9.3	F5			
16 00.2	+58 58	Σ 2006	AB	Dra	1830 1955	204 189		1.6	1.6	8.2	9.9	A3	9891		
--			AC		1830 1955	224 214		43.5	46.2		8.9		9891		
16 00.3	+54 31	Hu 154		Dra	1900 1921	270		1.4		7.9	11.9	A0	9889		
16 00.3	+73 56	M1r 349		UMi	1972	312		7.0		8.0	11.0	K0			
16 00.3	−16 32	β pm		Lib	1886 1907	301 306		162.7 155.1		5.5	11.1	F8			49 Lib
16 00.4	−43 49	h 4823		Nor	1913	229		34.0		7.6	13.0	K0			
16 00.9	+13 16	OΣ 303		SerCp	1846 1959	111 164		0.6	1.2	7.5	8.0	F5	9880		
16 00.9	+39 11	OΣ 304		CrB	1847 1926	175		10.7		6.7	10.8	A0	9887		
16 00.9	−40 26	I 1280	AB	Lup		240		1.0		6.2	10.7	A0			
--		CorO 190	ABxC		1889 1933	158		8.0			10.1				
16 01.0	+33 18	S 676		CrB	1825 1923	109 71		74.2 89.6		5.5	8.7	F8			15 ρ CrB
16 01.1	+26 10	H V 75		CrB	1841 1923	108 111		42.4 49.3		8.0	11.0	F2			
16 01.1	−46 32	See 254		Nor	1897 1959	194 204		1.6	1.0	8.1	9.1	F8			
16 02.1	−0 49	Bail 884		SerCp	1892					7.4					
16 02.1	−51 07	B 1796	AB	Nor	1933	256		5.2		6.8	14.5	B8			
--			AC		1931	296		11.6			12.0				
16 02.7	−29 08	I 1282		Sco	1925 1934	242		11.0		6.2	13.0	K0	9896		
16 02.8	−2 28	Rst 4557		SerCp	1938 1943	320		2.8		8.0	12.5	K2			
16 02.9	−25 01	β 38		Sco	1868 1945	353 348		4.4	4.4	7.4	9.4	A0	9899		
16 03.0	−8 03	Rst 3928		Oph	1938 1943	166		3.1		7.2	13.8	F2			
16 03.0	+14 00	Σ 2000		SerCp	1830 1954	228		2.6		8.4	9.2	F2	9904		
16 03.4	−63 22	B 859		TrA	1927	66		5.7		7.8	13.1	F8			
16 03.5	+61 21	Hu 915	AB	Dra	1905 1952	308		2.3		7.8	12.0	A3	9916		
--			AC		1913	186		75.0					9916		
--			AD		1913	19		77.6					9916		
16 03.5	−57 47	See 258	AB	Nor	1897 1960					5.3	5.5	A2		o	ι¹ Nor
--		h 4825	ABxC		1836 1946	252 246		10.1	10.8		8.1				
16 03.7	+11 26	Σ 2003		SerCp	1831 1910	171		14.4		7.3	11.3	K2	9908		
16 03.8	+4 59	Kui 69		SerCp	1958	276		30.8		6.1	13.9	K0			43 Ser
16 03.8	−33 04	Howe 82		Lup	1877 1954	346		2.6		7.8	7.9	F0			
16 04.4	−11 22	Σ 1998	AB	Sco	1825 1960					4.8	5.1	F8	9909	o	ξ Sco
--			AC		1825 1975	79 51		6.8	7.6		7.3		9909		
--			BC		1933 1942	61 57		6.9	7.2				9909		
16 04.4	−11 27	Σ 1999	AB	Sco	1831 1975	102 99		10.5	11.6	7.4	8.1	K0	9910		
--			AC		1910 1959	82 83		75.8	80.7				9910		
--			Aa		1866 1957	169		280.6					9910		a is AB of ADS 9909
16 04.6	−37 52	HdO 253		Lup	1902	146		40.0		6.0	13.0	F0			
16 04.7	−39 26	I 373		Lup	1897 1930	217		8.6		7.2	12.2	B9			
16 04.8	+60 03	OΣΣ 142		Dra	1875 1918	265		104.8		7.6	9.4	K0			
16 04.8	+70 16	OΣΣ 143		UMi	1875 1924	84		46.7		6.7	9.3	A0			
16 04.9	+39 09	Giclas		CrB	1965					6.7	15				
16 04.9	−40 44	I 1284		Lup	1928 1931	92 92		0.3	0.2	8.5	8.8	F8			
16 05.0	−26 57	B 2378		Sco		n		12.0		7.8	14.0	G5			
16 05.4	−19 48	β 947	AB	Sco	1880 1959	88 132		0.9	0.5	2.6	10.3	B1	9913		β¹ Sco; single 1958 36-in
--		H III 7	AC		1823 1976	24 21		13.6	13.6		4.9		9913		C is β² Sco
16 05.7	−6 17	β 948	AB	Oph	1879 1960	150 120		1.5	1.3	6.4	9.1	F5	9918		Mag. of A here is for Aa
--		Σ 2005 rej	AC		1879 1958	233		28.9			9.9		9918		
--		Σ 2005 rej	AD		1879 1924	195		52.7			10.3		9918		
--		φ 384	Aa		1964	143		0.1		7.2	7.2		9918		
--		Σ 2005 rej	CD		1910	165		35.4					9918		
16 05.7	−32 52	See 264	AB	Lup	1897 1960					8.4	9.0	G5		o	
--			AC		1897 1957	8		9.6			10.6				Spectrum of C may be M
16 05.8	−39 51	I 1285		Lup	1926 1935	271		1.7		8.2	9.7	A3			
16 06.0	−6 08	Kui 70		Oph	1958	63		9.3		6.4	12.2	K0			
16 06.0	+13 19	Σ 2007	AB	SerCp	1830 1958	328 323		32.0	36.6	6.9	8.4	K0	9922		A very fine pair (Webb)
--			AC		1908 1921	138		167.6			10.5		9922		
16 06.5	−40 27	Rst 1876		Lup	1932 1952	102 77		0.3	0.1	7.8	8.2	G0			
16 07.3	−33 43	h 4832	AB	Lup	1919	354		37.7		8.0	10.1	A0			
--			BC		1919	20		23.7			10.8				
16 07.3	−36 45	h 4831		Lup	1900 1919	358		40.7		5.7	11.7	F0			
16 07.6	+29 00	Σ 2011		CrB	1829 1941	67		2.4		7.8	10.4	A3	9930		
16 07.6	−12 45	β 39		Sco	1875 1958	257		3.3		5.6	9.9	A0	9924		11 Sco
16 07.7	−38 02	CorO 193		Lup	1875 1959	73 102		10.0	2.5	8.4	8.5	G5			
16 07.9	+14 25	A 1798		SerCp	1908 1958	111 63		0.3	0.2	8.4	8.9	F0	9931		

α 2000	δ 2000	Star Name		Const	Years	Pos Ang	Separation	m₁	m₂	Spec	ADS	Orb	Notes
16ʰ08ᵐ1	+17°03'	Σ 2010	AB	Her	1832 1958	10° 12°	31″2 28″4	5.3	6.5	G5 G5	9933		7 κ Her
--			AC		1912 1960	212	62.7		13.6		9933		
16 08.1	+45 24	β 355	AB	Her	1876 1960	279	0.3	8.1	8.3	K0	9935		
--			ABxC		1905 1926	316	26.6		11.6	K0	9935		
16 08.5	-10 06	β 949		Sco	1880 1959			7.3	7.6	F8	9932	o	
16 08.6	-39 06	See 265	AB	Sco	1897 1920	298	16.2	6.9	13.3	A0			
--		Δ 199	AC		1835 1954	184	44.1		7.1				Spectrum of C may be M
--		Rst 3930	CD		1936 1945	112 105	1.3 1.3		11.7				
16 09.0	+36 29	β 1087		CrB	1889 1958	169 186	3.1 2.2	4.8	13.1	K0	9939		16 τ CrB
16 09.0	+57 56	Es		Dra	1907 1958	140	12.3	6.3	10.6	A0	9944		
16 09.1	-56 03	Rst 1878		Nor	1932	215	8.7	6.5	13.5	K0			Spectrum composite K0+A2
16 09.4	-31 03	I 557		Sco	1913 1959	205 218	0.5 0.4	7.8	8.2	F0			
16 09.5	-32 39	BrsO 11		Sco	1890 1934	85	7.8	6.7	7.4	G5 G5			
16 09.7	-27 54	Glp 4		Sco	1890 1906	289	56.3	7.9	10.3	K2			
16 10.2	-40 08	I 1082	AB	Sco	1920 1935		0.1	7.7	8.1	B5			Measures uncertain
--		δ 145	ABxC		1922 1934	145	2.5		12.1				
16 10.3	-84 14	h 4798		Oct	1919	134	20.8	7.7	11.2	K0			
16 10.5	+47 48	OΣ 307 rej		Her	1851 1946	202	17.5	7.7	10.7	K0	9950		
16 10.5	-55 21	Rst 814		Nor	1929	114	4.6	7.6	13.3	B3			
16 11.2	+47 34	Σ 2025		Her	1830 1936	164	2.6	7.8	11.1	F0	9956		
16 11.5	+9 43	φ 354		Her	1959 1960	90	0.1	7.3	7.3	A3			
16 11.7	+33 21	OΣ 305	AB	CrB	1852 1941	263	5.4	6.3	10.3	K0	9958		
--			AC		1905	86	62.7				9958		
16 11.8	+36 25	Wils 21		CrB	1937 1940	121 311	0.1 0.1	5.6		K5			Single in 1949,50
16 11.8	+42 22	Σ 2024 rej		Her	1900 1958	44	23.6	5.9	9.6	K5	9962		H IV 115
16 12.0	-19 28	β 120	AB	Sco	1876 1955	3	0.9	4.3	6.8	B3	9951		14 ν Sco
--		H V 6	AC		1821 1955	337	41.1		6.4		9951		
--		Mh	CD		1846 1955	39 51	1.1 2.3		7.8	A	9951		
16 12.2	-23 55	B		Sco		sp	10.0	8.0	14.0	A0			
16 12.3	-28 25	h 4839		Sco	1834 1951	82 73	4.6 4.0	5.9	7.9	B9	9953		12 Sco
16 12.3	-42 23	See 267		Nor	1897 1926	191	5.0	8.0	13.0	F0			
16 12.7	+40 47	Σ 2030		Her	1831 1924	238	5.6	7.8	11.1	A0	9968		
16 12.8	+26 40	Σ 2022		CrB	1830 1958	130 146	2.8 2.5	6.4	10.0	F2	9966		
16 13.3	+13 32	Σ 2021	AB	Her	1829 1971	316 348	3.2 4.1	7.4	7.5	K0	9969		
--			AC		1854 1921	128 123	262.2 236.2		10.5		9969		
16 13.4	-56 25	CorO 195		Nor	1901 1946	132 129	3.2 2.9	7.9	9.6	F8			
16 13.5	-54 38	HdO 254		Nor	1897	210	17.2	4.9	12.8	G4			κ Nor
16 13.8	+28 44	Σ 2029		CrB	1830 1942	187	6.3	7.7	9.5	F2	9973		
16 13.8	-24 25	B 307		Sco	1926 1952	229	1.4	6.4	10.2	B8	9967		
16 14.3	-3 42	β pm		Oph	1879 1908	292 294	64.6 65.5	2.7	13.1	M			1 δ Oph
16 14.3	-10 24	Rst 3936	AB	Sco	1937 1958	284 265	0.2 0.2	8.2	8.4	F8	9971		
--		Σ 2019 rej	ABxC		1848 1938	154	22.2		9.5	F8	9971		
16 14.7	+33 52	Σ 2032	AB	CrB	1827 1976			5.6	6.6	G0	9979	o	17 σ CrB
--			AC		1851 1935	234 148	21.2 8.7		13.1		9979		
--			AD		1836 1933	89 85	43.8 71.0		10.6		9979		
--			BD		1911 1923	82 83	67.1 71.3				9979		
16 14.7	-39 08	See 268		Sco	1897 1947	165 174	1.9 1.7	8.4	8.6	F0			
16 15.4	-63 41	R 274		TrA		120	30.0	3.9	11.9	G0			δ TrA
16 15.6	-8 22	β pm		Sco	1910 1958	216 279	23.6 25.8	5.6	13.3	G0			18 Sco
16 15.6	-56 41	I 558		Nor	1872 1943	50	1.6	8.2	10.0	G5			
16 15.9	-49 18	Rst 5539		Nor	1948	68	0.2	8.8	8.8	A2			
16 16.0	+11 26	β pm	AB	Her	1887 1919	259 257	66.2 65.1	7.6	9.9	K0			
--			AC		1919	341	109.5		11.0				
--			BC		1907 1918	18	108.1						
16 16.1	-30 37	I 1586		Sco	1927 1959	14 299	0.3 0.3	8.0	8.5	F0			
16 16.3	-1 39	Σ 2031 rej	AB	SerCp	1901 1972	230	20.8	7.0	9.1	F8	9984		
--			AC		1922	21	93.4		11.1		9984		
16 16.4	-24 17	Rst 3010		Sco	1935 1940	311	1.3	6.6	11.6	B8			
16 16.7	+29 09	S,h 223	AB	CrB	1879 1911	29 29	56.0 54.5	5.8	11.8	A0	9990		18 υ CrB
--			AC		1823 1924	24 22	88.7 86.8		12.8		9990		
--			AD		1823 1924	55 52	126.4 123.6		11.8		9990		
--			AE		1898	18	74.0				9990		
--			CE		1879 1924	225	13.3				9990		
16 16.9	-29 45	See 270		Sco	1897 1899	139	8.8	7.6	13.4	B8	9983		
16 17.0	-53 42	I 987		Nor	1914 1960	275 123	0.5 0.3	6.8	8.0	K0		o	Direct motion
16 17.3	-53 05	See 269		Nor	1897 1933	7	21.0	6.3	12.9	A5			
16 17.4	-34 49	h 4840		Sco	1836 1935	298	4.9	8.5	9.1	F8			
16 17.5	+75 45	LDS 1844		UMi	1966	125	227.0	5.0	15	F0			21 η UMi
16 18.3	-4 42	β		Oph	1913 1922	246	110.6	3.2	12.3	K0			2 ε Oph
16 18.4	-36 34	B 2383		Sco		sp	8.0	7.8	14.0	F8			
16 19.2	+41 40	OΣ 309		Her	1846 1960	236 272	0.5 0.4	8.6	8.8	A5	10006		Spectrum composite A5+G
16 19.3	-35 29	See 272		Sco	1897 1960	280 274	0.3 0.2	7.4	7.6	F2			
16 19.3	-42 40	See 271		Nor	1897 1960	152 130	0.3 0.3	5.9	6.7	A2			λ Nor; direct motion
16 19.3	-64 34	R 275		TrA	1878 1937	351 349	4.1 3.8	8.3	9.1	A0			
16 19.5	-30 54	BrsO 12		Sco	1879 1898	320	22.8	5.4	6.9	F5 F8			

α 2000	δ 2000	Star Name		Const	Years	Pos Ang	Separation	m₁	m₂	Spec	ADS	Orb	Notes
16ʰ 19.7ᵐ	+46° 19′	β 1198		Her	1890 1958	146°	6″.7	3.9	14.6	B5	10010		22 τ Her
16 19.8	−47 13	h 4842	AxBC	Nor	1913	200	19.8	7.9		F0			
—		Rst 836	BC		1930 1948	43	0.2	10.2	10.2				
16 19.8	−50 09	h 4841		Nor	1900 1959	1 12	41.7 44.9	4.0	10.0	G8			γ² Nor
16 19.9	+39 43	Kui 72		Her	1935 1960	153 137	1.2 2.0	5.5	10.7	F2			
16 20.0	−64 39	I 15		TrA	1895 1937	317	1.2	6.8	9.0	A0			
16 20.1	−20 03	S,h 225		Sco	1823 1916	333	47.1	7.2	8.8	B9 B9			H V 134
16 20.3	−78 42	BrsO	AB	Aps	1872 1918	12	102.9	4.7	5.1	M5 M1			δ Aps
—			AC		1918	74	90.5		12.4				
16 20.5	−20 07	B 1808	AB	Sco	1929 1959	191 195	0.2 0.2	7.6	8.1	A0 A	10005		
—		S,h 226	ABxC		1826 1940	21	12.9		8.2		10005		H V 124
16 20.5	−39 26	I 91		Sco	1896 1959	296 293	10.8 15.1	6.1	10.1	A0			
16 20.6	−29 31	I 562		Sco	1913 1952	193 202	0.8 0.8	7.7	10.5	A0	10004		
16 21.0	−54 55	Rst 3012		Nor	1934	158	1.1	7.8	10.3	A3			
16 21.2	+22 59	Hu 481		Her	1902 1960	228 178	0.5 0.7	8.0	9.8	F8	10017		
16 21.2	−25 36	H IV 121		Sco	1783 1959	270 273	20.8 20.0	2.9	8.5	B1	10009		20 σ Sco; variable
16 21.4	−33 18	h 4843		Sco	1897 1933	267	12.4	7.3	11.8	F5			
16 21.9	+19 09	S,h 227	AB	Her	1821 1938	244 233	38.3 41.6	3.8	9.8	F0	10022		20 γ Her
—			BC		1910	298	84.7		12.2		10022		
16 22.1	+30 54	β pm		CrB	1909	59	185.1	4.9	12.3	K0			19 ξ CrB
16 22.2	−52 44	B 2384		Nor		170	7.0	7.7	11.5	B2			
16 22.4	+33 48	Σ I 29	AB	CrB	1835 1922	166 165	371.9 364.4	5.4	5.3	M K5			ν CrB; both deep ;y. (Webb)
—		H N 81	Aa		1879 1936	238	68.8		11.1				
—		Σ I 29	Ab		1913 1936	154 152	281.7 180.0		12.7				H VI 18?
—		H VI 18	Bb		1879 1921	16 16	104.6 102.4						
16 22.5	−43 55	Δ 200		Nor	1836 1914	198 196	42.8 40.7	5.9	9.5	G5			
16 22.9	+32 20	H V 38		CrB	1783 1914	21 19	36.4 34.7	6.3	8.8	A2	10031		
16 22.9	−23 07	β 624		Sco	1878 1959	320	1.1	7.7	9.2	A2	10024		
16 23.1	−0 51	A 692		Oph	1904 1925	223	3.1	7.3	14.8	A5	10028		
16 23.1	+13 50	Σ 2040		Her	1831 1943	313	6.7	8.0	10.0	F2	10030		
16 23.2	−48 38	I 563		Nor	1913 1950	356 5	0.6 0.6	8.1	9.3	K0			
16 23.8	+61 42	Σ 2054		Dra	1832 1958	7 355	0.9 1.0	6.0	7.2	G5	10052		
16 23.8	−41 15	h 4845		Sco	1836 1942	137 130	1.9 1.9	8.2	8.7	F0			
16 23.9	−33 12	h 4848	AB	Sco	1836 1952	155 153	6.2 6.2	7.1	7.6	A0			
—			AC		1846 1890	357	92.0		9.0				
16 24.0	+61 31	OΣ 312	AB	Dra	1843 1974	144 142	4.7 5.2	2.7	8.7	G5	10058		14 η Dra
—			ABxC		1923	241	564.9		7.6		10058		
16 24.2	+37 02	Σ 2044		CrB	1830 1954	342	8.4	8.7	8.9	K0	10044		
16 24.5	−37 34	B 868		Sco	1927 1947	130	0.1	6.1	6.2	B8			Measures uncertain
16 24.7	−29 42	H N 39		Sco	1835 1957	349 354	7.4 5.4	5.9	6.6	G0 G0	10035		
16 24.9	−45 21	I 1291		Nor	1927 1934	38	5.3	6.5	12.5	A2			
16 25.3	−49 09	CorO 197	AB	Nor	1895 1951	207 131	1.6 1.6	8.0	8.2	G5		o	
—			AC		1895 1959	108 95	10.2 15.7		12.0				
16 25.4	+14 02	β 625	AB	Her	1879 1960	177 223	1.9 1.0	4.6	11.6	A0	10054		24 ω Her
—			AC		1879 1957	104 96	33.9 28.4		11.1		10054		
—			AD		1925 1960	270 271	126.2 127.5				10054		
16 25.4	+17 18	Σ 2043		Her	1830 1904	87	9.8	8.0	11.3	G0	10057		
16 25.6	−23 27	H II 19	AB	Oph	1822 1959	3 344	4.1 3.1	5.3	6.0	B4	10049		5 ρ Oph
—			AC		1846 1925		151.0		7.9	A	10049		
—			AD		1846 1925	253	156.0		7.0	B3	10049		
—		β 1115	DE		1889 1959	26 358	0.9 0.6		8.2		10045		
16 25.7	−29 24	B 310		Sco	1927	206	3.8	7.7	13.6	B8	10048		
16 25.8	−29 55	I 94		Sco	1897 1946	204 194	0.7 0.9	8.1	9.3	G5	10050		
16 25.8	−33 34	See 276		Sco	1897 1900	153	14.9	8.0	14.5	G0			
16 26.8	−48 03	I 93		Nor	1897 1935	284	1.0	7.5	10.0	B8			
16 27.2	−27 11	Stone 32		Sco	1880 1931	343 345	9.2 10.1	7.8	11.2	F2	10062		
16 27.2	−47 33	h 4853		Nor	1836 1951	335	22.8	4.8	7.5	B5 A			ε Nor
16 27.2	−48 47	B 2386		Nor		f	10.0	7.5	13.0	K5			
16 27.8	−8 22	Rst 3949		Oph	1938 1958	50 95	0.8 1.0	4.6	7.8	A2			3 υ Oph
16 27.9	+25 59	Σ 2049		Her	1829 1959			7.1	8.1	A3	10070	o	
16 28.0	+51 36	Hu 663		Dra	1903 1919	235	2.9	7.5	12.3	K0	10076		
16 28.0	−64 03	Δ 201		TrA	1836 1918	25 16	24.6 19.6	5.3	10.3	F3			ι TrA
16 28.3	−16 13	Rst 3950		Oph	1937 1951	55 91	0.4 0.3	7.6	8.4	F5			
16 28.7	+51 24	Hu 748		Dra	1904 1958	82	5.9	6.3	12.9	K0	10079		
16 28.8	−8 08	Σ 2048	AB	Oph	1831 1951	303 299	4.7 5.3	6.5	9.2	F0	10072		
—			AC		1910 1960	299 301	132.0 131.9		11.8		10072		
16 28.9	+18 25	Σ 2052	AB	Her	1829 1960			7.7	7.8	K0	10075	o	
—			AC		1911 1925	27 29	147.5 143.3		11.1		10075		
16 29.4	+10 36	Σ 2051		Her	1832 1934	19	13.7	7.9	9.4	K0	10077		
16 29.4	−26 26			Sco	1847 1959	273 275	3.3 2.9	1.2	5.4	M1 B3	10074	o	21 α Sco, Antares [67]
16 30.2	+21 29	β pm		Her	1879 1909	274	256.2	2.8	10.1	K0			27 β Her
16 30.9	+1 59	Σ 2055	AB	Oph	1825 1960			4.2	5.2	A0	10087	o	10 λ Oph
—			ABxC		1912 1960	170 170	118.8 119.2		11.1		10087		
—			AD		1918 1921	246	313.8		9.9		10087		
16 30.9	+38 04	Σ 2059		Her	1829 1959	209 199	1.2 0.9	8.7	8.8	F5	10093		

α 2000	δ 2000	Star Name		Const	Years	Pos Ang		Separation		m₁	m₂	Spec	ADS	Orb	Notes
16ʰ31.0	−46°29′	Rst 5413	AB	Nor	1947	233°		4.″7		8.0	13.4	B3			
—		h 4857	AC		1872 1948	73	67	7.0	6.1		9.6				Spectrum of C may be M
16 31.1	−16 37	β 626	AB	Oph	1878 1916	36	37	32.5	34.4	4.3	12.8	K0	10086		8 ø Oph
—			AC		1916 1960	318	318	119.6	120.0		11.1		10086		
16 31.5	+8 18	S,h 233		Her	1823 1923	71		58.9		7.2	9.1	G5 G5			
16 31.5	+33 31	β 816		Her	1881 1924	223		5.2		6.7	12.2	A0	10100		
16 31.6	+5 26	Σ 2056		Her	1831 1945	315		6.3		7.9	9.0	A3	10094		
16 31.7	−41 49	δ 146	AB	Sco	1921 1934	131		8.6		5.5	12.5	B1			
—		Δ 202	AC		1836 1921	180		58.0			9.8				
16 31.8	−7 01	Σ 3105		Oph	1830 1960					8.1	8.2	A0	10092	o	
16 31.8	+45 36	Σ 2063		Her	1830 1958	195		16.4		5.7	8.2	A0	10105		
16 31.8	−62 17	I 336		TrA	1900 1946	198		1.2		8.1	8.3	B9			
16 32.6	+40 07	OΣ 313		Her	1847 1959	162	139	0.8	0.9	7.7	8.3	G5	10111		
16 32.6	−43 44	I 1587		Nor	1928	34		0.4		7.6	9.1	A2			
16 32.9	−10 34	Ho 407		Oph	1890 1910	217		14.2		6.8	11.8	A5	10104		
16 33.0	−33 32	I 95		Sco	1897 1952	359		1.6		7.7	9.7	A0			
16 33.1	−60 54	Δ 203		TrA	1836 1930	249	262	36.2	27.4	8.5	8.7	A0 F2			
16 33.2	+30 54	Σ 2061		Her	1829 1944	25		2.5		8.0	10.8	F2	10113		
16 33.5	+30 30	β 818		Her	1881 1926	34		3.8		6.7	13.9	F0	10116		
16 33.9	−48 07	I 1588	AB	Nor	1927 1936	269		2.9		6.7	12.5	B0			
—		h 4861	AC		1913 1934	359		36.0			13.1				
16 34.4	−65 02	B 1815		TrA	1930 1960	184	200	0.4	0.4	8.3	9.2	K0			
16 35.2	+37 21	Par		Her	1904	331		6.4		7.3		M	10121		W Her; variable
16 35.4	+17 03	Webb		Her	1914 1960	359	0	156.2	156.6	6.3	7.3	A0 A5			
16 36.1	−51 14	I 374	AB	Ara	1900 1936	299		2.7		7.5	10.8	B8			
—			AC		1900	358		22.9			13.0				
16 36.2	+52 55	Σ 2078	AB	Dra	1831 1958	116	108	3.7	3.4	5.4	6.4	A2	10129		17 Dra
—		Σ I 30	AC		1833 1956	194		90.3			5.5	A0	10129		C is 16 Dra
—			BC		1902 1908	195		90.0					10129		
—			CD		1879 1956	122	122	116.9	120.1		11.1		10129		
16 36.4	−2 19	β pm		Oph	1910	245		100.3		5.8	13.5	K0			12 Oph
16 36.4	+33 49	Σ 2069 rej		Her	1901 1922	73		27.7		7.2	10.8	K0	10127		
16 36.6	+69 48	β 953	AB	Dra	1879 1958					8.6	9.1	F5	10140	o	
—			ABxC		1911	152		72.2			9.8		10140		
—			ABxD		1842 1911	47		146.6			8.0		10140		
16 36.8	+26 32	Cou 489		Her	1970	332		0.7		7.8	9.8	F5			
16 37.0	+69 49	Baize	AB	Dra	1956 1959	296	299	0.8	0.7	8.0	11.0	F0			
—		LDS 1851	ABxC		1966	227		147.0			8.7				
16 37.1	+0 15	Rst 5414		Oph	1946	22		5.9		7.1	13.1	K0			
16 37.1	−52 23	B 2390		Ara		42		6.0		6.9	12.5	M			
16 37.4	−61 33	h 4862		TrA	1872 1917	179		11.0		8.5	9.1	F5			
16 37.7	+19 33	Σ 2070 rej		Her	1905 1911	140		28.4		8.0	9.8	K0			
16 38.4	−43 24	h 4867		Sco	1880 1934	294		16.3		5.9	9.1	B3			
16 38.6	+38 19	Σ 2080	AB	Her	1830 1958	29	16	5.6	1.3	9.2	13.0		10138		
—			AC		1910 1921	220		109.7			12.1		10138		
—			AD		1875 1910						8.2		10138		ADS has rectang. measures
16 38.7	+48 56	Σ 2082		Her	1828 1958	92	92	22.4	25.6	5.1	11.8	M	10144		42 Her
16 39.1	−37 13	ø 340	AB	Sco	1954 1959	147		0.1		6.9	6.9	A0			
—		h 4870	ABxC		1898 1919	10		30.6			11.0				
16 39.4	−3 06	β 820	AB	Oph	1881 1952	236		4.2		8.1	9.6	K0	10139		
—			AC		1911	257		162.5			12.8		10139		
16 39.6	+23 00	Σ 2079		Her	1831 1930	91		16.8		7.4	8.2	F0	10146		
16 39.9	−47 47	Slr 12	AB	Ara	1893 1959	180	165	1.3	1.3	8.1	8.1	F5			
—		h 4871	AC		1836 1914	47	45	27.8	30.0		10.9	A3			
16 40.2	−28 00	Vou 44	AB	Sco	1938 1943	60		0.1		9.1	9.1	G0	10143		
—		h 4878	ABxC		1879 1938	359		8.3			8.6		10143		
16 40.4	−39 55	See 283	AB	Sco	1897 1945	183	172	11.6	8.9	7.7	12.9	F2			
—			AC		1897	269		25.2			10				
16 40.5	−34 18	Howe 84		Sco	1898 1931	113		10.4		8.1	10.1	K0			
16 40.6	+4 13	Σ I 31	AB	Her	1835 1932	230		69.8		5.8	7.0	A0 A0	10149		37 Her
—		Σ 2074 rej	Ba		1904 1960	316	316	24.0	25.2		11.8		10149		
16 40.7	−47 40	CorO 198		Ara	1889 1960	98		2.6		8.4	9.3	A0			
16 40.7	−62 33	B 1816		TrA	1930	341		0.2		8.7	8.8	B5			
16 40.8	−60 27	B 1818		Ara	1930 1952	35		1.4		6.3	9.0	F8			
16 40.8	−72 18	HdO 255		Aps	1902 1931	246		2.2		6.7	10.7	K0			
16 40.9	+21 57	Ho 553		Her	1897 1906	183		11.8		7.9	12.4	F8	10155		
16 41.1	−2 51	β pm		Oph	1909	62		91.6		7.2	12.7	G0			
16 41.1	−47 45	Slr 21		Ara	1895 1942	320		1.6		7.5	9.1	B8			
16 41.3	+31 36	Σ 2084		Her	1826 1960					2.9	5.5	G0	10157	o	40 ζ Her
16 41.3	−48 46	MlbO 8	AB	Ara	1878 1938	14		1.6		5.6	8.9	O5			
—		Δ 206	AC		1836 1938	266		9.6			6.8	O5			
—		h 4876	AD		1878 1938	160		13.4			10.4				
—			AE		1900 1938	15		13.9			11.3				
—		I 96	AF		1900 1938	192		20.8			12.4				
16 41.9	+73 53	Mlr 198		Dra	1971	216		0.3		7.3	7.6	A2			

α 2000	δ 2000	Star Name		Const	Years	Pos Ang	Separation	m₁	m₂	Spec	ADS	Orb	Notes
16ʰ 41.9ᵐ	−19° 55′	Kui 73		Oph	1958	214°	22".5	5.6	13.6	F5			
16 42.1	+41 12	Σ 2091		Her	1830 1959	302 313	1.3 0.7	8.3	8.8	F0	10169		
16 42.4	+21 36	Σ 2085		Her	1830 1941	309	6.1	7.3	8.8	A0	10167		
16 42.5	−37 05	R 283		Sco	1881 1960	88 12	1.0 0.2	7.0	7.8	G5			
16 42.8	−62 33	HdO 256		TrA		240	20.0	8.0	13.0	K2			
16 42.9	+38 55	Σ 2093 rej		Her	1879 1911	262	113.5	3.5		K0			44 η Her
16 43.0	+3 27	Σ 2081		Oph	1901 1912	322	21.1	7.8	10.5	A5			
16 43.1	+77 31	Ku 1	AB	UMi	1889 1958	188	2.9	6.1	9.4	F2	10214		
—			AC		1914 1959	16 14	124.4 115.0		9.8		10214		
16 43.1	−77 31	h 4858		Aps	1900	72	51.1	4.2	12.0	G8			β Aps
16 43.5	+20 43	OΣΣ 149		Her	1875 1923	135	99.0	7.1	8.5	F8 A0			
16 43.6	+6 37	H V 127		Her	1840 1923	291	53.6	7.7	9.1	K0 F0			
16 43.7	+51 32	Hu 664		Dra	1904 1959	304	0.4	8.4	8.4	F5	10189		
16 43.8	−38 09	HdO 259		Sco		160	50.0	6.1	12.3	A0			
16 43.9	−41 07	CapO 70		Sco	1931	258	95.7	6.2	6.3	B8 A3			
16 44.2	+23 31	Σ 2094	AB	Her	1831 1959	83 77	1.6 1.3	7.4	7.7	F2	10184		
—			AC		1830 1954	312	24.9		11.0		10184		
16 44.2	−58 34	φ 250		Ara	1930 1943	171	0.5	8.7	8.8	F8			
16 44.3	−0 33	Σ 2086		Oph	1831 1914	157	13.9	7.5	10.0	A0	10180		
16 44.3	−27 27	β 1116	AB	Sco	1889 1943	356 8	2.0 2.0	6.6	10.6	A0	10173		
—			AC		1897 1933	197	24.2		14.0		10173		
16 44.4	−42 23	Δ 207		Sco	1836 1930	185	11.3	8.5	9.2	K A			
16 44.6	+13 37	Hu 1277		Her	1905 1935	9	2.9	7.9	12.4	A0	10186		
16 44.7	+2 20	Σ 2088 rej	AB	Oph	1905 1910	334	20.0	8.0	12.0	A2	10185		
—			AC		1909 1910	358	15.8		13.4		10185		
—			BC		1905 1910	105	8.2				10185		
16 44.7	−40 50	I 1592		Sco		240	8.0	5.7	12.0	B3			
16 44.7	−53 09	HdO 258		Ara		40	40.0	6.0	12.0	K0			
16 44.9	+9 57	Σ 2090 rej	AB	Her	1904	156	20.4	7.9	13.3	K5			
—			AC		1904	26	66.9		9.9				
—			AD		1904	32	92.9		9.6				
16 45.0	+6 05	OΣ	AB	Her	1854 1922	191	163.6	6.7	9.5	G5			
—			AC		1854 1923	244 247	175.9 155.0		10.0				
—			BC		1912	310	149.4						
—			Ba		1903 1907	346	62.8		13.0				
—			Ca		1903 1911	108	106.0						
16 45.0	+29 28	Cou 490		Her	1970	73	0.1	8.8	8.8	F5			
16 45.0	−28 31	Rst 1901		Sco	1934 1940	219	5.4	6.0	13.8	A2			
16 45.1	+28 21	Σ 2095		Her	1830 1937	162	5.1	7.3	9.3	F5	10194		
16 45.4	−71 50	B 2392		Aps		82	14.0	7.7	14.5	G0			
16 45.7	−0 46	A 1141		Oph	1905 1960			8.7	8.7	F8	10196	o	
16 45.7	+30 00	Σ 2098	AB	Her	1831 1915	146	14.3	8.7	9.4	G5	10201		
—			AC		1825 1915	135	64.6		10.2		10201		
—			AD		1909 1915	16	62.6		8.7		10201		
—			BC		1900	132	50.7				10201		
—			Dd		1910	271	5.6		14.0		10201		
16 45.8	+8 35	S,h 239	AB	Her	1852 1916	230	82.5	5.2	9.6	K2			43 Her
—			BC		1911 1960	133 133	98.6 98.1		12.4				
16 45.8	+35 38	Σ 2101		Her	1829 1960	60 52	4.3 4.2	7.4	10.1	G0	10203		
16 45.9	−61 40	Rst 5063		TrA	1942	147	1.4	7.0	12.5	A0			
16 46.0	+82 02	HdO 143		UMi	1879 1959	6 3	77.6 76.9	4.2	11.0	G5	10242		22 ε UMi; variable
16 46.6	−3 12	A 27		Oph	1899 1938	22	2.0	8.0	11.5	F0	10204		
16 46.6	−47 05	B 1825	AB	Ara	1931 1936	245	0.3	8.1	8.4	O			
—			ABxC			sp	30.0		12.0				
—			CD			190	1.0		12.5				
16 46.7	−67 07	HdO 257	AB	TrA		123	25.0	5.1	11.3	A0			
—		I	BC			np	6.0		11.8				
16 46.8	−13 20	Rst 3959		Oph	1937 1951	35 50	0.4 0.3	8.0	9.4	A2			
16 46.9	+2 15	β pm	AB	Oph	1887 1909	219	150.3	6.7	9.1	F0 B8			
—			BC		1909	173	75.8		12.6				
16 47.2	+2 04	Σ 2096	AB	Oph	1832 1969	93 89	22.2 23.4	6.1	9.4	A2	10207		19 Oph
—			AC		1911 1960	194 194	213.4 214.6		11.3		10207		
16 47.5	−45 28	HdO 260		Sco	1902 1942	357	0.4	8.5	8.7	B8			
16 47.5	−48 19	Δ 211	AB	Ara	1913	124	106.1	7.3	8.4	K5 A2			
—			BC		1880 1913	194	44.9		8.4	A2			
—		h 4885	CD		1880 1933	242	4.0		9.7				
16 47.8	+5 15	β pm		Her	1911 1959	84	122.7	5.2	10.4	A0			45 Her
16 47.8	−40 58	LDS 570		Sco		sf	26.0	8.1	11.0	F5			
16 47.8	−43 56	I 1595		Sco	1928 1960	119 114	0.3 0.3	8.8	8.8	A0			
16 48.2	−24 32	h 1294		Oph	1898 1935	131	25.1	7.5	12.5	K5			
16 48.2	−36 53	Δ 209		Sco	1836 1959	147 141	23.0 23.4	7.4	8.5	A5 A			
16 48.7	+35 55	Σ 2104		Her	1829 1968	19	5.8	7.3	9.1	F2	10224		
16 48.7	−55 26	Δ 210		Ara	1913 1938	352	75.1	8.2	8.7	K A0			
16 48.9	+59 30	OΣ 316 rej		Dra	1867 1946	349	47.3	8.1	9.1	G5			
16 49.2	+45 59	β 627	AD	Her	1881 1911	229 230	67.0 66.2	4.9	11.9	A2	10227		

α 2000	δ 2000	Star Name		Const	Years	Pos Ang	Separation	m_1	m_2	Spec	ADS	Orb	Notes
16 49.2	+45 59'	β 627	AE	Her	1881 1911	268° 269°	43.3 45.7		11.9		10227		
--			AxBC		1878 1958	309 10	1.8 1.8				10227		52 Her
--		A 1866	BC		1908 1960			9.5	9.6		10227	o	
16 49.3	-34 01	See 288	AB	Sco	1897	136	19.3	7.1	13.3	A2			
--			AC		1897	162	20.0		13.4				
16 49.5	-43 57	I 99		Sco	1900 1942	75 72	1.1 1.1	8.0	8.6	A0			
16 49.6	+13 16	Σ 2103	AB	Her	1830 1958	37 43	5.7 5.3	6.0	10.8	A0	10225		
--			AC		1911 1960	245 248	18.1 16.5		14.4		10225		
--			AD		1911 1960	141 139	40.5 40.6		13.6		10225		
16 49.6	-31 39	I 993		Sco	1897 1936	99	3.6	6.7	11.2	K0			
16 49.8	-59 02	Rst 5067		Ara	1942	120	25.7	3.8	13.6	K5			η Ara
16 50.6	-50 03	CorO 201		Ara	1889 1955	43	2.9	7.2	7.3	A5			
16 51.0	-37 31	h 4889		Sco	1835 1951	5	6.7	6.2	8.3	B9			
16 51.1	+9 24	Σ 2106		Oph	1827 1960			7.0	8.0	F8	10229	o	
16 51.4	+1 13	OΣ 315		Oph	1844 1960	173 128	0.9 0.5	5.7	7.6	A0	10230		21 Oph
16 51.4	-24 50	B 2397		Sco	1937 1959	27 358	0.2 0.1	8.2	8.3	F5			
16 51.4	-28 01	B 319		Sco	1926 1960	224	1.3	8.0	12.4	F2	10228		
16 51.8	+28 40	Σ 2107	AB	Her	1829 1960			6.8	8.2	F5	10235	o	
--			AC		1912 1925	309	82.6				10235		
16 52.0	-35 28	See 290		Sco	1897 1900	73	9.8	7.5	13.5	K0			
16 52.1	-40 41	I 1599		Sco	1934 1959	21	0.9	7.4	10.8	K0			
16 52.3	-25 36	See 291		Sco	1897 1933	5 2	2.5 2.7	6.9	11.0	A0	10232		
16 53.0	+31 42	β		Her	1913 1960	63 64	66.1 70.9	5.3	12.0	F0			53 Her
16 53.0	+44 24	OΣ 317	AB	Her	1846 1924	235 214	15.7 19.3	8.4	12.0	G5	10246		
--			AC		1874 1923	318 317	113.7 119.3		9.1		10246		
16 54.0	-41 48	B 1833	AB	Sco	1930 1960	71	0.4	5.6	7.1	B0			In open cluster NGC 6231
--			AC		1931	101	5.5		13.8				
--		See 293	AD		1897	128	18.2		13.7				
--			AE		1897	282	20.9		13.0				
--			AF		1847	21	56.6		7.3	B0			
--		See 294	FG		1897 1920	47	6.5		13.0				
--		h 4892	FH		1847 1933	300	8.6		10.3				
16 54.0	-46 55	h 4890		Ara	1836 1959	328 323	31.1 30.6	7.9	8.0	B8 A0			In a vacant field (h)
16 54.1	-41 44	See 296		Sco	1897	242	20.0	7.0	13.0				
16 54.2	-1 37	Rst 5071	AB	Oph	1943 1958	214	16.0	6.3	13.3	F0			
--		Kui 74	AC		1958	37	20.0		14.1				
16 54.2	-41 50	See 297	AB	Sco	1897	129	13.4	6.8	12.0	B			
--			AC		1897	120	24.0		12.				
16 54.3	-41 49	B 1834		Sco	1931	312	4.3	6.7	13.3	O			
16 54.4	-30 43	B 2399		Sco	1933 1936	287	6.0	8.0	13.0	K0			
16 54.4	-41 50	h 4893		Sco	1897 1934	52	7.3	8.0	10.0	B			
16 54.7	-43 34	B 1836		Sco	1931 1959	33 45	0.4 0.4	8.0	8.2	F8			
16 55.0	+25 44	Σ 2110 rej		Her	1841 1958	93	18.1	6.1	10.6	K0	10259		56 Her
16 55.0	-24 31	B 323		Oph	1926 1959	120 100	0.3 0.3	8.6	9.0	F5	10252		
16 55.0	-40 35	See 304		Sco	1897	245	18.8	7.7	13.0	K5			
16 55.0	-41 09	I 576		Sco	1913 1934	265	5.4	5.8	12.8	O			
16 55.1	-31 13	I 577		Sco	1911 1944	14	0.8	8.2	10.2	G5			
16 55.1	-31 24	HdO		Sco	1897 1944	130 127	3.5 3.5	6.7	11.1	A0			
16 55.1	-41 55	See 303		Sco	1897	164	21.8	6.6	12.0	B0			Error in position?
16 55.4	+18 26	β 954	AB	Her	1879 1958	175 183	2.6 2.5	5.4	12.7	K2	10262		54 Her
--		Fox	AC		1913	314	96.6				10262		
16 55.4	-63 16	HdO 261		Ara		315	7.0	6.1	13.0	A0			
16 55.5	-21 34	β 241		Oph	1874 1959	340 1	0.7 0.4	7.5	7.7	B8	10257		
16 55.6	-39 30	See 310		Sco	1897	126	12.3	7.2	13.8	K5			
16 56.0	-16 48	Kui 76		Oph	1958	310	21.4	6.5	13.8	K2			
16 56.2	-31 31	B 1328	AxBC	Sco		40	117.6	8.0		A3			
--			BC		1932 1934	245	0.4	10.6	11.0				
16 56.2	-37 11	R 286		Sco	1881	318	17.1	7.4	11.0	G5			
16 56.2	-46 51	h 4896		Ara	1836 1933	25	4.0	7.8	9.0	B8			
16 56.2	-65 19	Rst 3053		TrA	1929 1935	180	0.2	8.5	9.2	F8			
16 56.3	+74 17	A 1144		Dra	1905 1946	307 303	5.1 5.8	7.2	14.0	F5	10299		
16 56.3	-31 18	Rst 1913		Sco	1934 1944	140	0.8	7.3	10.3	K0			
16 56.4	+65 02	Σ 2118		Dra	1832 1960			7.1	7.3	F0	10279	o	20 Dra
16 56.7	+14 08	OΣ 318		Her	1847 1968	249	2.7	7.0	9.6	K0	10270		
16 56.8	-23 09	β 1117		Oph	1889 1959	264 294	0.6 0.8	6.2	6.5	A0	10265		24 Oph
16 56.9	-40 31	Rst 5421	AB	Sco	1945	129	7.2	7.3	13.1	B3			
--		CorO	AC		1897 1945	252	7.6		9.8				
--			AD		1897 1903	238	15.7		10.6				
16 57.0	-32 20	See 314		Sco	1897 1900	21	14.5	7.5	14.0	A2			
16 57.0	-38 26	See 313		Sco	1897 1942	230	2.4	8.1	9.5	F2			
16 57.0	-71 07	φ 300		Aps	1930	162	2.8	6.7	12.2	A2			
16 57.1	-19 32	S,h 240		Oph	1823 1951	231	4.7	6.3	8.3	B8	10266		
16 57.5	-60 36	Rst 5070		Ara	1942	335	0.2	7.2	8.2	B8			
16 57.9	+47 22	A 1874	AB	Her	1908 1971	50 60	2.9 4.6	8.3	11.6	K0	10288		
--		Σ I 32	AC		1834 1920	263	114.0		8.2	K0	10288		

α 2000	δ 2000	Star Name		Const	Years	Pos Ang	Separation	m₁	m₂	Spec	ADS	Orb	Notes
16ʰ57ᵐ9	−27°37′	h 4902		Oph	1879 1920	31° °	11″2 ″	7.6	10.3	A0	10271		
16 57.9	−38 00	Jsp 700		Sco	1929 1945	130	9.1	7.3	11.5	B5			
16 58.1	+15 08	OΣ 319		Her	1847 1960	64	0.9	8.2	9.2	K0	10277		
16 58.4	−67 31	Don 816		TrA	1929	82	8.9	7.9	15	A0			
16 58.9	−37 37	See 315		Sco	1897 1960			6.7	6.9	A3		o	
16 59.3	−16 55	Hu 162		Oph	1900 1959	236 220	0.4 0.6	8.6	8.9	F0	10287		
16 59.5	+9 42	β 1298	AB	Oph	1901 1959	88 116	0.3 0.5	7.9	9.2	F0	10295		
--		OΣΣ 150	ABxC		1874 1914	165	77.0		8.4	F5	10295		
--		Σ 2111 rej	CD		1901	164	24.0		12.4		10295		
16 59.9	−59 20	h 4900		Ara	1917 1959	12 9	26.1 29.1	7.2	11.0	F8			
16 59.9	−73 25	I 100		Aps	1902 1932	178	0.7	7.2	8.7	B9			
17 00.1	−54 36	HdO 262		Ara	1900	71	20.1	5.7	11.8	A2			
17 00.5	−48 39	See 316		Ara	1897 1943	180 177	0.7 0.7	6.4	7.5	G5			
17 00.9	−28 24	CorO 205		Oph	1912 1939	77	2.9	8.6	9.0	G0	10300		
17 01.0	+16 36	A 2085	AB	Her	1909 1927	351 355	5.6 5.4	7.2	13.7	A0	10308		
--			BC		1909 1927	327	1.5		14.2		10308		
17 01.0	+46 16	Es 1255		Her	1913	45	8.4	8.0	11.7	K5	10311		
17 01.1	−4 13	β pm		Oph	1887 1919	70 69	91.8 94.1	4.8	9.6	K0			30 Oph
17 01.1	−42 04	B 1841		Sco	1931 1952	346 359	0.2 0.2	7.7	8.5	B8			
17 01.1	−56 33	H1d 131		Ara	1895 1943	133	2.2	6.6	10.2	B9			
17 01.1	−58 51	h 4901		Ara	1872 1952	130	2.8	7.8	7.9	B8			
17 01.2	−20 26	h 4911		Oph	1909	357	98.7	7.2		A0			
17 01.2	−29 41	Ho 554	AB	Oph	1896 1904	357	9.8	7.9	12.9	G5	10304		
--			AC		1896 1904	352	36.0		10.6		10304		
17 01.5	+14 57	H IV 122		Her	1783 1958	239	19.0	6.3	11.1	A0	10310		Σ 2115
17 01.6	+42 44	Es 633		Her	1908	259	6.3	7.3	11.8	G5	10316		
17 01.7	−33 22	Ho 410	AB	Sco	1892 1932	347	9.6	7.2	12.8	A2			
--			AC		1892 1904	245	29.3		12.6				
17 01.8	−51 08	I 1306		Ara	1926 1959	196 215	0.3 0.1	7.1	7.4	A3			
17 01.9	−32 09	HdO	AB	Sco	1897 1959	161 158	27.1 23.8	5.0	12.5	B8			
--			AC		1897	109	43.0		13.1				
--			AD		1897	41	48.2		13.3				
17 02.0	+8 27	Σ 2114		Oph	1830 1959	136 181	1.3 1.2	6.6	7.8	A0	10312		
17 02.9	−36 52	δ 20		Sco	1914 1942	240	1.5	8.4	9.4	B5			
17 02.9	−50 10	CorO 206		Ara	1888 1933	234	8.0	7.4	8.4	A0			
17 03.1	−53 14	HdO 263		Ara		20	25.0	5.4	13.0	F5			ε² Ara
17 03.1	−58 33	I 997		Ara	1911 1943	161	0.6	8.3	8.6	F5			
17 03.4	−51 12	Rst 1928		Ara	1934	53	2.3	7.9	13.5	A3			
17 03.7	+13 36	Σ I 33	AB	Her	1835 1923	115 116	292.5 299.2	5.9	6.1	A0 K2			
--			Ba		1887 1918	267	152.2		10.0				
--			ab		1909 1919	209	98.6		10.5				
17 03.7	−47 10	HdO 264		Ara	1926 1934	247	6.2	6.3	12.3	A2			
17 03.9	+19 41	β 822	AB	Her	1881 1943	227	1.6	6.5	10.9	K5	10323		
--			AC		1914 1959	134	108.1		9.5		10323		
17 04.0	+53 13	OΣΣ 151	AB	Dra	1875 1922	173 172	78.2 81.3	8.3	9.5	K0			
--			Bb		1905 1909	359	18.3						
17 04.5	−12 41	Hu 164		Oph	1900 1937	341 345	1.8 1.7	7.0	12.7	K0	10324		
17 04.7	+19 36	Perry		Her	1881 1958	231	1.8	6.2	9.5	A0	10326		
17 04.8	+28 05	Σ 2120	AB	Her	1829 1958	11 234	3.8 16.3	7.3	10.1	K0	10332		
--			AC		1907 1924	174	145.7		10.5		10332		
17 04.8	−34 07	HdO 265		Sco			20.0	4.9	14.0	B1			
17 05.1	−5 04	Luy	AB	Oph	1922	123	185.0	7.8	10.0	K5 M			Wolf 531
--			AC		1922	87	292.6						
17 05.2	+69 47	A 1146		Dra	1905 1958	316 233	0.3 0.1	8.2	9.5	F5	10340		
17 05.3	+54 28	Σ 2130	AB	Dra	1828 1976			5.7	5.7	F5	10345	o	21 μ Dra
--		β 1088	AC		1925 1958	179 175	13.8 13.2		13.7		10345		
--			BC		1889 1958	191 186	12.3 13.6				10345		
17 05.4	+12 44	H V 133		Her	1783 1924	307 309	48.7 56.0	4.9	10.9	A3	10334		60 Her
17 05.4	−33 46	WNO 5		Sco	1877 1959	231 296	5.2 17.6	8.1	8.6	G0			Spectrum of B may be M
17 05.4	−58 05	Rst 3062		Ara	1934	76	5.8	8.0	11.0	M			
17 05.6	−41 06	H1n		Sco	1966	190	0.2	8.4	8.4				
17 06.2	−38 37	See 318		Sco	1897 1959	235 332	0.4 0.7	8.4	8.8	F2			
17 06.3	+22 05	Ho 556		Her	1897 1958	124 112	24.2 30.3	5.6	13.1	K2	10343		
17 06.3	−37 14	HdO 266	AB	Sco	1897 1931	82	5.9	6.0	11.4	A2			
--			AC		1900	187	43.3		12.9				
17 06.5	−13 56	Σ 2119		Oph	1831 1949	18 11	2.0 2.3	8.2	8.2	F8	10331		
17 06.5	−35 27	B 894		Sco	1927	356	2.8	6.1	12.6	B3			
17 06.9	−1 39	Σ 2122		Oph	1831 1958	280	20.4	6.3	8.5	A2	10347		
17 07.4	−44 27	CorO 208		Sco	1897 1934	139	5.1	7.2	9.3	A0			
17 07.4	−53 45	Rst 3067		Ara	1935	127	3.2	8.0	13.5	A2			
17 07.6	+35 57	h 264	AB	Her	1900 1926	197 204	12.5 12.9	9.0	11.0				
--					1907	100	245.8	9.0	6.1				Measured to A of Hu 1176
17 08.0	+31 12	OΣ 324		Her	1853 1945	220	3.8	6.6	11.1	K2	10356		
17 08.0	+35 56	Hu 1176	AB	Her	1905 1958			6.1	6.1	A5	10360	o	
--		Ho 412	AC		1892 1958	143 136	19.5 20.0		11.4		10360		

α 2000	δ 2000	Star Name		Const	Years	Pos Ang	Separation	m_1	m_2	Spec	ADS	Orb	Notes
17ʰ08.1ᵐ	−41°37′	I 407		Sco	1902 1959	186° 179°	0.″7 0.″3	7.7	8.2	A3			
17 08.2	−1 05	A 1145		Oph	1905 1952	241 155	0.4 0.2	6.2	8.2	A0	10355	o	Round 1955−59
17 08.3	+50 51	Es 77		Dra	1902 1959	273 274	16.8 18.3	6.3	11.0	B9	10369		
17 08.5	−20 13	h 4922	AB	Oph	1879 1919	310	21.8	7.5	10.7	A0			
—		β	AC			265	22.0		13.5				
17 09.9	−82 19	h 4884		Aps	1919 1940	8	34.9	7.2	9.2	A0			
17 10.1	−27 46	I 246		Oph	1902 1942	34 44	1.3 1.3	7.7	11.2	F2	10368		
17 10.2	−0 46	β 124		Oph	1875 1953	253 263	1.1 1.0	8.0	11.0	A0	10376		
17 10.3	−46 44	Δ 213		Ara	1836 1934	167	8.1	6.9	8.4	B2			
17 10.3	−75 23	h 4904		Aps	1872 1940	187	7.1	7.6	9.3	F2			
17 10.4	−15 43	β 1118	AB	Oph	1889 1960			3.0	3.5	A2	10374	o	35 η Oph
—			AC		1898 1911	142 143	93.4 94.6		12.0		10374		
—			AD		1898 1911	288			10.5		10374		Sep. >100″ (IDS)
17 10.7	−44 33	I 1312		Sco	1926 1934	127	13.2	5.1	13.0	G5			
17 10.8	−24 57	h 589		Oph	1879 1940	302	10.0	8.4	8.7	A3	10379		
17 11.3	−46 11	Rst 5541		Ara	1949	95	0.2	8.3	8.3	B5			
17 11.4	−68 38	Don 824		TrA	1928	214	2.2	8.0	12.0	A2			
17 11.5	+39 16	Hu 1178	AB	Her	1905 1957	10 15	0.3 0.3	8.8	9.1	F8	10391		
—		Σ 2136	ABxCD		1831 1930	114	15.5				10391		
—		Hu 1178	CD		1905 1957	81	1.1	9.0	13.5		10391		
17 11.5	−16 29	Hu 169		Oph	1900 1958	223 140	0.1 0.2	8.2	8.3	A3	10385		
17 11.6	−26 42	β 956	AB	Oph	1880 1942	164	0.8	8.1	9.1	F8	10384		
—			AC		1925	148	67.1		10.9		10384		
17 11.7	−4 38	Rst 4567		Oph	1938 1943		4.6	7.7	13.2	A2			
17 11.7	+49 45	Σ 2142	AB	Her	1830 1958	114	5.2	6.0	9.8	A2	10397		
—		Kui 78	AC		1958	105	28.8		14.5		10397		
17 12.1	+21 14	Σ 2135	AB	Her	1829 1968	166 188	6.7 8.0	7.4	8.7	K0	10394		
—			AC		1911	208	158.9		11.8		10394		
17 12.2	−27 03	β 125	AB	Oph	1877 1942	66	1.7	6.9	9.7	G5	10388		
—			AC		1892	90	77.7		13.0		10388		
17 12.5	+10 35	B		Oph	1958	34	30.3	5.6	14.3	K5			37 Oph
17 12.8	−33 22	CorO 209		Sco	1920	138	10.2	7.8	10.5	B5			
17 13.0	+7 45	OΣ 325		Oph	1857 1959	203 227	1.7 0.7	7.0	8.9	F0	10398		
17 13.0	−32 26	HdO		Sco		290	30.0	6.0	13.5	B2			
17 13.0	−58 36	h 4920		Ara	1872 1943	329 325	3.0 3.0	7.1	8.8	F2			
17 13.1	+54 08	Σ 2146	AB	Dra	1831 1934	225	2.8	7.2	9.2	F0	10410		
—			AC		1909	231	88.8		9.2	G5	10410		
17 13.3	−67 12	Δ 214		Ara	1835 1918	328 356	26.6 30.1	6.0	9.2				
17 13.6	−9 17	β 1247		Oph	1891 1944	343	1.5	7.6	9.9	A0	10400		
17 13.7	−47 57	Rst 906		Ara	1930 1949	54	2.3	8.0	12.5	G5			
17 13.9	−38 18	Howe 86		Sco	1880 1955	143	2.8	7.0	8.8	F5			
17 14.1	−8 24	Barnard 7		Oph	1892 1923	155 145	2.2 1.8	8.0	11.4	K0	10404		
17 14.1	+56 08	OΣ 327		Dra	1846 1960			8.5	8.8	F2	10425	o	
17 14.5	−39 46	h 4926	AB	Sco	1896 1933	334	14.4	6.3	10.5	K5			
—			AC		1896 1933	210	16.9		11.3				
17 14.6	+14 23	Σ 2140	AB	Her	1829 1968	118 107	4.6 4.7	3.5	5.4	M	10418	o	64 α Her, Rasalgethi
—		AGC	AC		1888 1934	336 330	23.5 21.1		15		10418		
—		Σ 2140	AD		1890 1960	39 39	84.8 81.2		11.1		10418		
17 15.0	+24 50	Σ 3127	AB	Her	1830 1958	174 236	25.8 8.9	3.1	8.2	A2	10424		65 δ Her; optical pair
—			AC		1921	352	158.9				10424		
—			AD		1921	95	190.9				10424		
17 15.3	−14 35	β 282		Oph	1875 1958	153	4.2	6.2	11.3	K0	10419		
17 15.3	−26 36	S,h 243	AB	Oph	1822 1976			5.1	5.1	K0 K0	10417	o	36 Oph; an orange pair
—			AC			nf	732.0		6.6		10417		
—			AD		1823 1905	286 315	195.0 208.0		8.1	K0	10417		
—			Aa		1898 1904	298	38.6		13.4		10417		
17 15.6	−10 18	β 957		Oph	1880 1959			8.0	8.1	F5	10421	o	
17 15.6	−38 36	ø 355		Sco	1959	142	0.1	6.7	6.8	F8			
17 15.7	−9 49	A 2592		Oph	1913 1959	343 273	0.3 0.4	7.6	8.1	F5	10423	o	
17 15.8	−33 44	See 322		Sco	1897 1954	1 307	0.2 0.2	7.3	7.6	B3			
17 16.1	+60 43	Σ 2155		Dra	1830 1955	114	9.8	6.8	10.1	F0	10448		
17 16.3	−42 20	I 408		Sco	1902 1943	168	1.7	7.0	9.0	B5			
17 16.3	−72 33	B 2046		Aps	1932	297	3.3	8.0	14.0	K2			
17 16.4	−35 45	HdO 267		Sco	1898	18	36.5	6.1	13.5	F8			
17 16.5	+1 13	h 854		Oph	1878 1958	358	20.4	5.9	13.0	B8	10428		U Oph; A is variable
17 16.5	+14 40	β 1200		Her	1890 1929	10	1.3	8.0	12.4	G5	10433		
17 16.6	−0 27	A 2984		Oph	1915 1959	298 346	0.5 1.0	4.8	7.8	K0	10429		41 Oph
17 16.9	+89 02	h 2985		UMi	1906	269	54.7	6.6	14.1	M			λ UMi
17 17.3	+33 06	OΣ 328		Her	1847 1943	60	4.4	4.8	10.2	B3	10449		68 u Her; A is variable
17 17.3	−30 10	β 1119		Sco	1889 1959	356 297	0.6 0.3	7.5	8.1	G5			
17 17.4	−7 38	Σ 2144 rej		Oph	1848 1915	183	25.5	8.2	9.2	K2			
17 17.4	−66 57	LDS 587		Ara		f	33.0	6.7	12.9	G5			
17 17.6	+28 55	Σ 2147		Her	1833 1924	94	6.2	7.1	11.0	M	10451		
17 17.7	−26 38	H I 35		Oph	1825 1937	331 335	6.5 5.7	7.1	8.6	A0	10436		
17 18.0	−24 17	H III 25		Oph	1782 1951	355	10.3	5.4	6.9	K2 F6	10442		39 o Oph

α 2000	δ 2000	Star Name		Const	Years		Pos Ang		Separation		m₁	m₂	Spec	ADS	Orb	Notes
17ʰ18.1ᵐ	+28°44′	S 686		Her	1825	1918	4°	4°	55.0″	52.4″	8.1	9.7	A2			
17 18.3	−32 33	HdO 268		Sco	1934		174		19.5		6.4	12.0	B6			
17 18.4	−29 52	B 337		Oph	1927	1933	203		0.6		8.1	10.1	A2	10444		
17 18.6	+21 47	OΣΣ 152		Her	1874	1919	49		51.9		8.2	10.1	K0			
17 18.9	−49 20	B 2407		Ara			f		8.0		7.5	12.0	K2			
17 19.0	+53 41	Fur	AB	Dra	1910		153		3.9		8.0	10.5				
—			BC		1910		144		1.8			11.0				
17 19.0	−34 59	M1bO 4	AB	Sco	1877	1959					6.1	7.6	K2		o	
—		h 4935	AC		1889	1948	129	136	30.5	30.8		10.0				Spectrum of C may be M
—		See	AD		1897	1952	86	328	55.4	14.6		12.9				
17 19.1	−46 38	BrsO 13	AB	Ara	1880	1959					5.5	8.6	K0		o	
—			AC		1900		279		41.8			12.5				
—			AD				30		47.0			14.0				
17 19.2	−59 42	HdO		Ara			280		20.0		6.0	13.0	K2			
17 19.3	−53 23	Δ 215	AB	Ara	1938		49		58.7		8.3	9.0	G5 G			
—			AC		1938		99		63.7			9.9	A			
			CD		1938		153		12.8			11				
17 19.4	+28 02	Cou 625		Her	1971		16		0.3		7.3	9.8	A0			
17 19.4	−44 13	HdO 269		Sco	1901	1947	15		0.3		7.3	7.6	A0			
17 19.5	−50 04	ø 356		Ara	1959		112		0.1		7.2	7.2	F0			
17 19.9	−17 45	β 126	AB	Oph	1875	1958	262		1.9		6.3	7.4	A0	10465		
—			AC		1879	1958	140		11.4			11.3		10465		
17 20.0	−8 01	β pm		Oph	1893	1908	223	223	105.5	102.9	8.2	10.2	G0			
17 20.0	−19 59	I 592		Oph	1914		258		10.8		8.0	13.2	F2	10466		
17 20.2	−70 03	I 104		Aps	1901	1934	137	132	1.9	1.5	6.6	9.3	G0			
17 20.6	−19 20	A 2241		Oph	1910	1926	76		5.1		6.5	14.0	G0	10476		
17 20.6	−59 26	h 4931		Ara	1871	1942	256		1.1		7.8	7.9	A0			
17 20.7	+32 28	Dorpat 544	AB	Her	1853	1923	328	335	162.6	230.0	5.4	9.7	G0	10488		72 Her
—			BC		1900	1911	216		8.7			12.9		10488		
17 20.8	−12 51	S,h 247		SerCd	1863	1959	32	28	48.1	46.3	4.3	8.3	A0	10481		53 ν Ser
17 20.9	+24 30	S 687	AB	Her	1879	1960	56	56	220.7	222.9	5.1	8.6	A0			70 Her
—			AC		1910		20		239.2			11.8				
—			BC		1911		315		146.7							
17 20.9	−10 42	Rst 3971		SerCd	1937	1943	82		8.5		6.5	14.3	F0			
17 20.9	−27 20	β 127		Oph	1876	1951	92		5.2		8.4	9.2	F5	10477		
17 21.0	+1 31	Σ 2150	AB	Oph	1832	1941	185	204	8.1	9.8	9.4	10.3		10486		
—			AC		1905	1925	57		163.7			7.9	F8	10486		
17 21.0	−21 07	Don 832		Oph	1931	1959	66	50	2.7	3.7	4.4	8.9	F2			40 ξ Oph
17 21.5	+28 45	Kui 80		Her	1934	1958	157		0.7		6.4	8.6	F8			
17 22.1	+5 00	β 959		Oph	1879	1924	257		3.4		6.7	11.6	G5	10498		
17 22.1	−70 07	ø 373		Aps	1960		119		0.1		6.2	6.2	B8			ι Aps
17 22.4	−30 12	Ho 413		Sco	1892	1924	281	297	7.4	10.4	8.0	11.8	F5			
17 22.7	−37 48	B 908	AB	Sco	1927		114		2.7		6.4	12.2	G			Spectrum composite G+B8p
—			AC				208		12.0			13.5				
17 22.7	−38 12	I 1317		Sco	1926	1947	174	190	0.3	0.3	8.3	9.2	A0			
17 22.8	−58 28	CorO 213		Ara	1917		283		9.4		6.9	9.8	A5			
17 22.9	−37 13	See 327		Sco	1897		222		20.2		5.9	14.3	K0			
17 22.9	−58 01	HdO 270		Ara	1900	1938	187		2.0		5.9	9.5	K0			
17 23.3	−47 28	HdO 271		Ara	1900		62		42.8		5.3	10.6	B3			ι Ara
17 23.5	+16 54	A 2183		Her	1910	1944	128		1.0		7.7	10.7	A0	10516		
17 23.6	+13 24	β 46		Oph	1875	1959	204		2.0		7.5	10.7	K5	10517		
17 23.7	+37 09	Σ 2161	AB	Her	1830	1958	307	316	3.6	4.1	4.6	5.6	A0	10526		75 ρ Her
—			AC		1911		223		119.5			13.4		10526		
17 23.7	+47 16	Σ 2164		Her	1829	1976	17	13	8.8	9.3	8.0	9.5	F8	10530		
17 24.0	−9 21	Rst 3972		Oph	1938	1958	278	332	0.2	0.1	8.6	8.6	G0		o	
17 24.0	+38 35	Hu 1179		Her	1905	1960	273	259	0.2	0.1	7.2	7.3	F8	10531		
17 24.0	−11 41	β 242	AB	SerCd	1875	1942	69	77	1.0	1.0	8.4	9.2	A2	10515		
—			AC		1876	1939	63		9.3			10.9		10515		
—			AD		1876	1937	64		47.9			10.2		10515		
17 24.4	+36 57	OΣ 329		Her	1867	1955	12	13	32.6	33.2	6.4	8.1	G5	10535		
17 24.5	−30 32	Hld 28		Sco	1881	1938	233		3.6		8.6	9.0	K0			
17 24.6	+15 36	Σ 2160		Her	1830	1958	66		3.9		6.2	10.7	B9	10528		
17 24.7	+38 02	Cou 1142		Her	1974		217		1.7		7.9	11.3	G5			
17 24.7	+39 12	S 689		Her	1825	1925	198		89.6		8.1	8.7	K5 G			
17 24.7	−21 26	Hld 134		Oph	1889	1933	146		4.3		6.0	12.0	G7	10522		
17 24.8	−59 13	I 385	AB	Ara	1900	1942	170	149	0.4	0.4	8.0	8.5	A0			
—			ABxC		1901		211		17.8			13.0				
17 24.9	+13 20	Σ 2159		Oph	1831	1935	326		26.4		8.6	9.0	F8			
17 25.1	−34 34	I 595		Sco	1914	1943	150		0.8		7.5	9.3	A0			
17 25.4	−56 23	h 4942	AB	Ara	1872	1934	328		17.9		3.3	10.3	B1			γ Ara
—			AC		1913		66		41.6			11.8				
17 25.9	+16 55	Kui 81	AB	Her	1958		217		8.5		6.0	14.6	M			
—			AC		1958		221		23.1			13.6				
17 26.0	+26 53	h 1299	AB	Her	1901	1924	20		50.0		6.4	11.4	A5			
—			AC		1901	1924	58		52.1			11.9				

α 2000	δ 2000	Star Name		Const	Years	Pos Ang		Separation		m₁	m₂	Spec	ADS	Orb	Notes
17ʰ26ᵐ.0	−50°38′	HdO	AB	Ara		180°	°	25″.0	″	5.2	14.0	K1			κ Ara
---			AC			95		30.0			13.5				
17 26.1	−27 36	I 1069		Oph	1913 1927	350		5.5		7.4	11.4	B9	10538		
17 26.2	+29 27	Σ 2165	AB	Her	1832 1956	46	61	6.7	9.2	7.7	9.2	F0	10553		
---			AC		1890 1956	252	250	98.2	95.5		9.5		10553		
---			Cc		1909 1922	180		49.7			11.8		10553		
17 26.3	+67 46	β 1201		Dra	1890 1955	340		0.4		8.8	8.8	A5	10573		
17 26.8	−26 20	β 128		Oph	1876 1934	324		4.0		7.5	9.7	B9	10547		
17 26.9	−45 51	h 4949	AB	Ara	1836 1953	267	256	2.9	2.2	6.0	6.7	B9			
---		Δ 216	AC		1836 1913	313		103.0			7.6	A0			
17 27.2	−50 38	See 328	AB	Ara	1900	159		76.1		6.1	10.8				
---			Aa			160		40.0			14.0	B9			
---			BC		1897 1937	274		4.2			11.6				
17 27.5	+16 27	A 2184		Her	1910 1960	358	20	0.9	1.3	7.3	10.6	F0	10560		
17 27.8	−12 11	Hu 234	AB	SerCd	1900 1922	167	172	1.0	1.0	8.0	12.0	F2	10559		
---			AC		1900 1945	308		5.4			9.8		10559		
17 27.9	+11 23	Σ 2166		Oph	1831 1949	283		27.3		7.1	8.9	A	10562		
17 27.9	−47 02	Rst 5543		Ara	1949	356		1.9		6.7	13.0	B2			
17 28.0	−52 18	HdO 272	AB	Ara		355		17.0		5.8	13.0	K0			
---			AC			330		25.0			14				
17 28.2	−31 12	PrO 162		Sco	1911	120		14.4		7.8	11.3	K2			
17 28.3	−20 58	A 2244		Oph	1910 1959					8.6	8.8	F8	10561	o	
17 28.6	−25 31	β 129		Oph	1877 1951	99	115	0.9	0.9	7.7	7.9	F0	10564		
17 28.6	−55 10	HdO 273		Ara	1900	151		37.0		6.0	12.5	K0			
17 28.7	−22 35	B		Oph		sp		12.0		8.0	12.0	K2			
17 28.7	−82 47	h 4912		Oct	1919	122		25.1		6.8	11.4	B8			
17 29.0	+50 52	Σ 2180		Dra	1831 1955	262		3.2		7.7	7.9	F0	10597		
17 29.0	−43 58	Δ 217		Sco	1836 1952	169		13.4		6.3	8.7	B9			
17 29.3	−10 00	Σ 2171		Oph	1830 1952	76	64	1.6	1.4	8.5	8.6	F2	10576		
17 29.4	−38 31	B 342		Sco	1926 1959	67	85	0.4	0.4	7.0	7.3	A2			
17 29.5	+34 56	Σ 2178		Her	1832 1934	130		10.7		7.5	9.1	K0	10594		H III 40 may be same star
17 29.5	−61 46	I 598		Ara	1911 1944	135		1.2		6.9	9.0	F0			
17 29.7	−8 01	Rst 3977		Oph	1938 1951	291		0.6		8.2	10.2	A0			
17 29.8	−5 55	β 1089		Oph	1888 1960	5	336	1.0	1.3	6.4	10.6	G5	10583		
17 29.9	+15 58	OΣ 330		Her	1848 1912	57		14.3		7.6	11.2	F5	10592		
17 30.1	−33 43	Howe 39	AB	Sco	1877 1934	321		4.4		6.8	9.6	B5			
---		Ho 646	AC		1893 1934	314		14.6			11.5				
---			AD		1881 1935	29		58.7			9.4	B8			
17 30.3	−4 22	Rst 5085		Oph	1943	333		10.8		6.6	11.5	F0			
17 30.4	−1 04	Σ 2173		Oph	1830 1960					6.0	6.1	G5	10598	o	
17 30.4	+52 18	β 1090	AB	Dra	1889 1934	13		4.2		2.8	13.8	G0	10611		23 β Dra
---			AC		1912 1960	157	156	115.6	117.4		12.5		10611		
17 30.6	−43 14	B 2410		Sco		p		8.0		6.9	12.0	F5			
17 30.7	−43 04	CapO 76		Sco	1901 1928	340		4.7		8.4	9.4	B9			
17 30.8	−37 26	B 912		Sco	1927 1959	259	40	0.2	0.2	7.6	7.9	A0		o	Direct motion
17 31.1	−60 41	h 4951		Ara	1917	313		47.4		3.6	11.6	B8			δ Ara
17 31.3	+19 01	Cou 499		Her	1970	88		0.2		8.7	8.7	A5			
17 31.3	−10 58	Hu 673		SerCd	1900 1924	186		5.4		7.8	13.1	F5	10605		
17 31.3	−39 01	Howe 87		Sco	1896 1951	232		3.1		7.5	8.8	F8			
17 31.4	+2 43	A 2386		Oph	1911 1955	324	3	0.1	0.1	6.4	6.4	G0	10607		Single 1957,59
17 31.5	+49 44	Fox		Her	1909	230		44.4		8.1	10.8	K0			
17 31.7	+30 19	Σ 2181 rej	AB	Her	1904 1926	306	313	26.8	27.2	7.2	9.4	K0			
---			AC		1926	210		47.8			10.4				
17 31.7	−41 02	Hld 136		Sco	1896 1937	109		1.1		7.8	8.1	B9			
17 31.7	−48 45	See 331		Ara	1897 1959	15	291	0.7	0.2	7.7	7.9	F0			
17 31.8	+56 55	Es		Dra	1918	208		13.6		8.3	9.5	K0	10626		
17 31.8	−46 02	I 40		Ara	1900 1933	210		18.0		6.3	10.5	G0			
17 31.8	−49 53	h 4955		Ara	1913	173		55.6		3.0	11.0	B3			α Ara
17 32.0	+2 49	OΣ 331	AB	Oph	1848 1959	326	349	0.8	1.4	7.9	9.4	B8	10614		
---			AC		1898 1902	241		51.0			11.2		10614		
---			DE		1903	166		2.6		11.5	11.5		10614		Faint pair 1′ p OΣ 331
17 32.0	+68 08	β		Dra	1913	345		170.0		5.1	11.3	K0			27 Dra
17 32.2	+55 11	Σ I 35		Dra	1833 1955	312		61.9		4.9	4.9	A5 A5	10628		ν Dra; grand obj. (Webb)
17 32.6	+34 45	Hu 1181		Her	1904 1960	331	190	0.2	0.1	8.4	8.7	G0	10624		Quadrant uncertain
17 32.8	+47 53	Σ 2189 rej	AB	Her	1901 1919	100		21.0		7.7	10.1	A0	10630		
---			AC		1901 1919			65.1			8.8		10630		
17 33.2	−45 31	I 603		Sco	1911 1939	80		1.2		7.3	8.8	B9			
17 33.5	−42 24	B 1862		Sco	1930 1945	334		0.3		7.8	8.4	B5			
17 33.6	−37 06	See 334	AB	Sco	1897	109		41.7		1.6	14.8	B2			35 λ Sco
---		Δ 218	AC		1897	331		94.9			11.9				
17 33.9	−44 05	See 333		Sco	1897	287		13.8		7.9	13.0	B8			
17 34.2	−35 44	Vou 71		Sco	1934	139		9.2		7.6	13.0	B8			
17 34.4	+13 10	Σ 2184	AB	Oph	1830 1924	72		21.5		6.7	11.6	G5	10633		
---			AC		1920	293		127.2					10633		
17 34.6	+9 35	Σ I 34	AB	Oph	1835 1949	191		41.2		5.8	8.5	A2	10635		53 Oph

α 2000	δ 2000	Star Name		Const	Years	Pos Ang		Separation		m₁	m₂	Spec	ADS	Orb	Notes
17ʰ34.ᵐ6	+9°35′	Σ I 34	AC	Oph	1912	345°		94″.0			10.8		10635		
---			AD		1912	223		91.4			10.8		10635		
17 34.7	-32 35	h 4962	AB	Sco	1877 1933	102		5.4		5.7	10.5	O5			
---		Ho 647	AC		1893 1907	83		13.3			10.5				
17 34.8	+6 01	Σ 2185	AB	Oph	1830 1934	5		27.7		8.0	11.0	F8	10638		
---			AC		1864 1949	190 225		97.1	80.8		8.5	G	10638		
---			AD		1912	202		89.6			9.1		10638		
---			AE		1909 1921	182 180		140.7	145.4		11.4		10638		
17 34.8	-11 15	h 4964		SerCd	1879 1959	226 225		54.0	54.6	5.7	8.7	B8			
17 34.9	+15 19	OΣ 332	AB	Her	1848 1947	112		10.3		8.0	11.1	K0	10641		
---			AC		1912	60		41.5			12.4		10641		
17 35.0	+61 52	β 962	AB	Dra	1879 1959					5.3	8.0	F8	10660	o	26 Dra
---		LDS 2736	ABxC		1963	162		(12′.3)			11.6	F8 M0	10660		
17 35.0	-62 07	B 1861		Ara	1931	265		1.4		7.4	11.4	A5			
17 35.4	+13 22	A 1879		Oph	1908 1958	36 86		0.3	0.3	7.8	10.0	A3	10648		
17 35.5	-5 56	Σ 2183 rej	AB	Oph	1879 1911	163		28.0		7.8	11.3	A0			
---			AC		1848 1914	13		28.4			10.7				
17 35.9	+1 00	Σ 2186		Oph	1831 1953	81		2.9		8.4	8.4	B8	10650		
17 35.9	-4 12	J 2668	AB	Oph	1943	255		50.0		8.0	13.5	A2			
---			BC		1943	225		4.0			14.0				
17 36.0	+20 34	h 2807		Her	1830	22		8.0		7.0	11.0				Probably BDS 8082, Σ 2190
17 36.0	+21 00	Σ 2190		Her	1829 1958	23		10.3		5.8	9.3	A2	10655		
17 36.5	+68 23	β pm		Dra	1877 1911	100 59		49.6	71.4	9.5	8.1	F5			
17 36.6	+7 22	A 1156		Oph	1905 1957	169		0.3		8.6	8.8	A3	10659		
17 36.7	-41 56	h 4963		Sco	1897 1940	316 312		6.9	6.9	8.0	10.5	K2			
17 36.8	+10 34	OΣ 333		Oph						7.9		G5			Very unlikely a close pair
17 36.8	-20 58	Hu 751		Oph	1903 1952	329 120		0.3	0.3	8.7	8.9	A2	10657		Direct motion
17 36.9	+68 45	Fox		Dra	1913 1922	276		72.3		4.9	13.2	F5			28 ω Dra
17 37.0	+67 07	Σ 2207		Dra	1832 1955	128 120		1.1	0.6	8.3	8.8	A2	10690		
17 37.3	-40 19	Wg 212		Sco	1903	341		12.7		7.2	9.7	B8			
17 37.3	-49 15	I 106		Ara	1900 1959	37		0.9		7.6	8.4	B9			
17 37.5	-37 47	B 915	AB	Sco	1927 1959	120 175		0.2	0.2	8.6	8.8	F0			
---		CorO 216	ABxC		1920 1940	194		14.0			9.0				
17 37.6	-15 24	Rst 5090		SerCd	1943 1960	81 79		24.5	25.1	3.5	12.9	A5			55 ξ Ser
17 37.8	-37 52	I 247		Sco	1902 1959	113 106		1.4	1.4	6.8	9.2	G5			
17 38.6	+55 46	Σ 2199		Dra	1830 1959	116 71		1.7	1.8	7.8	8.4	F5	10699		
17 38.8	-44 06	I 1331		Sco	1926 1959	53		0.6		8.5	8.7	F5			
17 38.9	-45 46	Don 851		Ara	1932 1948	249		1.8		7.5	12.2	G5			
17 39.1	+2 02	S,h 251	AB	Oph	1823 1923	328		111.2		6.3	7.4	K0 F0			
---			AC		1823 1912	18		132.4			12.3				
---			BC		1823	73		114.3							
17 39.5	+3 24	β 961	AB	Oph	1880 1925	141		8.0		6.8	11.4	K0	10688		
---			AC		1913	152		30.9			12.9		10688		
17 39.5	+46 00	β		Her	1913	49		116.0		3.8	12.1	B3			85 z Her
17 39.5	-13 39	Ho 418		SerCd	1892 1898	287		16.7		7.5	13.5	F0	10685		
17 39.8	-4 58	Σ 2191	AB	Oph	1831 1934	268		26.4		7.8	8.8	F2 F2	10693		
---		Ho 419	BC		1893 1902	33		8.1			12.8		10693		
17 40.0	-0 38	β 631		Oph	1879 1958	73 29		0.4	0.2	7.5	7.5	A0	10696		
17 40.1	+29 14	Σ 2192		Her	1833 1957	88 45		10.4	12.9	7.6	10.0	K0	10703		
17 40.3	+63 41	Σ 2218		Dra	1836 1959	355 326		2.5	1.7	7.1	8.3	F5	10728		
17 40.6	-35 39	δ 148		Sco	1922 1943	138		1.2		7.9	9.7	F5			
17 40.7	+31 17	OΣΣ 157		Her	1874 1924	111 109		112.9	115.0	6.4	8.5	K0 A0			
17 41.1	+24 31	Σ 2194	AB	Her	1831 1958	8		16.3		6.6	8.9	K0	10715		
---			AC		1880 1959	161 161		162.9	167.7		9.4		10715		
17 41.2	+16 29	A 2091		Her	1900 1945	274		0.6		8.1	10.1	A0	10714		
17 41.2	+41 39	Σ 2203		Her	1830 1959	334 306		0.7	0.7	7.6	7.9	A2	10722		OΣΣ 158 probably same
17 41.5	+37 52	Cou 996		Her	1973	321		1.3		7.8	11.6	G0			
17 41.5	-53 28	Pol 4		Ara	1887 1933	296		10.7		7.9	9.9	A3			
17 41.8	+21 30	Cou 114		Her	1966	20		0.3		6.9	8.1	F5			
17 41.9	+72 09	Σ 2241	AB	Dra	1832 1958	15		30.3		4.9	6.1	F5 F5	10759		31 ψ Dra
---			AC		1857 1923	128 120		101.5	89.7		11.4		10759		
---			AD		1905	84		100.5			12.9		10759		
---			CD		1908	19		67.6					10759		
17 42.0	+15 57	β 1251	AB	Her	1891 1948	79 4		1.4	0.6	5.6	11.1	F5	10723	o	Single 1958
---			AC		1910 1960	276 274		155.1	155.5		12.4		10723		
17 42.2	-48 39	h 4970	AB	Ara	1881 1938	69		7.9		8.0	8.8	F2			
---			AC		1933	233		18.2			10.5				
17 42.2	-69 01	Don 850		Aps	1928	319		3.2		7.6	13.6	A2			
17 42.5	+24 34	β pm		Her	1886 1909	260		159.6		5.5	9.1	K5			83 Her
17 42.6	+26 33	Σ 2198		Her	1829 1911	26		7.6		7.5	11.5	K0	10731		
17 43.2	+24 47	A 233		Her	1901 1927	235		3.2		7.9	12.7	G5	10737		
17 43.3	+17 41	HΣ		Her	1887 1925	50		15.6		8.0	11.0	A0	10738		
17 43.4	+33 57	Ho 560		Her	1894 1956	92 85		0.4	0.9	8.7	8.7	F5	10742		
17 43.5	-57 01	HdO 276		Ara	1902 1944	108		1.2		6.8	8.8	B9			
17 43.6	+34 46	OΣ 334		Her	1848 1947	356		15.2		8.2	9.6	G5	10747		

α 2000	δ 2000	Star Name		Const	Years		Pos Ang		Separation		m₁	m₂	Spec	ADS	Orb	Notes
17ʰ43ᵐ.9	+5°51′	Σ 2200	AB	Oph	1830	1943	165°		1″.7		8.4	9.2	A0	10741		
--			AC		1912		246		31.1			12.6		10741		
17 44.3	+14 25	h 1303		Her	1904	1959	150	151	37.9	39.7	6.2	9.3	A3	10749		
17 44.3	−72 13	HdO 275		Aps	1914	1959	88	336	0.9	0.6	7.3	7.6	F5		o	
17 44.4	+40 27	Es 1556		Her	1916		240		13.1		7.9	12.7	K5	10760		
17 44.5	+19 00	Σ 2206		Her	1830	1960	246		1.1		8.2	9.8	B9	10756		
17 44.5	−49 57	B 354		Ara	1927	1942	296		4.2		7.8	12.0	K0			
17 44.6	+2 35	Σ 2202	AB	Oph	1827	1968	93		20.6		6.2	6.6	A0 A0	10750		
--			AC		1912		28		95.9			12.5		10750		
17 44.6	−1 44	Schj 15	AB	Oph	1890	1932	355		55.2		8.2	9.3	K2			
--			AC		1890	1910	148		138.2			9.5				
17 44.7	−42 44	φ 341		Sco	1957	1959	6	27	0.1	0.1	6.9	7.0	A2			
17 44.8	+31 08	Σ 2213		Her	1836	1952	329		4.6		8.0	8.5	F8	10765		
17 44.9	−48 10	I 1602		Ara	1927	1944	18	9	0.2	0.2	8.6	8.9	A0			
17 44.9	−57 33	HdO 277	AB	Pav			265		30.0		6.1	12.0	G5			Mag. of A here is for Aa
--		H1n	Aa		1967		306		0.2		6.9	6.9				
17 45.0	−45 11	h 4973		Sco	1881	1951	27	26	12.9	13.4	8.4	9.2	A0			
17 45.0	−54 08	R 303		Ara	1880	1954	109		3.4		8.0	9.1	A0			
17 45.6	−3 30	A 32	AB	Oph	1899	1941	237		0.6		7.9	9.9	A0	10763		
--		h 4977	AC		1899	1910	143		22.0			13.8		10763		
17 45.6	−50 47	CapO 17		Ara	1891	1959	218	196	1.2	1.2	7.9	8.1	F8			
17 45.7	+16 50	A 2092		Her	1909	1945	337		0.8		8.1	11.1	K0	10770		
17 45.7	−46 55	I 248		Ara	1900	1959	125		1.8		8.0	9.5	A0			
17 45.8	+17 43	Σ 2205		Her	1830	1960	291	326	2.5	1.7	8.5	8.9	K0	10769		
17 45.9	+21 54	OΣ 335		Her	1846	1949	140		25.0		8.1	9.2	A0	10774		
17 46.0	+39 19	Σ 2224		Her	1831	1931	350		7.6		6.7	9.9	K0	10782		
17 46.3	−13 19	Σ 2204		SerCd	1830	1953	25		14.4		8.1	8.2	A0 A0	10771		
17 46.4	−70 20	I 610		Aps	1914	1930	72		2.2		8.0	11.0	G5			
17 46.5	+27 43	Σ 2220	AB	Her	1831	1970	241	247	29.9	33.8	3.4	10.1	G5	10786		86 μ Her; 2nd meas. AxBC
--		AC 7	AD		1921		4		256.1			11.1		10786		
--			BC		1857	1960						10.6		10786	o	
17 46.9	−43 30	CorO 220		Sco	1897	1920	335		13.5		7.6	9.7	B8			
17 47.1	+17 42	Σ 2215		Her	1831	1959	311	275	0.8	0.6	5.8	7.8	A0	10795		
17 47.1	−38 07	I 1336		Sco	1926	1959					7.1	7.3	B9		o	
17 47.5	−18 06	Hu 186		Sgr	1900	1952	340	331	1.0	0.9	7.5	11.6	F0	10788		
17 47.6	−40 08	See 338		Sco	1897	1900	95		37.5		3.0	12.9	F6			z¹ Sco
17 47.9	+34 17	β 632	AB	Her	1877	1926	343		5.6		6.6	12.8	B3	10807		
--			AC		1898	1921	349		32.0			13.3		10807		
--		OΣ 336 rej	AD		1843	1947	164	163	44.7	42.1		10.6		10807		
17 48.9	+4 58	Σ 2223 rej		Oph	1894	1924	211		18.3		7.3	8.6	F0	10813		
17 49.0	+37 04	Cou 1145		Her	1974		101		0.2		7.9	7.9	G0			
17 49.1	+62 48	LDS 1454		Dra	1963		212		17.0		8.0	17				
17 49.1	−11 20	Rst 3984		SerCd	1937	1943	58		2.9		8.0	13.7	K0			
17 49.3	−30 36	Rst 1990		Sco	1934	1944	242		2.0		6.8	12.5	F0			
17 49.6	−20 00	h 2810	AB	Sgr	1890	1918	189		40.8		7.7	10.5	A2			
--			AC		1909		270		297.0							C is the planetary NGC 6445
--			CD		1909		148		30.4							C, D are nebulous knots
--			CE		1909		302		32.7			10.7				
17 49.9	−37 03	See 340	AB	Sco	1897		201		26.4		3.2	14.9	K2			
--			AC		1897		103		41.7			14.7				
17 50.0	+41 57	Σ 2237		Her	1829	1932	8		20.4		8.0	10.3	F2	10829		
17 50.1	−35 09	B 2418		Sco	1932		32		6.0		7.8	12.7	K2			
17 50.2	+25 17	Σ 2232		Her	1830	1941	141		6.4		6.8	8.3	A2	10827		
17 50.2	−40 05	HdO 279		Sco	1900	1903	37		32.6		4.8	10.9	A3			z² Sco
17 50.4	−41 39	See 341		Sco	1897		67		19.7		7.2	14.0	K5			
17 50.5	+7 15	OΣ 337		Oph	1843	1960	308	197	0.5	0.3	8.3	8.6	F2	10828		
17 50.5	+36 51	Cou 1146		Her	1974		147		0.4		7.3	8.7	G0			
17 50.5	−48 17	h 4982		Ara	1837	1913	59		41.9		6.8	9.8	K0			
17 50.5	−53 37	h 4978		Ara	1834	1933	269		12.3		6.0	9.0	B3			
17 50.6	−49 30	B 2416		Ara			f		7.0		7.2	12.0	A5			
17 50.7	−29 17	B 359	AB	Sgr	1927		339		45.3		8.0	9.8	A2	10820		
--			BC		1927		225		4.2			13.8		10820		
--			BD		1927		118		7.3			13.1		10820		
17 51.2	+44 54	Σ 2242		Her	1830	1954	327		3.5		8.0	8.0	F0	10849		
17 51.2	−30 33	Pz		Sco	1877	1952	190		10.1		6.8	8.2	A0			
17 51.5	−40 46	HdO 281		Sco	1900	1959	194	187	27.8	23.6	6.0	12.5	M			
17 51.6	−60 10	HdO 278		Pav			350		30.0		5.8	13.0	K0			
17 51.9	+2 54	Σ 2233	AB	Oph	1832	1942	67		2.3		8.0	10.8	A0	10844		
--			AC		1905	1908	141		22.6			14.0		10844		
--			AD		1905	1908	237		30.4			13.6		10844		
17 51.9	+7 24	A 1164		Oph	1905	1957	38	35	0.2	0.3	8.4	8.8	A0	10846		
17 52.0	+15 20	OΣ 338	AB	Her	1845	1959	44	357	0.7	0.7	6.8	7.1	K0	10850		See BDS 8201
--			AC		1894		217		28.0			12.9		10850		
--			AD		1906		251		95.2			9.9		10850		
--		Lewis 16			1896	1908	353	342	0.8	0.9	10.0	10.5				

α 2000	δ 2000	Star Name		Const	Years	Pos Ang	Separation	m₁	m₂	Spec	ADS	Orb	Notes
h m	+1° 07′					° °	″ ″						
17 52.1	+1 07	S 694		Oph	1825 1923	237	81.8	6.8	7.1	K2 A0			
17 52.1	−60 24	h 4979		Pav	1887 1933	239	10.4	7.5	10.5	F0			
17 52.2	−28 30	Howe 40	AB	Sgr	1877 1937	9	7.0	8.0	10.5	G5	10838		
—			AD		1897 1898	357	12.5		12.0		10838		
—			AE		1898	4	20.2				10838		
—			BC		1889 1942	175 144	1.4 1.4		10.9		10838		
17 52.5	+27 12	h 1307		Her	1875 1910	323	33.2	7.9	11.9	A3			
17 53.0	−7 55	Σ 3128	AB	Oph	1834 1938	27 2	1.5 0.4	7.6	11.1	G5	10858		Companion not seen 1923,46
—			AC		1911	102	92.7		13.4		10858		
17 53.1	−13 39	Hu 189		SerCd	1900 1947	232 241	1.2 1.5	7.5	8.7	F5	10856		
17 53.3	−5 56	Ho 561		Oph	1897 1910	329	32.8	6.9	12.1	K0	10861		
17 53.3	+40 00	β 130		Her	1875 1958	123 116	1.8 1.6	5.2	8.5	K0	10875		90 Her
17 53.3	−34 44	β 1123		Sco	1889 1959	213 341	0.5 0.1	6.9	6.9	B9			
17 53.4	−34 54	See 342		Sco	1897 1959	286 242	0.4 0.4	6.4	6.5	K0			
17 53.5	+56 52	LDS 1457		Dra	1963	290	316.0	3.8	14.9				32 ξ Dra
17 53.6	+61 03	Es 1833		Dra	1920	261	7.7	7.8	12.6	F8	10887		
17 53.7	−39 53	Rst 1998		Sco	1935 1945	245	5.0	8.0	14.0	K0			
17 53.9	−34 45	B 1871		Sco	1929 1957	87 111	0.3 0.1	6.5	7.3	B9			
17 54.1	−11 20	h 4995		SerCd	1900 1920	155	28.6	6.6	11.6	K0	10872		
17 54.2	+11 08	ø		Oph	1963	28	0.1	7.1	7.3	F5			
17 54.9	−11 38	H1d 139	AB	SerCd	1888 1945	152	3.5	7.0	10.8	F2	10878		
—			AC		1909	94	65.6		12.5		10878		
17 55.6	+25 08	Cou 503		Her	1970	77	0.3	8.0	8.6	A3			
17 55.7	−30 34	I 1011		Sco	1912 1935	153	1.3	7.7	10.1	A0			
17 56.2	+39 26	β 417		Her	1877 1960	270 292	1.6 1.4	7.9	9.8	G0	10911		
17 56.3	+2 59	A 2189		Oph	1910 1958	326 11	0.3 0.1	8.5	8.8	A2	10899		Too close 1956,57,60
17 56.3	+62 37	OΣΣ 163	AB	Dra	1874 1924	37 42	59.4 57.2	7.2	7.7	K0 A2			
—			BC		1911 1919	339 339	183.8 184.7		10.6				
17 56.3	−15 49	h 2814	AB	SerCd	1878 1904	157	20.8	6.1	8.6	A0	10891		
—			AC		1904	349	33.7		11.5		10891		
17 56.3	−32 29	Jsp 748		Sco	1928 1944	168	13.5	6.6	12.4	B1			
17 56.4	+18 20	Σ 2245		Her	1829 1960	293	2.6	7.4	7.4	A2	10905		
17 56.6	+51 29	β 633	AB	Dra	1878 1934	151	20.9	2.2	13.2	K5	10923		33 γ Dra, Eltanin
—			AC		1878 1898	227 230	47.9 45.7		12.7		10923		
—			AD		1878 1915	14 12	56.7 57.6		12.7		10923		
—			AE		1898 1913	235 235	97.5 96.9		11.7		10923		
—			AF		1879 1914	116	125.3		11.0		10923		
—			AG		1898 1913	28 28	139.2 140.1		11.7		10923		
17 56.8	+48 37	Σ 2258		Her	1830 1937	222	2.5	8.7	8.9	G5	10924		
17 56.9	−39 56	CorO 222		Sco	1900 1952	124	3.5	7.6	9.2	F5			Spectrum composite F5+A2
17 57.1	+0 04	Σ 2244		Oph	1830 1959	273 286	1.0 0.3	6.7	6.9	A2	10912		
17 57.1	+45 51	Hu 235		Her	1900 1955	265 270	1.5 1.6	6.9	9.5	F5	10934		
17 57.2	+86 51	OΣ 340		UMi	1847 1925	237 226	31.5 32.1	8.5	9.0	F2	10437		
17 57.2	−55 23	Rmk 22		Ara	1880 1952	88 93	2.5 2.5	7.0	8.0	F8			
17 57.3	+35 41	Σ 2257		Her	1900 1925	150	21.6	7.8	11.0	K0	10932		
17 57.3	−38 47	Rst 2008		Sco	1935 1947	172 179	0.7 0.7	7.8	10.8	B9			
17 57.5	+10 57	β 1299	AB	Oph	1900 1960			8.8	8.8	K5	10916	o	
—			ABxC		1900 1920	63	27.0		11.5		10916		
17 57.5	−57 40	h 4992		Pav	1872 1951	10 23	4.8 3.5	8.3	8.9	G5			
17 57.7	−76 11	h 4974		Aps	1835 1918	120	25.5	6.1	14.0	K2			
17 57.9	−36 01	R 306	AB	Sco	1880 1947	16	3.6	7.0	9.5	A0			
—		See 343	AC		1896 1931	97 99	11.0 11.8		12.8				
17 58.0	−39 08	I 1013		Sco	1913 1959	178 151	0.7 0.7	6.6	8.1	A0			
17 58.1	+52 13	Σ 2261		Dra	1829 1931	262	9.5	7.7	9.7	A2	10953		
17 58.1	−45 27	Don 872		Sco	1931 1948	99	3.6	7.7	15	A3			
17 58.2	+27 24	Ho 425		Her	1891 1923	143	5.3	8.0	13.0	K2	10942		
17 58.7	+35 38	Cou 1002		Her	1973	167	0.6	7.2	9.1	K0			
17 58.9	−36 52	Δ 219		Sgr	1836 1935	265 259	47.1 50.1	5.8	8.1	G5 F0			
17 59.0	+2 02	Σ 2252	AB	Oph	1831 1954	24	4.0	8.5	8.8	A2	10945		
—			AC		1890 1931	164	94.2		8.7		10945		
17 59.1	+30 03	Σ 2259		Her	1831 1949	278	19.6	7.3	8.3	A0	10955		Spectrum composite A0+G0
17 59.1	−30 15	Pz	AB	Sgr	1836 1952	105	5.5	5.2	6.9	M0			
—			AC		1897 1919	239	26.2		13.0				
17 59.1	−38 50	I 1015		CrA	1913 1944	203	1.9	7.7	10.2	A0			
17 59.2	+64 09	Σ 2273	AB	Dra	1832 1930	284	21.1	7.5	7.8	F2 F2	10985		
—			AC		1905 1921	266	23.7		12.7		10985		
—			Aa		1921	341	18.9		14.7		10985		
17 59.4	−47 46	I 110		Ara	1902 1944	127	1.5	8.2	10.0	F8			
18 00.0	+24 49	Cou 115		Her	1966	115	0.3	8.8	8.8	K0			
18 00.0	−39 04	B 925		CrA	1927 1931	269	4.8	7.2	10.8	K0			
18 00.2	+8 51	OΣΣ 161	AB	Oph	1874 1923	77	62.8	7.1	9.1	G5 K0			
—			AC		1911	126	14.2		11.3				
18 00.2	+80 00	Σ 2308	AB	Dra	1832 1955	236 232	20.6 19.3	5.7	6.1	F5 F5	11061		41 Dra
—			AC		1932	128	221.6				11061		
18 00.3	+52 51	Σ 2271	AB	Dra	1831 1976	262 268	1.9 3.1	8.1	9.1	G0	10988		

α 2000	δ 2000	Star Name		Const	Years		Pos Ang		Separation		m₁	m₂	Spec	ADS	Orb	Notes
18ʰ00ᵐ.3	+52°51′	Σ 2271	AC	Dra	1913 1922	18°		167″.2				12.4		10988		
18 00.6	+2 56	β 1124	AB	Oph	1889 1934	196		6.6			4.0	13.8	B5	10966		67 Oph
—		H VI 2	AC		1823 1925	143		54.5				8.6		10966		
—		β 634	AE		1878 1925	179		45.8				11.0		10966		
—			CD		1878 1925	128		8.4				12.6		10966		
—			CE		1914	265		32.3				13.5		10966		
18 01.1	+45 21	Σ 3129		Her	1830 1904	169		30.8			7.6	10.5	B9			
18 01.2	−45 50	I 1112	AB	Ara	1914 1915	233		19.6			7.7	12.3	G5			
—			AC		1913	58		20.1				13.0				
—		h 5005	AD		1836 1915	27		26.6				10.4				
—			AE			75		30.0				13.5				
18 01.4	+40 19	Ho 77		Her	1884 1927	310		2.0			7.8	12.1	A0	10999		
18 01.4	+65 57	Σ 2284		Dra	1832 1953	194 190		3.7	3.5		7.8	9.4	F5	11016		
18 01.5	+21 36	Σ 2264		Her	1829 1974	262 258		6.1	6.3		5.0	5.1	A3 G5	10993		95 Her
18 01.6	+33 19	Ho 76	AB	Her	1884 1960	202 202		13.3	14.2		6.2	13.2	K5	10998		
—			AC		1895 1904	203		49.8				13.2		10998		
18 01.6	+40 11	Σ 2267		Her	1830 1959	234 255		1.4	0.8		8.6	8.6	A3	11001		
18 01.8	+1 18	β 1125		Oph	1889 1959	15 68		1.0	0.6		4.5	9.3	A2	10990		68 Oph
18 01.9	−19 06	I 621		Sgr	1914 1925						7.5		B3			Single 1925 (26-inch)
18 01.9	−22 47	β 283	AB	Sgr	1878 1933	238		8.2			5.8	12.9	B0	10983		
—			AC		1892	34		14.1				14.1		10983		
18 02.4	−23 02	H N 40	AB	Sgr	1866 1934	22		6.0			7.6	10.7	O7	10991		
—		H N 6	AC		1823 1956	212		10.8				8.7		10991		
—			CD		1866 1934	281		2.3				10.7		10991		
—			CE		1875 1890	191		6.1				12.6		10991		
—			CF		1875 1890	82		28.4				14.0		10991		
—			CG		1878 1890	212		29.6				13.4		10991		
18 02.5	+26 19	Ho 564	AB	Her	1897 1908	325 325		23.2	31.5		7.0	12.7	K0	11003		
—			AC		1908 1924	64 57		82.9	80.3			10.9		11003		
18 02.5	+44 14	β 1127		Her	1889 1960	145 94		0.8	0.9		7.4	9.3	F2	11010	o	Prob. Lewis 28, ADS 11011
18 02.6	−24 15	Rst 3149	AB	Sgr	1935 1943	95		1.5			6.9	12.4	B3			
—		Arg 31	AC		1879 1904	27		35.5				8.5				
—		β	AD			197		28.0				11				
18 02.8	+75 47	Σ 2302	AB	Dra	1833 1937	247		5.8			6.9	9.9	A0	11072		
—			AC		1833 1925	280		23.1				9.4		11072		
18 02.8	−27 05	H1d 32	AB	Sgr	1881 1940	101		5.1			8.4	9.4	G5	10995		
—		J 1592	AB		1948	348		6.2				12.9				
—			AC		1948	10		21.4				12.0				
—		Fox	AC		1909	68		43.3						10995		
18 02.9	+56 26	Σ 2278	AB	Dra	1831 1949	22 26		38.9	36.9		7.1	8.1	A5 A0	11035		
—			AC		1900 1945	33 35		35.0	34.3			8.5		11035		
—			AD		1880 1918	191		200.8				9.7		11035		
—			BC		1831 1949	146		6.1						11035		
—			Dd		1909	226		5.1				14.1		11035		
18 02.9	−27 50	B 2428		Sgr	1931	26		8.9			6.7	13.0	K0			
18 03.0	−43 28	See 345		CrA	1897 1913	141		16.0			7.5	12.0	K0			
18 03.1	−8 11	Σ 2262	AB	Oph	1836 1976						5.2	5.9	F0	11005	o	69 τ Oph
—			AC		1879 1959	127		100.3				9.3		11005		
18 03.1	+48 28	Σ 2277	AB	Her	1830 1972	118 125		27.6	26.8		6.4	8.3	A0	11028		
—			AC		1897 1959	284 292		88.0	94.0			9.8		11028		
—			BC		1901 1909	289 290		114.7	115.3					11028		
18 03.2	+7 55	OΣΣ 164		Oph	1875 1923	2		50.1			8.3	8.9	K0			
18 03.4	+2 31	A 2191		Oph	1910 1930	201		3.3			7.9	13.6	A0	11008		
18 03.9	+26 39	Ho 426		Her	1890 1922	193 204		12.4	12.0		7.9	13.6	F2	11031		
18 03.9	−22 50	B 2430		Sgr		5		8.0			7.2	12.8	B0			
18 04.2	+14 47	Σ 2269		Her	1830 1911	164 164		20.1	21.6		8.0	11.3	A0	11032		
18 04.2	−22 30	S 698	AB	Sgr	1825 1900	316		29.6			7.7	8.6	B0 B8			
—		B 1875	BC		1930	156		2.0				13.0				
—		Ara 1842			1922	111		12.5			10.6	12.9				
18 04.7	+27 07	Es 471	AB	Her	1907	271		18.2			7.2	14.2	A2	11040		
—			AC		1907	44		30.4				10.3		11040		
18 04.7	−59 13	h 5006		Pav	1915 1917	336		28.9			6.9	12.5	G0			
18 04.8	−35 54	HdO 283		Sgr	1900	290		12.5			5.8	11.0	K0			
18 04.8	−72 57	B		Aps		15		8.0			8.1	11.1	A3			
18 04.9	+33 16	Ho 78		Her	1884 1925	200		7.7			7.6	13.6	G0	11047		
18 04.9	+48 08	OΣ 343		Her	1846 1947	76		2.9			8.0	11.0	A2	11058		
18 05.0	+41 57	Roe 1		Her	1908 1909	187		23.4			6.3	12.3	F0	11054		
18 05.0	−29 35	See 346	AB	Sgr	1897 1899	234		32.9			5.1	13.5	F8	11029		W Sgr; A is variable
—			AC		1897 1899	100		47.8				13.5		11029		
18 05.3	+81 29	Σ 2326		Dra	1832 1940	202 196		15.6	16.4		7.7	10.0	A5	11156		
18 05.5	+2 30	Σ 2272	AB	Oph	1825 1976						4.2	6.0	K0	11046	o	70 Oph [68]
—			Aa		1878 1947	198 282		71.4	34.9			12.9		11046		
—			Ab		1905 1946	252 287		70.0	76.7			13.9		11046		
—			Ac		1905 1946	73 50		105.0	119.4			12.9		11046		
—			Ad		1878 1946	50 22		87.2	141.1			12.4		11046		

α 2000	δ 2000	Star Name		Const	Years	Pos Ang	Separation	m₁	m₂	Spec	ADS	Orb	Notes
18 05.5	+2 30′	Σ 2272	Ae	Oph	1905 1946	103° 80°	120.3 110.6		13.0		11046		
--			Af		1905 1945	329 334	142.4 184.5		12.9		11046		
--			Ag		1886 1946	224 246	165.9 138.6		11.7		11046		
--			Ah		1924	4	158.3				11046		
--			Ba		1899 1920	208 239	50.8 38.2				11046		
--			Bb		1900	247	69.3				11046		
--			Bc		1900	168	68.4				11046		
--			ge		1906	73	247.9				11046		
--			gh		1906	270	180.5				11046		
--			gx		1906	250	16.6		13.4		11046		
18 05.5	−63 12	Rst 959		Pav	1929	117	0.3	8.3	8.6	G0			
18 05.7	+12 00	Σ 2276	AB	Oph	1830 1967	257	6.9	7.0	7.4	A0	11056		
--			AC		1905 1932	306 305	63.1 61.1		12.0		11056		
18 05.7	−34 37	h 5012		Sgr	1919	191	23.4	8.5	9.1	K5 K2			
18 05.7	−36 35	Howe 88	AxBC	Sgr	1896 1934	4	3.2	7.7		B8			
--		B 1876	BC		1933	15	0.4	8.9	10.3				
18 05.8	+21 27	OΣ 341	AB	Her	1849 1960			7.4	8.5	G0	11060	o	
--			AC		1905 1971	172	28.2		9.5		11060		
--			AD		1905 1971	101	37.9		9.5		11060		
--			AE		1905 1971	38	63.1		9.5		11060		
--			AF		1905 1933	355	108.0		9.5		11060		
--			AG		1933	238	135.9		8.5		11060		
18 06.0	+4 33	OΣΣ 165		Oph	1874 1923	142	66.1	8.5	8.5	K2 A0			
18 06.0	−8 07	A 2595		SerCd	1913 1925	62	2.6	7.0	13.8	F8	11052		
18 06.1	−24 12	B 376		Sgr	1927	98	1.9	7.7	11.7	B9	11049		
18 06.1	−56 49	Rst 961		Ara	1928	220	1.5	8.0	11.8	K2			
18 06.3	+38 24	Hu 1186		Her	1905 1958	129 93	0.2 0.3	8.7	8.8		11071		Hu has p.a. 309° in 1905
18 06.3	−34 19	O1 47	AB	Sgr	1910 1943	116	1.7	8.3	9.6	B8			
--		CorO 223	AC		1922	244	27.0		10.3				
--		δ	CD			247	26.7		11.3				
18 06.4	−41 45	h 5011		CrA	1882 1919	350	29.6	7.7	9.1	A0			
18 06.5	+40 22	Σ 2282		Her	1831 1960	93 86	2.4 2.5	7.4	8.4	A0	11074		
18 06.7	−18 59	h 592		Sgr	1890 1918	207	36.4	7.8	11.3	A0			
18 06.8	+8 53	β pm	AB	Oph	1893 1918	240	131.5	7.8	10.1	F5			
--			BC		1908	38	80.4		13.1				
18 06.8	−43 25	h 5014		CrA	1836 1959			5.7	5.7	A3		o	
18 07.0	+30 34	AC 15	AB	Her	1859 1960			5.1	8.4	F8	11077	o	99 Her
--			AC		1912	53	93.4		10.5		11077		
18 07.0	+46 47	Es 1157	AB	Her	1912 1922	180	26.2	8.1	10.1	F2	11085		
--			AD		1922	243	47.2		10.7		11085		
--			BC		1912 1922	14	4.3		10.8		11085		
18 07.1	+49 43	OΣ 344		Her	1847 1958	155 145	2.2 2.3	6.3	10.4	A0	11090		
18 07.2	−49 49	I 1350a		Ara	1926 1944	193	0.4	8.4	8.9	A2			
18 07.3	+9 34	OΣ 342	AB	Oph				3.7		A3	11076		72 Oph; single 1952,53
--			ABxC		1890 1934	301 299	25.3 25.4		14.0		11076		
--			ABxD		1878 1911	168 166	51.7 54.4				11076		
18 07.5	+19 40	OΣ 524		Her	1853 1960	86 267	0.4 0.2	7.7	8.9	A2	11080	o	Single 1957 (24-inch)
18 07.8	+13 04	H V 74		Oph	1783 1960	129 138	40.9 42.3	6.6	9.6	A0	11086		
18 07.8	+26 06	Σ 2280	AB	Her	1831 1955	183	14.2	5.9	6.0	A3 A3	11089		100 Her
--			AC		1905	124	75.9		11.0		11089		
18 08.1	+2 13	β 636	AB	Oph	1878 1914	127	4.6	6.5	11.7	A2	11088		
--			AC		1898 1920	100 100	15.1 13.6		13.5		11088		
--			AD		1923	349	41.5		11.5		11088		
18 08.1	−7 07	Rst 3997		SerCd	1938 1943	80	1.3	7.5	13.2	K0			
18 08.5	−45 46	h 5015		Ara	1891 1935	258	3.9	6.2	9.8	B8			
18 08.6	+18 39	Hu 314		Her	1901 1958	147 119	0.4 0.3	8.6	8.7	A0	11098		
18 08.6	−27 52	β 244		Sgr	1876 1942	258 263	2.1 2.1	7.5	8.7	G5	11084		
18 08.8	+20 49	AGC 8		Her	1878 1934	136	23.4	4.4	11.9	B3	11102		102 Her
18 08.9	−25 28	WNO		Sgr	1877 1904	64	13.3	6.6	8.6	B8	11069		
18 09.2	−26 07	B 379	AB	Sgr	1927	339	4.7	7.6	13.0	G0	11096		
--			AC			195	6.0		14.5		11096		
18 09.6	+4 00	Σ 2281	AB	Oph	1831 1960			6.1	7.0	F2	11111	o	73 Oph
--			AC		1912 1960	191 193	67.1 67.9		12.6		11111		
18 09.6	+6 09	Σ 2283		Oph	1832 1959	92 74	1.2 0.7	8.1	8.6	F5	11110		
18 09.7	+50 24	Hu 674		Her	1904 1958	279 243	0.5 0.5	7.5	8.0	A2	11128		
18 09.9	+3 07	β 637	AB	Oph	1878 1934	194	7.3	5.7	11.7	F5	11113		
--			AC		1910 1924	238 238	104.9 103.5		10.0		11113		
18 10.0	+48 24	Σ 2293 rej		Her	1902 1925	83	13.1	8.1	10.3	G0	11130		
18 10.0	−59 02	HdO 285		Pav	sp		40.0	6.4	12.0	K5			
18 10.1	+16 29	Σ 2289		Her	1829 1959	243 224	1.2 1.2	6.5	7.2	F2	11123		Composite spectrum F2+A0
18 10.1	−23 46	Rst 5104		Sgr	1943 1949	188 172	0.2 0.2	8.8	8.8	B5			
18 10.1	−30 44	β 245		Sgr	1868 1951	352	4.0	5.6	8.6	K0			
18 10.4	+0 33	Σ 2286		Oph	1831 1946	316	2.6	8.0	10.7	A0	11119		
18 10.4	+32 21	Ho 81		Her	1883 1925	212	2.8	8.0	11.7	A2	11131		
18 10.6	−16 46	S 700	AB	Sgr	1879 1910	292	18.7	8.8	8.9	B			4th star said to lie south

α 2000	δ 2000	Star Name		Const	Years		Pos Ang		Separation		m_1	m_2	Spec		ADS	Orb	Notes
18ʰ10.6ᵐ	-16°46′	S 700	AC	Sgr	1825	1915	353°	°	28.7″	″		9.6					
—			Aa		1876		310		15.0			10.9					
18 10.6	-75 11	h 4999		Aps	1835	1940	173		13.1		8.0	9.3	A0	A			
18 10.8	-40 26	h 5023		CrA	1836	1951	276		8.7		8.1	8.4	A2				
18 11.1	-47 31	B 1879		Tel	1934		266		1.7		6.1	11.5	K0				
18 11.2	-19 51	β 132		Sgr	1875	1959	240	203	0.8	1.0	6.9	7.3	A2		11127		Same as ADS 11250, See 501
18 11.2	-45 57	HdO 286		Tel	1897		228		21.2		4.5	12.9	G5				ε Tel
18 11.3	-29 38	B 927		Sgr	1927		204		6.4		8.0	12.5	G5				
18 11.3	-75 53	HdO 282		Aps	1902		231		25.9		5.9	13.2	K5				
18 11.4	-5 12	Ho 267		SerCd	1889	1907	356	355	15.7	19.0	6.6	12.6	B9		11135		
18 11.7	+33 27	B 2545	AB	Her	1958	1960	233	242	0.1	0.1	6.4	6.7	A2		11149		
—		Ho 82	ABxC		1885	1960	207	218	0.6	0.8		9.8	A2		11149		
18 11.7	-23 42	h 5030		Sgr	1901	1907	287		42.1		5.0	10.7	K0		11133		1 Sgr
18 12.0	-56 26	h 5021		Tel	1900	1913	326		4.9		7.6	13.5	A0				
18 12.4	-7 18	A 36	AxBC	SerCd	1899	1915	42		65.6		7.6		G0		11165		
—			BC		1899	1925	196		1.3		11.0	11.3			11165		
18 12.5	-15 22	Lv 7	AB	SerCd	1892	1923	278		3.8		8.0	11.6	B5		11146		
—			AC		1912		240		33.6			11.6			11146		
18 12.6	+41 23	Σ 2298	AB	Her	1831	1953	186	177	2.4	2.0	8.6	8.8	F5		11174		
—			AC		1881	1924	40	40	78.0	76.4		10.4			11174		
18 12.6	-73 40	HdO 284		Aps	1901	1935	232	243	2.5	2.5	6.0	9.0	F5				
18 12.8	+5 50	OΣ 345	AB	Oph	1845	1943	64		1.3		8.1	11.1	B9		11160		
—			AC		1909		200		87.4						11160		
18 12.9	-50 34	HdO 288		Tel	1901	1943	161	149	2.3	1.8	6.8	10.0	F0				
18 13.0	+28 14	H V 93		Her	1880	1926	136		54.8		8.2	8.2	F8	F8			
18 13.0	-39 10	I 1018		CrA	1911	1943	272	283	1.4	1.4	7.3	10.8	G0				
18 13.4	-27 12	B 2435		Sgr			np		6.0		8.0	15	B9				
18 13.6	-15 37	β 131	AB	SerCd	1875	1953	278		2.8		7.3	9.3	F5		11166		
—			AC		1880	1944	279	289	7.1	8.4		11.5			11166		
—		Fox	AD		1913		36		37.0			12.0			11166		
—		β 131	BC		1933		296		4.8						11166		
18 13.6	-31 58	B 1353		Sgr	1930	1954	143	35	0.1	0.1	7.6	7.7	B8				
18 13.8	-21 04	H V 7	AB	Sgr	1836	1932	258		16.9		3.9	11.4	B8		11169		13 μ Sgr; variable
—		β 292	AC		1878	1901	119		25.8			13.4			11169		
—		h 2822	AD		1835	1925	312		48.5			9.8		B3	11169		
—			AE		1835	1925	115		50.0			9.3			11169		
—		Slv	BD		1968		331		40.6						11169		
—			BE		1968		106		64.5						11169		
18 14.0	+20 17	Ho 268		Her	1903		58		42.0		7.5	11.5	K0				
18 14.2	+64 45	Com		Dra	1903		281		299.1		7.3		K2				
18 14.2	-27 30	B 384		Sgr	1927		330		4.6		7.2	13.7	B9		11173		
18 14.2	-37 37	B 934	AB	CrA	1927		30		14.7		7.7	11.5	A0				
—			BC		1927		185		2.4			12.0					
18 14.6	+0 11	Σ 2294		Oph	1831	1960					8.6	8.7	F2		11186	o	
18 14.6	-36 54	B 935		Sgr	1927	1936	108	99	0.2	0.2	8.5	8.7	B9				
18 14.6	-47 22	h 5031		Tel	1882	1928	87	76	23.4	27.9	8.2	9.9	K0	G5			
18 14.7	+56 35	β 1274	AB	Dra	1892	1924	239		95.6		6.4	9.8	F0		11213		
—			BC		1892	1914	146		0.9			10.6			11213		
—			BD		1878	1924	7		5.4			10.4			11213		
18 14.8	-66 49	h 5019		Pav	1918		334		38.0		7.6	12.8	B8				
18 15.0	-50 18	I 429		Tel	1902	1952	146		0.3	0.1	8.0	8.2	A0				Too close after 1938
18 15.1	-57 51	h 5029		Pav	1871	1959	113	96	2.4	2.0	8.3	8.5	G5				
18 15.2	-20 23	β 286		Sgr	1878	1933	216		6.0		6.0	13.0	B1		11191		16 Sgr
18 15.3	-18 59	β 284	AB	Sgr	1890	1913	359		17.9		8.3	12.0	B0		11193		
—			AC		1891	1913	87		31.3			11.9			11193		
—			AD		1890	1913	192		57.5			10.1			11193		
—			AE		1890	1913	180		76.1			10.2			11193		
—			AF		1890	1914	268		138.5			10.8			11193		
—			Aa		1891	1914	199		12.0			11.5			11193		
—			Ab		1891	1913	67		22.0			11.9			11193		
—			Ad		1898		103		17.5			13.5			11193		
—			Bc		1891	1914	329	324	5.0	5.0		12.0			11193		
—		Ara 743	Cb		1916		302		13.0						11193		
18 15.4	+19 46	OΣ 346		Her	1847	1949	329		5.4		8.2	9.0	F2		11208		
18 15.4	-48 51	h 5033	AB	Tel	1902	1913	115		17.3		6.8	10.8	K0				
—			AC		1913		10		18.2			12.2					
—			AD		1913		64		27.9			10.7					
18 15.4	-76 40	Gli 241		Aps	1911	1925	76		2.0		8.5	8.8	F2				
18 15.5	-18 40	See 349		Sgr	1897	1899	123	130	11.1	10.1	6.1	11.8	A2		11196		Comp. not seen 1934, 36-in
18 15.7	-3 21	Σ 2296		SerCd	1829	1937	8		3.2		7.3	10.9	K0		11205		
18 15.7	-63 03	h 5024		Pav	1871	1917	8		41.9		5.6	11.2	K0				
18 16.1	-56 39	I 111		Tel	1900	1944	305	295	0.8	1.0	6.9	9.4	K0				
18 16.2	-31 10	h 5037		Sgr	1890	1919	227		37.2		7.7	9.9	B5				
18 16.2	-46 02	h 5034		Tel	1836	1959	90	97	2.4	2.4	7.5	8.8	A2				
18 16.4	-40 28	I 1020		CrA	1913	1944	303	294	0.4	0.4	8.2	8.5	B8				

α 2000	δ 2000	Star Name		Const	Years		Pos Ang		Separation		m_1	m_2	Spec	ADS	Orb	Notes
18^{h}16.5	+23°24′	β pm	AB	Her	1877	1908	3°	°	129″.2	″	7.6	10.9	K0			
—			BC			1908	246		17.9			11.6				
18 17.5	−18 28	Rst 3170		Sgr	1933	1940	170		8.4		6.5	13.1	O5			
18 17.6	−27 52	B 386		Sgr	1926		34		0.2		8.0	9.0	B9	11224		
18 17.6	−36 46	β 760	AB	Sgr	1879	1959	105		3.6		3.2	7.8	M4			η Sgr
—			AC		1896		276		33.3			13.0				
—			AD		1879	1896	303		93.2			10.0				
18 17.7	−19 40	β 246		Sgr	1875	1952	105	113	0.5	0.5	8.3	8.3	B2	11228		
18 17.8	−18 48	S,h 263	AB	Sgr	1823	1916	12		54.3		6.7	9.7	B5	11232		
—		β 299	Ae		1892		22		22.2			12.5		11232		
—			Af		1891		66		29.4			13.1		11232		
—			Ah		1891		328		22.0			13.1		11232		
—			Bc		1891		132		10.4			13.2		11232		
—			ed		1891		305		7.1			12.5		11232		
—			gh		1891		317		8.4		13.0			11232		
18 18.0	−68 14	HdO 289		Pav	1901		298		2.5		6.3	9.8	A0			
18 18.1	+23 18	β pm	AB	Her	1908	1959	222	222	37.8	44.9	6.7	11.5	K5			
—			AC		1878	1959	126	125	156.3	155.4		9.1				
18 18.7	−18 37	β 639	AB	Sgr	1878	1958	155	135	0.6	0.4	6.8	7.3	B0	11240		
—		S,h 264	AC		1823	1921	52		17.2			9.3		11240		
—		Fox	AE		1918		330		146.6			11.0		11240		
—		β 300	CD		1891	1934	326	325	8.3	7.5		14.8		11240		
18 19.2	+7 16	h 5494		Oph	1901	1911	70		39.6		5.4	11.2	K0	11254		
18 19.7	−4 57	Schj 16		SerCd	1874	1957	193	203	2.3	2.9	8.1	9.4	G0	11257		
18 19.7	−63 53	I 249		Pav	1900	1928	4	1	7.1	7.5	6.2	10.8	G0			
18 20.1	−7 59	Σ 2303		SerCd	1831	1959	216	236	3.2	2.1	6.6	9.1	F5	11262		
18 20.4	−49 08	I 1361		Tel	1926	1935	254		2.3		7.6	12.3	A2			
18 20.4	−80 14	Rst 3156		Aps	1934		30		2.6		7.3	11.0	K2			
18 20.5	+20 55	Cou 202		Her	1967		264		0.2		8.4	8.5	A0			
18 20.6	+22 48	Σ 2310		Her	1830	1935	236		5.1		6.8	10.1	B9	11273		
18 20.7	−36 47	B 1356		Sgr	1930		232		2.4		7.5	12.9	M			
18 20.8	+27 32	β 640		Her	1878	1957	346	320	2.4	1.9	7.0	11.7	G5	11275		
18 20.8	+71 20	OΣ 353	AB	Dra	1856	1960	64	304	0.6	0.2	4.4	6.1	A0	11311	o	43 φ Dra
—			AC		1912		115		70.8			12.7		11311		
18 20.9	+3 23	h 5495	AB	Oph	1878	1935	286		28.1		4.9	11.5	G5	11271		74 Oph
—			AC		1912	1925	80		57.8			11.9		11271		
18 21.0	−19 39	β 48		Sgr	1874	1959	0	354	2.3	2.3	8.1	10.1	F0	11268		
18 21.0	−29 50	See 350	AB	Sgr	1896		276		25.8		2.7	14.4	K2	11264		19 δ Sgr
—			AC		1896		165		40.1			14.9		11264		
—			AD		1896		221		58.1			12.9		11264		
18 21.1	+72 44	β pm	AB	Dra	1879	1908	26	19	144.6	149.2	3.6	11.9	F8			44 χ Dra
—			BC		1908		302		10.2			13.4				
18 21.3	−2 54	Σ II 8		SerCd	1836	1923	77	61	112.7	179.7	3.3	12.0	K0			58 η Ser
18 21.8	+21 30	β 641		Her	1880	1956	349	342	1.0	0.9	6.8	8.7	B8	11287		
18 22.1	+48 21	Hu 582		Dra	1902	1921	195		2.1		8.0	12.2	G5	11301		
18 22.2	−15 05	Σ 2306	AB	Sct	1831	1959	220	221	12.8	10.2	7.9	8.6	F5	11282		
—		Mh	AC		1936		218		10.1			9.0		11282		
—		Dem 18	BC		1848	1950	64	70	0.6	1.2				11282		
18 22.2	−55 34	Δ 220		Tel	1836	1938	178		31.1		8.1	8.5	G0			
18 23.0	−18 52	Stone 42		Sgr	1879		85		6.7		8.5	9.0				Not found (β)
18 23.0	−62 35	CapO 79		Pav	1901	1928	55		5.7		8.4	8.6	F5			
18 23.1	−60 45	Rst 5110		Pav	1942		216		3.2		7.0	12.0	A3			
18 23.2	−61 30	Gale 2		Pav	1895	1955	150	154	3.3	3.3	4.4	8.6	M1		o	ξ Pav
18 23.3	−42 47	CorO 224		CrA	1889	1935	137	137	3.5	3.2	3.6	8.9	F5			
18 23.5	−26 10	Ho 566		Sgr	1897	1952	150		0.5	0.1	8.4	8.9	F2	11296		Too close after 1942
18 23.7	+21 46	β pm		Her	1854	1909	321	321	204.8	221.8	3.8	10.3	K0			109 Her
18 23.7	+38 21	Ho 431		Lyr	1892	1925	0		21.2		6.8	12.3	A0	11317		
18 23.9	+58 48	Σ 2323	AB	Dra	1833	1975	6	351	3.1	3.8	5.0	8.0	A2	11336		39 Dra
—			AC		1834	1956	21		88.9			7.4		11336		
—			AE		1880	1910	66		198.0			10.9		11336		
—			AF		1910		80		150.3			11.2		11336		
—			AG		1918		254		36.0			14.2		11336		
—			AH		1918		86		41.3			14.2		11336		
—			BC		1909	1937	22		85.5					11336		
18 24.0	+1 30	h 858		SerCd	1904	1909	229		12.8		8.2	9.5	F5			
18 24.0	+38 44	Ho 432		Lyr	1892	1934	288		17.0		6.4	12.9	K2	11320		[69]
18 24.1	+20 27	Ho		Her	1903		350		39.0		6.6	11.6	F0			
18 24.2	−34 23	See 351		Sgr	1896		295		32.5		1.9	14.2	A0			20 ε Sgr
18 24.3	−44 07	Δ 221		CrA	1837	1914	164		75.3		5.3	9.9	B5			
18 24.4	−30 41	B 1359		Sgr	1928		77		3.1		8.0	14.5	A0			
18 24.5	−29 47	B 2446		Sgr			sp		7.0		8.0	14.0	A2			
18 24.7	−6 36	Σ 2313		Sct	1832	1951	198		6.1		7.5	8.8	G0	11318		
18 24.8	+7 14	OΣ 347		Oph	1843	1946	347		3.2		7.8	11.6	B9	11323		
18 24.9	−1 35	AC 11		SerCd	1854	1960					6.8	7.0	F5	11324	o	
18 25.0	+27 23	Σ 2315	AB	Her	1830	1960					6.5	7.5	A0	11334	o	Error in position?

α 2000	δ 2000	Star Name		Const	Years	Pos Ang	Separation	m₁	m₂	Spec	ADS	Orb	Notes
18ʰ25.0	+27°23'	Σ 2315	AC	Her	1898 1934	80° 80°	21".7 44".9		13.5		11334		1898 dist. prob. in error
18 25.3	+26 05	Σ 2317 rej	AB	Her	1904 1925	224	24.8	7.7	10.4	K0	11337		
—			AD		1904 1925	190	44.6		9.4		11337		
—			BC		1901 1954	323	1.1		10.6		11337		
18 25.3	+28 05	Ho 85		Her	1885 1925	195	4.8	8.0	12.0	B9	11338		
18 25.3	+46 53	β 134		Lyr	1861 1955	134	1.1	7.8	9.7	A0	11343		
18 25.3	+48 46	Hu 66	AB	Dra	1898 1960	310 271	0.3 0.4	7.9	8.1	G5	11344		Nearly an equilateral tri-
—		OΣ 351	AC		1846 1960	25 18	0.5 0.7		8.2		11344		ple in 1898; AB single in
—		Hu 66	BC		1898 1954	43 39	0.5 0.9				11344		1943,45 (VBs, 82-inch)
18 25.3	−20 32	Jc 6		Sgr	1851 1959	297 289	1.8 1.8	4.9	7.4	K0 A0	11325		21 Sgr
18 25.4	+26 00	Σ 2318		Her	1829 1903	254	20.8	8.1	10.3	K0			
18 25.5	−30 15	B 939		Sgr	1927 1931	191	4.9	7.4	12.5	B8			
18 25.8	−53 39	h 5041		Tel	1872 1943	260	2.8	7.3	9.3	F0			
18 26.1	+0 47	β 1203		SerCd	1890 1960	68 134	0.3 0.3	7.6	7.8	A3	11339		
18 26.7	+26 27	β 1326	AB	Her	1904 1958	105	5.5	6.5	12.7	B3	11356		
—			AC		1904 1924	61	62.0		8.7		11356		
18 26.7	+36 10	Es 2173		Lyr	1925	300	5.6	7.7	11.9	K0	11361		
18 26.9	−17 12	β 965		Sgr	1880 1921	106	1.5	7.9	11.6	A3	11345		
18 27.1	−8 02	h 5496		Sct	1823				6.6	K0			Not double (IDS)
18 27.2	+0 12	Σ 2316	AB	SerCd	1828 1958	314 318	4.0 3.8	5.3	7.6	A0	11353		59 d Ser; A is sp. binary
—			Aa		1951	290	0.1				11353		
18 27.5	−15 22	J 1631	AB	Sct	1942	265	20.0	6.8	13.9	F5			
—			AC		1941	155	120.0		10.3				
18 27.6	−54 02	B 390		Tel	1927	19	1.2	8.0	10.7	K0			
18 27.7	+19 17	Σ 2319	AB	Her	1830 1956	191	5.4	8.2	8.6	F5	11372		
—			AC		1829 1956	279 278	38.4 41.2		10.5		11372		
—			AD		1956	246	155.2				11372		
—			BC		1890 1909	283 282	39.3 40.3				11372		
18 27.7	−26 38	β 133		Sgr	1875 1959	265 251	1.8 1.3	6.9	7.0	A5	11354		
18 27.7	−32 07	I 1023		Sgr				8.5	8.7	A2			Measures uncertain
18 27.8	+24 42	Σ 2320		Her	1831 1956	11 3	1.8 1.3	7.0	8.9	B9	11373		
18 28.2	−52 30	B 1362		Tel	1930 1948	233 237	0.3 0.3	8.1	8.2	A2			
18 28.6	+4 51	OΣ 168	AB	SerCd	1875 1923	164	47.6	7.7	8.4	A0			
—			AC		1912	218	245.6						
18 28.9	−25 03	H N 125		Sgr	1889 1951	100 104	2.8 2.8	8.1	8.3	G0	11378		
18 29.0	−26 35	WNO 6		Sgr	1890 1905	182	41.9	6.5	8.2	A0			
18 29.2	+29 33	Ho 434		Lyr	1891 1925	186 186	11.5 13.7	7.7	12.6	G5	11393		
18 29.2	−29 16	B 2450		Sgr		np	8.0	6.8	12.0	F5			
18 29.3	−41 57	B 2449		CrA		p	8.0	8.0	12.0	K0			
18 29.5	+7 22	A 582		Oph	1903 1927	45	2.8	7.7	13.5	G0	11390		
18 29.9	−57 31	Rst 987	AB	Pav	1929	346	1.8	5.8	12.5	K0			
—		HdO 290	AC		1900 1913	120 120	34.5 33.9		10.5		11396		
18 30.0	+1 11	Σ 2321		SerCd	1830 1940	190	6.8	8.2	9.8	A0	11399		
18 30.1	+4 04	Σ 2322	AB	Oph	1828 1935	168	20.1	6.7	12.0	B5	11399		
—			AC		1912	70	66.9		13.0		11399		
—			AD		1912	194	85.2		13.1		11399		
18 30.5	−23 15	Rst 4013		Sgr	1935 1949	263	1.3	7.0	11.7	K0			
18 30.7	+4 31	OΣΣ 170		Oph	1875 1922	5	101.3	6.8	9.0	A2			
18 30.9	−48 01	h 5045		Tel	1897 1933	24	8.0	6.7	10.5	G0			
18 31.0	+1 23	Σ 2324	AB	SerCd	1829 1943	146	2.4	8.4	8.7	B8	11410		
—		Fox	AC		1923	238	33.0		12.0		11410		
18 31.1	+32 15	Σ 2333	AB	Lyr	1831 1952	334	6.4	7.8	8.4	A0	11424		
—			AC		1909 1952	35	163.8		12.5		11424		
—			AD		1952	339	85.0		13.2		11424		
—			BC		1880 1909	37	160.9				11424		
18 31.1	−32 59	Howe 43	AB	Sgr	1877 1942	200 193	3.2 3.2	5.3	9.8	A3			
—			AC		1897 1899	338	29.0		12.8				
18 31.2	+13 11	Σ 2330		Her	1829 1952	177 169	20.3 17.6	8.1	9.8	K0	11418		
18 31.2	+65 26	Kui 86		Dra	1958	36	26.6	6.3	10.5	A3			
18 31.2	−43 44	I 1369		CrA	1928 1944	238	4.2	7.9	12.2	B3			
18 31.3	+62 32	Hu 932		Dra	1904 1921	89	2.7	7.0	12.8	F2	11443		
18 31.4	+6 27	Σ 2329		Oph	1830 1956	45	4.3	8.2	9.5	B9	11420		
18 31.4	−10 48	Σ 2325		Sct	1829 1925	257	12.3	5.8	9.1	B3	11414		
18 31.4	−18 24	See 354		Sgr	1897 1934	183	25.2	5.2	14.1	A0	11411		In open cluster M25
18 31.9	−19 08	β 966	AB	Sgr	1879 1918	253	66.5	6.5	9.6	G0	11433		A is U Sgr
—			AE		1890 1940	282	145.0		9.1		11433		
—			AG		1893	254	430.7		8.4		11433		
—			AH		1890 1940	295	72.2		10.0		11433		
—			Aa		1890 1940	195	164.0		9.3		11433		
—			Ac		1890 1940	90	76.8		11.0		11433		
—			BC		1880 1952	122	0.7		10.1		11433		
—			Bb		1890 1940	200	110.4		10.4		11433		
—			CD		1879 1918	172	11.2		10.0		11433		
—			EF		1890 1940	291	25.2		9.6		11433		
—			EK		1890 1940	27	30.4		9.5		11433		

α 2000	δ 2000	Star Name		Const	Years	Pos Ang	Separation	m₁	m₂	Spec	ADS	Orb	Notes
18ʰ31.9ᵐ	−19°08′	β 966	EL	Sgr	1890 1940	123° °	21.0 ″		9.7		11433		
—			Ge		1890	63	61.5		9.1		11433		
—			HJ		1890 1940	311	44.0		9.8		11433		
—			JM		1890 1940	345	70.2		10.0		11433		
—			aC		1890 1940	264	93.6				11433		
—			dE		1890	142	131.3	8.3		B9	11433		
18 31.9	−28 31	Rst 2062		Sgr	1931 1951	311 330	0.5 0.7	8.2	9.8	G0			
18 31.9	−59 34	I 633		Pav	1912 1943	314 322	0.5 0.6	8.5	8.7	F5			
18 32.0	+6 47	OΣ 354		Oph	1846 1960	154 190	0.8 0.7	7.7	8.5	F5	11432		
18 32.0	+25 04	Pou 3419		Her	1901	238	11.3	7.7	12.1	A5			
18 32.2	−7 50	β 419		Sct	1877 1957	58 39	1.2 1.4	8.5	9.2	A0	11430		
18 32.2	−9 22	β 247		Sct	1875 1916	167	7.8	7.7	11.1	A0	11429		
18 32.8	+23 37	Wils		Her	1937			5.8		K5			Single (IDS)
18 32.8	+30 33	β 1253		Lyr	1891 1958	157	7.1	5.5	12.8	B8	11446		
18 32.9	+38 50	OΣΣ 171	AB	Lyr	1829 1925	319 322	141.6 144.0	6.9	8.2	F8 G5			
—			Aa		1911 1926	139 137	85.9 84.9		11.0				
18 33.2	+40 10	OΣ 356 rej	AB	Lyr	1843 1919	308 305	40.2 35.2	7.3	9.9	F5			
—			AC		1895 1918	4	52.2		9.7				
—			BC		1895 1918	46	45.6						
18 33.4	+8 16	OΣ 355 rej		Oph	1866 1946	248	38.7	6.4	9.7	B8	11448		
18 33.4	−38 44	Δ 222		CrA	1836 1936	359	21.4	5.9	6.6	B8 B9			
18 33.7	−36 48	Jsp 777		Sgr	1929	284	11.7	8.0	15	B9			
18 33.8	+17 44	Hu 322	AB	Her	1901 1959	86 59	0.2 0.1	7.7	7.9	F5	11454		Less than 0″.1 in 1960
—		Σ 2339	ABxC		1830 1960	272	2.3		8.0		11454		
18 33.9	+52 21	A 1377	AB	Dra	1906 1960			6.1	6.1	K0	11468	o	
—		Σ 2348	ABxC		1832 1969	272	25.7		8.8		11468		
18 34.2	−66 17	MlbO 5		Pav	1873 1942	293	4.6	7.0	9.5	G5			
18 34.4	−34 49	Stone 62		Sgr	1876 1955	140 135	2.1 2.1	7.6	7.9	F0			
18 34.6	+36 08	Hu 1293		Lyr	1905 1922	79	0.9	7.6	14.6	K0	11471		
18 34.9	−48 02	h 5047		Tel	1913	173	42.8	8.4	8.5	B8			
18 35.5	+23 36	OΣ 359		Her	1849 1959			6.3	6.5	K0	11479	o	
18 35.6	+4 56	β 643	AB	SerCd	1878 1924	338 330	8.9 10.5	6.5	12.3	A0	11477		
—		Σ 2342	AC		1830 1930	12 5	26.9 30.5		9.1		11477		
18 35.9	+16 59	OΣ 358	AB	Her	1845 1971			6.8	7.0	G0	11483	o	
—			ABxC		1923	234	201.3				11483		
18 36.0	+11 44	OΣ 357		Oph	1845 1960			8.1	8.2	A2	11484	o	
18 36.2	+41 17	Σ 2351		Lyr	1830 1953	340	5.2	7.7	7.7	A0	11500		
18 36.3	−38 09	I 1372		CrA	1926 1959	20 18	7.6 8.2	7.3	11.5	M	11474		
18 36.5	+4 57	h 864		SerCd	1876 1910	317 323	18.0 18.2	6.8	11.0	K2	11494		
18 36.6	+6 40	J 99	AB	Oph	1911 1934	323 327	26.8 28.9	5.5	14.5	F2	11496		
—			AC		1910 1960	89 83	72.4 74.8		11.5		11496		
18 36.6	+22 06	h 2834		Her	1879 1895	250	20.2	6.7	11.7	B9	11498		
18 36.6	+33 28	Σ 2349		Lyr	1830 1958	205	7.3	5.4	10.6	B8	11504		
18 36.7	−20 19	Ho 567		Sgr	1895 1943	163	1.2	7.4	10.4	B5	11487		
18 36.8	−26 17	Rst 3187		Sgr	1934 1959	231	0.2	8.4	8.5	G0			Too close in 1959
18 36.9	+38 47	H V 39	AB	Lyr	1836 1946	138 173	43.0 62.8	0.0	9.5	A0	11510		3 α Lyr, Vega
—		Σ II 9	AC		1864 1899	299 285	46.9 54.4		11.0		11510		
—			AE		1905 1921	39 39	124.5 118.5		9.5		11510		The components are optical
—			CD		1897	35	31.3				11510		
18 37.0	+34 52	Σ 2352	AB	Lyr	1830 1910	284	15.3	8.1	11.1	K0	11511		
—			AC		1881 1909	161	210.2		11.0		11511		
18 37.1	−19 13	See 355		Sgr	1897 1919	240	12.4	7.2	13.0	K2	11497		
18 37.3	+7 32	Σ 2346		Oph	1829 1925	283 292	15.4 23.6	7.7	9.1	G0	11506		Optical (IDS)
18 37.4	+2 38	β 1327		SerCd	1903	179	13.2	7.8	15	F5	11507		
18 37.4	+77 41	OΣ 363		Dra	1852 1960	20 128	0.6 0.2	7.6	7.8	F0	11584	o	
18 37.4	−47 04	h 5049		Tel	1913	264	19.9	7.0	11.5	F2			
18 38.1	−14 00	β 135	AB	Sct	1875 1947	186	2.4	6.4	11.2	B9	11512		
—			AC		1916	65	20.0		14.7		11512		
18 38.3	+8 50	Hu 198		Oph	1900 1960	195 150	0.2 0.4	5.9	8.6	M	11524	o	X Oph; A is variable
18 38.3	+28 42	Σ 2356		Lyr	1831 1958	47 61	1.0 1.0	8.4	9.4	F0	11529		
18 38.3	−32 41	ø 3		Sgr	1926 1931	268	4.7	7.2	14.0	F5			
18 38.4	−3 12	A 88	AB	SerCd	1900 1960			7.2	7.3	F8	11520	o	
—			ABxC		1909 1960	114 120	15.5 16.3		14.0		11520		
18 38.4	+36 03	Σ 2362		Lyr	1830 1950	183	4.2	7.5	8.8	A5	11534		
18 38.4	+60 43	h 2836	AB	Dra	1905	324	36.9	6.7	9.9	F2			
—			AC		1905	259	54.7		9.9				
18 38.4	+63 32	Σ 2377 rej		Dra	1900 1920	340	16.5	7.1	8.7	K0	11559		
18 38.6	+15 05	h 1331		Her	1893 1910	208	39.2	6.9	11.4	A0			
18 38.6	+16 32	Ho 87	AB	Her	1883 1960	79 186	0.3 0.3	8.4	8.4	G5	11530	o	
—			AC		1893	131	45.6		12.5		11530		
18 38.7	+4 51	OΣ 360		SerCd	1849 1951	293 285	1.1 1.5	6.6	10.1	K0	11526		
18 38.7	+24 38	h 1332		Her	1843 1910	226 229	25.1 26.2	8.1	10.5	F5			
18 38.9	+52 21	Σ 2368	AB	Dra	1831 1951	331 323	2.0 1.9	7.6	7.8	A3	11558		
—			AC		1901 1920	125	36.6		10.7		11558		
18 39.3	+20 56	Σ 2360		Her	1831 1950	6 357	2.5 2.4	7.8	9.0	B8	11546		

α 2000	δ 2000	Star Name		Const	Years	Pos Ang	Separation	m_1	m_2	Spec	ADS	Orb	Notes
18h 39.5	-45°45'	Hu 1527		Tel	1913 1929	311°	4.6"	7.4	10.8	K0			
18 39.9	+7 21	Σ 2355 rej	AB	Oph	1904 1960	139 133	25.0 23.1	6.3	11.3	K0	11555		
—			AC		1918 1960	40 38	107.3 108.0				11555		
—		Fox	AD		1912	224	50.4		12.4		11555		
18 40.0	-7 47	Σ 2350 rej		Sct	1880 1915	197 194	23.3 22.0	5.8	10.8	K0	11552		
18 40.1	+24 43	Σ 2364		Her	1831 1933	182 177	6.5 8.7	8.1	10.3	K0	11560		
18 40.2	+9 42	S 704		Oph	1825 1919	268	57.2	8.5	8.5	F0 F5			
18 40.6	+31 38	Ho 437	AB	Lyr	1892 1960	116 128	0.4 0.5	8.4	8.6	A5	11566		
—			AC		1893 1904	273	40.2		10.7		11566		
—			Aa		1906	290	22.5		13.9		11566		
—			CD		1893 1951	337 346	2.3 4.0		11.2		11566		
18 41.0	+24 50	A 2988		Her	1916 1960			8.6	8.6	A5	11574	o	
18 41.2	-42 10	I 250		CrA	1900 1951	133 120	1.0 1.0	7.6	8.7	A0			
18 41.3	+4 33	J 2143		SerCd	1942	315	10.0	7.1	12.1	F8			
18 41.3	+30 18	Σ 2367	AB	Lyr	1833 1960			7.5	8.0	G5	11579	o	
—			AC		1832 1968	193	14.2		8.9		11579		
—			AD		1891	95	22.9		12.2		11579		
—			AE		1880 1909	341	149.4		11.2		11579		
18 41.7	+51 08	Hu 754		Dra	1904 1923	92	1.4	7.9	15	A2	11599		
18 42.1	+34 45	Ball	AC	Lyr	1879 1910	65	90.9	6.1	11.4	B5	11593		Mag. of A here is for Aa
—		B 2546	Aa		1958	254	0.2	6.6	7.6		11593		
—		Σ 2372	AaxB		1829 1958	83	25.0		8.4	B5	11593		
18 42.2	-32 22	Rst 2071		Sgr	1934 1944	7	0.6	7.9	10.4	B8			
18 42.3	-9 03	Rst 4594	AB	Sct	1938 1943	46	15.2	4.7	12.2	F0	11581		δ Sct; variable
—		H V 36	AC		1879 1976	130	52.6		9.2		11581		
18 42.6	+55 32	H VI 37	AB	Dra	1879 1910	160	146.8	5.0	10.6	A0			46 Dra; fine contr. (Webb)
—			AC		1852 1869	341	333.8						
18 42.9	+44 56	Σ 2380		Lyr	1831 1951	9	25.8	7.4	8.9	G0	11616		
18 42.9	-39 17	I 1379		CrA	1926 1936	182	0.3	7.8	8.0	B8			
18 43.0	-71 26	h 5048		Pav	1872 1917	355 356	49.7 55.6	4.0	11.9	K0			ζ Pav
18 43.2	+19 28	β 645		Her	1877 1924	305	9.3	7.1	12.1	B9	11609		
18 43.3	+39 18	β pm		Lyr	1877 1910	191	60.2	6.6	10.4	K5			
18 43.3	-22 25	I 1381		Sgr	1925 1944	35	1.0	7.7	10.7	B9	11595		
18 43.4	-52 52	h 5055		Tel	1837 1952	76	7.5	8.6	9.0	A2		o	Retrograde motion
18 43.4	-55 46	B 398	AB	Tel	1926 1952	341 157	0.2 0.2	8.2	8.6	F5			
—		h 5053	ABxC		1835 1913	197	32.6		10.0				
18 43.5	-8 17	J 104	AB	Sct	1910 1934	97	13.6	5.1	14.6	G5	11601		ε Sct
—			AC		1910 1934	195	37.6		13.7		11601		
—		Kui 87	AD		1934	312	15.4		14.7		11601		
18 43.5	-33 55	I 638		Sgr	1929	131	0.7	8.1	10.6	F8			
18 43.7	+5 39	OΣ 361		SerCd	1848 1946	172	22.6	8.3	9.0	A2			
18 43.7	-22 40	Don 919	AB	Sgr	1928 1944	72	3.4	7.8	12.7	F5			
—			AC		1944	287	15.3		12.6				
18 43.8	+30 24	Σ 2376	AB	Lyr	1830 1949	63	22.4	8.5	9.2	A0	11624		
—			AC		1880 1934	207	81.7		11.2		11624		
18 43.8	-38 19	CorO 227	AB	CrA	1900 1953	214	29.2	5.1	9.7	A0 K0			λ CrA
—			AC		1900	57	40.0						
18 43.9	+2 37	Σ 2369		SerCd	1830 1959	98 86	1.5 0.6	8.2	8.7	G0	11617		
18 43.9	-0 13	A 859		Aql	1904 1953	14	0.3	8.5	8.9	A0	11614		
18 44.3	+39 40	Σ 2382	AB	Lyr	1831 1976	27 357	3.1 2.6	5.0	6.1	A3	11635	o	4 ε¹ Lyr
—		Σ I 37	ABxCD		1835 1955	173	207.7				11635		Famous double-double [70]
—			AD		1910	173	210.7			A5	11635		
—			AI		1863 1956	135 137	145.4 147.1				11635		
—			BC		1910	172	208.9			A3 A5	11635		
—			BD		1910	172	212.4				11635		
—			BI		1910	137	147.8				11635		
—		Σ 2383	CD		1831 1975	155 94	2.7 2.3	5.2	5.5		11635	o	5 ε² Lyr
—		Σ I 37	CI		1863 1956	37 36	129.1 123.1				11635		
—			DI		1910	36	126.5				11635		
—		S,h 277	EF		1878 1904	38 38	46.7 46.0	10.1			11635		
—			EG		1878 1904	247 245	42.6 43.6				11635		
—			GH		1878 1904	356 353	25.0 26.4				11635		
18 44.3	+61 03	Σ 2403		Dra	1832 1958	259 271	1.9 1.4	6.1	8.9	K0	11661		
18 44.3	-35 39	B 942		Sgr	1927 1931	313	9.4	4.9	12.7	B5			
18 44.5	+34 00	OΣΣ 172		Lyr	1875 1923	6	65.5	8.0	8.7	F8			
18 44.7	-29 59	B 2460		Sgr		sf	8.0	7.5	13.0	K2			
18 44.8	+37 36	β 968	AB	Lyr	1889 1934	50 50	27.0 25.5	4.3	15	A3	11639		7 ζ² Lyr
—			AC		1880 1923	275 272	43.4 46.1		13.3		11639		
—		Σ I 38	AD		1835 1955	150	43.7		5.9	A3	11639		
—		β 968	AE		1880 1960	304 301	61.7 61.8		11.5		11639		
18 44.8	-25 48	See 358		Sgr	1897 1953	30 46	1.6 1.4	8.3	8.7	F8	11622		
18 44.9	-33 23	O1 20		Sgr	1909 1959	352 336	0.7 0.5	8.2	8.7	G0			
18 45.0	+44 22	Es 1424		Lyr	1915	91	8.8	7.9	13.0	F5	11654		
18 45.1	+46 08	Σ 2395		Lyr	1831 1931	309	8.3	8.0	10.4	A0	11660		
18 45.1	-51 21	Rst 5452		Tel	1946	218	0.2	8.6	8.6	G5			

α 2000	δ 2000	Star Name		Const	Years		Pos Ang		Separation		m_1	m_2	Spec	ADS	Orb	Notes
18h 45m.2	+38° 19′	Σ 2393	AB	Lyr	1829	1933	22°	23°	10″.4	14″.6	7.7	10.4	K0	11656		
—			AC		1880	1921	300	300	178.2	179.6		11.3		11656		
18 45.5	+5 30	ø 332	AB	SerCd	1953	1960	136	136	0.2	0.1	6.9	6.9	A0	11640		Tweedledee (ø)
—		Σ 2375	ABxCD		1829	1974	108	118	2.2	2.5				11640		
—		ø 332	CD		1953	1960	135		0.1		7.9	7.9		11640		Tweedledum (ø)
18 45.7	+20 33	h 2839	AB	Her	1879	1935	96	73	44.7	48.2	4.2	12.9	F5	11658		110 Her
—			AC		1879	1935	92	75	61.2	63.8		10.9		11658		
—			Aa		1900	1935	197	209	23.7	12.6		13.9		11658		
—			Ab		1900	1935	28	20	27.1	37.6		13.6		11658		
18 45.7	-13 40	Ho 438		Sct	1889	1953	75		2.7		8.0	10.8	F2	11637		
18 45.8	+34 31	Σ 2390		Lyr	1830	1951	157		4.2		7.2	8.6	A5	11669		
18 45.9	-10 30	Σ 2373		Sct	1832	1953	338		4.2		7.2	8.2	F2	11642		
18 46.3	-22 24	See 360		Sgr	1897	1925	209	214	12.2	12.5	5.8	13.5	K4	11652		28 Sgr
18 46.5	-0 58	Σ 2379	AB	Aql	1832	1942	121		13.0		6.0	7.8	A0 A0	11667		5 Aql
—			AC		1880	1912	146	146	27.5	26.3		11.2		11667		
—			BC		1933		168		15.3					11667		
18 46.6	-8 07	Rst 4597		Sct	1939	1953	343	334	0.2	0.4	7.7	7.9	A0			
18 46.7	-10 07	A 1887		Sct	1908	1926	256		3.4		5.7	13.9	F5	11670		
18 46.8	-14 28	Rst 4596		Sct	1939	1943	159		6.0		7.1	12.0	F8			
18 47.0	+18 11	β pm	AB	Her	1886	1909	247	246	96.2	99.2	4.4	10.4	A3			111 Her
—			AC		1886	1909	258	256	118.3	121.3		9.9				
—			AD		1886	1909	265	264	131.2	133.5		10.4				
—			BC		1907		295		29.9							
—			CD		1907		312		19.8							
18 47.2	+31 24	Σ 2397		Lyr	1830	1951	267		3.9		7.2	9.5	G5	11685		
18 47.4	-49 38	h 5059	AB	Tel	1913		238		25.6		7.1	11.9	B9			
—			AC		1913		201		32.0			11.5				
18 47.5	+49 26	β 971	AB	Dra	1879	1954	355		0.5	0.2	7.4	9.4	F5	11698		A is a sp. binary (F5,A)
—		Fox	ABxD		1914		132							11698		
—		β 971	AC		1914		113		107.6			11.2		11698		
18 48.0	-16 47	h 2838		Sgr	1879	1914	333		33.5		7.1	11.1	K2			
18 48.0	-18 14	Rst 3198		Sgr	1933	1944	162	156	0.4	0.4	8.5	9.0	F2			Composite spectrum F2+A2
18 48.2	+10 39	OΣ 362	AB	Aql	1853	1922	340		7.7		8.0	12.1	G5	11693		
—			AC		1898		103		11.6			13.7		11693		
18 48.4	+15 24	J 2150		Her	1942		35		12.0		6.9	13.8	B8			
18 48.4	-31 52	Rst 2076	AB	Sgr	1934	1944	240		5.4		7.4	14.9	B8			
—			AC		1934	1944	131		10.2			13.5				
18 48.4	-37 01	h 5064		CrA	1919		269		18.6		6.9	13.2	K0			
18 48.5	+10 45	Σ 2396	AB	Aql	1829	1925	233	329	11.7	40.9	8.2	11.7	K0	11696		
—			AC		1891	1910	13	11	153.2	160.0		10.7		11696		
—			AD		1891	1910	50	48	173.5	176.5		11.0		11696		
18 48.7	-6 00	Σ 2391	AB	Sct	1829	1923	332		37.9		6.5	9.8	A2	11695		
—			BC		1907		107		12.6			14.3		11695		
18 48.7	-18 36	Kui 88		Sgr	1933	1959	195	182	0.3	0.3	7.0	7.7	A0			
18 48.8	-51 08	δ		Tel	1920	1929			3.6		8.0	12.0	K0			
18 48.9	-41 20	Rst 2075		CrA	1934	1945	96		9.1		7.6	13.8	G5			
18 49.0	-8 28	Σ 2388 rej	AB	Sct	1905		345		53.7		7.8	10.0	B3			
—			BC		1905		20		21.4			10.1				
18 49.0	+21 10	Σ 2401	AB	Her	1828	1951	38		4.3		7.1	8.7	B5	11715		
—			AC		1935		150		78.6					11715		
18 49.6	+10 41	Σ 2402		Aql	1830	1959	198	210	0.7	1.1	8.6	9.0	A3	11722		
18 49.7	-5 55	H VI 50	AB	Sct	1901	1934	0		23.3		6.1	12.2	K0	11719		
—			AC		1899	1923	171		113.7			8.6	K0	11719		
18 49.7	-20 19	See 362		Sgr	1897	1899	1		17.6		5.4	14.0	K1	11713		29 Sgr
18 49.7	-33 36	See 361		Sgr	1897	1959	230		1.3		8.4	8.5	F2			
18 49.7	-73 00	R 314		Pav	1880	1947	260	269	1.6	1.9	6.3	8.3	A0			
18 49.8	+32 49	H V 40	AB	Lyr	1879	1924	70	73	36.2	34.8	5.9	11.4	B2	11732		8 Lyr
—			AC		1854	1923	122		58.7			10.4		11732		
—			CD		1879	1905	213		18.1			11.6		11732		
18 49.9	+26 26	Σ 2406		Lyr	1830	1941	2		4.8		6.9	10.9	A2	11733		
18 49.9	+32 33	Ho 440		Lyr	1892	1934	177		19.0		5.3	12.8	A2	11737		9 ν Lyr
18 50.1	+29 49	h 1352		Lyr	1879	1918	243		11.9		7.5	11.7	B8	11741		
18 50.1	+33 22	Σ I 39	AB	Lyr	1835	1955	149		45.7		3.4	8.6	B8	11745		10 β Lyr; variable
—		β 293	AC		1878	1913	248		46.6			13.0		11745		
—			AD		1898		68		64.3			14.3		11745		
—			AE		1879	1929	318		66.9			9.9		11745		
—			AF		1879	1929	19		85.8			9.9		11745		
18 50.2	+11 31	β 265		Aql	1875	1960	236	231	1.5	1.4	7.3	9.3	A0	11735		
18 50.3	-7 54	β 969	AB	Sct	1880	1936	238		14.3		6.8	11.9	N	11726		S Sct; variable
—			AC		1913		102		44.1			12.5		11726		
18 50.6	-26 46	B 409		Sgr	1927		258		5.2		7.2	14.0	A2	11724		
18 50.8	+10 59	Σ 2404		Aql	1829	1957	183		3.6		6.9	8.1	K2	11750		
18 50.8	-22 10	β 1300		Sgr	1901		247		21.5		6.6	13.4	A6	11731		30 Sgr
18 51.0	-41 04	h 5066		CrA	1881	1933	85		10.1		6.5	9.5	B5			
18 51.1	-6 17	J 107	AB	Sct	1912	1915	191		5.8		8.0	12.4	B8	11752		

α 2000	δ 2000	Star Name		Const	Years	Pos Ang	Separation	m₁	m₂	Spec	ADS	Orb	Notes
18ʰ 51ᵐ.1	−6°17′	J 107	AC	Sct	1892 1915	140° 144°	10″.0 8″.1		13.2		11752		
--			AE		1901 1915	88	15.4		13.1		11752		
--			CD		1915	196	3.6		15		11752		
--			EF		1915	328	3.5		14.1		11752		
18 51.2	+59 23	Σ 2420	AB	Dra	1833 1949	346 326	30.3 34.2	4.8	7.8	K0	11779		47 o Dra
--			AC		1880 1960	327 326	107.8 139.4		11.4		11779		
18 51.4	−14 41	Rst 4604		Sct	1939 1943	90	0.8	8.1	11.0	B9			
18 51.6	−35 23	B 952		Sgr	1927 1930	70	4.6	7.8	12.4	K0			
18 51.6	−60 54	Rst 5126		Pav	1942	247	0.2	8.1	8.1	F0			
18 51.7	+13 32	Σ 2409	AB	Her	1829 1950	33 24	1.0 0.8	8.1	9.4	G5	11763		
--			AC		1900 1911	196	32.8		10.9		11763		
18 52.0	+37 31	Es 2026	AB	Lyr	1923	108	19.7	7.3	13.0	B3	11777		
--			BC		1923	340	5.0		13.5		11777		
18 52.0	−57 56	h 5065		Pav	1872 1913	22	22.4	8.2	10.0	A0			
18 52.2	+39 25	Weisse 33		Lyr	1925	152	99.9	8.6	8.6				
18 52.2	−41 43	HdO 291	AxBC	CrA	1901 1959	340 342	35.1 36.0	6.5		K2			
--		I 1385	BC		1926	292	0.6	10.2	12.2				
18 52.2	−62 11	h 5062		Pav	1917	206	63.1	4.2	12.2	B3			λ Pav
18 52.3	+14 32	Σ 2411		Her	1829 1937	95	13.5	6.6	9.4	K0	11773		
18 52.3	−35 38	B 953		Sgr	1927 1959	99 126	0.4 0.6	8.1	9.5	F2			
18 52.5	+14 00	Σ 2412		Her	1830 1959	56	1.4	8.6	8.7	K2	11778		
18 53.0	+83 54	OΣ 349		Dra	1846 1960	95 52	0.6 0.3	8.1	8.6	G5	11006		Apparently in rapid motion
18 53.0	−16 23	Rst 4022		Sgr	1937 1951	318	0.6	6.9	9.5	B9			
18 53.1	−18 38	Ho 569		Sgr	1894 1905	40	18.5	6.9	11.8	K0	11776		
18 53.1	−27 45	I 1031		Sgr	1913 1947	285 302	0.5 0.5	7.7	8.9	A5	11774		
18 53.2	−11 21	A 1889	AxBC	Sct	1908 1917	32	61.2	7.9		B9	11783		
--			BC		1908 1926	117	0.8	10.5	10.8		11783		
18 53.3	+25 23	OΣ 364		Her	1842	163	0.7	7.7	10.7	G5			Not double (IDS)
18 53.6	+34 37	Lewis 26		Lyr	1899	85	5.1	8.1	10.1	K0			No pair here or nearby
18 53.6	+75 47	Σ 2452		Dra	1832 1951	218	5.7	6.6	7.4	A0	11870		
18 53.7	+36 58	H VI 3		Lyr	1905	20	174.6	5.6	9.3	B3			11 δ¹ Lyr
18 53.9	+37 22	β 137	AB	Lyr	1875 1959	124 150	1.2 1.4	8.2	8.7	G0	11811		
--			ABxC		1924	144	18.7		11.1		11811		
--			AC		1880 1925	142 143	17.9 20.7				11811		
18 54.0	−47 16	I 112	AB	Tel	1900 1948	182 186	1.6 1.6	7.4	9.4	F5			
--		Δ 224	AC		1835 1949	63	84.2		7.6	A0			
18 54.1	−13 52	A 1891		Sct	1908 1945	267	0.3	8.7	8.7	B9	11803		
18 54.2	−22 45	β 1033	AB	Sgr	1888 1930	97	2.5	4.8	10.6	K2	11794		32 v¹ Sgr
--		h 5072	AC		1878 1930	58	28.2		10.6		11794		
18 54.5	+1 54	OΣΣ 176	AB	Aql	1874 1923	115	96.8	7.6	7.6	F5 K0			
--			AC		1918	5	150.7		9.3				
--			BD		1918	359	149.5		9.7				
18 54.5	+20 37	Σ 2415		Her	1831 1960	299 293	2.0 1.9	7.0	8.9	A0	11816		
18 54.5	+36 54	Es 2028	AB	Lyr	1923	349	86.2	4.5	11.2	M	11825		12 δ² Lyr
--			BC		1923	138	2.2		11.6		11825		
18 54.7	+22 39	β 646	AB	Her	1878 1924	32	35.4	4.5	11.0	G0	11820		113 Her; sp. compos. G0+A3
--			AC		1878 1924	25 23	40.7 37.8		11.0		11820		
--			BC		1877 1915	159 145	7.0 7.0				11820		
18 54.8	+48 52	β 1255	AB	Dra	1891 1937	88 95	1.6 1.5	5.8	12.5	F5	11846		
--			AC		1911 1960	250 259	34.7 30.6		13.0		11846		
18 54.9	+33 58	OΣ 525	AB	Lyr	1849 1958	128	1.7	6.0	10.2	G0	11834		
--		S,h 282	AC		1846 1935	350	45.4		7.7	A	11834		
--			AD		1880 1910	285	193.0		10.6		11834		
18 54.9	+41 36	Ho 270	AB	Lyr	1887 1934	307	8.2	5.4	12.4	K0	11840		
--			AC		1887 1934	40	23.6		11.4		11840		
18 55.3	−26 18	Smyth		Sgr	1837	244	309.0	2.0	9.4	B3			34 σ Sgr
18 55.4	+23 24	Pou 3591	AB	Her	1894 1905	46	24.1	8.4	9.4				Mag. of A here is for Aa
--		Cou 511	Aa		1970	200	0.4	8.8	9.6	G0			
18 55.8	+3 27	A 2192		SerCd	1910 1960			7.8	7.9	A2	11842	o	
18 55.9	+44 14	OΣ 365	AB	Lyr	1841 1898	168	0.5	7.2	8.3	A2	11863		Not seen double after 1841
--		Σ 3130	AC		1833 1958	262	2.7		10.6		11863		
--			AD		1880 1908	311	181.1		10.2		11863		
18 56.0	−25 03	B 413		Sgr	1927	252	5.4	7.4	14.2	G0	11832		[71]
18 56.0	−28 08	See 364		Sgr	1897 1942	96 109	0.4 0.8	8.3	9.0	F0	11828		
18 56.2	+4 12	Σ 2417	AB	SerCd	1830 1973	104 104	21.6 22.3	4.5	5.4	A5 A5	11853		63 θ Ser
--			BC		1927	56	414.1		7.9		11853		
18 56.3	−0 48	A 1171	AB	Aql	1905 1926	94	0.7	8.0	13.5	A2	11854		
--		Σ 2414	AC		1831 1926	278	17.2		11.1		11854		
18 56.4	+2 28	J 2681	AB	SerCd	1943	35		6.2	11.9	K0			Separation >100″ (IDS)
--			BC		1943	95	5.0		12.2				
18 56.4	+45 30	h 1356		Lyr	1873 1918	343	30.0	8.0	9.1	F8			
18 56.4	−11 19	Rst 4024		Sct	1937 1943	314	2.7	7.2	13.3	K5			
18 56.4	−61 49	h 5069	AB	Pav	1878 1942	82	0.7	7.8	8.0	F2			
--			ABxC		1871 1935	93	15.0		11.7				
18 56.7	−8 26	J 2156	AB	Sct	1948	1	52.4	7.6	12.2	A0			

α 2000	δ 2000	Star Name		Const	Years		Pos Ang		Separation		m₁	m₂	Spec	ADS	Orb	Notes
18ʰ56.ᵐ7	−8°26′	J 2156	BC	Sct	1948		306°		4.″6			12.2				
18 56.8	−14 51	h 5503		Sct	1903		79		38.0		7.2	10.8	B9			
18 56.9	+56 45	Σ 2433	AB	Dra	1834	1937	124		7.6		7.1	10.2	F2	11883		
			AC		1905	1919	127		36.1			11.9		11883		
18 56.9	−33 20	Glp		Sgr	1890		9		50.0		7.1	8.8	B9			
18 57.0	+2 28	h 5504		SerCd	1909		284		54.6		7.6	10.1	F8			
18 57.0	+32 54	β 648	AB	Lyr	1878	1960					5.4	7.5	G0	11871	o	
—			ABxC		1912	1921	285	287	54.1	55.1		12.1		11871		
—			ABxD		1912	1921	187	189	93.3	91.0		12.2		11871		
—			AE		1921		320		101.9					11871		
—			AF		1921		88		100.4					11871		
18 57.0	+45 51	h 1357		Lyr	1901	1913	212		26.8		7.3	9.8	A0			
18 57.1	+26 06	Σ 2422		Lyr	1832	1960	106	82	0.8	0.7	8.0	8.1	A0	11869		
18 57.2	+14 50	Hu 676		Her	1902	1959	80		1.4		7.2	10.0	G5	11867		
18 57.3	+62 24	Σ 2440	AB	Dra	1832	1936	123		17.0		6.5	9.0	K0	11901		
—			AC		1880	1908	62		160.0			10.5		11901		
18 57.5	+58 14	Σ 2438		Dra	1832	1960					7.1	7.4	A2	11897	o	
18 58.2	+17 22	Ho 91		Aql	1886	1959	132	141	6.3	6.8	5.4	11.1	F5	11884		FF Aql; variable
18 58.2	−20 25	Ho 271		Sgr	1889	1930	330		17.0		6.9	12.2	O8	11872		
18 58.2	−83 25	Rst 3188	AB	Oct	1934		320		0.2		7.4	7.4	A0			
—		h 5043	ABxC		1918		5		45.4			14.3				
18 58.4	+36 25	Σ 2429		Lyr	1829	1934	289		5.5		8.2	9.7	A3	11900		
18 58.5	+25 15	A 2990	AB	Vul	1916	1927	140		2.7		7.9	14.5	K5	11895		
—			AC		1916		44		14.5			14.0		11895		
18 58.8	+40 41	Σ 2431		Lyr	1829	1969	236		18.9		6.2	8.5	B5	11910		BDS 8981, H IV 93 is same
18 58.9	+32 41	AGC 9	AB	Lyr	1868	1958	297	304	13.8	13.4	3.2	12.0	A0	11908		14 γ Lyr
—			AC		1879	1904	21		176.9			10.5		11908		
18 58.9	−48 30	I 113		Tel	1900	1948	229		3.0		6.7	10.5	K5			
18 59.1	+13 37	Σ 2424	AB	Aql	1831	1957	242	286	18.7	17.5	5.2	8.7	F5	11902		11 Aql
—			AC		1911	1959	262	267	78.5	78.2		12.2		11902		
18 59.2	−39 32	h 5074		CrA	1903	1920	246		15.9		6.5	11.8	A0			
18 59.4	−12 50	Kui 89		Sgr	1934	1960	164	207	0.2	0.3	5.9	6.4	B5			
18 59.6	+15 04	β pm	AB	Aql	1887	1909	187	187	132.4	131.1	4.0	9.9	K0			ε Aql; fine field (Webb)
—			AC		1887	1909	163		148.6			10.0				
—			BC		1909		101		59.6							
19 00.0	+12 53	Σ 2426	AB	Aql	1829	1953	260		16.9		7.4	8.8	K5 F	11916		Spectrum composite F+K5
—			BC		1902	1905	167		3.7			13.3		11916		
19 00.6	+12 14	Hu 678		Aql	1902	1942	359		2.8		8.0	10.5	A2	11926		
19 00.8	−0 27	h 874	AB	Aql	1901	1910	2		9.7		7.2	13.4	G5	11927		
—			AC		1895	1901	304		22.5			11.8		11927		
19 00.8	+23 18	h 2850		Vul	1876	1947	275		2.6		8.4	9.2	A0	11934		
19 01.1	+20 09	Cou 319		Sge	1968		84		1.8		7.3	11.8	B8			
19 01.1	−37 04	BrsO 14		CrA	1837	1951	281		12.7		6.6	6.8	B8 B8			
19 01.4	+46 56	h 1362		Lyr	1879	1925	276	281	41.9	43.7	5.0	10.5	A5	11964		16 Lyr
19 01.4	−34 31	I 252		Sgr	1897	1959	7	26	1.0	0.9	8.6	8.9	F2			
19 01.6	−36 53	B 957	AB	CrA	1927		138		4.1		7.2	13.3	A0			
—			AC		1928		24		57.7				B2			C is variable
19 01.8	+12 32	Σ 2432	AB	Aql	1904	1932	87	93	14.6	14.8	6.7	9.2	B9	11952		
—			AC		1904		164		20.3			13.1		11952		
19 01.8	+19 11	Σ 2437		Sge	1830	1959	81	36	1.1	0.6	8.2	8.4	G5	11956		
19 01.8	+33 37	J 112		Lyr	1910	1934	178		25.4		6.4	13.5	B3	11965		
19 02.0	+19 07	h 2851	AB	Sge	1879	1926	120	131	17.2	15.3	6.8	11.1	G5	11957		
—			AC		1879	1926	303	298	43.2	44.2		9.2		11957		
—			Aa		1953		53							11957		Uncertain (IDS)
19 02.1	+14 26	Ho 93	AB	Aql	1883	1924	335	325	1.1	1.3	7.6	11.9	A2	11958		
—			AC		1892		210		39.2			12.4		11958		
19 02.1	+52 16	Σ 2450	AB	Dra	1832	1958	302		5.1		6.4	8.6	K2	11979		Here, mag. of B is for BC
—		Hu 757	BC		1904	1960	93	353	0.2	0.2	9.1	9.7		11979		
19 02.2	+8 46	Σ 2436	AB	Aql	1830	1925	309	313	34.6	31.9	8.3	9.1	K0 F8			
—			AC		1923		301		135.1			9.8	A0			
—			CD		1923	1924	11		10.5			11.2				
19 02.6	−29 53	HdO 150	AB	Sgr	1867	1959					3.2	3.4	A4	11950	o	38 ζ Sgr
—		H V 78	ABxC		1905		302		75.0			9.9		11950		
19 02.7	−0 42	Σ 2434	AB	Aql	1831	1955	147	106	25.6	24.3	8.5	8.5	G5 G5	11971		Mag. of B is for BC
—			BC		1831	1957	80	17	1.9	0.9	8.7	10.6		11971		
19 02.7	−36 06	h 5080		Sgr	1877	1951	247		5.5		7.9	8.8	A0			
19 02.9	−3 42	A 3105		Aql	1927	1954	162	40	0.2	0.1	6.2	6.2	A0			14 Aql; rapid changes
19 03.1	−19 15	h 5082	AB	Sgr	1878	1958	88		7.5		6.0	9.5	G5	11972		Elegantly triple (h)
—			AC		1878	1958	113		20.2			10.7		11972		
—			AD		1958		260		25.9			13.5		11972		
—			BC		1933		126		13.7					11972		
19 03.1	−45 43	R 317	AB	Tel	1881	1959	283		1.4		8.2	9.0	B9			
—		h 5078	AC		1881	1959	213		18.9			9.4				
—		B	CD		1935		270		6.1			13.5				
19 03.4	+26 03	Σ 2444		Lyr	1829	1918	322	316	24.8	23.6	8.2	9.9	A0			

α 2000	δ 2000	Star Name		Const	Years	Pos Ang		Separation		m₁	m₂	Spec	ADS	Orb	Notes
h m	° '					°	°	"	"						
19 03.5	−68 45	∅ 357		Pav	1959	16		0.1		6.6	6.6	G0		o	
19 03.6	+16 58	Σ 2442		Aql	1828 1925	208 208		23.0	16.6	8.1	9.6	K0	11992		
19 03.6	−39 27	B 421		CrA	1927	41		5.2		8.0	13.7	G5			
19 03.7	+35 45	Σ 2448		Lyr	1831 1948	193		2.4		8.5	8.5	A3	12002		
19 03.7	+57 27	Arg 33		Dra	1903 1919	58		10.5		8.5	9.0	A0	12019		
19 03.9	+34 09	β 1285	AB	Lyr	1899	295		11.1		7.1	13.3	B5	12003		[72]
—			AC		1899	208		39.8			10.5		12003		
19 04.0	+14 47	Σ 2443		Aql	1829 1956	312		6.6		8.6	9.0	F5	11999		
19 04.0	−51 01	HdO 292	AB	Tel	1926	40		10.8		5.9	13.4	K5			
—			AC		1900 1926	77 64		21.6	22.0		12.4				
19 04.1	−63 47	h 5075		Pav	1835 1947	109 113		1.6	1.6	8.1	8.1	A0			[73]
19 04.2	−22 54	H N 129		Sgr	1867 1951	308		8.0		6.9	8.4	A0	11987		
19 04.3	+43 53	Dawes 9		Lyr	1859 1935	178		2.0		6.8	10.4	B9	12016		
19 04.3	−21 32	H N 126		Sgr	1879 1959					7.7	7.9	G0	11989	o	
19 04.6	+23 20	Σ 2445	AB	Vul	1830 1951	263		12.6		7.2	8.9	B3	12010		
—			AC		1900	106		142.7			8.9		12010		
—			BC		1900	104		154.4					12010		
19 04.7	−21 45	See 369		Sgr	1897 1949	237 246		34.0	35.7	3.8	13.7	G8	11996		39 o Sgr
19 05.0	−4 02	S,h 286		Aql	1823 1959	207 209		35.6	38.4	5.5	7.2	K0 K0	12007		15 Aql
19 05.2	+49 55	Es		Dra	1911	130		12.3		6.6	11.9	K5	12034		
19 05.4	+13 52	β 287	AB	Aql	1878 1934	60 53		4.9	6.5	3.0	12.0	A0	12026		17 ζ Aql
—			AC		1879 1909	77 77		156.6	158.6		11.8		12026		Stream of stars np (Webb)
19 05.6	+27 17	Ho 95		Lyr	1885 1958	219 191		0.4	0.2	8.8	8.8	F5	12032		
19 05.7	−15 40	h 5507		Sgr	1900	63		46.9		5.9	12.1	A0			
19 05.8	+6 33	Σ 2446	AB	Aql	1831 1951	154 153		10.1	9.6	7.1	9.1	F5	12029		
—			AC		1905	341		34.5			10.2		12029		
19 05.8	+59 18	Mlr 217		Dra	1971	15		0.2		8.5	8.5	A2	12055		
19 06.0	+35 53	H V 103		Lyr	1840 1917	55 56		54.6	56.0	8.3	9.7	A2			
19 06.0	−63 45	Rst 1018		Pav	1930	230		0.2		7.9	8.7	A0			
19 06.4	+7 09	Σ 2449		Aql	1829 1951	291		8.0		7.2	7.9	F2	12037		
19 06.4	−11 54	Rst 4028		Sgr	1937 1943	238		1.5		7.3	13.4	F8			
19 06.4	−37 04	h 5084		CrA	1834 1959					4.8	5.1	F8 F8		o	γ CrA
19 06.6	−1 21	Σ 2447	AB	Aql	1829 1930	344		13.9		6.8	9.2	B8	12038		
—			Aa		1905	190		13.9			14.0		12038		
19 06.7	−15 28	Rst 5130		Sgr	1943	328		6.4		7.9	13.0	K0			
19 06.9	+22 10	Σ 2455	AB	Vul	1828 1960	144 40		4.9	6.6	7.4	8.5	F0	12050		
—			AC		1911 1921	22 22		92.5	93.5		11.9		12050		
19 06.9	−16 14	S 710		Sgr	1825 1958	2		6.4		5.9	9.9	B8	12039		
19 07.1	+22 35	Σ 2457		Vul	1828 1957	201		10.3		7.5	9.0	F0	12053		
19 07.1	+72 04	OΣ 369		Dra	1948 1958	43 20		0.7	0.7	7.6	7.9	F8	12113		
19 07.4	+32 30	Σ 2461	AB	Lyr	1830 1958	331 300		3.7	3.4	5.2	9.3	F0	12061		17 Lyr
—			AC		1894 1935	69 61		112.3	171.7		10.8		12061		Mag. of C may include C'
—			AD		1887 1956	296 293		124.3	131.8		9.4		12061		
—			AE		1891 1956	116 118		152.4	140.9		9.5		12061		
—			AF		1891 1956	358 354		168.2	159.7		9.1		12061		
—			Aa		1933	50		48.0			11.6		12061		
—			CE		1908 1933	173 186		122.6	149.7				12061		
—			CG		1908 1933	120 138		130.7	124.2		10.8		12061		
—			CH		1908 1910	29 29		163.1	160.1		11.6		12061		
—		Kui 90	CC'		1936 1960	80 91		0.3	0.3				12061		[74]
—		Σ 2461	Cc		1908 1910	52 52		63.2	59.9		12.5		12061		
—			EG		1908 1910	60		113.7					12061		
19 07.8	+38 56	Σ 2469		Lyr	1831 1956	123		1.4		7.8	8.9	A3	12075		
19 07.9	+29 48	Σ 2466	AB	Lyr	1831 1955	109 104		2.3	2.4	8.5	9.0	A0	12071		
—			AC		1915	142		99.2			9.6		12071		
19 08.0	+16 51	β pm		Sge	1893 1921	273 277		19.6	21.5	6.1	10.6	G5			
19 08.0	−34 29	B 964		Sgr	1927 1934	168		1.6		7.8	12.3	K2			
19 08.1	+27 05	Ho 98	AB	Lyr	1886 1959	164 102		0.3	0.2	8.8	8.8	A0	12079		
—			AC		1893	111		27.6			12.2		12079		
19 08.1	+49 05	A 1390		Dra	1906 1929	263 258		4.4	3.8	8.0	12.5	K2	12089		
19 08.1	−26 50	S 711		Sgr	1825 1933	124		45.4		7.0	8.6	G5 A3			
19 08.2	+8 59	h 876		Aql	1904 1914	5		30.2		7.8	10.9	K0			
19 08.2	−5 20	Rst 4618		Aql	1938 1952	323 322		0.4	0.2	8.6	8.8	A3			In 1958 160°, 0".2 (B)
19 08.4	+55 20	Dem 19	AB	Cyg	1874 1953	20 1		0.6	0.7	7.4	10.4	A3	12104		Elongated 1863–70 (Dem)
—		Σ 2479	ABxC		1832 1953	38 31		6.6	6.8		9.4		12104		
19 08.5	−30 58	h 5091		Sgr	1896 1947	207 210		9.0	9.0	7.9	9.5	G5			
19 08.6	+38 31	Ali 881		Lyr	1929	267		15.3		7.8	12.7	A5			
19 08.7	−23 11	B 2469		Sgr		np		10.0		6.5	15	G5			
19 08.7	−33 48	CorO 233	AB	Sgr	1890 1954	255		11.8		8.0	9.1	F5 F5			
—		LPO	BC		1920 1959	290 322		2.5	5.6		11.8				
19 08.8	+34 46	Σ 2470		Lyr	1829 1933	271		13.4		6.6	8.6	B3	12093		
19 09.0	+1 21	Rst 5460		Aql	1946	202		4.6		7.8	14.0	F5			
19 09.1	+34 36	Σ 2474		Lyr	1830 1974	259 262		17.3	16.2	6.7	8.8	G5	12101		
19 09.2	+11 51	Σ 2464		Aql	1830 1944	23		1.4		8.1	10.4	A0	12095		
19 09.4	+30 34	Ho 572		Lyr	1896 1902	315		18.5		6.7	12.4	B9	12110		

α 2000	δ 2000	Star Name		Const	Years	Pos Ang	Separation	m₁	m₂	Spec	ADS	Orb	Notes
19ʰ09ᵐ.4	−13°27′	Rst 4031		Sgr	1937 1943	48°	1".0	7.1	12.0	K0			
19 09.8	−19 48	B 427		Sgr	1925 1953	290 8	0.1 0.1	7.0	7.2	K0	12096	o	
19 09.8	−21 01	φ 311	AB	Sgr	1936 1939	150	0.1	3.7	3.7	F3			41 π Sgr
—			ABxC		1936 1939	122	0.4		5.9				
19 10.6	−60 03	h 5085		Pav	1857 1942	241	2.8	7.7	9.2	B9			
19 10.8	−12 09	Ho 100		Sgr	1884 1944	327	4.8	7.7	10.7	A0	12119		
19 10.9	+8 07	Σ 2471		Aql	1830 1909	122 124	7.6 8.3	7.5	10.3	A5	12129		
19 11.0	−7 26	A 95		Aql	1900 1959			7.3	7.7	G0	12126	o	
19 11.1	+38 47	Σ 2481	AB	Lyr	1830 1975	234 206	3.8 4.6	8.5	8.5	G5	12145		
		Se 2	BC		1858 1960				9.5		12145	o	Σ 2481's np component (B)
—		Tar	DE		1888 1925	201	7.6				12145		3′ south of ADS 12145
				Dra				7.8		F8			Not seen double in 1904
19 11.3	+63 13	Arg 34		Sge	1900 1957	126 144	0.5 0.6	7.9	9.1	A2	12140		
19 11.4	+21 16	A 151		Lyr	1916	254	34.2	7.4	11.9	K0	12150		
19 11.5	+29 53	J 1263	AB		1916	72	5.0		12.9		12150		
			BC										
19 11.7	+17 53	Hu 333		Sge	1901 1919	86	1.9	7.9	13.0	A0	12146		
19 11.7	+26 15	Σ 2480		Lyr	1829 1935	24 24	14.6 15.3	7.5	10.8	A3	12153		
19 12.1	+2 37	β 1204	AB	Aql	1890 1960	4	0.4	7.4	8.2	B9	12147		
—			AC		1890 1916	195	13.0		13.3		12147		
—			AD		1890 1916	160	21.3		14.1		12147		
—			AE		1890 1916	317	26.5		13.3		12147		
—			AF		1890 1916	292	27.7		13.3		12147		
—		Σ 2476	AG		1830 1906	214	31.5		10.4		12147		
19 12.1	+49 51	Σ 2486	AB	Cyg	1832 1976	225 210	10.5 7.9	6.6	6.8	G5 G5	12169		In a singular and beautiful field (Webb)
—			AC		1910 1924	7 13	55.8 47.1		13.3		12169		
—			AD		1881 1910	78 85	170.7 173.1		11.4		12169		
19 12.1	−21 56	See 371		Sgr	1897 1899	330	7.4	8.0	13.0	K0	12139		
19 12.2	+32 15	Hu 941		Lyr	1904 1958	144	1.2	7.5	10.5	A0	12161		
19 12.3	+21 13	Cou 320		Sge	1968	111	0.2	8.6	8.6	A2			
19 12.3	+47 23	h 1377		Lyr	1901 1912	357	36.2	6.7	12.6	K0			
19 12.4	+30 21	Σ 2483	AB	Lyr	1831 1951	318	9.9	8.1	9.2	A0	12162		
—			AC		1831 1934	236	71.2		10.6		12162		
—			BC		1900	228	70.7				12162		
19 12.4	−33 04	Ol 22		Sgr	1909 1959	67 172	0.7 0.2	8.7	8.8	G0			
19 12.4	−51 48	Δ 225	AB	Tel	1836 1938	252	70.0	7.1	8.5	K2 F8			
—		Hu	BC		1914	80	29.7		11.1				
19 12.6	+16 51	β 139	AB	Sge	1875 1958	140	0.7	6.7	8.0	B9	12160		
—		OΣΣ 177	AC		1874 1919	288 285	120.8 113.4		7.9	G5	12160		
—		β 139	AD		1899 1958	103	27.7 28.6		12.7		12160		
—			Aa		1899 1911	103	27.7		12.7		12160		
—			CD		1911 1917	261 261	125.0 127.6				12160		
				Dra	1879 1907	354 352	90.3 88.1	3.1	12.3	K0			57 δ Dra
19 12.6	+67 40	β pm		Vul	1898 1959	293 290	2.7 3.5	8.0	13.5	F5	12166		
19 12.7	+24 35	A 264	AB		1843 1930	58 35	8.7 3.3		12.5		12166		
—		Ma 7	AC		1901 1916	112 114	33.6 31.7		15		12166		
—		A 264	AD										
—			DE		1901 1916	119 119	5.4 4.5		15		12166		
19 12.7	−33 51	h 5094	AB	Sgr	1880 1959	215 191	15.2 23.6	7.3	7.9	A0 A0			
—		B 428	BC		1925 1950	20 12	2.6 2.6		12.9				
19 12.8	−50 29	B 2473		Tel		300	7.0	6.1	11.4	K0			
19 13.2	−40 45	I 1396		CrA	1926 1939	95 91	5.0 5.2	7.6	11.0	K0			
19 13.4	−14 27	β 138		Sgr	1875 1946	278 313	1.5 1.1	7.7	11.1	F5	12168		
19 13.5	+39 02	S,h 289		Lyr	1879 1925	57	39.4	8.2	9.2	A0			
19 13.6	+20 12	Hu 335		Sge	1901 1958	222 234	0.5 0.5	8.0	11.7	K0	12182		21 Aql
19 13.7	+2 18	h 879		Aql	1878 1912	287	36.2	5.2	12.4	B8	12197		20 η Lyr
19 13.8	+39 09	Σ 2487	AB	Lyr	1830 1969	82	28.1	4.4	9.1	B3	12197		
—			AC		1880 1909	151	161.3		11.0		12197		
19 13.9	−47 22	h 5092		Tel	1871 1953	351	17.7	8.5	8.7	F8			
19 14.2	+34 13	OΣ 366 rej		Lyr	1866 1946	230	21.8	7.7	10.3	B9	12211		
19 14.3	−8 43	A 98	AB	Aql	1925 1937	130 128	27.4 26.0	6.5	11.0	K0	12188		
—			AxBC		1900	128	28.8				12188		
—			BC		1900 1925	54	1.2		11.1		12188		
19 14.3	+19 04	Σ 2484		Sge	1831 1959	218 234	2.5 2.5	7.9	9.4	F8	12201	o	
19 14.5	+10 33	A 1175		Aql	1905 1926	38	2.8	7.9	13.9	F0	12203		
19 14.5	+34 34	OΣ 367 rej	AB	Lyr	1843 1935	228	33.4	7.1	9.6	F5	12215		
—		Ho 648	Aa		1881 1913	68 74	23.0 21.4		12.6		12215		
—		β 975	BC		1880 1957	222 244	0.8 1.3		9.7		12215		
19 14.6	−27 19	CorO 234	AB	Sgr	1897 1959	329 327	2.2 2.2	8.5	9.0	A0	12189		
—			AC		1911	55	63.0		10.0		12189		
19 14.8	−17 21	Howe		Sgr	1878 1906	78	27.9	7.4	10.9	A2			
19 14.8	−58 00	B 1891		Pav	1931 1942	185 195	0.5 0.5	7.4	9.4	G0			
19 15.1	+83 28	Σ 2572 rej		Dra	1905 1925	200	25.0	6.3	10.0	A2			
19 15.3	+15 05	OΣΣ 178		Aql	1875 1925	268	89.6	5.7	7.8	G5 A0			
19 15.3	+50 04	Σ 2496	AB	Cyg	1832 1958	78	2.3	6.3	10.1	G5	12240		
—			AC		1881 1909	242	185.6		10.0		12240		
19 15.3	−34 17	I 644		Sgr	1913 1928	173	8.1	7.9	10.9	K0			

α 2000	δ 2000	Star Name		Const	Years	Pos Ang		Separation		m₁	m₂	Spec	ADS	Orb	Notes
19ʰ15.5ᵐ	−25°15′	B 430		Sgr	1926 1959	°	°	″	″	5.6	5.7	F5	12214	o	42 ψ Sgr
19 15.5	−32 12	B 1380		Sgr	1929 1935	317		0.3		8.6	8.8	A2			
19 15.6	+18 31	β pm		Sge	1879 1909	165		136.4		6.4	11.0	K5			
19 15.9	+20 18	J 2959	AB	Sge	1944	190		140.0		8.1	10.0	K2			
−−			BC		1944	165		2.0			12.2				
19 15.9	+27 27	OΣ 371	AB	Lyr	1846 1958	154 161		0.8	0.8	7.0	7.1	B9	12239		
−−			AC		1851 1933	268		47.6			8.6		12239		
19 16.0	+16 10	OΣ 368	AB	Aql	1850 1955	216		1.0		7.3	8.5	F0	12236		
−−			AC		1878 1907	100		17.3					12236		
19 16.2	+21 23	h 2862	AB	Vul	1901 1924	13		39.1		4.8	11.8	B5	12243		1 Vul
−−			AC		1901 1924	158		43.6			13.0		12243		
19 16.4	+14 33	Σ 2489		Aql	1828 1928	348		8.2		5.6	8.6	A0	12248		
19 16.4	+38 08	S,h 292	AB	Lyr	1879 1924	71		99.8		4.4	9.1	K0			21 θ Lyr
−−			AC		1880 1908	127		99.9			10.9				
19 16.7	+11 51	J 2170	AB	Aql	1942	30		60.0		7.9	12.1	A2			
−−			BC		1942	100		3.0			13.1				
19 16.7	+33 08	Ho 102	AB	Lyr	1881	345		86.1		7.2	10.2	B5	12263		
−−			BC		1884 1902	57		1.9			11.5		12263		
19 16.8	−49 59	h 5099		Tel	1872 1916	37		13.2		7.9	10.8	F0			
19 16.9	+63 12	Σ 2509		Dra	1832 1959	353 330		0.5	1.7	7.2	8.3	F5	12296		
19 16.9	−10 58	β 140	AB	Aql	1891 1916	327 321		36.9	39.4	7.0	10.4	G0	12244		
−−			BC		1891 1916	209		7.5			10.6		12244		
19 17.0	+9 21	OΣ 370	AB	Aql	1846 1973	14		19.6		8.5	9.1	K0 A	12259		
−−			AC			78							12259		
−−			BC			128							12259		
19 17.1	+1 42	ø		Aql	1936	140		0.2		8.6	8.8				Nova Aql 1936
19 17.2	−66 40	Gale 3		Pav	1901 1959	40 259		0.8	0.3	6.1	6.5	A2			
19 17.4	−45 25	B 967		Tel	1927 1938	332 330		6.3	7.0	7.4	12.0	A5		o	Direct motion
19 17.7	+23 02	β 248	AB	Vul	1876 1953	127		1.8		5.4	9.2	B0	12287		2 Vul = ES Vul
−−		Ho	AC		1881	120		50.8			11.0		12287		
19 17.7	−15 58	S 715		Sgr	1825 1951	15		8.5		7.0	7.5	A3	12266		
19 17.8	−16 55	Rst 3220		Sgr	1934 1943	234		7.1		7.9	14.1	B5			
19 18.0	−0 57	h 5508		Aql	1903	96		19.5		7.2	11.7	A5	12281		
19 18.0	+20 12	Cou 321		Sge	1968	133		0.2		8.0	8.0	A5			
19 18.1	−5 25	h 881	AB	Aql	1895 1937	341		33.2		7.9	10.1	K2	12283		
−−		Ho 574	Aa		1895 1923	62		15.7			11.5		12283		
−−		h 881	BC		1895 1932	307		7.0			10.1		12283		
19 18.1	−15 57	S 716		Sgr	1825 1951	196		5.1		8.4	8.6	A0	12271		
19 18.2	−18 52	H V 77		Sgr	1783 1900	159		39.7		6.8	9.0	B9			
19 18.3	−55 27	ø 30		Tel	1926 1927	34		1.5		8.0	12.7	K0			
19 18.5	+1 05	Σ 2492	AB	Aql	1830 1958	11 5		3.4	3.1	5.3	9.3	K0	12289		23 Aql
−−			AC		1905 1958	61 66		12.5	11.3		13.5		12289		
19 18.7	−56 09	h 5100		Tel	1913 1915	152		19.4		6.8	11.8	B5			
19 18.8	+0 20	Σ I 40	AB	Aql	1835 1922	316		423.4		6.4	6.6	K0 F0			24 Aql
−−			Aa		1911	160		49.5			12.5				
−−			Bb		1911	302		59.8							
19 18.8	−5 46	Ho 575		Aql	1894 1929	11		5.8		7.9	11.9	A0	12290		
19 18.8	+53 58	A 1393		Cyg	1906 1957	240 252		0.6	0.8	8.1	9.7	A3	12315		
19 19.0	+39 16	Σ 2502		Lyr	1831 1935	205		1.7		7.8	9.7	A2	12310		
19 19.0	−15 32	h 2863		Sgr	1900 1909	16 18		45.8	48.6	6.1	11.6	K2			
19 19.0	−33 17	I 253		Sgr	1897 1959					7.6	7.7	G0		o	
19 19.1	−33 09	B 431		Sgr	1927 1959	140 148		0.9	1.3	8.4	8.7	G0			
19 19.4	+19 43	Σ 2500 rej	AB	Sge	1842 1911	25		20.1		8.0	10.5	A0	12309		Mag. given for B is for BC
−−		Cou 421	BC		1969	171		0.3		11.3	11.3		12309		
19 19.5	−36 40	I 646		Sgr	1914 1935	213		4.4		7.2	12.0	F5			
19 19.6	−51 34	I 1397		Tel	1926 1928	338		1.5		6.6	10.6	A0			
19 19.7	+12 22	S 717	AB	Aql	1856 1913	175		60.2		5.5	9.0	F0			28 Aql
−−			AC		1912	105					13.0				
−−			BC		1912	58									
19 19.8	+64 23	β pm		Dra	1909	181		111.9		6.3	9.8	B9			
19 19.8	−62 53	I 114		Pav	1900 1930	285		0.6		8.0	8.5	G5			
19 19.9	+35 32	Σ 2505	AB	Lyr	1831 1937	315		10.5		8.3	9.0	A0	12328		Spectrum composite A0+K
−−			AC		1905	224		19.3			12.8		12328		
19 20.0	+5 35	Σ 2497		Aql	1830 1935	357		30.0		8.0	9.1	G5			Fine object; fixed (Webb)
19 20.0	−6 38	Σ 2494 rej		Aql	1848 1932	76 85		26.6	25.9	7.1	10.1	K0	12314		
19 20.1	+26 39	OΣΣ 181	AB	Lyr	1875 1923	5 3		54.5	57.8	7.6	7.4	F5 K			
−−			Aa		1910	169		36.2			14.2				
−−			Bb		1910	42		93.7			13.1				
19 20.2	+4 03	Σ 2498	AB	Aql	1827 1949	66		12.1		8.4	9.0	K0	12322		
−−		Bot	AC		1958	101		171.6			9.5		12322		
19 20.4	−10 34	Hu 72		Aql	1899 1922	62		1.1		7.3	12.5	M3	12320		
19 20.5	−5 25	β pm		Aql	1909 1959	12 9		120.1 115.9		5.0	11.8	G5			26 Aql
19 20.5	+50 20	Σ 2511 rej	AB	Cyg	1904 1912	42		12.4		7.5	11.1	K5	12343		
−−			AC		1904 1912	116 116		78.0	79.3		9.8		12343		
19 20.6	+2 56	Howe 47		Aql	1890 1958	334 318		0.4	0.5	8.3	8.4	A2	12329		

α 2000	δ 2000	Star Name		Const	Years	Pos Ang		Separation		m₁	m₂	Spec	ADS	Orb	Notes
19ʰ20.ᵐ6	+6°39′	Ho 576	AB	Aql	1894 1943	180°	°	3".8	"	7.8	11.5	F8	12332		
--			AC		1906 1943	254		45.0			10.9		12332		
19 20.7	+14 25	OΣΣ 180		Aql	1874 1923	266		80.3		8.1	9.0	G0 A5			
19 20.9	-33 03	h 5107		Sgr	1896 1952	127		13.6		7.8	9.1	G0			
19 21.0	+19 09	Σ 2504		Sge	1830 1968	285		8.9		7.0	8.7	F5	12336		
19 21.2	+9 49	h 884	AB	Aql	1901 1906	299		44.6		7.8	11.2	K5			
--			BC		1901	238		10.5			10.9				
19 21.6	+52 23	β 1129		Cyg	1889 1960	344 324		0.3	0.2	7.7	7.7	A5	12366		
19 21.6	-19 14	β	AB	Sgr	1904	338		46.9		6.4	10.8	B8			
--		H VI 120	AC		1899 1916	309		90.8			10.0	K0			
19 21.7	+25 34	Es 483		Vul	1907 1914	358 359		12.6	13.5	7.0	12.0	B3	12352		A is Z Vul (variable)
19 21.7	-17 15	Ho 272	AB	Sgr	1888 1959	38 52		6.7	12.3	7.4	11.9	F8	12337		
--			AC		1897	301		4.9			12.4		12337		
19 21.9	+10 55	A 1178		Aql	1905 1926	331		4.2		7.3	13.4	K2	12351		
19 22.0	+22 30	β 141	AB	Vul	1875 1955	81		0.8		7.8	9.4	B5	12355		
--		h 2867	AC		1875 1916	332		28.0			11.6		12355		
--		β 141	AE		1877 1916	90		50.4			11.1		12355		
--			AF		1898 1916	213		50.0			12.6		12355		
--			Aa		1916	239		20.8			15		12355		
--			CD		1897 1916	181		5.5			12.8		12355		
19 22.1	-4 44	Σ 2501	AB	Aql	1829 1934	22		19.8		7.7	10.2	F5	12347		
--			AC		1901	108		9.6			13.9		12347		
19 22.6	-44 28	Δ 226		Sgr	1835 1953	77		28.3		4.0	7.1	B8 A3			
19 23.2	+47 12	OΣ 372	AB	Cyg	1849 1921	57		79.2		8.0	9.8	A0	12397		
--			BC		1847 1927	295		3.7			11.5		12397		
19 23.5	-69 24	Don 957		Pav	1929 1948	251		2.7		7.8	13.5	F8			
19 23.8	-17 12	Rst 3223		Sgr	1934 1943	206		5.3		6.9	13.0	B9			
19 24.1	+29 37	Wils		Cyg						5.0		B3			2 Cyg; single (Wils)
19 24.1	+46 26	OΣ 373		Cyg	1847 1959	231		1.7		7.6	10.5	A0	12412		
19 24.1	+61 12	Mlr 11		Dra	1969	305		2.3		7.8	10.5	K2			
19 24.4	+16 56	Σ I 41		Sge	1835 1922	79 78		336.2	340.6	6.3	7.1	A0 A0			2 Sge
19 24.7	+23 37	Ho 448		Vul	1890 1924			7.6		8.1	11.1	G5	12413		
19 25.0	+11 57	β pm	AB	Aql	1852 1909	4 343		142.6	105.6	5.2	8.7	G5 G0			31 Aql
--			AC		1909	322		113.7			10.6				
--			BC		1852 1907	247 254		42.6	42.2						
19 25.1	-23 03	Rst 3225		Sgr	1935 1951	34 61		1.5	1.2	7.5	12.5	F8			
19 25.1	-29 19	h 5113		Sgr	1867 1959	171 160		17.7	12.2	6.1	10.5	K0	12400		
19 25.3	-24 31	ø 327		Sgr	1951 1959	84		0.1		5.8	5.8	A5		o	47 χ¹ Sgr
19 25.4	+24 55	OΣ	AB	Vul	1866 1923	123 58		27.9	40.1	6.2	10.6	F8			
--			AC		1924	26		67.8							
--			AD		1923	281		88.4							
19 25.5	+3 07	β pm		Aql	1879 1907	272 271		101.3	108.9	3.4	10.9	F0			30 δ Aql
19 25.5	+19 48	h 2871	AB	Vul	1887 1957	108 100		26.2	18.9	5.2	9.9	K0	12425		4 Vul
--			AC		1908 1959	199 204		53.9	52.6		11.6		12425		
19 25.7	-62 11	I 115		Pav	1900 1937	55 52		2.9	2.9	7.7	10.7	F5			
19 25.8	+14 37	Σ 2518		Aql	1829 1933	358		5.2		8.0	10.9	F5	12429		
19 25.8	+28 46	Σ 2522		Cyg	1830 1946	339		4.4		8.2	9.7	A0	12437		
19 26.1	+15 33	A 1647	AB	Aql	1907 1944	34		1.2		7.9	12.4	F0	12435		
--			AC		1907 1917	85		10.6			13.9		12435		
--			AD		1907 1917	143 140		12.4	11.5		14.4		12435		
19 26.1	+38 49	Ho 450	AB	Lyr	1892 1944	267		0.9		8.5	9.2	F5	12446		
--			AC		1892 1925	73		30.1			12.2		12446		
19 26.3	+1 49	H VI 48		Aql	1904	174		153.5		8.1	10.5	K2			
19 26.5	+0 20	β		Aql	1901 1912	288		201.0		4.7	8.9	F0			32 ν Aql
19 26.5	+19 53	Σ 2521	AB	Vul	1829 1958	44 35		22.6	26.7	5.9	10.7	K5	12445		
--			AC		1908 1918	323		70.4			9.9		12445		
--			AD		1878 1918	64		149.6			9.9		12445		
--			Aa		1908 1934	74 71		25.0	26.5		14.4		12445		
19 26.6	+27 19	Σ 2525		Vul	1830 1959					8.1	8.4	F8	12447	o	
19 26.6	+32 21	Σ 2528	AB	Cyg	1831 1905	244		14.4		8.2	10.2	F2	12456		
--			AC		1902 1905	176		32.6					12456		
19 26.7	-26 19	See 375		Sgr	1897 1939	167		12.6		7.5	12.0	A2	12432		
19 26.8	+50 09	OΣΣ 182	AB	Cyg	1874 1956	307 300		71.8	73.4	7.3	8.5	F5 A5	12470		
--			Aa		1901	114		12.5			14.6		12470		
--			Ab		1904 1956	139 150		31.7	34.7		11.1		12470		
19 26.9	+12 04	A 1181		Aql	1905 1958	196 199		0.3	0.5	6.9	9.1	A0	12452		
19 26.9	+21 10	Σ 2523	AB	Vul	1830 1953	148		6.4		8.4	8.5	B8	12451		
--			AC		1928	144		250.6			9.0		12451		
19 26.9	+73 22	Σ 2550	AB	Dra	1832 1956	249		2.0		8.4	8.4	F2	12524		
--			AC		1901 1956	168 156		77.8	77.8		9.3		12524		
--			BC		1901	167		77.4					12524		
19 27.0	+21 06	Kru 8		Vul	1935	139		10.1		7.1	13.3	B5			
19 27.0	-15 57	Stone 46		Sgr	1880 1933	195		4.8		6.8	7.3				Not seen 1901 by β, 40-inch
19 27.5	-54 58	Rst 1035		Tel	1929 1948	284		0.3		8.6	8.6	A0			
19 27.7	+36 32	Σ 2534		Cyg	1830 1969	64		6.9		7.6	7.8	A0	12478		

α 2000	δ 2000	Star Name		Const	Years	Pos Ang	Separation	m₁	m₂	Spec	ADS	Orb	Notes
19ʰ27.8ᵐ	−54°20′	h 5114	AB	Tel	1837 1940	271° 248°	66.″0 70.″2	5.7	8.4	K2 G0			
−−			AD		1873	96	70.5		11.1				
−−		I	BC		1913 1932	191 195	11.8 8.3		12.9				
19 27.9	+2 10	β pm		Aql	1879 1909	253	115.2	7.8	10.9	K2			
19 27.9	−29 30	β 423		Sgr	1877 1959	125 129	1.3 1.2	8.4	8.9	G0	12453		
19 28.2	+54 47	Es 654		Cyg	1908	192	7.6	7.5	12.8	G5	12509		
19 28.2	−12 09	Schj 22		Sgr	1874 1960			8.0	8.3	G0	12469	o	
19 28.4	−43 53	h 5117		Sgr	1836 1951	266 261	6.0 6.0	8.1	9.3	F8			
19 28.6	+37 11	Hu 1303	AB	Cyg	1905 1956	312 305	0.8 0.7	7.9	9.3	A3	12504		
−−			AC		1905 1919	70	4.5		14.0		12504		
19 28.7	+24 40	Σ I 42		Vul	1835 1924	28 28	396.2 413.7	4.4	5.8	M K0			6 α Vul and 8 Vul
19 29.0	+12 28	A.G.		Aql	1894 1903	304	19.6	7.9	9.9	A5			
19 29.0	+15 14	A 1651		Aql	1907 1956	257	0.4	8.3	9.5	A0	12502		
19 29.4	−7 03	Kui 91	AB	Aql	1934 1947	228	1.5	6.6	11.7	F8	12503		U Aql; variable
−−		h 887	AC		1900 1912	348	35.2		13.2		12503		
19 29.4	−40 57	B 1385		Sgr	1930 1931	317	0.2	8.2	8.3	F5			
19 29.5	+3 05	A 2197	AB	Aql	1910 1957	242	2.5	8.0	13.0	B5	12508		
−−		Σ 2531	AC		1830 1927	30	31.4		9.9		12508		
19 29.5	+78 16	Σ 2571		Dra	1832 1949	23 20	11.5 11.3	7.6	8.3	F0	12608		
19 29.6	+12 24	A 1653		Aql	1907 1960	302 224	0.2 0.2	8.0	9.0	A3	12515		
19 29.6	−12 39	Hu 75		Sgr	1899 1960	202 257	0.5 0.7	7.4	7.9	F8	12505		
19 29.8	−67 18	h 5109	AB	Pav	1871 1940	141	24.7	7.7	9.5	K0			
−−			AC		1871 1940	14	36.4		10.0				
19 29.9	−26 59	H N 119		Sgr	1876 1951	142	7.8	5.6	8.6	K3	12506		
19 30.1	−0 27	Σ 2533		Aql	1831 1924	212	22.8	7.4	9.2	A3	12518		
19 30.2	+2 54	Σ 2532	AB	Aql	1829 1929	5 5	34.9 33.7	6.1	10.3	K5	12520		
−−			AD		1923	197	119.8				12520		
−−			BC		1911	135	34.0		13.6		12520		
19 30.2	+55 25	Arg		Cyg	1919	86	76.3	6.9	9.4	K0			
19 30.2	+56 39	A 712		Cyg	1904 1959	89 289	0.2 0.1	7.4	7.7	A0	12552	o	Single 1934−41
19 30.5	−48 26	I 254		Tel	1900 1947	213 216	0.8 0.8	8.8	8.8	F5			
19 30.6	+25 07	Cou 1027		Vul	1973	196	3.6	7.8	12.4	G0			
19 30.6	+28 17	β 651		Cyg	1878 1930	288	6.4	9.1	13.1	F2	12536		
19 30.7	+27 58	Σ I 43		Cyg	1832 1967	54 54	34.3 34.4	3.1	5.1	K0	12540		6 β Cyg, Albireo [75]
19 30.8	+36 14	J 117	AB	Cyg	1911 1934	160	27.7	6.3	14.0	A0	12545		
−−			AC		1911 1934	29	22.5		14.0		12545		
19 30.9	−6 31	Ho 578		Aql	1894 1959	111 90	21.5 29.6	7.3	12.3	G0	12529		
19 31.0	+50 12	OΣ 374 rej		Cyg	1867 1909	298	18.7	7.6	11.1	K0	12563		
19 31.1	+2 01	β pm	AB	Aql	1879 1909	286 287	113.1 110.9	6.8	10.6	K0			
−−			BC		1909	215	75.8		10.9				
19 31.1	+63 19	β 655	AB	Dra	1878 1924	331	2.0	7.4	12.2	K0	12586		
−−		Σ 2549	AC		1832 1949	291 286	21.1 25.8		9.3		12586		
−−			AD		1832 1924	279 276	47.5 50.7		8.0	A0	12586		
−−			DC		1832 1949	89 80	26.9 28.3				12586		
19 31.3	−2 07	Dem 20	AB	Aql	1869 1959	67	1.2	6.8	9.9	B8	12538		V822 Aql (var.)
−−		Σ 2535	AC		1831 1959	298 298	27.8 25.2		9.8		12538		
19 31.3	+47 29	A 713		Cyg	1904 1960	211 256	0.3 0.4	7.7	8.2	A3	12567		
19 31.4	+36 43	β 438	AB	Cyg	1879 1926	38	4.4	7.9	13.0	G0	12561		
−−		Σ 2538	AC		1830 1949	246	52.8		8.9		12561		
−−			AD		1862 1925	248	46.8		9.1		12561		
−−		β 438	AE		1878 1924	238	21.2		13.0		12561		
−−		Σ 2538	CD		1830 1968	52	6.0				12561		
19 31.7	+33 48	Goyal 17		Cyg	1933	216	22.2	7.5	10.0				
19 32.1	+28 16	β 652	AB	Cyg	1878 1924	323	5.1	8.0	13.1	A0	12572		
−−		Σ 2539	AC		1830 1926	3	5.4		10.0		12572		
−−		Fox	AD		1913	242	54.6		11.0		12572		
−−			DE		1913	225	8.3		13.8		12572		
19 32.2	+6 30	β 650	AB	Aql	1891 1927	146	6.5	7.6	11.1	A0	12569		
−−			AC		1891 1913	330	11.5		12.5		12569		
−−			AD		1891 1913	254	26.7		9.7		12569		
−−			AE		1913	274	88.6				12569		
−−		Fox	EF		1913	48	16.0				12569		
19 32.2	+9 20	β 976		Aql	1880 1953	105	2.2	6.9	10.7	A0	12568	o	
19 32.2	−60 16	I 117		Pav	1900 1947	183	1.0	7.6	8.1	A2			
19 32.4	+69 40	Com		Dra	1902	340	315.6	4.7		K0			61 σ Dra
19 32.4	−22 44	Don 961		Sgr	1932 1940	22 24	7.2 7.7	7.6	14.6	F2			
19 32.9	−46 47	I 118		Tel	1900 1959	131	1.1	7.5	9.7	G5			
19 33.2	+32 54	Hu 948	AB	Cyg	1904 1956	164 156	0.4 0.4	8.4	9.2	A0	12596		
−−		hz	ABxC		1966	64	62.7				12596		
19 33.2	+60 10	Σ I 44		Dra	1834 1924	287	76.2	6.3	8.3	K5 K0			
19 33.3	+20 25	Σ 2540	AB	Vul	1830 1951	147	5.1	7.3	8.8	A3	12594		
−−			AC		1912	221	147.1		12.4		12594		
−−			Aa		1920	120	0.4				12594		
19 33.3	+62 03	Σ 2553		Dra	1832 1957	80 113	1.1 1.0	8.3	9.1	G0	12626		
19 34.1	+7 23	β 653	AB	Aql	1878 1914	275 280	21.4 31.2	4.5	13.0	K0	12607		38 μ Aql

α 2000	δ 2000	Star Name		Const	Years	Pos Ang	Separation	m₁	m₂	Spec	ADS	Orb	Notes
19ʰ34ᵐ.1	+7°23′	β 653	AC	Aql	1878 1914	286° 292°	21″.2 31″.0		13.0		12607		
--			AD		1887 1922	350 344	59.6 66.4		12.5		12607		
--			AE		1887 1922	72 69	183.1 177.5		9.6		12607		
--			Aa		1903 1915	39 35	38.1 38.4		12.8		12607		
--			BC		1891 1915	196	5.1				12607		
19 34.1	+33 01	Sei 638		Cyg	1893	263	24.8	8.0	10.8	K0			
19 34.4	+71 36	Ku 2		Dra	1889 1960	271 252	1.4	6.9	8.9	F2	12690		
19 34.6	+18 08	OΣ 375		Sge	1847 1957	138 165	0.6 0.6	7.5	8.7	G5	12623		
19 34.6	+19 46	β 1130	AB	Vul	1889 1934	32	9.3	4.9	13.4	B8	12622		9 Vul
--			AC		1918 1923	318	108.3		12.5		12622		
19 34.8	+29 28	Wils		Cyg	1949 1951	121 141	0.2 0.1	5.9	6.4	F0			9 Cyg, Hynek 84
19 35.1	+34 12	OΣ 376		Cyg	1848 1955	233	2.7	8.1	10.8	A2	12638		
19 35.2	+74 23	Mlr 222		Dra	1971	198	7.5	7.1	11.8	K0			
19 35.4	+11 56	A.G.		Aql	1893 1903	312	27.7	8.1	9.7	K5			
19 35.5	+46 26	OΣΣ 187	AB	Cyg	1875 1924	287	64.3	8.3	9.0	A			
--			AC		1875 1918	255	129.2		9.0				
--			BC		1875 1924	50	82.0						
19 35.6	+40 01	A 1400		Cyg	1906 1944	290 337	0.3 0.3	8.1	10.1	A3	12662		
19 35.6	-23 19	See 384		Sgr	1897 1959	167 172	5.8 5.2	7.9	11.0	A5	12625		
19 36.0	+0 05	A 1188	AB	Aql	1905 1926	359	3.3	7.6	13.8	G5	12644		
--			AC		1909	228	109.1		11.6		12644		
19 36.0	+0 15	β pm		Aql	1909	266	67.6	7.1	11.6	A2			
19 36.0	-36 48	I 651		Sgr	1913 1959	122 136	0.5 0.6	8.3	8.7	A3			
19 36.1	+11 09	β 1257		Aql	1891 1933	176	3.8	6.5	12.9	A2	12660		
19 36.2	+6 00	Σ 2543		Aql	1830 1933	155	12.6	6.8	9.7	K0	12661		
19 36.4	+15 54	H V 104		Aql	1893 1926	127	38.8	7.3	9.8	F8			
19 36.4	+50 13	β 1131	AB	Cyg	1889 1934	44 53	3.6 4.2	4.5	12.8	F5	12695		13 θ Cyg
--			AC		1852 1923	186 182	29.9 48.4		10.5		12695		
19 36.5	+24 00	Cou 1033		Vul	1973	191	0.2	8.0	8.4	A0			
19 36.6	+41 00	OΣ 378		Cyg	1846 1946	288	1.4	8.2	10.0	A0	12687		
19 36.7	-1 17	J 118		Aql	1910	161	47.0	4.4	13.1	B5	12663		41 ι Aql
19 36.7	-24 53	β 654		Sgr	1878 1959	163 170	2.7 2.5	4.7	9.2	B9	12654		52 Sgr
19 36.9	+11 16	J 133	AB	Aql	1910 1934	57 61	17.2 16.4	6.0	13.8	G5	12670		
--			AC		1910 1934	315	20.1		12.3		12670		
--		Kui 92	Bb		1934	25	5.0		14.6		12670		
19 36.9	+30 20	A 369		Cyg	1902 1929	6	4.2	7.6	14.1	B9	12688		
19 37.2	+29 20	h 1423	AB	Cyg	1900 1924	128	20.8	6.4	11.1	B5	12696		Mag. of A here refers to Aa
--			AC		1924	340	28.0		13.0		12696		
--			AD		1924	357	35.1		12.1		12696		
--		Wils 23	AE		1949	289	25.0		9.7		12696		
--			AF		1949	18	71.5		9.0		12696		
--			AG		1949	88	76.0		8.7		12696		
--			Aa		1949 1951	119	0.1				12696		
19 37.3	+16 28	H VI 26	AB	Sge	1782 1949	82 81	91.9 89.2	5.7	8.0	K0	12693		4 ε Sge
--			AC		1912	280	99.4		12.5		12693		
--			AD		1899	342	160.8		9.5		12693		
19 37.3	+30 25	A 370		Cyg	1902 1929	268	4.8	7.9	14.4	A0	12700		
19 37.6	+63 44	Mlr 56	AB	Dra	1970	133	0.2	8.7	8.7				
--			ABxC		1972	35	17.5		13.0				
19 37.7	-9 58	Lv	AB	Aql	1903 1912	286	82.1	7.3	10.1	A2	12694		
--			BC		1888 1956	284	4.2		11.1		12694		
19 37.7	-41 28	Vou 34		Sgr	1936 1956	134	0.2	8.4	8.5	F2			
19 37.9	+19 21	Σ 2552		Sge	1828 1930	197	5.2	8.2	9.0	A2	12705		
19 37.9	+22 48	A 164		Vul	1900 1953	210 226	0.4 0.4	7.9	9.4	G0	12704		
19 37.9	+49 17	h 1428	AB	Cyg	1905	258	24.7	6.6	9.5	K0			
--			AC		1905	259	55.9		9.5				
--			CD		1905	274	10.5		11.9				
19 38.0	+33 53	OΣ 379 rej		Cyg	1866 1946	86	24.8	8.1	9.4	G5			
19 38.2	+17 15	J 138	AB	Sge	1910	138	21.2	6.7	13.4	A0	12711		
--			AC		1904 1910	130	37.7		11.4		12711		
19 38.4	+0 21	β 249	AB	Aql	1875 1952	142 124	1.3 0.9	7.4	9.5	A2	12708		
--			AC		1916	116	18.1		14.1		12708		
--		Lv	AD		1916	56	38.5		11.1		12708		
19 38.5	+17 15	β		Sge	1904 1914	333	12.4	7.5	11.1	B3	12723		
19 38.6	+23 03	A 165	AB	Vul	1900 1930	136 132	5.2 6.3	7.0	14.0	A0	12731		
--			AC		1930	343	18.6		13.5		12731		
19 38.7	-10 09	Σ 2545	AB	Aql	1829 1959	315 324	3.5 3.7	6.8	8.7	A5	12715		
--			AC		1886 1959	169 165	28.0 26.1		11.4		12715		
19 38.9	-10 20	Σ 2547	AB	Aql	1830 1932	331	20.8	8.5	8.9	A0	12725		
--			AC		1905 1932	143	50.8		11.4		12725		
19 39.0	+76 25	Mlr 224		Dra	1971	276	0.2	8.4	9.1	K0			
19 39.2	+5 24	h 2886		Aql	1903 1924	328	47.8	5.2	12.1	B3	12737		44 σ Aql; variable
19 39.2	-16 54	S 722		Sgr	1825 1940	236	10.2	7.1	7.6	A2	12728		
19 39.4	+16 34	H N 84		Sge	1796 1931	302	28.2	6.5	8.9	K5	12750		
19 39.4	+22 15	Σ 2556		Vul	1829 1960			7.8	8.3	F2	12752	o	

α 2000	δ 2000	Star Name		Const	Years	Pos Ang	Separation	m₁	m₂	Spec	ADS	Orb	Notes
19ʰ39.ᵐ6	+29°45′	Σ 2557	AB	Cyg	1831 1950	104° °	11″.2 ″	7.5	10.0	A0	12759		
--		β 54	AC		1878 1916	303	21.5		11.1		12759		
19 39.7	+33 59	Ho 111	AB	Cyg	1885 1958	4	1.1	6.1	10.6	A2	12765		
--			AC		1905 1958	170 168	24.3 24.6		11.7		12765		
19 39.7	−39 45	See 388		Sgr	1879 1959	198 193	2.5 2.5	7.7	9.7	G5			
19 39.8	−23 26	See 389		Sgr	1897 1959	323	0.2	6.9	7.1	A0	12741		53 Sgr; too close 1935,59
19 39.9	−14 51	Ho 109		Sgr	1868 1921	114	11.4	7.3	13.3	B8	12748		BDS 9497 is likely same
19 39.9	−22 04	I 656		Sgr	1926 1937	140	0.1	7.9	8.1	A0	12743		Measures uncertain
19 39.9	−66 41	δ		Pav	1916	242	19.7	6.4	12.5	A0			
19 40.0	+27 12	Cou 822		Vul	1972	143	0.2	8.8	8.8	A0			
19 40.0	−49 57	Rst 1043		Tel	1930	211	2.3	7.3	14.2	K2			
19 40.1	+18 01	J 121	AB	Sge	1910 1960	180 179	33.0 31.5	4.4	13.2	G0	12766		5 α Sge
--			AC		1910	249	35.8		14.9		12766		
19 40.2	+35 09	Sei 656		Cyg	1894 1933	287	11.0	8.1	11.1	A0	12776		
19 40.2	+60 30	Σ 2573		Dra	1832 1955	27	18.2	6.2	9.5	A2	12789		
19 40.5	+62 40	Σ 2574		Dra	1832 1959	129 230	1.0 0.3	8.1	8.1	F5	12803		
19 40.7	−0 37	h 2888		Aql	1900 1924	354	42.2	5.7	12.9	A0	12775		45 Aql
19 40.7	+23 43	Σ 2560 rej	AB	Vul	1892 1958	295	15.3	6.6	8.9	B3	12778		
--			AC		1910	67	118.7				12778		
19 40.7	−16 18	h 599	AB	Sgr	1878 1903	273 274	35.8 38.0	5.4	11.9	K0	12767		54 Sgr
--			AC		1878 1932	42	45.6		8.9	G0	12767		
19 40.8	+45 31	β pm		Cyg	1911 1960	144 149	84.8 88.0	5.1	11.8	F2			
19 41.1	+10 41	Σ 2558 rej		Aql	1905 1908	300	28.9	8.0	10.5	F8			
19 41.1	+13 49	Kui 93		Aql	1934 1959	137 150	0.2 0.1	6.7	6.9	B3			Too close in 1960
19 41.3	+30 43	β 145	AB	Cyg	1875 1943	267	0.8	7.3	10.0	G5	12786		Spectrum composite G5+A3
--			AC		1878 1925	29	9.0		13.4		12786		
--			AD		1878 1925	157	26.9		11.2		12786		
19 41.8	+50 32	Σ I 46		Cyg	1832 1976	136 134	37.3 39.3	6.0	6.1	G0 G0	12815		
19 41.9	+27 23	OΣ 382		Vul	1849 1957	354 333	0.5 0.4	7.2	7.7	B8	12798		
19 42.0	+40 15	Kui 94		Cyg	1937 1960	160 130	0.5 0.2	6.5	7.8	A3			
19 42.3	−52 57	CorO 238		Tel	1871 1953	50	3.3	7.6	8.7	A3			
19 42.5	−16 07	β 1288		Sgr	1889		0.2	5.9	5.9	F0			55 Sgr; single in 1900 (A)
19 42.6	+11 50	OΣ 380	AB	Aql	1844 1958	78 77	0.6 0.5	5.6	6.8	F5	12808		χ Aql; sp. compos. F5+A3
--			AC		1912	48	81.6		12.3		12808		
--		Hlm 26			1924	317	9.4	11.0	11.5		12808		
--		J 1858			1941	275	140.0	5.3	10.3	F5	12808		
19 42.6	+40 02	β pm		Cyg	1909	165	102.1	7.9	7.9	A0			
19 42.6	−59 01	I 119		Pav	1900 1947	180 163	1.2 1.7	7.9	8.8	G0			
19 42.8	+8 23	Σ 2562	AB	Aql	1829 1931	252	27.1	6.9	8.6	F5	12813		
--			AC		1908	294	81.7		12.2		12813		
--			AD		1904 1931	224	110.8				12813		
--			BC		1909	310	64.2				12813		
--			BD		1905 1909	216	87.2				12813		
19 42.8	+37 41	OΣΣ 188		Cyg	1875 1925	121	59.0	7.9	7.9	K2 F2			[76]
19 42.9	+40 43	OΣ 383	AB	Cyg	1845 1956	27 19	0.9 0.9	6.9	8.4	A0	12831		
--			AC		1903 1933	99	45.2				12831		
--			AD		1933	44	68.9				12831		
19 43.1	−8 18	Σ I 45		Aql	1835 1922	146	96.5	7.0	7.3	F5 F8			
19 43.2	+38 40	Sei 666		Cyg	1896	326	25.4	6.7	9.7	K0			
19 43.3	+4 10	OΣ 381 rej		Aql	1843 1927	6	15.5	8.0	11.2	A5	12825		
19 43.5	+0 40	h 2892		Aql	1896 1909	90	12.8	8.1	10.8	B5			
19 43.6	−15 28	β pm		Sgr	1910	129	81.8	5.5	11.7	F2			
19 43.8	+38 19	OΣ 384	AB	Cyg	1851 1959	195	1.0	7.6	7.9	B8	12851		
--			AC		1856 1933	297	58.8				12851		
19 43.9	+15 28	h 1432		Aql	1828	315	10.0	8.0	10				
19 43.9	+27 08	β 658		Vul	1878 1954	295 291	0.6 0.4	6.8	8.3	K0	12850		Spectrum composite K0+A0
19 43.9	+34 10	Ho 453	AB	Cyg	1892 1934	50	15.4	6.1	12.6	A0	12852		
--			AC		1892 1924	136	33.9		11.6		12852		
19 44.0	−66 18	h 5132		Pav	1835 1916	309	21.5	7.7	10.0	G0			
19 44.1	+12 22	Σ 2567		Aql	1829 1921	314	18.2	7.8	9.6	A2	12848		
19 44.5	+4 59	Σ 2566 rej		Aql	1879 1908	234	25.2	7.8	10.8	K5			
19 45.0	+10 46	AGC 10	AB	Aql	1878 1957	146 137	0.3 0.3	8.2	8.2	B3	12864		
--		Σ 2570	ABxC		1827 1955	277	4.1		9.5		12864		
19 45.0	+45 08	Σ 2579	AB	Cyg	1830 1960			2.9	6.3	A0	12880	o	18 δ Cyg
--			AC		1913 1960	66 66	68.2 65.7		11.9		12880		
19 45.0	+55 51	h 1439		Cyg	1902 1919	200	30.6	8.0	9.7	M			
19 45.2	−70 01	φ 273		Pav	1930	247	2.5	7.6	13.0	F5			
19 45.7	+36 05	Σ 2578	AB	Cyg	1831 1956	125	15.0	6.4	7.2	A0	12893		BDS 9591, H N 113 same
--			AC		1894 1908	358	45.9		11.4		12893		
--			AD		1900	71	95.2				12893		
--			AF		1880 1956	251	143.3		9.0		12893		
--			Cc		1908	75	11.6		13.4		12893		
19 45.8	+40 33	OΣ 385		Cyg	1845 1949	53	1.3	7.6	9.9	B8	12904		
19 45.9	+35 01	H V 137	AB	Cyg	1783 1956	33 27	35.0 38.7	6.2	9.2	K0	12900		
--		Bot	AC		1958	77	44.6		8.5		12900		

α 2000	δ 2000	Star Name		Const	Years		Pos Ang		Separation		m₁	m₂	Spec	ADS	Orb	Notes
19ʰ45ᵐ.9	+39°53′	A 1404	AB	Cyg	1906	1958	322°	72°	0″.2	0″.2	7.5	8.2	B9	12906		Too close 1935–43
––			ABxC		1906	1918	273		19.6			12.9		12906		
––			ABxD		1896	1918	79		21.8			11.9		12906		
19 46.0	+4 15	β 468		Aql	1876	1919	183		9.7		7.0	11.3	G0	12882		
19 46.3	+10 37	β pm		Aql	1879	1907	258		132.6		2.7	10.7	K2			50 γ Aql
19 46.4	+33 44	Σ 2580	AB	Cyg	1832	1970	73	69	25.8	26.0	5.0	9.2	F5 K5	12913		17 Cyg
––			AC		1887	1928	142	138	148.8	134.6		9.0		12913		
––			BC		1901	1912	151	150	137.4	133.5				12913		
19 46.4	–24 53	See 394		Sgr	1897	1959	287	287	0.4	0.9	8.1	9.1	F0	12883		
19 46.5	–21 31	β 467		Sgr	1879	1933	134		3.1		7.5	10.0	A5	12887		
19 46.6	–1 23	Rst 5143		Aql	1943	1953	132	138	0.2	0.2	8.4	8.4	A0			
19 46.6	+32 53	Ho 114	AB	Cyg	1886	1943	239	217	3.1	2.7	6.2	12.7	K2	12920		
––		β	AC		1901	1943	214		9.7			13.7		12920		
––		S 726	AD		1825	1933	206	197	33.4	30.7		9.2		12920		
19 47.1	–8 09	A 108		Aql	1900	1958	183	54	0.3	0.2	8.5	8.9	K0	12911	o	Round or too close 1936,48
19 47.4	+18 32	Wils		Sge	1937						3.8		M			7 δ Sge [77]
19 47.5	+38 24	h 601		Cyg	1904	1934	241		24.7		5.8	12.8	B9	12944		
19 47.5	–21 49	Stone 49		Sgr	1879	1953	9	352	1.5	1.5	8.5	9.0	G0	12917		
19 47.6	+1 05	β pm	AB	Aql	1893	1910	123	120	102.0	100.8	6.8	10.5	G5			
––			BC		1908		195		30.7			12.2				
19 47.9	+10 02	A.G.		Aql	1918		294		49.9		8.0	9.3	K5 G0			
19 48.2	+37 10	OΣ 386		Cyg	1846	1959	78	74	1.0	0.9	8.2	8.5	A2	12965		
19 48.2	+70 16	Σ 2603		Dra	1832	1957	354	15	2.8	3.1	3.8	7.4	K0	13007		63 ε Dra
19 48.6	+24 58	Σ 2586	AB	Vul	1830	1955	227		3.7		7.5	10.4	B9	12964		
––			AC		1912	1933	268		63.4					12964		
19 48.7	+11 49	Σ 2583	AB	Aql	1829	1960	121	110	1.5	1.4	6.1	6.9	F2	12962		π Aql. Sp. compos. F2+A2
––			AC		1886	1911	305		33.6			12.2		12962		
19 48.7	+15 04	A 1658		Aql	1907	1960					8.2	8.6	F5	12961	o	
19 48.7	+35 19	OΣ 387		Cyg	1944	1960					7.2	7.7	F5	12972	o	
19 49.0	+19 09	AGC 11	AB	Sge	1878	1959					5.5	6.2	A2	12973	o	8 ζ Sge
––		Σ 2585	ABxC		1831	1967	311		8.6			8.7		12973		
––			ABxD		1921	19711	248	247	74.7	76.0		11.0		12973		
19 49.0	+44 23	Σ 2588	AxBC	Cyg	1833	1932	160		9.6		7.2		B8	12986		
––		A 718	BC		1904	1959	52	48	0.3	0.3	8.6	9.1		12986		
19 49.1	–61 49	I 120	AB	Pav	1901	1959					7.9	8.1	G0		o	
––		h 5141	ABxC		1872	1933	343		13.9			10.0				
19 49.5	+38 43	Es 84	AB	Cyg	1902	1955	161		11.6		6.1	11.0	G5	12992		
––			AC		1902	1955	98		22.8			11.1		12992		
19 49.8	–11 25	Σ 2581 rej		Aql	1879	1918	282		38.2		7.2	8.8	F5			
19 49.8	–64 54	h 5140		Pav	1836	1952	89	80	2.0	1.5	8.0	8.0	G0			
19 50.0	+17 57	H IV 99	AB	Sge	1879	1917	85		24.8		8.3	10.3	A0			
––			AC		1879	1917	256		68.7			9.3				
19 50.1	+54 39	Σ 2598		Cyg	1832	1922	148		11.2		8.0	10.4	K5	13018		
19 50.2	–10 00	Rst 4643		Aql	1939	1941	277		1.3		6.9	11.0	G5			
19 50.2	–45 23	h 5148		Tel	1901	1914	318		14.2		7.2	12.0	K0			
19 50.4	+24 09	Cou 1034		Vul	1973		207		0.3		8.5	8.7	K0			
19 50.6	+38 43	h 603	AB	Cyg	1887	1923	102	106	53.9	54.6	5.1	10.1	M0	13014		19 Cyg
––			BC		1902	1907	13	11	21.8	21.5		10.8		13014		
19 50.7	–41 52	I 122		Sgr	1900	1938	339		5.1		7.6	10.4	A2			
19 50.7	–59 12	I 121		Pav	1900	1959	89	133	0.5	0.5	5.8	7.3	A2			
19 50.8	+8 52	Σ II 10	AB	Aql	1836	1925	322	301	152.4	165.2	0.8	9.5	A5	13009		53 α Aql, Altair
––			AC		1905	1959	94	100	238.4	247.2		10.0		13009		
19 50.8	–10 46	Ho 275		Aql	1887	1903	117	117	19.0	21.1	5.6	13.6	F0	13017		51 Aql
19 51.0	+10 25	J 124	AB	Aql	1910	1958	162	203	20.5	14.4	5.1	13.4	G0	13012		54 o Aql
––			AC		1910	1958	222		22.5			13.6		13012		
19 51.2	–36 54	h 5151		Sgr	1919	1935	358		7.7		8.4	9.4	F2			
19 51.2	–72 48	h 5137	AB	Pav	1917		201		29.6		7.4	11.4	F8			
––			AC		1917		313		42.8			10.9				
19 51.4	+4 05	Σ 2587		Aql	1828	1939	100		4.1		6.7	9.4	K0	13019		
19 51.6	+50 46	Hu 686		Cyg	1904	1921	148		4.4		7.9	12.9	K0	13049		
19 51.7	+31 08	Es 357	AB	Cyg	1906	1907	307		9.6		7.1	12.6	A5	13038		
––		OΣ 389	AC		1849	1946	183		12.5			9.5		13038		
19 52.0	+31 42	Es 355		Cyg	1906		295		13.0		7.8	13.4	A0	13042		
19 52.0	–10 21	β 148	AB	Aql	1875	1959	333	272	0.9	0.6	7.7	8.1	F2	13028		
––			AC		1891	1925	64		27.0			12.7		13028		
19 52.1	+11 38	β pm		Aql	1909	1924	160	156	93.9	90.5	5.8	11.3	G0			
19 52.2	+19 51	Hu 351		Vul	1901	1924	155		2.0		7.9	12.1	K0	13044		
19 52.3	+10 21	Σ 2590	AB	Aql	1830	1959	309		13.5		6.6	9.5	B5	13041		
––			ab		1909		272		6.6		11.6	12.2		13041		These lie 2′ np Σ 2590
19 52.4	+25 52	OΣ 388	AB	Vul	1848	1954	139		3.9		8.2	8.2	A0	13050		
––			ABxC		1907	1929	134		28.4			8.9		13050		
––			AC		1907	1950	135		31.1					13050		
––			BC		1850	1949	139	134	26.6	27.1				13050		
19 52.4	–16 53	Ara 300		Sgr	1917		358		14.2		7.9	11.6				
19 52.5	+22 28	Ho 580		Vul	1895	1959	268	273	0.6	0.8	8.7	8.8	M5 A1	13055		

α 2000	δ 2000	Star Name		Const	Years	Pos Ang		Separation		m₁	m₂	Spec	ADS	Orb	Notes
19ʰ52.ᵐ6	−54°58′	Δ 227		Tel	1826 1952	149°	°	22″.9	″	6.1	6.8	G5 A2			Yellow, pale green (h)
19 52.8	+64 11	Σ 2604		Dra	1831 1925	184		27.8		6.9	9.1	G5	13092		
19 53.0	+38 46	β pm		Cyg	1909	314		43.7		7.6	11.5	G5			
19 53.1	−25 28	B 454		Sgr	1926 1959	337 11		0.3	0.1	8.5	8.7	F8	13048		
19 53.4	+20 20	Σ I 48		Vul	1831 1928	147		42.2		7.2	7.5	A0 A0			
19 53.5	+24 05	Dju		Vul	1953 1960	243		0.8		4.6	7.8	A0			13 Vul
19 53.6	+59 43	OΣΣ 194		Cyg	1875 1923	0		75.3		6.1	9.4	A0			
19 53.7	+23 00	Σ 2599		Vul	1829 1955	51		3.9		8.2	9.9	B8	13076		
19 54.0	+15 18	Σ 2596		Aql	1831 1959	353 310		2.1	2.1	7.3	8.7	F8	13082		
19 54.0	+49 15	Es		Cyg	1911	305		16.9		8.0	11.0	K0	13100		
19 54.1	−8 34	h 900		Aql	1880 1901	77		46.5		5.8	11.9	K5			
19 54.2	+1 57	Hzg	AB	Aql	1907 1921	93		61.8		8.5	8.7	K0 K0			56 Aql
––		Abetti	BC		1921	38		77.2			11.5				
19 54.3	−23 56	h 2904		Sgr	1877 1959	142 49		18.3	28.9	6.2	10.2	K0	13072		
19 54.6	−8 14	Σ 2594		Aql	1833 1955	170		35.7		5.8	6.5	B3 B	13087		57 Aql
19 54.6	+24 35	Cou 1035		Vul	1973	34		1.1		8.0	14.0	F5			
19 54.7	+7 08	β 659		Aql	1878 1958	316 306		12.3	13.8	6.2	12.2	A0	13093		
19 54.9	+50 49	Hu 687		Cyg	1904 1960					8.5	8.5	A0	13135	o	
19 55.0	+41 52	Ho 581		Cyg	1895 1960					8.0	8.5	K0	13125	o	
19 55.0	−51 49	ø 31		Tel	1927	54		1.8		8.0	12.7	K0			
19 55.1	+30 12	OΣ 390	AB	Cyg	1849 1953	22		9.7		6.6	8.9	B9	13117		
––			AC		1849 1932	175		16.4			10.6		13117		
19 55.1	+59 27	M1r 230		Cyg	1971	236		0.9		7.9	12.7	A2			
19 55.3	+6 24	OΣ 532	AB	Aql	1852 1975	17 5		12.4	12.9	3.7	11.6	K0	13110		60 β Aql
––			AC		1879 1925	347 348		151.9	175.0		10.3		13110		
19 55.3	−6 44	Σ 2597		Aql	1826 1959	92 82		1.9	0.6	6.8	7.9	F2	13104		
19 55.6	+52 26	Σ 2605	AB	Cyg	1831 1958	185 178		3.3	3.2	4.9	7.4	A3	13148		24 ψ Cyg
––			AC		1908 1958	55 55		18.8	21.2		13.6		13148		
––			AD		1880 1908	62 61		163.2	165.4		10.2		13148		
19 55.7	+40 24	h 604		Cyg	1905 1919	94		70.4		7.2	9.3	B3			See note in BDS
19 56.0	+17 53	Ho 116	AB	Sge	1886 1937	23		4.1		7.9	12.6	B9	13140		
––		h 2908	AC		1886 1904	11		18.1			12.9		13140		
19 56.2	−38 20	See 398		Sgr	1896 1928	307		10.5		6.7	12.7	K5			
19 56.3	+35 05	β 980	AB	Cyg	1879 1958	208		7.4		3.9	11.9	K0	13149		21 η Cyg
––		h 1455	AC		1879 1924	327		46.0			10.4		13149		
––			AD		1879 1924	169		49.7			10.4		13149		
––		β 980	AE		1898 1924	247 246		61.7	60.2		11.4		13149		
19 57.2	+40 22	β		Cyg	1905 1919	317		64.6		5.5	8.5	B3			
19 57.7	+51 19	Hu 689		Cyg	1904 1959	17 5		0.4	0.2	8.6	9.0	A0	13191		BDS 9745 is same star
19 57.8	+44 23	OΣ 393		Cyg	1847 1953	226 228		21.8	18.9	8.1	9.3	K0 A0	13190		
19 57.9	−9 03	β	AB	Aql	1906 1937	112		10.1		7.7	10.3	K2	13163		
––		HdO 155	AC		1868 1919	119		13.6			8.0	K0	13163		
––		β	CD		1906	158		5.4			11.3		13163		
19 57.9	+11 25	β 266		Aql	1875 1975	167		15.8		7.4	11.5	A3	13168		
19 57.9	+27 15	AC 16	AB	Vul	1859 1955	236		0.4		8.5	9.0	A0	13176		
––			AC		1875 1921	136		93.2			8.4	A0	13176		
19 57.9	+42 16	OΣ 392	AB	Cyg	1844 1957	322 279		0.4	0.1	6.7	8.5	A2	13186		Single 1954 (M1r, 20-inch)
––		Σ 2607	ABxC		1831 1958	290		3.1			9.0		13186		
19 58.2	+16 30	β 149	AB	Sge	1893 1898	279 279		126.6	127.8	6.8	10.2	B8	13182		
––			BC		1893 1901	199		8.4			12.8		13182		
19 58.2	−51 54	Δ 229		Tel	1826 1938	243		80.5		7.5	8.1	F0 F8			
19 58.3	−2 14	AC 12		Aql	1854 1959	334 300		0.9	1.4	7.4	8.2	F5	13178		
19 58.4	+28 59	Es 495	AB	Vul	1907	245		40.4		7.8	10.8	A0	13193		
––			BC		1907 1927	312		4.2			11.5		13193		
19 58.5	+33 17	Σ 2606		Cyg	1832 1956	131 137		1.2	1.0	7.7	8.4	A5	13196		
19 58.6	+38 06	Σ 2609		Cyg	1831 1954	29 23		2.4	2.1	6.6	7.7	B5	13198		
19 58.8	+47 22	Σ 2611		Cyg	1831 1968	27		5.3		8.4	8.4	K0	13209		
19 59.1	+35 33	Σ 2610	AB	Cyg	1830 1947	296		4.2		8.4	8.9	B9	13204		
––		Ma	AC		1843 1933	203		12.6			11.8		13204		
––		G1p	AD		1894	15		62.1					13204		
19 59.1	+50 40	h 1464		Cyg	1902	32		13.4		7.4	12.0	K0	13221		
19 59.2	−21 55	LDS 697		Sgr		np		42.0		8.0	11.5	G0			
19 59.7	+30 55	A 379		Cyg	1902 1929	226		2.4		8.0	13.0	A0	13224		
19 59.7	+31 50	β 1133		Cyg	1889 1956	338		0.9		6.8	9.5	A0	13223		
19 59.8	−9 57	Ho 276		Aql	1887	173				6.7	6.7	F8			Single (IDS)
19 59.9	−34 42	I 1410	AB	Sgr		sp		30.0		5.3	10.8	A3			θ² Sgr
––			BC		1925	106		1.2			11.5				
20 00.1	+17 31	H IV 100	AB	Sge	1904 1924	256		23.8		10.3	10.3	K			These are A and X Sge
––			AC		1904 1924	297		114.0			5.6	M			
––		β	Cc		1878 1934	205 208		29.0	28.6		11.8		13230		C is 13 Sge = VZ Sge (var.)
20 00.1	+17 37	S 730	AB	Sge	1825 1921	15		114.6		7.1	8.7	K0 F5			
––			AC		1879 1904	338		79.9			9.5				
––			AD		1879 1904	198		39.8			9.9				
20 00.1	+25 57	h 1462		Vul	1905 1918	24		33.3		7.7	9.5	K2			
20 00.2	+11 12	h 1458		Aql	1875 1954	133		16.3		8.5	8.5	A A	13228		

α 2000	δ 2000	Star Name		Const	Years	Pos Ang	Separation	m₁	m₂	Spec	ADS	Orb	Notes
20ʰ00ᵐ.2	+36°25	OΣ 394		Cyg	1847 1933	294° °	11.0 ″ ″	7.1	9.9	K0	13240		Orange, blue (Es)
20 00.3	+26 11	Ho 584		Vul	1896 1924	226	2.3	6.5	12.0	K0	13237		
20 00.3	+29 55	β 1258		Cyg	1878 1937	160 149	1.5 1.4	7.4	11.4	A0	13236		
20 00.3	−37 42	HdO 293		Sgr	1896 1901	209	23.0	6.0	13.6	K0			
20 00.6	−0 12	H I 93	AB	Aql	1876 1955	295	2.0	7.5	8.1	A0	13259		
			AC		1906	5	28.0				13259		
20 00.7	+36 35	Webb	AB	Cyg	1900 1925	202	70.6	6.7	8.8	B9 G			
—			AC		1910	221	81.8		9.6				
—			BC		1896 1925	278	26.3						
—			CD		1900 1910	329	15.1		10.6	A2			
20 01.0	−12 15	Rst 4649		Sgr	1939 1941	224	6.9	7.7	13.9	G0			
20 01.2	−38 35	HdO 294		Sgr	1901 1959			8.1	8.9	F2		°	
20 01.4	+10 45	Σ 2613		Aql	1829 1967	351 353	4.7 3.9	7.6	7.8	F2	13256		
20 01.4	+50 06	H V 47	AB	Cyg	1875 1949	147	41.8	5.1	10.1	K0	13278		26 Cyg
—			AE		1899	347	167.4		10.3		13278		
—			BC		1878 1933	74 76	9.0 8.5		12.7		13278		
—			BD		1898	258	10.1		15		13278		
20 01.4	−47 24	I 256		Tel	1900 1942	202 194	1.0 1.0	7.2	9.2	F0			
20 01.5	+40 18	h 1468		Cyg	1874 1928	279	13.0	7.8	8.8	A0	13271		
20 01.5	+64 49	H VI 38		Dra	1905	172	193.8	5.4		M			64 Dra
20 01.8	−17 33	h 2918		Sgr	1879 1882	135	16.8	8.2	8.9	A2			
20 01.9	+40 52	OΣΣ 196	AB	Cyg	1873 1924	168	55.2	7.0	9.6	A0			
—			AC		1922	133	138.6		12.0				
20 02.0	+24 56	OΣ 395		Vul	1844 1959	79 115	0.6 0.8	5.8	6.2	F0	13277		16 Vul
20 02.2	+54 39	β 426	AB	Cyg	1877 1949	308	5.8	8.5	10.5	K0	13301		
—			AC		1877 1949	53 53	166.2 164.6		8.3	F0	13301		
—		β 427	CD		1877 1949	336	3.0		10.4		13301		
20 02.3	+64 38	OΣΣ 200	AB	Dra	1875 1921	336	96.8	6.6	8.7	G5			65 Dra
—			Aa		1905	192	26.7		11.9				
20 02.5	−11 50	Rst 4650		Sgr	1939 1941	332	5.4	7.9	15	K2			
20 02.7	+64 27	Σ 2632 rej		Dra				8.0	11.0				
20 02.8	+14 35	Σ 2616		Aql	1829 1955	265	3.3	6.8	9.7	K0	13290		
20 02.8	+67 52	LDS 2447	AB	Dra	1953	180	124.0	4.5	13.1				67 ρ Dra
—			AC		1953	225	145.0		16				
20 02.9	+8 24	Σ 2615		Aql	1828 1927	323 312	10.8 9.8	7.9	10.8	B9	13294		
20 03.0	+24 56	Lewis 33		Vul	1900	20	0.4	8.0	9.0				Not a real double? (IDS)
20 03.2	+50 28	Ho 454		Cyg	1889 1905	57	5.6	7.5	12.5	A0	13317		
20 03.2	−18 15	Rst 3246		Sgr	1935 1941	338	1.0	7.7	13.5	F5			
20 03.3	+18 30	OΣ 396		Sge	1866 1914	206	47.4	6.1	9.4	K2			
20 03.3	−42 55	I 1490		Sgr	1925 1959	130 149	0.6 0.6	7.9	9.1	F2			
20 03.5	+36 01	Σ 2624	AB	Cyg	1830 1957	179 173	2.0 1.7	7.2	7.8	O	13312		
—			AC		1831 1956	328	42.5		9.1		13312		
—			AD		1894 1899	172	29.0		11.0		13312		
—			BC		1902	329	44.0				13312		
20 03.5	−36 36	B 463	AB	Sgr	1926 1938	206	0.8	7.3	9.8	A2			
—		See 402	AC		1896 1929	310 309	21.0 22.3		12.5				
20 03.7	+36 26	Sei 830		Cyg	1896	357	28.6	8.1	10.9	B2			
20 03.7	+38 20	h 1470		Cyg	1892 1925	337	28.8	7.3	9.4	M	13318		
20 03.8	+14 59	β		Aql	1900 1910	170	28.9	7.0	9.0	F8	13310		
20 03.9	+23 51	Cou 520		Vul	1970	150	1.2	7.6	13.2	K0			
20 04.0	−78 47	h 5149		Oct	1835 1918	150	32.4	8.2	9.7	A0			
20 04.1	+17 04	β pm	AB	Sge	1886 1918	272 276	201.4 190.7	5.9	9.1	G0			15 Sge
—			AC		1886 1918	315 320	202.8 203.7		6.8	A2			
—			Aa		1908 1924	320 330	58.7 60.0		11.6	G0			
—			BC		1907	23	147.7						
—			Bb		1903 1918	231	183.4		8.9				
—			Cc		1908	184	93.4		11.6				
20 04.1	+54 28	Stein 2497		Cyg	1917	123	10.1	8.0	11.3	K0			
20 04.2	+11 48	Σ 2620	AB	Aql	1830 1935	290	1.8	8.2	9.3	B9	13320		
—			AC		1906	45	38.9		11.7		13320		
20 04.3	−32 03	h 5165		Sgr	1919	309	51.3	5.0	12.6	K2			
20 04.6	+9 04	Σ 2621		Aql	1829 1954	223	5.7	8.6	8.8	B9	13330		
20 04.6	+32 13	h 1471		Cyg	1893 1933	6 7	29.2 31.2	5.6	11.2	B0	13335		
20 04.7	+63 53	Σ 2640	AB	Dra	1832 1974	27 16	4.9 5.6	6.3	10.2	A2	13371		
			AC		1880 1924	81	153.4		11.1		13371		
20 04.8	+15 54	OΣ 397 rej		Aql	1845 1947	170 176	34.0 40.8	7.3	9.0	G0			
20 05.1	−4 19	β 56		Aql	1875 1952	162 176	1.6 1.5	8.0	9.0	F5	13334		
20 05.1	+16 41	Ptt		Sge	1916	190	6.3	8.1	8.8	A0	13341		
20 05.1	−11 36	H IV 3		Aql	1780	10	25.0	6.5		F5			
20 05.1	−63 04	h 5163		Pav	1872 1945	249	1.5	7.9	8.3	A2			
20 05.2	+38 29	β		Cyg	1906	230	12.4	6.2	12.8	G5	13348		
20 05.3	+0 27	h 2927	AB	Aql	1901	125	24.3	6.9	11.4	A0	13340		
—		β	BC		1901	185	4.8		12.6		13340		
20 05.4	+15 30	β 57		Aql	1875 1935	120	2.3	6.3	10.7	M	13344		
20 05.5	−33 00	See 404	AB	Sgr	1897 1949	70 86	0.5 0.5	6.9	7.9	B8			

α 2000 (h m)	δ 2000	Star Name		Const	Years	Pos Ang	Separation	m_1	m_2	Spec	ADS	Orb	Notes
20 05.5	−33 00′	See 404	ABxC	Sgr	1896 1949	181° 185°	21.4 21.4		12.6				
20 06.0	+35 47	β 440	AB	Cyg	1876 1943	61 65	6.5 6.9	6.8	11.8	O	13374		In open cluster NGC 6871
—		β 429	AC		1876 1976	26 28	7.8 11.4		10.8		13374		
—		S,h 314	AD		1876 1956	300	11.4		9.4		13374		
—		β 429	AE		1876 1956	107	28.1		11.3		13374		
—		S,h 314	AF		1876 1956	28	36.1		7.8		13374		
—		Doo	AH		1899	57	30.1		13.6		13374		
—		β 429	FG		1876 1933	110	10.2		12.1	B2	13374		
—		Doo	FH		1899	154	17.8				13374		
20 06.2	+53 10	β pm		Cyg	1893 1908	123 125	171.8 170.4	5.9	9.7	F5			
20 06.4	+16 15	A 1667		Sge	1907 1931	265	3.9	7.9	14.7	K0	13375		
20 06.4	+35 58	β pm	AB	Cyg	1909 1924	64 53	16.2 23.2	5.4	13.6	K0			27 Cyg
—			AC		1909 1924	154 144	40.2 36.2		11.5				
		Sei 878			1896	84	11.9	5.5	9.5	K0			
20 06.6	+7 35	OΣΣ 198	AB	Aql	1875 1921	186	65.1	7.1	7.7	A0 A	13379		
—			Aa		1901	174	37.3				13379		
20 06.6	+12 41	h 1477		Aql	1901 1924	271	20.0	7.6	10.2	K5	13377		
20 06.7	+12 56	β 428		Aql	1876 1955	344 353	0.6 0.8	7.5	8.8	F2	13384		
20 06.7	−28 22	See 405		Sgr	1897 1946	232 236	0.5 0.5	8.4	8.9	F2	13357		
20 06.8	−12 56	Σ 2625		Sgr	1827 1959	8	12.8	7.6	11.4	K0	13370		
20 06.9	−8 55	HdO 156	AB	Cap	1892 1946	88	6.2	7.8	10.5	F0	13380		
—			AC		1907 1946	58	8.4		12.6		13380		
20 07.0	+66 18	Σ 2650 rej		Dra	1905 1925	228	21.8	7.0	9.9	A0	13424		
20 07.3	+16 05	Σ 2629 rej		Aql	1874 1932	188	9.2	7.6	10.7	B9	13394		
20 07.3	−19 35	See 406		Cap	1897	1	2.7	7.9	10.8				
20 07.4	+35 43	OΣ 398	AB	Cyg	1846 1945	82	0.9	7.2	9.7	B0	13405		
—		A 280	AC		1901 1929	133	5.4		14.6		13405		
20 07.4	−29 43	h 5168		Sgr	1877 1919	80	18.7	6.7	10.7	K0			
20 07.5	+38 38	A 1416		Cyg	1906 1931	42	4.7	8.0	10.9	A0	13408		
20 07.7	+4 46	Σ 2627	AB	Aql	1829 1930	29	1.9	9.9	12.4	A5	13399		
—			AC		1912	262	80.4		7.9	K2	13399		
20 07.8	+9 24	Σ 2628		Aql	1830 1975	349 341	4.5 3.4	6.5	8.6	F5	13403		
20 07.9	−70 49	h 5162		Pav	1835 1917	292	6.7	8.0	10.5	K0			
20 08.0	+32 35	Σ 2633	AB	Cyg	1831 1905	103	11.7	7.9	10.9	B9	13410		
—		h 1483	AC		1905	263	46.9				13410		
—			AD		1905	249	72.0				13410		
—		Sei 900			1893	226	28.7	9.5	9.7				
20 08.0	+42 23	A 382		Cyg	1902 1953	85 94	1.3 1.6	7.0	10.4	K0	13415		
20 08.3	+38 07	h 606		Cyg	1900 1918	228	43.6	8.0	8.6	B A0			
20 08.6	−71 33	I 1042		Pav	1911 1928	95 83	0.6 0.6	8.5	9.1	A0			
20 08.9	+62 05	Σ 2652		Dra	1832 1958	280 238	0.3 0.3	7.2	7.5	A0	13449		
20 08.9	+77 43	Σ 2675	AB	Cep	1832 1954	122	7.4	4.4	8.4	B9	13524		1 κ Cep
—			AC		1895 1912	336	169.8		8.4		13524		
20 09.2	+35 29	Σ 2639	AB	Cyg	1830 1967	302	5.7	7.7	8.7	B3	13429		
—		Sei 923			1894 1896	208	27.9	9.7	10.1		13429		
20 09.5	+51 41	Σ 2645		Cyg	1831 1955	138	1.5	8.5	8.8	A0	13447		
20 09.6	+16 48	Σ 2634	AB	Sge	1830 1975	14 14	6.4 4.7	7.9	9.4	K0	13434		
—			AC		1911 1924	314 312	76.4 74.8		12.6		13434		
20 09.6	+33 25	h 1485		Cyg	1875 1937	277	4.7	8.3	9.0	A0	13441		
20 09.9	+20 55	Σ 2637	AB	Sge	1832 1951	327 325	11.4 11.9	6.5	9.0	F2	13442		
—			AC		1832 1949	227 223	70.7 83.9		7.4	K2	13442		
—			AD		1923	228	162.2				13442		
—			BC		1901	216	82.2				13442		
20 09.9	+21 01	S 737		Sge	1824 1919	129	100.8	8.2	9.6	K0			
20 10.1	+8 27	Σ 2635	AB	Aql	1828 1924	79	7.4	6.6	10.1	F8	13443		
—			AC		1911 1924	43 42	71.6 73.7		12.4		13443		
20 10.2	+43 57	OΣ 400		Cyg	1845 1960			8.0	8.5	G5	13461	o	
20 10.3	−79 07	h 5153		Oct	1918	125	41.4	7.4	12.8	G5			
20 10.4	+49 49	Σ 2648		Cyg	1831 1931	117	6.4	8.1	9.4	F5	13471		
20 10.5	+25 03	Cou 122		Vul	1966	336	0.5	7.3	10.0	A0			
20 10.6	+33 38	S 738	AB	Cyg	1824 1934	112 108	41.8 41.2	8.5	8.6	A0 A0	13463		
—		OΣ 541	BC		1840 1935	192 185	1.6 1.8		10.5		13463		
20 10.6	−46 44	I 123		Tel	1900 1954	175 184	10.7 8.2	6.8	11.1	K0			
20 10.8	+37 03	OΣ 399		Cyg	1846 1947	279	4.5	7.4	10.0	G5	13473		
20 10.9	−32 19	B 987		Sgr	1927 1947	25	5.1	7.6	11.8	A0			
20 11.1	+16 11	β pm	AB	Sge	1893 1908	153 152	144.0 152.0	7.3	9.5	K0			
—			AD		1907	293	86.6						
—			BC		1908	271	63.8		12.0				
—		Giclas			1962	96	108.0	8.3	15				
20 11.1	−57 31	HdO 295		Pav	1901 1952	228 256	0.6 0.6	6.8	7.7	A0			
20 11.2	−36 06	h 5173		Sgr	1880 1949	119 123	9.8 7.1	5.3	11.5	K4			
20 11.3	−0 08	S 735		Aql	1825 1923	206	55.3	7.1	8.5	F0 A0			
20 11.3	−0 49	H VI 27		Aql	1901	260	113.7	3.2	12.8	A0			65 θ Aql
20 11.4	+29 10	A 1200		Vul	1905 1929	198	4.9	7.3	13.7	K2	13487		
20 11.6	+6 25	LDS 1036		Aql	1936	192	44.0	8.3	8.6	G5 G			

α 2000	δ 2000	Star Name		Const	Years	Pos Ang	Separation	m₁	m₂	Spec	ADS	Orb	Notes
20 11.6	+38 53'	A 1418		Cyg	1896 1931	324°	2".9	7.7	12.0	K0	13495		
20 11.6	+62 05	LDS 2791		Dra	1966	250	93.0	5.8	15	F5			68 Dra
20 11.9	+36 12	β pm		Cyg	1878 1909	110	53.2	8.0	10.3	O			
20 12.0	+34 29	A 282	AB	Cyg	1901 1958	206 220	0.2 0.2	7.9	8.2	A2	13508		
		Ho 121	AC		1884 1929	18 17	21.4 23.0		12.2		13508		
--			AD		1889 1904	14 15	41.6 42.5		11.7		13508		
20 12.2	+38 27	OΣ 401		Cyg	1847 1929	60	13.8	7.3	10.6	G5	13515		
20 12.3	−8 05	β 1205		Aql	1890 1952	50 95	0.6 0.2	7.5	8.8	F8	13493		Single 1958 (82-inch)
20 12.3	+22 48	Cou 123		Vul	1966	245	0.2	8.6	8.9	K0			
20 12.3	+32 05	Σ 2649	AB	Cyg	1832 1951	152 151	26.1 22.8	8.0	9.3	A2	13513		
--			AC		1910 1922	284 285	129.2 130.5		10.7		13513		
--			AD		1921	38	175.3		9.7		13513		
20 12.3	+34 51	A.G. 249		Cyg	1902 1910	132	34.5	8.0	9.8	K2			
20 12.4	−12 37	β pm		Cap	1909	259	71.6	5.9	12.1	F5			2 ξ Cap
20 12.5	+51 28	AC 17	AB	Cyg	1859 1958	80	4.0	6.0	11.5	K2	13535		
--			AC		1913 1922	8	83.4		12.6		13535		
--			AD		1913 1922	81	74.2		12.0		13535		
--			AE		1922	149	130.7				13535		
20 12.6	+0 52	Σ 2644		Aql	1830 1958	208	3.0	6.9	7.2	A0	13506		
20 12.6	+25 39	Doo 15	AB	Vul	1900 1951	302 299	127.8 122.8	7.6	9.1	F2	13518		
--			BC		1900 1951	170	2.2		10.8		13518		
20 12.8	−3 00	Σ 2643		Aql	1830 1937	71 76	3.2 3.0	7.0	9.5	A0	13511		
20 13.1	+49 11	Es 502		Cyg	1907 1933	223 222	11.2 12.1	7.9	10.2	A0	13545		
20 13.2	+34 13	OΣ 203		Cyg	1876 1913	37	90.4	8.4	9.1	A2			
20 13.4	+38 44	β pm	AB	Cyg	1879 1909	355	132.4		7.1	P B2			A is variable
--			AC		1878 1909	106	55.9		9.3				
20 13.6	+46 44	h 1495	AB	Cyg	1878 1903	331	36.6	3.8	13.1	K0	13554		31 Cyg = V695 Cyg (var.)
--		Σ I 50	AC		1836 1926	173	107.0		6.7	B9	13554		
--			AD		1835 1926	323	337.5		4.8	A2	13554		
--		Wils	Aa		1949 1954	140 319	0.1				13554		Single 1953; 0".04 in 1954
--		Gui	DC		1901	150	433.6				13554		
--		Es 26	DE		1900 1912	252 251	32.9 34.3		13.0		13554		
20 13.7	+24 14	Σ 2653		Vul	1831 1958	255 270	2.4 2.6	6.6	9.7	A0	13543		
20 13.7	+43 23	β 660	AB	Cyg	1878 1958	319	9.6	6.1	12.6	K2	13555		
--			AC		1913	190	57.3		11.1		13555		
--		Fox	AD		1914	278	30.0		12.9		13555		
20 13.7	+53 08	Σ 2658	AB	Cyg	1831 1968	127 111	5.5 5.4	7.1	9.2	F5	13560		
--			AC		1832 1931	217 208	32.1 49.3		10.3		13560		
--			Aa		1910 1913	313	34.0		12.7		13560		
--			BC		1901 1911	217	45.9				13560		
--			Bb		1910	83	11.2		14.0		13560		
20 13.7	−34 07	h 5178		Sgr	1856 1955	10	2.7	7.0	8.5	G0			
20 13.8	+16 09	Σ 2651		Sge	1830 1957	279	1.4	8.5	8.5	F8	13542		
20 14.1	+22 13	Σ 2655	AB	Vul	1831 1954	3	6.2	7.9	7.9	A0	13553		
--			AC		1825 1932	150 154	57.3 60.0		9.6		13553		
20 14.2	+6 35	S 740		Aql	1824 1923	193	43.4	7.8	8.0	G5 G			
20 14.2	+35 22	Es 204		Cyg	1896 1933	231 257	11.8 12.6	7.7	10.5	F5	13562		
20 14.3	+31 29	A 1204		Cyg	1905 1951	134	0.3	8.6	8.9	F	13564		
20 14.4	−6 03	Σ 2646	AB	Aql	1829 1959	52 44	24.7 20.0	7.3	9.2	F0	13552		
--			BC		1909	106	26.5		12.5		13552		
20 14.4	+42 06	OΣ 403	AB	Cyg	1848 1958	173	0.8	7.4	7.6	B8	13572		
--		Σ 2657	AC		1848 1958	33	11.6		10.0		13572		
20 14.4	+80 32	Σ 2694		Dra	1832 1936	346	4.0	6.8	10.8	A0	13708		
20 14.5	+24 51	OΣ 402		Vul	1849 1907	35	15.4	8.0	11.5	B9	13566		
20 14.5	+36 40	β pm		Cyg	1878 1909	118	113.6	8.1	9.9	O			
20 14.5	+36 48	β pm	AB	Cyg	1881 1918	153	212.4	5.0	6.6	A0 K5			29 Cyg
--			AC		1887 1918	23 23	225.2 221.7		10.0				
--			Aa		1908 1960	341 333	38.8 36.9		12.2				
--			BD		1894 1918	120 120	216.5 217.4		10.2				
20 14.6	−64 26	h 5171	AB	Pav	1873 1927	306	17.3	7.0	10.0	A2			
--			AC		1873 1916	336	30.0		10.0				
20 14.9	−56 59	Rmk 25		Pav	1835 1940	29	7.5	8.0	8.1	F5			
20 15.1	+2 18	Bail 1551		Aql	1909	42	9.3	7.9	11.9	K0			
20 15.1	+41 18	A 387		Cyg	1902 1928	151	4.9	7.9	13.7	A0	13595		Faint 7" pair, 0'.3 sf
20 15.2	−3 30	Σ 2654		Aql	1831 1951	233	14.2	6.9	9.3	F0	13574		
20 15.2	+54 09	Ho 455	AB	Cyg	1889 1925	82	33.6	7.5	11.5	M	13610		
--			AD		1889 1925	257	32.3		11.5		13610		
--			AE		1889 1925	76 75	36.7 38.1		10.6		13610		
--			BC		1889 1925	190 182	2.7 3.7		11.5		13610		
--			BE		1904 1925	214 210	5.1 5.7				13610		
20 15.3	+25 36	β 983	AB	Vul	1879 1955	155 166	0.9 0.7	4.8	8.9	B3	13589		
--			AC		1912	83	115.8		9.7		13589		
20 15.3	+64 12	Mlr 60		Dra	1970	200	0.2	8.7	8.7	G5			
20 15.4	+33 44	A 283		Cyg	1901 1958	296 290	2.5 3.0	5.7	13.7	G5	13596		
20 15.5	+2 51	h 910	AB	Aql	1881 1925	320 323	13.6 15.0	7.6	12.6	F5	13586		

α 2000	δ 2000	Star Name		Const	Years	Pos Ang	Separation	m₁	m₂	Spec	ADS	Orb	Notes
20ʰ15ᵐ5	+2°51′	h 910	AC	Aql	1879 1925	248° 246°	26″8 29″8		12.3		13586		
20 15.5	+47 43	S 743		Cyg	1824 1902	175	208.9	4.0	9.5	K0			32 Cyg; sp. compos. K0+A3
20 15.6	+7 49	Σ 2656		Aql	1827 1910	234	9.5	7.2	11.9	A2	13590		
20 15.8	+52 30	OΣ 404 rej		Cyg	1867 1952	114	29.6	7.4	9.9	K5	13616a		
20 16.1	−7 44	Ho 589	AB	Aql	1895 1911	324	15.4	8.0	12.0	B9	13598		
—			AC		1906 1911	201	56.2		10.2		13598		
20 16.4	−12 20	β 294	AB	Cap	1891 1912	35	26.9	6.4	13.7	B9	13600		3 Cap
			BC		1891 1898	178 180	8.2 7.5		14.2		13600		
20 16.4	−36 27	h 5183	AB	Sgr	1919	229	38.0	6.1	11.2	M			
—			AC		1919	180	46.7		12.6				
20 16.5	+37 38	β 442	AB	Cyg	1876 1949	103	18.8	8.5	9.0	B	13626		In open cluster IC 4996
—		S1v	AC		1967	77	31.8		9.6		13626		
—		β 442	Aa		1876 1922	156	4.1		10.8		13626		
—			Ab		1888 1913	158	8.9				13626		
—			Ac		1876 1913	332	19.2		9.6		13626		
—			BC		1876 1949	49	17.1				13626		
—			Bd		1898 1914	128	3.7		14.5		13626		
—			Be		1876 1913	163	6.9		11.5		13626		
—			Cf		1879 1933	109	12.4		13.0		13626		
—			Cg		1879 1913	116	20.6		12.8		13626		
—			Ch		1898 1913	306	15.3		13.7		13626		
—					1896	3	19.8	8.1	11.1	B	13626		Unidentified measures in
—					1896	76	25.7	8.0	10.9		13626		IDS at this position
20 16.8	−3 30	Rst 4659		Aql	1939 1943	356	1.2	7.0	9.8	F5			
20 16.8	+39 41	Σ 2663		Cyg	1831 1952	324	5.4	8.1	8.6	A0	13636		
20 16.9	+31 30	Ho 588	AB	Cyg	1896 1949	297	51.1	7.0	8.8	A0	13630		
—			BC		1896 1925	16	8.5		12.5		13630		
20 16.9	+40 22	β 661	AB	Cyg	1878 1958	66	12.8	5.2	11.5	K5	13640		
—			AC		1912	218	111.7		12.3		13640		
20 16.9	−32 36	Stone 64		Sgr	1876 1959	303 299	2.2 2.2	8.2	8.4	G5			
20 17.0	+37 25	A 1423		Cyg	1906 1931	130	4.4	8.0	12.1	O	13641		
20 17.5	+29 09	β 441		Vul	1876 1958	66	5.9	6.2	10.7	K0	13648		
20 17.6	−12 30	β 295	AB	Cap	1891 1960	182 182	43.5 44.3	4.2	13.7	G0	13632		5 α¹ Cap
—		h 607	AC		1830 1932	220 221	44.1 45.4		9.2		13632		
—		Wirtz	DC		1905	290	29.3	13.9			13632		
20 17.7	+20 25	Cou 219		Sge	1967	115	0.3	8.3	9.1	K2			
20 17.8	−40 11	Δ 230		Sgr	1836 1952	116	9.8	7.7	8.0	F5			
20 18.0	+33 12	Barnard 11	AB	Cyg	1898 1944	200	0.4	8.2	8.7	B9	13660		
—		Ho 592	AC		1895 1921	256	2.9		11.4		13660		
20 18.1	+40 44	Σ 2666	AB	Cyg	1831 1958	245	2.7	5.8	8.0	B2	13672		
—			AC		1887 1908	208	34.1		8.5		13672		
20 18.1	−12 33	h 608	AB	Cap	1846 1959	144 172	6.4 6.6	3.6	11.0	G5	13645		6 α² Cap
—		AGC 12	AD		1879 1909	156	154.6		9.3		13645		
—			BC		1877 1959	240	1.2		11.3		13645		
—		Σ I 51			1835 1924	291 291	374.5 377.7	3.6	4.2	G5	13645		Measure of α², α¹ Cap
20 18.2	+18 20	Hu 357		Sge	1901 1925	196	1.8	7.8	12.9	A0	13662		
20 18.3	+25 39	β 985	AB	Vul	1880 1915	149	5.1	7.0	13.0	B3	13666		
—		h 1499	AC		1880 1915	356	21.6		9.9		13666		
—		β 985	CD		1893 1915	63	9.8		12.6		13666		
20 18.4	+55 24	Σ 2671	AB	Cyg	1831 1958	341 338	3.0 3.5	6.1	7.5	A0	13692		
—			AC		1911	55	83.9		12.6		13692		
—			CD		1911	119	5.4		14.2		13692		
20 18.7	+33 14	OΣ 405	AB	Cyg	1846 1955	150	0.7	8.4	9.4	A0	13682		
—		Barnard	ABxC		1917	256	2.8		14.0		13682		
20 18.9	+38 17	A 1425	AB	Cyg	1906 1960	297 280	0.2 0.2	8.5	8.5	B1	13686		
—			ABxC		1906 1917	310	8.8		13.2		13686		
20 19.0	+39 00	Ho 125		Cyg	1885 1941	194	3.1	6.3	10.6	K2	13690		
20 19.0	−14 17	Rst 4057		Cap	1937 1943	261	1.3	7.4	11.2	K0			
20 19.1	+36 45	β 1206		Cyg	1890 1938	1	2.0	7.6	10.6	F2	13693		
20 19.4	+14 22	Σ 2665	AxBC	Del	1829 1943	17	3.3	6.8		A0	13688		Spectrum composite A0+G
—		A 1672	BC		1907 1945	266 253	0.2 0.3	9.5	10.2		13688		
20 19.4	−19 07	H V 87		Cap	1782 1913	175 179	55.9 55.9	5.5	9.0	K4	13675		7 σ Cap = S,h 380
20 19.6	+13 00	Σ 2664		Del	1829 1915	322	27.7	8.4	8.9	K0			
20 19.7	+41 08	OΣΣ 205		Cyg	1875 1923	319	45.5	7.1	8.9	B9			
20 19.7	+47 54	β pm		Cyg	1878 1918	51	63.5		7.8	R8 G0			U Cyg; A is variable
20 19.8	+45 22	OΣ 406		Cyg	1845 1959			7.4	8.3	F5	13723	o	
20 19.9	−2 15	Σ 2661		Aql	1828 1922	341	24.3	8.3	9.5	A0			
20 20.1	+51 15	Es 800	AB	Cyg	1909	315	28.2	9.0	9.5	G0	13732		
—			AC		1909	104	40.3		10.4		13732		
—			AE		1909	97	112.6		8.9		13732		
—			CD		1909	150	2.2		12.4		13732		
20 20.2	−29 08	Glp		Sgr	1890	198	53.7	7.7	9.6	K0			
20 20.2	−34 35	I 1416		Sgr	1925 1954	300	0.3	7.1	7.5	G0		o	Close 1937,59; p.a. steady
20 20.3	+39 24	A 1427	AB	Cyg	1906 1958			6.3	8.9	A0	13728	o	
—		Σ 2668	ABxC		1831 1969	294 283	3.3 3.3		9.3		13728		

α 2000	δ 2000	Star Name		Const	Years		Pos Ang		Separation		m₁	m₂	Spec	ADS	Orb	Notes
20ʰ20.ᵐ5	+43°51′	β 1207		Cyg	1890	1919	217°	°	5.8″	″	6.9	14.5	O	13736		
20 20.5	+53 36	β 663	AB	Cyg	1891	1958	314	314	6.6	7.2	6.2	15	K5	13743		
—			AC		1885	1958	75	69	7.7	6.5		12.4		13743		
—			AD		1958		232		11.5			14.5		13743		
20 20.5	−29 12	h 5188	AB	Sgr	1876	1959	64	50	4.7	4.1	6.4	9.4	A0	13702		
—			AC		1876	1957	322	321	27.1	27.2		8.3	A0	13702		
—			AD		1951		32		106.3			10.2		13702		
—			DE		1876	1951	189	186	4.7	4.7		10.2	F5	13704		
20 20.6	−58 44	h 5185		Pav	1914	1916	61		18.6		7.3	11.1	A0			
20 20.7	−12 46	Lam 5		Cap	1877	1960	210	211	55.4	54.1	4.8	11.8	A0	13714		8 ν Cap
20 20.8	−7 44	Schj 25		Aql	1875	1953	217		2.7		8.4	9.2	G0	13721		
20 20.8	−14 47	Barnard 12		Cap	1884	1952	106	89	0.8	0.8	6.2	10.2	B9	13717		β¹ Cap
20 20.9	−45 33	See 415		Tel	1897	1946	97		11.1		7.3	12.0	F5			
20 21.0	−14 47	Σ I 52	AB	Cap	1835	1922	267		205.3		3.4	6.2	G0 B9			β²,β¹ Cap. Fine contrast
—			AC		1912		134		226.6			9.0	F8			
—			BC		1913		111		397.6							
—		h 2948			1891		294		111.7							β² Cap and h 2948A
—					1891	1913	322		6.4		13.0	13.4				
20 21.2	+43 35	h 1505		Cyg	1900	1925	110		16.9		7.9	10.1	A0	13753		
20 21.2	+63 59	β 1134		Dra	1889	1925	81		4.2		5.7	12.6	K5	13769		
20 21.6	+19 30	Cou 327	AB	Del	1968		72		0.2		8.7	8.7	F8			
—			ABxC		1968		126		4.0			12.5	F8			
20 21.9	+13 36	Hu 1197		Del	1905	1921	303		1.1		7.4	14.0	A0	13755		
20 22.2	+40 15	β 665	AxBC	Cyg	1878	1923	196		41.2		2.2		F8	13765		37 γ Cyg
—			BC		1878	1929	302		1.8		9.9	10.9		13765		
—			BCxD		1911		134		40.9			12.7		13765		
20 22.2	−16 47	H N 138		Cap	1878	1936	330		3.0		8.5	9.0	G5	13751		BDS 10151
20 22.5	+26 18	h 1504	AB	Vul	1904	1925	248		22.2		7.1	10.6	K2			
—			AC		1904		227		51.8			9.9				
20 22.5	−42 03	h 5190	AB	Sgr	1900	1914	307	307	29.9	31.7	5.6	12.6	A0			κ¹ Sgr
—			AC		1900	1914	277	278	50.7	52.0		11.6				
20 22.6	−54 16	h 5187		Tel	1914		323		17.8		8.0	13.5	K0			
20 22.7	+13 20	Σ 2673	AB	Del	1830	1955	331		2.5		8.6	10.1	F2	13767		
—			AC		1829	1955	103		76.0			8.7		13767		
—		Σ 2674	CD		1829	1935	1		15.6			11.5		1376?		
20 22.8	+53 25	Σ 2681	AB	Cyg	1831	1912	40		6.7		7.7	11.2	A0	13795		
—			AC		1831	1949	204	201	41.8	39.6		8.8		13795		
—			AD		1842	1913	172	170	44.4	42.9		11.4		13795		
—			CD		1830	1910	102		22.4					13795		
20 22.9	+42 59	Ho 128	AB	Cyg	1886	1957	35	12	1.0	1.2	6.6	11.3	K0	13786		
—		ΟΣΣ 207	AC		1876	1920	63	63	96.4	93.2		8.5	A0	13786		
—			AD		1908		164		57.4			13.6		13786		
20 23.0	+39 13	ΟΣ 206	AB	Cyg	1876	1949	256		43.0		6.7	9.2	B9	13783		
—			Aa		1905		262		23.4			10.9		13783		
—		Sei 1107	BC		1895		70		19.3			11.4	B			
20 23.1	−33 03	B 2483		Sgr			sf		12.0		7.6	11.0	K5			
20 23.2	+20 52	A 288		Del	1901	1958	352	49	0.3	0.2	8.7	8.9	F2	13777		
20 23.6	−26 44	Ho 456		Cap	1889	1928	215	208	14.4	14.1	7.3	11.7	K0	13770		
20 23.9	−42 25	β 763	AB	Sgr	1879	1952	204	234	1.3	0.8	6.0	6.9	A3			κ² Sgr
—		I	AC				270		15.0			14.9				
—		See	AD		1897		226		29.3			13.9				
20 24.1	+29 00	β 443	AB	Vul	1878	1915	134	138	13.0	14.0	7.2	11.2	A5	13807		
—			AC		1878	1915	89		35.2			11.7		13807		
20 24.4	+19 35	Σ 2679	AB	Del	1830	1955	80	78	21.9	23.5	7.9	9.2	A2	13808		
—			AC		1908	1910	152		39.1			12.0		13808		
20 24.4	+24 17	Cou 125		Vul	1966		115		0.4		7.4	8.3	A3			
20 24.6	+1 04	Σ 2677	AB	Aql	1946		14		17.1		6.2	13.2	A0	13811		
—			AC		1828	1927	29		32.7			10.7		13811		
20 24.6	+5 31	β 664		Del	1878	1926	288		8.9		7.2	12.7	B9	13810		A 12th-mag. * lies 0′.3 nf
20 24.6	+48 42	Es		Cyg	1911		321		11.4		7.8	10.3	B8	13831		
20 24.7	−8 46	Rst 4062		Cap	1938	1952			0.3		7.4	8.9	A3			
20 24.8	+25 11	Cou 521		Vul	1970		286		0.4		7.7	9.6	A0			
20 24.8	+40 09	Ho 278		Cyg	1886	1953	173	70	0.2	0.1	7.5	7.5	K0	13833		Single or round 1953-60
20 24.9	−55 56	φ 277		Tel	1931		32		0.2		8.8	8.8	K0			
20 25.1	+59 36	A 730		Cyg	1904	1959					7.1	7.4	A0	13850	o	
20 25.5	+40 06	Dem 22	AB	Cyg	1875	1959	140	155	2.8	3.1	8.1	9.2	G5	13847		
—			AC		1917	1926	95	97	17.4	15.6		13.9		13847		
20 25.5	+54 41	h 1516	AB	Cyg	1907	1912	148		47.0		7.4	10.8	B3			
—			AC		1912		204		49.4			12.1				
20 25.6	+26 42	β	AB	Vul	1910		73		75.6		8.2	9.6	F5	13843		
—			BC		1910		356		2.3			12.6		13843		
20 25.6	+55 08	A 1429		Cyg	1906	1959	197	190	0.7	0.7	8.3	9.3	A0	13857		
20 25.6	−56 44	h 5193	AB	Pav	1879		84		245.4		1.9	9.0	B3			α Pav
—			BC		1873	1914	332		17.6			10.3				
20 25.7	+60 11	A 731		Cep	1904	1932	214		2.1		7.4	12.1	A3	13864		

α 2000	δ 2000	Star Name		Const	Years	Pos Ang		Separation		m₁	m₂	Spec	ADS	Orb	Notes
20ʰ 26.3	+54°02′	Frk		Cyg	1917	239°		59″.5		8.7	8.7				
20 26.4	+56 38	Σ 2687		Cyg	1831 1950	118		26.6		6.2	9.0	A0	13870		
20 26.6	−39 55	Scott		Sgr	1903	320		30.8		8.5	9.0	K0 K2			
20 26.6	−47 57	I 1419	AB	Tel		sf		60.0		8.2	9.4	F0 F			
—			BC		1926 1930	342		3.0			12.2				
20 26.9	−37 24	R 321		Sgr	1880 1959					6.7	8.0	K1		o	
20 27.3	−18 13	β 60	AB	Cap	1846 1955	145 148		2.8	3.2	5.3	8.9	B8	13860		10 π Cap
—		β 296	AC		1898	44		38.1			14.2		13860		
20 27.5	−2 06	S 749	AB	Aql	1825 1920	189		59.9		6.6	8.0	F8 F8	13868		
—			AC		1891 1908	314 316		44.9	43.6		10.7		13868		
—			AD		1891 1908	266 266		100.4	97.4		10.8		13868		
20 28.2	+81 25	β pm	AB	Dra	1908 1960	13 12		109.5	109.8	5.5	11.0	K0			75 Dra
—			AC		1884 1960	285 283		198.4	197.7		6.8	K0			
20 28.3	+18 46	Ho 131	AB	Del	1881 1938	324		4.1		6.8	10.2	G5	13886		
—			AC		1903 1921	81 81		101.0	195.8		9.8		13886		
20 28.4	−3 21	Rst 4667		Aql	1939 1943	178		9.8		6.1	13.8	B9			68 Aql
20 28.5	+54 30	Σ 2693		Cyg	1830 1915	13		13.7		8.1	9.6	A0	13903		
20 28.9	−17 49	S,h 323	AB	Cap	1823 1958	177 158		4.0	0.5	5.0	10.0	F0	13887		11 ρ Cap
—		β 61	AC		1891 1901	151		55.2			13.2		13887		
—		S,h 323	AD		1823 1913	151 150		238.0	247.6		6.7	K0	13887		
—		Dob	DE		1913	106		59.4					13887		
20 29.1	+35 50	Ho 594		Cyg	1894 1925	209		18.2		7.3	13.0	K2	13905		
20 29.1	−51 05	I 664		Tel	1914 1959	287 280		0.8	1.0	8.0	9.2	F0			
20 29.2	+37 31	Weisse 35	AB	Cyg	1883 1949	213		4.0		8.5	9.0	F5	13909		
—			AC		1883 1949	100		87.1			9.5	B8	13909		
—			BC		1910 1921	98		88.0					13909		
—			CD		1883 1933	203		11.8			11.2		13909		
20 29.4	+56 04	Kui 97		Cyg	1938 1958	150 137		0.5	0.6	6.0	8.5	B9			
20 29.5	+81 05	β pm	AB	Dra	1851 1908	39		214.4		6.1	9.3	K0			74 Dra
—			BC		1908	306		107.3			12.6	G5			
20 29.6	−27 19	CorO 240		Cap	1882 1931	21		6.9		8.2	10.0	F0	13896		
20 29.6	−46 09	I 1422		Ind	1925 1936			0.1		8.2	8.3	A0			Measures uncertain
20 29.9	−18 35	S,h 324		Cap	1823 1955	239		21.9		6.1	6.6	A2 A3	13902		12 o Cap
20 30.1	+48 57	β 669	AB	Cyg	1878 1935	341		17.8		5.0	13.0	B3	13932		45 ω¹ Cyg
—			AC		1878 1918	86		56.3			9.5		13932		
20 30.2	+19 25	β 987	AB	Del	1880 1941	128		2.4		6.6	10.9	B9	13921		
—		S 752	AC		1824 1915	288		105.9			7.0	B9	13921		
—		Fox	AD		1913	296		56.4			10.9		13921		
—			Aa		1913 1924	71		22.1			10.5		13921		
20 30.3	+10 54	β 63	AB	Del	1874 1958	346		0.9		6.1	8.1	A0	13920		1 Del
—		β 297	AC		1898 1915	349		16.8			14.1		13920		
20 30.3	−69 04	h 5194		Pav	1871 1917	256		4.2		7.3	12.6	A2			
20 30.6	+2 24	Bail 2028		Aql	1909 1910					7.9					
20 30.9	+49 13	S 755	AB	Cyg	1825 1925	278		60.4		6.6	9.2	A2			
—		Es	Aa		1909	121		21.4			13.5				
20 31.0	+20 36	β 363	AB	Del	1878 1919	63 64		21.8	16.7	6.0	10.0	A2	13913		
—			AC		1892 1911	198 199		44.1	45.6				13913		
20 31.0	+36 56	AC 18	AB	Cyg	1859 1958	155 159		2.6	2.0	6.2	11.2	F8	13949		44 Cyg
—		Kui 98	AC		1935 1958	76		16.5			12.7		13949		
—		AC 18	AD		1913	166		63.0			12.2		13949		
20 31.1	+15 48	A 1675		Del	1907 1960					7.6	7.6	A2	13944	o	
20 31.1	−15 03	ø 336		Cap	1954 1959	117 129		0.1	0.1	6.9	6.9	G0			
20 31.2	+11 16	h 269	AD	Del	1878 1958	106		23.5		7.0	12.0	A0	13946		
—		Σ 2690	AxBC		1831 1958	256 256		14.2	16.7				13946		
—		Dawes 1	BC		1841 1960	210 244		0.5	0.2	7.9	8.0	A0	13946		
20 31.3	+49 13	S 756		Cyg	1825 1922	323		56.8		5.4	10.1	M			46 ω² Cyg
20 31.7	+62 27	β 671		Cep	1877 1958	336 316		0.5	0.5	8.4	8.9	A0	13989		
20 31.9	−40 54	Gli 259	AB	Mic	1876 1959	152 156		3.0	3.8	8.2	8.3	G0			
—		I	AC		1925 1959	14 356		6.7	10.3		12.5				
20 32.0	+25 48	Σ 2695		Vul	1831 1959	77 87		0.8	0.6	6.5	8.3	A2	13964		Spectrum composite A2+G
20 32.1	−5 15	A 170		Aql	1900 1941	217		1.6		6.7	11.0	K0	13956		
20 32.1	+15 11	A 1677		Del	1907 1937	172		1.1		8.0	11.0	G0	13965		
20 32.1	−45 21	Rst 5470	AB	Ind	1945	136		0.3		8.4	8.6	F0			
—		h 5204	ABxC		1836 1936	35		6.3			8.8				
20 32.2	−22 09	h 2973		Cap	1879 1934	129		39.3		7.8	8.4	F0 F0			
20 32.4	−9 51	β 668	AB	Cap	1878 1959	29		4.6	2.9	5.7	11.2	G5	13960		
—			AC		1908 1921	198 200		100.2	103.2		9.9		13960		
20 32.5	−16 37	See		Cap	1897 1954	120 110		0.4	0.1	7.9	7.9	F0	13961	o	[78]
20 32.7	−31 23	h 5206		Mic	1896 1918	192		16.4		7.7	12.2	A2			
20 33.0	+31 44	Es 365	AB	Cyg	1906	262		25.7		8.0	11.8		13995		
—			AC		1906	319		33.7			12.3		13995		
—			CD		1906	288		2.5			12.4		13995		
20 33.0	−68 22	h 5200		Pav	1835 1917	137		12.2		7.5	11.0	A0			
20 33.0	−71 19	I 1047	AB	Pav	1912	331		9.6		8.4	12.8	K2			
—		h	AC		1836	65		117.0			9.0				

α 2000	δ 2000	Star Name		Const	Years	Pos Ang		Separation		m₁	m₂	Spec	ADS	Orb	Notes
20ʰ33ᵐ.2	+33°24′	h 1535	AB	Cyg	1902 1924	246°		16″.8		8.1	11.7	F5	13998		
—			AC		1902 1924	150	165	18.4	19.1		11.7		13998		
—			AD		1905	232		34.3			11.1		13998		
20 33.3	−80 58	h 5182		Oct	1880 1918	357		26.8		5.8	11.5	K0			
20 33.4	−6 13	h 1529		Aql	1905 1920	108		36.6		7.4	9.6	M			
20 33.5	−22 14	h 2975		Cap	1877 1955	23		9.8		7.4	12.0	F8	13988		
20 33.6	+5 27	Σ 2696	AB	Del	1831 1959	302		0.8		8.3	8.7	A2	13997		
—			AC		1901	349		13.8			13.7		13997		
20 33.7	+38 35	A 1431		Cyg	1906 1950	35		0.8		8.3	9.4	A0	14007		
20 33.8	−40 33	Jc 18	AB	Mic	1846 1951	225		4.4		7.9	8.5	A0			
—			AC		1879	144		106.1			10.5				
20 33.9	+35 15	Wils	AB	Cyg	1949 1950	94	127	0.1	0.1	4.6		K5			47 Cyg, Hynek 94
—		β	AC		1880 1909	190		117.6			10.2				
20 33.9	+46 42	β		Cyg	1913 1921	17		117.7		5.8	9.9	B9			
20 34.0	+34 41	OΣ 408		Cyg	1846 1954	190		1.6		6.7	9.7	B8	14016		
20 34.3	−0 28	Σ 2697 rej		Aql	1848 1923	2		30.4		8.1	10.1	K2			
20 34.5	+6 53	β 1208		Del	1890 1923	333		3.0		7.0	11.8	B5	14017		
20 34.7	+32 30	Σ 2700		Cyg	1831 1951	285		23.9		7.0	8.8	K0	14027		
20 34.7	−63 19	Hu 1615		Pav	1914 1959	359	22	0.4	0.2	8.1	8.3	A0			
20 34.8	−33 55	h 5207		Mic	1896 1919	258		10.2		7.8	10.2	F5			
20 34.9	+46 51	OΣΣ 208		Cyg	1876 1925	241	241	76.4	78.5	8.1	8.8	K0 A0			
20 35.0	+32 25	Stone 51		Cyg	1879	268		32.8		8.5	9.0				β could not find this pair
20 35.5	+27 39	Cou 1171		Vul	1973	105		0.4		8.0	10.5	A0			
20 36.5	−45 33	I 41		Ind	1895 1959			1.9		7.8	8.9	A3			
20 36.6	−71 04	Δ 231		Pav	1826 1917	288		57.2		7.1	8.9	A0			
20 36.7	+50 53	Hld 39		Cyg	1881 1908	176		7.6		7.9	10.7	A0	14072		
20 36.9	−12 44	Σ 2699	AB	Cap	1829 1947	195		9.5		8.1	9.1	F0	14054		BDS 10334, H N 134 same
—			AC		1904	170		38.7					14054		
—			AD		1904	158		59.9					14054		
20 37.4	+75 36	hz 7		Cep	1974	223		0.6		7.3	10.2			o	VW Cep; A is variable [79]
20 37.5	+14 36	β 151	AB	Del	1873 1960					4.0	4.9	F5	14073	o	6 β Del
—		h	ABxC		1878 1924	116	121	27.7	22.8		12.9		14073		
—		Σ 2704	ABxD		1829 1931	344	327	32.5	39.1		10.8		14073		
—		Lbz	ABxE		1922 1960	270	271	102.7	106.5				14073		
20 37.5	+31 34	Σ I 53	AB	Cyg	1835 1927	176	176	178.1	180.8	6.9	7.0	A0 F0			48 Cyg
—			Aa		1911	1		71.2			12.9				
20 37.6	−47 17	h 5209	AB	Ind	1896 1914	199		66.0		3.1	11.9	G2			α Ind
—			AC		1900 1914	343		62.0			13.4				
20 37.7	+33 22	Σ 2705	AB	Cyg	1831 1976	261		3.1		7.4	8.4	K0	14078		
—			AC		1880 1909	341		183.0			11.2		14078		
—			AD		1953	3		53.8			12.9		14078		
—			AE		1953	42		74.7			12.1		14078		
20 37.8	+29 43	A 742	AxBC	Cyg	1904 1916	344		58.2		8.0		A2	14079		
—			BC		1904 1940	127	123	1.0	1.0	9.4	10.2		14079		
20 37.8	+60 45	Σ 2717	AB	Cyg	1832 1956	267	259	2.1	1.8	7.2	9.7	G0	14102		
—			AC		1898 1956	52	52	46.3	42.7		9.7		14102		
20 38.1	−76 54	h 5199		Oct	1918	210		28.9		7.7	12.7	G5			
20 38.3	−1 06	β 672		Aql	1878 1913	281	282	30.5	32.0	4.3	10.8	K0	14081		71 Aql
20 38.3	+48 04	Es 89		Cyg	1906 1912	200		16.3		6.6	11.5	G5	14100		
20 38.4	−21 00	Rst 5551a		Cap	1951	82		0.1		8.1	8.1	G5			Measures uncertain
20 38.7	+38 38	Σ 2708	AB	Cyg	1829 1926	355	327	10.8	34.2	7.0	8.7	G0			
—			AC		1878 1911	48	14	15.0	16.9		13.3				
—			AD		1880 1922	214	219	115.1	113.5		10.8				
20 38.9	+47 41	A 746		Cyg	1904 1944	147		2.1		7.6	13.0	A0	14119		
20 39.1	+10 05	OΣ 533	AB	Del	1852 1947	11	286	10.4	28.8	5.1	11.7	G5	14101		7 κ Del
—			AC		1865 1903	101		214.4					14101		
20 39.1	+15 50	β 288	AB	Del	1878 1934	168	159	7.9	6.8	6.0	12.5	B2	14106		
—			AC		1913	279		43.8			10.9		14106		
—			AD		1934	145		30.3			14.4		14106		
—			AE		1934	35		22.5			14.4		14106		
20 39.3	+29 06	Ho 136		Vul	1882 1939	6	357	2.5	2.4	8.0	11.5	F5	14117		
20 39.3	−14 57	Hu 200	AB	Cap	1900 1959					5.8	6.3	B5	14099	o	14 τ Cap
—			AC		1912	272		158.6			12.0		14099		
20 39.4	+0 29	h 2984	AB	Aqr	1879 1924	217	221	55.9	59.7	5.2	11.3	K0	14108		1 Aqr
—			AC		1879 1924	39	36	72.9	69.8		11.1		14108		
20 39.5	+38 23	Weisse 37		Cyg						7.9		A3			Single (IDS); see Σ 2708
20 39.6	+15 55	β 298	AB	Del	1891 1911	224		29.5		3.8	13.3	B8	14121		9 α Del
—		h 1554	AC		1878 1911	279		43.4			11.8		14121		
—		β 298	AD		1877 1959	150	155	48.0	46.5		12.8		14121		
—			AE		1891 1959	309	306	51.6	54.5		12.5		14121		
—			AF		1879 1911	114	115	80.7	79.4		10.6		14121		
20 39.6	+30 31	Σ 2711		Cyg	1831 1955	223		2.4		8.1	9.1	A0	14124		
20 39.6	+40 35	OΣ 410	AB	Cyg	1850 1959	23	10	0.6	0.8	6.8	7.1	B8	14126		
—			ABxC		1851 1939	70		69.0			8.9		14126		
—			ABxD		1923	327		104.6					14126		

α 2000	δ 2000	Star Name		Const	Years	Pos Ang		Separation		m₁	m₂	Spec	ADS	Orb	Notes
20ʰ39ᵐ7	+47°21′	A 748		Cyg	1904 1927	24°	°	1″.4	″	7.9	12.4	A0	14142		
20 39.9	+11 15	β pm		Del	1886 1908	9	8	194.6	191.9	6.4	10.4	F8			
20 40.0	−6 48	HdO 158		Aqr	1868	144		8.2		8.0	9.				
20 40.3	+3 26	OΣ 409 rej		Del	1866 1926	84		16.9		6.9	10.4	K0	14133		
—			AC		1915	334		65.3			8.4		14133		
20 40.3	−51 14	Hu 1617		Ind	1914 1959	124 124		0.5	0.2	8.5	9.0	F8			
20 40.6	+21 56	A 2795		Vul	1914 1958	219 295		0.2	0.1	7.8	8.0	B9	14148		Too close 1954,55,57,58
20 40.6	+29 48	Ho 137		Cyg	1885 1958	279 306		1.2	0.9	6.1	10.6	A0	14149		
20 40.7	−64 11	Hu 1616		Pav	1914 1937	83 76		0.8	0.8	8.1	9.3	K0			
20 40.9	−42 24	h 5211	AB	Mic	1880 1916	300		20.3		6.5	10.4	G5			
—		I 1121	BC		1916	248		6.3			12.8				
20 41.0	+32 18	Σ 2716	AB	Cyg	1830 1954	47		2.7		5.7	7.8	K0	14158		49 Cyg [80]
—			AC		1912	91		68.3			11.6		14158		
20 41.0	+39 05	h 612		Cyg	1902 1924	9		48.3		6.5	9.8	B9			
20 41.0	−14 31	Ho 135		Aqr	1883 1944	224		2.6		7.7	12.7	F0	14145		
20 41.4	+45 17	H N 73		Cyg	1879 1924	106		75.4		1.3	11.7	A2	14172		50 α Cyg, Deneb
20 41.7	−75 21	Δ 232		Oct	1826 1940	17		17.4		7.1	7.6	G5 G5			μ² Oct
20 41.8	+12 31	Σ 2715		Del	1830 1974	3	2	12.0	12.3	7.7	10.3	F8	14168		
20 41.8	+22 59	Ho 595		Vul	1896 1907	110		17.4		7.8	13.0	K0	14169		
20 41.9	+19 31	Cou 226	AB	Del	1967	4		0.3		8.0	8.3	A2			
—			AC		1967	340		13.5			11.5				
20 41.9	+20 43	β 673	AB	Vul	1878 1926	296		3.9		7.4	11.9	A3	14170		
—			AC		1901 1949	165 164		105.7	104.3		8.0	G5	14170		
20 42.2	+50 20	β 675	AB	Cyg	1878 1934	102		3.0		5.4	12.4	B3	14189		51 Cyg
—			AC		1878 1934	182		25.7			11.4		14189		
—			AD		1878 1924	329		32.8			11.4		14189		
—			AE		1916	240		72.3			9.4		14189		
20 42.3	+45 49	OΣ 411		Cyg	1845 1940	274 331		15.3	21.0	7.7	10.5	G0	14190		
20 42.3	+57 23	β 152		Cep	1876 1957	111 92		0.4	1.0	7.3	8.1	A3	14196		
20 42.6	+12 44	Σ 2718	AB	Del	1831 1954	86		8.5		8.0	8.0	F5 F5	14184		
—			AC		1889 1912	345		167.0			8.7		14184		
—			BC		1901	342		168.9					14184		
20 42.7	+52 38	A 1683		Cyg	1907 1957	32 40		0.2	0.1	8.5	9.2	B9	14204		
20 43.3	+23 00	h 1565		Vul	1892 1918	71		18.1		8.5	9.0	A2			
20 43.3	+25 50	β pm	AB	Vul	1910	228		16.9		7.0	14.5	F0			
—			AC		1910	208		39.7			12.5				
—			AD		1875 1910	103		387.1			9.1				
—			AE		1875 1910	286		442.8			9.4				
20 43.4	+25 36	Ho 138	AB	Vul	1881 1958	350 340		2.6	2.9	7.3	13.8	K5	14208		
—			AC		1881 1906	306		128.2			11.2		14208		
—			AxCD		1899	306		129.2					14208		
—			CD		1881 1958	329 353		2.7	3.1		11.8		14208		
20 43.5	+19 53	Σ 2721		Del	1830 1952	32 28		2.4	2.6	7.8	9.9	G5	14205		
20 43.6	+19 44	Σ 2722		Del	1830 1937	307		7.2		8.4	8.9	G5	14209		
20 43.6	−51 48	I 128		Ind	1900 1936	327		3.2		8.0	11.3	F0			
20 43.7	+46 19	Ho 140		Cyg	1885 1906	312		7.0		7.2	13.3	G5	14219		
20 43.8	+39 28	Sei 1230		Cyg	1896	128		25.5		7.8	10.6	K0			
20 44.0	−5 57	Rst 4679		Aqr	1938 1952	359		0.2		8.6	8.7	G5			
20 44.8	+19 43	Cou 424		Del	1969	239		5.1		8.0	13.0	K5			
20 44.8	−20 54	β 674		Cap	1877 1942	103 99		1.6	1.6	7.7	10.0	K0	14218		
20 44.9	+12 19	Σ 2723	AB	Del	1831 1960	86 116		1.5	1.2	6.9	8.7	A0	14233		Mag. of A is for Aa
—			Aa		1940 1953	334 43		0.1	0.1				14233		
20 45.1	+12 44	β 64	AB	Del	1876 1959					9.1	9.3	G	14238	o	
—		OΣΣ 209	AC		1874 1912	158		96.6			8.3	F	14238		Mag. given for C is for Cc
—		β 64	Aa		1891 1913	119		62.2			10.5		14238		
—		Franz	Ca		1899	198		62.9					14238		
—		Dom	Cc		1953 1957	55 24		0.3	0.2				14238		Measures uncertain
20 45.1	−50 30	I 17	AB	Ind	1895 1953	39		1.0		8.2	8.3	A0			
—		Δ 235	AC			sf		124.0			7.3				
20 45.2	−1 03	Leon 56		Aqr	1943	280		0.8		7.9	11.5	G5			
20 45.2	−35 10	See 428		Mic	1896 1936	185		4.4		6.9	10.9	G0			
20 45.3	+57 35	β pm		Cep	1894 1918	165 161		71.8	66.4	4.5	10.2	G0			
20 45.3	+61 50	β		Cep	1879 1957	34 66		100.5	51.7	3.4	11.1	K0	14276		3 η Cep; optical
20 45.4	−30 29	h 5218		Mic	1877 1935	192		9.8		6.6	11.0	F0			Very beautiful object (h)
20 45.5	+54 01	Hld 41	AB	Cyg	1881 1937	234		3.5		8.0	11.8	K0	14272		
—			AC		1881 1937	262		8.2			10.6		14272		
20 45.7	+30 43	Σ 2726		Cyg	1830 1955	57 67		6.6	6.0	4.2	9.4	K0	14259		52 Cyg
20 45.7	+50 40	OΣ 412 rej	AB	Cyg	1899 1901	282		25.5		7.1	13.1	F5	14273		
—		Hu	AC		1899	270		25.6					14273		
—		OΣ 412 rej	BC		1899 1901	186		5.0					14273		
20 46.2	+15 54	Σ 2725		Del	1829 1959	358 8		4.2	5.8	7.6	8.4	K0	14270	o	
20 46.2	+33 58	β 676	AB	Cyg	1878 1959	321 272		37.7	54.9	2.5	11.5	K0	14274		53 ε Cyg
—			AC		1959	265		78.1					14274		
20 46.5	−26 52	h 5220		Cap	1877 1951	354		18.0		7.0	9.2	F2	14258		
20 46.7	+16 07	Σ 2727		Del	1830 1976	274 268		11.9	9.6	4.5	5.5	G5	14279		12 γ Del

α 2000	δ 2000	Star Name		Const	Years	Pos Ang		Separation		m₁	m₂	Spec	ADS	Orb	Notes
20ʰ46.ᵐ7	+46°32′	Ho 143	AB	Cyg	1885	306°		1.″0		6.3	11.8	A2			Single in 1901 (Doo)
--		Kui 101			1937 1951	105		0.5		6.5	8.4	A2			Same as Ho 143? (IDS)
20 47.2	+34 22	β 677	AB	Cyg	1878 1934	121		9.9		4.9	9.9	K0	14290		T Cyg
--			AC		1890 1934	194 201		12.4	14.0		11.2		14290		
20 47.2	+42 25	OΣ 414	AB	Cyg	1848 1949	95		10.0		7.6	8.7	B9	14295		
--		Bot	AC		1958	16		105.6			9.7		14295		
20 47.2	+51 48	β 365		Cyg	1892 1902	285		14.5		7.9	11.2	F0	14302		
20 47.3	+45 35	Ho 280		Cyg	1888 1960	75 78		14.0	12.9	7.0	13.0	K5	14298		
20 47.3	−26 25	β 153		Cap	1876 1947	283 265		1.6	1.6	7.3	9.1	A2	14280		
20 47.4	+36 29	OΣ 413	AB	Cyg	1842 1960					4.8	6.1	B5	14296	o	54 λ Cyg
--		S 765	ABxC		1863 1955	105 105		85.6	84.6		9.9		14296		
20 47.4	−18 37	h 2995		Cap	1830	284		20.0		8.1	11.1	G0			
20 47.7	+42 04	β 268		Cyg	1875 1955	221 206		0.4	0.5	7.7	8.6	A0	14306		
20 47.8	+6 00	β 65		Del	1875 1958	186 194		1.6	1.6	5.6	9.2	A0	14293		13 Del
20 47.8	+47 50	Es 810		Cyg	1909 1935	42		18.3		5.6	14.6	K0	14318		
20 47.9	+52 24	Es 93		Cyg	1901 1935	273 302		7.4	12.2	6.3	11.4	G5	14322		
20 47.9	−27 45	Stone 53		Mic	1877 1885	177		17.3		7.5	10.5	G5	14288		
20 48.0	+27 27	β 66		Vul	1876 1959	159 168		1.2	1.3	8.5	9.0	F0	14312		
20 48.0	+39 17	A 1434	AB	Cyg	1934 1953	63 50		0.1	0.1	7.8	7.8	F5	14314		V367 Cyg; compos. sp. F5+A3
--			ABxC		1906 1960	255		2.3			13.7		14314		
20 48.2	+26 24	Σ 2728	AB	Vul	1831 1957	25 25		4.2	8.4	8.1	10.4	K5	14315		
--			AC		1905 1914	214		18.4			13.8		14315		
20 48.4	−18 12	S 763	AB	Cap	1878 1950	294		15.8		6.7	8.6	G5	14299		
--			CD		1903	159		9.4		10.3	10.5		14299		
20 48.5	+1 43	Rst 5471		Aqr	1946 1950	39		0.6		7.4	10.5	K0			
20 48.5	−43 59	Don 992		Mic	1932	271		4.3		5.1	15	A7			
20 48.7	+51 55	Σ 2732		Cyg	1831 1958	74		4.1		6.5	8.5	B2	14336		
20 48.7	−43 59	Don 993	AB	Mic	1932	330		11.6		7.7	15	F5			
--		h 5222	AC		1914 1932	277 275		27.2	27.5		12.5				
20 48.9	+46 07	h 1581		Cyg	1875 1934	174 174		20.5	21.3	4.8	10.8	B2	14337		55 Cyg
20 49.0	+39 48	Σ 2731	AB	Cyg	1830 1935	86		4.1		7.7	10.8	A0	14334		
--		Ho 596	AC		1895 1922	171		16.4			12.7		14334		
--			AD		1895 1922	249		18.5			12.9		14334		
20 49.2	+39 17	Ali 952		Cyg	1929	54		6.0		8.3	9.2				
20 49.3	+58 45	OΣΣ 211		Cep	1875 1923	262 264		115.2	107.7	6.7	7.9	B9 K0			
20 49.3	−18 02	h 3000		Cap	1905	244		31.6		6.4	12.9	K0			
20 49.6	+12 33	β	AB	Del	1901	21		65.8		6.0	14.1	F5			15 Del
--			AC		1890 1922	82		108.0			11.2				
--			AD		1922	280		180.7							
20 49.6	+50 08	Es 94	AB	Cyg	1901 1931	13		101.5		6.9	9.9	M	14345		
--			BC		1901 1928	79		2.5			10.4		14345		
20 49.9	+46 40	β 250		Cyg	1875 1934	8 7		20.3	19.2	6.3	11.3	B3	14350		
20 50.0	+5 33	OΣΣ 210		Del	1875 1960	122 125		81.8	79.9	6.3	8.8	K0			
20 50.0	−33 47	h 5224		Mic	1835 1933	166		20.5		5.0	10.0	G6			α Mic
20 50.1	+44 04	β pm		Cyg	1852 1907	45 46		86.0	75.5	5.0	11.0	A5			56 Cyg
20 50.1	−27 22	h 5226		Cap	1877 1950	68		18.6		7.0	9.0	K0 K	14335		
20 50.6	+30 25	OΣ 415		Cyg	1846 1955	235		3.4		8.2	10.2	F5	14356		
20 50.6	+30 55	β 67		Cyg	1875 1959	287 303		1.5	1.6	6.9	10.2	A2	14355		
20 51.1	+6 23	Σ 2730		Del	1830 1955	335		3.4		8.6	8.7	K0	14359		
20 51.1	+51 25	β 155	AB	Cyg	1876 1957	25 36		0.6	0.9	7.3	8.0	F0	14370		
--			AC		1885 1915	26 23		17.7	16.2		12.0		14370		
--			AD		1880 1909	48 48		199.8	198.3		10.9		14370		
20 51.4	−5 38	Σ 2729	AB	Aqr	1829 1959					6.4	7.2	F2	14360	o	4 Aqr
--			AC		1912	316		68.7			12.9		14360		
--			AD		1899 1921	329		131.3			9.4		14360		
20 51.6	−62 26	Rmk 26		Pav	1835 1959	101 87		2.6	2.3	6.5	6.5	A2			
20 51.7	−29 47	Rst 2164		Mic	1934 1945	317		1.0		7.3	11.3	K2			
20 51.7	−40 54	h 5228		Mic	1879 1919	104		32.1		7.2	9.5	K0			
20 52.1	−33 53	See 430		Mic	1896 1942	347		13.7		7.8	12.6	K5			
20 52.4	+20 08	Ho 144		Vul	1886 1959	168 352		0.4	0.3	8.1	8.1	F5	14379		Changing; quadrant in doubt
20 52.7	+7 21	Σ 2733		Del	1832 1975	145		40.2		8.5	8.7	A F			
20 52.8	−16 09	β 154		Cap	1875 1944	61		2.8		8.3	9.6	G5	14377		
20 53.0	+29 09	OΣ 417	AB	Vul	1847 1956	39 29		0.6	0.8	8.2	8.8	A0	14397		
--			AC		1847 1941	109		30.9			9.8		14397		
20 53.0	−23 47	h 3003		Cap	1876 1959	217 206		2.1	1.7	6.3	8.5	G5	14380		
20 53.2	+62 11	β pm	AB	Cep	1875 1910	247 270		98.7	85.3	8.6	9.0	K2 F8			
--			BC		1910	293		60.6			13.2				
20 53.5	+30 57	OΣΣ 212		Cyg	1875 1913	154		65.7		8.0	9.6	B9			
20 53.6	+19 36	Ho 597		Del	1895 1943	221		10.0		7.6	11.9	A3	14402		
20 53.6	+35 14	Ho 146		Cyg	1886 1957	56 52		0.4	0.3	8.7	8.7	B9	14404		
20 53.7	+21 16	J 2327	AB	Vul	1942	50		70.0		8.0	11.8	F0			
--			BC		1942			3.0			12.2				
20 53.7	+59 18	A 751		Cep	1904 1959					7.4	7.7	F2	14412	o	Spectrum composite F2+A2
20 53.9	+33 26	β pm		Cyg	1880 1909	58		187.3		5.5	9.5	K2			
20 54.0	+57 11	Es 135	AB	Cep	1902 1923	196		6.6		7.2	11.4	A0	14415		

α 2000	δ 2000	Star Name		Const	Years	Pos Ang	Separation	m₁	m₂	Spec	ADS	Orb	Notes
20ʰ54.ᵐ0	+57°11′	Es 135	AC	Cep	1923	117°	71.8″		11.5		14415		
20 54.1	+45 07	OΣ 422		Cyg	1851 1945	333	2.8	7.5	9.2	B9	14411		
20 54.4	+40 42	OΣ 420		Cyg	1848 1930	3	5.8	6.5	10.7	B8	14413		
20 54.4	+69 57	A 1216	AB	Cep	1904 1932	340 337	21.6 20.8	7.8	14.8	K0	14441		
--			AC		1904 1932	351 351	41.4 40.9		10.4		14441		
--			CD		1904 1932	91	0.8		11.4		14441		
20 54.7	+37 04	OΣ 419		Cyg	1847 1935	34	1.8	7.3	10.6	A0	14420		
20 54.8	+32 42	OΣ 418		Cyg	1842 1959	302 288	0.6 1.1	8.1	8.2	G0	14421		
20 54.8	-46 36	I 1429		Ind	1925 1959	140	0.4	8.0	8.4	F8			
20 54.8	-58 27	B 2487		Ind	1945	104	17.3	3.7	12.5	K2			β Ind
20 55.0	+28 05	β 367	AB	Vul	1876 1960			8.4	8.8	G5	14424	o	
--			AC		1875 1924	28 17	30.9 32.6		11.4		14424		
--			AD		1899 1916	93 90	30.9 28.6		13.4		14424		
--		A			1912	65	54.7	8.0	13.5		14424		
20 55.1	-52 07	I 18		Ind	1900 1935	2	4.2	7.0	10.2	G5			
20 55.3	+42 31	OΣ 423		Cyg	1853 1954	80	2.9	7.0	9.5	B9	14432		
20 55.6	+12 34	h 1592		Del	1879 1911	22 21	40.0 38.2	5.5	11.8	A2	14429		16 Del
20 55.7	+4 32	Σ 2735		Del	1829 1958	284	2.1	6.1	7.6	G0	14430		
20 55.9	+32 06	OΣ 421 rej	AB	Cyg	1867 1914	193	36.8	7.7	9.9	K0			
--		Hu	Aa		1898	78	30.6		12.8				
20 56.0	+27 35	Ho 460	AB	Vul	1892 1911	86	12.5	6.8	12.5	K0	14440		
--			AC		1892 1911	262	35.5		12.9		14440		
20 56.4	+50 44	β 1137		Cyg	1889 1934	344 348	6.9 7.2	5.8	13.5	F0	14460		
20 56.4	-59 16	I 129	AB	Ind	1900 1959	18 351	1.6 2.0	8.0	9.5	G0			
--			AC		1927	29	35.8		13.3				
20 56.7	+13 00	Σ 2736		Del	1830 1939	218	5.1	8.2	10.1	F2	14453		
20 56.7	+61 33	Σ 2740		Cep	1832 1934	329	4.3	8.1	10.4	F5	14478		
20 56.8	+42 54	β pm	AB	Cyg	1893 1920	355 349	116.4 110.7	7.6	9.8	K0			
--			AC		1894 1920	25 24	160.1 152.4		8.9				
--			BC		1908 1920	69 69	86.0 88.1						
--			CD		1893 1920	356 357	93.1 92.6		10.4				
20 56.9	-9 42	β 1034	AB	Aqr	1888 1940	165	2.1	5.7	11.4	K2	14449		7 Aqr
--		β			1912	67	176.6	5.7	9.2	K2	14449		
20 57.0	+23 40	A 175	AB	Vul	1898 1930	291	1.9	8.0	13.5	K2	14461		
--			AC		1891 1930	208	16.7		12.7		14461		
20 57.2	+0 28	Howe 55	AB	Aqr	1879 1928	72	26.2	6.1	9.8	K2	14457		
--		β	AC		1912 1959	119 116	36.2 35.4		12.3		14457		
20 57.3	+66 45	Milb 268		Cep	1922 1938	336 332	4.8 5.9	7.9	10.4	K	14496		
20 57.7	+58 50	A 756	AB	Cep	1904 1957	221	0.5	8.1	9.0	A0	14493		
--			AC		1918	197	55.1		9.3		14493		
20 57.7	+61 22	Σ 3133		Cep	1832 1968	101	3.6	8.1	9.6	G5	14495		
20 58.4	+10 50	β pm	AB	Del	1886 1908	162 161	197.4 198.2	5.5	10.1	K0			18 Del
--			BC		1907	223	67.2		10.9				
20 58.5	+16 26	Σ 2738	AB	Del	1830 1951	254	14.9	6.6	8.7	A0	14490		Spectrum composite A0+B9
--			AC		1889 1909	103	210.5		9.2	F5	14490		
--			BC		1901	101	223.9				14490		
20 58.5	+41 21	β pm		Cyg	1880 1909	352	168.7	7.2	10.3	K5			
20 58.5	+50 28	Σ 2741	AB	Cyg	1831 1958	36 28	1.9 1.9	5.9	7.2	B8	14504		
--			AC		1879 1925	356	139.4		10.6		14504		
20 58.7	-70 25	h 5231	AB	Pav	1836 1917	111 116	7.3 7.3	8.1	8.7	F2			
--		I 668	BC		1911 1927	322 325	1.0 1.0		9.2				
20 59.1	+4 18	Σ 2737	AB	Equ	1835 1960			6.0	6.3	F5	14499	o	1 ε Equ
--			ABxC		1833 1967	78 70	10.8 10.7		7.1		14499		
--			AD		1912 1924	279 280	77.1 74.8		12.4		14499		
20 59.2	+76 31	Hu 956	AB	Cep	1904 1952	104 94	0.9 0.9	9.7	10.7	G5	13296		
--			AC		1944	130			6.4	M	13296		
20 59.3	+15 34	OΣ 424	AB	Del	1846 1955	329 306	0.5 0.6	8.3	9.5	A3	14505		
--			AC		1891 1929	306	34.3		10.6		14505		
20 59.4	+36 26	Hu 764		Cyg	1904 1951	187 198	0.3 0.4	8.0	9.2	B9	14512		
20 59.8	+16 49	OΣΣ 213		Del	1875 1924	37	70.7	6.7	9.4	F2			
20 59.8	+47 31	Σ 2743	AB	Cyg	1831 1951	352	20.2	4.7	9.6	B0	14526		59 Cyg = V832 Cyg
--			AC		1842 1921	141	26.7		11.5		14526		
--			AD		1913	220	38.3		11.0		14526		
20 59.9	+40 16	β pm	AB	Cyg	1893 1908	217 217	132.7 137.1	6.6	9.9	F8			
--			AC		1908	145	208.4		10.3				
--			BC		1908	107	212.2						
--			Cc		1908	39	8.7		12.6				
21 00.1	+7 31	Kui 102		Equ	1934 1958			6.3	7.6	A5		o	
21 00.1	+48 41	β 1210	AB	Cyg	1890 1924	119	2.3	7.3	12.0	A0	14531		
--		OΣ 425	AC		1847 1924	28 27	12.3 14.9		10.6		14531		
--			AE		1898 1907	18	45.1				14531		
--			CD		1890 1924	134	4.3		11.0		14531		
21 00.2	+36 58	Ho 147	AB	Cyg	1885 1906	353	7.0	7.9	14.0	K5	14530		
--			AC		1897 1906	249	26.0		12.7		14530		
21 00.3	+61 30	h 1607	AB	Cep	1873 1935	93 82	8.4 11.5	7.5	10.7	K0	14544		

α 2000	δ 2000	Star Name		Const	Years	Pos Ang	Separation	m₁	m₂	Spec	ADS	Orb	Notes
21ʰ00ᵐ.3	+61°30′	h 1607	AC	Cep	1920	325°	30″.6		13.7		14544		
—			AD		1909	334	84.9		11.7		14544		
21 00.5	+19 20	β		Del	1880 1923	333 334	44.7 46.6	5.7	9.6	M			
21 00.7	−35 18	β 765		Mic	1891 1953	139 120	2.1 1.2	6.7	10.2	K0			
21 01.0	+40 00	A 1438		Cyg	1906 1958	224 241	0.2 0.3	8.7	8.8	A0	14543		
21 01.2	+46 09	OΣ 426		Cyg	1848 1958	167 162	2.5 2.5	5.4	9.6	B3	14549		60 Cyg
21 01.3	−32 15	Rst 5475		Mic	1945	93	26.2	4.7	13.7	G4			γ Mic
21 01.8	+39 16	Σ 2746		Cyg	1830 1959	276 310	0.9 1.0	8.0	8.6	F0	14558		
21 01.8	+44 11	Ho 600		Cyg	1896 1924	80	1.9	6.7	11.7	A2	14560		
21 01.9	+23 40	Cou 128		Vul	1966	129	0.2	7.5	7.7	F8			
21 01.9	+57 04	β 1139		Cep	1889 1898	140	2.0	6.7	13.2	B8	14574		
21 02.1	+56 40	Σ 2751		Cep	1831 1958	344 353	1.9 1.5	6.1	7.1	B9	14575		
21 02.2	+7 11	Σ 2742		Equ	1831 1955	225 218	2.6 2.8	7.4	7.4	F8	14556		
21 02.3	−43 00	Δ 236		Mic	1826 1951	73	57.4	6.5	6.9	G5 G5			
21 02.3	+39 31	H IV 113	AB	Cyg	1783 1933	300	18.5	6.5	10.6	K2	14567		
—			AC		1878 1905	250	25.9		11.8		14567		
21 02.4	+37 40	Σ 2747		Cyg	1830 1954	258 263	4.6 4.7	8.5	8.5	G5	14569		
21 02.5	+21 41	β 69	AB	Vul	1875 1957	315 329	1.0 0.7	8.3	9.1	F0	14570		
—			AC		1875 1915	238 240	78.4 77.3		7.7	K0	14570		
—		Ho 599	CD		1891 1915	155 158	19.5 18.8		13.7		14570		
21 02.8	+36 45	Hlm 38		Cyg	1924	262	11.8	7.5	11.0		14579		
21 02.8	+45 51	β 1138	AB	Cyg	1889 1959	189 180	0.3 0.2	7.2	7.2	B8	14585		
—			AC		1894 1908	313 315	167.8 165.6		8.3	K2	14585		
—		Kui 104	CD		1934 1958	104 97	2.8 3.1		13.7		14585		
—		β 1138	CE		1908 1958	68 250	17.0 3.9		13.3		14585		An optical pair (IDS)
21 03.0	+1 32	Σ 2744	AB	Aqr	1830 1959	190 137	1.5 1.3	6.7	7.2	F5	14573	o	
—			AC		1912 1924	102 102	87.6 89.1		12.5		14573		
21 03.1	−52 46	Rst 1080		Ind	1930 1948	52	4.6	7.8	13.7	G5			
21 03.2	+36 45	Sei 1380		Cyg	1896	261	11.2	8.1	11.4	A0			
21 03.2	−27 44	See 435		Mic	1897 1959	293	0.2	6.8	7.4	K0	14565		Too close after 1952
21 03.5	+24 00	Ho 281		Vul	1889 1924	299 300	13.1 13.7	7.4	13.4	M	14589		DY Vul (variable)
21 03.5	+29 06	β 445	AB	Cyg	1877 1925	110	4.8	7.0	11.5	K0	14590		
—			AC		1914	19	48.4		12.0		14590		
—			AD		1915	317	86.5		12.2		14590		
—			AE		1914	218	105.8		9.8		14590		
21 03.6	+53 58	Es		Cyg	1910	98	55.1	8.1	8.5	A2 A0			
21 03.7	+31 04	OΣ 427		Cyg	1846 1959	149	5.3	7.8	11.9	K2	14597		
21 03.8	−24 19	See 436		Cap	1897 1936	87 76	0.2 0.2	8.4	8.7	F5	14584		
21 03.9	+18 44	Brt 2489		Del	1904 1938	247	4.7	7.9	12.0	K0			
21 03.9	+41 38	OΣΣ 214	AB	Cyg	1875 1924	185	57.3	6.4	8.7	F2			
—			BC		1911	242	107.3		12.9				
21 04.0	−18 06	β 1211		Cap	1890 1944	344	0.6	8.5	9.1	F2	14588		
21 04.1	−5 49	Σ 2745		Aqr	1831 1955	192	2.8	5.9	7.3	F5 A3	14592		12 Aqr
21 04.1	−47 58	I 130		Ind	1900 1936	318	3.1	7.1	10.1	K0			
21 04.4	−19 51	φ 328		Cap	1951 1952	355	0.3	5.1	6.4	A4			22 η Cap
21 04.5	+7 46	β 269		Equ	1876 1950	253 230	1.1 1.0	8.1	10.1	F2	14600		
21 04.6	+5 30	β pm		Equ	1911	252	80.7	5.6	12.5	K2			3 Equ
21 04.6	+12 01	β 70	AB	Equ	1891 1899	239	78.8	7.6	9.8	K0	14601		
—			AC		1899	237	75.0		10.0		14601		
—			BC		1891 1899	96	5.3				14601		
21 04.6	+52 24	Σ 2757		Cyg	1831 1937	268	1.8	7.6	9.1	A0	14615		
21 04.9	+12 27	h 1608		Del	1879 1910	258	19.8	7.6	11.4	F2	14607		
21 05.0	+35 26	Σ 2753		Cyg	1831 1921	346 341	31.2 29.8	7.4	10.9	A0			
21 05.4	+5 57	β pm		Equ	1911 1924	275 278	35.6 34.9	5.9	12.3	F8			4 Equ
21 05.5	+62 10	Σ 2764	AB	Cep	1831 1937	301	7.0	7.8	8.3	A0	14634		
—		Hu 765	BC		1904 1957	43 35	0.5 0.6		8.8		14634		
21 05.7	+47 48	β 158		Cyg	1875 1943	315 310	10.4 11.1	7.5	12.0	A0	14627		
21 05.9	+34 59	Pop 24		Cyg	1970	244	7.1		2.5				
21 06.0	+21 37	Hall		Vul	1875	64		6.0	8.0				No 6th-mag. star here (IDS)
21 06.4	−41 23	See 437	AB	Mic	1897	56	44.1	5.6	14.5	K0			η Mic
—			AC		1879	88	132.9		8.4	F8			
21 06.4	−45 48	Hu 1333		Ind	1914 1936	319	1.7	7.7	10.4	K0			
21 06.6	+47 39	Es 32		Cyg	1899 1934	153	15.7	4.6	14.1	K5	14649		63 Cyg
21 06.7	−80 42	Gli 263		Oct	1890 1918	248	4.8	7.3	10.3	A5			
21 06.8	+34 08	Σ 2760	AB	Cyg	1829 1960	223 291	13.7 0.7	8.0	8.8	A2	14645		
—			AC		1903 1925	152	60.0		9.8		14645		
21 06.9	+38 45	Σ 2758	AB	Cyg	1830 1976			5.2	6.0	K5 K5	14636	o	61 Cyg; large p.m. [81]
—			AC		1907 1914	196 195	306.1 321.4				14636		
—			AD		1918	282	304.9		8.6		14636		
—			AE		1918	12	306.0		8.1		14636		
—			Aa		1918 1921	31 30	157.1 140.6		10.5		14636		
—			BC		1902 1914	198 200	277.7 337.2				14636		
21 07.0	+59 00	Σ 2766		Cep	1831 1937	249	5.0	8.6	8.8	F5	14669		
21 07.1	−25 00	See 439		Cap	1897 1933	186	26.2	4.6	11.7	M1	14632		24 Cap
21 07.2	+36 57	Sei 1407		Cyg	1895 1933	286 280	10.4 12.8	7.9	10.5	A0			Ali 438

α 2000	δ 2000	Star Name		Const	Years	Pos Ang	Separation	m_1	m_2	Spec	ADS	Orb	Notes
21 07.2	+44 40′	A 881		Cyg	1904 1937	218°	4″.2	7.4	11.9	B9	14664		
21 07.2	+47 44	Es 817		Cyg	1902	352	11.4	7.3	11.8	A0	14665		
21 07.2	−13 55	Σ 2752	AB	Aqr	1827 1975	145 171	5.2 4.7	7.2	11.2	K0	14638		
--		β 157	AC		1876 1959	82 301	21.4 11.7		12.5		14638		
21 07.4	−8 14	β 368	AB	Aqr	1876 1959	99 106	0.5 0.1	7.5	7.8	A0	14648		Very close after 1952
--			AC		1890 1898	29	12.1		13.5		14648		
--			CD		1890 1898	318 316	6.2 5.1		14.2		14648		
21 07.5	−0 10	Σ 2755		Aqr	1829 1931	83	24.3	6.8	10.4	M	14654		
21 07.5	+63 04	Mlr 14		Cep	1969	332	1.5	8.0	13.5	K0			
21 07.7	−5 34	β pm		Aqr	1908	296	42.0	7.6	12.5	G0			
21 07.9	−10 13	β 473	AB	Aqr	1877 1938	116	2.0	8.3	9.5	G0	14659		
--			AC		1898	357	25.4		11.7		14659		
21 07.9	−48 56	Rst 5477		Ind	1946	145	15.6	6.7	13.1	A0			
21 08.0	+5 09	OΣ 527		Equ	1846 1960	306 201	0.4 0.1	6.9	8.4	A2	14666	o	
21 08.4	−75 59	I 257		Oct	1898 1932	289 300	1.6 1.6	8.5	8.9	F8			
21 08.6	+30 12	Σ 2762	AB	Cyg	1829 1975	316 306	3.6 3.4	5.8	7.8	A0	14682		
--			AC		1865 1931	226	57.7		8.9		14682		
--			AD		1915	67	74.1				14682		
21 08.6	−21 12	h 3009	AB	Cap	1900 1909	66	67.0	5.3	11.0	A0			25 χ Cap
--			Aa		1900	111	39.2		15				
--			Bb		1900	89	18.2		12.2				
21 08.7	−37 15	B 1895		Mic	1930 1952	128	0.9	6.9	10.3	B8			
21 09.0	+47 16	Es 512		Cyg	1907	18	13.3	7.0	12.3	F2	14693		
21 09.0	−28 29	δ 153		Mic	1922 1941	280	3.5	7.0	11.7	G5	14674		
21 09.1	+38 44	S 779		Cyg	1824 1925	11 10	114.8 111.6	7.8	9.4	K2			
21 09.4	+5 02	OΣ 429 rej		Equ				7.9		K5			Not double (IDS)
21 09.4	−73 10	I 379	AB	Pav				6.5	6.5	K6			Measures uncertain
--		HdO 305	ABxC		1901 1931	136 131	8.1 7.9		13.4				
21 10.3	+10 08	Knott	AB	Equ	1867 1958	276 268	2.1 1.9	4.7	11.5	F0	14702		5 γ Equ
--		β 71	AC		1877 1925	10 5	41.3 47.7		12.5		14702		
--		Σ I 54	AD		1835 1922	153 153	366.2 352.5		5.9	A2	14702		
21 10.4	−54 34	h 5246		Ind	1856 1951	118 129	3.5 3.5	7.8	8.0	G0			
21 10.5	+19 58	Σ 2767		Peg	1830 1955	30	2.5	8.1	8.5	F5	14708		
21 10.5	+22 27	Σ 2769		Vul	1830 1954	300	17.9	6.9	7.7	A0 A0	14710		
21 10.5	+47 42	β 159	AB	Cyg	1876 1955	318 315	1.3 1.3	6.5	9.6	B5	14720		
--		OΣΣ 215	AC		1875 1921	190 189	134.1 135.0		7.9		14720		
--		Lv	AE		1916	337			14.5		14720		
--		β 159	Aa		1911 1934	147	14.5		13.3		14720		
--		Kui 105	Ab		1934		46.6		13.7		14720		
21 11.0	+9 33	Σ 2765	AB	Equ	1830 1955	86 81	3.0 2.8	8.4	8.6	A3	14715		
--			AC		1909	163	41.6		13.4		14715		
21 11.1	+36 18	Ho 283		Cyg	1887 1960	211 211	22.1 21.5	6.5	12.2	B1	14724		
21 11.4	−52 20	Hu 1626		Ind	1914 1959	208 152	0.7 1.2	7.3	8.5	F8			
21 11.6	−3 07	Σ 2770		Aqr	1828 1937	245	7.7	7.6	11.1	K0	14725		
21 11.6	+41 15	OΣ 431	AB	Cyg	1846 1946	117 124	3.2 3.1	8.5	8.9	F8	14733		
--			BC		1932	353	50.6		12.4		14733		
21 11.6	−23 07	h 5251		Cap	1876 1951	305	8.7	8.5	9.0	F5	14719		
21 11.8	+59 59	Σ 2780	AB	Cep	1831 1959	229 219	1.1 1.0	6.0	7.0	B2	14749		Webb notes pale ruby * nf
--			AC		1878 1909	212	120.6		8.7		14749		
21 12.0	−5 49	Σ 2768		Aqr	1829 1933	194	7.7	7.9	10.9	K0	14730		
21 12.0	+24 10	OΣ 430		Vul	1843 1959	216 206	1.3 1.3	8.2	10.2	F8	14738		
21 12.1	+45 42	β 160	AB	Cyg	1892 1915	154	57.4	7.9	11.4	B8	14745		
--			BC		1892 1915	114	6.2		11.6		14745		
--					1913	85	23.2	10.7	13.2		14745		
21 12.1	+76 19	OΣ 436		Cep	1848 1898	230	11.7	7.0	10.5	B9	14782		
21 12.1	−30 35	β 251		Mic	1876 1943	233	2.5	7.5	8.8	F5			
21 12.2	+10 39	β pm	AB	Equ	1892 1918	359 357	187.3 186.4	8.2	10.2	F8			
--			BC		1908	30	19.4		14.1				
21 12.2	+58 55	Arg		Cep	1918	194	36.6	8.1	9.3	B3			
21 12.3	−15 00	H I 47		Cap	1855 1957	324 313	2.9 3.7	8.0	8.0	G5	14736		
21 12.5	+38 34	Sei 1445		Cyg	1895 1921	24	27.3	7.3	10.6	B9			
21 12.6	−60 03	I 258		Pav	1900 1943	106 132	0.6 0.6	8.1	9.1	F0			
21 12.9	+30 14	β pm	AB	Cyg	1907	204	69.0	3.2	11.4	K0			64 ζ Cyg
--			AC		1907	300	90.4		11.1				
--			AD		1879 1907	67 66	101.2 102.8		12.4				
21 13.0	+2 48	β 1303		Equ	1900 1938	234	3.9	7.8	14.0	F5	14750		
21 13.3	+16 55	β 681		Peg	1878 1945	240	2.6	7.4	11.7	K2	14758		
21 13.4	+7 13	β 270	AB	Equ	1875 1958	355 5	0.6 0.3	7.4	9.7	F0	14759	o	
--			AC		1898 1916	32	32.5		12.8		14759		
--		S 781	AD		1824 1931	172	184.5		7.3	A2	14759		
21 13.5	+15 59	Hu 767		Peg	1904 1960			7.0	7.0	A5	14761	o	
21 13.7	+64 24	H I 48		Cep	1860 1959			7.0	7.2	G0	14783	o	
21 14.1	+58 18	Σ 2783		Cep	1831 1959	43 15	1.3 0.9	7.8	7.8	A0	14784		
21 14.2	+30 58	Cou 834		Cyg	1972	77	7.6	7.9	11.3				
21 14.3	+34 18	OΣΣ 216		Cyg	1875 1923	47	102.1	7.0	8.2	B8 A2			

α 2000	δ 2000	Star Name		Const	Years	Pos Ang	Separation	m₁	m₂	Spec	ADS	Orb	Notes
21ʰ14.3ᵐ	+41°09′	OΣ 432		Cyg	1847 1957	130° 119°	1″.2 1″.4	7.8	8.2	G5	14778		
21 14.5	+4 41	β 682	AB	Equ	1877 1943	103	5.6	7.5	12.0	K5	14774		
—			AC		1913	176	91.5		10.1		14774		
21 14.5	+10 00	OΣ 535	AB	Equ	1852 1960			5.2	5.3	F5	14773	o	δ Equ; followed by three
—		Σ 2777	AC		1833 1925	38 14	27.4 47.7		9.4		14773		small stars in line (Webb)
21 14.7	−0 50	A 883	AB	Aqr	1904 1958			8.1	8.3	A0	14775	o	
—		Σ 2775	AC		1825 1970	178	21.4		10.5		14775		
—			AD		1911	20	168.3		9.6		14775		
—			DC		1911	125	75.0				14775		
—			ab		1879	60	99.4	9.0	9.5		14775		
21 14.8	+38 03	AGC 13	AB	Cyg	1874 1959			3.8	6.4	F0	14787	o	65 τ Cyg
—			AC		1876 1914	260 228	15.7 29.4		13.3		14787		
—			AD		1851 1924	213 210	136.6 170.0		10.0		14787		
—			Aa		1912	288	71.7		12.3		14787		
—			Ab		1912 1921	185	89.9		11.7		14787		
—			Ac		1921	185	90.3				14787		
21 14.8	−25 55	h 3014		Cap	1856 1954	298	2.1	8.7	8.9	A3	14770		
21 15.0	−49 06	I 1437		Ind	1926 1942	352	1.2	7.0	10.0	K0			
21 15.2	+51 17	h 1631		Cyg	1902 1925	17	49.2	7.1	11.1	K0			
21 15.6	+37 15	Ho 285		Cyg	1888 1905	26	8.9	7.8	13.0	B8	14796		
21 15.6	+78 36	Σ 2796		Cep	1832 1951	43 43	24.6 25.9	7.2	10.3	A0	14845		
21 15.8	+5 15	Wils		Equ	1949			3.9		F8			8 α Equ [82]
21 15.8	−53 16	φ 329		Ind	1951 1959	82 138	0.2 0.1	6.6	6.6	A5		o	Retrograde motion
21 15.9	−83 16	h 5233		Oct	1900 1918	270	11.7	7.6	12.6	K0			
21 16.3	−9 13	Dem 24		Aqr	1876 1923	147 174	0.5 0.3	7.6	7.6	M			Duplicity doubtful (IDS)
21 16.7	−7 39	Σ 2781		Aqr	1828 1968	171	3.0	8.8	8.8	F2	14804		H I 90 probably same star
21 17.1	−80 21	I 670		Oct	1915 1933	11	0.5	7.9	8.5	B9			
21 17.3	+58 37	β 1140	AB	Cep	1889 1946	275	4.1	6.4	12.0	B3	14832		
—			AC		1912	271	69.9		12.6		14832		
21 17.4	+35 46	β 162	AB	Cyg	1875 1959	246	1.2	8.4	8.9	A2	14822		
—			AC		1880	138			12.9		14822		
21 17.6	+33 45	Ho 153		Cyg	1883 1959	111 124	0.8 0.9	8.1	9.1	A5	14829		
21 17.6	+39 45	Σ 2785		Cyg	1832 1937	235	2.7	8.0	9.9	F0	14828		
21 17.6	+82 31	Σ 2807		Cep	1837 1944	316 313	2.3 2.1	8.5	8.6	F8	14921		
21 17.9	+34 54	OΣ 433	AB	Cyg	1849 1958	220	15.1	4.4	10.0	B3	14831		66 υ Cyg
—			AC		1849 1958	177 181	21.2 21.5		10.0		14831		
—			BC		1946	137	13.6				14831		
21 18.1	+30 35	Ho 154		Cyg	1882 1933	207	3.6	8.1	11.3	A0	14835		
21 18.5	+80 21	Σ 2801		Cep	1832 1954	273 269	1.4 1.9	7.8	8.5	F8	14916		
21 18.6	+11 34	β 163	AB	Equ	1876 1959			7.2	8.7	G0	14839	o	
—			AC		1908	242	28.9		13.5		14839		
—			AD		1908	161	50.2		12.9		14839		
—			AE		1901 1908	9	82.6		10.8		14839		
21 18.6	+62 35	β	AB	Cep	1879 1907	23 22	209.2 206.8	2.4	10.2	A5	14858		5 α Cep
—			BxCD		1907	172	19.9				14858		
—			CD		1907	104	2.6	10.9	11.1		14858		
21 18.8	−26 37	See 443		Cap	1897 1954	280 289	1.8 1.4	7.8	10.3	F8	14834		
21 18.9	+41 02	Ho 601	AB	Cyg	1895 1924	179	17.3	6.2	12.7	A5	14849		
—			AC		1895	82	56.0		11.7		14849		
21 19.0	+39 45	OΣ 434 rej		Cyg	1866 1946	122	24.7	6.7	9.5	A0	14850		
21 19.1	+61 52	Es 137	AB	Cep	1902 1905	74	45.5	6.7	10.1	B0	14868		
—			BC		1902 1921	75	2.7		12.8		14868		
21 19.3	+58 37	Σ 2790	AB	Cep	1832 1967	45	4.5	5.7	10.0	K0 A0	14864		
—			AC		1898 1935	183	16.0		15		14864		
—			AD		1880	351	74.5		10.3		14864		
21 19.3	−13 03	Wils		Aqr	1949			6.7		A3			Checked by Wils; single
21 19.3	−47 03	h 5259		Ind	1913 1959	130 126	27.5 27.0	7.1	10.4	G5			
21 19.4	+38 14	Ho 286		Cyg	1886 1953	288 58	0.3 0.1	6.6	6.6	F2	14859		Rapid; single 1948,51,58
21 19.4	+52 19	Es 98	AB	Cyg	1901 1928	311	26.5	7.3	10.0	A0	14865		
—			AC		1901 1928	87	30.1		9.8		14865		
—			BC		1912	108	52.1				14865		
—			CD		1901 1907	256 262	4.9 7.3		14.3		14865		
21 19.7	+9 32	Σ 2786		Equ	1831 1955	185	2.5	7.2	8.3	A2	14856		
21 19.7	+33 15	Ho 155		Cyg	1884 1940	33	2.2	8.0	9.5	G0	14861		
21 19.7	+53 03	S 786		Cyg	1824 1913	301	48.5	6.8	8.9	K2			
21 19.8	−26 21	β 271	AB	Cap	1876 1959	227 255	2.2 3.2	6.6	9.5	G4	14847		
—			AC		1898 1909	73 72	74.6 81.7		12.0		14847		
—			AD		1917	62	247.2		12.0		14847		
—			AE		1918	37	179.4		12.0		14847		
21 19.9	+53 20	A 1695		Cyg	1907 1957	212 199	0.4 0.4	8.0	9.5	A0	14876		
21 19.9	−53 27	h 5258		Ind	1834 1957	305 275	3.5 6.0	4.5	7.0	A5			θ Ind
21 20.0	+52 59	Σ 2789	AB	Cyg	1832 1974	116 115	5.9 6.7	7.7	7.7	G5	14878		
—			AC		1913	172	126.6		11.8		14878		
21 20.1	−27 19	β 252		Cap	1876 1953	99 94	2.4 2.4	8.0	8.1	G5	14852		
21 20.2	+60 38	A 763		Cep	1904 1932	212	1.1	7.5	12.2	K5	14892		

α 2000	δ 2000	Star Name		Const	Years	Pos Ang	Separation	m₁	m₂	Spec	ADS	Orb	Notes
21ʰ20.4ᵐ	+60°41′	Es 138		Cep	1902 1906	265°	9.0″	6.7	13.0	F5	14898		
21 20.8	+32 27	OΣ 437	AB	Cyg	1845 1959	68 28	1.4 2.1	6.2	6.9	G5	14889		
--			AC		1913 1921	140	85.8		10.5		14889		
21 20.9	+3 08	β 838		Equ	1881 1958	90 130	1.3 1.6	8.1	10.0	F8	14880		
21 20.9	+41 27	Ho 156		Cyg	1885 1923	45 50	1.7 2.3	7.8	12.8	G5	14896		
21 20.9	-30 54	h 5263		Mic	1918	94	27.9	8.0	12.4	K2			
21 21.3	-84 19	h 5235		Oct	1872 1918	264	3.3	8.2	8.3	F0			
21 21.4	+2 54	OΣ 435		Equ	1848 1956	203 229	0.6 0.6	8.1	8.6	K0	14894		
21 21.4	+10 20	A 617		Equ	1903 1960			7.5	7.5	F5	14893	o	
21 21.8	+2 02	Σ 2787	AB	Aqr	1830 1931	20	22.7	8.0	9.3	A2	14901		
--			AC		1892 1908	94	70.5		10.6		14901		
21 21.8	+43 09	OΣ 438		Cyg	1847 1955	355	2.2	8.0	10.9	A0	14907		
21 22.1	+19 48	Σ II 11	AB	Peg	1835 1967	311	36.3	4.1	8.2	K0	14909		1 Peg
--			AC		1912 1921	20	75.0		11.9		14909		
21 22.2	-80 07	LDS 736		Oct		nf	164.0	7.6	10.6	G5			
21 22.4	-59 52	Rst 5160		Pav	1943	219	4.3	7.8	13.5	K2			
21 22.6	-55 47	Hu 1536		Ind	1913 1931	171	5.8	7.6	12.2	F2			
21 22.7	-87 03	I 337	AB	Oct	1900 1918	289 284	1.1 1.1	9.1	10.1	G5			
--		h 5192	ABxC		1871 1918	327 323	18.2 18.2		8.8				
21 22.9	+6 49	h 3023	AB	Equ	1878 1912	260 257	31.6 34.4	5.2	13.7	A0	14920		10 β Equ
--			AC		1877 1912	309 307	67.4 69.2		11.7		14920		
--			AE		1878 1912	276 274	86.3 89.2		12.2		14920		
--		β	CD		1877 1901	10 8	6.0 5.8		12.7		14920		
21 23.0	-51 57	I 132		Ind	1900 1948	298 287	1.4 1.2	7.8	9.6	G5			
21 23.6	+64 56	Σ 2798		Cep	1832 1921	146	6.4	8.2	10.1	B9	14952		
21 23.7	+55 18	A 1892		Cyg	1908 1959	347	0.7	7.9	9.9	A0	14945		
21 23.8	-6 35	S 788		Aqr	1824 1934	84 90	36.8 48.2	8.0	8.6	F8 K2			
21 23.8	+37 21	Σ I 55	AB	Cyg	1835 1925	303	365.4	6.6	6.6	K5 F8			
--			Aa		1880 1909	155	179.1		11.6				
--			Ab		1909	184	10.6		14.6				
--			Ac		1909	326	17.8		13.4				
21 23.8	+47 10	A 765	AB	Cyg	1904 1958	42 32	0.3 0.4	7.4	8.6	B2	14944		
--			ABxC		1904 1928	25 25	25.7 26.5		14.0		14944		
--			CD		1904 1928	330 339	6.4 7.4		14.5		14944		
21 24.0	+24 16	h 1641		Vul	1902 1923	313 309	51.8 53.9	5.7	10.5	F0			
21 24.1	+25 19	β 447	AB	Vul	1878 1934	330 320	8.5 9.3	6.2	12.2	A0	14943		
--			AC		1917	187	56.6		11.7		14943		
--			AD		1914 1960	82 82	57.1 56.6		11.8		14943		
--			AE		1914 1960	115 115	67.8 66.8		10.5		14943		
--			AF		1914 1960	216 217	80.5 82.4		10.7		14943		
21 24.2	-12 53	h 5517		Aqr	1877 1912	326 321	47.8 48.9	5.5	13.0	A5			18 Aqr
21 24.2	-25 33	β 1035		Cap	1888 1942	207	0.8	8.2	10.2	A0	14936		
21 24.4	-41 00	β 766	AB	Mic	1879 1959	300 267	0.9 0.5	6.4	7.0	A0			θ² Mic
--			AC		1879	66	78.4		10.5				
21 24.7	-34 58	B 528		Mic	1927 1933	170	6.9	7.0	13.0				
21 24.9	+49 19	Es		Cyg	1907 1924	162 164	19.2 20.3	6.6	12.1	A0	14962		
21 25.1	+9 23	β 164	AB	Equ	1875 1957	242 229	0.6 0.4	7.8	8.3	A2	14954		
--		Σ 2793	AC		1828 1968	242	26.6		8.5		14954		
21 25.2	+18 28	Cou 430		Peg	1969	239	0.5	7.5	8.4	A0			
21 25.3	+29 28	Cou 940		Cyg	1972	285	0.4	7.7	9.1	A0			
21 25.3	-71 48	h 5260		Pav	1917	271	44.6	6.1	12.0	K0			
21 25.5	+2 03	A 2289	AB	Aqr	1910 1958	304 273	0.2 0.1	7.9	8.9	A0	14960		
--		OΣ 439	ABxC		1850 1927	221	15.5		11.5		14960		
21 25.6	+58 14	β 1141		Cep	1889 1957	166	3.0	7.8	13.3	A0	14974		
21 25.8	+33 29	Goyal 56		Cyg	1934	4	13.0	8.0	10.5				
21 25.8	+36 40	S 790	AB	Cyg	1896 1923	30	33.0	5.9	10.3	B0	14969		69 Cyg
--			AC		1896 1922	98	54.0		9.0		14969		
21 26.5	+52 45	β 369	AB	Cyg	1891 1917	31	16.3	7.8	11.8	B9	14989		
--			AC		1905 1917	63	51.0		12.5		14989		
21 26.6	-46 04	I 1442	AB	Ind	1914	208	22.8	7.0	13.3	F8			
--		h 5267	AC		1914	182	44.0		10.0				
21 26.7	+13 41	Σ 2797		Peg	1830 1954	216	3.3	7.3	8.8	A2	14977		
21 26.7	+37 31	Sei 1516		Cyg	1895	339	4.1	8.1	9.6	K0	14990		
21 26.7	-22 25	See 446		Cap	1897 1933	13	21.3	3.7	12.3	G4	14971		34 ζ Cap
21 27.0	-42 33	MlbO 6		Mic	1879 1959	144 147	2.9 2.9	5.6	7.9	A3			
21 27.4	+59 45	OΣ 440		Cep	1848 1935	189 181	12.4 11.4	6.4	10.7	M	14998		
21 28.7	+70 34	Σ 2806		Cep	1832 1975	250 249	13.6 13.3	3.2	7.9	B1	15032		8 β Cep; a Cepheid var.
21 28.8	+77 56	Hu 771		Cep	1904 1921	195	2.6	7.3	11.6	K0	15054		
21 28.9	+11 05	Σ 2799	AB	Peg	1831 1959	333 276	1.4 1.6	7.5	7.5	F2	15007		
--			AC		1906 1912	338 336	137.1 136.2		9.3		15007		
21 29.0	+22 11	h 1647	AB	Peg	1892 1924	177	41.0	6.0	8.7	M3			
--			AC		1892 1926	126	41.2		10.4				
21 29.1	+44 55	β 448		Cyg	1876		2.0	7.0	11.0	B9			Single
21 29.3	+67 03	Hu 964		Cep	1904 1922	277	1.5	6.9	12.6	A2	15040		
21 29.9	+23 38	β 685		Peg	1878 1934	332	29.8	4.6	11.6	K5	15027		2 Peg

α 2000	δ 2000	Star Name		Const	Years		Pos Ang		Separation		m₁	m₂	Spec	ADS	Orb	Notes
21ʰ29ᵐ.9	+52°56'	Σ 2803		Cyg	1832	1914	290°	287°	23".2	24".8	7.2	10.3	A0	15035		
21 30.4	−60 12	h 5270		Ind	1914	1916	54		27.4		7.6	11.6	G5			
21 30.6	+42 13	OΣ 441		Cyg	1847	1922	321		6.8		7.9	10.6	A2	15045		
21 30.7	−38 38	B 530		Gru	1926	1942	36		1.0		7.6	10.6	F5			
21 31.0	−36 33	B 1008	AB	Gru	1927	1952	31	331	0.3	0.1	8.3	8.5	F2			
−−		h 5275	ABxC		1919		201		40.8			11.3				
21 31.4	+40 04	Ho 161		Cyg	1881	1943	358	6	2.8	2.9	7.4	11.4	F5	15057		Slow direct motion?
21 31.4	−19 14	See 449		Cap	1897	1943	197	190	1.8	2.0	7.3	14.1	F0	15046		
21 31.5	+48 17	A 771		Cyg	1904	1955	66		0.3		8.5	8.7	A0	15058		
21 31.6	−5 34	H V 76	AB	Aqr	1879	1947	319	321	34.3	35.4	2.9	10.8	G0	15050		22 β Aqr
−−		β 73	AC		1879	1915	185	186	54.5	57.2		11.4		15050		
21 31.8	+33 49	Σ 2802		Cyg	1830	1954	9		3.8		8.5	8.5	A5	15060		
21 32.0	+34 12	Ho 603	AB	Cyg	1896	1906	252	251	80.6	81.6	8.2	10.2	F0	15066		
−−			BC		1896	1933	272		3.4			11.7		15066		
21 32.2	−33 57	B 1009		PsA	1927		59		6.8		6.0	13.3	A2			6 PsA
21 32.5	+4 52	h 3032		Peg	1903	1929	93		17.3		7.3	12.0	F0	15067		
21 33.0	+20 43	Σ 2804	AB	Peg	1828	1974	314	350	2.9	3.1	7.6	8.3	F5	15076		
−−			AC		1912	1924	106		97.1			11.6		15076		
21 33.3	−80 02	h 5262		Oct	1836	1918	94		24.4		6.5	11.3	A0			
21 33.6	+60 20	β 166		Cep	1875	1938	259		1.3		7.9	10.7	A0	15094		
21 33.7	−16 12	A 2096		Cap	1909	1959	60	38	0.9	0.8	7.1	10.5	F5	15080		
21 34.0	+61 48	OΣ 442		Cep	1847	1959	11	350	0.6	0.5	8.6	8.8	A2	15103		
21 34.1	−4 22	Ho 288		Aqr	1887	1906	278	278	17.0	18.9	6.8	13.3	A2	15085		
21 34.2	+64 00	Es		Cep	1921		228		13.0		7.8	12.3	A0	15108		
21 34.4	+66 44	Σ I 57	AB	Cep	1836	1924	26		180.5		7.0	7.2	K0 K2			
−−			Aa		1909	1924	345		123.7			11.0				
21 34.5	+9 30	Σ 3112		Peg	1831	1952	238		7.0		7.9	9.7	G0	15092		
21 34.5	+59 06	Σ 2810		Cep	1831	1940	290		17.0		8.4	9.4	F0	15113		
21 34.7	−7 24	A 296		Aqr	1901	1925	58		2.7		8.0	14.0	F0	15093		
21 35.2	+21 24	β 74		Peg	1874	1956	320	329	1.5	1.3	7.6	9.5	F8	15109		
21 35.4	+24 27	Hu 371		Peg	1901	1960	163	277	0.2	0.2	6.7	7.0	A3	15115		
21 35.8	+31 01	h 939	AB	Cyg	1900	1904	173		14.8		8.0	12.0	K0	15120	o	
−−		Σ 2808 rej	AC		1879	1905	291		40.7			10.2		15120		
−−		h 939	CD		1903		344		15.3			12.7		15120		
21 36.2	+30 03	β 167		Cyg	1876	1916	89		2.1		6.5	10.9	K0	15126		
21 36.2	+50 30	h 1669		Cyg	1902		240		15.6		7.2	11.4	B9	15135		
21 36.2	−26 10	Rst 1098		PsA	1930	1940	8		18.4		5.7	13.9	A3			8 PsA
21 36.6	+54 49	A 1699		Cyg	1907	1932	97	97	4.2	4.5	7.9	14.9	A5	15140		
21 36.9	+50 32	Es 34	AB	Cyg	1899	1956	140	132	2.7	2.8	8.5	9.2	A0	15145		
−−			AC		1899	1956	70		39.6			9.0		15145		
21 37.1	−19 28	H VI 6		Cap	1877	1921	47		68.1		4.7	9.5	B5			39 ε Cap
21 37.5	+75 14	OΣ 449		Cep	1848	1946	122		1.3		8.1	10.1	A0	15180		
21 37.6	−0 23	Σ 2809		Aqr	1828	1923	163		31.1		6.3	8.7	A2	15142		
21 37.7	+6 37	Σ I 56	AB	Peg	1834	1934	349		39.2		6.0	8.3	A0	15147		3 Peg
−−			AC		1911	1959	118	119	91.0	88.3		13.2		15147		
21 38.0	+48 29	Es 102		Cyg	1908	1911	36		11.4		7.8	9.7	B8	15158		
21 38.5	+5 46	h 941		Peg	1878	1909	345	336	25.9	26.5	5.7	11.7	F0	15157		4 Peg
21 38.5	+43 17	Dawes 15		Cyg	1860	1935	72		1.5		7.9	10.8	K0	15167		
21 38.8	+55 48	β 687	AB	Cep	1878	1958	4		0.7		8.3	9.3	A0	15178		
−−			AC		1909	1913	204		54.4			9.8		15178		
21 39.0	+20 36	OΣ 444		Peg	1850	1916	276		8.0		8.1	11.1	F5	15171		
21 39.0	+57 29	β 1143	AB	Cep	1889	1935	324		1.6		5.6	13.3	O5	15184		
−−		Σ 2816	AC		1832	1969	121		11.7			7.7		15184		
−−			AD		1832	1969	339		19.9			7.8		15184		
−−		Gui	CD		1900		325		30.2					15184		
−−		Fle			1956		359		53.3		7.9	11.0		15184		
21 39.3	−30 18	See 451		PsA	1896	1928	257		12.0		7.5	12.5	M			
21 39.5	−0 03	β 1212	AB	Aqr	1890	1960					7.2	7.6	F8	15176	o	
−−			ABxC		1891	1959	141	160	44.5	37.6		10.8		15176		
21 39.5	+41 44	β 449	AB	Cyg	1876	1915	19	14	6.8	6.2	8.2	12.8	K0	15186		
−−		OΣ 447	AC		1783	1937	158	174	13.9	14.3		12.2		15186		
−−		β 449	AD		1876	1915	247		17.7			11.3		15186		
−−		OΣ 447	AE		1848	1923	45		29.0			8.8	K0	15186		
−−		Fox	EF		1920		47		43.5			10.8		15186		
21 39.5	+57 29	Stein 2582		Cep	1906		236		2.8		8.0	11.1	B3			
21 39.6	+2 15	β pm		Aqr	1911		46		132.6		5.3	11.6	K0			25 Aqr
21 39.6	+47 12	h 1676		Cyg	1907	1919	135		29.0		8.0	9.3	K0			
21 39.9	+39 31	A 1445	AB	Cyg	1906	1927	281		1.5		6.8	12.8	A3	15191		
−−		h 1675	AC		1828		263		15.0			12.8		15191		
21 39.9	+49 08	Es 825	AB	Cyg	1909		253		9.6		7.4	11.7	K2	15201		
−−			AC		1909		286		55.1			8.7		15201		
21 40.2	+43 16	AC 20	AB	Cyg	1860	1958	324		2.7		5.1	10.4	K5	15208		75 Cyg
−−		OΣΣ 221	AC		1875	1924	255	254	54.4	57.9		9.3		15208		
21 40.4	+57 35	Σ 2819		Cep	1832	1951	57		12.4		7.5	8.5	F5	15214		
21 40.7	+54 19	Es 35	AB	Cyg	1899		224		11.1		6.8	11.5	M	15220		A is RU Cyg (variable)

α 2000	δ 2000	Star Name		Const	Years	Pos Ang	Separation	m_1	m_2	Spec	ADS	Orb	Notes
21ʰ40ᵐ7	+54°19′	Es 35	AC	Cyg	1899	29°	18."6		10.2		15220		
21 41.0	+29 21	OΣ 448		Cyg	1845 1959	248 213	0.7 0.7	8.4	9.4	G0	15215		
21 41.1	+35 05	Ho 164	AB	Cyg	1882 1942	62 67	3.0 3.7	8.8	8.8	K0	15219		
--			AC		1893 1923	238	25.3		12.0		15219		
21 41.1	-57 44	B 2497		Ind		232	7.0	8.0	13.0	K0			
21 41.3	+39 28	β 274		Cyg	1875 1935	181	3.4	7.9	11.0	K0	15224		
21 41.5	-14 50	Ho 464		Cap	1893 1908	104	17.5	7.1	11.4	F0	15216		
21 41.6	+40 48	S 796	AB	Cyg	1879 1925	230 232	64.3 62.1	6.1	10.1	A0			76 Cyg
--			AC		1925	278	137.6						
21 42.0	+18 56	Σ 2818 rej		Peg	1899 1934	21 22	20.3 22.3	7.7	9.9	K0			
21 42.0	-23 16	See 454		Cap	1897 1954	199 205	5.2 5.5	5.3	11.5	G9	15223		41 Cap
21 42.3	+5 54	Hu 280		Peg	1900 1958	138 270	0.2 0.2	8.4	8.7	A5	15236	o	
21 42.4	+41 05	Kui 108		Cyg	1937 1960			6.4	6.6	A0		o	77 Cyg
21 42.5	-37 56	h 5288		Gru	1879 1952	60	19.7	7.6	9.3	F0			
21 42.6	+41 03	β 688	AB	Cyg	1878 1959	209 210	0.4 0.3	8.3	8.3	F0	15251		Round 1951-55
--			AC		1914	34	29.3		11.9		15251		
21 42.6	+42 26	Σ 2820 rej		Cyg	1903 1921	232	16.3	7.4	9.8	A0	15249		
21 43.1	+62 13	Es 1856		Cep	1920	52	5.4	8.0	12.0		15269		
21 43.4	+38 17	S 799	AB	Cyg	1824 1925	59 60	153.2 151.2	5.6	6.9	A0 A0			
--			AC		1904 1917	319	134.1		12.6				
--		Kui 109	Aa		1937 1948	151 159	1.4 1.5		11.0				
--		S 799	BD		1905	248	30.8						
21 43.5	+58 47	β 690	AB	Cep	1878 1946	260	19.5	4.1	12.3	M	15271		μ Cep; A is H's so-called
--			AC		1878 1924	299	41.2		12.7		15271		Garnet Star (variable)
21 43.9	+27 51	Ho 166		Peg	1886 1959			8.1	8.2	F5	15267	o	
21 44.0	-26 31	Howe 57		PsA	1876 1954	300 297	1.6 1.6	8.3	9.6	F0	15257		
21 44.0	-57 20	Jc	AB	Ind	1845	5	152.2	6.8	7.0	G0 G0			
--			AC		1845	221	205.4		7.9	K0			
21 44.1	+7 09	OΣΣ 222		Peg	1874 1922	258	87.8	7.9	9.2	F2 F5			
21 44.1	+26 32	β pm	AB	Peg	1910	319	30.9	8.2	12.9	G5			
--			AC		1875 1910	261 261	539.8 547.0		9.8				
--			AD		1875 1910	291 290	414.0 423.1		10.0				
--			CD		1875 1910	32	275.6						
21 44.1	+28 45	Σ 2822	AB	Cyg	1780 1971			4.8	6.1	F5	15270	o	H III 15
--			AC		1878 1924	263 277	35.3 48.6		11.5		15270		
--			AD		1823 1948	61 52	217.4 199.0		6.9	A5	15270		D is the A star of Es 521
--			BD		1902 1909	54 53	206.1 204.7			A5	15270		
21 44.2	+9 52	S 798	AB	Peg	1879 1910	325	81.8	2.4	11.2	K0	15268		8 ε Peg, Enif
--			AC		1874 1913	322 320	140.4 142.5		8.4		15268		
21 44.3	+28 47	Es 521		Cyg	1907 1938	270 276	14.2 16.2	6.9	13.3	A5	15275		
21 44.3	+60 55	Hu 969		Cep	1904 1922	325	2.6	7.5	12.5	G5	15288		
21 44.6	+25 39	β 989	AB	Peg	1880 1960			4.7	5.0	F5	15281	o	10 κ Peg
--		Σ 2824	ABxC		1831 1958	308 292	11.0 13.8		10.6		15281		
21 44.8	+3 00	β 689		Aqr	1878 1957	241	1.8	7.7	10.9	K2	15278		
21 44.9	+62 28	Mlr 16		Cep	1969	33	17.0	6.0	9.5	B2			
21 44.9	-33 02	I 1051		PsA		290	20.0	4.3	11.3	A0			9 z PsA
21 45.6	+33 07	Cou 1187		Cyg	1974	268	4.4	7.9	11.4	F5			
21 45.8	+15 46	OΣΣ 224		Peg	1875 1920	5	58.4	8.1	9.7	G5			
21 46.5	+6 32	OΣ 450 rej		Peg	1866 1914	249	42.6	7.8	10.6	F5			
21 46.5	+22 10	Ho 465	AB	Peg	1893 1923	246	42.6	7.1	9.1	A2	15311		
--			BC		1893 1923	80 79	3.6 4.7		10.9		15311		
21 46.9	+0 51	Σ 2825		Aqr	1827 1960	100 129	1.1 0.7	8.4	8.6	F2	15313		
21 47.0	+47 59	A 773		Cyg	1904 1927	200	3.1	7.7	12.1	F0	15329		
21 47.0	-16 08	h 3056	AB	Cap	1901	94	69.1	2.9	15	A5	15314		49 δ Cap
--			AC		1901 1910	302 302	115.2 118.9		12.6		15314		
21 47.4	-13 09	Howe 58	AB	Cap	1890 1933	105 94	0.6 0.7	8.3	9.4	F0	15322		
--		Σ 2826	AC		1829 1932	82 82	4.3 3.9		9.0		15322		
21 47.6	-17 18	β 1036		Cap	1888 1944	206	4.7	7.3	10.3	B9	15325		
21 47.7	-30 54	φ 330	AB	PsA	1951 1959	37 6	0.2 0.1	5.7	5.8	A2			10 θ PsA
--		h 5296	ABxC		1877 1918	339	35.6		11.2				
21 48.3	-47 18	BrsO 15		Gru	1836 1931	14 356	30.3 55.0	5.7	8.7	G5			
21 48.6	-6 15	h 1691		Aqr	1877 1936	277 270	14.7 13.5	7.9	10.8	A3	15337		
21 48.7	-65 30	I 19		Ind	1900 1947	342 320	1.3 1.3	7.3	8.6	F2			
21 49.1	+66 48	Σ 2836		Cep	1832 1926	153	11.8	6.5	9.5	F2	15366		
21 49.2	+50 31	Σ 2832	AB	Cyg	1832 1946	213	13.1	7.8	8.3	A0	15355		
--			AC		1912	319	45.9		9.4		15355		
--			BC		1912	334	51.1				15355		
21 49.4	+64 54	Stein 1054		Cep	1903	257	11.9	8.0	10.2				
21 49.6	+36 06	Ho 169		Peg	1882 1926	136	3.2	8.0	12.0	F8	15356		
21 49.7	+65 14	Mlr 74		Cep	1970	274	0.7	7.8	10.0	M			
21 50.1	+17 17	Cou 14		Peg	1959 1960	55	0.4	5.5	7.5	F2		o	13 Peg
21 50.1	+31 51	β 692	AB	Peg	1878 1934	10	2.7	7.3	10.8	K0	15364		
--		Fox	AD		1913	144	43.0				15364		
--			AE		1913	172	52.0				15364		
--		β 692	BC		1878 1912	299 296	36.9 37.6		10.8		15364		

α 2000	δ 2000	Star Name		Const	Years	Pos Ang		Separation		m₁	m₂	Spec	ADS	Orb	Notes
21ʰ 50ᵐ.1	+34° 49′	h 1697		Peg	1876 1928	258°	°	9″.1	″	7.7	11.4	K5	15367		
21 50.5	+39 25	Ho 170		Cyg	1886 1960	162 220		0.2	0.3	8.8	8.8	A3	15375		
21 50.7	+22 15	Ho 467	AB	Peg	1893 1959	181 210		1.0	2.1	8.1	10.3	K2	15373		
—			AC		1892 1905	339 338		39.6	40.1		12.0		15373		
21 50.8	−27 56	h 3059		PsA	1877 1930	253		25.2		7.4	10.5	B9	15365		
21 50.9	−20 32	See 460		Cap	1897 1959	115 144		0.5	0.5	8.2	8.6	F8	15371		
21 50.9	−82 43	h 5278		Oct	1835 1946	83 70		3.1	3.1	5.4	7.7	G0			λ Oct; sp. compos. G0+A3
21 51.0	+32 39	Es 382	AB	Peg	1906 1927	316		10.8		8.5	14.8	A0	15377		
—			AC		1906	320		60.2			8.5	K0	15377		
21 51.0	+61 37	OΣ 451	AB	Cep	1847 1967	221		4.3		7.5	8.5	A2	15390		
—			CD		1868	131		3.2		9.3	10.2		15390		2′ north of brighter pair
21 51.6	+19 50	h 947	AB	Peg	1879 1924	93 95		20.3	19.4	5.8	9.1	B9	15383		
—			AC		1879 1924	325 320		24.1	23.8		11.1		15383		
21 51.6	+65 45	Σ 2843	AB	Cep	1831 1958	134 145		2.4	1.5	7.1	7.3	A2	15407		
—			AC		1912 1939	276		56.0			9.9		15407		
21 51.7	+19 18	Σ 2834	AB	Peg	1830 1954	289 295		4.1	4.3	6.9	10.2	F2	15386		
—		β	AC		1908	121		26.0			12.5		15386		
21 51.8	+64 54	Σ 2844 rej		Cep	1906 1912	261		11.6		7.2	9.2	K0	15408		
21 51.9	+9 05	Σ 2833		Peg	1829 1920	338		9.0		8.0	10.8	A5	15392		
21 51.9	+42 21	Ho 172	AB	Cyg	1886 1976	87 83		10.5	10.6	7.8	12.3	G5	15400		
—			AC		1895 1911	52 50		51.7	55.0		12.3		15400		
—			AD		1911	36		189.7			9.0		15400		
21 52.0	+55 48	Σ 2840	AB	Cep	1832 1958	194 196		20.0	18.3	5.5	7.3	B3 A	15405		A splendid pair (Webb)
—			AC		1912	348		55.2			13.2		15405		
21 52.2	−85 50	h 5261		Oct	1870 1918	204 201		5.1	5.1	8.5	8.8	F5			
21 52.4	+63 05	Σ 2845		Cep	1832 1954	172		2.1		8.4	8.5	B3	15417		
21 53.1	+68 06	OΣΣ 226	AB	Cep	1876 1923	245		76.2		7.5	9.5	K0 A			
—			AD			176		45.0			14.3				
—			BC		1913 1922	149		85.1			15				
21 53.3	−46 50	φ 374		Gru	1960	135		0.1		8.2	8.2	F8			
21 53.4	−14 11	Rst 4092		Cap	1936 1943	28		1.1		8.0	11.5	A5			
21 53.4	−78 35	I 672		Oct		308		15.0		7.5	13.5	K2			
21 53.6	−10 19	φ 358		Cap	1959	129		0.1		7.0	7.5	B9			
21 53.8	+62 37	S 800	AB	Cep	1824 1970	146		62.6		7.0	8.7	B3 B8	15434		EM Cep (variable)
—		Es 144	Aa		1902	281		19.8			13.1		15434		
—			Bb		1902	43		22.4			13.6		15434		
21 53.8	−20 00	β 168	AD	Cap	1868	103		35.6		8.0	14.0		15419		
—			AxBC		1868 1953	72		5.7				F8	15419		
—		Hu 380	BC		1901 1958					10.2	10.3		15419	o	
21 54.0	+44 03	A 620		Cyg	1903 1958	247 271		0.2	0.2	8.6	8.7	G5	15435		
21 54.3	+19 43	Σ 2841	AB	Peg	1829 1958	110		22.3		6.4	7.9	K0	15431		The mag. of B is for BC
—		Cou 432	BC		1969	29		0.2		8.6	8.8		15431		
21 54.6	−3 18	Σ 2838		Aqr	1829 1959	185 184		21.6	17.6	6.3	9.1	F8	15432		Webb notes * stream, np
21 54.9	−11 59	h 616		Cap	1879 1918	274		30.9		8.1	10.3	A0			
21 55.2	−61 53	HdO 296		Ind	1901 1959					6.6	6.7	F0		o	
21 55.5	+10 53	β 75	AB	Peg	1875 1960					8.5	8.9	G5	15447	o	
—			ABxC		1912 1924	213 212		48.5	46.4		12.5		15447		
21 55.5	+52 31	OΣ 456	AB	Cyg	1847 1960	26 35		1.4	1.5	8.4	8.6	F2	15460		
—			AC		1881 1912	187		25.3			10.0		15460		
—			AD		1881 1912	283		35.2			9.2		15460		
21 55.5	+65 19	OΣ 457		Cep	1848 1954	247		1.4		5.9	8.1	B2	15467		
21 55.6	+35 05	Sei 1544		Peg	1894	216		21.1		7.9	10.9	A0			
21 55.6	−21 08	h 3065	AB	Cap	1904	133		39.6		7.2	11.5	M3			
—			AC		1904	149		57.6			10				
21 55.9	+59 50	OΣ 537		Cep	1856 1959	194 205		2.6	2.1	7.9	11.0	A0	15476		
21 56.0	−30 55	h 5307	AB	PsA	1918	161		33.3		8.0	11.4	F5			
—			BC		1918	159		20.2			12.8				
21 56.1	+24 20	OΣ 454	AB	Peg	1850 1955	278		7.2		8.2	10.2	A0	15461		
—			AC		1895 1915	265		43.2			11.9		15461		
21 56.2	−6 59	β 693		Aqr	1878 1959	54 48		0.9	0.9	7.5	10.0	B9	15459		
21 56.5	+59 48	OΣ 458	AB	Cep	1851 1960	349		0.8		6.9	8.4	A0	15481		
—			AC		1878 1905	31		22.7			12.1		15481		
21 56.5	−53 03	B 2498	AB	Ind	1933	251		2.4		8.0	13.0	A3			
—		h 5302	AC		1880 1934	349 352		12.7	12.2		11.0				
21 56.6	−31 23	B 1014		PsA	1928 1932	145		2.5		7.9	12.2	A0			
21 56.7	+63 38	Wils		Cep	1949 1950	147 174		0.1	0.1	4.9		M			VV Cephei
21 56.9	−35 22	B 1394		PsA	1930 1944	302		0.5		7.6	9.2	A0			
21 57.3	+61 18	β 275		Cep	1876 1958	183 174		0.3	0.4	7.8	7.8	B3	15499		
21 57.5	+4 09	OΣΣ 225		Peg	1875 1922	286		75.1		7.0	9.3	F5			
21 57.8	+61 01	Mlr 17		Cep	1968	290		0.2		8.3	8.3	G0			
21 57.8	−15 07	h 3071		Cap	1877 1959	319 323		20.3	18.0	7.2	10.9	A2	15489		
21 57.9	−55 00	φ 307		Ind	1931 1959					5.3	5.3	F0		o	δ Ind
21 58.0	+5 56	Σ 2848		Peg	1829 1933	56		10.7		7.2	7.5	A2	15493		
21 58.1	−3 30	Σ 2847		Aqr	1831 1959	296 309		1.2	0.9	8.4	8.8	F0	15494		
21 58.2	+82 52	Σ 2873	AB	Cep	1832 1975	77 69		13.8	13.7	7.0	7.3	F5	15571		

α 2000	δ 2000	Star Name		Const	Years		Pos Ang		Separation		m₁	m₂	Spec	ADS	Orb	Notes
21ʰ58.2	+82°52′	Σ 2873	AC	Cep	1921		205°	°	145.1	″				15571		
21 58.4	+27 40	h 950		Peg	1820		10		10.0		8.0	10.0				Not found; no BD star here
21 58.6	+6 01	S,h 336	AB	Peg	1823	1908	226	225	105.9	99.9	7.9	9.7	A2			
—			AC		1908		102		85.3			11.9				
21 58.7	+29 49	β pm		Peg	1909		134		29.8		6.9	14.5	F5			
21 59.2	+73 11	β		Cep	1910	1930	174	173	134.3	131.4	5.0	11.8	F5			16 Cep; optical (Kui)
21 59.5	−29 03	h 5311	AB	PsA	1919		292		40.6		7.1	10.6	K0			Very nearly an equilateral
—			AC		1919		227		48.6			10.9				triangle (h, 1830's)
21 59.6	+74 18	A 879		Cep	1904	1932	140		4.1		8.0	13.3	K0	14557		
21 59.6	−27 38	Stone 56		PsA	1878	1935	36		11.7		7.5	10.5	F0	15509		
21 59.7	+49 08	Hu 774		Cyg	1904	1953	151	256	0.2	0.2	8.2	8.2	A0	15530		Single 1957–60 (82-inch)
21 59.8	+23 56	Σ 2850		Peg	1830	1943	261		2.9		7.0	11.0	M3	15525		
21 59.8	+67 58	Σ 2853		Cep	1832	1925	188		3.7		8.0	10.5	F5	15542		
22 00.6	−13 45	Hu 282		Aqr	1900	1959	32	38	0.7	0.4	7.5	8.8	F0	15534		
22 00.7	+62 29	Kui 110		Cep	1937	1952	225		0.7		6.7	8.7	B0			
22 00.8	−28 27	β 276		PsA	1876	1955	115		1.7		5.8	6.8	B8	15536		12 η PsA
22 00.9	+43 38	Ho 175	AB	Lac	1885	1957	308		1.1		7.7	10.7	A5	15547		
—			AC		1895		284		32.7			12.6		15547		
22 01.1	+13 07	h 289		Peg	1878	1925	326	324	51.1	54.7	5.7	11.1	F2	15543		20 Peg
22 01.1	+39 15	Sei 1549	AB	Lac	1894		78		14.2		7.1	10.8		15549		Mag. of A here refers to Aa
—		A 1451	Aa		1906	1934	352		0.1		7.9	7.9	A0	15549		Too close 1933,46,57,58
22 01.2	+26 50	A 306		Peg	1901	1929	306		1.2		7.8	13.8	B9	15548		
22 01.5	−15 37	Howe 59	AB	Aqr	1877	1937	271		9.1		7.1	10.6	G5	15546		
—		h 5524	AC		1901	1918	292		102.6			10.2		15546		
—			AD		1913	1918	313		178.9			8.8		15546		
22 01.8	−9 52	Rst 4095		Aqr	1937	1951	194	191	0.3	0.2	8.8	8.8	G0			P.a.'s discordant 1950 [83]
22 01.9	+4 46	OΣΣ 228	AB	Peg	1875	1927	28	26	73.5	78.3	7.9	9.6	G5	15559		
—			Aa		1901	1925	158	160	25.7	22.4		12.4		15559		
22 02.3	+36 59	Ho 177		Lac	1886	1932	110		8.1		7.2	13.7	M	15566		
22 02.3	+61 06	β 695	AB	Cep	1878	1943	146		3.0		8.2	12.5	A0	15574		
—		Wak 5	AC		1970		205		13.9			12.0		15574		
22 02.4	−16 58	S 802	AB	Aqr	1866	1951	244		3.7		7.2	7.4	A2	15562		29 Aqr
—			AC		1913		291		143.0			11.7		15562		
22 02.8	+67 16	Mlb		Cep	1923		312		18.8		8.2	9.4	A2			
22 02.9	+39 34	OΣ 459		Lac	1845	1928	196		10.8		7.9	10.6	A0	15576		
22 02.9	+44 39	β 694	AB	Cyg	1878	1960	352	4	0.5	0.8	5.6	8.1	A0	15578		
—			AC		1918		309		66.8			10.5		15578		
—							277		27.1			13.5		15578		
—							328		24.1			13.0		15578		
22 03.1	−27 03	B 547		PsA	1927	1935	294		2.3		7.8	14.2	A2	15570		
22 03.1	−76 07	h 5306		Oct	1917		72		34.6		6.0	10.2	F2			
22 03.2	+60 51	Σ 2860	AB	Cep	1832	1941	251	258	3.3	9.4	7.9	9.5	K0	15590		An optical pair (IDS)
—			AC		1904		99		27.4			13.0		15590		
22 03.6	−55 59	LDS 771		Ind			sp		43.0		7.0	14.8	G0			
22 03.8	+64 38	Σ 2863	AB	Cep	1831	1974	289	277	5.6	7.7	4.4	6.5	A3 G	15600	o	17 ξ Cep
—			AC		1912	1925	199	200	95.3	96.8		12.6		15600		
22 03.9	+59 49	OΣ 461	AB	Cep	1848	1947	298		11.1		6.7	11.4	B5	15601		In a fine field (Webb)
—			AC		1876	1923	39		89.9			9.7		15601		
—			AD		1876	1921	72		183.9			7.9	A	15601		
—			AE		1876	1918	37		236.7			7.0	A0	15601		
—			Aa		1905		334		18.3			14.3		15601		
—			DE		1876	1918	347		136.1					15601		
—			EF		1876	1921	34		192.4			8.0	A0	15601		
22 04.4	+13 39	Σ 2854		Peg	1830	1975	83	83	3.1	2.0	7.6	7.9	F5	15596		
22 04.5	+15 52	β 696	AB	Peg	1877	1956	355		0.5		8.8	8.8	G5	15599		
—			AC		1900	1925	318		59.9			9.4		15599		
—			AD		1912	1925	28		84.8			9.3		15599		
—			AE		1915		0		120.3			9.6		15599		
22 04.5	−25 53	B 549		PsA	1926	1959	213		3.6		7.0	13.0	K0	15592		
22 04.6	+32 57	h 953		Peg	1903	1934	105		21.4		6.4	11.3	G5	15602		
22 05.1	+62 17	β 697	AB	Cep	1878	1958	94		19.8		5.1	11.1	O5	15624		19 Cep
—			AC		1913		271		60.4			10.1		15624		
22 05.7	+29 54	h 1721		Peg	1879	1973	278	268	8.3	11.2	8.0	9.3	M	15620		
22 05.7	+57 08	Stein 2618		Cep	1911		78		9.8		7.6	13.1	F5			
22 05.8	+4 52	Σ 2856		Peg	1830	1942	200		1.2		8.2	8.8	G5	15614		BDS 11412 probably same
22 05.8	−0 19	β pm		Aqr	1879	1907	42	41	114.6	113.0	3.0	12.0	G0			34 α Aqr
22 06.0	+20 48	Σ 2861		Peg	1830	1955	221		7.1		8.3	8.8	A3	15627		
22 06.0	−56 28	I 381		Ind	1901	1942	112	102	1.8	1.8	8.1	10.3	G0			
22 06.2	+10 06	Σ 2857		Peg	1828	1976	113		19.8		7.4	9.1	A2	15630		
22 06.2	+82 40	Mlr 257		Cep	1971		223		0.2		8.3	8.6	A0			
22 07.0	+25 21	β pm		Peg	1852	1907	213	220	92.1	103.7	3.8	11.2	F5			24 ι Peg
22 07.0	+36 06	OΣ 462	AB	Lac	1848	1959	334	320	1.4	1.1	7.9	9.7	A3	15645		
—			AC		1850	1954	32		7.5			11.3		15645		
22 07.1	+0 34	Σ 2862		Aqr	1828	1959	104	98	2.3	2.5	8.2	8.6	G0	15639		
22 07.1	+34 31	Ho 612		Peg	1895	1927	68	69	26.5	22.0	7.0	12.2	F5	15646		

α 2000	δ 2000	Star Name		Const	Years	Pos Ang		Separation		m₁	m₂	Spec	ADS	Orb	Notes
22ʰ 07ᵐ.7	+26°22′	Cou 537		Peg	1970	339°		0″.3		8.6	8.8	G0			
22 07.7	−28 04	Howe 60		PsA	1876 1957	152 143		2.6	3.1	8.1	9.5	G0	15647		
22 08.2	−46 58	Rst 5483		Gru	1947	149		28.4		1.7	11.8	B5			α Gru
22 08.3	+49 48	β		Lac	1910	273		27.5		6.6	13.0	K5	15659		
22 08.5	+39 22	Ho 470		Lac	1892 1907	352		11.8		7.8	13.8	A0	15662		
22 08.6	+59 17	Σ 2872	AD	Cep	1878 1924	153		117.6		6.6	10.1	A0	15670		
—			AxBC		1832 1925	316		21.6					15670		
—			BC		1833 1957	334 307		0.5	0.8	8.0	8.0	A0	15670		
22 08.9	+67 44	h 1742		Cep	1904 1923	344 333		23.0	24.0	8.4	9.3	A0			
22 09.2	+33 10	β	AB	Peg	1877 1934	314 323		27.4	27.7	5.6	11.9	K0	15672		27 Peg
—			AC		1879 1923	262 263		72.8	70.2		10.1		15672		
—			AD		1880 1923	90 89		185.2	189.2		10.6		15672		
—			AE		1909	299		116.4			12.5		15672		
22 09.3	+44 51	h 1735	AB	Lac	1901 1921	110		27.1		6.9	8.4	B9	15679		
—			AC		1961			43.5					15679		
—			BC		1901 1913	164		22.7					15679		
—					1922	286		110.3					15679		
22 10.0	+23 08	Cou 136		Peg	1966	72		0.2		7.6	7.6	F5			
22 10.1	+7 58	Σ 2867	AB	Peg	1831 1951	209		10.5		8.3	9.4	G0	15685		
—			BC		1904	253		8.9			13.9		15685		
22 10.1	+32 21	Ho 641		Peg	1900 1906	131		13.1		7.9	12.1	K0	15689		
22 10.2	+44 05	A 2493		Lac	1912 1928	334		3.7		8.0	14.0	K2	15692		
22 10.4	+14 38	Σ 2869		Peg	1829 1934	254 254		22.7	21.5	6.3	12.3	K0	15690		
22 10.5	−33 50	δ 156		PsA	1922 1959	298 317		1.3	1.3	7.8	10.8	G0			
22 10.6	+47 55	Es		Lac	1908 1918	308		22.1		7.7	10.0	F5	15698		
22 10.6	+70 08	Σ 2883		Cep	1833 1958	254		14.6		5.6	7.6	F2	15719		
22 10.9	+57 56	β 436	AB	Cep	1876 1903	328		19.6		7.4	11.4	A5	15706		Ho suspected close compan.
—		Ho 290	AC		1889 1903	100		19.4			12.9		15706		
—		β 436			1903	178		23.2		8.0	16	A5	15706		
22 11.0	+63 24	Σ 2879	AB	Cep	1834 1955	229		0.8		8.8	8.8	B5	15712		
—			AC		1916	260		29.5					15712		
22 11.2	+50 49	h 1741	AB	Lac	1876 1923	319 303		23.2	27.6	5.4	10.4	A2	15708		
—			AC		1923	274		51.3					15708		
—			AD		1924	272		73.7					15708		
22 11.3	+40 11	OΣ 464		Lac	1847 1960	54 118		0.8	0.3	8.6	8.8	A2	15707		
22 11.5	+32 05	Ho 178		Peg	1881 1938	224		3.3		7.8	12.5	F0	15709		
22 11.6	−34 28	β 769		PsA	1891 1952	354		0.7		7.0	8.5	F2			
22 11.8	+56 50	β pm	AB	Cep	1892 1918	36 34		79.6	72.9	5.2	10.3	F8			
—		Giclas	AC		1967	126		56.0			13.5				
22 11.8	+59 43	Σ 2880		Cep	1833 1941	352		4.2		7.8	9.7	K0	15729		
22 11.9	+6 54	β 698	AB	Peg	1878 1926	338		10.5		7.0	11.8	F0	15716		
—			AC		1913	289		108.4			10.6		15716		
22 12.0	+37 39	Σ 2876		Lac	1829 1931	68		11.8		7.8	9.3	F8	15723		
22 12.0	+50 12	OΣ 465	AB	Lac	1848 1924	322		14.8		7.3	10.8	F0	15727		
—			AC		1908	235		18.0			13.3		15727		
22 12.0	+60 05	β 376		Cep	1876 1931	149		3.6		7.7	10.9	K5	15737		
22 12.0	+60 46	H I 49		Cep	1783	4				5.4		K0			Not found (IDS)
22 12.0	−38 18	h 5319		Gru	1836 1955	110 127		2.1	2.1	7.6	7.7	F2			
22 12.2	+63 44	Σ 2884		Cep	1833 1914	149		2.1		8.0	9.5	A0	15742		
22 12.6	−8 01	β 475		Aqr	1891 1957	228 203		1.5	1.3	7.1	9.9	F2	15725		
22 12.9	+49 22	Ho 291		Lac	1888 1910	198		9.2		7.9	13.4	K5	15747		
22 12.9	+73 18	Σ 2893		Cep	1833 1967	348		28.9		6.2	8.3	G5	15764		
22 13.0	−49 03	CorO 250		Gru	1889 1933	353		5.5		7.6	10.3	G5			
22 13.4	−37 29	B 2056		Gru	1934 1959	161 49		0.2	0.2	8.5	8.4	A3			
22 13.6	+52 34	β 991		Lac	1880 1956	144		0.6		8.8	8.8	B9	15756		
22 13.7	+7 43	β 699	AB	Peg	1878 1953	183		2.4		7.9	12.0	K2	15750		
—			AC		1913	319		52.4			10.9		15750		
22 13.9	+39 43	h 1746	AB	Lac	1879 1927	179 184		27.2	28.4	4.5	10.5	K2	15758		
—			AC		1904	188		69.8			13.5		15758		
22 13.9	+72 28	A 895		Cep	1904 1930	178		1.1		8.1	11.3	K0	15781		
22 14.3	+17 11	Σ 2877	AB	Peg	1828 1940	316 13		7.6	16.0	6.5	9.7	K2	15763		An optical pair (IDS)
—			AC		1923 1933	45		86.9					15763		
—			AD		1923 1933	307 306		106.4	105.2				15763		
—			BC		1910 1926	300 300		101.0	99.6				15763		
22 14.3	−21 04	H N 56	AB	Aqr	1823 1959	120 114		5.0	5.0	5.6	7.1	G8	15753		41 Aqr
—			AC		1916 1950	43		212.1			9.0	F5	15753		
—		β 171	CD		1878 1959	259 256		11.5	12.0		11.7		15760		
22 14.5	+7 59	Σ 2878	AB	Peg	1830 1958	131 122		1.4	1.5	6.8	8.3	A0	15767		
—			AC		1843 1937	126 122		62.6	66.0		9.7		15767		
—			AD		1843 1937	274		124.3			11.1		15767		
22 14.6	+29 34	Σ 2881		Peg	1830 1960	111 85		1.8	1.4	7.6	8.1	F5	15769		
22 14.7	+49 21	Σ 2886		Lac	1832 1922	109 109		19.3	20.3	7.6	9.8	F8	15778		
22 14.8	+22 31	OΣ 467 rej		Peg	1843 1914	273		23.0		6.7	10.7	K0	15771		
22 15.0	+57 03	β pm		Cep	1891 1918	20 15		132.6	127.8	4.2	9.5	F0			23 ε Cep
22 15.1	+51 29	Ho 614		Lac	1897 1927	175		4.7		7.8	10.3	A0	15783		

α 2000	δ 2000	Star Name		Const	Years	Pos Ang	Separation	m₁	m₂	Spec	ADS	Orb	Notes
22ʰ15ᵐ.7	+43°54′	Ho 180		Lac	1886 1959	222° 233°	0″.5 0″.7	8.2	8.2	B9	15794		
22 15.9	+54 40	β 377	AB	Lac	1891 1911	66 65	63.9 59.1	7.7	10.3	G5	15797		Near open cluster NGC 7245
---			Aa		1899 1911	113 116	33.5 31.9		12.6		15797		
---			Ab		1899 1911	264 263	34.9 37.4		12.6		15797		
---			Ac		1899	261	40.1		13.6		15797		
---			Ad		1911	358	57.8		12.2		15797		
---		Giclas	Ae		1967	106	78.0		15		15797		
---		β 377	BC		1891 1917	303 300	7.0 5.9		11.2		15797		
---			bc		1899	242	6.1				15797		
22 16.1	−57 34	B 554		Tuc	1927 1930	110	4.3	7.2	12.4	M			
22 16.3	+33 44	OΣ 468		Peg	1854 1907	164	12.8	7.2	11.4	F8	15798		
22 16.5	+37 08	Sei 1561		Lac	1894	124	8.6	7.5	10.0				
22 17.5	+49 08	Es		Lac	1907	292	21.2	6.6	11.6	K0	15814		
22 18.0	−62 49	I 20		Tuc	1901 1959	342 243	0.6 0.3	7.6	8.1	F5			
22 18.3	−53 38	HdO 298		Gru	1900 1959	16 32	2.3 3.4	5.4	9.9	F7			
22 18.8	−68 19	HdO 297		Ind	1901	89	26.8	7.1	11.5	G5			
22 18.9	+37 46	Σ 2894	AB	Lac	1831 1955	194	15.6	6.1	8.3	F0	15828		
---			AC		1905	248	43.7		13.0		15828		
---			AD		1908 1918	176	221.7		9.4		15828		
---			BD		1880 1918	174	206.6				15828		
22 19.5	−60 48	h 5323		Tuc	1871 1952	204	26.5	8.4	8.7	G0 G0			
22 20.4	+11 02	h		Peg	1901	305	39.9	7.9	10.9	K0			
22 20.4	+35 07	OΣ 469	AB	Peg	1843 1947	280 287	32.2 28.7	7.5	9.1	F0	15848		An optical pair (IDS)
---			AC		1908	52	53.1		12.3		15848		
22 20.5	+5 47	h 962	AB	Peg	1867 1958	19	6.2	5.4	10.7	B5	15847		30 Peg
---			AC		1876 1958	221 225	9.9 14.8		11.8		15847		
22 21.0	+31 19	β 1217		Peg	1890 1946	222	0.6	7.7	10.6	K0	15858		
22 21.0	+46 32	h 1755		Lac	1879 1925	9	48.2	4.6	10.8	B5	15862		2 Lac
22 21.0	+66 58	OΣ 470		Cep	1850 1937	354 354	3.7 4.2	7.4	9.9	A5	15870		
22 21.3	+28 20	Ho 615	AB	Peg	1893 1923	127	72.6	4.8	9.1	B8	15863		32 Peg
---			AD		1895 1924	307	42.3		11.8		15863		
---			AE		1893 1924	116	60.3		11.8		15863		
---			BC		1895 1924	18	2.4		10.8		15863		
22 21.4	+41 48	A 411		Lac	1902 1957	201 214	0.3 0.3	8.5	9.2	A0	15867		
22 21.7	−1 23	h 3106		Aqr	1838 1959	126 140	49.5 37.4	3.8	12.0	A0	15864		48 γ Aqr
22 21.8	+66 42	Σ 2903		Cep	1832 1955	96	4.3	6.7	6.7	F5 A2	15881		Sp. compos. in HD cat.
22 21.9	+40 40	h 1756	AB	Lac	1901 1924	286	22.1	6.7	10.6	K2	15874		
---			AC		1901 1925	325	22.4		13.2		15874		
---			AD		1912 1925	76	58.0		11.3		15874		
22 22.0	−67 05	Don 1025		Ind	1928	349	0.7	8.5	8.9	F5			
22 22.1	−34 29	B 557		PsA	1926 1952	344 321	0.3 0.3	8.0	8.1	F2			
22 22.5	+30 21	Ho 474	AB	Peg	1892 1906	111	44.8	7.6	11.6	G0	15883		
---			BC		1892 1915	36	4.2		11.6		15883		
---			BD		1904	254	11.6		13.6		15883		
---			CD		1906	255	15.6				15883		
22 22.6	+6 28	Σ 2899 rej		Peg	1904 1926	32	18.6	8.0	11.2	A3	15880		
22 22.6	+9 56	OΣΣ 231		Peg	1875 1922	110	91.0	8.0	9.1	F0			
22 22.6	−56 09	I 134		Gru	1900 1947	308 296	0.6 0.4	7.8	8.4	A3			
22 22.7	−45 57	I 135		Gru	1900 1956	201	2.7	6.6	10.8	S G0			π¹ Gru
22 23.1	−45 56	I 382		Gru	1901 1953	207 214	4.6 4.6	5.8	11.3	F0			π² Gru
22 23.2	+5 39	Ho 292		Peg	1887 1935	64	4.1	7.5	11.0	A2	15889		
22 23.6	+45 21	Σ 2902	AB	Lac	1833 1942	89	6.4	7.6	8.5	G5	15900		
---			AC		1914	49	12.1		12.1		15900		
---			CD		1914	80	5.0		12.9		15900		
22 23.7	+20 51	Σ 2900	AB	Peg	1832 1958	181 173	2.5 0.3	6.2	9.4	F5	15896		33 Peg
---			AC		1832 1956	343 312	56.6 78.4		8.9		15896		
---		Cou 139			1966	74	0.2	8.7	8.7	F0			
22 23.8	−72 48	h 5325	AB	Ind	1879 1917	268	18.9	8.3	8.9	A0 A0			
---			AC		1917	99	32.4		11.5				
22 24.1	−4 50	β 172	AB	Aqr	1875 1960	20 324	0.5 0.5	6.5	6.5	A0	15902	o	51 Aqr
---			ABxD		1898 1917	191 191	113.7 116.0		10.1		15902		
---			AC		1898 1917	342	54.4		10.2		15902		
---			AE		1898 1917	133	132.4		8.6		15902		
22 24.6	−72 15	φ 285		Ind	1928 1959		0.1	6.0	6.1	G0			ν Ind; measures uncertain
22 24.7	−41 26	Jc 19		Gru	1845 1952	83 74	44.0 24.8	6.7	8.4	F5 F5			
22 24.9	−57 48	I 383		Tuc	1914	237	81.1	5.3	12.9	G0			
22 25.3	+73 36	h 3119		Cep	1906	93	35.9	8.0	12.2	F0			
22 25.5	+22 34	Ho 616		Peg	1895 1924	358	19.1	7.6	12.9	A2	15924		
22 25.8	−20 14	S 808		Aqr	1825 1951	152	6.8	7.1	8.3	F8	15926		
22 25.9	−45 07	I 136		Gru	1900 1958	269 276	1.7 1.7	7.7	8.7	K0			
22 25.9	−75 01	Δ 238		Oct	1836 1917	83 81	18.1 20.1	6.1	8.7	G0			
22 26.4	+54 06	A 1464		Lac	1906 1929	98	4.7	7.6	14.1	A0			
22 26.6	+4 24	β 290	AB	Peg	1875 1937	219 224	2.6 3.5	5.8	12.3	G0	15935		34 Peg
---			AC		1911 1924	272 272	101.2 103.3		12.8		15935		
22 26.6	−16 45	S,h 345	AB	Aqr	1823 1976	303 334	10.0 3.1	6.4	6.6	G0 G0	15934		53 Aqr

α 2000	δ 2000	Star Name		Const	Years		Pos Ang		Separation		m₁	m₂	Spec	ADS	Orb	Notes
22ʰ 26ᵐ.6	−16° 45′	S,h 345	BC	Aqr	1901		339°		46″.7			12.9		15934		
−−		β 1307	CD		1901		101		1.8			13.9		15934		
22 26.7	+22 49	h 3115	AB	Peg	1892	1902	263		32.8		7.8	11.2	A2			
−−			AC		1902		330		41.3			12.2				
−−			AD		1892	1902	95		57.8			11.6				
22 26.8	+37 27	Σ 2906		Lac	1832	1937	2		4.3		6.5	10.1	B3	15942		
22 26.8	+49 42	β 380	AB	Lac	1876	1917	323		24.9		8.2	12.9	A	15951		
−−		OΣΣ 234	AC		1874	1922	134		36.3			8.6		15951		
−−		β 380	Aa		1905	1912	88		23.5			14.4		15951		
−−			CD		1877	1912	243		21.4			13.4		15951		
−−			Cc		1912		27		17.8			14.5		15951		
22 27.3	−64 58	h 5334		Tuc	1836	1928	282		6.9		4.5	9.0	B9			δ Tuc
22 27.4	−58 00	I 137		Tuc	1901	1943	335		2.0		7.1	10.3	A2			
22 27.9	+4 42	β pm	AB	Peg	1891	1908	208	210	101.5	98.3	4.8	9.8	K0			35 Peg
−−			AC		1892	1908	241		181.5			9.7				
−−			BC		1907		269		109.5							
22 28.1	+12 15	β 701	AB	Peg	1877	1959	283	227	1.2	0.9	7.2	10.2	K0	15962		
−−			AC		1910	1924	129		119.2			11.1		15962		
22 28.1	+44 07	Es 1467		Lac	1915		228		7.9		6.8	14.0	K5	15968		
22 28.1	+57 42	Kr 60	AB	Cep	1890	1960					9.8	11.3	M2 M5	15972	o	Famous red-dwarf pair [84]
−−			AC		1890	1941	56	62	26.6	73.2		10.1		15972		
−−			AD		1900	1912	21	34	21.3	30.8		14.9		15972		
−−			AE		1900	1915	99	93	67.9	78.5		12.5		15972		
−−			AF		1901	1915	275	290	39.6	30.1		13.9		15972		
−−			AG		1912		191		61.2			13.7		15972		
−−			AH		1912		236		90.0			13.8		15972		
−−			AI		1873	1918	152	140	195.4	201.5		8.5	B9	15972		
−−			BC		1915	1921	59	60	48.2	53.7				15972		
−−			CE		1907	1918	130		46.0					15972		
22 28.2	+17 16	Σ 2908		Peg	1828	1937	116		9.0		7.7	9.4	K0	15967		
22 28.5	+48 33	A 187		Lac	1900	1944	128		2.0		7.7	12.7	K0	15973		
22 28.5	−51 47	h 5338		Gru	1914	1916	183		30.3		7.0	10.8	K5			
22 28.7	−39 08	See 473		Gru	1896	1904	116		27.3		5.5	12.5	G4			ν Gru
22 28.8	−0 01	Σ 2909		Aqr	1781	1975					4.3	4.5	F2	15971	o	55 ζ Aqr
22 29.2	+58 25	β 702	AB	Cep	1878	1934	285	284	19.4	20.4		13.0	G0	15987		27 δ Cep; A is variable
−−		Σ I 58	AC		1835	1972	191		41.0			7.5	A0	15987		
22 29.3	−43 30	I 1054		Gru	1935		228		5.6		4.0	12.8	G5			δ¹ Gru
22 29.4	−28 40	See 474	AB	PsA	1896	1955			0.1		7.9	8.2	K0	15978		Duplicity doubtful (IDS)
−−		H N 34	ABxC		1906	1959	297		33.3			11.0		15978		
22 29.5	+29 59	Cou 735		Peg	1971		200		0.6		8.0	10.5	A3			
22 29.8	−43 45	Δ 239		Gru	1836	1953	216	212	61.3	60.7	4.1	9.0	M4			δ² Gru
22 29.9	+52 25	Leon 53	AB	Lac	1928		312		13.9		6.6	11.5	G5	15993		
−−		OΣ 472 rej	AC		1867	1933	6		15.8	14.2		12.6		15993		
22 29.9	−46 04	B 562		Gru	1927				4.1		7.2	14.0	G5			
22 30.0	+4 26	Σ 2912		Peg	1831	1960					5.8	7.1	F5	15988	o	37 Peg
22 30.1	+49 21	Frk		Lac	1918		94		63.9		6.5	9.7	K0			
22 30.2	+0 08	β 1264		Aqr	1891	1944	22		4.1		7.7	13.2	A0	15990		
22 30.2	+22 28	Hu 388		Peg	1901	1958					8.5	9.0	F0	15992	o	
22 30.2	+40 58	Es 1696	AB	Lac	1917		344				7.5	11.3	F5	15996		
−−			BC		1917		141		2.3			11.7		15996		
22 30.3	+57 14	OΣ 473		Cep	1848	1922	357		14.8		6.8	10.1	A2	16001		Spectrum composite A2p+G
22 30.5	−8 07	Σ 2913		Aqr	1830	1954	329		8.2		7.7	8.7	F0	15994		
22 30.5	+53 31	Σ 2917	AB	Lac	1832	1967	71		4.7		8.6	8.6	F0	16008		
−−			AC		1911		186		89.5			11.3		16008		
22 30.5	+61 37	Hu 981		Cep	1904	1958	254	223	0.1	0.3	7.6	7.8	A0	16011		
22 30.7	+17 58	Cou 234		Peg	1967		77		0.2		8.4	8.4	F8			
22 31.0	−49 26	HdO 299		Gru	1901		284		19.8		6.8	12.0	G0			
22 31.2	+50 52	Σ 2918		Lac	1834	1936	246		1.4		7.8	9.5	A2	16020		
22 31.3	+50 17	β 703		Lac	1878	1925	299	294	30.2	36.3	3.8	11.8	A0	16021		7 α Lac; optical (IDS)
22 31.5	−32 21	Pz		PsA	1826	1952	172		30.3		4.4	7.9	A0			
22 32.0	−42 44	I 1455		Gru	1925	1935	358	352	0.3	0.3	8.3	8.6	K0			
22 32.4	+39 47	Roe 47	AB	Lac	1910	1911	158		43.1		5.8	9.8	A3	16031		
−−			AC		1910	1911	344		32.4			10.1		16031		
−−			AD		1910		216		105.7			9.4		16031		
−−			DE		1910	1928	175		6.6			9.8		16031		
22 32.6	+34 14	h 1779		Peg	1904	1926	217		22.0		7.9	9.3	A5	16034		
22 32.9	+49 23	Hu 1320		Lac	1904	1957					8.4	9.1	F5	16046	o	
22 33.0	+69 55	Σ 2924	AB	Cep	1831	1959					6.5	7.0	F2	16057	o	Spectrum composite F2+A5
−−			AC		1910		193		114.4			10.6		16057		
−−			AD		1910		194		178.3			10.3		16057		
22 33.1	+68 30	β 706	AB	Cep	1881	1938	15		2.8		7.7	12.3	A3	16058		
−−			AC		1881	1913	253		29.8			11.3		16058		
22 33.3	+70 22	Σ 2923	AB	Cep	1833	1958	46		9.4		6.4	8.7	A0	16062		
−−			AC		1911		146		97.0			11.3		16062		
22 33.6	+40 49	β 705		Lac	1878		158		1.5		6.8	12.3	K0			May be single (IDS)

α 2000	δ 2000	Star Name		Const	Years	Pos Ang	Separation	m₁	m₂	Spec	ADS	Orb	Notes

α 2000	δ 2000	Star Name		Const	Years	Pos Ang		Separation		m_1	m_2	Spec	ADS	Orb	Notes
22 33.9	+65 50'	Hu 983		Cep	1904 1960	154° 188°		0".2	0".2	8.2	8.5	K0	16072		
22 34.0	−1 34	β	AB	Aqr	1912 1959	298 298		95.9	98.1	5.9	10.3	K0			60 Aqr
--			AC		1912 1959	274 275		124.0	125.6						
22 34.2	+54 05	A 1468		Lac	1906 1953	256		0.3		8.3	8.3	A2	16073		
22 34.5	+4 13	Σ 2920	AB	Peg	1829 1936	144		13.6		8.1	9.2	A0	16069		
--			AC		1878 1907	63		21.3			14.7		16069		
22 34.5	+40 46	h 1786		Lac	1904	226		43.9		7.3	8.7	B5			
22 34.9	+41 34	h 1788		Lac	1902 1972	299 298		3.4	3.7	8.4	8.7	F2	16084		
22 35.0	+22 18	Ho 617		Peg	1895 1907	53		17.5		7.9	13.4	K0	16081		
22 35.0	+30 48	h 966	AB	Peg	1882 1926	269		13.3		7.4	11.4	K2	16085		
--			AC		1906	276		36.8			12.4		16085		
22 35.2	+14 37	Hu 982		Peg	1904 1956	215		0.7		7.7	11.2	K2	16087		
22 35.6	−38 43	h 5344		Gru	1889 1955	168		5.0		7.9	9.9	F2			
22 35.7	+54 13	Mlr 3		Lac	1953 1957	168		0.4		8.5	9.1	K0			
22 35.7	+56 52	h 1791	AB	Lac	1894 1926	61		17.5		7.9	9.2	G0	16097		
--			AC		1910 1925	150		85.0			10.7		16097		
22 35.9	+39 38	Σ 2922	AB	Lac	1831 1969	186		22.4		5.7	6.5	B3 B5	16095		8 Lac
--		A 1469	AC		1903 1956	169		48.8			10.5		16095		
--			AD		1902 1968	144		81.8			9.3		16095		
--			AE		1928	239		336.6			7.8		16095		
--		Σ 2922	BC		1830 1969	155		28.2					16095		
--		A 1469	BD		1830 1969	131		66.5					16095		
--			BE		1928	242		323.7					16095		
--			CD		1902 1906	116		42.2					16095		
--			Cc		1906 1932	255		1.4			14.6		16095		
--			Dd		1906 1935	226		9.5			13.3		16095		
22 35.9	−20 56	H V 96		Aqr	1910	247		50.9		7.5	9.5	K2			
22 36.1	+35 35	OΣ 474 rej		Peg					6.2		K0			Probably single (IDS)
22 36.1	+72 53	β 1092	AB	Cep	1889 1959	237 210		0.3	0.3	8.3	8.3	F5	16111		Close in 1954,55
--		h 3133	ABxC		1889 1926	264 261		29.2	32.6		12.8		16111		
--		OΣΣ 236	ABxD		1783 1926	137		42.2			8.5		16111		
22 36.6	+27 47	Ho 186		Peg	1881 1910	24		7.2		7.1	12.3	K0	16103		
22 36.6	−31 40	Δ 241		PsA	1826 1918	31		89.5		5.8	7.6	K0 K0			A fine object (h)
22 37.0	−40 35	β 771		Gru	1891 1934	263		2.7		5.8	11.0	A2			
22 37.8	−4 14	h 5529		Aqr	1896 1925	246 247		103.1	98.3	5.0	8.5	K0			63 κ Aqr
22 37.8	−39 51	I 138		Gru	1896 1934	278		3.4		6.8	11.0	F8			
22 38.0	−33 28	B 568		PsA	1927 1935	310		4.9		7.6	13.4	A3			
22 38.4	−7 54	A 2695		Aqr	1921 1959	102 116		0.3	0.3	6.4	8.2	G0	16130		[85]
22 38.5	+2 18	Ho 479		Aqr	1893 1957	232 152		0.6	0.5	8.2	9.7	G0	16131		
22 38.6	+56 48	h 1796		Lac	1902 1958	11 10		29.5	30.9	5.5	10.8	M	16140		
22 38.6	−14 04	h 5355	AB	Aqr	1905 1919	294 294		76.8	78.0	7.6	8.9	A3 F8			
--			AC		1905 1919	2		107.1			10.1				
--			BC		1909	45		105.5							
22 38.8	+44 19	Ho 295		Lac	1887 1960					7.6	7.6	G0	16138	o	
22 38.8	−53 44	B 569		Gru	1927 1959	185 178		1.7	1.7	7.9	12.0	F0			
22 38.9	+19 31	Kui 113		Peg	1935 1948	236		1.8		5.8	11.4	G5			40 Peg
22 39.1	+37 23	OΣ 475		Lac	1847 1932	73		15.5		6.8	10.8	B3	16143		
22 39.1	−52 42	h 5349		Gru	1914 1916	118		33.8		6.7	11.4	F0			
22 39.3	+39 03	S 813		Lac	1825 1974	49 49		60.4	62.3	4.9	8.4	O5	16148		10 Lac
22 39.6	+37 36	Ho 187		Lac	1883 1934	286 285		18.3	19.5	6.0	12.9	G5	16154		
22 39.7	−28 20	H VI 119	AB	PsA	1783 1951	160		86.6		6.3	7.3	G0 F5	16149		
--		H N 117	BC		1837 1959	58 67		3.2	3.2		8.1		16149		
22 40.0	−19 12	h 3128		Aqr	1877 1934	227		10.9		7.3	10.6	F8	16155		
22 40.2	+37 31	Ho 188		Lac	1885 1960	43 163		0.4	0.2	8.7	8.7	F8	16164		
22 40.8	−3 33	Kui 114		Aqr	1934 1958	128 124		0.3	0.2	6.9	7.4	G0		o	Too close 1954,55
22 40.9	+14 33	Ho 296	AB	Peg	1903 1960					6.3	6.8	G5	16173	o	
--			AC		1910 1924	234 235		68.2	72.2		11.5		16173		
22 41.3	+72 43	Σ 2940		Cep	1832 1953	140		2.8		8.3	9.5	A3	16191		
22 41.5	+10 50	β		Peg	1879 1925	139		64.3		3.4	11.4	B8	16182		42 ζ Peg
22 41.5	+40 14	S 815		Lac	1825 1925	16 16		72.0	69.0	5.3	9.3	B2			12 Lac
22 42.1	−5 06	H N 140	AB	Aqr	1901 1911	263		75.1		7.1	9.7	M			
--			AC		1890 1911	182		162.4			10.3				
22 42.4	+54 15	Es 1028		Lac	1910	243		5.7		7.5	10.0	B9	16203		
22 42.6	−47 13	CorO 252		Gru	1897 1943	133 129		7.8	7.8	6.0	10.0	G0			
22 43.0	+1 13	Σ 2936		Aqr	1832 1946	49		4.7		7.0	10.0	A3	16201		
22 43.0	+30 13	β 1144	AxBC	Peg	1824 1926	339		90.4		2.9		G0	16211		44 η Peg
--			BC		1889 1949	83		0.2		9.9	9.9		16211		
--			BCxD		1916	320		62.9			13.8		16211		
--			DE		1916	181		5.7			15		16211		
22 43.1	−8 19	Σ 2935	AB	Aqr	1831 1959	313 308		2.6	2.3	6.9	7.9	A2	16208		
--			AC		1913 1924	357		70.2			10.7		16208		
--			AD		1913	277		77.2			12.2		16208		
22 43.1	+47 10	OΣ 476	AxBC	Lac	1847 1958	335 315		0.5	0.5	6.4		B9	16214		
--		Hu 91	BC		1898 1958	218		0.2		8.0	10.0		16214		
22 43.5	+46 02	OΣ 477	AB	Lac	1846 1953	123 250		9.6	11.6	7.4	11.3	F5	16220		

α 2000 δ 2000		Star Name		Const	Years	Pos Ang	Separation	m₁	m₂	Spec	ADS	Orb	Notes
22ʰ43.ᵐ5	+46°02′	OΣ 477	AC	Lac	1880 1953	350° 346°	176″.2 179″.4		11.2		16220		
--			CD		1908 1953	126	10.3		12.7		16220		
22 43.5	−41 25	I 1457		Gru	1925	282	14.2	4.9	14.0	G5			ρ Gru
22 43.8	+78 31	OΣ 481		Cep	1855 1954	268	2.4	7.5	9.3	A0	16243		
22 44.1	+39 28	Σ 2942	AB	Lac	1831 1953	280	2.8	6.1	8.3	K5	16228		
--		β 450	AC		1878 1933	235	10.6		11.5		16228		
22 44.1	+41 49	OΣ 479		Lac	1849 1958	129	14.6	5.1	10.5	K0	16227		13 Lac
22 44.6	−63 57	HdO 300		Tuc		290	20.0	7.1	13.0	A0			
22 44.7	−60 07	h 5358		Tuc	1916	91	31.3	7.8	9.5	G5 G0			
22 44.8	+24 23	A.G. 286		Peg				7.3		K0			Single (IDS)
22 45.3	−9 39	Σ 2939		Aqr	1831 1909	62	11.1	7.4	10.4	A5	16238		
22 45.3	+51 28	Hu 783		Lac	1904 1957	133 161	0.2 0.2	8.8	8.8	B9	16249		
22 45.6	+30 27	Che		Peg	1906	324	10.6	6.4	10.8	K0	16248		
22 45.6	−53 30	Cru1s		Gru	1879 1934	185	24.4	4.9	11.5	G8			η Gru
22 45.9	+19 14	Σ 2941		Peg	1830 1954	270 262	8.7 12.5	8.1	10.8	K0	16254		An optical pair
22 46.1	+52 06	Ho 619		Lac	1897 1905	8	18.7	8.0	13.0	F0	16259		
22 46.1	+58 04	OΣ 480		Cep	1845 1924	117	30.9	8.0	9.0	F8	16260		
22 46.1	+65 18	β pm		Cep	1879 1909	240 240	145.0 146.7	7.6	9.5	G5			
22 46.7	+4 54	Bail 2986		Peg	1910			7.2					Error in position?
22 46.7	+12 10	h 301	AB	Peg	1866 1975	118 100	12.2 11.5	4.2	12.2	F5	16261		46 ξ Peg
--			AC		1879 1924	22 15	127.3 145.0		11.0		16261		
22 46.7	−46 56	h 5362		Gru	1856 1937	141	10.5	6.7	9.7	A5			
22 46.8	−42 32	I 682		Gru	1925 1931	4	3.2	8.0	11.8	K0			
22 46.8	−48 18	I 304		Gru	1900 1945	359	4.5	8.2	9.7	K0			
22 47.0	+44 46	A 189	AB	Lac	1900 1959	205	0.9	8.7	8.8	A0	16266		
--			AC		1900	127	24.0		14.1		16266		
22 47.3	−16 09	Hu 291		Aqr	1900 1945	6 349	2.1 2.4	7.4	10.1	G0	16265		
22 47.5	+83 09	OΣ 482		Cep	1850 1940	33	3.5	4.7	9.4	K0	16294		
22 47.6	−65 34	h 5361		Tuc	1916	41	80.6	6.5	10.8	K0			
22 47.7	−14 03	Σ 2943		Aqr	1783 1959	110 121	35.6 23.7	5.8	9.0	B9	16268		69 Aqr
22 47.8	−4 14	Σ 2944	AB	Aqr	1832 1976	247 282	4.1 2.3	7.5	8.0	G0 G0	16270		
--			AC		1833 1955	157 106	55.6 49.7		8.5		16270		
--			AD		1921	342	201.1				16270		
--			BC		1900 1913	127 122	49.5 49.6				16270		
22 48.5	+31 06	β 1146		Peg	1889 1958			7.6	8.5	B9	16278	o	
22 49.0	+68 34	Σ 2947	AB	Cep	1832 1975	76 58	3.0 4.3	7.2	7.2	F5 F5	16291		
--			AC		1910 1925	206 206	109.3 111.2		10.2		16291		
--			BC		1925	207	113.7				16291		
22 49.6	+66 33	Σ 2948		Cep	1832 1949	5	2.7	7.2	8.9	B9	16298		
22 49.6	−13 36	H VI 97		Aqr	1825 1913	293	133.4	4.0	8.5	K5			71 τ Aqr
22 49.7	+31 19	Σ 2945		Peg	1832 1953	295	4.0	8.8	8.8	F0	16292		
22 49.9	+75 10	h 1826	AB	Cep	1905	356	21.5	7.8	12.2	A0			
--			AC		1905	192	29.5		10.8				
22 50.0	−32 48	HdO 301		PsA	1925 1959			7.0	7.2	A5		o	
22 50.6	+40 30	Σ 2946		Lac	1831 1967	253 260	5.0 5.3	8.3	8.3	F8	16293		
22 51.3	+13 57	β pm	AB	Peg	1853 1908	343 338	200.6 198.9	8.0	10.7	K0 G5			
--			BC		1908	193	106.7		12.4				
22 51.4	+26 23	Ho 482	AB	Peg	1893 1960			7.7	7.7	A3	16314	o	
--			AC		1893 1971	198	50.5		9.2		16314		
22 51.4	+61 42	Σ 2950	AB	Cep	1832 1960	319 295	2.0 1.7	6.1	7.4	G0	16317		
--			AC		1910 1959	354	39.3		10.7		16317		
22 51.6	−1 49	Rst 4711		Psc	1939 1943	128	6.4	7.9	13.0	A0			
22 51.8	+41 19	h 1823	AB	Lac	1874 1921	259	19.2	7.1	12.8	A0	16321		
--			AC		1874 1923	338	82.1		8.8	A0	16321		
--			AE		1885 1921	263	118.3		9.2	A0	16321		
--			CD		1886 1917	139	4.9		11.0		16321		
22 51.8	+53 11	Es 1032	AB	Lac	1910	134	24.6	7.9	11.6	A5	16323		
--			BC		1910	191	2.3		12.2		16323		
22 51.8	−35 53	h 5365	AB	PsA	1896 1933	278	4.8	7.4	12.4	A3			
--			AC		1919	35	55.6		11.0				
22 52.0	+43 19	β 451	AB	Lac	1888 1958	128 144	29.6 25.7	4.9	11.9	K5	16325		15 Lac
--			AC		1908	310	102.9		11.8		16325		
--			AD		1908	228	127.7		12.2		16325		
--		Fox	AE		1921	355	59.0				16325		
22 52.0	+57 43	A 632	AB	Cep	1903 1958			8.6	9.1	K0	16326	o	Mag. of A is for Aa
--		Dom	Aa		1952 1953	232 212	0.3 0.3						Not seen in 1957 (Dom)
22 52.2	−63 11	I 340		Tuc	1900 1949	20 359	1.2 0.9	6.1	9.1	K0			
22 52.5	−32 53	h 5367		PsA	1835 1957	276 262	4.2 4.2	4.5	8.0	A0			22 γ PsA
22 52.6	−42 47	h 5366	AB	Gru	1879 1950	252	14.8	8.5	9.2	A2			
--			CD		1914	111	13.0	13.7	13.9				
22 52.7	+67 59	OΣΣ 238	AB	Cep	1875 1972	280	69.1	7.4	8.1	F2			
--			BC		1876	250	25.0						
22 52.8	+60 55	Σ 2953		Cep	1832 1916	138	8.3	7.8	9.8	A0	16334		
22 53.6	+30 46	Ho 191	AB	Peg	1881 1907	88	3.4	7.5	13.5	A2	16339		
--			AC		1881 1923	280	24.3		10.6		16339		

α 2000	δ 2000	Star Name		Const	Years	Pos Ang		Separation		m₁	m₂	Spec	ADS	Orb	Notes
22ʰ53ᵐ.7	+44°45'	β 382	AB	Lac	1876 1960	°	°	"	"	5.8	7.8	A0	16345	o	
--		h 1828	ABxC		1876 1925	354	356	26.4	28.0		10.7		16345		
22 54.2	+28 01	Σ 2952 rej		Peg	1879 1925	136		17.6		7.8	10.3	G0	16352		
22 54.2	+76 20	Σ 2963		Cep	1832 1956	355		2.3		7.9	8.6	A3	16371		
22 54.7	+7 15	Σ 2955 rej		Psc	1879 1908	337		22.9		7.8	11.2	K2			
22 54.7	-67 36	Don 1039		Ind	1928	256		1.5		7.6	11.0	F0			
22 55.0	-40 56	I 1460		Gru	1925 1942	347		0.3		8.6	8.7	F2			
22 55.1	+9 28	β 383	AB	Peg	1877 1914	119		2.6		8.0	12.7	K2	16364		
--			AC		1877 1914	240		15.6			12.4		16364		
22 55.2	-4 59	β 178		Aqr	1879 1958	325	312	0.9	0.4	5.9	7.9	G5	16365		In 1938-57, sep. small
22 55.3	-48 30	I 22	AB	Gru	1900 1959	185	166	0.4	0.3	7.6	7.9	G0			Direct motion
--			AC			s		94.0				F5			Direct motion
--		B 2506	CD		1944 1956	320	101	0.1	0.1	7.4	7.6				
22 55.7	+36 21	h 975		Lac	1875 1923	243	243	49.6	51.0	5.6	9.5	B9	16376		
22 55.9	-32 32	Howe 91		PsA	1877 1953	236	244	5.0	5.0	4.2	9.2	G3			23 δ PsA
22 56.1	+66 43	β 849		Cep	1881 1954	127	108	3.7	5.3	8.0	11.9	A0	16385		
22 56.2	+72 50	OΣ 484	AB	Cep	1846 1958					7.8	8.7	A2	16393	o	
--			AC		1855 1971	256	255	31.9	30.7		11.3		16393		
--		S 820	AD		1825 1918	279		120.8			8.8	A0	16393		BDS 11988 same star
22 56.4	+22 57	Cou 240		Peg	1967	290		0.6		7.4	7.8	F0			
22 56.4	+41 36	Σ 2960	AB	Lac	1831 1923	344		27.6		5.6	11.6	B3	16381		16 Lac = EN Lac
--			AC		1831 1923	48		62.8			8.7		16381		
--			CB		1904	252		55.9					16381		
22 56.5	+62 52	Σ 2961		Cep	1833 1954	348		1.9		8.6	8.6	B5	16394		
22 56.7	+78 30	Σ 2971		Cep	1832 1967	4		5.4		8.0	9.2	G5	16407		
22 56.8	+62 44	Mlr 266		Cep	1971	177		2.9		7.7	14.1	B0			
22 56.9	+11 51	Σ 2958		Peg	1831 1958	7	13	3.9	3.8	6.6	8.9	A3	16389		
22 57.1	-3 15	Σ 2959	AB	Psc	1832 1959	97	115	15.7	11.0	6.6	10.6	A0	16392		
--		β 713	AC		1914	102		22.5			13.1		16392		
--			BC		1891 1959	94	97	10.1	11.2				16392		
22 57.4	+43 01	β 452	AB	Lac	1880 1933	256		6.7		6.9	11.0	K5	16401		
--			AC		1911	16		9.8			14.9		16401		
--			AD		1911	213		85.6			13.7		16401		
22 57.6	+62 22	Hu 990		Cep	1904 1958	288	292	1.0	0.9	8.1	11.1	A2	16404		
22 57.8	-26 06	h 5371		PsA	1876 1950	343		8.9		7.4	8.9	G0	16400		
22 57.9	+13 37	Hu 989		Peg	1904 1957	76	70	0.4	0.4	7.9	10.4	F2	16403		Single 1924-28
22 58.4	+67 05	h 1838		Cep	1879 1922	266		2.1		8.4	9.4	G0	16423		
22 58.6	+9 21	OΣ 536	AB	Peg	1852 1959					7.1	7.3	G0	16417	o	
--			AC		1903 1924	85	84	241.5	233.8		9.2		16417		
--					1924	112		174.4					16417		Unidentified measures in
--					1909	218		115.6		9.6	11.0		16417		IDS at this position
22 58.6	-45 31	Hu 1335		Gru	1929 1956					8.1	8.5	G5		o	
22 59.0	+27 45	Σ 2967		Peg	1831 1927	6		6.7		8.0	9.6	F5	16424		
22 59.2	+11 44	OΣ 483		Peg	1845 1960					6.1	7.4	F0	16428	o	52 Peg
22 59.2	+80 21	β		Cep	1913	11		49.8		7.4	10.4	K2			
22 59.5	+0 58	Barnard 18		Psc	1889 1958	94	85	3.8	3.8	5.4	13.1	K0	16431		2 Psc
23 00.0	-46 30	Rst 3314		Gru	1935 1947	291		3.3		8.0	13.4	F2			
23 00.3	+14 00	β 850		Peg	1881 1945	119		3.0		8.1	10.6	A2	16437		
23 00.7	+31 05	Σ 2968		Peg	1832 1941	90		3.4		6.6	9.1	A0	16443		
23 01.0	+76 07	β 851		Cep	1881 1924	160		2.0		7.5	13.0	G5	16454		
23 01.0	-58 26	Hu 1643		Tuc	1914 1943	4	12	1.8	1.8	7.2	11.5	A3			
23 01.2	+26 46	Σ 2969		Peg	1831 1955	34		4.0		8.1	10.0	A5	16449		
23 01.2	+80 47	OΣ 487		Cep	1890 1955	208	162	0.2	0.1	6.9	8.7	A3	16469		Single 1956,57
23 01.7	+67 48	Hu 993		Cep	1904 1928	220		1.9		8.0	10.3	F8	16459		
23 02.0	+48 00	A 194		And	1900 1960	98	102	0.2	0.2	8.6	8.6	A0	16457		Quad. reversal about 1928?
23 02.1	+62 31	Es 1861		Cep	1920	290		4.3		8.0	12.2	A5	16466		
23 02.3	-64 18	Δ 244		Tuc	1836 1916	100	96	48.9	47.2	7.6	10.0	F0			
23 02.4	+64 13	Mlr 69		Cep	1970	108		0.3		8.4	9.2	G5			
23 02.6	+42 45	β 1147	AB	And	1889 1958	318	25	0.3	0.5	5.1	8.8	A2	16467		2 And
--			AC		1911 1912	192		90.4			13.7		16467		
23 02.6	-18 32	β 384		Aqr	1877 1959	65		1.1		6.8	8.8	A0	16465		
23 02.6	-36 25	β 1011		Gru	1881 1936	302	298	2.1	2.1	6.5	9.5	K0			
23 02.7	+55 14	OΣ 485 rej	AB	Cas	1866 1958	52	50	21.8	20.1	6.6	9.8	B9	16474		
--			AC		1907 1924	260		56.6			10.0		16474		
--			AD		1915	95		39.8					16474		
23 02.8	+44 04	Σ 2973		And	1831 1958	41		7.4		6.4	9.6	B3	16472		
23 03.2	-62 23	B 583		Tuc	1927 1930	308		3.0		8.0	12.0	A0			
23 03.4	+60 27	OΣ 486 rej		Cep	1845 1920	276		33.9		6.7	9.3	B5	16481		
23 03.8	+28 05	h 1842	AB	Peg	1879 1924	208	211	98.4	108.5	2.4	11.6	M	16483		53 β Peg
--			AC		1852 1924	95	98	264.2	253.1		9.4		16483		
23 04.8	+64 05	Mlr 70		Cep	1970	259		0.4		8.1	8.9	F0			
23 05.0	+33 22	Σ 2974		Peg	1831 1953	162		2.7		8.1	8.1	F0	16496		
23 05.1	-43 02	I 1464		Gru	1918 1931	262		1.0		7.7	11.7	K0			
23 05.2	-7 42	A 417	AB	Aqr	1902 1959					6.2	6.3	F0	16497	o	83 Aqr
--		Σ I 59	ABxC		1835 1923	147		262.1			7.5	K0	16497		

α 2000	δ 2000	Star Name		Const	Years	Pos Ang	Separation	m₁	m₂	Spec	ADS	Orb	Notes
h m 23 05.2	−17° 05′	h 3164		Aqr	1901	130°	55.1″	6.3	11.4	K0			
23 05.5	+46 43	A 196		And	1900 1956	324	0.5	8.5	9.0	A0	16505		
23 06.2	+41 47	Ho 194		And	1885 1945	60	0.4	7.7	10.0	G5	16516		Spectrum composite G5+A3
23 06.4	−47 47	Hu 1547		Gru	1913 1935	155	2.9	8.0	11.0	K0			
23 06.5	+46 55	OΣΣ 242		And	1876 1924	31	79.8	8.0	8.0	B3 B9			
23 06.5	+61 26	Σ 2977		Cep	1832 1955	335 350	2.2 1.9	6.7	10.6	F5	16514		
23 06.6	+42 38	Es 1596	AB	And	1916	22	35.8	7.8	11.9	K5	16513		
—			BC		1916	76	5.7		12.9		16513		
23 06.7	−23 45	B 588		Aqr	1925	83	2.9	4.5	14.5	G9	16511		86 Aqr
23 06.9	−38 54	β 773		Gru	1920 1948	211	1.1	5.7	8.0	A0			υ Gru
23 06.9	−43 31	Jc 20	AB	Gru	1846 1959	11 75	3.0 1.1	4.5	7.0	F4			θ Gru
—			AC		1846 1904	293	160.0		8.1	G0			
23 07.2	+60 49	β 180	AB	Cep	1875 1957	177 155	0.6 0.6	8.1	8.6	A2	16518		
—			AC		1875 1915	106	34.3		10.7		16518		
23 07.2	−50 41	Δ 246		Gru	1835 1952	261 256	8.1 8.6	6.1	6.8	F5 F5			
23 07.4	+20 35	OΣ 488 rej		Peg	1865 1925	334	14.1	6.7	10.4	K0	16520		
23 07.4	+70 40	Σ 2984	AB	Cep	1832 1924	295	4.5	7.7	10.2	K0	16525		
—			AC		1911	81	74.8		13.1		16525		
23 07.5	+32 50	Σ 2978		Peg	1830 1969	145	8.4	6.3	7.5	A2	16519		
23 07.7	+46 23	h 1849		And	1879 1925	346 347	47.2 48.2	5.3	11.7	K5	16526		4 And
23 07.8	+12 40	β 1025	AB	Peg	1891 1957	269 304	0.8 0.8	7.9	10.7	F0	16524		
—			AC		1891 1925	84	22.1		11.7		16524		
23 07.8	+39 48	Σ 2979		And	1831 1945	218 224	3.1 3.1	7.7	9.7	F2	16527		
23 07.8	+63 38	Hu 994		Cep	1904 1960	306 141	0.2 0.1	6.8	7.3	B3	16530		Single 1958 (82-inch)
23 07.9	+31 28	β 78	AB	Peg	1879 1915	55 55	17.2 18.2	7.3	11.1	A2	16528		
—			AC		1879 1915	62	48.4		11.6		16528		
23 07.9	+75 23	OΣ 489	AB	Cep	1846 1960			4.6	6.6	G5	16538	o	33 π Cep
—		h 1852	AC		1906 1911	240	58.6		12.0		16538		
23 08.6	−59 44	Δ 245		Tuc	1836 1933	291	13.7	7.6	9.8	F2			
23 08.6	−63 52	B 590		Tuc	1926 1928	355	0.1	8.0	8.4	G0			
23 08.8	+10 57	A 1238	AB	Peg	1905 1959			8.3	8.5	F5	16539	o	
—			ABxCD		1905 1926	294	70.6				16539		
—			CD		1905 1944	120	1.2	11.6	11.6		16539		
23 09.4	+18 44	Ho 487		Peg	1892 1924	117 119	17.4 18.5	6.6	12.4	B9	16547		
23 09.5	+8 41	Σ 2982		Peg	1831 1923	198	32.6	5.1	9.7	M	16550		57 Peg
23 09.7	+59 20	β	AB	Cas	1904 1934	322	19.9	5.6	13.0	A3	16556		2 Cas
—		S 823	AC		1824 1916	163	167.2		8.4	A	16556		
—			AD		1899 1912	276 297	162.7 162.7		8.9		16556		
—		Kui 115	Aa		1934	41	10.6		14.7		16556		
—			Ab		1934	278	28.9		15		16556		
—		S 823	CD		1912	310	276.0				16556		
23 09.9	−22 27	Rst 3320		Aqr	1935 1959	38 7	0.4 0.4	5.1	5.9	G2			89 Aqr; sp. compos. G2+A2
23 10.0	+14 26	Σ 2986		Peg	1829 1967	273	31.6	7.8	10.6	G0	16553		
23 10.0	+36 51	S 825	AB	And	1825 1925	319	66.4	7.7	8.2	K2 K2	16558		
—		β	BC		1904 1922	334	12.4		14.7		16558		
23 10.0	+47 58	Σ 2985		And	1832 1955	254	15.4	6.6	8.6	G5 G5	16557		
23 10.0	−42 52	Don 1042		Gru	1930 1956	164 204	0.9 0.9	5.8	9.2	F8			
23 10.2	+57 27	OΣ 490	AB	Cas	1846 1955	308 299	1.4 1.3	7.3	9.3	G5	16560		
—			AC		1909	351	82.5				16560		
23 10.3	+32 29	β 385	AB	Peg	1876 1959	136 101	0.4 0.6	7.4	8.2	B9	16561		
—		h 5532	AC		1876 1951	77	58.1		9.1		16561		
23 10.4	+49 01	Σ 2987		And	1832 1950	166 156	3.4 4.1	7.3	10.2	G0	16562		
23 10.7	+26 31	β 852	AxBC	Peg	1881 1913	283	58.5	7.2		A2	16567		
—			BC		1881 1954	11 347	1.2 1.2	10.8	11.3		16567		
—			BD		1898 1914	208 210	18.0 19.3		14.0		16567		
23 10.8	+45 31	h 1853		And	1905	281	31.4	7.2	10.4	K0			[86]
23 11.2	+5 00	h 980	AB	Psc	1902	184	73.4	6.9	10.6	M			
—			Aa		1876	120	40.0						
—			BC		1902	200	9.1		11.2				
23 11.5	+38 13	Ho 197	AB	And	1885 1958	111 3	0.4 0.1	8.3	8.6	F5	16576	o	
—			AC		1885 1925	330 327	42.6 40.3		8.7		16576		
—			AD		1894 1925	281 281	47.3 50.5		8.7		16576		
23 11.8	+26 51	β pm	AB	Peg	1909 1924	240 240	86.7 82.9	6.3	11.2	K0			60 Peg
—			AC		1853 1909	292 294	238.4 231.5		9.6				
23 11.8	−44 43	I 1466		Gru	1925 1950	226 206	0.4 0.6	8.2	8.7	F8			
23 12.0	−11 56	Σ 2988		Aqr	1830 1967	279	3.6	7.8	7.8	K0	16579		
23 13.0	−32 19	Howe 92		Scl	1877 1951	266	7.0	7.6	10.2	K0			
23 13.1	+40 00	Σ 2992	AB	And	1830 1928	286	14.2	7.6	9.3	A3	16599		
—			AC		1910	343	51.1		13.4		16599		
23 13.2	−49 37	I 1467		Gru	1926 1936	9	0.5	6.8	8.4	G5			
23 13.3	+57 10	β pm		Cas	1852 1908	140 208	102.8 106.6	5.6	9.4	K2			Bradley 3077
23 13.3	−60 35	Hu 1645		Tuc	1914 1942	73	1.2	8.5	9.0	A3			
23 13.4	+11 04	Σ 2991 rej		Peg	1883 1959	359	33.6	5.9	9.2	K0	16603		
23 13.8	−13 24	β 181	AB	Aqr	1876 1959	309	1.4	7.4	10.7	K2	16608		
—			AC		1878 1959	235 243	18.8 18.9		12.2		16608		

α 2000	δ 2000	Star Name		Const	Years	Pos Ang	Separation	m₁	m₂	Spec	ADS	Orb	Notes
23ʰ14.ᵐ0	−8°56′	Σ 2993	AB	Aqr	1830 1969	176°	25.″2	8.3	9.4	G0 G	16611		
—		S 826	AC		1824 1954	109 120	158.2 103.0		11.4		16611		
—			BC		1909 1942	104 106	112.4 96.7				16611		
23 14.0	+24 14	Ho 299		Peg	1887 1935	75	1.1	8.0	10.2	A0	16612		
23 14.1	−2 38	β 714		Psc	1878 1958	146 121	0.6 0.4	7.5	10.5	A2	16613		
23 14.4	+29 46	h 1859		Peg	1905 1924	121	35.2	6.4	9.9	F5			
23 14.5	+20 27	h 982	AB	Peg	1902	176	41.2	7.2	11.9	K2			
—			AC		1902	218	46.0		11.5				
23 14.7	−10 41	β 715	AB	Aqr	1877 1958	256	3.5	6.4	10.9	K5	16618		
—		Fox	AC		1913	227	37.0		12.5		16618		
—			AD		1913	133	120.0		11.0		16618		
23 15.5	−59 42	I 143		Tuc	1900 1943	9	3.1	7.7	11.0	K2			
23 15.9	−9 05	β 1220	AD	Aqr	1877 1924	275 274	63.0 80.4	4.5	13.5	K0	16633		
—		Σ 12	AxBC		1836 1938	312	49.6			K	16633		
—		β 1220	BC		1890 1958	101 105	0.2 0.3	10.8	11.5	K	16633		
—			BE		1877 1924	34 341	18.4 19.7		14.3		16633		
23 15.9	+56 41	Stein 2965		Cas	1907	249	12.9	7.9	12.4	A0			
23 16.4	+64 07	β 992		Cep	1880 1958	170 67	0.4 0.2	8.2	8.4	F0	16638	o	
23 16.6	−1 35	Σ 2995		Psc	1830 1950	31	5.0	8.1	8.7	G5	16642		
23 16.7	+53 13	β pm		Cas	1909 1930	319 320	124.0 129.3	5.5	12.8	F8			Bradley 3084
23 16.7	−44 29	h 5390		Gru	1913 1933	44	22.5	5.9	10.7	K0			
23 17.5	+16 52	H1d 171	AB	Peg	1884 1948	32	2.4	8.4	9.4	F5	16648		Mag. of B here is for BC
—		Hu 497	BC		1901 1958	241 187	0.4 0.2	9.9	10.4		16648		
23 17.5	−45 14	Don 1046		Gru	1929 1947	58 67	0.4 0.4	8.0	9.5	G0			
23 17.6	+18 18	Hu 400		Peg	1901 1958	249 168	0.3 0.4	6.9	8.5	F0	16650	o	
23 17.7	+49 01	β 717	AB	And	1878 1934	161	7.6	4.9	12.9	M	16656		8 And
—			AC		1880 1908	131	219.4		10.2		16656		
—		Fox	AD		1916	232	54.9		11.5		16656		
—		Kui 116	Aa		1934	173	14.0		15		16656		
—			Ba		1934	185	6.4				16656		
23 18.0	−61 00	Δ 247		Tuc	1870 1959	281 291	40.6 46.6	6.7	7.8	G5 A3			
23 18.4	+47 16	A 202		And	1900 1932	257	2.5	7.8	10.1	B9	16661		
23 18.6	+68 07	Σ 3001	AB	Cep	1832 1970			4.9	7.1	G5	16666	o	34 o Cep
—			AC		1912	4	45.6		12.7		16666		
23 18.6	−58 18	h 5392		Tuc	1835 1959	19 322	7.0 35.1	7.5	9.5	G0			
23 18.9	+48 30	OΣ 493		And	1847 1911	25	8.4	7.5	10.5	A5	16673		
23 19.0	−9 37	Ho 199	AB	Aqr	1884 1959	224 174	1.2 1.5	5.0	11.0	A0	16671		95 ψ³ Aqr
—			AC		1912	230	130.1		13.1		16671		
23 19.1	+48 55	Es		And	1881 1918	231	49.4	7.7	9.2	A2			
23 19.1	−13 28	Σ 2998		Aqr	1830 1976	345 350	13.4 12.7	5.3	7.3	G5	16672		94 Aqr
23 19.4	−5 07	h 5394		Aqr	1869 1958	24 20	9.8 10.6	5.6	10.5	F2	16676		96 Aqr
23 19.7	+48 23	OΣΣ 244	AB	And	1875 1920	305 300	78.9 85.8	6.4	9.7	K0			
—			AC		1910 1925	104 105	136.4 134.3		11.3				
23 19.7	+73 55	h 1870		Cep	1905	294	15.6	7.8	11.4	F0	16683		
23 19.7	−33 43	HdO 302		Scl	1901	307	43.8	6.4	12.5	K0			
23 19.7	−46 19	Rst 3323		Gru	1932 1947	191	0.4	8.2	9.9	G5			
23 19.8	+57 15	β 229		Cas	1876 1915	36	17.5	7.2	11.9	K2	16681		
23 19.9	+34 44	Cou 742		Peg	1971	32	0.3	8.6	8.8	G0			
23 20.6	+62 13	β 278	AB	Cas	1890 1934	175	12.7	6.4	11.6	K5	16690		[87]
—			AC		1913 1922	259	58.5		11.3		16690		
—			AD		1913 1922	185	74.4		11.3		16690		
23 20.7	+44 07	Σ 3004		And	1833 1958	178	13.2	6.1	9.6	A3	16685		
23 20.8	+2 27	Σ 3002	AB	Psc	1831 1947	201	4.1	7.9	10.1	F5	16687		
—		Ho 488	BC		1890 1947	215	0.8		11.0		16687		
23 20.8	+21 57	OΣ 494		Peg	1850 1959	82	3.2	8.3	9.0	F0	16686		
23 20.8	+62 25	Es 220	AB	Cas	1905 1922	83	36.9	7.6	11.1	A0	16692		
—			BC		1905 1921	76	6.0		12.1		16692		
23 20.8	−50 18	Rst 5560	AB	Gru	1947	243	0.5	6.3	8.3	F5			
—		Δ 248	AC		1836 1952	211	16.8		8.9	A3			
23 20.9	+38 11	β pm	AB	And	1890 1918	256 256	119.0 121.1	5.8	9.2	F5			12 And
—			BC		1907	214	155.8						
23 21.0	−18 32	h 3184		Aqr	1869 1953	283	5.4	7.2	9.2	G5	16688		
23 21.2	−8 40	Rst 4721		Aqr	1939 1943	62	3.7	7.4	14.6	A2			
23 21.9	+31 49	β 718	AB	Peg	1878 1958	88 133	0.5 0.5	5.3	9.0	B8	16702		64 Peg
—			AC		1912	147	112.9		13.3		16702		
23 22.2	+56 05	h 1873		Cas	1904 1908	55	8.2	8.2	10.2	A0	16705		
23 22.5	+25 55	Es		Peg	1907	264	20.0	6.3	12.6	K2	16707		
23 22.7	−15 02	Hu 295		Aqr	1900 1959			5.6	6.4	A3	16708	o	97 Aqr
23 22.8	+20 34	Σ 3007	AB	Peg	1829 1976	79 91	5.7 5.9	6.6	9.6	G0	16713		
—			AC		1894 1956	322 311	75.8 88.2		8.9		16713		
23 23.1	+12 19	Ho 300		Peg	1906 1954	304 56	0.2 0.1	5.9	5.9	K0	16715		66 Peg; round in large ap.
23 23.5	+0 17	h 3189		Psc	1879 1909	136	41.7	6.3	10.3	K2			
23 23.8	−8 28	Σ 3008		Aqr	1830 1960	273 176	7.5 4.0	7.2	8.2	K0	16725		An optical pair
23 23.9	−53 49	Δ 249		Gru	1826 1951	212	26.5	6.5	7.3	A5 A5			
23 24.0	−27 17	Howe 63		Scl	1877 1934	267	6.5	8.0	11.0	B9	16728		

α 2000	δ 2000	Star Name		Const	Years	Pos Ang	Separation	m₁	m₂	Spec	ADS	Orb	Notes
23ʰ24.1ᵐ	+57°32′	OΣ 495		Cas	1846 1960	130° 111°	0″.6 0″.2	7.5	7.7	B5	16731		
23 24.1	−21 46	I 1058		Aqr	1913	233	1.9	6.6	10.3	F5	16727		Comp. prob. not real (IDS)
23 24.3	+3 43	Σ 3009		Psc	1829 1944	230	7.0	6.8	8.8	K2	16730		
23 24.3	+86 25	Ho 200		Cep	1885 1912	142	2.3	6.6	12.1	F0	16759		
23 24.4	+14 29	β 719	AB	Peg	1877 1954	11 350	1.1 1.3	8.1	11.1	F8	16735		
—			AC		1895 1916	108 106	106.4 107.6		11.1		16735		
23 24.8	+62 17	H VI 24	AB	Cas	1879 1917	226	98.6	5.2	7.7	K5			4 Cas
—			AC		1879 1917	258	215.0		8.6				
—			CD		1879 1912	36	9.8		9.6				
23 25.9	+27 42	Ho 489	AB	Peg	1889 1956	241 235	0.4 0.6	8.6	8.6	F8	16748		
—		OΣΣ 245	AC		1875 1918	195	63.2		8.9		16748		
23 26.2	+70 41	β 386	AB	Cep	1876 1925	313	19.9	6.7	12.1	A2	16754		
—			AC		1907 1910	203	46.5		12.3		16754		
23 26.5	−63 14	Hu 1648		Tuc	1914 1934	283	1.9	7.8	12.5	A3			
23 26.6	+23 42	Cou 338		Peg	1968	25	0.2	8.5	8.9	A5			
23 26.6	−21 44	See 485		Aqr	1897 1933	131	5.2	6.6	12.0	K0	16753		
23 26.9	+1 15	S 830	AB	Psc	1824 1921	345 344	150.1 163.4	4.9	11.9	A2			8 κ Psc; optical (IDS)
—			BC		1907	154	76.1		13.1				
23 26.9	−15 15	Hu 297		Aqr	1900 1951	312 292	0.4 0.5	7.1	9.1	A3	16758		
23 27.2	+1 07	Weisse 40		Psc				6.3		K0			9 Psc; no companion found
23 27.2	−50 17	Δ 250		Phe	1826 1954	94 86	51.8 33.4	7.7	8.7	K2 K2			
23 27.3	+65 37	β 1148		Cep	1889 1948	75	2.1	7.0	12.9	K0	16764		
23 27.6	+16 38	Σ 3012	AB	Peg	1831 1953	191	2.8	8.2	9.3	G0 G0	16766		
—			AC		1831 1953	66	52.4		9.0		16766		
—		Σ 3013	BC		1909 1953	64	54.1				16766		
—			BD		1909 1953	62	51.1		9.8		16766		
—			CD		1831 1974	274 275	2.9 3.2				16766		
23 27.7	+74 07	Σ 3017	AB	Cep	1832 1955	35 24	2.4 1.6	7.5	8.6	F0	16775		
—			AC		1913	141	89.0		13.4		16775		
23 27.8	−50 21	I 1059		Phe	1913 1928	210 188	1.4 1.1	7.5	11.3	G5			
23 28.2	−56 26	I 23		Phe	1900 1942	322 334	0.8 0.8	7.8	9.0	F2			
23 28.3	+6 05	H V 48		Psc	1903 1917	3	88.4	8.2	9.7	F2			
23 28.8	−0 50	A 896		Psc	1904 1944	69	0.8	7.6	10.1	F8	16781		
23 29.3	−85 43	Jsp 851		Oct	1929	64	3.0	7.7	10.9	K0			
23 29.5	+51 41	h 1885	AB	Cas	1905	248	21.1	8.0	12.0	M			
—			AC		1905	245	45.6		12.5				
23 29.8	+31 42	Cou 544		Peg	1970	261	1.0	7.7	11.0	A3			
23 30.0	+58 33	OΣ 496	AB	Cas	1851 1947	337 347	1.5 1.1	4.9	9.3	B3	16795		BDS 12349, H VI 25, AR Cas
—		S,h 355	AC		1841 1922	269	75.7		7.1	A0	16795		
—		h 1888	AE		1870 1918	114	43.4		8.9		16795		
—			AF		1870 1905	338	67.3		8.9		16795		
—			AG		1914 1918	348	67.0		9.1		16795		
—		Dawes 2	CD		1841 1956	222	1.4		8.9		16795		
—		h 1886	CH		1841 1880	341 337	23.4 26.9		12.9		16795		
—		h 1887	FG		1870 1918	74	10.6				16795		
23 30.4	+30 50	β 1266	AB	Peg	1891 1958			8.1	8.1	F5	16800	o	
—		Σ 3018	ABxC		1830 1969	204	18.9		9.6		16800		
—		Lv	AD		1916	8	36.8		13.9		16800	o	
23 30.7	+5 15	Σ 3019		Psc	1832 1951	185	10.7	7.6	8.6	A3	16803		
23 30.7	+64 20	β 774		Cep	1880 1958	7 350	0.5 0.5	8.5	8.9	A3	16806		
23 31.0	−69 05	h 5402		Tuc	1870 1916	198	36.3	7.2	9.3	G0			
23 31.1	+18 47	Σ 3020		Peg	1831 1955	111 104	1.7 2.0	7.6	9.6	A3	16808		
23 31.2	+62 05	Hu 1000		Cas	1904 1947	186 192	0.9 0.8	8.1	10.6	K0	16814		
23 31.3	+52 25	OΣ 498 rej		Cas	1866 1923	244	17.2	7.4	10.2	F5	16810		
23 31.4	+16 13	Σ 3021	AB	Peg	1830 1951	308	8.6	7.9	9.1	F8	16812		
—			AC		1913 1931	24 24	118.0 118.7		10.1		16812		
23 32.0	+38 18	h 1889	AB	And	1900	241	43.9	7.6	11.9	K2			
—			AC		1900	57	55.8		11.6				
23 32.2	+7 05	Hu 298		Psc	1900 1960			7.5	7.7	F5	16819	o	
23 32.4	+17 24	Σ 3023		Peg	1831 1955	280	1.9	7.2	9.9	F2	16821		
23 32.7	−45 07	See 488		Phe	1897	337	18.9	7.0	12.0	A2			
23 33.2	+57 25	OΣ 499	AxBC	Cas	1847 1929	78	9.5	7.4		G5	16828		
—		A 641	BC		1903 1958	147 160	0.4 0.3	9.0	10.8		16828		
23 33.3	−20 55	B 1900		Aqr	1929 1959	126 126	0.6 0.9	4.8	7.1	A0			101 Aqr
23 33.6	+17 49	Ho		Peg	1894	210	13.8	6.7	12.5	G0	16830		
23 33.6	+60 28	Σ I 60	AB	Cas	1835 1923	211 209	247.2 240.8	7.4	7.3	K0 K0			
—			Aa		1910	136	135.1		11.2				
—			Bb		1910	309	82.5		12.6				
23 33.8	−36 16	See 489		Scl	1896 1899	145	19.9	7.2	11.5	F0			
23 33.9	−61 39	Rst 3326		Tuc	1936	142	1.3	8.1	11.0	G0			
23 34.0	+31 20	β 720		Peg	1878 1960			5.7	5.8	K2	16836	o	72 Peg
23 34.8	+38 01	β 388		And	1876 1908	338 333	21.8 20.2	6.2	11.7	K5	16843		
23 35.1	−42 37	B 603		Phe	1925 1930	270	6.7	4.7	12.7	A2			z Phe
23 35.2	−64 41	h 5403		Tuc	1836 1916	45	37.7	7.4	9.8	F0			
23 35.3	−57 30	I 25		Phe	1895 1947	57 42	0.8 0.8	8.1	8.2	F2			

α 2000	δ 2000	Star Name		Const	Years		Pos Ang		Separation		m₁	m₂	Spec	ADS	Orb	Notes
23ʰ 35.7	−27° 29′	See 492		Scl	1897	1956	°	°	″	″	7.0	8.0	F8	16850	o	
23 36.3	−7 07	β 721	AB	Aqr	1878	1958	138	143	0.5	0.1	8.7	8.7	A3	16858		Too close 1936–39
			AC		1898	1899	302		21.7			11.4		16858		
23 36.9	−36 48	I 693		Scl	1914	1956	4	54	0.9	0.9	8.1	9.1	F5			
23 37.1	−31 52	Howe 93		Scl	1877	1937	251		5.5		6.5	9.8	K0			
23 37.5	+44 26	OΣ 500	AB	And	1845	1959	299	352	0.4	0.5	6.3	7.2	B9	16877		
			AC		1912		334		115.8			10.5		16877		
23 37.6	+46 28	β pm	AB	And	1908		199		47.5		3.8	13.1	K0			16 λ And
			AC		1868	1908	93	89	222.3	217.6		10.3				
			CD		1908		158		68.5			11.3				
23 37.7	+16 50	β pm	AB	Peg	1887	1909	36	34	100.1	98.3	6.2	10.7	A0			74 Peg
			BC		1907	1909	239		25.7			12.2				
23 37.7	−13 04	h 316		Aqr	1877	1909	93		33.1		5.7	9.6	G5	16878		
23 37.9	+18 24	Kui 117		Peg	1958		246		27.6		5.4	11.6	A0			75 Peg
23 38.0	+52 50	Es		Cas	1911	1918	142		19.5		8.2	9.5	K5			
23 38.4	+42 31	β 722	AB	And	1878	1917	348		7.4		6.8	12.5	B9	16890		
			AC		1916		219		38.4			11.3		16890		
			AD		1916		247		45.1			10.9		16890		
23 38.6	+35 02	Σ 3028		And	1829	1957	205	201	19.5	16.0	7.1	9.6	A2	16894		
23 38.6	+44 41	Webb	AB	And	1919		306		125.4		8.5	8.6	F5 A0			
			AC		1919		248		117.2			9.8				
			BC		1919		183		114.2							
			BD		1919		253		104.4			9.9				
			CD		1919		312		129.2							
23 38.8	+62 08	h 1896		Cas	1902	1912	116		16.3		6.6	11.0	A2	16898		
23 39.3	+45 43	A 643		And	1903	1959	264	195	0.2	0.2	8.4	8.6	A2	16904	o	
23 39.5	−46 38	Δ 251		Phe	1835	1959	269	275	4.0	4.0	6.6	7.2	A3			
23 39.9	+5 38	β pm		Psc	1879	1909	288	294	54.0	69.9	4.1	12.8	F8			17 ι Psc
23 39.9	+63 44	OΣ 502		Cep	1848	1916	223		3.6		6.9	10.6	A2	16911		
23 39.9	+64 19	Es 149	AB	Cep	1902	1922	119		5.7		8.7	8.9	F8	16910		
			AC		1910		177		50.5			10.6		16910		
23 40.0	+37 39	OΣ 501		And	1847	1958	164		14.9		6.5	9.9	F0	16913		
23 40.0	−47 20	HdO 303		Phe	1901	1942	65		2.0		7.0	10.0	G5			
23 40.4	+44 20	h 1898	AB	And	1879	1923	189	194	46.6	46.8	4.1	11.1	A0	16916		19 κ And
			AC		1879	1923	295	294	103.2	107.0		11.1		16916		
23 40.5	+24 42	Pou 5842		Peg	1905		54		11.6		7.8	13.1	F5			
23 40.5	+67 33	β 857		Cep	1881	1958	297		1.4		8.4	8.8	F8	16921		
23 40.7	+0 25	β 723		Psc	1878	1960	168	164	3.8	3.7	7.7	12.0	K0	16919		
23 41.1	+46 13	Mlr 4		And	1953	1960	151	259	0.1	0.2	7.8	8.1	F5		o	
23 41.3	+32 34	β 858	AB	Peg	1881	1959	277	238	0.5	0.8	7.4	8.9	A0	16928		
		β 389	AC		1881	1909	51	52	23.7	23.3		12.3		16928		
23 41.5	−41 35	Hu 1550		Phe	1913	1959	182	192	0.7	0.7	8.4	8.7	F5			
23 41.8	−17 49	h 5413	AB	Aqr	1892	1919	6	6	123.2	120.1	4.9	7.7	G0			104 Aqr
			BC		1913		76		90.2			11.7				
23 42.1	+20 18	OΣ 503	AB	Peg	1848	1958	132		1.4		8.1	8.7	F8	16937		
			AC		1911		234		37.9			14.2		16937		
23 42.3	+64 31	β 993		Cas	1880	1955	280	275	2.7	2.7	6.6	11.0	M	16940		
23 42.5	+61 41	Mlr 23		Cas	1969		217		8.0		6.5	13.0	K2			
23 42.6	+18 40	OΣ 504		Peg	1849	1920	175		7.7		7.4	10.2	K0	16942		
23 42.7	−14 33	β 279		Aqr	1875	1958	86		5.7		4.5	10.5	A0	16944		105 ω² Aqr
23 42.8	−11 20	β 725		Aqr	1877	1933	240		4.3		7.3	11.3	K0	16946		
23 43.1	−46 19	δ 28	AB	Phe	1914	1933	74		3.8		6.7	11.5	G5			
		h 5416	AxBC		1914		215		45.2							
		I 1607	CD		1927	1934	171		0.5		10.7	11.5				
23 43.3	−70 49	h 5415		Tuc	1917		126		38.2		7.5	11.2	G5			
23 43.4	+58 05	β pm	AB	Cas	1894	1918	212	213	91.5	106.3	7.1	9.6	G0			
			AC		1894	1918	134	141	115.6	118.9		9.6				
			AD		1894	1918	188	190	158.3	171.3		9.3				
			AE		1894	1918	120	126	160.5	159.6		9.6				
			AF		1894	1918	190	192	211.6	225.3		9.2				
23 43.5	+25 06	β 994		Peg	1880	1955	306		1.4		8.1	11.2	K0	16956		
23 43.9	+7 15	Σ 3033		Psc	1832	1948	6		3.3		8.8	8.8	F2	16958		
23 44.0	+29 22	AGC 14		Peg	1876	1959	192	235	1.4	1.0	5.0	8.1	K0	16957		78 Peg
23 44.5	+46 23	Σ 3034		And	1831	1934	103		5.4		7.6	9.8	A0	16965		
23 44.5	−26 15	h 5417		Scl	1856	1952	320		8.5		6.3	9.0	F5	16963		
23 44.9	−38 20	B 613		Scl	1926	1956	145	133	0.3	0.2	8.6	8.8	F0			
23 45.5	+20 25	OΣ 505		Peg	1849	1943	60		2.2		6.8	10.0	K0	16970		
23 45.6	−58 17	B 1020		Tuc	1927	1949	213		0.3		7.8	8.4	A3			
23 46.0	+46 25	Wils	AB	And	1908		218		24.8		5.0	14.5	K0			20 ψ And; single (Wils)
		β	AC		1908		17		62.0			13.0				
			AD		1879	1908	152		184.0			9.0				
23 46.0	−18 41	H II 24		Aqr	1823	1971	143	136	5.5	6.6	5.7	6.7	A5	16979		107 Aqr
23 46.1	+50 40	OΣΣ 248	AB	Cas	1876	1923	140		52.7		7.7	9.8	K0	16980		
			AC		1901	1915	339		23.3			12.4		16980		
23 46.1	+60 28	Σ 3037	AB	Cas	1832	1956	213		2.7		7.1	8.6	K0	16982		

α 2000	δ 2000	Star Name		Const	Years		Pos Ang		Separation		m₁	m₂	Spec	ADS	Orb	Notes

α 2000	δ 2000	Star Name		Const	Years		Pos Ang		Separation		m_1	m_2	Spec	ADS	Orb	Notes
23ʰ46.1ᵐ	+60°28′	Σ 3037	AC	Cas	1832	1956	186°		29.2″			9.0		16982		
—			AD		1915	1956	228	229	50.4	52.6		9.2		16982		
—			AE		1956		242		110.7					16982		
—			AF		1956		325		125.1					16982		
—			BC		1899		183		26.7					16982		
23 46.3	−13 22	Rst 4133		Aqr	1936	1943	156		2.9		8.0	12.5	G5			
23 46.9	+28 25	Σ 3039	AB	Peg	1830	1923	36	33	30.3	33.1	7.4	9.8	M0			
—			AC		1875	1927	94		709.0			8.2	A2			
23 47.4	+63 13	Dem 27	AB	Cas	1877	1940	359	354	1.6	1.7	8.1	10.7	B0	17001		
—			AC		1877	1940	144		10.4			10.6		17001		
23 47.4	−71 18	I 695	AB	Tuc	1914		24		8.3		7.1	13.5	K0			Mag. of A is for Aa here
—		φ 375	Aa		1960		131		0.1		7.9	7.9				
23 47.5	+25 35	β 727		Peg	1878	1915	315		16.9		7.1	12.6	K0	17005		
23 47.5	+49 18	β 390		Cas	1880	1958	234	231	18.0	15.4	7.5	11.0	B9	17004		
23 47.6	+46 50	β 995		And	1880	1958	241		0.8		6.1	8.1	B3	17006		
23 47.6	−60 31	CorO 261		Tuc	1900	1930	100		5.6		8.5	9.0	K0			
23 47.7	−0 46	A 899		Psc	1904	1930	38		3.6		7.2	14.2	K0	17007		
23 47.9	−2 46	S 835	AB	Psc	1824	1918	285		172.7		5.5	9.9	K0 G2			20 Psc
—			BC		1911		280		52.2			12.8				
23 47.9	+17 03	Σ 3041	AB	Peg	1900	1936	351	352	64.4	61.0	8.0	7.8	A0 G	17009		
—			AC		1900	1921	351	353	68.1	66.4		9.2		17009		
—			AxBC		1832	1923	348	352	71.1	64.6				17009		
—			BC		1832	1955	3	357	3.3	3.2				17009		
23 48.1	+63 49	β 1152	AxBC	Cas	1889	1913	136		74.4		7.6		A0	17010		
—			BC		1889	1957	102		0.6		10.0	10.0	F5	17010		
23 48.1	−50 53	I 305		Phe	1900	1936	123		3.1		7.0	11.0	K0			
23 48.5	+36 08	Cou 944		And	1972		108		0.1		8.8	8.8				
23 48.6	+36 16	B 2547	AB	And	1958	1960	346		0.3		7.6	8.6	G0	17019		
—		OΣ 506 rej	AC		1868	1958	80	81	17.9	19.5		10.6		17019		
23 48.6	+64 53	OΣ 507	AB	Cas	1843	1959	218	297	0.4	0.8	6.9	7.6	A0	17020	o	
—			AC		1847	1959	354	351	48.8	50.4		8.6		17020		
—		Fox	AD		1916		109		51.0			13.1		17020		
23 48.8	+62 13	OΣ 508	AB	Cas	1854	1958	193		1.6		5.5	8.0	A2	17022		6 Cas
—			AC		1912		309		62.4			10.5		17022		
—		Gui			1924		20		12.2					17022		Unidentified meas. by Gui
—					1923		130		69.2					17022		Another?
23 48.9	−28 08	β 1013	AB	Scl	1881	1956	228	243	3.3	3.9	4.5	11.5	A0	17021		δ Scl
—		h 3216	AC		1881	1929	297		74.3			9.3	G	17021		
23 49.3	+25 21	Weisse 42		Peg							8.3	9.0	K0			Single
23 49.8	+27 41	A 424		Peg	1902	1960					7.6	7.8	F0	17030	o	
23 49.8	−25 20	h 5423		Scl	1877	1944	307		13.5		6.4	11.6	A0	17029		
23 50.2	−29 24	B 620		Scl	1927		9		2.5		7.7	14.7	G5			
23 50.4	+51 37	Gui	AB	Cas	1915		163		21.6		6.4		F2	17032		
—			AC		1915		281		46.2					17032		
23 50.4	+57 10	Stein 3046		Cas	1906		218		10.0		7.9	13.0	B9			
23 50.5	+43 25	OΣ 509		And	1854	1941	104		5.5		7.8	9.7	A0	17037		
23 50.6	+54 12	Es 700		Cas	1908		35		14.6		7.2	11.2	F5	17038		
23 50.6	−51 42	Slr 14		Phe	1891	1959					8.1	8.4	G5		o	
23 51.4	+47 45	A 796		And	1904	1956	30	14	0.5	0.6	7.7	10.2	B9	17047		
23 51.5	−32 26	See 496		Scl	1896	1927	192		16.8		7.8	12.5	F0			
23 51.6	+42 05	OΣ 510	AB	And	1848	1959	348	312	0.4	0.4	7.7	8.0	A5	17050		
—		h 1911	ABxC		1847	1925	345		21.3			9.0		17050		
23 51.9	+37 53	Σ 3042		And	1832	1968	89	87	4.2	5.4	7.8	7.8	F5	17054		
23 52.1	+43 30	β 728		And	1878	1957	353	3	1.1	1.1	8.7	8.7	F8	17063		
23 52.4	+75 33	β 996	AB	Cep	1880	1958	65	95	5.5	4.6	6.5	11.4	K2	17062		
—			AC		1914		132		161.9			9.2		17062		
23 52.7	+60 42	β 1153	AB	Cas	1889	1946	318		0.4		9.7	9.9		17073		
—			ABxC		1889	1913	340		13.9			10.1		17073		
—			ABxD		1889	1911	66		176.5			6.9	K5	17073		
23 52.8	−72 24	B 1901		Tuc	1931		297		5.4		7.6	12.6	F2			
23 52.9	−3 09	φ 359		Psc	1959		42		0.1		6.7	6.7	K0			24 Psc
23 53.0	+11 55	Σ 3044		Peg	1830	1951	282		19.0		7.3	7.9	F0 F0	17079		
23 53.0	+41 21	Arg		And	1905	1917	145		51.2		7.3	9.6	M			
23 53.1	+60 42	OΣ 511	AB	Cas	1848	1913	35		10.5		6.7	10.9	K5	17080		
—			AC		1911		39		36.0			13.9		17080		
—			AD		1911		130		69.3			10.9		17080		
23 53.6	+51 31	OΣΣ 251	AB	Cas	1875	1920	197	201	42.4	44.1	7.0	9.4	K0			
—			AC		1910		129		50.1			11.8				
—			BD		1905		171		14.6			13.4				
23 53.7	−29 24	h 5429		Scl	1918	1955	224	226	27.4	28.8	7.5	10.6	K5			
23 53.7	−65 57	h 5428		Tuc	1900	1916	114		12.3		6.7	12.5	F8			
23 54.1	+39 17	Ho 205	AB	And	1885	1906	180		4.6		6.7	12.7	F8	17087		
—			AC		1913		250		92.5			11.3		17087		
—			AD		1913		211		102.3			9.1		17087		
23 54.4	−27 03	Lal 192		Scl	1826	1950	270		6.6		6.9	7.5	A2	17090		

α 2000	δ 2000	Star Name		Const	Years	Pos Ang	Separation	m_1	m_2	Spec	ADS	Orb	Notes
23ʰ 54.8	+74° 25′	h 3226		Cep	1905	7	45.1	6.6	12.1	B9			
23 54.9	+29 29	OΣΣ 252		Peg	1875 1923	143	111.5	6.6	7.6	B9 K0			
23 55.1	−37 22	I 146		Scl	1900 1947	223	0.9	8.2	9.4	F5			
23 55.2	−21 23	HCWils 29		Aqr	1885 1954	194 193	46.0 44.0	7.3	10.0	K0			
23 55.4	−17 50	β 729		Aqr	1877 1916	344	11.5	7.7	11.7	F5	17101		
23 55.7	+56 47	Stein 3057		Cas	1906	317	8.7	7.8	12.6				
23 56.1	−62 52	B 626		Tuc	1927	258	7.5	7.3	13.3	A3			
23 56.8	+4 43	A 2100		Psc	1909 1957	289 229	0.2 0.4	7.4	7.9	F0	17111	o	
23 56.9	+55 50	β 1224	AB	Cas	1890 1951	203	4.0	7.0	13.7	K0	17114		
—			AC		1912	357	77.2		12.1		17114		
23 57.3	+61 02	OΣ 512	AB	Cas	1853 1936	292 293	4.6 3.0	6.7	11.0	M	17119		
—			AC		1874	84	363.5		9.6		17119		
—		Arg 99	CD		1869 1936	318	4.8		10.0		17119		
—			CE		1936	86	21.5				17119		
23 57.4	+72 51	A 900		Cep	1904 1958	104 124	0.3 0.3	8.3	8.8	B5	17118		
23 57.8	+11 28	h 321		Peg	1901 1911	133	20.7	6.6	10.3	A0	17125		
23 58.4	+35 01	OΣ 513		And	1851 1939	22	3.6	6.8	9.3	A3	17136		
23 58.6	−14 08	Rst 4136	AB	Cet	1936 1958	18 14	0.2 0.2	8.0	8.0	F0			
—		B	ABxC		1958	220	11.2		14.7	F0			
23 58.7	−3 33	β 730		Psc	1878 1958	266 292	1.4 1.3	4.9	10.2	K0	17137		27 Psc
23 59.0	+55 45	Σ 3049	AB	Cas	1833 1958	326	3.0	5.0	7.1	B2	17140		8 σ Cas
—			AC		1879 1909	67	109.9				17140		Glorious wide field (Webb)
23 59.1	+74 50	β 1154		Cep	1889 1959	310 322	1.0 1.1	8.6	8.8	A3	17143		
23 59.4	+54 41	A 1498		Cas	1906 1958	68 75	0.4 0.4	8.7	8.9	F5	17151		Spectrum composite F5+A3
23 59.5	+33 43	Σ 3050	AB	And	1832 1971			6.6	6.6	F8	17149	o	
—			AC		1909 1960	289 293	82.2 81.2		12.9		17149		
23 59.5	+57 40	H1m		Cas	1901	76	18.6	7.4	11.4	A0			
23 59.5	−26 31	Lal 193		Scl	1877 1951	170	10.6	8.2	8.5	F0 F0	17150		
23 59.7	+37 48	Es 2443	AB	And	1930	232	75.0	7.8	11.3	B9			
—			BC		1930	150	7.3		11.0				
23 59.8	+11 16	β		Peg	1901	122	25.7	6.6	11.1	A2			

SUPPLEMENTARY NOTES
(Listed by number in brackets)

1. Σ 3. A faint star was seen in field: 1831.33, 133°0, 4".57 (h). An illusion?

2. A 1803. No certain change. Too close 1932, '46, '47.

3. β, AC. 1909 and 1921 measures disagree. BDS 494 is the same star.

4. H V 16. Examined by β and not seen double.

5. h. Probably some error in h position. No such pair here (IDS).

6. Ho 215. Single in 1903, '06 (Doo); also 1907 (A).

7. OΣ 27 rej. Single. A measure at same position: 1920.66, 295°3, 38".49 (Gui).

8. Kui 7. During 1934-49, measured at about 160°, 0".25; 1949-58, 170° to 190°, 0".12.

9. γ Ari. One of the earliest doubles recognized, by R. Hooke in 1664. The southern component is an αCV-type variable.

10. γ And. Splendid double star, orange and emerald green (Smyth), or yellow and blue (H, Σ). Discovered by C. Mayer in 1778. Σ split B into the close BC pair in 1842. In 1960, L. A. Maestre and J. A. Wright found the B star to be a spectroscopic binary as well, consisting of two similar B9.5 components, period 2.67 days.

11. Polaris AB. The companion could be seen by W. Kitchiner with a 1¾-inch refractor, and many years later β succeeded with a 1.6-inch. The primary is itself an astrometric binary, period 30.5 years, but attempts to resolve it directly have failed.

12. HdZ. Marked double in Harvard Zones. Single in 1901 with Yerkes 40-inch.

13. Kui 9. This is HR 775 in Cetus. Some scatter in measured p.a. Single in 1958.

14. η Per. Called "orange, smalt blue" by Smyth. His contemporary Peter Barlow likened this star to a miniature representation of Jupiter and several satellites.

15. Edg. Spectrum of A component is composite, G0+A5; star was checked for duplicity and found single (Wils).

16. Σ 317. BDS mentions a third star 15ˢ following at 102°4.

17. Σ 346. Very close after 1934, and measures uncertain. In 1954, 316°0, 0".1 (VBs).

18. See 23. Motion direct; has gone through more than one revolution.

19. OΣ 60. Probably not double (IDS). Single in 1865, '78, '83, '84, '98, '99.

20. Atlas, in the Pleiades. After 1842, not seen double again until 1936; also reported single in 1937 and 1953. Yet on January 11, 1876, E. Hartwig at Strasbourg noted that the star's disappearance behind the Moon was not instantaneous. Position angles are scattered; distances range from "single" to 0".6. The star is a spectroscopic binary that was resolved in photoelectric diffraction-pattern traces at four lunar occultations in 1968-73; this recent work cannot explain Hartwig's observation, however, and casts doubt on the visual reports of duplicity.

21. β 544. Spectrum composite, F5+A.

22. I 270. Duplicity of AB doubtful.

23. 51 Tau. In 1957, A was not seen double (B).

24. H VI 101. There is a wide faint pair, noticed by h, at p.a. 320°.

25. Σ I 10. Separated 5'.6, these stars are a good naked-eye test. Garrett P. Serviss noted their contrasting colors in an opera glass. The primary, θ² Tauri, is the brightest true member of the Hyades cluster.

26. Lewis 4. Doubtful; often appeared single with Lick 36-inch refractor (A).

27. Lewis 5. May be BDS 2302 = β 1044 with 10° error in declination.

28. A 2424. Too close 1929, '34, '51, '53.

29. Σ 586 rej. Same position and in field of ADS 3442.

30. Dem 5. Single or not seen in 1950, '51, '54, '56.

31. J 307, BC. Single in 1951, '53, '57.

32. OΣ 89. Single in 1956 (Mlr).

33. Rigel. B is a spectroscopic binary, period 9.86 days. But the reality of the visual pair BC is a lingering puzzle. In 1871, β suspected elongation with his 6-inch Clark refractor, and he made the 1878 measure with the 18½-inch Dearborn instrument on two nights. Others also reported splitting the star: "When best defined clearly divided; distance not more than ¼ second" (R, 1878, 11½-inch, 600x and 800x). But β's repeated attempts in 1889-91 with the 36-inch failed, even using powers as high as 2,600x. Recent interferometric observations place an upper limit of 0".1 on the separation, if any.

34. Sei 136. Is B the diffuse nebula Ced 42? Single in 1934 (Kui); "component does not exist" (VBs).

35. Capella. The primary has been known since 1899 as a spectroscopic binary, period 104 days. Visual scrutiny by A and Hu with the Lick 36-inch showed no trace of elongation in 1900, but visual measures with the Greenwich 28-inch in 1900-01 suggested the p.a. decreases with time. This was confirmed by JAn, and separations of 0".041 to 0".054 measured, in 1919-21 with an eyepiece interferometer on the Mount Wilson 100-inch reflector. In 1971, A. Labeyrie first resolved Capella by speckle interferometry, using the Palomar 200-inch.

36. Σ 672 rej. Class IV. No double star

found near here (BDS); may be h 3269 = BDS 2568.

37. GΣ. Measure is said to be of AE, but it must be of another pair.

38. HCWils 3. No pair of this description nearby. May be BDS 2804 = β 1048.

39. Orion Trapezium. Christiaan Huygens distinguished A, B, and C in 1656, and glimpsed D in 1684. Σ found E in 1826; h added F in 1832. AGC spotted G with the 36-inch in 1888; Barnard saw and measured the faint pair HH′ with the same telescope. Both A and B are eclipsing binaries.

40. β 1007. Close binary in rapid motion. No elongation in 1918, '21, '36, '45.

41. Betelgeuse. Doo measures a star in direction 23°.1 from primary. 1903.44, 173°.2, 8″.09, 11.0 to 11.8.

42. 1 Gem. Spectroscopic binary. Composite according to *Henry Draper Catalogue*: *G8 + G0* and *K0 + F8*? Evidently both visual components are giants and one of them, probably the later-type component, is itself a close binary (period 9.59 days).

43. β 1192. BC round in 1955 (VBs and φ), but it is unclear whether the measures belong to BC or AA′.

44. Sirius. B is the white-dwarf companion discovered visually by AGC while testing the 18½-inch Dearborn objective on January 31, 1862. The companion's existence had been deduced by F. W. Bessel in 1844 from irregularities in Sirius' proper motion. The 50-year period, eccentricity, and periastron date were rather well known before AGC's observation, which then supplied the missing information about the orbit's size. The C component was suspected by Fox in 1920, and also reported by I, B, and φ, while other observers have failed to confirm the duplicity of B. Single in 1945, '46, '49, '50. Furthermore, a study by I. W. Lindenblad of the B star's motion, 1965-72, produced strong astrometric evidence that neither A nor B can have a close companion.

45. OΣΣ 79. A faint star lies almost exactly between A and B.

46. B 719. This multiple star appears involved in a small nebulosity, "shaped like the tail of a comet" (IDS). Companions C to F may be knots in the nebula.

47. Castor. Reported to be double by J. Bradley and J. Pound in 1718, and the first star that H argued must be a true physical binary (1803). Although Castor has now been followed through three-quarters of a revolution, the orbit is still poorly determined.

48. Procyon. Discovered November 14, 1896, by J. M. Schaeberle with the Lick 36-inch refractor, when the AB separation was 4″.7. As in the case of Sirius, the existence of the companion was known from astrometry before the visual discovery, and it has proven to be a white dwarf.

49. A 2132. Round in 1950, '51, '53, '55, '56, '57. Must have short period (VBs).

50. Ho 252. Rapid binary. Not separated 1906, '40, '47, '50.

51. Wils 16. Closer than 0″.1, by interferometer (1954).

52. Wils 17. Less than 0″.1 in 1954 (φ).

53. B 2529. In 1936 and '38, p.a. 139° to 122°, sep. 3″.0. Not resolved in 1948 (B).

54. Hu 1128. Companion not seen 1941-58 (VBs, 82-inch).

55. OΣ 215. The following measure may refer to this star: 1909.36, 209″.3, 1″.68.

56. A.G. Star seemed elongated in 1869 and 1903. No companion seen in 1893.

57. η Car, the puzzling novalike star that in 1843 became almost as bright as Sirius.

58. R 155. Companion invisible after 1942.

59. Dubhe. Faint distant pair (β), 1881.12, 203°.5, 384″.9.

60. H VI 13. All stars in the field are much more than 90″ away.

61. Dom. In 1952, 48° or 71°; 1954.2, 210°; 1954.3, not seen (Dom).

62. β. This is a measure of the galaxy NGC 4966 with respect to ADS 8777A.

63. 42 α Com. The 25.9-year orbit of this pair is almost exactly edge on to our view, so the p.a. remains near either 11° or 191°. The parallax of the system is 0″.051, according to S. L. Lippincott, who notes that eclipses lasting two hours could occur if the orbital inclination is within 1′ of 90° (*Astronomical Journal, 66*, 272, 1961). The uncertainty of the various published inclination values makes such eclipses possible, though none has yet been observed.

64. Mizar. This star and Alcor form the familiar naked-eye double in the handle of the Big Dipper. Mizar AB was the first double discovered telescopically, in about 1650 by the Italian astronomer G. B. Riccioli. The three stars have the same parallax and move through space together as members of the Ursa Major moving cluster. Each is now known to be a spectroscopic binary as well, and there is an unrelated 3.7-year oscillation in the position of B, making this a septuple system.

65. h 2659. Single in 1878. Nothing nearer in 1901 than a 13th-magnitude star at 202°.3, 38″.7.

66. α Cen. Duplicity first noted by Jean Richaud while observing a comet in 1689 from Pondicherry, India. This pairing of 1st-magnitude stars is " . . . beyond all comparison the most striking object of the kind in the heavens" (h). The red-dwarf star Proxima Centauri, discovered by I in 1911 to share the proper motion of α Cen, is the nearest star to the Sun. In 1978, K. W. Kamper and A. J. Wesselink found the distance of Proxima as 1.295 parsecs (4.22 light-years), and the distance of the AB pair as 1.333 parsecs (4.35 light-years).

67. Antares. The bright star has been called "fiery red" (Smyth). Dawes believed the companion's green tint was real, rather than a contrast effect, when he saw it emerge ahead of the primary from behind the dark limb of the Moon in 1856. Spectral types *M1Ib, B2.5V*; Skylab confirms cool star, hot companion (Yale *Bright Star Catalogue*).

68. 70 Oph. Orbit computers have long had difficulty with the AB pair, even though two full 88-year revolutions have now been observed. As early as 1841, when J. Mädler could not reconcile the 1804-23 measures with later ones, he concluded that "the law of gravity does not hold good in this system." See ascribed the apparent departure from Kepler's law to a possible "dark satellite" of B that made it weave slightly in and out of the ellipse of best fit, and marginal supporting evidence appeared in a 1976 radial-veloci-

ty study. But A. H. Batten's continued spectroscopic work through 1979 has yielded a smooth radial-velocity curve that strengthens the case against a third body.

69. Ho 432. Ho also notes an 11th-magnitude star at 100″ in p.a. 44° from A.

70. ε Lyr, the well-known double-double. Keen eyesight distinguishes ε¹ Lyr 3′.5 north of ε² Lyr without optical aid; a 3-inch glass easily resolves these as AB and CD, respectively.

71. Σ 3130. Lewis measured a pair 1ᵐ following and 1′ south: 1897, 37°, 3″.1, magnitudes 10 and 11.

72. β 1285. Concerning the brighter star, β noted "possibly a close pair."

73. h 5075. " . . . the nearest star is one of a small triangle; seems strange h did not say so" (R).

74. Kui 90. Ca: single 1946-56 (VBs, 82-inch); 1954, 301°, 0″.33 (Wils).

75. Albireo. "One of the finest double stars in the heavens" (Webb). Topaz yellow, sapphire blue (Smyth). The spectrum of A is composite, *K5 + B*; its close companion was first distinguished by H. A. McAlister in 1977 using speckle interferometry, and then by C. E. Worley visually in 1979 (magnitude difference 1.5, separation 0″.4, 26-inch).

76. OΣΣ 188. There is a 13th-magnitude star between A and B. From A, 28″.73; from B, 30″.48.

77. δ Sge. Checked for duplicity but found single. Spectrum composite, *Ma + A0*.

78. See. In 1922: 150°, 0″.18 (A). The 1897 measure does not seem consistent with motion since 1922. Single in 1958 (82-inch).

79. hz 7. The primary is VW Cep, an unresolved eclipsing binary. The existence of a more distant third star was established astrometrically by J. L. Hershey in 1974; the visual discovery followed quickly (hz, Sproul 24-inch).

80. Σ 2716. The A star was examined for duplicity by Wils; found single. Composite spectrum, *K0 + A*.

81. 61 Cyg. This is "Piazzi's flying star" (proper motion 5″.22 per year), and the first star whose parallax was successfully determined (by F. W. Bessel, 1838). It is also the most observed double star, with over 34,000 photographic images and 2,000 visual measures amassed through 1983. Several tentative astrometric studies suggest one, two, or three unseen companions with periods of 5 to 12 years.

82. α Equ. Spectrum composite. Wils checked for duplicity; single.

83. Rst 4095. Discordant p.a.'s in 1950: 220°, five nights (IDS).

84. Kr 60. Located 1ᵐ.1 preceding and 43′ south of δ Cep. The AC pair was discovered by Kr, while β first detected B with the Lick 36-inch. Photographs made by Barnard in 1908-20, showing the rapid orbital motion of AB, have frequently appeared in astronomy texts. Both components are red dwarfs, and B is a flare star.

85. A 2695. In 1913.7, measured at 117°, 0″.22 (A). Many discordant measures.

86. h 1853. Compare the star at 23ʰ 12ᵐ.1, +45° 17′ (2000.0), which fits the h description better.

87. β 278. There is a measure by OΣ (1876.45, 351°.9, 14″.76, one night).

Visual Binary Stars

Visual Binary Stars

Column Headings

ADS — The binary star's number in R. G. Aitken's *New General Catalogue of Double Stars* (1932).

Star Name — The discoverer's designation, and under this the Flamsteed number, Bayer letter, and common name, if any. (A key to discoverers appears in Table I of the Introduction.) In the case of multiple systems, the traditional Roman letters identify the components for which the orbit has been calculated. Thus, "AB" means the orbital elements describe the motion of the B component relative to A; "AB × C" means the orbit is for C relative to a close AB pair.

α 2000 and **δ 2000** — Right ascension and declination, referred to the 2000.0 equinox.

m_1 and **m_2** — Visual magnitudes of the two components. Immediately underneath is the combined magnitude of the pair. Values listed only to tenths are approximate, but those with two decimal places are based on photoelectric **V** magnitudes. Often, for very close pairs, an accurate combined magnitude has been measured photoelectrically but the relative brightnesses of the components are only roughly known. When the individual magnitudes are given to two decimal places, either the stars are wide enough to have been separately measured photoelectrically, or the combined magnitude is photoelectric and the magnitude difference has been determined by a special photometric technique.

Also listed here are the spectral types of the components, when known, and the measured trigonometric parallax, if any. The reciprocal of the parallax is the binary's distance from the Sun in parsecs. Obtained from various sources, the information in this column often differs somewhat from that in *Sky Catalogue 2000.0*, Vol. 1; the values given here are preferred.

Orbital Elements — These describe the angular size and shape of the true (unforeshortened) orbit of the companion relative to the primary star, and also the orientation of this orbit in space:

P. Period of revolution in years.

T. Time of a periastron passage, expressed as a year with decimals.

a. Semimajor axis of the true orbit, in seconds of arc.

e. Eccentricity of the true orbit.

i. Inclination, in degrees, of the true orbit plane to the plane of the sky. The inclination can range from 0° to 180°. If *i* is 90°, the orbit appears edge on to our view. For *i* less than 90°, the motion of the secondary star around the primary appears direct (counterclockwise on the sky); for *i* greater than 90° the motion is retrograde (clockwise).

ω. Longitude of periastron, in degrees. This is the angle, in the plane of the true orbit, measured from the ascending node Ω to the periastron point and in the direction of motion.

Ω. Position angle of the ascending node, in degrees, measured on the plane of the sky counterclockwise from north. The ascending node is the node at which the radial velocity of the companion, relative to the primary, is positive (receding). When radial-velocity measurements are lacking, the node lying between 0° and 180° is taken as the ascending one. Immediately under Ω is the equinox to which the node is referred, when known.

Under each set of elements are the orbit computer(s) and date of publication. Sometimes, when several alternative orbits were given originally, the name is followed by a digit to distinguish which set of elements has proven more reliable by recent observations and is the one given here.

The orbits have been graded by either P. Couteau or C. E. Worley on a scale of 1 to 5. Grades 1 (definitive) and 2 (reliable) are reserved for those based on many observations through at least one complete revolution; grades 3 and 4 are progressively less certain, though at least *P* and *a* are roughly known; grade 5 orbits, based upon only a small observed arc, are indeterminate but may serve for rough predictions.

Finally, the year of discovery or earliest useful measurement is listed.

In a few cases, when two alternative orbits seem to satisfy the existing measures well, both are listed. Only future observation can settle which interpretation is correct.

Const — Constellation.

Ephemeris — This brief tabulation shows the character of the companion's current motion. For the *beginning* of selected years are given the position angle (PA) and separation (Sep) of the companion relative to the primary, calculated from the orbital elements. The position angles are in degrees from north, counterclockwise through east, south, and west, and have been corrected to the equinox of date. Separations are in seconds of arc.

ADS	Star Name	α 2000	δ 2000	m₁	m₂	Orbital Elements P / T	a / e	i / ω	Ω / equinox	Const	Ephemeris Date	PA	Sep
17175	β 733	0ʰ 02.2ᵐ	+27°05′	5.84	8.88	26.27ʸ	0.83″	50.0°	288.6°	Peg	1985	287°	0.7″
	85 Peg			comb. 5.78		1910.11	0.38	94.4	2000		1987	314	0.5
				G3V		R. G. Hall, 1949					1989	29	0.3
				par. 0.081		grade 1 first meas. 1878					1991	93	0.6
											1993	118	0.7
102	Σ 2	0 09.3	+79 43	6.6	6.9	300	0.686	113.2	169.9	Cep	1940	59	0.3
				comb. 6.0		1888.2	0.57	330.1			1960	32	0.5
				A7IV		W. D. Heintz, 1952					1980	19	0.6
						grade 3 first meas. 1830					2000	12	0.8
											2020	7	0.9
148	β 1026	0 12.1	+53 37	7.3	8.1	72.0	0.22	25.1	17.5	Cas	1985	12	0.2
				comb. 6.9		1919.18	0.77	122.4			1990	78	0.1
				A7V	F2V	P. Baize, 1954					1995	248	0.1
						grade 2 first meas. 1888					2000	278	0.2
											2005	292	0.3
161	OΣ 2 AB	0 13.4	+26 59	6.7	7.5	693.5	1.018	119.1	182.4	And	1940	16	0.4
				comb. 6.30		1963.77	0.84	266.5			1960	334	0.1
				G0III	F2IV	S. Arend, 1954					1980	180	0.3
						grade 4 first meas. 1851					2000	165	0.5
											2020	156	0.6
207	Σ 13	0 16.2	+76 57	7.0	7.3	1600	1.26	135.7	81.0	Cep	1940	70	0.8
				comb. 6.4		1830.0	0.50	304.0	2000		1960	63	0.8
				B8Vnn		W. D. Heintz, 1960					1980	57	0.9
						grade 5 first meas. 1831					2000	51	0.9
											2020	46	0.9
221	OΣ 4	0 16.7	+36 30	8.2	8.9	112.5	0.40	153.9	118.7	And	1970	184	0.6
				comb. 7.7		1908.1	0.56	106.3			1980	172	0.5
				F6V	F9V	P. Muller, 1957					1990	158	0.5
						grade 2 first meas. 1854					2000	140	0.4
											2010	111	0.3
246	Grb 34 AB	0 18.0	+44 00	8.07	11.04	2600	41.15	61.4	45.3	And	1940	59	37.7
				comb. 8.00		1745	0.0	0.0	2000		1960	61	37.0
				M3V	M6V	S. L. Lippincott, 1972					1980	62	36.3
				par. 0.278		grade 5 first meas. 1860					2000	64	35.5
											2020	66	34.6
283	h 1018	0 21.0	+67 40	8.4	8.9	163.4	1.24	88.5	84.9	Cep	1970	86	1.3
				comb. 7.9		1943.0	0.96	230.4			1980	86	1.4
				G5		P. Muller, 1957					1990	86	1.5
						grade 3 first meas. 1897					2000	87	1.6
											2010	87	1.6
293	OΣ 6	0 21.4	+67 00	7.7	8.3	240	0.46	103.0	147.3	Cep	1940	167	0.2
				comb. 7.23		1927.0	0.80	184.3			1960	155	0.4
				B8V	A1V	P. Muller, 1954					1980	152	0.6
						grade 3 first meas. 1849					2000	150	0.7
											2020	148	0.8
	B 1909	0 28.4	−20 20	7.2	7.2	5.625	0.134	59.7	119.0	Cet	1985	108	0.2
				comb. 6.4		1982.95	0.60	174.7			1986	119	0.2
				G2IV		W. H. van den Bos 1, 1956					1987	131	0.2
						first meas. 1929					1988	169	0.1
											1989	24	0.0
						11.25	0.214	69.8	119.0	Cet	1985	109	0.2
						1985.88	0.0	0.0			1986	120	0.2
						W. H. van den Bos 2, 1956					1987	133	0.2
						first meas. 1929					1988	159	0.1
											1989	236	0.1
	I 260	0 31.6	−62 58	4.8	6.0	44.43	0.385	142	40.3	Tuc	1985	305	0.5
	β² Tuc			comb. 4.5		1923.50	0.80	282.1			1987	301	0.5
				A2V	A7V	O. J. Eggen 1, 1965					1989	297	0.6
				par. 0.023		grade 3 first meas. 1900					1991	294	0.6
											1993	290	0.6
434	OΣ 12	0 31.8	+54 31	5.5	5.8	640	0.586	47.7	174.4	Cas	1940	167	0.6
	14 λ Cas			comb. 4.73		1958.0	0.0	0.0	2000		1960	175	0.6
				B8Vn		W. D. Heintz, 1963					1980	183	0.6
				par. 0.027		grade 5 first meas. 1845					2000	191	0.6
											2020	200	0.5
490	Ho 212	0 35.2	−3 36	5.6	6.3	6.94	0.20	30.4	346.3	Cet	1985	274	0.3
	13 Cet			comb. 5.1		1960.10	0.73	97.8	1950		1986	289	0.3
				F8V		A. L. Behall, 1961					1987	314	0.2
				par. 0.058		grade 2 first meas. 1887					1988	145	0.1
											1989	224	0.2

ADS	Star Name	α 2000	δ 2000	m_1	m_2	Orbital Elements				Const	Ephemeris		
						P	a	i	Ω				
						T	e	ω	equinox		Date	PA	Sep
520	β 395	0ʰ 37ᵐ.3	−24° 46′	6.3	6.4	25ʸ.00	0″.670	78°.0	112°.0	Cet	1985	97°	0″.5
				comb.	5.6	1924.00	0.22	142.0			1987	104	0.7
				G5V		W. H. van den Bos, 1937					1989	109	0.8
				par.	0″.070	grade 1		first meas. 1886			1991	113	0.8
											1993	119	0.6
588	OΣ 18 AB	0 42.4	+4 10	7.8	9.4	545.8	1.67	34.6	78.5	Psc	1940	169	1.3
				comb.	7.6	1840.0	0.14	9.0			1960	186	1.4
				F8V	G7V	P. Baize, 1958					1980	200	1.5
				par.	0.013	grade 4		first meas. 1845			2000	213	1.6
											2020	225	1.7
673	β 495	0 48.8	+18 42	8.4	8.4	128	0.54	109.8	30.8	Psc	1970	134	0.1
				comb.	7.6	1850.34	0.59	349.3	2000		1980	27	0.2
				G0		C. E. Worley, 1973					1990	301	0.1
						grade 3		first meas. 1878			2000	242	0.3
											2010	229	0.5
671	Σ 60	0 49.1	+57 49	3.44	7.51	480	11.99	34.8	278.4	Cas	1940	281	9.3
	24 η Cas			comb.	3.41	1889.6	0.50	268.6	1900		1960	296	10.8
				G0V	M0V	K. Aa. Strand, 1969					1980	307	12.0
				par.	0.169	grade 2		first meas. 1832			2000	317	12.9
											2020	326	13.5
684	β 232	0 50.4	+50 38	8.4	8.8	218.6	0.59	37.0	81.8	Cas	1940	200	0.4
				comb.	7.8	1913.2	0.65	350.3			1960	224	0.7
				F5V	F7V	P. Baize, 1964					1980	236	0.8
						grade 3		first meas. 1876			2000	246	0.9
											2020	253	1.0
732	A 2307	0 53.3	+4 05	7.4	8.4	60.02	0.23	69.8	36.6	Psc	1985	43	0.3
				comb.	7.0	1955.29	0.49	198.9	2000		1990	48	0.3
				F0		R. L. Walker, 1975					1995	54	0.2
						grade 4		first meas. 1911			2000	66	0.2
											2005	99	0.1
						60.9	0.251	73.0	44.2	Psc	1985	44	0.4
						1956.6	0.43	185.8	2000		1990	48	0.3
						W. D. Heintz, 1975					1995	53	0.3
						grade 4		first meas. 1911			2000	59	0.2
											2005	74	0.1
746	OΣ 20	0 54.6	+19 11	6.23	6.89	360.4	0.700	131.5	120.1	Psc	1940	276	0.5
	66 Psc			comb.	5.76	1902.75	0.30	149.1			1960	251	0.5
				A1Vn		P. Couteau, 1965					1980	224	0.5
				par.	0.005	grade 3		first meas. 1847			2000	199	0.5
											2020	180	0.6
755	Σ 73	0 55.0	+23 38	6.0	6.4	164.7	1.014	46.4	171.5	And	1970	218	0.6
	36 And			comb.	5.4	1957.15	0.31	4.65			1980	259	0.6
				K1		P. Muller, 1957					1990	292	0.8
						grade 2		first meas. 1836			2000	313	0.9
											2010	327	1.1
784	β 1099	0 56.8	+60 22	6.0	6.7	83.4	0.25	53.5	170.1	Cas	1985	319	0.2
				comb.	5.5	1952.9	0.23	342.0	2000		1990	331	0.3
				B9IVn		W. D. Heintz, 1978					1995	340	0.3
						grade 2		first meas. 1889			2000	348	0.3
											2005	357	0.3
819	A 1902	0 59.3	−0 40	8.7	8.8	85.9	0.29	50.4	105.7	Cet	1985	185	0.3
				comb.	8.0	1961.50	0.80	280.0	2000		1990	193	0.3
				F5		M. Miranian, 1973					1995	200	0.3
						grade 3		first meas. 1908			2000	206	0.3
											2005	212	0.3
862	OΣ 21	1 03.0	+47 23	6.7	8.0	450	0.816	87.0	176.7	And	1940	173	0.4
				comb.	6.4	1902.0	0.80	180	2000		1960	174	0.7
				A9IV		W. D. Heintz, 1966					1980	175	0.9
				par.	0.003	grade 4		first meas. 1847			2000	175	1.0
											2020	176	1.2
918	A 1516	1 07.2	+38 39	8.3	8.3	33.60	0.16	30	95.0	And	1985	56	0.2
				comb.	7.57	1930.0	0.23	104.1	2000		1987	71	0.2
				F8V	F8V	P. J. Morel, 1970					1989	87	0.2
				par.	0.024	grade 4		first meas. 1907			1991	105	0.2
											1993	128	0.1
	Rst 3352	1 07.8	−41 29	5.8	6.2	28.5	0.209	72.0	138.2	Phe	1985	265	0.1
	υ Phe			comb.	5.2	1949.66	0.51	324.3			1987	286	0.1
				A3V		W. S. Finsen, 1967					1989	296	0.2
				par.	0.010	grade 3		first meas. 1936			1991	303	0.2
											1993	307	0.3

ADS	Star Name	α 2000	δ 2000	m₁	m₂	Orbital Elements P / T	a / e	i / ω	Ω / equinox	Const	Ephemeris Date	PA	Sep

ADS	Star Name	α 2000	δ 2000	m₁ m₂	P / T	a / e	i / ω	Ω equinox	Const	Date	PA	Sep
940	OΣ 515 42 φ And	1ʰ09.5ᵐ	+47°15′	4.6 5.5 comb. 4.2 B7Ve	371.6ʸ 1889.9 P. Baize, 1958 grade 3	0."447 0.21 first meas. 1851	139°.8 259°.7	148°.2	And	1940 1960 1980 2000 2020	172° 155 140 127 114	0."4 0.4 0.5 0.5 0.5
974	A 655	1 11.2	+41 13	8.5 8.9 comb. 7.9 G5 par. 0.004	197.6 1927.49 J. M. Costa-Morales, 1978 grade 4	0.31 0.10 first meas. 1904	31.2 39.4	155.0	And	1970 1980 1990 2000 2010	288 305 321 335 348	0.3 0.3 0.3 0.3 0.3
999	β 1100	1 14.8	+60 56	8.0 8.0 comb. 7.2 F7IV F3V	75 1947.0 P. Muller, 1955 grade 3	0.36 0.88 first meas. 1889	129.0 244.7	77.4	Cas	1985 1990 1995 2000 2005	24 20 15 10 3	0.5 0.5 0.4 0.4 0.3
					150.0 1908.0 D. J. Zulevic, 1972 grade 3	0.58 0.0 first meas. 1889	108.0 0.0	27.0	Cas	1970 1980 1990 2000 2010	218 209 202 192 174	0.5 0.6 0.6 0.5 0.3
	I 27 CD κ Tuc	1 15.0	−68 49	7.8 8.2 comb. 7.2 G5 par. 0.042	86.2 1924.0 W. D. Heintz, 1978 grade 2	1.12 0.04 first meas. 1897	33.5 159.0	148.9 2000	Tuc	1985 1990 1995 2000 2005	193 215 240 266 290	1.0 1.0 0.9 0.9 1.0
1081	φ 337 BC 42 Cet	1 19.8	−0 31	7.4 7.6 comb. 6.7 A7V	21.5 1961.5 W. D. Heintz, 1975	0.136 0.03 first meas. 1958	25.0 0.0	108.0 2000	Cet	1985 1987 1989 1991 1993	141 176 213 247 277	0.1 0.1 0.1 0.1 0.1
1097	β 4	1 21.3	+11 32	7.4 8.0 comb. 6.9 F0	180 1943 P. Muller, 1954 grade 3	0.364 0.85 first meas. 1877	133.8 253.9	157.4	Psc	1970 1980 1990 2000 2010	120 114 108 104 100	0.4 0.4 0.4 0.5 0.5
	I 263	1 22.0	−69 43	7.5 8.0 comb. 7.0 F2	214.8 1922.67 M. Klerk, 1973 grade 4	0.575 0.36 first meas. 1900	58.7 76.7	53.3	Tuc	1940 1960 1980 2000 2020	207 234 250 266 286	0.3 0.5 0.6 0.6 0.5
1123	β 1163	1 24.3	−6 55	6.6 6.8 comb. 5.9 F3V F4V	16.14 1972.73 W. S. Finsen, 1973 grade 1	0.188 0.92 first meas. 1890	116.2 349.6	29.2	Cet	1985 1986 1987 1988 1989	209 207 204 199 307	0.3 0.3 0.2 0.1 0.0
	Rst 33	1 25.9	−47 54	8.3 9.0 comb. 7.8 G5	171.6 1952.64 J. L. Newburg, 1966 grade 3	0.968 0.83 first meas. 1928	117.7 64.4	145.7 2000	Phe	1970 1980 1990 2000 2010	312 306 301 297 293	0.8 1.0 1.1 1.1 1.1
1158	β 1164 95 Psc	1 27.7	+5 21	7.6 7.9 comb. 7.0 G0V	63.8 1945.70 W. H. van den Bos, 1950 grade 2	0.311 0.86 first meas. 1890	110.8 222.0	174.9	Psc	1985 1990 1995 2000 2005	153 150 145 135 104	0.4 0.3 0.3 0.2 0.1
1183	A 1910	1 29.7	+22 50	7.4 7.7 comb. 6.8 A2V A4V par. 0.018	74.5 1928.1 W. D. Heintz, 1976 grade 2	0.248 0.87 first meas. 1908	115.9 78.6	38.7 2000	Psc	1985 1990 1995 2000 2005	119 106 89 62 210	0.2 0.1 0.1 0.1 0.1
	δ 31 ÀB	1 35.0	−29 55	7.8 7.9 comb. 7.1 K3V par. 0.052	4.559 1932.613 N. Wieth-Knudsen, 1956 grade 1	0.171 0.316 first meas. 1920	25.25 41.64	68.12 2000	Scl	1985 1986 1987 1988 1989	285 335 62 202 262	0.2 0.2 0.1 0.1 0.2
	β 1000 ABxC			7.1 11.5 comb. 7.1 K3V M2V par. 0.052	111.8 1960.1 J. L. Newburg, 1969 grade 3	1.419 0.21 first meas. 1881	29.3 64.6	141.8 2000	Scl	1970 1980 1990 2000 2010	257 297 325 348 10	1.1 1.3 1.5 1.6 1.6

ADS	Star Name	α 2000	δ 2000	m_1	m_2	Orbital Elements				Const	Ephemeris		
						P	a	i	Ω		Date	PA	Sep
						T	e	ω	equinox				
	h 3447	$1^h36^m.1$	−29°54′	6.0	7.1	1876^y	12″.39	77°.8	138°.0	Scl	1940	111°	1″.7
	τ Scl			comb.	5.7	1968.60	0.98	85.39	1900		1960	124	1.1
				F4		S. Arend, R. Mourao, 1961					1980	331	1.5
						grade 5		first meas. 1837			2000	338	2.2
											2020	342	2.7
	Kui 7	1 37.6	−9 24	6.8	7.2	28.3	0.19	107.0	158.0	Cet	1985	185	0.1
				comb.	6.2	1952.3	0.60	173.0			1987	174	0.2
				F7V		W. D. Heintz, 1968					1989	168	0.2
				par.	0.026	grade 4		first meas. 1934			1991	165	0.3
											1993	162	0.3
	Δ 5	1 39.8	−56 12	5.79	5.83	483.7	7.817	142.8	13.1	Eri	1940	206	9.8
	p Eri			comb.	5.06	1813.49	0.53	18.37	2000		1960	201	10.5
				K0V	K5V	G. B. van Albada, 1956					1980	195	11.1
				par.	0.148	grade 4		first meas. 1835			2000	191	11.5
											2020	186	11.7
1359	β 870	1 44.3	+57 32	6.4	7.8	573.5	1.07	180.0	0.0	Cas	1940	29	1.1
				comb.	6.1	1985.76	0.0	0.0			1960	16	1.1
				A3V		G. M. Popovic, 1972					1980	4	1.1
						grade 5		first meas. 1880			2000	351	1.1
											2020	339	1.1
1394	h 3461	1 45.6	−25 03	5.4	8.6	1192	4.652	180.0	0.0	Scl	1940	41	4.7
	ε Scl			comb.	5.3	2076.2	0.0	0.0	1900		1960	35	4.7
				F1V		R. Mourao, 1969					1980	29	4.7
				par.	0.030	grade 5		first meas. 1836			2000	23	4.7
											2020	17	4.7
1411	OΣ 34	1 49.9	+80 53	7.8	8.1	395.0	0.635	67.0	119.3	Cep	1940	206	0.2
				comb.	7.2	1941.0	0.16	90.5	2000		1960	258	0.3
				A0		W. D. Heintz, 1962					1980	280	0.4
						grade 4		first meas. 1843			2000	292	0.6
											2020	300	0.6
1522	Σ 183	1 55.1	+28 48	7.7	8.4	368.0	0.54	124.1	172.5	Tri	1940	334	0.3
				comb.	7.2	1966.6	0.57	296.9			1960	264	0.1
				F2		P. Couteau, 1973					1980	179	0.3
						grade 2		first meas. 1832			2000	152	0.3
											2020	132	0.4
1538	Σ 186	1 55.9	+1 51	6.76	6.76	170.3	1.05	73.6	40.4	Cet	1970	52	1.5
				comb.	6.01	1893.35	0.71	220.7			1980	55	1.4
				F9V		R. Mourao, 1976					1990	57	1.3
				par.	0.025	grade 2		first meas. 1831			2000	61	1.1
											2010	65	0.9
1598	β 513	2 02.0	+70 54	4.70	6.37	60.44	0.653	22.8	64.2	Cas	1985	218	0.8
	48 Cas			comb.	4.49	1964.78	0.345	4.5	2000		1990	234	0.9
				A4V		W. D. Heintz, 1969					1995	248	0.9
				par.	0.032	grade 1		first meas. 1878			2000	263	0.9
											2005	279	0.8
1615	Σ 202	2 02.0	+2 46	4.18	5.21	933.05	4.0	120.9	23.3	Psc	1940	304	2.3
	113 α Psc			comb.	3.82	2098.64	0.696	225.4	2000		1960	295	2.1
				A2	A2	M. Scardia, 1983					1980	284	2.0
						grade 5		first meas. 1831			2000	272	1.8
											2020	257	1.7
1631	Σ 208	2 03.7	+25 56	5.90	7.27	309.0	1.34	49.1	22.3	Ari	1940	221	0.6
	10 Ari			comb.	5.63	1931.48	0.59	162.5	1900		1960	283	0.7
				dF4		A. Kranjc, L. Pigoni, 1960					1980	320	0.9
				par.	0.024	grade 3		first meas. 1833			2000	339	1.3
											2020	350	1.6
1630	OΣ 38 BC	2 03.9	+42 20	5.5	6.3	61.1	0.296	111.1	104.15	And	1985	107	0.6
	57 γ² And			comb.	4.84	1952.1	0.93	171.15	1950		1990	106	0.5
				B8V	A0V	P. Muller, 1957					1995	105	0.5
				par.	0.005	grade 1		first meas. 1842			2000	103	0.4
											2005	101	0.3
1709	Σ 228	2 14.0	+47 29	6.60	7.08	144.7	0.908	63.4	97.6	And	1970	257	0.9
				comb.	6.06	1898.2	0.26	320.9	1900		1980	266	1.0
				F5V		W. D. Heintz, 1954					1990	274	1.1
				par.	0.030	grade 2		first meas. 1831			2000	282	1.0
											2010	292	0.8
1733	Hst 1	2 15.8	−18 14	8.3	9.3	173	2.01	52.5	155.7	Cet	1970	83	1.6
				comb.	7.9	2027.0	0.38	69.0	2000		1980	101	1.6
				K3V		W. D. Heintz, 1978					1990	119	1.7
						grade 4		first meas. 1879			2000	135	1.7
											2010	153	1.5

ADS	Star Name	α 2000	δ 2000	m_1	m_2	Orbital Elements P / T	a / e	i / ω	Ω / equinox	Const	Ephemeris Date	PA	Sep
1778	Joy 1 68 o Cet Mira	$2^h 19^m.3$	$-2°59'$	2.0v comb. 2.0v gM7e par. 0.013	9.5v	400^y 2001.5 P. Baize, 1980 grade 5	0.85 0.66	$111°.3$ 106.0	$118°.5$ first meas. 1923	Cet	1940 1960 1980 2000 2020	129° 123 112 9 296	0.8 0.7 0.4 0.1 0.4
	β 738	2 23.2	$-29\ 52$	7.6 comb. 7.0 F8	7.9	110.5 1960.55 W. S. Finsen, 1969	0.640 0.48	100.7 94.0	28.3 first meas. 1879	For	1970 1980 1990 2000 2010	210 202 193 178 137	0.5 0.5 0.5 0.3 0.2
1833	Σ 257	2 25.7	$+61\ 33$	7.6 comb. 7.1 B8	8.1	581.9 1934.92 A. Valbousquet, 1980 grade 4	0.68 0.61	47.3 170.2	141.2 2000 first meas. 1830	Cas	1940 1960 1980 2000 2020	325 7 45 71 87	0.3 0.3 0.3 0.4 0.5
1860	Σ 262 AaxB z Cas	2 29.0	$+67\ 24$	4.62 comb. 4.49 A3Vp par. 0.020	6.87 F5	840.0 1550.0 W. D. Heintz, 1962 grade 5	2.27 0.40	132.0 299.0	6.3 2000 first meas. 1829	Cas	1940 1960 1980 2000 2020	243 239 234 230 226	2.3 2.4 2.4 2.5 2.6
	Plq 32 268 G. Cet GC 3121	2 36.1	$+6\ 53$	5.84 comb. 5.84 K3V par. 0.128	12.0	60.0 1937.0 G. Martin, P. A. Ianna, 1975 grade 4	0.257 0.45	70.3 115.5	120.5 first meas. 1912	Cet	1985 1990 1995 2000 2005	118 129 186 295 314	0.3 0.2 0.1 0.2 0.2
1990	Hu 1216	2 37.1	$-11\ 12$	8.6 comb. 8.4 G0	10.1	161.8 2007.53 V. Erceg, 1970 grade 4	0.50 0.27	46.2 326.0	132.9 first meas. 1900	Cet	1970 1980 1990 2000 2010	336 357 31 79 116	0.5 0.4 0.3 0.3 0.3
	φ 312 83 ε Cet	2 39.6	$-11\ 52$	5.8 comb. 5.0 F5IV-V par. 0.059	5.8	2.667 1956.55 W. S. Finsen, 1970 grade 1	0.108 0.27	31.9 279.1	30.0 first meas. 1951	Cet	1985 1986 1987 1988 1989	172 341 110 204 39	0.1 0.1 0.1 0.1 0.1
2034	OΣ 43	2 40.7	$+26\ 37$	7.9 comb. 7.6 F7V	9.1	475.0 1835.0 W. D. Heintz, 1962 grade 4	1.21 0.38	122.5 276.5	51.0 2000 first meas. 1848	Ari	1940 1960 1980 2000 2020	28 18 8 357 345	1.1 1.1 1.0 1.0 0.9
	φ 333	2 43.4	$-66\ 43$	7.1 comb. 6.3 F5V	7.1	28.5 1962.2 W. D. Heintz, 1978 grade 3	0.28 0.84	89.9 344.5	35.5 2000 first meas. 1955	Hor	1985 1987 1989 1991 1993	215 215 215 36 215	0.4 0.3 0.2 0.0 0.1
2081	Σ 296 13 θ Per	2 44.2	$+49\ 14$	4.13 comb. 4.12 F7V par. 0.077	9.87 M1V	2720 1613.0 J. Hopmann, 1958 grade 5	22.29 0.13	75.4 100.6	128.0 1900 first meas. 1783	Per	1940 1960 1980 2000 2020	302 303 304 305 306	18.7 19.2 19.6 20.0 20.4
2122	Σ 305	2 47.5	$+19\ 22$	7.4 comb. 7.0 F9V par. 0.036	8.2	720.0 1790.0 W. Rabe, 1961 grade 5	2.913 0.77	114.3 55.67	145.6 1900 first meas. 1830	Ari	1940 1960 1980 2000 2020	312 311 309 308 306	3.4 3.6 3.6 3.7 3.7
2200	β 524 AB 20 Per	2 53.7	$+38\ 20$	5.6 comb. 5.40 F4IV par. 0.010	6.7	62.3 1976.93 A. da Silva , 1970 grade 2	0.197 0.01	143.7 346.9	127.1 2000 first meas. 1880	Per	1985 1990 1995 2000 2005	98 69 34 0 333	0.2 0.2 0.2 0.2 0.2
2242	β 741	2 57.2	$-24\ 58$	8.1 comb. 7.36 G5	8.2	137.0 1874.76 W. H. van den Bos, 1956 grade 3	1.548 0.67	83.5 260.4	162.3 first meas. 1879	For	1970 1980 1990 2000 2010	311 326 333 339 350	0.5 0.7 0.9 1.0 0.4
2236	A 2413	2 57.2	$+1\ 53$	8.2 comb. 7.5 F9V	8.4	150.0 1930.6 P. Muller, 1952 grade 3	0.70 0.55	69.8 287.6	2.3 first meas. 1912	Cet	1970 1980 1990 2000 2010	53 78 105 126 141	0.4 0.4 0.4 0.4 0.5

ADS	Star Name	α 2000	δ 2000	m₁	m₂	Orbital Elements				Const	Ephemeris		
						P / T	a / e	i / ω	Ω / equinox		Date	PA	Sep
2246	β 1173 AB	2ʰ 58.6ᵐ	+24° 08′	8.3	8.4	317.5ʸ	0.285″	61.2°	155.8°	Ari	1940	27°	0.2″
				comb. 7.6		1887.0	0.38	144.9	2000		1960	65	0.1
				K0		W. D. Heintz, 1965					1980	95	0.2
						grade 4		first meas. 1890			2000	114	0.2
											2020	126	0.3
2253	β 525	2 58.9	+21 37	7.5	7.5	241.9	0.44	60.1	92.5	Ari	1940	218	0.3
				comb. 6.7		1846.95	0.22	317.4			1960	242	0.4
				A3		J. M. Costa-Morales, 1978					1980	256	0.5
						grade 4		first meas. 1877			2000	268	0.5
											2020	279	0.5
2402	h 3555 α For	3 12.1	−28 59	4.0	7.0	314	4.37	81.5	117.7	For	1940	113	1.3
				comb. 3.9		1947.0	0.76	42.0	2000		1960	291	1.6
				F7IV	G7V	W. D. Heintz, 1978					1980	297	3.8
						grade 3		first meas. 1836			2000	299	5.1
											2020	301	5.8
2390	Σ 360	3 12.2	+37 13	8.1	8.3	616.9	2.122	105.4	112.2	Per	1940	130	2.3
				comb. 7.4		1625.4	0.61	128.4	1900		1960	128	2.4
				G0		J. Hopmann, 1965					1980	127	2.5
						grade 5		first meas. 1831			2000	125	2.6
											2020	124	2.7
	Jc 8 AB	3 12.4	−44 25	6.6	6.9	43.87	0.425	151.7	17.5	Eri	1985	190	0.6
				comb. 6.0		1932.53	0.87	28.26			1987	186	0.6
				F6III		S. Wierzbinski, 1954					1989	183	0.7
						grade 2		first meas. 1897			1991	181	0.7
											1993	178	0.8
2377	OΣ 50 AB	3 12.6	+71 33	8.4	8.4	344.9	1.102	125.2	18.9	Cas	1940	187	1.3
				comb. 7.6		2117.31	0.26	24.2			1960	178	1.2
				F8		G. M. Popovic, 1972					1980	168	1.1
						grade 5		first meas. 1847			2000	154	0.9
											2020	133	0.7
2436	OΣ 52	3 17.5	+65 40	6.8	7.3	330.0	0.386	180.0	0.0	Cam	1940	96	0.5
				comb. 6.3		2075.0	0.31	65.0	2000		1960	83	0.5
				A3V		W. D. Heintz, 1963					1980	69	0.4
						grade 5		first meas. 1846			2000	52	0.4
											2020	31	0.4
2446	OΣ 53	3 17.7	+38 38	7.8	8.3	118.2	0.57	131.4	116.3	Per	1970	272	0.8
				comb. 7.3		1929.0	0.76	47.4	1900		1980	266	0.9
				G0		W. Rabe, 1948					1990	260	0.8
						grade 3		first meas. 1845			2000	253	0.8
											2010	245	0.7
2459	AC 2 95 Cet	3 18.4	−0 56	5.6	7.5	217.2	0.974	72.9	69.4	Cet	1940	227	0.8
				comb. 5.4		1837.68	0.37	309.9	1900		1960	236	1.1
				K1IV	G8V	T. Jastrzebski, 1961					1980	243	1.2
				par. 0.006		grade 4		first meas. 1854			2000	250	1.1
											2020	259	0.8
2524	A 2909 AB	3 24.4	−15 39	8.3	8.3	25.0	0.148	53.5	2.2	Eri	1985	48	0.1
				comb. 7.5		1949.0	0.18	254.7			1987	76	0.1
				G1V	G1V	P. Muller, 1955					1989	110	0.1
						grade 3		first meas. 1918			1991	140	0.1
											1993	162	0.1
2531	A 829	3 26.2	+12 29	8.3	9.8	120.6	0.30	146.1	37.0	Tau	1970	304	0.3
				comb. 8.1		1991.6	0.47	211.7			1980	265	0.2
				G0		P. Baize, 1980					1990	203	0.2
						grade 4		first meas. 1904			2000	116	0.2
											2010	70	0.3
2538	A 980	3 28.4	+60 15	6.8	7.8	259.9	0.370	143.6	158.1	Cam	1940	132	0.3
				comb. 6.4		1968.29	0.22	90.0			1960	91	0.2
				B9V		D. J. Zulevic, 1969					1980	38	0.2
						grade 4		first meas. 1905			2000	359	0.3
											2020	333	0.4
	B 52	3 33.9	−31 05	6.8	7.1	19.40	0.225	84.4	141.1	For	1985	320	0.3
				comb. 6.2		1958.40	0.34	8.8			1986	321	0.3
				F5V		W. S. Finsen, 1963					1987	322	0.3
						grade 2		first meas. 1926			1988	323	0.3
											1989	324	0.3
2616	Σ 412 7 Tau	3 34.4	+24 28	6.6	6.7	568.2	0.650	155.2	5.6	Tau	1940	39	0.3
				comb. 5.94		1911.18	0.71	228.4			1960	18	0.5
				A3V		S. Vlaicu, M. Vasile, 1961					1980	7	0.6
				par. 0.003		grade 3		first meas. 1830			2000	359	0.7
											2020	353	0.8

ADS	Star Name	α 2000	δ 2000	m₁	m₂	Orbital Elements P / T	a / e	i / ω	Ω / equinox	Const	Date	PA	Sep
2612	Σ 400	3ʰ35ᵐ.0	+60°02′	6.8	7.6	287.7 ʸ	1.245″	67.6	82.4	Cam	1940	106°	0.3″
	comb. 6.43					1938.4	0.70	32.5	1900		1960	247	0.7
	F4V					P. Baize, 1952					1980	259	1.3
						grade 3	first meas. 1829				2000	264	1.6
											2020	268	1.8
2644	Σ 422	3 36.8	+0 35	5.88	8.83	2101	8.023	32.1	92.3	Tau	1940	257	6.5
	comb. 5.81					1900.0	0.18	152.0	1900		1960	261	6.6
	G9V K6V					J. Hopmann, 1964					1980	266	6.6
	par. 0.028					grade 5	first meas. 1832				2000	270	6.7
											2020	274	6.7
2756	β 1003	3 45.3	−27 52	8.2	11.8	425.0	2.761	48.0	50.8	For	1940	62	2.2
	comb. 8.2					1983.0	0.28	77.58	1900		1960	83	1.8
	K5V M3V					J. Hopmann, 1960					1980	116	1.4
	par. 0.031					grade 4	first meas. 1881				2000	162	1.4
											2020	195	1.8
2755	β 536	3 46.3	+24 11	8.3	9.3	1000	1.761	99.0	3.6	Tau	1940	194	0.4
	comb. 7.9					1962.70	0.72	193.1	1900		1960	183	0.5
	A6V					S. Wierzbinski, 1956					1980	167	0.3
						grade 5	first meas. 1878				2000	56	0.2
											2020	22	0.5
2799	OΣ 65	3 50.3	+25 35	5.81	6.21	62.28	0.430	83.2	26.0	Tau	1985	209	0.5
	comb. 5.24					1937.80	0.62	349.5	1900		1990	212	0.4
	A3V A5V					S. Wierzbinski, 1957					1995	224	0.1
						grade 2	first meas. 1846				2000	25	0.2
											2005	102	0.0
2959	Σ 483	4 04.1	+39 31	7.5	8.6	394.7	2.101	110.6	10.3	Per	1940	155	0.8
	comb. 7.2					1909.10	0.65	140.1			1960	113	0.7
	G5V					P. Couteau, 1958					1980	77	0.9
	par. 0.018					grade 3	first meas. 1830				2000	58	1.2
											2020	47	1.6
2980	A 1710	4 06.5	+43 25	8.2	8.2	87.0	0.398	119.9	112.2	Per	1985	317	0.5
	comb. 7.4					1947.2	0.61	324.0	2000		1990	313	0.5
	G9V G9V					W. D. Heintz, 1969					1995	309	0.6
						grade 3	first meas. 1907				2000	305	0.6
											2005	301	0.6
2995	OΣ 531	4 07.6	+38 04	7.4	8.9	705.9	3.599	112.8	161.8	Per	1940	82	1.0
	comb. 7.2					1998.24	0.50	166.9	1900		1960	33	1.0
	K2V					W. Rabe, 1961					1980	3	1.4
	par. 0.029					grade 4	first meas. 1852				2000	346	1.8
											2020	333	1.8
2963	Σ 460	4 10.0	+80 42	5.54	6.29	415.1	1.00	44.8	90.8	Cep	1940	78	0.9
	comb. 5.10					2012.1	0.24	70.0	1900		1960	93	0.9
	G8III A6V					P. Baize, 1958					1980	112	0.8
	par. 0.004					grade 4	first meas. 1828				2000	137	0.6
											2020	173	0.5
3041	A 2801	4 10.7	−4 52	8.1	8.1	20.0	0.145	62.3	172.7	Eri	1985	28	0.2
	comb. 7.3					1911.5	0.76	42.0			1986	33	0.1
	G2IV G2IV					P. Muller, 1954					1987	40	0.1
						grade 3	first meas. 1914				1988	51	0.1
											1989	69	0.1
						40.0	0.197	67.6	10.0	Eri	1985	215	0.1
						1959.40	0.0	0.0			1987	234	0.1
						D. J. Zulevic, 1972					1989	271	0.1
						grade 3	first meas. 1914				1991	314	0.1
											1993	339	0.1
3064	A 1938	4 13.6	+7 43	6.0	6.1	7.18	0.135	66.1	318.0	Tau	1985	203	0.1
	46 Tau					1954.72	0.29	147.9			1986	280	0.1
	comb. 5.3					W. S. Finsen, 1962					1987	304	0.2
	F3V					grade 2	first meas. 1908				1988	317	0.2
	par. 0.025										1989	333	0.1
	Rst 2338	4 14.2	−46 08	7.3	7.7	18.60	0.250	62.8	115.9	Hor	1985	195	0.2
	comb. 6.7					1954.45	0.68	251.2	1950		1986	208	0.2
	G0					W. H. van den Bos, 1965					1987	223	0.2
						grade 2	first meas. 1934				1988	239	0.2
											1989	256	0.2
3082	OΣ 77 AB	4 15.9	+31 42	8.0	8.1	200.0	0.585	56.7	72.4	Per	1970	265	0.8
	comb. 7.33					1887.6	0.46	35.60	1950		1980	270	0.8
	G0V					P. Muller, 1956					1990	275	0.7
						grade 3	first meas. 1846				2000	281	0.7
											2010	289	0.6

ADS	Star Name	α 2000	δ 2000	m₁	m₂	Orbital Elements P / T	a / e	i / ω	Ω / equinox	Const	Ephemeris Date	PA	Sep
3102	Ho 328	4ʰ17ᵐ.0	+19°41′	7.7	8.1	63ʸ.3	0″.36	130°.0	81°.8	Tau	1985	12°	0″.4
				comb.	7.1	1972.4	0.98	263°.0	2000		1990	9	0.4
				F4V	F6V	W. D. Heintz, 1978					1995	7	0.4
						grade 2		first meas. 1890			2000	4	0.5
											2005	2	0.5
3098	Σ 511	4 17.9	+58 47	7.5	7.9	254.0	0.480	138.0	130.5	Cam	1940	171	0.3
				comb.	6.9	1951.0	0.18	333.4	2000		1960	136	0.4
				A0		W. D. Heintz, 1969					1980	107	0.4
						grade 3		first meas. 1829			2000	75	0.4
											2020	40	0.4
3114	Σ 520	4 18.3	+22 49	8.2	8.3	690.5	1.115	71.1	115.2	Tau	1940	128	0.6
				comb.	7.5	1972.43	0.47	83.74	1900		1960	149	0.3
				F5		A. da Silva, 1961					1980	224	0.2
						grade 4		first meas. 1837			2000	272	0.4
											2020	286	0.6
3135	OΣ 79	4 19.9	+16 31	7.2	8.2	91.04	0.561	52.8	64.3	Tau	1985	124	0.2
	55 Tau			comb.	6.8	1897.58	0.60	131.3	1900		1990	231	0.2
				F9V		S. Wierzbinski, 1955					1995	272	⁺0.3
				par.	0.028	grade 2		first meas. 1846			2000	302	0.3
											2005	326	0.4
3159	β 744 AB	4 21.5	−25 44	6.6	6.9	76.92	0.500	35.0	0.0	Eri	1985	334	0.6
				comb.	6.04	1924.00	0.58	106.0	1900		1990	349	0.5
				F1V	F3V	W. H. van den Bos, 1951					1995	13	0.3
				par.	0.049	grade 2		first meas. 1891			2000	86	0.2
											2005	179	0.3
3169	OΣ 82	4 22.7	+15 03	7.3	8.5	255.5	1.185	138.0	12.0	Tau	1940	33	1.1
				comb.	7.0	1891.6	0.29	230.3	2000		1960	14	1.3
				F9V	G1V	W. D. Heintz, 1969					1980	359	1.4
				par.	0.029	grade 3		first meas. 1848			2000	345	1.4
											2020	330	1.3
3182	Hu 304	4 23.9	+9 28	5.8	5.9	51.6	0.18	0.0	0.0	Tau	1985	103	0.1
	66 Tau			comb.	5.1	1937.4	0.70	215.0			1990	265	0.1
				A3V	A4V	G. Van Biesbroeck, 1951					1995	343	0.2
						grade 2		first meas. 1901			2000	5	0.2
											2005	17	0.3
	φ 342 AB	4 25.6	+15 56	7.1	7.1	13.15	0.13	132.6	141.4	Tau	1985	158	0.1
	70 Tau			comb.	6.3	1962.84	0.07	97.1			1986	139	0.1
				F7V		W. S. Finsen, 1978					1987	119	0.1
				par.	0.041	grade 2		first meas. 1959			1988	90	0.1
											1989	48	0.1
3211	Hu 609	4 26.2	+34 43	8.6	9.0	165.0	0.276	135.7	138.6	Per	1970	187	0.3
				comb.	8.0	1915.5	0.24	163.2	2000		1980	171	0.3
				F5		W. D. Heintz, 1967					1990	159	0.3
						grade 4		first meas. 1902			2000	149	0.3
											2010	139	0.3
3230	β 311	4 26.9	−24 05	6.8	7.0	175.7	1.25	79.7	163.4	Eri	1970	104	0.4
				comb.	6.1	1917.5	0.87	107.0			1980	117	0.5
				A2		E. Horeschi, 1957					1990	125	0.6
						grade 3		first meas. 1877			2000	131	0.7
											2010	136	0.8
3248	Hu 1080	4 29.0	+16 10	7.0	7.7	40.4	0.425	92.6	77.7	Tau	1985	260	0.5
				comb.	6.5	1917.80	0.41	297.4	1950		1987	259	0.5
				F7V	G0V	W. H. van den Bos, 1956					1989	258	0.5
				par.	0.008	grade 2		first meas. 1904			1991	258	0.4
											1993	257	0.3
3264	Σ 554	4 30.1	+15 38	5.70	7.98	189.5	1.01	108.0	12.8	Tau	1970	20	1.7
	80 Tau			comb.	5.57	1888.0	0.83	162.0			1980	19	1.8
				F0V		P. Baize, 1977					1990	18	1.8
				par.	0.022	grade 2		first meas. 1831			2000	17	1.8
											2010	16	1.7
	B 2092	4 34.0	−55 03	3.8	4.3	13.00	0.212	54.1	118.7	Dor	1985	169	0.2
	α Dor			comb.	3.3	1962.60	0.84	229.7			1986	178	0.2
				A0III		G. Knipe, 1966					1987	194	0.1
				par.	0.011	grade 3		first meas. 1956			1988	236	0.1
											1989	106	0.1
3358	β 1295 AB	4 40.0	+53 28	5.8	7.4	26.25	0.18	115.0	161.0	Cam	1985	118	0.1
	2 Cam			comb.	5.6	1965.1	0.54	207.0	2000		1987	85	0.1
				F0V	F9V	W. D. Heintz, 1962					1989	23	0.1
				par.	0.016	grade 2		first meas. 1901			1991	337	0.1
											1993	228	0.0

ADS	Star Name		α 2000	δ 2000	m_1	m_2	Orbital Elements				Const	Ephemeris		
							P / T	a / e	i / ω	Ω / equinox		Date	PA	Sep
3358	Σ 566 2 Cam	ABxC	$4^h40^m.0$	+53°28'	5.6 comb. 5.4 F5V par.	7.3 0".016	425ʸ.0 2035.0 W. D. Heintz, 1962 grade 4	1".38 0.30 first meas. 1829	131°.5 331°.5	115°.0 2000	Cam	1940 1960 1980 2000 2020	268° 253 229 194 156	1".3 1.0 0.8 0.7 0.8
	h 3683		4 40.3	−58 57	7.2 comb. 6.5 G5V par.	7.3 0.059	552.3 1923.61 S. Wierzbinski, 1956 grade 3	3.164 0.99 first meas. 1836	140.1 134.7	48.1 1900	Dor	1940 1960 1980 2000 2020	101 96 94 93 92	1.3 2.2 2.9 3.4 3.9
3390	Σ 577		4 42.2	+37 31	8.6 comb. 7.8 F8	8.6	654.6 2028.0 E. Hock, 1966 grade 4	1.55 0.25 first meas. 1829	143.8 88.3	59.9 1900	Per	1940 1960 1980 2000 2020	49 37 22 4 342	1.3 1.2 1.1 1.0 0.9
3434	Hu 612		4 47.8	+53 18	7.0 comb. 6.8 F2	8.8	165.0 1871.7 P. Baize, 1958 grade 3	0.38 0.25 first meas. 1902	58.3 317.0	163.3	Cam	1970 1980 1990 2000 2010	332 340 349 0 17	0.4 0.4 0.4 0.4 0.3
3465	A 2621		4 49.6	+2 12	8.4 comb. 7.6 A0	8.4	225.8 1830.87 V. Erceg, 1978 grade 4	0.18 0.27 first meas. 1913	41.8 77.6	129.0	Ori	1940 1960 1980 2000 2020	18 43 70 95 121	0.2 0.2 0.2 0.2 0.2
3475	β 883		4 51.2	+11 04	7.5 comb. 6.8 F6V par.	7.6 0.033	16.30 1955.95 W. D. Heintz, 1969 grade 1	0.202 0.44 first meas. 1879	20.2 69.8	156.4 2000	Ori	1985 1986 1987 1988 1989	101 120 147 191 254	0.2 0.2 0.2 0.1 0.1
3483	β 552	AB	4 51.8	+13 39	6.5 comb. 6.3 F6V par.	8.1 0.017	100.9 1885.6 W. D. Heintz, 1962 grade 3	0.705 0.55 first meas. 1874	50.0 315.6	139.8 2000	Ori	1970 1980 1990 2000 2010	327 7 142 199 240	0.7 0.3 0.4 0.4 0.6
3588	β 314	AB	4 59.0	−16 23	5.94 comb. 5.66 F3V par.	7.28 F9V 0.020	54.62 1980.9 E. van Dessel, 1978 grade 2	0.507 0.81 first meas. 1876	107.3 8.9	142.8 1900	Lep	1985 1990 1995 2000 2005	337 329 326 324 322	0.3 0.6 0.8 0.9 0.9
3596	OΣ 93		5 00.5	+5 06	8.5 comb. 8.0 G5 par.	9.0 0.022	531.4 1934.03 J. M. Costa, 1962 grade 4	1.39 0.56 first meas. 1847	127.1 105.7	68.5	Ori	1940 1960 1980 2000 2020	292 255 237 225 213	0.5 0.8 1.0 1.1 1.1
3614	Hu 445		5 01.7	+20 50	8.6 comb. 8.0 G7V	8.9 G8V	82.0 1933.73 P. Baize, 1957 grade 2	0.66 0.87 first meas. 1901	72.0 274.8	154.4	Tau	1985 1990 1995 2000 2005	278 285 293 300 308	0.4 0.4 0.4 0.4 0.4
3608	A 1844	AB	5 01.7	+26 40	7.1 comb. 6.81 G2V	8.4	25.0 1951.42 P. Baize, 1959 grade 2	0.225 0.29 first meas. 1908	27.0 85.0	137.3	Tau	1985 1987 1989 1991 1993	5 24 43 62 81	0.3 0.3 0.3 0.3 0.2
3686	A 3009		5 05.9	−13 55	8.1 comb. 8.0 G5	10.4	217.0 1916.0 W. D. Heintz, 1973 grade 4	1.162 0.18 first meas. 1926	44.0 7.0	86.8 2000	Lep	1940 1960 1980 2000 2020	141 191 228 250 268	0.8 0.8 1.1 1.3 1.4
3701	A 3010 104 Tau		5 07.5	+18 39	5.6 comb. 4.8 G4V par.	5.6 0.056	1.19 1911.37 O. J. Eggen 1, 1956 grade 5	0.18 0.90 first meas. 1912	73.0 90.3	122.3	Tau	1985 1986 1987 1988 1989	89 65 42 18 354	0.1 0.1 0.1 0.1 0.1
							2.38 1910.60 O. J. Eggen 2, 1956 grade 5	0.10 0.0 first meas. 1912	0.0 0.0	0.0	Tau	1985 1986 1987 1988 1989	94 245 37 188 339	0.1 0.1 0.1 0.1 0.1

ADS	Star Name	α 2000	δ 2000	m₁	m₂	Orbital Elements				Const	Ephemeris		
						P / T	a / e	i / ω	Ω / equinox		Date	PA	Sep
3711	OΣ 98 14 Ori	5ʰ07.9ᵐ	+8°30'	5.78 comb. 5.34 A0	6.52	198.9ʸ 1974.8 P. Baize, 1969	1.04″ 0.17	141.0° 59.3°	88.7° 1900	Ori	1970 1980 1990 2000 2010	49° 21 349 322 301	0.8″ 0.7 0.7 0.8 0.9
						first meas. 1844							
3728	A 2636	5 08.9	+3 13	7.1 comb. 6.6 B8V	7.6	150.0 1950.0 P. Muller, 1963 grade 3	0.252 0.74	59.5 91.0	116.0	Ori	1970 1980 1990 2000 2010	326 339 351 2 13	0.2 0.2 0.2 0.2 0.2
						first meas. 1913							
3730	β 1047 BC	5 09.8	+28 02	9? comb. 8.65 dG7	9.5?	32.1 1969.1 W. D. Heintz, 1976 grade 3	0.217 0.93	139.5 239.0	128.7 2000	Tau	1985 1987 1989 1991 1993	77 75 72 69 66	0.3 0.3 0.3 0.3 0.3
						first meas. 1889							
3799	OΣ 517 AB	5 13.5	+1 58	6.9 comb. 6.2 A5V	7.1 G0	312.0 1926.0 W. H. van den Bos, 1960 grade 3	0.383 0.85	23.3 267.5	174.4	Ori	1940 1960 1980 2000 2020	205 227 237 244 249	0.2 0.4 0.5 0.5 0.6
						first meas. 1854							
3841	JAn Aa 13 α Aur Capella	5 16.7	+46 00	0.6 comb. 0.1 G5III	1.1 G0III	0.28481 1936.458 H. A. McAlister, 1983 grade 1	0.055 0.0	136.64 0.0	220.2 2000	Aur	1985.00 1985.02 1985.04 1985.06 1985.08	57 38 19 356 326	0.053 0.055 0.052 0.046 0.041
						first meas. 1919							
3841	St 3 HL			10.5 comb. 10.1 dM1	13.0 dM5	388 2010 W. D. Heintz, 1974 grade 5	3.72 0.0	65.0 0.0	168.5 2000	Aur	1940 1960 1980 2000 2020	126 144 156 165 173	2.1 2.8 3.4 3.7 3.7
						first meas. 1935							
3959	A 2641	5 22.6	+2 37	8.4 comb. 8.3 K3V	10.9	83.9 1957.75 P. Baize, 1980 grade 3	1.06 0.07	113.7 179.3	158.3	Ori	1985 1990 1995 2000 2005	190 176 166 158 151	0.7 0.9 1.1 1.1 1.1
						first meas. 1913							
3991	A 847 BC	5 23.9	−0 52	7.8 comb. 7.1 F8V	7.9 F8V	48.0 1968.02 P. Baize, 1973 grade 3	0.34 0.10	90.0 115.0	141.8	Ori	1985 1987 1989 1991 1993	322 322 142 142 142	0.1 0.0 0.0 0.1 0.2
						first meas. 1904							
3956	Σ 677	5 24.7	+63 23	7.9 comb. 7.3 G0	8.2	370.3 2086.2 W. D. Heintz, 1962 grade 4	1.637 0.14	123.8 44.6	97.0 2000	Cam	1940 1960 1980 2000 2020	214 187 159 136 119	1.1 1.0 1.0 1.2 1.4
						first meas. 1831							
4020	A 848	5 25.5	−0 33	6.7 comb. 6.2 B9	7.3	194.3 1792.68 V. Erceg, 1978 grade 4	0.17 0.24	23.3 155.5	42.4	Ori	1970 1980 1990 2000 2010	149 179 209 236 262	0.1 0.1 0.1 0.1 0.1
						first meas. 1904							
4115	Σ 728 32 Ori	5 30.8	+5 57	4.49 comb. 4.21 B5IV par. 0.001	5.82	586.0 2005.0 L. Siegrist, 1951 grade 5	1.293 0.14	101.6 345.6	36.3	Ori	1940 1960 1980 2000 2020	61 51 45 41 37	0.5 0.7 0.9 1.1 1.1
						first meas. 1830							
4076	A 1034	5 32.2	+70 49	8.5 comb. 7.9 G5 par. 0.007	8.9	400 1890 W. D. Heintz, 1976 grade 3	0.705 0.50	119.0 172.0	112.8 2000	Cam	1940 1960 1980 2000 2020	191 160 146 137 131	0.3 0.4 0.6 0.7 0.8
						first meas. 1905							
4200	Σ 742	5 36.4	+22 00	7.2 comb. 6.7 F8	7.8	2959 2115.5 J. Hopmann, 1973 grade 5	5.57 0.21	56.1 173.2	110.6	Tau	1940 1960 1980 2000 2020	266 269 272 274 277	3.8 3.9 4.0 4.1 4.2
						first meas. 1837							
4229	β 1240 AB 26 Aur	5 38.6	+30 30	6.0 comb. 5.40 G5III	6.3 F0V	53.2 1922.0 P. Baize, 1956 grade 2	0.135 0.56	136.2 320.2	147.5	Aur	1985 1990 1995 2000 2005	57 31 15 3 353	0.1 0.1 0.2 0.2 0.2
						first meas. 1892							

ADS	Star Name	α 2000	δ 2000	m_1	m_2	P T	a e	i ω	Ω equinox	Const	Ephemeris Date	PA	Sep
4241	β 1032 AB	$5^h38^m.7$	$-2°36'$	4.0	6.0	$170^y.0$	0".25	165°.0	124°.5	Ori	1970	184°	0".2
	48 σ Ori			comb.	3.73	1970.0	0.07	299°.0	2000		1980	160	0.2
				O9.5V		W. D. Heintz, 1974					1990	137	0.2
				par.	0".001	grade 3		first meas.	1888		2000	115	0.2
											2010	94	0.2
4263	Σ 774	5 40.8	−1 57	1.88	4.02	1509	2.728	72.0	155.5	Ori	1940	159	2.5
	50 ζ Ori			comb.	1.74	2070.6	0.07	47.3	1900		1960	161	2.5
				O9.5IbB0III		J. Hopmann, 1967					1980	163	2.4
				par.	0.022	grade 5		first meas.	1830		2000	165	2.3
											2020	168	2.2
4265	β 1007	5 41.3	+16 32	5.3	5.9	78.45	0.19	72.4	56.3	Tau	1985	260	0.1
	126 Tau			comb.	4.8	1921.37	0.48	358.1			1990	290	0.1
				B3IV		P. Baize, 1961					1995	24	0.1
				par.	0.009	grade 2		first meas.	1881		2000	57	0.1
											2005	89	0.1
4299	A 494	5 42.9	−6 48	6.5	6.9	20.10	0.201	65.2	99.3	Ori	1985	127	0.2
				comb.	5.97	1958.75	0.36	269.1			1987	151	0.1
				F5V	F6V	P. Couteau, 1958					1989	190	0.1
						grade 2		first meas.	1903		1991	228	0.1
											1993	253	0.2
4505	OΣ 122	5 55.8	+36 56	7.6	8.3	320.0	0.257	39.4	109.9	Aur	1940	176	0.2
				comb.	7.1	1990.0	0.36	168.9			1960	215	0.2
				A5		P. Muller, 1957					1980	261	0.2
						grade 3		first meas.	1847		2000	302	0.2
											2020	340	0.2
	Hu 1399 AB	6 00.4	−31 02	8.9	9.7	72.0	0.94	101.8	128.2	Col	1985	316	0.9
				comb.	8.5	1930.37	0.50	284.5			1990	312	0.9
				K5V	K5V	P. Baize, 1952					1995	306	0.7
				par.	0.044	grade 3		first meas.	1915		2000	284	0.2
											2005	136	0.4
4617	A 2715 AB	6 02.4	+9 39	4.4	6.0	17.5	0.276	95.1	26.3	Ori	1985	25	0.3
	61 μ Ori			comb.	4.2	1929.25	0.76	223.7			1986	25	0.4
				A2V		H. L. Alden, 1942					1987	24	0.4
				par.	0.029	grade 2		first meas.	1914		1988	23	0.4
											1989	23	0.4
	Kui 23	6 04.1	+23 16	4.7	5.1	13.17	0.19	57.3	181.0	Gem	1985	149	0.2
	1 Gem			comb.	4.1	1955.60	0.32	192.0	2000		1986	165	0.2
				G8III		W. D. Heintz, 1962					1987	175	0.2
				par.	0.026	grade 2		first meas.	1948		1988	183	0.2
											1989	192	0.2
	Δ 23	6 04.8	−48 28	7.2	7.4	463.5	2.94	56.8	130.7	Pup	1940	83	2.0
				comb.	6.5	2049.5	0.15	38.8	2000		1960	99	2.2
				G6V		W. D. Heintz, 1962					1980	113	2.5
				par.	0.011	grade 4		first meas.	1836		2000	124	2.6
											2020	135	2.5
4603	OΣ 121	6 05.3	+74 00	7.6	8.8	163.0	0.41	115.4	149.5	Cam	1970	269	0.2
				comb.	7.3	1952.0	0.77	120.0	2000		1980	242	0.2
				F8		W. D. Heintz, 1978					1990	223	0.3
						grade 2		first meas.	1843		2000	210	0.3
											2010	201	0.4
	Rst 3442	6 09.8	−22 46	6.4	6.5	18.2	0.203	35.4	91.3	Lep	1985	346	0.2
				comb.	5.7	1961.37	0.35	117.7			1986	1	0.2
				F5V	F5V	W. S. Finsen, 1964					1987	14	0.2
						grade 2		first meas.	1935		1988	27	0.2
											1989	38	0.2
4841	β 1008	6 14.9	+22 30	3.3v	8.8	473.7	1.08	142.7	84.5	Gem	1940	272	1.4
	7 η Gem			comb.	3.3v	1819.7	0.54	26.2			1960	266	1.5
				M3III		P. Baize, 1980					1980	261	1.5
				par.	0.013	grade 4		first meas.	1882		2000	257	1.6
											2020	252	1.6
	Rst 5225	6 15.9	+1 10	7.1	7.1	28.1	0.165	32.0	133.3	Ori	1985	247	0.2
				comb.	6.3	1965.7	0.35	253.0	2000		1987	266	0.2
				F5		W. D. Heintz, 1975					1989	288	0.2
						grade 2		first meas.	1946		1991	316	0.1
											1993	359	0.1
4890	φ 331 Aa	6 17.1	+9 57	6.1	6.1	8.7	0.13	69.6	130.9	Ori	1985	332	0.1
	75 Ori			comb.	5.40	1965.10	0.19	105.2			1986	15	0.1
				A2V	A2V	J. Dommanget, 1977					1987	86	0.1
				par.	0.013	grade 2		first meas.	1954		1988	114	0.1
											1989	129	0.1

ADS	Star Name	α 2000	δ 2000	m₁	m₂	Orbital Elements P / T	a / e	i / ω	Ω / equinox	Const	Ephemeris Date	PA	Sep
4890	φ 331 Aa	6ʰ17.1ᵐ	+9°57'	6.1	6.1	9.0 y	0.11"	59.6°	135.2°	Ori	1985	303°	0.1"
	75 Ori			comb.	5.40	1967.27	0.23	173.6°			1986	338	0.1
				A2V	A2V	W. S. Finsen, 1978					1987	44	0.1
				par.	0.013	grade 2	first meas. 1954				1988	99	0.1
											1989	121	0.1
4929	β 895 AB	6 20.0	+28 26	7.9	7.9	108.0	0.28	63.8	143.5	Aur	1970	54	0.1
				comb.	7.1	1889.0	0.0	0.0			1980	110	0.2
				A3V		D. J. Zulevic, 1971					1990	133	0.3
						grade 2	first meas. 1879				2000	148	0.3
											2010	167	0.2
4971	A 2667	6 21.4	+2 16	6.8	7.4	120.0	0.448	58.4	104.8	Mon	1970	138	0.4
				comb.	6.3	1932.0	0.45	261.1			1980	154	0.4
				A3V	A6V	P. Muller, 1963					1990	174	0.4
						grade 3	first meas. 1913				2000	198	0.3
											2010	222	0.3
5042	OΣ 139	6 25.6	+22 27	7.9	9.5	514.3	0.75	47.5	154.2	Gem	1940	200	0.2
				comb.	7.7	1927.0	0.87	304.0	2000		1960	233	0.4
				A3		W. D. Heintz, 1962					1980	248	0.5
						grade 3	first meas. 1847				2000	258	0.6
											2020	264	0.7
	R 65 AB	6 29.8	−50 14	6.0	6.1	52.9	0.48	133.0	124.8	Pup	1985	274	0.7
				comb.	5.3	1969.1	0.96	46.5	2000		1990	272	0.8
				F2V		W. D. Heintz, 1978					1995	269	0.8
				par.	0.020	grade 3	first meas. 1879				2000	267	0.8
											2005	264	0.7
5197	Σ 932	6 34.3	+14 44	8.1	8.3	2360	3.210	124.1	4.2	Gem	1940	321	2.0
				comb.	7.4	2215.0	0.20	119.7	1900		1960	316	1.9
				F5		J. Hopmann, 1960					1980	311	1.8
						grade 5	first meas. 1830				2000	306	1.7
											2020	299	1.6
5212	Ho 234	6 34.5	−11 14	8.1	8.2	169.0	0.38	38.8	36.5	CMa	1970	326	0.3
				comb.	7.4	1823.8	0.18	351.8			1980	357	0.3
				F0		P. Baize, 1958					1990	24	0.3
						grade 4	first meas. 1888				2000	48	0.3
											2010	73	0.3
	φ 19	6 35.9	−36 05	6.8	7.3	29.0	0.30	99.7	156.5	Col	1985	345	0.3
				comb.	6.3	1964.5	0.49	287.1			1987	342	0.3
				G0V		W. S. Finsen, 1977					1989	338	0.3
						grade 2	first meas. 1925				1991	332	0.2
											1993	266	0.0
5234	OΣ 149	6 36.4	+27 17	7.14	8.71	114.8	0.86	112.2	75.0	Gem	1970	344	0.5
				comb.	6.91	1922.8	0.71	278.3	2000		1980	324	0.6
				G2V		W. D. Heintz, 1967					1990	308	0.7
				par.	0.027	grade 2	first meas. 1848				2000	295	0.7
											2010	284	0.8
5280	OΣ 150	6 39.3	+42 00	8.6	9.5	394.1	0.26	121.7	347.8	Aur	1940	89	0.1
				comb.	8.2	1962.08	0.27	299.9			1960	35	0.1
				A0		J. M. Costa-Morales, 1972					1980	6	0.2
						grade 4	first meas. 1847				2000	349	0.2
											2020	336	0.2
5423	AGC 1	6 45.1	−16 43	−1.46	8.49	50.09	7.500	136.53	44.57	CMa	1985	35	8.2
	9 α CMa			comb.	−1.46	1894.13	0.592	147.27	1950		1990	5	4.5
	Sirius			A1V	DA	W. H. van den Bos, 1960					1995	232	3.1
				par.	0.374	grade 1	first meas. 1862				2000	150	4.6
											2005	111	6.7
5400	Σ 948 AB	6 46.2	+59 27	5.37	5.95	699.0	1.66	180.0	0.0	Lyn	1940	99	1.7
	12 Lyn			comb.	4.87	1740.0	0.03	154.8			1960	89	1.7
				A3V		P. Brosche, 1957					1980	80	1.7
						grade 4	first meas. 1831				2000	70	1.7
											2020	60	1.7
5447	OΣ 156	6 47.4	+18 12	6.81	7.02	1057	1.536	113.6	134.7	Gem	1940	276	0.6
				comb.	6.16	1858.52	0.61	147.8	1900		1960	260	0.5
				A2V		J. Dommanget, 1953					1980	243	0.5
				par.	0.008	grade 4	first meas. 1844				2000	226	0.6
											2020	212	0.6
5455	OΣ 157	6 47.8	+0 20	7.5	7.8	299.0	0.56	143.0	176.2	Mon	1940	303	0.5
				comb.	6.9	1991.0	0.34	339.5	2000		1960	267	0.4
				A2		W. D. Heintz, 1973					1980	218	0.3
						grade 3	first meas. 1847				2000	174	0.4
											2020	136	0.4

ADS	Star Name	α 2000	δ 2000	m₁	m₂	Orbital Elements P / T	a / e	i / ω	Ω / equinox	Const	Ephemeris Date	PA	Sep

ADS	Star Name	α 2000	δ 2000	m₁ m₂		Orbital Elements				Const	Ephemeris		
						P	a	i	Ω		Date	PA	Sep
						T	e	ω	equinox				

5469	A 2731	$6^h 48.7^m$	$+7°37'$	8.4 9.0 comb. 7.9 G0V par. 0.012		180.0^y 1934.7 P. Muller, 1957 grade 3	0.77 0.81	42.7 141.9	108.2 first meas. 1913	Mon	1970 1980 1990 2000 2010	54° 61 66 70 74	0.8 0.9 1.1 1.1 1.2
5514	Σ 963 14 Lyn	6 53.1	+59 27	5.7 6.9 comb. 5.4 G0I A2		480.0 1950.0 W. D. Heintz, 1963 grade 4	0.736 0.53	56.4 130.0	58.8 2000 first meas. 1830	Lyn	1940 1960 1980 2000 2020	169 226 251 270 290	0.2 0.3 0.4 0.4 0.4
5559	Σ 982 38 Gem	6 54.6	+13 11	4.74 7.68 comb. 4.67 F0V G4V par. 0.040		3190 1636.0 J. Hopmann, 1949 grade 5	9.550 0.48	148.4 310.5	185.6 1900 first meas. 1829	Gem	1940 1960 1980 2000 2020	154 151 148 145 142	6.7 6.9 7.0 7.1 7.3
	I 65	6 57.3	−35 30	6.9 7.1 comb. 6.2 F8IV par. 0.064		16.74 1958.85 W. S. Finsen, 1963 grade 1	0.218 0.43	35.0 241.6	128.5 1950 first meas. 1898	Pup	1985 1986 1987 1988 1989	197 208 222 237 255	0.3 0.2 0.2 0.2 0.2
5625	A 2681	6 57.4	+2 53	8.5 8.7 comb. 7.8 A5		250 1950.0 W. D. Heintz 1, 1975 grade 4	0.268 0.25	126.0 284.4	128.0 2000 first meas. 1913	Mon	1940 1960 1980 2000 2020	235 164 133 112 92	0.1 0.2 0.2 0.2 0.2
						85 1943.0 W. D. Heintz 2, 1975 grade 4	0.180 0.80	130.5 282.1	55.0 2000 first meas. 1913	Mon	1985 1990 1995 2000 2005	307 302 296 290 284	0.2 0.2 0.2 0.2 0.2
5871	Σ 1037	7 12.8	+27 13	7.2 7.2 comb. 6.4 F6V par. 0.025		116.53 1920.85 M. Scardia, 1983 grade 1	0.931 0.946	131.50 257.26	32.54 2000 first meas. 1827	Gem	1970 1980 1990 2000 2010	325 321 317 313 308	1.2 1.2 1.2 1.1 1.0
5925	β 575	7 14.8	−15 29	8.0 8.0 comb. 7.27 F8		512.3 1593.9 R. F. Mourao, 1966 grade 4	0.727 0.0	0.0 0.0	0.0 1900 first meas. 1878	CMa	1940 1960 1980 2000 2020	243 258 272 286 300	0.7 0.7 0.7 0.7 0.7
	I 7	7 17.5	−46 59	7.1 7.8 comb. 6.6 K2V par. 0.077		84.0 1957.58 W. H. van den Bos, 1961 grade 3	0.609 0.90	111.1 36.9	40.9 1950 first meas. 1896	Pup	1985 1990 1995 2000 2005	210 209 207 206 205	0.9 1.0 1.0 1.0 0.9
5983	Σ 1066 55 δ Gem	7 20.1	+21 59	3.55 8.18 comb. 3.53 F0IV K3V		1200 1437.0 J. Hopmann, 1960 grade 5	6.975 0.11	63.28 57.19	18.4 1900 first meas. 1829	Gem	1940 1960 1980 2000 2020	215 218 222 226 230	6.7 6.4 6.1 5.8 5.4
6126	Σ 1104 AB	7 29.4	−15 00	6.4 7.5 comb. 6.1 F7V par. 0.021		1090.9 1885.0 J. Hopmann, 1969 grade 5	3.684 0.36	49.4 176.5	145.2 1900 first meas. 1831	Pup	1940 1960 1980 2000 2020	351 2 14 27 41	2.2 2.2 2.1 2.0 2.0
6175	Σ 1110 AB 66 α Gem Castor	7 34.6	+31 53	1.94 2.92 comb. 1.57 A1V A2Vm par. 0.066		511.3 1950.65 P. Muller, 1956 grade 3	7.37 0.36	112.9 239.8	41.7 1950 first meas. 1826	Gem	1940 1960 1980 2000 2020	202 167 98 68 55	3.7 2.2 2.3 4.0 5.6
						420.1 1965.30 W. Rabe, 1958 grade 3	6.295 0.33	115.9 261.4	40.5 1900 first meas. 1826	Gem	1940 1960 1980 2000 2020	202 168 95 62 48	3.8 2.2 2.2 3.8 5.2
6251	Schb 10 α CMi Procyon	7 39.3	+5 13	0.35 10.3 comb. 0.35 F5IV par. 0.283		40.65 1927.60 K. Aa. Strand, 1951 grade 1	4.548 0.40	35.7 269.8	284.3 2000 first meas. 1896	CMi	1985 1987 1989 1991 1993	356 6 16 26 37	5.1 5.2 5.2 5.2 5.1

ADS	Star Name	α 2000	δ 2000	m₁	m₂	Orbital Elements				Const	Ephemeris		
						P	a	i	Ω		Date	PA	Sep
						T	e	ω	equinox				
6276	OΣ 177	7^{h}41^m.7	+37°26′	8.0	9.0	180^y	0″.38	157°.7	116°.8	Lyn	1970	195°	0″.3
				comb.	7.6	1952.0	0.74	153°.2			1980	179	0.4
				A3		P. Muller, 1977					1990	170	0.5
						grade 3		first meas. 1845			2000	163	0.5
											2010	157	0.6
6315	Hu 710	7 43.0	−17 04	7.1	7.6	115.9	0.309	142.8	113.9	Pup	1970	103	0.3
				comb.	6.6	1952.6	0.52	257.5	2000		1980	83	0.3
				G5III	G5III	W. D. Heintz, 1965					1990	68	0.4
						grade 3		first meas. 1902			2000	54	0.4
											2010	40	0.4
6347	Ho 247	7 46.1	+21 07	8.0	8.4	169.0	0.386	46.0	101.5	Gem	1970	216	0.3
				comb.	7.4	2022.0	0.32	247.0	2000		1980	234	0.3
				F2		W. D. Heintz, 1962					1990	253	0.3
						grade 4		first meas. 1887			2000	272	0.3
											2010	294	0.3
6354	Hu 1247	7 48.0	+60 18	7.8	7.8	18.80	0.20	129.0	61.8	Cam	1985	277	0.2
				comb.	7.0	1954.48	0.50	294.7			1986	270	0.2
				F9V	F9V	P. Baize, 1961					1987	262	0.2
						grade 2		first meas. 1900			1988	254	0.2
											1989	244	0.2
6420	β 101	7 51.8	−13 54	5.6	6.2	23.18	0.58	77.8	103.3	Pup	1985	114	0.1
	9 Pup			comb.	5.16	1915.71	0.69	67.7			1987	283	0.4
				G1V		R. Woolley, L. Symms, 1937					1989	290	0.6
				par.	0.065	grade 1		first meas. 1875			1991	294	0.6
											1993	299	0.6
	φ 325	7 52.8	−5 26	6.4	6.7	30.8	0.30	76.5	165.4	Mon	1985	183	0.3
				comb.	5.8	1970.20	0.68	235.2	2000		1987	187	0.3
				F5IV		W. D. Heintz, 1978					1989	193	0.2
						grade 3		first meas. 1952			1991	202	0.2
											1993	220	0.1
6483	OΣ 185	7 57.3	+1 08	7.0	7.1	57.04	0.440	79.0	9.0	CMi	1985	84	0.1
				comb.	6.3	1945.62	0.77	252.0			1990	129	0.1
				F7V		W. H. van den Bos, 1949					1995	162	0.2
						grade 2		first meas. 1847			2000	182	0.2
											2005	8	0.2
6532	Σ 1175	8 02.5	+4 09	8.6	9.9	1590	3.426	60.3	25.0	CMi	1940	243	1.5
				comb.	8.3	1901.5	0.38	217.1	1900		1960	256	1.3
				G5		J. Hopmann, 1964					1980	272	1.2
				par.	0.029	grade 5		first meas. 1831			2000	290	1.2
											2020	307	1.2
6549	OΣ 187	8 04.1	+33 02	7.1	7.7	217.4	0.28	148.1	40.3	Gem	1940	86	0.1
				comb.	6.6	1935.5	0.65	266.2	2000		1960	18	0.2
				A1.5V		P. J. Morel, 1970					1980	355	0.3
						grade 3		first meas. 1844			2000	340	0.4
											2020	328	0.4
6623	Σ 1187	8 09.5	+32 13	7.1	8.0	4390	6.52	124.9	26.4	Cnc	1940	33	2.5
				comb.	6.7	1869.40	0.63	317.6			1960	28	2.6
				F2		J. Hopmann, 1971					1980	24	2.7
						grade 5		first meas. 1829			2000	20	2.8
											2020	17	2.9
6650	Σ 1196 ABxCc	8 12.2	+17 39	5.05	6.20	1150	7.96	144.6	256.7	Cnc	1940	100	5.7
	16 ζ Cnc			comb.	4.73	1960.0	0.26	163.9	1900		1960	90	5.8
				F8V	G0V	C. Gasteyer, 1954					1980	81	5.9
						grade 5		first meas. 1841			2000	72	6.0
											2020	64	6.0
6650	Σ 1196 AB			5.63	6.02	59.7	0.884	172.0	58.0	Cnc	1985	239	0.7
	16 ζ Cnc			comb.	5.05	1930.0	0.32	233.0	1900		1990	182	0.6
				F8V	G0V	C. Gasteyer, 1954					1995	125	0.7
				par.	0.040	grade 1		first meas. 1826			2000	86	0.8
											2005	59	1.0
6719	A 337	8 17.3	−5 22	8.4	8.7	99.5	0.57	109.0	53.1	Hya	1985	93	0.4
				comb.	7.8	1907.0	0.83	91.0	2000		1990	86	0.4
				F2		W. D. Heintz 1, 1978					1995	78	0.4
						grade 3		first meas. 1902			2000	68	0.3
											2005	48	0.1
						161.5	0.39	131.8	42.9	Hya	1970	116	0.4
						1892.0	0.34	104.3	2000		1980	101	0.4
						W. D. Heintz 2, 1978					1990	87	0.4
						grade 3		first meas. 1902			2000	74	0.4
											2010	62	0.4

ADS	Star Name	α 2000	δ 2000	m₁	m₂	Orbital Elements P / T	a / e	i / ω	Ω / equinox	Const	Ephemeris Date	PA	Sep
6762	Σ 1216	8ʰ21ᵐ.3	−1°36′	7.1	7.4	435ʸ.0	0″.654	45°.0	133°.2	Hya	1940	242°	0″.5
				comb. 6.5		1842.9	0.20	0°.0			1960	261	0.5
				A2Vpn		B. Ekenberg, 1945					1980	276	0.6
						grade 4	first meas. 1831				2000	287	0.7
											2020	297	0.7
	Rst 321	8 25.1	−49 10	8.4	8.7	25.6	0.277	41.3	126.6	Vel	1985	150	0.3
				comb. 7.8		1923.95	0.24	236.7	2000		1987	167	0.3
				G5		C. E. Worley, 1981					1989	186	0.3
						grade 2	first meas. 1928				1991	209	0.2
											1993	236	0.2
6796	Hu 856	8 25.4	+37 24	8.4	8.9	86.0	0.24	46.7	21.2	Lyn	1985	250	0.3
				comb. 7.9		1952.3	0.60	72.5			1990	258	0.3
				F5		P. Couteau, 1963					1995	266	0.3
						grade 3	first meas. 1904				2000	275	0.3
											2005	285	0.3
6811	A 1746 BC	8 26.7	+24 32	8.6	8.6	21.82	0.19	44.9	126.0	Cnc	1985	342	0.2
	24 Cnc			comb. 7.81		1955.84	0.10	95.0			1987	14	0.2
				F7V		P. Baize, 1958					1989	51	0.2
						grade 2	first meas. 1908				1991	84	0.2
											1993	110	0.2
	B 1606	8 27.0	−52 42	7.2	7.4	14.5	0.158	52.7	105.9	Vel	1985	50	0.1
				comb. 6.5		1952.79	0.28	180.4			1986	73	0.1
				F6V		W. S. Finsen, 1963					1987	87	0.2
						grade 2	first meas. 1929				1988	97	0.2
											1989	106	0.2
6825	A 550	8 27.7	−4 25	7.5	7.6	40.0	0.193	106.4	170.8	Hya	1985	185	0.2
				comb. 6.8		1918.50	0.18	96.1			1987	179	0.2
				F0		W. H. van den Bos, 1952					1989	174	0.2
						grade 2	first meas. 1903				1991	169	0.2
											1993	162	0.2
6828	A 551 AB	8 28.5	−2 31	7.1	7.2	53.0	0.321	84.1	62.0	Hya	1985	95	0.1
				comb. 6.42		1948.40	0.45	229.4			1990	216	0.1
				F0		W. H. van den Bos, 1953					1995	236	0.1
						grade 2	first meas. 1903				2000	245	0.2
											2005	49	0.1
6851	Hu 716	8 31.6	+34 58	7.5	9.1	31.0	0.34	132.2	139.2	Lyn	1985	177	0.1
				comb. 7.3		1952.84	0.70	232.4			1987	143	0.3
				G5		P. Baize, 1972					1989	131	0.4
						grade 2	first meas. 1902				1991	123	0.4
											1993	116	0.5
6871	β 205	8 33.1	−24 36	6.9	7.0	159.2	0.568	133.7	176.1	Pyx	1970	22	0.5
				comb. 6.2		1948.31	0.29	65.4			1980	3	0.6
				A8IV		J. L. Newburg, 1964					1990	349	0.6
				par. 0.009		grade 3	first meas. 1877				2000	337	0.6
											2010	325	0.6
	φ 335	8 34.5	−32 36	7.1	7.2	13.9	0.16	55.7	150.5	Pyx	1985	146	0.2
				comb. 6.4		1961.24	0.51	140.4			1986	154	0.2
				G5IV-V		W. S. Finsen, 1977					1987	167	0.1
						grade 3	first meas. 1955				1988	199	0.1
											1989	303	0.1
6914	β 208	8 39.1	−22 40	5.3	6.7	145.0	1.700	81.7	209.3	Pyx	1970	220	1.1
				comb. 5.0		2011.0	0.13	167.7	2000		1980	239	0.5
				G6		W. D. Heintz, 1967					1990	356	0.4
				par. 0.058		grade 3	first meas. 1877				2000	20	1.0
											2010	27	1.4
	I 314	8 39.4	−36 36	6.5	7.6	66.5	0.527	102.0	55.7	Pyx	1985	233	0.5
				comb. 6.2		1925.7	0.86	341.0	2000		1990	225	0.2
				F3IV		W. D. Heintz, 1968					1995	271	0.1
						grade 3	first meas. 1900				2000	249	0.4
											2005	245	0.6
	B 1624	8 42.1	−52 45	8.2	9.2	76.1	0.400	69.7	94.9	Vel	1985	19	0.1
				comb. 7.8		1997.4	0.24	0.6	2000		1990	69	0.2
				G1V		W. D. Heintz, 1969					1995	88	0.3
						grade 4	first meas. 1929				2000	103	0.3
											2005	123	0.2
	Rst 3603	8 43.2	−12 26	8.5	8.6	70.5	0.32	136.5	122.0	Hya	1985	349	0.3
				comb. 7.8		1966.7	0.29	0.0	2000		1990	331	0.3
				F5		W. D. Heintz, 1978					1995	317	0.4
						grade 3	first meas. 1938				2000	306	0.4
											2005	295	0.4

ADS	Star Name	α 2000	δ 2000	m₁	m₂	Orbital Elements				Const	Ephemeris		
						P	a	i	Ω				
						T	e	ω	equinox		Date	PA	Sep
6993	Σ 1273 ABxC	8ʰ46ᵐ.8	+6°25′	3.4	6.8	890ʸ.0	4″.536	42°.0	55°.5	Hya	1940	257°	3″.1
	11 ε Hya			comb.	3.4	1933.0	0.29	203°.1	2000		1960	271	2.9
				G0III		W. D. Heintz, 1963					1980	286	2.8
				par.	0″.013	grade 4		first meas. 1830			2000	302	2.7
											2020	318	2.7
6993	Sp AB			3.8	4.7	15.05	0.238	49.3	109.3	Hya	1985	204	0.3
	11 ε Hya			comb.	3.4	1961.05	0.67	264.0	2000		1986	214	0.2
				G0III		W. D. Heintz, 1963					1987	226	0.2
				par.	0.013	grade 1		first meas. 1888			1988	238	0.2
											1989	253	0.2
7012	A 2552	8 48.7	+0 57	8.5	8.5	90.0	0.22	138.2	25	Hya	1985	141	0.2
				comb.	7.7	1966.0	0.30	140.0			1990	119	0.2
				F5		P. Baize, 1976					1995	100	0.2
						grade 3		first meas. 1913			2000	84	0.2
											2005	71	0.2
7054	A 1584	8 53.1	+54 58	8.1	8.1	74.1	0.397	28.8	61.9	UMa	1985	209	0.2
				comb.	7.3	1987.4	0.72	221.0	2000		1990	7	0.2
				G2V	G2V	W. D. Heintz, 1964					1995	51	0.3
						grade 3		first meas. 1907			2000	66	0.5
											2005	75	0.5
	ø 316	8 53.8	-47 31	6.1	6.1	7.24	0.10	33.2	8.6	Vel	1985	332	0.1
				comb.	5.3	1964.36	0.20	33.3			1986	32	0.1
				A9IV-V		W. S. Finsen, 1973					1987	109	0.1
				par.	0.012	grade 2		first meas. 1951			1988	163	0.1
											1989	197	0.1
7082	A 2131 AB	8 54.9	+26 12	7.0	8.1	45.0	0.335	40.6	155.8	Cnc	1985	190	0.4
				comb.	6.7	1973.5	0.46	264.2	2000		1987	199	0.4
				G2V	K0V	P. Baize, 1980					1989	208	0.4
				par.	0.010	grade 2		first meas. 1910			1991	217	0.4
											1993	225	0.4
7114	h 2477 AxBC	8 59.2	+48 02	3.14	10.2	817.9	9.09	57.8	4.8	UMa	1940	9	6.0
	9 ι UMa			comb.	3.14	1993.90	0.79	129.7			1960	18	4.1
				A7IV	dM1	J. Hopmann, 1973					1980	50	1.7
				par.	0.066	grade 4		first meas. 1841			2000	177	2.0
											2020	214	3.0
	Kui 37	9 00.6	+41 47	4.1	6.2	21.85	0.619	134.8	22.8	Lyn	1985	336	0.6
	(10 UMa)			comb.	4.0	1950.10	0.15	210.0	2000		1987	302	0.5
				F5V		W. D. Heintz, 1967					1989	261	0.5
				par.	0.071	grade 2		first meas. 1936			1991	225	0.5
											1993	194	0.5
7158	A 1585	9 03.6	+47 09	4.2	4.4	70.1	0.27	105.8	106.6	UMa	1985	275	0.2
	12 κ UMa			comb.	3.5	1973.2	0.04	152.4	2000		1990	258	0.1
				A1Vn		P. J. Morel, 1970					1995	202	0.1
				par.	0.010	grade 2		first meas. 1907			2000	139	0.1
											2005	121	0.2
7203	Σ 1306	9 10.4	+67 08	4.85	8.16	1067	6.20	146.2	99.7	UMa	1940	50	1.7
	13 σ² UMa			comb.	4.80	1917.7	0.81	331.5	1900		1960	21	2.5
				F7IV		P. Baize, 1948					1980	5	3.2
				par.	0.052	grade 4		first meas. 1832			2000	355	3.9
											2020	348	4.6
	ø 347	9 12.3	+15 00	7.2	7.2	2.65	0.126	124.7	144.4	Cnc	1985	329	0.1
	81 Cnc			comb.	6.4	1963.86	0.31	189.0			1986	157	0.1
				G9V		W. S. Finsen, 1966					1987	99	0.1
				par.	0.062	grade 3		first meas. 1959			1988	254	0.1
											1989	141	0.2
7251	Σ 1321	9 14.4	+52 41	7.61	7.71	975.0	16.73	21.0	173.8	UMa	1940	75	18.5
				comb.	6.91	2260.0	0.28	44.0	2000		1960	81	18.1
				M0V	M0V	K. Chang, 1972					1980	87	17.6
				par.	0.161	grade 5		first meas. 1832			2000	93	17.2
											2020	99	16.8
	ø 363	9 17.3	-68 41	6.1	6.1	3.2	0.124	123.6	156.1	Car	1985	85	0.1
				comb.	5.3	1961.73	0.51	294.2			1986	14	0.1
				F4V		W. S. Finsen, 1964					1987	325	0.1
				par.	0.051	grade 3		first meas. 1960			1988	105	0.1
											1989	24	0.1
7284	Σ 3121	9 17.9	+28 34	7.9	8.0	34.20	0.660	77.0	25.0	Cnc	1985	213	0.5
				comb.	7.2	1878.30	0.35	130.0			1987	223	0.4
				K3V	K3V	W. H. van den Bos, 1938					1989	243	0.3
				par.	0.055	grade 1		first meas. 1832			1991	290	0.2
											1993	338	0.3

ADS	Star Name	α 2000	δ 2000	m₁	m₂	Orbital Elements P a i Ω / T e ω equinox	Const	Ephemeris Date PA Sep

ADS	Star Name	α 2000	δ 2000	m₁	m₂	P / T	a / e	i / ω	Ω / equinox	Const	Date	PA	Sep
7307	Σ 1338	9ʰ21.ᵐ0	+38°11′	6.50	6.74	219ʸ.7	1″.459	47°.8	91°.5	Lyn	1940	201°	1″.3
				comb.	5.86	1996.77	0.61	252°.9			1960	224	1.2
				F3V		G. A. Starikova, 1966					1980	256	0.9
						grade 4	first meas. 1829				2000	16	0.4
											2020	95	1.2
7334	A 1342	9 22.8	−9 50	7.3	7.3	25.4	0.28	75	119.7	Hya	1985	34	0.1
				comb.	6.5	1947.7	0.96	92.0			1987	42	0.1
				A2		W. S. Finsen, 1976					1989	50	0.1
						grade 3	first meas. 1906				1991	58	0.1
											1993	66	0.1
7341	A 2477	9 24.5	+18 08	7.3	8.8	243.4	0.350	0.0	0.0	Leo	1940	273	0.4
				comb.	7.1	1755.73	0.0	0.0			1960	302	0.4
				G0III		D. J. Zulevic, 1970					1980	332	0.4
						grade 4	first meas. 1912				2000	2	0.4
											2020	31	0.4
7390	Σ 1356	9 28.5	+9 03	5.87	6.53	118.227	0.880	66.05	325.69	Leo	1970	343	0.5
	2 ω Leo			comb.	5.40	1959.40	0.557	302.65	1900		1980	14	0.5
				F9V		E. van Dessel, 1976					1990	53	0.5
				par.	0.028	grade 1	first meas. 1825				2000	84	0.6
											2010	103	0.7
	--	9 30.7	−40 28	4.1	4.6	33.99	0.795	58.5	287.2	Vel	1985	131	0.9
	ψ Vel			comb.	3.6	1935.75	0.44	48.4			1987	139	0.9
				F2IV		O. J. Eggen, 1967					1989	148	0.8
				par.	0.059	grade 1	first meas. 1897				1991	161	0.7
											1993	177	0.6
	B 780	9 40.7	−57 59	6.0	6.1	10.72	0.142	130.0	90.4	Car	1985	61	0.1
				comb.	5.3	1952.50	0.34	17.0			1986	348	0.1
				A3IV		W. H. van den Bos, 1959					1987	304	0.1
				par.	0.036	grade 1	first meas. 1927				1988	285	0.2
											1989	271	0.2
	ø 326	9 44.2	−27 46	5.4	5.6	18.83	0.127	117.8	176.5	Ant	1985	184	0.1
	θ Ant			comb.	4.7	1965.54	0.42	312.6			1986	165	0.1
				F7V		W. S. Finsen, 1966					1987	146	0.1
				par.	0.045	grade 3	first meas. 1952				1988	124	0.1
											1989	97	0.1
7545	OΣ 208	9 52.1	+54 04	5.30	5.38	105.5	0.36	28.0	118.4	UMa	1970	55	0.3
	30 ø UMa			comb.	4.59	1987.7	0.46	48.0	2000		1980	100	0.2
				A3IV		W. D. Heintz, 1971					1990	188	0.2
				par.	0.025	grade 2	first meas. 1843				2000	264	0.3
											2010	294	0.4
7555	AC 5	9 52.5	−8 06	5.58	6.07	75.60	0.385	143.3	36.7	Sex	1985	72	0.6
	γ Sex			comb.	5.05	1958.10	0.70	151.4	2000		1990	67	0.6
				A1V		W. H. van den Bos, 1960					1995	61	0.6
				par.	0.009	grade 2	first meas. 1854				2000	56	0.6
											2005	51	0.6
	I 173	10 06.2	−47 22	5.3	6.9	232.0	0.64	35.7	20.6	Vel	1940	271	0.2
				comb.	5.1	1932.8	0.72	178.3			1960	336	0.5
				K0IV		D. Jones, 1960					1980	354	0.8
				par.	0.011	grade 4	first meas. 1901				2000	3	0.9
											2020	11	1.0
7662	A 2145	10 09.3	+20 20	7.4	7.4	64.7	0.24	115.8	161.2	Leo	1985	277	0.1
				comb.	6.6	1981.75	0.95	104.5			1990	253	0.1
				Am	F5III	W. S. Finsen, 1977					1995	242	0.2
						grade 3	first meas. 1910				2000	235	0.2
											2005	229	0.2
7685	OΣ 213	10 13.2	+27 25	8.4	10.4	157.5	0.845	124.7	60.0	Leo	1970	132	0.7
				comb.	8.2	1939.5	0.95	111.3	2000		1980	127	0.8
				G5V		W. D. Heintz, 1962					1990	124	0.9
						grade 3	first meas. 1844				2000	121	1.0
											2010	118	1.0
	Hu 1597	10 16.1	−59 54	6.9	7.0	124.0	0.435	53.5	81.7	Car	1970	280	0.4
				comb.	6.2	1973.0	0.12	220.5	2000		1980	315	0.3
				A5V		W. D. Heintz, 1967					1990	10	0.2
						grade 3	first meas. 1914				2000	49	0.3
											2010	70	0.4
7704	OΣ 215	10 16.3	+17 44	7.2	7.5	552.0	1.299	131.0	117.6	Leo	1940	193	1.2
				comb.	6.6	1829.0	0.82	122.3	1900		1960	188	1.3
				A9IV		S. Wierzbinski, 1956					1980	183	1.4
						grade 4	first meas. 1844				2000	179	1.5
											2020	176	1.6

ADS	Star Name	α 2000	δ 2000	m₁	m₂	Orbital Elements				Const	Ephemeris		
						P / T	a / e	i / ω	Ω / equinox		Date	PA	Sep
7706	I 851	10ʰ16ᵐ.3	-28°59'	8.4 comb. 7.7 F5	8.5	89ʸ.6 1956.2 W. D. Heintz, 1969 grade 3	0".295 0.75	118°.2 48°.5 first meas. 1911	83°.9 2000	Ant	1985 1990 1995 2000 2005	246° 243 240 237 233	0".4 0.4 0.4 0.4 0.4
7724	Σ 1424 41 γ Leo	10 20.0	+19 51	2.22 comb. 1.92 K0III par. 0.019	3.47 G7III	618.6 1743.32 W. Rabe, 1958 grade 5	2.505 0.84	36.4 162.5 first meas. 1831	143.2 1900	Leo	1940 1960 1980 2000 2020	120 121 123 125 127	4.0 4.2 4.3 4.4 4.5
7744	OΣ 216	10 22.7	+15 21	7.5 comb. 7.4 G5	9.6	306.5 1941.7 W. D. Heintz, 1978 grade 3	1.74 0.54	147.0 39.5 first meas. 1845	48.7 2000	Leo	1940 1960 1980 2000 2020	21 295 256 238 226	0.8 0.9 1.4 1.9 2.2
7775	OΣ 217	10 26.9	+17 13	7.9 comb. 7.4 F6V	8.4	139.6 1966.9 W. D. Heintz, 1975 grade 4	0.424 0.97	53.5 169.5 first meas. 1844	157.5 2000	Leo	1970 1980 1990 2000 2010	132 142 145 147 148	0.1 0.4 0.6 0.7 0.7
7780	Hu 879 31 β LMi	10 27.9	+36 42	4.40 comb. 4.20 G9III par. 0.021	6.12	37.2 1955.5 P. Baize, 1976 grade 3	0.32 0.70	78.6 344.6 first meas. 1904	47.5	LMi	1985 1987 1989 1991 1993	231 233 238 266 49	0.4 0.3 0.2 0.0 0.1
7846	β 411	10 36.1	-26 40	6.7 comb. 6.3 F6V	7.5	210.1 1948.23 J. L. Newburg, 1966 grade 3	0.982 0.80	126.1 35.8 first meas. 1877	144.8 2000	Hya	1940 1960 1980 2000 2020	204 340 322 315 310	0.3 0.6 1.1 1.4 1.5
	See 119 p Vel	10 37.3	-48 14	4.2 comb. 3.8 F3IV par. 0.033	5.1 A6V	16.30 1953.03 W. S. Finsen, 1968 grade 1	0.340 0.73	129.4 293.8 first meas. 1897	38.3	Vel	1985 1986 1987 1988 1989	201 40 357 334 317	0.1 0.1 0.2 0.2 0.3
7871	OΣ 224	10 39.7	+8 51	7.8 comb. 7.5 F6 par. 0.019	9.0	214.1 1808.91 P. Baize, 1955 grade 3	0.51 0.09	142.2 56.9 first meas. 1843	149.7	Leo	1940 1960 1980 2000 2020	238 202 171 141 106	0.4 0.4 0.5 0.5 0.4
7896	A 2768	10 42.6	+3 35	6.7 comb. 6.6 F5	8.8	84.4 1976.8 W. D. Heintz, 1978 grade 3	0.38 0.55	143.0 340.0 first meas. 1914	45.0 2000	Sex	1985 1990 1995 2000 2005	331 301 282 270 261	0.2 0.3 0.4 0.4 0.5
	R 155 μ Vel	10 46.8	-49 25	2.7 comb. 2.7 G5III	6.4 G2V	116.2 1953.16 W. H. van den Bos, 1966 grade 3	1.437 0.79	63.1 188.6 first meas. 1880	57.5 2000	Vel	1970 1980 1990 2000 2010	45 51 55 58 61	1.5 2.0 2.3 2.5 2.6
7982	β 1076 55 Leo	10 55.7	+0 44	6.1 comb. 5.9 F2III par. 0.015	8.0	128.3 1917.19 P. J. Morel, 1970 grade 3	0.823 0.68	119.6 310.5 first meas. 1889	214.2 2000	Leo	1970 1980 1990 2000 2010	72 65 59 53 48	0.9 1.0 1.1 1.1 1.1
8035	β 1077 50 α UMa Dubhe	11 03.7	+61 45	1.86 comb. 1.79 K0III par. 0.031	4.80	44.66 1921.00 P. Couteau, 1958 grade 2	0.77 0.45	126.9 71.0 first meas. 1889	167.5	UMa	1985 1987 1989 1991 1993	301 292 283 272 260	0.8 0.8 0.7 0.7 0.6
	φ 47 χ¹ Hya	11 05.3	-27 18	5.6 comb. 4.94 F4V par. 0.033	5.7	7.40 1945.90 W. H. van den Bos, 1957 grade 2	0.140 0.28	94.7 180.0 first meas. 1927	45.0	Hya	1985 1986 1987 1988 1989	51 47 44 41 4	0.1 0.2 0.2 0.1 0.0
8086	β 220 ψ Crt	11 12.5	-18 30	6.7 comb. 6.1 A1V par. 0.127	7.1	233.0 1966.5 W. D. Heintz, 1968 grade 4	0.383 0.64	115.5 161.4 first meas. 1875	142.5 2000	Crt	1940 1960 1980 2000 2020	121 15 275 192 171	0.3 0.1 0.1 0.2 0.3

ADS	Star Name	α 2000	δ 2000	m₁	m₂	Orbital Elements				Const	Ephemeris		
						P / T	a / e	i / ω	Ω / equinox		Date	PA	Sep
8094	Σ 1517	11^{h}13^m.7	+20°08′	7.7	7.7	4050^y	2″.41	116°.2	120°.4	Leo	1940	212°	0″.2
				comb.	6.9	1964.80	0.83	309°.0			1960	157	0.3
				G0		J. Hopmann, 1970					1980	132	0.4
						grade 5		first meas. 1829			2000	120	0.5
											2020	111	0.6
8119	Σ 1523 AB	11 18.2	+31 32	4.32	4.79	59.84	2.530	122.65	101.59	UMa	1985	91	2.3
	53 ξ UMa			comb.	3.78	1935.17	0.414	127.53	1900		1990	60	1.3
	Alula Aus.			G0V	G0V	W. D. Heintz, 1967					1995	317	1.1
						grade 1		first meas. 1826			2000	273	1.8
											2005	243	1.7
8128	Σ 1527	11 19.0	+14 16	7.0	8.1	1148	4.590	80.1	16.3	Leo	1940	21	2.7
				comb.	6.7	2006.5	0.45	78.98	1900		1960	24	2.2
				F9V		J. Hopmann, 1960					1980	30	1.5
						grade 5		first meas. 1829			2000	46	0.8
											2020	115	0.4
	I 879	11 21.0	−54 29	4.3	5.0	39.18	0.277	59.6	98.9	Cen	1985	122	0.3
	π Cen			comb.	3.8	1936.56	0.86	236.5			1987	125	0.4
				B5V		J. L. Newburg, 1967					1989	128	0.4
						grade 2		first meas. 1912			1991	131	0.4
											1993	133	0.4
8148	Σ 1536	11 23.9	+10 32	4.03	6.70	192.0	1.92	130.5	52.2	Leo	1970	178	1.1
	78 ι Leo			comb.	3.94	1948.47	0.55	140.0			1980	152	1.3
				F2IV		P. Baize, 1980					1990	131	1.5
				par.	0.047	grade 3		first meas. 1832			2000	117	1.7
											2010	105	2.0
	BrsO 5	11 24.7	−61 39	7.6	8.6	421.5	5.760	48.5	77.8	Cen	1940	190	2.1
				comb.	7.2	1918.74	0.68	16.39	2000		1960	224	3.9
				K7V	M0	W. H. van den Bos, 1965					1980	237	5.4
				par.	0.088	grade 3		first meas. 1838			2000	245	6.7
											2020	250	7.6
8166	Hu 462	11 27.2	−15 39	8.2	8.6	48.50	0.450	159.0	169.2	Crt	1985	316	0.5
				comb.	7.6	1960.44	0.10	33.1			1987	304	0.5
				K0		P. Couteau, 1962					1989	291	0.5
						grade 2		first meas. 1902			1991	278	0.5
											1993	264	0.4
8189	OΣ 234	11 30.8	+41 17	7.6	7.9	86.44	0.41	56.7	337.1	UMa	1985	135	0.4
				comb.	7.0	1967.14	0.37	26.3	1950		1990	145	0.5
				F6V		P. Couteau, 1965					1995	153	0.5
				par.	0.037	grade 2		first meas. 1844			2000	160	0.5
											2005	166	0.5
8197	OΣ 235	11 32.3	+61 05	5.75	7.07	72.87	0.813	47.7	81.0	UMa	1985	256	0.5
				comb.	5.47	1981.50	0.398	130.7	2000		1990	287	0.6
				F6V		W. D. Heintz, 1972					1995	315	0.6
						grade 1		first meas. 1844			2000	341	0.6
											2005	4	0.7
8311	β 603	11 48.6	+14 17	6.0	8.3	122.0	0.68	148.3	149.0	Leo	1970	359	0.9
				comb.	5.9	1939.0	0.67	352.5			1980	350	1.0
				A6V		W. D. Heintz, 1963					1990	342	1.1
				par.	0.047	grade 3		first meas. 1879			2000	335	1.1
											2010	329	1.1
8337	β 794	11 53.7	+73 45	7.2	8.4	77.15	0.388	33.7	24.6	Dra	1985	226	0.2
				comb.	6.9	1987.80	0.48	245.0	2000		1990	303	0.2
				F7V		W. D. Heintz, 1963					1995	358	0.3
				par.	0.030	grade 2		first meas. 1881			2000	23	0.4
											2005	39	0.4
8344	A 75	11 54.1	+71 55	8.0	8.8	78.6	0.291	109.0	24.0	Dra	1985	199	0.3
				comb.	7.6	1914.6	0.11	241.0			1990	187	0.2
				F5V		P. Baize, 1953					1995	160	0.1
						grade 3		first meas. 1900			2000	87	0.1
											2005	48	0.2
	I 215	12 01.8	−34 39	7.4	8.1	108.0	0.79	118.0	40.0	Hya	1970	16	0.6
				comb.	6.9	1953.0	0.20	325.0			1980	336	0.4
				G0		A. V. Bespalov, 1961					1990	280	0.5
				par.	0.041	grade 4		first meas. 1897			2000	250	0.7
											2010	234	0.9
	See 143	12 03.6	−39 01	7.2	7.4	109.2	0.67	150.0	22.8	Cen	1970	101	0.9
	89 Cen			comb.	6.5	1913.77	0.58	98.5			1980	89	0.9
				G0V		O. J. Eggen 1, 1965					1990	75	0.9
				par.	0.032	grade 3		first meas. 1897			2000	59	0.7
											2010	35	0.6

ADS	Star Name	α 2000	δ 2000	m₁	m₂	Orbital Elements				Const	Ephemeris		
						P / T	a / e	i / ω	Ω / equinox		Date	PA	Sep
8419	Σ 3123	12^{h}06.0	+68°42'	7.9	7.9	115.5	0".31	135°.8	85.0	Dra	1970	75°	0".2
				comb. 7.1		1978.53	0.60	106°.1			1980	307	0.1
				F5		P. Baize, 1974					1990	240	0.2
						grade 2	first meas. 1832				2000	213	0.3
											2010	192	0.3
8446	Σ 1606	12 10.8	+39 53	7.3	8.0	326.8	0.774	155.5	26.6	CVn	1940	314	0.8
				comb. 6.8		1996.37	0.73	218.9	1900		1960	299	0.6
				A3		R. van der Wiele, 1974					1980	265	0.4
						grade 4	first meas. 1831				2000	134	0.2
											2020	51	0.5
8481	β 920	12 15.8	−23 21	6.8	7.9	175.0	1.00	57.0	92.0	Crv	1970	337	0.5
				comb. 6.5		1995.0	0.26	335.0			1980	26	0.5
				F5		A. V. Bespalov, 1961					1990	65	0.6
						grade 4	first meas. 1879				2000	88	0.7
											2010	107	0.7
8539	Σ 1639	12 24.4	+25 35	6.76	7.79	678.0	1.30	161.0	105.7	Com	1940	332	1.0
				comb. 6.40		1891.0	0.95	334.0			1960	328	1.3
				F0V		R. M. Aller, 1951					1980	325	1.5
						grade 4	first meas. 1836				2000	323	1.7
											2020	321	1.8
8569	OΣ 251	12 29.1	+31 23	8.3	9.9	600.0	0.61	37.3	12.0	CVn	1940	38	0.4
				comb. 8.1		1898.0	0.84	257.0			1960	47	0.5
				F8V		P. Baize, 1957					1980	53	0.6
						grade 4	first meas. 1844				2000	58	0.7
											2020	62	0.7
8573	β 28	12 30.1	−13 24	6.5	8.6	180.0	1.51	32.3	104.0	Crv	1970	307	1.5
				comb. 6.4		1944.50	0.75	65.4			1980	316	1.8
				G0V		P. Muller, 1977					1990	323	2.0
				par.	0.042	grade 3	first meas. 1875				2000	329	2.2
											2010	334	2.3
8575	Σ 1647	12 30.6	+9 43	8.5	8.8	4273	4.36	67.0	41.5	Vir	1940	232	1.4
				comb. 7.9		1864.30	0.69	161.8			1960	237	1.4
				F2		J. Hopmann, 1970					1980	242	1.3
						grade 5	first meas. 1830				2000	247	1.3
											2020	253	1.2
	h 4539 γ Cen	12 41.5	−48 58	2.9	2.9	84.50	0.930	112.9	2.4	Cen	1985	355	1.5
				comb. 2.1		1931.22	0.79	187.8			1990	353	1.4
				A0III		W. H. van den Bos, 1936					1995	351	1.2
				par.	0.010	grade 1	first meas. 1835				2000	347	1.0
											2005	341	0.7
8630	Σ 1670 29 γ Vir Porrima	12 41.7	−1 27	3.48	3.50	171.37	3.746	146.05	31.78	Vir	1970	303	4.6
				comb. 2.74		1836.433	0.881	252.88	1900		1980	297	3.9
				F0V	F0V	K. Aa. Strand, 1937					1990	287	3.0
				par.	0.094	grade 1	first meas. 1825				2000	267	1.8
											2010	44	0.9
	R 207 β Mus	12 46.3	−68 06	3.7	4.0	383.1	1.735	61.3	161.8	Mus	1940	6	1.4
				comb. 3.1		1872.29	0.53	98.32	1900		1960	18	1.4
				B2V		R. F. Mourao, 1964					1980	30	1.3
						grade 4	first meas. 1880				2000	43	1.3
											2020	57	1.3
8695	Σ 1687 AB 35 Com	12 53.3	+21 14	5.06	7.23	510	1.42	35.0	230.1	Com	1940	118	1.0
				comb. 4.92		1931.0	0.15	243.0	2000		1960	141	1.0
				G8III	F6	W. D. Heintz, 1973					1980	163	1.1
				par.	0.022	grade 4	first meas. 1829				2000	182	1.2
											2020	198	1.3
	I 83	12 56.7	−47 41	7.4	7.6	294.0	0.695	42.6	105.0	Cen	1940	186	0.3
				comb. 6.7		1923.88	0.87	308.7			1960	210	0.6
				F5		W. H. van den Bos, 1953					1980	220	0.7
						grade 3	first meas. 1897				2000	227	0.9
											2020	232	1.0
8739	β 1082 78 UMa	13 00.7	+56 22	5.05	7.43	115.7	1.256	51.5	269.2	UMa	1970	26	1.2
				comb. 4.94		1921.74	0.44	301.0			1980	44	1.3
				F2V		P. Baize, 1946					1990	57	1.5
				par.	0.030	grade 3	first meas. 1889				2000	69	1.5
											2010	82	1.4
8804	Σ 1728 AB 42 α Com	13 10.0	+17 32	5.05	5.08	25.87	0.662	89.92	11.6	Com	1985	191	0.6
				comb. 4.31		1911.65	0.500	278.9	1900		1987	191	0.5
				F6V		H. Haffner, 1948					1989	193	0.0
						grade 1	first meas. 1827				1991	11	0.4
											1993	11	0.5

ADS	Star Name	α 2000	δ 2000	m_1	m_2	Orbital Elements P / T	a / e	i / ω	Ω / equinox	Const	Ephemeris Date	PA	Sep
	ø 305	$13^h11^m.7$	−26°33′	7.2	7.3	$19^y.0$	0".118	148°.9	10°.4	Hya	1985	113°	0".1
				comb.	6.5	1962.25	0.93	92°.5			1986	109	0.2
				A5V		W. S. Finsen, 1968					1987	106	0.2
						grade 3	first meas. 1933				1988	104	0.2
											1989	101	0.2
	See 170 AB	13 12.3	−59 55	5.0	5.7	27.0	0.185	68.3	114.6	Cen	1985	120	0.2
				comb.	4.5	1941.1	0.32	168.5			1987	127	0.2
				B8V		W. S. Finsen, 1964					1989	139	0.1
						grade 4	first meas. 1897				1991	174	0.1
											1993	256	0.1
	I 1567	13 16.9	−34 36	8.6	9.0	64.72	0.252	180.0	0.0	Cen	1985	189	0.3
				comb.	8.0	1954.25	0.0	0.0			1990	161	0.3
				G5		J. L. Newburg, 1969					1995	133	0.3
						grade 3	first meas. 1897				2000	105	0.3
											2005	78	0.3
	See 179	13 31.0	−39 24	4.5	4.7	78.7	0.165	152.9	153.2	Cen	1985	106	0.2
				comb.	3.8	1955.9	0.46	250.7			1990	95	0.2
				G8III		W. S. Finsen, 1964					1995	85	0.2
				par.	0.007	grade 3	first meas. 1897				2000	74	0.2
											2005	63	0.2
8939	OΣ 269	13 32.8	+34 54	7.3	7.8	54.8	0.18	68.6	44.0	CVn	1985	248	0.1
				comb.	6.8	1934.6	0.80	2.0	1900		1990	70	0.0
				A6III		P. Baize, 1952					1995	207	0.1
						grade 2	first meas. 1844				2000	215	0.2
											2005	219	0.3
8949	Σ 1757	13 34.3	−0 19	7.8	8.7	334.0	2.35	58.0	97.1	Vir	1940	95	2.7
				comb.	7.4	1723.0	0.24	140.8	1900		1960	104	2.6
				K1		W. D. Heintz, 1956					1980	114	2.2
				par.	0.015	grade 4	first meas. 1831				2000	131	1.6
											2020	166	1.1
8954	β 932	13 34.7	−13 13	6.5	6.9	197.7	0.563	70.5	114.1	Vir	1970	39	0.3
				comb.	5.9	1922.1	0.93	105.6	2000		1980	46	0.3
				A0Vp		W. D. Heintz, 1969					1990	51	0.4
						grade 3	first meas. 1879				2000	56	0.4
											2010	60	0.4
	I 365	13 37.2	−61 42	6.2	6.5	34.63	0.489	114.3	71.8	Cen	1985	151	0.4
				comb.	5.6	1932.45	0.81	93.4			1987	142	0.4
				F6V		A. Abrami, 1952					1989	134	0.4
						grade 2	first meas. 1900				1991	126	0.4
											1993	118	0.4
8974	Σ 1768	13 37.5	+36 18	5.01	6.91	240.0	1.091	144.0	67.0	CVn	1940	115	1.6
	25 CVn			comb.	4.84	1863.95	0.83	137.9	1900		1960	109	1.7
				A7III		S. Wierzbinski, 1955					1980	104	1.8
				par.	0.029	grade 3	first meas. 1831				2000	99	1.8
											2020	94	1.8
8987	β 612	13 39.6	+10 45	6.3	6.3	22.35	0.208	47.8	35.4	Boo	1985	213	0.3
				comb.	5.5	1929.69	0.527	359.9			1987	221	0.3
				F0V		A. Danjon, 1956					1989	230	0.3
				par.	0.008	grade 1	first meas. 1879				1991	241	0.2
											1993	262	0.2
9019	Σ 1781	13 46.1	+5 07	7.8	8.2	312.5	1.091	43.7	168.8	Vir	1940	311	0.8
				comb.	7.2	1974.7	0.63	249.3	2000		1960	345	0.6
				F8V	F0	W. D. Heintz, 1976					1980	96	0.3
						grade 3	first meas. 1830				2000	162	0.7
											2020	181	1.0
9031	Σ 1785	13 49.1	+26 59	7.59	8.03	155.0	2.423	46.8	155.1	Boo	1970	152	3.2
				comb.	7.04	1916.90	0.44	200.9	2000		1980	159	3.4
				K6		K. Aa. Strand, 1955					1990	167	3.4
				par.	0.062	grade 2	first meas. 1830				2000	174	3.3
											2010	182	3.0
	Howe 28	13 53.5	−35 40	6.3	6.5	257.7	1.268	72.3	113.8	Cen	1940	109	0.8
				comb.	5.54	1958.9	0.71	96.2	2000		1960	252	0.2
				F4V		W. D. Heintz, 1969					1980	305	0.8
						grade 4	first meas. 1889				2000	319	0.8
											2020	333	0.8
9053	Σ 1788 AB	13 55.0	−8 04	6.5	7.7	2613	5.00	50.9	108.0	Vir	1940	86	3.1
				comb.	6.2	1823.8	0.35	293.5			1960	90	3.2
				F8V		J. Hopmann, 1970					1980	94	3.4
				par.	0.021	grade 5	first meas. 1831				2000	97	3.5
											2020	101	3.6

ADS	Star Name	α 2000	δ 2000	m₁	m₂	Orbital Elements				Const	Ephemeris		
						P / T	a / e	i / ω	Ω / equinox		Date	PA	Sep
9089	A 1097	14ʰ 02.0ᵐ	+57° 13′	8.4	8.7	333.3ʸ	0.518″	39.3°	30.0°	UMa	1940	142°	0.3″
				comb.	7.76	1951.70	0.34	134.2			1960	189	0.3
				F5		V. Janova, 1960					1980	221	0.4
						grade 4		first meas. 1905			2000	247	0.4
											2020	271	0.4
9094	β 1270	14 03.7	+8 29	8.3	8.4	38.65	0.194	15.0	149.8	Boo	1985	69	0.1
				comb.	7.6	1988.90	0.43	0.0	2000		1987	105	0.1
				F5		W. D. Heintz, 1973					1989	152	0.1
				par.	0.017	grade 2		first meas. 1892			1991	199	0.1
											1993	234	0.1
9159	OΣ 278	14 12.2	+44 11	8.4	8.6	215	0.36	134.0	146.0	Boo	1940	26	0.3
				comb.	7.7	1994.0	0.24	229.0	2000		1960	352	0.3
				F2IV		W. D. Heintz, 1976					1980	319	0.3
						grade 3		first meas. 1843			2000	269	0.2
											2020	201	0.2
9182	Σ 1819	14 15.3	+3 08	7.8	7.9	213.4	1.10	148.0	2.5	Vir	1940	317	1.1
				comb.	7.1	2002.94	0.24	180.0	1900		1960	284	0.9
				G0V		P. Baize, 1974					1980	239	0.8
						grade 3		first meas. 1830			2000	189	0.8
											2020	140	?0.8
9229	Σ 1834	14 20.3	+48 30	8.05	8.28	321.0	0.898	76.9	110.5	Boo	1940	100	0.7
				comb.	7.41	1902.5	0.88	168.2	1900		1960	102	1.0
				F9V		W. H. van den Bos, 1936					1980	104	1.2
						grade 3		first meas. 1831			2000	105	1.4
											2020	106	1.5
9247	β 1111 BC	14 23.4	+8 27	7.6	7.7	39.50	0.228	37.5	39.5	Boo	1985	50	0.3
				comb.	6.86	1957.98	0.254	149.9			1987	62	0.2
				A9V		P. Couteau, 1958					1989	76	0.2
				par.	0.008	grade 1		first meas. 1889			1991	96	0.2
											1993	123	0.2
9264	A 2069	14 26.7	+16 25	8.9	9.1	44.1	0.20	145.1	52.5	Boo	1985	302	0.2
				comb.	8.2	1933.85	0.35	9.4			1987	283	0.2
				F8		P. Baize, 1954					1989	270	0.2
						grade 2		first meas. 1909			1991	259	0.2
											1993	250	0.2
	Rst 4529	14 31.0	−5 48	8.4	8.4	43.08	0.322	58.6	178.6	Vir	1985	1	0.3
				comb.	7.6	1972.00	0.07	68.0			1987	9	0.3
				G5		P. J. Morel, 1968					1989	18	0.3
						grade 3		first meas. 1938			1991	30	0.3
											1993	45	0.2
9301	A 570	14 32.3	+26 41	6.5	6.7	30.0	0.210	148.0	29.0	Boo	1985	315	0.2
				comb.	5.8	1924.36	0.15	66.0			1987	277	0.2
				A7IV		W. H. van den Bos, 1949					1989	247	0.2
				par.	0.010			first meas. 1903			1991	223	0.2
											1993	204	0.2
9324	A 347	14 36.9	+48 13	8.4	8.6	322.6	0.704	134.0	56.2	Boo	1940	341	0.4
				comb.	7.7	1910.9	0.46	0.0	1950		1960	300	0.5
				F2		U. Guntzel-Lingner, 1956					1980	277	0.6
						grade 4		first meas. 1902			2000	264	0.8
											2020	254	0.9
	-- α Cen Rigil Kent.	14 39.6	−60 50	−0.04	1.17	79.92	17.515	79.24	204.87	Cen	1985	212	21.2
				comb.	−0.35	1955.56	0.516	231.56	2000		1990	215	19.7
				G2V	K0V	W. D. Heintz, 1960					1995	218	17.3
				par.	0.754	grade 1		first meas. 1752			2000	222	14.1
											2005	230	10.5
9343	Σ 1865 30 ζ Boo	14 41.1	+13 44	4.52	4.55	123.4	0.595	142.0	130.0	Boo	1970	307	1.1
				comb.	3.78	1897.59	0.96	1.47	1900		1980	305	1.1
				A2III		S. Wierzbinski, 1956					1990	303	1.0
				par.	0.007	grade 2		first meas. 1830			2000	300	0.8
											2010	295	0.6
9378	OΣ 285	14 45.5	+42 23	7.9	8.2	88.40	0.325	172.9	72.5	Boo	1985	143	0.3
				comb.	7.30	1971.52	0.53	171.6	1950		1990	127	0.4
				F6V		P. Couteau, 1973					1995	115	0.4
						grade 2		first meas. 1845			2000	105	0.4
											2005	97	0.5
	φ 309 17 Lib	14 46.2	−21 11	7.1	7.2	25.95	0.297	56.1	134.0	Lib	1985	284	0.2
				comb.	6.4	1918.38	0.04	292.4	1900		1987	303	0.3
				F7V		A. da Silva, J. Balca, 1972					1989	318	0.3
						grade 3		first meas. 1935			1991	335	0.3
											1993	3	0.2

ADS	Star Name	α 2000	δ 2000	m_1	m_2	P / T	a / e	i / ω	Ω / equinox	Const	Date	PA	Sep
9380	Σ 1879	$14^h 46^m.3$	+9°39'	7.76	8.38	226.0	0".99	121°.0	73°.5	Boo	1940	103°	1".2
				comb.	7.27	1866.82	0.68	150°.1	1900		1960	95	1.4
				G2V		S. Wierzbinski, 1957					1980	90	1.5
						grade 3		first meas. 1829			2000	85	1.5
											2020	79	1.5
9392	Σ 1883	14 48.9	+5 57	7.6	7.6	227.9	0.835	107.4	100.3	Vir	1940	142	0.3
				comb.	6.8	1964.38	0.62	41.7	2000		1960	99	0.3
				F6V		B. H. Feierman, 1969					1980	298	0.4
						grade 3		first meas. 1830			2000	283	0.8
											2020	277	1.1
9413	Σ 1888	14 51.4	+19 06	4.72	6.97	151.505	4.904	140.04	348.1	Boo	1970	340	7.2
	37 ξ Boo			comb.	4.59	1909.361	0.512	203.92	2000		1980	333	7.2
				G8V	K4V	R. Wielen, 1962					1990	326	7.0
				par.	0.145	grade 1		first meas. 1836			2000	318	6.6
											2010	308	6.0
9418	OΣ 287	14 51.4	+44 55	8.4	8.6	400.0	0.934	64.4	167.6	Boo	1940	335	1.0
				comb.	7.7	2118.0	0.25	319.5	2000		1960	340	1.1
				G0		W. D. Heintz, 1962					1980	346	1.1
						grade 4		first meas. 1843			2000	351	1.0
											2020	358	0.9
9425	OΣ 288	14 53.4	+15 42	6.8	7.5	215.4	1.09	112.5	4.7	Boo	1940	182	1.7
				comb.	6.3	1819.3	0.60	0.0	1900		1960	177	1.6
				F9V		W. D. Heintz, 1956					1980	171	1.2
						grade 4		first meas. 1845			2000	159	0.8
											2020	89	0.3
	h 4707	14 54.2	-66 25	7.6	7.9	288.0	1.386	119.4	60.5	Cir	1940	53	1.0
				comb.	7.0	1932.0	0.30	354.2			1960	23	0.8
				G0		R. Woolley, B. Mason, 1948					1980	333	0.6
						grade 4		first meas. 1837			2000	289	0.9
											2020	268	1.2
	I 227 AB	14 56.6	-34 38	8.0	8.2	40.0	0.256	126.2	85.7	Cen	1985	87	0.4
				comb.	7.3	1955.50	0.86	162.8	1950		1987	85	0.4
				F8		O. P. Lategan, 1961					1989	82	0.3
						grade 2		first meas. 1897			1991	77	0.3
											1993	69	0.2
9453	β 239	14 58.7	-27 39	6.3	6.6	339.3	0.93	73.0	138.2	Hya	1940	330	0.8
	59 Hya			comb.	5.7	1880.0	0.0	153.3	1900		1960	343	0.5
				A4V		R. F. Mourao, 1972					1980	15	0.3
						grade 5		first meas. 1878			2000	80	0.3
											2020	112	0.5
9494	Σ 1909	15 03.8	+47 39	5.3	6.2v	225.0	3.77	83.9	57.8	Boo	1940	249	2.4
	44 i Boo			comb.	4.9	2021.0	0.43	38.8	2000		1960	268	1.0
				G0V	G2	W. D. Heintz, 1978					1980	33	0.9
				par.	0.083	grade 3		first meas. 1832			2000	53	2.2
											2020	62	1.8
9504	A 689	15 07.1	-2 17	8.1	8.8	61.19	0.20	110.1	148.2	Lib	1985	176	0.1
				comb.	7.6	1926.40	0.65	17.3			1990	64	0.0
						W. H. van den Bos, 1966					1995	344	0.1
						grade 3		first meas. 1904			2000	335	0.2
											2005	330	0.3
9505	A 2385	15 07.3	+18 26	6.7	6.8	8.0	0.10	122.6	41.3	Boo	1985	128	0.1
				comb.	6.0	1939.97	0.45	71.4			1986	89	0.1
				A2V		O. J. Eggen, 1946					1987	54	0.1
						grade 4		first meas. 1910			1988	337	0.0
											1989	227	0.1
	See 219	15 08.8	-45 17	4.6	4.9	72.88	0.422	79.6	29.1	Lup	1985	196	0.4
	λ Lup			comb.	4.0	1929.62	0.82	290.7			1990	199	0.4
				B3V		W. H. van den Bos, 1962					1995	203	0.3
						grade 3		first meas. 1897			2000	209	0.2
											2005	37	0.1
9532	B 2351 Aa	15 12.2	-19 47	5.1	5.6	22.35	0.107	161.0	2.3	Lib	1985	341	0.1
	24 ι Lib			comb.	4.53	1972.00	0.35	187.6	2000		1987	323	0.1
				B9IV		W. D. Heintz, 1964					1989	300	0.1
				par.	0.023	grade 3		first meas. 1940			1991	268	0.1
											1993	219	0.1
9557	Σ 3091	15 16.0	-4 54	7.9	8.0	156.4	0.37	98.7	44.3	Lib	1970	229	0.5
				comb.	7.2	2052.7	0.40	335.4	2000		1980	227	0.5
				F8V		P. Laques, P. J. Morel, 1971					1990	225	0.5
								first meas. 1832			2000	224	0.5
											2010	221	0.4

ADS	Star Name	α 2000	δ 2000	m_1	m_2	Orbital Elements P a i Ω / T e ω equinox	Const	Ephemeris Date PA Sep

9578 Σ 1932 AB — 15^{h}18.3 — +26°50 — m_1 7.33 m_2 7.36; comb. 6.59; G0V

Orbital Elements: P 203^y.0, a 1″.216, i 58°.2, Ω 64°.0; T 1941.42, e 0.65, ω 51°.0, equinox 2000; W. D. Heintz, 1965; grade 2 — first meas. 1830

Const: CrB

Ephemeris:
Date	PA	Sep
1940	85°	0″.4
1960	234	0.9
1980	250	1.4
2000	259	1.6
2020	267	1.6

9617 Σ 1937, 2 η CrB — 15 23.2 — +30 17 — m_1 5.61 m_2 5.88; comb. 4.98; G2V; par. 0.063

Orbital Elements: P 41.56, a 0.839, i 58.9, Ω 24.2; T 1892.39, e 0.28, ω 219.6, equinox 1900; A. Danjon, 1938; grade 1 — first meas. 1826

Const: CrB

Date	PA	Sep
1985	9	0.8
1987	17	0.9
1989	24	1.0
1991	30	1.0
1993	36	1.0

— h 4757, γ Cir — 15 23.4 — −59 19 — m_1 5.1 m_2 5.5; comb. 4.5; B5IV

Orbital Elements: P 180.0, a 1.16, i 103.0, Ω 182.0; T 2036.0, e 0.48, ω 250.0; A. V. Bespalov, 1961; grade 5 — first meas. 1836

Const: Cir

Date	PA	Sep
1970	92	0.4
1980	55	0.4
1990	33	0.6
2000	20	0.7
2010	11	0.8

9626 Σ 1938 BC, 51 μ² Boo — 15 24.5 — +37 21 — m_1 6.98 m_2 7.63; comb. 6.50; G1V; par. 0.030

Orbital Elements: P 260.1, a 1.463, i 135.4, Ω 174.9; T 1865.0, e 0.59, ω 338.5, equinox 1900; P. Baize, 1952; grade 3 — first meas. 1826

Const: Boo

Date	PA	Sep
1940	33	1.7
1960	23	2.0
1980	15	2.2
2000	8	2.3
2020	2	2.2

9643 A 1120 — 15 27.3 — +9 42 — m_1 8.5 m_2 9.1; comb. 8.0; G0

Orbital Elements: P 50.0, a 0.189, i 69.7, Ω 166.6; T 1930.2, e 0.80, ω 319.6; P. Muller, 1955; grade 3 — first meas. 1905

Const: SerCp

Date	PA	Sep
1985	273	0.1
1987	295	0.1
1989	306	0.1
1991	312	0.1
1993	317	0.2

9645 A 2074 — 15 27.4 — +17 38 — m_1 8.2 m_2 8.9; comb. 7.7; F8

Orbital Elements: P 59.0, a 0.206, i 59.9, Ω 117.9; T 1981.9, e 0.90, ω 305.2; P. Baize, 1976; grade 3 — first meas. 1909

Const: SerCp

Date	PA	Sep
1985	200	0.1
1990	233	0.1
1995	245	0.2
2000	252	0.2
2005	257	0.3

9689 See 238 BC — 15 33.2 — −24 29 — m_1 7.8 m_2 8.0; comb. 7.12; F0Vo

Orbital Elements: P 57.32, a 0.24, i 25.3, Ω 0.6; T 1938.08, e 0.62, ω 338.1, equinox 1900; S. Wierzbinski, 1960; grade 2 — first meas. 1897

Const: Lib

Date	PA	Sep
1985	200	0.3
1990	228	0.2
1995	327	0.1
2000	76	0.1
2005	113	0.2

9701 Σ 1954 AB, 13 δ Ser — 15 34.8 — +10 32 — m_1 4.2 m_2 5.2; comb. 3.8; F0; par. 0.015

Orbital Elements: P 3168, a 6.02, i 112.6, Ω 166.8; T 1700.0, e 0.31, ω 274.6; J. Hopmann, 1973; grade 5 — first meas. 1833

Const: SerCp

Date	PA	Sep
1940	181	3.9
1960	179	4.1
1980	178	4.3
2000	176	4.4
2020	174	4.6

— h 4786, γ Lup — 15 35.1 — −41 10 — m_1 3.5 m_2 3.6; comb. 2.8; B3V

Orbital Elements: P 147.0, a 0.59, i 95.6, Ω 92.8; T 1887.0, e 0.49, ω 301.0, equinox 1900; W. D. Heintz, 1956; grade 3 — first meas. 1835

Const: Lup

Date	PA	Sep
1970	279	0.6
1980	277	0.6
1990	275	0.7
2000	274	0.7
2010	272	0.6

9716 OΣ 298 — 15 36.0 — +39 48 — m_1 7.45 m_2 7.60; comb. 6.77; K2V K5; par. 0.052

Orbital Elements: P 55.88, a 0.785, i 62.6, Ω 2.6; T 1938.26, e 0.591, ω 19.3, equinox 1950; P. Couteau, 1966; grade 1 — first meas. 1843

Const: Boo

Date	PA	Sep
1985	235	0.4
1990	309	0.3
1995	27	0.3
2000	151	0.4
2005	170	0.8

9769 Σ 1989, π² UMi — 15 39.6 — +79 59 — m_1 7.4 m_2 8.2; comb. 7.0; F2V; par. 0.004

Orbital Elements: P 150.7, a 0.349, i 139.6, Ω 0.0; T 1902.46, e 0.90, ω 144.8, equinox 1950; M. A. Giannuzzi, 1956; grade 3 — first meas. 1832

Const: UMi

Date	PA	Sep
1970	30	0.6
1980	27	0.6
1990	24	0.6
2000	21	0.6
2010	18	0.6

9744 Hu 580, 21 ι Ser — 15 41.6 — +19 40 — m_1 5.2 m_2 5.3; comb. 4.5; A1V; par. 0.005

Orbital Elements: P 11.07, a 0.117, i 69.1, Ω 70.0; T 1942.0, e 0.80, ω 180.0, equinox 1950; W. H. van den Bos 1, 1964; grade 2 — first meas. 1902

Const: SerCp

Date	PA	Sep
1985	86	0.1
1986	163	0.0
1987	43	0.0
1988	58	0.1
1989	63	0.2

Orbital Elements: P 22.14, a 0.21, i 84.3, Ω 70.0; T 1936.47, e 0.0, ω 0.0, equinox 1950; W. H. van den Bos 2, 1964; grade 2 — first meas. 1902

Const: SerCp

Date	PA	Sep
1985	84	0.1
1987	224	0.0
1989	244	0.1
1991	248	0.2
1993	252	0.2

9747 A 2176 — 15 42.0 — +0 27 — m_1 8.0 m_2 8.0; comb. 7.2; A5V

Orbital Elements: P 56.1, a 0.19, i 45.6, Ω 87.5; T 1932.55, e 0.77, ω 281.7; P. Couteau, 1960; grade 3 — first meas. 1910

Const: SerCp

Date	PA	Sep
1985	255	0.1
1990	86	0.1
1995	136	0.1
2000	156	0.2
2005	170	0.2

ADS	Star Name	α 2000	δ 2000	m₁	m₂	Orbital Elements P / T	a / e	i / ω	Ω / equinox	Const	Date	PA	Sep
9757	Σ 1967 8 γ CrB	15ʰ 42ᵐ.7	+26°18′	4.08 comb. 3.82 A0IV par. 0″.026	5.51	91ʸ.0 1840.5 P. Baize, 1953 grade 2	0″.74 0.42	96°.0 106°.0 first meas. 1826	111°.0 1900	CrB	1985 1990 1995 2000 2005	121° 118 115 114 112	0″.5 0.6 0.7 0.8 0.7
9806	Hu 912	15 49.3	+60 32	8.5 comb. 7.8 F5	8.7	140.0 2000.0 W. D. Heintz, 1966 grade 4	0.269 0.44	39.0 253.5 first meas. 1905	101.6 2000	Dra	1970 1980 1990 2000 2010	236 258 289 351 66	0.3 0.2 0.2 0.1 0.2
9831	A 2080	15 54.1	+16 59	8.5 comb. 7.7 F2	8.5	257.0 1943.11 D. J. Zulevic, 1973 grade 3	0.30 0.14	50.2 270.0 first meas. 1909	115.1	SerCp	1940 1960 1980 2000 2020	16 69 99 119 137	0.2 0.2 0.3 0.3 0.3
9836	I 977	15 55.7	−26 45	7.8 comb. 7.3 F8	8.3	124.0 1984.0 W. D. Heintz, 1965 grade 4	0.476 0.42	58.0 105.5 first meas. 1914	111.7 2000	Sco	1970 1980 1990 2000 2010	122 175 275 302 323	0.4 0.2 0.3 0.4 0.4
9842	Σ 1985	15 55.9	−2 10	7.0 comb. 6.7 G0	8.1	2331 1533.80 J. Hopmann, 1973 grade 5	6.95 0.38	31.7 132.6 first meas. 1831	107.8 1900	SerCp	1940 1960 1980 2000 2020	343 346 349 352 355	6.0 6.0 6.1 6.2 6.3
	See 258 ι¹ Nor	16 03.5	−57 47	5.3 comb. 4.6 A5V par. 0.015	5.5	26.93 1937.24 W. H. van den Bos, 1961 grade 1	0.366 0.52	161.5 311.8 first meas. 1897	33.5 1950	Nor	1985 1987 1989 1991 1993	218 199 164 85 7	0.4 0.3 0.2 0.2 0.2
9909	Σ 1998 AB ξ Sco	16 04.4	−11 22	4.8 comb. 4.16 F5IV par. 0.036	5.1	45.69 1951.14 P. Baize, 1942 grade 1	0.72 0.74	36.9 348.2 first meas. 1825	201.7 1884	Sco	1985 1987 1989 1991 1993	29 34 39 47 61	1.0 0.9 0.8 0.7 0.5
	See 264 AB	16 05.7	−32 52	8.4 comb. 7.9 K0V par. 0.042	9.0	134.0 1900.10 W. D. Heintz, 1971 grade 3	0.82 0.15	143.5 129.0 first meas. 1897	109.1 2000	Lup	1970 1980 1990 2000 2010	148 130 112 92 68	0.9 0.9 0.8 0.8 0.7
9932	β 949	16 08.5	−10 06	7.3 comb. 6.7 F7V	7.6	55.0 1903.0 R. H. Wilson, 1940	0.510 0.88	84.9 299.6 first meas. 1880	22.8 1910	Sco	1985 1990 1995 2000 2005	193 195 196 197 198	0.5 0.5 0.6 0.5 0.5
9969	Σ 2021 AB	16 13.3	+13 32	7.39 comb. 6.68 K0 par. 0.050	7.48	5229 1850.0 J. Hopmann, 1970 grade 5	11.93 0.68	60.9 153.3 first meas. 1829	155.0	Her	1940 1960 1980 2000 2020	342 346 350 354 358	4.1 4.2 4.2 4.2 4.2
9979	Σ 2032 AB 17 σ CrB	16 14.7	+33 52	5.58 comb. 5.22 G0V par. 0.042	6.59 G1V	1000 1828.0 W. Rabe, 1958 grade 4	6.599 0.78	33.3 84.35 first meas. 1827	7.7 1900	CrB	1940 1960 1980 2000 2020	225 229 233 236 239	5.7 6.2 6.7 7.1 7.5
	I 987	16 17.0	−53 42	6.8 comb. 6.5 G5V	8.0	86 1953.9 M. Miranian, 1973 grade 3	0.36 0.20	18.1 74.6 first meas. 1914	9.1	Nor	1985 1990 1995 2000 2005	227 242 257 272 287	0.4 0.4 0.4 0.4 0.4
10017	Hu 481	16 21.2	+22 59	8.0 comb. 7.8 F8	9.8	141.4 1865.94 G. M. Popovic, 1969 grade 4	0.491 0.43	136.4 161.6 first meas. 1902	175.3	Her	1970 1980 1990 2000 2010	157 142 113 54 355	0.6 0.4 0.3 0.2 0.3
	CorO 197AB	16 25.3	−49 09	8.0 comb. 7.3 G5	8.2	311.2 2073.6 R. F. Mourao, 1976 grade 4	1.66 0.0	180 0.0 first meas. 1895	0.0	Nor	1940 1960 1980 2000 2020	155 131 108 85 61	1.7 1.7 1.7 1.7 1.7

ADS	Star Name	α 2000	δ 2000	m_1	m_2	P / T	a / e	i / ω	Ω / equinox	Const	Date	PA	Sep
10075	Σ 2052	16h 28.9m	+18°25′	7.7	7.8	236.1	2.234	107.9	94.1	Her	1940	222	0.6
				comb.	7.0	1921.07	0.77	130.4	1900		1960	159	0.9
				K2V		L. Siegrist, 1952					1980	135	1.4
				par.	0.057	grade 2		first meas.	1829		2000	124	2.0
											2020	118	2.4
10074	--	16 29.4	-26 26	1.2v	5.4	878	2.90	90	93.7	Sco	1940	274	3.1
	21 α Sco			comb.	1.2v	1461	0.10	0.0			1960	273	3.0
	Antares			M1Ib	B3	P. Baize, 1978					1980	273	2.8
				par.	0.019	grade 5		first meas.	1847		2000	273	2.6
											2020	273	2.3
10087	Σ 2055	16 30.9	+1 59	4.16	5.22	129.87	0.970	26.8	52.5	Oph	1970	1	1.1
	10 λ Oph			comb.	3.81	1939.54	0.618	158.9	1950		1980	13	1.3
				A1V		C. Finzi, M. Giannuzzi, 1955					1990	22	1.5
				par.	0.001	grade 2		first meas.	1825		2000	30	1.5
											2010	37	1.5
10092	Σ 3105	16 31.8	-7 01	8.1	8.2	365	0.465	130.0	23.2	Oph	1940	358	0.2
				comb.	7.4	1959.5	0.58	109.0	2000		1960	262	0.1
				A0		W. D. Heintz, 1974					1980	201	0.3
				grade 3				first meas.	1830		2000	177	0.3
											2020	159	0.4
10140	β 953 AB	16 36.6	+69 48	8.6	9.1	202.1	0.35	118.6	142.3	Dra	1940	134	0.4
				comb.	8.00	1895.40	0.48	243.8			1960	122	0.4
				F5		P. Baize, 1953					1980	110	0.4
						grade 3		first meas.	1879		2000	94	0.3
											2020	69	0.2
10157	Σ 2084 AB	16 41.3	+31 36	2.90	5.53	34.49	1.36	132.9	229.2	Her	1985	110	1.4
	40 ζ Her			comb.	2.80	1967.80	0.46	290.9	1900		1987	99	1.5
				G0IV	G7V	P. Baize, 1976					1989	89	1.6
				par.	0.102	grade 2		first meas.	1826		1991	80	1.6
											1993	70	1.6
10229	Σ 2106	16 51.1	+9 24	7.0	8.0	1080	1.25	105.9	161.0	Oph	1940	222	0.3
				comb.	6.6	1945.0	0.35	282.5	2000		1960	193	0.4
				F6IV		W. D. Heintz, 1963					1980	180	0.5
				par.	0.004	grade 4		first meas.	1827		2000	173	0.7
											2020	168	0.8
10235	Σ 2107	16 51.8	+28 40	6.8	8.2	261.8	1.012	27.1	52.7	Her	1940	60	1.0
				comb.	6.5	1895.51	0.56	242.1	1900		1960	76	1.2
				F4V		W. Rabe, 1927					1980	88	1.4
				par.	0.021	grade 3		first meas.	1829		2000	98	1.4
											2020	108	1.5
10279	Σ 2118	16 56.4	+65 02	7.1	7.3	729.3	1.38	97.0	66.2	Dra	1940	73	0.8
	20 Dra			comb.	6.4	1853.50	0.47	220.3	1950		1960	70	1.1
				F2V		M. A. Giannuzzi, 1955					1980	69	1.3
				par.	0.025	grade 4		first meas.	1832		2000	67	1.4
											2020	66	1.6
	See 315	16 58.9	-37 37	6.7	6.9	37.9	0.181	32.5	69.0	Sco	1985	353	0.1
				comb.	6.0	1948.6	0.12	300.0	2000		1987	20	0.1
	B 885			A3V		W. D. Heintz, 1967					1989	43	0.2
						grade 2		first meas.	1897		1991	63	0.2
											1993	81	0.2
10340	A 1146	17 03.6	+69 48	8.2	9.5	155	0.403	105.7	114.8	Dra	1970	147	0.2
				comb.	7.9	1958.0	0.12	258.2	2000		1980	127	0.3
				F5		W. D. Heintz, 1975					1990	119	0.4
				par.	0.003	grade 3		first meas.	1905		2000	112	0.4
											2010	106	0.4
10345	Σ 2130 AB	17 05.3	+54 28	5.65	5.70	482.0	3.330	143.4	267.4	Dra	1940	101	2.2
	21 μ Dra			comb.	4.92	1964.0	0.37	204.3	2000		1960	74	2.1
				F7V	F6	W. D. Heintz, 1965					1980	42	1.9
				par.	0.043	grade 3		first meas.	1828		2000	8	1.9
											2020	338	2.2
10360	Hu 1176 AB	17 08.0	+35 56	6.1	6.1	8.159	0.116	126.0	114.3	Her	1985	137	0.1
	c Her			comb.	5.37	1950.69	0.38	206.4			1986	117	0.1
				A5		B. Cester, 1964					1987	103	0.2
				par.	0.016	grade 2		first meas.	1905		1988	88	0.1
											1989	64	0.1
10355	A 1145	17 08.2	-1 05	6.2	8.2	137	0.42	130.2	22.3	Oph	1970	58	0.3
				comb.	6.0	1967.0	0.27	297.5	2000		1980	26	0.3
				A1V		W. D. Heintz, 1979					1990	3	0.4
						grade 4		first meas.	1905		2000	340	0.3
											2010	315	0.3

ADS	Star Name	α 2000	δ 2000	m₁	m₂	Orbital Elements P / T	a / e	i / ω	Ω / equinox	Const	Ephemeris Date	PA	Sep
10374	β 1118 35 η Oph	17ʰ10ᵐ4	−15°43′	3.0 comb. 2.5 A2V par. 0″.047	3.5	84ʸ.3 1936.80 G. Knipe, 1959 grade 2	1″.057 0.89 first meas. 1889	97°.3 277°.1	38°.8 1900	Oph	1985 1990 1995 2000 2005	253° 247 242 237 234	0″.4 0.5 0.6 0.6 0.6
						88.0 1936.80 G. Van Biesbroeck, 1960 grade 2	0.86 0.84 first meas. 1889	100.4 279.8	41.4	Oph	1985 1990 1995 2000 2005	260 253 248 243 239	0.4 0.5 0.6 0.6 0.6
10425	OΣ 327	17 14.1	+56 08	8.5 comb. 7.9 F2	8.8	90.2 1969.2 W. D. Heintz, 1976 grade 2	0.28 0.62 first meas. 1846	59.0 8.3	148.5 2000	Dra	1985 1990 1995 2000 2005	307 315 320 323 327	0.3 0.3 0.4 0.4 0.4
10418	Σ 2140 AB 64 α Her Rasalgethi	17 14.6	+14 23	3.5? comb. 3.3? M5Ib	5.4 G5III	3600 3635 P. Baize, 1978 grade 5	4.68 0.0 first meas. 1822	155.8 180.0	119.6	Her	1940 1960 1980 2000 2020	110 108 106 104 102	4.7 4.7 4.7 4.6 4.6
10417	S,h 243 36 Oph	17 15.3	−26 36	5.05 comb. 4.31 K0V par. 0.183	5.08 K1V	548.7 1643.48 P. Brosche 2, 1960 grade 5	13.91 0.90 first meas. 1822	99.2 90.0	93.6	Oph	1940 1960 1980 2000 2020	172 163 154 146 139	4.3 4.4 4.6 4.9 5.2
10421	β 957	17 15.6	−10 18	8.0 comb. 7.3 F5V	8.1	106.0 1934.60 W. H. van den Bos, 1953 grade 3	0.351 0.58 first meas. 1880	98.7 356.6	21.5 1950	Oph	1970 1980 1990 2000 2010	205 203 201 200 197	0.5 0.5 0.6 0.5 0.4
10423	A 2592	17 15.7	−9 49	7.6 comb. 7.1 F7V	8.1	157.0 1891.0 W. D. Heintz, 1978 grade 4	0.37 0.35 first meas. 1913	140.0 105.0	177.5 2000	Oph	1970 1980 1990 2000 2010	248 233 220 207 193	0.4 0.4 0.4 0.4 0.4
	MlbO 4 AB	17 19.0	−34 59	6.1 comb. 5.9 K3V par. 0.137	7.6	42.18 1933.77 R. Wielen 4, 1962 grade 1	1.734 0.57 first meas. 1877	128.4 247.2	313.0 2000	Sco	1985 1987 1989 1991 1993	296 290 283 277 271	1.9 2.0 2.1 2.1 2.0
	BrsO 13	17 19.1	−46 38	5.53 comb. 5.47 G8V par. 0.125	8.65 M0V	693.2 1907.18 N. Wieth-Knudsen, 1957 grade 4	10.41 0.78 first meas. 1880	35.6 333.4	131.8 2000	Ara	1940 1960 1980 2000 2020	213 234 245 253 259	4.5 6.2 7.8 9.2 10.5
						2205 1907.76 R. Wielen 4, 1962 grade 4	23.90 0.90 first meas. 1880	44.9 331.8	137.0 2000	Ara	1940 1960 1980 2000 2020	212 233 244 252 257	4.4 6.2 7.9 9.6 11.2
	Rst 3972	17 24.0	−9 21	8.6 comb. 7.8 G0	8.6	31.38 1949.84 W. H. van den Bos, 1963 grade 3	0.207 0.10 first meas. 1938	46.5 127.9	102.4	Oph	1985 1987 1989 1991 1993	282 300 320 345 13	0.2 0.2 0.2 0.2 0.2
10561	A 2244	17 28.3	−20 58	8.6 comb. 7.9 F5	8.8	45.0 1925.2 P. Muller, 1955 grade 2	0.205 0.54 first meas. 1910	35.0 165.4	120.7	Oph	1985 1987 1989 1991 1993	88 94 99 104 109	0.3 0.3 0.3 0.3 0.3
10598	Σ 2173	17 30.4	−1 04	6.00 comb. 5.30 G8IV par. 0.050	6.10	46.08 1916.06 R. Duncombe, J. Ashbrook, 1950 grade 1	1.02 0.174 first meas. 1830	99.16 327.52	152.66	Oph	1985 1987 1989 1991 1993	338 336 334 332 330	1.0 1.1 1.2 1.2 1.1
	B 912	17 30.8	−37 26	7.6 comb. 7.0 A0	7.9	111.9 1928.57 M. Klerk, 1973 grade 4	0.242 0.33 first meas. 1927	17.2 142.9	72.5	Sco	1970 1980 1990 2000 2010	11 29 46 63 83	0.3 0.3 0.3 0.3 0.3

ADS	Star Name	α 2000	δ 2000	m₁	m₂	Orbital Elements P / T	a / e	i / ω	Ω / equinox	Const	Ephemeris Date	PA	Sep
10660	β 962	17^h 35.0^m	+61° 52'	5.34	7.95	76.0^y	1".52	105.7	152.0	Dra	1985	349°	1".2
	26 Dra			comb.	5.25	1950.40	0.16	322.0	1900		1990	341	1.5
				G0V		P. Baize, 1965					1995	335	1.7
				par.	0".064	grade 2 first meas. 1879					2000	330	1.7
											2005	324	1.5
10723	β 1251 AB	17 42.0	+15 57	5.6	11.1	122.7	1.00	137.4	49.4	Her	1970	170	0.8
				comb.	5.6	1948.43	0.40	136.1	1900		1980	139	0.9
				F5		J. Dommanget, 1975					1990	116	1.0
				par.	0.032	grade 3 first meas. 1891					2000	98	1.1
											2010	84	1.2
	HdO 275	17 44.3	−72 13	7.3	7.6	101.3	0.65	159.0	121.4	Aps	1970	303	0.6
				comb.	6.7	1987.0	0.25	268.0	2000		1980	257	0.5
				F8V		W. D. Heintz, 1978					1990	194	0.5
						grade 3 first meas. 1914					2000	141	0.6
											2010	104	0.7
	I 1336	17 47.1	−38 07	7.1	7.3	34.0	0.157	55.6	179.0	Sco	1985	161	0.1
				comb.	6.4	1948.72	0.62	240.6			1987	179	0.1
				B9		W. H. van den Bos, 1960					1989	189	0.2
						grade 3 first meas. 1926					1991	196	0.2
											1993	202	0.2
10828	OΣ 337	17 50.5	+7 14	8.3	8.6	500.0	0.622	119.0	158.2	Oph	1940	229	0.3
				comb.	7.7	2000.0	0.55	28.0	2000		1960	198	0.3
				F2		W. D. Heintz, 1963					1980	171	0.3
						grade 4 first meas. 1843					2000	144	0.3
											2020	85	0.2
11006	OΣ 349	17 53.0	+83 54	8.2	8.6	180.0	0.505	75.5	107.0	Dra	1970	53	0.3
				comb.	7.6	1909.0	0.80	111.0	2000		1980	63	0.3
				G5		W. D. Heintz 1, 1962					1990	69	0.3
						grade 3 first meas. 1846					2000	74	0.4
											2010	78	0.4
10871	A 235	17 53.3	+24 59	8.6	8.8	37.4	0.270	72.0	84.6	Her	1985	82	0.4
				comb.	7.9	1929.83	0.46	177.0			1987	85	0.4
				G8V		O. J. Eggen, 1965					1989	87	0.4
				par.	0.015	grade 2 first meas. 1901					1991	90	0.4
											1993	93	0.3
11010	β 1127	18 02.5	+44 14	7.4	9.3	336.6	0.981	130.0	36.7	Her	1940	104	0.9
				comb.	7.2	1802.10	0.41	119.8			1960	91	1.0
				F2		G. M. Popovic, 1970					1980	79	1.1
						grade 4 first meas. 1889					2000	69	1.1
											2020	60	1.2
11005	Σ 2262	18 03.1	−8 11	5.24	5.94	280.0	1.494	59.3	63.0	Oph	1940	267	2.1
	69 τ Oph			comb.	4.78	1829.0	0.72	49.78	1900		1960	271	2.0
				F4IV	F3	S. Wierzbinski, 1959					1980	277	1.9
				par.	0.053	grade 3 first meas. 1836					2000	283	1.7
											2020	292	1.4
11046	Σ 2272 AB	18 05.5	+2 30	4.21	6.00	88.13	4.545	121.15	301.7	Oph	1985	287	2.1
	70 Oph			comb.	4.02	1984.05	0.50	13.2	2000		1990	224	1.5
				K0V		W. D. Heintz, 1973					1995	168	2.5
				par.	0.193	grade 1 first meas. 1825					2000	148	3.8
											2005	138	4.9
11060	OΣ 341 AB	18 05.8	+21 27	7.4	8.5	20.1	0.27	76.0	98.0	Her	1985	91	0.5
				comb.	7.1	1938.2	0.97	160.0			1987	92	0.5
				G0V	G8V	G. Van Biesbroeck, 1948					1989	93	0.5
				par.	0.018	grade 1 first meas. 1849					1991	93	0.5
											1993	94	0.4
	h 5014	18 06.8	−43 25	5.7	5.7	191.2	1.062	145.2	49.2	CrA	1970	19	1.4
				comb.	4.9	1841.68	0.52	190.4	1900		1980	11	1.3
				A5V		S. Wierzbinski, 1958					1990	0	1.1
						grade 3 first meas. 1836					2000	345	0.9
											2010	323	0.7
11077	AC 15	18 07.0	+30 34	5.10	8.45	55.8	1.000	32.0	218.7	Her	1985	9	1.3
	99 Her			comb.	5.05	1941.8	0.74	300.6	2000		1990	23	1.0
				F7V		W. D. Heintz, 1972					1995	57	0.5
				par.	0.058	grade 2 first meas. 1859					2000	250	0.5
											2005	295	0.9
11080	OΣ 524	18 07.5	+19 40	7.7	8.9	271	0.302	141.8	62.0	Her	1940	355	0.2
				comb.	7.4	1958.0	0.44	133.0	2000		1960	275	0.2
				A3V		W. D. Heintz, 1974					1980	223	0.2
						grade 3 first meas. 1853					2000	188	0.2
											2020	161	0.3

ADS	Star Name	α 2000	δ 2000	m₁	m₂	Orbital Elements				Const	Ephemeris		
						P / T	a / e	i / ω	Ω / equinox		Date	PA	Sep
11111	Σ 2281 73 Oph	18ʰ09ᵐ.6	+4°00′	6.07 comb. 5.69 F2V par. 0″.022	7.03	270ʸ.0 1913.0 W. D. Heintz, 1968 grade 3	1″.16 0.57	103°.0 308°.1 first meas. 1831	70°.7 2000	Oph	1940 1960 1980 2000 2020	56° 27 327 290 276	0″.6 0.4 0.3 0.6 0.8
11123	Σ 2289	18 10.1	+16 29	6.5 comb. 6.0 B9V	7.2 F2	3040 1551.0 J. Hopmann, 1964 grade 5	1.500 0.24	142.3 141.4 first meas. 1829	71.8	Her	1940 1960 1980 2000 2020	226 223 220 217 214	1.2 1.2 1.2 1.2 1.2
11186	Σ 2294	18 14.6	+0 11	8.6 comb. 7.9 F5	8.7	278.0 1889.7 W. J. Luyten 2, 1934 grade 3	0.67 0.95	102.7 160.7 first meas. 1831	88.3	Oph	1940 1960 1980 2000 2020	96 94 94 93 92	0.8 1.0 1.1 1.2 1.2
11260	Hu 197	18 19.7	+10 17	8.6 comb. 8.2 G1V	9.4 G6V	118.4 1899.61 P. Baize, 1972 grade 3	0.45 0.06	127.1 34.0 first meas. 1900	54.9	Oph	1970 1980 1990 2000 2010	183 141 99 71 50	0.3 0.3 0.3 0.4 0.4
11311	OΣ 353 43 φ Dra	18 20.8	+71 20	4.4 comb. 4.2 A0 par. 0.008	6.1	271.7 1720.75 D. Olevic, 1975 grade 4	0.39 0.44	119.0 201.3 first meas. 1856	72.3	Dra	1940 1960 1980 2000 2020	0 306 263 221 133	0.2 0.2 0.2 0.2 0.2
11324	AC 11	18 24.9	−1 35	6.8 comb. 6.1 A9III	7.0	240.0 1870.0 W. D. Heintz, 1958 grade 4	0.59 0.34	93.3 5.0 first meas. 1854	176.0 1900	SerCd	1940 1960 1980 2000 2020	358 357 356 355 353	0.5 0.7 0.8 0.8 0.7
11334	Σ 2315	18 25.0	+27 23	6.5 comb. 6.1 A2V	7.5	775.0 1865.0 W. D. Heintz, 1960 grade 4	0.90 0.30	115.9 237.7 first meas. 1830	111.9 2000	Her	1940 1960 1980 2000 2020	150 137 129 123 119	0.4 0.6 0.7 0.8 0.9
11468	A 1377 AB	18 33.9	+52 21	6.1 comb. 5.35 K0III par. 0.003	6.1	169.5 1914.62 P. Baize, 1976 grade 3	0.243 0.33	48.6 251.2 first meas. 1906	61.0 1950	Dra	1970 1980 1990 2000 2010	86 98 110 124 140	0.3 0.2 0.2 0.2 0.2
11479	OΣ 359	18 35.5	+23 36	6.3 comb. 5.6 G9III	6.5	210.9 1929.50 L. Symms, 1964 grade 3	0.416 0.82	118.7 192.9 first meas. 1849	6.9	Her	1940 1960 1980 2000 2020	34 15 9 5 2	0.2 0.5 0.6 0.7 0.7
11483	OΣ 358	18 35.9	+16 59	6.8 comb. 6.1 G2V par. 0.014	7.0 G2V	292.0 1784.0 W. D. Heintz, 1954 grade 4	1.358 0.48	134.2 19.43 first meas. 1845	16.6 1900	Her	1940 1960 1980 2000 2020	179 171 161 147 127	1.9 1.8 1.6 1.3 1.0
11484	OΣ 357	18 36.0	+11 44	8.1 comb. 7.4 A2	8.2	256.0 1731.0 A. Florsch, 1954 grade 4	0.43 0.21	123.7 0.0 first meas. 1845	87.4	Oph	1940 1960 1980 2000 2020	177 127 96 70 34	0.2 0.3 0.3 0.3 0.2
11584	OΣ 363	18 37.4	+77 41	7.6 comb. 6.9 F0	7.8	251.5 1750.65 D. J. Zulevic 1, 1975 grade 4	0.38 0.48	65.5 216.8 first meas. 1842	4.9 1900	Dra	1940 1960 1980 2000 2020	68 116 161 194 308	0.2 0.2 0.2 0.2 0.1
						434.6 1846.36 D. J. Zulevic 2, 1975 grade 4	0.54 0.0	73.4 0.0 first meas. 1842	14.9 1900	Dra	1940 1960 1980 2000 2020	66 118 156 172 180	0.2 0.2 0.2 0.3 0.4
11524	Hu 198 X Oph	18 38.3	+8 50	5.9v comb. 5.9v K1III	8.6	485.3 1873.6 P. Baize, 1980 grade 4	0.34 0.72	150.6 139.9 first meas. 1900	72.9	Oph	1940 1960 1980 2000 2020	156 146 138 132 127	0.3 0.4 0.4 0.4 0.5

ADS	Star Name	α 2000	δ 2000	m₁	m₂	P / T	a / e	i / ω	Ω / equinox	Const	Date	PA	Sep
11520	A 88	18ʰ38.4ᵐ	−3°12′	7.2	7.3	12ʸ.18	0″.196	117°.45	174°.2	SerCd	1985	1°	0″.2
				comb. 6.5		1922.17	0.26	78°.8			1986	347	0.2
				F9V	F9V	W. H. van den Bos, 1953					1987	334	0.2
				par.	0″.024	grade 1	first meas. 1900				1988	318	0.2
											1989	292	0.1
11568	Σ 2384	18 38.4	+67 08	8.3	8.6	137	0.587	85.5	128.9	Dra	1970	311	0.8
				comb. 7.7		2005.5	0.70	352.0	2000		1980	312	0.6
				G5		W. D. Heintz, 1975					1990	314	0.4
				par.	0.008	grade 3	first meas. 1832				2000	32	0.0
											2010	138	0.1
11530	Ho 87 AB	18 38.6	+16 32	8.4	8.4	112.5	0.28	23.0	41.4	Her	1970	256	0.1
				comb. 7.6		1972.0	0.65	254.0	2000		1980	35	0.2
				G5		W. D. Heintz, 1978					1990	68	0.3
						grade 3	first meas. 1883				2000	84	0.4
											2010	96	0.4
11579	Σ 2367 AB	18 41.3	+30 18	7.5	8.0	90.0	0.25	116.5	243.5	Lyr	1985	90	0.1
				comb. 6.94		1889.0	0.90	345.0	1900		1990	83	0.2
				G5III	K0III	P. Baize, 1950					1995	79	0.3
						grade 2	first meas. 1833				2000	76	0.3
											2005	75	0.4
	B 398 AB	18 43.4	−55 46	8.2	8.6	47.7	0.27	106.0	160.4	Tel	1985	212	0.1
				comb. 7.6		1975.2	0.18	187.8	2000		1987	189	0.1
				F5		W. D. Heintz, 1978					1989	178	0.2
						grade 2	first meas. 1926				1991	172	0.2
											1993	168	0.3
11635	Σ 2382	18 44.3	+39 40	5.00	6.10	1165	2.78	138.0	29.0	Lyr	1940	4	2.9
	4 ε¹ Lyr AB			comb. 4.66		1152.4	0.19	165.7	1950		1960	0	2.8
				A4V	F1V	U. Guntzel-Lingner, 1956					1980	355	2.7
				par.	0.015	grade 5	first meas. 1831				2000	350	2.6
											2020	344	2.5
11635	Σ 2383	18 44.4	+39 37	5.23	5.47	585.0	2.95	120.5	17.4	Lyr	1940	110	2.2
	5 ε² Lyr CD			comb. 4.59		1644.5	0.49	88.0	1950		1960	100	2.2
				A8Vn	F0Vn	U. Guntzel-Lingner, 1956					1980	91	2.3
				par.	0.015	grade 5	first meas. 1831				2000	82	2.3
											2020	73	2.4
11842	A 2192	18 55.8	+3 27	7.8	7.9	225.0	0.350	138.0	37.5	SerCd	1940	189	0.3
				comb. 7.1		1933.0	0.20	199.0	2000		1960	140	0.2
				B9V	F0III	W. D. Heintz 2, 1963					1980	93	0.3
						grade 4	first meas. 1910				2000	64	0.4
											2020	44	0.4
11871	β 648	18 57.0	+32 54	5.36	7.46	61.20	1.24	115.2	48.1	Lyr	1985	36	1.2
				comb. 5.21		1910.67	0.25	279.3	1900		1990	20	1.0
				G0V		G. von Schrutka, 1939					1995	356	0.8
				par.	0.052	grade 2	first meas. 1878				2000	317	0.7
											2005	279	0.8
11897	Σ 2438	18 57.5	+58 14	7.1	7.4	259.1	0.601	127.2	148.0	Dra	1940	11	0.7
				comb. 6.5		1874.38	0.89	315.0	1900		1960	6	0.8
				A2IV		T. Jastrzebski, 1959					1980	2	0.9
						grade 3	first meas. 1832				2000	359	0.9
											2020	355	1.0
11950	HdO 150	19 02.6	−29 53	3.2	3.4	21.138	0.532	110.6	74.5	Sgr	1985	72	0.4
	38 ζ Sgr			comb. 2.5		1921.39	0.205	1.4	1950		1987	46	0.3
				A2III	A2V	W. H. van den Bos, 1960					1989	327	0.2
				par.	0.019	grade 1	first meas. 1867				1991	279	0.4
											1993	264	0.6
	ø 357	19 03.5	−68 45	6.6	6.6	13.58	0.131	158.8	78.1	Pav	1985	52	0.2
				comb. 5.8		1962.1	0.38	172.0			1986	35	0.1
				F8V		W. S. Finsen, 1969					1987	11	0.1
						grade 3	first meas. 1959				1988	335	0.1
											1989	281	0.1
11989	H N 126	19 04.3	−21 32	7.7	7.9	665.0	2.256	118.1	70.4	Sgr	1940	236	1.0
				comb. 7.0		1912.7	0.62	142.5			1960	215	1.0
				G0		K. Gottlieb, 1946					1980	193	1.0
						grade 4	first meas. 1879				2000	171	1.0
											2020	152	1.1
12040	Σ 2454 AB	19 06.2	+30 26	8.5	9.7	687	1.25	55.8	132.5	Lyr	1940	264	1.0
				comb. 8.2		1638.2	0.25	310.5			1960	272	1.1
				K0		P. Baize, 1976					1980	278	1.2
						grade 4	first meas. 1831				2000	284	1.3
											2020	289	1.3

ADS	Star Name	α 2000	δ 2000	m₁	m₂	Orbital Elements P / T	a / e	i / ω	Ω / equinox	Const	Ephemeris Date	PA	Sep
	h 5084 γ CrA	19ʰ06.ᵐ4	−37°04′	4.84 comb. 4.20 F8V par.	5.08 4.20 F8V 0″.048	120ʸ.42 1998.80 W. D. Heintz, 1963 grade 1	1″.907 0.313	149°.0 350°.0	53°.0 2000 first meas. 1834	CrA	1970 1980 1990 2000 2010	188° 157 109 55 6	1″.9 1.5 1.3 1.3 1.4
12096	B 427	19 09.8	−19 48	7.0 comb. 6.3 K3III	7.2 6.3 K2III	2.68 1931.81 N. Voronov, 1934 grade 5	0.129 0.531	45.6 86.1	45.0 first meas. 1925	Sgr	1985 1986 1987 1988 1989	21 257 327 77 280	0.1 0.1 0.1 0.1 0.1
12126	A 95	19 11.0	−7 26	7.3 comb. 6.7 F9III	7.7 6.7 G1III	120.0 1944.0 W. H. van den Bos, 1961 grade 3	0.256 0.69	126.7 251.9	116.8 1950 first meas. 1900	Aql	1970 1980 1990 2000 2010	89 79 69 59 48	0.3 0.3 0.3 0.3 0.3
12201	Σ 2484	19 14.3	+19 04	7.9 comb. 7.7 F8	9.4 7.7	2316 730.5 J. Hopmann, 1973 grade 5	2.64 0.33	58.7 58.3	6.3 first meas. 1831	Sge	1940 1960 1980 2000 2020	231 233 235 237 240	2.3 2.2 2.2 2.1 2.1
12214	B 430 42 ψ Sgr	19 15.5	−25 15	5.6 comb. 4.9 F5 par.	5.7 4.9 F5 0.001	18.75 1953.77 W. S. Finsen, 1976 grade 2	0.15 0.47	83.3 167.9	111.9 first meas. 1926	Sgr	1985 1986 1987 1988 1989	113 115 116 120 130	0.2 0.2 0.1 0.1 0.0
	Gale 3	19 17.2	−66 40	6.1 comb. 5.5 A5V	6.5 5.5	157.0 1951.0 W. D. Heintz, 1973 grade 3	0.56 0.58	47.0 100.3	85.0 2000 first meas. 1901	Pav	1970 1980 1990 2000 2010	289 307 322 336 348	0.4 0.5 0.5 0.6 0.6
	I 253	19 19.0	−33 17	7.6 comb. 6.9 G0	7.7 6.9	60.0 1937.50 W. H. van den Bos, 1954 grade 3	0.510 0.78	92.1 133.1	138.2 1950 first meas. 1897	Sgr	1985 1990 1995 2000 2005	139 138 137 315 165	0.6 0.5 0.2 0.1 0.0
	φ 327 47 χ¹ Sgr	19 25.3	−24 31	5.8 comb. 5.0 Am par.	5.8 5.0 0.028	10.8 1958.3 W. S. Finsen, 1965 grade 3	0.123 0.0	90.0 0.0	85.1 first meas. 1951	Sgr	1985 1986 1987 1988 1989	265 265 265 265 85	0.1 0.1 0.1 0.0 0.1
12447	Σ 2525	19 26.6	+27 19	8.10 comb. 7.48 F9 par.	8.38 7.48 0.009	990.0 1887.9 F. Job, T. Tamburini, 1967 grade 4	1.97 0.93	133.7 8.5	103.1 1950 first meas. 1830	Vul	1940 1960 1980 2000 2020	299 296 293 291 289	1.2 1.6 1.8 2.1 2.3
12469	Schj 22	19 28.2	−12 09	8.0 comb. 7.4 dG7	8.3 7.4	172 1989.0 W. D. Heintz, 1975 grade 3	1.070 0.48	36.0 353.0	163.5 2000 first meas. 1874	Sgr	1970 1980 1990 2000 2010	61 107 163 215 256	0.7 0.6 0.6 0.6 0.7
12515	A 1653	19 29.6	+12 24	8.0 comb. 7.6 A5I	9.0 7.6	240.0 1868.47 W. D. Heintz, 1963 grade 4	0.233 0.0	180.0 0.0	0.0 2000 first meas. 1907	Aql	1940 1960 1980 2000 2020	253 223 193 163 133	0.2 0.2 0.2 0.2 0.2
12552	A 712	19 30.2	+56 39	7.4 comb. 6.8 A0	7.7 6.8	125.0 1965.0 W. D. Heintz, 1965 grade 3	0.172 0.25	110.0 212.5	105.3 2000 first meas. 1904	Cyg	1970 1980 1990 2000 2010	258 168 125 113 105	0.1 0.1 0.1 0.2 0.2
12557	Σ 2536	19 31.7	+17 48	8.4 comb. 8.3 G2V	11.4 8.3	730 2000 W. D. Heintz, 1960 grade 4	1.88 0.0	0.0 0.0	120.0 2000 first meas. 1831	Sge	1940 1960 1980 2000 2020	91 100 110 120 130	1.9 1.9 1.9 1.9 1.9
12656	Hu 679	19 35.0	+50 37	8.6 comb. 8.2 F8	9.4 8.2	97.4 1977.6 W. D. Heintz, 1978 grade 3	0.28 0.55	0.0 0.0	114.6 2000 first meas. 1904	Cyg	1985 1990 1995 2000 2005	199 226 243 255 264	0.2 0.2 0.3 0.3 0.4

ADS	Star Name	α 2000	δ 2000	m₁	m₂	Orbital Elements P / T	a / e	i / ω	Ω / equinox	Const	Ephemeris Date	PA	Sep
12752	Σ 2556	19ʰ 39ᵐ.4	+22° 15′	7.8 comb. 7.3 F6IV	8.3	342ʸ.5 2003.06 P. Baize, 1972 grade 4	0″.54 0.29	128°.1 21°.5	21°.4 1900 first meas. 1829	Vul	1940 1960 1980 2000 2020	93° 61 35 11 340	0″.3 0.4 0.4 0.4 0.3
12880	Σ 2579 18 δ Cyg	19 45.0	+45 08	2.91 comb. 2.86 B9III	6.33	827.6 1885.8 P. Baize, 1973 grade 4	3.20 0.49	147.0 134.0	98.7 1900 first meas. 1830	Cyg	1940 1960 1980 2000 2020	260 245 232 221 211	2.0 2.2 2.3 2.5 2.7
12889	Σ 2576	19 45.6	+33 36	8.3 comb. 7.6 K3V par. 0.043	8.4	224.7 1945.24 W. Rabe, 1948 grade 3	2.048 0.76	152.7 132.1	94.8 1900 first meas. 1831	Cyg	1940 1960 1980 2000 2020	38 208 177 162 152	0.6 1.2 2.0 2.6 3.0
12911	A 108	19 47.1	−8 09	8.5 comb. 7.9 K0	8.9	192.0 1946.0 W. D. Heintz, 1973 grade 3	0.268 0.64	39.0 276.0	46.2 2000 first meas. 1900	Aql	1970 1980 1990 2000 2010	78 92 103 112 121	0.2 0.3 0.3 0.3 0.3
12961	A 1658	19 48.7	+15 04	8.3 comb. 7.7 F5V	8.6 F6V	89.8 1903.3 P. Couteau, 1963 grade 2	0.225 0.07	172.9 182.3	2.9 first meas. 1907	Aql	1985 1990 1995 2000 2005	217 195 172 149 126	0.2 0.2 0.2 0.2 0.2
12972	OΣ 387	19 48.7	+35 19	7.2 comb. 6.7 F5V par. 0.019	7.7 F7V	156.5 1858.34 P. Baize, 1961 grade 3	0.624 0.08	131.4 73.3	157.2 1900 first meas. 1844	Cyg	1970 1980 1990 2000 2010	185 167 151 133 107	0.6 0.6 0.6 0.5 0.4
12973	AGC 11 8 ζ Sge	19 49.0	+19 09	5.5 comb. 5.04 A3V par. 0.008	6.2	22.8 1933.35 W. S. Finsen, 1937 grade 1	0.146 0.85	135.0 180.0	163.7 first meas. 1878	Sge	1985 1987 1989 1991 1993	174 169 166 162 159	0.2 0.3 0.3 0.3 0.3
	I 120 AB	19 49.1	−61 49	7.9 comb. 7.2 G0	8.1	61.45 1932.01 S. Wierzbinski, 1957 grade 3	0.302 0.67	154.6 131.1	110.6 1900 first meas. 1910	Pav	1985 1990 1995 2000 2005	109 73 285 220 196	0.3 0.2 0.1 0.2 0.3
13135	Hu 687	19 54.9	+50 49	8.5 comb. 7.7 A0	8.5	112.5 1971.9 P. Baize, 1959 grade 3	0.17 0.16	49.8 158.4	128.5 first meas. 1904	Cyg	1970 1980 1990 2000 2010	288 317 349 32 73	0.1 0.1 0.1 0.1 0.1
13125	Ho 581	19 55.0	+41 52	8.0 comb. 7.5 G8V par. 0.012	8.5 K0V	25.69 1937.06 G. Van Biesbroeck, 1927 grade 2	0.285 0.52	39.2 245.0	34.6 first meas. 1895	Cyg	1985 1987 1989 1991 1993	177 215 308 18 41	0.2 0.2 0.1 0.2 0.3
	HdO 294	20 01.2	−38 35	8.1 comb. 7.7 F2	8.9	474.7 1922.46 J. Dommanget 2, 1978 grade 5	1.13 0.76	36.9 114.4	136.0 1900 first meas. 1901	Sgr	1940 1960 1980 2000 2020	339 8 23 34 42	0.5 0.7 0.9 1.0 1.2
13461	OΣ 400	20 10.2	+43 57	8.0 comb. 7.44 G4V par. 0.041	8.5 G8V	86.16 1970.83 W. D. Heintz, 1963 grade 2	0.455 0.48	116.7 340.2	142.2 2000 first meas. 1845	Cyg	1985 1990 1995 2000 2005	39 10 355 346 340	0.2 0.3 0.4 0.5 0.6
13723	OΣ 406	20 19.8	+45 22	7.4 comb. 7.0 F5IV	8.3	113.5 1914.5 W. D. Heintz, 1976 grade 2	0.34 0.85	128.7 151.5	101.3 2000 first meas. 1845	Cyg	1970 1980 1990 2000 2010	121 117 113 109 103	0.6 0.6 0.6 0.5 0.4
	I 1416	20 20.2	−34 35	7.1 comb. 6.5 G0	7.5	18.60 1957.80 W. H. van den Bos, 1961 grade 2	0.152 0.71	99.8 0.0	125.5 first meas. 1925	Sgr	1985 1986 1987 1988 1989	306 305 304 303 302	0.3 0.3 0.3 0.2 0.2

ADS	Star Name	α 2000	δ 2000	m₁	m₂	Orbital Elements P / T	a / e	i / ω	Ω / equinox	Const	Ephemeris Date	PA	Sep
13728	A 1427 AB	20ʰ20.ᵐ3	+39°24′	6.33	8.87	69ʸ.86	0″.28	62°.7	3°.9	Cyg	1985	116°	0″.3
				comb.	6.23	1875.61	0.95	278°.7			1990	120	0.3
				A1V		J. Dommanget, 1978					1995	125	0.2
						grade 3	first meas. 1906				2000	130	0.2
											2005	136	0.2
13850	A 730	20 25.1	+59 36	7.1	7.4	83.72	0.172	136.5	75.0	Cyg	1985	316	0.2
				comb.	6.5	1932.09	0.77	281.3			1990	310	0.2
				A3V	A5V	P. Baize, 1955					1995	303	0.2
						grade 2	first meas. 1904				2000	295	0.2
											2005	285	0.2
	R 321	20 26.9	−37 24	6.7	8.0	232.8	1.23	133.4	150.2	Sgr	1940	5	0.5
				comb.	6.4	1949.0	0.73	222.1	1900		1960	181	0.6
				K2IV–V		J. Dommanget, 1979					1980	151	1.2
				par.	0.034	grade 3	first meas. 1880				2000	140	1.6
											2020	132	1.8
13944	A 1675	20 31.1	+15 48	7.6	7.6	48.25	0.166	114.7	147.0	Del	1985	199	0.1
				comb.	6.8	1948.95	0.47	66.3			1987	185	0.1
				A3V	A3V	P. Baize, 1955					1989	173	0.1
						grade 2	first meas. 1907				1991	163	0.1
											1993	152	0.1
13961	See	20 32.5	−16 37	7.9	7.9	42.5	0.23	115.0	87.6	Cap	1985	120	0.3
				comb.	7.1	1957.1	0.93	118.3	2000		1987	118	0.3
				F0		W. D. Heintz, 1978					1989	116	0.3
						grade 3	first meas. 1897				1991	114	0.2
											1993	111	0.2
	hz 7 ABxC	20 37.4	+75 36	7.3v	10.2	30.5	0.51	29.2	0.9	Cep	1985	269	0.7
	VW Cep			comb.	7.2v	1966.48	0.60	75.5	2000		1987	278	0.7
				K0V		J. L. Hershey, 1975					1989	289	0.6
				par.	0.041	grade 2	first meas. 1942				1991	303	0.5
											1993	321	0.4
14073	β 151	20 37.5	+14 36	4.0	4.9	26.65	0.475	63.6	178.6	Del	1985	45	0.3
	6 β Del			comb.	3.6	1962.72	0.35	346.4	1950		1987	115	0.2
				F5IV		P. Couteau, 1962					1989	167	0.3
				par.	0.026	grade 1	first meas. 1873				1991	194	0.3
											1993	236	0.2
14099	Hu 200	20 39.3	−14 57	5.8	6.3	200	0.31	75.0	96.0	Cap	1970	110	0.3
	14 τ Cap			comb.	5.3	1913.0	0.57	251.2	2000		1980	113	0.3
				B6III		W. D. Heintz, 1979					1990	118	0.3
						grade 3	first meas. 1900				2000	123	0.2
											2010	131	0.2
	Kui 99	20 39.6	+4 58	8.4	9.1	42.35	0.84	85.4	127.5	Del	1985	106	0.2
				comb.	7.9	1964.35	0.07	105.5			1987	118	0.4
				K5V		P. Baize, 1968					1989	122	0.6
				par.	0.054	grade 2	first meas. 1935				1991	124	0.7
											1993	126	0.8
14270	Σ 2725	20 46.2	+15 54	7.55	8.38	2851	7.36	66.0	20.9	Del	1940	6	5.5
				comb.	7.13	1658	0.18	276.3	1900		1960	7	5.7
				K0		J. Hopmann, 1973					1980	9	5.9
				par.	0.017	grade 5	first meas. 1829				2000	10	6.1
											2020	12	6.2
14296	OΣ 413	20 47.4	+36 29	4.85	6.07	391.3	0.777	133.8	138.6	Cyg	1940	37	0.7
	54 λ Cyg			comb.	4.54	1795.0	0.45	298.4	1900		1960	25	0.8
				B5V		W. Rabe, 1948					1980	15	0.8
				par.	0.005	grade 4	first meas. 1842				2000	6	0.9
											2020	358	0.9
14360	Σ 2729	20 51.4	−5 38	6.40	7.22	147.2	0.797	70.3	171.9	Aqr	1970	9	0.9
	4 Aqr			comb.	5.98	1894.32	0.39	39.8	1950		1980	17	0.7
				F5V	F8V	K. H. Hintze, 1959					1990	30	0.5
				par.	0.025	grade 2	first meas. 1829				2000	57	0.4
											2010	106	0.3
14412	A 751	20 53.7	+59 18	7.4	7.7	58.8	0.19	134.0	13.0	Cep	1985	163	0.2
				comb.	6.8	1918.0	0.51	109.0			1990	137	0.2
				F2V	F5V	W. D. Heintz, 1956					1995	115	0.2
				par.	0.008	grade 2	first meas. 1904				2000	96	0.2
											2005	80	0.2
14424	β 367	20 55.0	+28 05	8.4	8.8	102.0	0.407	67.2	121.1	Vul	1970	109	0.4
				comb.	7.8	1935.25	0.27	186.3	2000		1980	118	0.5
				G1V	G3V	W. D. Heintz, 1962					1990	127	0.5
						grade 2	first meas. 1876				2000	137	0.4
											2010	157	0.3

ADS	Star Name	α 2000	δ 2000	m₁	m₂	Orbital Elements P / T	a / e	i / ω	Ω / equinox	Const	Ephemeris Date	PA	Sep

ADS	Star Name	α 2000	δ 2000	m_1	m_2	P T	a e	i ω	Ω equinox	Const	Date	PA	Sep
14499	Σ 2737 AB ε Equ	20ʰ 59.1ᵐ	+4° 18′	6.0 comb. 5.42 F5IV par. 0″.016	6.3	101ʸ.4 1920.21 W. H. van den Bos, 1933 grade 1 first meas. 1835	0″.656 0.702	92°.8 339°.3	105°.2 1900	Equ	1970 1980 1990 2000 2010	286° 286 285 284 283	1″.0 1.1 1.0 0.8 0.5
	Kui 102	21 00.1	+7 31	6.3 comb. 6.0 A5V	7.6 F4V	52.5 1959.0 P. J. Morel, 1970 grade 2 first meas. 1934	0.27 0.22	127.1 128.9	3.0 2000	Equ	1985 1990 1995 2000 2005	41 22 6 347 313	0.3 0.3 0.3 0.3 0.2
14573	Σ 2744	21 03.0	+1 32	6.74 comb. 6.20 F5V par. 0.035	7.21	1532 2060.0 G. M. Popovic, 1962 grade 5 first meas. 1830	2.56 0.57	137.0 26.0	91.0	Aqr	1940 1960 1980 2000 2020	147 136 125 113 100	1.4 1.3 1.3 1.2 1.2
14636	Σ 2758 AB 61 Cyg	21 06.9	+38 45	5.23 comb. 4.81 K5V par. 0.293	6.05 K7V	653.3 1676.94 E. De Caro, G. Veca, 1948 grade 4 first meas. 1830	24.31 0.40	55.0 147.0	171.4 1880	Cyg	1940 1960 1980 2000 2020	137 142 146 150 153	25.9 27.5 29.0 30.3 31.2
14666	OΣ 527	21 08.0	+5 09	6.9 comb. 6.7 A2V	8.4	177.0 1987.0 W. D. Heintz, 1976 grade 3 first meas. 1846	0.30 0.59	126.2 35.8	119.2 2000	Equ	1970 1980 1990 2000 2010	174 134 69 339 316	0.2 0.1 0.1 0.1 0.3
14759	β 270 AB	21 13.4	+7 13	7.4 comb. 7.25 F0	9.7	113 1938.0 W. D. Heintz, 1979 grade 3 first meas. 1875	0.35 0.83	115.2 342.0	163.0 2000	Equ	1970 1980 1990 2000 2010	358 355 352 350 347	0.5 0.6 0.6 0.6 0.6
14761	Hu 767	21 13.5	+15 59	7.0 comb. 6.2 A7V	7.0 A7V	34.44 1943.62 P. Baize, 1961 grade 2 first meas. 1904	0.227 0.63	69.0 118.2	165.2	Peg	1985 1987 1989 1991 1993	47 74 96 111 122	0.1 0.1 0.1 0.2 0.2
14783	H I 48	21 13.7	+64 24	7.0 comb. 6.3 G2IV	7.2 G2IV	84.4 1922.0 P. Baize, 1950 grade 2 first meas. 1860	0.61 0.85	82.4 38.0	63.0 1900	Cep	1985 1990 1995 2000 2005	252 255 259 277 53	0.6 0.4 0.3 0.1 0.1
14773	OΣ 535 7 δ Equ	21 14.5	+10 00	5.2 comb. 4.49 F7V par. 0.053	5.3 F7V	5.70 1912.77 W. Luyten, E. Ebbighausen, 1934 grade 1 first meas. 1852	0.26 0.42	100.0 169.0	23.0	Equ	1985 1986 1987 1988 1989	19 2 201 50 29	0.3 0.1 0.2 0.1 0.3
14775	A 883 AB	21 14.7	-0 50	8.1 comb. 7.46 A1V	8.3 A2V	78.35 1950.48 P. Baize, 1955 grade 3 first meas. 1904	0.166 0.17	133.8 275.2	111.7	Aqr	1985 1990 1995 2000 2005	36 12 350 331 314	0.1 0.1 0.1 0.2 0.2
14787	AGC 13 65 τ Cyg	21 14.8	+38 03	3.82 comb. 3.73 F0IV par. 0.046	6.42	49.9 1939.6 W. D. Heintz, 1970 grade 1 first meas. 1874	0.88 0.25	134.2 119.0	159.7 2000	Cyg	1985 1987 1989 1991 1993	102 72 39 12 353	0.5 0.5 0.5 0.6 0.7
	φ 329	21 15.8	-53 16	6.6 comb. 5.8 A7V	6.6	17.0 1960.0 W. D. Heintz, 1973 grade 2 first meas. 1951	0.11 0.63	144.0 207.0	109.5 2000	Ind	1985 1986 1987 1988 1989	90 84 79 72 65	0.2 0.2 0.2 0.2 0.1
14839	β 163	21 18.6	+11 34	7.2 comb. 7.0 F7V	8.7 G6V	76.1 1985.5 W. D. Heintz, 1969 grade 2 first meas. 1876	0.520 0.88	99.8 170.2	253.2 2000	Equ	1985 1990 1995 2000 2005	102 265 260 258 257	0.0 0.2 0.5 0.6 0.8
14893	A 617	21 21.4	+10 20	7.5 comb. 6.7 F8V	7.5 F8V	12.20 1954.10 P. Baize, 1958 grade 2 first meas. 1903	0.184 0.05	113.5 56.6	99.0	Equ	1985 1986 1987 1988 1989	235 178 131 111 98	0.1 0.1 0.1 0.2 0.2

ADS	Star Name	α 2000	δ 2000	m₁	m₂	Orbital Elements P / T	a / e	i / ω	Ω / equinox	Const	Ephemeris Date	PA	Sep
14926	A 764	21ʰ22.ᵐ3	+57°34'	8.4 comb. G5 grade 4	9.6 8.14	203ʸ.3 1830.0 W. D. Heintz, 1961	0″.760 0.12	56°.9 302°.4	172°.4 2000 first meas. 1904	Cep	1940 1960 1980 2000 2020	323° 343 1 26 87	0″.7 0.8 0.8 0.5 0.4
15115	Hu 371	21 35.4	+24 27	6.7 comb. A4V grade 4	7.0 6.1	173.9 1863.9 P. Baize, 1961	0.293 0.29	60.5 287.0	134.9 first meas. 1901	Peg	1970 1980 1990 2000 2010	285 295 304 312 323	0.3 0.3 0.3 0.3 0.3
15176	β 1212 24 Aqr	21 39.5	−0 03	7.2 comb. F7V par. grade 2	7.6 6.6 F8V 0.029	48.7 1923.01 A. Danjon, 1942	0.423 0.86	55.2 295.0	139.8 first meas. 1890	Aqr	1985 1987 1989 1991 1993	249 254 258 261 264	0.4 0.4 0.5 0.5 0.5
15236	Hu 280	21 42.3	+5 54	8.4 comb. A5 grade 2	8.7 7.8	77.84 2013.61 P. J. Morel, 1970	0.260 0.73	63.2 285.7	1.5 2000 first meas. 1900	Peg	1985 1990 1995 2000 2005	140 147 154 161 170	0.3 0.3 0.3 0.3 0.2
	Kui 108 77 Cyg	21 42.4	+41 05	6.4 comb. B9V grade 2	6.6 5.7 A0V	27.2 1975.9 P. J. Morel, 1970	0.16 0.33	155.0 206.6	32.6 2000 first meas. 1937	Cyg	1985 1987 1989 1991 1993	39 25 12 358 343	0.2 0.2 0.2 0.2 0.2
15267	Ho 166	21 43.9	+27 51	8.1 comb. F6V grade 2	8.2 7.4 F6V	80.5 1936.14 P. Couteau, 1958	0.303 0.29	144.8 179.3	103.2 first meas. 1886	Peg	1985 1990 1995 2000 2005	85 72 56 35 6	0.4 0.3 0.3 0.3 0.2
15270	Σ 2822 78 μ Cyg	21 44.1	+28 45	4.78 comb. F6V par. grade 4	6.09 4.50 0.045	507.5 1962.5 W. D. Heintz, 1966	4.278 0.58	76.5 160.0	109.6 2000 first meas. 1780	Cyg	1940 1960 1980 2000 2020	228 283 298 320 18	0.6 1.6 1.8 1.2 0.8
15281	β 989 10 κ Peg	21 44.6	+25 39	4.7 comb. F5IV grade 1	5.0 4.1	11.558 1979.13 P. J. Morel, P. Couteau, 1972	0.255 0.288	108.4 126.9	111.1 2000 first meas. 1880	Peg	1985 1986 1987 1988 1989	133 123 115 108 95	0.3 0.3 0.3 0.3 0.2
	Cou 14 13 Peg	21 50.1	+17 17	5.5 comb. F2III par. grade 3	7.5 5.3 0.012	31.0 1964.60 W. D. Heintz, 1972	0.400 0.29	69.5 90.0	51.3 2000 first meas. 1959	Peg	1985 1987 1989 1991 1993	24 35 44 53 66	0.3 0.4 0.4 0.4 0.3
	HdO 296	21 55.2	−61 53	6.6 comb. F0IV grade 2	6.7 5.9	27.5 1965.9 W. S. Finsen, 1969	0.240 0.59	81.4 208.1	107.1 first meas. 1901	Ind	1985 1987 1989 1991 1993	118 124 150 264 287	0.2 0.1 0.1 0.1 0.1
15447	β 75	21 55.5	+10 53	8.5 comb. G5 grade 3	8.9 7.9	163.6 1968.28 P. Baize, 1974	0.77 0.62	60.6 192.7	29.0 2000 first meas. 1875	Peg	1970 1980 1990 2000 2010	227 330 3 14 20	0.3 0.3 0.6 0.8 1.0
	φ 307 δ Ind	21 57.9	−55 00	5.3 comb. F0IV par. grade 3	5.3 4.5 0.015	12.0 1954.7 W. S. Finsen, 1956	0.16 0.12	65.4 307.2	102.3 first meas. 1931	Ind	1985 1986 1987 1988 1989	259 273 284 297 323	0.1 0.2 0.2 0.1 0.1
15530	Hu 774	21 59.7	+49 08	8.2 comb. A0 grade 3	8.2 7.4	87 1960.5 W. D. Heintz, 1979	0.14 0.54	23.5 254.0	94.9 2000 first meas. 1904	Cyg	1985 1990 1995 2000 2005	136 145 153 161 169	0.2 0.2 0.2 0.2 0.2
15600	Σ 2863 17 ξ Cep	22 03.8	+64 38	4.4 comb. A3m par. grade 5	6.5 4.2 dF7 0.029	3800 1750.0 G. Zeller, 1965	11.5 0.24	109.0 114.0	85.0 1950 first meas. 1831	Cep	1940 1960 1980 2000 2020	278 277 276 274 273	7.4 7.7 8.0 8.2 8.5

ADS	Star Name	α 2000	δ 2000	m₁	m₂	Orbital Elements				Const	Ephemeris		
						P / T	a / e	i / ω	Ω / equinox		Date	PA	Sep
15902	β 172 51 Aqr	22ʰ24.ᵐ1	−4°50′	6.5 comb. A0V	6.5 5.78	190ʸ 1997.0 W. D. Heintz, 1975 grade 4	0″.490 0.46 first meas. 1875	137°.0 329°.5	160°.9 2000	Aqr	1970 1980 1990 2000 2010	305° 277 223 171 134	0″.4 0.3 0.2 0.3 0.3
15972	Kr 60 AB	22 28.1	+57 42	9.80 comb. dM4	11.46 9.6 dM6	44.6 1925.64 S. L. Lippinccott, 1953 grade 1	2.412 0.41 first meas. 1890	164.5 217.8	161.1 2000	Cep	1985 1987 1989 1991 1993	153 145 137 130 123	3.1 3.2 3.3 3.4 3.4
15971	Σ 2909 AB 55 ζ Aqr	22 28.8	−0 01	4.31 comb. F2IV	4.51 3.65	856.0 1957.6 R. S. Harrington, 1968 grade 3	5.055 0.49 first meas. 1781	131.2 55.12	310.2 2000	Aqr	1940 1960 1980 2000 2020	289 263 227 192 170	2.5 2.0 1.8 2.1 2.7
15988	Σ 2912 37 Peg	22 30.0	+4 26	5.77 comb. F5IV par.	7.14 5.50 0.026	140.0 1908.0 G. Knipe, 1960 grade 2	0.75 0.51 first meas. 1831	89.0 202.3	117.2 1900	Peg	1970 1980 1990 2000 2010	117 117 118 118 119	1.1 1.0 0.9 ?0.7 0.5
15992	Hu 388	22 30.2	+22 28	8.5 comb. F2V	9.0 8.0 F5V	304.1 1933.96 P. Baize, 1976 grade 2	0.36 0.74 first meas. 1901	12.3 194.8	75.4	Peg	1940 1960 1980 2000 2020	330 34 53 63 71	0.1 0.3 0.4 0.5 0.5
16046	Hu 1320	22 32.9	+49 23	8.4 comb. F4V	9.1 7.9 F5V	62.6 1956.55 P. Couteau, 1972 grade 2	0.21 0.62 first meas. 1904	28.0 132.0	5.3 1950	Lac	1985 1990 1995 2000 2005	316 324 333 342 354	0.3 0.3 0.3 0.3 0.3
16057	Σ 2924	22 33.0	+69 55	6.5 comb. A9III	7.0 6.0	225.6 1975.8 W. D. Heintz, 1956 grade 3	0.78 0.27 first meas. 1831	79.0 6.7	82.6 1900	Cep	1940 1960 1980 2000 2020	33 74 86 110 232	0.2 0.5 0.5 0.2 0.3
16111	β 1092	22 36.1	+72 53	8.3 comb. F5	8.3 7.56	52.0 1952.0 P. Baize, 1972 grade 2	0.23 0.80 first meas. 1889	66.5 190.0	40.0	Cep	1985 1990 1995 2000 2005	48 51 57 74 301	0.4 0.3 0.2 0.1 0.0
16138	Ho 295	22 38.8	+44 19	7.6 comb. F9V	7.6 6.8 F9V	30.0 1950.0 D. L. Harris, 1947 grade 3	0.353 0.30 first meas. 1887	90.0 84.3	153.3	Lac	1985 1987 1989 1991 1993	333 333 333 333 333	0.3 0.3 0.3 0.2 0.2
	Kui 114	22 40.8	−3 33	6.9 comb. F6V	7.4 6.4	28.0 1953.6 P. Baize, 1976 grade 3	0.17 0.90 first meas. 1934	80.0 180.0	128.6	Aqr	1985 1987 1989 1991 1993	124 126 127 127 128	0.2 0.2 0.3 0.3 0.3
16173	Ho 296	22 40.9	+14 33	6.26 comb. G3V par.	6.81 5.75 G8V 0.033	20.93 1942.00 P. Baize, 1957 grade 2	0.30 0.72 first meas. 1896	131.4 204.2	69.8	Peg	1985 1987 1989 1991 1993	113 81 70 63 57	0.1 0.3 0.4 0.5 0.5
16278	β 1146	22 48.5	+31 06	7.6 comb. B9	8.5 7.2	161.8 1929.9 P. J. Morel, 1970 grade 3	0.20 0.15 first meas. 1889	135.3 284.3	152.3 2000	Peg	1970 1980 1990 2000 2010	130 111 91 68 44	0.2 0.2 0.2 0.2 0.2
	HdO 301	22 50.0	−32 48	7.0 comb. F2IV	7.2 6.3	26.9 1962.36 W. S. Finsen, 1964 grade 2	0.212 0.54 first meas. 1925	148.3 88.6	31.9	PsA	1985 1987 1989 1991 1993	60 32 319 226 192	0.2 0.1 0.1 0.1 0.2
16314	Ho 482	22 51.4	+26 23	7.7 comb. A1V	7.7 6.94 F5III	243.3 1923.56 P. J. Morel, 1970 grade 3	0.422 0.45 first meas. 1893	127.1 345.1	169.3 2000	Peg	1940 1960 1980 2000 2020	135 76 39 21 9	0.2 0.2 0.3 0.5 0.5

ADS	Star Name	α 2000	δ 2000	m_1	m_2	Orbital Elements P T	a e	i ω	Ω equinox	Const	Ephemeris Date	PA	Sep
16326	A 632	22^{h}52^m.0	+57°43′	8.6	9.1	98^y.25	1″.096	103°.2	20°.0	Cep	1985	164°	0″.7
				comb. 8.1		1931.8	0.83	69°.8	2000		1990	159	0.7
				K5III		W. D. Heintz, 1962					1995	151	0.6
						grade 3	first meas. 1903				2000	141	0.5
											2005	125	0.4
16345	β 382	22 53.7	+44 45	5.8	7.8	104.5	0.647	43.0	202.3	Lac	1970	195	0.9
				comb. 5.6		1939.8	0.54	199.3			1980	205	1.0
				A3V	F6V	P. Muller, 1954					1990	215	1.0
				par. 0.016		grade 2	first meas. 1876				2000	224	0.9
											2010	236	0.8
16393	OΣ 484	22 56.2	+72 50	7.8	8.7	128.6	0.24	128.7	100.8	Cep	1970	114	0.4
				comb. 7.38		1918.0	0.64	168.7			1980	109	0.4
				A2		P. Muller, 1955					1990	104	0.4
						grade 2	first meas. 1846				2000	98	0.4
											2010	92	0.3
16417	OΣ 536	22 58.6	+9 21	7.1	7.3	26.5	0.39	90.0	165.7	Peg	1985	166	0.4
				comb. 6.4		1951.9	0.55	260.3			1987	166	0.3
				G2V	G4V	B. Cester, 1962					1989	166	0.2
				par. 0.033		grade 2	first meas. 1852				1991	166	0.1
											1993	166	0.0
	Hu 1335	22 58.6	−45 31	8.1	8.5	44.6	0.35	62.7	82.4	Gru	1985	70	0.4
				comb. 7.5		1972.7	0.39	197.5	2000		1987	75	0.4
				G5		W. D. Heintz, 1973					1989	79	0.5
						grade 3	first meas. 1929				1991	83	0.5
											1993	87	0.5
16428	OΣ 483 52 Peg	22 59.2	+11 44	6.1	7.4	286.0	0.88	44.6	6.3	Peg	1940	252	0.8
				comb. 5.8		2041.0	0.33	42.6			1960	274	0.7
				F0V		U. Guntzel-Lingner, 1956					1980	301	0.7
						grade 4	first meas. 1845				2000	330	0.7
											2020	359	0.6
16467	β 1147 2 And	23 02.6	+42 45	5.1	8.8	76.6	0.277	48.9	159.6	And	1985	340	0.4
				comb. 5.1		1875.0	0.62	8.2			1990	345	0.4
				A3Vn		P. Baize, 1974					1995	350	0.4
						grade 3	first meas. 1889				2000	355	0.4
											2005	1	0.4
16497	A 417 83 Aqr	23 05.2	−7 42	6.2	6.3	21.84	0.200	44.7	18.2	Aqr	1985	356	0.1
				comb. 5.5		1917.68	0.356	269.8			1987	28	0.2
				F2V		W. P. Hirst, 1944					1989	51	0.2
				par. 0.018		grade 1	first meas. 1902				1991	72	0.2
											1993	95	0.2
16538	OΣ 489 33 π Cep	23 07.9	+75 23	4.6	6.6	147.0	0.92	32.0	84.0	Cep	1970	322	1.0
				comb. 4.4		1935.3	0.56	100.0			1980	335	1.1
				G2III		G. Van Biesbroeck, 1951					1990	346	1.2
				par. 0.002		grade 2	first meas. 1846				2000	357	1.2
											2010	6	1.2
16539	A 1238	23 08.8	+10 57	8.3	8.5	72.0	0.260	143.9	109.1	Peg	1985	193	0.2
				comb. 7.60		1961.2	0.28	132.5			1990	174	0.3
				F6V	F7V	P. Muller, 1955					1995	157	0.3
						grade 3	first meas. 1905				2000	142	0.3
											2005	129	0.3
16576	Ho 197 AB	23 11.5	+38 13	8.3	8.6	166.75	0.35	110.1	114.5	And	1970	336	0.2
				comb. 7.7		2018.57	0.38	251.5			1980	319	0.3
				F5		J. M. Costa-Morales, 1978					1990	308	0.3
						grade 4	first meas. 1885				2000	297	0.3
											2010	284	0.2
16638	β 992	23 16.4	+64 07	8.2	8.4	480.0	0.430	128.3	26.5	Cep	1940	100	0.2
				comb. 7.5		1983.0	0.46	0.0	2000		1960	62	0.2
				F0		W. D. Heintz, 1960					1980	31	0.2
						grade 4	first meas. 1888				2000	1	0.2
											2020	325	0.2
16649	β 79	23 17.6	−1 31	8.5	9.6	388.3	1.99	108.1	8.8	Psc	1940	46	1.2
				comb. 8.2		2045.0	0.49	75.8	2000		1960	34	1.4
				dG4		W. D. Heintz, 1962					1980	24	1.5
				par. 0.010		grade 4	first meas. 1876				2000	16	1.5
											2020	6	1.2
16650	Hu 400	23 17.6	+18 18	6.9	8.5	172.0	0.38	141.0	37.0	Peg	1970	154	0.4
				comb. 6.7		1874.0	0.25	55.7	2000		1980	137	0.4
				A8		W. D. Heintz, 1962					1990	118	0.3
						grade 4	first meas. 1902				2000	99	0.3
											2010	78	0.3

ADS	Star Name	α 2000	δ 2000	m_1	m_2	Orbital Elements P / T	a / e	i / ω	Ω / equinox	Const	Ephemeris Date	PA	Sep
16666	Σ 3001	23ʰ 18.6ᵐ	+68° 07′	4.86	7.13	796ʸ.2	2″.991	58°.2	37°.9	Cep	1940	208°	2″.9
	34 o Cep			comb. 4.73		2134.38	0.17	268°.7	1900		1960	213	2.9
				K0III	F6V	S. Wierzbinski, 1956					1980	217	2.9
				par.	0″.023	grade 4	first meas. 1832				2000	223	2.8
											2020	228	2.6
16665	β 80	23 18.9	+5 24	8.5	9.1	91.80	0.768	30.7	5.9	Psc	1985	327	0.7
				comb. 8.0		1904.97	0.77	97.4	1950		1990	343	0.5
				K0V		P. Couteau, 1960					1995	33	0.3
				par.	0.018	grade 2	first meas. 1875				2000	196	0.3
											2005	228	0.6
16708	Hu 295	23 22.7	−15 02	5.6	6.4	63.16	0.420	78.0	99.0	Aqr	1985	114	0.3
	97 Aqr			comb. 5.2		1942.00	0.12	178.0			1990	143	0.1
				A2V	A7V	W. H. van den Bos, 1953					1995	246	0.1
				par.	0.016	grade 2	first meas. 1900				2000	269	0.3
											2005	278	0.4
16800	β 1266 AB	23 30.4	+30 50	8.1	8.1	49.4	0.20	148.0	83.4	Peg	1985	88	0.3
				comb. 7.38		1910.3	0.43	173.0			1987	82	0.3
				F7V	F7V	S. Arend, 1971					1989	77	0.3
						grade 2	first meas. 1891				1991	71	0.3
											1993	64	0.3
16819	Hu 298	23 32.2	+7 05	7.5	7.7	30.53	0.203	43.0	146.7	Psc	1985	140	0.2
				comb. 6.8		1926.62	0.25	42.0			1987	168	0.1
				F7V	F8V	B. Cester, 1963					1989	208	0.1
						grade 2	first meas. 1900				1991	256	0.1
											1993	289	0.2
16836	β 720	23 34.0	+31 20	5.7	5.8	241.2	0.447	35.6	123.8	Peg	1940	39	0.4
	72 Peg			comb. 5.0		1857.45	0.28	129.0			1960	61	0.5
				K3IIIb		P. Baize, 1976					1980	80	0.5
				par.	0.006	grade 3	first meas. 1878				2000	97	0.5
											2020	114	0.5
16850	See 492	23 35.7	−27 29	7.0	8.0	76.1	0.577	51.5	75.2	Scl	1985	306	0.5
				comb. 6.6		1967.8	0.53	105.7	2000		1990	323	0.5
				G0IV		W. D. Heintz, 1966					1995	339	0.5
						grade 3	first meas. 1897				2000	354	0.5
											2005	7	0.6
16904	A 643	23 39.3	+45 43	8.4	8.6	292.0	0.232	180.0	0.0	And	1940	218	0.2
				comb. 7.7		1825.17	0.0	0.0	2000		1960	194	0.2
				A2		W. D. Heintz, 1967					1980	169	0.2
						grade 4	first meas. 1903				2000	144	0.2
											2020	120	0.2
	Mlr 4	23 41.1	+46 13	7.8	8.1	20.0	0.157	55.4	106.1	And	1985	297	0.1
				comb. 7.2		1968.4	0.30	294.3			1986	312	0.1
				F7V	F8V	P. Muller, 1974					1987	340	0.1
						grade 2	first meas. 1953				1988	34	0.1
											1989	77	0.1
17020	OΣ 507 AB	23 48.6	+64 53	6.9	7.6	565.8	0.74	52.5	125.0	Cas	1940	288	0.7
				comb. 6.4		1699.83	0.0	0.0			1960	296	0.7
				A0		D. J. Zulevic, 1977					1980	304	0.7
				par.	0.009	grade 4	first meas. 1847				2000	312	0.7
											2020	320	0.7
17030	A 424	23 49.8	+27 41	7.6	7.8	113.0	0.22	70.0	62.0	Peg	1970	88	0.2
				comb. 6.9		2045.0	0.32	269.0	2000		1980	112	0.1
				F1V	F2V	P. J. Morel, 1970					1990	157	0.1
						grade 3	first meas. 1902				2000	198	0.1
											2010	219	0.2
	Slr 14	23 50.6	−51 42	8.1	8.4	135.0	0.84	164.5	10.3	Phe	1970	242	0.6
				comb. 7.5		1974.0	0.24	145.0	2000		1980	198	0.7
				G5		W. D. Heintz, 1978					1990	160	0.7
						grade 3	first meas. 1891				2000	129	0.8
											2010	104	0.9
17111	A 2100	23 56.8	+4 43	7.4	7.9	83.0	0.34	134.0	139.3	Psc	1985	169	0.1
				comb. 6.9		1988.0	0.87	84.0			1990	309	0.1
				F0		W. D. Heintz, 1975					1995	281	0.2
				par.	0.000	grade 3	first meas. 1909				2000	270	0.3
											2005	262	0.4
17149	Σ 3050	23 59.5	+33 43	6.6	6.6	355	3.125	65.0	349.8	And	1940	247	1.6
				comb. 5.8		2023.0	0.56	46.3	2000		1960	277	1.5
				G0V		W. D. Heintz, 1974					1980	309	1.5
						grade 3	first meas. 1832				2000	335	1.7
											2020	5	1.2

Spectroscopic Binary Stars

Spectroscopic Binary Stars

Column Headings

HD — The star's number in the *Henry Draper Catalogue* or *Henry Draper Extension*. Sometimes "ftr" follows the number, indicating that the spectroscopic binary is the fainter companion of the HD star. In a few cases, a *Bonner Durchmusterung* number (preceded by " + ") is given when the star has no HD number.

Immediately under this number may appear the code "eb" (eclipsing binary) or "vb" (visual binary); this signifies that the spectroscopic orbit refers to the same two components that produce eclipses, or for which a visual orbit exists.

Star Name — The Flamsteed or Bayer designation of the star, if any. Otherwise (or immediately underneath) a common name, variable star designation, or other identification may be given. The prefix "HR" denotes a star's number from the *Harvard Revised Photometry* (1908), as used in the Yale *Bright Star Catalogue* (1982). In the case of a visual double or multiple system, the same capital Roman letter that is used by double star observers is given here to identify which of the visual components is the spectroscopic binary. In the case of a spectroscopic triple, small Roman letters in parentheses distinguish the orbits referred to.

The old designations 5 Cet, 6 Per, and 10 UMa are still sometimes encountered today. They are placed in parentheses because these stars lie outside the official 1930 boundaries of the constellations for which they are named (see the last column).

α 2000 and δ 2000 — Right ascension and declination, referred to the 2000.0 equinox.

Mag — The apparent magnitude of the star. Magnitudes listed to hundredths are photoelectric **V** (visual) values in the **UBV** system, while those given only to tenths are rougher visual magnitudes determined by other methods. Exceptions to this rule are flagged by qualifying letters: "B" means the magnitude is a photoelectric **B** (blue) value, and "p" means the magnitude was measured using blue-sensitive photographic plates. If the star is variable, the magnitude at maximum brightness is given on the upper line, and that at minimum on the lower line.

Spec — The spectral types of the components. That of the primary star is given on the upper line, and that of the secondary (when known) on the lower line. These spectral types should be considered more reliable than those given in *Sky Catalogue 2000.0*, Vol. 1.

Orbital Elements — These describe the shape and orientation of the orbit of one star around the other, insofar as this information can be deduced from spectroscopic observations. Often the spectrum of only one component is visible, and the orbit describes the motion of this primary star around the center of mass of the system. (When two spectra are seen, separate orbital elements are of-

ten derived for the primary and secondary stars. Only those of the primary are given fully here.)

P. Period of revolution, usually expressed in days. Exceptions are followed by "y," which signifies that the period is in years.

T. The date and time of the orbital epoch, usually expressed as a Julian date and fraction of a day. If followed by "y," the epoch is a calendar year and fraction instead. The epoch is the instant of a periastron passage, unless followed by a qualifying letter. For near-circular orbits periastron is not well determined and it is customary to choose some other epoch. The letter "e" means the epoch is a mid-eclipse; "n" means the epoch is that of mean longitude zero (that is, when the star is passing through its ascending node and has maximum positive radial velocity); "a" denotes an arbitrary epoch.

ω. The angle in the plane of the orbit measured from the ascending node to the periastron point, in the direction of motion.

e. Eccentricity of the true orbit.

K_1 and K_2. The half-amplitude of the radial-velocity curve, in kilometers per second. That of the primary is always given on the upper line; that of the secondary, when known, is on the lower line.

V_0 — The radial velocity of the barycenter of the system, in kilometers per second, relative to the Sun. A positive value means the system is receding from our solar system. In some cases, a value has been derived separately from the spectra of the two components. That from the secondary is given on the second line.

f(m) or $m \sin^3 i$ — For a single-spectrum binary, $f(m) = m_2^3 \sin^3 i / (m_1 + m_2)^2$, the mass function, is given and the second line is left blank. For a two-spectrum system, the so-called minimum masses are given; the upper line is $m_1 \sin^3 i$ and the lower line $m_2 \sin^3 i$. In either case the units are solar masses. Sometimes an exponential notation is used, where the value in parentheses is a power of 10; thus, 8.89(− 3) is 0.00889 solar mass.

a sin *i* — The projection of the semimajor axis of the orbit on the plane of the sky, in kilometers. These values are expressed in exponential notation; for example, 6.22(+ 8) is 622,000,000 km. In the case of a two-spectrum system, the upper line gives $a_1 \sin i$, and the lower line $a_2 \sin i$. Since the inclination *i* cannot be determined from radial-velocity measurements, the semimajor axis itself, *a,* usually remains unknown too.

Author — The name of the orbit computer (first author only in the case of several), and the date of publication.

Quality — The orbits have been graded by A. H. Batten, J. M. Fletcher, and P. J. Mann. Grade A orbits are the most reliable, usually based on more than one series of observations; secular changes of the elements are either ruled out or quantitatively established. Grades B, C, D, or E are those rated good, average, poor or preliminary, and very poor, respectively. When there was not enough information to rate an orbit, it has been graded I.

Const — The constellation in which the star lies.

HD	Star Name	α 2000	δ 2000	Mag	Spec	P / T	ω / e	K₁ / K₂	V₀	f(m) or m sin³i	a sin i	Author Quality Const
224930	85 Peg	0ʰ 02.2ᵐ	+27° 05'	5.75	G2V	26.27 y 1910.11 y	274.4° 0.38	5.1	−37.9	0.105	6.22(+8)	A. B. Underhill, 1963 grade D Peg
28	33 Psc	0 05.3	−5 42	4.61	K1III	72.93 2422530.330	337.7 0.27	16.4	−6.6	0.0298	1.58(+7)	W. E. Harper, 1926 grade A Psc
352	(5 Cet)	0 08.2	−2 27	6.1	K0	96.41 2420006.84	222.1 0.12	23.8	+1.1	0.132	3.13(+7)	W. H. Christie, 1933 grade C Psc
358	21 α And Alpheratz	0 08.4	+29 05	2.17	A0p	96.6960 2442056.32	77.1 0.52	30.1	−11.6	0.171	3.42(+7)	G. C. L. Aikman, 1976 grade A And
232121 eb	SX Cas	0 10.7	+54 53	8.92 9.70	A6e gG6	36.567 2418015.33 a	10 0.50	50.0	−9.7	0.308	2.18(+7)	O. Struve, 1944 grade D Cas
1061 eb	35 Psc UU Psc	0 15.0	+8 49	6.01 7.62	F1IV-V F1IV-V	0.8417 2436094.121 n	0	89.9 96.8	+2.6	0.295 0.274	1.04(+6) 1.12(+6)	B. Cester, 1959 grade C Psc
1337 eb	AO Cas	0 17.7	+51 26	6.09 6.24	O9III O9III	3.5236 2435957.058	20.8 0.04	223.4 174.8	−31.1	10.1 12.9	1.08(+7) 8.46(+6)	K. D. Abhyankar, 1958 grade B Cas
1976	HR 91	0 24.3	+52 01	5.56	B5IV	27.8 2435018.9 n	140 0.2	30	−18	0.0733	1.12(+7)	A. Blaauw, 1963 grade E Cas
2261	α Phe	0 26.3	−42 18	2.40	K0III	3848.83 2416201.85	19.8 0.34	5.8	+75.2	0.0649	2.89(+8)	J. Lunt, 1924 grade C Phe
2421	HR 104	0 28.2	+44 24	5.2	A2Vs	3.9558 2418841.59	233.2 0.15	41.7	+2.0	0.0288	2.24(+6)	S. Udick, 1912 grade C And
3196	13 Cet A	0 35.2	−3 36	5.20	F8V	2.0819 2424548.60	280 0.01	38.4		0.0122	1.10(+6)	W. J. Luyten, 1933 grade C Cet
3369	29 π And	0 36.9	+33 43	4.35	B5V B5V	143.6065 2427898.567	349.0 0.56	47.5 117.4	+8.35	27.1 11.0	7.77(+7) 1.92(+8)	J. A. Pearce, 1936 grade A And
3627	31 δ And	0 39.3	+30 52	3.22	K3III	41 y 1901.50 y	356.1 0.34	4.0	−8.5	0.111	1.04(+9)	G. A. Bakos, 1976 grade E And
4089	ρ Tuc	0 42.5	−65 28	5.38	F5	4.8202 2419299.110	269.3 0.02	26.1	+14.1	8.89(−3)	1.73(+6)	F. J. Neubauer, 1929 grade A Tuc
4058	20 π Cas	0 43.5	+47 01	4.94	A5V	1.9642 2427535.740 n	0.0	120.5 122.1	+12.9	1.47 1.45	3.25(+6) 3.30(+6)	G. Mannino, 1955 grade A Cas
4161 eb	21 Cas YZ Cas	0 45.7	+74 59	5.6 6.0	A1V	4.4672 2423962.947	132 0.00	73.0	+8.1	0.180	4.48(+6)	C. L. Perry, 1966 grade A Cas
4502 eb	34 ζ And	0 47.3	+24 16	4.06	K1II	17.7692 2432756.059 n	0.00	26.3	−23.7	0.0336	6.43(+6)	L. Gratton, 1950 grade B And
4382	23 Cas	0 47.8	+74 51	5.4	B8III	33.75 2420577.41	269.7 0.41	16.3	−4.1	0.0115	6.90(+6)	R. K. Young, 1915 grade C Cas
4676	64 Psc	0 49.0	+16 56	5.07	F8V	13.8208 2440012.71	196 0.21	58.5 60.3	+4.9 +8.6	1.14 1.11	1.09(+7) 1.12(+7)	H. A. Abt, 1976 grade C Psc
4614	24 η Cas	0 49.1	+57 49	3.45	G0V	9.209 2439065.14	287 0.45	2.2	+9.2	7.25(−6)	2.49(+5)	H. A. Abt, 1976 grade E Cas
4727	35 ν And	0 49.8	+41 05	4.53	B5V	4.2828 2418155.661	0.00	75.6 104	−23.9	1.49 1.08	4.45(+6) 6.12(+6)	F. C. Jordan, 1910 grade C And
5015	HR 244	0 53.1	+61 07	4.79	F8IV	127.951 2414164.5	218 0.55	1.6	+22.5	3.17(−5)	2.35(+6)	H. A. Abt, 1976 grade C Cas
5516	38 η And	0 57.2	+23 25	4.4	G8III-IV G8III-IV	115.71 2430119.11	166.4 0.01	17.9 19.8	−10.3	0.338 0.306	2.85(+7) 3.15(+7)	K. C. Gordon, 1946 grade B And
5679 eb	U Cep	1 02.3	+81 53	6.95 9.11	B7V G8III-IV	2.4930 2438270.939	0.0	120 180	−2	4.19 2.80	4.11(+6) 6.17(+6)	A. H. Batten, 1974 grade E Cep
6118	69 σ Psc	1 02.8	+31 48	5.5	B9V	81.12 2431308.667	345.2 0.90	54.3 60.0	+10.4	0.547 0.495	2.64(+7) 2.92(+7)	L. P. Belserene, 1947 grade C Psc
6582	30 μ Cas	1 08.3	+54 55	5.15	G5VI	8393 2434753	178 0.30	2.8	−98.0	0.0166	3.08(+8)	T. F. Worek, 1977 grade E Cas
6882 eb	ζ Phe	1 08.4	−55 15	3.91	B8V	1.6698 2432772.217 e	0.0	130.6 201.0	+19.6 +17.1	3.83 2.49	3.00(+6) 4.62(+6)	D. M. Popper, 1970 grade B Phe
	SMC X-1	1 17.2	−72 39	13.2	B0I	3.8924 2443066.335 e	299.5 0.0	+180 19		0.779 12.3	1.60(+7) 1.02(+6)	F. Primini, 1976 grade B Tuc
8374	47 And	1 23.7	+37 43	5.5	F2m F2m	35.371 2421894.691	320.1 0.63	39.0 40.5	+13.3	0.440 0.424	1.47(+7) 1.53(+7)	J. M. Fletcher, 1967 grade A And
8556	HR 404	1 24.3	−6 55	5.91	F4V	16.14 y 1972.742 y	0.0 0.92	40.1		2.38	1.27(+9)	J. M. Fletcher, 1973 grade D Cet

HD	Star Name	α 2000	δ 2000	Mag	Spec	P / T	ω / e	K_1 / K_2	V_0	f(m) or $m \sin^3 i$	a sin i	Author / Quality Const
8634	HR 407	1^h 25.6^m	+23° 31′	6.2	F5III	5.4291 / 2433243.762	322°.5 / 0.38	14.5	−15.9	1.36(−3)	1.00(+6)	K. O. Wright, 1954 / grade C Psc
9053	γ Phe	1 28.4	−43 19	3.4	K5Ib	193.79 / 2419544.92	/ 0.0	16.0	+26.0	0.0824	4.26(+7)	W. J. Luyten, 1936 / grade B Phe
9021	38 Cas	1 31.2	+70 16	5.8	F6V	134.078 / 2429000.433	188.2 / 0.31	19.9	+1.1	0.0943	3.49(+7)	K. O. Wright, 1954 / grade C Cas
9826	50 υ And	1 36.8	+41 24	4.09	F8V	197.9 / 2439001.0	340 / 0.21	1.8	−27.7	1.12(−4)	4.79(+6)	H. A. Abt, 1976 / grade D And
10308	HR 484	1 41.3	+25 45	6.18	F2III F2III	4.4347 / 2421940.987	295.6 / 0.11	81.5 88.6	+4.6	1.16 1.07	4.94(+6) 5.37(+6)	R. F. Sanford, 1919 / grade C Psc
10516	ø Per	1 43.7	+50 41	4.06	B2Vpe	126.696 / 2424473.5?	285 / 0.15	42.5 77.5	0.0	14.2 7.78	7.32(+7) 1.33(+8)	E. M. Hendry, 1976 / grade D Per
11353	55 ζ Cet Baten Kaitos	1 51.5	−10 20	3.72	K2III	1652 / 2414377.6	85 / 0.59	3.3	+9.2	3.25(−3)	6.05(+7)	H. S. Jones, 1928 / grade E Cet
11291	2 Per	1 52.2	+50 48	5.6	B9p	5.6270 / 2440281.3	208 / 0.02	26.5	+11.4	0.0109	2.05(+6)	J. F. Heard, 1975 / grade B Per ?
11443	2 α Tri	1 53.1	+29 35	3.40	F6IV	1.7365 / 2439068.26	110 / 0.06	12.4	−20.0	3.42(−4)	2.96(+5)	H. A. Abt, 1976 / grade D Tri
11636	6 β Ari Sheratan	1 54.6	+20 48	2.60	A5V	106.9973 / 2440208.398	20.0 / 0.90	37.1	−4.0	0.0470	2.38(+7)	W. L. Gorza, 1971 / grade B Ari
11529	46 ω Cas	1 56.0	+68 41	4.99	B8III	69.92 / 2420426.02	50.0 / 0.30	29.6	−24.8	0.163	2.71(+7)	R. K. Young, 1915 / grade C Cas
11909	8 ι Ari	1 57.4	+17 49	5.10	K1p	1567.66 / 2420961.1	94.0 / 0.36	10.8	−4.9	0.167	2.17(+8)	K. C. Gordon, 1946 / grade B Ari
12211 eb	X Tri	2 00.5	+27 53	8.82 9.66	A5V G0V	0.9715 / 2422772.565 n	110 / 0.0		−5	0.134	1.47(+6)	O. Struve, 1946 / grade C Tri
12111 vb	48 Cas	2 02.0	+70 54	4.7 6.4	A4V	55.16 y / 1925.01 y	356.1 / 0.34	4.0	−8.5	0.0830	7.76(+8)	H. A. Abt, 1965 / grade E Cas
12534	57 γ² And B	2 03.9	+42 20	4.84	B9.5V B9.5V	2.67 / 2436122.18	175.2 / 0.29	141 112.5	+2.5	1.76 2.20	4.95(+6) 3.95(+6)	J. A. Maestre, 1960 / grade D And
12869	12 κ Ari	2 06.6	+22 39	5.04	Am	15.2938 / 2421844.121	358.3 / 0.61	34.5 35.4	+11.5	0.137 0.133	5.75(+6) 5.90(+6)	R. B. Jones, 1931 / grade B Ari
—	TT Ari	2 06.9	+15 17	9.4	sd	0.1376 / 2442004.710	0.0 / 0.0	65.4	+4.6	4.00(−3)	1.24(+5)	A. P. Cowley, 1975 / grade E Ari
13161	4 β Tri	2 09.5	+34 59	3.00	A5III	31.3884 / 2432004.255	318.4 / 0.53	33.3 69.2	+15.2	1.45 0.695	1.22(+7) 2.53(+7)	E. G. Ebbighausen, 195 / grade B Tri
13480	6 Tri A	2 12.4	+30 18	5.5	G5III	14.732 / 2422243.157	5.4 / 0.04	56.5 57.0	−19.1	1.12 1.11	1.14(+7) 1.15(+7)	W. E. Harper, 1921 / grade C Tri
13530	(6 Per) HR 645	2 13.6	+51 04	5.32	K0III	1650 / 2425200	270 / 0.75	13.5	+27.3	0.122	2.03(+8)	W. H. Christie, 1936 / grade E And
13974	8 δ Tri	2 17.1	+34 13	4.87	G0V	10.0201 / 2418911.3	61 / 0.01	9.4	−5.9	8.64(−4)	1.30(+6)	H. A. Abt, 1976 / grade B Tri
14214	HR 672	2 18.0	+1 45	5.56	F9V	93.50 / 2423389.995	101.0 / 0.45	19.4	+25.8	0.0505	2.23(+7)	W. E. Harper, 1930 / grade D Cet
15064	HR 706	2 24.6	−40 50	6.17	G5IV	142.33 / 2437159.06	188.2 / 0.29	17.8	+0.7	0.0731	3.33(+7)	E. S. Barker, 1967 / grade C Phe
15138	66 And	2 27.9	+50 34	6.1	F4V F4V	10.9903 / 2437005.979	271.3 / 0.18	46.5 50.0	−5.3	0.506 0.470	6.91(+6) 7.43(+6)	R. J. Northcott, 1965 / grade C And
8890	1 α UMi Polaris	2 31.8	+89 16	2.02	F8Ib A5:	30.46 y / 1928.48 y	307.2 / 0.64	4.1	−16.4	0.0361	4.82(+8)	E. Roemer, 1965 / grade B UMi
16620	83 ε Cet	2 39.6	−11 52	4.84	F5IV-V	975.9 / 2437661.8	76 n / 0.28	2.2	+16.4	9.55(−4)	2.83(+7)	H. A. Abt, 1976 / grade E Cet
16920	ζ Hor	2 40.7	−54 33	5.20	F2V F5V	12.9274 / 2433207.361	78.6 / 0.25	58.1 66.1	+5.8	1.24 1.09	1.00(+7) 1.14(+7)	J. Sahade, 1964 / grade C Hor
16739	12 Per	2 42.2	+40 12	4.92	F9V	331.0 / 2415019.0	267.7 / 0.67	21.4 24.8	−21.8	0.744 0.642	7.23(+7) 8.38(+7)	A. Colacevich, 1941 / grade C Per
16769	HR 791	2 44.8	+67 49	5.8	A5III	2.5364 / 2425319.979	/ 0.0	55.1	+4.3	0.0441	1.92(+6)	W. J. Luyten, 1936 / grade C Cas
17094	87 μ Cet	2 44.9	+10 07	4.26	F0IV	1202.2 / 2417111.5	165.2 / 0.46	14.4	+32.5	0.261	2.11(+8)	H. A. Abt, 1965 / grade E Cet

HD	Star Name	α 2000	δ 2000	Mag	Spec	P / T	ω / e	K_1 / K_2	V_0	f(m) or m sin³i	a sin i	Author / Quality Const
17206	1 τ¹ Eri	2ʰ 45.1ᵐ	−18° 34′	4.48	F6V	958	180°	3.0	+3.0	1.91(−3)	3.53(+7)	H. A. Abt, 1976
						2439391.9	n 0.45					grade E Eri
17138 eb	RZ Cas	2 48.9	+69 38	6.21 / 7.55	A2V	1.1953		68	−40	0.0390	1.12(+6)	H. G. Horak, 1952
						2416886.393	n 0.0					grade B Cas
17543	42 π Ari	2 49.3	+17 28	5.21	B6IV	3.854	78.3	24.8	+7.8	6.09(−3)	1.31(+6)	R. K. Young, 1917
						2420370.259	0.04					grade D Ari
17904	20 Per	2 53.7	+38 20	5.40	F4V	1269	306	4.8	+2.4	9.65(−3)	7.30(+7)	H. A. Abt, 1976
						2440023.3	n 0.49					grade E Per
17878/9	18 τ Per	2 54.3	+52 46	3.95	G4III A4V	1515.6	234.6	19.0	+2.2	0.345	2.71(+8)	A. Colacevich, 1941
						2415693.4	0.73					grade A Per
18256	46 ρ Ari	2 56.4	+18 01	5.58	F6V	3507	149	6.7	+12.1	0.0951	3.08(+8)	H. A. Abt, 1976
						2419459	n 0.30					grade D Ari
18925/6	23 γ Per	3 04.8	+53 30	2.94	G8III A3V	5350	344	12.7	+2.5	4.87 / 2.82	6.48(+8) / 1.12(+9)	D. B. McLaughlin, 1948
						2432263	0.72	21.9				grade I Per
+67 244 eb	RX Cas	3 07.8	+67 34	8.99 / 9.3	gG3 gA5e	32.315		36	−24	0.157	1.60(+7)	O. Struve, 1944
						2431138.0	n 0.0					grade E Cas
19356	26 β Per Algol (abc)	3 08.2	+40 57	2.12	B8V Am	1.8613y	313	12.0	+3.7	3.91 / 1.48	1.09(+8) / 2.88(+8)	G. Hill, 1971
						1952.007 y	0.23	31.6				grade C Per
19356 eb	26 β Per Algol (ab)	3 08.2	+40 57	2.12	B8V	2.8673	62	44.0	var.	0.0253	1.73(+6)	G. Hill, 1971
						2428482.739	0.02					grade B Per
18778	HR 906	3 11.7	+81 28	5.93	Am	11.665	286.0	4.5	−7.4	9.68(−5)	6.91(+5)	H. A. Abt, 1961
						2436842.346	n 0.29					grade E Cep
19820 eb	CC Cas	3 14.1	+59 34	7.11	O9IV O9IV	3.3690	300.8	141.7 / 291.8	−4.2	18.9 / 9.18	6.53(+6) / 1.35(+7)	J. A. Pearce, 1927
						2424426.705	0.10					grade C Cas
20320	13 ζ Eri	3 15.8	−8 49	4.80	Am	17.922	90.5	18.8	−4.5	0.0106	4.40(+6)	H. A. Abt, 1961
						2418893.160	0.31					grade D Eri
20210	HR 976	3 16.0	+34 41	6.24	A9m	5.5435	72	62.7	24.2	0.142	4.78(+6)	H. A. Abt, 1976
						2426482.66	n 0.03					grade B Per
21120	1 o Tau	3 24.8	+9 02	3.58	G8III	1654.9	155.6	4.4	−20.1	0.0132	9.67(+7)	E. S. Jackson, 1957
						2429974.34	0.26					grade B Tau
21754	5 Tau	3 30.9	+12 56	4.11	K0II-III	960.0	326.3	8.3	+14.2	0.0439	1.00(+8)	W. E. Harper, 1924
						2414889.565	0.40					grade C Tau
21629	GK Per	3 31.2	+43 55	12.88 / 13.44	sdBe K2IVp	0.685		140 / 34	+29	0.0732 / 0.302	1.32(+6) / 3.20(+5)	B. Paczynski, 1965
						2438051.830	n 0.0					grade D Per
21912	IW Per	3 33.6	+39 54	5.8	A5m	0.9172		95.7	+2.7	0.0835	1.21(+6)	W. E. Harper, 1935
						2400000.518	n 0.0					grade B Per
22203	19 τ⁵ Eri	3 33.8	−21 38	4.26	B8V	6.2236	313	107 / 103	+15.0	2.76 / 2.87	8.97(+6) / 8.64(+6)	O. Struve, 1925
						2424446.548	n 0.20					grade E Eri
22468	HR 1099	3 36.8	+0 35	5.88	G5V K0V	2.8378	356	50.2 / 61.6	−14.0	0.227 / 0.185	1.96(+6) / 2.40(+6)	B. W. Bopp, 1976
						2442763.909	n 0.03					grade C Tau
22805	11 Tau	3 40.8	+25 20	6.0	A2V	20.4870	323	34.2	−5.2	0.0423	7.63(+6)	J. A. Pearce, 1975
						2421876.0	0.61					grade D Tau
23817	β Ret	3 44.2	−64 48	3.84	K0IV	1911.5	13.8	5.2	+51.1	0.0261	1.34(+8)	H. S. Jones, 1928
						2420086.1	0.21					grade C Ret
23180	38 o Per	3 44.3	+32 17	3.83	B1III	4.4192	344	109.3 / 159.4	+19.8	5.26 / 3.61	6.63(+6) / 9.67(+6)	C. R. Lynds, 1960
						2427325.619	0.05					grade A Per
23302	17 Tau Electra	3 44.9	+24 07	3.69	B6III	100.46	0.0	26.0	−0.3	0.114	3.07(+7)	H. A. Abt, 1965
						2424472.86	n 0.52					grade D Tau
23466	29 Tau	3 45.7	+6 03	5.34	B3V	2.4079		23	+17	3.04(−3)	7.62(+5)	A. Blaauw, 1963
						2435000.7	n 0.0					grade D Tau
23850	27 Tau Atlas	3 49.2	+24 03	3.62	B8III	1254.68	213	14.5	−0.4	0.386	2.48(+8)	H. A. Abt, 1965
						2415870.6	n 0.14					grade E Tau
23277	HR 1138	3 49.2	+70 52	5.43	A0	15.5132	107.4	22.2 / 24.7	+16.0	0.0813 / 0.0730	4.62(+6) / 5.14(+6)	R. M. Petrie, 1940
						2429186.817	0.22					grade B Cam
23848	42 Per	3 49.5	+33 05	5.10	A2V	1.7653	260.6	34.4	−11.9	7.37(−3)	8.32(+5)	C. L. Morbey, 1974
						2437640.030	0.09					grade D Per
+16 516 eb	V471 Tau	3 50.2	+17 15	9.6 / 9.71	wd K2V	0.5212		147	+40	0.172	1.05(+6)	A. Young, 1976
						2440609.935	e 0.00					grade C Tau
24534	X Per	3 55.4	+31 03	6.07	Ope	580.7		59	−50	12.4	4.71(+8)	J. B. Hutchings, 1975
						2442084	n 0.0					grade E Per

HD	Star Name	α 2000	δ 2000	Mag	Spec	Orbital Elements P / T	ω / e	K₁ / K₂	V₀	f(m) or m sin³i	a sin i	Author Quality	Const
24546	43 Per A Per	3ʰ 56.6ᵐ	+50° 42'	5.27	F5IV	30.4380 2440873.134	27.1 0.63	51.9 54.4	+25.1	0.910 0.868	1.69(+7) 1.77(+7)	G. Wallerstein, 1973 grade B	Per
24769	33 Tau	3 57.1	+23 11	5.97	B9.5IV	1.5919 2431410.50	240 0.37	56.9	+25.0	0.0244	1.16(+6)	J. A. Pearce, 1975 grade E	Tau
25267	36 τ⁹ Eri	3 59.9	−24 01	4.64	ASi	5.9537 2417600.950	151 0.10	37.6	+24.9	0.0324	3.06(+6)	J. Sahade, 1950 grade B	Eri
25204 eb	35 λ Tau	4 00.7	+12 29	3.41	B3V A4IV	3.9540 2439137.623	142.0 0.12	55.4	+15.2	0.0683	2.99(+6)	C. Casini, 1968 grade C	Tau
25487 eb	RW Tau	4 03.9	+28 08	8.0 p 11.59	A0 G	2.7688 2426315.884	36.5 0.29	53.3	−20.2	0.0382	1.94(+6)	W. A. Hiltner, 1940 grade D	Tau
25823	41 Tau	4 06.6	+27 36	5.19	A0p	7.2274 2421944.74	121 0.18	16.6	+2.3	3.27(−3)	1.62(+6)	H. A. Abt, 1973 grade C	Tau
26591	HR 1300	4 11.6	−20 21	5.78	Am Am	3.6587 2440610.125 n	0.0	104.5 104.9	+32.8	1.75 1.74	5.26(+6) 5.28(+6)	H. A. Abt, 1977 grade C	Eri
26609 eb	YY Eri	4 12.2	−10 29	8.5 p 9.2 p	G5 G5	0.3215 2427364.360 n	0.0	130 200	−20	0.727 0.473	5.75(+5) 8.84(+5)	O. Struve, 1947 grade D	Eri
26630	51 μ Per	4 14.9	+48 25	4.12	G0Ib	283.272 2430518.418	266.5 0.06	20.7	+7.7	0.260	8.05(+7)	H. L. Johnson, 1946 grade A	Per
26673/4	52 Per	4 14.9	+40 29	4.71	G5Ib A2	1576.44 2425927.40	66.7 0.41	18.1	−4.5	0.737	3.58(+8)	K. Osawa, 1957 grade A	Per
27376	41 Eri	4 17.9	−33 48	3.55	B8.5V	5.0105 2417562.266	124.3 0.01	63.8 64.9	+17.8	0.559 0.550	4.40(+6) 4.47(+6)	G. F. Paddock, 1915 grade B	Eri
26961	b Per (abc)	4 18.2	+50 18	4.59	A2V	701.76 2440143.4	263 0.24	11.4	+21	0.0988	1.07(+8)	G. Hill, 1976 grade D	Per
26961	b Per (ab)	4 18.2	+50 18	4.59	A2V	1.5274 2440001.58	111 0.02	39.4	var.	9.70(−3)	8.27(+5)	G. Hill, 1976 grade C	Per
27176	51 Tau	4 18.4	+21 35	5.65	A8V G0	4035 2439075	311 0.34	9.1	+38.3	0.263	4.75(+8)	A. J. Deutsch, 1971 grade D	Tau
27295	53 Tau	4 19.4	+21 09	5.35	B9Vp	4.4521 2441316.857 n	81 0.06	9.6	+12.5	4.07(−4)	5.87(+5)	M. M. Dworetsky, 1972 grade B	Tau
27483	HR 1358	4 20.9	+13 52	6.15	F6V F6V	3.0591 2431706.09	193.8 0.02	71.0 73.9	+36.1	0.493 0.473	2.99(+6) 3.11(+6)	R. J. Northcott, 1952 grade B	Tau
27628	60 Tau	4 22.1	+14 05	5.72	Am	2.1433 2436851.869 n	307.0 0.04	26.6	+41.7	4.18(−3)	7.83(+5)	H. A. Abt, 1961 grade D	Tau
27697	61 δ¹ Tau	4 22.9	+17 33	3.76	K0III	529.8 2434356.5	335 0.42	3.0	+39.5	1.11(−3)	1.98(+7)	R. F. Griffin, 1977 grade D	Tau
27749	63 Tau	4 23.4	+16 47	5.63	Am	8.4178 2419818.843	189.2 0.10	37.6	+35.0	0.0458	4.33(+6)	W. E. Harper, 1935 grade B	Tau
28052	71 Tau	4 26.3	+15 37	4.49	F0V	5200 2422211	286.0 0.24	15.1	+40.8	1.70	1.05(+9)	H. A. Abt, 1965 grade E	Tau
28319	78 θ² Tau	4 28.7	+15 52	3.41	A7III	140.728 2436489.792	49.1 0.75	31.0	+39.6	0.126	3.97(+7)	E. G. Ebbighausen, 195 grade B	Tau
28204	HR 1401	4 33.5	+72 32	5.9	A5	4.195 2426034.645 n	0.0	31.3	+9.0	0.0134	1.81(+6)	W. J. Luyten, 1936 grade C	Cam
28910	86 ρ Tau	4 33.8	+14 51	4.65	F0V	488.5 2417962.4	55.0 0.09	18.5	+30.3	0.317	1.24(+8)	H. A. Abt, 1965 grade E	Tau
29140	88 d Tau	4 35.7	+10 10	4.24	Am	3.5712 2419735.889	0.0	76.3	+28.7	0.165	3.75(+6)	R. E. Wilson, 1913 grade B	Tau
29094/5	58 e Per	4 36.7	+41 16	4.22	G8III B	28.7 y 1921.21 y	197.0 0.65	12.8	+5.9	1.00	1.40(+9)	R. F. Sanford, 1953 grade B	Per
29365 eb	HU Tau	4 38.3	+20 41	5.7	B8V	2.0563 2438805.169 n	0.0	65.4	−2.3	0.0597	1.85(+6)	A. Mammano, 1967 grade C	Tau
29479	91 σ¹ Tau	4 39.2	+15 48	5.07	Am	251.205 2420304.785 n	282.0 0.33	5.0	+17.0	2.74(−3)	1.63(+7)	H. A. Abt, 1961 grade E	Tau
29317	3 Cam	4 39.9	+53 05	5.05	K0III	121 2421137.55	285 0.02	28.2	−40.5	0.282	4.69(+7)	J. B. Cannon, 1918 grade C	Cam
29763	94 Tau	4 42.2	+22 57	4.3	B3V	2.9565 2436425.247	126.7 0.05	53.6 179	+12.3	2.96 0.887	2.18(+6) 7.27(+6)	R. M. Petrie, 1961 grade C	Tau
30211	57 μ Eri	4 45.5	−3 15	4.02	B5IV	7.3589 2416392.46	150 0.26	19.4	+23.3	5.02(−3)	1.90(+6)	G. Hill, 1969 grade C	Eri

HD	Star Name	α 2000	δ 2000	Mag	Spec	Orbital Elements P / T	ω / e	K_1 / K_2	V_0	f(m) or m sin³i	a sin i	Author Quality Const
30453	HR 1528	4ʰ49ᵐ.3	+32°35′	5.9	A8m	7.0507 2426327.626	289.7 0.03	57.8	+20.3	0.141	5.60(+6)	W. E. Harper, 1953 grade B Aur
30836	3 π⁴ Ori	4 51.2	+5 36	3.68	B2III	9.5191 2418275.65 n	165 0.03	25.8	+23.3	0.0170	3.38(+6)	W. J. Luyten, 1936 grade C Ori
31109	61 ω Eri	4 52.9	−5 27	4.39	A9IV	3057 2419973	227 n 0.46	18.1	−8.3	1.32	6.76(+8)	H. A. Abt, 1965 grade D Eri
31237	8 π⁵ Ori	4 54.3	+2 26	3.69	B2III	3.7005 2417921.64 n	0.0	57.9	+24.2	0.0746	2.95(+6)	O. J. Lee, 1950 grade C Ori
31278	7 Cam	4 57.3	+53 45	4.43	A1V	3.8845 2418686.714 n		35.8	−9.5	0.0185	1.91(+6)	L. B. Lucy, 1971 grade C Cam
31964 eb	7 ε Aur	5 02.0	+43 49	2.98	F0Iap B	9890 2433464	346.4 0.20	15.0 17.0	−1.4	16.8 14.8	2.00(+9) 2.27(+9)	K. O. Wright, 1970 grade D Aur
32068/9 eb	8 ζ Aur	5 02.5	+41 05	3.75	K4Ib B6V	972.162 2434585.74	336.0 0.41	24.6 31.4	+12.9	7.54 5.91	3.00(+8) 3.83(+8)	K. O. Wright, 1970 grade C Aur
32357	12 Cam	5 06.2	+59 01	6.2	K0III	80.1745 2421174.114	72.5 0.35	22.7	+0.6	0.0801	2.34(+7)	H. A. Abt, 1969 grade C Cam
32537	9 Aur	5 06.7	+51 36	4.99	F0V	391.7 2418619 n	84 0.37	5.8	−7.4	6.36(−3)	2.90(+7)	H. A. Abt, 1965 grade D Aur
32964	66 Eri	5 06.8	−4 39	5.10	B9.5V B9.5Vp	5.5227 2441384.13 n	161 0.10	103.8 100.7	+32.6	2.38 2.45	7.84(+6) 7.61(+6)	A. Young, 1976 grade B Eri
32990	103 Tau	5 08.1	+24 16	5.50	B2V	58.31 2424221.266	273.9 0.19	36.7	+16.2	0.283	2.89(+7)	S. N. Hill, 1929 grade C Tau
33254	16 Ori	5 09.3	+9 50	5.43	Am	155.83 2439801.8	264 0.67	8.2	+42.9	3.65(−3)	1.30(+7)	P. S. Conti, 1969 grade D Ori
33856	17 ρ Ori	5 13.3	+2 52	4.46	K3III	1031.40 2426182.46	17.9 0.10	8.7	+40.5	0.0695	1.23(+8)	F. C. Bertiau, 1957 grade B Ori
—	19 β Ori B Rigel B	5 14.5	−8 12	6.7	B9	9.860 2429633.196	10 0.1	25.0 32.6	+19.1	0.109 0.0837	3.37(+6) 4.40(+6)	R. F. Sanford, 1942 grade D Ori
33959	14 Aur	5 15.4	+32 41	4.94	A9V	3.789 2421207.866 n	0.0	23.0	−9.8	4.79(−3)	1.20(+6)	L. B. Lucy, 1971 grade C Aur
34029	13 α Aur Capella	5 16.7	+46 00	0.06	G5III G0III	104.0204 2442119.352 n	292.4 0.01	26.1	+29.5	0.192	3.73(+7)	A. H. Batten, 1975 grade A Aur
34334	16 Aur	5 18.2	+33 22	4.54	K3III	434.8 2418690	40 0.1	14.8	−27.5	0.144	8.80(+7)	W. H. Christie, 1936 grade D Aur
34364	17 Aur AR Aur	5 18.3	+33 46	6.11	B9V B9V	4.1346 2428102.679 n	0.0	107.2 116.0	+25.1	2.48 2.29	6.09(+6) 6.60(+6)	A. B. Wyse, 1936 grade B Aur
34790	HR 1752	5 21.2	+29 34	5.7	A2V	2.1517 2421140.396 n	0.0	113.3 129.8	−19.7	1.71 1.50	3.35(+6) 3.84(+6)	W. E. Harper, 1926 grade C Aur
34759	20 ρ Aur	5 21.8	+41 48	5.22	B5V	35.5 2430005 n	0.0	28	+19	0.0809	1.37(+7)	A. Blaauw, 1963 grade E Aur
35411 eb	28 η Ori A (ab)	5 24.5	−2 24	3.35	B1V	7.9841 2415839.717 n	0.0	145.2	+35.9	2.54	1.59(+7)	L. B. Lucy, 1971 grade C Ori
35411 eb	28 η Ori A (abc)	5 24.5	−2 24	3.35	B1V	9.2 y 1900.0 y	270 0.1	17.5	+19.5	1.84	8.05(+8)	A. Pogo, 1928 grade D Ori
35588	HR 1803	5 25.8	+0 31	6.15	B3V	2.8884 2433202.34	198.8 0.12	71.2	+16.8	0.106	2.81(+6)	M. Duflot, 1953 grade C Ori
35715	30 ψ Ori	5 26.8	+3 06	4.59 6.7	B2IV B2IV	2.5260 2429189.107	93.0 0.07	142.8 235.1	+15.8	8.74 5.31	4.95(+6) 8.15(+6)	J. A. Pearce, 1953 grade C Ori
39780 eb	TZ Men	5 30.2	−84 47	6.19 6.87	A1III B9V	8.5690 2439193.87 n	235 0.06	59.8 119.4	−1	3.39 1.70	7.03(+6) 1.40(+7)	D. H. P. Jones, 1969 grade C Men
36486 eb	34 δ Ori Mintaka	5 32.0	−0 18	2.24	O9.5II	5.7324 2420024.212	358.7 0.10	101.0	+20.1	0.604	7.92(+6)	R. H. Curtiss, 1914 grade B Ori
36371	25 χ Aur	5 32.7	+32 12	4.75	B5Iab	655.16 2420629.78	135.5 0.17	20.5	−0.2	0.561	1.82(+8)	R. K. Young, 1916 grade D Aur
37297	28 Dor	5 33.0	−64 14	5.24	Kp	180.8757 2423108.418	330.0 0.51	22.4	+9.8	0.134	4.79(+7)	J. Lunt, 1924 grade B Dor
36695	VV Ori (abc)	5 33.5	−1 09	5.35	B1V	119.088 2419810.3 n	359.7 0.29	13.5	+3.3	0.0267	2.12(+7)	H. W. Duerbeck, 1975 grade E Ori
36695 eb	VV Ori (ab)	5 33.5	−1 09	5.38	B1V B5-B9V	1.4854 2442041.683 e	0.0	128.6 285.6	+22.1	7.56 3.40	2.63(+6) 5.83(+6)	H. W. Duerbeck, 1975 grade D Ori

HD	Star Name	α 2000	δ 2000	Mag	Spec	P / T	ω / e	K_1 / K_2	V_0	f(m) or m sin³i	a sin i	Author Quality	Const
36822	37 φ¹ Ori	5ʰ 34.8ᵐ	+9°29′	4.41	B0IV	8.4 y / 1908.3 y	105° / 0.22	13.3	+33.2	0.696	5.47(+8)	O. Struve, 1926 grade E	Ori
37021 eb	41 θ¹ Ori B BM Ori	5 35.3	−5 23	7.95 8.52	B3V A	6.4705 / 2440265.343 e	0.0	52.8 / 171	+20.7 +23.6	5.76 1.78	4.70(+6) 1.52(+7)	D. M. Popper, 1976 grade C	Ori
37041	43 θ² Ori A	5 35.4	−5 25	5.07	O9.5V	20.9672 / 2430641.416	179.8 / 0.19	90.5	+32.1	1.53	2.56(+7)	G. C. L. Aikman, 1974 grade C	Ori
37043	44 ι Ori	5 35.4	−5 55	2.76	O9III O9III	29.1351 / 2429999.610	116.9 / 0.76	115.2 / 195.8	+26.6	15.7 9.26	3.00(+7) 5.10(+7)	J. A. Pearce, 1953 grade B	Ori
37202	123 ζ Tau	5 37.6	+21 09	2.99	B2IVp	132.91 / 2415771.78	318.5 / 0.16	8.9	+21.8	9.36(−3)	1.61(+7)	A. B. Underhill, 1952 grade E	Tau
37507	49 Ori	5 38.9	−7 13	4.79	A4IV	445.74 / 2420498.19	200.7 / 0.55	28.6	−1.5	0.631	1.46(+8)	H. A. Abt, 1965 grade D	Ori
37438	125 Tau	5 39.7	+25 54	5.15	B2V	27.864 / 2420471.607	335 / 0.55	25.5	+14.8	0.0280	8.16(+6)	J. B. Cannon, 1916 grade D	Tau
37756	HR 1952	5 40.8	−1 08	4.93	B3III	27.1546 / 2429991.098	84.6 / 0.73	88.5 / 137.9	+36.2	6.36 4.08	2.26(+7) 3.52(+7)	J. A. Pearce, 1953 grade C	Ori
39357	136 Tau	5 53.3	+27 37	4.54	B9.5V	5.969 / 2420147.25 n	0.0	48.9 / 71	−17.2	0.633 0.436	4.01(+6) 5.83(+6)	W. J. Luyten, 1936 grade C	Tau
39698	57 Ori	5 54.9	+19 45	5.90	B2V	7.9969 / 2416805.89	151 / 0.01	70 / 176	+21	8.84 3.52	7.70(+6) 1.94(+7)	G. Hill, 1969 grade D	Ori
39220 eb	31 Cam TU Cam	5 55.0	+59 53	5.12 5.29	A2V	2.9333 / 2423443.050 n	0.0	78.0	−4.0	0.145	3.15(+6)	L. B. Lucy, 1971 grade C	Cam
40372	59 Ori	5 58.4	+1 50	5.88	A5	2.7405 / 2432141.143	183.0 / 0.02	55.6	+45.3	0.0489	2.09(+6)	P.-H. Nadeau, 1952 grade C	Ori
40536	2 Mon	5 59.1	−9 34	5.04	Am	9.3553 / 2419673.815	35.4 / 0.21	57.1	+22.2	0.169	7.18(+6)	C. T. Elvey, 1924 grade C	Mon
40183 eb	34 β Aur Menkalinan	5 59.5	+44 57	1.90 2.83	A2IV A2IV	3.9600 / 2431075.759 n	0.0	107.5 / 111.5	−17.1	2.20 2.12	5.85(+6) 6.07(+6)	B. Smith, 1948 grade A	Aur
40932	61 μ Ori A	6 02.4	+9 39	4.12	Am	4.4476 / 2436095.405 n	54 / 0.01	28.4	var.	0.0106	1.74(+6)	C. D. Scarfe, 1967 grade C	Ori
40932 vb	61 μ Ori AB	6 02.4	+9 39	4.12	Am	17.5 y / 1911.75 y	43 / 0.76	14.9	+43.3	0.603	8.51(+8)	P. Bourgeois, 1929 grade E	Ori
41116	1 Gem	6 04.1	+23 16	4.16	G8III–IV	9.5966 / 2440443.129 n	0.0	51.7	+31.7	0.138	6.82(+6)	R. F. Griffin, 1976 grade B	Gem
41511	17 Lep	6 05.0	−16 29	4.92	B9V M1III	260.0 / 2439135.8	185.5 / 0.13	21.0	+15.1	0.244	7.44(+7)	A. P. Cowley, 1967 grade C	Lep
41357	40 Aur	6 06.6	+38 29	5.35	A4m	28.28 / 2420468.197	178.4 / 0.56	51.4 / 62.5	+16.9	1.35 1.11	1.66(+7) 2.01(+7)	R. K. Young, 1917 grade D	Aur
41753	67 ν Ori	6 07.6	+14 46	4.40	B3V	131.211 / 2436475.852	6.6 / 0.64	33.3	+24.1	0.228	4.62(+7)	E. G. Ebbighausen, 195 grade A	Ori
42933 eb	δ Pic	6 10.3	−54 58	4.66 4.88	B0.5III B	1.6725 / 2438500.79 n	90 / 0.05	167.0	+30.6	0.806	3.84(+6)	A. D. Thackeray, 1966 grade B	Pic
42083	HR 2172	6 11.8	+52 39	6.17	A2m	106 / 2440918.8	114.8 / 0.63	40.5 / 43.1	+13.5	1.55 1.46	4.58(+7) 4.88(+7)	C. V't Veer-Menneret 1 grade D	Aur
42995	7 η Gem Propus	6 14.9	+22 30	3.28	M3III	2983 / 2426570	168 / 0.53	8.8	+17.6	0.129	3.06(+8)	D. B. McLaughlin, 1944 grade C	Gem
44402	1 ζ CMa Furud	6 20.3	−30 04	3.02	B3V	675 / 2416508.0	207 / 0.57	13.5	+32.2	0.0957	1.03(+8)	A. Colacevich, 1941 grade D	CMa
43905	45 Aur	6 21.8	+53 27	5.36	F5III	6.5013 / 2423634.166	330.6 / 0.02	31.7	−1.5	0.0215	2.83(+6)	W. E. Harper, 1925 grade C	Aur
44762	δ Col	6 22.1	−33 26	3.84	G5	868.78 / 2419915.02	117.1 / 0.70	10.6	−2.6	0.0391	9.04(+7)	H. S. Jones, 1928 grade C	Col
44691 eb	RR Lyn	6 26.4	+56 17	5.5 5.9	A3m	9.9451 / 2471458 n	0.0	65.7 / 87.2	−11.9 −14.4	2.11 1.59	8.98(+6) 1.19(+7)	D. M. Popper, 1971 grade B	Lyn
46792	HR 2410	6 31.2	−61 53	6.14	B3V	2.9723 / 2431998.871	39 / 0.10	119	+34	0.512	4.84(+6)	J. Sahade, 1950 grade D	Pic
46052 eb	WW Aur	6 32.5	+32 27	5.87 6.56	A4Vm A5Vm	2.5250 / 2432945.539 e	0.0	115.6 / 127.7	−8.7	1.98 1.79	4.01(+6) 4.43(+6)	M. Kitamura, 1976 grade A	Aur
47129	V640 Mon	6 37.4	+6 08	6.05	O8e	14.3961 / 2423039.94	22.4 / 0.01	205.2	+24.9	12.9	4.06(+7)	K. D. Abhyankar, 1959 grade C	Mon

HD	Star Name	α 2000	δ 2000	Mag	Spec	P / T	ω / e	K₁ / K₂	V₀	f(m) or m sin³i	a sin i	Author / Quality / Const
48915	9 α CMa	6 45.1	−16 43	−1.47	A1V	50.04 y	145.7	2.4	−7.6	0.0138	4.87(+8)	R. G. Aitken, 1918
vb	Sirius				wdA5	1894.133 y	0.59					grade A CMa
46588	HR 2401	6 46.2	+79 34	5.47	F8V	60.0	112	1.7	+16.2	2.14(−5)	1.25(+6)	H. A. Abt, 1976
						2439545.2 n	0.46					grade E Cam
50337	A Car	6 49.9	−53 37	4.39	G5	195.26		24.9	+25.5	0.313	6.69(+7)	R. E. Wilson, 1918
						2420415.81 n	0.0					grade C Car
50310	τ Pup	6 49.9	−50 37	2.92	K0III	1066.0	64.0	4.1	+36.4	7.54(−3)	5.99(+7)	H. S. Jones, 1928
						2420992.8	0.09					grade B Pup
55719	HR 2727	7 12.3	−40 30	5.30	Ap	46.314	209.7	48.5	−4.4	0.535	3.06(+7)	W. K. Bonsack, 1976
						2441674.94	0.13					grade C Pup
57061	30 τ CMa	7 18.7	−24 57	4.39	O9III	154.90	107.7	49.5	+43	1.69	1.01(+8)	O. Struve, 1954
						2425206.5	0.30					grade C CMa
57060	29 CMa	7 18.7	−24 34	4.95	O7f	4.3934	42.2	222.5	+13.5	4.96	1.34(+7)	O. Struve, 1958
eb	UW CMa				O7	2436189.134	0.09					grade C CMa
57167	R CMa	7 19.5	−16 24	5.74	F1V	1.1359		31.5	−42.5	3.69(−3)	4.92(+5)	O. Struve, 1950
eb				6.33		2422030.254 n	0.0					grade D CMa
56986	55 δ Gem	7 20.1	+21 59	3.52	F2IV	2238.6	214.6	27.1	+4.1	3.80	7.81(+8)	H. A. Abt, 1965
	Wasat					2415466	0.35					grade D Gem
57103	19 Lyn	7 22.9	+55 17	5.6	A0V	2.2596	126.1	106.4	+4.2	4.32	3.30(+6)	J. A. Pearce, 1932
					A0V	2419031.632	0.08	199.1		2.31	6.17(+6)	grade C Lyn
58728	63 Gem	7 27.7	+21 27	5.25	F5IV-V	1.9327	42	96.2	+22.5	1.05	2.56(+6)	H. A. Abt, 1976
						2423429.68 n	0.03	115.9	+23.6	0.867	3.08(+6)	grade B Gem
58972	4 γ CMi	7 28.2	+8 56	4.31	K3III	389.0	107.4	18.6	+46.8	0.223	9.46(+7)	W. H. Christie, 1934
						2400388.53	0.31					grade D CMi
59717	σ Pup	7 29.2	−43 18	3.24	K5III	257.8	349.3	18.6	+87.3	0.165	6.50(+7)	R. E. Wilson, 1917
					G5V	2420418.6	0.17					grade A Pup
60414/5	KQ Pup	7 33.8	−14 31	4.88	M2Iabpe	9752	202.6	17.1	+34.4	3.54	2.04(+9)	A. P. Cowley, 1965
				5.17	B	2431811	0.46					grade D Pup
60179	66 α² Gem	7 34.6	+31 53	1.9	A1V	9.2128	266.4	12.9	+5.2	1.33(−3)	1.42(+6)	J. Vint.-Hansen, 1940
	Castor A					2427543.938	0.50					grade A Gem
60178	66 α¹ Gem	7 34.6	+31 53	2.9	Am	2.9283	94.7	31.9	−1.2	9.87(−3)	1.28(+6)	J. Vint.-Hansen, 1940
	Castor B					2427501.703	0.00					grade A Gem
+32 1582	YY Gem	7 34.6	+31 53	9.21	dM1e	0.8143		121.0	+0.9	0.580	1.35(+6)	B. W. Bopp, 1974
eb	Castor C			9.6	dM1e	2424989.117 e	0.0	119.0		0.589	1.33(+6)	grade B Gem
61421	10 α CMi	7 39.3	+5 13	0.35	F5IV	40.65 y	87.7	1.7	−4.1	6.15(−3)	3.24(+8)	J. M. Fletcher, unpubl.
vb	Procyon					1927.20 y	0.36					grade B CMi
61859	HR 2962	7 42.7	+34 00	6.0	F7V	31.50	44.0	45.2	−12.1	1.53	1.91(+7)	W. E. Harper, 1926
					F7V	2423884.45	0.21	52.4		1.32	2.22(+7)	grade C Gem
62044	75 σ Gem	7 43.3	+28 53	4.25	K1III	19.605		34.2	+45.8	0.0814	9.22(+6)	W. J. Luyten, 1931
						2418962.43 n	0.0					grade C Gem
62623	3 Pup	7 43.8	−28 57	3.96	A2Ia	137.767	247.3	3.6	+25.4	5.85(−4)	6.53(+6)	H. L. Johnson, 1947
						2430278.777	0.29					grade B Pup
64096	9 Pup	7 51.8	−13 54	5.16	G1V	8467	67.7	4	−21.2	0.0213	3.37(+8)	H. A. Abt, 1976
vb						2420757	0.69					grade E Pup
64440	A Pup	7 52.2	−40 35	3.71	G5III	2660	190	11.8	+24.0	0.349	3.96(+8)	W. H. Christie, 1936
						2417165	0.4					grade E Pup
64511	U Gem	7 55.1	+22 00	9.93	sdBe	0.1769		143	+40	0.0537	3.48(+5)	J. Smak, 1976
eb				15.0		2442061.480 e	0.0					grade D Gem
65818	V Pup	7 58.2	−49 15	4.30	B1Vp	1.4545			+19		7.12(+6)	H. O. Frieboes, 1962
eb					B2	2428649.131 e		356				grade D Pup
68520	ε Vol	8 07.9	−68 37	4.34	B5V	14.1683		66.7	+9.7	0.437	1.30(+7)	R. F. Sanford, 1914
						2419457.104	0.00					grade A Vol
68273	γ² Vel	8 09.5	−47 20	1.82	WC7	78.5		43.1	−18.0	0.625	4.58(+7)	K. S. Ganesh, 1967
					O7	2439125.28 n	0.17					grade D Vel
68351	15 ψ Cnc	8 13.1	+29 39	5.63	A0p	585.4154	87	12.1	+17.8	0.0985	9.46(+7)	H. A. Abt, 1973
						2421962.65	0.24					grade C Cnc
69142	h² Pup	8 14.0	−40 21	4.43	K0	930	140	10.8	+13.5	0.0937	1.27(+8)	W. H. Christie, 1936
						2418060	0.4					grade E Pup
70958	1 Hya	8 24.6	−3 45	5.59	F1	1.5630	123.9	30.3	+71.3	4.50(−3)	6.50(+5)	R. F. Sanford, 1922
						2422650.082	0.05					grade C Hya

HD	Star Name	α 2000	δ 2000	Mag	Spec	Orbital Elements P / T	ω / e	K₁ / K₂	V₀	f(m) or m sin³i	a sin i	Author Quality	Const

HD	Star Name	α 2000	δ 2000	Mag	Spec	P	ω	K_1	V_0	f(m) or	a sin i	Author	
						T	e	K_2		$m \sin^3 i$		Quality	Const
—	Z Cam	8 25.2	+73 07	13.5 B	G	0.2898	193	−38		0.492	7.69(+5)	R. P. Kraft, 1969	
				13.9 B	sdBe	2438470.841 a	0.0	144	−37	0.660	5.74(+5)	grade E	Cam
72945	HR 3395	8 35.8	+6 37	5.99	F5	14.296	220.8	22.7	+24.2	0.0154	4.28(+6)	A. H. Joy, 1919	
						2421599.474	0.28					grade C	Cnc
71973	HR 3352	8 36.8	+74 43	6.2	A5	4.285	101.7	63.5	−7.4	0.112	3.72(+6)	W. E. Harper, 1930	
						2424971.150	0.11					grade B	Cam
+19 2068	TX Cnc	8 40.0	+19 00	10.02	F8V	0.3829		117.3	+26.6	0.712	6.18(+5)	J. A. J. Whelan, 1973	
eb				10.4		2438011.391 e	0.0	189.8		0.440	9.99(+5)	grade C	Cnc
75747	RS Cha	8 43.2	−79 04	6.04	A8IV	1.6699		136.1	+16.1	1.82	3.13(+6)	J. Andersen, 1975	
eb				6.7	A8IV	2438380.526 e	0.0	138.9	+15.6	1.78	3.19(+6)	grade C	Cha
74874	11 ε Hya AB	8 46.8	+6 25	3.38	G0III	5492	86.8	7.3	+36.4	0.110	4.37(+8)	B. Adams, 1939	
vb						2420839	0.61					grade C	Hya
75759	HR 3525	8 50.3	−42 05	5.98	B0V	33.311	85.2	121.3	+23.4	18.0	4.31(+7)	A. D. Thackeray, 1966	
						2438848.28	0.63	150.4		14.5	5.35(+7)	grade C	Vel
76805	H Vel	8 56.3	−52 43	4.68	B5V	0.9147	44.3	46.3	+22.2	9.19(−3)	5.77(+5)	F. J. Neubauer, 1930	
						2417967.119	0.13					grade C	Vel
76644	9 ι UMa A Talitha	8 59.2	+48 02	3.14	A7V	4028	33	6.0	+6.8	0.0734	3.10(+8)	H. A. Abt, 1965	
						2416342	0.36					grade E	UMa
77258	w Vel	9 00.1	−41 15	4.45	F8IV	74.1469	90.0	17.8	−7.4	0.0433	1.81(+7)	J. Lunt, 1924	
						2422728.629	0.05					grade C	Vel
76943	HR 3579 (10 UMa)	9 00.6	+41 47	3.97	F5V	7980.7	30	4.0	+27.2	0.0513	4.34(+8)	H. A. Abt, 1976	
vb						2433319	0.15					grade D	Lyn
77581	Vela X−1	9 02.1	−40 33	6.88	B0.5Ib	8.966	146	273	−7.2	18.5	3.34(+7)	S. Rappaport, 1976	
						2446412.01 n	0.13					grade B	Vel
77350	69 ν Cnc	9 02.7	+24 27	5.43	B9p	1401.4	264	7.7	−14.3	0.0546	1.39(+8)	H. A. Abt, 1973	
						2419687	0.35					grade D	Cnc
78316	76 κ Cnc	9 07.7	+10 40	5.24	B8p	6.3933	157	67.4	+24.5	0.198	5.88(+6)	G. C. L. Aikman, 1976	
						2440001.95	0.13					grade A	Cnc
78418	75 Cnc	9 08.8	+26 38	6.00	G5IV	19.4589	252.5	20.2	+12.3	0.0156	5.28(+6)	R. F. Sanford, 1922	
						2422426.634	0.21					grade C	Cnc
78515	77 ξ Cnc	9 09.4	+22 03	5.14	K0III	1700.76	301.1	4.4	−7.7	0.0150	1.03(+8)	E. S. Jackson, 1957	
						2428876.86 n	0.06					grade B	Cnc
78362/3	14 τ UMa	9 10.9	+63 31	4.66	Am	1062.4	349.4	3.9	−8.7	4.42(−3)	5.00(+7)	M. C. Bretz, 1961	
						2425721.6	0.48					grade A	UMa
79351	a Car	9 11.0	−58 58	3.43	B2IV	6.7447	113.0	21.5	+23.3	6.63(−3)	1.96(+6)	W. Buscombe, 1960	
						2416534.215	0.18					grade C	Car
79193	21 Hya	9 12.4	−7 07	6.11	Am	7.7504	233	73.2	−4	1.59	7.75(+6)	M. T. Chauville, 1975	
						2441405.09 n	0.11	82.8		1.41	8.77(+6)	grade C	Hya
79028	16 UMa	9 14.3	+61 25	5.17	dF9	16.2397	143	35.3	−14.6	0.0733	7.85(+6)	H. A. Abt, 1976	
						2423048.47 n	0.09					grade B	UMa
79910	23 Hya	9 16.7	−6 21	5.23	K2III	922	92.3	10.0	−7.7	0.0839	1.21(+8)	H. S. Jones, 1928	
						2418549.21	0.29					grade C	Hya
79763	HR 3676	9 17.5	+46 49	5.7	A1V	15.986	355.2	63.3	−13.1	1.49	1.21(+7)	W. E. Harper, 1916	
						2419408.027	0.50	73.6		1.28	1.40(+7)	grade D	UMa
81188	κ Vel	9 22.1	−55 01	2.50	B2IV	116.65	96.2	46.5	+21.9	1.15	7.32(+7)	H. D. Curtis, 1907	
						2416459.00	0.19					grade C	Vel
81809	HR 3750	9 27.8	−6 04	5.38	G2V	917.1	73	2.2	+56.3	9.48(−4)	2.71(+7)	H. A. Abt, 1976	
						2439728	0.21					grade E	Hya
81858	2 ω Leo	9 28.5	+9 03	5.40	dF8	116.85 y	124.6	2.2	−6.0	0.0268	1.07(+9)	H. A. Abt, 1976	
vb						1959.55 y	0.56					grade E	Leo
82328	25 θ UMa	9 32.9	+51 41	3.18	F6IV	371.0	270	4.3	+13.0	2.78(−3)	2.12(+7)	H. A. Abt, 1976	
						2439040	0.25					grade E	UMa
83808/9	14 o Leo	9 41.2	+9 54	3.50	A2	14.4980		54.1	+27.1	1.30	1.08(+7)	H. C. Plummer, 1908	
					F6III	2414660.301 n	0.0	63.1		1.12	1.26(+7)	grade C	Leo
83950	W UMa	9 43.8	+55 57	8.5 p	F8V	0.3336		131.3	−42.3	1.18	6.02(+5)	S. P. Worden, 1973	
eb				9.2 p	F8V	2441372.749	0.0	243.0		0.637	1.11(+6)	grade C	UMa
85040	20 Leo	9 49.8	+21 11	6.09	A8IV	4.1467		101.2	+27.2	1.80	5.77(+6)	F. C. Fekel, 1977	
						2442094.111 n	0.0	101.7		1.79	5.80(+6)	grade D	Leo
85217	4 Sex	9 50.5	+4 21	6.23	F6V	3.0546		100.3	+17.4	1.41	4.21(+6)	D. M. Popper, 1948	
						2426435.252 n	0.0	105.4		1.35	4.43(+6)	grade C	Sex

HD	Star Name	α 2000	δ 2000	Mag	Spec	P / T	ω / e	K₁ / K₂	V₀	f(m) or m sin³i	a sin i	Author Quality Const
85622	m Vel	9ʰ 51.7ᵐ	−46° 33′	4.57	G5	329.30 / 2423596.5 n	° / 0.0	14.1	+11.0	0.0959	6.38(+7)	W. J. Luyten, 1936 grade C Vel
86146	19 LMi	9 57.7	+41 03	5.13	F5V	9.2847 / 2439162.3	88 / 0.07	17.8	−8.8	5.40(−3)	2.27(+6)	H. A. Abt, 1976 grade C LMi
88215	HR 3991	10 10.1	−12 49	5.30	F5V	28.098 / 2440271.3	177 / 0.07	10.1	+7.5	2.98(−3)	3.89(+6)	H. A. Abt, 1976 grade D Hya
88284	41 λ Hya	10 10.6	−12 21	3.61	K0III	1585.8 / 2418795.1	238.9 / 0.14	3.7	+19.4	8.10(−3)	7.99(+7)	H. S. Jones, 1928 grade C Hya
89758	34 μ UMa	10 22.3	+41 30	3.04	M0III	230.089 / 2425577.03 n	236.4 / 0.06	7.4	−20.4	9.63(−3)	2.34(+7)	E. S. Jackson, 1957 grade B UMa
89822	30 UMa	10 24.1	+65 34	4.93	A0p	11.5791 / 2418468.175	171.0 / 0.26	38.9 64.8	−2.6	0.754 0.453	5.98(+6) 9.96(+6)	K. Nariai, 1976 grade C UMa
90569	45 Leo	10 27.6	+9 46	6.03	A2p	34.66 y / 1892.11 y	2 / 0.51	11.0	−7.8	1.11	1.65(+9)	H. A. Abt, 1973 grade E Leo
90537 vb	31 β LMi	10 27.9	+36 42	4.20	G8III−IV	37.9 y / 1917.0 y	5.0 / 0.66	3.2	+5.7	0.0208	4.76(+8)	A. B. Underhill, 1963 grade D LMi
91312	HR 4132	10 33.2	+40 26	4.72	A7IV	292.56 / 2419108	311 / 0.30	14.5	+9.0	0.0804	5.56(+7)	H. A. Abt, 1965 grade E UMa
91636 eb	49 Leo TX Leo	10 35.0	+8 39	5.66	A2V	2.4451 / 2427160.738	295.9 / 0.06	62.6	+16.7	0.0620	2.10(+6)	C. Chamberlin, 1957 grade B Leo
92139/40	p Vel A	10 37.3	−48 14	4.2	F3IV F0V	10.2104 / 2416461.175 n	185.0 / 0.51	42.3 53.6	+21.2	0.333 0.263	5.11(+6) 6.47(+6)	D. S. Evans, 1969 grade A Vel
92214	φ Hya	10 38.6	−16 53	4.90	K0III	1200 / 2420760	270 / 0.1	4.0	+16.2	7.86(−3)	6.57(+7)	W. H. Christie, 1936 grade D Hya
92168	38 LMi	10 39.1	+37 55	5.85	F8V	7.7991 / 2420165.164	285.6 / 0.02	24.1	+6.1	0.0113	2.58(+6)	N. Ginestet, 1974 grade C LMi
94334	45 ω UMa	10 54.0	+43 11	4.68	A1V	15.8307 / 2435185.246	27.3 / 0.31	22.2	−18.7	0.0155	4.59(+6)	E. G. Ebbighausen, 1959 grade A UMa
95689 vb	50 α UMa Dubhe	11 03.7	+61 45	1.79	K0III F	44.0 y / 1865.9 y	174 / 0.35	2.0	−8.7	0.0110	4.14(+8)	H. S. Jones, 1937 grade E UMa
96314	χ² Hya	11 06.0	−27 17	5.64	B8III−IV B8.5V	2.2677 / 2439925.545 e	123.3 / 0.0 168.9	+30.6 +28.0	3.40 2.48	3.84(+6) 5.27(+6)	J. Andersen, 1975 grade B Hya
98088	SV Crt	11 17.0	−7 08	6.14	A3Vp A8V	5.9051 / 2441847.78	316.9 / 0.17	74.7 98.5	−8.1	1.73 1.32	5.98(+6) 7.88(+6)	S. C. Wolff, 1974 grade B Crt
98230	53 ξ UMa B	11 18.2	+31 32	4.9	G0	3.9805 / 2425000 a	5.0 / 0.0	5.0	−15.9	5.17(−5)	2.74(+5)	L. Berman, 1931 grade C UMa
98231	53 ξ UMa A	11 18.2	+31 32	4.4	G0	669.18 / 2418582.0	320.0 / 0.53	8.0	−15.0	0.0217	6.24(+7)	W. H. van den Bos, 1928 grade C UMa
98353	55 UMa	11 19.1	+38 11	4.76	A2V	2.5 / 2421487.014	173.4 / 0.11	38.5 54.5	−3	0.120 0.0849	1.32(+6) 1.86(+6)	F. Henroteau, 1919 grade E UMa
98991	13 λ Crt	11 23.4	−18 47	5.08	F5IV	1940 / 2420136	90 / 0.62	6.9	+13.0	0.0320	1.44(+8)	H. A. Abt, 1976 grade D Crt
—	Cen X-3	11 23.5	−60 52	13.25 13.45	B0Ib	2.0871 / 2441132.083 e	415.1 / 0.0			15.5	1.19(+7)	E. Schreier, 1972 grade B Cen
99028 vb	78 ι Leo	11 23.9	+10 32	3.93	F2IV	192.0 y / 1948.47 y	140.7 / 0.54	2.5	−11.0	0.0678	2.03(+9)	H. A. Abt, 1976 grade E Leo
102509	93 Leo	11 48.0	+20 13	4.54	A G5III−IV	71.70 / 2418029.255	331.2 / 0.08	23.7	−0.2	0.0982	2.33(+7)	J. B. Cannon, 1910 grade D Leo
102660	HR 4535	11 49.2	+16 15	6.02	A3m	2.7818 / 2423521.231	61.8 / 0.02	31.0	−24.2	8.60(−3)	1.19(+6)	R. M. Petrie, 1926 grade C Leo
102713	HR 4536	11 49.7	+34 56	5.73	F5IV	32.864 / 2422059.507	223.6 / 0.09	37.3	−8.3	0.175	1.68(+7)	W. E. Harper, 1935 grade B UMa
103578	95 Leo	11 55.7	+15 39	5.50	A3V	6.6254 / 2424941.115	4.1 / 0.02	57.6	−20.4	0.131	5.25(+6)	O. Struve, 1927 grade C Leo
104321	8 π Vir	12 00.9	+6 37	4.64	A4V	282.69 / 2417164.7 n	312 / 0.27	26.2	−10.4	0.471	9.81(+7)	H. A. Abt, 1965 grade D Vir
104337	HR 4590	12 00.9	−19 40	5.3	B1.5V	2.9631 / 2426378.885	68.8 / 0.06	120.5 225	+3	8.22 4.40	4.90(+6) 9.15(+6)	R. N. Van Arnam, 1932 grade D Crv
104671	θ¹ Cru	12 03.0	−63 19	4.32	Ap	24.4828 / 2419453.347	358.9 / 0.61	46.1 56.1	−2.8	0.741 0.609	1.23(+7) 1.50(+7)	F. C. Moore, 1931 grade A Cru

HD	Star Name	α 2000	δ 2000	Mag	Spec	P / T	ω / e	K₁ / K₂	V₀	f(m) or m sin³i	a sin i	Author / Quality	Const
104841	θ² Cru	12ʰ 04.3ᵐ	−63° 10′	4.71	B2IV	3.4280 / 2419604.367	/ 0.0	51.3	+16.1	0.0481	2.42(+6)	D. S. Grattin, 1926 grade I	Cru
105981	4 Com	12 11.9	+25 52	5.66	K2V	462.8 / 2424665.79	235.3 / 0.17	14.3	+21.3	0.134	8.93(+7)	W. E. Harper, 1930 grade C	Com
— eb	CC Com	12 12.1	+22 32	11.31 12.17		0.2207 / 2442467.831 e	0.0 /	122.0 235.9	−10.2	0.693 0.358	3.70(+5) 7.16(+5)	S. M. Rucinski, 1977 grade D	Com
106112	HR 4646	12 12.2	+77 37	5.10	Am	1.2710 / 2420685.265 n	0.0 /	63.2	+0.3	0.0333	1.10(+6)	O. J. Lee, 1916 grade C	Cam
106760	HR 4668	12 16.5	+33 04	4.99	K1III	1300 / 2421750	300 / 0.3	6.8	−41.5	0.0368	1.16(+8)	W. H. Christie, 1936 grade D	Com
107259	15 η Vir Zaniah	12 19.9	−0 40	3.88	A2V	71.9 / 2417644.41	191.6 / 0.34	30.5 43.7	+5.3	1.49 1.04	2.84(+7) 4.06(+7)	W. E. Harper, 1935 grade C	Vir
107700	12 Com	12 22.5	+25 51	4.83	F8:p	396.49 / 2423885.847	/ 0.60	25.3	+0.5	0.341	1.10(+8)	J. Vint.–Hansen, 1940 grade B	Com
108248	α¹ Cru Acrux	12 26.6	−63 06	1.41	B1IV	75.769 / 2417641.1	18 / 0.48	32.8	+7.5	0.187	3.00(+7)	A. D. Thackeray, 1974 grade C	Cru
108722	18 Com	12 29.4	+24 07	5.47	F5IV	17.954 / 2439243.5 n	286 / 0.42	11.5	−42.5	2.12(−3)	2.58(+6)	H. A. Abt, 1976 grade E	Com
109358	8 β CVn Chara	12 33.7	+41 21	4.29	G0V	2429.9 / 2413397	307 / 0.49	2.6	+6.5	2.94(−3)	7.57(+7)	H. A. Abt, 1976 grade E	CVn
109510	24 Com	12 35.1	+18 23	5.03	K2III F1V	7.3366 / 2428423.039	287.6 / 0.21	66.3 77.4	+4.1	1.14 0.975	6.54(+6) 7.63(+6)	R. M. Petrie, 1937 grade B	Com
110317	HR 4821	12 41.3	−13 01	6.1	F2	1.4605 / 2428707.080	86.8 / 0.09	88.2 100	−14.6	0.531 0.468	1.76(+6) 2.00(+6)	R. F. Sanford, 1942 grade D	Crv
110318	HR 4822	12 41.3	−13 01	6.0	F5	44.4137 / 2429089.062	238.8 / 0.25	25.9	−11.6	0.0727	1.53(+7)	R. F. Sanford, 1942 grade C	Crv
110951	32 Vir	12 45.6	+7 40	5.20	Am	38.324 / 2434039.463 n	210.0 / 0.07	48.1	−10.6	0.440	2.53(+7)	F. C. Bertiau, 1957 grade B	Vir
112014	32 Cam	12 49.1	+83 25	5.83	B9V	3.2866 / 2424226.669	211.1 / 0.04	108.3 128.9	−0.1	2.47 2.07	4.89(+6) 5.82(+6)	J. S. Plaskett, 1926 grade D	Cam
112486	HR 4917	12 56.3	+54 06	5.82	A5m	5.1274 / 2440092.515	22 / 0.05	69.9 64.0	+7.3	0.609 0.665	4.92(+6) 4.51(+6)	R. Margoni, 1969 grade D	UMa
112985	δ Mus	13 02.3	−71 33	3.61	K2III	847 / 2421790	0 / 0.4	7.8	+36.5	0.0321	8.33(+7)	W. H. Christie, 1936 grade E	Mus
113791	ξ² Cen	13 06.9	−49 54	4.26	B2IV	7.6497 / 2418077.493	308.6 / 0.35	38.8	+14.3	0.0381	3.82(+6)	F. J. Neubauer, 1931 grade C	Cen
113904	θ Mus	13 08.1	−65 18	5.69	WC6 O9.5	18.341 / 2440663.193 n	173 / 0.0	6		0.366 10.6	4.36(+7) 1.51(+6)	A. F. J. Moffat, 1977 grade D	Mus
114519 eb	RS CVn	13 10.6	+35 56	7.93 9.0	F4IV-V K0IV	4.7978 / 2433015.620 n	89.7 / 0.0	86	−15	1.32 1.38	5.92(+6) 5.67(+6)	D. M. Popper, 1961 grade C	CVn
114911	η Mus	13 15.2	−67 54	4.79	B8V	20.0052 / 2420606.722	120 / 0.12	56.5	−8.1	0.367	1.54(+7)	W. Buscombe, 1961 grade D	Mus
116656	79 ζ UMa A Mizar A	13 23.9	+54 56	2.27	A2V	20.5386 / 2436997.212	104.2 / 0.54	68.8 67.6	−5.6	1.60 1.63	1.64(+7) 1.61(+7)	C. Fehrenbach, 1961 grade A	UMa
116657	79 ζ UMa B Mizar B	13 23.9	+54 55	3.95	Am	175.55 / 2437294.4	4.3 / 0.46	6.3	−9.3	3.19(−3)	1.35(+7)	F. Gutmann, 1965 grade B	UMa
116658	67 α Vir Spica	13 25.2	−11 10	0.97	B1V B3V	4.0145 / 2440284.78	142 / 0.18	120 189	0 −2	7.16 4.55	6.52(+6) 1.03(+7)	R. R. Shobbrook, 1972 grade C	Vir
118216	BH CVn	13 34.8	+37 11	4.97	F2IV	2.6132 / 2417022.522	214.8 / 0.04	9.5	+6.4	2.32(−4)	3.41(+5)	W. E. Harper, 1938 grade B	CVn
— eb	UX UMa	13 36.7	+51 54	12.7 p 13.7 p	sdB3	0.1967 / 2427341.222 n	250 / 0.0		0.0	0.319	6.76(+5)	O. Struve, 1948 grade E	UMa
119756	1 Cen	13 45.7	−33 03	4.23	F2III	9.9448 / 2422737.382	137.7 / 0.25	6.0	−23.9	2.02(−4)	7.94(+5)	H. Spencer Jones, 1928 grade E	Cen
119834	M Cen	13 46.7	−51 26	4.64	gG9	437.00 / 2424162.96	58.6 / 0.13	12.3	−5.6	0.0823	7.33(+7)	H. Spencer Jones, 1928 grade C	Cen
120064	3 Boo	13 46.7	+25 42	5.95	F6IV-V	36.04 / 2422014.483	258.2 / 0.49	54.0 65.8	+6.5	2.34 1.92	2.33(+7) 2.84(+7)	R. M. Petrie, 1926 grade C	Boo
120307	ν Cen	13 49.5	−41 41	3.40	B2V	2.6252 / 2418642.950 n	0.0 /	20.6	+9.1	2.38(−3)	7.44(+5)	L. B. Lucy, 1971 grade D	Cen

HD	Star Name	α 2000	δ 2000	Mag	Spec	Orbital Elements P / T	ω / e	K₁ / K₂	V₀	f(m) or m sin³i	a sin i	Author / Quality Const
120955	4 h Cen	13 53.2	−31 56	4.72	B5IV	6.927 / 2418733.25	147.2 / 0.23	21.4	+5.2	6.50(−3)	1.98(+6)	G. F. Paddock, 1916 / grade C Cen
121370	8 η Boo Muphrid	13 54.7	+18 24	2.69	G0IV	494.173 / 2428136.19	326.3 / 0.26	8.4	+1.0	0.0274	5.51(+7)	F. C. Bertiau, 1957 / grade A Boo
121263	ζ Cen	13 55.5	−47 17	2.54	B2IV	8.024 / 2429798.46	290 / 0.5	110.7 / 159.4	+6.5	6.29 / 4.37	1.06(+7) / 1.52(+7)	D. M. Popper, 1942 / grade D Cen
122223	υ² Cen	14 01.7	−45 36	4.34	F7I-II	1025 / 2423960	140 / 0.4	8.2	−0.5	0.0452	1.06(+8)	W. H. Christie, 1936 / grade D Cen
123299	11 α Dra Thuban	14 04.4	+64 23	3.66	A0III	51.420 / 2417352.596	24.5 / 0.38	46.9	−13.0	0.436	3.07(+7)	J. A. Pearce, 1956 / grade B Dra
124547	4 UMi	14 08.8	+77 33	4.87	K3III	605.8 / 2438901.7	311.8 / 0.14	12.7	+5.9	0.125	1.05(+8)	C. D. Scarfe, 1971 / grade B UMi
123515	HR 5296	14 09.6	−51 30	5.96	B9III	26.005 / 2437744.71	150.8 / 0.20	39.6	+5.2	0.158	1.39(+7)	D. S. Evans, 1967 / grade C Cen
123999	12 d Boo	14 10.4	+25 06	4.82	F8IV	9.6045 / 2417680.052	290 / 0.19	67.4 / 66.5	+9.1	1.13 / 1.14	8.74(+6) / 8.62(+6)	H. A. Abt, 1976 / grade B Boo
124425	HR 5317	14 13.7	−0 51	5.93	F6IV	2.6960 / 2422744.103 n	0.0	24.3	+17.6	4.02(−3)	9.01(+5)	J. C. Duncan, 1921 / grade C Vir
124570	14 Boo	14 14.1	+12 58	5.5	dF6	726.6 / 2439271	155 / 0.55	2.9	−39.2	1.07(−3)	2.42(+7)	H. A. Abt, 1976 / grade D Boo
125351	A Boo	14 18.0	+35 31	4.80	K0III	212.085 / 2440286.002	224.9 / 0.57	20.1	−21.6	0.0992	4.82(+7)	C. D. Scarfe, 1975 / grade A Boo
125248	CS Vir	14 18.6	−18 43	5.89	A0p	4.4 y / 1945.05 y	82.4 / 0.21	7.5	−10.9	0.0658	1.62(+8)	M. S. Hockey, 1969 / grade C Vir
125337	100 λ Vir	14 19.1	−13 22	4.52	Am	1.9302 / 2436744.445 n	31.1 / 0.10	29.5 / 33.5	−5.4	0.0263 / 0.0231	7.79(+5) / 8.85(+5)	H. A. Abt, 1961 / grade D Vir
126983	HR 5413	14 30.3	−49 31	5.36	A2V A2V	11.82 / 2441440.40	41.8 / 0.33	76.6 / 70.9	+6.3	1.59 / 1.72	1.18(+7) / 1.09(+7)	J. P. Kaufmann, 1973 / grade D Lup
128620/1 vb	α Cen	14 39.6	−60 50	−0.04 1.17	G2V K0V	81.18 y / 1875.71 y	52 / 0.53	5.0 / 5.0	−21.6	0.939 / 0.939	1.73(+9) / 1.73(+9)	J. Lunt, 1918 / grade D Cen
131041ftr	39 Boo B	14 49.7	+48 43	6.9	F5	12.822 / 2422379.490	97.1 / 0.39	58.3 / 72.2	−28.2	1.28 / 1.03	9.47(+6) / 1.17(+7)	W. E. Harper, 1922 / grade C Boo
132813	RR UMi	14 57.6	+65 56	6.2 p 6.5 p	gM7	750 / 2422252.5 n	0.0	6.7	+6.9	0.0234	6.91(+7)	R. K. Young, 1927 / grade C UMi
132742 eb	19 δ Lib	15 01.0	−8 31	4.91	B9.5V g	2.3273 / 2422854.108 n	153.4 / 0.05	78.3	−36.4	0.116	2.50(+6)	J. Sahade, 1963 / grade B Lib
133640ftr eb	44 i Boo B	15 03.8	+47 39	6.2 6.8	G2V G2V	0.2678 / 2428635.368 n	0.0	115.4 / 231.1	+3.4	0.772 / 0.385	4.25(+5) / 8.51(+5)	D. M. Popper, 1943 / grade D Boo
134759 vb	24 ι Lib	15 12.2	−19 47	4.53	B9IVp	22.35 y / 1949.65 y	7.6 / 0.35	3.4	−14.2	0.0274	3.58(+8)	W. D. Heintz, 1966 / grade E Lib
134687	e Lup	15 12.8	−44 30	4.81	B3III	0.9014 / 2435000.366 n	0.03	22	+13.5	9.95(−4)	2.73(+5)	W. Buscombe, 1962 / grade D Lup
135240	δ Cir	15 16.9	−60 57	5.08	O9V O	3.9025 / 2439542.08	295.8 / 0.09	158.3	+9.2	1.59	8.46(+6)	A. D. Thackeray, 1969 / grade C Cir
136403	HR 5702	15 19.5	+32 31	6.14	A7m F1	3.5773 / 2441347.95	39.3 / 0.09	59.4 / 74.6	−22.2	0.492 / 0.391	2.91(+6) / 3.65(+6)	A. H. Batten, 1976 / grade B CrB
136504	ε Lup	15 22.7	−44 41	3.36	B3IV B3V	4.5598 / 2439370.68	330 / 0.26	56.1 / 64.8	+7.9	0.404 / 0.350	3.40(+6) / 3.92(+6)	A. D. Thackeray, 1970 / grade C Lup
137107/8 vb	2 η CrB	15 23.2	+30 17	4.98	G0V	41.56 y / 1933.829 y	218.0 / 0.27	4.5	−6.0	0.128	9.04(+8)	Y. C. Chang, 1929 / grade E CrB
137052	31 ε Lib	15 24.2	−10 19	4.93	F5V	226.95 / 2414785.116	339.5 / 0.68	14.0	−9.7	0.0255	3.20(+7)	R. B. Jones, 1931 / grade A Lib
137909	3 β CrB Nusakan	15 27.8	+29 06	3.66	F0p	10.496 y / 1938.20 y	185.4 / 0.41	9.2	−18.0	0.235	4.42(+8)	F. J. Neubauer, 1944 / grade B CrB
138213	HR 5752	15 28.7	+47 12	6.0	A5m	105.95 / 2424232.1 n	0.0	10.8	−17.1	0.0139	1.57(+7)	L. B. Lucy, 1971 / grade D Boo
139006 eb	5 α CrB Alphecca	15 34.7	+26 43	2.24	A0V	17.3599 / 2436803.695	313.9 / 0.40	35.8	+1.5	0.0637	7.83(+6)	E. G. Ebbighausen, 1976 / grade B CrB
139892	7 ζ² CrB	15 39.4	+36 38	5.1	B7V	12.5842 / 2423866.601	40 / 0.01	126.1 / 119.4	−14.3	9.40 / 9.93	2.18(+7) / 2.07(+7)	K. D. Abhyankar, 1966 / grade D CrB

HD	Star Name	α 2000	δ 2000	Mag	Spec	P / T	ω / e	K₁ / K₂	V₀	f(m) or m sin³i	a sin i	Author / Quality	Const

HD	Star Name	α 2000	δ 2000	Mag	Spec	P T	ω e	K_1 K_2	V_0	f(m) or m sin³i	a sin i	Author Quality	Const
140008	4 ψ² Lup	15ʰ 42.7ᵐ	−34° 43′	4.74	B6V	12.26 ∠438252.97 n	82°.8 0.19	63.3 66.4	+3.9	1.35 1.28	1.05(+7) 1.10(+7)	A. D. Thackeray, 1964 grade B	Lup
140873	25 Ser	15 46.1	−1 48	5.40	B8III	38.937 2419528.565	206.8 0.80	53.6	−12.6	0.134	1.72(+7)	R. M. Petrie, 1949 grade B	SerCp
141004	27 λ Ser	15 46.4	+7 21	4.43	G0V	1837 2440152	288 0.55	2.8	−6.7	2.44(−3)	5.91(+7)	H. A. Abt, 1976 grade D	SerCp
141556	5 χ Lup	15 51.0	−33 38	3.94	B9IVHg	15.2565 2438429.538 n	166 0.00	135.6	−16.3	3.95	2.84(+7)	M. M. Dworetsky, 1972 grade C	Lup
141544	HR 5882	15 51.5	−47 04	6.00	K1IV	137.55 2430086.203	156.6 0.14	21.2	+30.3	0.132	3.97(+7)	D. S. Evans, 1961 grade I	Nor
142926	4 Her	15 55.5	+42 34	5.75	B7IV-Ve	46.194 2441473.8	5 0.34	12.0	−19.3	6.89(−3)	7.17(+6)	J. F. Heard, 1975 grade E	Her
143018	6 π Sco	15 58.9	−26 07	2.89	B1V B1V	1.5701 2436266.5 n	0.0	131 197	−4	3.46 2.30	2.83(+6) 4.25(+6)	C. Hetzler, 1959 grade D	Sco
143454	T CrB recurr. nova	15 59.5	+25 56	2.0 10.8	gM3 Q	227.6 2432046.0	90 0.06	24.0 33.5	−27.0	2.60 1.87	7.50(+7) 1.05(+8)	R. P. Kraft, 1958 grade D	CrB
143333	49 Lib	16 00.3	−16 32	5.46	F8V	3100 2419783 n	100 0.13	9.0	−25.7	0.229	3.80(+8)	H. A. Abt, 1976 grade D	Lib
144284	13 θ Dra	16 01.9	+58 34	4.01	F8IV-V	3.0708 2439277.02 n	244 0.01	25.3	−8.0	5.16(−3)	1.07(+6)	H. A. Abt, 1976 grade B	Dra
144208/9	HR 5983	16 03.3	+36 38	5.8	dF9 A0	108.075 2429765.532 n	0.0	19.6 23.2	−1.9	0.477 0.403	2.91(+7) 3.45(+7)	R. M. Petrie, 1942 grade B	CrB
144069/70 vb	ζ Sco	16 04.4	−11 22	4.16	F6IV	44.70 y 1905.39 y	343.6 0.75	3.7	−29.4	0.0249	5.49(+8)	Y. C. Chang, 1929 grade E	Sco
144217	8 β¹ Sco Graffias	16 05.4	−19 48	2.63	B0.5V	6.8281 2435845.228	37.9 0.28	129.0 215.2	−1.0 +39.2	16.0 9.59	1.16(+7) 1.94(+7)	K. D. Abhyankar, 1959 grade B	Sco
145389	11 φ Her	16 08.8	+44 56	4.24	B9p	560.5 2440525.2	357 0.47	2.4	−16.8	5.53(−4)	1.63(+7)	G. C. L. Aikman, 1976 grade C	Her
145849	HR 6046	16 11.8	+36 25	5.5	K3III	2150 2424290	340 0.6	16.0	−30.6	0.468	3.78(+8)	W. H. Christie, 1936 grade D	CrB
145502	14 ν Sco	16 12.0	−19 28	4.01	B2IV-V	5.9222 2435188.282 n	0.0	26.4	+1.4	0.0113	2.15(+6)	L. B. Lucy, 1971 grade E	Sco
146361	17 σ² CrB	16 14.7	+33 52	5.8	F8	1.1398 2423869.105 n	94 0.02	60.1 68.2	−11.9	0.133 0.117	9.42(+5) 1.07(+6)	R. W. Tanner, 1949 grade B	CrB
—	Sco X-1	16 20.0	−15 38	12.25 13.3	sdBe	0.7874 2442565.741 n	0.0	58.2	−138.5	0.0161	6.30(+5)	A. P. Cowley, 1975 grade D	Sco
147165	20 σ Sco	16 21.2	−25 36	2.88	B1III	34.23 2434895.2	308 0.36	34.0	+2.5	0.113	1.49(+7)	O. Struve, 1961 grade E	Sco
147869	21 Her	16 24.2	+6 57	5.84	A2pSr	4.951 2421773.086	355.9 0.51	16.3	−34.4	1.42(−3)	9.55(+5)	W. E. Harper, 1928 grade D	Her
147971	ε Nor	16 27.2	−47 33	4.46	B3V	3.2617 2438825.931	271.5 0.13	122.5 132.9	−12.5	2.86 2.64	5.45(+6) 5.91(+6)	A. D. Thackeray, 1966 grade B	Nor
148367	3 υ Oph	16 27.8	−8 22	4.64	Am	27.218 2438914.84	333.7 0.74	34.9 41.1	−33.6	0.204 0.173	8.79(+6) 1.03(+7)	F. Gutmann, 1966 grade B	Oph
147787	ı TrA	16 28.0	−64 03	5.3	F0	39.8880 2423236.454	87.2 0.28	38.7	−5.6	0.212	2.04(+7)	H. S. Jones, 1928 grade C	TrA
147584	ζ TrA	16 28.5	−70 05	4.90	G0V	12.9762 2418103.642	274.5 0.06	7.4	+7.6	5.43(−4)	1.32(+6)	H. S. Jones, 1928 grade C	TrA
148856	27 β Her Kornephoros	16 30.2	+21 29	2.78	G8III	410.575 2415500.374	24.6 0.55	12.8	−25.5	0.0521	6.04(+7)	H. C. Plummer, 1908 grade B	Her
— eb	CM Dra	16 34.4	+57 09	12.87 13.65	dM4e dM4e	1.2684 2442893.932 n	0.0	70.0 80.1	−118.6	0.238 0.208	1.22(+6) 1.40(+6)	C. H. Lacy, 1977 grade C	Dra
150680 vb	40 ζ Her	16 41.3	+31 36	2.80	G0IV KOV	34.417 y 1898.755 y	113.8 0.44	3.8	−69.9	0.0519	5.90(+8)	L. Berman, 1941 grade C	Her
150682	39 Her	16 41.6	+26 55	5.93	F2III F5	2.3076 2435964.771 n	0.0	100.6 108.7	−9.0	1.14 1.06	3.19(+6) 3.45(+6)	A. Abrami, 1959 grade A	Her
151613	HR 6237	16 45.3	+56 47	4.85	F2V	363.57 2415232.4	80.7 0.35	6.0	−2.0	6.70(−3)	2.81(+7)	H. A. Abt, 1965 grade D	Dra
153751 eb	22 ε UMi	16 46.0	+82 02	4.23	G5III	39.4809 2433083.47	323.5 0.04	31.8	−10.6	0.132	1.73(+7)	J. L. Climenhaga, 1951 grade A	UMi

HD	Star Name	α 2000	δ 2000	Mag	Spec	P / T	ω / e	K₁ / K₂	V₀	f(m) or m sin³i	a sin i	Author Quality / Const
151676 eb	V1010 Oph	16ʰ 49.5ᵐ	−15° 40′	6.1 / 7.0	A5V	0.6614 / 2441090.716 e	° / 0.0	101	−41	0.0708	9.19(+5)	E. F. Guinan, 1977 grade D Oph
151769	20 Oph	16 49.8	−10 47	4.64	F5IV-V	1290 / 2440154	320 / 0.21	3.2	+0.2	4.10(−3)	5.55(+7)	H. A. Abt, 1976 grade E Oph
151890 eb	μ¹ Sco	16 51.9	−38 03	3.03	B1.5V B	1.4463 / 2428282.335 n	185 / 0.0	185 / 280	0	9.09 / 6.01	3.68(+6) / 5.57(+6)	O. Struve, 1940 grade C Sco
152248		16 54.2	−41 50	6.12	O7f	5.97 / 2440815.6 n	191 / 0.0	191 / 206	−44 / −8	20.1 / 18.7	1.57(+7) / 1.69(+7)	G. Hill, 1974 grade D Sco
152830	V644 Her	16 55.3	+13 37	6.1	F2II	11.848 / 2426576.200	267.7 / 0.31	28.5	−3.7	0.0245	4.41(+6)	W. E. Harper, 1932 grade B Her
153597	19 h Dra	16 56.0	+65 08	4.88	F6V	52.1089 / 2439983.57	339 / 0.21	17.6	−20.9	0.0276	1.23(+7)	H. A. Abt, 1976 grade A Dra
152667 eb	V861 Sco	16 56.6	−40 49	6.16 / 6.4	B0Iae	7.8483 / 2439652.16	348.0 / 0.08	84.5	−34.6	0.487	9.09(+6)	E. N. Walker, 1970 grade D Sco
—	HZ Her Her X-1	16 57.8	+35 21	13.0 p / 14.5 p	A	1.7002 / 2441329.577 e	0.0	169.2		0.855	3.96(+6)	H. Tananbaum, 1972 grade B Her
153808	58 ε Her	17 00.3	+30 56	3.92	A0V	4.0235 / 2417947.242 n	138 / 0.02	70.7 / 112.0	−24.2	1.56 / 0.985	3.91(+6) / 6.20(+6)	W. J. Luyten, 1936 grade C Her
153890	V923 Sco	17 03.8	−38 09	5.91	F3IV-V F3V	34.8189 / 2430028.332	287.1 / 0.42	53.6 / 50.5	−14.8	1.48 / 1.57	2.33(+7) / 2.19(+7)	N. W. W. Bennett, 1963 grade C Sco
153919		17 03.9	−37 51	6.56	O6f	3.4126 / 2441453.14 e	324 / 0.20	20.6	−64.5	2.91(−3)	9.47(+5)	J. B. Hutchings, 1974 grade E Sco
154732	HR 6363	17 04.8	+48 48	6.09	K0	786 / 2426275	0 / 0.3	10.0	+12.8	0.0709	1.03(+8)	W. H. Christie, 1936 grade E Her
155937 eb	AK Her	17 14.0	+16 22	8.8 B / 9.3 B	F8V	0.4215 / 2422977.149 n	0.0	78	−13	0.0208	4.52(+5)	R. F. Sanford, 1934 grade D Her
156015	64 α Her B	17 14.7	+14 23	5.4	G0II-III	51.578 / 2434791.026 n	67.5 / 0.02	36.1	−38.4	0.252	2.56(+7)	A. J. Deutsch, 1956 grade B Her
156247 eb	U Oph	17 16.5	+1 13	5.88 / 6.56	B5V B5V	1.6773 / 2428054.881	306 / 0.01	180.0 / 199.4	−10.8	5.00 / 4.51	4.15(+6) / 4.60(+6)	J. A. Pearce, 1960 grade B Oph
156633 eb	68 u Her	17 17.3	+33 06	4.77 / 5.4	B2.5V B5	2.0510 / 2437801.92	259 / 0.06	95.6 / 263	−30.0	7.16 / 2.60	2.69(+6) / 7.40(+6)	B. J. Kovachev, 1975 grade B Her
157978/9 (abc)	HR 6497	17 26.3	+7 36	6.05	A0 G	1170 / 2433085	222.3 / 0.45	19.5 / 27.0	−12.1	5.05 / 3.65	2.80(+8) / 3.88(+8)	D. B. McLaughlin, 1962 grade I Oph
157978/9 (ab)	HR 6497	17 26.3	+7 36	6.05	A0 A0	3.7581 / 2432020.509 n	110 / 0.0	110 / 110	var.	2.08 / 2.08	5.68(+6) / 5.68(+6)	D. B. McLaughlin, 1962 grade I Oph
157950	HR 6493	17 26.6	−5 05	4.53	F3V	26.2765 / 2418411.524	14.5 / 0.49	47.5 / 50.7	+0.4	0.884 / 0.828	1.50(+7) / 1.60(+7)	T. H. Parker, 1915 grade B Oph
158261	HR 6506	17 26.8	+34 42	5.9	A0V	5.9182 / 2423585.527	35.7 / 0.03	25.1	−22.7	9.71(−3)	2.04(+6)	W. H. Christie, 1925 grade B Her
158614 vb	HR 6516	17 30.4	−1 04	5.30	G8IV-V G8IV-V	46.08 y / 1916.06 y	327.5 / 0.17	5.4 / 5.4	−76.4	1.05 / 1.05	1.23(+9) / 1.23(+9)	A. H. Batten, 1971 grade D Oph
159082	HR 6532	17 32.2	+11 56	6.1	A0III	6.7984 / 2422878.154	116.3 / 0.07	50.2	−12.8	0.0887	4.68(+6)	J. W. Campbell, 1922 grade C Oph
159560	25 ν² Dra	17 32.3	+55 10	4.89	A4m	38.5958 / 2415916.386 n	329.9 / 0.27	12.4	−16.9	6.82(−3)	6.34(+6)	H. A. Abt, 1960 grade D Dra
159176	HR 6535	17 34.7	−32 35	5.68	O7V O7V	3.3666 / 2438536.953 n	213.8 / 0	213.8 / 206.3	9.1 / 8.6	12.7 / 13.2	9.90(+6) / 9.55(+6)	W. Seggewiss, 1976 grade C Sco
160922	28 ω Dra	17 36.9	+68 45	4.8	F5V	5.2798 / 2440052.6	35 / 0.04	35.4 / 44.6	−14.4	0.156 / 0.124	2.57(+6) / 3.24(+6)	H. A. Abt, 1976 grade A Dra
159876	55 ξ Ser	17 37.6	−15 24	3.54	F0IV	2.2923 / 2419210.191 n	0.0	19.4	−42.8	1.74(−3)	6.12(+5)	R. K. Young, 1911 grade A SerCd
161321 eb	V624 Her	17 44.3	+14 25	6.18 / 7.73	Am Am	3.894 / 2424701.985	74.3 / 0.04	96.6 / 108.1	−32.4	1.83 / 1.63	5.17(+6) / 5.78(+6)	R. M. Petrie, 1928 grade C Her
161701	HR 6620	17 47.6	−14 44	6.07	B9III	12.4520 / 2439034.786 n	252.4 / 0.06	52.5	−18.5	0.186	8.97(+6)	D. P. Hube, 1969 grade C SerCd
161783	V539 Ara	17 50.5	−53 37	5.9	B3V B4V	3.170 / 2433114.719 n	29 / 0.04	154 / 181	−6	6.67 / 5.68	6.71(+6) / 7.88(+6)	J. Sahade, 1952 grade B Ara
162724	V906 Sco	17 53.9	−34 45	5.96	B9V B9V	2.7754 / 2439698.22	221 / 0.18	121 / 153	−32	3.15 / 2.49	4.54(+6) / 5.74(+6)	H. A. Abt, 1970 grade D Sco

HD	Star Name	α 2000	δ 2000	Mag	Spec	Orbital Elements			V_0	$f(m)$ or $m \sin^3 i$	$a \sin i$	Author Quality Const
						P T	ω e	K_1 K_2				
166865	40 Dra	18 00.1	+80 00	6.04	K2V	10.5217 2421764.648	256.8 0.31	46.2 51.5	+2.9	0.462 0.414	6.36(+6) 7.08(+6)	S. L. Boothroyd, 1920 grade C Dra
165341 vb	70 Oph AB	18 05.5	+2 30	4.02	K0V K5V	88.13 y 1984.05 y	13.2 0.50	3.4	−7.1	0.0853	1.30(+9)	A. H. Batten, 1976 grade D Oph
--- eb	DQ Her	18 07.5	+45 52	14.2	sdBe	0.1936 2434954.945 e	0.0	149	−20	0.0665	3.97(+5)	J. L. Greenstein, 1959 grade E Her
166126 eb	W Ser	18 09.8	−15 33	9.0 9.9	F5III	14.1567 2426625.361	31 0.37	66	+18.6	0.339	1.19(+7)	J. Sahade, 1957 grade E SerCd
166937 eb	13 μ Sgr	18 13.8	−21 04	3.85	B8Iap	180.45 2423632.21	79.2 0.40	56.8	−2.7	2.64	1.29(+8)	O. Kohl, 1932 grade B Sgr
167647 eb	RS Sgr	18 17.6	−34 06	6.0 p 6.8 p	B4IV	2.4157 2423480.06	41 0.09	85.1	+6.0	0.153	2.82(+6)	J. Sahade, 1949 grade C Sgr
168532	105 Her	18 19.2	+24 27	5.27	K4II	478 2423540.65	234.5 0.40	16.1	−14.4	0.159	9.70(+7)	W. E. Harper, 1925 grade D Her
170000	43 φ Dra	18 20.8	+71 20	4.18	A0p	26.768 2438853.6	171 0.39	26.6	−20.8	0.0408	9.02(+6)	H. A. Abt, 1973 grade E Dra
168913	108 Her	18 20.9	+29 52	5.62	A6Vm	5.5146 2419551.742	0.0	70.1 101.7	−20.2	1.72 1.18	5.32(+6) 7.71(+6)	Z. Daniel, 1914 grade C Her
170153	44 χ Dra	18 21.1	+72 44	3.58	F7V	280.531 2422440.160	122.6 0.45	18.0	+32.5	0.121	6.20(+7)	J. Vint.-Hansen, 1942 grade A Dra
168339	ξ Pav	18 23.2	−61 30	4.36	M1III	2214 2418076.27	187.2 0.26	17.9	+12.4	1.19	5.26(+8)	H. S. Jones, 1928 grade D Pav
169156	ζ Sct	18 23.7	−8 56	4.68	K0III	2373.7911 2418278.392	242.1 0.10	5.8	−5.2	0.0474	1.88(+8)	J. Grobben, 1969 grade C Sct
169981	HR 6917	18 26.0	+29 50	5.88	A2V	9.6120 2422048.711	326.4 0.47	28.5	+7.5	0.0159	3.32(+6)	R. K. Young, 1919 grade B Her
169985/6	59 d Ser A (bc)	18 27.2	+0 12	5.3	A6V A6V	1.8505 2430172.495 n	0.0	90 100		0.694 0.624	2.29(+6) 2.54(+6)	E. C. Tilley, 1943 grade C SerCd
169985/6	59 d Ser A (abc)	18 27.2	+0 12	5.3	G0III A6V	386.0 2429332.71	277 0.47	28 19	−23.3	1.16 1.71	1.31(+8) 8.90(+7)	E. C. Tilley, 1943 grade D SerCd
170200	HR 6928	18 28.0	+6 12	5.7	B8III	3.4941 2440806.838	0.06	38.5	−14.7	0.0206	1.85(+6)	D. P. Hube, 1976 grade D SerCd
170465	δ¹ Tel	18 31.8	−45 55	4.95	B6IV	18.8456 2435003.693	78 0.51	65	+7	0.342	1.45(+7)	W. Buscombe, 1962 grade D Tel
170523	δ² Tel	18 32.0	−45 45	5.07	B6II-III	21.7056 2435216.669	12.7 0.22	34.9	−7.6	0.0889	1.02(+7)	W. Buscombe, 1956 grade C Tel
234677	BY Dra	18 33.9	+51 44	8.41	K7Ve	5.9760 2441147.250	211.1 0.49	30.6 37.3	−25.7	0.0707 0.0580	2.19(+6) 2.67(+6)	B. W. Bopp, 1973 grade D Dra
172044	HR 6997	18 36.6	+33 28	5.41	B8IIp	1675 2420438.5	120 0.16	3.2	−26.9	5.48(−3)	7.28(+7)	H. A. Abt, 1973 grade E Lyr
171978	HR 6993	18 37.6	−0 19	5.77	A2V A2V	14.674 2432304.628	224.9 0.21	38.6 38.1	+11.4	0.319 0.323	7.62(+6) 7.52(+6)	R. M. Petrie, 1948 grade C SerCd
173524	46 Dra	18 42.6	+55 32	5.05	A0 A0	9.8107 2440003.22	173 0.20	25.1 29.5	−31.0	0.0843 0.0717	3.32(+6) 3.90(+6)	G. C. L. Aikman, 1976 grade B Dra
173648	6 ζ¹ Lyr	18 44.8	+37 36	4.35	Am	4.2999 2418110.722 n	0.0	51.2	−26.0	0.0599	3.03(+6)	F. C. Jordan, 1909 grade B Lyr
175286	50 Dra	18 46.4	+75 26	5.35	A0 A0	4.1175 2420293.519	107.6 0.01	79.1 83.9	−8.8	0.953 0.899	4.48(+6) 4.75(+6)	W. E. Harper, 1919 grade B Dra
173764	β Sct	18 47.2	−4 45	4.22	G5II	834 2422480.9	33.9 0.35	16.7	−21.9	0.332	1.79(+8)	R. K. Young, 1927 grade C Sct
174107	V603 Aql	18 48.9	+0 35	11.95	sdBe	0.1385 2437858.78 n	0	37.5	−23	7.58(−4)	7.14(+4)	R. P. Kraft, 1964 grade E Aql
174638/9 eb	10 β Lyr Sheliak	18 50.1	+33 22	3.38 4.29	B8pe	12.9349 2442260.922 n	0.0	184.0	−17.8	8.37	3.27(+7)	A. H. Batten, 1975 grade B Lyr
175306	47 o Dra	18 51.2	+59 23	4.67	K0II-III	138.420 2419258.16	274.3 0.11	23.5	−19.5	0.183	4.45(+7)	R. K. Young, 1920 grade C Dra
174933	112 Her	18 52.3	+21 25	5.48	B7V A3V	6.3624 2424589.683	195.5 0.12	17.7	−19.6	3.59(−3)	1.54(+6)	W. F. Meyer, 1927 grade B Her
175426	11 δ¹ Lyr	18 53.7	+36 58	5.75	B2.5V	88.352 2428406.613	191.3 0.37	39.7	−17.2	0.460	4.48(+7)	E. H. Richardson, 1957 grade C Lyr

HD	Star Name	α 2000	δ 2000	Mag	Spec	Orbital Elements P / T	ω / e	K_1 / K_2	V_0	f(m) or m sin³i	a sin i	Author Quality Const
175492/3	113 Her	18 54.7	+22 39'	4.56	G4III	245.3	171°	16.0	−23.3	0.102	5.36(+7)	W. J. Luyten, 1936
					A6V	2418709.73 n	0.12					grade B Her
176318	HR 7174	18 58.0	+38 16	5.70	B6V	2.9116	92.5	76.5	−31.1	0.130	3.02(+6)	W. Gorza, 1971
						2432721.793	0.17					grade B Lyr
176155	FF Aql	18 58.2	+17 22	5.38	F4Ia	1435	6.3	3.5	−17.4	6.39(−3)	6.91(+7)	H. A. Abt, 1959
						2425610.7	0.01					grade D Aql
175813 eb	ε CrA	18 58.7	−37 06	4.74 5.00	F0V	0.5914		25.7	+61.9	1.04(−3)	2.09(+5)	S. Tapia, 1976
						2441583.688	0.02					grade D CrA
178125	18 Aql Y Aql	19 07.0	+11 04	5.10	B7V	1.3023		27.6	−18.7	2.84(−3)	4.94(+5)	F. C. Jordan, 1914
						2418157.502 n	0.00					grade C Aql
178449	17 Lyr	19 07.4	+32 30	5.21	dA7	49.09	223	11.9	−35.2	5.80(−3)	7.05(+6)	H. A. Abt, 1976
						2439360.5 n	0.48					grade E Lyr
178428	HR 7260	19 08.0	+16 51	6.07	G5V	21.998	92.7	12.8	+14.4	4.48(−3)	3.79(+6)	V. Albitzky, 1933
						2426206.422	0.21					grade D Sge
179094	HR 7275	19 08.4	+52 26	5.81	K1IV	28.59	330.1	40.3	+5.2	0.194	1.58(+7)	R. K. Young, 1944
						2431048.179	0.04					grade B Cyg
178322	HR 7257	19 10.0	−41 54	5.87	B6V B5V	12.47 2438237.45 n	19.1 0.05	79.7 78.9	+13.3	2.56 2.59	1.36(+7) 1.35(+7)	A. D. Thackeray, 1965 grade C CrA
179950 vb	42 ψ Sgr AB (abc)	19 15.5	−25 15	4.86	F2+G5 F0III-IV	7319.0 2442418.795	2.6 0.51	10.0 13.8	−26.8	3.78 2.74	8.66(+8) 1.19(+9)	F. C. Fekel, 1975 grade D Sgr
179950	42 ψ Sgr B (ab)	19 15.5	−25 15	5.7	F2II-III G5III-IV	10.7786 2442226.01	181.5 0.47	72.5 90.6	−43.0	1.86 1.48	9.48(+6) 1.19(+7)	F. C. Fekel, 1975 grade B Sgr
—	PSR 1913+16	19 15.5	+16 05	22.5 ?		0.3230 2442321.433	179 0.62	199		0.128	6.93(+5)	R. A. Hulse, 1975 grade A Aql
181182 eb	U Sge	19 18.8	+19 37	6.51 9.13	B8.5IV-V G3III-IV	3.3806 2417131.417	194.4 0.03	69.7	−10.1	0.119	3.24(+6)	D. H. McNamara, 1951 grade D Sge
181470	HR 7338	19 19.0	+37 27	6.2	A0III	10.3932 2423570.622	198.7 0.52	60.3 84.0	−14.9	1.18 0.845	7.36(+6) 1.03(+7)	W. E. Harper, 1928 grade C Lyr
181391	26 f Aql	19 20.5	−5 25	5.00	G8III-IV	266.544 2433420.207	152.7 0.83	29.9	−18.0	0.128	6.11(+7)	K. L. Franklin, 1952 grade B Aql
181615/6 eb	46 υ Sgr	19 21.7	−15 57	4.61	B8p+F2pe	137.9567 2419643.99	16.8 0.06	49.1	+13.3	1.69	9.30(+7)	F. L. Seydel, 1929 grade C Sgr
182490	2 Sge	19 24.4	+16 56	6.25	Am	7.390 2421047.175 n	0.0	52.8 77.6	+11.2	1.01 0.689	5.37(+6) 7.89(+6)	W. J. Luyten, 1936 grade C Sge
183056	4 Cyg	19 26.2	+36 19	5.11	B8p	35.0225 2438929.1	290 0.45	5.7	−11.0	4.80(−4)	2.45(+6)	H. A. Abt, 1973 grade E Cyg
183007	HR 7392	19 29.4	−43 27	5.69	A3p	164.64 2437104.6	79.7 0.12	11.8	−30.9	0.0275	2.65(+7)	D. S. Evans, 1967 grade B Sgr
184035	201 G. Sgr HR 7422	19 34.1	−40 02	5.9	A3III	4.625 2436702.433	315 0.09	72.5	+11.9	0.181	4.59(+6)	W. Buscombe, 1961 grade D Sgr
184759	9 Cyg	19 34.8	+29 28	5.39	F5	1717.3 2420553 n	135 0.71	15.7	−17.1	0.241	2.61(+8)	H. A. Abt, 1976 grade D Cyg
184552	51 Sgr	19 36.0	−24 43	5.66	A8m	8.1158 2418110.796 n	260.2 0.14	20.9	−30.3	7.47(−3)	2.31(+6)	E. N. Walker, 1969 grade C Sgr
185912 eb	V1143 Cyg	19 38.7	+54 58	5.92 6.42		7.6408 2439339.616	48.3 0.54	87.1 89.8	−15.9	1.33 1.29	7.70(+6) 7.94(+6)	M. S. Snowden, 1969 grade B Cyg
185507 eb	44 σ Aql	19 39.2	+5 24	5.18	B3V B3V	1.9503 2428200.615 n	0.0	164.2 208	−4.5	5.84 4.61	4.40(+6) 5.58(+6)	W. J. Luyten, 1939 grade B Aql
185734	12 φ Cyg	19 39.4	+30 09	4.66	G8III-IV G8III-IV	434.086 2430837.64	216.5 0.52	26.8 27.9	+5.0	2.35 2.25	1.37(+8) 1.42(+8)	R. A. Rach, 1961 grade B Cyg
185936 eb	QS Aql	19 41.1	+13 49	5.95 6.05	B5V	2.4968 2423963.599	95.7 0.06	47.3	−14.2	0.0273	1.62(+6)	S. N. Hill, 1930 grade E Aql
187076/7	7 δ Sge	19 47.4	+18 32	3.82	M2II A0V	3725 2414525	270 0.32	7.5	+1.9	0.139	3.64(+8)	D. B. McLaughlin, 1952 grade D Sge
187879 eb	V380 Cyg	19 50.6	+40 36	5.68	B1IV	12.4256 2437455.08	127.2 0.22	92.7 163.7	−2.9	12.9 7.30	1.55(+7) 2.73(+7)	A. H. Batten, 1962 grade A Cyg
187949 eb	V505 Sgr	19 53.1	−14 36	6.48 7.51	A2V F8:IV:	1.1829 2425501.080 n	0.0	104	−2	0.138	1.69(+6)	D. M. Popper, 1949 grade D Sgr
188727	10 Sge S Sge	19 56.0	+16 38	5.36	F6Ib-G5Ib	676.2 2429089.7	202.3 0.25	15.0	−10.0	0.215	1.35(+8)	G. H. Herbig, 1952 grade A Sge

HD	Star Name	α 2000	δ 2000	Mag	Spec	Orbital Elements P / T	ω / e	K_1 / K_2	V_0	f(m) or $m \sin^3 i$	a sin i	Author Quality	Const
188728	61 φ Aql	19^h 56.2^m	+11° 25′	5.28	A1V	3.3207 2423210.628 n	0.0	37.2	−28.0	0.0178	1.70(+6)	L. B. Lucy, 1971 grade C	Aql
226868	Cyg X-1	19 57.8	+38 42	8.89	O9.7Iab	5.5998 2401556.46	330 0.06	72.2	−1.7	0.218	5.55(+6)	C. T. Bolton, 1975 grade C	Cyg
189103	θ¹ Sgr	19 59.7	−35 17	4.35	B3IV	2.1051 2411140.645 n	0.0	15.9	+0.9	8.79(−4)	4.60(+5)	R. E. Wilson, 1921 grade B	Sgr
191110	HR 7694	20 08.5	−10 04	6.17	B9.5III	9.3464 2441584.693 n	220 0.01	51.5 57.1	−8.4	0.654 0.589	6.62(+6) 7.34(+6)	M. M. Dworetsky, 1974 grade C	Cap
191747	18 Vul	20 10.6	+26 54	5.51	A3III	9.316 2421103.17 n	0.0	78.5 86.3	−13.0	2.27 2.06	1.01(+7) 1.11(+7)	W. J. Luyten, 1936 grade C	Vul
191692	65 θ Aql	20 11.3	−0 49	3.24	B9III B9III	17.1243 2431636.344	34.5 0.61	51.0 63.7	−27.9	0.742 0.594	9.52(+6) 1.19(+7)	C. U. Cesco, 1946 grade A	Aql
192577/8 eb	31 o¹ Cyg V695 Cyg	20 13.6	+46 44	3.80	K4Ib B4V	3784.3 2437169.73	201.1 0.22	14.0 20.8	−7.7 −12.3	9.19 6.19	7.11(+8) 1.06(+9)	K. O. Wright, 1970 grade B	Cyg
192909/10 eb	32 o² Cyg V1488 Cyg	20 15.5	+47 43	3.98	K5Iab B4IV-V	1147.8 2433141.80	218.2 0.30	17.0 34	−5.7	9.15 4.58	2.56(+8) 5.12(+8)	K. O. Wright, 1970 grade C	Cyg
192713	22 Vul	20 15.5	+23 31	5.17	G2Ib	251.0 2423333.24	140 0.05	26.8	−23.7	0.500	9.24(+7)	W. J. Luyten, 1936 grade C	Vul
193370	35 Cyg	20 18.6	+34 59	5.17	F5Ib	2440.0 2427388.8	342.7 0.51	9.6	−17.3	0.143	2.77(+8)	K. Osawa, 1957 grade B	Cyg
193576 eb	V444 Cyg	20 19.5	+38 44	8.00	WN5 B1	4.2124 2428770.159 n	305 0.0 120	9.48 24.1	+10	1.77(+7) 6.95(+6)	G. Munch, 1950 grade D	Cyg
193964	71 Dra	20 19.6	+62 15	5.6	B9V	5.2981 2441140.485	291.2 0.04	49.7	−7.8	0.0674	3.62(+6)	D. P. Hube, 1973 grade D	Dra
—— eb	V Sge	20 20.2	+21 06	9.5 13.9	WN5	0.5142 2437889.915 n	320 0.0 85	−5 +10	0.745 2.80	2.26(+6) 6.01(+5)	G. H. Herbig, 1965 grade E	Sge
193495/6	9 β Cap Dabih (abc)	20 21.0	−14 47	3.08	F8V B8	1374.126 2421521.26	119.1 0.42	21.9 20.0	−18.9	3.75 4.10	3.76(+8) 3.43(+8)	R. F. Sanford, 1939 grade A	Cap
193495/6	9 β Cap Dabih (bc)	20 21.0	−14 47	3.08	B8	8.6780 2428383.898	343.2 0.36	37.9		0.0398	4.22(+6)	R. F. Sanford, 1939 grade D	Cap
194215	HR 7801	20 25.4	−28 40	5.84	K3V	377.60 2430279.9	0.0 0.07	11.2	−7.3	0.0547	5.80(+7)	B. W. Bopp, 1970 grade D	Sgr
193924	α Pav	20 25.6	−56 44	1.93	B3IV	11.753 2417547.68 n	0.0	7.2	+2.0	4.56(−4)	1.16(+6)	W. J. Luyten, 1936 grade C	Pav
194184	HR 7799	20 25.8	−40 48	6.15	K3III	117.776 2437063.4	288.9 0.24	22.6	+42.1	0.129	3.55(+7)	D. S. Evans, 1967 grade B	Sgr
195725	2 θ Cep	20 29.6	+63 00	4.21	Am	840.0 2416214.5 n	83.7 0.03	13.9	−6.4	0.234	1.61(+8)	H. A. Abt, 1961 grade E	Cep
197433 eb	VW Cep	20 37.4	+75 36	7.27 7.7	K0V G5	0.2783 2431984.665 n	0.0	75 230	−35	0.618 0.202	2.87(+5) 8.80(+5)	D. M. Popper, 1948 grade E	Cep
196524 vb	6 β Del	20 37.5	+14 36	3.62	F5IV	26.65 y 1938.38 y	190 0.48	7.6	−24.1	0.300	8.92(+8)	H. A. Abt, 1976 grade D	Del
196544	5 z Del	20 37.8	+11 23	5.44	A2V	11.039 2422139.862	61.8 0.23	26.0	−4.9	0.0186	3.84(+6)	W. E. Harper, 1935 grade B	Del
196574	71 Aql	20 38.3	−1 06	4.30	G8III	205.2 2423358.0	0.0	9.8	−5.9	0.0201	2.77(+7)	L. B. Lucy, 1971 grade B	Aql
198743	6 μ Aqr	20 52.7	−8 59	4.72	Am	1782 2417116.97 n	43.1 0.24	5.4	−11.4	0.0267	1.28(+8)	H. A. Abt, 1961 grade E	Aqr
199081	57 Cyg	20 53.2	+44 23	4.77	B5V B5V	2.8548 2441571.275 n	159.6 0.15	111.9 126.0	−21.1	2.04 1.81	4.34(+6) 4.89(+6)	R. W. Hilditch, 1973 grade B	Cyg
199579	HR 8023	20 56.6	+44 56	5.96	O6	48.608 2422892.612	66.8 0.10	42.2	−5.8	0.374	2.81(+7)	J. S. Plaskett, 1922 grade C	Cyg
199870	HR 8035	20 58.3	+44 28	5.53	G8III	635.1 2439186.1	148.1 0.44	6.4	−22.6	0.0125	5.02(+7)	G. A. Radford, 1975 grade C	Cyg
199603 eb	DV Aqr	20 58.7	−14 29	6.00	F0IV	1.5755 2426160.500 e	0.0	95.5	+10.3	0.143	2.07(+6)	W. Paffhausen, 1976 grade D	Aqr
199766	1 Equ	20 59.1	+4 18	5.22	F5IV	2.0313 2440051.36	51 0.17	15.8	+7.8	7.96(−4)	4.35(+5)	H. A. Abt, 1976 grade D	Equ
199532	α Oct	21 04.7	−77 01	5.14	F4III F5III	9.073 2435302.404	276.2 0.39	47.0 47.0	+45.0	0.306 0.306	5.40(+6) 5.40(+6)	W. Buscombe, 1960 grade D	Oct

HD	Star Name	α 2000	δ 2000	Mag	Spec	Orbital Elements P / T	ω / e	K_1 / K_2	V_0	f(m) or m sin³i	a sin i	Author Quality	Const
202275	7 δ Equ	21ʰ 14.5ᵐ	+10° 00′	4.49	F7V	5.7 y / 1970.11	193.3° y / 0.46	24.9	-15.2	2.34	6.33(+8)	M. M. Dworetsky, 1971 grade C	Equ
202447/8	8 α Equ Kitalpha	21 15.8	+5 15	3.92	G2III A5V	98.81 / 2441548.828	35.2 / 0.03	15.7 / 12.5	-17.8	0.102 / 0.128	2.1$_{5}$(+7) / 1.70(+7)	D. J. Stickland, 1976 grade D	Equ
203439	HR 8169	21 21.4	+32 37	6.0	A2V	20.30 / 2424363.558	219.7 / 0.44	45.7 / 79.0	-4.0	1.88 / 1.08	1.15(+7) / 1.98(+7)	W. E. Harper, 1926 grade C	Cyg
203858	HR 8194	21 24.1	+25 19	6.2	A2V	6.9463 / 2430007.986 n	0.0	71.4 / 81.5	-19.5	1.37 / 1.20	6.82(+6) / 7.78(+6)	C. G. Patten, 1942 grade B	Vul
204188	IK Peg	21 26.4	+19 23	6.08	A8m	21.724 / 2423307.986 n	0.0	41.5	-12.4	0.161	1.24(+7)	W. E. Harper, 1927 grade C	Peg
205021	8 β Cep Alfirk	21 28.7	+70 34	3.19	B2III	10.893 / 2433555.4	338 / 0.52	3.1	-3.1	2.10(-5)	3.97(+5)	W. S. Fitch, 1968 grade E	Cep
205767	23 ζ Aqr	21 37.7	-7 51	4.68	A7V	8016 / 2413270	42.1 / 0.54	11.3	-21.0	0.716	1.05(+9)	H. A. Abt, 1965 grade E	Aqr
206267	HR 8281	21 39.0	+57 29	5.62	O6.5V O9	3.7098 / 2441819.64	332.3 / 0.09	103.7 / 298	-6.4	18.3 / 6.37	5.27(+6) / 1.51(+7)	D. Crampton, 1975 grade D	Cep
206155 eb	EE Peg	21 40.0	+9 11	6.9 7.6	A4V F5:V:	2.6282 / 2429486.408	357.7 / 0.03	86.2	-13.4	0.175	3.11(+6)	G. A. Bakos, 1965 grade C	Peg
205478	ν Oct	21 41.5	-77 23	3.75	K0III	1020 / 2418525	80 / 0.4	8.0	+34.0	0.0418	1.03(+8)	W. H. Christie, 1936 grade D	Oct
206301	42 Cap	21 41.5	-14 03	5.18	G2IV	13.1736 / 2422069.255	175.4 / 0.16	23.0	-1.8	0.0160	4.11(+6)	R. F. Sanford, 1942 grade B	Cap
206672	80 π¹ Cyg	21 42.1	+51 11	4.67	B3V	26.33 / 2431306.5	0.0	16.5	-8.2	0.0123	5.97(+6)	C. Fehrenbach, 1948 grade E	Cyg
206644	77 Cyg	21 42.4	+41 05	5.74	A0V	1.7290 / 2424255.750 n	0.0	110.0 / 110.0	-25.5	0.956 / 0.956	2.62(+6) / 2.62(+6)	W. J. Luyten, 1936 grade D	Cyg
206697	SS Cyg	21 42.7	+43 34	8.2 12.1	sdBe dG5	0.2695 / 2430267.619 a	0.0	123 / 116	-9 -3	0.185 / 0.197	4.56(+5) / 4.30(+5)	J. Smak, 1969 grade E	Cyg
206546	HR 8293	21 43.2	-19 37	6.22	Am	6.3702 / 2425489.752 n	0.0	82.1 / 86.3	-25.5	1.62 / 1.54	7.19(+6) / 7.56(+6)	R. F. Sanford, 1931 grade C	Cap
206901	10 κ Peg	21 44.6	+25 39	4.15	F5IV	5.9715 / 2419054.957	148 / 0.03	41.7		0.0449	3.42(+6)	W. J. Luyten, 1934 grade D	Peg
207330	81 π² Cyg	21 46.8	+49 19	4.24	B3III	72.0162 / 2428410.6	238.1 / 0.34	7.8	-12.3	2.95(-3)	7.26(+6)	S. Taffara, 1939 grade D	Cyg
207098 eb	49 δ Cap	21 47.0	-16 08	2.83	Am	1.0228 / 2436498.8	141 / 0.01	70.8	-0.2	0.0377	9.96(+5)	A. H. Batten, 1961 grade B	Cap
207650	14 Peg	21 49.8	+30 10	5.09	A0V	5.3047 / 2429117.474	302.7 / 0.53	37.0 / 40.4	-23.9	0.0813 / 0.0745	2.29(+6) / 2.50(+6)	R. M. Petrie, 1940 grade C	Peg
207757	AG Peg	21 51.0	+12 38	8.37	M3III	820 / 2440928 n	0.0	5.1	-16.3	0.0113	5.75(+7)	J. B. Hutchings, 1975 grade E	Peg
208095	HR 8357	21 52.0	+55 48	5.68	B7V B9	17.3263 / 2424415.682	269.9 / 0.22	108.3 / 165.8	-6.5	20.8 / 13.6	2.52(+7) / 3.85(+7)	J. A. Pearce, 1938 grade C	Cep
208816 eb	VV Cep	21 56.7	+63 38	4.90	M2Iape B9	7430.5 / 2438461.0	59.2 / 0.35	19.4 / 19.1	-20.2 -18.5	18.0 / 18.2	1.86(+9) / 1.83(+9)	K. O. Wright, 1977 grade C	Cep
209481	14 Cep LZ Cep	22 02.1	+58 00	5.56	O9Vn O9Vn	3.0705 / 2437182.42	171. / 0.03	98.0 / 207.4	-7.1	6.16 / 2.91	4.14(+6) / 8.75(+6)	R. W. Hilditch, 1974 grade C	Cep
209791	17 ξ Cep A Kurhah	22 03.8	+64 38	4.29	Am F2III-IV	810.9 / 2438529.8	106 / 0.46	7.1 / 19.9	-10.7	0.855 / 0.305	7.03(+7) / 1.97(+8)	C. R. Vickers, 1976 grade C	Cep
209625	32 Aqr	22 04.8	-0 54	5.28	Am	7.8327 / 2421808.627 n	0.0	6.8	+20.5	2.56(-4)	7.32(+5)	R. B. Jones, 1932 grade C	Aqr
210027	24 ι Peg	22 07.0	+25 21	3.76	F5V	10.2130 / 2427364.96	240 / 0.01	49.1	-4.6	0.126	6.90(+6)	H. A. Abt, 1976 grade A	Peg
210334 eb	AR Lac	22 08.7	+45 45	6.11	F8 K2III	1.9832 / 2426623.844 n	0.0	116.1 / 115.6	-33.7	1.28 / 1.28	3.17(+6) / 3.15(+6)	R. F. Sanford, 1951 grade B	Lac
211416	α Tuc	22 18.5	-60 16	2.85	K3III	4197.7 / 2418666.4	48.5 / 0.39	7.2	+42.2	0.127	3.83(+8)	H. S. Jones, 1928 grade C	Tuc
212120	2 Lac	22 21.0	+46 32	4.57	B6IV B6V	2.6164 / 2427700.8	97.4 / 0.04	79.5 / 100.0	-8.9	0.873 / 0.694	2.86(+6) / 3.59(+6)	R. W. Hilditch, 1974 grade C	Lac
213235	37 Peg	22 30.0	+4 26	5.50	F5IV	372.4 / 2440042	208 / 0.48	8.2	0.0	0.0144	3.68(+7)	H. A. Abt, 1976 grade D	Peg

HD	Star Name	α 2000	δ 2000	Mag	Spec	Orbital Elements P T	ω e	K_1 K_2	V_0	f(m) or m sin³i	a sin i	Author Quality Const
213420	6 Lac	22h 30.5m	+43° 07′	4.48	B2 IV	880 2416300	190° n 0.30	9	−11.9	0.0578	1.04(+8)	W. R. Beardsley, 196 grade E Lac
214419 eb	CQ Cep	22 36.8	+56 54	8.8	WN6 O7	1.6410 2431038.23	n 0.0	295	−75	4.38	6.66(+6)	W. A. Hiltner, 1944 grade E Cep
215182	44 η Peg Matar	22 43.0	+30 13	2.96	G2 II−III	818.0 2415288.7	5.6 0.15	14.2	+4.3	0.235	1.58(+8)	R. T. Crawford, 1901 grade B Peg
216489	IM Peg	22 53.0	+16 50	5.65	K1 III	24.65 2422237.154	n 0.0	33.2	−12.8	0.0937	1.13(+7)	W. E. Harper, 1920 grade B Peg
216494	74 Aqr	22 53.5	−11 37	5.81	B9p	3.4298 2441989.07	66 n 0.05	95.3 115.1	−7.1 −7.5	1.81 1.50	4.49(+6) 5.42(+6)	R. J. Wolff, 1974 grade C Aqr
216598 eb	SW Lac	22 53.7	+37 55	8.65	G3 G3	0.3207 2423373.037	n 0.0	202.5 172.5	−22.5	0.808 0.948	8.93(+5) 7.61(+5)	O. Struve, 1949 grade D Lac
216916	16 Lac EN Lac	22 56.4	+41 36	5.61	B2 IV	12.097 2433841.3	104 0.04	23.0	−13.0	0.0152	3.82(+6)	W. S. Fitch, 1969 grade C Lac
217792	π PsA	23 03.5	−34 45	5.10	F0V	178.3177 2435319.73	2.6 n 0.53	21.3	−6.0	0.109	4.43(+7)	B. W. Bopp, 1970 grade C PsA
218658	33 π Cep A	23 07.9	+75 23	4.40	G2 III	556.2 2414126.33	5.7 0.28	23.0	−19.6	0.622	1.69(+8)	W. E. Harper, 1925 grade C Cep
218670	ι Gru	23 10.4	−45 15	3.88	K0 III	409.614 2416115.569	240.8 0.66	13.6	−4.2	0.0454	5.75(+7)	H. S. Jones, 1928 grade B Gru
+36 5017 eb	AB And	23 11.6	+36 54	10.4 p 11.2 p	G5 G5	0.3319 2425497.397	n 0.0	165 265	−45	1.69 1.05	7.53(+5) 1.21(+6)	O. Struve, 1950 grade E And
219113 eb	SZ Psc	23 13.4	+2 41	7.2 7.7	K1 IV F8V	3.9653 2435744.085	119 0.04	81.5 110.5	+8.8	1.67 1.23	4.44(+6) 6.02(+6)	S. Jakate, 1976 grade D Psc
219815 eb	9 And AN And	23 18.4	+41 46	6.0 6.2	A7m	3.2196 2436094.876	353.1 0.03	71.6	−3.8	0.123	3.17(+6)	B. Cester, 1959 grade B And
219834	94 Aqr	23 19.1	−13 28	5.21	G5 IV K2V	2323.6 2429308.4	225.7 n 0.08	5.5	+10.8	0.0398	1.75(+8)	M. B. K. Sarma, 1961 grade D Aqr
221253 eb	AR Cas	23 30.0	+58 33	4.88	B3V	6.0663 2440087.193	31.4 0.25	56.7	−13.4	0.104	4.58(+6)	W. L. Gorza, 1971 grade B Cas
221700 eb	Y Psc	23 34.4	+7 55	9.44 12.23	A3V K0 IV	3.7659 2425495.635	103 0.12	37	+6	0.0194	1.90(+6)	O. Struve, 1946 grade D Psc
221950	16 Psc	23 36.4	+2 06	5.66	F6V F6V	45.459 2439777.392	103.0 n 0.37	40.1 41.5	+48.4	1.05 1.01	2.33(+7) 2.41(+7)	G. C. de Strobel, 19 grade C Psc
222107	16 λ And	23 37.6	+46 28	3.88	G8 III−IV	20.5212 2429202.389	313.6 0.04	6.6	+6.8	6.11(−4)	1.86(+6)	E. C. Walker, 1944 grade A And
224113	HR 9049 AL Scl	23 55.3	−31 55	6.11	B5 IV	2.4445 2434990.964	167 n 0.07	85.8	+13.0	0.159	2.88(+6)	S. Archer, 1958 grade C Scl
224151	V373 Cas	23 55.6	+57 25	6.00	B0.5 II	13.4187 2420801.379	0 0.1	117.5 144	−24.3	13.5 11.0	2.16(+7) 2.64(+7)	R. F. Sanford, 1936 grade D Cas
224355	HR 9059	23 57.1	+55 42	5.6	dF3 dF3	12.1562 2439327.11	216.7 0.31	71.8 72.6	+10.3	1.64 1.62	1.14(+7) 1.15(+7)	M. Imbert, 1977 grade B Cas
224617	28 ω Psc	23 59.3	+6 52	4.01	F4 IV	2.158	185 0.35	6.5	−6.0	5.06(−5)	1.81(+5)	W. R. Beardsley, 196 grade E Psc

Variable Stars

Variable Stars

Column Headings

Star Name — Variable star designation according to the standard system described in the Introduction.

HD — The star's number in the *Henry Draper Catalogue* or *Henry Draper Extension*.

α 2000 and **δ 2000** — Right ascension and declination, referred to the 2000.0 equinox.

Type — Type of variability. Among the most common types are the Mira variables (M), the α^2 Canum Venaticorum stars (αCV), semiregular variables (SR), slow irregulars (L), classical Cepheids (Cδ), and eclipsing binaries of the Algol type (EA). A key to all types appears in the Introduction.

Max and **Min** — The magnitudes of the star at maximum and minimum brightness. These are the *extreme* values that have been recorded and may not represent the normally observed amplitude. For many stars, the *average* values at maximum and minimum are given in the notes following this section. Sometimes the amplitude is listed within parentheses in place of the minimum brightness. In cases where the values for maximum and minimum could not be found, the photoelectric **V** magnitude from the Yale *Bright Star Catalogue* (1982) was placed in the maximum-brightness column and labeled with the letter "Y." The symbol ">" means "fainter than." Here and in later columns, uncertain values or those qualified in some way are followed by a colon.

The magnitude system is indicated by a letter after the minimum column (and also occasionally the maximum), as follows: a photoelectric magnitude on the **UBV** system (U, B, or V), visual (v), photographic (p), yellow (y), blue (b), infrared magnitude whose system is unknown (i), and infrared photoelectric magnitude (H, I, J, K, or R). In general, both magnitudes for a given star are on the same system. If they are not, the system is implied by the number of decimal places shown. For example, if "V" is indicated, an entry to hundredths is photoelectric while one given only to tenths is either visual or an approximate photoelectric value (unless labeled otherwise). A similar convention is used when "B" is given.

Epoch — The Julian date of an observed maximum or minimum. Epochs specified to within a small fraction of a day are *heliocentric*; the correction needed for geocentric predictions, which never exceeds 0.0058 day, is described in the Introduction. For eclipsing variables and RV Tauri stars, the epoch of least light is given; otherwise the epoch of maximum is listed. The notes should always be consulted for exceptions to this rule.

Period — Period of variation in days. The average period is listed for Mira and semiregular variables.

F — The fraction of the period either taken up by the star's rise from minimum to maximum brightness, if it is an intrinsic variable, or spent in eclipse, if it is an eclipsing variable.

Spec — Spectral type of the variable.

Notes — The common name or Flamsteed designation of the star. For a nova or supernova, the year of outburst (or the most recent flare in the case of a recurrent nova) is listed.

Star Name	HD	α 2000	δ 2000	Type	Max	Min		Epoch	Period	F	Spec	Notes
Z Peg	224709	0^{h}00^m.1	+25°53'	M	7.7	13.6	v	2442410	325^d.47	0.50	M6e-M8e	
CG And	224801	0 00.7	+45 15	αCV	6.32	6.42	V	2440101.6502	3.73975		A0Vp	
WZ Cas	224855	0 01.3	+60 21	SRb	9.4	11.4	p		186		C9,2JLi(N1p)	
YY Psc	224935	0 02.0	-6 01	Lb?	4.35	4.41	V				M3III	30 Psc
W Cet	224960	0 02.1	-14 41	M	7.1	14.8	v	2438578	351.31		S6,3e-S9,2e	
TW And	—	0 03.3	+32 51	EA/SD	8.8	10.86	V	2439020.4104	4.122774	0.13	F0V + K0	
Y Cas	225082	0 03.4	+55 41	M	8.7	15.3	v	2444506	413.48	0.43	M6e-M8.5e	
SV And	225192	0 04.3	+40 07	M	7.7	14.3	v	2442887	316.21	0.38	M5e-M7e	
SU And	225217	0 04.6	+43 33	Lc	8.0	8.5					C6,4(Nb)	
V567 Cas	225289	0 05.1	+61 19	αCV	5.71	5.81	V	2440482.444	6.4322	0.50	B9pHgMn	
SW Scl	—	0 06.2	-32 49	SR	9.4	11.7	B	2425550	144		M1Ib-IIe - M4e	
TT Peg	—	0 06.5	+27 05	SRa	9.3	11.2	v	2437178	154		M6e	
XY Scl	178	0 06.6	-32 36	SRb	8.44	9.00	V		60:		M6III	
KU And	—	0 06.9	+43 05	M	6.5	10.5:	I	2442400	750:		M10I-III:	
α And	358	0 08.4	+29 05	αCV	2.02	2.06	V	2441862.126*	0.966222		B8IVpHgMn	21 And
YY Scl	393	0 08.5	-25 59	Lb	8.69	8.91	V				M6III	
V Scl	409	0 08.6	-39 14	M	8.7	15.0	v	2441263	296.15	0.48	M4e-M6e	
β Cas	432	0 09.2	+59 09	δSctc	2.25	2.31	V	2438991.876	0.10430	0.4	F2III-IV	11 Cas
SS Cas	499	0 09.6	+51 34	M	8.8	13.3	v	2444208	140.57	0.48	M3e-M8e	
SX Cas	—	0 10.7	+54 53	EA/GS	8.96	9.83	V	2439009.525	36.56375	0.098	B7IIIe + K3III	
AC Cet	672	0 11.0	-18 34	Lb	7.96	8.29	V				M5III	
WW Cet	—	0 11.4	-11 29	UG	9.3	16.8	p		31.2:		P(UG)	
γ Peg	886	0 13.2	+15 11	βC	2.80	2.87	V	2436168.371*	0.157495	0.50:	B2IV	88 Peg
UW And	—	0 14.2	+29 03	M	9.4	>14.0	v	2442777	244.2		M5	
AD Cet	1014	0 14.5	-7 47	Lb?	4.9	5.16	V				M3III	
FM Cas	—	0 14.5	+56 15	Cδ	8.82	9.47	V	2442817.713	5.809284	0.30	F7-G0I	
AE Cet	1038	0 14.6	-18 56	Lb?	4.26	4.46	V				M1III-M3III	7 Cet
UU Psc	1061	0 15.0	+8 49	E11	6.0	6.03	V	2439765.175	0.841678	0.18	F0IV + F0IV	35 Psc
S Scl	1115	0 15.4	-32 03	M	5.5	13.6	v	2442343	365.32	0.48	M3e-M8e	
AF Phe	1198	0 16.1	-48 33	Lb	7.57	7.82	V				Mb	
X And	1167	0 16.2	+47 01	M	8.3	15.2	v	2443429	346.18	0.37	S2,9e-S5,5e	
AO Cas	1337	0 17.7	+51 26	E11/KE	6.07	6.24	v	2440855.4016	3.523487		O9III + O9III	
V377 Cas	1479	0 19.2	+59 42	δSct?	7.78	7.83	V		0.03		F0	
TV Cas	1486	0 19.3	+59 08	EA/SD	7.22	8.22	V	2444602.4534	1.8125956	0.18	B9V + F7IV	
VX And	1546	0 19.9	+44 43	SRa	7.8	9.3	v	2425558	369		C4,5(N7)	
T Cet	1760	0 21.8	-20 03	SRc	5.0	6.9	v	2440562	158.9		M5-M6SIIe	
T And	1795	0 22.4	+27 00	M	7.7	14.5	v	2442959	280.76	0.46	M4e-M7.5e	
S Tuc	1925	0 23.1	-61 40	M	8.2	15.0	v	2441432	240.71	0.44	M3e-M5e	
T Cas	1845	0 23.2	+55 48	M	6.9	13.0	v	2444160	444.83	0.56	M6e-M9.0e	
R And	1967	0 24.0	+38 35	M	5.8	14.9	v	2443135	409.33	0.38	S3,5e-S8,8e(M7e)	
S Cet	1987	0 24.1	-9 20	M	7.6	14.7	v	2442650	320.45	0.47	M3e-M6.5e:	
TZ Cep	—	0 25.2	+73 54	SRd	9.0	11.0	v	2425840	83.0	0.48	G6-K2e(M2)	
B Cas	—	0 25.3	+64 09	SNI?	-4.0	>19:	v	2295550:				Tycho's SN 1572
TU Cas	2207	0 26.3	+51 17	Cep	6.88	8.18	V	2441704.839	2.139298	0.31	F3II-F5II	
V379 Cas	—	0 26.6	+60 48	Cδs	8.88	9.23	V	2444883.47	4.30575	0.45	F6-G0	
AG Phe	2320	0 26.9	-39 53	EB	8.2	8.8	p	2444170.7948	0.75533809		A3	
η Scl	2429	0 27.9	-33 00	Lb	4.7	4.90	V				M4III	
TV Psc	2411	0 28.0	+17 54	SR	4.65	5.42	V		70		M3IIIv	47 Psc
AG Cet	2438	0 28.0	-11 40	SRb	6.99	7.45	V		90:		M3	
GR And	2453	0 28.5	+32 26	αCV	6.87	6.95	V				A2pSrCrEu	
YZ Scl	2489	0 28.5	-35 43	Lb	8.45	8.78	V				M5III	
T Scl	2585	0 29.2	-37 55	M	8.5	13.5	v	2441378	201.70	0.49	M3	
DL Cas	—	0 30.0	+60 13	Cδ	8.63	9.26	v	2442780.334	8.000669	0.33	F5Ib-G2Ib	
GN And	2628	0 30.1	+29 45	δSctc	5.18	5.22	v		0.0696		A7III(A5-F0)	28 And
T Phe	2725	0 30.4	-46 25	M	8.7	14.6	v	2441350	281.23	0.38	M5e	
T Psc	—	0 32.0	+14 36	SR	9.2	12.3	v	2424848	260		M5	
TU And	2890	0 32.4	+26 02	M	7.8	13.1	v	2443820	316.77	0.48	M5e	
κ Cas	2905	0 33.0	+62 56	αCyg	4.22	4.30	B		0.09028		B1Iaeα	15 Cas
θ Tuc	3112	0 33.4	-71 16	δSct?	6.06	6.15	V		0.052:		A7IV	
ZZ Scl	3287	0 35.8	-24 53	Lb	8.01	8.33	V				M5III	
AA Scl	3373	0 36.5	-30 20	Lb	8.63	8.99	V				M5III	
AG Psc	3379	0 36.8	+15 14	βC	6.1	6.12	v		0.09165:		B2.5IV	53 Psc
Y Cep	3344	0 38.4	+80 21	M	8.1	16.0	v	2444134	332.57	0.40	M5e-M8.2e	
BB Cet	3580	0 38.5	-20 18	αCV	6.63	6.64	V	2442620.629	1.4788	0.30	B8pSi	
Z Scl	3735	0 40.0	-33 58	cst	6.68		V				F8V	
α Cas	3712	0 40.5	+56 32	cst?	2.20	2.27	V				K0IIIa	Schedar
ξ Phe	3980	0 41.8	-56 30		5.89 Y				4:		ApSrEuCr	
AI Cet	3953	0 42.0	-9 39	SRb	8.48	9.00	V		47:		M8	
V486 Cas	3950	0 42.6	+52 20	E?	6.91	(0.04)	V	2439012.27	5.551		B1III	
S And	3969	0 42.7	+41 16	SN	5.7	>16	v	2409772:			P	SN 1885 in M31
AB Scl	4226	0 44.5	-33 39	Lb	8.40	8.85	V				M5III	
EG And	4174	0 44.6	+40 41	Z And	7.08	7.8	V				M2IIIep	
o Cas	4180	0 44.7	+48 17	γC	4.50	4.62	V				B2-B5eIV	22 Cas
YZ Cas	4161	0 45.7	+74 59	EA/DM	5.71	6.12	B	2428733.4218	4.467224	0.15	A2V + F2V	21 Cas
U Cas	4350	0 46.4	+48 15	M	8.0	15.7	v	2444621	277.19	0.44	S3,5e-S8,6e	

Star Name	HD	α 2000	δ 2000	Type	Max	Min		Epoch	Period	F	Spec	Notes
XX Psc	4490	0^{h}47^m.2	+19°35'	δSct?	5.99	6.03	V	2441602.241	0^d.1040		F0Vn	59 Psc
ζ And	4502	0 47.3	+24 16	EB/GS/RS	3.92	4.14	V	2442321.05	17.7693		K1II-III	34 And
RW And	4489	0 47.3	+32 41	M	7.9	15.7	v	2443078	430.30	0.36	M5e-M10e(S6,2e)	
V And	4779	0 50.1	+35 39	M	9.0	15.2	v	2443001	257.73	0.44	M2e-M3e	
RX Cep	4499	0 50.1	+81 58	SRd?	7.2	8.2	v		55:		G5	
AZ Phe	4849	0 50.1	-43 24	δSct	5.20	5.25	V		0.055		A9-F0III	
GO And	4778	0 50.3	+45 00	αCV	6.14	(0.03)	V	2438365.03	2.156		A0p	
ρ Phe	4919	0 50.7	-50 59	δSct	5.17	5.27	V		0.10	0.50:	F2III	
V526 Cas	4818	0 51.0	+51 30	δSctc	6.34	6.37	V				F5III	
RR And	4895	0 51.4	+34 23	M	8.4	15.6	v	2443390	328.15	0.52	S6.5,2e	
RV Cas	5016	0 52.7	+47 25	M	7.3	16.1	v	2444313	331.68	0.38	M4.5e-M9.5e	
BQ Tuc	5276	0 53.6	-62 52	Lb?	5.60	5.80	V				M4III	
BM Cas	--	0 54.8	+64 05	EB/GS	8.78	9.31	V	2433859.9	197.28		A5Ia-F0Iabe	
W Cas	5235	0 54.9	+58 34	M	7.8	12.5	v	2444209	405.57	0.46	C7,1e	
AC Scl	5473	0 56.3	-25 33	SRb	7.92	8.26	V		47:		M4III	
γ Cas	5394	0 56.7	+60 43	γC(X)	1.6	3.0	v				B0.5IVpe	27 Cas
U Tuc	5774	0 57.2	-75 00	M	8.0	14.8	v	2441467	259.45	0.46	M3e-M5e	
BC Cet	5601	0 57.5	-10 29	αCV	7.64	7.66	V		1.11		A0pSi	
WW Psc	5820	0 59.8	+6 29	Lb	5.95	6.12	V				M2III	
AK Cet	--	1 00.5	-12 12	Lb	7.68	7.94	V				M3	
V551 Cas	--	1 00.6	+60 27	αCV	8.43	8.77	V	2441206	69.0		A0pSrCr	
U Cep	5679	1 02.3	+81 53	EA/SD	6.75	9.24	V	2444541.6031	2.4930475	0.15	B7Ve + G8III-IV	
CC Tuc	6311	1 02.7	-65 27		6.21 Y	(0.10)	V				M2III	
WX Psc	--	1 06.4	+12 36	M	1.0	3.0	K	2440460	650	0.4	M	
Z Cet	6592	1 06.8	-1 29	M	8.4	14.2	v	2441679	184.81	0.48	M1e-M6.5e	
V487 Cas	6474	1 07.0	+63 46	SRd	7.44	7.74	V		134		G0-G4Ia	
BS Tuc	6870	1 08.1	-61 52	δSct	7.43	7.57	V		0.065		A5III	
ζ Phe	6882	1 08.4	-55 15	EA	3.92	4.42	V	2441643.6890	1.6697671	0.12	B6V + B9V	
AL Cet	6816	1 08.6	-17 04	Lb	8.75	9.07	V				M5III	
AI Phe	6980	1 09.6	-46 16	EA	8.2	9.0	p	2443410.6885	24.5923		G0	
AM Cet	7122	1 11.3	-13 30	SRb	6.84	7.12	V		70:		M5III	
U Scl	--	1 11.6	-30 07	M	8.3	15.2	v	2441430	333.73	0.39	M5e	
RU Cas	6972	1 11.7	+65 01	cst?	5.50	5.60	V				B9IV	32 Cas
AI Scl	7312	1 12.8	-37 51	δSct?	5.93	5.98	V		0.05		F0III	
RT Psc	7307	1 13.8	+27 08	SRb	8.2	10.4	p		70		M	
UY Psc	--	1 14.2	+7 19	SRd	8.75	9.15	V		80:		K	
AN Cet	7421	1 14.2	-2 11	SRb	8.37	8.74	V		20:		M4	
U And	7482	1 15.5	+40 43	M	8.9	15.0	v	2442982	346.55	0.41	M6e	
Z Psc	7561	1 16.1	+25 46	SRb	8.8	10.1	p		144		C7,3	
VV Scl	7676	1 16.1	-34 09	EB	8.4	8.6	v	2436498.450	2.47962		A5p	
UZ And	--	1 16.2	+41 45	M	9.1	15.6	v	2442663	314.30	0.39	M7e-M9e	
VZ Cas	--	1 16.4	+56 24	M	9.3	15.0	v	2444676	169.24	0.46	M0e-M3e	
AY Cet	7672	1 16.6	-2 30	RS	5.35	5.58	V				G5III-IVe	39 Cet
UV Psc	--	1 16.9	+6 49	EB	9.4	10.4	p	2443406.5225	0.8610482		G5	
WX Cet	--	1 17.1	-17 56	UG	9.5	18	p		450:		P(UG)	
S Psc	7773	1 17.6	+8 56	M	8.2	15.3	v	2442322	405.37	0.42	M5e-M7e	
V538 Cas	--	1 18.1	+61 43	Isb	9.4	10.6	p				K5III	
V465 Cas	7733	1 18.2	+57 48	SRb	7.7	8.9	p		60		M5	
AA Cas	7861	1 19.4	+56 20	Lb	8.3	9.4	v				M6III	
AK Phe	8106	1 19.6	-47 18	Lb	7.40	7.64	V				Mb	
S Cas	7769	1 19.7	+72 37	M	7.9	16.1	v	2443870	612.43	0.43	S3,4e-S5,8e	
W Phe	8166	1 19.9	-55 55	M	8.1	14.4	v	2441452	331.20	0.42	M6e	
RU Cep	--	1 21.2	+85 08	SRd	8.2	9.8	v		109		G6-M3.5III	
XZ Cas	--	1 22.9	+61 11	Lb	9.0	9.6	v				M0	
AV Cet	8511	1 24.0	-8 00	δSctc	6.20	6.22	V		0.070		F0V	44 Cet
HN And	8441	1 24.3	+43 09	αCV	6.67	6.76	V	2441244.0	69.5	0.40:	A2pSrCrEu	
AL Phe	8729	1 25.2	-45 56	Lb?	8.33	8.7	V				Mb	
RX Psc	--	1 25.6	+21 23	M	8.8	14.6:	v	2439388	280.54	0.44	M1e	
δ Cas	8538	1 25.8	+60 14	EA?	2.68	2.76	V	2420161	759		A5V	37 Cas
R Scl	8879	1 27.0	-32 33	SRb	9.1	12.8	p	2440587	370		C6II	
γ Phe	9053	1 28.4	-43 19	?	3.39	3.49	V				M0IIIa	
WZ Scl	9065	1 28.7	-33 46	δSct	6.56	6.59	V	2440401.872	0.090		F0IV	
XX Scl	9133	1 29.4	-33 19	δSct	8.87	8.90	V	2440401.870	0.0458		A7	
RZ Per	--	1 29.7	+50 51	M	8.7	14.0	v	2442310	353.82	0.47	S4,9e	
VX Psc	9100	1 29.9	+18 21	δSct	5.90	5.92	V		0.131		A4III	97 Psc
R Psc	9203	1 30.6	+2 53	M	7.1	14.8	v	2439731	344.04	0.44	M3e-M6e	
RR Cet	9356	1 32.1	+1 21	RRab	9.10	10.10	V	2433181.404	0.55302814	0.12	A7-F5	
IZ Per	9234	1 32.1	+54 01	EA	7.8	9.0	p	2441332.440	3.687661	0.12	B8	
AE Phe	9528	1 32.6	-49 32	EW	7.7	8.4	v	2440857.8140	0.36237459		G0	
V636 Cas	9250	1 32.7	+63 36	Cδs	7.09	7.26	V	2444001.355	8.377	0.5	G0Ib	
WW Cas	--	1 33.5	+57 45	Lb	9.1	11.7	v				C5,5(N1)	
SX And	--	1 33.6	+46 31	M	8.7	>13.5	v	2442905	333.27		M6.5e	
KK And	9531	1 34.3	+37 14	αCV?	5.91	(0.012)	V	2440187.755	0.6684		B8VpSi	
RW Cas	--	1 37.2	+57 46	Cδ	8.62	9.76	V	2435575.227	14.7949	0.37	F6-G5	
V539 Cas	--	1 37.9	+62 48	Lb?	8.36	8.72	V				M2	

Star Name	HD	α 2000	δ 2000	Type	Max	Min		Epoch	Period	F	Spec	Notes
									d			
GY And	9996	1^{h}38^m.5	+45°24'	αCV	6.27	6.41	V				B9VpCrEu	
UV Cet	--	1 38.8	-17 58	UV	6.8	12.95	V				M5.5Ve	
Y And	10112	1 39.6	+39 21	M	8.2	15.1	v	2442640	220.53	0.48	M3e-M4.5e	
V557 Cas	10221	1 42.3	+68 03	αCV	5.55	5.64	V	2440974.88	3.1848	0.50	B9pSiSrCr	43 Cas
V595 Cas	--	1 43.0	+56 31	Lc	8.75	9.05	V				M2Ib	
∅ Per	10516	1 43.7	+50 41	γC	4.03	4.11	V		19.5		B2Vep	
UZ Psc	10783	1 45.7	+8 34	αCV	6.43	6.65	V	2439757.91	4.1327	0.05	A2p	
V589 Cas	--	1 46.1	+61 00	Lc	8.75	9.22	V				M3Iab-Ib	
VY Psc	10845	1 46.6	+17 25	δSct	6.54	6.59	V		0.2:		A9III	
AQ Cet	11193	1 49.9	-4 52	Lb	8.49	9.00	V				Mc	
TT Per	11094	1 50.5	+53 45	SRb	9.2	10.6	p		82		M5II-III	
VV Ari	11285	1 51.2	+20 31	δSctc	6.8	6.82	V		0.0764		F0	
V436 Per	11241	1 52.0	+55 09	E?	5.46	5.68	y				B1.5V	1 Per
η¹ Hyi	11733	1 52.6	-67 57	cst?	6.7		p				A0	
γ² Ari	11503	1 53.5	+19 18	αCV	4.62	(0.04)	V	2440998.99	2.6095	0.50	A1pCrSiSr	
ψ Phe	11695	1 53.6	-46 18	SR	4.3	4.5	V		30		M4III	
WX Cas	--	1 54.1	+61 07	Lc	9.4	11.5	v				M2Iab-Ib	
RR Ari	11763	1 55.9	+23 35	EA?	6.42:	6.84:	p	2436493.2	47.9:	0.08:	K0III	7 Ari
X Cas	--	1 56.6	+59 16	M	9.45	13.2	V	2443922	422.84	0.55	C5,4e(N1e)	
AA Cet	12180	1 59.0	-22 55	EW	6.2	6.7	p	2441268.689	0.53616996		F2	
U Per	12025	1 59.6	+54 49	M	7.4	12.3	v	2442197	321.03	0.46	M5e-M7e	
YY Cet	--	2 00.2	-18 13	EB	9	9.4	p	2427333.500	1.117455		A7	
AR Cet	12292	2 00.4	-8 31	SR?	5.40	5.61	V				M3III	
X Tri	12211	2 00.6	+27 53	EA	8.9 p	11.89	B	2441499.133	0.97153097	0.22	A3 + G3	
V Per	12244	2 01.9	+56 44	N	4.0:	>17.5	p	2410500			P(Q)	N 1887
V598 Cas	12208	2 02.0	+61 54	Lb	7.38	7.54	V				M3-M4III	
V393 Cas	12160	2 02.7	+71 18	SRa	7.0	8.0	v	2443210	393		M0	
V540 Cas	12288	2 03.5	+69 35	αCV	7.73	7.75	V	2440854.7	34.9		A0pCr	
ν For	12767	2 04.5	-29 18		4.69 Y (0.035)		V		1.89		B9.5pSi	
WZ Psc	12872	2 06.2	+8 15	SR	6.20	6.38	V		20:		M2III:	
V558 Cas	--	2 07.5	+65 29	E	9.0	9.6	B				A-F	
VX Per	--	2 07.8	+58 27	Cδ	8.99	9.69	V	2436812.137	10.89364	0.51	F6-G1	
SS For	--	2 07.9	-26 52	RRab	9.45	10.60	V	2438668.951	0.495432	0.14	A3-G0	
BX And	13078	2 09.1	+40 48	EW/DW?	8.9	9.57	p	2436528.7777	0.61011534		F2V	
V351 Per	13051	2 09.4	+57 00	βC	8.69	8.72	V		0.3746		B1III-IVe	
KK Per	13136	2 10.3	+56 34	Lc	6.6	7.78	V				M1-M3.5 Iab-Ib	
RV And	--	2 11.0	+48 57	SRa	9.0	11.5	v	2439247	171.65		M4e	
TZ Tri	13480	2 12.4	+30 18		4.94 Y (0.075)		V				G5III + F5V	6 Tri
V353 Per	13544	2 13.9	+53 55	βC	8.85	8.89	B		0.3908		B0.5IV	
V354 Per	13745	2 15.8	+56 00	βC	8.01	8.05	B		0.45039		O9.7II	
R Ari	13913	2 16.1	+25 03	M	7.4	13.7	v	2444505	186.78	0.45	M3e-M6e	
V357 Per	13866	2 17.0	+56 43	βC	7.44	7.52	V		0.28619:		B2Ib-II:p	
V358 Per	13890	2 17.1	+56 46	γC	8.50	8.59	v		1.241:	0.40:	B1IIIpe	
PP Per	--	2 17.1	+58 32	Lc?	9.2	10.30	V				M0Iab-M1.5Ib	
W And	14028	2 17.6	+44 18	M	6.7	14.6	v	2443504	395.93	0.42	S6,1e-S9,2e(M4-M1)	
AS Cet	14284	2 18.1	-14 08	Lb	7.90	8.09	V				M2	
o Cet	14386	2 19.3	-2 59	M	2.0	10.1	v	2444839	331.96	0.38	M5e-M9e	Mira
VZ Cet	--	2 19.4	-2 58	unq	9.5	12	v	2433300:	4750:	0.5	Beq	
T Per	14142	2 19.4	+58 58	SRc	8.2	9.1	v	2429300	260:		M2Iab	
V605 Cas	--	2 20.4	+59 40	Lc	8.22	8.48	V				M2Iab	
SU For	14729	2 21.6	-37 13	EA	9.5	11.0	p	2414869.805	2.4346597	0.12	A2	
SU Per	14469	2 22.1	+56 36	SRc	9.0	10.5	p		533		M3.5Iab	
S Per	14528	2 22.9	+58 35	SRc	7.9	11.5	v	2432200			M3Iae-M6Iae	
V440 Per	14662	2 23.9	+55 22	Cδ	6.28 Y (0.123)		V		7.572		F7Ib	
TZ Hor	15379	2 25.4	-66 30		6.4	6.51	V				M5III	
V559 Cas	14817	2 25.7	+61 33	EA	7.01	7.23	V	2441357.560	1.58064	0.12	B9V	
R Cet	15105	2 26.0	-0 11	M	7.2	14	v	2443768	166.24	0.43	M4e-M9	
DM Per	14871	2 26.0	+56 06	EA	7.7	8.4	p	2441920.4543	2.7277427	0.16	B6V	
AB Per	15144	2 26.0	-15 20	αCV?	5.71	5.88	V	2433226.69	2.997814	0.50	A5VpSrCr	
VW Ari	15165	2 26.8	+10 34	δSct?	6.64	6.76	V		0.149		F0IV	
S Tri	--	2 27.3	+32 44	M	8.7	>12.4	v	2435847	247.37	0.39	M2e·	
RS Cet	15329	2 28.0	+0 13	SR?	7.9	8.6	v				G	
RR Per	15186	2 28.5	+51 16	M	8.1	15.1	v	2442208	390.14	0.45	M6e-M7e	
ι Cas	15089	2 29.1	+67 24	αCV	4.45	4.53	V	2437248.313	1.74050	0.35	B9pCrSr	
R For	--	2 29.3	-26 06	M	7.5	13.0	v	2438053	387.85	0.52	Ce	
V528 Cas	--	2 29.6	+60 39	βC?	8.69	8.86	B				B5V	
V529 Cas	--	2 29.7	+60 41	γC	8.44	8.86	V				B5Veα	
TY For	15634	2 30.2	-25 11		6.49	6.51	V		0.05		dA9n	
TV Hor	15793	2 30.3	-57 49	SRb	6.6	6.81	V	2440892:	30:		M5III	
UU Ari	15550	2 30.6	+19 51	δSctc	6.10	6.15	V		0.080		A9III	26 Ari
α UMi	8890	2 31.8	+89 16	Cδ	1.92	2.07	V	2439253.01	3.969778	0.50	F7:Ib-IIv	Polaris
UX And	--	2 33.5	+45 39	SRb	8.2	9.9	v		400:		M6III	
U Cet	15971	2 33.7	-13 09	M	6.8	13.4	v	2442137	234.76	0.44	M2e-M6e	
V362 Per	15752	2 34.2	+58 24	βC?	8.22	8.24	B		0.25942:		B0III	
CC Eri	16157	2 34.4	-43 48	BY	8.70	9.03	V		1.56145		K7Ve	

Star Name	HD	α 2000	δ 2000	Type	Max	Min		Epoch	Period	F	Spec	Notes
CO Eri	16308	2^{h}35^m.6	−45°04′	EA	9.0	9.6	p	2428776.575	5^d.7836		G0	
R Tri	16210	2 37.0	+34 16	M	5.4	12.6	v	2442014	266.48	0.44	M4IIIe	
CS Eri	16456	2 37.1	−42 58	RRc	8.76	9.31	V	2438417.087	0.311331	0.47	A2	
DI Eri	16554	2 37.8	−45 37	Lb	8.76	8.98	V				K8	
UY And	16326	2 38.3	+39 11	Lb	7.4	12.3	v				C5,4(N3)	
δ Cet	16582	2 39.5	+0 20	βC	4.05	4.10	V	2438338.4763	0.16113668		B2IV	82 Cet
V482 Cas	16429	2 40.8	+61 17	βC	8.27	8.31	B		0.37822*		O9.5I-II	
DO Cas	16506	2 41.4	+60 33	EB/KE	8.39	9.01	V	2433926.4573	0.6846661		A4V	
W Tri	16682	2 41.5	+34 31	SRc	8.5	9.7	p		108		M5II	
AT Cet	16896	2 42.0	−22 36	SRb	8.08	8.32	V		60:		M5	
X For	17004	2 43.0	−26 07	SRb	9.5	10.8	p		76:		M3	
RR Cep	--	2 43.3	+81 08	M	9.0	15.5	v	2443601	384.18	0.41	M5e-M8.8e	
ST For	17166	2 44.4	−29 12	SRa	8.7	10.2	p	2426328	277		Mb(e)	
UV Ari	17093	2 45.0	+12 27	δSctc	5.18	5.22	V		0.0355		A7IV	38 Ari
RY Per	17034	2 45.7	+48 09	EA	8.5	10.7	p	2427070.708	6.8635663	0.12	B8III-V + G8IV	
TW Cas	16907	2 45.9	+65 44	EA	8.32	8.98	V	2442008.3873	1.4283240	0.16	B9V + A0:	
CU Eri	17387	2 47.0	−13 20	EW	8.0	8.6	v		0.633798		G5	
Z Eri	17491	2 47.9	−12 28	SRb	7.0	8.6	p		80		M4III	
T Ari	17446	2 48.3	+17 31	SRa	7.5	11.3	v	2443830	316.6	0.49	M6e-M8e	
SU Ari	--	2 48.5	+17 22	N	9.5	>14.5	v					N 1854
SS Cet	--	2 48.6	+1 46	EA/SD	9.4	13.0	v	2442451.329	2.973976	0.13	A0	
VY Ari	17433	2 48.7	+31 07	BY+UV	6.83	7.01	V	2442035.108	7.854		K0	
RZ Cas	17138	2 48.9	+69 38	EA/SD	6.18	7.72	V	2443200.3063	1.195247	0.17	A3V	
W Per	237008	2 50.6	+56 59	SRc	8.7	11.8	v		465		M3Ia-Iab – M5Ia:	
SU Cas	17463	2 52.0	+68 53	Cδs	5.70	6.18	V	2438000.598	1.949319	0.40	F5Ib-II – F7Ib-II	
RR Eri	17895	2 52.2	−8 16	SRb	7.4	8.6	p		97		M5III	
R Hor	18242	2 53.9	−49 53	M	4.7	14.3	v	2441490	403.97	0.40	M7IIIe	
RZ Ari	18191	2 55.8	+18 20	SRb	5.62	6.01	V		30		M6III	45 Ari
LT Eri	18296	2 57.3	+31 56	αCV	5.03	5.14	V	2439837.7*	2.88422		B9pSi	21 Per
XY Cet	18597	2 59.6	+3 31	EA/DM	8.65	9.54	V	2438372.949	2.780712	0.10	Am + Am	
CV Eri	18760	3 00.8	−2 53	Lb?	6.11	6.29	V				M2III	
T Hor	18949	3 00.9	−50 39	M	7.2	13.7	v	2441530	217.66	0.46	M5IIe	7 Eri
α Cet	18884	3 02.3	+4 05	Lb?	2.45	2.54	V				M2III	Menkar
V Hor	19285	3 03.5	−58 56	SRb	8.7	9.8	p				Mb	
CW Eri	19115	3 04.0	−17 44	EA	8.1	8.61	V	2441267.6752	2.7283737	0.10	F0	
ρ Per	19058	3 05.2	+38 50	SRb	3.30	4.0	V		50:		M4II	25 Per
V383 Per	19216	3 06.5	+33 38	αCV?	7.84	7.87	V	2439860.7	7.7		B9V	
UW Ari	19374	3 07.4	+17 53	βC	6.10	(0.13)	V	2438379.296	0.15275		B1.5V	53 Ari
V400 Per	--	3 07.6	+47 08	Nb	8	19.5	p	2442315			P(Q)	N 1974
RX Cas	--	3 07.8	+67 35	EB/GS	8.64	9.49	V	2445341.359	32.322593		K1III + A5IIIe	
β Per	19356	3 08.2	+40 57	EA(X)	2.12	3.40	V	2440953.4657	2.8673075	0.14	B8V + G5IV + Am	Algol
X Ari	19510	3 08.5	+10 27	RRab	8.97	9.95	V	2437583.570	0.6511426	0.13	A8-F4	
ST Ari	--	3 10.1	+13 26	SRb	9.0	10.6	v	2428434	99.0		M4	
U Ari	19737	3 11.0	+14 48	M	7.2	15.2	v	2442835	371.13	0.40	M4e-M9.5e	
V623 Cas	--	3 11.4	+57 54	Lb?	9.0	9.8	v				C4,5J(R5)	
SS Eri	--	3 11.8	−11 53	M	9.4	>13.0	v	2434383	314.4		M5?	
SX Ari	19832	3 12.2	+27 15	SX Ari	5.67	5.81	V	2437667.728*	0.7278925	0.25	B7pSiHeI	56 Ari
V368 Cas	19644	3 12.6	+59 55	EA	8.45	9.2	B	2444166.469	4.451642	0.12	B3III	
TW Hor	20234	3 12.6	−57 19	SRb	5.25	5.95	V		158:		C5II	
LX Per	--	3 13.3	+48 06	EA/RS	8.2	9.2	v	2427033.120	8.038207	0.07	G0V + K0IV	
CC Cas	19820	3 14.1	+59 34	EB/D	7.06	7.30	V	2443818.166	3.368753		O9IV + O9IV	
TV Cet	20173	3 14.6	+2 45	EA/DM	8.60	9.32	V	2441275.962	9.1032884	0.03	F2	
AA Per	--	3 15.2	+46 35	SRa	9.0	10.0	v	2424890.2	130.4		M6	
V423 Per	20210	3 16.0	+34 41	E11	6.16	6.28	V		5.54349		C5II	
X Cet	20646	3 19.4	−1 04	M	8.4	13.0	v	2441953	177.14	0.49	M2e(S)-M6e	
τ⁴ Eri	20720	3 19.5	−21 45	Lb?	3.59	3.72	V				gM3	16 Eri
BK Cam	20336	3 20.0	+65 39	γC	4.78	4.89	V		1640		B2Ve	
UZ Per	--	3 20.1	+32 01	SR	9.2	11.5	p		927		M5II-III	
W Ari	--	3 20.8	+28 57	N?	9.5	>20	v					N 1855
VW Eri	--	3 21.5	−21 27	SRd	9.2	11.0	p		83.4		K5	
UX Ari	21242	3 26.6	+28 43	RS	6.26	6.62	V		6.43791		G5V + K0IVeα	
Y Per	21280	3 27.7	+44 11	M	8.1	11.1	v	2440499	252.61	0.48	C4,3e	
R Per	21567	3 30.1	+35 40	M	8.1	14.8	v	2442416	209.97	0.49	M2e-M5e	
TU Hor	21981	3 30.6	−47 23	E?	5.91	6.04	V				A1V	
GK Per	21629	3 31.2	+43 54	Na(X)	0.2	14.0	v	2415439.4			K2IVp + WD	N 1901
V396 Per	21699	3 32.1	+48 01	αCV	5.45	5.53	V	2441619.956	2.4761	0.5	B8IIIpMn	
AS Eri	21985	3 32.4	−3 19	EA	8.33	9.00	V	2428538.066	2.664152	0.09	A2	
KP Per	21803	3 32.7	+44 51	βC	6.37	6.47	V	2438671.996	0.201753	0.50	B2IV	
IW Per	21912	3 33.6	+39 54	E11	5.80	5.85	V	2433617.317	0.9171877		A5m	
RT Eri	22228	3 34.2	−16 10	M	8.5	12.9	p	2429360	370.8	0.46	M7e	
BT Eri	223364	3 34.5	−39 25	EA	8.5	9.2	p	2438400.741	2.112299	0.13	A0	
IX Per	22124	3 35.0	+32 01	E11	6.6	6.62	v	2429146.901	1.326363		F2III-IV	
EG Eri	22470	3 36.3	−17 28		5.23 Y	(0.06)	V		1.93		B9p	20 Eri
V711 Tau	22468	3 36.8	+0 35	E/RS(X)	5.70	5.81	V		2.83782		G5IV + K1IV	
V486 Tau	22374	3 37.0	+23 13	αCV	6.65	6.78	V	2441252.12	10.61		A2p	

Star Name	HD	α 2000	δ 2000	Type	Max	Min		Epoch	Period	F	Spec	Notes
									d			
RR Hor	23126	3^{h}40^m.2	−52°44'	cst	9.0		p				G	
VY Eri	23001	3 41.3	−10 45	SRb	9.2	10.6	p		102.5:		M5	
BD Cam	22649	3 42.2	+63 13	Lb	5.04	5.17	V				S5,3(M4III)	
AF Per	278890	3 42.4	+36 31	SR	9.5	10.7	v		89		M6	
δ Per	22928	3 42.9	+47 47	αCV?	2.98	3.03	V		1:		B5III	39 Per
V624 Tau	23156	3 43.7	+24 22	δSct	8.50	8.51	V		0.024		A7V	
o Per	23180	3 44.3	+32 17	E11	3.79	3.85	V	2436459.0*	4.419171		B1III + B1V	38 Per
CH Tau	--	3 45.0	+9 55	SRb	9.4	10.7	v	2428733*	97		M6.5	
π Eri	23614	3 46.1	−12 06	Lb?	4.38	4.44	V				M2IIIab	26 Eri
S For	23686	3 46.2	−24 24	cst?	5.6	8.5	v				F8	
V647 Tau	23607	3 47.3	+24 08	δSct	8.10	8.11	V		0.049		A7V	
V650 Tau	23643	3 47.4	+23 41	δSct	7.77	7.8	V		0.031		A3V	
RX Ret	24308	3 47.7	−66 42	SRd	9.1	11.2	v				K0	
BR Eri	23937	3 48.8	−7 01	SRb	8.7	10.0	p		175.5		M5	
V376 Per	23728	3 49.1	+43 58	δSct	5.8	5.91	V		0.091		A9IV	
BU Tau	23862	3 49.2	+24 08	γC	4.77	5.50	V				B8Vpe	Pleione
BU Eri	24082	3 49.3	−20 53	EA	8.5	9.0	p	2432244.33			A3	
V467 Per	23848	3 49.5	+33 05		5.11 Y						A3V	42 Per
BE Cam	23475	3 49.5	+65 32	Lc	4.35	4.48	V				M2II	
SS Cep	22689	3 49.5	+80 19	SRb	8.0	9.1	p		90		M5III	
XY Per	275877	3 49.6	+38 59	Ina	9.2	10.6	p				B6 + A2II	
U Eri	24220	3 50.5	−24 57	M	8.5	14.9	v	2440110	274.91	0.48	M4e	
V766 Tau	24155	3 51.3	+13 03		6.30 Y (0.06)		y		2.53		B9pSi	
RX Per	279135	3 51.3	+33 02	M	9.0	>13.0	v	2438817	421.5		M3e	
SU Eri	24244	3 51.5	−1 22	SRb	9.5	10.7	p		112		M4III	
BV Eri	24327	3 51.9	−10 32	EB?	8.6	9.0	p	2431180.49			F2	
U Hor	24607	3 52.8	−45 50	M	7.8	>15.1	p	2434638	348.4		M6IIIe	
X Tau	24400	3 53.2	+7 46	cst	8.4		p				F5	
RW Cam	--	3 54.4	+58 39	Cδ	8.20	9.10	V	2437389.57	16.41437	0.34	F8Ib(F5-G1) + A:	
V479 Tau	24550	3 54.5	+5 10	δSct	7.41	7.60	V		0.076		F3II-III	
T Eri	24754	3 55.2	−24 02	M	7.4	13.2	v	2439046	252.24	0.45	M3e-M5e	
DO Eri	24712	3 55.3	−12 06	αCV	5.97	6.00	V	2440578.0*	12.448	0.5	A5p	
X Per	24534	3 55.4	+31 03	γC(X)	6.07	7.0	V				O9.5ep	
DL Eri	24832	3 56.6	−9 45	δSct	6.17	6.24	V		0.1562		F1V	
γ Eri	25025	3 58.0	−13 31		2.95 Y (0.03)		V				M0.5IIICaCr	34 Eri
V386 Per	24809	3 58.1	+34 49	δSct	6.50	6.58	V		0.052		A8V	
IQ Per	24909	3 59.7	+48 09	EA	7.72	8.27	V	2440222.5974	1.7435673	0.12	B8Vp: + A0Vnp:	
τ⁹ Eri	25267	3 59.9	−24 01		4.66 Y						B6V + B9.5V	36 Eri
XY Dor	25470	4 00.3	−51 34		6.51 Y (0.09)		V				M1III	
λ Tau	25204	4 00.7	+12 29	EA	3.3	3.80	p	2435089.204	3.952955	0.15	B3V + A4IV	35 Tau
γ Ret	25705	4 00.9	−62 10	Lb?	4.42	4.64	V		25		M4III	
WW Tau	281505	4 01.7	+30 15	SRd	9.0	12.9	p	2428039	116.4		G2e-K2	
V380 Per	25354	4 03.2	+38 03	αCV	7.71	7.79	V	2437315.14*	3.9007		A3p	
CX Eri	25761	4 03.4	−39 23	SRb	8.89	9.43	V	2440666	97		Ma	
DP Eri	25675	4 03.5	−24 28	SRb	7.18	7.34	V		30:		M5III	
√RW Tau	25487	4 03.9	+28 08	EA	7.98	11.47	V	2442776.9313	2.7688396	0.14	B8Ve + K0IV	
V Eri	25725	4 04.3	−15 43	SRc	8.8	10.4	p		97		M6II	
RX Cam	25361	4 05.0	+58 40	Cδ	7.30	8.07	V	2442766.583	7.912024	0.28	F6Ib-G2Ib	
UV Cam	25408	4 05.9	+61 48	SRb	7.5	8.1	v		294:		C5,3(R8)	
CY Eri	25921	4 05.9	−10 18	SRb	6.95	7.20	V		25		gM4	
GS Tau	25823	4 06.6	+27 36	αCV	5.15	5.22	V	2421944.74*	7.227424	0.50	B9pSi	41 Tau
AG Per	25833	4 06.9	+33 27	EA	6.71	7.00	V	2441673.385	2.02870517	0.12	B4 + B5V	
CZ Eri	26231	4 07.4	−39 30	SRb	8.75	8.97	V		50		Ma	
SZ Cam	25638	4 07.9	+62 20	EA/DM	7.0	7.29	B	2441665.2516	2.6985439	0.17	O9.5V + B0	
TV Tau	--	4 08.5	+26 51	SR	9.3	11.9	v	2425870	120	0.54	M6	
XX Cam	25878	4 08.6	+53 22	RCB?	8.09	9.8	B				G1I(C0-2,0)	
MX Per	25940	4 08.7	+47 43	γC	4.00	4.10	V		55:		B3Ve	48 Per
DD Eri	26258	4 08.9	−8 06	SRb	8.60	8.9	v		100:		Mc	
VW Hyi	--	4 09.1	−71 18	UG	8.4	14.4	v		27.8:			
U Men	--	4 09.6	−81 51	M	8.0	10.9	p	2436065	407		Me	
IM Tau	26322	4 10.8	+26 29	δSct	5.37	5.44	V		0.144923		F2IV-V	44 Tau
DE Eri	26535	4 11.1	−20 03	SRb	8.91	9.06	V		150:		Mb	
W Eri	26601	4 11.5	−25 08	M	7.5	14.5	v	2440345	376.63	0.40	M7e	
o¹ Eri	26574	4 11.9	−6 50	δSct	4.00	4.05	V		0.0815		F2II-III	38 Eri
BZ Eri	--	4 12.2	−6 02	EA	9.0	10.0	p	2425558.445	0.6641704	0.12	F2	
YY Eri	26609	4 12.2	−10 28	EW	8.80	9.50	B	2439187.1216	0.32149510		G5 + G5	
DF Eri	26832	4 13.1	−36 33	SRb	8.21	8.33	V	2440643	36		Ma	
BM Eri	26750	4 13.5	−10 23	EA	9.0	9.8	p	2431140	>20000	<0.02	gM6	
WW Dor	27002	4 13.6	−50 35	SRb	8.66	8.87	V	2440625	52		Ma	
WX Dor	27199	4 15.1	−53 01	SRb	8.76	8.91	V		80:		Mb	
DY Eri	26965	4 15.3	−7 39	UV	4.43 Y						K1V + M4Ve	40 Eri
V774 Tau	26923	4 15.5	+6 11		6.31 Y						G0IV	
γ Dor	27290	4 16.0	−51 29	EW?	4.23	4.27	V				F2-F5 IV-V	
SY Per	--	4 16.6	+50 38	SRa	9.5	12.5	v	2430525	476.2		C6,4e	
ZZ Cam	26763	4 17.7	+62 21	Lb	8.7	9.3	p				M0-M5	

Star Name	HD	α 2000	δ 2000	Type	Max	Min		Epoch	Period	F	Spec	Notes
RS Eri	—	4ʰ17ᵐ.9	−18°31′	M	9.2	>12.8	p	2438048	296ᵈ.00		Me	
b Per	26961	4 18.2	+50 18	E11	4.59	4.65	V	2422780.433	1.527365		A2V	
V724 Tau	27309	4 19.6	+21 46		5.38 Y	(0.07)	V		2.7098		A0pSi	56 Tau
V483 Tau	27397	4 20.0	+14 02	δSct	5.56	5.59	V		0.054		F0IV	57 Tau
DQ Eri	27498	4 20.3	−2 38	SRb	6.84	7.05	V		30:		M4III	
V696 Tau	27459	4 20.6	+15 06	δSct	5.20	5.26	V		0.036		F0V	58 Tau
DG Eri	27598	4 20.7	−16 50	SRc	6.94	7.14	V	2440623:	82		M5II	
TW Cam	—	4 20.8	+57 26	RVb	8.98	10.27	V	2438607.8	87.22		F8Ib−G8Ib	
V469 Per	27396	4 21.6	+46 30		4.85 Y	(0.01)	V		0.303:		B4IV	53 Per
T Tau	284419	4 22.0	+19 32	InT	8.4	13.5	v				dGe−K1e(T)	
RY Tau	283571	4 22.0	+28 27	InT	9.3	13.0	p				dF8e−dG5e(T)	
V775 Tau	27628	4 22.1	+14 05		5.72 Y						A3m	60 Tau
DH Eri	27957	4 23.7	−27 50	SRb	8.13	8.33	V		60		Mb	
V776 Tau	27962	4 25.5	+17 56		4.29 Y	(0.02)	U		57.25		A2IV	68 Tau
V777 Tau	28052	4 26.3	+15 37		4.49 Y						F0V	71 Tau
DF Tau	—	4 27.0	+25 42	I	9.2	15.0	p				dM1e(T)	
W Tau	28236	4 28.0	+16 03	SRb	9.1	13.0	v	2412857	260.65	0.53	M5	
R Tau	28309	4 28.3	+10 10	M	7.6	14.7	v	2442357	323.72	0.41	M5e−M9e	
θ² Tau	28319	4 28.7	+15 52	δSct?	3.40 Y	(0.03)	V		0.07		A7III	78 Tau
DU Eri	28497	4 29.1	−13 03		5.60 Y	(0.08)	V				B1Vne	
RV Cam	28257	4 30.7	+57 25	SRb	9.3	10.6	p	2428861	101		M4II−III − M6	
RY Cam	28168	4 30.8	+64 27	SRb	8.9	11.0	p	2439238	135.75		M3III	
DZ Eri	28843	4 32.6	−3 13	δSct?	5.81 Y	(0.08)	y		1.374*		B9III	
R Ret	29383	4 33.5	−63 02	M	6.5	14.0	v	2440585	278.28	0.48	M4e−M6.5e	
ρ Tau	28910	4 33.8	+14 51		4.65 Y						A8V	86 Tau
EH Eri	29009	4 33.9	−6 44		5.72 Y	(0.045)	V		3.82		B9pSi	46 Eri
α Dor	29305	4 34.0	−55 03		3.27 Y	(0.03)	V		2.95		A0IIISi	
DV Eri	29064	4 34.2	−8 14		5.11 Y	(0.02)	p				M3III	47 Eri
α Tau	29139	4 35.9	+16 31	Lb	0.75	0.95	V				K5III	Aldebaran
v Eri	29248	4 36.3	−3 21	βC	3.4	3.6	p	2433629.277	0.17790414		B2III	48 Eri
R Dor	29712	4 36.8	−62 05	SRb	4.8	6.6	v		338:	0.30	M8IIIq:e	
SZ Tau	29260	4 37.2	+18 33	Cδ	6.37	6.71	V	2434628.57	3.14873	0.45	F5Ib−F9.5Ib	
RX Tau	29411	4 38.2	+8 20	M	9.1	14.8	v	2441618	335.06	0.47	M7e	
HU Tau	29365	4 38.3	+20 41	EA	5.92	6.7	V	2442412.456	2.056302	0.16	A8V	
VY Tau	—	4 39.3	+22 48	IsT?	9.0	14.5	p				e(T)	
SV Pic	29906	4 40.0	−52 24	SRb	8.99	9.17	V	2440666	60		Mb	
T Cam	29147	4 40.1	+66 09	M	7.3	14.4	v	2443433	373.20	0.47	S4,7e−S8.5,8e	
DM Eri	29755	4 40.4	−19 40	SRb	4.28	4.36	V		30		M4III	54 Eri
R Cae	29844	4 40.5	−38 14	M	6.7	13.7	v	2440645	390.95	0.41	M6e	
DW Eri	30020	4 43.6	−8 48	δSct?	6.82 Y	(0.025)	V				F4IIIpSr	
RZ Eri	30050	4 43.8	−10 41	EA/RS	7.79	8.71	V	2423854.306	39.28244	0.05	A5−F5Vm + sgG8	
DX Eri	30076	4 44.1	−8 30		5.90 Y	(0.09)	V				B2Ve	56 Eri
X Cam	29384	4 45.7	+75 06	M	7.4	14.2	v	2444679	143.56	0.49	K8−M8e	
R Pic	30551	4 46.2	−49 15	SRa	6.7	10.0	v	2438091	164.2		M1IIe−M4IIe	
T Cae	30593	4 47.3	−36 13	SR	9.0	10.8	p	2427840	156		C6,4(N4)	
SU Dor	—	4 47.5	−55 41	M	8.5	>14.0	v	2440741	235.86			
AW Per	30282	4 47.8	+36 43	Cδ	7.04	7.85	V	2437040.488	6.46342	0.25	F6−G0	
KS Per	30353	4 48.9	+43 17	SR	7.60	7.85	V		30:		A5Iaep	
V473 Tau	30466	4 49.3	+29 34	αCV	7.22	7.4	V	2439870.63	1.39:		B9p	
SV Per	—	4 49.8	+42 17	Cδ	8.58	9.36	V	2439076.24	11.12875	0.39	F6−G1	
Y Cae	31036	4 50.8	−40 13	Lb	8.74	8.86	V				M3(III)	
ST Cam	30243	4 51.2	+68 10	SRb	9.2	12.0	p		300:		C5,4(N5)	
V480 Tau	30780	4 51.4	+18 50	δSct	5.09	5.11	V		0.042		A7IV−V	97 Tau
V Tau	30868	4 52.0	+17 32	M	8.5	14.6	v	2442102	169.80	0.45	M0e−M4e	
o¹ Ori	30959	4 52.5	+14 15	Lb	4.65	4.88	V		30		S3.5	4 Ori
Z Cae	31311	4 52.6	−43 04	SR	7.84	7.99	V	2440667	52		M2III	
π⁵ Ori	31237	4 54.3	+2 26	E11	3.70	3.77	V	2417922.565	3.70045		B3III + B0V	8 Ori
R Eri	31444	4 55.3	−16 25	cst?	5.72		v				gG4	
SX Eri	—	4 55.4	−6 55	M	9.5	>13.6	v	2438325	282.80			
AB Aur	31293	4 55.8	+30 33	Ina	6.90	8.4	V				B9IV−Vneq	
SU Aur	—	4 56.0	+30 34	Inbs	9.3	11.8	p				F5−G2IIIne	
V429 Ori	—	4 56.2	−3 32	?	9	10	v				G4	
R Ori	31798	4 59.0	+8 08	M	9.1	13.4	v	2441391	378.06	0.40	Ne(C8,1e)	
R Lep	31996	4 59.6	−14 48	M	5.5	11.7	v	2440800	432.13	0.55	C6IIe	
S Eri	32045	4 59.9	−12 32	δSct?	4.78	4.80	V				F0IV	64 Eri
RX Aur	31913	5 01.4	+39 58	Cδ	7.28	8.02	V	2439075.63	11.623515	0.49	F6−G2	
ε Aur	31964	5 02.0	+43 49	EA/GS	2.92	3.83	V	2435629	9892	0.08	A8Ia−F2Iaep	7 Aur
ζ Aur	32068	5 02.5	+41 05	EA/GS	3.70	3.97	V	2427692.825	972.160	0.041	K5II + B7V	8 Aur
X Cae	32846	5 04.4	−35 42	δSctc	6.28	6.39	V	2439395.570	0.1352227	0.50	F1III−F2IV	
UX Ori	293049	5 04.5	−3 47	Isa	8.7	12.8	p				A2e	
V1032 Ori	32549	5 04.6	+15 24		4.68 Y	(0.03)	y		4.63		A0pSi	11 Ori
T Lep	32803	5 04.8	−21 54	M	7.4	13.5	v	2434887	368.13	0.47	M6e−M8e	
BF Aur	32419	5 05.1	+41 17	EB	8.79	9.51	V	2441752.456	1.583217		B5Vn + B5Vn	
W Ori	32736	5 05.4	+1 11	SRb	8.6	11.1	p		212		N5(C5,4)	
V Ori	—	5 06.1	+4 06	M	8.9	14.7	v	2438820	267.71	0.49	M3e−M5e	

Star Name	HD	α 2000	δ 2000	Type	Max	Min		Epoch	Period	F	Spec	Notes
HZ Aur	32633	5ʰ06ᵐ1	+33°55′	αCV	7.02	7.13	V	2439499.9	6ᵈ42929	0.20	B9pSiCrEu	
BM Cam	32357	5 06.2	+59 01	RS	6.18	(0.140)	V	2444288.8	82.8		K0III	12 Cam
V430 Ori	––	5 06.6	+0 33	SRb	9.4	10.8	p		104.5		M3–M5	
WZ Dor	33684	5 07.6	–63 24	SRb	5.1	5.28	V		40		M3III	
λ Eri	33328	5 09.1	–8 45	βC	4.27 Y	(0.09)	V		0.24		B2IVne	69 Eri
TX Aur	33016	5 09.1	+39 00	Lb	8.5	9.2	v				C5,4(N3)	
TT Aur	33088	5 09.7	+39 35	EB/DM	8.59	9.5	B	2441682.3969	1.3327333		B2Vn + B5:	
U Dor	271044	5 10.2	–64 19	M	9.2	>15.0	p	2429850	394.4		M8IIIe	
UX Cam	32730	5 10.8	+68 40	Lb	9.5	10.5	p				M6	
S Pic	33894	5 11.0	–48 30	M	6.5	14.0	v	2439595	426.61	0.36	M6.5e–M8(II–III)e	
NV Aur	––	5 11.3	+52 52	M	3.6	6.2	H	2441023	635		M10	
RX Lep	33664	5 11.4	–11 51	Lb	5.0	7.0	v				M6III	
SX Aur	33357	5 11.7	+42 10	EB/KE?	8.38	9.14	V	2442403.4045	1.2100774		B3V + B5V	
BN Cam	32650	5 12.4	+73 57	αCV	5.3	(0.05)	v	2441252.91	0.7325		A0pSi	
SY Aur	––	5 12.7	+42 50	C5	8.75	9.38	V	2436843.52	10.14452	0.42	F5–F8	
μ Lep	33904	5 12.9	–16 12	αCV?	2.97	3.36	V		2:		B9pHgMn	5 Lep
T Pic	34448	5 15.1	–46 55	M	7.9	14.4	v	2439478	200.61	0.49	M6IIIe	
KW Aur	33959	5 15.4	+32 41	δSctc+E11	4.95	5.08	V		0.088088	0.50	A9IV	14 Aur
IM Aur	33853	5 15.5	+46 24	EA	7.90	8.51	V	2440515.5465	1.247296	0.20	B7V	
V431 Ori	––	5 15.9	+11 58	SRb	9.3	11.1	v	2430033	122:		N(C5,5)	
UX Aur	33877	5 16.0	+49 33	SRc	9.5	10.6	p		90:		M4II	
AE Aur	34078	5 16.3	+34 19	Ina	5.78	6.08	V				O9.5V	
α Aur	34029	5 16.7	+46 00	RS	0.06	(0.15)	V		104.023		G0III + G5III	Capella
R Aur	34019	5 17.3	+53 35	M	6.7	13.9	v	2444004	457.51	0.51	M6.5e–M9.5e	
CD Tau	34335	5 17.5	+20 08	EA	7.27	7.9	B	2437253.345	3.435137	0.08	F7V + F7V	
S Dor	35343	5 18.2	–69 15	S Dor	8.6	11.7	p				A0eq	
AR Aur	34364	5 18.3	+33 46	EA/DM	6.15	6.82	V	2438402.1832	4.134695	0.07	ApHgMn + B9V	17 Aur
EO Aur	34333	5 18.3	+36 38	EA/DM?	7.56	8.13	V	2421190.7414	4.06563724	0.12	B3V + B3V	
PU Aur	34269	5 18.3	+42 48	Lb?	5.64	(0.10)	V				M4III	
RZ Lep	34738	5 18.8	–22 13	SRb	8.33	8.44	V		10:		M0	
IQ Aur	34452	5 19.0	+33 45	αCV	5.35	5.43	V	2437295.88*	2.4660	0.22	B9VpSi	
T Col	34897	5 19.3	–33 42	M	6.6	12.7	v	2441973	225.84	0.50	M3e–M6e	
XX Tau	––	5 19.4	+16 43	Na	6.0	16.5	p	2425154.7			P(Q)	N 1927
MZ Aur	34626	5 20.5	+36 38	SX Ari?	8.1	8.19:	B		0.5:		B2Vnpeα	
UV Aur	––	5 21.8	+32 31	M	7.4	10.6	v	2441062	394.42		C6,2–C8,2Jep(Ne)	
SW Col	35515	5 23.4	–39 41	Lb?	5.71	(0.34)	V				M1III	
η Ori	35411	5 24.5	–2 24	EB	3.14	3.35	B		7.98926		B1V + B2e	28 Ori
R Oct	40857	5 26.1	–86 23	M	6.4	13.2	v	2441172	405.57	0.44	M5.5e	
DH Ori	290428	5 26.2	–0 17	SRa	9.4	11.7	v	2426727	165.3		M0e	
ψ Ori	35715	5 26.8	+3 06	E11	4.31	4.34	B		2.52588:		B2IV	30 Ori
W Aur	281118	5 26.9	+36 54	M	8.0	15.3	v	2443165	274.27	0.41	M3e–M8e	
TY Men	37909	5 26.9	–81 35	EW	7.7	8.2	p	2441353.9862	0.461667		A3	
S Aur	35556	5 27.1	+34 09	SR	8.2	13.3	v	2442000	590.1	0.59	C4–5,4–5(N3)	
IU Aur	35652	5 27.9	+34 47	EB/SD	8.19	8.83	V	2438448.4063	1.81147536		B0p + B1Vp	
QZ Aur	––	5 28.7	+33 19	Na	6.0	18.0	p	2438440			P(Q)	N 1964
S Ori	36090	5 29.0	–4 42	M	7.5	13.5	v	2438062	419.20	0.48	M6.5e–M8e	
CI Ori	36167	5 29.7	–1 06	cst	4.70		V				K5III	31 Ori
LY Aur	35921	5 29.7	+35 23	EB/SD?	6.66	7.35	V	2439061.4640	4.0024943		O9.5 + O9.5II–IIIeα	
AS Cam	35311	5 29.8	+69 30	EA/DM	8.57	9.19	V	2440204.5137	3.4309714	0.08	B8V + B9	
UX Men	37513	5 30.0	–76 15	EA	8.8	9.2	p	2428778.650	4.181100		G1V + G1V	
TZ Men	39780	5 30.2	–84 47	EA	6.2	6.9	p	2438196.370	8.569	0.04	B9.5IV–V	
CK Ori	36217	5 30.3	+4 12	SR?	5.9	7.1	v		120:		K2IIIv	
V451 Ori	––	5 31.5	+11 01	γC	8.5	9.5	p				B9e	
δ Ori	36486	5 32.0	–0 18	EA?	1.94	2.13	B	2419068.20	5.732476	0.13	B0III + O9V	Mintaka
T Aur	36294	5 32.0	+30 27	Nb+EA	4.1	15.5	B	2412088			P(Q)	N 1891
CE Tau	36389	5 32.2	+18 36	SRc	6.1	6.5	p		165		M2Iab–Ib	119 Tau
VV Ori	36695	5 33.5	–1 09	EB	5.1	5.5	p	2440545.8994	1.48537769		B1IV + A0:	
β Dor	37350	5 33.6	–62 29	Cδ	3.46	4.08	V	2435206.44	9.84200	0.49	F4Ia–G4Iab	
CM Tau	––	5 34.5	+22 01	SN(X)	–6.0: v	15.9	p	2106216:			CONT	SN 1054 (Crab)
WX Men	37993	5 34.7	–73 44		5.78 Y	(0.15)	V				M3III	
V372 Ori	36917	5 34.8	–5 34	Ina	7.94	8.05	V				A0V	
KX Ori	36958	5 35.1	–4 44	Ina?	6.9	8.1	p				B2IV	
LP Ori	36982	5 35.2	–5 28	Inas?	7.8	9.2	p				B1.5Vp	
V1016 Ori	37020	5 35.3	–5 23	EA	6.73	(0.8)	V	2443144.62	65.4325		O7	θ¹ Ori A
BM Ori	37021	5 35.3	–5 23	EA	7.95	8.52	V	2440265.343	6.470525	0.12	B2V + A5:	θ¹ Ori B
V1046 Ori	37017	5 35.4	–4 30		6.56 Y						B1.5V	
NU Ori	37061	5 35.5	–5 16	Inas?	6.83	6.93	V				B1V	
V361 Ori	37062	5 35.5	–5 25	Inas	8.16	8.24	V				B4V	
NV Ori	––	5 35.5	–5 33	Inbs	8.7	11.3	v				F0III–IV – F6V	
RR Cam	––	5 35.5	+72 28	SRa	9.5	11.3	v	2437750	123.88	0.44	M6	
V359 Ori	37058	5 35.6	–4 50	Inas	7.25	7.36	V				B3Vp	
ι Men	38602	5 35.6	–78 49		6.05 Y	(0.01)	V		2.644		B8III	
RV Col	––	5 35.7	–30 50	SRd	9.3	10.3	p	2427800	105.7	0.40:	G5	
T Ori	––	5 35.8	–5 29	Inas	9.5	12.6	v				B8–A3epV	
CQ Tau	36910	5 36.0	+24 45	Inas	8.7 p	12.55	B				A1–F5IVe	

Star Name	HD	α 2000	δ 2000	Type	Max	Min		Epoch	Period	F	Spec	Notes
									d			
ε Ori	37128	5ʰ36ᵐ.2	−1°12'	I?	1.68	1.71	V				B0Iae	Alnilam
V380 Ori	—	5 36.4	−6 43	InT	8.2	10.81	B				B8−A2eq(T)	
BN Ori	245465	5 36.5	+6 50	Inas	9.0	13.7	p				A9pe	
ζ Tau	37202	5 37.6	+21 09	γC	2.90	3.03	V				B4IIIpe	123 Tau
ET Tau	245523	5 37.7	+27 16	EA	9.1	10.1	p	2429362.476	5.996879	0.18	B8	
V1030 Ori	37479	5 38.8	−2 36	E	6.65 Y (0.10)		y		1.19081		B2Vp	
ω Ori	37490	5 39.2	+4 07	γC	4.40	4.59	V				B3IIIe	47 Ori
HH Aur	37386	5 39.5	+29 50	Inbs?	9.2	10.5	p				G6IV	
RU Aur	—	5 40.1	+37 38	M	9.0	16.0	v	2442520	466.47	0.40	M7e−M9e	
NO Aur	37536	5 40.7	+31 55	Lc	6.10	6.30	V				M2SIab	
V1051 Ori	37808	5 40.8	−10 25		6.52 Y (0.03)		v		1.099		B9.5IIIpSi	
V901 Ori	37776	5 40.9	−1 30	βC?	6.82	6.85	B		0.37968:		B2V	
S Cam	36972	5 41.0	+68 48	SRa	7.7	11.6		2443360	327.26	0.51	C7,3e(R8e)	
FX Ori	246473	5 41.9	+14 50	SR	8.2	10.4	v		720		M3	
U Aur	37724	5 42.1	+32 02	M	7.5	15.5	v	2443579	408.09	0.39	M7e−M9e	
V356 Aur	—	5 42.5	+29 00	δSct	8.01	8.12	V		0.18916	0.40	F4IIIp	
R Men	39247	5 42.7	−75 15	SRb	9.5	12.0	p	2430150			Mb	
V731 Tau	37967	5 43.3	+23 12		6.21 Y						B2.5Ve	
V351 Ori	38238	5 44.3	+0 09	Inas	8.3	11.6	p				A7III	
ST Tau	38262	5 45.1	+13 35	CWb	7.79	8.56	V	2436916.891	4.034269	0.30	F5−G5	
TU Tau	38218	5 45.2	+24 25	SR?	5.9	8.6	v		190:		N3(C5,4)	
EU Tau	38321	5 45.7	+18 39	Cδ	7.98	8.31	V	2441324.22	2.10250	0.40	G5	
S Col	—	5 46.9	−31 42	M	8.9	14.2	v	2440559	325.85	0.46	M6e−M8	
RY Lep	38882	5 48.1	−20 01	EA	8.2	9.1	v	2438315.595			F0	
SU Tau	247925	5 49.1	+19 04	RCB	9.1	16.0	v				G0ep(C1,0)	
R Col	39324	5 50.5	−29 12	M	7.8	15.0	v	2442016	327.62	0.39	M3e−M7	
RS Lep	248515	5 51.8	+15 53	cst	9.0		v				F5	
Z Tau	—	5 52.4	+15 48	M	9.2	14.2	v	2440645	490.7	0.41	S7.5,1e	
V593 Tau	39340	5 53.1	+26 27	γC	8.13	8.32	V				B3Vev	
λ Col	39764	5 53.1	−33 48	E11?	4.85	4.92	V		0.640	0.35	B5V	
TU Cam	39220	5 55.0	+59 53	EB/DM	5.12	5.29	V	2438051.375	2.933241		A0IV−V	31 Cam
BH Cam	—	5 55.0	+64 59	Lb	2.37	2.53	K				M8	
α Ori	39801	5 55.2	+7 24	SRc	0.40	1.3	V		2110		M1−M2 Ia−Iab	Betelgeuse
U Ori	39816	5 55.8	+20 10	M	4.8	12.6	v	2442280	372.40	0.38	M6.5IIIe	
TW Aur	39783	5 57.1	+45 31	SRb	9.1	10.6	p		150:		M5III	
V1004 Ori	40372	5 58.4	+1 50	δSct	5.88	5.89	v		0.054		A5m	59 Ori
V474 Mon	40535	5 59.0	−9 23	δSct	5.93	6.36	V	2440593.526	0.13494		F2IV	1 Mon
RS Lep	40640	5 59.3	−20 13	EA	9.3	10.9	p	2436191.148	1.2885439	0.17	A2V	
V642 Ori	40490	5 59.5	+9 15	EA	8.8	9.1	p	2427126.360	2.757240	0.17	A0	
β Aur	40183	5 59.5	+44 57	EA/D	1.89	1.98	V	2431076.719	3.9600421	0.06	A2IV + A2IV−V	34 Aur
θ Aur	40312	5 59.7	+37 13	αCV	2.62	2.70	V		1.3735		B9.5pSi	37 Aur
π Aur	40239	5 59.9	+45 56	Lc	4.24	4.34	V				M3.5II	35 Aur
V529 Ori	—	6 00.1	+20 17	Nr?	6.0:	>11.0	v					N 1894?
CO Aur	40457	6 00.5	+35 19	SRd	7.46	8.08	V				F5Ib	
V352 Ori	40913	6 01.8	−2 21	Lb	8.5	10.0	p				M7ep	
Z Aur	—	6 01.8	+53 18	SRd	9.2	11.7	v			0.46	G0e−G6e	
SW Pic	41586	6 02.2	−60 06	Lb?	6.39	6.51	V				M4III	
V Cam	39741	6 02.5	+74 30	M	7.7	16.0	v	2443402	522.45	0.31	M7e	
RR Ori	250795	6 03.0	+16 23	M	9.1	>14.7	v	2436905	251.50		M6.5e	
SS Lep	41511	6 05.0	−16 29	Z And?	4.82	5.06	V				A2e(Shell)	17 Lep
CF Pup	—	6 05.3	−49 08	EA	9.4	12.6	p	2428821.58	7.64556	0.08:		
S Lep	41698	6 05.8	−24 12	SRb	7.1	8.9	p		90		M6	
V916 Ori	252214	6 08.3	+13 58	βC?	8.11	8.13	B		0.39912:		B2V	
V917 Ori	252248	6 08.5	+13 56	βC	8.77	8.90	V		0.4033		B3Vn	
SS Gem	41870	6 08.6	+22 37	RV	9.3	10.7	p	2434365	89.31	0.19	F8Ib−G5Ib	
δ Pic	42933	6 10.3	−54 58	EB	4.65	4.90	V	2441695.336	1.672541	0.18	B3III + O9V	
TU Gem	42272	6 10.9	+26 01	SRb	9.4	12.5	p		230		C4,6	
GQ Ori	42532	6 11.2	+9 37	Cδ	9.0	9.8	p	2427866.7	8.61566	0.33	G0	
V638 Mon	42657	6 11.7	−4 40		6.18 Y (0.02)		y		0.724		B9pHgMn	
TV Gem	42475	6 11.8	+21 52	SRc	8.7	9.5	p		182		M1Iab	
WY Gem	42474	6 11.9	+23 12	Lc	8.94	9.8	B				M2epIab + B2V	
RR Aur	42311	6 12.1	+43 10	M	8.2	15	v	2443031	307.87	0.47	M3e−M7e	
X Aur	42212	6 12.2	+50 14	M	8.0	13.6	v	2444604	163.79	0.50	M3e−M7e	
BU Gem	42543	6 12.3	+22 54	Lc?	5.74 V	7.5	v				M1−M2 Ia−Iab	6 Gem
QR Aur	42616	6 13.7	+41 42	αCV	7.10	7.16	V	2438450.7	17.0		A0pCrSrEu	
η Gem	42995	6 14.9	+22 30	SRb(E)	3.2	3.9	v	2437725*	232.9	0.05	M3III	Propus
V345 Ori	—	6 16.0	−1 04	M	9.5	>14	p	2435507	332			
GK Ori	—	6 17.7	+8 31	SR	9.5	11.0	v	2426002	236	0.5:	N(C5,4)	
UW Lyn	42973	6 17.9	+61 31	Lb?	4.93	5.04	V				M3IIIab	1 Lyn
LT Gem	254699	6 18.4	+23 34	E	8.96	9.19	V		1.0748:		B1V	
LU Gem	43818	6 19.3	+23 28	βC	7.21	7.24	B		0.21909		B0II	
UZ Lyn	43378	6 19.6	+59 01	δSct	4.48 Y						A2Vs	2 Lyn
KS Mon	44096	6 19.7	−5 17	SRb	9.5	10.7	p				M	
SV Mon	44320	6 21.4	+6 28	Cδ	7.61	8.88	V	2436437.070	15.2321	0.38	F6−G4	
FR CMa	44458	6 21.4	−11 46	γC	5.46	5.64	V				B1Vpe	

Star Name	HD	α 2000	δ 2000	Type	Max	Min		Epoch	Period	F	Spec	Notes
RS Ori	44415	6^{h}22^m.2	+14°41'	Cδ	8.03	8.83	V	2435912.305	7^d.56681	0.32	F5Ib-G1Ib	
V Mon	44639	6 22.7	-2 12	M	6.0	13.7	v	2440231	333.80	0.46	M5e-M8e	
β CMa	44743	6 22.7	-17 57	βC	1.93	2.00	V	2441296.175	0.25003		B1II-III	Mirzam
IM Mon	44701	6 23.0	-3 17	EB	6.40	6.49	B	2433383.225	1.190243		B5n + B8n	
μ Gem	44478	6 23.0	+22 31	Lb?	2.76	3.02	V				M3.0IIIae	13 Gem
V Aur	44388	6 24.0	+47 42	M	8.5	13.0	v	2443435	353.00	0.52	C6,2e(N3e)	
ψ¹ Aur	44537	6 24.9	+49 17	Lc	4.75	5.02	V				K5-M0 Iab-Ib	46 Aur
T Mon	44990	6 25.2	+7 05	Cδ	5.59	6.60	V	2436137.090	27.0205	0.27	F7Iab-K1Iab	
BL Ori	44984	6 25.5	+14 43	Lb	8.5	9.7	p				Nb(C6,2)	
RR Lyn	44691	6 26.4	+56 17	EA	5.64	6.03	V	2438046.8402	9.945070	0.04	A3Vm	
SW Mon	—	6 27.0	+5 23	SRb	9.2	10.8	v	2430670	112		M4III	
W Col	—	6 27.8	-40 06	M	9.3	>11.5	p	2415411	327		M6e	
FS CMa	45677	6 28.3	-13 03	unq	7.55	8.58					B2IVep	
RT Aur	45412	6 28.6	+30 30	Cδ	5.00	5.82	V	2442361.155	3.728115	0.25	F4Ib-G1Ib	48 Aur
V Lyn	—	6 29.7	+61 33	SRb	9.5	12.0	p				M5III-IV	
AX Mon	45910	6 30.5	+5 52	unq	6.59	6.87	V	2438386	232.5		B1IVeq + K0III	
DW Gem	—	6 31.0	+27 27	Lb	8.5:	11.9	p				M4	
RX Pic	46919	6 31.5	-63 32	EA	9.0	10.5	p				A0	
SX Col	46431	6 31.6	-36 56	Lb?	6.28	(0.13)	V				M1III	
ξ¹ CMa	46328	6 31.9	-23 25	βC	4.33	4.36	V	2441296.0514	0.2095755		B0.5IV	4 CMa
V578 Mon	—	6 32.0	+4 51	E	8.54	?	V		2.420		B0V	
WW Aur	46052	6 32.5	+32 27	EA/DM	5.79	6.54	V	2441399.305*	2.52501922	0.10	A3m: + A3m:	
RW Mon	259986	6 34.8	+8 50	EA	9.26	11.43	V	2437004.6781	1.9060913	0.16	B9V	
W Gem	46595	6 35.0	+15 20	Cδ	6.54	7.36	V	2437136.473	7.91413	0.30	F5-G1	
RR Pic	—	6 35.6	-62 38	Nb	1.2: p	12.42	V	2424310			P(Q)	N 1925
UU Aur	46687	6 36.5	+38 27	SRb	7.83	10.0	B		234		C5,3-C7,4(N3)	
SU Cam	—	6 38.2	+73 55	M	8.9	12.6	v	2444265	285.03		M5	
SY Gem	—	6 40.6	+31 11	N?	9.2	>13.0	v				M7e-M8e	N 1866
U Lyn	—	6 40.8	+59 52	M	8.8	15.0	v	2440319	435.94	0.42	M4-M5pe	
V340 Car	—	6 40.9	-52 26	SR	9.5	10.5	p					
S Mon	47839	6 41.0	+9 54	Ia?	4.62	4.67	V				O7Ve	15 Mon
SV Cam	—	6 41.3	+82 16	EA/DW/RS	8.40	(0.71)	V	2442594.61518	0.59306995	0.17	G5V + G3V	
VW Gem	47883	6 42.2	+31 27	Lb	8.1	8.5	v				C3,9	
BT Mon	—	6 43.8	-2 01	Na	4.5	>17	p	2429515			P(Q)	N 1939
DM Gem	48328	6 44.2	+29 57	Na	4.8	16.5	p	2416179			P(Q)	N 1903
FT CMa	48917	6 44.5	-31 04	γC	5.13	5.44	V				B2Ve	10 CMa
S Lyn	47929	6 44.6	+57 55	M	8.5	14.8	v	2440225	300.44	0.45	M6e:	
CH Pup	—	6 45.2	-36 32	M	7.9	>14	v	2440815	494.0:	0.19	Me	
V505 Mon	48914	6 45.8	+2 30	EB?	7.2	7.7	p				B5Ib	
HK CMa	49333	6 47.0	-21 01	αCV	6.06	6.09	y	2442818.88	2.181		B7IIIn	12 CMa
X Gem	48912	6 47.1	+30 17	M	7.5	13.6	v	2440604	263.72	0.49	M5e-M6e	
AW Cam	48049	6 47.5	+69 38	EB/KE	8.22	8.66	V	2438738.452	0.7713468		A0V	
V613 Mon	49368	6 48.4	+5 32	SRb?	7.64	7.76	V				M2(S5,1)	
IS Gem	49380	6 49.7	+32 36	SRd	6.6	7.3	p		47:		K3II	
κ CMa	50013	6 49.8	-32 31	γC	3.78	3.97	V				B2Vne	13 CMa
RX Gem	49521	6 50.2	+33 14	EA	9.2	11.2	p	2440555.782	12.2086588	0.08	A3IIIe	
V592 Mon	49976	6 50.7	-8 02	αCV	6.16	6.31	V		2.976		A2pSrCrEu	
GP CMa	—	6 52.9	-12 11	Lb?	9	12.1	v				Ceα:	
OX Aur	50018	6 53.0	+38 52	δSct	5.94	6.14	V		0.154412	0.5	A7n-F3IV	59 Aur
GY Mon	50436	6 53.2	-4 35	Lb	9.4	10.6	p				N3(C6,3)	
EY CMa	50707	6 53.5	-20 13	βC	4.79	4.84	V	2441296.1640	0.184557	0.50	B1III-IV	15 CMa
o¹ CMa	50877	6 54.1	-24 11	Lc	3.78	3.99	V				K2.5Iab	16 CMa
EZ CMa	50896	6 54.2	-23 56	E?/WR	6.71	6.95	V	2443200.47	3.763		WN5	
NP Pup	51208	6 54.4	-42 22	Lb	6.25	6.37	V				C3II	
AU Mon	50846	6 54.9	-1 23	EA	8.2	9.5	v	2442801.3752	11.1130371	0.15	B5 + F0	
DN Gem	50480	6 54.9	+32 09	Nb	3.6 p	15.76	B	2419476			P(Q)	N 1912
V352 Aur	50420	6 55.2	+43 55	δSctc	6.03	6.06	y		0.17		A9III	
ι CMa	51309	6 56.1	-17 03	βC	4.36	4.40	V		0.08:		B3II-III	20 CMa
BG Mon	—	6 56.4	+7 04	SRb	9.2	10.4	v	2431881	30:		N(C5,2)	
Y Mon	51189	6 56.9	+11 15	M	8.6	14.9	v	2440637	230.19	0.49	M4e-M5.5e	
X Mon	51478	6 57.2	-9 04	SRa	6.9	10.0	v	2441370	155.70	0.51	gM3e-M4pe	
HH CMa	51630	6 57.2	-22 12	βC	6.59	6.66	V		0.19		B2III	
UW Aur	—	6 57.3	+41 07	SRa	9.5	11.6	v	2442800	560.7	0.55	C4,5J(R6p/N3)	
V523 Mon	51725	6 58.1	-9 02	SRb	6.96	7.45	V		45:		Mb	
V383 Car	—	6 59.1	-58 31	δSctc	8.84	(0.06)	V				Fm	
NY Aur	51418	6 59.3	+42 19	αCV	6.60	6.77	V	2441241.654	5.4379	0.50	A0pEuSrCr	
SW Gem	267341	6 59.5	+26 03	SRa	8.9	10.3	v	2432360*	680		M5III	
FU CMa	52437	7 00.3	-22 07	γC	6.48	6.60	V				B3IV-Vne	
V614 Mon	52432	7 01.0	-3 15	SRb	7.25	7.6	V		60:		R5(C4,5J)	
R Lyn	51610	7 01.3	+55 20	M	7.2	>14.5	v	2442135	378.66	0.44	S2.5,5e-S6,8e:	
σ CMa	52877	7 01.7	-27 56	Lc	3.43	3.51	V				K7Ib	22 CMa
GU CMa	52721	7 01.8	-11 18	γC	6.49	6.72	V				B2Vne	
V526 Mon	52610	7 01.9	-1 08	Cep	9.0	9.4	p	2433304.284	2.6751834	0.48	G0	
NP Gem	52554	7 02.4	+17 45	Lb?	5.89	6.02	V				M1	
ω Gem	52497	7 02.4	+24 13		5.18 Y	(0.086)	V		0.7282		G5Ib-IIa	42 Gem

Star Name	HD	α 2000	δ 2000	Type	Max	Min		Epoch	Period	F	Spec	Notes
FZ CMa	52942	7^{h}02^m.7	-11°27'	EA/DM	8.05	8.44	V	2441742.324	1^d.27306	0.16	B2.5IV-Vn	
V637 Mon	52918	7 02.9	-4 14	βC	4.99 Y (0.04)		V		0.17		B1V	19 Mon
Z CMa	53179	7 03.7	-11 33	Ina	8.8	11.2	p				B8peq	
ζ Gem	52973	7 04.1	+20 34	Cδ	3.66	4.16	V	2436791.922	10.15082	0.50	F7Ib-G3Ib	Mekbuda
R Vol	--	7 05.6	-73 01	M	8.8	13.9	v	2442281	450.30	0.48	Me	
FM CMa	53756	7 05.7	-12 19	EB/DM	7.28	7.50	V		2.7888		B1V	
V569 Mon	53755	7 05.8	-10 40	βC	6.42	6.53	V		0.267:		B0.5V + F5III	
FN CMa	53974	7 06.7	-11 18	βC	5.38	5.42	V		0.12377		B0III	
AC Car	54795	7 06.8	-58 23	SRb	9.1	10.3	p		99:		M7III	
RY Mon	--	7 06.9	-7 33	SRa	7.7	9.2	v	2423743	466	0.43	N3(C5,5)	
V CMi	53847	7 07.0	+8 53	M	7.4	15.1	v	2442737	366.10	0.39	M4e-M10	
TW Gem	53792	7 07.3	+22 31	cst?	8.49		V				M0III	
R Gem	53791	7 07.4	+22 42	M	6.0	14.0	v	2440725	369.81	0.36	S2,9e-S8,9e	
AM Gem	--	7 07.4	+28 18	M	9.5	v>16.5	p	2424880	355.1		M10	
FV CMa	54309	7 07.4	-23 50	γC	5.64	5.94	V				B2IV-Ve	
W CMa	54361	7 08.1	-11 55	Lb	6.35	7.9	V				C6,3(N)	
R CMi	54300	7 08.7	+10 01	M	7.25	11.6	V	2441323	337.78	0.48	C7,1Je(Sep)	
S CMa	--	7 09.5	-32 56	cst	9.0		p				A5	
AP Mon	--	7 09.6	-6 44	Lb	8.74	9.5	V				M3	
FF CMa	55173	7 10.6	-30 39	EB/KE	7.38	7.74	V	2428847.465	1.213375		B2V + B2V	
V571 Mon	55057	7 11.4	-0 18	δSct	5.43	5.50	V		0.0999081		A8Vn-F3Vn	21 Mon
VW CMa	--	7 12.4	-25 31	EB/KE	9.0	9.2	p	2427924.229	0.720831			
HI CMa	55538	7 12.8	-15 30	γC	7.8	(0.4)	p				B2IIIe	
OU Pup	56022	7 13.2	-45 11		4.89 Y (0.03)		y		0.90		A0pSi	L¹ Pup
BQ Gem	55383	7 13.4	+16 10	SRb	6.63	7.01	B		50:		M4IIIab	51 Gem
UY Lyn	54895	7 13.4	+51 26		5.47 Y (0.14)		V				M3III	
L² Pup	56096	7 13.5	-44 39	SRb	2.6	6.2	v	2440813	140.42	0.40	M5IIIe	
GY CMa	55857	7 13.6	-27 21	βC?	6.12	(0.04)	V		0.112		B0.5V	
GG CMa	55958	7 13.8	-31 05	E11+βC?	6.55	6.61	V				B2IV	
V536 Mon	55708	7 13.9	-2 55	E	9.1	10.1	p				A	
EW CMa	56014	7 14.3	-26 21	γC	4.42	4.82	V				B3IVe	27 CMa
ω CMa	56139	7 14.8	-26 46	γC	3.60	4.18	V				B2IV-Ve	28 CMa
PR Pup	56455	7 14.8	-46 51		5.72 Y (0.04)		y		2.24*		A0pSi	
AA Cam	54587	7 14.9	+68 48	Lb	9.0	9.6	p				M5(S)	
GZ CMa	56429	7 16.3	-16 43	EA/DM	8.1	8.7	p	2438814.273	4.801052	0.07	A0	
RY CMa	56450	7 16.6	-11 29	Cδ	7.71	8.45	V	2436416.937	4.67825	0.24	F6-G0Ib	
RR Mon	56567	7 17.5	+1 06	M	8.4	15.3	v	2440285	393.22	0.39	S7,2e:	
NV Pup	57150	7 18.3	-36 44		4.66 Y (0.18)		V				B2V + B3IVne	
NW Pup	57219	7 18.6	-36 45	βC/E11	5.11 Y				1		B2IVne	
UW CMa	57060	7 18.7	-24 34	EB/KE?	4.84	5.33	V	2442424.014	4.393407		O7Ia:fp + OB	29 CMa
R CMa	57167	7 19.5	-16 24	EA/SD	5.70	6.34	V	2444289.361	1.1359405	0.15	F1V	
T CMa	--	7 21.4	-25 27	SR	9.0	11	v	2423692	309			
RU Cam	56167	7 21.7	+69 40	CWa	8.10	9.79	V		22	0.35	C0,1-C3,2e(K0-R0)	
CW CMa	57802	7 21.9	-23 48	EA/DM	8.58	8.98	V	2442090.1657	2.11797737	0.094	A0V	
GH CMa	57890	7 22.4	-20 30	SRb	6.82	7.19	V		20:		M6III	
VY CMa	58061	7 23.0	-25 46	unq	6.5	9.6	v				M5Ibep(C6,3)	
V Gem	57770	7 23.2	+13 06	M	7.8	14.9	v	2442142	275.07	0.45	M4e(Se)-M5e	
FW CMa	58343	7 24.7	-16 12	γC	5.00	5.50	V				B3Ve	
BX Mon	--	7 25.4	-3 36	M	9.5	13.4	p	2430345	1374	0.39	M4ep	
TT Mon	--	7 25.7	-5 51	M	8.9	>13.5	p	2438323	322.97		M6e	
SS CMa	--	7 26.1	-25 16	Cδ	9.26	10.36	V	2441109.19	12.361	0.45	F6-G2	
GI Mon	58756	7 26.8	-6 41	Na	5.2:	15.1	p	2421595			P(Q)	N 1918
FY CMa	58978	7 27.0	-23 05	γC	5.54	5.69	V				B0IVpe	
FX CMa	58881	7 27.1	-11 43	SRb	8.56	8.86	V		40:		S3,9(C1)	
β CMi	58715	7 27.2	+8 17	γC	2.84	2.92	V		0.09:		B8Ve	Gomeisa
RY Gem	58713	7 27.4	+15 40	EA	8.5	11.3	p	2440383.674	9.300525	0.09	A2Ve + K2	
Y Lyn	58521	7 28.2	+45 59	SRc	7.8	10.3	p		110		MSIb-II	
S Vol	--	7 29.8	-73 23	M	7.7	13.9	v	2438790	395.83	0.54	M4e	
U Mon	59693	7 30.8	-9 47	RVb	6.1	8.1	p	2437395	92.26	0.22	F8e-K0Ib:p	
VZ Cam	55966	7 31.1	+82 25	SR	4.80	4.96	V		23.7		M4IIIa	
MQ Pup	60099	7 31.3	-38 00	EB	8.0	8.9	p	2428789.650	1.468565		B9	
PS Pup	60168	7 31.7	-35 53	E11	6.61 Y (0.03)		v		0.6711		A0V	
NQ Gem	59643	7 31.9	+24 30	SR?	7.4	7.80	V				C1-C6.5,3ev	
Z Pup	60218	7 32.6	-20 40	M	7.2	14.6	v	2437985	499.67	0.38	M4e-M9e	
VX Pup	60219	7 32.6	-21 56	Cδ	7.73	8.51	V	2436237.953	3.01172	0.53	F5-F8	
S CMi	59950	7 32.7	+8 19	M	6.6	13.2	v	2443911	332.94	0.49	M6e-M8e	
X Pup	60266	7 32.8	-20 55	Cδ	7.82	9.24	V	2436279.500	25.9610	0.14	F6-G2	
KQ Pup	60415	7 33.8	-14 31	unq	4.88	5.17	V				M2Iabpe + B2V	
OW Pup	60606	7 33.8	-36 20		5.5	5.60	B				B3Vne	
T CMi	--	7 34.0	+11 44	M	9.5	15.1	v	2442045	328.3	0.47	M4Se-M8	
AI CMi	--	7 35.7	+0 15	L?	8.8	10.6	p				G5Iab	
PT Pup	61068	7 36.7	-19 42		5.74 Y						B2III	
BN Gem	60848	7 37.1	+16 54	γC	6.0	6.6	p				O8Vspe	
PU Pup	61429	7 38.3	-25 22	EW	4.70 Y (0.06)		V		2.57895		B8IV	
MY Pup	61715	7 38.3	-48 36	Cδ	5.54	5.76	V	2441043.72	5.6952	0.37	F4Iab	

Star Name	HD	α 2000	δ 2000	Type	Max	Min		Epoch	Period	F	Spec	Notes
VZ Pup	--	7^{h}38^m.6	−28°30'	Cδ	8.92	10.32	V	2434888.699	231^d.1640	0.17	F5−G2	
OV Cep	51802	7 40.5	+87 01	SR	5.00	5.07	V				M2IIIa	
R Pup	62058	7 40.9	−31 40	SRd?	6.56	6.64	V				G2 0−Ia	
U CMi	61789	7 41.3	+8 23	M	8.0	14.0	v	2443150	413.88	0.52	M4e	
WX Pup	--	7 42.0	−25 53	Cδ	8.80	9.37	V	2435042.184	8.93825	0.31	F6−G1	
NZ Gem	61913	7 42.1	+14 13		5.56 Y (0.10)		p				M3II−III	
S Gem	62045	7 43.0	+23 27	M	8.2	14.7	v	2442112	293.23	0.42	M4e−M7e	
σ Gem	62044	7 43.3	+28 53	RS	4.28 Y (0.12)		V		19.603		K1III	75 Gem
AZ CMi	62437	7 44.1	+2 24	δSctc	6.44	6.51	V	2440886.0713	0.09526		F0III	
YZ CMi	--	7 44.7	+3 34	BY+UV	8.6	12.93	B	2441355.178*	2.780964		M4.5Ve	
RZ Pup	63033	7 45.1	−39 51	cst?	8.8		p				K2	
PV Pup	62863	7 45.5	−14 41	E	6.89 Y (0.41)		y		1.660		A8V	
W Pup	63218	7 46.0	−42 12	M	7.3	13.6	v	2441246	120.10	0.47	M3e	2 Pup B
BC Cam	62140	7 46.5	+62 50	αCV	6.43	6.48	V	2441254.08	4.285		F0pSrCrEu	49 Cam
S Pup	63451	7 46.7	−48 07	cst	7.2		v				A2	
OX Pup	63401	7 47.1	−39 20		6.31 Y (0.06)		V		2.41		B8III	
T Pup	63640	7 48.1	−40 39	cst	6.14		V				M2III	
ZZ Pup	63482	7 48.4	−19 18	EA	9.5	11.5	p	2426783.107	6.33811	0.10	A2	
W CMi	--	7 48.8	+5 24	SRb	8.72	9.04	V		95:		C7,2(R6)	
T Gem	63334	7 49.3	+23 44	M	8.0	15.0	v	2442120	287.79	0.50	S1.5,5e−S9,5e	
BC CMi	64052	7 52.1	+3 17	SRb	6.14	6.42	V		35:		M4III	
V372 Car	64722	7 52.5	−54 22	βC	5.69	(0.027)	V		0.1160		B2III	
NQ Pup	64332	7 53.1	−11 38	Lb	7.55	7.68	V				M2p(S6,3)	
TU Mon	--	7 53.3	−3 03	EA	9.0	10.9	p	2420930.57	5.049025	0.15	B5 + A5	
HS Pup	--	7 53.4	−31 39	Nb	8.0	>20.0	p	2438385				N 1963
U Gem	64511	7 55.1	+22 00	UG(E/X)	8.2	14.9	v		103:		M4.5 + WD	
MW Pup	65293	7 56.0	−44 55	EA	8.8	9.2	p	2434336.450	2.398735		A0	
SU Pup	--	7 56.2	−44 09	M	9.3	>14.0	v	2431385	339.8		Me	
PX Pup	65183	7 56.4	−30 17		6.33 Y (0.42)		V				M6III	
χ Car	65575	7 56.8	−52 59	βC	3.46	(0.015)	V		0.101		B2IV	
V341 Car	65750	7 56.8	−59 08	L	6.2	7.10	V				M1−M3 II−III	
ZZ Cnc	--	7 57.1	+11 00	EA/DS	9.4	10.9	p	2426770.350	25.5950	0.08	A8IV−V: + K0III:	
FW Mon	65259	7 57.6	−7 11	EA	9.4	10.6	p	2427562.220	3.8735833	0.13	B5 + F2	
AP Pup	65592	7 57.8	−40 07	Cδ	7.11	7.78	V	2435243.926*	5.0843102	0.51	F5−F8	
V373 Car	--	7 57.8	−60 50	γC	8.9	9.2	v				B9pe	
V356 Car	65987	7 58.1	−60 37	EA?	7.59	8.01	V				B9.5IVpSi	
V Pup	65818	7 58.2	−49 15	EB	4.7	5.2	p	2428648.3048	1.4544877		B1Vp + B3IV:	
AQ Pup	65589	7 58.4	−29 08	Cδ	8.12	9.39	V	2435939.500	29.8568	0.17	F5Ib−G2Ib	
V374 Car	66194	7 58.8	−60 49	γC	5.72	5.84	V				B2IV−Vpne	
UX Mon	65607	7 59.3	−7 30	EA	8.0	8.9	v	2433328.849	5.90450	0.17	A6p + G2IVp	
PY Pup	66255	8 00.5	−48 52		6.12 Y (0.04)		v		6.82		A0pSi	
U Pup	65940	8 00.8	−12 51	M	8.5	14.8	p	2437900	316.99	0.41	M5e−M8e	
V645 Mon	65953	8 01.2	−1 24		4.68 Y (0.02)		V		5		K4III	28 Mon
AX Cam	65339	8 01.7	+60 19	αCV	5.95	6.08	V	2441701.41	8.0278*	0.4	A2pSrCrEu	53 Cam
UU Cnc	--	8 02.5	+15 11	EB/GS	8.68	9.35	V	2441072.03	96.71		K4III	
AR Pup	--	8 03.0	−36 36	RVb	8.7	10.9	p		75		cF0−cF8	
HZ Pup	--	8 03.4	−28 29	Nb	7.7	18.5	p	2438048			P(Q)	N 1963
SV Lyn	66175	8 03.7	+36 21	SRb	8.2	9.1	p		70:		gM5	
MZ Pup	66888	8 04.3	−32 41	Lc?	5.2	5.44	V				M1Ib	
V375 Car	67536	8 04.7	−62 50	βC?	6.29	(0.04)	V		0.264:		B4Vn	
BL Cnc	66875	8 06.3	+22 38	Lb	5.97	6.04	V				M3III	9 Cnc
YY CMi	67110	8 06.6	+1 56	EB	8.33	9.13	V	2428023.147	1.0940197		F6Vn	
ρ Pup	67523	8 07.5	−24 18	δSct	2.7	2.8	v	2435555.911*	0.14088143	0.40	F6IIp	15 Pup
VZ Gem	--	8 07.8	+30 51	N?	8.7	16.0:	v					N 1856
BH Pup	67889	8 08.3	−42 02	EA	8.4	9.1	p	2421692.523	1.915854		B9	
PQ Pup	67888	8 08.6	−37 41		6.37 Y						B4Ve	
Z Lyn	66976	8 08.7	+57 50	cst	8.8		p				A2	
RT Mon	67650	8 08.7	−10 47	SRb	8.2	10.3	v	2433315	115.3	0.49	M3III	
γ² Vel	68273	8 09.5	−47 20		1.6	1.8	V				WC8 + O7.5e	
XY Pup	67862	8 09.6	−11 59	EA	9.2	11.4	v	2426417.805	13.77830	0.10	A3e	
AS Pup	--	8 09.7	−38 10	M	9.0	12.8	p	2435467	328.05	0.50	M7e	
AX Vel	68556	8 10.8	−47 42	Cδ	7.9	8.5	V	2441023.43	2.59285	0.25	F8	
HU Hya	68178	8 11.0	−10 37	EA	9	9.5	p	2427120.550	2.516440		A2	
NS Pup	68553	8 11.4	−39 37	Lc	4.4	4.5	V				K3Ib	
CP Pup	--	8 11.8	−35 21	Na	0.5	>17.0	p	2430675			P(Q)	N 1942
AH Vel	68808	8 12.0	−46 39	Cep	5.50	5.86	V	2434866.28	4.22717	0.49	F7Ib−II	
AT Pup	--	8 12.4	−36 57	Cδ	7.54	8.40	V	2436422.395	6.6650	0.22	F8−G0	
BM Cnc	68351	8 13.1	+29 39	αCV	5.53	5.65	V	2439482.9	4.116		B9pSiCr	15 Cnc
RS Pup	68860	8 13.1	−34 35	Cδ	6.53	7.62	V	2435734.426	41.3876	0.24	F9−G7	
XZ Pup	68884	8 13.5	−23 57	EA	8.0	10.9	p	2442412.1946	2.1923631	0.21	A0	
MX Pup	68980	8 13.5	−35 54	γC	4.60	4.88	V				B1.5IIIe	
DY Pup	--	8 13.8	−26 34	N	7.0	>16.0	p	2416073				N 1902
OS Pup	69081	8 14.0	−36 19		4.60	4.71	V				B1.5IV	
AI Vel	69213	8 14.1	−44 34	δSct	6.4	7.1	v		0.11157396		A2p−F2p	
RX Cnc	68775	8 14.7	+24 44	SRb	9.2	11.3	p		120:		M8	

Star Name	HD	α 2000	δ 2000	Type	Max	Min		Epoch	Period	F	Spec	Notes
R Cnc	69243	8^{h}16^m.6	+11°44'	M	6.07	11.8	V	2444231	361^d.60	0.47	M6e-M9e	
W Lyn	--	8 16.8	+40 08	M	8.8	14.0	v	2442050	295.2		M6	
AU Pup	69951	8 17.7	-41 42	EB	8.50	9.40	V	2439237.985	1.126411		A0	
AI Hya	--	8 18.8	+0 17	EA	9.0	9.5	v	2428935.46	8.289676		F0 + F5	
SW Pup	70196	8 18.8	-42 45	EA	9.3	10.4	p	2419282.068	2.7473405	0.14	F0	
HQ Hya	69997	8 19.3	-10 10	δSct	6.29	6.33	V		0.097:		F3IIIp	
FZ Hya	--	8 21.6	+4 58	Lb	9.5	10.5	p				M6	
V Cnc	70276	8 21.7	+17 17	M	7.5	13.9	v	2443485	272.13	0.46	S0e-S7,9e	
R Cha	71793	8 21.8	-76 21	M	7.5	14.2	v	2442006	334.58	0.41	M4e-M7e	
Z Cnc	70421	8 22.4	+15 00	SRb	9.4	10.7	p	2437026	104:		M6III	
AC Pup	--	8 22.7	-15 55	Lb	8.9	10.1	v				N(C5,4)	
FK Hya	70938	8 24.5	-8 31	Lb	9.0	10.1	p				Mb	
X Lyn	--	8 25.5	+35 24	M	9.5	16.0	v	2440233	320.8		M5e	
NO Pup	71487	8 26.3	-39 04	EA	6.7	7.14	v	2441361.7635	1.25689059	0.13	B9V	
BP Cnc	71250	8 26.7	+12 39	SRb	5.41	5.75	V		40:		M3III	27 Cnc
VV Pyx	71581	8 27.6	-20 51	E	6.56 Y	(0.3)	p		4.56960		A2V	
GU Vel	71935	8 27.6	-53 05	δSct	5.09 Y	(0.015)	V		0.07:		A9-F0 III-IV	
AS Vel	71872	8 28.3	-38 58	EA	8.8	9.4	p	2426454.440	1.55788874	0.12	A3	
CX Cnc	71496	8 28.6	+24 09	δSctc	6.10	(0.025)	V		0.096		F0Vn	28 Cnc
V Car	72275	8 28.7	-60 07	Cδ	7.08	7.82	V	2437454.023	6.69668	0.30:	F6-G2 Ib-II	
TW Cnc	71780	8 29.6	+12 27	EA	8.50	8.97	V	2431854.76	70.760	0.03	G8III + A8	
RT Hya	71887	8 29.7	-6 19	SRb	7.0	11.0	v		253	0.46	M6e-M7	
U Pyx	72085	8 29.9	-30 19	SR	8.6	9.4	p		345:		K5	
TZ Lyn	71866	8 31.2	+40 13	αCV	6.65	6.79	V	2437311.12*	6.79760	0.56	A0p	
AL Vel	--	8 31.2	-47 40	EA	8.60	9.13	V	2436263.51	96.1116	0.09	K0III + A3	
X Car	72698	8 31.3	-59 14	EB/KE	7.90	8.65	V	2428857.146	1.0826310		A0V + A0V	
VZ Hya	72257	8 31.7	-6 19	EA	8.97	9.69	V	2421925.825	2.9042998	0.07	F5 + F5	
FY Vel	72754	8 32.4	-49 36	EB?	6.84	7.06	v		33.72		B2Ibpe	
FX Vel	--	8 32.6	-37 59	EB	9.2	10.4	p	2434302.525	1.052565		Be	
HV Hya	72968	8 35.5	-7 59	αCV	5.66	5.76	V	2440619.8	5.57		A1pSrCrEu	3 Hya
U Cnc	72863	8 35.8	+18 54	M	8.5	15.5	v	2440627	304.78	0.40	M2e	
HV Vel	73340	8 35.9	-50 58		5.80 Y	(0.025)	y		2.67		B8Si	
RZ Vel	73502	8 37.0	-44 07	Cδ	6.44	7.64	V	2434906.43	20.3969	0.26	G8	
BR Cnc	73175	8 37.7	+19 31	δSctc	8.26	(0.02)	V		0.038		F0Vn	
GO Vel	73588	8 37.7	-40 26	Lb?	6.61	6.98	V		75		M5III	
T Vel	73678	8 37.7	-47 22	Cδ	7.70	8.34	V	2434890.37	4.63974	0.30	F6-F9	
CY Cnc	--	8 38.6	+19 59	δSctc	8.14	(0.02)	V		0.1:		F0V	
UV Cnc	73375	8 38.8	+21 10	Lb	9	10.5	p				M4	
RZ Cnc	73343	8 39.1	+31 48	EA/GS/RS	8.67	10.03	V	2418702.531	21.642998	0.15	K2III + K4III	
BS Cnc	73450	8 39.2	+19 36	δSctc	8.50	(0.02)	V		0.051		A9Vn	
HW Vel	74071	8 39.4	-53 26		5.48 Y	(0.02)	V		0.26145		B5V	
RV Hya	73766	8 39.7	-9 35	SRc	8.7	10.0	p		116		M5II	
BU Cnc	73576	8 39.7	+19 16	δSctc	7.67	(0.03)	V		0.071		A7Vn	
BT Cnc	73575	8 39.7	+19 47	δSctc	6.66	(0.06)	V		0.10228		F0III	
AK Hya	73844	8 39.9	-17 18	SRb	6.33	6.91	V		112:		M4III	
o Vel	74195	8 40.3	-52 55	βC	3.56	3.67	V		0.131977		B3IV	
BQ Cnc	73729	8 40.4	+20 11	δSctc	8.19	(0.02)	V		0.074		F2Vn	
BV Cnc	73746	8 40.5	+19 12	δSctc	8.65	(0.02)	V		0.21		F0V	
V343 Car	74375	8 40.6	-59 46	βC?	4.3	4.33	V				B1.5III	
BN Cnc	73763	8 40.7	+19 14	δSctc	7.80	(0.03)	V		0.0388205		A8V	
X UMa	73507	8 40.8	+50 08	M	8.1	14.8	v	2442200	248.84	0.43	M4e	
VZ Cnc	73857	8 40.9	+9 49	δSct	7.18	7.91	V	2441304.364*	0.17836415	0.26	A7III-F2III	
BW Cnc	73798	8 40.9	+20 16	δSctc	8.48	(0.01)	V		0.072		F0Vn	
BX Cnc	74028	8 42.1	+19 25	δSctc	7.96	(0.02)	V		0.053		A7V	
BY Cnc	74050	8 42.2	+18 56	δSctc	7.91	(0.01)	V		0.058		A7Vn	
HX Vel	74455	8 42.3	-48 06		5.51 Y	(0.037)	b		0.563		B1.5Vn	
HY Vel	74560	8 42.4	-53 07		4.86 Y	(0.015)	b		0.0284		B3IV	
η Hya	74280	8 43.2	+3 24		4.30 Y	(0.015)	v		0.16:		B3V	7 Hya
RS Cha	75747	8 43.2	-79 04	EA+δSct	6.02	6.68	V	2442850.7688	1.669870	0.15	A5V + A7V	
SW Vel	74712	8 43.6	-47 24	Cδ	7.44	8.96	V	2435015.461	23.4744	0.25	K2	
S Cnc	74307	8 43.9	+19 02	EA/DS	8.29	10.25	V	2436985.029	9.4845516	0.08	B9V + G8IV	
BI Cnc	74521	8 44.8	+10 05	αCV	5.58	5.71	V	2441616.50	4.2359		A0pSiCr	49 Cnc
SX Vel	74884	8 44.9	-46 21	Cδ	7.97	8.65	V	2435245.3*	9.54993	0.48	G5	
V344 Car	75311	8 46.7	-56 46	γC	4.4	4.51	V				B3Vne	
TT Pyx	75322	8 48.5	-26 10	EA	8.8	9.4	p	2425622.57	1.515769	0.30	B9	
KX Hya	75333	8 49.4	-3 27		5.31 Y	(0.02)	y		6:		B9pHgMn	14 Hya
HZ Vel	75654	8 49.9	-39 09	δSct	6.39 Y				0.087		A5III	
RS Cam	74110	8 50.8	+78 58	SRb	7.9	9.7	v	2427143	88.6	0.45	M4III	
RZ Pyx	75920	8 52.1	-27 29	EB	8.83	9.72	B	2438431.474	0.656273		B7V	
BO Cnc	75716	8 52.5	+28 16	Lb?	5.9	6.37	V				M3III	53 Cnc
V Pyx	76181	8 53.4	-34 49	SR	9.0	11.5	p				K2	
S Hya	76011	8 53.6	+3 04	M	7.4	13.3	v	2441683	256.45	0.49	M4e-M6.5e	
X Cnc	76221	8 55.4	+17 14	SRb	5.6	7.5	v	2443631	195:		C5,4(N3)	
T Hya	76400	8 55.7	-9 08	M	6.7	13.2	v	2441990	289.21	0.49	M3e-M9e	
AC UMa	--	8 55.9	+64 58	EA	9.2	13	p	2442521.58	6.85493	0.07	A2	

Star Name	HD	α 2000	δ 2000	Type	Max	Min		Epoch	Period	F	Spec	Notes
T Cnc	—	8ʰ56ᵐ.7	+19°51'	SRb	7.6	10.5	v		482ᵈ	0.35	C3,8–C5,5(R6–N6)	
V376 Car	77002	8 57.0	−59 14	βCs	4.91	4.96	V		0.0208		B2IV–V	
RT Cnc	76734	8 58.3	+10 51	SRb	7.12	8.6	V		60:		M5III	
FZ Vel	77140	8 58.9	−47 14	δSct	5.15	5.17	V		0.065		Am	
CQ Vel	—	8 58.9	−53 20	Na	9.0	>16.5	p	2429739				N 1940
TY Pyx	77137	8 59.7	−27 49	E/RS	6.87	7.47	V	2443187.230	3.198584		G5 + G5	
CV Vel	77464	9 00.6	−51 33	EA	6.5	7.3	p	2442048.6689	6.889494	0.10	B2V + B2V	
GP Vel	77581	9 02.1	−40 33	E(X)	6.76	6.93	V	2441738.257	8.962		B0.5Ibe	Vela X-1
CW Vel	77756	9 02.3	−52 51	EA	9.5	10.6	p	2429037.307	2.360917	0.17	B9	
UX Lyn	77443	9 03.8	+38 45	SRb?	6.60	6.78	V				M6III	
T Pyx	—	9 04.7	−32 23	Nr	6.3	14.0	v	2439501	7000:		P	
S Pyx	78000	9 05.1	−25 05	M	8.0	14.2	v	2442533	206.44	0.45	M3e–M5e	N 1966
TT UMa	—	9 05.2	+60 17	Lb	8.9	9.5	v				M6III	
V345 Car	78764	9 05.6	−70 32	γC	4.67	4.78	V				B2Vne	
κ Cnc	78316	9 07.7	+10 40	αCV?	5.22	5.27	V	2439633.5	5.0035		B9IIIpHgMn	76 Cnc
λ Vel	78647	9 08.0	−43 26	Lc	2.14	2.22	V				K4Ib–II	
BG Vel	78801	9 08.3	−51 26	Cδ	7.43	7.91	V	2434918.8	6.92357	0.32	G5	
W Cnc	78585	9 09.9	+25 15	M	7.4	14.4	v	2443896	393.22	0.40	M6.5e–M9e	
RS Cnc	78712	9 10.6	+30 58	SRc?	6.2	7.7	p		120:		M6eIb–II(S)	
V357 Car	79351	9 11.0	−58 58	E?	3.41	3.44	V		6.751154		B2IV–V	a Car
GX Vel	79186	9 11.1	−44 52		5.00 Y (0.07)		V				B5Ia	
KM Hya	79193	9 12.4	−7 07		6.11 Y						A3m	21 Hya
SY Vel	79402	9 12.4	−43 47	SRb	7.60	8.06	V		63		M5	
GG Vel	79459	9 12.8	−43 29	EA	8.0	8.5	p	2438379.575	1.475205		A0	
WX Hya	—	9 13.4	−14 23	M	9.5	>12.5	v	2438735	235.78		M3e	
IQ Hya	—	9 13.5	−23 24	M	5.0	6.2	I				Ne	
DD UMa	79439	9 16.2	+54 01	δSct?	4.83 Y (0.03)		V		0.13:		A5V	18 UMa
UZ Hya	—	9 16.8	−4 36	M	8.8	14.0	v	2438370	261.13*		M4e	
RT UMa	—	9 18.4	+51 24	Lb	8.6	9.6	v				C4,4	
RW Car	81055	9 19.6	−68 46	M	8.5	15.0	v	2438991	318.62	0.47	M4e–M7e	
RW Vel	80837	9 20.3	−49 31	M	8.9	>13.1	p	2430170	451.7		M7III(II)e	
IN Hya	80567	9 20.6	+0 11	SRb	6.27	6.75	V		45:		gM4	
QV Car	81099	9 21.3	−58 41	SRb	8.51	8.76	V	2440689:	75		M5III	
CG UMa	80390	9 21.7	+56 42	Lb	5.87	5.95	V				M4IIIa	
WY Vel	81137	9 22.0	−52 34	Z And	8.8	10.2	p				M3Ib:ep + B	
V Vel	81222	9 22.3	−55 58	Cδ	7.20	7.93	V	2434909.719	4.370991	0.30	F6–F9	
KU Hya	81009	9 22.8	−9 50		6.53 Y (0.05)		V		3.41		A5pSrCrEu	
DF Leo	81028	9 23.5	+7 43	SRb	6.74	6.95	V		70:		gM4	
RS Vel	81432	9 23.9	−48 52	M	9.1:	>12.6	p	2429810	409.5		M7e	
RZ Hya	—	9 24.8	−6 48	M	9.2	>12.5	v	2439882	332.54		M4e–M5	
IL Hya	81410	9 24.8	−23 50	SR	7.45	7.95	V		25.4		K2IV	
GI Vel	81576	9 25.0	−45 23	SRb	7.88	8.05	V		120:		Mb	
GK Vel	81575	9 25.1	−43 59	SRb	6.26	6.5	V		120:		M5III	
V377 Car	—	9 25.5	−57 22	EA/DM	8.05	8.34	V		2.242624		B4V	
GL Vel	81922	9 26.7	−53 31	SRb	7.31	8.01:	V	2440644	117		Mb	
IW Car	82085	9 26.9	−63 38	RVb	7.9	9.6	p	2429401	67.5		F7–F8	
Y Vel	—	9 29.0	−52 11	M	8.1	14.2	v	2440875	444.61	0.40	M8e–M9.5	
N Vel	82668	9 31.2	−57 02	L?	3.10	3.16	V				K5III	
R Car	82901	9 32.2	−62 47	M	3.9	10.5	v	2442000	308.71	0.48	M4e–M8e	
S Ant	82610	9 32.3	−28 38	EW	6.4	6.92	V	2435139.929	0.648345		A9Vn	
S Vel	82829	9 33.2	−45 13	EA	7.74	9.50	V	2438921.9020	5.9336663	0.10	A5Ve + K5IIIe	
U Vel	82850	9 33.2	−45 31	SR	7.87	8.19	V		37		M5	
T Ant	—	9 33.8	−36 37	Cδ?	8.89	9.76	V	2436120.578	5.89771	0.22	F6Iab	
ζ Cha	83979	9 33.9	−80 56		5.06	5.17	V				B5IV	
DK UMa	82210	9 34.5	+69 50		4.56 Y (0.058)		V		0.9202		G4III–IV	24 UMa
X Hya	83048	9 35.5	−14 42	M	8.0	13.6	v	2438077	301.44	0.42	M7e–M7.5e	
RR Ant	83199	9 35.7	−39 54	Lb	9.4	10.9	p				M5–M5.5II	
IM Vel	83368	9 36.4	−48 45		6.17 Y (0.05)		V		1.428		ApSrEuCr	
ST Hya	—	9 37.9	−20 39	M	8.8	14.4	v	2440282	304.75		Me	
Y Dra	83114	9 42.4	+77 51	M	6.24 V	15.0	v	2441213	325.49	0.45	M5e	
IP Vel	84400	9 43.5	−51 14	E	6.15 Y (0.2)		p				B6V	
W UMa	83950	9 43.8	+55 57	EW	7.9	8.63	V	2441004.3977	0.33363696		dF8p + F8p	
RR Hya	84474	9 45.0	−24 01	M	8.6	14.5	v	2438184	343.35	0.50	M4e	
ZZ Car	84810	9 45.2	−62 30	Cδ	3.28	4.18	V	2440736.9	35.53584	0.26	F6Ib–K0Ib	1 Car
R LMi	84346	9 45.6	+34 31	M	6.3	13.2	v	2442462	371.93	0.41	M6.5e–M9e	
CS UMa	84335	9 46.5	+57 08		5.2 Y (0.16)		p				M3IIIab	
R Leo	84748	9 47.6	+11 26	M	4.4	11.3	v	2441688	312.43	0.43	M8IIIe	
GM Vel	85008	9 47.7	−46 39	SRb	8.83	9.10	V		120		M6	
XX Ant	85207	9 49.4	−38 20	EB/DM	8.7	9.2	p	2438441.425	8.1070		F2	
DG Leo	85040	9 49.8	+21 11	δSct	6.08	6.12	V	2440654.6465	0.08184505	0.50	G5Iab–Ib	20 Leo
SZ Vel	—	9 50.1	−44 39		9.4	10.9	p		150:		M5e	
υ UMa	84999	9 51.0	+59 02	δSct	3.77	3.86	V	2439149.823	0.133		F2IV	29 UMa
Y Hya	85405	9 51.1	−23 01	SRb	8.3	12.0	p		302.8		N3p(C5,0)	
Z Vel	298731	9 52.9	−54 11	M	7.8	14.8	v	2440742	421.56	0.46	M9e	
S LMi	85597	9 53.7	+34 55	M	7.9	14.3	v	2441312	233.76	0.42	M4e–M7e	

Star Name	HD	α 2000	δ 2000	Type	Max	Min		Epoch	Period	F	Spec	Notes
QX Car	86118	9^{h}54^m.6	−58°25′	EA/DM	6.60	7.21	V	2440701.3715	4^d.47804	0.06	B5V + B5V	
V366 Car	--	9 54.8	−57 19	Z And	5.92	8.60	J				Pe	
GX Car	--	9 55.4	−58 26	Cδ	8.94	9.77	V	2440741.13	7.19673	0.29	F8II−K0	
SY UMa	85795	9 55.7	+49 49	cst?	5.23		V				A3III	31 UMa
V367 Car	86441	9 56.8	−57 39	EB/DM	7.49	7.59	V	2442468.79	5.73		B6V	
RR Car	86655	9 58.1	−58 52	SRb	9.1	10.4	p				M6.5SII−III	
V Leo	86608	10 00.0	+21 16	M	8.4	14.6	v	2441335	273.35	0.44	M5e	
YY Ant	87023	10 01.5	−38 05	SRb	8.74	9.0	V		120:		Mb	
RY Leo	--	10 04.3	+13 59	SRb	9.5	12.0	p		155	0.47	M2e	
HW Hya	87555	10 05.3	−21 21	Lb	8.39	8.59	V				Mb	
R Vel	87816	10 06.1	−52 11	cst	6.49		V				K1III	
IO Hya	87870	10 07.5	−22 29	SRb	6.87	7.02	V		80:		gM4	
CM Vel	88028	10 07.5	−53 16	SRc	8.7	11.0	p	2428780	780		M0	
FS Hya	--	10 07.6	−16 41	SR	9.5	11.5	p		166.7		M	
S Car	88366	10 09.4	−61 33	M	4.5	9.9	v	2442112	149.49	0.51	K5e−M6e	
R Ant	88262	10 09.8	−37 44	cst	7.9		p				A0	
V368 Car	88647	10 11.6	−58 50	Lb?	6.1	6.40	V				M5III	
QY Car	88661	10 11.8	−58 04	γC	5.63	5.83	V				B2IVpne	
AB Ant	88539	10 11.9	−35 19	Lb?	6.79	6.89	V				C6,3(N0)	
RT Sex	88517	10 12.3	−10 19	SRb	8.0	8.5	V		96		M6	
V347 Car	89143	10 13.6	−73 20	EA/DM	8.5	9.0	p	2438474.425	5.72555		A3m	
U UMa	88651	10 15.1	+59 59	cst?	6.25		V				M0IIIv	
W Vel	299011	10 15.3	−54 29	M	8.3	14.0	v	2440699	394.72	0.44	M8IIIe	
RW LMi	--	10 16.1	+30 34	M	6.9	>10.3	I	2439508:			C	
GY Vel	89273	10 16.7	−51 12		6.30 Y (0.32)		V				M4−M5III	
V337 Car	89388	10 17.1	−61 20	Lc	3.36	3.44	V				K3II	q Car
WZ Vel	89356	10 17.6	−47 57	SRb	9.0	10.0	p		130:		M3	
GZ Vel	89682	10 19.6	−55 02		6.37	6.44	V				F0IV	
HP Car	89714	10 19.6	−57 24	EA	8.85	9.3	V	2424348.182	1.6004464	0.12	B2IIIn	
AD Leo	--	10 19.7	+19 52	UV	9.41	10.94	B				M4.5Ve	
VY Hya	89639	10 20.3	−23 09	EA	9.0	11.3	p	2423535.601	2.00119519	0.09	A3	
EV Car	89845	10 20.4	−60 27	SRc	9.2	10.6	p	2429475	347:		M4.5Ia	
RY Vel	89841	10 20.7	−55 19	Cδ	7.89	8.72	V	2440246.67	28.1270	0.22	F6−G3	
RS Sex	89688	10 21.0	+2 17	cst?	6.64	6.68	V				B2.5IV	23 Sex
V Ant	--	10 21.2	−34 48	M	9.2	>12.5	p	2428608	302.76		M7IIIe	
AQ Car	89991	10 21.4	−61 04	Cδ	8.55	9.15	V	2436188.449	9.76896	0.50	F8−G0Ib	
HR Car	90177	10 22.9	−59 37	S Dor	8.2	9.6	p				B2eq	
U Leo	--	10 24.1	+14 00	N?	9.5	>14	v					N 1855
CK Car	90382	10 24.4	−60 11	SRc	9.2	10.5	p	2429500	525:		M3.5Iab	
HS Hya	90242	10 24.6	−19 06	EA	8.5:	9.2	p	2441374.5954	1.568035		F3−F4	
DE Leo	90254	10 25.3	+8 47	SRb?	5.60	5.67	V				M3IIIab	44 Leo
RX Sex	90386	10 26.2	+3 56	δSct	6.6	6.62	v	2441303.024	0.0799		A3V	
X Oct	91620	10 26.2	−84 21	SRa	8.7	12.7	p	2428885	205		M3e−M6IIIe	
UW Car	--	10 26.8	−59 40	Cδ	8.98	9.86	V	2434897.059	5.345773	0.28	G0	
V348 Car	90707	10 27.0	−57 41	EB	8.55	8.93	V	2439637.3	5.562107		B1III + B	
CX Leo	90569	10 27.6	+9 46	αCV	5.97	6.15	V				A0pSiCr:	45 Leo
YZ Car	90912	10 28.3	−59 21	Cδ	8.24	9.08	V	2434907.04	18.1631	0.41	G5	
UX Car	91039	10 29.2	−57 37	Cδ	7.81	8.67	V	2434906.805	3.682246	0.26	F5−G1II	
UY Leo	--	10 29.4	+23 04	Lb	9.5	11	p				gM7	
V349 Car	91093	10 29.6	−57 58	Lc	7.76	8.31	V				M2Iab	
PP Car	91465	10 32.0	−61 41	γC	3.27	3.37	V				B5Vne	p Car
UY Car	--	10 32.1	−61 47	Cδ	8.54	9.33	V	2434890.645	5.543726	0.27	G	
ρ Leo	91316	10 32.8	+9 18		3.85 Y (0.07)		V				B1Ib	47 Leo
Y Car	91595	10 33.2	−58 30	Cδ	7.53	8.48	V	2441041.39	3.639760	0.29	F3	
V369 Car	91619	10 33.4	−58 11	αCyg	6.11	(0.1)	V		20:		B6Iae	
QS Car	91908	10 34.3	−71 10	EA/DM	9.0	9.4	p	2428656.315	9.3208		F2−F5	
S Sex	91637	10 34.9	−0 20	M	8.2	13.5	v	2440620	261.0	0.50	M3e−M5e	
TX Leo	91636	10 35.0	+8 39	EA	5.66	5.75	v	2438844.3055	2.4450566	0.08	A2V	49 Leo
U Ant	91793	10 35.2	−39 34	Lb	8.1	9.7	p				C5,3(Nb)	
RZ Car	92090	10 35.6	−70 43	M	8.8	>15.0	v	2439586	272.77	0.42	M4e−M8e	
V361 Car	--	10 35.7	−58 15	Lc	7.09	7.57	V				M1.5Iab−Ib	
V379 Car	--	10 36.0	−58 14	βC+E?	8.21	9.34	V		0.1753		B1III	
V380 Car	--	10 36.0	−58 15	βC?	8.95	(0.02)	V		0.236:		B0.5III	
UZ Car	--	10 36.3	−61 01	Cδ	9.00	9.62	V	2434894.69	5.20466	0.29	G0	
HX Hya	92017	10 37.1	−23 54	SRb	8.63	8.82	V	2440689	50		Mb	
V370 Car	92207	10 37.4	−58 44	αCyg	5.45	5.52	V				A0Iae	
U Hya	92055	10 37.6	−13 23	SRb	7.0	9.2	v		450:		N2(C7,3)	
FF Hya	92096	10 37.9	−12 01	SRb	8.2	10.3	p	2440662	85		Mb	
HW Car	92490	10 39.3	−61 09	Cδ	9.0	9.8	p	2424404.94*	9.2002	0.50	G2−G3 Ib−II	
V364 Car	92664	10 40.2	−65 06	αCV	5.48	5.52	V	2442428.81	1.668		B8IIIpSi	
RX LMi	92620	10 42.2	+31 42	SRb	5.98	6.16	V		150:		M2III	
RZ Cha	93486	10 42.4	−82 02	EA/DM	8.2	9.1	p	2441401.7711	2.832084	0.11	F5 + F5V	
QZ Car	93206	10 44.4	−60 00	EB	6.16	6.49	V	2443192.4	5.9981		O9III	
R UMa	92763	10 44.6	+68 47	M	6.7	13.4	v	2442587	301.68	0.39	M3e−M9e	
VY Car	93203	10 44.6	−57 34	Cδ	6.87	8.05	V	2410009.58	18.990	0.36	F6−G4 Iab−Ib	

Star Name	HD	α 2000	δ 2000	Type	Max	Min		Epoch	Period	F	Spec	Notes
RT Car	--	10^h44.^m8	-59°25'	Lc	8.2	9.9	v		d		M2 0-Ia	
SV Vel	93247	10 44.9	-56 17	Cδ	7.93	9.07	V	2436195.125	14.09707	0.34	F6-G5	
VY UMa	92839	10 45.1	+67 25	Lb	5.89	6.5	V				C5II	
η Car	93308	10 45.1	-59 41	S Dor	-0.8	7.9	v				Pe	
TX UMa	93033	10 45.3	+45 34	EA	7.06	8.76	V	2442961.0934	3.0632253	0.13	B8 + F2	
BO Car	93420	10 45.8	-59 29	Lc	7.18	8.5	V				M4Ib	
SX Car	93444	10 46.1	-57 33	Cδ	8.66	9.47	V	2435074.336	4.8600	0.30	F5-G2 Ib-II	
AC Vel	93468	10 46.3	-56 50	EB	8.5	9.0	p	2429342.594	4.5622426		B6	
VV Leo	--	10 48.8	+8 40	SR	9.5	11	p		181.5		M7	
IX Car	94096	10 50.4	-59 59	SRc	9.0	10.0	p	2428900	400		M2Iab	
RS Hya	94103	10 51.3	-28 38	M	9.2	14.4	v	2439161	334.90	0.45	M6e	
W Leo	94362	10 53.6	+13 43	M	8.9	14.8	v	2438925	385.41	0.35	M6e-M8.8	
BZ Car	94613	10 54.1	-62 03	SRc	8.9	10.8	p		97		M3Ib	
T Car	94776	10 55.3	-60 31	cst?	7.00		B				K0III	
WZ Car	94777	10 55.3	-60 56	Cδ	8.65	10.01	V	2444143.17	23.0132	0.17	F8	
VY Leo	94705	10 56.0	+6 11	Lb?	5.69	6.03	V				M5.5IIIv:	56 Leo
AG Car	94910	10 56.2	-60 27	S Dor	7.1	9.0	p				B0I-A2Ieq	
XX Car	310331	10 57.1	-65 08	Cδ	8.67	9.89	V	2436221.730	15.71624	0.33	G0	
U Car	95109	10 57.8	-59 44	Cδ	5.72	7.02	V	2437320.055	38.7681	0.21	F6-G7Iab	
VW UMa	94902	10 59.0	+69 59	SR	6.85	7.71	V		125		M2	
AM Leo	--	11 02.2	+9 54	EW	8.2	8.9	p	2439936.8337	0.36579720		F8	
XY Car	--	11 02.3	-64 16	Cδ	8.82	9.77	V	2436190.230	12.43483	0.39	G5	
XZ Car	--	11 04.2	-60 59	Cδ	8.05	9.13	V	2436205.754	16.6499	0.37	K5	
GV Car	96368	11 05.6	-58 44	EA/DM	8.92	9.32	V	2423828.433	4.294621	0.10	A0	
RW Crt	96297	11 06.0	-9 09	SRb	8.71	8.94	V	2440683	77		Mb	
χ² Hya	96314	11 06.0	-27 17	EB	5.6	5.9	p	2442848.6107	2.267701		B8III-IV + B8.5V	
V385 Car	96548	11 06.3	-65 31	E11?+WR	7.70	(0.04)	V	2442000.0	4.762		WN8	
V815 Cen	96616	11 07.3	-42 38	αCV	5.14	(0.03)	V		2.433		A3pSr	
RW Cen	96650	11 07.3	-55 07	SRa	8.6	10.3	v	2438134	185.2	0.53	N3	
CI Cha	--	11 07.4	-81 47	Lb?	8.82	8.95	V				C	
RS Car	96830	11 08.1	-61 56	Na	7.0	>15.8	p	2413285:			P(Q)	N 1895
CU Cha	--	11 08.1	-77 39	Ina	8.38	8.48	V				B9-A0Vpe	
V382 Car	96918	11 08.6	-58 59	Cδ?	3.84	4.02	V				F8-G2Ia	
V371 Car	96919	11 08.6	-61 57	αCyg	5.12	5.19	V		20 :		B9.5Iae	
CO UMa	96813	11 09.3	+36 19	Lb?	5.79	5.95	V				M3.5IIIab	
ER Car	97082	11 09.7	-58 50	Cδ	6.58	7.13	V	2440277.88	7.71855	0.34	F8-G1 Iab-Ib	
V353 Car	97151	11 10.0	-60 06	γC	7.59	7.78	V				B2Ve	
GH Car	--	11 10.7	-60 45	Cδs	9.00	9.35	V	2435069.395	5.72557	0.44	G0	
S Leo	--	11 10.8	+5 28	M	9.0	14.5	v	2438442	189.41	0.48	M3e-M6e:	
EM Car	97484	11 12.1	-61 06	EA/DM	8.73	9.0	B	2429551.574	3.41427	0.14	O8V + O8V	
IT Car	97485	11 12.2	-61 45	Cδ	7.90	8.29	V	2437299.766	7.53320	0.36	F8-K2 Iab-Ib	
U Crt	--	11 12.8	-7 18	M	9.5	>13.5	v	2428269	169	0.45	M0e	
TT Hya	97528	11 13.2	-26 28	EA	7.5	9.5	p	2424615.388	6.9534124	0.10	A3e + dG6p	
GI Car	97746	11 14.0	-57 55	Cδs	8.10	8.47	V	2434924.602	4.43061	0.46	F3-F8 Iab-Ib	
HY Hya	97754	11 14.6	-26 04	Lb	7.47	7.60	V				Mb	
SY Crt	97918	11 15.7	-12 36	Lb?	6.34	6.62	V				M4III	
SV Crt	98088	11 17.0	-7 08	αCV	6.32	6.35	B	2440373.62	5.90513		A8IVpSrCrSi	
RX Crt	98218	11 17.8	-22 09	SRb	7.3:	7.7:	V		300:		M3	
ξ UMa	98230	11 18.2	+31 32	RS	4.87		V		3.9805		G0V	53 UMa B
RS Cen	98678	11 20.5	-61 52	M	7.75	14.1	V	2441966	164.37	0.46	M1Ibe-M5(III)e	
AY Cen	99325	11 25.1	-60 44	Cδ	8.58	9.16	V	2436733.520	5.30975	0.32	G1I-G5	
AZ Cen	99355	11 25.2	-61 22	Cδs	8.41	8.80	V	2435223.36	3.21068	0.40	F7	
HZ Hya	99448	11 26.3	-25 45	SRb	7.63	8.18	V	2440689:	95		Mb	
V771 Cen	99619	11 27.0	-61 22	SRc?	6.87	(0.2)	V				M2Ib-II	
ST UMa	99592	11 27.8	+45 11	SRb	7.7	9.5	B		81		M4III	
AF Leo	99635	11 27.9	+15 09	SRb	9.5	11	p		107	0.43	M5	
MN Cen	99769	11 28.1	-61 25	EA/DM	8.6	9.0	p	2424918.58	3.48916	0.17	B2-B3V	
V808 Cen	99953	11 29.3	-63 33	αCyg	6.42	6.47	V				B2Iaeα	
AW UMa	99946	11 30.1	+29 58	EW	6.84	7.10	V	2443948.7928	0.43872687		F0-F2	
V419 Cen	100148	11 30.9	-56 54	Cδs	7.98	8.36	V	2440760.28	5.50691	0.42	F7II-K0	
TU Mus	100213	11 31.2	-65 44	EB	8.17	8.75	V	2441699.8270	1.3872833		B3 + B3	
CZ Leo	100141	11 31.3	-4 18	SRb	8.51	9.01	V	2440680:	115		M5	
V809 Cen	100198	11 31.3	-61 17	αCyg	6.31	6.36	V				A3Iae	
RR Crt	--	11 31.7	-12 23	SRb?	9	10.5	p	2431273:			M5	
o¹ Cen	100261	11 31.8	-59 27	SRd	5.8	6.6	B		200:		G3 0-Ia	
o² Cen	100262	11 31.8	-59 31	αCyg	5.12	5.22	V		46.3:		A2Iaeα	
SU Crt	100363	11 32.8	-12 02	δSctc	8.62	8.65	V		0.055		F2V	
V763 Cen	100733	11 35.2	-47 22	SRb	5.55	5.80	V		60:		M3III	
V785 Cen	--	11 35.6	-47 10	Lb?	7.6	(0.31)	V				M3II-III	
SS Crt	100766	11 35.7	-17 56	SRb	8.54	8.82	V	2440692	65			
BF Cen	100915	11 36.3	-61 28	EA/DM?	8.5	9.4	p	2424262.28	3.69334	0.20	B7	
V646 Cen	100987	11 37.0	-53 13	EA/SD	9.0	11.7	p	2443916.1946	2.24657322	0.15	B8IV	
LW Cen	--	11 37.5	-63 21	EB/KE	8.90	9.65	V	2424824.462	1.0025674		B1.5V	
V816 Cen	101065	11 37.6	-46 43	δSct?	7.996	8.020	V	2443400.00090	0.00843060		F8p	
ω Vir	101153	11 38.5	+8 08	Lb	5.23	5.37	V				M4III	1 Vir

Star Name	HD	α 2000	δ 2000	Type	Max	Min		Epoch	Period	F	Spec	Notes
V420 Cen	--	11^{h}39^m.8	-47°58'	CWa	9.37	10.60	V	2425350.67	24^d.7678	0.27		
AI Leo	101441	11 40.5	+11 12	Lb	8.44	10.5	V				M5	
AK Leo	101487	11 40.8	+13 05	Lb	8.4	9.4	v		60:		M5	
UZ Leo	101602	11 41.0	-62 42	Cep	8.30	9.12	V	2440746.1	3.33434	0.27	F3Ib-II	
RU UMa	101605	11 41.7	+38 29	M	8.5	14.0	v	2440634	252.46		M3.5e-M5e	
V772 Cen	101712	11 41.8	-63 25	Lc?	7.80	8.36	V				M2Ibep + B	
V346 Cen	101837	11 42.8	-62 26	EA	8.48	8.9	B	2421963.674	6.32227	0.11	B3II-III	
V644 Cen	--	11 43.1	-60 44	γC?	8.7	10.2	p				B2IIIe	
V810 Cen	101947	11 43.5	-62 29	SRd	4.95	5.12	V		130:		F5-G0 0-Ia + B1Iab	
MT Cen	--	11 44.0	-60 33	Na	8.5	>15	p	2426473				N 1931
RT Mus	310831	11 44.5	-67 18	Cδ	8.58	9.31	V	2434957.758	3.08608	0.33	F8	
TV UMa	102159	11 45.6	+35 54	SRb	8.3	9.2	p		50.38		M5III	
CX Mus	102545	11 47.7	-68 59	EA	8.7	9.3	p	2428687.275	5.90322		B9	
SV Cen	102552	11 48.0	-60 34	EB/KE?	8.71	9.98	V	2444061.0600	1.658500		B1V + B6.5II-III	
V801 Cen	102567	11 48.0	-62 12	γC?(X)	8.93	9.39	V				B1Vne	
μ Mus	102584	11 48.2	-66 49	Lb	4.6	4.8	V				K4III	
II Hya	102620	11 48.8	-26 45	SRb	4.85	5.12	V	2440684	61		M4III	
X Cen	102681	11 49.2	-41 45	M	7.0	13.8	v	2441709	315.1	0.41	M5e-M6.5e	
LZ Cen	102893	11 50.6	-60 48	EB	8.10	8.50	B	2426096.384	2.757717		B2III	
RU Crt	102946	11 51.1	-11 12	Lb?	8.5	9.5	p				M3	
TY Vir	103036	11 51.8	-5 46	SRd	8.00	8.32	V		50:		G3Ibp	
UU Mus	103137	11 52.3	-65 24	Cδ	9.22	10.25	V	2436208.270	11.63641	0.50	G0	
VZ Cen	103146	11 52.5	-61 31	EA	8.34	8.6	B	2429125.519	4.9287012	0.2	B2III-IV	
β Hya	103192	11 52.9	-33 54		4.2	4.24	B		1.94		B9IIIpSi	
CF UMa	--	11 53.0	+37 43	UV?	8.5	12	v					real?
AD Cen	--	11 53.2	-59 19	Lc?	9.4	11.4	p				K3(II)-M3e	
W Cen	103513	11 55.0	-59 15	M	7.60	13.7	V	2441786	201.57	0.47	M3e-M8(III)e	
DN UMa	103483	11 55.1	+46 29		6.54 Y (0.09)		B		0.87:		A3Vn	65 UMa A
Z UMa	103681	11 56.5	+57 52	SRb	7.9	10.8	p	2439368	196	0.05	M5IIIe	
AE Cru	104012	11 58.6	-61 10	EA/D	9.0	9.7	p	2430399.114	3.4781475	0.12	B7III	
DZ Mus	104191	11 59.9	-69 53	EA	8.3	8.9	p	2418093.728	3.247619		A0	
GK Com	104207	12 00.1	+19 25	SRb	6.84	7.13	V		50		M4III	
SV Vir	--	12 00.4	-10 12	M	9.3	>12.5	v	2438380	295.60		M4e	
AG Vir	104350	12 01.1	+13 00	EW	8.4	8.98	V	2439946.7472	0.64265057		A2+A2	
X Vir	--	12 01.9	+9 04	?	7.3	11.2	v				FIV-Vp	
DP UMa	104513	12 02.1	+43 03	δSct	5.21 Y (0.02)		V		0.06		A7m	67 UMa
R Com	104785	12 04.0	+18 49	M	7.1	14.6	v	2443539	362.82	0.38	M5e-M8ep	
θ² Cru	104841	12 04.3	-63 10	βC?	4.70	4.74	V		0.0889		B3IV	
TZ Vir	104851	12 04.6	+2 37	SRb	9.4	11.1	p		134		M5	
RX Vir	104886	12 04.8	-5 46	SR?	8.7	9.1	p		200:		K0	
SU Vir	104959	12 05.2	+12 22	M	8.4	14.5	v	2442190	209.95	0.48	M2e-M5.5e	
RW Vir	105266	12 07.2	-6 46	Lb	8.6	9.1	p				M5III	
δ Cen	105435	12 08.4	-50 43	γC	2.51	2.65	V				B2IVne	
V817 Cen	105521	12 08.9	-41 14	γC	5.47	5.58	V				B3IVeα	
V788 Cen	105509	12 08.9	-44 20	EA/D	5.74	5.93	V	2441370.496	4.966377	0.08	A2m(A5-F2)	
RU Cen	105578	12 09.4	-45 25	RV	8.7	10.7	p	2428015.51	64.727		A7Ib-G2pe	
W Cru	--	12 12.0	-58 47	EB/GS	9.04	10.38	B	2440731.6	198.53		G2Iabe	
GM Com	106103	12 12.4	+27 23	δSctc	8.06	8.14	V		0.208:		F5V	
S Mus	106111	12 12.8	-70 09	Cδ	5.90	6.44	V	2435837.992	9.66011	0.49	F6Ib	
FG Vir	106384	12 14.3	-5 43	δSct	6.53	6.58	V		0.07		dF3	
AH Vir	106400	12 14.3	+11 49	EW	9.0	9.61	V	2435245.6522	0.40752189		K0V + K0V	
T Vir	106430	12 14.6	-6 02	M	9.0	14.8	v	2440277	339.46	0.36	M6e	
V369 Cen	106474	12 15.0	-54 49	SRb	8.02	(0.51)	V		70:		M5II	
δ Cru	106490	12 15.1	-58 45	βC	2.78	2.84	V		0.151038	0.40	B2IV	
DK Dra	106677	12 15.7	+72 33	RS	6.29 Y (0.15)		V		64.44		K0III + K0III	
BH Cru	--	12 16.3	-56 17	M	7.2	10.0	V	2440858	421		SC4.5,8e-SC7,8e	
AB Cru	106871	12 17.6	-58 10	EA/DM	8.56	9.2	B	2429235.019	3.4132987	0.19	O8Vne	
ε Mus	106849	12 17.6	-67 58	SRb?	3.99	4.31	V		40		M5III	
AO Cru	106873	12 17.8	-63 37	Lc	8.5	10.0	p				M0Ia-Iab	
FM Com	107131	12 19.0	+26 00	δSctc	6.40	6.48	V		0.0551		A5-A7mIV-V	
R Crv	107199	12 19.6	-19 15	M	6.7	14.4	v	2442781	317.03	0.41	M4.5e-M9:e	
CH Vir	107317	12 20.4	-9 00	Lb	9.5	10.6	p				Mc	
RY UMa	107397	12 20.5	+61 19	SRb	6.68	8.5	V	2440810	311		M2-M3IIIe	
T Cru	107447	12 21.4	-62 17	Cδ	6.32	6.83	V	2434541.340	6.73331	0.34	F6-G2Ib	
AS Dra	107760	12 22.2	+73 15	RS	8.00	8.06	V	2435926.055	5.414905		G3V + K0V	
TT Crv	107814	12 23.3	-11 49	SR	6.47	6.57	V		11.5		M3III	
R Cru	107805	12 23.6	-61 38	Cδ	6.40	7.23	V	2434514.629	5.82575	0.28	F6-G2 Ib-II	
AI CVn	107904	12 23.8	+42 33	δSct	5.89	6.15	V		0.2085		F3IV	4 CVn
FK Vir	107937	12 24.2	+5 58	SRb	7.50	7.78	V		40		gM4	
XZ Cen	107913	12 24.2	-35 38	M	7.8	10.7	v	2430136	290.7		M5e	
GN Com	107966	12 24.3	+26 06	αCV?	5.15	5.18	V				A2V(Am)	13 Com
S Cen	107957	12 24.6	-49 26	SR	9.2	10.7	p		65		C4,5(Nbp)	
SS Vir	108105	12 25.3	+0 48	M	6.0	9.6	v	2440653	354.66	0.48	Ne(C5,3e)	
SS Dra	108345	12 26.3	+68 41	SRb	8.4	10.4	v	2427667	51.5	0.48	M5	
BL Cru	108396	12 27.5	-59 00	SR?	5.43	(0.35)	V				M4-M5III	

238

Star Name	HD	α 2000	δ 2000	Type	Max	Min		Epoch	.Period	F	Spec	Notes
FT Vir	108506	12^{h}27^{m}9	−4°37′	δSct	6.20	6.23	V		0^d.05		A8n	
AI Com	108662	12 28.9	+25 55	αCV+δSct?	5.23	5.40	V	2439586.07	5.0633		A0pCrEuSr	17 Com
CQ Dra	108907	12 30.1	+69 12	Lb?	4.95	5.04	V				M3IIIa	4 Dra
T CVn	108833	12 30.2	+31 30	M?	7.6	12.6	v	2442784	290.09	0.42	M6.5e	
BK Vir	108849	12 30.4	+4 25	SRb	7.28	8.8	V		150:		M7III	
UU Com	108945	12 31.0	+24 34	αCV+δSctc	5.41	5.46	V	2440334.194	2.1953		A3pSrCrEu	21 Com
BG Cru	108968	12 31.7	−59 25	Cδs	5.34	5.58	V	2440393.66	3.3428	0.47	F5Ib−G0p	
κ Dra	109387	12 33.5	+69 47	EB?	3.5	3.9	V		30:		B6IIIpe	5 Dra
U Cen	109231	12 33.5	−54 40	M	7.0	14.0	v	2439941	220.28	0.47	M3II:e−M5IIe	
Y Vir	−−	12 33.9	−4 25	M	8.3	15.0	v	2442532	218.43	0.46	M3e−M5e	
BO Mus	109372	12 34.9	−67 45	Lb	6.0	6.7	v					
TU Crv	109585	12 36.0	−20 32	δSctc	6.53	(0.025)	B		0.082		Mb	
T UMa	109729	12 36.4	+59 29	M	6.6	13.4	v	2442024	256.54	0.41	F0III	
BO Cru	−−	12 36.7	−61 41	M?	7.7	12.0	I				M4e−M7e	
α Mus	109668	12 37.2	−69 08		2.17	2.24	p		0.09:		M0−M10(S)e:	
											B2IV−V	
RV Dra	−−	12 37.6	+65 34	M	8.4	14.2	v	2442265	208.02	0.35	M1e−M3e	
RV Crv	109796	12 37.7	−19 35	EB	8.60	9.16	V	2441029.384	0.7472521		F0 + G0:	
FW Vir	109896	12 38.4	+1 51		5.71 Y	(0.1)	V		0.15		M3IIICa	
R Vir	109914	12 38.5	+6 59	M	6.0	12.1	v	2442512	145.64	0.50	M4.5IIIe	
RS UMa	110064	12 39.0	+58 29	M	8.3	14.8	v	2442363	258.97	0.42	M4.5e−M6e	
AX CVn	110066	12 39.3	+35 57	αCV	6.32	6.55	V				A0pSrCrEu	
FH Mus	110020	12 39.9	−66 31	E11?	6.26 Y	(0.025)	V		0.3:		B8V	
SX Crv	110139	12 40.2	−18 48	EW/KW	8.99	9.25	V	2441017.4557	0.3166386		F8	
AG Cru	110258	12 41.4	−59 48	Cδ	7.73	8.58	V	2434908.770	3.83728	0.27	F8Ib−II	
GG Vir	110377	12 41.6	+10 26	δSct	6.19 Y	(0.02)	V				A7Vn	27 Vir
R Mus	110311	12 42.1	−69 24	Cδ	5.93	6.73	V	2440896.13	7.47665	0.27	F7Ib	
UW Cen	−−	12 43.3	−54 32	RCB	9.1	>14.5	v				K	
S UMa	110813	12 43.9	+61 06	M	7.0	12.4	v	2442170	226.02	0.47	S0.5,9e−S5,9e	
TX CVn	−−	12 44.7	+36 46	Z And	9.2	11.8	p				B1−B9Veq + K0III−M4	
Y CVn	110914	12 45.1	+45 26	SRb	7.4	10.0	p		157		C5,4J(N3)	
FM Vir	110951	12 45.6	+7 40	δSct?	5.20	5.23	V		0.07		F0IIIm	32 Vir
BR Cru	−−	12 46.4	−56 30	αCV	8.70	8.82	V	2443014.000	2.8730	0.50	Ap	
X Cru	110945	12 46.4	−59 08	Cδ	8.10	8.70	V	2434939.562	6.21997	0.31	F6−G2Ib	
EP Vir	111133	12 47.0	+5 57	αCV	6.29	6.39	V	2440640.2	16.31		A0pSrCrEu	
RU Vir	111166	12 47.3	+4 09	M	9.0	14.2	v	2437827	436.52	0.49	C8,1e	
U CVn	111223	12 47.3	+38 23	M	8.8	>12.5	p	2442491	345.65		M7e	
β Cru	111123	12 47.7	−59 41	βC	1.23	1.31	V		0.2365072		B0.5III−IV	
SV Crv	111499	12 49.8	−15 05	SRb	6.78	7.6	V		70:		M5III	
U Vir	111691	12 51.1	+5 33	M	7.5	13.5	v	2442472	206.80	0.47	M2e−M8e	
BV Cru	−−	12 53.7	−60 21	βC	8.77	(0.05)	B		0.16		B0.5III	
BU Cru	111934	12 53.7	−60 21	E?	6.80	6.90	V	2443228.61			B1.5Ib	
ε UMa	112185	12 54.0	+55 58	αCV	1.76	1.79	V	2426437.0*	5.0887		A0pCr	Alioth
EF Mus	111953	12 54.3	−70 02	Lb?	8.0	8.8	p				K0	
ψ Vir	112142	12 54.4	−9 32	Lb	4.7	4.8	V				M3IIICa−1	40 Vir
S Cru	112044	12 54.4	−58 26	Cδ	6.22	6.92	V	2434973.520	4.68997	0.34	F6−G1 Ib−II	
μ² Cru	112091	12 54.6	−57 10	γC	4.99	5.18	V				B5Vne	
λ Cru	112078	12 54.7	−59 09	βC?	4.62	(0.02)	V	2441779.081	0.3951		B4Vne	
TU CVn	112264	12 54.9	+47 12	SRb	5.55	6.6	V		50		M5III	
α² CVn	112413	12 56.0	+38 19	αCV	2.84	2.98	V	2439012.61	5.46939	0.6	A0pSiEuHg	Cor Caroli
RY Dra	112559	12 56.4	+66 00	SRb	9.4	11.4	p		172.5:		C3,4	
V823 Cen	112381	12 57.0	−54 35	αCV	6.7	(0.005)	V		2.84		A0pSiCr	
V377 Cen	112455	12 57.3	−48 04	EA/DS	8.4	9.1	p	2430393.956	8.251658	0.06	A2V	
CN Vir	112737	12 58.8	+8 13	SRb	8.17	9.0	V		60:		M3	
BZ Vir	−−	13 00.7	−17 39	M	9.5	>13	p	2429701	150.92		M5e	
UY Vir	113158	13 01.9	−19 46	EA	8.5	9.2	p	2430020.667	1.9945051	0.14	A7V	
RZ Cen	−−	13 01.9	−64 38	EB/KE?	8.95	9.60	V	2429342.942	1.8759517		B2V	
RT Vir	113285	13 02.6	+5 11	SRb	9.0	10.3	p		155:		M8III	
CO Vir	−−	13 03.9	+7 04	SRb	8.49	9.54	V		70:		M5	
V789 Cen	113523	13 04.8	−41 12	Lb?	6.22	(0.09)	V				M3−M4III	
FS Com	113866	13 06.4	+22 37	SRb	5.30	6.1	V		58:		M5III	40 Com
θ Mus	113904	13 08.1	−65 18		5.51 Y						B0Ia + WC5:	
WW Cen	−−	13 09.4	−60 15	SRb	8.8	11.6	v		304		M5−M7	
RS CVn	114519	13 10.6	+35 56	EA/AR/RS	7.93	9.14	v	2441825.542	4.79781	0.11	F4IV−V + K0IVe	
V824 Cen	114365	13 11.0	−52 34	αCV	6.3	(0.035)	V		1.272		A0pSi	
SS Cen	114720	13 13.6	−64 09	EA/SD	9.4	11.0	p	2429552.475	2.4787192	0.14	B8V	
SW Vir	114961	13 14.1	−2 48	SRb	6.85	7.88	V	2440709	150:		M7III	
η Mus	114911	13 15.2	−67 54	E?	4.80 Y	(0.1)	V				B8V	
UW Vir	115122	13 15.4	−17 29	EA	8.90	12.20	V	2442545.481	1.810730	0.14	A2	
FH Vir	115322	13 16.4	+6 30	SRb	6.92	7.45	V	2440740	70:		M6III	
DK Vir	115308	13 16.4	−1 23	δSct	6.67	6.72	V	2439176.981	0.121		F1IV	
UY Cen	115236	13 16.5	−44 42	SR	9.22	11.2	B		114.6:		SC	
AO CVn	115604	13 17.5	+40 34	δSctc	4.70	4.75	V		0.12168	0.50	F3IIIp	20 CVn
HH Com	115708	13 18.6	+26 22	αCV	7.77	7.85	V	2441731.0	5.07		A2pSrCrEu	
V378 Cen	115514	13 19.0	−62 23	Cδs	8.27	8.68	V	2434917.113	6.45930	0.32	F5Iab−Ib − G5	
V819 Cen	−−	13 19.4	−58 12	RR?	9.00	(0.07)	V	2442460.59	0.6755:	0.6	A2	

239

Star Name	HD	α 2000	δ 2000	Type	Max	Min		Epoch	Period	F	Spec	Notes
V CVn	115898	13^{h}19^m.5	+45°32'	SRa	6.52	8.56	V	2443929	191^d.89	0.50	M4e-M6eIIIa:	
T Mus	115673	13 21.2	-74 27	SR	9.4	11.3	p		93		Np	
AV CVn	--	13 21.3	+43 59	Lb	9.5	12.1	p				S2.5,9	
V790 Cen	116087	13 22.6	-60 59	βC?	6.16	6.27:	V				B2.5Vn	
U Oct	115486	13 24.5	-84 13	M	7.1	14.1	v	2440366	302.57	0.47	M4e-M6 II-IIIe	
α Vir	116658	13 25.2	-11 10	E11(βC)	0.97	1.04	V	2419530.49	4.01454		B1III-IV + B2V	Spica
V379 Cen	116507	13 25.3	-59 47	EA/SD	8.8	9.6	p	2428402.23	1.874685	0.18	B5V	
RR UMa	--	13 25.9	+62 23	M	8.6	14.2	v	2442244	230.58	0.43:	M4e	
RR Cha	--	13 26.4	-82 20	Na	7.1	>15	p	2434478				N 1953
V Vir	117045	13 27.8	-3 10	M	8.1	15.0	v	2440664	249.63	0.42	M3e-M6e	
V743 Cen	116994	13 28.4	-51 17	δSct	8.57	8.82	V	2439243.6436	0.10225435	0.36	A0V	
EZ Mus	116890	13 28.8	-69 38		6.2	6.29	v		4.5		ApSi	
R Hya	117287	13 29.7	-23 17	M	3	11		2441676	389.61	0.48	M7IIIe	
FO Vir	117362	13 29.8	+1 06	RR? ✓	6.5	6.8	V		0.6:		A2	
SS Hya	117408	13 30.5	-23 39	EA?	7.88	8.1	B				B9	
FK Com	117555	13 30.8	+24 13	FK Com	8.14	8.33	V	2442192.345	2.400	0.40:	G2IIIpneα + K3V	
V701 Cen	117470	13 31.6	-51 46	EB/KE	8.8	9.3	v	2439243.2661	0.738447		B9V	
V659 Cen	117399	13 31.6	-61 35	Cδ	6.45	6.71	V	2440348.77	5.62180	0.43	F7Ib	
S Vir	117833	13 33.0	-7 12	M	6.3	13.2	v	2442414	377.43	0.45	M7IIIe	
S Cha	117360	13 33.3	-77 34	cst	6.51		V				F5V	
CW Vir	118022	13 34.1	+3 40	αCV	4.91	4.99	V	2434816.9	3.7220		A1pSrCrEu	78 Vir
T UMi	118556	13 34.7	+73 26	M	8.1	15.0	v	2442389	313.86	0.45	M5.5e	
BH CVn	118216	13 34.8	+37 11	RS	4.94	5.01	V	2443639.52	2.6131738		F2IV + KIV	
FP Vir	118289	13 35.9	+8 18	SRb	6.72	7.35	V		55:		gM4	
V764 Cen	118238	13 36.1	-33 29	SRd	8.84	9.13	V				K2III	
KN Cen	--	13 36.5	-64 33	Cδ	9.28	10.36	V	2436238.172	34.0457	0.21		
TV Hya	118412	13 37.2	-23 37	E?	8.0	8.2	p				A3	
RV Cen	118322	13 37.5	-56 29	M	7.0	10.8	v	2440960	446.0	0.56	N3e	
BZ Boo	118743	13 38.4	+27 17	δSctc	8.2	(0.03)	v		0.37:		A5	
V UMi	119227	13 38.7	+74 19	SRb	8.8	9.9	p		72.0		M5III	
V765 Cen	118781	13 39.7	-39 45	Lb?	6.2	(0.08)	V				M4	
ε Cen	118716	13 39.9	-53 28	βC	2.29	2.31	V	2441040.965	0.169608		B1III	
Z Cen	118843	13 40.0	-31 38	SNI	8.0	>20.5	p	2413383:			P	SN 1895
V744 Cen	118767	13 40.0	-49 57	SRb	5.14	6.55	V		90:		M8III	
XX Cen	118769	13 40.3	-57 37	Cδ	7.30	8.31	V	2440366.24	10.954348	0.49	F6-G4(F7-F8II)	
CQ UMa	119213	13 40.4	+57 12	αCV	6.24	6.26	V	2441450.74			A2VpSrCrEu	
T Cen	119090	13 41.8	-33 36	SRa	5.5	9.0	v	2443242	90.44	0.47	K0:e-M4II:e	
T Cir	119123	13 43.4	-65 28	EA/SD	9.3	10.6	p	2429095.586	3.2984345	0.13	B6-B8 II-III	
V827 Cen	119419	13 44.3	-51 01	αCV	6.46	(0.025)	V		2.605		A0pSi	
SX Hya	119592	13 44.6	-26 47	EA	8.9	11.9	p	2442540.415	2.895697	0.11	A3 + K5:	
CR UMa	120198	13 46.6	+54 26	αCV	5.64	5.67	V		>1		B9pEuCr	84 UMa
V766 Cen	119796	13 47.2	-62 35	S Dor?	6.17	7.50	V				G8 0-Ia	
RT Cen	--	13 48.4	-36 52	M	8.1	13.6	v	2442098	255.02	0.47	M6II:e	
R CVn	120499	13 49.0	+39 33	M	6.5	12.9	v	2443586	328.53	0.46	M5.5e-M9e	
W Hya	120285	13 49.0	-28 22	SRa	7.7	11.6	p	2440336	397.35	0.50	M7.5e-M9pe	
V806 Cen	120323	13 49.4	-34 27	SRb	4.16	4.26	V		12:		M5III	2 Cen
μ Cen	120324	13 49.6	-42 28	γC	2.92	3.47	V				B2IV-Ve	
SZ Cen	120359	13 50.6	-58 30	EA/D	8.3	8.9	p	2441386.7466	4.107983	0.15	A6III	
V381 Cen	120400	13 50.7	-57 35	Cep	7.32	8.01	V	2436201.953	5.07878	0.28	F6-G7(F8Ib-II)	
VX Cen	120460	13 51.2	-60 25	SR	9.5	12.8	v	2429800	307.8		S8,5e(M4-M8 II-III)	
CU Dra	121130	13 51.4	+64 43		4.46	4.94	V				M3.5III	10 Dra
RX Cen	--	13 51.4	-36 57	M	8.7	>15	v	2442114	327.90	0.38	M5e	
AW CVn	120933	13 51.8	+34 27	SR?	4.72	4.81	V				K5III	
V757 Cen	120734	13 51.9	-36 37	EW/KW	8.3	8.7	V	2442308.69312	0.34316929			
DL Vir	120902	13 52.6	-18 43	EA	7.0	7.5	v	2438796.475	1.315475	0.14	G8III + A	
V758 Cen	120738	13 52.7	-55 32	EW/KE	8.8	9.40	B	2444403.2797	0.58078556		B9IV	
V774 Cen	120958	13 53.5	-39 03	γC	7.63	7.72	V				B3Vne	
XZ Boo	--	13 53.9	+17 17	Lb	8.8	9.6	v				M5	
V767 Cen	120991	13 53.9	-47 08	γC	5.86	6.26	V				B2IIIe	
T Aps	--	13 55.9	-77 48	M	8.4	15.0	v	2441096	261.03	0.43	M3e	
ZZ Boo	121648	13 56.2	+25 55	EW/KW	5.8	6.40	V	2443692.4521	4.991775		G2V + G2V	
CP Vir	121713	13 57.0	+6 34	SRb	8.38	8.91	V		70:		M7	
V412 Cen	121518	13 57.5	-57 43	Lb	7.1	9.6	B				M3Iab-Ib - M7	
TW Cen	121714	13 57.7	-31 04	M	8.8	>12.6	p	2429310	269.27	0.43	M4e-M8II:	
V828 Cen	122532	14 03.5	-41 25	αCV	6.10	(0.045)	V		1.837		A0pSi	
AT Cir	122314	14 03.4	-66 44	EA/DM	8.4	8.8	p	2415221.517	3.257494	0.15	A5IV-V	
β Cen	122451	14 03.8	-60 22	βC	0.61	(0.045)	V		0.157		B1III	
AQ Cen	--	14 05.0	-35 30	M	9.3	>13.5	p	2429719	387.5		Me	
θ Aps	122250	14 05.3	-76 48	SRb	6.4	8.6	p		119		M7III	
χ Cen	122980	14 06.0	-41 11	βCs	4.15	(0.020)	B		0.035		B2V	
Z Boo	123304	14 06.5	+13 29	M	8.2	15.0	v	2441096	281.14	0.39	M5e-M6e	
ER Vir	123214	14 06.7	-14 12	SRb	6.45	6.63	V	2440752	55		Mb	
AB Boo	--	14 07.1	+20 45	N?	4.5							N 1877
BY Boo	123657	14 07.9	+43 51	Lb?	4.98	5.33	V				M4-M4.5III	
DM Vir	123423	14 07.9	-11 09	EA	9.0	9.7	p	2426087.505	4.669430	0.12	F6IV + F6IV	

Star Name	HD	α 2000	δ 2000	Type	Max	Min		Epoch	Period	F	Spec	Notes
CF Boo	123782	14h08m.3	+49°27'	Lb	5.2	5.3	v		d		M2IIIab	13 Boo
ES Vir	123576	14 08.6	-8 52	Lb	8.15	8.33	V				Mb	
FR Vir	123598	14 09.0	-19 15	Lb?	7.02	7.19	V				M3III	
V759 Cen	123732	14 10.7	-47 46	EW/KW	7.4	7.56	V	2442196.09732	0.39395129		F9V	
ET Vir	123934	14 10.8	-16 18	SRb	4.80	5.00	V	2440697:	80		M2III-IIIa	
RU Hya	--	14 11.6	-28 53	M	7.2	14.3	v	2438518	333.19	0.35	M6e	
CU Vir	124224	14 12.3	+2 25	αCV	4.98	5.05	V	2439995.4413	0.52067688		A0VpSi	
V760 Cen	--	14 12.5	-59 26	UV?	8.1	13	p					
EV Vir	124304	14 13.2	-13 52	SRb	6.74	7.09	V	2440726	120		Mb	
κ² Boo	124675	14 13.5	+51 47	δSctc	4.50	4.58	V		0.076242	0.50	A8IV	17 Boo
V716 Cen	124195	14 13.7	-54 38	EB/KE	5.96	6.52	V	2438524.4069	1.490096		B5V	
V820 Cen	--	14 13.9	-38 06	RV	8.42	>10.10	V	2443325	150		K0e	
FS Vir	124681	14 14.9	+3 20	Lb?	6.37	6.52	V				M4IIIab	
V795 Cen	124367	14 14.9	-57 05	γC	4.97	5.10	V				B4Vne	
CN Boo	124953	14 16.1	+18 55	δSctc	6.24	(0.03)	B		0.04:		A9V	
R Cen	124601	14 16.6	-59 55	M	5.3	11.8	v	2441942	546.2		M4e-M8IIe	
V636 Cen	124784	14 17.0	-49 57	EA/DM?	8.7	9.2	v	2434540.340	4.28398	0.04	G0V	
RR Cen	124689	14 17.0	-57 51	EW	7.27	7.68	V	2424231.0981	0.60569071		A9-F0V	
U UMi	125556	14 17.3	+66 48	M	7.4	12.7	v	2442113	326.51	0.50	M6e-M8e	
R Cam	127226	14 17.8	+83 50	M	6.97	14.4	V	2443978	270.22	0.45	S2,8e-S8,7e	
CS Vir	125248	14 18.6	-18 43	αCV	5.73	5.93	V	2433103.95	9.2954	0.50	A0pCrEu	
EY Vir	125356	14 19.2	-13 26	SRb	8.2:	8.6:	V	2440765	>100		Mc	
--	--	14 20	-60	SN	-6:		v	1788970:				SN of A.D. 185
EZ Vir	125624	14 21.1	-19 00	SRb	7.77	7.97	V	2440710:	39		Ma	
V418 Cen	125332	14 21.3	-64 14	Lc	8.7	9.5	p				K4II	
CQ Vir	125753	14 21.4	+6 27	SRb	8.99	9.64	V				M3	
V339 Cen	125465	14 21.8	-61 33	Cδ	8.40	9.17	V	2440768.15	9.4660	0.50	F7II-G5	
Y Boo	125920	14 22.0	+19 48	cst?	7.94		V				K0III	
CI Boo	126009	14 22.2	+29 22	Lb	6.47	6.81	V				M3III	
UV Boo	126030	14 22.5	+25 33	cst?	8.11	8.16	V				F5V	
S Boo	126289	14 22.9	+53 49	M	7.8	13.8	v	2444116	270.73	0.44	M3e-M6e	
V761 Cen	125823	14 23.0	-39 31	SX Ari	4.38	4.43	V	2442807.75	8.8171		B2V-B8IIIpHeSi	
BS Cir	125630	14 23.5	-66 39	αCV	6.7	(0.14)	V		2.205		A2pSiCr	
RX Boo	126327	14 24.2	+25 42	SRb	8.6	11.3	p		340:		M6.5e-M8IIIe	
FF Vir	126515	14 25.9	+1 00	αCV	7.07	7.13	V				A2p	
τ¹ Lup	126341	14 26.1	-45 13	βC	4.36	4.43	B	2435250.483	0.177365		B2IV	
V745 Cen	--	14 27.2	-62 04	EB/SD	9.3	10.3	p	2433792.235	3.025101		B6-B8 Ib-II	
RS Vir	126753	14 27.3	+4 41	M	7.0	14.4	v	2442565	352.80	0.37	M6e-M7e	
V Boo	127335	14 29.8	+38 52	SRa	7.0	12.0	v	2444780	258.01	0.49	M6e	
Y Cen	127233	14 31.0	-30 06	SRb?	8.9	10.0	p		180:		M4e-M7	
γ Boo	127762	14 32.1	+38 19	δSctc?	3.02	3.07	V		0.2903137	0.5:	A7III	Seginus
V Cen	127297	14 32.5	-56 53	Cδ	6.43	7.21	V	2440308.60	5.493839	0.26	F5Ib-II - G0	
CP Boo	127986	14 33.3	+36 58	δSctc?	6.9	(0.02)	B				F8IVw	
TU Cen	--	14 34.0	-31 42	M	9.5	14.5	p	2428214	293.90		M4e-M7e	
CH Boo	128333	14 34.7	+49 22	Lb	5.74	(0.14)	V				M1IIIab	
V798 Cen	--	14 34.8	-60 58	Lb?	8.7	9.3	p				M6III	
CK Boo	--	14 35.1	+9 07	EW/KW	8.99	9.26	v	2442896.8759	0.3551501		F8	
η Cen	127972	14 35.5	-42 09	γC	2.30	2.41	V				B1.5Vne	
BW Boo	128661	14 37.1	+35 56	EA/DM	7.13	7.46	V	2440362.9026	3.332821	0.06	F0V	
R Boo	128609	14 37.2	+26 44	M	6.2	13.1	v	2444518	223.40	0.46	M3e-M8e	
V737 Cen	128037	14 37.2	-62 01	Cδ	7.5	8.0	p	2428656.350	7.06585	0.39	G2Ib	
RV Boo	129004	14 39.3	+32 32	SRb	7.9	9.88	B		137	0.50	M5e-M7e	
V Lib	--	14 40.4	-17 39	M	9.0	15.0	v	2438443	255.26	0.42	M5e	
RW Boo	129355	14 41.2	+31 34	SRb	8.0	9.5	p		209		M5	
α Lup	129056	14 41.9	-47 23	βC	2.28	2.31	V		0.259864		B1.5III	
DL Dra	129798	14 42.1	+61 16	δSct	6.25 Y	(0.14)	V		0.0825:		F2V	
X Cir	128982	14 43.0	-65 15	Nb	6.5	>16.5	p	2424762			P(Q)	N 1926
W Boo	129712	14 43.4	+26 32	SRb?	4.73	5.4	V		450:		M2-M4III	34 Boo
UV Dra	130082	14 44.0	+56 06	SRa	8.6	9.8	v	2424436.4	77.4	0.43	M5	
V553 Cen	129981	14 46.6	-32 11	CWb	8.23	8.80	V	2441124.28	2.06051	0.41	F4-K0Ia(C)	
BP Cir	129708	14 46.7	-61 28	Cep	7.54	(0.33)	V		2.3984	0.42	F2-F3II	
BG Vir	130254	14 47.1	+4 53	SRb?	9.5	10.7	p		50:		M5	
RR Boo	--	14 47.1	+39 19	M	8.3	13.9	v	2443047	194.70	0.48	M2e-M6e	
V768 Cen	130328	14 48.6	-36 38	SRb	5.93	6.15	V				M3III	
RY Boo	130818	14 49.7	+23 02	cst?	7.12	7.16	V				F5III-IV	
AV Cir	130233	14 50.5	-67 30	Cδ	8.0	8.6	p	2438206.05	3.0651	0.40	F7II	
ξ Boo	131156	14 51.4	+19 06	BY?	4.52	4.67	V		10.137		G8V + K4V	37 Boo
AX Cir	130701	14 52.6	-63 49	Cδ	5.65	6.09	V	2438199.54	5.273268	0.32	F2-G2II + B4	
S Lup	131169	14 53.4	-46 37	M	7.8	13.5	v	2438021	342.72	0.52	Se	
θ Cir	131492	14 56.7	-62 47	γC	5.02	5.44	V				B3Vne	
EN TrA	131356	14 57.0	-68 50	Cδ	8.7	9.1	p	2438207.0	36.9	0.32	G5	
RR UMi	132813	14 57.6	+65 56	SR?	6.1	6.5	p		40:		M5III	
FY Lib	132112	14 57.8	-12 26	SRb	7.06	7.46	V	2440670	45:		gM5	
R Aps	131109	14 57.9	-76 40	cst?	5.35		V				K4III-M0III	
Y Lup	132125	14 59.6	-54 57	M	8.2	15.2	v	2441917	401.14	0.37	M7e	

Star Name	HD	α 2000	δ 2000	Type	Max	Min		Epoch	Period	F	Spec	Notes
BX Boo	133029	15^{h}00^m.6	+47°17'	αCV	6.33	6.41	V	2438544.72	2^d.8881		B9VpSiCrSr	
δ Lib	132742	15 01.0	−8 31	EA	4.92	5.90	V	2442937.4236	2.327374	0.23	B9.5V	19 Lib
BF Cir	132461	15 02.6	−64 57	EB/DM?	8.8	9.2	p	2438199.345	6.458997		B5V	
		15 02.8	−41 57	SN	−8:		v					SN 1006
i Boo	133640	15 03.8	+47 39	EW	6.5	7.1	v	2439370.4222	0.2678160		G2V + G2V	44 Boo
σ Lib	133216	15 04.1	−25 17	SRb	3.20	3.36	V		20		M3IIIa	20 Lib
GM Lup	133220	15 04.7	−40 52	Lb?	6.3	6.46	V				M6III	
RT Lib	133710	15 06.4	−18 44	M	8.2	14.6	v	2438447	251.74	0.45	M3pe−M5.5e	
YY Lib	—	15 08.2	−21 10	M	9.0	>12.0	p		229.53		Me	
T TrA	133638	15 09.6	−68 43	cst?	6.8		v				A0	
Y Lib	134739	15 11.7	−6 01	M	7.6	14.7	v	2440703	275.05	0.41	M5e	
BV Dra	153421	15 11.9	+61 51	EW	8.40	9.00	B	2442858.0687	0.35006571		G0	
FL Ser	134943	15 12.1	+18 59	Lb	5.79	6.02	V				M4IIIab	
X TrA	134453	15 14.3	−70 05	Lb	8.1	9.1	p				C5,5	
Z Ser	—	15 16.0	+2 10	SRa	9.4	10.9	p		87		M5	
ES Lib	135681	15 16.8	−13 02	EB	7.10	7.57	V	2440329.4669	0.8830356		A2−A3V	
RT Boo	—	15 17.2	+36 22	M	8.3	13.9	v	2442722	273.86	0.49	M6.5e−M8e	
U CrB	136175	15 18.2	+31 39	EA/SD	7.66	8.79	V	2444372.492	3.45220133	0.14	B6V + F8III−IV	
GG Lup	135876	15 18.9	−40 47	EB	5.4	6.0	p	2434532.325	2.164175		B5 + A0	
FZ Lib	136140	15 19.4	−9 09	Lb?	6.87	7.11	V				gM4	
R TrA	135592	15 19.8	−66 30	Cδ	6.39	6.93	v	2434930.633	3.389287	0.32	F6−G0	
γ UMi	137422	15 20.7	+71 50	δSct?	3.05 Y	(0.03)	V				A3II−III	Pherkad
BR Cir	—	15 20.7	−57 10	X	7.21	11.38	K				eα	Cir X−1
S CrB	136753	15 21.4	+31 22	M	5.8	14.1	v	2444604	360.26	0.35	M6e−M8e	
S Lib	136458	15 21.4	−20 23	M	7.5	13.0	v	2437797	192.37	0.49	M2e	
δ Lup	136298	15 21.4	−40 39	βC	3.21	3.24	V	2441045.172	0.16547	0.50	B1.5IV	
S Ser	136695	15 21.7	+14 19	M	7.0	14.1	v	2442100	368.59	0.43	M5e−M6e	
UU CrB	—	15 22.9	+31 33	unq	8.59	8.64	v	2444381.753			F8	
RS Lib	136986	15 24.3	−22 55	M	7.0	13.0	v	2438862	217.65	0.48	M7e−M8e	
GH Lup	136739	15 24.7	−52 51	Cep	8.6	9.05	B	2438202.145	9.285	0.50	G5−K2	
HP TrA	136828	15 26.1	−63 26	E	8.2	8.6	p	2438196.250	2.75815		B8	
β CrB	137909	15 27.8	+29 06	αCV	3.65	3.72	V	2440335.0*	18.487		F0IIIpSrCrEu	Nusakan
GO Lup	137597	15 28.2	−37 22	SRb	6.98	7.21	V				M4III	
BP Oct	129723	15 28.3	−88 08	δSct	6.47	6.52	V		0.08		Am	
IL Nor	137677	15 29.4	−50 35	Nb	5.5	>16.3	p	2412638			P(Q)	N 1893
S UMi	139492	15 29.6	+78 38	M	7.7	12.9	v	2442284	326.25	0.50	M7e−M9e	
κ¹ Aps	137387	15 31.5	−73 23	γC	5.43	5.61	V				B3IVe	
GG Lib	138344	15 32.2	−23 53	SRb?	6.83	6.93	V				M5:III	
RU Lib	138547	15 33.3	−15 20	M	7.2	14.2	v	2441840	316.56	0.46	M5e−M6e	
TW Dra	139319	15 33.9	+63 54	EA	8.2	10.5	p	2438539.4457	2.8068352	0.15	A5V + K0III	
EI Lib	138672	15 34.4	−23 00	EA	9.5	10.5	p	2430869.310	1.98691	0.13	A2	
α CrB	139006	15 34.7	+26 43	EA/DM	2.21	2.32	B	2423163.770	17.359907	0.03	A0V + G5V	Alphecca
δ Ser	138918	15 34.8	+10 32	δSct	4.20	4.25	V		0.134		F0IV	
R Nor	138743	15 36.0	−49 30	M	6.5	13.9	v	2440863	492.74		M3e−M6II	
τ⁴ Ser	139216	15 36.5	+15 06	Lb	7.5	8.9	p				M5IIb−IIIa	17 Ser
IM Nor	—	15 39.5	−52 19	Nb	9.0	21:	p	2422512				N 1920
SW CrB	140155	15 40.8	+38 43	SRb	7.8	8.5	v		100:		M0	
RR CrB	140297	15 41.4	+38 33	SRb	8.4	10.1	p		60.8	0.50	M5	
χ Ser	140160	15 41.8	+12 51	αCV	5.33	5.36	v	2434134.06	1.59584		A0pSr	20 Ser
U Lib	—	15 42.1	−21 11	M	9.0	15.0	v	2437735	226.40	0.44	M3e	
U Nor	139717	15 42.3	−55 19	Cδ	8.68	9.64	V	2434926.566	12.64133	0.46	F7−G5	
γ CrB	140436	15 42.7	+26 18	δSctc	3.80	3.86	V		0.030		B9IV + A3V	8 CrB
BP Boo	140728	15 42.8	+52 22	αCV	5.34	(0.02)	U	2440725.336	1.30488		B9pSiCr	
FQ Lup	140145	15 43.6	−37 10	L?	9.5	10.5	p					
T Nor	140041	15 44.1	−54 59	M	6.2	13.6	v	2440978	242.56	0.39	M3e−M6	
CT Ser	—	15 45.6	+14 23	Na	6.0	>16.0	p	2432600			P(Q)	N 1948
R CrB	141527	15 48.6	+28 09	RCB	5.71	14.8	V				C0,0(F8pep)	
X CrB	141678	15 48.9	+36 15	M	8.5	14.2	v	2443719	241.17	0.46	M5e−M7e	
IR Nor	141080	15 48.9	−44 06	EA	9.0	9.5	p					
V CrB	141826	15 49.5	+39 34	M	6.9	12.6	v	2443763	357.63	0.41	C6,2e(N2e)	
R Ser	141850	15 50.7	+15 08	M	5.16	14.4	v	2442315	356.41	0.41	M7IIIe	
ST Her	142143	15 50.8	+48 29	SRb	8.8	10.3	p	2434879	148.0		M6SIIIa	
FP Ser	142500	15 54.7	+8 35	δSct	6.20	6.23	V		0.250:		A7Vn	40 Ser
SY Nor	—	15 54.7	−54 34	Cδ	8.99	9.88	V	2434920.617	12.6452	0.35		
Z CrB	142927	15 56.1	+29 14	M	8.8	15.5	v	2442514	250.68	0.42	M4e−M5e	
RR Lib	142641	15 56.4	−18 18	M	7.8	15.0	v	2438502	276.96	0.47	M4e−M5.5e	
CL Dra	143466	15 57.8	+54 45	δSct	4.95	5.00	V		0.063		F0IV	
FX Lib	142983	15 58.2	−14 17	γC	4.79	4.96	V				B5IIIpe	48 Lib
RS CrB	143347	15 58.5	+36 01	SRa	8.7	11.6	p	2434825	332.2	0.47	M7	
V913 Sco	142990	15 58.6	−24 50		5.43 Y						B5IV	
T CrB	143454	15 59.5	+25 55	Nr	2.0	10.8	v	2431860	29000:		M3III + P(Q)	N 1946
S TrA	142941	16 01.2	−63 47	Cδ	6.06	6.81	V	2434910.570	6.32344	0.32	F8II	
AG Dra	—	16 01.7	+66 48	Z And	8.8	11.8	p				Gep	
X Her	144205	16 02.7	+47 14	SRb	7.5	8.6	p		95.0	0.50	M6e	
RR Her	144578	16 04.2	+50 30	SRb	8.8	13.5	p		239.7		K5−N0e(C6,4)	

Star Name	HD	α 2000	δ 2000	Type	Max	Min		Epoch	Period	F	Spec	Notes
TV Nor	143654	16^{h}04^m.2	−51°33′	EA	8.7	9.3	p	2425832.231	8^d.524406	0.025	A0	
RZ Sco	144018	16 04.6	−24 06	M	8.0	12.8	v	2441092	155.6	0.45	M3e−M4e	
Z Sco	144311	16 06.0	−21 44	M	8.7	13.4	v	2439350	349.93	0.50	M5.5e:−M7e	
R Her	144622	16 06.2	+18 22	M	7.76	15.0	V	2442224	318.38	0.40	M5e	
U Ser	144782	16 07.3	+9 56	M	7.8	14.7	v	2442326	237.85	0.47	M4e:−M6e	
U TrA	143999	16 07.3	−62 55	Cep	7.47	8.25	V	2419722.284	2.56842	0.28	F5−F7	
SX Her	144921	16 07.5	+24 55	SRd	8.6	10.9	p	2434218	102.90	0.46	G3ep−K0(M3)	
FS Ser	145002	16 08.5	+8 32		5.73 Y	(0.04)	V				M3.5IIIa	47 Ser
W UMi	150265	16 08.5	+86 12	EA	8.7	9.78	V	2439758.846	1.7011576	0.23	A3	
FQ Ser	145050	16 08.6	+8 37	Lb	6.31	6.60	V				gM4	
V856 Sco	144667	16 08.6	−39 06		6.64	8.00	V		30:		A7IIIe	
EQ TrA	−−	16 10.0	−66 09	EA	8.9	9.5	p	2438228.285	2.7095	0.20	A5	
RU Her	145459	16 10.2	+25 04	M	6.8	14.3	v	2441058	485.49	0.43	M6e−M9e	
LQ Her	145713	16 11.6	+23 30	Lb?	5.58	5.83	V				M4.5IIIa	10 Her
V718 Sco	145718	16 13.2	−22 29	EA	9.0	10.4	p	2442630.4			A2	
TZ CrB	146361	16 14.7	+33 52	RS+δSct?	5.69	(0.05)	V	2423869.561	1.139789		G0VeCa	17 CrB
W CrB	146560	16 15.4	+37 48	M	7.8	14.3	v	2444192	238.40	0.45	M2e−M5e	
VZ Aps	−−	16 16.3	−74 02	M	8.2	17.5	p	2436845	385		M5e	
T Sco	146417	16 17.0	−22 58	N	6.8	>12.0	v	2400552				N 1860
AT Dra	147232	16 17.3	+59 45	Lb	6.8	7.5	p				M4IIIa	
S Nor	146323	16 18.9	−57 54	Cδ	6.12	6.77	V	2437805.32	9.75411	0.48	F8−G0Ib	
X Ser	−−	16 19.3	−2 30	Nb	9.0	18.3	p					N 1903
δ¹ Aps	145366	16 20.3	−78 42	Lb?	4.66	4.87	V				M4−M5III	
σ Sco	147165	16 21.2	−25 36	βC	2.94	3.06	B	2438566.155	0.2468406	0.55	B2III + O9.5V	20 Sco
W Oph	−−	16 21.4	−7 42	M	9.3	14.9	v	2438079	331.10	0.41	M(6)e	
U Sco	−−	16 22.5	−17 53	Nr	8.8	19	p	2444048	13400:			N 1979
RY CrB	−−	16 23.1	+30 51	SRb	9.2	10.4	v	2431344	90	0.52	M10III	
V760 Sco	147683	16 24.7	−34 54	EA	7.3	7.7	p	2438230.250	1.7309	0.17	B5	
ω Her	148112	16 25.4	+14 02		4.57 Y	(0.04)	V		2.951		B9pCr	24 Her
U Her	148206	16 25.8	+18 54	M	6.5	13.4	v	2442540	406.05	0.40	M7IIIe	
V Oph	148182	16 26.7	−12 26	M	7.3	11.6	v	2438538	297.99	0.48	N3e(C6,3e)	
χ Oph	148184	16 27.0	−18 27	γC	4.18	5.0	V				B2IV:pe	7 Oph
V2105 Oph	148349	16 27.7	−7 36		5.23 Y	(0.06)	V				M2.5III	
PQ Nor	148013	16 28.3	−57 13	E?	7.3	7.6	p				A0	
g Her	148783	16 28.6	+41 53	SRb	5.7	7.2	p		70:		M6III	30 Her
OZ Nor	148259	16 28.8	−44 49	γC?	7.5	7.67	B				B2Ve	
α Sco	148478	16 29.4	−26 26	SRc	0.88	1.80	V	2408600	1733		M1.5Iab−Ib + B4Ve	Antares
QU Nor	148379	16 29.7	−46 15		5 30	5.35	V				B1.5Iape	
R UMi	149683	16 30.0	+72 17	SRa	8.8	11.0	v	2434656	324.40	0.50	M7e	
WW Aps	−−	16 31.7	−74 59	M	9.0	16.8	p	2427946	267		M4(Ib)e	
GN Her	−−	16 31.9	+38 52	Lb	9.0	11.5	p				gM4	
ω Oph	148898	16 32.1	−21 28	αCV?	4.45	4.51	V		1.5	0.17	A7p	9 Oph
UV Oct	146008	16 32.5	−83 55	RRab	8.92	9.79	V	2434328.396	0.542625		A6−F6	
R Dra	149880	16 32.7	+66 45	M	6.7	13.0	v	2442059	245.47	0.45	M5e−M9e	
SS Her	−−	16 32.9	+6 51	M	8.5	13.7	v	2441090	107.20	0.48	M0e−M5e	
UY Her	149431	16 33.2	+38 04	cst	8.8		p				A2	
T Oph	−−	16 33.7	−16 08	M	8.8	>14.2	v	2441874	366.69	0.36	M6.5e	
μ Nor	149038	16 34.1	−44 03		4.94 Y						B0Ia	
S Oph	−−	16 34.3	−17 10	M	8.9	14.7	v	2442247	233.49	0.45	M5e	
TX Dra	150077	16 35.0	+60 28	SRb	7.9	10.2	p		78	0.50	M4e−M5	
W Her	149749	16 35.2	+37 21	M	7.6	14.4	v	2442580	280.37	0.45	M3e−M5e	
UU Her	−−	16 35.9	+37 58	SRd	8.5	10.6	p				F2Ib−cF8	
V918 Sco	149404	16 36.4	−42 52		5.47 Y						O9Iae	
X Ara	149234	16 36.4	−55 24	M	9.0	14.5	p	2436705	175.78	0.51	M5e−M7e II−III	
Y Her	149805	16 36.9	+7 06	cst?	7.3		p				B9	
V600 Her	149881	16 37.0	+14 29	βC?	7.0	7.03	V		0.205797		B0.5III	
ζ Oph	149757	16 37.2	−10 34		2.56 Y						O9.5Vn	13 Oph
WW Dra	150708	16 39.1	+60 42	EA/RS	8.29	9.49	V	2428020.3693	4.629583	0.12	sgG2 + sgK0	
V349 Her	149573	16 39.4	−60 58	EB/DM	8.6	8.8	v	2438229.310	1.13837		A5V	
R Ara	149730	16 39.7	−57 00	EA/DM?	6.0	6.9	p	2425818.028	4.42507	0.09	B9IV−V	
AZ Dra	151481	16 40.7	+72 40	Lb	8.0	8.9	p				M2	
OY Ara	149990	16 40.8	−52 26	Nb	6.2	17.5	p	2418764:			P(Q)	N 1910
V502 Oph	150484	16 41.4	+0 30	EW	8.34	8.84	V	2441174.229	0.45339345		G2V + F9V	
V449 Her	151056	16 42.7	+48 24	Lb	8.9	9.9	p				M6	
UV Her	151204	16 45.6	+12 08	M	9.5	15.0	p	2436439	342.0		M6e−M6.5e	
ε UMi	153751	16 46.0	+82 02	EA/RS	4.22	4.28	V	2433077.75	39.4809	0.04	G5III + dA8−F0	22 UMi
V636 Her	151732	16 47.3	+42 14	Lb	5.83	6.03	V				M4III−IIIa	
RW UMi	−−	16 47.7	+77 02	Na	6.0	>21.0		2435741			P(Q)	N 1956
AH Dra	152152	16 48.3	+57 49	SRb	8.5	9.3	p	2430520	158		M7	
RR Oph	151592	16 49.0	−19 28	M	8.0	14.9	v	2438502	293.27	0.46	M3e	
V637 Her	152107	16 49.2	+45 59	αCV	4.78	4.83	V		0.96		A2VpSrCrEu	52 Her
V1010 Oph	151676	16 49.5	−15 40	EB	6.1	7.0	v	2438937.7715	0.66142629		A5V	
KQ Sco	−−	16 51.7	−45 26	Cep?	9.3	10.3	v	2435281.4	28.6896	0.35	K5	
S Her	152276	16 51.9	+14 56	M	6.4	13.8	v	2441987	307.44	0.47	M2e−M8.8e	
μ¹ Sco	151890	16 51.9	−38 03	EB	2.80	3.08	B	2432001.0451	1.44026907		B1.5V + B6.5V	

Star Name	HD	α 2000	δ 2000	Type	Max	Min		Epoch	Period	F	Spec	Notes
W TrA	--	16^{h}51^m.9	-67°58'	M	9.4	12.5	p	2411441	248^d.7		Me	
V610 Ara	--	16 52.2	-57 16	EB/SD?	8.8	9.2	p	2436689.430	1.484060		F0V	
V919 Sco	151932	16 52.3	-41 51		6.51	6.61	V		3.5		WN7a	
V900 Sco	152235	16 54.0	-42 00		6.29	6.34	V		2.631:		B1Iae	
ζ¹ Sco	152236	16 54.0	-42 22		4.73 Y (0.09)		V				B1Iape	
V840 Oph	--	16 54.7	-29 38	Na	5.5:	>17.0	p	2421344				N 1917
AK Sco	152404	16 54.7	-36 53	Inbs	8.8	10.3	p				F5Vp	
V645 Her	152896	16 55.0	+29 02	δSct?	7.23	7.37	V				A5	
V644 Her	152830	16 55.3	+13 37	δSct	6.32	6.36	V	2440770.7548	0.11505449	0.50	F3Vs	
RS Sco	152476	16 55.6	-45 06	M	6.2	13.0	v	2442134	320.06		M5e-M8e	
V380 Sco	152699	16 56.1	-30 19	SR	9.2	10.6	v	2427682	187.17	0.29	M1	
AI Dra	153345	16 56.3	+52 42	EA	7.05	8.09	V	2437544.5095	1.19881520	0.18	A1	
RR Sco	152783	16 56.6	-30 35	M	5.0	12.4	v	2442037	279.42	0.47	M6II-IIIe - M8(II)e	
V861 Sco	152667	16 56.6	-40 49	EB	6.07	6.69	V		7.84818		B0.5Iae	
κ Oph	153210	16 57.7	+9 22	cst?	3.14	3.20	V				K2III	27 Oph
SS Oph	153167	16 57.9	-2 46	M	7.8	14.5	v	2438945	180.03	0.47	M5e	
V883 Sco	152901	16 57.9	-38 00	EB	7.34	7.66	V	2438228.315	1.294745		B2.5Vn	
RV Sco	153004	16 58.3	-33 37	Cδ	6.61	7.49	V	2436046.875	6.06133	0.30	F5-G1	
V841 Oph	--	16 59.5	-12 54	Nb	4.2	13.5	v	2396146:				N 1848
WZ Dra	--	16 59.9	+52 20	M	8.5	14.0	v	2437724	398.0		M6e	
RV Her	--	17 00.5	+31 13	M	9.0	15.5	v	2440842	205.31	0.44	M2e	
V451 Her	153882	17 01.5	+14 57	αCV	6.26	6.34	V	2436724.60*	6.0075	0.65	B9pCrEu	
SY Her	--	17 01.5	+22 29	M	8.4	14.0	p	2442185	116.81	0.49	M1e-M6e	
RT Sco	153858	17 03.5	-36 55	M	7.0	16.0:	v	2439252	449.04	0.39	M6e-M7e	
V923 Sco	153890	17 03.8	-38 09	E	5.91 Y (0.35)		V		34.8269:		F4IV + F2V	
V884 Sco	153919	17 03.9	-37 51	EB(X)	6.51	6.60	V	2441453.14	3.4120		O6.5Iaf	
UX Oph	--	17 05.0	-12 12	M	9.4	13.6	v	2438879	116.71		M4e	
BF Oph	154365	17 06.1	-26 35	Cδ	6.93	7.63	V	2436055.672	4.06784	0.30	F6-G2	
V616 Ara	154339	17 07.0	-47 00	EB/GS	8.2	8.6	p	2428716.400	4.99525		B3II-III	
R Oph	154721	17 07.8	-16 06	M	7.0	13.8	v	2442175	302.57	0.45	M4e-M6e	
RT Her	155481	17 10.8	+27 04	M	8.5	15.5	v	2437960	298.49	0.40	M4e	
V620 Her	155514	17 11.1	+24 14	δSct	6.19	6.23	V		0.0797	0.48	A8V	63 Her
AH Sco	155161	17 11.3	-32 20	SRc	8.1	12.0	p	2429769	713.6	0.46	M4e-M5 Ia-Iab	
V463 Her	155526	17 11.6	+16 25	unq	8.1	8.15	v	2436029*	54.0		K0III:	
V447 Oph	--	17 12.0	+8 20	SRb?	9.5	13.5	p	2430257	21:		gM4ev	
FV Sco	155550	17 13.7	-32 51	EA	7.9	8.6	p	2442954.0424	5.72786	0.13	B6V	
AK Her	155937	17 14.0	+16 21	EW	8.83	9.32	B	2438531.4318	0.42152309		F2 + F6	
UW Her	156163	17 14.4	+36 22	SRb	8.6	9.5	p		100		M5e	
V915 Sco	155603	17 14.5	-39 46		6.60 Y (0.15)		V				G5Ia	
α¹ Her	156014	17 14.6	+14 23	SRc	3	4	v				M5Ib-II	Rasalgethi
V438 Oph	--	17 14.7	+11 04	SRa	9.3	11.6	p	2437866	154.4	0.42	M7e	
RW Sco	155734	17 14.9	-33 26	M	8.8	15.0	v	2439480	389.69	0.40	M5e	
V727 Sco	155852	17 15.5	-33 06	Lb	9.0	9.7	p				M1	
U Oph	156247	17 16.5	+1 13	EA	5.88	6.58	V	2436727.424	1.6773460	0.17	B5Vnn + B5V	
VW Dra	156947	17 16.5	+60 40	SRd	6.0	6.5	v		170:		K1.5IIIb	
V360 Her	--	17 16.6	+24 27	N	6.3	>15.5	p					N 1892
u Her	156633	17 17.3	+33 06	EB	4.6	5.3	p	2444069.386	2.0510264		B1.5Vp + B5III	68 Her
TX Her	156965	17 18.6	+41 53	EA	8.54	9.31	V	2441499.6687	2.0598133	0.10	A5 + F0	
V2112 Oph	156697	17 18.9	+6 05	δSct	6.51 Y (0.03)		V		0.188		F0-F2 IV-Vn	
Z Oph	156801	17 19.5	+1 31	M	7.6	14.0	v	2442238	348.75	0.40	K4ep-M7.5e	
V656 Her	157049	17 20.3	+18 03		5.00 Y (0.16)		p				M2IIIab	
RY Ara	--	17 21.1	-51 07	RV	9.2	12.1	p	2430220	143.5		G5-K0	
RS Her	157330	17 21.7	+22 55	M	7.0	13.0	v	2442554	219.65	0.47	M4e-M8e	
θ Oph	157056	17 22.0	-25 00	βC	3.25	3.29	V	2440324.230	0.140531		B2IV	42 Oph
V635 Sco	156957	17 22.4	-41 45	Lb	9.5	10.7	p				S7,6:	
V636 Sco	156979	17 22.8	-45 37	Cδ	6.04	6.92	V	2440364.39	6.79671	0.34	G5	
DW Aps	156545	17 23.5	-67 56	EA/SD?	7.9	9.1	p	2439209.502	2.312950	0.14	B6III	
V640 Her	157967	17 25.9	+16 55	Lb	6.12	6.29	V				M4IIIab	
V906 Oph	--	17 26.4	-21 53	Na	8.4	>15	p	2434240			P(Q)	N 1952
V750 Ara	157832	17 27.9	-47 02	γC	6.64	6.70	V				B2-B5Vne	
V499 Sco	158155	17 29.0	-33 00	EB	8.8	9.3	p	2434598.3094	2.3332977		B5	
V843 Oph	--	17 30.6	-21 29	SN	-2.5	>19.0	v	2307214:				Kepler's SN 1604
V482 Sco	158443	17 30.8	-33 37	Cδ	7.63	8.30	V	2440392.37	4.52786	0.28	F5	
V648 Her	159223	17 32.4	+26 26	δSct	6.83	6.87	v		0.29		A5	
RU Oph	--	17 32.9	+9 25	M	8.6	14.2	v	2439033	202.29	0.49	M3e	
λ Sco	158926	17 33.6	-37 06	βC	1.59	1.65	V	2440380.1225	0.2137015	0.5	B2IV + B	Shaula
V642 Her	159354	17 33.7	+14 50	SRb	6.41	6.56	V		12:		M4IIIa	
V701 Sco	317844	17 34.4	-32 30	EB	8.2	8.67	B	2443574.8358	0.76187547		B5	
RV Oph	--	17 34.6	+7 15	EA	9.4	11.4	v	2423997.3833	3.6871222	0.10	A0	
V972 Oph	--	17 34.8	-28 10	Nb	8.0	>16.5	p	2436161			P(Q)	N 1957
RW Ara	158830	17 34.8	-57 09	EA/SD	8.85	11.45	V	2441861.8801	4.367215	0.13	A1IV + K3III:	
TY Dra	160540	17 37.0	+57 44	Lb	8.8	9.9	v				M5-M8	
V449 Sco	159595	17 37.0	-32 08	?	7.0	7.6	p				A2V	
V535 Ara	159441	17 38.1	-56 49	EW/DW?	7.17	7.75	V	2439292.9353	0.62929677		A8V	
V728 Sco	--	17 39.2	-45 29	N	5.0	>11.0	v					N 1862

Star Name	HD	α 2000	δ 2000	Type	Max	Min		Epoch	Period	F	Spec	Notes
V862 Sco	160202	17ʰ40ᵐ0	−32°12′	γC?	6.63	6.76	V		d		B1ne−B8	
V551 Oph	160408	17 40.9	−27 24	Lb	9.0	9.4	p				M2	
BM Sco	160371	17 41.0	−32 13	SRd	6.8	8.7	p		850:		K2.5Ib	
o Ser	160613	17 41.4	−12 53	δSct	4.20	4.26	V		0.053		A2V	56 Ser
V626 Ara	160342	17 42.1	−50 31	Lb	6.2	6.57	V				M3III	
V703 Sco	160589	17 42.3	−32 31	δSct	7.82	8.50	B	2437186.365	0.11521790	0.40	F0−F5	
V2024 Oph	−−	17 42.4	−24 59	N	9.5:	>18.0	p				P(Q)	N 1967
RU Sco	160496	17 42.4	−43 45	M	7.8	13.7	v	2440727	369.20	0.54	M7II−IIIe	
V721 Sco	−−	17 42.5	−34 40	N	9.5	>18.0	p	2433528			P(Q)	N 1950
κ Sco	160578	17 42.5	−39 02	βC	2.39	2.42	V	2440690.062	0.19987		B1.5III	
V Pav	160435	17 43.3	−57 43	SRa	9.3	11.2	p	2438560	225		C6,4	
XX Oph	161114	17 43.9	−6 16	Ia	9.1	11.1	p				Beqp	
V624 Her	161321	17 44.3	+14 25	EA	6.18	6.36	V	2440321.0049	3.894977	0.097	A3m	
KP Sco	−−	17 44.3	−35 44	Na	9.4	>16.5	p	2425419				N 1928
CF Her	−−	17 45.0	+21 30	M	9.0	>13.0	v	2434902	306.2		M0	
V620 Ara	161160	17 47.2	−56 06	EA/SD?	9.0	9.8	p	2428686.375	1.554965		B9III	
X Sgr	161592	17 47.6	−27 50	Cδ	4.24	4.84	V	2436968.852	7.01225	0.36	F7 II	3 Sgr
V885 Sco	161562	17 48.0	−37 38	EA	8.7	9.0	p	2428786.300	3.119975		A0	
SV Sco	161652	17 48.3	−35 42	M	8.7	14.9	v	2439601	256.20	0.47	M3e	
V3894 Sgr	161756	17 48.5	−26 59		6.27	6.39	V				B4IVe	
V500 Sco	−−	17 48.6	−30 29	Cδ	8.40	9.13	V	2436239.324	9.311665	0.49	K0	
V3888 Sgr	−−	17 48.7	−18 46	N	8:	>13	v	2442327:			P(Q)	N 1974
V1274 Sgr	−−	17 48.8	−17 52	N	9.5	>13.0	p	2434985			P(Q)	N 1954
V393 Sco	161741	17 48.8	−35 03	EA	7.7	8.6	p	2428321.118	7.71249	0.12	B9	
V825 Sco	−−	17 49.9	−33 33	N	8.0	>13.0	p	2438380			P(Q)	N 1964
V744 Her	162732	17 50.1	+48 24		6.68 Y (0.07)		V		86.7207		Bep(Shell)	88 Her
V723 Sco	−−	17 50.1	−35 24	Na	9.0	22	B	2434236.4				N 1952
RS Oph	162214	17 50.2	−6 43	Nr	5.3	12.3	p	2439791			Ocp + M2ep	N 1967
V1172 Sgr	−−	17 50.4	−20 41	N	9.0	?	p	2433713			P(Q)	N 1951
W Pav	161567	17 50.4	−62 25	M	8.1	14.6	v	2440420	283.35	0.40	M4e−M7e	
V539 Ara	161783	17 50.5	−53 37	EA/DM	5.66	6.18	V	2439314.342	3.169128	0.14	B2V + B3V	
RY Sco	162102	17 50.9	−33 42	Cδ	7.52	8.44	V	2440365.129	20.3157	0.37	F6−G2	
V697 Sco	−−	17 51.3	−37 25	Na	8.0	>16.5	p	2430000			P(Q)	N 1941
V720 Sco	162287	17 51.9	−35 22	Na	7.5	>18.0	p	2433500			P(Q)	N 1950
V382 Sco	−−	17 51.9	−35 25	Na	9.0:	>16.5	p	2415632				N 1901
Y Oph	162714	17 52.6	−6 09	Cep	5.92	6.38	V	2439853.30	17.12413	0.44	F8Ib−G3Ibv	
V533 Oph	162812	17 53.1	−2 35	SR	8.3	9.3	p		32:		M6	
V696 Sco	−−	17 53.2	−35 50	Na	7.2	>16.5	p	2431226			P(Q)	N 1944
V771 Sgr	162718	17 53.5	−24 47	γC	8.9	9.4	p				B0ne	
U Ara	162398	17 53.6	−51 41	M	7.7	14.1	v	2440289	225.21	0.44	M3IIep−M5e	
V906 Sco	162724	17 53.9	−34 45	E?	5.96 Y (0.26)		V				B9V + B9V	
UX Her	163175	17 54.1	+16 57	EA	9.05	10.21	V	2440022.4194	1.54885307	0.15	A3	
TX Her	162980	17 55.3	−34 14	cst	7.7		p				A2n	
V441 Her	163506	17 55.4	+26 03	SRd	5.34	5.48	V		70:		F2Ibe	89 Her
V732 Sgr	316633	17 56.1	−27 22	Na	6.0:	>16.0	p	2428280			P(Q)	N 1936
V764 Sco	162985	17 56.1	−45 10	EB	8.6	9.1	p	2438233.210	6.8084		A2	
V2052 Oph	163472	17 56.3	+0 40	βC	5.81	5.84	V	2441442.048	0.1398903	0.5	B2IV−V	
V453 Sco	163181	17 56.3	−32 29	EB	6.36	6.73	V	2441762.58	12.0061		B0.5Iae	
AI Sco	320921	17 56.3	−33 49	RVb	9.5	12.7	p		71.0		G0−K2	
SV Oph	−−	17 56.4	+3 23	M	9.4	13.0	v	2438674	216.47		M2e	
T Dra	−−	17 56.4	+58 13	M	7.2	13.5	v	2440556	421.22	0.44	C6,2e−C8,3e	
RT Oph	−−	17 56.5	+11 10	M	8.6	15.5	v	2437475	425.60	0.36	M7e	
OP Her	163990	17 56.8	+45 21	Lb	7.7	8.3	p				M5IIb−IIIa	
V566 Oph	163611	17 56.9	+4 59	EW	7.5	7.96	V	2443281.5034	0.40964660		F4V	
UW Dra	164345	17 57.5	+54 40	Lb?	7.0	8.0	v				K5p	
Z Her	163930	17 58.1	+15 08	EA/RS	7.3	8.1	p	2413086.348	3.9928012	0.11	F4IV−V + K0IV	
V Dra	−−	17 58.2	+54 52	M	9.5	14.7	v	2442233	277.97	0.50	M4e	
V3889 Sgr	−−	17 58.4	−28 22	Na	8.4	20:	p	2442607			ea	N 1975
v Her	164136	17 58.5	+30 11	δSct?	4.41 Y (0.05)		V				F2II	94 Her
V771 Her	164429	17 58.9	+45 29		6.48 Y (0.04)		U		0.517436		B9pSiSr	
V1275 Sgr	−−	17 59.1	−36 19	Na	7.5	>13.0	p	2434928			P(Q)	N 1954
V1647 Sgr	163708	17 59.2	−36 56	EA	7.0	7.1	p	2441829.6951	3.282797	0.06	A0	
RW CrA	163726	17 59.3	−37 53	EA/SD	9.3	10.3	p	2431017.297	1.6835995	0.18	A0	
RY Her	164307	17 59.8	+19 29	M	8.3	14.1	v	2441004	221.48	0.44	M4e−M6e	
V1944 Sgr	−−	17 59.9	−27 18	N	7.0:	13.0	p				P(Q)	N 1960
V787 Sgr	316917	18 00.0	−30 30	Na	7.2	>16.5	p	2428680			P(Q)	N 1937
V999 Sgr	163982	18 00.1	−27 33	Nb	7.5:	16.6	p	2418680			P(Q)	N 1910
V540 Sgr	163869	18 00.1	−35 56	Lc	9.4	10.8	p				M5Iab	
V2048 Oph	164284	18 00.3	+4 22	γC	4.55	4.85	V				B2Ve	66 Oph
DZ Ser	−−	18 00.3	−10 34	N	8.0:	16.9	p				P(Q)	N 1960
V394 CrA	−−	18 00.4	−39 01	Na	7.5	>13.5	p	2432999			P(Q)	N 1949
FL Sgr	−−	18 00.5	−34 36	Na	8.3	>13.0	p	2423960				N 1924
V1946 Sgr	−−	18 02.1	−23 33	M?	9.00	13.30	I	2436705	440:			
V1950 Sgr	−−	18 02.7	−23 22	M	7.00	14.20	I	2436454	400:		M	
V1951 Sgr	−−	18 03.5	−23 00	M	8.30	14.80	I	2437180	510			

Star Name	HD	α 2000	δ 2000	Type	Max	Min		Epoch	Period	F	Spec	Notes
V986 Oph	165174	18^{h}04^m.6	+1°55'	βC?	6.10	6.15	V	2436338.53	0^d.284653	0.50	B0IIIn	
W Sgr	164975	18 05.0	−29 35	Cδ	4.30	5.08	V	2437678.578	7.594710	0.32	F4−G1Ib	
W Dra	166407	18 05.6	+65 57	M	8.9	15.4	v	2442176	278.6	0.43	M3e−M8e	
V1012 Sgr	—	18 06.2	−31 44	Na	8.0	>17.0	p	2420356				N 1914
X Dra	—	18 06.9	+66 09	M	8.9	15.8	v	2441099	257.54	0.44	M5e	
V737 Sgr	—	18 07.1	−28 45	N	9.5:	>13.0	p	2427250				N 1933
o Her	166014	18 07.5	+28 46	γC	3.81	3.90	B				B9.5V	103 Her
DQ Her	—	18 07.5	+45 51	Nb(EA)	1.3	v 15.6		2427794			sdBe + P(Q)	N 1934
V927 Sgr	—	18 07.7	−33 21	Na	7.5:	>16.5	p	2431197			P(Q)	N 1944
V566 Her	166253	18 07.9	+41 43	SR?	8.6	9.3	p		400:		gM4	
VX Sgr	165674	18 08.1	−22 13	SRc	6.5	12.5	v	2436493	732	0.44	M4Iae−M9.5	
S Oct	158413	18 08.7	−86 48	M	7.3	14.0	v	2439970	258.94	0.42	M4IIe−M5e	
V630 Sgr	321353	18 08.8	−34 20	Na	4.0	14.4	p	2428445			P(Q)	N 1936
V3792 Sgr	165814	18 08.9	−25 28	EB	6.43	6.88	V	2441879.349	2.24815		B3III	
T Her	166382	18 09.1	+31 01	M	6.8	13.9	v	2442330	165.01	0.47	M2e−M8e	
V1148 Sgr	—	18 09.1	−26 00	N	8.0	>16.0	p	2430956			P(Q)	N 1943
V1015 Sgr	—	18 09.1	−32 28	Na	7.1	>12.0	p	2417054				N 1905
W Ser	166126	18 09.8	−15 33	E	8.42	10.20	V	2440048.9	14.16245	0.20	cF5ep	
R Cep	—	18 11.1	+88 59	cst	9.1		p				G2V	(in UMi)
NQ Her	166801	18 11.6	+18 19	EA?	8.0	8.6	p	2426894.433	0.870218	0.15	A0	
V669 Her	167006	18 11.9	+31 24		4.97 Y (0.14)		p		6.3		M3III	104 Her
V4045 Sgr	166469	18 12.0	−28 54		6.51 Y (0.013)		v		2.90		A0p	
R Pav	165961	18 12.9	−63 37	M	7.5	13.8	v	2439247	229.82	0.48	M4e−M5e	
AP Sgr	166767	18 13.0	−23 07	Cδ	6.59	7.41	V	2440390.31	5.05793	0.30	F6−G1	
V692 CrA	166596	18 13.2	−41 20	SX Ari	5.46	(0.05)	V		1.67		B3pSi	
μ Sgr	166937	18 13.8	−21 04	EA	3.79	3.92	V	2429051	180.45	0.11	B8Iape	13 Sgr
V849 Oph	167276	18 14.1	+11 37	Nb	6.0	>17.5	p	2422268			P(Q)	N 1919
V533 Her	—	18 14.3	+41 51	Na	3.0:	14.9	p	2438060			P(Q)	N 1963
V1175 Sgr	—	18 14.3	−31 07	N	7.0	>12.0	p	2434064			P(Q)	N 1952
TV Her	—	18 14.7	+31 49	M	9.0	14.6	v	2441866	303.72	0.37	M4e	
W Lyr	167740	18 14.9	+36 40	M	7.3	13.0	v	2442117	196.54	0.48	M2e−M8e	
V1583 Sgr	—	18 15.5	−23 54	Na	8.9	>16.5	p	2425422				N 1928
TZ Lyr	—	18 15.8	+41 07	EB	9.4	10.4	v	2440737.8285	0.52882613		K0	
V2509 Sgr	167231	18 15.8	−35 38	EB	7.6:	8.0	p	2438233.325	1.086970		A0	
RY Oph	167766	18 16.6	+3 42	M	7.5	13.8	v	2442348	150.53	0.46	M3e−M6e	
WZ Sgr	167660	18 17.0	−19 05	Cδ	7.45	8.53	V	2435506.629	21.849708	0.34	F8−K1	
FM Sgr	—	18 17.3	−23 38	Na	8.2:	16.5	p	2424727				N 1926
NV Tel	167405	18 17.5	−48 36	EA	8.8	9.5	v	2428006.400	3.54495		A0	
RS Sgr	167647	18 17.6	−34 06	EA	6.0	6.9	p	2420586.387	2.4156832	0.17	B3V + A	
η Sgr	167618	18 17.6	−36 46	Lb	3.08	3.12	V				M3.5III	
IQ Her	168198	18 17.9	+17 59	SRb	6.99	7.47	V	2430496	75	0.50	M4	
V1149 Sgr	—	18 18.5	−28 17	N	7.4	>16.5	p	2431500				N 1945
RS Tel	—	18 18.9	−46 33	RCB	9.3	>13.0	p				R8	
V928 Sgr	—	18 19.0	−28 06	N	8 5	>16.5	p	2432316			P(Q)	N 1947
RU Dra	—	18 19.4	+59 32	M	9.4	>13.6	v	2434214	297.1	0.40:	M5e	
FO Ser	168227	18 19.4	−15 37	L	8 45	8.80	V				R6(C4.5J)	
V726 Sgr	315532	18 19.6	−26 54	Na	9.0:	>16.5	p	2428290			P(Q)	N 1936
V1016 Sgr	—	18 20.0	−25 11	Na	7.0:	15.0	p	2414880				N 1899
φ Dra	170000	18 20.8	+71 20		4.22 Y (0.05)		V		1.716		A0pSi:	43 Dra
Y Sgr	168608	18 21.4	−18 52	Cδ	5.40	6.10	V	2436230.180	5.77335	0.34	F8I	
V4028 Sgr	168574	18 21.5	−24 55		6.25 Y						M5III	
XZ Sgr	168710	18 22.1	−25 14	EA	8.82	10.93	V	2441150.348	3.2755357	0.11	A3V + G5IV:	
V441 Sgr	315574	18 22.1	−25 29	Na	8.2:	16.0	p	2426230			P(Q)	N 1930
V1977 Sgr	—	18 22.4	−16 17	M	9.5	14.6	i	2437875	467			
V4050 Sgr	168733	18 22.9	−36 40		5.34 Y (0.01)		v				A0	
GR Sgr	—	18 23.0	−25 35	N	7.5:	16.6	p	2423906				N 1924
TW Lyr	—	18 23.9	+39 35	M	9.5	15.0	p	2442285	376.63			
XX Sgr	169315	18 24.7	−16 48	Cδ	8.41	9.47	V	2435308.449	6.4243198	0.30	F6−G2	
V655 CrA	—	18 24.8	−37 00	N	8:	17:	p				P(Q)	N 1967
AW Her	348635	18 25.6	+18 18	EA	9.5	10.9	v	2425719.42	8.80086	0.07	K2 + G4	
V909 Sgr	—	18 25.9	−35 02	Na	6.8	>16.0	p	2430172			P(Q)	N 1941
EG Ser	169691	18 26.0	−1 41	EA	8.7	9.5	p	2426487.525	4.97362	0.05	A0	
SV Her	—	18 26.4	+25 02	M	9.1	15.1	v	2441598	239.32	0.46	M5e	
RZ Sct	169753	18 26.6	−9 12	EA	7.34	8.84	V	2419261.1025	15.1902079	0.17	B3Ib	
V988 Oph	169931	18 26.9	+3 55	SR	9.0	10.2	p	2431291	63.2		M7e	
BS Sgr	—	18 26.9	−27 07	N1	9.2	>16	p	2421427				N 1917
d Ser	169986	18 27.2	+0 12	?	4.9	5.9					G0III + A6V	59 Ser
RV Sgr	169831	18 27.9	−33 19	M	7.2	14.8	v	2440425	317.51	0.47	M5e−M9	
V2349 Sgr	170097	18 28.4	−16 42	EA	8.4	9.0	p	2426916.650	5.02565	0.12	B1Vne	
T Ser	—	18 28.8	+6 18	M	9.1	15.5	v	2441117	340.76	0.47	M7e	
V451 Oph	170470	18 29.2	+10 54	EA	7.9	8.5	p	2444757.4807	2.196522	0.12	A0 + A2	
V4031 Sgr	170235	18 29.4	−25 15		6.59 Y (0.10)		V				B2IVpe	
V432 Sct	170397	18 29.8	−14 35		5.96 Y (0.01)		v		2.21		B9pSiCr	
AC Her	170756	18 30.3	+21 52	RVa	7.43	9.74	B	2435052	75.4619	0.23	F2Ibp−K4e	
RX Her	170757	18 30.7	+12 37	EA	7.2	7.8	p	2433170.398	1.7785724	0.14	A0 + A0	

Star Name	HD	α 2000	δ 2000	Type	Max	Min		Epoch	Period	F	Spec	Notes
									d			
V3890 Sgr	--	18ʰ30ᵐ.7	-24°01'	Na	8.4:	17.2	p	2437818				
FH Ser	--	18 30.8	+2 37	Na	4.41	16:	V	2440636			P(Q)	N 1962
V Sgr	170656	18 31.4	-18 16	cst	8.1		v				F0	N 1970
V3508 Sgr	170682	18 31.4	-19 09	γC	7.73	7.92	V	2440580:			B6IIIe	
V2572 Sgr	--	18 31.6	-32 36	Na	6.5	p 14.83	B	2440413			P	N 1969
U Sgr	170764	18 31.9	-19 07	Cδ	6.34	7.08	V	2438084.230	6.744925	0.27	F5Ib-G1.5Ib	
V1017 Sgr	--	18 32.1	-29 24	Z And?	6.2	p 14.73	B				G5IIIep	
T Lyr	--	18 32.3	+37 00	Lb	7.8	9.6	v				R6(C5,3)	
V668 CrA	170625	18 32.4	-42 19	δSctc	8.70	8.76			0.088		A4-A5V	
SV Dra	--	18 33.6	+49 22	M	9.1	15.0	v	2442170	256.20	0.49	M7e	
BY Dra	--	18 33.9	+51 43	BY	8.07	8.48	V	2441127.228	3.813378	0.40	K7Ve	
FV Sct	--	18 34.5	-12 55	N	7.0:	21.0	p				P(Q)	N 1960
FR Ser	171586	18 35.6	+4 56	αCV	6.34	6.52	V	2441460.79	2.1436		A2p	
V3645 Sgr	--	18 35.8	-18 41	Nb	8:	18	p	2440580:			P(Q)	N 1969
V3877 Sgr	171451	18 36.4	-35 27	SRb	6.84	7.30	V	2440760	50:		M5III	
RZ Her	172008	18 36.8	+26 03	M	9.0	15.5	v	2437934	329.01	0.44	M5e	
α Lyr	172167	18 36.9	+38 47		0.03	Y (0.02)	V		0.07		A0Va	Vega
RS Dra	--	18 37.6	+74 20	SRa	9.0	12.0	v	2415120	282.72	0.44	M5e	
V681 CrA	--	18 37.7	-42 57	EA/DM	7.6	8.1	p	2428748.350	2.163925		B9.5V	
EW Sct	171955	18 37.9	-6 48	Cep?	7.77	8.24	V	2441525.0	10:		K0	
XY Lyr	172380	18 38.1	+39 40	Lc	7.3	7.8	p				M4-M5 Ib-II	
X Oph	172171	18 38.3	+8 50	M	5.9	9.2	v	2441478	334.39	0.53	M6IIIe + K1III	
AM CrA	172321	18 41.2	-37 29	SR	8.6	12.7	v		187.5:		M3e	
V679 Oph	172804	18 41.9	+6 49	Lb	8.96	9.3	V				M2(S4.5,8)	
RU Sct	172730	18 41.9	-4 07	Cδ	8.87	10.02	V	2423236.0	19.69767	0.36:	F4-G5	
V693 CrA	--	18 42.0	-37 31	Na	6.5	>19	v	2444697			P(Q)	N 1981
FI Lyr	--	18 42.1	+28 58	SR	9.5	10.5	v	2427580	146		M	
HK Aql	172829	18 42.3	+0 09	Lb?	8.9	10.9	p				K5III	
δ Sct	172748	18 42.3	-9 03	δSct	4.98	5.16	B	2427991.786	0.193770		F2IIIp	
HK Lyr	173291	18 42.8	+36 58	Lb	9.5	11.6	p				Nb(C6,2)	
Z Sct	--	18 42.9	-5 49	Cep	9.1	10.1	v	2436247.160	12.9014	0.40	F8-G4	
V3879 Sgr	172816	18 42.9	-19 17	SRb	6.05	6.58	v		50:		M4III	
RW Tel	172605	18 43.4	-45 47	SRb	9.4	11.8	p		127.35		M6II-IIIe	
SS Sct	173058	18 43.7	-7 44	Cδ	7.97	8.43	V	2435315.625	3.671253	0.37	F6-G0	
V1331 Aql	173198	18 44.2	-1 33	EB/KE?	7.7	8.05	V	2442610.070	1.364209		B1V	
RY Lyr	--	18 44.9	+34 41	M	9.0	>15.6	v	2441764	326.31	0.40	M5e-M6e	
V350 Sgr	173297	18 45.3'	-20 39	Cδ	7.08	7.82	V	2435317.227	5.15424	0.28	F5-G2	
V535 Her	173650	18 45.6	+21 59	αCV	6.41	6.58	V	2437127.8*	9.9748		B9pSiCr:	
V368 Sct	--	18 45.7	-8 33	Na	6.9	17	v	2440797			P(Q)	N 1970
CX Dra	174237	18 46.7	+52 59		5.88	Y (0.10)	V		10:		B2.5Ve	
KO Aql	173847	18 47.2	+10 46	EA/SD?	8.3	9.50	B	2441887.4724	2.864055	0.13	A0V-A3V	
R Sct	173819	18 47.5	-5 42	RVa	4.45	8.20	V	2432078.3	140.05		G0Iae-K0Ibpv	
V CrA	173539	18 47.5	-38 09	RCB	8.3	>16.5	v				C(R0)	
DH Her	343047	18 47.6	+22 51	EA	9.4	12.0	v	2426575.456	4.779173	0.15	A5	
V356 Sgr	173787	18 47.9	-20 16	EA	7.00	7.87	B	2433900.827	8.89610	0.12	B3:V + A2II	
V603 Aql	174107	18 48.9	+0 35	Na/E(X)	-1.4	12.03	V	2421755			P(Q) + CONT	N 1918
YZ Sgr	174089	18 49.5	-16 43	Cδ	7.02	7.76	V	2435514.301	9.55345	0.54	F6-G2	
β Lyr	174639	18 50.1	+33 22	EB	3.34	4.34	V	2445342.39	12.93578		B7Ve + A8p	Sheliak
S Sct	174325	18 50.3	-7 54		6.80	Y			148.0		C5II	
HS Her	174714	18 50.8	+24 43	EA	8.50	8.97	V	2440146.6080	1.637438	0.12	B6III	
BB Sgr	174383	18 51.0	-20 18	Cδ	6.69	7.29	V	2436053.535	6.63699	0.34	F6-G1	
o Dra	175306	18 51.2	+59 23	RS	4.66		V		138.420		G9IIIb	47 Dra
λ Pav	173948	18 52.2	-62 11	γC	3.4	4.3	V				B2II-IIIe	
DI Her	175227	18 53.4	+24 17	EA	8.3	8.9	p	2438287.5843	10.5501609	0.04	B4III + B5III	
HR Lyr	175268	18 53.4	+29 14	Na	6.5	15.4	p	2422299			P(Q)	N 1919
FN Sgr	--	18 53.9	-19 00	Z And	9.0	13.9	p				P	
V913 Aql	--	18 54.5	+10 38	SRa	9.2	10.5	p	2429735	50		M5II	
δ² Lyr	175588	18 54.5	+36 54	SRc?	4.22	4.33	V				M4II	12 Lyr
UX Sgr	175188	18 54.9	-16 31	SRb	8.9	9.6	p		100:		Mb	
R Lyr	175865	18 55.3	+43 57	SRb	3.88	5.0	V	2435920	46.0		M5III	13 Lyr
V1182 Aql	175514	18 55.4	+9 21	EB/KE?	8.5	8.65	v	2439651.720	1.621924		O8Vnn	
V373 Sct	--	18 55.5	-7 43	Na	6.0	18.5	p	2442541			P(Q)	N 1975
EL Aql	--	18 56.0	-3 19	Na	5.5:	19.0	p	2425043:			P(Q)	N 1927
EU Sct	--	18 56.2	-4 12	N	8.0	17.0	p	2433135			P(Q)	N 1949
V686 CrA	175362	18 56.7	-37 21	αCV	5.25	5.41	V	2442254.500	3.670		B8IVSi	
κ Pav	174694	18 56.9	-67 14	CWa	3.94	4.75	V	2440858.53	9.088	0.46	F5I-II	
V446 Her	--	18 57.4	+13 14	Na	3.0	18.8	p	2436998			P(Q)	N 1960
FF Aql	176155	18 58.2	+17 22	Cδs	5.18	5.68	V	2441576.428	4.470916	0.48	F5Ia-F8Ia	
ε CrA	175813	18 58.7	-37 06	EW	4.74	5.00	V	2439707.6619	0.5914264		F2V	
V1286 Aql	176232	18 58.8	+13 54	αCV	5.83	5.93	V	2441517.4	6.05		A4pEuCrSi	10 Aql
AD Aql	--	18 59.1	-8 10	RVa	8.83	13.42	V	2427628	65.4		Fp(R)	
AR Sgr	--	18 59.7	-23 42	RV	9.1	13.5	p	2421113.1	87.87		F5e-G6	
ST Sgr	176592	19 01.5	-12 46	M	7.6	16.0	v	2440460	395.12	0.44	S4,3e-S9,5e	
V1059 Sgr	176654	19 01.8	-13 10	Na	4.5:	16.5	p	2414360			P(Q)	N 1898
V604 Aql	176779	19 02.1	-4 27	Na	7.6:	18	p	2417072:			P(Q)	N 1905

Star Name	HD	α 2000	δ 2000	Type	Max	Min		Epoch	Period	F	Spec	Notes
MT Tel	176387	19^{h}02.^{m}2	−46°39′	RRc ✓	8.68	9.28	V	2438479.332	0.^{d}316897	0.37	A0	
V599 Aql	176853	19 02.6	−10 43	EB/KE?	6.67	6.75	V	2421836.539	1.849084		B2V + B8	
LT Vul	177392	19 03.7	+21 16	δSct	6.58	6.62	V	2440720.7916	0.10901407	0.47	F2III	
BH Dra	178001	19 03.7	+57 28	EA	8.38	9.27	V	2440019.7982	1.81723857	0.10	A2V + Ap	
SU Sgr	177017	19 03.7	−22 43	SRb	8.1	8.8	v	2416620	88		M6	
V337 Aql	177284	19 04.2	−2 02	EB/DM	8.57	9.27	V	2441168.401	2.733826		B0.5Vp + B2V	
V Aql	177336	19 04.4	−5 41	SRb	6.6	8.4	v		353		C5,4−C6,4(N6)	
SZ Aql	177441	19 04.7	+1 18	Cδ	7.92	9.26	V	2435528.937	17.137939	0.37	F7−K1	
V805 Aql	177708	19 06.3	−11 39	EA/DM	7.58	8.22	V	2442324.266	2.4082337	0.11	A2 + A7	
R Aql	177940	19 06.4	+8 14	M	5.5	12.0	v	2443458	284.2	0.42	M5e−M9e	
BL Tel	177300	19 06.6	−51 25	EA	7.72	9.82	B	2434692.6	778.1	0.12	cF8 + M	
Y Aql	178125	19 07.0	+11 04	E/KE	5.02	(0.04)	B	2438607.445	1.30227		B8III−V	18 Aql
RX Tel	177456	19 07.0	−45 58	SRc	8.9	10.4	p	2429960	349.6		M	
V525 Sgr	177768	19 07.2	−30 10	EB	7.9	8.8	p	2429662.4593	0.7051220		A2	
TT Aql	178359	19 08.2	+1 18	Cδ	6.46	7.70	V	2437236.10	13.7546	0.34	F6−G5	
V398 Lyr	178770	19 08.2	+39 09	Lb	7.3	7.6	V				M6	
V4024 Sgr	178175	19 08.3	−19 17		5.54 Y	(0.2)	V				B2Ve	
V496 Aql	178287	19 08.4	−7 26	Cδs	7.59	7.98	V	2436017.062	6.80703	0.33	G5	
V3880 Sgr	−−	19 08.9	−22 14	?	1.8	2.8	K		550		M7e	
V Lyr	178876	19 09.1	+29 40	M	8.2	15.7	v	2441476	373.53	0.33		
FM Aql	178695	19 09.3	+10 33	Cδ	7.89	8.66	V	2435151.723	6.11423	0.30	F5−F9Ia	
U Dra	180050	19 10.0	+67 17	M	9.1	14.6	v	2441142	316.42	0.47	M6e−M8	
V363 Sgr	−−	19 11.3	−29 52	Na	8.4:	>16.0	p	2425096			P(Q)	N 1927
RT Vul	179370	19 11.5	+22 23	cst	7.5		p				B8	
V471 Lyr	179527	19 11.8	+31 17		5.98 Y						B9pSi	19 Lyr
V1344 Aql	179315	19 12.0	+4 21	Cδ	7.65	8.00	V	2443398.071	7.47803	0.35	G1Ib	
FL Lyr	179890	19 12.1	+46 19	EA	8.7	9.3	v	2438221.5525	2.1781544	0.07	G5	
FN Aql	179494	19 12.8	+3 33	Cδs	7.96	8.75	V	2436804.603	9.48151	0.49	F8−G2	
X Lyr	337821	19 13.1	+26 47	Lb	8.6	9.8	v				gM3.5	
SS Lyr	180162	19 13.3	+46 59	M	8.4	13.0	v	2438852	350.48		M5IIIe	
TW Sgr	179451	19 13.4	−21 34	M	9.1	>13.5	v	2436465	220.63		M2e−M3e	
V1288 Aql	179761	19 13.7	+2 18	αCV	5.06	5.16	V	2444099.23	1.73		B8II−IIIpHg:	21 Aql
RW Sgr	179604	19 13.9	−18 52	SRa	9.0	11.5	v	2438206	189.53	0.47	M4II−IIIe − M6III:e	
BQ Sgr	179482	19 14.2	−36 15	EA	9.5	12.5	p	2422224.378	8.019537	0.10	A2n	
RX Sgr	179769	19 14.5	−18 49	M	9.0	14.4	v	2442170	334.07	0.49	M5e	
XY Aql	−−	19 14.9	+4 14	M	9.5	17:	v	2438663	423.4		M8	
W Aql	−−	19 15.4	−7 03	M	7.3	14.3	v	2439116	490.43	0.37	S3,9e−S6,9e	
V473 Lyr	180583	19 16.0	+27 56	Cep	6.11	6.22			1.49107		F6Ib−II	
T Sgr	180196	19 16.3	−16 59	M	7.6	12.9	v	2442125	392.35	0.47	S4.5,8e−S5.5,8e	
RY Sgr	180093	19 16.5	−33 31	RCB	6.0	>15.0	v				G0Ipe(C1,0)	
R Sgr	180275	19 16.7	−19 18	M	6.7	12.8	v	2442298	268.81	0.46	M4e−M6e	
V342 Aql	180639	19 17.1	+9 21	EA	9.5	12.9	p	2443723.3861	3.390614	0.14	A4II	
V356 Aql	180621	19 17.2	+1 43	Nb	7.0	17.7	p	2428444:			P(Q)	N 1936
RS Vul	180939	19 17.7	+22 26	EA	6.9	7.6	p	2432808.257	4.4776635	0.14	B5V + A2	
ES Vul	180968	19 17.7	+23 02	βC?	5.4	5.46	v	2436338.5	0.6096	0.50	O8IV−B0.5IV	2 Vul
TY Sgr	180491	19 17.7	−23 56	M	9.1	15.0	v	2440704	325.41	0.44	M3e	
V889 Aql	181166	19 18.8	+16 15	EA/DM	8.52	9.1	V	2438242.334	11.120879	0.03	B9	
U Sge	181182	19 18.8	+19 37	EA	6.58	9.18	v	2440774.4638	3.3806260	0.14	B8III + K:	
V1942 Sgr	180953	19 19.2	−15 54	Lb	6.74	7.10:	V				C4,6	
V528 Aql	−−	19 19.3	+0 38	Na	7.2	18.1	p	2431695			P(Q)	N 1945
W Sge	181332	19 19.5	+17 12	M	8.8	13.2	v	2442200	278.12		M4e	
V1208 Aql	181333	19 19.7	+12 22	δSctc	5.51	5.56	V		0.149663		F0III	28 Aql
Z Sgr	181060	19 19.7	−20 56	M	8.4	16.0	v	2437890	450.25	0.47	M4e−M6e(Se)	
SW Sgr	180958	19 19.9	−31 43	M	9.4	14.5	v	2439565	290.34	0.51	M5e	
U Lyr	−−	19 20.1	+37 53	M	8.3	13.5	v	2440688	455.60	0.52	N0e(C4,5e)	
V606 Aql	181419	19 20.4	−0 08	Na	5.5:	17.3:	p	2414754:			P(Q)	N 1899
UX Dra	183556	19 21.6	+76 34	SRa	5.94	7.1	V		168:		C7,3	
Z Vul	181987	19 21.7	+25 34	EA	7.38	9.20	B	2443831.251	2.45492679	0.18	B4V + A2−A3III	
υ Sgr	181615	19 21.7	−15 57	EB	4.3	4.4	p	2433134	137.939		B2Vpe + A2Ia(Shell)	46 Sg
ρ¹ Sgr	181577	19 21.7	−17 51	δSct	3.90	3.93	V		0.050		F0IV−V	44 Sgr
BF Cyg	−−	19 23.9	+29 40	Z And	9.3	13.4	p				Bep + M5III	
CH Cyg	182917	19 24.5	+50 14	Z And	6.4	8.7	V		97		M7IIIab + B	
V1229 Aql	−−	19 24.7	+4 15	Na	6.7	18:	B	2440687			P(Q)	N 1970
RR Lyr	182989	19 25.5	+42 47	RRab	7.06	8.12	V	2442995.405	0.566867	0.19	A8−F7	
V368 Aql	−−	19 26.6	+7 36	Na	5.0	15.7	v	2428438.4			P(Q)	N 1936
TT Lyr	−−	19 27.6	+41 42	EA	9.34	11.43	V	2438605.2644	5.243727	0.14	A0	
U Aql	183344	19 29.4	−7 03	Cδ	6.08	6.86	V	2434922.31	7.02393	0.30	F5I−II − G1	
V374 Aql	−−	19 30.2	−0 50	SRa	8.8	11.7	v	2438314	456.50		C7,3(Ne)	
AF Cyg	184008	19 30.2	+46 09	SRb	7.4	9.4	p		94.1		M5e	
V923 Aql	183656	19 30.5	+3 27	γC	6.04	(0.12)	V				B5.5IIIpe−B8V	
V822 Aql	183794	19 31.3	−2 07	EB/DM	6.87	7.44	V	2442577.333	5.294950		B5 + B8:V	
DO Aql	−−	19 31.4	−6 26	Nb	8.6	17.8	p	2424450:			P(Q)	N 1925
V1125 Cyg	184128	19 31.8	+31 52	L	9.0	9.9	p				A6−F6	
XZ Cyg	239124	19 32.5	+56 23	RRab	9.00	10.16	V	2441453.3856	0.4664731			
WY Sge	−−	19 32.7	+17 45	N	5.4	19.5	p					N 1783

Star Name	HD	α 2000	δ 2000	Type	Max	Min		Epoch	Period	F	Spec	Notes
V1293 Aql	184201	19^{h}33^m.1	+5°02'	SRb	8.3	9.0	p		d		M5III	
V3790 Sgr	184077	19 33.3	-16 37	SRb	7.24	7.54	V				M2	
PW Tel	183806	19 33.4	-45 16		5.61 Y	(0.02)	v		2.85		ApCrEuSr	
V1294 Aql	184279	19 33.6	+3 46	γC	6.82	7.23	V				B0.5IV	
V450 Aql	184313	19 33.8	+5 28	SRb	6.30	6.65	V	2431320	64.20	0.52	M5III-M8III	
TY Cyg	184524	19 33.9	+28 20	M	9.0	15.0	v	2438099	350.02	0.48	M6e	
AQ Sgr	184283	19 34.3	-16 22	SRb	9.1	10.9	p		199.6		C5,1	
V1264 Cyg	184905	19 34.7	+43 57	αCV	6.48	6.67	V	2440829.81	2.17	0.30	A0p	
Z Pav	183847	19 35.5	-62 45	SRa	9.1	10.5	p	2429850	135.5		Me	
U Vul	185059	19 36.6	+20 20	Cδ	6.78	7.51	V	2436410.812*	7.990676	0.33	F8Iab-G2	
R Cyg	185456	19 36.8	+50 12	M	6.1	14.2	v	2442021	426.44	0.35	S3,9e-S6,8e	
RT Aql	185293	19 38.0	+11 43	M	7.6	14.5	v	2443290	327.11	0.42	M6e-M8e(S)	
V1143 Cyg	185912	19 38.7	+54 58	EA	5.85	6.37	V	2439385.6881	7.6407543	0.022	F5V + F5V	
BG Cyg	—	19 38.9	+28 31	M	9.0	12.8	v	2438333	291.28	0.5:	M7e-M8e	
σ Aql	185507	19 39.2	+5 24	EB/DM	5.14	(0.2)	V	2422486.797	1.95026		B3V + B3V	44 Aql
RV Aql	185821	19 40.7	+9 56	M	8.1	15.0	v	2442618	218.60	0.47	M2e-M7:e	
BR Cyg	—	19 40.9	+46 47	EA	9.4	10.6	v	2441539.4654	1.33256415	0.19	A5V + F0V	
QS Aql	185936	19 41.1	+13 49	EA/SD	5.93	6.06	V	2440443.489	2.513294	0.17	B5V	
V1351 Cyg	186532	19 42.1	+55 29	Lb	6.33	6.55	V				gM5	
V1276 Cyg	186357	19 42.8	+29 20	δSct	6.44	6.46	V		0.088		F1III	
RT Cyg	186686	19 43.6	+48 47	M	6.4	12.7	v	2442303	190.28	0.44	M2e-M8e	
PS Vul	186518	19 43.9	+27 08	E?	6.28 Y	(0.076)	V				G4III	
SU Cyg	186688	19 44.8	+29 16	Cδ	6.45	7.18	V	2436237.609	3.845507	0.37	F2I	
V973 Cyg	186776	19 44.8	+40 43	SRb	6.10	6.62	V		40:		M3III	
LU Vul	—	19 45.6	+28 35	Na	9.2	>21.0	p	2440057			P(Q)	N 1968
TU Cyg	187159	19 46.2	+49 04	M	8.7	15.5	v	2441892	219.43	0.49	M3e-M5e	
CN Dra	187764	19 46.7	+68 26	δSct	6.29	6.35	V		0.100		F0III	
δ Sge	187076	19 47.4	+18 32		3.82 Y						M2II + A0V	7 Sge
CK Vul	—	19 47.4	+27 19	Nb	2.7	>17.0:	v	2331186:				N 1670
LV Vul	—	19 48.0	+27 10	Na	5.17	16.90	B	2439964			P(Q)	N 1968
S Vul	—	19 48.4	+27 17	SRd	8.75	9.30	V	2439755.3	68.40	0.29	G0-K2(M1)	
V697 Cyg	235013	19 49.3	+52 47	Lb	8.9	10.8	p				M2	
V1339 Aql	187567	19 50.3	+7 54	γC?	6.33	6.52	V				B2.5IVe	
χ Cyg	187796	19 50.6	+32 55	M	3.3	14.2	v	2442143	406.93	0.41	S6,2e-S10,4e	
V1509 Cyg	187849	19 50.6	+38 43		5.12 Y	(0.12)	V				M2IIIa	19 Cyg
V380 Cyg	187879	19 50.6	+40 36	EA	5.5	5.6	p	2445332.1727	12.425703	0.11	B1III + B3V	
T Pav	186484	19 50.7	-71 46	M	7.0	14.0	v	2439186	243.97	0.43	M4e	
NZ Pav	186786	19 51.0	-65 36	δSct	6.04	6.06	v		0.080		F2III-IV	
X Aql	187757	19 51.5	+4 28	M	8.3	15.5	v	2441487	347.04	0.46	M6e	
SV Vul	187921	19 51.5	+27 28	Cδ	6.73	7.76	V	2438268.9	45.035	0.19	F7Iab-K0Iab	
V3961 Sgr	187474	19 51.8	-39 52		5.28	5.33	V		2445.2		ApCrSi:	
HO Tel	187418	19 52.0	-46 52	EA	7.9	8.35	p	2438986.3496	1.6131409	0.15	A2	
V1162 Aql	187826	19 52.3	-11 22	Cδs	8.6	9.3	p	2425803.400	5.3761	0.50:	G5	
CU Cyg	—	19 52.4	+55 20	M	9.5	14.1	v	2437958	215.7	0.43	M6e	
η Aql	187929	19 52.5	+1 00	Cδ	3.48	4.39	V	2436084.656	7.176641	0.32	F6Ib-G4Ib	55 Aql
V500 Aql	—	19 52.5	+8 29	Na	6.1:	17.8:	p	2430845:			P(Q)	N 1943
V819 Cyg	188439	19 53.0	+47 48	βC?	6.30	6.33	V	2436338.65	0.3775:	0.70	B0.5IIIn	
V505 Sgr	187949	19 53.1	-14 36	EA	6.48	7.51	V	2440087.3367	1.18287141	0.20	A0V + F8IV	
V1291 Aql	188041	19 53.3	-3 07	αCV	5.61	5.67	V	2432323*	224.5		A5pSrCrEu	
V449 Cyg	188344	19 53.3	+33 57	Lb	7.4	9.0	p				M1-M4	
S Pav	187835	19 55.2	-59 12	SRc	6.6	10.4	v	2439051*	386.26	0.49	M7IIe	
RR Sgr	188378	19 55.9	-29 11	M	5.6	14.0	v	2441133	334.58	0.43	M5e-M7e	
S Sge	188727	19 56.0	+16 38	Cδ	5.28	6.04	v	2436082.168	8.382173	0.31	F6Ib-G5Ibv	10 Sge
V548 Cyg	189371	19 57.0	+54 48	EA	8.9	9.7	p	2442279.3927	1.8052186	0.19	A0	
X Vul	339279	19 57.5	+26 33	Cδ	8.47	9.22	V	2435309.977	6.319422	0.35	F8Ia-G1	
RR Aql	188915	19 57.6	-1 53	M	7.8	14.5	v	2441764	394.78	0.30	M6e-M9	
PX Aql	—	19 58.0	-9 13	SR	9.4	11.7	v	2427658	154.8		M5	
V1357 Cyg	226868	19 58.4	+35 12	EB?(X)	8.78	8.93	V	2441166.22	5.60125		O9.7Iabpev	Cyg X-1
V476 Cyg	—	19 58.4	+53 37	Na	2.0	16.2	v	2422561			P(Q)	N 1920
RU Sgr	188813	19 58.7	-41 51	M	6.0	13.8	v	2439497	240.31	0.43	M3e-M6e	
V393 Cyg	189548	19 58.8	+43 18	EA?	8.8	9.7	p				A0	
RS Aql	189191	19 59.1	-7 53	M	8.7	15.4	v	2439193	410.12	0.48	M5e-M8	
VZ Sge	189577	20 00.1	+17 31	Lb?	5.27	5.57	V				M4IIIa	13 Sge
NT Vul	189849	20 01.1	+27 45		4.64 Y	(0.03)	V		14.0		A4III	15 Vul
V485 Cyg	189918	20 01.2	+33 56	Lb	8.8	9.8	p				M5III	
Z Cyg	190163	20 01.4	+50 03	M	7.4	14.7	v	2442247	263.69	0.45	M5e-M6e	
CF Cyg	235081	20 01.4	+52 05	cst	8.9		v				A2	
NU Pav	189124	20 01.7	-59 23	SRb	4.91	5.26	v		80:		M6III	
V3872 Sgr	189763	20 02.7	-27 43	L?	4.45	4.61	V				M4III	62 Sgr
V1295 Aql	190073	20 03.0	+5 44	unq	7.87	(0.02)	V				A0ep	
V1359 Aql	—	20 03.7	-3 23	Cδs?	8.83	9.02	V		3.7317	0.60:	G5	
V1362 Cyg	190467	20 03.7	+36 26	E?	8.0	8.16	V		7:		B5II:n	
RR Tel	—	20 04.2	-55 43	Z And	6.5	16.5	p				F5ep	
CD Cyg	—	20 04.4	+34 07	Cδ	8.35	9.52	V	2437480.293	17.0751	0.28	F8Ib-K0Ib	
AA Cyg	190629	20 04.5	+36 49	SRb	9.4	12.4	p		212.7		S7,5	

Star Name	HD	α 2000	δ 2000	Type	Max	Min		Epoch	Period	F	Spec	Notes
X Sge	190606	20^{h}05^m.1	+20°39'	SR	7.0	8.36	V	2425648	196^d	0.56	N(C5,5)	
V477 Cyg	190786	20 05.5	+31 58	EA	8.50	9.34	V	2440769.7137	2.3469846	0.07	A3 + F5	
V448 Cyg	190967	20 06.2	+35 23	EB	8.0	8.8	p	2416361.095	6.5197334		B1Ib-II + O9	
V453 Cyg	227696	20 06.6	+35 44	EA	8.29	8.72	V	2439340.0988	3.8898128	0.15	B0.5IV:	
V1943 Sgr	190643	20 06.9	-27 13	Lb	9.0	11.0	p				M8	
SW Cyg	191240	20 07.0	+46 18	EA	9.24	11.83	V	2438602.6009	4.573116	0.11	A2e + K0	
SY Aql	190970	20 07.1	+12 57	M	8.3	15.4	v	2442651	355.92	0.37	M5e-M7e	
WZ Sge	--	20 07.6	+17 42	Nr(E)	7.0	15.5	p	2432001	11900:		P(Q)	N 1978
V395 Cyg	191546	20 08.7	+44 04	SRd?	7.8	8.4	v		40.5		F8Ib	
V1624 Cyg	191610	20 09.4	+36 50		4.93 Y (0.05)		V				B2.5Ve	28 Cyg
SZ Cep	--	20 09.9	+77 11	M	8.6	15.5	v	2442624	326.59		S3.5,8e-S4,8:e	
V346 Aql	191515	20 10.0	+10 21	EA/SD	9.0	10.1	p	2442959.4707	1.10636313	0.19	A0V	
V1300 Aql	--	20 10.4	-6 16	M?	2.2	3.2	K		680		M:	
RY Cyg	191783	20 10.4	+35 57	Lb	8.5	10.0	v				N(C4,8-C6,4)	
RX Cyg	192035	20 10.8	+47 49	cst	8.0		v				B3	
R Cap	--	20 11.3	-14 16	M	9.4	14.9	v	2439014	345.13	0.44	Ne	
S Aql	--	20 11.6	+15 37	SRa	8.9	12.8	v	2443855	146.45	0.48	M3e-M5.5e	
X Pav	191171	20 11.8	-59 56	SRa	9.2	11.1	p	2439527	199.19	0.48	Mc	
RW Aql	355077	20 11.9	+16 04	cst?	9.0		p				F3n	
FG Sge	--	20 11.9	+20 20	unq	9.5	13.7	B				B4Ieq-F6Iep	
V1042 Cyg	192103	20 11.9	+36 12	Ia	8.06	8.16	V				WC8(+OB)	
V1357 Aql	--	20 12.1	+15 21	SX Ari	7.9	(0.04)	V	2444084.32	19.5		B5p	
RU Aql	192081	20 12.7	+13 00	M	8.7	14.8	v	2439171	274.24	0.42	M5e	
RS Cyg	192443	20 13.4	+38 44	SRa	6.5	9.3	v	2438300*	417.39		N0pe(C8,2e)	
V695 Cyg	192577	20 13.6	+46 44	EA	3.77	3.88	V	2441470.0	3784.3	0.017	K2II + B3V	31 Cyg
V1372 Cyg	192678	20 13.6	+53 40	αCV	7.33	7.39	V		18:		A4p	
DR Vul	339770	20 13.8	+26 45	EA	8.65	9.19	v	2445335.801	2.2509169	0.15	B8	
R Sge	192388	20 14.1	+16 44	RVb	9.46	11.46	B	2423627.0	70.594		G0Ib-G8Ib	
NU Vul	192518	20 14.2	+28 42	δSct	5.18 Y (0.07)		V		16.478		A7IVn	21 Vul
TU Cap	--	20 14.2	-15 54	SRb	9.20	11.29	V		90:		M5	
V1644 Cyg	192640	20 14.5	+36 48		4.97 Y (0.03)		V				A2V	29 Cyg
R Tel	192016	20 14.7	-46 58	M	7.6	14.8	v	2439439	461.88	0.43	M5IIe-M7e	
R Del	192502	20 14.9	+9 05	M	7.6	13.8	v	2442355	284.88	0.45	M5e-M6e	
Z Aql	--	20 15.2	-6 09	M	8.2	14.8	v	2441938	129.226	0.47	M3e	
V1488 Cyg	192909	20 15.5	+47 43	EA	4.11	4.14	V	2441256.96	1147.4	0.02	K3Ib + B3V	32 Cyg
SX Cyg	192788	20 15.6	+31 04	M	8.2	15.2	v	2438719	411.02	0.41	M7e	
MW Vul	192913	20 16.5	+27 47	αCV	6.62	6.70	V	2441617.165	16.478		A0p	
RT Cap	192737	20 17.1	-21 19	SRb	8.9	11.7	p		393		C6,4(N3)	
SU Pav	--	20 17.6	-60 04	M	8.6	>14.2	v	2439910	245.79		M4(II)e - M6II-IIIe	
RT Sgr	192702	20 17.7	-39 07	M	6.0	14.1	v	2439315	305.31	0.47	M5e-M7e	
P Cyg	193237	20 17.8	+38 02	S Dor	3.0	6.0	v				B2pe	34 Cyg
CN Cyg	--	20 17.9	+59 48	M	7.3	14.0	v	2441972	198.48	0.44	M2e-M7e	
AU Cyg	193345	20 18.5	+34 23	M	9.5	15.3	p	2439301	435.63		M6e-M7e	
WX Cyg	193368	20 18.6	+37 27	M	8.8	13.2	v	2440480	410.45	0.48	N3e(C8,2e)	
AE Cap	193104	20 19.0	-15 51	SR	9.0	11.1	B		200		M4	
V470 Cyg	228911	20 19.4	+40 53	EA	8.7	8.8	p	2431741.815	1.87292	0.18	B2 + B2	
V444 Cyg	193576	20 19.5	+38 44	EA	8.3	8.6	p	2441164.332	4.212424	0.20	O6-O8III: + WN5.5	
V478 Cyg	193611	20 19.6	+38 20	EA	8.9	9.3	p	2418552.648	2.880891	0.17	B0V + B0V	
U Cyg	193680	20 19.6	+47 54	M	5.9	12.1	v	2442233	462.40	0.48	Npe(C7,2e-C9,2e)	
DE Dra	193964	20 19.6	+62 15	EA	5.72 Y (0.15)		V		5.2984		B9V	71 Dra
V1584 Cyg	193722	20 19.9	+46 50		6.50 Y (0.06)		V		1.132854		B9pSi	
MY Cyg	193637	20 20.1	+33 57	EA	8.7	9.4	p	2441561.5978	4.00518908	0.15	A7:m + A5:m	
AC Dra	194258	20 20.1	+68 53	?	7.0	7.3	p				M5IIIa	
V Sge	--	20 20.3	+21 06	N1(E)	9.5	13.9	v				P	
BI Cyg	--	20 21.4	+36 56	Lc	8.4	9.9	v				M4Iab	
V1322 Cyg	229221	20 23.8	+37 30	γC	8.77	9.6	V				B0pe	
V865 Aql	--	20 23.9	+0 57	M	9.5	13.9	p	2438593	364.8		M6-M7(S7.5e:)	
BR Tel	--	20 24.0	-52 52	SRd	9.5	11.4	p	2428686*	142		G5	
T Mic	194676	20 27.9	-28 16	SRb	7.7	9.6	p		344		M6e	
U Mic	194814	20 29.2	-40 25	M	7.0	14.4	v	2439540	334.29	0.39	M6e	
KN Aql	195275	20 30.3	+1 53	SRb	8.52	9.35	V		70	0.41	M5e	
AF Dra	196502	20 31.5	+74 57	αCV	5.16	5.22	V	2433901.7*	20.2728		A0pSrCrEu	73 Dra
AI Cyg	195691	20 31.8	+32 31	SRb	9.2	11.8	p		197.3		M6	
ST Cyg	196070	20 32.5	+54 57	M	9.4	14.5	v	2440417	336.63	0.48	M6e	
RU Cap	--	20 32.6	-21 41	M	9.2	15.2	v	2438891	347.37	0.36	M9e	
Z Del	195763	20 32.7	+17 27	M	8.3	15.3	v	2442308	304.44	0.48	S5,2e-S7,2e	
SZ Cyg	196018	20 32.9	+46 36	Cδ	8.96	9.84	V	2437489.629	15.10989	0.42	F9Ib-G5Ib	
CZ Del	195876	20 33.6	+9 31	SRb	9.0	10.2	p		123		M5	
AK Cap	--	20 34.1	-23 15	Lb	9.0	9.7	p				M2	
FG Vul	--	20 34.6	+28 17	SR?	8.97	9.51	V		80:		M5II?	
V Vul	340667	20 36.5	+26 36	RVa	8.06	9.35	V	2414871.1	75.72		G4e-K3(M2)	
BD Vul	--	20 37.3	+26 29	M	9.3	12.7	v	2425758	430	0.53	C6,3	
GO Cyg	196628	20 37.3	+35 26	EB	8.3	8.9	p	2433930.4056	0.71776383		B9n + A0n	
VW Cep	197433	20 37.4	+75 36	EW/KW	7.23	7.68	V	2444157.4131	0.27831460		G5 + K0Ve	
ρ Pav	195961	20 37.6	-61 32	δSct	4.88 Y (0.03)		V		0.11		F5	

Star Name	HD	α 2000	δ 2000	Type	Max	Min		Epoch	Period	F	Spec	Notes
EU Del	196610	20ʰ37ᵐ9	+18°16′	SRb	5.8	6.9	v	2435794	59ᵈ5		M6IIIFe	
RU Vul	196792	20 38.9	+23 15	M	8.1	12.2	v	2438631	156.33	0.47	M3.5e	
FF Cyg	---	20 38.9	+37 53	M	8.2	14.2	v	2441891	324.59	0.43	M4e	
DM Del	---	20 39.6	+14 26	EA	8.6	8.9	v	2444501.3913	0.8446733	0.29	A3	
R Mic	196717	20 40.0	-28 47	M	8.3	13.8	v	2439579	138.66	0.46	M4e	
V Cyg	---	20 41.3	+48 09	M	7.7	13.9	v	2441091	421.43	0.46	Npe(C6-C7,4e)	
α Cyg	197345	20 41.4	+45 17	αCyg	1.25 Y						A2Iae	Deneb
AV Mic	196829	20 41.4	-42 08	Lc?	6.25	6.35	V				M3II	
Y Del	---	20 41.7	+11 53	M	8.8	16.5:	v	2440373	468.55	0.43	M8e	
S Cap	---	20 41.7	-19 03	cst?	9.2		p				G5	
HR Del	---	20 42.3	+19 10	Nb	3.70	12.38	V	2439837			P(Q)	N 1967
V568 Cyg	197419	20 42.4	+35 27	γC	6.40	6.68	V				B2IV-Ve	
S Del	197420	20 43.1	+17 05	M	8.3	12.4	v	2442330	277.21	0.52	M5e-M8e	
X Cyg	197572	20 43.4	+35 35	Cδ	5.87	6.86	V	2435915.918	16.3866	0.35	F7Ib-G8Ibv	
δ Del	197461	20 43.5	+15 04	δSct	4.39	4.49	V		0.158		A7IIIp	11 Del
DR Cyg	---	20 43.7	+38 09	M	9.3	>15.5	p	2439419	313.88		M5e	
Y Aqr	---	20 44.4	-4 50	M	8.4	15.0	v	2441212	382.34	0.43	M6.5e-M9	
KP Del	197753	20 45.2	+18 18	Lb	7.70	7.86	v				M5	
AU Mic	197481	20 45.2	-31 20	BY?	8.59	8.96	V	2441054*	4.865	0.55	M0Ve	
T Del	197772	20 45.3	+16 24	M	8.5	15.2	v	2442283	332.14	0.45	M3e-M6e	
U Del	197812	20 45.5	+18 05	SRb	7.6	8.9	p		110:		M5II-III	
W Aqr	---	20 46.4	-4 05	M	8.4	14.9	v	2439128	381.10	0.42	M6-M8e	
V1489 Cyg	---	20 46.4	+40 07		0.38	1.10	K		1280		M6II-III	
MW Pav	197070	20 46.5	-71 57	EW	8.49	9.00	V	2440862.608	0.79498855		A5	
V Aqr	197942	20 46.8	+2 26	SRa	7.6	9.4	V	2434275	244.0		M6e	
T Cyg	198134	20 47.2	+34 22	Lb?	5.0	5.5	v				K3III	
EN Aqr	198026	20 47.7	-5 02	Lb	4.41	4.45	V				M3III	
V Del	198136	20 47.8	+19 20	M	8.1	16.0	v	2441065	533.51	0.42	M4e-M6e	3 Aqr
V367 Cyg	198288	20 48.0	+39 17	EB	7.38	7.98	B	2437390.96	18.5972		A7Iape + A9:	
FI Vul	198315	20 48.9	+23 00	Lb	8.6	9.4	p				M3	
V1661 Cyg	198478	20 48.9	+46 07		4.84 Y (0.05)		V				B3Iae	55 Cyg
VY Mic	198103	20 49.1	-33 44	EA	8.4	8.7	p	2438295.265	4.4358	0.10	A3	
T Aqr	198373	20 49.9	-5 09	M	7.2	14.2	v	2443360	202.10	0.48	M2e-M5.5e	
TX Del	---	20 50.2	+3 39	CW	8.85	9.54	V	2439431.902	6.1629	0.33	G0-G5	
T Vul	198726	20 51.5	+28 15	Cδ	5.44	6.06	V	2435934.758	4.435572	0.31	F5Ib-G0Ib	
Y Cyg	198846	20 52.1	+34 39	EA	7.30	7.90	V	2409534.3195	2.9963331	0.10	B0IV + B0IV	
V1330 Cyg	---	20 52.7	+35 59	Na	7.5	18.1	p	2440722			P(Q)	N 1970
BW Vul	199140	20 54.4	+28 31	βC	6.2	6.4	p	2440841.576	0.2010353	0.61	B2IIIv	
X Del	199170	20 54.9	+17 39	M	8.2	14.8	v	2441551	281.04	0.42	M4e-M6e	
EM Aqr	199124	20 55.1	-1 22	δSctc	6.55	(0.03)	V		0.054		A9IV-V - F0V	
UX Cyg	199252	20 55.1	+30 25	M	9.0	16.5:	v	2439352	578.0	0.40	M4e-M6e	
U Pav	198735	20 55.5	-62 42	M	8.6	>12.0	v	2417139	289.7		M4e	
X Cep	201305	20 56.2	+83 03	M	8.1	16.0	v	2443834	535.19	0.42	M4.5e:-M7e	
S Ind	199003	20 56.4	-54 19	M	7.9	17.0	p	2439779	399.95	0.41	M6e-M8IIe	
S Equ	199454	20 57.2	+5 05	EA	8.0	10.08	V	2442596.7435	3.4360969	0.13	B8V + F0V:	
SV Equ	199465	20 57.3	+5 50	EW	9.0	9.20	V	2439382.427	0.881		A0	
VX Cyg	---	20 57.3	+40 11	Cep	9.5	10.6	v	2437361.262	20.13183	0.31	cF5e-cG2	
DQ Cep	199908	20 57.8	+55 29	δSct	7.22	7.32	V	2433924.8404	0.07886444	0.50	F4III	
DV Aqr	199603	20 58.7	-14 29	EB/DS	5.89	6.25	V	2426160.500	1.575531		F0IV	
KZ Pav	199005	20 58.7	-70 25	EA	7.8	8.4	p	2430448.617	0.949880		F2	
V450 Cyg	---	20 58.8	+35 56	Nb	7.0:	>17.0	p	2430510			P(Q)	N 1942
AO Cap	199728	20 59.6	-19 02	αCV	6.2	(0.050)	V		2.25		B9pSi	20 Cap
V832 Cyg	200120	20 59.8	+47 31	γC	4.49	4.88	V				B1ne	59 Cyg
TX Cyg	---	21 00.1	+42 36	Cδ	8.79	10.01	v	2437366.965	14.70825	0.38	F5Ib-G6Ib	
RR Cap	200128	21 02.3	-27 05	M	7.8	15.5	v	2439096	277.54	0.40	M5e-M8.5	
ER Vul	200391	21 02.4	+27 48	EW/RS	7.27	7.49	V	2440182.2662	0.6980960		G0V + G5V	
YZ Cyg	200528	21 02.7	+41 17	cst?	8.2		p				A0	
DY Vul	200563	21 03.5	+24 00	Lb	9.0	9.7	p				M3	
R Vul	200687	21 04.4	+23 49	M	7.0	14.3	v	2442166	136.36	0.49	M3e-M7e	
V1059 Cyg	---	21 04.6	+42 27	SRa	9.3	10.3	p	2430550	372:		Me	
RV Aqr	---	21 05.9	-0 13	M	8.7	>15	v	2438590	453.5		C6-7,2-4(Ne)	
TW Cyg	---	21 06.0	+29 24	M	8.9	15.0	v	2436111	341.32	0.48	M6.5-M9ep	
DT Cyg	201078	21 06.5	+31 11	Cδ	5.63	5.92	v	2436074.469	2.499140	0.48	F7.5Ib-IIv	
RS Cap	200994	21 07.2	-16 25	SRb	8.3	10.3	p		340		M4	
V Cap	201015	21 07.6	-23 55	M	8.2	14.4	v	2440395	275.72	0.42	M5e-M8.2	
V389 Cyg	201433	21 08.6	+30 12	unq	5.5	5.7	p				B9V	
σ Oct	177482	21 08.7	-88 57		5.47 Y						F0III	
T Cep	202012	21 09.5	+68 29	M	5.2	11.3	v	2444177	388.14	0.54	M5.5e-M8.8e	
AM Peg	---	21 10.1	+12 28	SRa	9.0	11.0	v	2438732	136.5	0.44	M1e-M3e	
γ Equ	201601	21 10.3	+10 08	αCV	4.58	4.77	V		314		F0pIII-V	5 Equ
Z Cap	358906	21 10.6	-16 10	M	8.6	15.0	v	2442017	181.48	0.49	M2e-M7.0	
RS Aqr	---	21 11.0	-4 02	M	9.5	14.4	v	2440419	214.62	0.49	Me	
V1425 Cyg	202000	21 11.0	+55 20	EB	7.9	8.36	p	2440400.944	1.252387		B9 + A0	
V Ind	201484	21 11.5	-45 04	RRab	9.12	10.48	V	2439792.276	0.4796030	0.11	A5-G3:	
V1500 Cyg	---	21 11.6	+48 09	Na	2.22	>21	B	2442655			P(Q)	N 1975

Star Name	HD	α 2000	δ 2000	Type	Max	Min		Epoch	Period	F	Spec	Notes
EW Aqr	201707	21^{h}11^m.7	−14°28′	δSctc	6.41	6.48	V	2441534.104	0^d.097		F0III−IV	
VY Aqr	−−	21 12.2	−8 50	Nr	8.0	16.6	p	2445667				N 1983
R Equ	202051	21 13.2	+12 48	M	8.7	15.0	v	2437179	260.70	0.44	M3e−M4e	
T Oct	200286	21 14.0	−82 06	M	9.1	14.8	v	2439504	218.50	0.51	M2e−M4II:e	
W Ind	201866	21 14.4	−53 02	SRc	9.4	11.5		2430090	198.8		M4IIe−M5IIe	
τ Cyg	202444	21 14.8	+38 03	δSct?	3.65	3.84	V				F2III−IV	65 Cyg
RV Equ	−−	21 14.9	+9 00	Isb	9.0	9.7	p				K0	
RR Aqr	202306	21 15.0	−2 54	M	9.1	14.4	v	2442664	182.45	0.44	M2e−M4e	
V344 Cep	−−	21 17.7	+56 00	SRa	9.1	(3.5)	R	2442190	500	0.5		
υ Cyg	202904	21 17.9	+34 54		4.43 Y						B2Vne	66 Cyg
CE Cyg	203154	21 18.9	+47 00	Isb?	9.0	10.6	p				K5	
V1334 Cyg	203156	21 19.4	+38 14	Cep	5.77	5.93	V	2440124.533	3.3336	0.50	F1II	
FZ Cep	203378	21 19.7	+55 27	SR	8.5	9.1	p				M5	
T Ind	202874	21 20.2	−45 01	SRb	7.7	9.4	p		320:		N(C7,3)	
RY Aqr	203069	21 20.3	−10 48	EA/SD?	8.8	10.1	v	2442960.1496	1.9665904	0.13	A3	
V532 Cyg	−−	21 20.6	+45 28	Cδ	8.88	9.26	V	2436817.694	3.283154	0.44	F5	
θ¹ Mic	203006	21 20.8	−40 49	αCV	4.77	4.87	V	2440345.32	2.1219		ApCrEuSr	
X Peg	−−	21 21.0	+14 27	M	8.8	14.4	v	2441616	200.82	0.49	M2e−M5e	
V836 Cyg	203470	21 21.4	+35 44	EB	8.59	9.30	B	2426547.5224	0.65341090		A0	
T Cap	203349	21 22.0	−15 10	M	8.4	14.3	v	2439267	269.28	0.44	M2e−M8.2	
SW Peg	−−	21 22.5	+22 00	M	8.0	14.0	v	2438750	396.33		M4e	
V1070 Cyg	203712	21 22.8	+40 56	SR?	6.7	7.7	v				M7III	
RW Aqr	−−	21 23.1	+0 50	M	8.7	13.6	v	2439044	139.57	0.38	M2e−M4e	
Y Pav	203133	21 24.3	−69 44	SRb	8.6	10.3	p	2430060	233.3		C7,3	
V1073 Cyg	204038	21 25.0	+33 42	EW	8.0	8.38	V	2438672.5816	0.7858597		A3Vm	
IK Peg	204188	21 26.4	+19 23	δSct	6.07 Y	(0.01)	V		0.044		A8m	
S Mic	204045	21 26.7	−29 51	M	7.8	14.3	v	2440693	208.93	0.43	M3e−M5.5	
TV Peg	−−	21 26.8	+16 35	M	9.0	>14.0	v	2438554	246.64		M0e	
AX Cep	−−	21 26.9	+70 13	M	9.5	13.0	v	2437023	395.0	0.46	C(N)	
EI Cep	205234	21 28.5	+76 24	EA/DM	7.54	8.06	V	2436820.4665	8.439334	0.06	Am + F1	
β Cep	205021	21 28.7	+70 34	βC	3.16	3.27	V	2440444.625	0.1904881	0.50	B2IIIev	Alfirk
SX Pav	203881	21 28.7	−69 30	SRb	5.43	5.97	V		50:		M5III	
X Ind	204438	21 30.5	−53 57	M	9.0	>13.0	p	2430043	225.85	0.49	M4e−M5IIe	
GK Cep	205372	21 31.0	+70 49	EB/KE	6.89	7.37	V	2438694.7063	0.936157		A2V + A2V	
S Cep	206362	21 35.2	+78 37	M	7.4	12.9	v	2443787	486.84	0.55	C7,4e(N8e)	
W Cyg	205730	21 36.0	+45 22	SRb	6.8	8.9	p	2438659.73	126.26	0.50	M5IIIae	
UU Cap	205526	21 36.2	−13 52	SRb	9.3	10.4	p		100:		M4	
ε Cap	205637	21 37.1	−19 28	γC	4.48	4.72	V				B3IV−Vpeq	39 Cap
CP Cyg	205939	21 37.5	+44 42	cst	6.20		V				A7III	
V337 Cep	206165	21 37.9	+62 05	αCyg	4.69	4.78	V				B2Ib	9 Cep
SV PsA	205819	21 38.7	−33 21	EA	9.0	9.5	p				G5	
UU Cyg	206223	21 39.5	+43 17	cst	8.9		p				A2	
EE Peg	206155	21 40.0	+9 11	EA	6.9	7.6	v	2440286.4329	2.6282284	0.10	A3Vm + F4:	
RU Cyg	206483	21 40.6	+54 19	SRa	9.2	11.6	p	2441244	233.85	0.5	M6e	
EK Cep	206821	21 41.4	+69 42	EA/DM	7.99	9.32	B	2439002.7240	4.4277926	0.06	A0V	
Q Cyg	−−	21 41.7	+42 50	Na	3.0	15.2	v	2406583			P(Q)	N 1876
V460 Cyg	206570	21 42.0	+35 31	Lb	5.6	7.0	v				N1(C6,5)	
V1339 Cyg	206632	21 42.1	+45 46	SRb?	5.9	7.1	v		35:		M4III:	
SS Cyg	206697	21 42.7	+43 35	UG(X)	8.2	12.4	v		50.1:		A1−dGep	
RS Gru	206379	21 43.1	−48 11	δSct	7.93	8.49	V	2441599.9991	0.14701121	0.26	A6−F0	
μ Cep	206936	21 43.5	+58 47	SRc	3.43	5.1	V		730		M2Iae	
ε Peg	206778	21 44.2	+9 52	?	0.7	3.5	V				K2Ib	Enif
HN Peg	206860	21 44.5	+14 46		5.94 Y	(0.02)	V		24.90		G0V	
RR Peg	206890	21 44.5	+25 00	M	8.5	14.6	v	2442384	263.87	0.43	M4e−M6e	
TU Peg	−−	21 45.1	+12 42	M	8.2	13.8	v	2442230	321.85		M7e−M8e	
AG Cap	207005	21 46.3	−9 17	SRb	5.90	6.14	V		25:		M3III	47 Cap
EP Aqr	207076	21 46.5	−2 13	SRb	6.37	6.82	V		55:		M8III	
δ Cap	207098	21 47.0	−16 08	EA	2.81	3.05	V	2435656.913	1.0227688	0.08	A7IIIm	49 Cap
DX Peg	207265	21 47.2	+23 51	Lb	9.4	10.4	v				M6	
RY PsA	−−	21 47.5	−36 12	M	8.8	>14.0	v	2437120	224.39		Me	
AP Cap	207188	21 47.6	−17 18	αCV	7.60	7.65	V	2442619.827	2.67		B9pSi	
R Gru	207192	21 48.5	−46 55	M	7.4	14.9	v	2440957	331.89	0.42	M7II−IIIe	
WY Cyg	−−	21 48.7	+44 15	M	9.5	17.0	p	2441161	304.50	0.44	M5e−M6e	
GH Peg	207741	21 50.9	+15 15	EA	8.8	9.28	V	2426647.354	2.556136	0.08	A3	
AG Peg	207757	21 51.0	+12 38	Z And	6.0	9.4	v		830.14		WN6 + M1−M3II−III	
V1619 Cyg	207857	21 51.1	+39 32		6.17 Y	(0.01)	V		20.70		B9pHgMn	
VZ Cyg	−−	21 51.7	+43 08	Cδ	8.62	9.28	v	2437103.945	4.8645598	0.28	F5−G0	
HO Peg	207932	21 52.3	+21 16		6.89 Y	(0.4)	B				M4III	
AW Peg	207956	21 52.3	+24 01	EA	7.8	9.2	p	2436783.5617	10.62249	0.10	A3−5Ve − F5IV	
EM Cep	208392	21 53.8	+62 37	EW/KE	7.02	7.17	V	2444290.6075	0.8061876		B1IVe	
DF Peg	−−	21 54.7	+14 33	EA	9.5	11.1	p	2433505.62	14.6987	0.04	A2	
VV Cep	208816	21 56.7	+63 38	EA/GS+SRc	4.80	5.36	V	2443360	7430	0.078	M2Ia−Iabep + B8:Ve	
IR Cep	208960	21 57.9	+61 01	Cδ	7.58	7.98	V	2441696.582	2.114124	0.40	G0	
PR Cep	−−	21 58.0	+56 44	Lb?	1.48	1.86	K				M8	
MR Cyg	−−	21 59.0	+47 59	EA	8.76	9.66	V	2434650.8278	1.6770345	0.22	A0 + F7	

Star Name	HD	α 2000	δ 2000	Type	Max	Min		Epoch	Period	F	Spec	Notes
CM Lac	209147	22^{h}00^m.1	+44°33'	EA	8.20	9.15	V	2427026.316	1.^{d}6046916	0.10	A2 + A8	
BG Lac	--	22 00.4	+43 27	Cδ	8.53	9.16	V	2435315.273	5.331908	0.33	F7-G4	
V Peg	209127	22 01.0	+6 07	M	7.0	15.0	V	2442390	302.31	0.44	M3e-M6e	
RT Lac	209318	22 01.5	+43 53	EB/RS	8.84	9.89	V	2440382.891	5.074015		G9IV + K1IV	
LZ Cep	209481	22 02.1	+58 00	E11	5.56	5.66	B	2441931.868	3.070510		O8.5III + O9Vn	14 Cep
DX Aqr	209278	22 02.4	-16 58	EA/KE?	6.37	6.78	V	2442687.697	0.9450132	0.15	A2V	29 Aqr
o Aqr	209409	22 03.3	-2 09	γC	4.68	4.89	V				B7IV-Ve	31 Aqr
TT PsA	209336	22 03.3	-31 27	Lb	7.00	7.49	V				M7III	
S PsA	209400	22 03.8	-28 03	M	8.0	14.5	v	2440690	271.66	0.42	M3e-M5IIe	
MO Cep	209772	22 03.9	+63 07	Lb?	5.13	5.33	V				M5III	18 Cep
TW Peg	209598	22 04.0	+28 21	SR	7.0	9.2	v	2430370	956.4		M6-M7	
RT Peg	209641	22 04.2	+35 07	M	9.4	15.4	v	2440443	215.50	0.44	M3e-M6e	
IV Cep	--	22 04.6	+53 30	Na	7.0	19.3	B	2441139			P(Q)	N 1971
HK Lac	209813	22 04.9	+47 14	RS?BY?	6.77	6.91	V	2440100.8	25.3:		K0III + F1V	
HT Lac	209857	22 05.3	+46 45	Lb?	6.09	6.23	V				M4IIIab	
SV Peg	209872	22 05.7	+35 21	SRb	9.2	11.0	p		144.6		M7	
RZ Peg	209890	22 05.9	+33 33	M	7.6	13.6	v	2439105	439.40	0.44	C9,1e	
CX Lac	--	22 07.8	+40 06	SR	9.5	10.6	p		130	0.55	K5	
DM Cep	210615	22 08.3	+72 46	Lb	8.4	9.6	p				M4	
AR Lac	210334	22 08.7	+45 45	EA/RS	6.11	6.77	V	2439376.4955	1.9831987	0.17	G2IV + K0III	
T Peg	210251	22 08.9	+12 32	M	8.7	15.4	v	2440984	373.15	0.49	M6e-M7e	
Y Lac	235739	22 09.0	+51 03	Cδ	8.80	9.45	V	2435347.676	4.323788	0.34	F5-G0	
RS Peg	210749	22 12.3	+14 33	M	8.2	14.7	v	2442260	412.27	0.44	M6e-M9e	
RU Peg	--	22 14.0	+12 42	UG	9.0	13.1	v		67.8		sdBe + G8IVn	
ε Cep	211336	22 15.0	+57 03	δSctc	4.15	4.21	V		0.041242		F0IV	23 Cep
CP Lac	--	22 15.7	+55 37	Na	2.1	15.6	p	2428340.0			P(Q)	N 1936
R PsA	211493	22 18.0	-29 36	M	8.5	14.7	v	2440812	292.79	0.38	M4(II)-M5IIe	
UW Peg	--	22 18.2	+2 44	SR	8.7	9.9	v		106		M5	
TX Peg	211647	22 18.3	+13 37	SR	9.3	10.6	p		120	0.50	M5e	
GP Cep	211853	22 18.7	+56 08	E/WR+E?	8.96	9.07	V	2431256.602	6.6883		WN6+O8-B0III:+O+O	
X Aqr	211610	22 18.7	-20 54	M	7.5	14.8	v	2439241	311.65	0.42	S6,3e:(M4e-M6.5e)	
SS Aqr	--	22 19.9	-14 24	M	8.6	13.2	v	2438316	192.6	0.55	M2e	
ε Oct	210967	22 20.0	-80 26	SRb	4.96	5.36	V		55:		M5III	
π¹ Gru	212087	22 22.7	-45 57	SRb	5.41	6.70	V		150:		S5	
RW Cep	212466	22 23.1	+55 58	SRd	8.6	10.7	p		346:		K0 0-Ia	
RT Aqr	212243	22 23.2	-22 03	M	8.8	13.1	v	2434986	246.3		M5e-M6e	
FI Aqr	212432	22 24.6	-23 22	αCV	7.49	7.52	V	2442620.799	4.689		B9p(Si)	
π Aqr	212571	22 25.3	+1 23	γC	4.42	4.70	V				B1Ve	52 Aqr
RV Peg	212678	22 25.6	+30 28	M	9.0	15.5	v	2442049	389.97	0.38	M6e	
T Gru	212537	22 25.7	-37 34	M	7.8	12.3	p	2441259	136.53	0.48	M1Iae-M2Ib	
S Gru	212539	22 26.1	-48 26	M	6.0	15.0	v	2440613	401.37	0.43	M8IIIe	
T PsA	212617	22 26.2	-29 05	cst	8.1		p				F0	
S Lac	213191	22 29.0	+40 19	M	7.6	13.9	v	2442350	241.80	0.46	M5e-M8e	
δ Cep	213306	22 29.2	+58 25	Cδ	3.48	4.37	V	2436075.445*	5.366341	0.25	F5Ib-G1Ib	27 Cep
δ² Gru	213080	22 29.8	-43 45		4.11 Y (0.11)		V				M4.5IIIa	
V350 Lac	213389	22 30.1	+49 21	E11/RS	6.34	6.47	V	2441119.6	17.755		K2III	
WX Cep	--	22 31.3	+63 31	EA/DM	8.7	9.29	V	2425088.537	3.3784535	0.134	A2 + A5:	
GX Peg	213534	22 31.6	+29 33	δSct	6.30	6.32	v		0.056		A5m	
XZ Cep	--	22 32.4	+67 09	EB/DM?	8.0	8.83	V	2443297.811	5.0972267		O9.5V	
KY Cep	--	22 32.8	+57 36		4:	13:	p	2440797.333			CONT	
ν Tuc	213442	22 33.0	-61 59		4.81 Y						M4III	
SS Peg	213837	22 34.0	+24 34	M	9.2	>14.5	p	2442209	416.35		M7e	
EE Aqr	213863	22 34.7	-19 51	EB	8.3	8.94	p	2440828.7804	0.50899555		F0	
RZ Lac	214197	22 35.7	+52 45	cst?	8.8		p				B7e	
DI Lac	214239	22 35.8	+52 43	Na	4.3	14.9	p	2419000.5			sdBe + P(Q)	N 1910
R Ind	213797	22 36.0	-67 17	M	8.2	14.6	v	2440505	216.26	0.47	M2e-M4(II)e	
W Cep	214369	22 36.5	+58 26	SRc	7.02	9.2	V				K0ep-M2epIa + B0-B1	
CQ Cep	214419	22 36.9	+56 54	EB/DM/WR	8.63	9.12	V	2432456.668	1.641249		WN5.5 + O7	
AB Aqr	--	22 38.6	-14 02	Lb	9.5	11.5	p				M7	
GQ Cep	215038	22 39.4	+75 40	αCV	8.11	8.29	V	2438350.57	2.037638	0.45	A0pSi	
T Tuc	214575	22 40.6	-61 33	M	7.7	13.8	v	2440731	250.76	0.46	M3IIe-M6IIe	
Z Lac	214975	22 40.9	+56 50	Cδ	7.94	8.85	V	2435152.859	10.88583	0.43	F6Ib-G6Ib	
RR Lac		22 41.4	+56 26	Cδ	8.43	9.22	V	2437175.363	6.41619	0.30	F6-G2	
DD Lac	214993	22 41.5	+40 14	βC	4.9	5.1	p	2421914.200	0.19308858		B2III	12 Lac
BC Peg	--	22 41.8	+21 10	SRb	9.3	10.3	v		125	0.4:	M6	
β Gru	214952	22 42.7	-46 53	Lc?	2.0	2.3	V				M3-M5 II-III	
R Lac	215254	22 43.3	+42 22	M	8.5	14.8	v	2441226	299.86	0.41	M5e-M5.5e	
GL Lac	215441	22 44.1	+55 35	αCV	8.73	8.90	V	2436864.88	9.4871	0.57	B8p	
ZZ Cep	215661	22 45.1	+68 08	EA/DM	8.60	9.55	V	2427928.451	2.141800	0.12	B7 + F0V	
EV Lac	--	22 46.9	+44 20	UV	8.28	11.83	B				dM4.5e	
DH Cep	215835	22 46.9	+58 04	E11	8.58	8.62	V	2432759.279	2.111040		O5.5 + O6.5IV-Vn	
AH Cep	216014	22 47.9	+65 04	EB/DM	6.78	7.07	V	2434989.4026	1.7747505		B0.5Vne + B0.5V	
FM Aqr	215874	22 48.5	-10 33	δSctc	6.16	6.19	V		0.087		A9III-IV	70 Aqr
V Lac	240073	22 48.6	+56 19	Cδ	8.42	9.39	V	2437128.836	4.983468	0.25	F5-G0	
X Lac	216105	22 49.0	+56 26	Cδ	8.20	8.60	V	2437129.648	5.44499	0.32	F6-G0	

Star Name	HD	α 2000	δ 2000	Type	Max	Min		Epoch	Period	F	Spec	Notes
DK Lac	—	22ʰ49ᵐ8	+53°17′	Na	5.0:	15.5	p	2433304	d		P(Q)	N 1950
RX Lac	216151	22 49.9	+41 03	SRb	9.1	10.8	p		174	0.32	M6	
V360 Lac	216200	22 50.4	+41 57	E11	5.92 Y (0.05)		V		5*		B3IV:e	14 Lac
AF Peg	—	22 51.4	+18 07	SRb	8.8	9.8	v	2426317	65	0.48	M5II-III	
AR Cep	217158	22 51.6	+85 03	SRb	7.0	7.9	v				M4III	
λ Aqr	216386	22 52.6	−7 35	Lb	3.70	3.80	V				M2.5IIIa	73 Aqr
MX Cep	216533	22 52.7	+58 48	αCV	7.81	7.96	V	2440785.2	17.22		A2pSrCrMn	
IM Peg	216489	22 53.0	+16 50	RS	5.64 Y (0.2)		V				K1-K2 II-IIIe	
SW Lac	—	22 53.7	+37 56	EW	8.51	9.33	V	2444499.5272	0.3207186		G3p + G3p	
ST PsA	—	22 54.3	−34 23	M	9.5	>13.0	p	2436006	179		Me	
HR Peg	216672	22 54.6	+16 56		6.34 Y (0.2)		V		50:		S4	
GO Peg	216724	22 55.0	+19 33	Lb	8.6	9.3	p				M4	
V PsA	216692	22 55.3	−29 37	SRb	9.3	10.5	p	2428370	148		Mb	
SX Lac	—	22 56.0	+35 12	SRd	9.0	10.0	p		190.0	0.47	K2	
EN Lac	216916	22 56.4	+41 36	βC	5.3	5.4	p	2433505.765	0.169165		B2IV	16 Lac
TW PsA	216803	22 56.4	−31 34		6.44	6.49	V		10		K4V	
KZ Cep	217035	22 56.5	+62 52	βC	8.20	8.22	B		0.24544		B0V	
EW Lac	217050	22 57.1	+48 41	γC	5.0	5.3	p				B4IIIep	
S Aqr	216907	22 57.1	−20 21	M	7.6	15.0	v	2439224	279.27	0.39	M4e-M6e	
GT Cep	217224	22 57.8	+68 24	EA/SD	8.2	9.1	v	2425628.250	4.908756	0.14	B3V	
TV PsA	217005	22 57.8	−26 10	SR	8.56	8.95	V		45:		M5:III	
TV And	—	22 58.1	+42 44	SRa	8.3	11.5	v	2440878	113.8	0.40	M4e-M5e	
NY Cep	217312	22 58.7	+63 05	EA	7.40	7.55	V	2441903.8136	15.275727	0.19	B0IV + B0IV	
SZ And	—	22 59.6	+42 51	M	9.5	15.8	v	2441165	343.38	0.40	M2e	
V509 Cas	217476	23 00.1	+56 57	SRd	4.75	5.5	V				F8e-K 0-Ia + B1V	
o And	217676	23 01.9	+42 20	γC	3.58	3.78	V				B6IIIpe + A2p	1 And
NN Cep	217796	23 02.1	+62 31	EA/DM	8.2	8.58	V	2444507.4033	2.058305	0.12	A5	
V638 Cas	217833	23 02.7	+55 14	αCV	5.7	(0.10)	U	2444115.24	5.36		B9IIIpHe	
LN And	217811	23 02.8	+44 04	βCs	6.38 V (0.02)		b		0.0196		B2V	
AK Peg	—	23 03.2	+11 21	SRa	8.9 p	12.0	B	2439037	193.6	0.40	M5e	
π PsA	217792	23 03.5	−34 45	cst	5.11		V				F0V + F3V	
β Peg	217906	23 03.8	+28 05	Lb	2.31	2.74	V				M2.5II-III	Scheat
CW Cep	218066	23 04.0	+63 24	EA/DM	7.60	8.04	V	2435373.4492	2.7291396	0.13	B0.5 + B0.5IV-Vea	
RW Peg	217949	23 04.2	+15 18	M	8.8	14.6	v	2442321	208.43	0.48	K3e:-M6.5e	
ER Aqr	218074	23 05.4	−22 29	Lb	7.14	7.81	V				M3	
R Peg	218292	23 06.6	+10 33	M	6.9	13.8	v	2442086	378.02	0.44	M6e-M9e	
KX And	218393	23 07.1	+50 12	γC	6.93	7.05	V				B3pe + K1III	
AF Scl	218348	23 07.4	−25 36	SRb	8.78	9.15	V		32:		M4:III	
Y Scl	218541	23 09.1	−30 08	SRb	8.7	10.3	p				Mb	
KY And	218674	23 09.3	+49 39	γC	6.71	6.90	V				B3Vne	
GZ Peg	218634	23 09.5	+8 41	Lb?	5.0	5.16	V				M4SIII + A2V	57 Peg
KZ And	218738	23 10.0	+47 57	BY	7.89	8.03	V				K0Ve	
RT And	—	23 11.2	+53 02	EA/DW/RS	8.55	9.47	V	2441141.88902	0.628929513	0.17	F8V	
V Cas	218997	23 11.7	+59 42	M	6.9	13.4	v	2444605	228.83	0.48	M5e-M8.5e	
SZ Psc	219113	23 13.4	+2 40	EA/RS	8.02	8.69	B	2443498.5020	3.9658663	0.11	K1IV-V + F8V	
TY And	219346	23 14.7	+40 48	SRb	8.8	10.5	v		260:		M5e-M6e	
χ Aqr	219576	23 16.8	−7 44	Lb	4.90	5.06	V				M3III	92 Aqr
EZ Peg	—	23 16.9	+25 43	UG?	9.5 p	10.52	B				G5Ve	
ET And	219749	23 17.9	+45 29	αCV	6.48	6.50	V	2438284.67	2.604		B9VpSi	
AN And	219815	23 18.4	+41 46	EB/DM	6.0	6.16	p	2436095.726	3.2195665		A7Vm	9 And
W Peg	219946	23 19.8	+26 17	M	7.9	13.0	v	2442107	344.92	0.46	M6e-M8e	
S Peg	220033	23 20.6	+8 55	M	7.1	13.8	v	2442427	319.22	0.47	M5e-M8.5e	
τ Peg	220061	23 20.6	+23 44	δSct	4.60	4.62	V		0.05433		A5V	62 Peg
RY Cep	—	23 21.2	+78 58	M	8.6	13.6	v	2444533	149.06	0.51	Ke-M0e	
SV Aqr	—	23 22.7	−10 49	Lb	9.44	11.06	V				Mb	
RU Aqr	220515	23 24.4	−17 19	SRb	8.5	10.1	v		68.7	0.50	M5e(Ne)C6.3e	
κ Psc	220825	23 26.9	+1 15	αCV	4.91	4.96	V	2437198.48*	0.5805		A0pCrSi:Sr:	8 Psc
HV Peg	220933	23 27.7	+25 10		5.98 Y (0.012)		V		6.97		A0pHgMn	69 Peg
CG Tuc	221006	23 29.0	−63 07		5.68 Y (0.06)		v		2.32		ApSi	
AR Cas	221253	23 30.0	+58 33	EA/DM	4.82	4.96	V	2435792.8948	6.0663309	0.061	B3IV-V	
V Phe	221433	23 32.4	−45 59	M	8.5	14.4	v	2440815	257.00	0.47	M4e	
V436 Cas	221568	23 32.8	+57 54	αCV	7.39	7.84	V	2438666	159.0		A0pSrCrEu	
HW Peg	221615	23 33.5	+22 30		7.0	7.30	p				M5IIIa	71 Peg
Z And	221650	23 33.7	+48 49	Z And	8.0	12.4	p				M2III + B1eq	
AN Phe	221621	23 33.9	−45 03	SR	8.83	9.54	V		70:		Me	
Y Psc	221700	23 34.4	+7 55	EA	9.0	12.0	v	2441225.473	3.765859	0.10	A3 + K0	
ι Phe	221760	23 35.1	−42 37	αCV	4.70	4.75	V		12.5:		A2VpSrCrEu	
V629 Cas	—	23 36.0	+52 38	αCV	9.0	(0.074)	V		0.63195		A0pSi	
SU Peg	—	23 37.0	+32 42	SRa	8.1	11.5	v	2426610	198.4		M3e	
GG And	—	23 37.3	+47 07	SRb?	8.4	8.9	v				M5:	
λ And	222107	23 37.6	+46 28	RS	3.69	3.97	V	2443886.0	54.20		G8III-IV	16 And
XX Cep	222217	23 38.3	+64 20	EA/SD	9.13	10.28	p	2444839.8022	2.3373266	0.14	A8V	
ST And	222241	23 38.8	+35 46	SRa	7.7	11.8	v	2438976	328.34	0.52	C4,3e-C6,4e(R3e)	
SV Cas	222293	23 39.0	+52 16	SRa	9.1	12.5	v	2437964	264.5	0.42	M6.5	
WY And	—	23 41.5	+47 36	SRd	9.5	10.9	p	2430377	108	0.43	G2e-K2(M3)	

Star Name	HD	α 2000	δ 2000	Type	Max	Min		Epoch	Period	F	Spec	Notes
R Aqr	222800	23h43m.8	−15°17'	M(Z And)	5.8	12.4	v	2442398	386d.96	0.42	M5e–M8.5e + P	
Z Cas	222914	23 44.5	+56 35	M	8.5	15.4	v	2442376	495.71	0.39	M7e	
TX Psc	223075	23 46.4	+3 29	Lb	6.9	7.7	p				N0(C6,2)	19 Psc
SX Phe	223065	23 46.5	−41 35	δSct	6.78	7.51	V	2438636.6178	0.05496438		A2V	
V566 Cas	223385	23 48.8	+62 13	αCyg	5.34	5.45	V		46		A3eq 0–Ia	6 Cas
TZ And	223608	23 50.9	+47 30	SRb	9.4	10.8	p				M6	
HH Peg	223637	23 51.4	+9 19	Lb?	5.74	6.0	V				M3IIIa	80 Peg
ET Aqr	223640	23 51.4	−18 55	αCV	5.16	5.21	V	2440900.80	3.730	0.31	B9pSiCrSr	108 Aqr
RY Cas	—	23 52.1	+58 44	Cδ	9.38	10.39	V	2437344.602	12.13726	0.41	F5–G3Ib	
Z Aqr	223737	23 52.2	−15 51	SRa	9.5	12.0	p	2442693	135.5	0.33	M1e–M7IIIe	
HT Peg	223781	23 52.6	+10 57	δSct	5.30 Y	(0.01)	V		0.06		A4Vn	82 Peg
EQ Cas	—	23 52.9	+55 01	RVa	9.3	13.4	v	2435155.11	58.34		Fp(R)	
TZ Cas	—	23 52.9	+61 00	Lc	8.86	10.5	V				M2Iab	
ρ Cas	224014	23 54.4	+57 30	SRd	4.1	6.2	v		320		F8Iap–K0 0–Iap	7 Cas
XZ Psc	224062	23 54.8	+0 07	Lb	5.55:	5.97	V				M5III	
AL Scl	224113	23 55.3	−31 55	E	6.06	6.30	V		2.445094		B6V	
RS And	224126	23 55.4	+48 38	SRa	8.7	10.8	p	2438803	136		M7–M10	
V373 Cas	224151	23 55.6	+57 25	E?/GS	5.9	6.3	v	2436491.237*	13.4192		B0.5II + B0.5II	
RR Cas	—	23 55.8	+53 43	M	9.5	14.7	v	2444470	300.07	0.48	M5e	
V Cep	224309	23 56.5	+83 11	cst?	6.56		V				A3V	
R Phe	224269	23 56.5	−49 47	M	7.5	14.4	v	2439210	267.86	0.49	M3e–M4e	
R Tuc	224379	23 57.4	−65 23	M	8.5	15.2	v	2439746	286.23	0.41	M5e	
V Cet	224442	23 57.9	−8 58	M	8.6	14.8	v	2438792	257.82	0.45	M3e	
R Cas	224490	23 58.4	+51 24	M	4.7	13.5	v	2444463	430.46	0.40	M6e–M10e	
LQ And	224559	23 58.8	+46 25	βC?	6.4	6.46	B		0.238:		B3IV:ne	
RR Phe	—	23 58.8	−39 27	M	9.4	>14.0	p	2430423	426.55			
S Phe	224583	23 59.1	−56 35	SR	8.6	10.6	p	2430060	141		M6IIIe	

SUPPLEMENTARY NOTES
(By star name)

Abbreviations: *mag.*, magnitude; *max. II*, secondary maximum; *min. II*, secondary minimum; *sp.*, spectroscopic (binary). Often, a statement that a star has a secondary minimum is followed by its magnitude at that time.

The magnitudes given here in brackets, [ ], are the average values at maximum and minimum light, while those in the main tabulation are the extreme observed values. Generally, magnitudes given to two decimals are photoelectric **V** values, and those to one place are visual. Sometimes qualifying letters are appended in parentheses: **B** and **V** refer to photoelectric magnitudes in the **UBV** system; "b" and "p" are photographic (blue-light) values; "v" means visual.

Other abbreviations are listed in the Introduction.

α And. Sp. binary; epoch of maximum given.
ζ And. Sp. binary; depth of min. II is 0.08 (p).
λ And. Sp. binary; brighter component is variable.
o And. Spectrum varies; rapid rotator (vsini = 300 km/sec).
R And. [6.9, 14.3]
S And. Supernova in galaxy M31; the first extragalactic supernova to be observed. It appeared 16" west of the nucleus in August, 1885, and remained visible in large telescopes until early 1890 (*Sky and Telescope*, **35**, February, 1968, page 97).
T And. [8.5, 13.8]
U And. [9.9, 14.3] Period varies; mean elements given.
V And. [9.5, 14.4]
W And. [7.4, 13.7]
X And. [9.0, 14.8]
Y And. [9.2, 14.2]
Z And. Prototype of Z

And symbiotic stars; 4-mag. outbursts every 10-20 years.
RR And. [9.1, 15.1]
RW And. [8.7, 14.8]
ST And. [8.8, 11.1] Period varies.
SV And. [8.7, 13.7] Maximum magnitude may vary with 930-day and 5,400-day periods.
TU And. [8.5, 12.5]
TV And. Amplitude varies strongly.
TW And. Period varies; min. II, 8.94.
TZ And. Period uncertain.
AN And. Min. II, 6.09; may be a triple system.
BX And. ADS 1671A; min. II, 9.15.
CG And. Period decreased from 3.7430 days in 1950 to 3.7394 days in 1963.
EG And. Short-period (40-day) and long-period variations seen.
ET And. Intense Si II lines in spectrum.
GO And. Subdwarf?
GY And. Sp. binary. Spectrum, color, and magnetic field possibly vary with a 23-year period.
HN And. Sp. binary.
KK And. *Bright Star Catalogue* gives spectral type B9IV.
LN And. Double; amplitude 0.02 (b).
LQ And. Shell star.

S Ant. Period probably varies; min. II.
T Ant. Period may vary.

κ¹ Aps. Shell star.
T Aps. [9.1, 14.7]

R Aqr. [6.5, 10.3] Strong emission lines; symbiotic star with matter expelled in a jet. Embedded in a nebula that may be the remnant of a novalike outburst seen by Japanese astronomers, A.D. 930.
S Aqr. [8.3, 14.1]
T Aqr. [7.7, 13.1]
W Aqr. [8.9, 14.2]
X Aqr. [8.3, 14.4]

Y Aqr. [9.4, 14.8]
RR Aqr. [9.5, 13.9]
RS Aqr. [10.0, 14.0]
RY Aqr. Period varies; min. II, 8.9.
SV Aqr. Cycle of 200-300 days is possible.
VY Aqr. Outbursts recorded in 1907, 1962, 1973, and 1983 (*Sky and Telescope*, **67**, February, 1984, page 125).
DV Aqr. Min. II, 6.10.
DX Aqr. Min. II.
EE Aqr. Min. II, 8.48.
EM Aqr. Period and amplitude vary.

η Aql. Period has changed twice in the course of 170 years.
σ Aql. Min. II.
R Aql. [6.1, 11.5] Sudden period changes. Source of radio flares?
U Aql. Period varies.
W Aql. [8.3, 14.0] Possible companion of spectral type F5-F8.
X Aql. [8.9, 14.9]
Y Aql. Min. II, 5.05; light curve may vary.
Z Aql. [9.0, 13.9]
RR Aql. [9.0, 13.9]
RS Aql. [9.7, 15.2]
RT Aql. [8.4, 14.0]
RU Aql. [9.4, 14.0]
RV Aql. [9.0, 14.2]
SY Aql. [9.5, 14.4]
SZ Aql. Period varies.
TT Aql. Period varies.
FF Aql. Sp. binary.
FN Aql. Period varies.
HK Aql. Possibly type Is.
KO Aql. Period varies; min. II, 8.4.
QS Aql. Visual binary (mag. 6.5, 6.7); period varies.
V337 Aql. Period varies; min. II.
V342 Aql. Period varies.
V346 Aql. Light curve varies; depth of min. II approximately 0.10 (p).
V599 Aql. Depth of min. II is 0.06 (**B**).
V603 Aql. Sp. binary.
V805 Aql. Min. II.

V822 Aql. Period of radial-velocity variations given; min. II.
V889 Aql. Min. II; apsidal rotation period is 720 years.
V923 Aql. Light variations probably related to rotation and nonuniform surface brightness.
V1182 Aql. Min. II, 8.63.
V1291 Aql. Elements from magnetic field variations; epoch of minimum magnetic field strength given.
V1295 Aql. Period of about 100 days possible; infrared excess.

R Ara. Min. II, 6.2.
U Ara. [8.4, 13.6]
X Ara. [10.9,]
RW Ara. Depth of min. II is 0.02 (p).
V349 Ara. 2.67-day period possible; min. II.
V535 Ara. Min. II.
V539 Ara. Min. II, 6.09.
V616 Ara. Deep min. II.
V620 Ara. Min. II?

γ² Ari. Southern component of close binary ADS 1507.
R Ari. [8.2, 13.2]
T Ari. [8.3, 10.9] Period varies, mean value given; amplitude varies strongly.
U Ari. [8.1, 14.6]
W Ari. No BD star brighter than mag. 12 at this position.
RR Ari. Min. II, 6.75.
RZ Ari. Cepheid-like variations?
ST Ari. Mean elements given.
SX Ari. Epoch of minimum given. Prototype of SX Ari class.
UW Ari. Period may be greater than 40 days.
UX Ari. Sp. binary; light curve varies.

β Aur. Sp. binary 'ADS 4556A.
ζ Aur. Light variations de-

pend strongly on wavelength.
θ Aur. ADS 4566B; combined magnitude given.
R Aur. [7.7, 13.3] Period varies.
S Aur. Possible 20-year cycle.
T Aur. Resembles DQ Her; mean magnitude between flares is 14.92 (**V**). Eclipsing binary epoch of minimum, 2437614.0088; period, 0.20437851 day.
U Aur. [8.5, 14.0]
V Aur. [9.2, 12.1] Minima became fainter between 1945 and 1965.
W Aur. [9.2, 14.6]
X Aur. [8.6, 12.7]
Z Aur. Period varies, 110-day variations noted in 1965.
RR Aur. [9.4, 13.7]
RT Aur. Period may vary.
RU Aur. [9.6,]
RX Aur. Period probably varies.
SX Aur. Period probably varies; min. II.
SY Aur. Period varies.
TT Aur. Min. II.
TW Aur. Mean magnitude varies with 1,370-day period.
UU Aur. Mean magnitude probably varies with 3,500-day period.
WW Aur. Epoch of primary minimum given; min. II, 6.28; possible apsidal motion.
AB Aur. Illuminates a nebula; infrared excess.
AE Aur. Near the nebula IC 405.
AR Aur. Min. II.
BF Aur. Min. II.
CO Aur. Superimposed 10-day and 40-day oscillations?
EO Aur. Min. II.
HZ Aur. Small secondary brightness oscillations of 0.07390-day period; radial velocity varies.
IM Aur. Min. II, 8.05.
IQ Aur. Epoch of minimum given.

IU Aur. Period varies; Min. II, 8.62.

KW Aur. Sp. binary; period and light curve may vary.

LY Aur. Near the open cluster NGC 1907; min. II, 7.26.

MZ Aur. Light variation may be due to the rotation of a star with nonhomogeneous surface brightness.

γ Boo. Light curve varies.

κ² Boo. ADS 9173A; sp. binary.

i Boo. ADS 9494B. Min. II, 6.99. The period, light curve, and min. II vary. · The orbital period of the AB pair is 225 years.

R Boo. [7.2, 12.3]
S Boo. [8.4, 13.3]
V Boo. Period may vary; amplitude varies strongly.
Z Boo. [9.3, 14.8]
RR Boo. [8.8, 12.7]
RT Boo. [9.0, 13.5]
ZZ Boo. Min. II.
BW Boo. Sp. binary.
BX Boo. Period varies.
CN Boo. *Bright Star Catalogue* gives spectral type A8III.

R Cae. [7.9, 13.1]

R Cam. [8.3, 13.2] Period varies; mean elements given.
T Cam. [8.0, 13.8]
V Cam. [9.9, 15.4]
X Cam. [8.1, 12.6] Period varies; mean elements given.
RS Cam. Mean brightness varies.

RU Cam. Period and light curve vary; the amplitude decreased sharply between 1964 and 1966. Unusual spectrum.

RW Cam. Period varies; spectral peculiarities may indicate hot companion.

SZ Cam. ADS 2984B. Component A is NSV 1458. Min. II, 7.25; possible member of NGC 1502.

TU Cam. Min. II, 5.22.
TW Cam. Infrared excess.
XX Cam. Enhanced carbon abundance.
AS Cam. Min. II.
AX Cam. Period may be 1.14229 days. Binary?
BE Cam. In a reflection nebula.
BK Cam. Optical double. Sp. binary; shell star.

κ Cnc. Sp. binary; period of light variations may be 6.91 days.
R Cnc. [6.8, 11.2] Period varies, mean elements given.
S Cnc. Near open cluster NGC 2632. Period varies; depth of min. II is 0.05 (v).
U Cnc. [9.9, 14.6] Near open cluster NGC 2632.
V Cnc. [7.9, 12.8] ADS 6763A.
W Cnc. [8.2, 14.1]
RS Cnc. Mean magnitude varies with 1,700-day period.
RT Cnc. Mean magnitude varies with 540-day period.
RZ Cnc. Min. II, 9.21.
TW Cnc. Period may be 70.745 days; min. II.
UV Cnc. Near open cluster NGC 2632.
VZ Cnc. Epoch of maximum given; light curve varies with 0.716290-day beat period.
BM Cnc. Sp. binary; elements of B − V maximum given.

BN Cnc. Member of Praesepe open cluster, M44.
BQ Cnc. In Praesepe.
BR Cnc. In Praesepe.
BS Cnc. In Praesepe.
BT Cnc. In Praesepe.
BU Cnc. In Praesepe.
BV Cnc. In Praesepe.
BW Cnc. In Praesepe.
BX Cnc. In Praesepe.
BY Cnc. In Praesepe.

α² CVn. ADS 8706A; B component's mag. is 5.6 (v). Prototype of αCVn class.
R CVn. [7.7, 11.9] Period varies.
T CVn. [9.6, 11.9] Semiregular variable?
U CVn. [9.8,]
V CVn. Amplitude varies strongly.
Y CVn. "La Superba," a very red star. Irregularities in light variation; 0.5-mag. oscillations, 2,000 to 2,200 days?
RS CVn. Period and light curve vary; min. II; probable starspot activity. Prototype of RS class of variables.
TX CVn. P Cyg profiles in spectrum.
AI CVn. Beat period 0.855 day; light curve varies.
AO CVn. Amplitude probably varies.
BH CVn. Sp. binary.

β CMa. Beat periods of 49.198, 5.442, and 4.900 days.
ι CMa. Possible member of the cluster Collinder 121.
κ CMa. Shell spectrum; *Bright Star Catalogue* gives spectral type B1.5IVne.
ξ¹ CMa. Secondary variations.
o¹ CMa. Member of cluster Collinder 121?
R CMa. Period varies; min. II, 5.78.
Z CMa. In a nebula.
UW CMa. Min. II; P Cyg profiles.
VW CMa. Depth of min. II is 0.1 (p).
VY CMa. The A component of the multiple system ADS 6033; in nebula near open cluster NGC 2362.
CW CMa. Depth of min. II is 0.1 (p).
EW CMa. Visual and sp. binary. *Bright Star Catalogue* has spectral type B3IIIe. Common extended envelope?
EY CMa. Period and radial velocity vary; superimposed 0.2033-day and 0.1690-day fluctuations.
EZ CMa. Member of open cluster Collinder 121; radial velocity varies.
FF CMa. Min. II, 7.6.
FM CMa. Min. II, 7.4.
FS CMa. Infrared excess; dust shell.
FW CMa. Radial velocity varies with 31.9045-day period.
FZ CMa. In CMa OB1 association, and the H II region Sharpless 295.
GG CMa. Possibly sp. binary with βC component.
HK CMa. In open cluster NGC 2287.

R CMi. [8.0, 11.0]
S CMi. [7.5, 12.6] Period varies, mean elements given.
U CMi. [8.8, 13.0]
V CMi. [8.7, 14.9]
YY CMi. Min. II, 8.9 (v).
YZ CMi. Epoch of minimum given. Period varies;

has outbursts like UV-type variables.

AI CMi. Possibly semiregular or RCB type.

δ Cap. Period derived from spectral observations; light curve probably varies.
ε Cap. Shell star.
T Cap. [9.5, 13.9]
V Cap. [9.2,]
Z Cap. [9.5, 14.0]
RR Cap. [9.3, 14.5]
RS Cap. Light variation with 3,360-day period also observed.
RU Cap. [9.7, 15.1]
AE Cap. Superimposed 2,300-day variations.

η Car. Massive young star embedded in large nebula; currently 6.21 (V). This star's chief outburst occurred in the 19th century. It was of 1st mag. or brighter during 1835-58, peaking at −0.8 in 1843.
χ Car. *Bright Star Catalogue* gives spectral type B3IVp.
l Car. See ZZ Car.
R Car. [4.6, 9.6]
S Car. [5.7, 8.5]
U Car. Period changed in 1945 from 38.750 days.
V Car. Light curve probably varies.
X Car. Min. II.
Y Car. Two components; light curve varies.
RR Car. Period and amplitude vary.
RT Car. Spectral type uncertain; near open cluster Trumpler 15.
RW Car. [9.3, 15.0]
RZ Car. [10.0, 15.4]
ZZ Car. Variable star designation of l Car. Period may vary.
AG Car. Periods of activity; P Cyg profiles in spectrum; lies in a ring nebula.
BO Car. B8III companion, 11.05 (V).
BZ Car. Mean magnitude probably varies in a cycle of about 1,800 days.
EM Car. Min. II, 8.9.
ER Car. Near open cluster NGC 3532.
GH Car. Near open cluster Trumpler 18.
GV Car. Depth of min. II is 0.15 (p); near open cluster NGC 3532.
HP Car. Min. II.
HR Car. May be associated with nebula.
HW Car. Epoch of minimum given.
IW Car. Mean magnitude varies with a period of about 1,500 days; infrared excess.
IX Car. In center of a bright, round nebula.
QS Car. Min. II, 9.2?
QZ Car. Near η Car.
V341 Car. In the reflection nebula IC 2220.
V343 Car. Sp. binary.
V345 Car. *Bright Star Catalogue* has spectral type B2IVe.
V347 Car. Type EB possible; deep min. II.
V348 Car. Min. II, 8.9. Member of open cluster IC 2581?
V357 Car. a Car; sp. binary.
V370 Car. Member of open cluster NGC 3324.
V372 Car. *Bright Star Catalogue* gives spectral type B1.5IV.

V374 Car. In cluster NGC 2516.
V375 Car. Double; period may be 0.359 day.
V376 Car. Double.
V382 Car. Member of NGC 3532? *Bright Star Catalogue* gives spectral type G4 0-Ia.

β Cas. Sp. binary.
γ Cas. ADS 782A; companion mag. 11.0. Embedded in H II region; X-ray source. Prototype of γC class.
δ Cas. Elements unconfirmed.
κ Cas. Sp. binary.
o Cas. Sp. binary; shell star. *Bright Star Catalogue* gives spectral type B5IIIe.
ϱ Cas. Usual minimum magnitude is 5.1 (v), but reached 6.2 in 1946. Long-period eclipsing binary?
B Cas. The supernova observed by Tycho Brahe. It appeared in November, 1572, and was visible in daytime. The star waned gradually and was lost after 16 months. Now only an optical supernova remnant and X-ray source.
R Cas. [7.0, 12.6]
S Cas. [9.7, 14.8]
T Cas. [7.9, 11.9]
U Cas. [8.4, 14.8]
V Cas. [7.9, 12.2]
W Cas. [8.8, 11.8]
Y Cas. [9.8, 14.5]
Z Cas. [10.0, 14.7]
RV Cas. [9.4, 15.2]
RW Cas. Period varies spontaneously, with a tendency to decrease.
RX Cas. Period and maximum magnitude vary; min. II.
RY Cas. Period varies spontaneously.
RZ Cas. Period varies; min. II, 6.26.
SS Cas. [9.8, 13.1]
SU Cas. Period probably varies.
TU Cas. Light curve and amplitude vary with 5.23026-day period.
TV Cas. Period varies; min. II, 7.32.
TW Cas. Light curve varies; period may vary. Depth of min. II is 0.08.
WZ Cas. Mean magnitude varies; HD 224869 lies within 1'.
YZ Cas. Min. II, 5.77 (p).
AA Cas. Mean magnitude varies with 850-day period; waves of 70 to 115 days.
AO Cas. Spontaneous period changes; apsidal motion with possible 70-year period.
AR Cas. ADS 16795A. Min. II, 4.86; apsidal motion with period 1,500 to 1,800 years.
CC Cas. Min. II.
DL Cas. Period varies. Member of open cluster NGC 129?
DO Cas. Period varies; min. II.
V373 Cas. Sp. binary; epoch of minimum given.
V482 Cas. Period may be 0.28292 day. Member of Per OB1 association.
V509 Cas. Similar to ϱ Cas.
V566 Cas. Binary; shell star. *Bright Star Catalogue* gives spectral type A3Iae.

β Cen. Sp. binary.

δ Cen. Member of Sco-Cen moving cluster.
ε Cen. Also 0.17696- and 0.2150-day periods.
η Cen. Shell star; double?
μ Cen. Member of Sco-Cen moving cluster.
R Cen. [5.8, 11.1] Double maxima and minima; epoch of main minimum given.
U Cen. [8.2, 13.4]
W Cen. [8.5, 13.2]
X Cen. [8.0, 13.4]
Z Cen. Supernova in galaxy NGC 5253.
RR Cen. Period varies; min. II.
RS Cen. [8.6, 13.4] X-ray source?
RT Cen. [9.0, 12.7] Period varies, mean elements given.
RV Cen. [7.7, 10.3]
RW Cen. [9.1, 10.0]
RX Cen. [9.4,]
SS Cen. Min. II, 9.5.
SV Cen. Period decreased from 1.66141 days in 1894 to 1.65904 days in 1969; min. II, 9.45.
SZ Cen. Min. II, 8.7.
UZ Cen. Light curve may vary; spectral type may be F2.
VZ Cen. Depth of min. II is 0.15 (p).
BF Cen. Min. II, 8.8; member of open cluster NGC 3766?
LZ Cen. Min. II.
MN Cen. Min. II, 8.7.
V346 Cen. Min. II.
V379 Cen. Min. II, 8.9.
V419 Cen. Period probably varies.
V636 Cen. Min. II, 8.8.
V646 Cen. Min. II, 9.1.
V701 Cen. Min. II.
V716 Cen. Min. II, 6.21.
V743 Cen. Amplitude varies.
V757 Cen. Min. II.
V758 Cen. Min. II, 9.15.
V759 Cen. Min. II.
V761 Cen. Spectrum, magnetic field vary periodically.
V766 Cen. Visual binary; infrared excess; dust shell.
V768 Cen. Possible period 60-80 days.
V772 Cen. Near cluster IC 2944.
V801 Cen. Ejecting shell; 292-second X-ray periodicity. X-ray flares every 188 days?
V810 Cen. Member of open cluster Stock 14.

β Cep. ADS 15032A, sp. binary. Period and radial velocity vary; elements for 1969 are given. Prototype of βC class.
δ Cep. Period varies; epoch of maximum given.
μ Cep. Herschel's Garnet star; very red. Possible 13.5-year variation in mean brightness.
S Cep. [8.3, 11.2]
T Cep. [6.0, 10.3] Period varies; mean elements given.
U Cep. ADS 830A. Component B mag. 11.2; C mag. 12.2. Period varies; A is surrounded by a gaseous disk.
W Cep. Superimposed 2,000-day and 350-day oscillations.
X Cep. [9.4, 15.7]
Y Cep. [9.6, 15.1]
RR Cep. [10.2, 14.7]
RW Cep. Infrared excess; dust shell.
SS Cep. Cycles of variation in mean brightness from

hundreds to thousands of days have been observed.

VV Cep. Superimposed 13.7-year, 0.15-mag. and 349-day, 0.3-mag. oscillations.

VW Cep. Visual binary. Period and light curve vary; min. II.

XX Cep. Period varies; min. II, 9.26. Apsidal motion possible.

XZ Cep. Min. II.

ZZ Cep. Min. II, 8.74.

AH Cep. Min. II.

CQ Cep. Min. II, 9.03. May have undergone a rapid 0.1-mag. fluctuation in 1969.

CW Cep. Period varies; min. II, 8.01. Member of the Cep OB3 association.

DH Cep. Period of radial-velocity variations is given; 2.6-year apsidal rotation period; near NGC 7380.

DQ Cep. Amplitude varies from 0.06 to 0.10 mag.

EI Cep. Min. II.

EM Cep. ADS 15434A. Light curve varies; min. II, 7.13; near open cluster NGC 7160.

GK Cep. Period varies; min. II, 7.35.

IV Cep. Close 16th-mag. companion.

KY Cep. Continuous spectrum; flares.

KZ Cep. Member of Cep OB3 association.

LZ Cep. Min. II.

MX Cep. Possible period 17.98 days.

NY Cep. Faint companion, 10.33 (V). Min. II. Member of Cep OB3 association.

o Cet. [3.45, 9.07] Mira, first seen to fade from view in 1596 by H. Fabricius. Prototype of M class. This star is ADS 1778A. Component B is the variable VZ Cet.

R Cet. [8.1, 13.0]
S Cet. [8.2, 14.2]
U Cet. [7.5, 12.6]
V Cet. [9.4, 14.3]
W Cet. [7.6, 14.4]
X Cet. [8.8, 12.3]
Z Cet. [8.9, 13.5]
RR Cet. Period varies.
SS Cet. Period varies.
TV Cet. Min. II.

UV Cet. Variable is the fainter member of a visual binary. Brighter member has spectral type dM5.5e. Prototype of the UV class.

VZ Cet = o Cet B = ADS 1778B. Rapid small fluctuations, perhaps due to interaction with o Cet A.

WW Cet. Rapid (approximately 10-minute) light fluctuations; spectrum resembles that of SS Cyg in light minimum.

XY Cet. Min. II, 9.12.

AA Cet. ADS 1581A. Min. II.

AB Cet. Sp. binary.

AD Cet. Has a companion, mag. 10.

R Cha. [8.5, 13.6]
S Cha. h 4590A.
RS Cha. Period may vary; min. II, 6.52.
RZ Cha. Min. II, 8.8.

T Cir. Min. II, 9.4.
AX Cir. HD 130702?
BF Cir. Min. II, 8.9.
BR Cir. X-ray source Cir X-1; mag. 22.5 (B). There are 16.6-day and millisecond

X-ray variations. Black-hole candidate.

R Col. [8.9, 14.3]
S Col. [9.3, 13.8]
T Col. [7.5, 11.9]

R Com. [8.5, 14.2]
UU Com. Member of Coma Berenices cluster; 31-minute, 0.01-mag. light fluctuations have also been observed.
FK Com. Min. II, 8.3. Prototype of FK Com class.
FM Com. Member of Coma Berenices cluster.
GM Com. Member of Coma Berenices cluster.

ε CrA. Min. II, 4.95.
RW CrA. Min. II, 9.4.
V686 CrA. Radial velocity varies with a 7.34-day period.

α CrB. Physical light variability also possible; min. II.

β CrB. Epoch of minimum given. Sp. binary; also a visual binary with an orbital period of 10.496 years.

γ CrB. ADS 9757A.

R CrB. Infrared excess. Prototype of RCB class.

S CrB. [7.3, 12.9]

T CrB. Outbursts have occurred in 1866 and 1946, with smaller ones in 1963 and 1975. Type M giant? Orbital period 227.6 days.

U CrB. Irregular period changes; min. II varies.
V CrB. [7.5, 11.0]
W CrB. [8.5, 13.5]
X CrB. [9.1, 13.6]
Z CrB. [10.0, 14.6]
RR CrB. Superimposed 377-day, 0.6-mag. oscillation.
RS CrB. Superimposed 69.5-day oscillation?
TZ CrB. Component A of multiple ADS 9979.

R Crv. [7.5, 13.8]
RV Crv. Min. II, 8.90.
SX Crv. Light curve varies; min. II, 9.23.

RR Crt. Possible 293- or 146-day period.
SV Crt. ADS 8115A. Sp. binary; period of light variations same as orbital period.

β Cru. Binary. Period, amplitude and light curve vary; many superimposed oscillations.

δ Cru. Expanding circumstellar shell.

θ² Cru. Sp. binary.
AB Cru. Min. II.
AE Cru. Depth of min. II is 0.03.
BH Cru. Two maxima and two minima, as in light curves of R Cen and R Nor.

α Cyg. Deneb. Spectrum varies. Prototype of αCyg class.

o² Cyg. Period varies; new designation is V1488 Cyg.

τ Cyg. Close binary ADS 14787; period varies from 2 to 3 hours.

υ Cyg. Sp. binary ADS 14831A; expanding circumstellar shells.

P Cyg. Mag. 3 in the year 1600; mag. 4.6-5.6 in 18th century.

R Cyg. [7.5, 13.9]

T Cyg. Multiple system, with a 10th-mag. physical companion and a 12th-mag.

optical companion.

U Cyg. [7.2, 10.7] Period varies.

V Cyg. [9.1, 12.8]

W Cyg. Secondary oscillation may exist.

X Cyg. Period varies.

Y Cyg. Period varies; min. II, 7.75; apsidal motion observed.

Z Cyg. [8.7, 13.3]

RS Cyg. [7.2, 9.0] Epoch of minimum given. Light curve varies; double maxima are sometimes seen.

RT Cyg. [7.3, 11.8]

RU Cyg. ADS 15220A; probably forms a physical system with the visual component C.

RY Cyg. Near the open cluster NGC 6883.

SS Cyg. Sp. binary. Intense emission lines seen in light minimum; 9-second and 6.6-hour optical, and 9-second X-ray, oscillations reported.

ST Cyg. [9.9, 13.9]
SU Cyg. Period varies.
SW Cyg. Period varies; min. II, 9.30.
SX Cyg. [9.0, 14.3]
TU Cyg. [9.4, 14.2]
TW Cyg. [10.0, 14.5]
TX Cyg. Period varies.
TY Cyg. [9.5, 14.6]
UU Cyg. ADS 15209A.
UX Cyg. Period may vary.
VX Cyg. Period varies.
WX Cyg. [9.7, 12.6]

XZ Cyg. Period and amplitude vary; strong Blazhko effect.

AF Cyg. Period sometimes twice that listed; mean magnitude varies.

BF Cyg. In a nebula. Mean brightness is gradually diminishing.

BI Cyg. Infrared excess.
BR Cyg. Min. II, 9.58.
CD Cyg. Period may vary.
CH Cyg. Period varies. Mean brightness varies with 4,700-day period; small outbursts.

DT Cyg. Period varies.
GO Cyg. Period varies; min. II, 8.49.
MR Cyg. Min. II, 9.12.
MY Cyg. Depth of min. II is 0.05.
V367 Cyg. Min. II, 7.78. Surrounded by diffuse envelope. Spectra may be gF0 and dF0.

V380 Cyg. Period varies. Double minimum. Apsidal motion with a period of approximately 1,490 years.

V389 Cyg. Sp. binary ADS 14682A. Two periods alternate every 1 to 3 weeks.

V444 Cyg. Period varies; min. II, 8.44.

V448 Cyg. Min. II, 8.42. Near open cluster NGC 6871.

V450 Cyg. Resembles DQ Her.

V453 Cyg. Min. II, 8.67; apsidal-motion period 71 years. Near open cluster NGC 6871.

V470 Cyg. Depth of min. II, 0.05.

V477 Cyg. Period varies; apsidal rotation period 349 or 396 years.

V478 Cyg. Min. II.
V532 Cyg. Period varies.
V548 Cyg. Min. II, 9.01.
V568 Cyg. No emission lines observed in spectrum before 1944.

V695 Cyg. Spectroscopic

and eclipsing binary ADS 13554A.

V819 Cyg. Sp. binary?
V832 Cyg. ADS 14526A; mag. of B component is 9.0.
V836 Cyg. Min. II, 8.81.
V1042 Cyg. In H II region Sharpless 109.
V1073 Cyg. Mean brightness varies; min. II, 8.34.
V1143 Cyg. Min. II, 6.06; Bright Star Catalogue has spectral type F6Va.
V1264 Cyg. Periods of 1.84 and 1.855 days are also possible.
V1322 Cyg. Star MWC 344; near open cluster NGC 6913.
V1334 Cyg. Visual binary ADS 14859; radial velocity varies.
V1357 Cyg. X-ray source Cyg X-1, with millisecond X-ray variations. Sp. binary. Unseen component is possibly a black hole. Supernova in 1408? (Sky and Telescope, 58, October, 1979, page 323.)
V1362 Cyg. Type Ell possible.
V1425 Cyg. Min. II, 8.21.
V1488 Cyg. Period varies. Former name is o² Cyg.
V1489 Cyg. Infrared source NML Cyg; dust shell.
V1509 Cyg. Double.
V1624 Cyg. Sp. binary.
V1661 Cyg. Optical double; member of NGC 457?

δ Del. Sp. binary. Pulsation period of main component is 0.158 day, and of secondary component 0.134 day.

R Del. [8.3, 13.3]
S Del. [8.8, 12.0] Double maxima are sometimes observed.
T Del. [9.3, 14.8]
U Del. Mean brightness varies with period of about 1,100 days.
V Del. [10.1, 15.5]
X Del. [9.0, 14.1]
Y Del. [9.9, 14.0]
Z Del. [8.8, 14.5]
TX Del. Period varies.
EU Del. Amplitude varies strongly.
HR Del. Outburst in 1967 from mag. 12 (p) to 3.3 (V). Small nebula discovered in 1981.

α Dor. Binary; combined magnitude is given.

β Dor. Period varies.

γ Dor. Period between 0.33 and 1.00 day.

R Dor. Deep min. II frequently observed.

S Dor. Extremely luminous star in nebulous cluster NGC 1910 in Large Magellanic Cloud. Prototype of S Dor class.

SU Dor. [9.3, 14.5]

x Dra. Sp. binary; shell star.

φ Dra. Visual and sp. binary.

R Dra. [7.6, 12.4]
T Dra. [9.6, 12.3] ADS 10937A; component B is UY Dra, mag. 10.8 (p), possibly constant.
U Dra. [9.5, 13.8] Period varies; mean elements given.
V Dra. [9.9, 14.2]
W Dra. [9.6, 14.4] Period varies strongly.
X Dra. [11.0, 14.7]
Y Dra. [9.2, 14.5]
RS Dra. Mean elements

given; light curve and amplitude vary strongly.

RV Dra. [9.2, 13.7]
SV Dra. [9.7, 14.3]
TW Dra. ADS 9706A; companion mag. 9.5. Period varies; min. II, 8.3.
WW Dra. ADS 10152A; mag. of component B is 9.7 (p). Min. II, 8.9.
WZ Dra. Period may be half of that listed.
AF Dra. Epoch of maximum given. Spectrum, radial velocity vary.
AG Dra. Composite spectrum; radial velocity varies.
AI Dra. Min. II, 7.16.
AS Dra. Binary; epoch of conjunction (G star behind) is given. Star has large space velocity.
BH Dra. ADS 12019A; min. II, 8.58.
BV Dra. ADS 9537A; min. II, 9.0: Component B is BW Dra type, also an EW type, with max. 9.2 (B), min. 9.6 (B), period 0.292298 day.
BY Dra. Sp. binary; variability due to rotation of star with nonuniform surface brightness. Prototype of BY class.
CX Dra. Shell star.
DE Dra. ADS 12214; sp. binary.
DK Dra. Sp. binary.
DL Dra. Binary; amplitude 0.14 (v?). Also a 0.0837-day period.

γ Equ. Magnetic field may vary with a 1,786-day period.
R Equ. [9.3, 14.5].
S Equ. Period and radial velocity vary; min II, 8.11.
SV Equ. Min. II, 9.16.

λ Eri. Shell star.
ν Eri. Superimposed 0.1779- and 0.1735-day oscillations; eclipsing, sp. binary?
o¹ Eri. Superimposed 0.1291-day period.
τ⁴ Eri. ADS 2472A.
τ⁹ Eri. Sp. binary.
T Eri. [8.0, 12.8]
U Eri. [9.4, 14.8]
V Eri. Superimposed 1,209-day cycle.
W Eri. [8.6, 13.8]
Z Eri. Superimposed 746-day cycle.
RZ Eri. Min. II, 8.0.
VW Eri. Alternating 83.4- and 93.3-day periods.
YY Eri. Period slowly increasing; min. II, 9.42.
AS Eri. One component may be a physical variable; min. II, 8.42.
BM Eri. Also 30-day, 0.2-mag. fluctuations at maximum brightness.
CC Eri. Sp. binary.
CO Eri. Min. II, 9.1.
CU Eri. Min. II.
CW Eri. Min. II, 8.41.
DL Eri. Superimposed 0.0921-day variation.
DM Eri. ADS 3380.
DO Eri. Epoch of minimum given.
DU Eri. Expanding circumstellar shell.
DY Eri. ADS 3093; component C is the variable, with min. 11.18 (V).
DZ Eri. Period may be 3.67 day.

R For. [8.9, 12.2]
S For. Close pair. Only one visual outburst, on March 6, 1899. Though reported by

three independent observers, it was not supported by concurrent photographs (*Sky and Telescope*, **18**, June, 1959, page 427).

— ζ Gem. Period varies.
η Gem. Eclipsing, sp. binary ADS 4841A. Epoch of minimum given; radial velocity varies with 2,984-day period.
σ Gem. Sp. binary.
ω Gem. Barium star.
R Gem. [7.1, 13.5]
S Gem. [9.0, 14.2]
T Gem. [8.7, 14.0]
U Gem. As an eclipsing binary, epoch of minimum is 2437638.82645 and period 0.17690617 day. Also, 25-second X-ray oscillations are reported. Prototype of UG class.
V Gem. [8.5, 14.2] Period may vary; mean elements given.
W Gem. Period varies.
X Gem. [8.2, 13.2]
RX Gem. Period varies; depth of min. II, 0.05. Is the main component surrounded by a gaseous ring?
RY Gem. Depth of min. I is 2.35 (**V**), and of min. II 0.06 (**V**).
SS Gem. Min. II, 10.0; light curve varies.
SW Gem. Epoch of minimum given.
WY Gem. May be an eclipsing variable similar to VV Cep.
BU Gem. May be an eclipsing binary with 32-year period.
LT Gem. Period may be 2.1497 days; member of Gem OB 1 association.
LU Gem. Member of Gem OB 1 association.

δ² Gru. Optical companion.
π¹ Gru. Visual binary; companion mag. 10.9, *G*0V.
R Gru. [8.3, 14.6]
S Gru. [7.7, 14.4]
T Gru. [8.6, 11.5]

α¹ Her. ADS 10418A; companion mag. 5.4. Superimposed 6-year light variation of 0.5-mag. amplitude.
ω Her. Double-wave light curve.
u Her. Min. II, 4.86; minima are asymmetrical.
R Her. [8.8, 14.6]
S Her. [7.6, 12.6] Period varies; mean value given.
T Her. [8.0, 12.8]
U Her. [7.5, 12.5]
W Her. [8.3, 13.5]
RS Her. [7.9, 12.5]
RT Her. [9.4, 15.0]
RU Her. [8.0, 13.7] Composite spectrum at light minimum; companion probable.
RV Her. [10.1, 14.8]
RY Her. [9.0, 13.8]
RZ Her. [9.4, 14.9]
SS Her. [9.2, 12.4] Period varies; mean value given.
SV Her. [9.8, 14.4]
TV Her. [9.7, 14.5]
TX Her. Period probably varies in a cycle of about 31 years.
UX Her. Period varies; min. II, 9.11.
AC Her. Maximum magnitude varies with 1.6-mag. amplitude.
AK Her. ADS 10408A; companion mag. 12. Period

varies; min. II, 9.20.
DH Her. Depth of min. II is 0.07.
DI Her. Min. II, 8.88; apsidal rotation observed.
DQ Her. Maximum between outbursts is normally 14.1 (p). As an eclipsing binary of the EA type, epoch of minimum is 2434954.9438 and period 0.193620897 day; amplitude 1.5 mag.
HS Her. Period varies; min. II, 8.63; apsidal rotation period about 130 years.
IQ Her. Superimposed variations of short period and small amplitude.
NQ Her. Elements not confirmed by other observations.
V360 Her. Recorded July 8, 1892, on a plate of the *Astrographic Catalogue*, Paris zone. Not visible on a Harvard patrol plate made 25 days earlier.
V441 Her. Radial velocity varies with a period of 70-72 days.
V451 Her. Epoch of minimum given.
V463 Her. Epoch of maximum given. There is a bump on descending branch of light curve.
V535 Her. Epoch of minimum given.
V600 Her. Sp. binary.
V624 Her. Sp. binary; min. II, 6.35.
V637 Her. ADS 10227A. Components B and C are of mag. 9.5 and 9.6.
V644 Her. Sp. binary.
V645 Her. Period changes from 0.125 to 0.145 day in a 40-day cycle; amplitude varies.
V744 Her. Sp. binary; similar to BU Tau (Pleione).

R Hor. [6.0, 13.0]
T Hor. [8.2, 13.2]
U Hor. [9.6,]
TU Hor. Possibly δSct type.

χ² Hya. Min. II, 5.8.
R Hya. [4.5, 9.5] ADS 8920A; period has decreased from 500 days since 18th century.
S Hya. [7.8, 12.7]
T Hya. [7.8, 12.6] Period may vary.
W Hya. Possibly Mira type; period and light curve vary strongly.
X Hya. [8.4, 12.8]
RR Hya. [9.3, 14.4]
RS Hya. [10.0, 14.1] Period varies, mean elements given.
RT Hya. [7.5, 9.2] Amplitude and light curve vary strongly.
RU Hya. [8.4, 14.0]
SS Hya. ADS 8923A; companion mag. is 12.0.
UZ Hya. Period may be 152 days.
VZ Hya. Min. II, 9.47
AI Hya. Min. II; period of light variation may not agree with that of radial-velocity variation.
FS Hya. Resembles S Vul.
HQ Hya. Period and light curve vary; brightness was constant in early 1967.
HS Hya. Elements uncertain.
HV Hya. Magnetic field may vary with 4.606-day period.
KM Hya. Sp. binary.

KU Hya. Binary; combined magnitude given.

VW Hyi. Brighter maxima lasting 17 days occur with a mean cycle of 179.6 days. Not a member of the Large Magellanic Cloud.

R Ind. [8.4, 14.3]
S Ind. [8.2 (v), 15 (v)]
X Ind. [9.4,]

R Lac. [9.1, 14.4]
S Lac. [8.2, 13.0] Period varies; mean value given.
X Lac. Period varies.
RT Lac. Light curve and depth of minima vary; min. II, 9.62.
SW Lac. Period, light curve, and depth of minima vary; min. II, 9.31.
AR Lac. *G*2IV component is physically variable, with amplitude 0.24 mag. Period varies; min. II, 6.43.
DD Lac. Superimposed 8.87601-day variations in amplitude, radial velocity and color index.
EN Lac. ADS 16381A; companion mag. 12.0. Superimposed 0.170845-day, 0.03-mag. variation.
EV Lac. There is an optical companion, mag. 12.00 (**V**), plus an unseen component.
EW Lac. Light variations may not be periodic; star surrounded by a diffuse envelope.
GL Lac. Period may vary from 9.47 to 9.49 days in a cycle of about 2.9 years; strong magnetic field.
HK Lac. Sp. binary. Period and light curve may vary; similar to BY Dra and λ And.
V350 Lac. Min. II, 6.45; strong Ca II emission lines.
V360 Lac. Period may be 20 days.

ϱ Leo. Binary; combined magnitude given.
— R Leo. [5.8, 10.0] Period varies; mean value given.
S Leo. [10.1, 13.9]
V Leo. [9.1, 13.7]
W Leo. [9.8, 14.2]
TX Leo. ADS 7837A; companion mag. is 8.0.
AM Leo. Min. II, 8.86.
CX Leo. Sp. binary.
DE Leo. Period of 1.2-2.0 days is possible.
DG Leo. Close binary; companion mag. 6.9.

R LMi. [7.1, 12.6]
S LMi. [8.6, 13.9]
RW LMi. Period more than 1 year; existence of a blue component is possible.

R Lep. [6.8, 9.8] Hind's Crimson star. Maximum magnitude varies from 5.5 to 6.5 in a 40-year cycle. Period varies; mean value given.
S Lep. Superimposed 875- to 890-day oscillation.
T Lep. [8.3, 12.9]
SS Lep. Many small outbursts.

δ Lib. Min. II, 5.01.
S Lib. [8.4, 12.0]
U Lib. [9.6, 14.4]
V Lib. [9.7, 14.7]
Y Lib. [8.6, 14.1]
RR Lib. [8.6, 14.2]
RS Lib. [7.5, 12.0]

RT Lib. [9.0, 14.3]
RU Lib. [8.1, 14.0]
ES Lib. Min. II, 7.33.
FX Lib. Member of Sco-Cen moving cluster.

α Lup. Period varies; value for 1973-74 given. In Sco-Cen moving cluster.
δ Lup. Member of Sco-Cen moving cluster.
S Lup. [8.6, 13.0] Companion of X Lup, which has a maximum mag. of 10.4 (p).
Y Lup. [9.8, 15.1]
GG Lup. Min. II, 5.8.

R Lyn. [7.9, 13.8]
S Lyn. [9.6, 14.3] Period varies.
U Lyn. [9.5, 14.4]
V Lyn. Alternating 55- and 87-day periods; resembles UU Her.
Y Lyn. ZrO bands in spectrum.
RR Lyn. Min. II, 5.90.
TZ Lyn. Epoch of minimum given.

β Lyr. Prototype of EB class of eclipsing binaries. Common envelope; period and light curve vary. Min. II, 3.84.
δ² Lyr. Member of open cluster Stephens 1.
U Lyr. [9.5, 12.0] Near open cluster NGC 6791; probably not a member.
V Lyr. [9.7, 14.8]
W Lyr. [7.9, 12.2] Double maximum. Period varies; mean value given.
RR Lyr. Strong Blazhko effect. Period and light curve change in 40.812289-day cycle; maximum magnitude varies. Prototype of RRab class.
RY Lyr. [9.8, 14.7]
TT Lyr. Min. II, 9.44.
TZ Lyr. ADS 11219A. Period varies; min. II, 9.58.
V473 Lyr. If Population I, this is the shortest-period classical Cepheid known in our galaxy.

R Men. Sometimes 115- or 240-day periods observed.
U Men. Visual binary; companion mag. is 10.4.
TY Men. Min. II, 8.06.
UX Men. Min. II, 9.2:

θ¹ Mic. Max. II.
R Mic. [9.2, 13.4]
S Mic. [9.0, 13.8]
T Mic. Period is sometimes 173.5 days.
U Mic. [8.8, 14.0]
AU Mic. Epoch of minimum given. Flares observed. Sp. binary?

S Mon. ADS 5322A; companion mag. is 7.7. Member of open cluster NGC 2264.
T Mon. Spontaneous period changes.
— U Mon. Mean magnitude varies with period 2,320 days.
V Mon. [7.0, 13.1]
— X Mon. [7.4, 9.1]
Y Mon. [9.1, 13.9] Period varies.
RR Mon. [9.4, 15.0]
RW Mon. Period varies.
TT Mon. [9.5,]
UX Mon. Rapid light fluctuations of amplitude up to 0.10 mag.; period varies. A6 component is a δSct variable.
AX Mon. Double system with envelope; rapid light fluctuations.

BX Mon. Composite spectrum.
FW Mon. Depth of min. II is 0.04.
IM Mon. Min. II, 6.47.
V474 Mon. Multiple periodicities; amplitude varies in 7.7455-day cycle.
V505 Mon. Period probably less than 1 day.
V536 Mon. A period of 31.035 days (or integral multiple) is possible.
V569 Mon. The A component of triple system ADS 5782.
V571 Mon. Amplitude varies; 0.3008269-day beat period.
V578 Mon. Near open cluster NGC 2244.

η Mus. Visual and sp. binary.
θ Mus. Sp. binary.
R Mus. Period varies.
S Mus. Period varies; sp. binary.
T Mus. Mean magnitude varies with a 1,020-day period.
TU Mus. Light curve varies; min. II, 8.65.
DZ Mus. Period may be twice as great.
FH Mus. Period 7 or 14 hours.

μ Nor. In cluster NGC 6169.
R Nor. [8.2, 13.2] Visual binary; component mag. is 14. Double maxima and minima; elements are given for the brighter maximum.
S Nor. In center of open cluster NGC 6087; period varies.
T Nor. [7.4, 13.2]
SY Nor. Companion mag. 12.8.
PQ Nor. Possible period 5.6 days. Variable of EW type?

R Oct. [7.9, 12.4]
S Oct. [8.4, 13.5]
T Oct. [9.5, 14.3]
U Oct. [7.9, 13.6]
UV Oct. Light curve varies with 80-day period.

ζ Oph. Runaway star from Sco OB2 association; variable spectrum.
θ Oph. Member of Sco-Cen moving cluster; sp. binary.
χ Oph. Member of Sco-Cen moving cluster.
ω Oph. Period may be 2.99 days.
R Oph. [7.6, 13.3]
S Oph. [9.5, 14.5]
T Oph. [9.8, 14.0]
U Oph. Period varies; min. II, 6.48.
V Oph. [7.5, 10.2]
W Oph. [9.9, 14.5]
X Oph. [6.8, 8.8] The A component of visual binary ADS 11524; companion mag. is 8.6.
Y Oph. Light curve is not typical for a Cepheid of this period; possibly sp. binary.
Z Oph. [8.1, 12.7] Period varies; mean value given.
RR Oph. [8.9, 14.6]
RS Oph. Previous outbursts observed in 1898, 1933 and 1958; red component may be a semiregular variable.
RT Oph. [9.6, 15.1]
RU Oph. [9.3, 13.8]
RV Oph. Light curve var-

ies; min. II, 9.56.
RY Oph. [8.2, 13.2]
SS Oph. [8.7, 13.5]
UX Oph. Period varies.
XX Oph. Shell star.
V451 Oph. Min. II, 8.31.
V502 Oph. Period varies; min. II, 8.81.
V566 Oph. Min. II, 7.91.
V843 Oph. Kepler's supernova of 1604. Rivaled Jupiter for several weeks and faded in 2 years.
V986 Oph. Period varies from 0.284653 to 0.2907 day.
V1010 Oph. Period varies.
V2048 Oph. Sp. binary; irregular brightness variations and rapid flares observed.
V2112 Oph. In a reflection nebula.

α Ori. Betelgeuse. Periodicity sometimes disappears; cyclical 200- to 400-day waves superimposed on main variation. Rapid decreases in radial velocity are followed by sudden 0.5-mag. dimming. (See L. Goldberg's review article, *Publications* of the Astronomical Society of the Pacific, **96**, 366, 1984.)
δ Ori. Min. II. Irregular light variations observed.
ε Ori. In Orion's Belt. Mass ejection.
η Ori. Min. II, 3.23. Period is from radial-velocity observations.
ψ Ori. May be βC type, with period 0.30806 day.
ω Ori. In a reflection nebula.
R Ori. [9.6, 13.1]
S Ori. [8.4, 12.9] Period varies strongly; mean value given.
U Ori. [6.3, 12.0]
V Ori. [9.4, 14.1]
W Ori. Mean magnitude varies with period 2,450 days.
UX Ori. Reversed P Cyg effect is sometimes observed in Hα line. Infrared excess.
VV Ori. Min. II, 5.31.
BM Ori. Component B (northernmost member) of Orion Trapezium, ADS 4186; min. II, 7.99.
BN Ori. Composite spectrum?
FX Ori. Superimposed 30-to 60-day irregular light variations, with an amplitude of up to 0.5 mag.
KX Ori. May not be variable.
LP Ori. May not be variable.
NU Ori. May not be variable; infrared excess.
NV Ori. Infrared excess.
V359 Ori. Weak helium lines.
V361 Ori. Infrared excess.
V372 Ori. Infrared excess.
V380 Ori. ADS 4209A. The companion is now unseen; may have been a bright knot in the nebula NGC 1999.
V429 Ori. Spectral type not consistent with that of an RR variable. Amplitude was less than 0.09 mag. in 1968-69.
V529 Ori. Nova Ori 1667, supposedly discovered by Hevelius. J. Ashbrook has made a strong case that this object does not exist (*Sky and Telescope*, **26**, August, 1963, page 81).
V916 Ori. Period of

0.28297 day is possible.
V1004 Ori. Sp. binary.
V1016 Ori. This is component A (westernmost of the four bright members) of the visual and sp. multiple ADS 4186, the Orion Trapezium. Variability discovered in 1973.
V1030 Ori. Binary.
V1046 Ori. Sp. binary; in northern Orion Sword region.

κ Pav. Period varies.
λ Pav. Member of Sco-Cen moving cluster.
R Pav. [8.5, 13.0]
S Pav. [7.3, 9.2] Epoch of minimum given; sometimes a double maximum is observed.
T Pav. [8.0, 13.8]
V Pav. Mean magnitude varies with 3,735-day period.
W Pav. [9.0, 14.1]
KZ Pav. Visual binary h 5231; double minimum.
MW Pav. Min. II, 8.94.

γ Peg. Epoch of maximum given; radial velocity varies with same period.
ε Peg. ADS 15268ABC; flare observed in 1972.
τ Peg. Superimposed 0.04895-day oscillation; amplitude varies with beat period 0.4943 day.
R Peg. [7.8, 13.2]
S Peg. [8.0, 13.0]
T Peg. [8.9, 14.3] Period may vary; mean value is given.
V Peg. [8.7, 14.4]
W Peg. [8.2, 12.7]
X Peg. [9.4, 13.8]
Z Peg. [8.4, 13.2]
RR Peg. [9.2, 14.1]
RS Peg. [9.3, 14.3]
RT Peg. [9.9, 14.5]
RU Peg. Sp. binary. Companion mag. 13.4. Wide (14.0-day) and narrow (5.8-day) maxima interchange.
RV Peg. [9.9, 14.6] Maximum varies by more than 3 mags.
RW Peg. [9.7, 14.0]
RZ Peg. [8.8, 12.8]
TW Peg. Superimposed 90-day oscillations of small amplitude.
AG Peg. Nature of light changes and spectrum is complex; period of radial-velocity variations given.
AW Peg. Min. II, 7.96.
DF Peg. Depth of min. II is 0.04.
EE Peg. Period may vary. Min. II, 7.10.
EZ Peg. Possible flare on November 16, 1943.
GH Peg. Min. II, 9.23.
GX Peg. Sp. binary.
GZ Peg. ADS 16550A. Sp. binary. Companion mag. 10.
HN Peg. Light variation due to starspots?
HW Peg. Sp. binary. *Bright Star Catalogue* gives mag. 5.32 (**V**).
IK Peg. Sp. binary.
IM Peg. Sp. binary.

β Per. Algol, the Demon star. Variability recognized by G. Montanari in 1667. Period varies. Min. II, 2.19. Infrared excess; X-ray and radio flares. Prototype of EA class of eclipsing binaries.
δ Per. Intensity of spectral

lines varies strongly; radial velocity also varies.
o Per. Epoch of light minimum given; period is from radial-velocity observations. Min. II.
ρ Per. Slower mean brightness variations with 1,100-day period are possible.
φ Per. Amplitude of 19.5-day oscillation varies with 127-day period.
b Per. Triple system in open cluster NGC 1545; orbital period of close pair around third star is 700 days.
R Per. [8.7, 14.0]
S Per. [8.6, 10.6] Two superimposed 0.22-mag. oscillations with 825- and 940-day periods. Member of h and χ Per cluster.
T Per. In h and χ Per cluster.
U Per. [8.1, 11.3] Period varies; mean value given. Depth of minimum varies strongly.
W Per. Near h and χ Per cluster.
X Per. Member of ζ Persei moving cluster. X-ray period 835 seconds; very rapid rotation.
Y Per. [8.4, 10.3]
RR Per. [9.2, 14.4]
RY Per. Depth of min. II is 0.02 mag.
RZ Per. [9.4, 13.7]
SU Per. In h and χ Per cluster.
SY Per. Near open cluster NGC 1528.
UZ Per. Oscillations with 91-day period are also observed.
VX Per. Near h and χ Per cluster.
XY Per. ADS 2788; probably both components vary. Infrared excess.
AG Per. ADS 2990A; companion mag. is 9.00. Period varies. Min. II. Member of ζ Per moving cluster.
AW Per. Sp. binary.
GK Per. Visual brightness varies within range 11-14 between outbursts; transient X-ray source.
IQ Per. Min. II, 7.88.
KK Per. In h and χ Per cluster.
KP Per. Period and light curve vary; amplitude changes with 10.8-day period.
KS Per. Sp. binary; light variations are not related to the orbital motion.
LT Per. Epoch of light minimum given. Member of α Per moving cluster.
MX Per. Radial velocity may vary with 16.934615-day period.
PP Per. In h and χ Per cluster.
V351 Per. Member of Per OB1 association.
V353 Per. In Per OB1.
V354 Per. In Per OB1.
V357 Per. In Per OB1.
V358 Per. In Per OB1.
V362 Per. In Per OB1.
V376 Per. Two periods, 0.097 and 0.067 day. Amplitude varies.
V380 Per. Epoch of minimum given.
V383 Per. Spectral-line strengths are not typical for this variable type.
V386 Per. Sp. binary.
V396 Per. Near α Persei cluster; period may be 1.6857 or 2.4928 days.

V423 Per. Sp. binary ADS 2433A.
V436 Per. Sp. binary.
V467 Per. Sp. binary.

γ Phe. Sp. binary.
ζ Phe. The A component of visual binary Rst 1205; companion mag. 6. Min. II, 4.22. Apsidal rotation period 32.5 years.
R Phe. [8.0, 14.1]
T Phe. [9.4, 14.2]
V Phe. [9.2, 14.0]
W Phe. [8.9, 14.0]
SX Phe. Period, light curve, and amplitude vary. There is a 0.19283425-day beat period, also a 21.0-day secondary period. Max. II, 7.10; min. II, 7.42.
AE Phe. Min. II, 8.33.
AG Phe. Deep min. II.

δ Pic. Min. II, 4.83.
R Pic. Period and light curve vary.
S Pic. [8.1, 13.8]
T Pic. [8.4, 13.9]
RR Pic. Short-period brightness variations; maximum between outbursts is normally mag. 12.04 (**V**).

κ Psc. Epoch of minimum given; period may vary.
R Psc. [8.2, 14.3]
S Psc. [9.6, 15.0]
Y Psc. Period varies.
RT Psc. Superimposed 533-day oscillation also observed.
SZ Psc. Period may vary; physical variability of one of the components is possible. Min. II, 8.10.
TV Psc. Cycles of 49-day duration are replaced by those of 70- to 85-day duration.
UU Psc. ADS 191A; min. II, 6.02. Period of apsidal motion about 3 years.
UV Psc. Min. II, 9.5.
UZ Psc. Magnetic variable; magnetic field varies with 4.134-day period. Sp. binary.
WX Psc. Large infrared excess. Surrounded by dust shell.
AG Psc. Period not confirmed.

π PsA. Sp. binary.
R PsA. [9.2, 14.7]
S PsA. [9.0,]
RY PsA. [9.5,]

ρ Pup. Epoch of maximum given; period is from radial-velocity observations.
R Pup. Member of open cluster NGC 2439; lies above Cepheid instability strip in HR diagram.
U Pup. [9.8, 14.1]
W Pup. [8.4, 12.4]
X Pup. Period varies.
Z Pup. [8.1, 14.5]
RS Pup. Period varies; associated with a faint reflection nebula.
SW Pup. Min. II, 9.4.
VX Pup. Period and light curve vary strongly; Blazhko effect with 7- to 10-day period.
XY Pup. Min. II, 9.3.
XZ Pup. Min. II, 8.12.
AP Pup. Epoch of maximum given.
AQ Pup. Period varies.
AR Pup. Mean magnitude varies with 1,200-day period; infrared excess.
AS Pup. [9.5, 12.7]

AT Pup. Near open cluster NGC 2546; period may vary.
AU Pup. Period varies.
CH Pup. [9.6,]
KQ Pup. *M*2 component is sp. binary; radial velocity varies with 27-year period. Resembles VV Cep.
MQ Pup. Depth of min. II is 0.2 mag.
MW Pup. Min. II.
NO Pup. Visual binary; companion mag. 7.3. Min. II, 6.82.
NV Pup. Shell star.
NW Pup. Ell, amplitude 0.045 (**V**), 1-day period; βC, amplitude 0.02 (**V**), 3-hour period?
OW Pup. Shell star.
OX Pup. Near cluster NGC 2451.
PR Pup. Period may be 1.72 days.
PU Pup. m Pup; binary.
PV Pup. Binary.

S Pyx. [9.0, 13.9]
T Pyx. Previous outbursts in 1890, 1902, 1920, and 1944.
V Pyx. Similar to AB Aur, BH Cep, and BO Cep; probably rapid irregular.
RZ Pyx. Min. II, 9.68.
TT Pyx. Period may be twice as large.
TY Pyx. Min. II about 7.45; H- and K-line emission out of eclipse.

R Ret. [7.6, 13.3]

δ Sge. Sp. binary; may be associated with a reflection nebula.
R Sge. Infrared excess.
S Sge. Period varies; radial velocity varies with 675-day period.
U Sge. Period and light curve vary. Near but probably not a member of the cluster Collinder 399.
V Sge. Sudden 3-mag. outbursts plus 0.6-mag., 1-year fluctuations; spectrum resembles that of a late-stage nova. As an eclipsing binary, epoch of minimum is 2441978.8020 and period is 0.514195 day.
WY Sge. J. D'Agelet's nova, observed July 26, 27, and 29, 1783. Recovered as a very blue, 19th-mag. variable star by H. Weaver in 1951.
WZ Sge. Previous outbursts in 1913 and 1946. Also an eclipsing system with light curve resembling that of EW class; epoch of minimum 2443857.4632, period 0.057213 day.
FG Sge. Spectrum changed from *B*4I to *F*6Ip, 1955-72. Brightened from 13.7 (p) to 9.45 (**B**), 1890-1968. Planetary with recent shell ejection?

μ Sgr. Irregular brightness fluctuations are also observed. Member of Sco-Cen moving cluster.
R Sgr. [7.3, 12.5]
T Sgr. [8.0, 12.6]
U Sgr. Member of open cluster IC 4725 (M25); possibly sp. binary.
Z Sgr. [8.6, 16.0]
RR Sgr. [6.8, 13.2]
RS Sgr. Period varies; min. II, 6.2.
RT Sgr. [7.0, 13.3]
RU Sgr. [7.2, 12.8]
RV Sgr. [7.8, 14.1]

RW Sgr. [9.4, 10.8]

RX Sgr. [9.7, 13.8]

RY Sgr. Sudden drops in brightness; also waves in brightness and radial velocity with 38.6-day period.

ST Sgr. [9.0, 15.2]

SU Sgr. Period not confirmed by some observations.

SW Sgr. [10.0,]

TY Sgr. [9.8, 15.0]

UX Sgr. Superimposed 1,000-day oscillations of small amplitude.

VX Sgr. Amplitude varies strongly; large infrared excess.

WZ Sgr. Member of Sgr OB1 association.

XZ Sgr. Period probably varies. Min. II, 8.95; G component probably an intrinsic variable.

BB Sgr. Nearby star HD 174403 is not physically related to Cepheid.

BQ Sgr. Period varies.

FN Sgr. Composite spectrum; outbursts observed in 1924-26 and 1936-41.

V350 Sgr. May be a sp. binary.

V356 Sgr. Min. II, 7.39; apsidal rotation possible.

V505 Sgr. Period varies. Min. II, 6.63.

V732 Sgr. Light curve resembles that of DQ Her.

V1017 Sgr. Evidence in spectrum of hot companion. Outbursts in 1901, 1919, and 1973.

V1148 Sgr. Symbiotic star; at edge of globular cluster NGC 6553.

V1942 Sgr. Rapid 0.4-mag. oscillations superimposed on 800- to 2,000-day ones of 0.8-mag.; near cluster Ruprecht 147.

V2349 Sgr. Min. II, 8.8 (p).

V2509 Sgr. Depth of min. II is 0.3.

V3508 Sgr. Near the open cluster IC 4725 (M25).

V3792 Sgr. Min. II, 6.81.

V3890 Sgr. It is uncertain whether the star at this position is the nova or its close companion.

V3961 Sgr. Sp. binary.

V4031 Sgr. Shell star; spectrum is variable.

V4050 Sgr. Sp. binary.

α Sco. Antares; member of Sco-Cen moving cluster.

ζ¹ Sco. Possible member of NGC 6231. Shell star.

χ Sco. Light curve varies; 0.20544-day secondary period causes modulation with a period of 14.74 days.

λ Sco. Sp. binary; 0.1068518-day secondary period causes modulation with a period of 10.15 days.

μ¹ Sco. Member of Sco-Cen moving cluster; min. II, 2.91.

σ Sco. ADS 10009A; sp. binary. Period, amplitude, and light curve vary; elements are for 1962-65. There is an 8.252-day beat period.

T Sco. Near the center of globular cluster NGC 6093 (M80).

U Sco. Outbursts in 1866, 1906, 1936, and 1979; late-type K-star?

Z Sco. [9.2, 13.4] Period may vary; mean value given.

RR Sco. [5.9, 11.8]

RS Sco. [7.0, 12.2]

RT Sco. [8.2,]

RU Sco. [9.0, 13.0] Period varies; mean value given.

RW Sco. [9.6,]

RY Sco. Period varies.

RZ Sco. [8.8, 12.2] Period varies strongly. The amplitude decreased sharply between 1957 and 1959.

SV Sco. [9.8, 14.8]

AH Sco. Amplitude changes from 1.6 to 3.3 mag. with a period of about 25 years.

AI Sco. Superimposed oscillations of 960 to 970 days.

AK Sco. Brightness varies with a 5.1480-day period. The spectral type based on metallic lines is given; that from hydrogen lines is A7-F0.

BM Sco. Near open cluster NGC 6405. Strong CN bands in spectrum.

FV Sco. Min. II, 8.0.

V382 Sco. Near open cluster NGC 6475.

V453 Sco. Min. II, 6.70.

V635 Sco. Superimposed rapid, irregular fluctuations.

V636 Sco. Sp. binary.

V701 Sco. In open cluster NGC 6383. Min. II.

V703 Sco. Regular period; amplitude and light curve variations similar to those of AI Vel. Near open cluster NGC 6405 (M6).

V718 Sco. Period may be as long as 5 days.

V720 Sco. Near the open cluster NGC 6475.

V723 Sco. Very similar to CP Lac (Nova Lac 1936).

V728 Sco. Recorded in early October, 1862, at 4th or 5th mag. by John Tebbutt while following Comet 1862 III.

V856 Sco. Same proper motion as HD 144668 (separation 44″), but radial velocity differs by 46 km/sec.

V861 Sco. Sp. binary; in open cluster NGC 6231.

V862 Sco. Outburst observed in 1965. In open cluster NGC 6405 (M6).

V883 Sco. Deep min. II.

V884 Sco. The X-ray source 1700-377. Min. II, 6.59; strong emission lines. Near open cluster NGC 6281.

V885 Sco. Possible EB Lyr type; noticeable min. II.

V900 Sco. In cluster NGC 6231. Period may be 6.28 or 6.85 days.

V906 Sco. Sp. binary in open cluster NGC 6475.

V913 Sco. Period may be 0.976 day.

V915 Sco. Multiple star.

V919 Sco. Member of cluster NGC 6231.

V923 Sco. Sp. binary in cluster NGC 6281.

R Scl. *Bright Star Catalogue* gives mag. as 5.79 (V). Double system LDS 2199.

S Scl. [6.7, 12.9]

T Scl. [9.2, 13.0]

U Scl. [9.8, 15.1]

V Scl. [9.9, 14.6]

Y Scl. Two periods of approximately 100 and 300 days.

AL Scl. Sp. binary.

δ Sct. Light curve varies with 5.24774-day period; second overtone present. Prototype of δSct class.

R Sct. Many irregularities. Period varies; emission lines are present at maximum.

S Sct. ADS 11726A.

RU Sct. Period varies.

RZ Sct. Min. II, 7.47 (V). Common-envelope binary.

EW Sct. Period needs confirmation.

o Ser. Sp. binary.

d Ser. ADS 11353A; sp. triple.

R Ser. [6.9, 13.4]

S Ser. [8.7, 13.5]

T Ser. Period varies; mean value given. Near open cluster NGC 6633.

U Ser. [8.5, 13.4]

W Ser. Common-envelope binary; orbital period increasing.

X Ser. Fluctuations in brightness, with a period of several minutes, are suspected.

FR Ser. Periods of 2.1308 and 2.1565 days are possible.

S Sex. [9.1, 13.4]

RS Sex. Probably not variable; may have been confused with the variable RX Sex.

α Tau. Aldebaran. Star surrounded by infrared-emission envelope.

ζ Tau. Common-envelope sp. binary.

θ² Tau. Visual and sp. binary; member of Hyades.

λ Tau. Triple system; min. II, 3.39.

ϱ Tau. Sp. binary; member of Hyades.

R Tau. [8.6, 14.2]

T Tau. Embedded in Burnham's nebula with NGC 1555 (Hind's variable nebula) and NGC 1554 nearby. Complex light curve; infrared companion.

V Tau. [9.2, 13.7]

W Tau. Period varies, mean elements given.

Z Tau. [9.8, 13.9] Period varies.

RW Tau. Period varies. There is a 12.5-mag. visual companion.

RX Tau. [9.6, 14.0]

RY Tau. Infrared excess.

ST Tau. Period varies.

SZ Tau. Period varies; near open cluster NGC 1647.

TU Tau. Close double; lines of A2III component are seen in spectrum.

VY Tau. Spectrum of T Tau type. Near a dark cloud. Flares of UG or Z And type.

WW Tau. The spectral type from TiO bands is M3.

BU Tau. Pleione, in the Pleiades. Shell star, and probably a long-period sp. binary.

CD Tau. Min. II.

CH Tau. Epoch of minimum given.

CM Tau. Lying in the Crab nebula, this 33-millisecond pulsar has been detected at radio, optical, X-ray, and gamma-ray wavelengths. Period increasing; the pulsar is the southwest component of a close double.

CQ Tau. Probably binary; superimposed 10-day, 1,000-day, and semiregular variations.

ET Tau. Min. II, 9.3.

GS Tau. Epoch of minimum given. Sp. binary.

HU Tau. Period may vary. Min. II, 6.1.

IM Tau. Periods of 0.126535 and 0.126569 day also possible.

V473 Tau. Periods of 3.5 and 7.1 days possible.

V480 Tau. Probable member of Hyades.

V483 Tau. Binary; member of Hyades.

V624 Tau. Member of Pleiades.

V647 Tau. In Pleiades.

V650 Tau. In Pleiades.

V696 Tau. In Hyades.

V711 Tau. Visual and sp. binary ADS 2644; similar to RS CVn. Transient X-ray source.

V775 Tau. Sp. binary; member of Hyades.

V776 Tau. Visual and sp. binary; in Hyades.

V777 Tau. Visual and sp. binary; massive unseen component. In Hyades.

R Tel. [8.6, 14.8]

RR Tel. Outburst from mag. 14 to 7 in 1944. Remained near maximum until 1949, when brightness began decreasing. There are cyclic variations spanning 350 to 410 days.

RS Tel. Infrared excess.

RX Tel. [9.2, 9.9]

BL Tel. Also a 0.1-mag. physical light variability with a period of about 2 months.

BR Tel. Epoch of minimum given; RV type?

MT Tel. Period may vary.

NV Tel. Possible EB type; noticeable min. II.

R Tri. [6.2, 11.7]

X Tri. Period varies. Visual binary.

TZ Tri. Visual and sp. binary.

HP TrA. Deep min. II.

θ Tuc. Period and amplitude vary.

R Tuc. [9.8, 15.1]

S Tuc. [9.3, 14.5]

T Tuc. [8.1, 13.2]

U Tuc. [8.6, 14.1] Near but not a member of the Small Magellanic Cloud.

CG Tuc. Spectrum variable.

ε UMa. Member of UMa moving cluster. Epoch of maximum given; radial velocity constant.

υ UMa. Light curve and amplitude vary.

R UMa. [7.5, 13.0]

S UMa. [7.8, 11.7] Period varies; mean value given.

T UMa. [7.7, 12.9]

W UMa. Period varies suddenly. Light curve varies. Min. II, 8.54.

X UMa. [9.7, 14.4]

Z UMa. Light curve of RV type, with deep secondary minima. Mean mag. varies with 1,560-day period and 0.2-mag. amplitude.

RS UMa. [9.0, 14.3]

ST UMa. Mean mag. may vary with a 590-day period.

TX UMa. Period varies; min. II, 7.13.

AC UMa. Southern component of visual binary BDS 4825.

CF UMa. Groombridge 1830B; may not exist.

CQ UMa. Amplitude 0.08 mag. in B band.

DD UMa. Sometimes constant.

DN UMa. Component of multiple ADS 8347.

DP UMa. Possible beat phenomena.

α UMi. Polaris. The A component of the visual triple ADS 1477; B component is of mag. 8.20 (V). Period varies; mean radial velocity varies in a 30-year cycle.

γ UMi. Shell star.

ε UMi. Sp. binary; min. II, 4.25.

R UMi. [9.1, 10.4]

S UMi. [8.4, 12.0] Period may vary; mean value is given.

T UMi. [9.2, 14.0] Period may vary; mean value given.

U UMi. [8.2, 12.0] Period may vary; mean value given.

V UMi. Mean mag. varies with 760-day period.

RR UMi. Sp. binary.

γ² Vel. Visual and sp. multiple in Gum nebula; 154-second period may indicate a close, collapsed component.

o Vel. Member of open cluster IC 2391.

S Vel. Period varies.

W Vel. [8.8, 13.6]

Y Vel. [9.5, 13.8]

Z Vel. [9.0, 14.3]

SW Vel. Period varies.

SX Vel. Light curve has a double maximum; epoch of max. II is given.

SY Vel. Slow variations with 1,400-day period are suspected.

WY Vel. Large infrared excess.

WZ Vel. Cycle of very long duration is also possible.

AH Vel. Light curve not characteristic of this period.

AI Vel. Light curve and period vary strongly. There is a 0.08620767-day secondary oscillation; 0.379188-day beat period.

AX Vel. Light curve varies; 0.22-mag. fluctuations are superimposed on the main variation.

CW Vel. Min. II, 9.6.

FX Vel. Very deep min. II.

FY Vel. Min. II, 7.03 (V).

GG Vel. Shallow min. II.

GP Vel. X-ray source Vel X-1; 8.97-day and 283-second X-ray oscillations observed. Light curve varies; min. II, 6.90.

HV Vel. In cluster IC 2391.

HW Vel. Probable member of cluster IC 2391. Beats observed?

HX Vel. In cluster IC 2395.

HY Vel. Probable member of cluster IC 2391.

IP Vel. Component A of visual double.

α Vir. Spica. Exhibits apsidal rotation with a 130-year period. Min. II.

R Vir. [6.9, 11.5] Period may vary; mean value given.

S Vir. [7.0, 12.7] Period may vary; mean value given.

T Vir. [9.6, 14.2]

U Vir. [8.2, 13.1]

V Vir. [8.9, 14.3]

Y Vir. [9.4, 13.6]

RS Vir. [8.1, 13.9]

RU Vir. [10.0, 13.3]

SS Vir. [6.8, 8.9] Period may vary; mean value given.

(Continued on page 362)

Suspected Variable Stars

Suspected Variable Stars

Column Headings

NSV — The number of the star in the *New Catalogue of Suspected Variable Stars* (1982), by B. V. Kukarkin and colleagues.

α 2000 and **δ 2000** — Right ascension and declination, referred to the 2000.0 equinox.

Const — The constellation in which the star lies.

Type — Type of variability. This is usually a mere guess, selected for consistency with the star's reported behavior. The notation is the same as that used in the Variable Stars section; a key to all the types appears in the Introduction.

Max and **Min** — The magnitudes of the star at the supposed maximum and minimum brightness. Generally speaking, one value is the star's *usual* brightness. The other is the brightness reported on one or more rare occasions; it is usually easy to tell which value is the exceptional one by looking at the supposed variable type or consulting other sources. The symbol ">" means "fainter than." The magnitude system is indicated by a letter after the minimum column, as follows: photoelectric magnitudes on the **UBV** system (U, B, or V), visual (v), photographic (p), or infrared photoelectric magnitude (I, K, or R).

Spec — The spectral type of the star.

Publ — The year in which a report of the star's suspected variability was first published. In some cases, the observation on which the report is based was made many years earlier.

HD — The star's number in the *Henry Draper Catalogue* or *Henry Draper Extension*.

SAO — The star's number in the *Smithsonian Astrophysical Observatory Star Catalog* (1966).

ADS — The star's number in R. G. Aitken's *New General Catalogue of Double Stars* (1932). If a particular component of a double or multiple system is the suspected variable, the capital Roman letter used by double star observers is appended.

BD — The star's designation in the *Bonner Durchmusterung* (BD), or else one of the southern extensions, the *Cordoba Durchmusterung* (CoD) or *Cape Photographic Durchmusterung* (CPD).

Notes — The common name, Bayer or Flamsteed designation, or other information. The following words are abbreviated: component (comp.), magnitude (mag.), spectroscopic (sp.), visual (vis.). An asterisk in this column signifies that there is a longer note about the star on page 272.

NSV	α 2000	δ 2000	Const	Type	Max	Min		Spec	Publ	HD	SAO	ADS	BD	Notes
14802	0ʰ01ᵐ8	+65°27'	Cas		7.9	8.5	p	F5	1907	224919	21014		BD +64°1887	
3	0 02.7	+2 08	Psc		9	11	v		1962				BD +01 4820B	
7	0 03.9	+27 59	Peg		9.2	10.9	v		1967			9B		
21	0 05.5	+50 31	Cas		7.0	8.0	v	K0	1970	37		24C	BD +49 4329	
24	0 05.9	−14 26	Cet		6.5	9.0	v	G0	1891		147080	49A	BD −15 6542	
110	0 16.1	+58 10	Cas		9.5	10.0	v		1926				BD +57 45	
145	0 23.0	+47 54	Cas	Cep?	10.2	11.1	p		1944				BD +47 76	
152	0 23.8	−9 28	Cet		6?	10.4	v	F8	1872	1960			BD −10 62	Borrelly's star*
216	0 35.4	−21 45	Cet		8.5	>9.5	v	K0	1892	3233	166385		BD −22 96	
224	0 36.8	+48 57	Cas	Lb	9.8	10.8	p	M5	1969	3332	36508		BD +48 183	
286	0 45.5	+61 00	Cas		9.6	10.3	p	K2	1966		11372		BD +60 97	
305	0 49.2	+57 05	Cas	Lb	8.5	9.0	p	M2	1958	4647	21738		BD +56 131	
332	0 52.9	+49 42	Cas	Lb	9.6	10.3	p	M6	1958	5033	36738		BD +48 266	
349	0 56.4	+23 03	And	EA?	10.0	11.1	v		1970					
360	0 59.3	+59 43	Cas		9	11?	v		1962			812B	BD +58 146B	
363	1 00.6	+47 19	And		8.0	9.2?	v		1962	5842	36840	829B	BD +46 229	
389	1 04.8	+41 17	And		11.0	12.5	v		1950					Near RX And
387	1 04.9	+61 43	Cas	Lb	10.0	10.5	p	M1	1968		11559		BD +60 160	
401	1 07.1	+49 25	Cas		2.91	3.63	K	M9	1969					
422	1 10.6	+2 27	Cet	L?	7.1	9.0	p	K4	1954	7014	109715		BD +01 221	33 Cet
421	1 10.9	+59 32	Cas	Lb	9.9	10.45	B	M2	1968		22065		BD +58 185	
436	1 13.2	+57 35	Cas	Lb	8.4	9.1	p	M0	1966	7177	22099		BD +56 223	
447	1 14.7	+35 37	And	N?	10?	>13	p		1876					
476	1 20.6	−8 25	Cet		10	10.7	p	M2	1975		129240		BD −09 260	
482	1 21.5	+9 41	Psc		9.5	>13.5	p		1880				BD +08 215	*
504	1 25.9	−3 56	Cet	E?	6.5	7.8	p	K0	1879	8713	129292		BD −04 207	142 G. Cet
518	1 27.9	−23 02	Cet		9.8	10.3	p		1972				CoD−23 526	
525	1 30.5	+62 44	Cas	E?	9.9	10.8	p	B0	1944				BD +61 277	
534	1 31.8	+30 38	Tri		8.03	8.63	V	K0	1948	9269	54758		BD +29 256	
557	1 34.4	−64 31	Hyi	Lb	10	10.5	p	M2	1959	9858			CoD−65 74	
549	1 34.8	+65 48	Cas	Lb	10.2	11.0	p	M5	1958	9455	11838		BD +65 179	
562	1 35.9	−17 32	Cet		7.16	7.66	V	G2	1959	9847	147900		BD −18 266	
589	1 41.6	+36 07	And		9.6	10.5	v	K0	1928				BD +35 323	
602	1 43.2	+48 31	And	Lb?	8.3	9.0	p	M2.5II	1966	10465	37456		BD +47 485	
650	1 54.6	+68 09	Cas	Ia?	6.9	7.7	p	B8	1933	11395	12030		BD +67 168	
665	1 55.3	−35 48	For		9.9	10.4	p	M0	1967		193382		CoD−36 732	
669	1 56.9	+55 47	Per		10	>11.5	p		1968					South of BD +55 440
681	1 58.0	+28 31	Tri		7.7	8.2	v	A0	1967	11962	75052		BD +27 311	
695	2 00.7	−13 50	Cet	S	10	11	p		1932				BD −14 376	
728	2 07.5	−57 52	Eri	E?	10	10.5	p		1963				CoD−58 432	
756	2 11.8	−71 29	Hyi		9.6	10.5	p		1901	13927			CoD−72 100	
768	2 16.3	−9 49	Cet		7.2	7.7	v	G2	1959	14044	148304		BD −10 462	
770	2 17.6	+57 05	Per		8.7	10.0	v	B5	1895				BD +56 481	Near cluster NGC 869
813	2 23.0	+33 02	Tri	I	10.0	10.9	p		1963		55501		BD +32 430	
828	2 27.0	−19 33	Cet	S	10	10.5	p		1933				BD −20 457	
840	2 27.8	−69 31	Hyi		7.8	8.8	p	M6	1901	15701	248570		CoD−70 124	
837	2 29.6	+52 26	Per	N?	8.9	>14	v		1907					Possible Nova Per 1873*
845	2 31.1	+35 08	Tri		9.3	>14.0	v		1922				BD +34 447	
857	2 35.8	+65 10	Cas		1.58	3.22	K	M9	1969					
878	2 37.3	−26 58	For		5.84	7.00	I	M9	1969					
908	2 44.2	−31 43	For		9.2	9.9	v		1897		193885		CoD−32 989	
969	2 51.1	−38 05	For	E	9.3	10.0	p	A0	1907	17886			CoD−38 935	
1039	3 10.5	+79 20	Cep		9.5	>13	R	M5	1943					
1063	3 11.4	+57 54	Cas	Lb	9.0	9.8	p	R5	1958	19557	23858		BD +57 702	
1074	3 12.1	−28 59	For		6	8	v		1927	20010B		2402B	CoD−29 1177B	α For B
1077	3 14.1	+48 17	Per	Lb	10.0	10.8	p	M0	1958		38660		BD +47 785	
1137	3 26.6	+41 55	Per		8.0	>12.8	p		1947					
1162	3 27.3	−61 16	Ret		7.9	8.4	p	F5	1964	21765	248810		CoD−61 632	
1151	3 28.1	+33 29	Per		6.29	6.88	I	M6	1969					
1166	3 29.3	−11 40	Eri		9.36	9.98	V	M0	1972				BD −12 662	
1168	3 31.4	+58 32	Cam	Lb?	9.8	10.5	p	K6	1893					
1203	3 35.0	−48 25	Hor		8.57	9.27	V	K7	1970	22496	216392		CoD−48 1011	Sp. binary
1194	3 35.1	+27 56	Tau		9.4	10.0	p	A6	1933				BD +27 524	
1199	3 36.4	+52 56	Cam	Lb	8.5	9.2	p	K5	1958	22135	24120		BD +52 703	
1214	3 37.7	−55 24	Ret		8.0	8.9	p	M6	1898	22868	233190		CoD−55 731	
1237	3 42.6	−14 51	Eri	L	9	9.5	p		1933				BD −15 642	
1282	3 45.2	−16 25	Eri		9.1	9.6	p	F0	1965	23517	149147		BD −16 696	
1280	3 45.8	+23 09	Tau	Is?	6.5	7.3	p	A0	1952	23410	76156	2748A	BD +22 545	Sp. binary, in Pleiades
1381	3 50.7	+1 34	Tau		7.3	7.8	p	F0	1910	24133	111453		BD +01 667	
1391	3 53.6	+25 41	Tau		6.34	7.84	V	A2	1958	24368	76305		BD +25 641	Sp. binary
1411	3 54.8	+20 41	Tau	M?	9.7	>15	p		1935					
1415	3 55.0	−14 54	Eri		4.97	5.47	I	M0	1969	24693	149250		BD −15 687	
1417	3 55.1	−15 01	Eri		9.6	10.5	v		1926				BD −15 688	
1431	4 01.7	+56 11	Cam	L	9.8	10.5	p	G0	1959	237206	24349		BD +55 840	
1484	4 11.7	+59 54	Cam	UV?	6	>12	p		1927					*

SUSPECTED VARIABLE STARS

NSV	α 2000	δ 2000	Const	Type	Max	Min		Spec	Publ	HD	SAO	ADS	BD	Notes
1510	4^{h}13^m.1	+14°33'	Tau		9.6	10.1	p	A0	1958	285621	93818		BD +14°666	
1529	4 15.4	+24 05	Tau	Lb?	9.4	10.1	p	M2	1956	26816	76523		BD +23 654	
1525	4 15.7	+47 25	Per	I	10	10.7	p		1966				BD +47 961	
1477	4 16.4	+82 28	Cep	E?	10.2	10.8	p	M	1958		660		BD +82 111	
1512	4 16.5	+74 06	Cam	Lb	10.0	10.6	p	Me	1959	26290			BD +73 216	
1591	4 18.0	-80 13	Men		6.53	7.13	B	Kp	1964	28525	258372		CoD-80 146	δ Men
1561	4 20.8	+15 09	Tau		10.2	11.3	p	K7	1958	285701			BD +14 684	
1578	4 22.0	-22 41	Eri		5.93	6.43	I	M6	1969					
1597	4 25.9	+18 52	Tau		7.53	8.28	V	G3	1973	27989	93926	3210	BD +18 636	
1607	4 27.6	-25 14	Eri	EA	10	10.5	p		1936				CoD-25 1890	
1612	4 29.0	+21 55	Tau		8.27	9.17	V	M1	1968	28343	76626		BD +21 652	
1675	4 33.0	-78 07	Men	L	10	11	p		1963				CoD-78 171	
1702	4 43.9	+22 57	Tau		6.6	8.0	v	B9	1967	29935	76729		BD +22 743	
1710	4 45.2	-23 52	Eri		2.35	3.78	K	M8	1969					
1794	4 59.1	-23 03	Lep	E?	10	10.5	p		1936				CoD-23 2304	
1887	5 13.7	-49 33	Pic	SR?	9.5	10.5	p		1949				CoD-49 1626	Visual binary*
1881	5 14.2	-21 42	Lep	EA	10	10.5	p		1936					
1873	5 14.3	+40 01	Aur		9.8	10.3	p	OB	1951				BD +39 1217	
1927	5 19.4	-7 21	Ori		7.5	8.1	B	Ap	1963	34736	131980		BD -07 1036	
1910	5 19.8	+63 16	Cam		1.69	3.09	K	M9	1969					
1937	5 20.4	-3 31	Ori		8.8	9.6	B	Ap	1963	34859	132000		BD -03 1065	
1944	5 21.3	-7 29	Ori		8.1	8.8	B	A2	1963	34992	132020		BD -07 1049	
1960	5 24.2	+1 52	Ori		9.0	12.5	p		1947					
1962	5 24.3	-6 49	Ori		8.4	8.9	B	F0	1963	35413	132069		BD -06 1166	
2002	5 27.7	-8 20	Ori		8.5	9.0	B	F0	1963	35929	132136		BD -08 1128	
2008	5 28.2	-20 46	Lep		7.0	11.0	v		1962		170457	4066B	BD -20 1096B	β Lep B
2025	5 30.0	+22 04	Tau		10.0	11.2	p	A2	1975	244206			BD +21 864	
2044	5 30.7	+21 52	Tau		10.0	11.0	p	F5	1975	244311			BD +21 869	
2068	5 31.8	+21 38	Tau		10.0	11.9	p	K0	1975	244495			BD +21 880	
2079	5 31.9	+11 12	Ori		9.5	10.6	p	K5	1951	244591	94622		BD +11 837	
2258	5 35.3	+8 42	Ori	Lb	8.5	9.3	p	M5	1958	36914	112927		BD +08 1005	
2426	5 37.1	+26 55	Tau		3.5	5.78	B	B9	1967	37098	77322	4208AB	BD +26 870	
2468	5 37.6	-1 25	Ori		8.5	10.5?	v		1962	37321B	132376B	4222B	BD -01 982B	
2477	5 37.9	+6 19	Ori		8.5	>10.4	v		1912				BD +06 975	
2548	5 38.5	-59 04	Pic		9.6	10.1	p	Fe	1951		234069		CoD-59 1105	Pec. variable spectrum
2501	5 38.8	+16 56	Tau		6.07	6.69	I		1969					
2533	5 39.7	-8 10	Ori	M?	2.07	2.95	K	C	1969					
2537	5 40.6	+31 22	Aur		4.0	6.07	v	B7	1964	37519	58319		BD +31 1048	Flares (1-2 magnitudes)
2571	5 40.8	-23 36	Lep	E?	10	11.5	p		1933				CoD-23 2981	
2557	5 41.2	+12 17	Ori		6.39	7.00	I	N	1969					
2560	5 41.3	+16 32	Tau		6.2	7.2	v		1961	37711B	94759	4265B	BD +16 841B	126 Tau B
2576	5 41.5	-4 24	Ori		9.3	10.5	v	M0	1932				BD -04 1215	
2587	5 42.6	+7 09	Ori	EA?	10.1	12.0	p	A0	1930					
2563	5 43.6	+66 33	Cam		8.8	9.8	p	A2	1837	37419	13577	4267	BD +66 405	
2622	5 45.4	-4 14	Ori		6.10	6.69	I	M4	1969					
2659	5 50.6	+22 40	Ori		9.7	10.3	p	A0	1975	248189			BD +22 1056	
2682	5 51.8	-3 38	Ori	UV?	4.5	9.0	v	M0	1926	294421	132598		BD -03 1216	
2698	5 52.9	-11 04	Lep		10	10.5	p		1967				BD -11 1310	
2688	5 53.3	+40 44	Aur	Lb	10.1	10.7	p	M0	1968		40680		BD +40 1443	
2706	5 53.6	-6 16	Ori	Lb	9.4	10.0	p	Ma	1958	39634	132619		BD -06 1344	
2699	5 53.6	+22 15	Ori		8.4	10.4	p	B8	1975	248792			BD +22 1083	
2721	5 54.4	-1 05	Ori	Lb	8.6	9.2	p	M3	1958	39732	132630		BD -01 1059	
2705	5 54.8	+45 21	Aur		8.1	8.86	B	A0	1940	39414	40692		BD +45 1194	
2736	5 56.9	+20 17	Ori		5.97	6.51	I		1969					
2756	5 59.7	+37 13	Aur		6.5	9.0	v		1962	40312B	58636	4566B	BD +37 1380B	37 θ Aur B
2832	6 00.3	-80 36	Men		9.8	10.3	p	F5	1964	43013	258430		CoD-80 208	
2813	6 03.1	-43 02	Pic	Lb	10	10.5	p	M4	1949	41440	217692		CoD-43 2215	
2794	6 03.3	+42 12	Aur	Lb	9.9	10.5	p		1968		40820		BD +42 1474	
2830	6 03.8	-69 43	Dor	Lb	10	10.5	p	M3	1949	42288			CoD-69 356	
2837	6 08.4	+22 43	Gem		10.0	10.7	R	C	1974					
2839	6 08.8	+25 39	Gem	E?	8.4	9.31	B	M5Ib	1958	41890	78029		BD +25 1131	
2841	6 09.1	+34 54	Aur	Lb	8.2	8.9	p	M6	1958	41849	58789		BD +34 1272	
2849	6 09.2	-14 04	Lep	S	8.5	12.0?	v		1962	42283B	151167	4755B	BD -14 1347B	
2846	6 09.7	+23 07	Gem		7.0	10.0	v		1962	42087B	78050	4751B	BD +23 1226B	3 Gem B
2869	6 13.1	+20 38	Ori		6.88	7.62	I		1969					
2882	6 15.5	+17 45	Ori	Lb	9.5	10.4	p	M2	1958	43151	95448		BD +17 1187	
2922	6 16.5	-74 27	Men		9.9	10.4	p		1964				CoD-74 299	
2938	6 21.9	-3 52	Mon		2.98	3.69	K	M10	1969					
2944	6 23.8	-19 47	CMa		7.5	9.0	v		1962	44953B	151453B	5023B	BD -19 1435B	
2965	6 24.0	-68 36	Dor		10.0	10.6	p	M5	1964	45819			CoD-68 397	
2956	6 25.0	-23 28	CMa		6.22	6.75	I	M5	1969					
2969	6 26.9	-23 16	CMa	EA?	9.5	10.0	p		1933				CoD-23 3867	
2942	6 28.1	+74 45	Cam	S	10.2	10.8	p	M2	1955		5852		BD +74 285	
3018	6 28.5	-74 10	Men		10	13	p		1938					
2999	6 31.2	+7 46	Mon		9.5	11.0	v		1962			5158B		

NSV	α 2000	δ 2000	Const	Type	Max	Min		Spec	Publ	HD	SAO	ADS	BD	Notes
3001	6ʰ31ᵐ.3	+6°21'	Mon	EW?	9.2	9.7	p	A5	1937	258916	113990		BD +06° 1273	
3002	6 31.6	+19 38	Gem	Lb	10.2	11.0	p	K7	1938	258810				
3015	6 31.8	−14 55	CMa		6.6	7.2	I	M8	1969					
3020	6 34.6	+60 57	Lyn		5.4	8.53	I	M9	1969					
3171	6 43.1	+46 54	Aur		9.9	10.6	p	K0	1937	47880			BD +47 1331	
3260	6 52.2	−56 31	Car		8.2	10.0?	v	F9	1927		234758		CoD−56 1669	Visual binary
3255	6 52.4	−20 06	CMa	SR	9.5	>12.5	p		1958					
3262	6 54.0	+4 11	Mon	EA?	9.3	10.7	p	A2	1952	266040			BD +04 1496	
3261	6 54.4	+21 10	Gem		9.5	>12	v		1962	50482B		5553B	BD +21 1426B	
3290	6 56.5	+6 29	Mon		9.3	11.0	v		1955	266734B		5604B	BD +06 1444B	
3281	6 56.5	+40 05	Aur		8.60	9.12	V	K5	1973		41448		BD +40 1758	
3299	6 57.2	−8 46	Mon		9.7	11.4	v		1955				BD −08 1642	
3320	6 58.6	−17 09	CMa	Lb	9	10.5	p	K0	1963	51920	152186		BD −16 1686	
3313	6 58.7	+17 02	Gem	N	7		v		1906					
3226	6 59.4	+83 29	Cam	E	10.2	10.8	p	G2	1958		1077		BD +83 168	Seen by E. E. Barnard*
3323	6 59.5	+9 19	Mon	E?	9.6	10.1	p	B8	1946	267564			BD +09 1467	
3365	7 04.1	−25 48	CMa	Lb?	9.5	10.6	p	K2	1907	53460	172892		CoD−25 3986	
3379	7 05.2	−35 56	Pup		8.0	8.6	p	Mc	1901	53917	197549		CoD−35 3334	
3412	7 08.1	+7 30	CMi		2.8	4.4	K	M9	1968					
3419	7 08.4	−13 10	CMa		9.94	11.70	V	K3+B5	1974				BD −12 1805	Sp. binary (K3II + B5IV)
3399	7 09.6	+65 40	Cam		10.2	11.0	p		1939					
3408	7 10.1	+65 58	Cam		2.52	3.36	K	M8	1969	53469	14094			
3439	7 10.3	−16 16	CMa		6.94	7.68	I	M7	1969					
3474	7 13.7	−41 42	Pup		10	10.5	p		1964				CoD−41 2894	
3473	7 14.3	−16 05	CMa	E	10.2	10.7	p	G?	1936		152597		BD −15 1713	
3497	7 14.9	−59 16	Car		9.5	10.0	p	F7	1965	56785	235039		CoD−59 1542	
3487	7 15.6	+5 55	CMi	Lb	9	9.5	p	M5	1934	56033	115160		BD +06 1587	
3516	7 16.6	−51 00	Car	SR?	10	11	p		1949				CoD−50 2665	
3524	7 18.9	+14 42	Gem		9.5	10.7	v	A1	1952				BD +14 1631	
3532	7 19.3	−10 54	Mon		5.78	6.48	I	M2	1969	57055	152715		BD −10 1983	
3543	7 20.6	−16 42	CMa		10.0	10.5	p		1936					
3552	7 21.3	−29 18	CMa	M	6.75	7.49	I	Me	1969					
3575	7 24.2	−2 35	Mon	S	9.7	10.2	p	M4	1957		134624		BD −02 2092	
3598	7 26.7	−44 34	Pup		9.8	10.3	p		1973				CoD−44 3424	
3608	7 28.3	−36 43	Pup	EA?	10	11	p		1949				CoD−36 3644	
3615	7 29.6	−17 34	Pup		8.6	9.5	p	Ma	1961	59494	152953		BD −17 1998	
3628	7 31.9	−12 47	Pup	EA?	9.9	>11.0	p	F1	1943				BD −12 2011	
3640	7 32.7	−30 16	Pup		9.4	10.1	p	G0	1975		198108		CoD−30 4669	
3633	7 33.8	+48 00	Lyn	EW?	10.98	11.48	V		1971					
3641	7 33.9	+30 31	Gem		7.51	9.01	I	M9	1969					
3646	7 34.1	−13 02	Pup	EA	8.3	8.8	p	B9	1974	60476	153079		BD −12 2032	
3670	7 38.6	−10 09	Mon	UV?	9	11	p		1955				BD −09 2150	
3687	7 39.8	−53 38	Car		9.7	10.2	p	A1	1971	62177	235404		CoD−53 1951	
3685	7 40.7	−1 09	Mon	E?	9	9.5	p	A5	1934		134968		BD −00 1785	
3697	7 43.1	+8 46	CMi	L?	10.0	10.5	p	G8	1916				BD +09 1752	
3726	7 47.3	+47 20	Lyn	L?	8.3	8.9	p	K0	1949	62668	41995		BD +47 1484	
3739	7 48.1	+5 29	CMi	Is?	9.7	10.2	v		1904					
3748	7 48.2	−42 30	Pup		7.74	8.5	B	A0	1900	63641	219009		CoD−42 3541	
3756	7 49.2	−35 15	Pup		5.87	6.37	B	A0	1964	63786	198480		CoD−34 3970	Sp. binary
3812	7 54.8	−33 03	Pup		9.7	10.3	p		1965				CoD−32 4556	
3822	7 55.7	−44 46	Pup	EA	9	9.5	p	A1	1949	65212	219156		CoD−44 3869	
3815	7 56.0	−0 45	Mon		10.2	>12.0	p		1950					
3841	7 58.4	−51 27	Car	E?	5.2	7.2	p	F+K	1970	65867	235646		CoD−51 2784	
3835	7 59.0	+2 36	CMi	S?	10.1	11.2	p		1934				BD +02 1838	Red star
3846	7 59.0	−43 49	Pup	N?	3	>12	v		1926				CoD−43 3805?	Seen in 1673*
3867	8 02.6	+1 43	CMi	EW?	9.5	10	p		1934				BD +02 1855	
3888	8 04.5	+13 25	Cnc	Lb	9.4	9.9	p	K0	1962	66529	97532		BD +13 1828	
3902	8 07.6	+24 40	Cnc		10.0	10.7	p		1938				BD +25 1850	
3916	8 08.9	+32 49	Cnc	UV?	10.2	11.4	v	Me	1954				BD +33 1646	
3933	8 10.2	−43 57	Vel		10.5	>16	p		1916					Near cluster IC 2391*
3959	8 14.8	−1 20	Hya		9.8	11.0	v		1955					B comp. of binary J 2055
3943	8 15.0	+68 08	UMa		10.2	11.2	p		1955					Period 200 days or less?
4031	8 23.0	+45 27	Lyn	EA?	8.0	8.8	v	G5	1963	70271	42321		BD +45 1570	
4051	8 24.9	+50 19	Lyn		9.1	10.0	p	F8	1938	233526	42341		BD +50 1536	
4098	8 27.5	−61 06	Car	EA	10	11	p		1949				CoD−60 2190	
4144	8 34.4	−54 40	Vel		8.2	8.7	p	B6	1965	73169	236094		CoD−54 2269	
4151	8 35.7	−52 19	Vel	L	8.7	9.2	p	M3	1949	73341	236108		CoD−51 3102	
4133	8 36.8	+74 43	Cam	L	9.0	>12	v		1962			6872B	BD +75 342B	
4189	8 40.4	−14 49	Hya		2.29	3.47	K	M8	1969					
4172	8 42.3	+72 14	UMa		10.2	10.7	p		1939					
4218	8 42.9	−49 24	Vel		10.0	10.5	p		1964				CoD−48 4047	
4215	8 43.4	+5 58	Hya	RR?	10	10.7	p	F9	1937				BD +06 2022	
4225	8 43.9	−25 36	Pyx		6.84	7.76	I	C	1969					
4240	8 44.0	−71 28	Vol	L	10	10.5	p		1963				CoD−71 497	
4230	8 44.7	−29 07	Pyx	L	8.5	10	p	M1	1931	74709	176398		CoD−28 6471	

SUSPECTED VARIABLE STARS

NSV	α 2000	δ 2000	Const	Type	Max	Min		Spec	Publ	HD	SAO	ADS	BD	Notes
4210	8^{h}45^m.3	+69°53'	UMa		10.2	11.8	p		1939					Near a 9th-mag. star
4295	8 55.4	+49 29	UMa		10.0	11.0	p	A1	1910				BD +50 1590	
4341	8 57.8	−73 06	Vol		9.7	10.4	p		1964				CoD−72 494	
4331	8 58.8	+14 14	Cnc		7.5	9.0	v	K0	1967	76793	98270	7117A	BD +14 2007	
4340	9 01.0	+41 51	Lyn		10	13.7	v		1939					
4345	9 02.0	+59 31	UMa		9.5	10.4	p	K2	1937		27091		BD +60 1165	
4384	9 07.1	−43 12	Vel		9.8	10.3	p		1973				CoD−42 4977	
4385	9 07.6	−26 14	Pyx		6.64	7.47	I	M6	1969					
4390	9 08.1	−27 18	Pyx		2.81	3.35	K	M1	1969	78542	177004		CoD−26 6752	
4409	9 11.0	−43 16	Vel		7.1	7.7	p	A0	1964	79154	220926		CoD−42 5038	
4429	9 14.3	−55 34	Vel		6.25	8.1	B	G8	1960	79846	236749		CoD−55 2590	
4441	9 17.6	+16 42	Cnc	E?	8.3	8.8	p	A0	1924	79992	98474		BD +17 2053	
4447	9 18.2	−33 02	Pyx		9.4	10.2	p	M6	1973				CoD−32 6252	
4458	9 22.0	+59 07	UMa		10.2	11.4	p	G5	1939				BD +59 1232	
4469	9 22.7	−18 11	Hya	EA	9.5	11	p		1933				BD −17 2838	
4478	9 24.7	+26 11	Leo		9.5	11.5	v		1962			7351B		
4485	9 25.8	−24 01	Hya		1.82	3.12	K	M9	1969					
4490	9 26.9	+14 18	Leo		7.7	8.2	p	K2	1929	81595	98574		BD +14 2095	
4495	9 27.1	−34 54	Pyx		9.8	10.6	p		1913				CoD−34 5884	
4499	9 28.0	−52 10	Vel		10	14	p		1938					
4513	9 30.7	−40 28	Vel		4.5	5.1	v		1955	82434B	221234B		CoD−39 5580B	ψ Vel B
4534	9 33.7	−34 43	Ant		10.0	10.6	v		1913				CoD−34 5985	
4539	9 35.0	+4 46	Hya		9	10	p	A2	1934	82908			BD +05 2200	
4550	9 37.3	+15 15	Leo		6?	8.0	v	F3	1878	83225	98676		BD +15 2083	Seen in 1612*
4561	9 37.4	−63 49	Car	E?	10	10.5	p	A5	1949	309731				
4577	9 39.3	−61 20	Car		4.42	4.92	B	B9	1963	83944	250653		CoD−60 2736	m Car
4582	9 41.5	+23 50	Leo	Lb	10.0	10.5	p	M6	1961				BD +24 2117	
4609	9 44.5	−45 46	Vel		10.11	10.61	V	M4	1953				CoD−45 5378	*
4617	9 45.5	−37 44	Ant		7.1	8.1	v	K0	1956	84610	200704		CoD−37 6041	*
4626	9 46.0	−72 13	Car		9.3	10.2	p	M6	1909	85140			CoD−71 584	
4629	9 50.3	+74 58	Dra	I	9.7	10.3	p	G	1962	84482	6972		BD +75 394	
4663	9 52.4	−53 59	Vel	S?	9.4	10.2	v		1951				CoD−53 3210	
4688	9 57.8	−61 56	Car		6.3	7.06	v	A0	1972	86675	250762		CoD−61 2429	
4696	10 00.1	+24 33	Leo		7.75	8.45	V	G5	1973	86590	81134		BD +25 2191	Sp. binary
4724	10 04.3	−55 12	Vel		10.0	10.8	p	F2	1909	300460			CoD−54 3226	
4732	10 05.6	−12 36	Hya	SR	10	11	p	M5	1933				BD −11 2793	
4746	10 07.2	−55 02	Vel	EA	10	10.5	p	B8	1949	300454			CoD−54 3257	
4771	10 10.8	−57 40	Car		10.0	10.5	p		1938					
4773	10 11.7	−11 01	Sex		8.2	9.3	v		1922				BD −10 3017	
4784	10 13.5	−1 14	Sex		9	10	p		1934				BD −00 2310	
4819	10 18.2	−58 40	Car		8.7	10.05	B	O9	1956	302686			CoD−58 3153	In small blue nebula
4836	10 22.8	+15 21	Leo		9.5	12.0	v		1962		99091B	7744B	BD +16 2116B	
4833	10 23.2	+61 37	UMa		9.8	10.6	p	G0	1958				BD +62 1122	
4848	10 24.8	−18 39	Hya		10.3	10.9	p	Pe	1968	90255	155965		BD −17 3140	In planetary NGC 3242
4871	10 28.5	−49 28	Vel		9.6	10.2	p		1971				CoD−48 5661	
4874	10 29.8	+29 31	LMi		9.2	9.87	B	F8	1954				BD +30 2024	
4878	10 30.2	−62 29	Car		8.68	9.18	B	B9	1964	91218	250987		CoD−61 2676	
4915	10 38.0	−20 11	Hya	SR?	9.5	10	p		1930				BD −19 3066	
4931	10 40.1	−47 20	Vel		10.2	11.0	p		1971				CoD−46 6381	
4945	10 42.6	+3 35	Sex		6.5	8.0	v	F6	1962	92749	118443	7896A	BD +04 2375	34 Sex
4950	10 42.8	−72 59	Car	E	10	10.5	p		1949				CPD−72 1017	
4956	10 45.9	+72 17	UMa		8.0	8.5	p	M4	1932	92880	7199		BD +73 504	
4971	10 47.4	−38 29	Ant		10.0	10.5	p		1969				CoD−37 6807	
4977	10 48.6	+36 18	LMi		5.64	6.18	I		1969					
4996	10 51.9	−66 20	Car	EA	10	10.5	p		1949				CoD−65 1011	
5035	10 59.0	+55 01	UMa		7.4	8.8	p	G9	1937	95001	27855		BD +55 1432	
5056	11 01.4	−37 10	Ant	EA	10	11	p		1949				CoD−36 6877	
5069	11 02.7	−59 36	Car		9.0	9.9	p		1927					
5085	11 06.5	+14 16	Leo		9.5	12.0	v		1962		99440B	8058B	BD +15 2288B	
5115	11 10.6	−28 54	Hya		9.3	9.9	p		1964				CoD−28 8693	
5121	11 11.3	−54 33	Cen		8.9	9.4	p	F3	1964	97317	238844		CoD−53 3911	
5133	11 12.1	−71 26	Car		9.8	11.2	v		1927				CoD−70 816	A comp. of binary h 4416
5162	11 17.5	−51 34	Cen		10.2	11.6	p		1969				CoD−50 5865	
5173	11 21.4	−55 46	Cen		9.7	10.4	p	N3	1907	98767				
5201	11 26.2	−51 22	Cen	Lb?	10	10.5	p		1949				CoD−50 6004	
5218	11 28.9	−53 43	Cen	Lb	10	10.5	p		1949				CoD−53 4065	
5219	11 29.2	−17 21	Crt		7.0	9.0	v	G0	1962	99877	156706	8179B	BD −16 3258	
5256	11 36.2	+81 18	Cam	E	10.1	10.6	p	F6	1962		1900		BD +82 338	
5266	11 36.6	−46 30	Cen	L?	9.9	10.4	p	A4	1949				CoD−45 7127	
5269	11 37.5	+58 27	UMa	I	10.0	10.7	p	F8	1933		28078		BD +59 1392	
5288	11 40.3	−30 16	Hya		7.06	7.84	I	M7	1969					
5327	11 46.8	−15 20	Crt	Lb?	9.5	10.5	p	M	1930				BD −14 3403	
5336	11 47.2	−76 37	Cha	Lb	9	9.5	p	M6	1963	102506	256874		CoD−75 552	
5344	11 48.6	+14 17	Leo		7.5	9.5	v		1967		99800	8311B	BD +15 2381B	
5374	11 53.0	+37 43	UMa		6.6	7.20	B	Gp	1974	103095	62738		BD +38 2285	Groombridge 1830*

NSV	α 2000	δ 2000	Const	Type	Max	Min		Spec	Publ	HD	SAO	ADS	BD	Notes
5381	11ʰ54ᵐ1	−54°10′	Cen		9.8	10.5	p		1906				CoD−53° 4281	
— 5394	11 56.1	+45 33	UMa	RR?	9.2	9.7	p	F8	1837	103626	43950	8354	BD +46 1759	Period 0.765 day?
5398	11 57.8	+45 18	UMa	Is?	9.4	10.2	p	G5	1933	103888	43971		BD +46 1764	
5401	11 58.3	+81 44	Cam	L	10.2	10.7	p	F8	1958		1959		BD +82 352	
5418	12 00.8	−40 21	Cen		8.0	9.5	p	A2	1939	104328	223166		CoD−39 7419	
5420	12 01.0	+64 08	UMa		10.2	11.2	p		1939					
5449	12 05.5	+28 30	UMa	EA	8.9	10.6	v	K3	1959	105020	82126		BD +29 2252	
5458	12 06.1	+27 33	Com		9.5	10.0	v	G5	1959	105101	82138		BD +28 2077	
— 5462	12 07.1	+27 40	Com	RR?	9.6	10.3	v		1959					
5465	12 07.4	+26 04	Com	EW?	10.0	10.8	v		1959					
5487	12 11.0	−50 41	Cen		9.8	>10.8	p		1971				CoD−49 6858	
5495	12 11.6	+36 05	CVn		8.83	9.65	V	A3	1837	105945	62887	8451A	BD +36 2246	*
5500	12 12.4	+27 21	Com	EW?	9.8	10.3	v		1959					
5503	12 13.2	−34 08	Hya		6.35	6.85	V	M4	1966	106198	203245		CoD−33 8252	
5524	12 16.7	+27 01	Com	EA?	9.8	10.3	v		1959					
5567	12 21.1	+25 43	Com	EW?	8.05	8.85	V	K0	1959	107468	82261		BD +26 2332	
5570	12 21.3	+25 41	Com	EW?	9.7	10.7	v		1959				BD +26 2333	
5579	12 22.5	+5 18	Vir		8.0	9.48	v	K5	1933	107705B	119360B	8531B	BD +06 2599B	17 Vir B
5598	12 24.3	+26 03	Com	EA?	9.7	10.6	p		1959					
5596	12 24.4	−53 34	Cen	E	8.8	10	p	K0	1927	107918	239924		CoD−52 5031	
5604	12 24.9	+23 09	Com	E?/I?	9.9	10.9	v		1959					
5608	12 25.1	+25 40	Com	RR?	9.8	10.5	v		1959					
5611	12 25.3	+35 59	CVn		10.37	11.40	V	B3	1973				BD +36 2268	*
— 5615	12 25.9	+26 46	Com	RR	9.3	9.8	v		1959					
— 5632	12 28.0	+7 56	Vir	RR	9.8	10.6	p		1933				BD +08 2605	
5654	12 32.6	−87 26	Oct		10.0	10.5	p		1939				CPD−86 258	
5717	12 33.4	+7 57	Vir		7.58	8.5	V	K5	1952	109270	119465		BD +08 2616	
5740	12 34.8	+26 50	Com	EW	10.11	10.61	V	F5	1959		82388		BD +27 2151	
5741	12 35.1	−28 47	Hya	Lb	10	11	p	M2	1936	109467	180925		CoD−28 9580	
5755	12 35.5	+23 29	Com	EW?	9.5	10.5	v		1959					
5783	12 37.3	−56 47	Cru		8.66	9.26	B	B5	1972	109724	240093		CoD−56 4556	
5800	12 38.1	+26 23	Com	RR?	9.5	10.0	v		1959					
5825	12 39.0	+16 30	Com	I?	10.1	10.9	p	G5	1884		100190		BD +17 2510	
5858	12 41.9	−59 41	Cru	EA?	4.91	>7.5	V	Be	1899	110335	240161		CoD−58 4692	*
— 5885	12 42.8	+24 30	Com	RR	10.2	11.5	v		1959					
5884	12 43.4	−68 13	Mus		10.0	>12.5	p		1971					
5902	12 43.5	+41 16	CVn	Lb	8.6	9.4	p	M3	1958	110687	44300		BD +42 2334	
6006	12 52.7	−26 00	Hya	SR?	9.5	10.8	p	M5	1933		181187		CoD−25 9487	
6008	12 53.4	−60 20	Cru		5.70	6.80	V	B9Ia	1958	111904	252069		CoD−59 4455	In open cluster NGC 4755
6021	12 55.2	−68 54	Mus	SRc?	9.4	10.0	p	M2Iab	1907	112094			CoD−68 1149	
6038	12 57.1	+70 30	Dra		10.0	>12.5	p		1939					
6044	12 58.8	−36 58	Cen	EA	9	10	p	G0	1949	112669	204000		CoD−36 8231	
6054	13 00.4	+35 45	CVn		9.57	10.58	V	Op+F2	1973	113001		8734	BD +36 2328	
6053	13 00.6	−3 22	Vir		8.0	11.0	v		1962		139096B	8732B	BD −02 3609B	
— 6058	13 00.7	+56 22	UMa		7.4	10.5	v		1957		28601B	8739B	BD +57 1408B	78 UMa B*
6061	13 01.9	−50 41	Cen		10	10.6	p		1964				CoD−50 7457	
6068	13 02.8	−23 58	Hya	EB?	9	10	p		1933				CoD−23 10870	
6075	13 03.9	−64 06	Cen		9.8	10.5	p	M6	1907	113282			CoD−63 803	
6078	13 04.6	−61 40	Cen	EA	10	10.5	p	B	1949	312256			CoD−60 4471	
6095	13 06.4	−4 51	Vir		7.6	8.3	v	K0	1956	113816	139157		BD −04 3419	
6123	13 10.4	+52 26	UMa		9.2	9.8	v	G0	1938	114536	28662		BD +53 1604	
6135	13 12.4	−57 00	Cen		9.8	10.4	p	S5	1901				CoD−56 4829	
6148	13 14.1	−16 33	Vir	cst	6.7?	8.2	v	K0	1974	114944	157805		BD −15 3621	Olbers' star M*
6161	13 15.5	+53 32	UMa		9.0	9.5	p	A5	1938	234024	28693		BD +54 1583	
6191	13 19.7	+47 47	CVn		10	11	v	M2	1968		44566	8862B	BD +48 2108B	
6192	13 22.0	−78 10	Cha		9.1	9.7	p	M5	1972	115637			CoD−77 593	
6264	13 28.2	+44 49	CVn	S?	9.2	13.3	p		1931					
6263	13 28.9	−32 00	Cen	Lb?	9.5	10.5	p	M6	1949				CoD−31 10386	
6269	13 29.2	−44 14	Cen	Lb	10	11	p		1949				CoD−43 8317	
6279	13 30.5	−44 08	Cen		10	>11	p		1971				CoD−43 8394	
6286	13 31.7	−58 23	Cen		10	10.7	p		1964				CoD−57 5052	
6347	13 37.2	+52 36	UMa		2.58	3.86	K	K5	1969	118668	28818		BD +53 1637	
6362	13 38.1	+73 00	UMi		9.9	10.8	v		1972					
6357	13 39.2	−5 41	Vir		11.0	13.5	v		1952		139467		BD −04 3527	B comp. of binary J 2660
6393	13 41.3	+56 04	UMa		10.1	10.6	p	G5	1937	238272	28850		BD +56 1674	
6396	13 42.4	−19 24	Vir		5.9	6.4	I	M2	1969	119231	158102		BD −18 3665	
6452	13 49.2	−47 46	Cen	L	10	10.5	p		1949				CoD−47 8633	
6471	13 50.7	+64 55	Dra		9	>13	R	M6	1947					
6473	13 52.4	−36 43	Cen		9	13?	p		1969					
6492	13 55.7	−44 39	Cen		8.7	9.2	p	A2	1964	121291	224547		CoD−44 8979	
6502	13 56.6	+27 30	Boo	I	6.1	6.6	p	K3	1951	121710	83084		BD +28 2278	9 Boo
6506	13 57.4	−1 41	Vir		9.32	10.38	V	K2	1967				BD −00 2767	
6520	13 58.0	+79 35	Cam	E	10.2	11.0	p	F8	1958				BD +80 426	
6507	13 58.3	−56 21	Cen	Lb?	9.8	10.5	p	Na	1907	121658	241334		CoD−55 5435	
6519	14 00.4	+37 52	CVn	Lb	9.4	10.29	B	M8	1954	122316	63879		BD +38 2501	

NSV	α 2000	δ 2000	Const	Type	Max	Min		Spec	Publ	HD	SAO	ADS	BD	Notes
6567	14ʰ08ᵐ7	+33°41′	Boo		9.5	12.0	v		1956		63952B	9126B	BD +34°2494B	
6583	14 12.0	+19 07	Boo		9.5	>13	p		1899				BD +19 2764	Near Arcturus*
6588	14 13.5	−29 55	Hya		6.47	9.20	I	M7	1969					
6589	14 14.1	−53 56	Cen		9.6	10.5	p	Nb	1907	124268	241556		CoD−53 5490	
6599	14 14.7	+19 08	Boo		9.5	>13	p		1899				BD +19 2773	Near Arcturus*
6640	14 20.3	+66 54	UMi	Lb	7.7	8.5	v	K2	1909	126048	16344		BD +67 832	
6634	14 20.4	+19 32	Boo	EB?	9.8	10.3	v		1923					Period 2.17 days?
6638	14 21.6	−2 23	Vir	Lb	9.8	10.4	p	M5	1961	125754	139885		BD −01 2942	
6654	14 24.0	+8 15	Boo	EA?	5.7	6.2	p	A3	1951	126200	120433		BD +08 2857	
6657	14 25.4	−30 13	Cen	EA	10	10.5	p	F8	1949	126312			CoD−29 11053	
6687	14 27.5	+75 42	UMi		4.0	4.8	v	K4	1969	127700	8024	9286A	BD +76 527	5 UMi
6676	14 29.0	−38 54	Cen	L	10	11	p		1949				CoD−38 9405	
6691	14 30.6	−3 33	Vir		9.4	11.8	p		1914					
6693	14 31.6	−50 24	Lup	EA	9.7	10.2	p	A5	1973	127197	241766		CoD−49 8813	
6692	14 32.0	−63 30	Cen		10.2	>16	p		1915					
6708	14 34.8	−39 33	Cen		9.7	10.5	p		1964				CoD−39 9021	
6728	14 36.9	−58 16	Cen		9.6	10.2	p		1964				CoD−57 5640	
6737	14 38.7	−63 43	Cen		9.5	14	p		1915					
6776	14 44.1	−56 22	Cir		9.5	10.0	B	A0	1964	129328	241953		CoD−55 5799	
6786	14 46.7	−61 28	Cir		8.1	8.6	p	F2	1958	129708	252879		CoD−60 5320	
6799	14 47.6	−49 19	Lup	Lb?	9.8	10.5	p	M5	1907	130036			CoD−48 9338	
6798	14 47.8	−60 25	Cir		9.2	9.7	B	B5	1965	129935	252890		CoD−59 5398	
6828	14 51.2	−24 54	Lib		5.48	6.10	I	M6	1969					
6844	14 51.6	+59 31	Dra		9.5	>13	R	K5	1947					
6870	14 57.1	+55 05	Dra		9.0	9.6	p	K0	1937	132465	29352		BD +55 1725	
6878	14 57.9	+66 08	UMi		8.5	>13	R	M7	1947					
6900	15 02.5	+31 41	Boo	Lb	8.2	8.9	p	M6	1960	133254	64471		BD +32 2537	
6903	15 04.1	−41 08	Lup		10.0	>12.5	p		1971					
6917	15 04.9	−37 58	Lup		9.6	10.1	p		1971				CoD−37 9917	
6933	15 06.8	−35 05	Lup		9.9	10.4	p	A0	1965	133674	206301		CoD−34 10176	
6972	15 09.5	+53 19	Boo	S	9.9	10.4	p	M7	1955				BD +53 1767	
6956	15 11.5	−73 26	Aps		8.4	9.0	p	A9	1964	133766	257246		CoD−72 1092	
7029	15 19.9	+16 28	SerCp	Is?	9.5	10.2	p	K5	1957	136377	101490		BD +16 2767	
7081	15 25.3	+59 01	Dra		9.7	10.2	p	K7	1937	238411	29523		BD +59 1655	
7086	15 28.3	−45 08	Lup		7.97	8.47	B	Be	1964	137518	225814		CoD−44 10140	
7098	15 28.8	+3 49	SerCp		8.79	10.12?	I	M9	1969					
7105	15 30.8	−59 26	Nor	L	10.29	10.79	B	OB	1959				CoD−58 6056	Visual binary?
7138	15 32.3	+57 58	Dra		9.9	10.7	p	K0	1937	238420	29563		BD +58 1574	
7118	15 32.5	−52 52	Nor		8.9	9.4	p	B9	1964	138141	242667		CoD−52 6743	
7135	15 32.9	+31 22	CrB		5.5	7.8	v		1971	138749B	64769B		BD +31 2750B	4 θ CrB B
7134	15 32.9	+31 22	CrB		4.12	4.82	V	Be	1971	138749A	64769A		BD +31 2750A	4 θ CrB A*
7187	15 38.7	+57 31	Dra		8.5	9.3	p	K0	1937	140024	29605		BD +57 1598	
7192	15 39.2	+57 55	Dra	Is	7.3	8.3	p	K1	1960	140117	29609		BD +58 1583	
7190	15 42.4	−65 39	TrA	Lb	11	11.5	p		1949					Near HK TrA
7264	15 43.0	+80 07	UMi		7.8	8.3	v	F0	1949	142297	2597		BD +80 489	
7275	15 43.4	+81 19	UMi		8.1	8.7	v	F8	1949	142653	2599	9798A	BD +81 530	
7244	15 46.2	+14 41	SerCp	EA?	8.8	9.4	v	A5	1949		101724		BD +15 2910	
7261	15 47.9	+0 41	SerCp		5.76	6.26	I	M5	1969	141209			BD +01 3133	
7294	15 51.5	−15 39	Lib	S?	10.0	10.7	v		1961					δSct type?
7340	15 54.9	+42 46	Her		9.9	11.0	p	A2	1936					
7349	15 58.5	−62 04	TrA		8.6	9.1	p	B9	1964	142493	253361		CoD−61 5175	
7378	15 59.0	+26 08	CrB	RV?	9.1	9.9	v	F2	1965	143352	84121		BD +26 2763	
7377	16 00.3	−45 08	Nor		8.6	9.3	p	F2	1907	143085	226433		CoD−44 10556	
7394	16 01.5	+26 24	CrB		7.9	8.6	v	F8	1968	143808	84151		BD +26 2769	
7391	16 03.1	−54 46	Nor		10.2	10.9	I	M8	1974					
7412	16 05.3	−53 38	Nor		10.2	10.7	I	M9	1974					
7418	16 05.6	−53 51	Nor		9.4	11.4	I	M6	1974					
7457	16 06.0	+50 11	Her	RR?	9.7	10.4	p	G0	1955				BD +50 2255	
7427	16 06.6	−54 45	Nor		8.9	11.4	I	M3	1974					
7440	16 07.4	−53 33	Nor		8.4	8.9	I	M9	1974					
7459	16 08.7	−54 57	Nor		8.6	9.1	I	Mc	1974					
7472	16 09.9	−54 02	Nor		9.9	10.7	I	M6	1974					
7583	16 10.0	+81 09	UMi		7.7	8.3	v	G5	1949	147620	2675		BD +81 543	
7550	16 10.9	+71 18	UMi		8.3	8.9	p	F5	1933	146508	8443		BD +71 768	
7526	16 11.7	+12 04	Her		10.2	11.1	p	Pc	1907	145649			BD +12 2966	In planetary NGC 4593
7521	16 13.1	−53 45	Nor		8.3	8.8	I	M6	1974					
7532	16 13.8	−53 38	Nor		9.5	10.9	I	M8	1974					
7553	16 15.5	−54 39	Nor		8.5	9.0	I	M7	1974					
7570	16 15.6	−28 33	Sco		10	>13	v		1965				CoD−28 12005	
7574	16 16.7	−53 49	Nor		4.9	5.47	v	M2	1965	146003	243526		CoD−53 6533	
7589	16 18.1	−54 03	Nor		8.5	10.7	I	M7	1974					
7607	16 18.8	−28 45	Sco		7.30	8.71	I		1969					
7645	16 22.1	−50 28	Nor	Lb	9.6	10.1	p	K2	1901	147088			CoD−50 10442	
7675	16 23.4	−26 35	Sco		10	10.5	p		1964				CoD−26 11312	
7707	16 25.3	+19 14	Her		6.8	7.3	v	M4	1951	148128	102152		BD +19 3096	

NSV	α 2000	δ 2000	Const	Type	Max	Min		Spec	Publ	HD	SAO	ADS	BD	Notes
7704	16ʰ26ᵐ6	−51°40′	Nor	UV?	10.0	12.0	p		1935				○	
7956	16 29.4	+86 26	UMi	UG?	9?	11.5?	v		1913					South of BD +86 252
7761	16 29.9	−12 02	Oph	SR?	9.5	10.2	p	K5	1908	148633	159950		BD −11 4154	
7851	16 35.5	+34 05	Her	Lb?	10	10.5	p		1968		65415		BD +34 2815	
7841	16 35.8	−25 18	Sco		10.2	11.5	p	A0e	1975	149437			CoD−25 11557	
7847	16 37.0	−45 19	Sco		7.9	8.4	p	B5	1964	149450	226961		CoD−45 10787	
7857	16 37.5	−36 19	Sco		10	11	p		1971				CoD−36 10857	
7894	16 41.4	−55 18	Ara	Lb?	8.9	9.6	p	M6	1910	150024	244063		CoD−55 6850	
7969	16 47.4	+23 13	Her		9.00	9.80	V	R0	1973		84606		BD +23 2998	
8006	16 52.2	−12 57	Oph		5.72	7.84	I	M8	1969					
8015	16 53.8	−41 19	Sco		7.88	9.13	V	B0	1966	152198	227367		CoD−41 11015	
8098	17 00.2	−10 37	Oph		8.44	10.29?I		M	1969					
8106	17 00.4	+5 01	Oph	E?	10	10.5	p	M2	1935				BD +05 3302	
8149	17 04.1	−14 56	Oph	L?	9.5	10.0	p	A2	1958	154129	160252		BD −14 4531	
8172	17 06.6	−57 43	Ara	Cep?	9.5	10.3	p	G5I	1910				CoD−57 6722	
8199	17 06.9	−16 18	Oph		10.1	11.9	v		1925					
8191	17 07.1	−35 46	Sco	I	8.0	8.6	p	Be	1943	154450	208426		CoD−35 11320	
8316	17 11.5	+48 50	Her	S	9.9	10.9	p	M1	1932		46557		BD +49 2601	
8322	17 13.0	−10 35	Oph		4.90	5.51	I	M7	1969					
8331	17 13.7	−31 51	Sco		7.97	9.07	I		1969					
8448	17 17.4	−29 15	Oph		10.0	>13.0	p		1967					
8523	17 18.7	+43 37	Her	SRb?	10.2	11.2	p	Mb	1959	157010	46630		BD +43 2716	Rapid light variations
8918	17 29.8	−5 55	Oph		8.5	11.0	v		1962		141691B	10583B	BD −05 4450B	
9118	17 31.9	+17 45	Her	SR?	2.38	3.24	K	M2	1969					
9257	17 36.7	+15 29	Her		10.0	10.5	p	K0	1952				BD +15 3226	
9246	17 37.6	−40 49	Sco		8.1	8.7	p	F5Ib	1964	159654	228211		CoD−40 11648	
9405	17 39.7	−32 06	Sco		8.77	10.2	V	A0	1972	318102			CoD−32 13076	In cluster NGC 6405
9424	17 40.2	−32 14	Sco		8.2	9.83	v		1972				CPD−32 4716	In cluster NGC 6405
9593	17 42.3	+50 56	Dra		9.7	10.4	p	G0	1938	234475	30527		BD +50 2451	
9596	17 43.6	+13 54	Oph	Lb	10	10.5	p	M0	1966				BD +13 3439	
9573	17 45.1	−60 31	Pav		8.5	9.3	p	M4	1910	160673	254017		CoD−60 6774	
9663	17 47.2	−33 11	Sco	N?	11	>13	p		1952					Seen on April 18, 1952
9708	17 49.3	−38 11	Sco	EA	10	10.5	p	G0	1949	324381			CoD−38 12214	
9725	17 49.4	−7 47	Oph	EW?	9.0	9.7	p	K0	1944	162059	141892		BD −07 4508	
9731	17 49.6	−2 14	Oph	Lb	10.0	10.5	p	M3	1951	162115	141894		BD −02 4461	
9741	17 50.3	−22 24	Sgr		6.00	6.57	I	M6	1969					
9764	17 51.2	−8 01	Oph	M?	8.91	10.58?I		M9	1969					OH emission
9797	17 52.9	−18 09	Sgr		6.70	7.20	I	M6	1969					
9818	17 54.1	−28 20	Sgr		7.01	10.97?I		M8	1969					
9833	17 54.3	−4 13	Oph	Lb	9.5	10.0	p	K5	1951	163036	141953		BD −04 4371	
9813	17 56.3	−70 22	Aps		9.5	10.9	p		1970				CoD−70 1564	
9950	17 59.2	+18 52	Her	Is?	10.1	11.6	p	M2	1935				BD +18 3516	
10013	18 02.4	−23 02	Sgr		7.30	8.21	V	O7	1960	164492	186145	10991A	CoD−23 13804	In open cluster NGC 6514
10081	18 03.7	−24 23	Sgr		9.1	10.0	v		1868	164740			CoD−24 13806	In Lagoon nebula (M8)
10164	18 06.0	−47 31	Ara		10.0	10.5	p		1969				CoD−47 12046	
10240	18 06.6	+46 16	Her		7.1	7.8	v	F2	1936	166067	47224		BD +46 2426	
10254	18 08.9	−26 16	Sgr		6.87?	7.80	I	M	1969					
10270	18 09.3	−17 22	Sgr		2.59	3.21	K	M7	1969					
10306	18 10.4	−10 34	SerCd		7.45	9.31?I		M8	1969					
10363	18 11.8	+33 27	Her		8.5	>11	v		1962		66733B	11149B	BD +33 3044B	
10371	18 12.8	+4 09	Oph	Lb	10	10.5	p	M5	1967	166929	123253		BD +04 3649	
10352	18 13.3	−49 55	Tel	Lb?	9.8	10.5	p	M4	1939	166456			CoD−49 11985	
10370	18 13.6	−34 30	Sgr		9.7	>13.0	p		1965				CoD−34 12602	
10404	18 14.1	+5 21	Oph	Lb	10.0	10.7	p	M5	1958	167243			BD +05 3653	
10548	18 18.7	−13 45	SerCd		9.53	10.08	V	O−F	1973				BD −13 4927	
10696	18 21.1	+33 56	Lyr	Lb	10.2	10.8	p	K2	1928				BD +33 3084	
10634	18 21.4	−56 24	Tel	E	10	10.5	p		1963				CoD−56 7292	
10756	18 24.0	+38 44	Lyr	I	7.9	8.7	p	K2	1966	169646	66936	11320A	BD +38 3160	
10780	18 24.3	+52 15	Dra		9.8	10.5	p	A	1938					
10724	18 24.3	−44 07	CrA		5.05	5.55	B	B3	1964	168905	228982		CoD−44 12569	
10869	18 27.8	+31 12	Lyr		5.31	5.91	I	M5	1969	170375	66998		BD +31 3272	
10846	18 28.1	−13 03	Sct		6.75	7.52	I	M5	1969					
10929	18 31.4	−42 36	CrA		9.3	10.0	p	K3	1900	170417			CoD−42 13342	
10970	18 32.2	−24 57	Sgr		7.62	8.60	I	M	1969					
11055	18 33.7	+49 11	Dra	UV?	9.5	>13	p		1948					This is not V1902 Sgr ∗
11077	18 36.4	−23 53	Sgr		2.84	3.37	K		1969					Near the globular M22
11099	18 37.6	−41 51	CrA		10	10.5	p	A4	1963	171576			CoD−41 12968	
11126	18 37.9	−6 52	Sct		5.71	6.80	I		1969					Near EW Sct
11129	18 38.1	−12 22	Sct		6.39	6.94	I	M6	1969					
11159	18 39.1	+1 42	SerCd		7.20	8.95	I	M8	1969					
11165	18 39.7	−4 48	Sct	Lb	9.5	10.0	p	G5	1958		142477		BD −04 4537	
11180	18 40.3	−5 43	Sct		7.01	10.72	I	Ms	1969					OH emission
11201	18 41.3	−6 22	Sct		7.28?	8.35	I	M?	1969					
11202	18 41.4	−4 21	Sct		6.73	7.24	I	M?	1969					
11225	18 41.9	+17 41	Her		1.23	2.58	K	C	1969					

NSV	α 2000	δ 2000	Const	Type	Max	Min	Spec	Publ	HD	SAO	ADS	BD	Notes
11233	18^{h}42^m.5	−2°18′	Aql		7.93	9.37?I	C	1969				BD −07 4668	*
11235	18 42.6	−7 38	Sct	RV?	10	10.5? p	K2	1953	172810	142524			
11263	18 43.6	+13 58	Her		1.60	2.90 K	M7	1969					
11276	18 44.3	−3 48	Aql		6.70	8.06 I	M7	1969					Spectrum M3III−M7?
11321	18 45.1	+40 11	Lyr	E?	10	10.5 p		1966				BD +40 3480	
11357	18 46.7	+54 17	Dra		10.2	10.9 p	K5	1938	234710	31167		BD +54 2037	
11345	18 47.3	+15 36	Her	E?	10.0	10.6 p	F8	1963	229491			BD +15 3562	
11355	18 48.2	−1 58	Aql		9.48	11.9 I	M6	1974					Spec. varies, M5.5−M7.8
11397	18 48.9	+50 49	Dra		9.6	10.2 p	F5	1938	234716	31181		BD +50 2671	
11386	18 49.9	+0 32	Aql		8.9	9.5 v	F0	1921	174299	123949		BD +00 4028	
11353	18 50.3	−65 27	Pav		9.8	11 p		1917					
11404	18 50.4	+12 41	Her		10.0	14.0 v		1955			11742B		
11483	18 52.1	+55 14	Dra		9.7	10.2 p	K2	1938		31228		BD +55 2123	
11461	18 52.9	−2 48	Aql		2.92	3.93 K	M9	1969					
11536	18 55.5	+6 37	Aql		2.82	3.36 K	K0	1969	175515	124050		BD +06 3978	
11563	18 55.9	+30 09	Lyr		7.42	8.55 I	M9	1969					
11597	18 56.5	+54 04	Dra		10.2	10.9 p	M2	1938		31278		BD +53 2155	
11617	18 59.1	+10 24	Aql		6.93	8.56 I		1969					
11640	19 01.1	−23 18	Sgr		9.1	11.5 v		1922				CoD−23 14946	
11689	19 03.3	+7 30	Aql		8.37?	11.05?I	C	1969					
11696	19 04.1	−19 24	Sgr		6.76	7.40 I	M6	1969					
11708	19 04.9	−5 14	Aql	EA?	9.8	10.8 p	F5	1945				BD −05 4861	
11737	19 06.7	−17 06	Sgr	EW?	10.6	11.3 p	K2	1944	177762	162196		BD −17 5470	Period 1.024 days
11774	19 09.0	+37 09	Lyr	I?	10.2	11.3 p	F2	1938		67878		BD +36 3419	
11769	19 09.8	−21 01	Sgr		1.75	2.38 K	F2II	1969	178524	187756		BD −21 5275	41 π Sgr, vis. binary
11764	19 10.2	−47 42	Tel		10.2	10.8 p		1971				CoD−47 12766	
11811	19 13.9	−46 00	Tel	Lb	10	10.5 p	M2	1949	179236			CoD−46 12875	
11847	19 15.0	+22 05	Vul		6.81	8.70 I	M9	1969					
11869	19 16.3	+32 32	Lyr		10	12 U		1939					
11872	19 17.0	+10 04	Aql		2.78	3.41 K	M8	1969					
11880	19 18.4	−19 04	Sgr	E?	9.4	10.1 v		1888				BD −19 5389	
11912	19 20.3	−8 02	Aql		1.63	2.61 K	C	1969				BD −17 5597	
11929	19 21.5	−17 13	Sgr		8.6	9.5 v	K2	1922	181537	162508		CoD−46 12968	
11941	19 23.0	−46 21	Tel		10	10.6 p		1971					
11963	19 23.2	−2 36	Aql		2.73	3.81 K	M9	1969					
12020	19 24.7	+59 24	Dra		9.7	10.3 p	G0	1937	239081	31642		BD +59 2028	
11986	19 24.8	−20 36	Sgr	EW?	9.0	11.5 p	K0	1944	182301	188091		BD −20 5531	
12009	19 25.0	+36 02	Lyr		2.77	3.32 K		1969					OH emission
12006	19 25.5	+2 48	Aql		10.9	>16 p		1949					Nova Aql 1949*
12022	19 26.3	+16 41	Sge		2.28	3.01 K	M9	1969					
12051	19 28.8	−42 07	Sgr		6.5?	8.5 v	Ma	1900	182929	229706		CoD−42 14230	
12107	19 32.1	−28 13	Sgr		7.5	8.1 p	B9	1964	183764	188244		CoD−28 15917	
12161	19 33.1	+51 44	Cyg		9.9	10.7 p	K7	1938	234929			BD +51 2618	
12165	19 34.2	+28 04	Cyg		2.20	3.35 K	C	1969					
12159	19 35.0	−34 10	Sgr	Lb?	10	10.5 p	Mb	1949	184259			CoD−34 13779	
12186	19 35.7	+6 40	Aql		9.8	10.3 I	M6	1959					
12220	19 36.4	+56 46	Cyg		9.0	9.5 p	K0	1914	185499	31818		BD +56 2273	
12218	19 36.6	+50 13	Cyg		10.8	14.7 v		1903					Near R Cyg
12247	19 37.7	+50 41	Cyg	Lb	8.6	9.1 p	M4	1962	185695	31831		BD +54 2187	
12207	19 37.7	−16 27	Sgr	E?	9.8	10.5 p	K0	1958	185006	162828		BD −16 5381	
12244	19 38.8	+8 26	Aql	Lb?	10	11 p		1964				BD +08 4173	
12344	19 43.1	+38 20	Cyg	S	10.2	11.4 p	A	1938					OH emission
12342	19 43.7	+3 45	Aql		7.12	8.80 I	M9	1969					In planetary NGC 6826
12382	19 44.8	+50 32	Cyg		9.6	10.3 p	Pd	1968	186924	31951		BD +50 2869	
12380	19 45.0	+45 08	Cyg		6.3	>8.5 v		1837		48796B	12880B	BD +44 3234B	18 δ Cyg B
12394	19 47.0	−2 04	Aql	S	10	10.5 p	A6	1935					
12439	19 49.1	+32 48	Cyg	SR	7.3	8.2 v	K5	1961	187503	68902		BD +32 3578	
12441	19 49.4	+18 51	Sge	Lb	9	9.5 p	K7	1964	350668	105311		BD +18 4258	
12470	19 50.1	+47 48	Cyg		9.3	9.8 v	A2	1949	187877	48885/6		BD +47 2933	
12467	19 51.2	−10 12	Aql	I	10.2	10.9 p	G6	1958	187622	163041		BD −10 5201	
12499	19 51.3	+51 58	Cyg		10.0	10.5 p	K0	1938	235024			BD +51 2698	
12518	19 54.0	−30 15	Sgr	RR?	9.4	9.9 p	F5	1973		211637		CoD−30 17453	
12495	19 55.4	−71 33	Pav	Lb	9.5	10 p	M5	1959	187336	257747		CoD−71 1578	
12612	19 56.4	+59 44	Cyg	L	10.2	10.9 p	M5	1959	189346	32132		BD +59 2149	
12759	20 03.0	+57 05	Cyg		7.2	7.9 p	A0	1937	190625	32238		BD +56 2344	
12794	20 06.7	−7 01	Aql	EA?	8.2	11 v	G5	1939	190754	144072		BD −07 5169	
12819	20 07.4	+36 34	Cyg	S	8.0	9.5 p	M2	1954	191226	69449		BD +36 3883	
12814	20 07.7	+6 03	Aql		2.49	3.47 K	M9	1969					
12808	20 07.9	−30 44	Sgr		8.3	10.2 v		1903				CoD−31 17327	
12817	20 09.0	−49 41	Tel		10	10.8 p		1971				CoD−50 12825	
12823	20 09.3	−42 07	Sgr	Cep?	9.5	10 p		1963				CoD−42 14707	
12861	20 10.3	+29 20	Vul		2.69	3.30 K	C4	1969					
12884	20 11.2	+40 17	Cyg		9.6	>15 v		1955				BD +39 4072	Bright during 1951−68
12912	20 12.2	+59 26	Cyg		9.7	10.4 p	K2	1937	239349	32358		BD +58 2072	
12901	20 12.5	+33 23	Cyg		2.70	3.42 K	M7	1969					

NSV	α 2000	δ 2000	Const	Type	Max	Min		Spec	Publ	HD	SAO	ADS	BD	Notes
12969	20ʰ16.ᵐ4	+31°47′	Cyg		8.5	12.6	p		1947				∘	
13007	20 19.1	+37 46	Cyg		8.61	9.53	V	Bp	1970	193516	69824		BD +37 3881	One of reports doubted
13033	20 20.2	+56 31	Cyg		9.7	10.5	p	F8	1937	239338	32487		BD +56 2398	
12990	20 20.2	−51 50	Tel	Lb	9.5	10	p	M3	1949	192976			CoD−52 9458	
13056	20 22.3	+62 53	Dra		7.22	8.04	I	M9	1969					
13053	20 22.9	+42 59	Cyg		8.5	11.0	v		1962		49550B	13786B	BD +42 3721B	
13082	20 25.0	+59 17	Cyg		9.6	10.2	p	G5	1937	239404	32557		BD +58 2109	
13083	20 25.2	+57 13	Cyg		9.8	10.5	p	G0	1937	239403	32560		BD +56 2415	
13122	20 27.7	+77 34	Cep	L	10.0	10.5	p	K2	1959		9787		BD +77 779	
13150	20 34.3	+19 32	Del		7.0	8.2	v	M7	1969	196036	106259		BD +19 4450	
13181	20 36.4	+56 05	Cyg	UVn?	9.9	10.6	p	K2	1937	239457	32727		BD +55 2436	
13180	20 36.9	+37 53	Cyg		8.03	9.16	I	M9	1969					
13190	20 39.2	−37 56	Mic	Lb	10	11	p		1949				CoD−38 14151	
13266	20 39.7	+80 30	Dra	Lb	10.1	11.1	p	Mb	1958	198327	3443		BD +79 683	
13234	20 41.6	+17 25	Del	L?	9.0	10.2	v	K0	1931	352828	106389		BD +16 4346	
13233	20 41.9	−5 38	Aqr		6.59	7.16	I	M7	1969					
13228	20 42.0	−19 04	Cap	L	9.8	10.7	v		1854				BD −19 5893	
13249	20 43.0	+22 03	Vul		6.87	7.71	I	M7	1969					
13284	20 46.6	−0 54	Aqr		6.48	7.98	I	M8	1969					
13313	20 48.3	+24 03	Vul	I?	9.7	10.6	v		1907				BD +23 4157	
13377	20 52.1	+44 04	Cyg		10.1	10.9	p	Be	1971					
13378	20 52.2	+44 26	Cyg		8.3	9.0	p	Be	1971	198931	50153		BD +43 3747	
13430	20 57.2	+16 15	Del	Lb?	8.7	9.2	p	M7	1929	199498	106690		BD +15 4297	
13425	20 58.1	−48 04	Ind		9.4	10.2	p	K0	1910	199303			CoD−48 13729	
13475	21 02.4	+23 19	Vul	I?	7.2	7.8	v	B9	1916	200352	89391		BD +22 4278	
13494	21 02.9	+46 16	Cyg		9.0?	10.5?	v	Be	1973		50393		BD +45 3377	
13503	21 04.7	+3 32	Equ		9.1	10.3	v		1953	200660B	126523	14602B	BD +02 4298B	
13517	21 04.7	+46 14	Cyg		10.2	10.8	v	Be	1973					
13581	21 10.6	+41 11	Cyg	SR?	9.9	11.0	p	G0	1959		50533		BD +40 4430	
13609	21 12.8	+60 06	Cep	Lb?	8.6	10.2	p	M2 Ib	1874	202380	33232		BD +59 2342	
13613	21 13.7	+45 32	Cyg	Lb	8.2	8.87	B	K3	1892	202348	50593		BD +44 3744	
13635	21 16.5	+55 23	Cep	E?	10.0	10.5	p	F5	1960		33275		BD +54 2502	
13637	21 17.0	+40 20	Cyg	I	9.5	10.0	p	A0	1962		50661		BD +39 4506	
13653	21 19.2	+23 28	Vul	Lb	8.5	9.0	p	M5	1961	203049	89615		BD +22 4364	
13668	21 21.4	+2 53	Equ		7.50	8.00	V	K1	1962	203323	126724	14894	BD +02 4346	
13681	21 25.3	−64 55	Pav		10.2	11.1	p		1970				CoD−65 2748	
13701	21 25.5	+47 01	.Cyg		9.5	10.0	p		1933				BD +46 3306	
13721	21 27.4	+36 42	Cyg	M?	2.37	3.33	K	M9	1969					OH emission
13743	21 27.5	+71 49	Cep		1.77	2.81	K	M5−M9	1969					
13739	21 28.4	+49 27	Cyg	L	10.2	11.4	p	G8Ib	1938				BD +48 3394	
13737	21 29.7	−50 21	Ind		7.7	8.3	p	A2	1966	204370	247049		CoD−50 13373	
13781	21 32.8	+54 19	Cyg		5.42	5.94	I		1969					
13806	21 36.3	+32 31	Cyg		1.55	2.38	K	M9	1969					
13846	21 37.0	+82 55	Cep		8.1	8.7	v	G5	1949	207146	3606	15229	BD +82 657	
13835	21 41.4	−38 25	Gru	Lb	10	10.5	p		1963					
13837	21 41.8	−51 10	Ind	SR?	10.0	10.5	p		1973				CoD−51 13027	
13852	21 42.1	+54 07	Cyg		8.8	9.6	p	B8	1936				BD +53 2686	
13857	21 43.1	+41 09	Cyg	Lb	6.3	7.08	B	M2	1959	206749	51221		BD +40 4623	
13865	21 43.8	+58 59	Cep		9.2	10.1	p	A2	1937	239751	33699		BD +58 2318	
13887	21 48.8	+39 57	Cyg		2.84	3.41	K	M8	1969					
13889	21 49.8	+12 43	Peg	S	9.6	10.1	p	M	1960				BD +12 4694	
13891	21 50.1	+17 17	Peg		5.32	6.16	V	F2	1966	207652	107425		BD +16 4612	13 Peg, visual binary
13933	21 53.5	+45 27	Cyg		8.2	9.8	p	G1	1938		51404		BD +44 3974	
13906	21 55.1	−76 56	Oct	I?	9.5	10	p	Mb	1963	207576			CoD−77 1093	
13967	21 56.9	−12 13	Cap	Lb	10	10.5	p	K5	1934	208505			BD −12 6130	
13969	21 57.0	−14 07	Cap	L	9	10.5	p	M4	1933	208519	164760		BD −14 6170	
13977	21 58.6	−45 45	Gru		8.3	8.8	p	A2	1971	208614	230922		CoD−46 14144	
13995	21 59.9	+54 04	Cyg		8.0	8.7	p	A0	1938	209179	33954		BD +53 2758	
14010	22 03.5	+45 34	Lac	Lb	11	11.5	p	Na	1966	209596				Visual binary*
14030	22 06.9	−19 23	Aqr	L?	9.7	10.4	v		1952				BD −20 6359	
14037	22 07.3	+11 54	Peg		1.53	2.08	K	M7	1943					
14039	22 07.5	+18 00	Peg		6.3	7.7?	v	M1	1977	210090	107676		BD +17 4693	
14126	22 20.1	+62 11	Cep		6.84	7.70	I		1969					
14145	22 23.5	+52 48	Lac	S	9.5	10.6	p	A0	1958				BD +52 3198	
14141	22 23.5	−20 19	Aqr	Lb	9.5	10.0	p	M2	1933	212294	191106		BD −21 6213	
14157	22 26.7	−48 22	Gru		9.6	10.4	p		1965		231140		CoD−48 14210	
14170	22 28.8	+10 43	Peg	Lb	10	10.5	p	M6	1961				BD +09 5050	
14184	22 31.2	−23 00	Aqr		8.3	8.8	p	G0	1964	213379	191201		CoD−23 17470	
14198	22 32.6	+45 32	Lac	Lb	10	10.5	p		1966				BD +44 4159	
14213	22 33.7	+56 38	Lac	L	5.6	6.8	v	G8	1892	213930	34574		BD +55 2769	
14231	22 35.5	−17 15	Aqr	EA	9.2	10.0	p	B9	1913	213985	165175		BD −18 6151	
14245	22 36.0	+50 36	Lac	I?	9.9	10.5	p	M0	1963		34606		BD +49 3905	
14229	22 36.0	−65 07	Tuc	L	10	11	p		1949					
14282	22 36.3	+86 40	Cep		8.3	9.0	v	G5	1949	215878	3773		BD +85 390	
14254	22 38.9	−60 39	Tuc		8.7	9.5	p	F6	1970		255270		CoD−61 6728	

NSV	α 2000	δ 2000	Const	Type	Max	Min		Spec	Publ	HD	SAO	ADS	BD	Notes
14347	22^{h}53^m.4	+36°30'	Lac		7.62	9.06	I		1969				°	
14381	22 57.1	+83 03	Cep		8.1	8.9	v	K0	1949	217585	3822		BD +82 707	
14392	23 02.1	+10 36	Peg		8.28	9.29?I		M9	1969					
14418	23 03.0	+81 51	Cep		8.0	8.7	v	K2	1949	218273	3835		BD +81 806	
14427	23 06.5	+61 28	Cep	EW?	9.6	10.1	p	A0	1938					
14460	23 13.8	−13 24	Aqr		8.5	11.0	v		1962		165571B	16608B	BD −14 6424B	
14470	23 15.6	−24 31	Aqr	E?	10.5	11	p		1931				CoD−25 16357	Period 0.99607 day
14489	23 18.5	+58 33	Cas		2.88	3.56	K	S	1969					
14492	23 19.0	−9 37	Aqr		9.0	>12	v		1962			16671B		
14501	23 19.4	+62 44	Cas		9.02	9.6	B	K5Ib	1935	219978	20558		BD +61 2423	
14522	23 20.5	+81 51	Cep		8.2	8.9	v	F8	1949	220361	3893		BD +81 818	
14515	23 20.7	−5 54	Aqr		5.7	7.4	v	G5	1959	220035	146652		BD −06 6191	
14566	23 24.3	+86 25	Cep		6.6	7.1	p	F0	1912	221142	3904		BD +85 399	
14580	23 27.6	+60 27	Cas	UV?	10.8	13.1	p		1954					Close binary
14582	23 28.0	+60 51	Cas		9.0	11.8	p	A2Iab	1951				BD +60 2546	
14602	23 32.4	−43 37	Phe		8.0	8.7	v	K2	1900	221432	231653		CoD−44 15273	
14611	23 33.8	+6 18	Psc		6.17	6.72	I	M7	1969					
14623	23 34.5	+43 33	And		2.05	2.87	K	C	1969					
14642	23 37.0	+35 48	And		7.8	8.7	p	K5	1937	222032	73374		BD +34 4966	
14643	23 37.4	−33 36	Scl	Lb	9.5	10	p	M1	1959	222024	214662		CoD−34 16103	
14654	23 39.2	−2 26	Psc	Lb?	9.5	10	p	K−M	1934				BD −03 5680	
14697	23 46.0	−35 35	Scl	L?	10	10.5	p		1963					
14708	23 47.4	+26 08	Peg	Lb	9.0	9.5	p	M6	1961	223203	91488		BD +25 5005	
14716	23 48.6	+26 27	Peg		9.5	10.0	p		1944					
14722	23 49.7	−43 17	Phe	EA	10	11	p		1949				CoD−43 15514	
14731	23 52.1	+61 49	Cas		0.92	1.54	K	M9	1969					
14747	23 54.2	+14 29	Peg	Lb	10.0	10.5	p	M6	1961					
14766	23 56.5	+44 39	And		10.0	13.0	v		1955			17109B		
14769	23 57.1	−52 14	Phe		10.1	11.0	p		1907				CoD−52 10546	
14785	23 58.7	−3 33	Psc		9.0	11.5	v		1962		147008B	17137B	BD −04 5996B	
14784	23 58.7	+46 44	And		9.3	9.8	v	M0	1970		53537		BD +45 4378	

SUPPLEMENTARY NOTES
(By NSV number)

152. In 1871, A. Borrelly recorded this star at magnitude 6 or 7 on November 3rd, 8 on November 8th, and 10 on November 24th. It has remained constant ever since (*Sky and Telescope*, **60,** July, 1980, page 21).

482. One of the "missing" BD stars (*Sky and Telescope*, **59,** May, 1980, page 389).

837. Nova Per 1873? Observed with a meridian circle on July 16th and 18th that year, but not seen thereafter.

1484. Seen by E. Hertzsprung on a Harvard patrol plate taken with double images on December 15, 1900 (*Sky and Telescope*, **34,** December, 1967, page 382). There is a star of photographic magnitude 15 at this position. But B. E. Schaefer believes Hertzsprung's object is an unusual defect on the plate (*Publications* of the Astronomical Society of the Pacific, **95,** 1019, 1983).

1887. Visual binary. Due to a misprint in the *Cordoba Durchmusterung*, this star is sometimes called CoD − 49°1726.

3133. Observed near Venus by E. E. Barnard, using the Lick 36-inch refractor on the morning of August 13, 1892 (*Sky and Telescope*, **15,** June, 1956, page 356). Not seen before or since.

3633. Variability supposedly confirmed by photographic and visual observations.

3846. J. Richer measured a 3rd-magnitude star at this location in 1673, on both January 12th and 21st (*Sky and Telescope*, **32,** August, 1966, page 79).

3933. Near the cluster IC 2391. This star was bright on March 29, 1895.

4550. On March 13, 1612, C. Scheiner observed a starlike object near Jupiter, calling it "very bright and large, like any of Jupiter's satellites." He recorded it on seven more nights through April 8th as it steadily faded, but he never saw it again (though he tried). The star is probably not BD + 15°2083, but the positions match closely (*Sky and Telescope*, **42,** December, 1971, page 344).

4609. Proper motion 0".75 per year, toward p.a. 217°.

4617. Period 370 days? Barium II star.

5374. This star is Groombridge 1830, with the exceptional proper motion of 7".05 per year. The 0.6-magnitude increase was noted on a plate taken with the Thaw 30-inch refractor April 27, 1939. Another flare was photographed February 27, 1968, with the Sproul 24-inch, but at a point 2".2 from Groombridge 1830 in position angle 166°. W. R. Beardsley *et al.* have suggested either that the star itself flared in 1939, or that the supposed companion (which has been named CF UMa) was then in line with it (*Astrophysical Journal*, **194,** 637, 1974). W. D. Heintz has recently questioned the existence of the companion.

5495. Component B = NSV 5496 = ADS 8451B (V = 8.53-8.82, *F*2).

5611. Not variable? Probable misprint in source article (*Astrophysical Journal*, **144,** 496, 1966).

5858. Did not appear on objective-prism plate of stars brighter than magnitude 7.5 taken May 22, 1896.

6058. Member of UMa moving cluster.

6148. This star was marked M by W. Olbers on a chart he prepared while observing a comet on April 1, 1796. He described the star as much brighter than any of the other neighbors of 53 Vir, but was startled 11 months later to find it faint (*Sky and Telescope*, **48,** September, 1974, page 163), where it has remained ever since.

6583 and 6599. Two well-documented BD stars near Arcturus. They were noticed "missing" by V. Safarik in 1885, and neither has been seen again with any certainty (*Sky and Telescope*, **39,** February, 1970, page 87).

6903. Same star as NSV 6904? Position of 6904 is given 3' north of 6903, and the magnitude range 15.7 to 17.5 (photographic).

7134 and 7135. Visual binary. For NSV 7135, the combined magnitude is given.

7427. Possibly the same as NSV 7428, whose position is given as 1'.6 east, with a photographic magnitude of 16.5 to fainter than 17.5 (W. J. Luyten, 1938).

7607. Same as NSV 7606? Positions agree within 1'; nothing reported about 7606 except a light amplitude of 1.5 magnitudes (photographic).

11055. Stellar image seen only on one plate in 1965.

11235. Period 34.88 days; mean brightness varies with period of about 900 days.

12006. Seen only on one plate taken May 2, 1949; a nova, a U Gem or UV Cet star, or a plate defect?

14010. Visual binary; magnitudes 11.0 and 11.8, spectral types *C*4:,3 and *F*8III-V.

Open Clusters

Open Clusters

Column Headings

Name — The cluster name or its number from J. L. E. Dreyer's *New General Catalogue of Nebulae and Clusters of Stars* (NGC) or *Index Catalogue* (IC). If none, one of the following designations is given: Antalova, Barkhatova (Bark), Basel, Berkeley (Berk), Biurakan (Biur), Blanco, Bochum, Collinder (Cr), Czernik, Dolidze (Do), Dolidze-Dzimselejsvili (DoDz), Feinstein (Fein), Frolov, Haffner, Harvard, Hogg, Iskudarian (Isk), King, Loden, Lynga, Markarian (Mrk), Melotte (Mel), Pismis, Roslund, Ruprecht (Ru), Sher, Stephenson (Steph), Stock, Tombaugh, Trumpler (Tr), Upgren, van den Bergh (vdB), van den Bergh-Hagen (vdB-Ha), Waterloo, Westerlund (Westr). A cross index of cluster names follows this section.

α **2000** and δ **2000** — Right ascension and declination, referred to the 2000.0 equinox.

Const — Constellation in which the cluster is located.

Diam — Apparent diameter in minutes of arc.

d(pc) — Approximate distance in parsecs. To convert to light-years, multiply by 3.26.

Tr Type — Descriptive type based on the system of R. J. Trumpler (*Lick Observatory Bulletin*, **14**, 154, 1930). This is a three-part code that characterizes the cluster's degree of concentration, the range in brightness of its stars, and the richness, as follows:

Concentration
   I. Detached; strong concentration toward center.
   II. Detached; weak concentration toward center.
   III. Detached; no concentration toward center.
   IV. Not well detached from surrounding star field.

Range in brightness
   1. Small range in brightness.
   2. Moderate range in brightness.
   3. Large range in brightness.

Richness
   p Poor (less than 50 stars).
   m Moderately rich (50 to 100 stars).
   r Rich (more than 100 stars).

The letter "n" following the Trumpler type means that there is nebulosity associated with the cluster.

N* — Approximate number of stars in the cluster, according to G. Lynga's 1983 examination of the National Geographic Society-Palomar Observatory Sky Survey plates.

Mag — In most cases, this is the cluster's total apparent **V** magnitude from **UBV** photometry, according to B. A. Skiff (1983). An asterisk means that he has computed **V** from RGU photometry, which is a system devised by W. Becker. Photographic magnitudes are indicated by the letter "p." Superimposed field stars are not included in the total magnitudes.

Br* — Apparent **V** magnitude of the brightest star that was used to compute the total brightness in the previous column. A value followed by the letter "p" is the approximate photographic magnitude of the brightest cluster star.

Spec — Earliest (hottest) spectral type observed among all cluster members.

Age — Approximate age in millions of years.

RV — Radial velocity in kilometers per second. Positive means receding (traveling away from the solar system).

Notes — Common names, Messier numbers, membership in OB associations (according to the *Catalogue of Star Clusters and Associations,* or R. M. Humphrey's 1978 study), and other information are placed in this column. Clusters belonging to the Large and Small Magellanic Clouds are flagged "LMC" and "SMC," respectively. The term "asterism?" is used for sparse groupings suspected of being mere chance alignments of stars, rather than true physical systems.

Name	α 2000	δ 2000	Const	Diam	d(pc)	Tr Type	N*	Mag	Br*	Spec	Age	RV	Notes
Berk 58	0^{h}00^m.2	+60°58′	Cas	8′		IV 2 p	30	9.7	15p				
Berk 59	0 02.6	+67 23	Cep	10		III 2 p n	40		11p				
Blanco 1	0 04.3	−29 56	Scl	90	250	III 2 m	30	4.5	5.02	B5	50		ζ Scl cluster
Stock 19	0 04.4	+56 02	Cas	3.0		II 2 p	6		8p	B5			
NGC 103	0 25.3	+61 21	Cas	5	3000	II 2 p	30	9.8	12.25	B3	38	−11	In Cas OB4
NGC 129	0 29.9	+60 14	Cas	21	1600	IV 2 p	35	6.5	8.57	B5	150	−14	Contains DL Cas
NGC 133	0 31.2	+63 22	Cas	7		IV 1 p	5	9.4p		A4			Asterism?
NGC 136	0 31.5	+61 32	Cas	1.2	4400	II 2 p	20		13p				
King 14	0 31.9	+63 10	Cas	7	2600	III 2 p	20	8.5	11.29	B2	16		
NGC 146	0 33.1	+63 18	Cas	7	2900	IV 3 p	20	9.1	11.56	B3	13		
NGC 189	0 39.6	+61 04	Cas	3.7	1080	III 2 p	15	8.8*	10.95	A0	20		
Stock 24	0 39.7	+61 57	Cas	4.0	1900	IV 2 p	20	8.8*	11.05	B9			
Do 12	0 40.8	+60 51	Cas	18									Asterism?
NGC 225	0 43.4	+61 47	Cas	12	630	III 1 p n	15	7.0	9.26	B8	140		
King 16	0 43.7	+64 11	Cas	3.0	2300	II 3 p	35	10.3*	12.49	OB			Asterism?
NGC 188	0 44.4	+85 20	Cep	14	1550	II 2 r	120	8.1	12.09	F2	5000	−49	
Berk 4	0 45.5	+64 24	Cas	5		I 2 p	25	10.6*	12.57				Asterism?
Do 13	0 50.0	+64 08	Cas	12		III 1 p	30						
NGC 281	0 52.8	+56 37	Cas	4.0		n		7.4p	9p	OB			
NGC 330	0 56.2	−72 29	Tuc	1.9				9.6					SMC
NGC 346	0 59.1	−72 11	Tuc	5.2		n		10.3					SMC
Berk 62	1 01.0	+63 57	Cas	10		III 2 p	50	9.3	10.91				
NGC 371	1 03.3	−72 05	Tuc	7.5		n							SMC, neb. Henize N76
NGC 366	1 06.4	+62 14	Cas	3.0		II 3 p	30		10p	B0			
NGC 381	1 08.3	+61 35	Cas	6		III 2 p	50	9.3p	10p	A2			In Cas OB1?
Stock 3	1 12.3	+62 20	Cas	2.0		IV 1 p	8		11p	OB			
NGC 433	1 15.3	+60 08	Cas	2.5		III 2 p	15		9p	OB			
NGC 436	1 15.6	+58 49	Cas	6	2200	I 3 m	30	8.8*	11.12	B5	79		
NGC 457	1 19.1	+58 20	Cas	13	2800	I 3 r	80	6.4	8.59	B2	25		ø Cas is a member, _not incl. in mag._
NGC 559	1 29.5	+63 18	Cas	4.4	900	II 2 m	60	9.5*	10.58	B7	1300		
NGC 581	1 33.2	+60 42	Cas	6	2600	III 2 p	25	7.4*	10.55	B2	22	−37	M103, in Cas OB8
Tr 1	1 35.7	+61 17	Cas	4.5	2200	I 3 p	20	8.1*	9.55	B2	26	−65	
NGC 609	1 37.2	+64 33	Cas	3.0	3100	II 3 r		11.0	14.43		200		
NGC 637	1 42.9	+64 00	Cas	3.5	2100	I 3 p	20	8.2*	9.97	B0		−48	
NGC 654	1 44.1	+61 53	Cas	5	2500:	II 3 m	60	6.5	7.36	B0	15	−31	In Cas OB8?
NGC 659	1 44.2	+60 42	Cas	5	2100	III 1 p	40	7.9*	10.43	B5	20		
NGC 663	1 46.0	+61 15	Cas	16	2200	III 2 m	80	7.1	8.42	B1	22	−32	In Cas OB8
Cr 463	1 48.4	+71 57	Cas	36	600	III 2 p	40	5.7	8.49		150		
Cr 21	1 50.1	+27 15	Tri	6		IV 2 p	20	8.2p					Asterism?
IC 166	1 52.5	+61 50	Cas	4.5	3300	III 1 r	120	11.7	14.88		1600		
Stock 4	1 52.8	+57 04	Per	20		III 1 p	15		11p	B5			
NGC 752	1 57.8	+37 41	And	50	400	III 1 m	60	5.7	8.96	A2	1100	−4	
NGC 744	1 58.4	+55 29	Per	11	1500	IV 2 p	20	7.9	10.44	B9	39		
NGC 743	1 58.7	+60 11	Cas	5		II 1 p	12		10p				
Stock 5	2 04.5	+64 26	Cas	15		IV 2 p	25		7p	OB			
Stock 2	2 15.0	+59 16	Cas	60	320	III 1 m	50	4.4	8.18	B8	100		
Basel 10	2 18.8	+58 19	Per		2600		12	9.9	10.23				
NGC 869	2 19.0	+57 09	Per	30	2200	I 3 r	200	4.3p	6.55	B0	5.6	−22	h Per, in Per OB1
NGC 884	2 22.4	+57 07	Per	30	2300	I 3 r	150	4.4p	8.05	B0	3.2	−21	χ Per, in Per OB1
Stock 6	2 23.7	+63 52	Cas	20		IV 2 p	20		11p	A2			
Mrk 6	2 29.6	+60 39	Cas	4.5	600	IV 2 p	6	7.1	8.44	B2			
NGC 956	2 32.4	+44 39	And	8		IV 1 p	30	8.9p	9p				
IC 1805	2 32.7	+61 27	Cas	22	2100	III 3 p n	40	6.5	7.87	O5	1.3	−34	In Cas OB6
Czernik 8	2 33.0	+58 44	Per	7	2400	II 3 m	10	9.7	9.87	B5			
NGC 957	2 33.6	+57 32	Per	11	2200	III 2 p	30	7.6	9.49	B1	15	−35	
King 4	2 35.7	+59 00	Cas	3.0	2200	III 1 p	20	10.5	12.85				
Tr 2	2 37.3	+55 59	Per	20	600	III 2 p	20	5.9	7.38	B9	78		
Berk 65	2 39.0	+60 25	Cas	5	3300	I 2 p	20	10.2	10.79				
NGC 1039	2 42.0	+42 47	Per	35	440	II 3 m	60	5.2	7.33	B8	190	−10	M34
NGC 1027	2 42.7	+61 33	Cas	20	1000	III 2 p n	40	6.7	9.33	B3	350		
Czernik 13	2 44.7	+62 21	Cas	6		II 2 p	10	10.4	12.84				Asterism?
DoDz 1	2 47.4	+17 12	Ari	12		III 2 p	12			A0			
IC 1848	2 51.2	+60 26	Cas	12	2200	IV 3 p n	10	6.5	7.10	O7	1.0	−17	
Cr 33	2 59.3	+60 24	Cas	40		n	25	5.9p					
Cr 34	3 00.9	+60 25	Cas	25		I 3 p		6.8p					
NGC 1193	3 05.8	+44 23	Per	1.5		II 3 m	40	12.6p	14p				
NGC 1220	3 11.7	+53 20	Per	1.6		II 2 p	15	11.8p	13p				
Tr 3	3 11.8	+63 15	Cas	23		III 3 p	30	7.0p		B5			
NGC 1245	3 14.7	+47 15	Per	10	2300	III 1 r	200	8.4	11.16	B9	1100		
Stock 23	3 16.3	+60 02	Cam	15		III 3 p n	25			B6			
Mel 20	3 22:	+49:	Per	185	170	III 3 m	50	1.2	2.88	B2	51	−2	α Per moving cluster _(mag. does not in-clude α Per)_
King 6	3 28.1	+56 27	Cam	7		IV 2 p	35		10p	B8			
NGC 1342	3 31.6	+37 20	Per	14	550	III 3 p	40	6.7	8.75	A2	300		
Berk 10	3 39.4	+66 32	Cam	12		II 3 p	50		14p				
IC 348	3 44.5	+32 17	Per	7	390	IV 2 p n	20	7.3	8.53	B6	130	+25	

Name	α 2000	δ 2000	Const	Diam	d(pc)	Tr Type	N*	Mag	Br*	Spec	Age	RV	Notes
Pleiades	3^{h}47^m.0	+24°07'	Tau	110'	125	I 3 r n	100	1.2	2.87	B5	78	+5	M45, NGC 1432 and 1435
Tombaugh 5	3 47.8	+59 03	Cam	17	1800	III 2 m	60	8.4	11.62				
NGC 1444	3 49.4	+52 40	Per	4.0	1000	IV 1 p		6.6	6.77	B0	160	+2	In Cam OB1?
NGC 1496	4 04.4	+52 37	Per	6		II 1 p	10	9.6p	12p				
Do 14	4 06.6	+27 26	Tau	12		IV 2 p	18						
NGC 1502	4 07.7	+62 20	Cam	8	950	II 3 p	45	5.7	6.93	B0	20	−16	In Cam OB1?
NGC 1513	4 10.0	+49 31	Per	9	820	II 1 m	50	8.4*	11.19	B8	430		
NGC 1528	4 15.4	+51 14	Per	24	800	II 2 m	40	6.4	8.75	A0	270		
IC 361	4 19.0	+58 18	Cam	6		II 1 r	60	11.7:	14.55				
Berk 11	4 20.6	+44 55	Per	6	2200	II 3 m	35	10.4:	11.75		30		
NGC 1545	4 20.9	+50 15	Per	18	800	II 2 p	20	6.2	7.13	B8	190		
Hyades	4 27:	+16:	Tau	330	46	II 3 m		0.5	3.40	A2	660	+43	Taurus mov. cl. (Aldeb-
NGC 1582	4 32.0	+43 51	Per	37		IV 2 p	20	7.0p	9p	B8			aran not a member)
NGC 1605	4 35.0	+45 15	Per	5	2800	III 1 m	40	10.7*	12.52	B3			
NGC 1624	4 40.4	+50 27	Per	1.9		I 2 p n	12	10.4	11.77	O			
Berk 68	4 44.5	+42 04	Per	12	3200	IV 2 p	60	9.8*	13.68	B8			
NGC 1647	4 46.0	+19 04	Tau	45	550	II 2 m	200	6.4	8.61	B8	210	−5	
Ru 148	4 46.5	+44 44	Per	3.0		IV 2 p	15	9.5*	9.91				
NGC 1662	4 48.5	+10 56	Ori	20	400	I 2 p	35	6.4	8.34	A0	300		
NGC 1664	4 51.1	+43 42	Aur	18	1200	III 1 p		7.6	10.56	B8	300	+35	
NGC 1755	4 55.0	−68 11	Dor	2.0				9.93					LMC
Czernik 19	4 57.0	+28 47	Aur	18		III 2 m	50						
NGC 1746	5 03.6	+23 49	Tau	42	420	III 1 p	20	6.1p	8p	B5			
NGC 1750	5 03.9	+23 39	Tau						8p				Part of NGC 1746
NGC 1818	5 04.2	−66 24	Dor	2.5				9.85					LMC
Do 15	5 04.6	+34 50	Aur	18		IV 1 p							Asterism?
NGC 1778	5 08.1	+37 03	Aur	7	1350	III 2 p	25	7.7	10.06	B6	160		
NGC 1850	5 08.5	−68 46	Dor	3.4		n		9.36					LMC, neb. Henize N103
NGC 1807	5 10.7	+16 32	Tau	17		II 2 p	20	7.0	8.60	A2			Asterism?
NGC 1817	5 12.1	+16 42	Tau	16	1750	III 1 m	60	7.7	11.17	A0	790		
NGC 1866	5 13.5	−65 28	Dor	5.1				9.89					LMC
Do 16	5 14.6	+32 43	Aur	12		III 2 p n	10						
NGC 1910	5 18.1	−69 13	Dor	0.9									LMC, with S Dor
Czernik 20	5 20.1	+39 28	Aur	18		II 2 r	12						
NGC 1857	5 20.2	+39 21	Aur	6	1900	II 2 m	40	7.0	7.38	A			
Berk 17	5 20.6	+30 36	Aur	14		III 1 r	100		16p				
Berk 18	5 22.2	+45 24	Aur	20		II 1 r	300		16p				
Do 17	5 22.4	+7 07	Ori	12		IV 2 p							Asterism?
Cr 62	5 22.5	+41 00	Aur	28		IV 3 p		4.2p					
IC 410	5 22.6	+33 31	Aur	6		I 2 p n			9p				In Aur OB2
NGC 1893	5 22.7	+33 24	Aur	11	4000	II 2 m n	60	7.5	9.31		1.0	−10	In Aur OB2
King 22	5 22.9	+45 28	Aur	14		III 3 r			15p				
Cr 464	5 22:	+73:	Cam	120		IV 3 p	50	4.2p					
Do 19	5 23.7	+8 11	Ori	24		IV 1 p							
DoDz 2	5 23.9	+11 28	Ori	12			12						
Berk 19	5 24.1	+29 36	Aur	7	4000	IV 2 p	40	11.4	14.70		3000		
Do 18	5 24.1	+33 18	Aur	12		IV 2 p n	15						
Berk 70	5 25.7	+41 54	Aur	12		IV 3 m	40		15p				Asterism?
NGC 1883	5 25.9	+46 33	Aur	2.5		II 1 p	30	12.0p	14p				
Cr 65	5 26:	+16:	Ori	220		II 3 p		3.0p		B3			
Do 21	5 27.4	+7 04	Ori	12		IV 2 p							
Stock 8	5 27.6	+34 25	Aur	5		I 2 p n	40		9p	OB			In nebula IC 417
NGC 1907	5 28.0	+35 19	Aur	7	1380	II 1 m n	30	8.2	11.26	B3	440		
Do 20	5 28.6	+33 47	Aur	12		IV 1 p	10						Asterism?
NGC 1912	5 28.7	+35 50	Aur	21	1320	III 2 m	100	6.4	9.53	B4	220		M38, in Aur OB1
NGC 2004	5 30.6	−67 17	Dor					9.86					LMC
NGC 1931	5 31.4	+34 15	Aur	1.0		n		11.3:	11.49				Contains the triple ADS
DoDz 3	5 33.7	+26 29	Tau	15		IV 2 p	10			B8			4112. In Aur OB1?
Cr 69	5 35.1	+9 56	Ori	65	500	II 3 p n	20	2.8p		O		+30	λ Ori cluster
NGC 1981	5 35.2	−4 26	Ori	25	400	III 2 p n	20	4.6	6.26	B1			In Ori OB1
Trapezium	5 35.4	−5 23	Ori	48	450	n		3.7		O6			In Orion nebula M42
DoDz 4	5 35.9	+25 57	Tau	28		IV 1 p	15						and Ori OB1
NGC 1960	5 36.1	+34 08	Aur	12	1270	II 3 m	60	6.0	8.86	B2	25	−4	M36, in Aur OB1
Cr 70	5 36:	−1:	Ori	150	430	II 3 m n	100	0.4	1.69	B0			Orion's Belt stars
NGC 2070	5 38.6	−69 05	Dor	5.0		n		8.27					30 Dor cluster in LMC
Stock 10	5 39.0	+37 56	Aur	25		IV 3 p	15			B7			
NGC 2100	5 42.0	−69 14	Dor	2.3				9.60					LMC
Basel 4	5 48.5	+30 13	Aur	8	5900	II 1 p	15	9.1*	12.18	B2			
King 8	5 49.4	+33 38	Aur	8	4150	I 3 m	30	11.2	13.48	B0	1300		
Berk 21	5 51.7	+21 47	Ori	7	12000	II 3 m	40	11.1	14.80		30		
NGC 2099	5 52.4	+32 33	Aur	24	1350	II 1 r	150	5.6	9.21	B9	300		M37
NGC 2112	5 53.9	+0 24	Ori	11		II 3 m n	50	9.1p	10p				
Basel 11B	5 58.2	+21 58	Ori	10	1500	II 2 m	12	8.9*	11.48				
NGC 2129	6 01.0	+23 18	Gem	7	2000	III 3 p	40	6.7	7.36	B1	16	+18	
NGC 2126	6 03.0	+49 54	Aur	6	1400	II 1 p	40	10.2p	13p				

Name	α 2000	δ 2000	Const	Diam	d(pc)	Tr Type	N*	Mag	Br*	Spec	Age	RV	Notes
NGC 2141	6^h03^m.1	+10°26'	Ori	10'	4400	II 3 r	100	9.4	13.33		4000		
IC 2157	6 05.0	+24 00	Gem	7	2000	III 2 p	20	8.4*	11.08	OB			
NGC 2158	6 07.5	+24 06	Gem	5	4900:	II 3 r		8.6	12.40	F0	3200		
NGC 2169	6 08.4	+13 57	Ori	7	1100	I 3 p n	30	5.9	6.94	B1	50	+16	In neb. Ced 78
NGC 2168	6 08.9	+24 20	Gem	28	870	III 2 m	200	5.1	8.18	B3	110	−5	M35
NGC 2175	6 09.8	+20 19	Ori	18	1950	IV 3 p n	60	6.8	7.55	O6			In Gem OB1?
NGC 2175S	6 10.9	+20 36	Ori			n	20						Small group nf NGC 2175
NGC 2186	6 12.2	+5 27	Ori	4.0	1800	II 2 p	30	8.7	9.82				
NGC 2194	6 13.8	+12 48	Ori	10	1600	III 1 r	80	8.5*	12.07				
NGC 2192	6 15.2	+39 51	Aur	6		III 1 p	45	10.9p	14p				
NGC 2204	6 15.7	−18 39	CMa	13	4450	III 3 m	80	8.6	12.20		3000		
Cr 89	6 18.0	+23 38	Gem	35	1300	IV 2 p n	15	5.7p		OB		+18	
NGC 2215	6 21.0	−7 17	Mon	11	1000	II 2 p	40	8.4	10.52	B9	350		
Cr 91	6 21.7	+2 22	Mon	17		IV 2 p	20	6.4p					Asterism?
Cr 92	6 22.9	+5 07	Mon	11				8.6p					Asterism?
Do 22	6 23.3	+4 39	Mon	18		IV 1 p	10						Asterism?
Bochum 1	6 25.5	+19 46	Gem					7.9	8.42				
NGC 2232	6 26.6	−4 45	Mon	30	400	IV 3 p	20	3.9	5.03	B3	22	+15	
NGC 2236	6 29.7	+6 50	Mon	7	3400	III 2 p	50	8.5*	10.97	A0			
NGC 2243	6 29.8	−31 17	CMa	5	4600	I 2 r	100	9.4	11.79		3900		
Cr 96	6 30.3	+2 52	Mon	8	1100	IV 2 p	15	7.3	8.75				
Cr 95	6 30.5	+9 56	Mon	19		IV 2 p n	10						
NGC 2239	6 31.0	+4 57	Mon	16		II 3 p n	40						
Cr 97	6 31.3	+5 55	Mon	21		IV 3 p	15	5.4p		B0			
NGC 2244	6 32.4	+4 52	Mon	24	1700	II 3 p n	100	4.8	5.84	O5	3.0	+34	In Rosette nebula
NGC 2250	6 32.8	−5 02	Mon	8		IV 2 p	10	8.9p	12p				
NGC 2251	6 34.7	+8 22	Mon	10	1550	III 2 p	30	7.3	9.10	B3	300	+8	
NGC 2252	6 35.0	+5 23	Mon	20		IV 2 p n	30	7.7p	9p	A			
NGC 2254	6 36.0	+7 40	Mon	4.0	2200	I 2 p	50	9.7	11.85	B9			
Ru 1	6 36.4	−14 11	CMa	11		III 1 p	15		11p				
Cr 104	6 36.5	+4 49	Mon	22		IV 1 p n	15	9.6p					
Basel 7	6 36.6	+8 21	Mon				15	8.9	10.35				
Tr 5	6 36.7	+9 26	Mon	8	2400	II 3 r n	150	10.9p	17p				
Cr 106	6 37.1	+5 57	Mon	45		III 3 p	20	4.6p		B0			
Cr 107	6 37.7	+4 44	Mon	35	1700	IV 3 p	15	5.1	7.14	OB		+21	Asterism?
NGC 2262	6 38.4	+1 11	Mon	3.5		I 2 p	35	11.3p					
Cr 110	6 38.4	+2 01	Mon	12		III 1 m	70	10.5p					
NGC 2259	6 38.6	+10 53	Mon	4.5		II 2 p n	25	10.8p	14p				
Cr 111	6 38.7	+6 54	Mon	3.2				7.0p					Asterism?
Ru 2	6 41.0	−29 33	CMa	7		IV 3 p	10		12p				
NGC 2264	6 41.1	+9 53	Mon	20	750	IV 3 p n	40	3.9	5p	O7	20	+21	Near Cone nebula, in
Ru 3	6 42.1	−29 27	CMa	2.8		II 1 p	15		11p				| Mon OB1
Do 23	6 43.2	0 00	Mon	12		IV 2 p	20						
NGC 2266	6 43.2	+26 58	Gem	7	3400	II 2 m	50	9.5p	11p				
NGC 2269	6 43.9	+4 34	Mon	4.0	1400	II 2 p	12	10.0:	11.61				
Do 24	6 44.2	+1 36	Mon	18		III 1 p	40						
Do 25	6 45.1	+0 18	Mon	24	5200	IV 2 p n	50	7.6	8.87				
Cr 115	6 46.5	+1 46	Mon	7		III 2 p	50	9.2p					
NGC 2287	6 47.0	−20 44	CMa	38	700	II 3 m	80	4.5	6.91	B3	190	+34	M41
NGC 2286	6 47.6	−3 10	Mon	15	1300	IV 3 m	50	7.5	9.71	B4			Asterism?
Bochum 2	6 48.9	+0 23	Mon				10	9.7	10.86				
NGC 2281	6 49.3	+41 04	Aur	15	500	I 3 p	30	5.4	7.30	A0	300	+21	
NGC 2301	6 51.8	+0 28	Mon	12	750	I 3 m	80	6.0	8.01	B8	110		
NGC 2302	6 51.9	−7 04	Mon	2.5	1100	II 2 p	30	8.9	10.24				
Biur 10	6 52.2	+2 56	Mon	4.0	5000	I 3 p	20	10.4:	10.69	O			Asterism?
Cr 121	6 54.2	−24 38	CMa	50	700	III 3 p	20	2.6	3.79	B3	13	+33	o CMa
NGC 2304	6 55.0	+18 01	Gem	5		II 1 p	30	10.0p					
NGC 2309	6 56.2	−7 12	Mon	3.0		II 2 m	40	10.5p	13p				
NGC 2311	6 57.8	−4 35	Mon	7		III 2 p	50	9.6p	12p				
Biur 8	6 58.1	+6 26	Mon	6	9000	II 2 m	70		14p	B8			
NGC 2323	7 03.2	−8 20	Mon	16	910	II 3 m	80	5.9	7.85	B6	78		M50
Bochum 3	7 03.4	−5 04	Mon				25	9.9	11.19				
NGC 2324	7 04.2	+1 03	Mon	8	2900	II 2 r	70	8.4	10.35	B9	660		
NGC 2335	7 06.6	−10 05	Mon	12	1000	III 3 m n	35	7.2	9.50	B9	160		In CMa OB1
NGC 2331	7 07.2	+27 21	Gem	18		IV 1 p	30	8.5p	9p				
Cr 465	7 07.2	−10 37	Mon	9		IV 2 p		10.1p					Asterism?
Cr 466	7 07.3	−10 49	Mon	4.0		III 2 p n	25	11.1p					
Ru 11	7 07.4	−20 48	CMa	2.9		III 2 p	20		11p				
NGC 2343	7 08.3	−10 39	Mon	7	1000	III 3 p n	20	6.7	8.39	A0	100		In CMa OB1?
NGC 2345	7 08.3	−13 10	CMa	12	1800	I 3 m	70	7.7	9.87	A2	79		
Haffner 23	7 09.4	−16 57	CMa	11		III 2 m	40		13p				
NGC 2354	7 14.3	−25 44	CMa	20	1850	III 2 m	100	6.5	9.13	A0	180		
Cr 132	7 14.4	−31 10	CMa	95		III 3 p	25	3.6	5.33	B3			
NGC 2353	7 14.6	−10 18	Mon	20	1100	II 2 p	30	7.1	9.19	B0	13	+33	In CMa OB1
NGC 2355	7 16.9	+13 47	Gem	9		II 2 p	40	9.7p	13p				

OPEN CLUSTERS

Name	α 2000	δ 2000	Const	Diam	d(pc)	Tr Type	N*	Mag	Br*	Spec	Age	RV	Notes
Cr 135	7^{h}17^m.0	−36°50'	Pup	50'		IV 2 p		2.1	2.71	B3			
Basel 11A	7 17.1	−13 58	CMa	9	1500		30	8.2	10.90		200		
NGC 2360	7 17.8	−15 37	CMa	13	1630	II 2 m	80	7.2	10.36	B8	1300		
NGC 2362	7 18.8	−24 57	CMa	8	1550	I 3 p n	60	4.1	4.39	O8	25	+33	
Haffner 6	7 20.1	−13 08	CMa	4.0	1100	IV 3 p n	60	9.2	11.11		790		
NGC 2367	7 20.1	−21 56	CMa	3.5	2000	IV 3 p	30	7.9	9.39	B3			
NGC 2368	7 21.0	−10 23	Mon	5		IV 2 p	15	11.8p					
Ru 16	7 23.2	−19 27	CMa	11		IV 2 p	15		13p				Asterism?
Haffner 8	7 23.4	−12 20	CMa	4.2	1700	IV 3 m	30	9.1	11.09		500		
Cr 140	7 23.9	−32 12	CMa	42	300	III 3 p	30	3.5	5.36	B3	22	+18	
NGC 2374	7 24.0	−13 16	CMa	19	1300	II 3 p	25	8.0	10.74		320		
NGC 2383	7 24.8	−20 56	CMa	6	2000	I 3 m	40	8.4:	9.76	A3			Near cluster NGC 2384
Ru 18	7 24.8	−26 13	CMa	4.0		III 2 p	40	9.4	10.98				
NGC 2384	7 25.1	−21 02	CMa	2.5	2000	IV 3 p	15	7.4:	8.58	B0		+38	Near cluster NGC 2383
Tr 6	7 26.1	−24 18	CMa	6		III 2 p		10.0p					
Mel 66	7 26.3	−47 44	Pup	10	2500	II 2 m	200	7.8	11.41	A0	6300		
Ru 20	7 26.7	−28 53	CMa	10		III 2 m	30	9.5	11.59				Asterism?
NGC 2395	7 27.1	+13 35	Gem	12	1200	III 1 p	30	8.0	9.96	B4	50		Asterism?
Ru 21	7 27.1	−31 11	Pup	11		III 1 m			11p				
Tr 7	7 27.3	−24 02	Pup	5	1600	II 3 p n	30	7.9	9.13				
NGC 2396	7 28.1	−11 44	Pup	10		III 3 p	30	7.4p	11p				
Czernik 29	7 28.3	−15 24	Pup	8		IV 2 p	40	10.3	12.22				
Haffner 10	7 28.6	−15 23	Pup	1.6		III 2 p	40	11.5:	12.68				Asterism?
NGC 2401	7 29.4	−13 58	Pup	2.0		II 3 p	20	12.6p					
Do 26	7 30.1	+11 54	CMi	24		IV 1 p							Asterism?
Bochum 5	7 30.9	−17 04	Pup				12	7.0	7.78				
Bochum 4	7 31.0	−16 57	Pup			n	30	7.3	8.04				
Ru 24	7 31.9	−12 45	Pup	2.0		III 1 p	15		11p				
Bochum 6	7 32.0	−19 26	Pup			n	40	9.9	10.65				
NGC 2414	7 33.3	−15 27	Pup	4.0	2500	I 3 m	35	7.9	8.21				
NGC 2421	7 36.3	−20 37	Pup	10	1900	I 2 m	70	8.3	10.45				
NGC 2422	7 36.6	−14 30	Pup	30	480	III 2 m	30	4.4	5.68	B3	78	+9	M 47
NGC 2423	7 37.1	−13 52	Pup	19	870	IV 2 m	40	6.7	9.02	B5	350		
Mel 71	7 37.5	−12 04	Pup	9	2800	II 1 m	80	7.1	10.18		420		
Ru 27	7 37.5	−26 36	Pup	18		II 2 m	30		12p				
NGC 2425	7 38.3	−14 52	Pup	3.3		III 2 p	30		14p				
Mel 72	7 38.4	−10 41	Mon	9		II 1 p	40	10.1p					
NGC 2420	7 38.5	+21 34	Gem	10	2500	I 2 r	100	8.3	11.06	F	4000	+115	
Bochum 15	7 40.1	−33 33	Pup			n	12	6.3	7.65				
Haffner 13	7 40.5	−30 07	Pup	15		II 3 p	15		8p	B9			
NGC 2439	7 40.8	−31 39	Pup	10	1610	II 3 m	80	6.9	8.90	B2	66	+72	Asterism due to uneven | absorption?
NGC 2432	7 40.9	−19 05	Pup	8		II 1 p	50	10.2p					
Ru 151	7 41.3	−16 15	Pup	15		IV 3 m	30		12p				
NGC 2437	7 41.8	−14 49	Pup	27	1660	III 2 m	100	6.1	8.68	B9	300	+42	M46, containing planet- | ary nebula NGC 2438
Ru 30	7 42.2	−31 28	Pup	4.0		II 2 p	30		11p				
Ru 31	7 42.7	−35 35	Pup	2.0		I 3 p	15		11p				
NGC 2447	7 44.6	−23 52	Pup	22	1100	IV 1 p	80	6.2:	8.20	B9	98		M93
Ru 32	7 45.0	−25 31	Pup	6	4100	III 2 p n	30	8.4	9.55				
Haffner 15	7 45.3	−32 47	Pup	3.5	2500	II 3 p	35	9.4	10.45				
NGC 2451	7 45.4	−37 58	Pup	45	260	II 2 p	40	2.8	3.60	B7	36	+26	
Ru 34	7 45.9	−20 23	Pup	4.0		III 2 p	35	9.5	11.19				Asterism?
Berk 39	7 46.7	−4 36	Mon	12		II 3 r	120		16p				
NGC 2453	7 47.8	−27 14	Pup	5	1500	I 2 p	30	8.3	9.46	OB	200		
Ru 36	7 48.5	−26 18	Pup	4.0		I 1 p	30	9.6	10.28				
NGC 2455	7 49.0	−21 18	Pup	8		III 2 p	50	10.2p	12p				
Haffner 16	7 50.3	−25 27	Pup	1.1	3900:	I 1 p	30	10.0:	11.62				
NGC 2477	7 52.3	−38 33	Pup	27	1300	I 3 r	160	5.8	9.81	B8	710		
Haffner 18	7 52.5	−26 22	Pup	1.0		I 3 p n	15	9.3	9.91				
NGC 2467	7 52.6	−26 23	Pup	16	3400:	n	50	7.1p		O5	250		In Pup OB1
Haffner 19	7 52.7	−26 15	Pup	1.8		I 3 p n	30	9.4	10.84				
NGC 2482	7 54.9	−24 18	Pup	12	800	III 1 m	40	7.3	10.04				
NGC 2479	7 55.1	−17 43	Pup	7		III 1 m	45	9.6p					
Tr 9	7 55.3	−25 56	Pup	6	2200	II 2 p	20	8.7	10.09				
NGC 2483	7 55.9	−27 56	Pup	10	2900		30	7.6	9.26		10		Asterism?
NGC 2489	7 56.2	−30 04	Pup	8	1200	II 2 m	45	7.9	11.11	B8	240		
Haffner 20	7 56.3	−30 24	Pup	1.8		II 3 p	20	11.0	13.13				
NGC 2516	7 58.3	−60 52	Car	30	400	I 3 r	80	3.8	7p	B3	110	+21	
Ru 43	7 58.7	−28 55	Pup	14		III 2 p	25		12p				
Ru 44	7 59.0	−28 35	Pup	5	6600	III 1 p	40	7.2	9.38				
Ru 45	7 59.6	−16 18	Pup	11		IV 2 p	35		13p				Asterism?
NGC 2506	8 00.2	−10 47	Mon	7	2200	I 2 r	150	7.6	10.76	A	4000		
NGC 2509	8 00.7	−19 04	Pup	8		II 1 p	70	9.3p					
Haffner 21	8 01.2	−27 10	Pup	1.1		II 1 p	20	10.3	12.14				
Ru 46	8 02.1	−19 28	Pup	2.0		II 3 p	15	9.1	9.43				
Ru 47	8 02.3	−31 06	Pup	5		I 1 p	20	9.6:	10.90				

Name	α 2000	δ 2000	Const	Diam	d(pc)	Tr Type	N*	Mag	Br*	Spec	Age	RV	Notes
Ru 49	8^{h}03^m.1	-26°47'	Pup	2.5		II 3 p	10	9.6	10.11				
Cr 173	8 04:	-46:	Pup	370		IV 2 p		0.6p				+36	
NGC 2527	8 05.3	-28 10	Pup	22	600	III 1 p	40	6.5	8.59		1000		
NGC 2533	8 07.0	-29 54	Pup	3.5	1700	III 1 p	60	7.6	8.99	B8	180		
NGC 2539	8 10.7	-12 50	Pup	22	1280	II 1 m	50	6.5	9.15	A0	660		
NGC 2547	8 10.7	-49 16	Vel	20	400	II 2 p n	80	4.7	6.47	B3	74	+16	
Ru 53	8 10.8	-27 01	Pup	18		IV 2 p	40		10p				
Ru 55	8 12.3	-32 36	Pup	17	4400	IV 2 p	12	7.8	8.56				Asterism?
NGC 2546	8 12.4	-37 38	Pup	41	1000	III 2 m	40	6.3	8.22	B	42		
Ru 56	8 12.6	-40 28	Pup	42		IV 2 p	40		9p				Asterism?
NGC 2548	8 13.8	-5 48	Hya	54	610	I 2 m	80	5.8	8.23	A0	300	-7	M48
vdB-Ha 23	8 14.4	-36 24	Pup	13			15						
Pismis 1	8 17.8	-37 05	Pup	4.6	2400	I 3 p	30	10.7	12.76	B7	85		
NGC 2567	8 18.6	-30 38	Pup	10	1700	III 2 m	40	7.4	10.10	B8	68		
NGC 2571	8 18.9	-29 44	Pup	13	2100	IV 1 p	30	7.0	8.82	B8	22		
Ru 59	8 19.1	-34 27	Pup	5		III 1 p	20	9.0	10.17				Asterism?
NGC 2579	8 21.1	-36 11	Pup	10	1000	IV 1 p n	20	7.5	9.51	B5	13		Asterism?
NGC 2580	8 21.6	-30 19	Pup	8		II 2 m	50	9.7p					
Cr 185	8 22.5	-36 10	Pup	9	1500	III 2 p	35	7.8	10.08		79		
NGC 2588	8 23.2	-32 59	Pup	2.0		II 1 p	20	11.8p					
NGC 2587	8 23.5	-29 30	Pup	9		II 1 p	40	9.2p					
Cr 187	8 24.2	-29 09	Pup	7		III 1 p	20	9.6p					
Ru 157	8 29.8	-19 06	Pyx	17		IV 2 p	30		11p				
vdB-Ha 34	8 31.3	-44 29	Vel	13		n	20						
Ru 62	8 32.5	-19 39	Pyx	7		IV 2 p	20		11p				Asterism?
Pismis 4	8 34.5	-44 16	Vel	18	600	IV 1 p n	45	5.9	7.32				
NGC 2627	8 37.3	-29 57	Pyx	11		III 2 m	60	8.4p	11p				
Ru 64	8 37.4	-40 06	Vel	67		II 3 p	80		9p				
Pismis 5	8 37.5	-39 40	Vel	2.0		III 2 p n	10	9.9	10.45				Asterism?
NGC 2635	8 38.5	-34 46	Pyx	3.0		I 3 p	15	11.2	12.45				Asterism?
Ru 65	8 39.1	-44 02	Vel	11		IV 1 p	20		13p				Asterism?
Pismis 6	8 39.3	-46 13	Vel	1.5	1600	II 2 p	15	7.2	8.85		32		
NGC 2632	8 40.1	+19 59	Cnc	95	160	II 2 m	50	3.1	6.30	A0	660	+33	M44. Praesepe, Beehive
IC 2391	8 40.2	-53 04	Vel	50	180	II 3 p	30	2.5	3.63	B3	36	+14	o Vel cluster
Waterloo 6	8 40.4	-46 09	Vel	2.2	1900		8	8.4	9.18		32		
IC 2395	8 41.1	-48 12	Vel	8	850	II 3 p	40	4.6	5.53	B5	16		
Pismis 8	8 41.5	-46 17	Vel	2.0	1400	II 3 p	25	9.5	10.54		32		
Ru 67	8 41.8	-43 23	Vel	6		IV 3 p	35	9.1	10.67				Asterism?
NGC 2660	8 42.2	-47 09	Vel	4.0	2100	I 3 m	70	8.8	11.68				
NGC 2659	8 42.6	-44 57	Vel	2.7		III 3 m	80	8.6:	9.69	B9			In Vel OB1?
vdB-Ha 47	8 42.6	-48 07	Vel	13									
NGC 2658	8 43.4	-32 39	Pyx	12		II 2 m	80	9.2p	12p	F8			
Cr 197	8 44.7	-41 22	Vel	17	1000	IV 2 p n	40	6.7	7.35				
Bochum 7	8 44.8	-45 59	Vel				9	6.8	7.55				
NGC 2669	8 44.9	-52 58	Vel	12	1000	II 3 p	40	6.1	7.64	B9	63		
Cr 196	8 45.0	-31 38	Pyx	5		IV 3 p	12	10.5p					Asterism?
Cr 198	8 45.3	-31 46	Pyx	6				11.2p					Asterism?
NGC 2670	8 45.5	-48 47	Vel	9	1000	II 2 p	30	7.8	9.31	B5	95		
NGC 2671	8 46.2	-41 53	Vel	4.0		I 3 p	40	11.6p		B9			
Tr 10	8 47.8	-42 29	Vel	15	420	II 2 p	40	4.6	6.42	B3	47		
Ru 71	8 49.5	-46 48	Vel	7		III 2 p	30		11p				
NGC 2682	8 50.4	+11 49	Cnc	30	800	II 2 m	200	6.9	9.69	B8	3200	+33	M67
vdB-Ha 56	8 57.3	-43 13	Vel	12		n	35						
Mrk 18	9 00.6	-48 59	Vel	2.0	1600	I 3 p	30	7.8	9.32	O			
Pismis 10	9 02.6	-43 38	Vel	2.5					10p				Asterism?
NGC 2818	9 16.0	-36 37	Pyx	9	3200	II 2 m n	40	8.2	11.32		1000		May contain plan. neb.
Pismis 12	9 19.9	-45 08	Vel	4.5		III 2 p	20	9.7	11.56				Asterism?
Pismis 13	9 22.3	-51 08	Vel	2.0		II 2 p	30	10.2	12.11				
Ru 76	9 24.2	-51 44	Vel	6		III 2 p	20	10.8	12.50				
Ru 77	9 27.1	-55 07	Vel	2.0		II 3 p	35	10.4*	12.48				
IC 2488	9 27.6	-56 59	Vel	15		II 2 m	70	7.4p	10p	B8			
NGC 2910	9 30.4	-52 54	Vel	5	1300	I 2 p	30	7.2	9.25	B8	40		
NGC 2925	9 33.7	-53 26	Vel	12	800	III 1 p	40	8.3p		B5	79		
NGC 2972	9 40.3	-50 20	Vel	4.0	1200	I 1 p	25	9.9	11.40				
Ru 79	9 41.0	-53 50	Vel	11	2000	III 3 p	20	9.2	11.47				
Ru 80	9 42.3	-44 02	Vel	12		III 1 p			13p				
Ru 82	9 45.6	-53 59	Vel	3.6		II 3 p	20	8.1*	10.82				
NGC 3033	9 48.8	-56 25	Vel	5	1100	II 3 p	50	8.8	10.59				
Ru 83	9 49.2	-54 34	Vel	3.4		I 3 p	30	9.8*	11.29				
Ru 84	9 49.2	-65 16	Car	3.6		II 1 p	20		11p				
Pismis 16	9 51.1	-53 11	Vel	1.5		I 3 p	12	8.0	8.69				
Cr 213	9 54.7	-50 43	Vel	17		IV 2 p		9.2p					
NGC 3105	10 00.8	-54 46	Vel	2.0		I 3 p	20	9.7	12.36				
Ru 86	10 01.6	-59 28	Car	12		III 2 m			12p				
NGC 3114	10 02.7	-60 07	Car	35	900	II 3 r		4.2	7.31	B9	110		

Name	α 2000	δ 2000	Const	Diam	d(pc)	Tr Type	N*	Mag	Br*	Spec	Age	RV	Notes
Tr 11	10^{h}05^m.0	−61°37′	Car	6′		II 3 m		10.8p					
Tr 12	10 06.4	−60 19	Car	4.0		I 3 p		10.5p					Asterism due to uneven
Ru 161	10 08.8	−61 12	Car	33		III 2 p			11p				absorption?
vdB−Ha 90	10 11.9	−58 05	Car	4.0				10.3	12.09				
Ru 87	10 15.5	−50 43	Vel	2.2		II 3 p	20		11p				
NGC 3228	10 21.8	−51 43	Vel	18	500	I 1 p	15	6.0	7.90	B9	42		
Tr 13	10 23.8	−60 05	Car	5		II 2 p	40	11.3p					
Westr 2	10 23.9	−57 45	Car	1.5	5000	I 3 p n	12	10.5	11.44	O			
NGC 3247	10 25.9	−57 56	Car	7	1400	II 2 p	20	7.6*	10.00		50		
NGC 3255	10 26.5	−60 40	Car	2.0		I 3 m	30	11.0	12.44				Asterism?
IC 2581	10 27.4	−57 38	Car	8	2000:	I 3 m n	25	4.3	4.64	B0	10	−3	In Car OB1?
Cr 223	10 30.5	−59 49	Car	9		II 2 p	35	9.4p					
Ru 90	10 31.0	−58 14	Car	9		III 3 m	15		12p				
NGC 3293	10 35.8	−58 14	Car	6	2600	I 3 r n		4.7	6.52	B0	10	−14	In Car OB1
Bochum 9	10 35.8	−60 08	Car			n	30	6.3	7.87				Asterism?
NGC 3324	10 37.3	−58 38	Car	6	3300	I 3 r n		6.7	8.21		2.2		
vdB−Ha 99	10 37.9	−59 12	Car	15		n	40						
NGC 3330	10 38.6	−54 09	Vel	7	1390	II 2 p	30	7.4	8.78	B8	50		
Mel 101	10 42.1	−65 06	Car	14	2100	II 3 m	50	8.0:	9.73	B6			
Bochum 10	10 42.2	−59 09	Car			n	40	6.2	8.59				
Cr 228	10 43.0	−60 01	Car	15	2600			4.4	6.28				
IC 2602	10 43.2	−64 24	Car	50	150	II 3 m	60	1.9	2.77	B0	36	+24	θ Car. At edge of
Tr 14	10 43.9	−59 34	Car	5	1660:			5.5	6.97	O	10		Sco−Cen association
vdB−Ha 103	10 44.3	−64 20	Car	60									
Tr 15	10 44.8	−59 22	Car	3.0	1500	n	20	7.0	8.36	O	6.0		
Cr 232	10 44.8	−59 34	Car	4.0				6.8p		B			
Tr 16	10 45.1	−59 43	Car	10	2500:	n		5.0	6.18	O	10		η Car nebula. Mag.
Cr 234	10 45.3	−59 45	Car	4.0				7.5p					includes η Car
Bochum 11	10 47.3	−60 06	Car			n	20	7.9	8.40				
Ru 91	10 47.7	−57 29	Car	1.7		II 2 p	15		10p				
Ru 92	10 53.9	−61 44	Car	2.2	2700	I 3 p	15	8.6*	10.92				
Tr 17	10 56.2	−59 13	Car	5	1400	II 2 p	30	8.4	10.30	B1	35		
Cr 236	10 57.0	−61 02	Car	8		III 2 p	20	7.7p		A0			
Bochum 12	10 57.4	−61 44	Car				20	9.7	11.43				
Hogg 9	10 58.4	−59 03	Car	1.5			10	10.6	11.54				Asterism?
NGC 3496	10 59.8	−60 20	Car	9	1100	III 1 m	60	8.2	11.78	B8	230		
Sher 1	11 00.0	−60 22	Car			II 2 p	8	8.8	9.45				
Pismis 17	11 01.1	−59 49	Car	0.6	4200	III 2 p		9.4	10.44				
Ru 93	11 04.4	−61 22	Car	4.0		III 2 p	30	7.7*	11.17				
Fein 1	11 06.0	−59 49	Car			IV 2 p	40	4.7	6.55	OB			
NGC 3532	11 06.4	−58 40	Car	55	410	II 1 m	150	3.0	7.08	B5	270		
Loden 282	11 10.4	−59 02	Car				20	7.7*	9.56				
NGC 3572	11 10.4	−60 14	Car	7	2300	I 2 m n	35	6.6	7.89	B0	13	−10	In Car OB2
Hogg 10	11 10.7	−60 22	Car	3.0	2200	I 3 p		6.9	7.11				In Car OB2
Cr 240	11 11.2	−60 17	Car	25		III 1 p n	30	3.9	4.60	B2			
Tr 18	11 11.4	−60 40	Car	12	2500	III 2 p	30	6.9	8.08	B2	250		In Car OB2
Hogg 11	11 11.5	−60 22	Car	1.5	2300		10	8.1	8.32				In Car OB2
Hogg 12	11 12.3	−60 45	Car	3.0		I 3 p	10	8.8	10.20				Asterism?
NGC 3590	11 12.9	−60 47	Car	4.0	1900	II 1 p	25	8.2	10.27	B3	50		In Car OB4
Stock 13	11 13.1	−58 55	Car	3.0	2700	I 2 p n	15	7.0	8.53	OB			In Car OB3
Tr 19	11 14.3	−57 35	Car	10		III 3 m	40	9.6p					
NGC 3603	11 15.1	−61 15	Car	2.5	3500	I 1 p n	30	9.1:	11.25				
IC 2714	11 17.9	−62 42	Car	12	1200	II 3 m	100	8.2p	10p	B9			
Mel 105	11 19.5	−63 30	Car	4.0	2100	I 2 p	70	8.5	11.13				
NGC 3680	11 25.7	−43 15	Cen	12	800	I 2 p	30	7.6	10.06		1800		
Ru 94	11 30.4	−63 27	Cen	21		II 1 m	60		10p				
NGC 3766	11 36.1	−61 37	Cen	12	1700	I 1 p	100	5.3	7.18	B0	22		
IC 2944	11 36.6	−63 02	Cen	15	2100	II 1 p n	30	4.5	6.38	O8	10	+2	λ Cen cluster, in
Stock 14	11 44.0	−62 30	Cen	4.0	2600	II 1 p	10	6.3	8.36	OB			Cen OB2
NGC 3960	11 50.9	−55 42	Cen	7		I 2 m	45	8.3	11.50				
Ru 97	11 57.3	−62 39	Cru	3.5	4000	II 2 m	20	9.1	12.00		1000		
Ru 98	11 58.0	−64 29	Cru	10	400	II 2 p	50	7.0	8.86				
NGC 4052	12 01.9	−63 12	Cru	8	1900	II 1 p	80	8.8p		B3	140		
Cr 285	12 03:	+58:	UMa	1400	23	p		0.4	2p	A0	160	−10	UMa moving cluster
NGC 4103	12 06.7	−61 15	Cru	7	1200	I 3 m	45	7.4p	10p	OB	22		
Stock 15	12 06.9	−59 30	Cru	12		IV 1 p			10p	B5			
NGC 4230	12 17.3	−55 08	Cen	6		IV 2 p	15	9.4p					
NGC 4337	12 23.9	−58 08	Cru	3.5		II 3 p		8.9:	10.53				Asterism?
NGC 4349	12 24.5	−61 54	Cru	16	1700	I 2 m	30	7.4	10.88	B8	220		
Mel 111	12 25:	+26:	Com	275	80	II 3 p	80	1.8	4.35	A0	400	0	Coma Berenices
NGC 4439	12 28.4	−60 06	Cru	4.0	1600	II 1 p		8.4	10.26	OB	63		
Hogg 14	12 28.6	−59 49	Cru	3.0		II 3 p		9.5	10.49				Asterism?
Ru 165	12 28.7	−56 28	Cru	22		IV 1 p	35		7p				Asterism?
Harvard 5	12 29.0	−60 46	Cru	6		II 3 p		7.1	8.44				
NGC 4463	12 30.0	−64 48	Mus	5		I 3 p	30	7.2	8.28*	B0			

Name	α 2000	δ 2000	Const	Diam	d(pc)	Tr Type	N*	Mag	Br*	Spec	Age	RV	Notes
Ru 105	12^{h}34$.^m$1	−61°34′	Cru	13′		III 2 p			12p				
Upgren 1	12 35.0	+36 18	CVn	15	140	IV 2 p	10			F3			
Harvard 6	12 37.9	−68 28	Mus	5		II 2 r	100	10.7p					
Tr 20	12 39.7	−60 36	Cru	8		II 2 m		10.1p					
NGC 4609	12 42.3	−62 58	Cru	5	1510	II 1 p	40	6.9	9.02	B4	36		Near Coalsack
Hogg 15	12 43.6	−63 06	Cru	2.0	4200	II 1 p	15	10.3	12.10		7.9		
NGC 4755	12 53.6	−60 20	Cru	10	2340	I 3 r		4.2	5.75	B2	7.1	−18	κ Cru cl., Jewel Box
NGC 4815	12 58.0	−64 57	Mus	3.0		I 3 m	100	8.6	9.63	O			Asterism?
NGC 4852	13 00.1	−59 36	Cen	11		II 2 p	60	8.9p					
Harvard 8	13 18.8	−67 12	Mus	4.0		I 2 p	30	9.5	11.57				
Stock 16	13 19.1	−62 34	Cen	3.0	2300	IV 2 p n	20		11.52	OB	28	−40	
Ru 107	13 20.6	−64 57	Mus	5		II 2 p	20	9.7	11.18				
Cr 269	13 22.6	−66 07	Mus	15		IV 2 p		9.2p					Asterism?
Loden 807	13 24.9	−62 26	Cen				35	7.9	9.90				
Ru 166	13 26.1	−63 25	Cen	2.8		III 3 p	30	10.8:	11.29				Asterism?
NGC 5138	13 27.3	−59 01	Cen	8	1400	II 2 p	40	7.6	10.30		320		
Basel 18	13 28.3	−62 22	Cen	4.0	1556		20	8.2*	10.99		160		
Hogg 16	13 29.3	−61 12	Cen	4.0	2100	II 3 p	20	8.4	10.10				
Cr 271	13 29.7	−64 11	Cen	7	1600	III 2 p	20	8.7:	9.50				
Cr 272	13 30.6	−61 16	Cen	9	2900	III 1 m	40	7.7*	10.50		2.0		
NGC 5168	13 31.2	−60 56	Cen	4.0	1400	I 3 p	50	9.1	10.36	OB	390		
Ru 108	13 32.2	−58 29	Cen	12	700	III 2 p	15	7.5	8.47				
Tr 21	13 32.2	−62 47	Cen	4.0	1100	I 3 p	20	7.7	8.94	B3			
Cr 275	13 34.6	−60 07	Cen	7		IV 2 p	20	10.2p					
Pismis 18	13 36.6	−62 09	Cen	4.0		II 2 p	35	9.7:	10.85				Asterism?
NGC 5281	13 46.6	−62 54	Cen	5	1300	I 1 p	40	5.9	6.61	OB	51		
NGC 5288	13 48.7	−64 41	Cir	4.0		II 2 p	25	11.8p					
Cr 277	13 48.7	−66 05	Mus	16		III 1 p	30	9.2p					
NGC 5316	13 53.9	−61 52	Cen	14	1120	III 1 p	80	6.0	7.76	B5	190		
vdB−Ha 155	13 57.5	−59 35	Cen	11									
NGC 5381	14 00.6	−59 34	Cen	14		II 2 p	50		12p				Asterism?
Ru 110	14 06.4	−67 34	Cir	28		III 1 p	35		10p				
NGC 5460	14 07.6	−48 19	Cen	25	500	II 3 m	40	5.6	8.01	B8	110		
Ru 167	14 18.2	−58 58	Cen	14		IV 2 p	40		11p				
Lynga 2	14 24.0	−61 24	Cen	12	1100	IV 2 p	30	6.4	7.68	B	32		
NGC 5606	14 27.8	−59 38	Cen	3.0	1700	I 1 p	15	7.7	7.89	OB	13		
NGC 5617	14 29.8	−60 43	Cen	10	1200	I 3 m	80	6.3	8.81	OB	46		
Pismis 19	14 30.7	−60 59	Cen	2.2		II 2 p	60		12p				
Tr 22	14 31.2	−61 10	Cen	7	1700	II 2 p	50	7.9	10.05	OB	110		
Hogg 17	14 33.7	−61 23	Cen	7	1700	II 3 p	10	8.3	9.55	B7	180		Asterism?
NGC 5662	14 35.2	−56 33	Cen	12		II 3 m	70	5.5	7.02	B8			
NGC 5715	14 43.4	−57 33	Cir	6		II 2 m	30	9.8p	11p				
vdB−Ha 164	14 44.1	−66 24	Cir	30									
NGC 5749	14 48.9	−54 31	Lup	8	900	IV 1 p	30	8.8p		B8	91		
Hogg 18	14 50.7	−52 15	Lup	3.0	1100	I 3 p	15	8.0	8.84	B3	50		
NGC 5764	14 53.6	−52 41	Lup	2.0	1000	II 2 p	12	12.6p					
NGC 5822	15 05.2	−54 21	Lup	40	550	II 1 r	150	6.5p	10p	B8	890		
NGC 5823	15 05.7	−55 36	Cir	10	700	III 2 m	100	7.9	9.71	F0	200		
Pismis 20	15 15.4	−59 04	Cir	4.5	4400	I 3 p		7.8	8.20	O9			
NGC 5925	15 27.7	−54 31	Nor	15		III 1 m	120	8.4p		A1			
Lynga 4	15 33.3	−55 13	Nor	3.0		IV 2 p	30	11.4	12.48				Asterism?
Cr 292	15 50.7	−57 40	Nor	16		III 2 m	50	7.9p					
NGC 5999	15 52.2	−56 28	Nor	5		I 3 m	40	9.0p	12p	A0			
NGC 6005	15 55.8	−57 26	Nor	3.0		I 2 p	35	10.7:	11.77				Asterism?
Ru 113	15 57.2	−59 28	Nor	45		III 1 m	20		9p				
Tr 23	16 00.5	−53 31	Nor	5		II 2 p	40	11.2p					
NGC 6025	16 03.7	−60 30	TrA	12	840	II 2 p	60	5.1	7.30	B3	110		
Lynga 6	16 04.8	−51 55	Nor	5	1600			9.5	10.68	B	40		
NGC 6031	16 07.6	−54 04	Nor	2.0	3200	I 2 p	20	8.5	10.85	B3	22		In Nor OB1?
NGC 6067	16 13.2	−54 13	Nor	13	2100	I 2 r	100	5.6	8.25	B2	78	−43	
Cr 299	16 18.4	−55 07	Nor	20		III 2 p		6.9p		B3			
NGC 6087	16 18.9	−57 54	Nor	12	900	I 2 p	40	5.4	7.93	B5	55		
Harvard 10	16 19.9	−54 59	Nor	30		II 2 p	30			B5			
Ru 116	16 23.6	−52 00	Nor	5		II 2 p			9p				
Ru 118	16 24.6	−51 58	Nor	3.4		I 2 p		9.8	10.90				
NGC 6124	16 25.6	−40 40	Sco	29	490	II 3 m	100	5.8:	8.67	B8	51		
Cr 302	16 26:	−26:	Sco	500		III 3 p		1.0p					Antares moving cluster
DoDz 5	16 27.4	+38 04	Her	27		III 1 p				F2			
NGC 6134	16 27.7	−49 09	Nor	7	790	II 3 m		7.2	9.33	B8	630		
Ru 119	16 28.3	−51 31	Nor	8		II 1 p		8.8	10.90				Asterism?
NGC 6152	16 32.7	−52 37	Nor	30	1030	II 2 m	70	8.1p	11p	B5			
NGC 6169	16 34.1	−44 03	Nor	7	1100		40	6.6p		O			μ Nor cluster. Aster-
NGC 6167	16 34.4	−49 36	Nor	8	1200	II 3 m		6.7	7.44	B			ism? In Ara OB1
Cr 307	16 35.2	−50 58	Ara	6		III 2 p		9.2	11.10				Asterism?
NGC 6178	16 35.7	−45 38	Sco	4.0		I 3 p	12	7.2	8.41	B5			

Name	α 2000	δ 2000	Const	Diam	d(pc)	Tr Type	N*	Mag	Br*	Spec	Age	RV	Notes
Do 27	16ʰ36.ᵐ5	−8°57′	Oph	24′		III 1 p n	15						
NGC 6192	16 40.3	−43 22	Sco	8		I 2 p	60	8.5p	11p	A2			
NGC 6193	16 41.3	−48 46	Ara	15	1350	II 3 p n		5.2	5.66	OB	1.0	−26	In Ara OB1
NGC 6200	16 44.2	−47 29	Ara	12	2400	III 2 m	40	7.4	9.20				In Ara OB2
DoDz 6	16 45.3	+38 17	Her	17		IV 2 p	5			F5			
NGC 6204	16 46.5	−47 01	Ara	5	2600	I 2 p	45	8.2	9.76	OB	13	−38	In Ara OB2
Hogg 22	16 46.7	−47 06	Ara	1.5	2800		8	6.7	7.29				
NGC 6216	16 49.4	−44 44	Sco	4.0		II 2 p	40	10.1p	12p				
NGC 6208	16 49.5	−53 49	Ara	16	1000	II 1 m	60	7.2	9.96		1000		
NGC 6231	16 54.0	−41 48	Sco	15	1800	I 3 p n		2.6	4.71	O8	3.2	−22	In Sco OB1
Lynga 14	16 55.2	−45 19	Sco	2.0	2300		15	9.7	11.16				
Cr 316	16 55.5	−40 50	Sco	105		I 2 m		3.4p	14p				
NGC 6242	16 55.6	−39 30	Sco	9	1200	I 3 m		6.4	7.28	B5	51		
Tr 24	16 57.0	−40 40	Sco	60	1600	IV 2 p n		8.6p		O7		−4	Near nebula IC 4628
NGC 6249	16 57.6	−44 47	Sco	6		II 1 p	30	8.2	9.78				
NGC 6250	16 58.0	−45 48	Ara	8	1020	IV 3 p	60	5.9	7.62	B8	14		
NGC 6253	16 59.1	−52 43	Ara	5		I 3 m	30	10.2p	13p				
NGC 6259	17 00.7	−44 40	Sco	10	770	III 2 m	120	8.0	11.64	F8	220		
NGC 6268	17 02.4	−39 44	Sco	6	1100	II 2 p		9.5p		B2			
NGC 6281	17 04.8	−37 54	Sco	8	600	II 2 p n		5.4	7.94		220		
Harvard 13	17 05.4	−48 11	Ara	15		II 2 p	15						
DoDz 7	17 10.6	+15 32	Her	20		III 1 p	6			F0			
Bochum 13	17 17.3	−35 33	Sco				35	7.2	8.04				
NGC 6318	17 17.8	−39 27	Sco	4.0		III 2 p		11.8p	12p				
NGC 6322	17 18.5	−42 57	Sco	10	1200	I 2 p	30	6.0	7.50	B1	10		
Ru 123	17 23.4	−37 56	Sco	9		II 3 p	50		10p				
IC 4651	17 24.7	−49 57	Ara	12	780	II 3 m	80	6.9	8.95		2400		
Pismis 24	17 25.3	−34 21	Sco	4.0	2100:	IV 2 p n	15	9.6	10.43				
NGC 6352	17 25.4	−48 28	Ara	2.0				10.7p	12p	A5			
DoDz 8	17 26.2	+24 11	Her	14		IV 2 p	6			A5			
Tr 26	17 28.5	−29 29	Oph	7	2800	III 2 p	40	9.5p					
Antalova 2	17 29.7	−32 30	Sco				8	8.8	9.70				Asterism?
Ru 125	17 29.7	−40 30	Sco	14		II 3 p	30		11p				
Cr 332	17 30.8	−37 05	Sco	10		IV 1 p	12	8.9p					
Cr 333	17 31.3	−34 05	Sco	5		III 2 p	30	9.8p					
Harvard 16	17 31.4	−36 51	Sco	15		IV 2 p	70						
NGC 6374	17 32.3	−32 36	Sco	2.5		IV 1 p		9.0	10.19				Asterism?
NGC 6383	17 34.8	−32 34	Sco	5	1380	IV 3 p n	40	5.5	5.64	O7	4.5	−4	
Tr 27	17 36.2	−33 29	Sco	7	2100	I 2 p	35	6.7	8.39		10		
Tr 28	17 36.8	−32 29	Sco	8	1500:	II 2 p n	30	7.7	9.84				
Ru 127	17 37.7	−36 16	Sco	8	1500	IV 2 p	20	8.8	10.35				
NGC 6396	17 38.1	−35 00	Sco	3.0		II 3 p	30	8.5	9.79				
Cr 338	17 38.2	−37 34	Sco	25		III 2 p	40	8.0p					
NGC 6404	17 39.6	−33 15	Sco	5		III 3 m	50	10.6p					
NGC 6405	17 40.1	−32 13	Sco	15	600	III 2 p	80	4.2	6.17	B5	51		M6, Butterfly cluster
NGC 6400	17 40.8	−36 57	Sco	8		II 2 m	60	8.8p	9p				
Tr 29	17 41.6	−40 06	Sco	9		II 3 p	30	7.5p					
NGC 6416	17 44.4	−32 21	Sco	18	800	IV 1 p	40	5.7	8.43	A0			
Cr 345	17 44.6	−33 45	Sco	6				10.9p					Asterism?
IC 4665	17 46.3	+5 43	Oph	41	430	III 2 p	30	4.2	6.86	B4	36	−12	
Cr 347	17 46.4	−29 18	Sgr	4.0	1500	III 2 p n	40	8.8	10.65				In nebula Sh2-16
NGC 6425	17 46.9	−31 32	Sco	8	800	I 1 p	35	7.2	10.15				
Cr 350	17 48.1	+1 18	Oph	45		IV 2 p	20	6.1p					
Ru 131	17 49.0	−29 13	Sgr	10		III 1 p	15		11p				
Cr 351	17 49.4	−28 44	Sgr	9		IV 2 p	30	9.3p					
NGC 6444	17 49.5	−34 49	Sco	12		III 2 m	8		11p				Ru 132
NGC 6451	17 50.7	−30 13	Sco	8	570	II 1 p n	80	8.2p	12p	A			
NGC 6469	17 52.9	−22 21	Sgr	12	1600	III 2 p	50	8.2p		B8			
NGC 6475	17 53.9	−34 49	Sco	80	240	II 2 r	80	3.3	5.60	B8	220	−14	M7
Tr 30	17 56.5	−35 19	Sco	10		III 2 p	20	8.8p					
NGC 6494	17 56.8	−19 01	Sgr	27	660	III 1 m	150	5.5	9.21	B9	220		M23
Ru 135	17 58.2	−11 41	SerCd	11		I 1 p	20						
NGC 6507	17 59.6	−17 24	Sgr	7		IV 2 p	35	9.6p	12p				
Tr 31	17 59.9	−28 11	Sgr	8	1000	III 1 p	25	9.8p		B9			
Mel 186	18 01:	+3:	Oph	240	200	IV 3 m		3.0p		B3			
Bochum 14	18 02.0	−23 42	Sgr			n	10	9.3	10.25				
NGC 6514	18 02.3	−23 02	Sgr	28	1600	n		6.3	7.29	O5			In Trifid nebula M20
NGC 6520	18 03.4	−27 54	Sgr	6	1650	I 2 m n	60	7.6p	9p	O	54	−26	| and Sgr OB1
NGC 6531	18 04.6	−22 30	Sgr	13	1300	I 3 m	70	5.9	7.25	B0	4.6	−9	M21, in Sgr OB1
NGC 6530	18 04.8	−24 20	Sgr	15	1600	II 2 m n		4.6	6.87	O5	2.0	−9	In Lagoon nebula M8
NGC 6540	18 06.3	−27 49	Sgr	0.8		III 1 p n	10	14.6p					
NGC 6546	18 07.2	−23 20	Sgr	13	830	III 2 m	150	8.0	10.62		250		
DoDz 9	18 08.8	+31 32	Her	34		III 2 p	15			F0			
Cr 367	18 09.6	−23 59	Sgr	37		IV 3 p n	30	6.4p		B1			
NGC 6568	18 12.8	−21 36	Sgr	13		III 1 m	50	8.6p					

Name	α 2000	δ 2000	Const	Diam	d(pc)	Tr Type	N*	Mag	Br*	Spec	Age	RV	Notes
NGC 6583	18^{h}15^m.8	−22°08′	Sgr	2.8		II 1 m	35	10.0p					
Cr 469	18 16.4	−18 13	Sgr	5	1970	II 2 p	10	9.1*	11.21	B2	20		
NGC 6595	18 17.0	−19 53	Sgr	11			30	7.0p					Asterism?
NGC 6605	18 17.1	−14 58	SerCd					6.0p					No cluster
NGC 6596	18 17.5	−16 40	Sgr			III 2 m n	30						
NGC 6604	18 18.1	−12 14	SerCd	2.0	700	I 3 p n	30	6.5	7.48	O9	4.0		In Ser OB2?
NGC 6603	18 18.4	−18 25	Sgr	5	2880	I 1 r n	100	11.1p	14p	B9			In star cloud M24
NGC 6611	18 18.8	−13 47	SerCd	7	2500	II 3 m n		6.0	8.21	O6		5.5	−20	M16, in Ser OB1
NGC 6613	18 19.9	−17 08	Sgr	9	1200	II 3 p n	20	6.9	8.65	B3	32		M18
NGC 6618	18 20.8	−16 11	Sgr	11	1500	III 2 p n	40	6.0	9.28	B0			In Omega nebula M17
Ru 140	18 21.8	−33 15	Sgr	3.5		III 2 p	15		11p				
NGC 6625	18 23.2	−12 03	Sct			n	30	9.0p					No cluster
Tr 33	18 24.8	−19 41	Sgr	7	1300	II 3 p	20	8.8p	9.74	B5	13		
Do 28	18 25.4	−14 39	Sct	12		IV 2 p	20						
NGC 6631	18 27.2	−12 02	Sct	5		II 2 m	30	11.7p					
NGC 6633	18 27.7	+6 34	Oph	27	320	III 2 m	30	4.6	7.57	B6	660	−23	
Ru 141	18 31.3	−12 19	Sct	7		III 2 m	20		12p				
Do 29	18 31.4	−6 38	Sct	18									Asterism?
NGC 6647	18 31.5	−17 21	Sgr					8.0p					No cluster
IC 4725	18 31.6	−19 15	Sgr	32	580	I 2 p	30	4.6	6.70	B5	89	+4	M25, containing U Sgr
NGC 6645	18 32.6	−16 54	Sgr	10		III 1 m	40	8.5p	12p	B9			
Do 30	18 32.9	−6 02	Sct	18									Asterism?
NGC 6649	18 33.5	−10 24	Sct	6	1300	II 2 m	50	8.9	11.56		25		With Cepheid V367 Sct
Do 31	18 34.9	−6 51	Sct	18		IV 1 p							Asterism?
NGC 6664	18 36.7	−8 13	Sct	16	1300	III 2 m	50	7.8	10.15	B3	140	+23	Contains Cepheid EV Sct
IC 4756	18 39.0	+5 27	SerCd	52	400	III 2 m	80	5.4p	8.67	B7	580		
Tr 34	18 39.8	−8 29	Sct	8		II 2 m	40	9.4*	11.17				
Do 32	18 40.4	−4 06	Sct	12		I 2 p	40						Asterism?
NGC 6683	18 42.2	−6 17	Sct	11	1250	I 2 p n	20	9.5p	11.71	B4			
Tr 35	18 42.9	−4 08	Sct	9	1610:	II 2 p	35	9.2	11.41	B4	42		
NGC 6694	18 45.2	−9 24	Sct	15	1550	I 1 m	30	8.0	10.30	B5	89	+4	M26
Basel 1	18 48.2	−5 51	Sct	9	1460	I 2 m	15	8.9*	12.58	B4	58		
Isk 1	18 48:	+37:	Lyr	110	250								
Czernik 38	18 49.7	+4 56	SerCd	14		II 3 r	80						
Ru 145	18 50.5	−18 05	Sgr	35		III 1 m			10p				
NGC 6704	18 50.9	−5 12	Sct	6	1810	I 3 m	30	9.2	12.20	B2	20		
NGC 6705	18 51.1	−6 16	Sct	14	1720	I 2 r		5.8	8.00	B8	220	+22	M11
NGC 6709	18 51.5	+10 21	Aql	13	950	III 2 m	40	6.7	9.07	B5	78	−15	
Steph 1	18 53.5	+36 55	Lyr	20	320	III 3 p	15	3.8	4.30	B5			δ Lyr cluster
Cr 394	18 53.5	−20 23	Sgr	22		IV 2 m		6.3p					
NGC 6716	18 54.6	−19 53	Sgr	7	600	IV 1 p	20	6.9	8.28	B5	160		
NGC 6738	19 01.4	+11 36	Aql	15		IV 2 p		8.3p					
NGC 6755	19 07.8	+4 14	Aql	15	1500	IV 2 m	100	7.5	10.23	B2	35		
NGC 6756	19 08.7	+4 41	Aql	4.0	1650	I 2 m	40	10.6p	13p	B3	47		
Ru 147	19 16.7	−16 17	Sgr	48		III 2 m	20		9p				
NGC 6791	19 20.7	+37 51	Lyr	16	5100	II 3 r	300	9.5	13.00	F2	6300	−70	
Cr 399	19 25.4	+20 11	Vul	60	130	III 2 p	40	3.6	5.19		200		Brocchi's cluster
NGC 6800	19 27.2	+25 08	Vul			III 2 p	20		10p	A			
NGC 6802	19 30.6	+20 16	Vul	3.2	990	III 1 m	50	8.8	12.93	B6	1700		
Stock 1	19 35.8	+25 13	Vul	60		IV 2 p	40	5.3:*	7.0:	A0			
NGC 6811	19 38.2	+46 34	Cyg	13	900	IV 3 p	70	6.8	9.88	A0	540		
Cr 401	19 38.4	+0 20	Aql	1.0		IV 2 m		7.0p					
NGC 6819	19 41.3	+40 11	Cyg	5	2200	I 1 r		7.3	11.49	A0	3500		
NGC 6823	19 43.1	+23 18	Vul	12	2700:	I 3 p n	30	7.1	8.81	O7	2.0	+2	In Vul OB1 with nebula
NGC 6830	19 51.0	+23 04	Vul	12	1470	II 2 p	20	7.9	9.88	B0	100	+22	NGC 6820
NGC 6834	19 52.2	+29 25	Cyg	5	2300	II 2 m	50	7.8	9.65	B2	79		
Harvard 20	19 53.1	+18 20	Sge	7		III 2 p	15	7.7	9.80				
Do 36	20 02.5	+42 06	Cyg	14		IV 1 p							Asterism?
NGC 6866	20 03.7	+44 00	Cyg	7	1200	II 2 m	80	7.6	10.66	A2	230		
Roslund 4	20 04.9	+29 13	Vul			IV 2 p n	30	10.0	11.58				In nebula IC 4954-5
DoDz 10	20 05.7	+40 32	Cyg	20		IV 1 p n	12			A0			
NGC 6871	20 05.9	+35 47	Cyg	20	1650:	IV 3 p n	15	5.2	6.83	O6	10	−14	In Cyg OB3?
Basel 6	20 06.8	+38 21	Cyg	14	2100	n	40	7.7*	10.18	B4			
Biur 1	20 07.5	+35 41	Cyg	15		III 2 p n	15						
Biur 2	20 09.2	+35 29	Cyg	13		III 2 p	10	6.3	7.87				
Roslund 5	20 10.0	+33 46	Cyg	45	300	IV 3 p n	15					−15	
IC 1311	20 10.3	+41 13	Cyg	9		II 3 r n	60	13.1p	17p				Behind γ Cyg nebula
Be 50	20 10.4	+34 46	Cyg	4		II 1 p n	12		14p				IC 1310?
NGC 6883	20 11.3	+35 51	Cyg	15	1380	I 3 p n	30	8.0p		B3	15		
NGC 6882	20 11.7	+26 33	Vul	18	590	II 2 p		8.1:	9.87	B5		−22	Near cluster NGC 6885
NGC 6885	20 12.0	+26 29	Vul	7	590	III 2 p	30	5.7p	6p	B5			With 20 Vul
Mel 227	20 12.1	−79 19	Oct	50		II 2 p	40	5.3p		B9			
Do 3	20 15.7	+36 47	Cyg	15		III 2 p n	40			O			
Do 39	20 16.4	+37 52	Cyg	13		III 1 m n	40						
IC 4996	20 16.5	+37 38	Cyg	6	1620	I 3 p n	15	7.3	8.51	B0	10	−22	

Name	α 2000	δ 2000	Const	Diam	d(pc)	Tr Type	N*	Mag	Br*	Spec	Age	RV	Notes
vdB 130	20^{h}17^m.7	+39°19'	Cyg	~4'		n	15	9.3	10.30				
Cr 419	20 18.1	+40 43	Cyg	4.5		IV 2 p		5.4p		B2			
Do 40	20 18.2	+37 50	Cyg	12		III 2 p n	12						
Do 41	20 19.3	+37 44	Cyg	11		IV 1 p							
Do 42	20 19.7	+38 08	Cyg	11		IV 2 p n	20						
Berk 86	20 20.4	+38 42	Cyg	8		I 3 p n	30	7.9	9.50				
Berk 87	20 21.7	+37 22	Cyg	12		IV 2 p	30		13p	O			
Do 43	20 21.7	+39 57	Cyg	20		IV 2 p							Asterism?
NGC 6910	20 23.1	+40 47	Cyg	8	1650	I 2 p n	50	7.4	9.61	O8	10	−33	In γ Cyg nebula and	Cyg OB9
Cr 421	20 23.3	+41 42	Cyg	6		III 1 p n	22	10.1p					
NGC 6913	20 23.9	+38 32	Cyg	7	1250	III 3 p n	50	6.6	8.59	B0	10	−28	M29, in Cyg OB1
Do 10	20 26.3	+40 07	Cyg	16		IV 2 p n	8						Asterism?
Do 44	20 29.7	+41 43	Cyg	12		IV 2 p n	15						Asterism?
NGC 6939	20 31.4	+60 38	Cep	8	1250	I 1 m	80	7.8	11.91	B8	1800		
NGC 6940	20 34.6	+28 18	Vul	31	800	III 2 m	60	6.3	9.31	B8	1100	+5	
Ru 173	20 41.8	+35 33	Cyg	50		III 3 p	20		8p				
Ru 175	20 45.2	+35 30	Cyg	9		III 2 p	30		11p				
DoDz 11	20 51.0	+35 57	Cyg	13		IV 1 p	12			A0			
Bark 1	20 53.7	+46 02	Cyg	20		III 2 m	12						
NGC 6996	20 56.5	+44 38	Cyg	7	500	III 2 p n	40	10.0p					
Berk 53	20 56.6	+51 02	Cyg	12		II 3 p	40		16p				
NGC 6994	20 59.0	−12 38	Aqr	2.8		IV 1 p	4	8.9p	10p				M73. Asterism?
NGC 7023	21 00.5	+68 10	Cep	5		n		7.1p					
Cr 428	21 03.2	+44 35	Cyg	14	480	III 2 p	20	8.7p					
NGC 7031	21 07.3	+50 50	Cyg	5	1000	IV 1 p	50	9.1	11.31	B5	56		
Do 45	21 09.0	+37 36	Cyg	18		II 2 p n	35						
NGC 7039	21 11.2	+45 39	Cyg	25	700	III 2 p	50	7.6	11.26	B5	1000		
IC 1369	21 12.1	+47 44	Cyg	4.0	1500	I 1 m	40	6.8	12.14		1300		
NGC 7062	21 23.2	+46 23	Cyg	7	1900	III 1 p	30	8.3	10.10	A1	100		
NGC 7067	21 24.2	+48 01	Cyg	3.0	3500:	II 2 p	20	9.7	11.17	B0	13		
NGC 7063	21 24.4	+36 30	Cyg	8	660	III 2 p	12	7.0	8.89	B8	140		
NGC 7082	21 29.4	+47 05	Cyg	25	1400	IV 2 p		7.2	9.90		1600		
NGC 7086	21 30.5	+51 35	Cyg	9	1200	II 2 m	50	8.4	10.19	A5	85		
NGC 7092	21 32.2	+48 26	Cyg	32	270	III 2 p	30	4.6	6.83	A0	270	−28	M39
IC 1396	21 39.1	+57 30	Cep	50	800	II 3 m n	50	3.5	3.82	O			In Cep OB2
NGC 7129	21 41.3	+66 06	Cep	2.7		n	10	11.5p					
NGC 7127	21 43.9	+54 37	Cyg	2.8	1300	IV 2 p	12						
NGC 7128	21 44.0	+53 43	Cyg	3.1	2600	II 3 m	35	9.7	11.50	B2	10		
NGC 7142	21 45.9	+65 48	Cep	4.3	1000	II 2 r	100	9.3	12.12	F3	4000		
IC 5146	21 53.4	+47 16	Cyg	9	1000	IV 2 p n	20	7.2	9.64	B1	230		Cocoon nebula
NGC 7160	21 53.7	+62 36	Cep	7	900	II 3 p	12	6.1	7.04	B1	10	−25	In Cep OB2
NGC 7209	22 05.2	+46 30	Lac	25	900	III 1 p	25	6.7	9.02	B9	300		
Cr 471	22 07:	+72:	Cep	130		III 2 p							
IC 1434	22 10.5	+52 50	Lac	8		II 1 p?	40	9.0p	12p				
NGC 7226	22 10.5	+55 25	Cep	1.8	2200	I 1 p	25	9.6*	10.80				
NGC 7235	22 12.6	+57 17	Cep	4.0	3800:	III 2 p	30	7.7	8.80	B0	2.0		
NGC 7243	22 15.3	+49 53	Lac	21	880	IV 2 p	40	6.4	8.47	B7	110	+3	
NGC 7245	22 15.3	+54 20	Lac	5	1900	II 1 p	3	9.2*	12.75				
IC 1442	22 16.5	+54 03	Lac		1800	II 2 m	20	9.1*	11.43				
NGC 7261	22 20.4	+58 05	Cep	6	900	III 1 p	30	8.4	9.61	B2	40		
Berk 94	22 22.7	+55 51	Cep	4.0	1600	I 1 p n	10	8.7:	9.65				
NGC 7281	22 24.7	+57 50	Cep			IV 2 p	20						Asterism?
NGC 7296	22 28.2	+52 17	Lac	4.0		III 2 p	20	9.7p	10p				
Berk 97	22 39.5	+59 01	Cep	6		III 1 p	12		11p				
NGC 7380	22 47.0	+58 06	Cep	12	3600	III 3 p n	40	7.2	8.58	O6	3.8	−38	In Cep OB1?
NGC 7419	22 54.3	+60 50	Cep	2.0	1920	II 3 r	40	13.0p	10p				
King 10	22 54.9	+59 10	Cep	3.0		II 3 m	40		11p				
NGC 7429	22 55.9	+59 59	Cep		1920	III 2 p	15		11p				
King 19	23 08.3	+60 31	Cep	7	1350	II 2 m	25	9.2	10.37	B6	40		
NGC 7510	23 11.5	+60 34	Cep	4.0	3160	II 2 m n	60	7.9	9.68	O9	10	−66	
Mrk 50	23 15.3	+60 28	Cep	5	2250	I 2 p n	5	8.5	9.83	B0	10		With binary HD 219460
Do 46	23 21.9	+55 46	Cas	12		IV 1 p							Asterism?
NGC 7654	23 24.2	+61 35	Cas	13	1600	I 2 r	100	6.9	8.22	B6	35	−35	M52
Czernik 43	23 25.8	+61 19	Cas	14		III 1 r	15						
NGC 7686	23 30.2	+49 08	And	15	1000	IV 1 p	20	5.6	6.17	A0			Asterism?
Stock 11	23 32.9	+55 29	Cas	10		IV 2 p			8p	A0			Asterism?
Stock 12	23 37.2	+52 26	Cas	20		IV 2 p			8p	B9			
NGC 7762	23 49.8	+68 02	Cep	11	1020	II 2 p	40	10.0p	11p				
King 21	23 49.9	+62 43	Cas	2.5		III 3 m	20		10p	OB			
King 12	23 53.0	+61 58	Cas	2.0		I 2 p	15		10p	OB			
Harvard 21	23 54.1	+61 46	Cas	4.0		IV 2 p	6	9.0p					
NGC 7788	23 56.7	+61 24	Cas	9	2410	I 2 p	20	9.4p		B4	16		In Cas OB5?
NGC 7789	23 57.0	+56 44	Cas	16	1900	II 1 r	300	6.7	10.70	B9	1600		
Frolov 1	23 57.4	+61 38	Cas					9.2	10.60				
NGC 7790	23 58.4	+61 13	Cas	17	3200	III 2 p	40	8.5	10.87	B4	78		Contains CE Cas, CF Cas

Open Cluster Cross Index

Open Cluster Cross Index

Column Headings

This listing supplements the main Open Clusters section (pages 273-284). The cluster name given there, as adopted by G. Lynga, is the one that is preferred or in wide use. But if a cluster in the main section also has some other designation(s) among the types tabulated here, it is entered. This index does not include objects for which the name given in the main section is the only name (for example, most of the Stock clusters). The order of this cross index is by right ascension (equinox 2000.0).

NGC — The number in J. L. E. Dreyer's *New General Catalogue of Nebulae and Clusters of Stars* (1888).

IC — The number from one of Dreyer's supplements to the NGC. Objects numbered 1 to 1529 inclusive are from the *Index Catalogue* (1895), comprising discoveries made in the years 1888 to 1894. Numbers 1530 through 5386 are from the *Second Index Catalogue* (1908) of discoveries made in 1895 to 1907.

M — The Messier number of the cluster. A complete list of Charles Messier's clusters and nebulae is given in the Introduction.

Be — Berkeley number. These are designated "Berk" in the main Open Clusters section.

Cr — The cluster's number assigned by the Swedish astronomer P. Collinder (d. 1975).

H — Harvard number. In the main section, "Harvard" is used.

K — Ivan King's number. In the main section, "King" is used.

Mel — The number assigned by P. J. Melotte (1880-1961).

St — The number of Jurgen Stock. In the main section, "Stock" is used.

Tr — The number of Robert J. Trumpler (1886-1956).

Other — Certain additional names or designations.

NGC	IC	M	Be	Cr	H	K	Mel	St	Tr	Other
103	--	--	--	1	--	--	--	--	--	
129	--	--	--	2	--	--	--	--	--	
133	--	--	--	3	--	--	--	--	--	
136	--	--	--	4	--	--	--	--	--	
146	--	--	--	5	--	--	--	--	--	
189	--	--	--	462	--	--	--	--	--	
225	--	--	--	7	--	--	--	--	--	
188	--	--	--	6	--	--	2	--	--	
281	1590	--	--	8	--	--	--	--	--	
366	--	--	--	9	--	--	--	--	--	
381	--	--	--	10	--	--	--	--	--	
433	--	--	--	--	--	--	--	22	--	
436	--	--	--	11	--	--	6	--	--	
457	--	--	--	12	--	--	7	--	--	
559	--	--	--	13	--	--	--	--	--	
581	--	103	--	14	--	--	8	--	--	
--	--	--	--	15	--	--	--	--	1	
609	--	--	--	16	--	3	--	--	--	
637	--	--	--	17	--	--	--	--	--	
654	--	--	--	18	--	--	9	--	--	
659	--	--	--	19	--	--	10	--	--	
663	--	--	--	20	--	--	11	--	--	
752	--	--	--	23	--	--	12	--	--	
744	--	--	--	22	--	--	--	--	--	
869	--	--	--	24	--	--	13	--	--	h Per
884	--	--	--	25	--	--	14	--	--	χ Per
--	--	--	--	--	--	--	--	7	--	Mrk 6
956	--	--	--	27	--	--	--	--	--	
--	1805	--	--	26	--	--	15	--	--	
957	--	--	--	28	--	--	--	--	--	
--	--	--	--	29	--	--	--	--	2	
1039	--	34	--	31	--	--	17	--	--	
1027	--	--	--	30	--	--	16	--	--	
--	1848	--	--	32	--	--	--	--	--	
1193	--	--	--	35	--	--	--	--	--	
1220	--	--	--	37	--	--	--	--	--	
--	--	--	--	36	1	--	--	--	3	
1245	--	--	--	38	--	--	18	--	--	
--	--	--	--	39	--	--	20	--	--	
1342	--	--	--	40	--	--	21	--	--	
--	348	--	--	41	--	--	--	--	--	
--	--	45	--	42	--	--	22	--	--	Pleiades
1444	--	--	--	43	--	--	--	--	--	
1496	--	--	--	44	--	--	--	--	--	
1502	--	--	--	45	--	--	--	--	--	
1513	--	--	--	46	--	--	--	--	--	
1528	--	--	--	47	--	--	23	--	--	
--	361	--	--	48	--	--	24	--	--	
1545	--	--	--	49	--	--	--	--	--	
--	--	--	--	50	--	--	25	--	--	Hyades
1582	--	--	--	51	--	--	--	--	--	
1605	--	--	--	52	--	--	--	--	--	
1624	--	--	--	53	--	--	--	--	--	
1647	--	--	--	54	--	--	26	--	--	
1662	--	--	--	55	--	--	--	--	--	
1664	--	--	--	56	--	--	27	--	--	
1746	--	--	--	57	--	--	28	--	--	
1778	--	--	--	58	--	--	--	--	--	
1807	--	--	--	59	--	--	29	--	--	
1817	--	--	--	60	--	--	--	--	--	
1857	--	--	--	61	--	--	32	--	--	
1893	--	--	--	63	--	--	33	--	--	
1883	--	--	--	64	--	--	--	--	--	
--	--	--	--	458	--	--	--	--	--	Do 21
1907	--	--	--	66	--	--	35	--	--	
1912	--	38	--	67	--	--	36	--	--	
1931	--	--	--	68	--	--	--	9	--	
1981	--	--	--	73	--	--	--	--	--	
1960	--	36	--	71	--	--	37	--	--	
2099	--	37	--	75	--	--	38	--	--	
2112	--	--	--	76	--	--	--	--	--	
2129	--	--	--	77	--	--	--	--	--	
2126	--	--	--	78	--	--	39	--	--	
2141	--	--	--	79	--	--	--	--	--	
--	2157	--	--	80	--	--	--	--	4	
2158	--	--	--	81	--	--	40	--	--	
2169	--	--	--	83	--	--	--	--	--	
2168	--	35	--	82	--	--	41	--	--	
2175	--	--	--	84	--	--	--	--	--	
2186	--	--	--	85	--	--	--	--	--	
2194	--	--	--	87	--	--	43	--	--	
2192	--	--	--	86	--	--	42	--	--	
2204	--	--	--	88	--	--	44	--	--	
2215	--	--	--	90	--	--	45	--	--	
2232	--	--	--	93	--	--	--	--	--	
2236	--	--	--	94	--	--	--	--	--	
2243	--	--	--	98	--	--	46	--	--	
2244	--	--	--	99	--	--	47	--	--	
2250	--	--	--	100	--	--	--	--	--	
2251	--	--	--	101	--	--	--	--	--	
2252	--	--	--	102	--	--	--	--	--	
2254	--	--	--	103	--	--	--	--	--	
--	--	--	--	105	--	--	--	--	5	
2262	--	--	--	109	--	--	--	--	--	
2259	--	--	--	108	--	--	48	--	--	
2264	--	--	--	112	--	--	49	--	--	
2266	--	--	--	113	--	--	50	--	--	
2269	--	--	--	114	--	--	--	--	--	
2287	--	41	--	118	--	--	52	--	--	
2286	--	--	--	117	--	--	--	--	--	
2281	--	--	--	116	--	--	51	--	--	
2301	--	--	--	119	--	--	54	--	--	
--	--	--	28	--	--	--	--	--	--	Biur 10
2304	--	--	--	120	--	--	55	--	--	
2309	--	--	--	122	--	--	56	--	--	
2311	--	--	--	123	--	--	--	--	--	
--	--	--	32	--	--	--	--	--	--	Biur 8
2323	--	50	--	124	--	--	58	--	--	
2324	--	--	--	125	--	--	59	--	--	
2335	--	--	--	127	--	--	60	--	--	
2331	--	--	--	126	--	--	--	--	--	
2343	--	--	--	128	--	--	--	--	--	
2345	--	--	--	129	--	--	61	--	--	
2354	--	--	--	131	--	--	--	--	--	
2353	--	--	--	130	--	--	62	--	--	
2355	--	--	--	133	--	--	63	--	--	
2360	--	--	--	134	--	--	64	--	--	
2362	--	--	--	136	--	--	65	--	--	
2367	--	--	--	137	--	--	--	--	--	
2368	--	--	--	138	--	--	--	--	--	
2374	--	--	--	139	--	--	--	--	--	
2383	--	--	--	141	--	--	--	--	--	
2384	--	--	--	143	--	--	--	--	--	
--	--	--	--	145	--	--	--	--	6	
--	--	--	--	147	--	--	66	--	--	
2395	--	--	--	144	--	--	--	--	--	
--	--	--	--	146	--	--	--	--	7	
2396	--	--	--	148	--	--	--	--	--	
2401	--	--	--	149	--	--	--	--	--	
2414	--	--	--	150	--	--	--	--	--	
2421	--	--	--	151	--	--	67	--	--	
2422	--	--	--	152	--	--	68	--	--	
2423	--	--	--	153	--	--	70	--	--	
--	--	--	--	155	--	--	71	--	--	
--	--	--	--	156	--	--	72	--	--	
2420	--	--	--	154	--	--	69	--	--	
2439	--	--	--	158	--	--	74	--	--	
2432	--	--	--	157	--	--	73	--	--	
2437	--	46	--	159	--	--	75	--	--	
2447	--	93	--	160	--	--	76	--	--	
2451	--	--	--	161	--	--	--	--	--	
--	--	--	38	--	--	--	--	--	--	Ru 34
2453	--	--	--	162	--	--	--	--	--	
2455	--	--	--	163	--	--	77	--	--	
2477	--	--	--	165	--	--	78	--	--	
2467	--	--	--	164	--	--	--	--	--	
2482	--	--	--	166	--	--	--	--	--	
2479	--	--	--	167	--	--	--	--	8	
--	--	--	--	168	2	--	--	--	9	
2489	--	--	--	169	--	--	79	--	--	

NGC	IC	M	Be	Cr	H	K	Mel	St	Tr	Other
2516	—	—	—	172	—	—	82	—	—	
2506	—	—	—	170	—	—	80	—	—	
2509	—	—	—	171	—	—	81	—	—	
2527	—	—	—	174	—	—	—	—	—	
2533	—	—	—	175	—	—	—	—	—	
2539	—	—	—	176	—	—	83	—	—	
2547	—	—	—	177	—	—	84	—	—	
2546	—	—	—	178	—	—	—	—	—	
2548	—	—	—	179	—	—	85	—	—	
2567	—	—	—	180	—	—	86	—	—	
2571	—	—	—	181	—	—	—	—	—	
2579	—	—	—	182	—	—	—	—	—	
2580	—	—	—	183	—	—	—	—	—	
2588	—	—	—	186	—	—	—	—	—	
2587	—	—	—	184	—	—	—	—	—	
2627	—	—	—	188	—	—	87	—	—	
2635	—	—	—	190	—	—	89	—	—	
2632	—	44	—	189	—	—	88	—	—	Praesepe
—	2391	—	—	191	—	—	—	—	—	
—	2395	—	—	192	—	—	—	—	—	
2660	—	—	—	193	—	—	92	—	—	
2659	—	—	—	194	—	—	91	—	—	
2658	—	—	—	195	—	—	90	—	—	
2669	—	—	—	202	3	—	—	—	—	
2670	—	—	—	200	—	—	93	—	—	
2671	—	—	—	201	—	—	—	—	—	
—	—	—	—	203	—	—	—	—	10	
2682	—	67	—	204	—	—	94	—	—	
—	—	—	—	205	—	—	—	—	—	Mrk 18
2818	—	—	—	206	—	—	96	—	—	
—	2488	—	—	208	—	—	97	—	—	
2910	—	—	—	209	—	—	—	—	—	
2925	—	—	—	210	—	—	—	—	—	
2972	—	—	—	211	—	—	—	—	—	
3033	—	—	—	212	—	—	—	—	—	
3105	—	—	—	214	—	—	—	—	—	
3114	—	—	—	215	—	—	98	—	—	
—	—	—	—	216	—	—	—	—	11	
—	—	—	—	217	—	—	—	—	12	
3228	—	—	—	218	—	—	—	—	—	
—	—	—	—	219	—	—	—	—	13	
3247	—	—	—	220	—	—	—	—	—	
3255	—	—	—	221	—	—	—	—	—	
—	2581	—	—	222	—	—	—	—	—	
3293	—	—	—	224	—	—	100	—	—	
3324	—	—	—	225	—	—	—	—	—	
3330	—	—	—	226	4	—	—	—	—	
—	—	—	—	227	—	—	101	—	—	
—	2602	—	—	229	—	—	102	—	—	
—	—	—	—	230	—	—	—	—	14	
—	—	—	—	231	—	—	—	—	15	
—	—	—	—	233	—	—	—	—	16	
—	—	—	—	235	—	—	—	—	17	
—	—	—	—	—	—	—	—	—	—	Hogg 9
3496	—	—	—	237	—	—	—	—	—	
3532	—	—	—	238	—	—	103	—	—	
3572	—	—	—	239	—	—	—	—	—	
—	—	—	—	241	—	—	—	—	18	
3590	—	—	—	242	—	—	—	—	—	
—	—	—	—	243	—	—	—	—	19	
3603	—	—	—	244	—	—	—	—	—	
—	2714	—	—	245	—	—	104	—	—	
—	—	—	—	246	—	—	105	—	—	
3680	—	—	—	247	—	—	106	—	—	
3766	—	—	—	248	—	—	107	—	—	
—	2944	—	—	249	—	—	—	—	—	
3960	—	—	—	250	—	—	108	—	—	
4052	—	—	—	251	—	—	—	—	—	
4103	—	—	—	252	—	—	109	—	—	
4230	—	—	—	253	—	—	—	—	—	
4337	—	—	—	254	—	—	—	—	—	
4349	—	—	—	255	—	—	110	—	—	
—	—	—	—	256	—	—	111	—	—	Coma Ber.
4439	—	—	—	259	—	—	—	—	—	
—	—	—	—	257	5	—	—	—	—	

NGC	IC	M	Be	Cr	H	K	Mel	St	Tr	Other
4463	—	—	—	260	—	—	—	—	—	
—	—	—	—	261	6	—	—	—	—	
—	—	—	—	262	7	—	—	—	20	
4609	—	—	—	263	—	—	—	—	—	
4755	—	—	—	264	—	—	114	—	—	
4815	—	—	—	265	—	—	—	—	—	
4852	—	—	—	266	—	—	116	—	—	
—	—	—	—	268	8	—	—	—	—	
5138	—	—	—	270	—	—	—	—	—	
5168	—	—	—	273	—	—	—	—	—	
—	—	—	—	274	—	—	—	—	21	
5281	—	—	—	276	—	—	120	—	—	
5288	—	—	—	278	—	—	—	—	—	
5316	—	—	—	279	—	—	122	—	—	
5460	—	—	—	280	—	—	123	—	—	
5606	—	—	—	281	—	—	—	—	—	
5617	—	—	—	282	—	—	125	—	—	
—	—	—	—	283	—	—	—	—	22	
5662	—	—	—	284	—	—	127	—	—	
5715	—	—	—	286	—	—	128	—	—	
5749	—	—	—	287	—	—	—	—	—	
5764	—	—	—	288	—	—	—	—	—	
5822	—	—	—	289	—	—	130	—	—	
5823	—	—	—	290	—	—	131	—	—	
5925	—	—	—	291	—	—	—	—	—	
5999	—	—	—	293	—	—	137	—	—	
6005	—	—	—	294	—	—	138	—	—	
—	—	—	—	295	—	—	—	—	23	
6025	—	—	—	296	—	—	139	—	—	
6031	—	—	—	297	—	—	—	—	—	
6067	—	—	—	298	—	—	140	—	—	
6087	—	—	—	300	—	—	141	—	—	
6124	—	—	—	301	—	—	145	—	—	
6134	—	—	—	303	—	—	146	—	—	
6152	—	—	—	304	—	—	—	—	—	
6169	—	—	—	306	—	—	—	—	—	
6167	—	—	—	305	11	—	—	—	—	
6178	—	—	—	308	—	—	—	—	—	
6192	—	—	—	309	—	—	149	—	—	
6193	—	—	—	310	—	—	—	—	—	
6200	—	—	—	311	—	—	—	—	—	
6204	—	—	—	312	—	—	—	—	—	
6216	—	—	—	314	—	—	152	—	—	
6208	—	—	—	313	—	—	—	—	—	
6231	—	—	—	315	—	—	153	—	—	
6242	—	—	—	317	—	—	155	—	—	
—	—	—	—	318	12	—	—	—	24	
6249	—	—	—	319	—	—	—	—	—	
6250	—	—	—	320	—	—	—	—	—	
6253	—	—	—	321	—	—	156	—	—	
6259	—	—	—	322	—	—	158	—	—	
6268	—	—	—	323	—	—	—	—	—	
6281	—	—	—	324	—	—	161	—	—	
6318	—	—	—	325	—	—	166	—	—	
6322	—	—	—	326	—	—	—	—	—	
—	4651	—	—	327	—	—	169	—	—	
6352	—	—	—	328	—	—	170	—	—	
—	—	—	—	331	15	—	—	—	26	
6374	—	—	—	334	—	—	—	—	—	
6383	—	—	—	335	—	—	—	—	—	
—	—	—	—	336	—	—	—	—	27	
—	—	—	—	337	—	—	—	—	28	
6396	—	—	—	339	—	—	—	—	—	
6404	—	—	—	340	—	—	—	—	—	
6405	—	6	—	341	—	—	178	—	—	
6400	—	—	—	342	—	—	177	—	—	
—	—	—	—	343	17	—	—	—	29	
6416	—	—	—	344	—	—	—	—	—	
—	4665	—	—	349	—	—	179	—	—	
6425	—	—	—	384	—	—	—	—	—	
6451	—	—	—	352	—	—	181	—	—	
6469	—	—	—	353	—	—	182	—	—	
6475	—	7	—	354	—	—	183	—	—	
—	—	—	—	355	18	—	—	—	30	
6494	—	23	—	356	—	—	184	—	—	

NGC	IC	M	Be	Cr	H	K	Mel	St	Tr	Other
6507	--	--	--	358	--	--	--	--	--	
--	--	--	--	357	--	--	--	--	31	
--	--	--	--	359	--	--	186	--	--	
6514	--	20	--	360	--	--	--	--	--	
6520	--	--	--	361	--	--	187	--	--	
6531	--	21	--	363	--	--	188	--	--	
6530	--	--	--	362	--	--	--	--	--	
6540	--	--	--	364	--	--	--	--	--	
6546	--	--	--	365	--	--	--	--	--	
6568	--	--	--	369	--	--	--	--	--	
6583	--	--	--	370	--	--	--	--	--	
6595	--	--	--	371	--	--	--	--	--	
6604	--	--	--	373	--	--	--	--	--	
6603	--	24	--	374	--	--	197	--	--	
6611	--	16	--	375	--	--	198	--	--	
6613	--	18	--	376	--	--	--	--	--	
6618	--	17	--	377	--	--	--	--	--	
--	--	--	--	378	--	--	--	--	33	
6631	--	--	--	379	--	--	--	--	--	
6633	--	--	--	380	--	--	201	--	--	
--	4725	25	--	382	--	--	204	--	--	
6645	--	--	--	383	--	--	205	--	--	
6649	--	--	--	384	--	--	206	--	--	
6664	--	--	--	385	--	--	--	--	--	
--	4756	--	--	386	--	--	210	--	--	
--	--	--	--	387	--	--	--	--	34	
--	--	--	--	388	--	--	--	--	35	
6694	--	26	--	389	--	--	212	--	--	
6704	--	--	--	390	--	--	--	--	--	
6705	--	11	--	391	--	--	213	--	--	
6709	--	--	--	392	--	--	214	--	--	
6716	--	--	--	393	--	--	--	--	--	
6738	--	--	--	396	--	--	--	--	--	
6755	--	--	--	397	--	--	--	--	--	
6756	--	--	--	398	--	--	--	--	--	
6791	--	--	46	--	--	--	--	--	--	
6802	--	--	--	400	--	--	--	--	--	
6811	--	--	--	402	--	--	222	--	--	
6819	--	--	--	403	--	--	223	--	--	
6823	--	--	--	405	--	--	--	--	--	
6830	--	--	--	406	--	--	224	--	--	
6834	--	--	--	407	--	--	225	--	--	
6866	--	--	--	412	--	--	229	--	--	
6871	--	--	--	413	--	--	--	--	--	
--	1311	--	--	414	--	--	--	--	36	
6883	--	--	--	415	--	--	--	--	--	
6882	--	--	--	416	--	--	--	--	--	
6885	--	--	--	417	--	--	--	--	--	
--	--	--	--	411	--	--	227	--	--	
--	4996	--	--	418	--	--	--	--	--	
6910	--	--	--	420	--	--	--	--	--	
6913	--	29	--	422	--	--	--	--	--	
6939	--	--	--	423	--	--	231	--	--	
6940	--	--	--	424	--	--	232	--	--	
6996	--	--	--	425	--	--	--	--	--	
6994	--	73	--	426	--	--	--	--	--	
7023	--	--	--	429	--	--	--	--	--	
7031	--	--	--	430	--	--	--	--	--	
7039	--	--	--	431	--	--	--	--	--	
--	1369	--	--	432	--	--	--	--	--	
7062	--	--	--	434	--	--	--	--	--	
7067	--	--	--	436	--	--	--	--	--	
7063	--	--	--	435	--	--	--	--	--	
7086	--	--	--	437	--	--	--	--	--	
7092	--	39	--	438	--	--	236	--	--	
--	1396	--	--	439	--	--	--	--	37	
7129	--	--	--	441	--	--	--	--	--	
7128	--	--	--	440	--	--	--	--	--	
7142	--	--	--	442	--	--	--	--	--	
--	5146	--	--	470	--	--	--	--	--	
7160	--	--	--	443	--	--	--	--	--	
7209	--	--	--	444	--	--	238	--	--	
--	1434	--	--	445	--	--	239	--	--	
7226	--	--	--	446	--	--	--	--	--	
7235	--	--	--	447	--	--	--	--	--	

NGC	IC	M	Be	Cr	H	K	Mel	St	Tr	Other
7243	--	--	--	448	--	--	240	--	--	
7245	--	--	--	449	--	--	241	--	--	
7261	--	--	--	450	--	--	--	--	--	
7296	--	--	--	451	--	--	--	--	--	
7380	--	--	--	452	--	--	--	--	--	
7419	--	--	--	453	--	--	--	--	--	
7510	--	--	--	454	--	--	--	--	--	
7654	--	52	--	455	--	--	243	--	--	
7686	--	--	--	456	--	--	--	--	--	
7762	--	--	--	457	--	--	244	--	--	
7788	--	--	--	459	--	--	--	--	--	
7789	--	--	--	460	--	--	245	--	--	
7790	--	--	--	461	--	--	--	--	--	

Globular Clusters

Globular Clusters

Column Headings

NGC — Number of the cluster in J. L. E. Dreyer's *New General Catalogue of Nebulae and Clusters of Stars* (1888).

Name — The cluster's alternate designation, usually consisting of the name of an observer, survey, or institution followed by a serial number. The following abbreviations are used: Messier (M), *Index Catalogue* (IC), Arp-Madore (AM), Dunlop (Dun), William Herschel (H), Haute Provence (HP), Lacaille (Lac), Palomar (Pal), Terzan (Ter), Tonantzintla (Ton), or United Kingdom Schmidt (UKS). Two globular clusters, 47 Tuc and ω Cen, are easily visible to the unaided eye and still carry the names given them on early star atlases.

α 2000 and **δ 2000** — Right ascension and declination of the cluster center, referred to the 2000.0 equinox.

Const — Constellation in which the cluster is located.

Diam — Apparent diameter, in minutes of arc, on the homogeneous system of B. V. Kukarkin (1974). It is roughly the size beyond which the cluster's gravity is insufficient to prevent stars from escaping. The listed diameter is based on star counts and may differ somewhat from that observed visually.

V — The cluster's total visual **V** (yellow) magnitude in the **UBV** photometric system.

B – V — Total color index in the **UBV** system.

Spec — Composite spectral type of the cluster. In general, the earlier (hotter) the spectral type, the lower the average abundance of heavy elements in the cluster stars.

d(kpc) — Distance to the cluster in kiloparsecs; to convert to light-years, multiply by 3,260.

RV — Radial velocity in kilometers per second. Positive means receding (traveling away from the solar system). An asterisk in this column indicates that discordant values were noted among the sources.

M_v — Total absolute visual magnitude. This would be the cluster's brightness if it were located at the standard distance of 10 parsecs (32.6 light-years) from our Sun.

$(B – V)_0$ — Intrinsic color index; that is, the index after the reddening effect of interstellar dust on the cluster's light has been removed. Compare this value to the observed color index in the **B – V** column.

Class — Concentration class on the system of H. Shapley and H. B. Sawyer (*Harvard Observatory Bulletin*, No. 824, 1927). The values range from 1 to 12; the smaller the number, the more highly concentrated the stars toward the center of the cluster.

[Fe/H] — Metallicity index, defined as the logarithm (to the base 10) of the iron-to-hydrogen abundance ratio in the cluster stars relative to that in the Sun. For example, a value of −2 means that, on the average, there is only one-hundredth as much iron in each globular member as in the Sun.

Notes — The abbreviations LMC and SMC indicate globulars that belong to the Large and Small Magellanic Clouds, respectively. X-ray sources are flagged by the code "XRS." Also appearing in this column are any special names by which the cluster is known or comments regarding the object's appearance.

NGC	Name	α 2000	δ 2000	Const	Diam	V	B–V	Spec	d(kpc)	RV	M_v	$(B-V)_0$	Class	[Fe/H]	Notes
104	47 Tuc	$0^h24^m.1$	$-72°05'$	Tuc	30.9	4.03	0.89	G3	4.6	–14	–9.43	0.85	3	–0.71	XRS
121		0 26.8	–71 32	Tuc	1.5	10.6	0.78			+144				–1.51	SMC
288	H 20⁶	0 52.8	–26 35	Scl	13.8	8.10	0.66		8.3	–48	–6.60	0.63	10	–1.40	
339		0 57.7	–74 29	Tuc	2.2	11.9	0.69								SMC
361		1 02.2	–71 33	Tuc	1.5	11.8	0.76								SMC, br. foregr. star
362	Dun 62	1 03.2	–70 51	Tuc	12.9	6.58	0.76	F8	9.0	+232	–8.32	0.72	3	–1.27	
416		1 08.1	–72 21	Tuc	1.1	11.0	0.77			+158				–1.59	SMC
419		1 08.3	–72 53	Tuc	2.6	10.0	0.67			+181					SMC
1261	Pal 1	3 12.3	–55 13	Hor	6.9	8.38	0.70	F8	13.4	+55	–7.32	0.68	2	–1.31	
	Pal 1	3 33.2	+79 35	Cep	1.8				46:	+3:			12		
1466		3 44.5	–71 41	Hyi	2.3	11.4	0.61:			+155					LMC
	AM–1	3 55.0	–49 36	Hor	1.7	15.89	0.68		116		–4.44	0.68		–1.69:	
		4 24.8	–21 11	Eri	1.0	15.3:	0.8:		100:	–44	–4.2:	0.7:		–1.43:	Eridanus cluster
	Pal 2	4 46.1	+31 23	Aur	1.9		1.9:		35.0	–133:		0.7:	9		
1786		4 59.1	–67 45	Dor	1.2	10.1	0.76			+269			2		LMC, br. foregr. star
1835		5 05.2	–69 24	Dor	1.2	9.8	0.72			+161:			3		LMC
1851	Dun 508	5 14.1	–40 03	Col	11.0	7.30	0.78	F7	10.8	+319	–8.10	0.71	2	–1.36	XRS
1904	M79	5 24.5	–24 33	Lep	8.7	8.00	0.63	F6	13.3	+185	–7.65	0.62	5	–1.69	
1978		5 28.6	–66 14	Dor	2.7	9.9	0.79			+324			5		LMC, elongated
2121		5 48.0	–71 28	Men	1.8	11.2	0.86						10		LMC, elongated
2210		6 11.5	–69 08	Dor	1.7	10.2	0.71			+258			3		LMC
2257		6 30.4	–64 19	Dor	3.4	13.5	0.68			+294					LMC
2298		6 49.0	–36 00	Pup	6.8	9.40	0.73	F7	12.3	+44	–6.40	0.62	6	–1.85	
2419		7 38.1	+38 53	Lyn	4.1	10.37	0.66	F5	93.1	–20	–9.57	0.63	2	–2.10:	
2808		9 12.0	–64 52	Car	13.8	6.30	0.93	F8	9.2	+104*	–9.22	0.71	1	–1.37	
	UKS0923–545	9 24.6	–54 43	Vel										–0.29:	Highly obscured
	Pal 3	10 05.5	+0 04	Sex	2.8	14.7:			96:	+22:	–5.3:		12	–1.78:	
3201	Dun 445	10 17.6	–46 25	Vel	18.2	6.75	0.98		5.0	+494	–7.40	0.76	10	–1.61	
	Pal 4	11 29.3	+28 58	UMa	2.1	14.20	0.80		93.3	+168:	–5.65	0.80	12	–2.2:	
4147	H 19¹	12 10.1	+18 33	Com	4.0	10.26	0.60	F2	17.5	+182	–6.02	0.58	6	–1.80:	
4372		12 25.8	–72 40	Mus	18.6	7.80	0.97:		4.9	+83	–7.10	0.52:	12	–2.08	
4590	M68	12 39.5	–26 45	Hya	12.0	8.20	0.63	F2	9.6	–117*	–6.81	0.60	10	–2.09	
4833	Lac I–4	12 59.6	–70 53	Mus	13.5	7.35	0.96		5.5	+217	–7.55	0.58	8	–1.86	
5024	M53	13 12.9	+18 10	Com	12.6	7.72	0.64	F4	17.2	–79	–8.62	0.59	5	–2.04	
5053		13 16.4	+17 42	Com	10.5	9.80	0.64		15.2		–6.20	0.61	11	–2.58:	
5139	ω Cen	13 26.8	–47 29	Cen	36.3	3.65	0.79	F7	5.2	+228	–10.27	0.68	8	–1.59:	
5272	M3	13 42.2	+28 23	CVn	16.2	6.35	0.69	F7	9.9	–147	–8.65	0.68	6	–1.66	
5286	Dun 388	13 46.4	–51 22	Cen	9.1	7.62	0.87	F8	8.9	+49	–7.99	0.60	5	–1.79	
5466	H 9⁶	14 05.5	+28 32	Boo	11.0	9.10	0.71		14.5	+120	–6.86	0.66	12	–2.22:	
5634	H 70¹	14 29.6	–5 59	Vir	4.9	9.57	0.67	F5	21.6	–63	–7.33	0.60	4	–1.82	
5694		14 39.6	–26 32	Hya	3.6	10.20	0.69	F3	32.3	–184*	–7.60	0.61	7	–1.92	
	IC 4499	15 00.3	–82 13	Aps	7.6	10.60	0.88		18.6		–6.52	0.64	11	–1.50:	
5824		15 04.0	–33 04	Lup	6.2	9.00	0.75	F5	23.7	–58	–8.32	0.61	1	–1.87	
	Pal 5	15 16.1	–0 07	SerCp	6.9	11.75	0.70		21.4		–5.00	0.67	12	–1.47:	
5897	H 19⁶	15 17.4	–21 01	Lib	12.6	8.55	0.75		12.1	+10	–7.05	0.69	11	–1.68	
5904	M5	15 18.6	+2 05	SerCp	17.4	5.75	0.71	F5	7.6	+52	–8.76	0.68	5	–1.40	
5927		15 28.0	–50 40	Lup	12.0	8.33	1.31	G2	7.4	–78	–7.77	0.76	8	–0.30	
5946		15 35.5	–50 40	Nor	7.1	9.6:	1.24		9.6:	+115	–7.0:	0.68	9	–1.37	
5986	Dun 552	15 46.1	–37 47	Lup	9.8	7.12	0.90	F8	10.2	–35:	–8.78	0.63	7	–1.67	
	Pal 14	16 11.1	+14 57	Her	2.1	14.8			75:	+81	–4.7			–1.47:	
6093	M80	16 17.0	–22 59	Sco	8.9	7.20	0.85	F7	8.3	+13	–8.08	0.64	2	–1.68	
6121	M4	16 23.6	–26 32	Sco	26.3	5.93	1.03		2.1	+64	–6.80	0.68	9	–1.33	
6101		16 25.8	–72 12	Aps	10.7	9.30	0.68:		12.3	+160*	–6.40	0.60:	10	–1.81	
6144		16 27.3	–26 02	Sco	9.3	9.12	1.01	G0	8.1	+130*	–6.57	0.65	11	–1.75	
6139		16 27.7	–38 51	Sco	5.5	9.2:	1.39	F8	8.9:	+8	–7.8:	0.71	2	–1.65	
	Ter 3	16 28.7	–35 22	Sco											
6171	M107	16 32.5	–13 03	Oph	10.0	8.13	1.14	G0	5.9	–60*	–6.90	0.77	10	–0.99	
6205	M13	16 41.7	+36 28	Her	16.6	5.86	0.69	F5	7.2	–248	–8.49	0.67	5	–1.65	Hercules cluster
6229	H 50⁴	16 47.0	+47 32	Her	4.5	9.43	0.71	F7	31.2	–154	–8.07	0.70	4	–1.54	
6218	M12	16 47.2	–1 57	Oph	14.5	6.60	0.82	F7	5.5	–44	–7.70	0.63	9	–1.61	
6235		16 53.4	–22 11	Oph	5.0	10.2:	1.04		10:	+85*	–6.2:	0.66	10	–1.40	
6254	M10	16 57.1	–4 06	Oph	15.1	6.57	0.92	F8	4.4	+70	–7.48	0.66	7	–1.60	
6256	Ter 12	16 59.5	–37 07	Sco			1.68								
	Pal 15	16 59.9	–0 32	Oph	4.2	14.2			70:		–5.4				
6266	M62	17 01.2	–30 07	Oph	14.1	6.60	1.17	F8	6.0	–61	–8.78	0.71	4	–1.28	
6273	M19	17 02.6	–26 16	Oph	13.5	7.15	1.00	F5	10.6	+121	–9.20	0.62	8	–1.68	
6284	H 11⁶	17 04.5	–24 46	Oph	5.6	9.0:	0.95	F8	10:	+22	–6.9:	0.68	9	–1.40	
6287	H 195²	17 05.2	–22 42	Oph	5.1	9.2:	1.20:		9.0:	–211:	–6.7:	0.84:	7	–2.05	
6293	H 12⁶	17 10.2	–26 35	Oph	7.9	8.2:	0.98	F4	7.3:	–73	–7.2:	0.64	4	–1.92	
6304	H 147¹	17 14.5	–29 28	Oph	6.8	8.42	1.33	G4	5.4	–98	–7.08	0.75	6	–0.59:	
6316	H 45¹	17 16.6	–28 08	Oph	4.9	9.0:	1.27		12:	+68	–8.0:	0.79	3	–0.47	
6341	M92	17 17.1	+43 08	Her	11.2	6.52	0.62	F2	7.8	–121	–7.98	0.61	4	–2.24	
6325		17 18.0	–23 46	Oph	4.3	10.70	1.54:		19.4	–5	–6.00	0.74:	4	–1.44	
6333	M9	17 19.2	–18 31	Oph	9.3	7.9:	0.94	F3	6.9:	+225	–7.4:	0.58	8	–1.78	
6342		17 21.2	–19 35	Oph	3.0	9.9:	1.29		15:	+83	–7.6:	0.80	4	–0.62	

GLOBULAR CLUSTERS

NGC	Name	α 2000	δ 2000	Const	Diam	V	B−V	Spec	d(kpc)	RV	M_v	$(B-V)_0$	Class	[Fe/H]	Notes
6356	H 48[1]	17^{h}23^m.6	−17°49′	Oph	7.2	8.40	1.11	G5	17.2	+32	−8.67	0.83	2	−0.62	
6355		17 24.0	−26 21	Oph	5.0	9.6:	1.46		6.8:	−160	−7.0:	0.70		−1.50	
6352		17 25.5	−48 25	Ara	7.1	8.15	1.06		5.4	−121	−6.32	0.81	11	−0.51	
	Ter 2	17 27.5	−30 48	Sco	1.5								9	−0.47:	XRS
6366		17 27.7	−5 05	Oph	8.3	10.0:	1.47		4.0:		−5.1:	0.82	11	−0.99:	
	Ter 4	17 30.6	−31 35	Sco										−0.21:	
	HP 1	17 31.1	−29 59	Oph	2.9										
6362	Dun 225	17 31.9	−67 03	Ara	10.7	8.30	0.85		7.1	−6	−6.35	0.73	10	−1.08	
	Grindlay 1	17 32.0	−33 50	Sco					10:						XRS, highly obscur
	Liller 1	17 33.4	−33 23	Sco					9:					−0.21:	XRS, highly obscur
6380		17 35.4	−39 04	Sco	3.9				8.7					−1.00:	
	Ter 1	17 35.8	−30 28	Sco	2.8									+0.24:	XRS?
6388		17 36.3	−44 44	Sco	8.7	6.85	1.17	G3	14.5	+84	−9.98	0.85	3	−0.74	
	Ton 2	17 36.6	−39 04	Sco	3.4										
6402	M14	17 37.6	−3 15	Oph	11.7	7.56	1.28	F8	10.2	−123	−9.34	0.70		−1.39	
6401		17 38.6	−23 55	Oph	5.6	9.5:	1.58		6.8:	−70	−7.2:	0.79	8	−1.13	
6397	Lac III−11	17 40.7	−53 40	Ara	25.7	5.65	0.75	F5	2.2	+19	−6.65	0.57	9	−1.91	
	Pal 6	17 43.7	−26 13	Oph	7.2	13.60			2.9:				11	−0.74:	
6426		17 44.9	+3 00	Oph	3.2	11.20	1.03		16.0	−155	−6.10	0.63	9	−2.20	
	Ter 5	17 48.1	−24 47	Sgr	2.1	13.50	2.77					1.0:		+0.24:	XRS
6440		17 48.9	−20 22	Sgr	5.4	9.65	1.98	G5	3.7	−84	−6.75	0.87	5	−0.26	XRS?
6441		17 50.2	−37 03	Sco	7.8	7.42	1.28	G3	10.3	+14	−9.08	0.83	3	−0.59	XRS?
	Ter 6	17 50.7	−31 17	Sco	1.2										
6453		17 50.9	−34 36	Sco	3.5	9.9:	1.28:		7.1:	−78	−6.5:	0.61:	4	−1.53	
	UKS1751−241	17 54.5	−24 09	Sgr					10:					−1.18:	Highly obscured
6496		17 59.0	−44 16	Sco	6.9	9.2:	0.93		9.0	−79*	−5.8:	0.86	12	−0.48	
6517		18 01.8	−8 58	Oph	4.3	10.3:	1.79		7.8:	−37	−7.8:	0.65	4	−1.34	
	Ter 9	18 01.8	−26 52	Sgr										−0.38:	
	Ter 10	18 03.6	−26 05	Sgr											
6522	H 49[1]	18 03.6	−30 02	Sgr	5.6	8.60	1.22	F8	6.4	+8	−7.04	0.72	6	−1.44	
6535		18 03.8	−0 18	SerCd	3.6	10.6:	0.97		11:	−126	−5.8:	0.61	11	−1.75	
6539		18 04.8	−7 35	SerCd	6.9	9.6:	1.89		2.3:	−39	−6.1:	0.67	10	−0.66	
6528		18 04.8	−30 03	Sgr	3.7	9.50	1.45		7.3	+160	−6.90	0.80	5	+0.12:	
6544		18 07.3	−25 00	Sgr	8.9	8.25	1.36	F9	4.6	−12	−7.10	0.73		−1.56	
6541	Dun 473	18 08.0	−43 42	CrA	13.1	6.64	0.77	F6	6.9	−153	−7.96	0.64	3	−1.83	
6553		18 09.3	−25 54	Sgr	8.1	8.25	1.62		5.9	−33	−8.15	0.83	11	−0.29	
6558		18 10.3	−31 46	Sgr	3.7		1.09		9.2:	−87*		0.69		−1.44	
	IC 1276	18 10.7	−7 12	SerCd	7.1		1.74		12.9			0.82	12		Pal 7
	Ter 11	18 12.6	−22 45	Sgr											
6569		18 13.6	−31 50	Sgr	5.8	8.7:	1.34		7.8:	−29	−7.8:	0.71	8	−0.86	
6584	Dun 376	18 18.6	−52 13	Tel	7.9	9.2:	0.79	F7	15:	+180	−7.0:	0.68	8	−1.54	
6624	H 50[1]	18 23.7	−30 22	Sgr	5.9	8.32	1.10	G3	8.5	+69	−7.13	0.85	6	−0.35	XRS
6626	M28	18 24.5	−24 52	Sgr	11.2	6.9:	1.10	F8	6.1:	+2	−8.1:	0.77	4	−1.44	
6638	H 51[1]	18 30.9	−25 30	Sgr	5.0	9.2:	1.16	G2	8.0	−14	−6.5:	0.80	6	−1.15	
6637	M69	18 31.4	−32 21	Sgr	7.1	7.70	0.99	G5	10.3	+50	−7.90	0.82	5	−0.59	
6642		18 31.9	−23 29	Sgr	4.5	8.80	1.08		6.2	−90*		0.72		−1.29	
6652		18 35.8	−32 59	Sgr	3.5	8.9:	0.92	G2	15.0	−87*	−7.3:	0.81	6	−0.89	
6656	M22	18 36.4	−23 54	Sgr	24.0	5.10	0.99	F5	3.1	−153	−8.45	0.64	7	−1.75	
	Pal 8	18 41.5	−19 49	Sgr	4.7		1.19		30.8	−38		0.89	10	−0.48	
6681	M70	18 43.2	−32 18	Sgr	7.8	8.08	0.71	G0	10.8	+198	−7.32	0.64	5	−1.51	
6712	H 47[1]	18 53.1	−8 42	Sct	7.2	8.21	1.16	G1	7.6	−124	−7.30	0.81	9	−1.01	XRS
6717	Pal 9	18 55.1	−22 42	Sgr	3.9		0.93		16:	+17		0.75	8	−1.32	
6715	M54	18 55.1	−30 29	Sgr	9.1	7.70	0.85	F7	21.5	+131	−9.41	0.71	3	−1.42	
6723	Dun 573	18 59.6	−36 38	Sgr	11.0	7.32	0.75	G2	8.7	−90	−7.48	0.72	7	−1.09	
6749		19 05.1	+1 47	Aql	6.3	11.07	1.76:					0.80:			
6752	Dun 295	19 10.9	−59 59	Pav	20.4	5.40	0.65	F6	4.2	−32	−7.80	0.62	6	−1.54	
6760		19 11.2	+1 02	Aql	6.6	9.10	1.66		4.0	−21	−6.80	0.75	9	−0.52	
6779	M56	19 16.6	+30 11	Lyr	7.1	8.25	0.86	F5	9.5	−138	−7.35	0.64	10	−1.94	
	Ter 7	19 17.7	−34 40	Sgr											
	Pal 10	19 18.2	+18 34	Sge	3.5				9.0:				12		
	Arp 2	19 28.2	−30 21	Sgr	3.7										
6809	M55	19 40.0	−30 58	Sgr	19.0	6.95	0.69		5.2	+167	−6.85	0.62	11	−1.82	
	Ter 8	19 41.7	−34 00	Sgr											
	Pal 11	19 45.3	−8 02	Aql	3.2				11:	−68			11	−0.7	
6838	M71	19 53.8	+18 47	Sge	7.2	8.30	1.13	G3	4.0	−19	−5.60	0.85		−0.58	
6864	M75	20 06.1	−21 55	Sgr	6.0	8.55	0.87	F8	18.2	−195	−8.30	0.70	1	−1.32	
6934	H 103[1]	20 34.2	+7 24	Del	5.9	8.88	0.74	F7	14.7	−379	−7.34	0.62	8	−1.54	
6981	M72	20 53.5	−12 32	Aqr	5.9	9.35	0.72	G0	17.3	−278	−6.94	0.69	9	−1.54	
7006	H 52[1]	21 01.5	+16 11	Del	2.8	10.60	0.74	F4	34.7	−385	−7.52	0.69	1	−1.59	
7078	M15	21 30.0	+12 10	Peg	12.3	6.35	0.68	F3	9.4	−112	−8.91	0.56	4	−2.15	XRS
7089	M2	21 33.5	−0 49	Aqr	12.9	6.50	0.67	F3	11.3	−6	−8.95	0.61	2	−1.62	
7099	M30	21 40.4	−23 11	Cap	11.0	7.50	0.58	F3	8.2	−172	−7.10	0.57	5	−2.13	
	Pal 12	21 46.5	−21 14	Cap	2.9		0.90		19.0	+9:	−4.30	0.88	12	−1.14	
	Pal 13	23 06.7	+12 46	Peg	1.8	14.50	0.69:		24.4	−28:	−2.60	0.64:	12	−1.79:	
7492		23 08.4	−15 37	Aqr	6.2	11.50	0.48		21.9	−189	−5.20	0.48	12	−1.82:	

Bright Nebulae

Bright Nebulae

Column Headings

Name — The nebula's number in J. L. E. Dreyer's *New General Catalogue of Nebulae and Clusters of Stars* (NGC) or *Index Catalogue* (IC). When not so numbered, the designation is that of one of the following authors: Cederblad (Ced); Gum; Minkowski (M1); Rodgers, Campbell, and Whiteoak (RCW); Sharpless (Sh2); van den Bergh (vdB); van den Bergh and Herbst (vdBH).

α 2000 and **δ 2000** — Right ascension and declination, referred to the 2000.0 equinox.

Const — Constellation in which the nebula is located, according to the coordinates given. Some nebulae may span more than one constellation.

Type — Nebula type, either emission (E), reflection (R), or a combination of the two (E + R).

Dim — Approximate maximum and minimum angular dimensions, in minutes of arc.

Description — Five observational characteristics of the nebula are listed, as follows:

1. *Visibility,* as given in the NGC or IC: extremely faint (eeF or eF), very faint (vF), faint (F), pretty faint (pF), pretty bright (pB), bright (B), very bright (vB), or extremely bright (eB). These descriptions, supplied by many different 19th-century observers using a wide variety of telescopes, are not homogeneous but may serve as a rough guide. It is important to note that the description for an NGC object is always the *visual* impression through a telescope (of perhaps 6- to 30-inch aperture), while for an IC object it is often based on the photographic appearance.

2. *Brightness* of the nebula on red-sensitive photographic plates. The scale ranges from 1 (brightest) to 6 (faintest), as in B. T. Lynds's 1965 catalogue.

3. *Color* based on the nebula's appearance on red- and blue-sensitive photographic plates. The color may be called very red (VR), red (R), moderately red (MR), intermediate (I), moderately blue (MB), blue (B) or very blue (VB). Intermediate means equally conspicuous on red and blue plates. Sometimes a range of color is noted, as for the Trifid nebula.

4. *Shape,* whether irregular (Ir), elliptical (E), or circular (C).

5. *Structure,* either amorphous (A), filamentary (F), or a combination of the two (AF).

HD — Henry Draper number, or other designation, of the star illuminating the nebula. Where appropriate, the variable star name may be listed instead. When a cluster is responsible for the nebular excitation, the NGC, IC, Collinder (Cr), or Stock (St) designation is given.

*Mag — Illuminating star's photoelectric visual **V** magnitude, if listed to two decimal places, or its visual magnitude, if listed to one. The letter "v" indicates that the brightness of the star varies.

*Spec — Spectral type of the illuminating star.

Notes — Common names, Messier numbers, and other information are listed here. Many words are abbreviated: association (assoc.), Barnard dark nebula (B), cluster (cl.), following or east of (f), Large Magellanic Cloud (LMC), nebula (neb.), north (n), preceding or west of (p), reflection (refl.), Small Magellanic Cloud (SMC), south (s), supernova remnant (SNR), Trumpler open cluster (Tr). If the illuminating star has a common name or a variable star designation not listed earlier, it appears here within brackets.

Name	α 2000	δ 2000	Const	Type	Dim		Description			HD	*Mag	*Spec	Notes
NGC 7822	0ʰ03ᵐ.6	+68°37′	Cep	E	60′	30′	eeF 3 R						Arc of nebulosity n of Ced 214
Ced 214	0 04.7	+67 10	Cep	E+R	50	40	2 R	IrAF		225216?	5.67	gK1	Surrounded by faint nebulous arcs; this
vdB 1	0 11.0	+58 46	Cas	R	9		2 VB			627	8.6	A	is often confused with NGC 7822
NGC 281	0 52.8	+56 36	Cas	E	35	30	F 1 R	IrAF		5005	7.76	O6	
IC 59	0 56.7	+61 04	Cas	E+R	10	5	pF 2 B			5394	2.47v	B0IVe	[γ Cas]
NGC 346	0 59.1	−72 11	Tuc	E	14	11	B	IrAF					Largest diffuse nebula in SMC
IC 63	0 59.5	+60 49	Cas	E+R	10	3	pF 1 I			5394	2.47v	B0IVe	[γ Cas]
NGC 456	1 14.4	−73 17	Tuc	E	15	15	pF	Ir					Several bright knots in SMC
Sh2−188	1 30.6	+58 22	Cas	E	10	3	1 R	IrF					Crescent-shaped
NGC 896	2 24.8	+61 54	Cas	E	27	13	eF 1 R					10.5	Nebulous complex, IC 1795 12′ nf
IC 1805	2 33.4	+61 26	Cas	E	60	60	3 R	IrF	IC 1805				
IC 1848	2 51.3	+60 25	Cas	E	60	30	F 2 R	IrAF	IC 1848				
vdB 8	2 51.6	+67 52	Cas	R	6		1 B			17443	8.5	A0	
IC 1871	3 03.2	+60 29	Cas	E	4	4	4 R	IrAF			10	K3	
vdB 16	3 28.3	+29 43	Ari	R	11	5	3 MB				9.1	A5	
vdB 14	3 29.2	+59 57	Cam	R	46		4 B			21291	4.23	B9Ia	
NGC 1333	3 29.3	+31 25	Per	R	9	7	F 3 B				9.5	B8p	Near dark nebula B205
vdB 15	3 30.1	+58 54	Cam	R	54		3 B			21389	4.58	A0Ia	
IC 348	3 44.6	+32 09	Per	R	10	10	pB 2 B			281159	8.53	B5V	Near dark nebula B4
Ced 19c	3 44.8	+24 17	Tau	R	16	16	3 VB			23288	5.46	B7IV	[Celaeno]
vdB 20	3 44.9	+24 07	Tau	R	20	16	3 VB			23302	3.70	B6III	[Electra]
Ced 19e	3 45.2	+24 28	Tau	R	21	21	3 VB			23338	4.30	B6IV	[Taygeta]
NGC 1432	3 45.8	+24 22	Tau	R	30	30	eF 2 VB			23408	3.88	B7III	[Maia]
Ced 19h	3 45.9	+24 33	Tau	R	15	15	3 VB			23432	5.76	B8V	[Sterope]
NGC 1435	3 46.1	+23 47	Tau	R	30	30	vF 2 VB			23480	4.18	B6IVnn	[Merope] Tempel's nebula
IC 349	3 46.3	+23 56	Tau				eF						Very small, 36″ sf Merope
vdB 23	3 47.5	+24 06	Tau	R	27	27	3 VB			23630	2.87	B7III	[Alcyone]
Ced 19o	3 49.2	+24 03	Tau	R	11	11	3 VB			23850	3.63	B8III	[Atlas]
Ced 19p	3 49.2	+24 08	Tau	R	10	10	3 VB			23862	5.09	B8Vpe	[Pleione]
vdB 24	3 49.6	+38 59	Per	R	9		3 MB			275877	8.8 v	A2II+B6	[XY Per]
IC 353	3 55.0:	+25 29:	Tau	R?	180	30	vF 3 B			24368	6.36	A0	
Sh2−205	3 56.1	+53 12	Cam	E	100	80	5 R	IrAF		24431	6.73	O9IV-V	Cam OB1
NGC 1499	4 00.7:	+36 37:	Per	E	145	40	vF 1 R	IrAF		24912	4.04	O7e	California nebula. [ξ Per, Per OB2]
NGC 1491	4 03.4	+51 19	Per	E	3	3	vB 1 R	IrAF			11	B0	Has a fainter extended envelope
IC 360	4 13.0:	+25 38:	Tau	E	180	100	vF 4 R						
vdB 26	4 13.6	+10 13	Tau	R	11		3 VB			26676	7.2	B8	
IC 359	4 19.0	+28 12	Tau	R	15	10	eeF 5 I				12	K8	
NGC 1554-5	4 21.8	+19 32	Tau	R	Var		vF 3 R	IrAF		284419	9.4 v	dG5e	Hind's variable nebula. [T Tau]
Ced 33	4 27.1	+26 06	Tau	R	5	2	4 R			DG Tau	10 v		
Ced 34	4 27.2	+22 57	Tau	R	10	6	2 B			28149	5.53	B7V	
NGC 1579	4 30.2	+35 16	Per	R	12	8	pB 4 B	IrAF			12		Dark lanes
IC 2087	4 40.0	+25 44	Tau	R	4	4	eeF 3 R						In dark nebula B22
NGC 1624	4 40.5	+50 27	Per	E	5	5	F 2 R	C AF	NGC 1624				
vdB 29	4 48.4	+29 47	Tau	R	14		4 B			30378	6.5	B9	
vdB 31	4 55.7	+30 33	Aur	R	9		3 B			31293	6.8 v	A0pe	[AB Aur]
NGC 1763	4 56.8	−66 24	Dor	E	25	20	vB	IrF					Has a bright nucleus. In LMC
NGC 1788	5 06.9	−3 21	Ori	R	8	5	B 1 VB			293815	10.12	B9V	
IC 2118	5 06.9:	−7 13:	Eri	R	180	60	F 3 B			34085	0.12	B8Iae:	[Rigel]
IC 405	5 16.2	+34 16	Aur	R	20		2			34078	5.96v	O9.5Ve:	Reflection part of IC 405, sf AE Aur
IC 405	5 16.2	+34 16	Aur	E	30	19	pB 2 R	IrAF		34078	5.96v	O9.5Ve:	Flaming Star nebula. [AE Aur]
Sh2−223	5 17.2	+42 12	Aur	E	70	10	4 VR	IrAF					SNR
vdB 37	5 18.2	+13 24	Ori	R	15		3 VR			34454	8.2	gM5	
Sh2−276	5 20:	−4:	Ori	E	400	40	3 R	AF					Barnard's Loop, p part. [ζ Ori?]
vdB 38	5 21.7	+8 27	Ori	R	39	30	3 I			34989	5.77	B1V	
NGC 1929,34-6	5 22.0	−67 58	Dor	E	20	15	pF	IrAF					OB assoc. + neb. in LMC, with IC 2126-7
IC 410	5 22.6	+33 31	Aur	E	40	30	2 R	IrAF	NGC 1893				
	5 26.8	−67 30	Dor	E	20	20		IrAF					Several knots and cl. NGC 1955 in LMC
NGC 1966	5 26.8	−68 49	Dor	E	13	12	pB	IrAF					In LMC
Sh2−224	5 27.3	+42 59	Aur	E	20	3	5 VR	IrF					SNR
IC 417	5 28.1	+34 26	Aur	E	13	10	2 R	IrAF	St 8				
NGC 2018	5 30.6	−71 04	Men	E	25	18	pB	IrF					Contains SNR, in LMC
NGC 1931	5 31.4	+34 15	Aur	E+R	3	3	vB 1 R	IrAF	NGC 1931				
IC 423	5 33.4	−0 37	Ori	R	6	4	vF 2 B						Annular. A planetary?
	5 33.4	−67 38	Dor	E	30	20		IrAF					Knots, filaments, cl. NGC 2014 in LMC
NGC 1952	5 34.5	+22 01	Tau	E	6	4	vB 1	E F			16 v		Crab nebula M1, SNR. [PSR 0531+21]
NGC 1973	5 35.1	−4 44	Ori	E+R	5	5	1 B			36958	7.36	B3V	
NGC 2048	5 35.2	−69 46	Dor	E	18	12	vF	IrAF					Lies sp 30 Dor, in LMC
NGC 1975	5 35.4	−4 41	Ori	E+R	10	5	B? 1 B				10.9		
NGC 1976	5 35.4	−5 27	Ori	E+R	66	60	eB 1 R	IrAF	37020-3	2.9	OB	Orion nebula M42. [θ¹ Ori]
NGC 1980	5 35.4	−5 54	Ori	E	14	14	vF 3			37043	2.77	O9III	[ι Ori]
NGC 1977	5 35.5	−4 52	Ori	E+R	20	10	1 B			37018	4.59	B1V	[42 Ori]
NGC 1982	5 35.6	−5 16	Ori	E+R	20	15	vB 1 R			37061	6.85	B1V	M43
Sh2−264	5 35:	+10:	Ori	E	270	240	4 R	C A	36861-2	3.39	O8e+B0.5V	[λ Ori]
NGC 1990	5 36.2	−1 12	Ori	E+R	50	50	3 B			37128	1.70	B0Iae	[ε Ori]
NGC 1999	5 36.5	−6 42	Ori	E+R	16	12	1 I?				9.3	B8	With IC 427-8 nf. [V380 Ori]

Name	α 2000	δ 2000	Const	Type	Dim		Description			HD	*Mag	*Spec	Notes
IC 426	5ʰ36ᵐ.8	−0°15′	Ori	R	5′	5′	vF 2 B			37140	8.54	B9	Fan-shaped
IC 430	5 38.5	−7 05	Ori	R	11	11	2			37507	4.80	A4V	Fan-shaped. [49 Ori]
NGC 2070	5 38.7	−69 06	Dor	E	40	25	vB 1 VR	IrAF					30 Dor nebula (Tarantula) in LMC
Sh2-240	5 39.1:	+28 00:	Tau	E	200	180	4	E F					SNR Simeis 147. Position of the center
vdB 49	5 39.2	+4 10	Ori	R	6		2 VB			37490	4.50	B3IIIe	[ω Ori]	is given
Sh2-231	5 39.4	+35 56	Aur	E	10	5	4 R	IrA					
NGC 2074,81	5 39.6	−69 27	Dor	E	16	10	pB	IrAF					Part of 30 Dor nebula in LMC
IC 431	5 40.3	−1 27	Ori	E	5	3	2 VB			37674	7.67	B5n	
NGC 2077-80	5 40.4	−69 38	Dor	E	15	11	B	IrAF					In LMC, sf 30 Dor. NGC 2083-4 near
NGC 2024	5 40.7	−2 27	Ori	E	30	30	B 2 R	IrAF		37742	2.05	O9.5Ibe	[ζ Ori]
IC 432	5 40.9	−1 29	Ori	R	8	4	1 B			37776	6.98	B2V	
IC 434	5 41.0	−2 24	Ori	E	60	10	1 R			37742	2.05	O9.5Ibe	Behind the Horsehead neb. [ζ Ori]
Sh2-235	5 41.1	+35 52	Aur	E	10		2	IrAF					
	5 41.5	−3 09	Ori	E	50	6	2 VR						An extension of IC 434 southward
NGC 2023	5 41.6	−2 14	Ori	E+R	10	10	1 B			37903	7.82	B1.5V	
IC 435	5 43.0	−2 19	Ori	R	5	3	1 VB			38087	8.30	B3n	
	5 43.2	−67 54	Dor	E	7			C F					Faint ring, in LMC
Ced 59	5 45.3	+9 04	Ori	E+R	3	2	2 R			FU Ori	12 v	F2p	Size varies. In dark nebula B35
NGC 2064	5 46.3	0 00	Ori	R	12	2	eF 2 I						
NGC 2067	5 46.5	+0 06	Ori	R	8	3	F 2 I			38563	10.49	B5	
NGC 2068	5 46.7	+0 03	Ori	R	8	6	B 1 I			38563	10.49	B5	M78
NGC 2071	5 47.2	+0 18	Ori	R	4	3	vF 2 I			290861	9.5	F8	
Sh2-276	5 48:	+1:	Ori	E	600	30	3 R			37742	2.05	O9.5Ibe	Barnard's Loop, f part. [ζ Ori?]
	5 56:	−6:	Ori	E	600	20	3 VR						Southern extension of Barnard's Loop
NGC 2149	6 03.5	−9 44	Mon	R	3	2	F 2 B				9.3		
Sh2-241	6 04.1	+30 15	Aur	E+R	10		2	IrAF					
NGC 2170	6 07.5	−6 24	Mon	R	2	2	vF 1 MB				9.5	B1	
Sh2-247	6 08.5	+21 37	Gem	E	10	10	3 VR	C A		41690	7.71	B1V	
Sh2-261	6 08.9	+15 49	Ori	E	30	15	3 R			41997?	8.42	O8	Lower's nebula
Ced 62	6 09.2	+18 41	Ori	R	3	2	2 B				13		Two symmetrical fans extending n-s
NGC 2182	6 09.5	−6 20	Mon	R	3	3	1 B			42261	9.0	B4	
NGC 2174-5	6 09.7	+20 30	Ori	R	40	30	eF 1 R	IrAF		42088	7.55	O6	With IC 2159, sf. Contains cluster
NGC 2183	6 10.8	−6 13	Mon	R	1	1	eF 3 B				13		
NGC 2185	6 11.1	−6 13	Mon	R	3	3	vF 3 B				12		
IC 2162	6 12.9	+17 59	Ori	E	15	4	vF 2 R	C AF			10		Also Sh2 256-8, with p-f extensions
NGC 2195	6 14.4	+17 39	Ori				F						Small and faint. IC 2162 is 30′ np
IC 443	6 16.9	+22 47	Gem	E	50	40	F 2	E F		43078	8.78	B0IV	SNR, with PSR 0611+22 near center
IC 444	6 20.4	+23 16	Gem	R	8	4	3 VB			43836	7.5	B9II	[12 Gem]
IC 446	6 31.0	+10 27	Mon	R	5	4	3 B				10.5	B1	Same as IC 2167
IC 2169	6 31.2	+9 54	Mon	R	25	20	F 3 B						Same as IC 447
NGC 2237-9,46	6 32.3	+5 03	Mon	E	80	60	pB 1 R			NGC 2244			Rosette nebula
IC 448	6 32.7	+7 19	Mon	R	15	10	3 VB			46300	4.48	A0Ib	[13 Mon]
NGC 2245	6 32.7	+10 10	Mon	R	5	3	1 I				10.8	B1p	
NGC 2247	6 33.2	+10 20	Mon	R	4	3	eF 2 I			259431	8.74	B6pe	
Sh2-282	6 38.0	+1 31	Mon	E	40	15	3 R	IrAF		47432	6.21	O9.5III	
NGC 2261	6 39.2	+8 44	Mon	E+R	2	1	B 1 I			R Mon	10.0 v		Hubble's variable neb. (fan-shaped)
NGC 2264	6 40.9	+9 54	Mon	E	60	30	1 R	IrAF		47839	4.66v	O7Ve	Cone nebula. [S Mon]
NGC 2282	6 46.9	+1 19	Mon	R	3	3	F 3 I			289120	9.3	B3	Same as IC 2172
NGC 2316	6 59.7	−7 46	Mon	E+R	4	3	pF 2 B						
Gum 1	7 04.3	−10 28	Mon	E+R	20	20	2 R	IrAF		53367	6.96	B0IVe?	RCW 2
IC 2177	7 05.1	−10 42	Mon	E	120	40	3 R			54662	6.21	O7III	Gum 2, RCW 1, Sh2-296
Ced 90	7 05.2	−12 20	CMa	E+R	10	10	1 I	IrAF		53623	8.5	B9	Gum 3, RCW 1
IC 466	7 08.6	−4 19	Mon	E+R	1	1	vF 1 R	IrAF					
Sh2-301	7 09.8	−18 29	CMa	E	8	7	1 R	E AF			10		Gum 5, RCW 6
NGC 2362	7 13.1:	−24 35:	CMa	E	300	90	5 VR	IrAF		57061	4.40	O9Ib	[τ CMa]
Sh2-294	7 16.6	−9 26	Mon	E	8	7	2 R	IrAF					
IC 468	7 17.5	−13 05	CMa	E	20	20	vF 3 R						
NGC 2359	7 18.6	−13 12	CMa	E	8	6	vF? 2 R	IrF		56925		WN4	Gum 4, RCW 5, ring-shaped
vdB 96	7 19.6	−23 58	CMa	R	10	5	3 B			57281	9.2	B9	Several components
Sh2-302	7 31.6	−16 58	Pup	E	20	20	4 R	E A		59934	7.8	B5.5V	Gum 6, RCW 7
vdB 97	7 32.6	−16 54	Pup	R	2		3 B				9.9		Near H II region Sh2-302
Sh2-307	7 35.5	−18 46	Pup	E	4	4	3 VR	IrAF					Gum 7, RCW 12
vdB 98	7 36.4	−25 20	Pup	R	15	15	3 VB			61071	7.3	B5	
NGC 2467	7 52.5	−26 24	Pup	E	8	7	pB 1 R	IrAF		64315	9.2	B	Gum 9, RCW 16. Faint streamers and cl.
NGC 2579	8 20.9	−36 13	Pup	E	2						10.2		Gum 11, RCW 20
Gum 12	8 30:	−45:	Vel	E	1200	720	4	C F					Gum nebula. Vela SNR with PSR 0833-45
NGC 2626	8 35.6	−40 40	Vel	E+R	5		pB 1				10.02	B1V	
Gum 15	8 44.6	−41 17	Vel	E	20		3						RCW 32. Brighter knot in Gum nebula
Gum 17	8 50.5	−42 07	Vel	E	100	65	3 I						RCW 3. Bright knot, dark lane in Gum
Gum 23	8 59.7	−47 27	Vel	E	20	10	2						RCW 38. Part of Gum nebula
Sh2-312	9 00:	−26:	Pyx	E	720		3	IrAF					
Gum 25	9 02.4	−48 42	Vel	E	7	6	2						RCW 40
RCW 47	10 05.2	−58 57	Car	E	25	20	3						
NGC 3199	10 17.1	−57 55	Car	E	22	22	vB 3			89358	11.1	WC	RCW 48. A partial ring
NGC 3247	10 24.8	−57 53	Car	E	6		2						RCW 49

Name	α 2000	δ 2000	Const	Type	Dim		Description	HD	*Mag	*Spec	Notes
RCW 50	10^{h}26^m.4	−57°09'	Car	E	12'	12'	3				
NGC 3293	10 34.0	−58 06	Car	E+R	40		B? 3 R−B	91969	6.51	B0Ib	Part of RCW 53
NGC 3324	10 37.7	−58 40	Car	E+R	16	14	pB 2 R−B	92206	7.7	O6	Part of RCW 53. Includes IC 2599
NGC 3372	10 43.8	−59 52	Car	E	120	120	1 R	93308	6.21v		η Car nebula. Contains Keyhole neb.
Gum 32	10 46.3	−58 39	Car	E	7		2		10.0		RCW 52
NGC 3503	11 01.3	−59 51	Car	E+R	3		vF 2 MR		10.46	B0Ve	
RCW 58	11 06.3	−65 34	Car	E	7	7	3	96548	7.70	WN8	Ring-shaped
NGC 3572	11 09.9	−60 06	Car	E	20		3	97151?	7.74	B2Ve	RCW 54
NGC 3576,79	11 12.0	−61 12	Car	E	20	15	F 2				Part of RCW 57 with NGC 3581−2,84,86
NGC 3603	11 15.1	−61 12	Car	E	12	10	2	97950	9.2	WN5+O	Part of RCW 57
Gum 39	11 28.9	−62 41	Cen	E	20	10	3 R	99897?	8.6	O6	Part of RCW 60
IC 2872	11 29.0	−62 57	Cen	E	12	4	4 R	99898?	9.3	B5:III:	Part of RCW 60
Gum 41	11 30.4	−63 50	Cen	E+R	15	15	3 VR−B	100099	8.08	O9III	RCW 61
RCW 59	11 37:	−57:	Cen	E	180	150	3 C				
IC 2944,8	11 38.3	−63 22	Cen	E	75	50	2 R	100841	3.13	B9III	RCW 62. [λ Cen]
NGC 3882	11 46.1	−56 22	Cen	R?			vF	102277	9.0	A2:V	
IC 2966	11 50.4	−64 54	Mus	R	3		2 I		11.47	B0.5V	
Ced 122	13 25.4:	−64 01:	Cen	E	150	150	3	113904?	5.51	B0Ia	Large nebula f Coalsack. [θ Mus]
NGC 5367	13 57.7	−39 59	Cen	R	4	3	vB 2				Double nucleus, includes IC 4347
vdBH 63	14 49.4	−65 14	Cir	R	1		2 B	130079	10.49	B9Vp?	Near dark nebula Bernes 145
vdBH 65a	15 01.1	−63 17	Cir	R			1 R		13.86v?		Cometary
Sh2−1	15 58.9	−26 09	Sco	E+R	90	10	4 VB IrAF	143018	2.89	B1V+B2V	[π Sco] Nebula brightest s of star
IC 4592	16 12.0:	−19 28:	Sco	R	150	60	3 R−MB	145501−2	4.00	B2IV−V	[ν Sco] Outer region is red due to
RCW 103	16 17.1	−51 07	Nor	E	5	3	2				SNR | light from Antares
RCW 102	16 17.8	−51 55	Nor	E	12	8	2	145492	9.3	B2Ib	
IC 4601	16 20.0	−20 02	Sco	R	20	10	3 VB	147009	8.08	B9.5V	Several components
RCW 106	16 20.8	−50 55	Nor	E	35	20	3				
Sh2−9	16 21.1	−25 35	Sco	E+R	60	50	3 VR−B IrAF	147165	2.89v	B2III+O9.5V	[σ Sco]
RCW 104	16 24.0	−51 31	Nor	E	20	20	3	147419	10.5	WN4	Ring-shaped
IC 4604	16 25.6	−23 26	Oph	R	60	25	2 VB	147933−4	4.61	B2V+B2V	[ρ Oph]
IC 4603	16 25.6	−24 28	Oph	R	20	10	eF 3 MB	147889	7.89	B2V	Near ρ Oph
vdB 107	16 29.2	−26 27	Sco	R	85	80	3 VR	148478−9	0.92	M1Ib+dB4	[Antares] Ced 132, but not IC 4606
IC 4605	16 30.2	−25 06	Sco	R	30	30	eF 3 VB	148605	4.78	B2V	[22 Sco]
NGC 6188,93	16 40.5	−48 47	Ara	E+R	20	12	F 2	NGC 6193			RCW 108. Near a dark nebula
NGC 6231	16 52.5:	−42 05:	Sco	E	240		4	151804	5.22	O8Ifp	RCW 113
IC 4628	16 57.0	−40 20	Sco	E	90	60	F 2	152723	7.31	O6k	Near cluster Tr 24
NGC 6281	17 02.8	−38 07	Sco	E	60		3 IrAF	153919	6.55	O7	Gum 57, RCW 119
vdBH 81	17 04.0	−51 05	Ara	R	6		2 MB	153772	8.32	B2V	
Sh2−3	17 12.3	−38 29	Sco	E	12		2 E F		10		Gum 58, RCW 120
RCW 126	17 16.9	−36 21	Sco	E	16	4	3 R Ir				
Sh2−5	17 18.6	−38 25	Sco	E	100	50	3 IrAF	157038	6.41	B4Ia	Gum 59, RCW 123
vdB 111	17 19.0	+6 05	Oph	R	12		3 B	156697	6.49	F0n	
NGC 6334	17 20.5	−35 43	Sco	E	40	30	pF 2 R IrAF	156369	8.9	A2V	Gum 61−64, RCW 127
NGC 6357	17 24.6	−34 10	Sco	E	50	40	F 2 R IrF		10	B	Gum 66, RCW 131
Sh2−13	17 29.1	−31 33	Sco	E	40		3 E AF	158186	7.00	B0V	Gum 68, RCW 133
NGC 6383	17 34.7	−32 35	Sco	E	80	30	3 R C AF	159176	5.70	O6V+O6V	Gum 67, RCW 132
Sh2−16	17 46.6	−29 18	Sgr	E	12	12	3 VR IrAF	Cr 347			Gum 70
NGC 6514	18 02.6	−23 02	Sgr	E+R	29	27	vB 1 R−VB IrF	164492	7.63	O5e	Trifid nebula M20
NGC 6526	18 02.6	−23 35	Sgr	E	40		F 3 C AF	164971	7.3	B0	Gum 74, RCW 145
NGC 6523	18 03.8	−24 23	Sgr	E	90	40	vB 1 R IrF	NGC 6530			Lagoon nebula M8. Faint f ext. IC 4678
Sh2−46	18 06.1	−14 10	SerCd	E	30	20	3 R E AF	165319	7.94	B0Ia	Gum 80, RCW 158
IC 4684	18 09.1	−23 25	Sgr	R	3	2	3 B	165872	9.3	B9	IC 4681 is 12' p
IC 4685	18 09.3	−23 59	Sgr	R	10	8	3	165921	7.28	O7.5	With S-shaped dark nebula B303
IC 1274	18 09.5	−23 44	Sgr	E	9	8	1 R C AF	166033	9.9	B5	Adjacent to dark nebula B91
IC 1275	18 10.0	−23 50	Sgr	E	10	6	1 R C A	166107	7.96	B5V	Adjacent to dark nebula B91
NGC 6559	18 10.0	−24 06	Sgr	E	8	5	vF 1 R IrAF		9.8	B1	Gum 75, RCW 146
Sh2−35	18 15.9	−20 15	Sgr	E	10	7	4 R IrA	167264	5.38	B0Ia	Gum 77b, RCW 151
NGC 6589	18 16.3	−19 48	Sgr	R	5	4	eF 1 MB	167638	9.4	B5	
IC 4701	18 16.5	−16 44	Sgr	E	60	40	4 R IrAF	NGC 6596			Gum 79, RCW 157, with IC 4715 f
NGC 6590,95	18 17.0	−19 53	Sgr	R	4	2	pF 1 MB		10.2	B6	Same as IC 4700
IC 1283−4	18 17.8	−19 40	Sgr	E+R	17	15	2 R IrAF	167815	7.56	B2	Gum 78, RCW 153
NGC 6604	18 17.9	−11 44	SerCd	E	60	30	1 R IrF	NGC 6604			Gum 84−85, RCW 167
NGC 6611	18 18.8	−13 47	SerCd	E	35	28	1 R IrF	NGC 6611			Eagle nebula and cluster M16. (IC 4703
IC 4715	18 19.9	−17 59	Sgr	E	30	5	2 VR				Star cloud | refers to nebula only.)
NGC 6618	18 20.8	−16 11	Sgr	E	46	37	B 1 R IrF	NGC 6618			Omega nebula M17. IC 4706−7 10' n
Sh2−53	18 25.2	−13 13	Sct	E	15		3 IrAF				
vdB 123	18 30.5	+1 11	SerCd	R	10	5	3 MB	170634	9.1	A0	
IC 1287	18 31.3	−10 50	Sct	R	44	34	3 VB	170740	5.5	B2V	
Sh2−64	18 31.6	−1 55	SerCd	E	20	8	2 R IrAF				
Sh2−55	18 32.2	−11 46	Sct	E	20	20	6 B IrAF				
NGC 6726−7	19 01.7	−36 53	CrA	R	2	2	F	176386	7.20	A0V	Double, near dark nebula Bernes 157
NGC 6729	19 01.9	−36 57	CrA	E+R	1			R CrA	9.7 v	F5pe	Cometary, with varying size
Sh2−72	19 03.8	+2 19	Aql	E	25		5 IrAF				
IC 4812	19 05.2	−37 03	CrA	R?			2	176269	6.69	B9V	
vdB 126	19 27.1	+22 43	Vul	R	7	5	2 VB	182918	8.3	A2	

Name	α 2000	δ 2000	Const	Type	Dim		Description		HD	*Mag	*Spec	Notes
Sh2-82	19^{h}30^m.3	+18°16'	Sge	E+R	7'	7'	3	IrA	231616	10.7	B0	
Sh2-91	19 35.6	+29 37	Cyg	E	120	2	4 R	IrF				Long thin filament. SNR
M1-92	19 36.3	+29 33	Cyg	R	0.2	0.1	2					Footprint nebula
NGC 6813	19 40.4	+27 18	Vul	E	3		vF 3	Ir				
NGC 6820	19 43.1	+23 17	Vul	E	40	30	F 3 R	C AF	NGC 6823			
Sh2-88	19 46.0	+25 20	Vul	E	18	6	3 R	IrAF	338916		B	Has two small, brighter knots
Sh2-84	19 49.0	+18 24	Sge	E	15	3	3 R	IrAF	187323	8.0	B5.5V	
Sh2-90	19 49.3	+26 52	Vul	E	8	3	2 R	IrAF				
Sh2-101	20 00.0	+35 17	Cyg	E	16	9	2 R	IrAF	227018	8.99	O7	Near dark nebula B144
Ced 174	20 02.8	+36 58	Cyg	E	15	5	3 R					
vdB 128	20 04.6	+32 15	Cyg	R	8		4 I		190603	5.63	B1.5Ia	
IC 4954-5	20 04.8	+29 15	Vul	R	25		2 B					Three bright components
IC 1311	20 10.8	+41 11	Cyg	E	60	20	eF 2 R	IrAF	194093	2.20	F8Ib	
NGC 6888	20 12.0	+38 21	Cyg	E	20	10	F 1 R	E F	192163	7.4	WN6	Crescent nebula, an oval ring
IC 1318	20 14.3	+39 54	Cyg	E	40	40	F 4 R	IrAF	194093	2.20	F8Ib	
IC 1318	20 16.4	+41 49	Cyg	E	45	25	1 R	IrAF	194093	2.20	F8Ib	
Sh2-104	20 17.8	+36 44	Cyg	E	7	7	2 R	C AF				Cyg OB1
Sh2-108	20 19.1	+39 21	Cyg	E	60	30	2 R	IrAF	194093	2.20	F8Ib	
IC 1318	20 19.3	+40 44	Cyg	E	40	20	2 R	IrAF	194093	2.20	F8Ib	Parts of γ Cyg nebula
Sh2-108	20 23.4	+39 46	Cyg	E	10	5	3 R	IrAF	194093	2.20	F8Ib	
NGC 6914a	20 24.3	+42 18	Cyg	R	6	6	1 VB			9.1	B7	
NGC 6914	20 24.7	+42 29	Cyg	R	13	12	vF					Connected with NGC 6914a and 6914b?
NGC 6914b	20 24.8	+42 23	Cyg	R	4	4	1 VB			8.7	B6	
IC 1318	20 26.2	+40 30	Cyg	E	50	30	1 R	IrAF	194093	2.20	F8Ib	With cl. NGC 6910
IC 1318	20 27.9	+40 00	Cyg	E	45	20	1 R	IrAF	194093	2.20	F8Ib	
vdB 133	20 30.7	+36 56	Cyg	R	10	10	2 MB		195593	6.19	F5Iab	[44 Cyg]
Sh2-112	20 33.9	+45 39	Cyg	E	9	7	2 R	C AF		8.8		
Sh2-115	20 34.5	+46 52	Cyg	E	30	20	3 R	IrAF				
NGC 6960	20 45.7	+30 43	Cyg	E	70	6	pB 2 R-B	IrF				Filamentary neb. at 52 Cyg
	20 48.5	+31 09	Cyg	E	45	30	3 R-B	IrF				Pickering's triangular wisp
IC 5068	20 50.8	+42 31	Cyg	E	80	30	vF 2 R					
IC 5070	20 50.8	+44 21	Cyg	E	80	70	F 3 R					Pelican neb. with IC 5067
IC 5076	20 55.9	+47 25	Cyg	R	9	6	vF 1 VB		199478	5.69	B8Ia	
IC 1340	20 56.2	+31 04	Cyg	E	25	20	3 R-B	IrF				
NGC 6992	20 56.4	+31 43	Cyg	E	60	8	eF 2 R-B	IrF				Network nebula — Veil nebula (SNR)
NGC 6995	20 57.1	+31 13'	Cyg	E	12		F 2 R-B	IrF				
NGC 7000	20 58.8:	+44 20:	Cyg	E	120	100	F 1 R	IrAF	195799?	5.96	O6V	North America nebula
NGC 7023	21 01.8	+68 12	Cep	R	18	18	eF 1 B	IrAF	200775	6.8	B5e	One of the brightest refl. nebulae
Sh2-129	21 11.8	+59 57	Cep	E	100	65	5 R	IrAF	202214	5.64	BOII	Cep OB1
vdB 140	21 17.5	+58 36	Cep	R	16	13	3 VB		203025	6.41	B2IIIe	
NGC 7076	21 26.3	+62 53	Cep	E	2		vF	C				Planetary?
vdB 142	21 37.1	+57 29	Cep	R	1		3 B		239710	8.8	B3	Near elephant-trunk shape in IC 1396
vdB 143	21 37.1	+68 12	Cep	R	8	8	2 VB		206135	8.3	A0	
IC 1396	21 39.1:	+57 30:	Cep	E	170	140	4 R	C AF	IC 1396			Cep OB2
vdB 145	21 43.7	+48 55	Cyg	R	9		3 MB?		206887	7.4	F2	
NGC 7129,33	21 44.4	+66 10	Cep	R	8	7	pF 1 B		NGC 7129			
IC 5146	21 53.5	+47 16	Cyg	E	12	12	pB 1 I	E AF		10.0	B1	Cocoon nebula, with a sparse cluster
Ced 199	22 11.6	+58 45	Cep	R	13	5	3 R		210839	5.04	O6If	[λ Cep]
vdB 152	22 13.6	+70 18	Cep	R	12	6	1 B			8.8	B9.5V	
Sh2-132	22 18.7	+56 08	Cep	E	30	20	2 R	IrF	211853	9.0	Ob	Cep OB1
NGC 7380	22 47.6	+58 04	Cep	E	25	30	1 R	IrF	NGC 7380			Cep OB1
Sh2-155	22 56.8	+62 37	Cep	E	50	30	2 R	IrAF	217086	7.65	O5	Cave nebulae, Cep OB3
IC 1470	23 05.2	+60 15	Cep	E	15	1	vF 2	C AF				Planetary?
NGC 7538	23 13.5	+61 31	Cep	E	10	5	vF 1 R	E AF				
Sh2-157	23 16.1	+60 02	Cas	E	60	50	3 I	IrF	219287	8.91	B0Ia	3' core is Sh2-157; rest Lynds Group 11
NGC 7635	23 20.7	+61 12	Cas	E	15	8	vF 1 R	IrF	220057	6.93	B2IV	Bubble nebula
Ced 211	23 43.8	-15 17	Aqr	E	2	1		IrF	222800	6.36v	M7IIIpev+B2	[R Aqr, symbiotic star]
vdB 150	23 55.1	+76 37	Cep	R	11		3 VB		210806	8.4	A0	

Dark Nebulae

Dark Nebulae

Column Headings

Name — The nebula's number in one of the following catalogues: Lynds, 1962 (LDN), Bernes, 1977 (Be), Sandqvist, 1977 (Sa), or Sandqvist and Lindroos, 1976 (SL).

α 2000 and **δ 2000** — Right ascension and declination, referred to the 2000.0 equinox.

Const — Constellation in which the nebula is located, according to the coordinates given. Some nebulae may span more than one constellation.

Area — The nebula's area in square degrees, as measured with a planimeter on photographs.

Description — A three-part code characterizing the object on photographs. First the opacity is given on a scale of 1 to 6. where

6 is the most opaque. Next, the shape is characterized as either irregular (Ir), elliptical (E), circular (C), cometary (Co), kidney-shaped (K), or S-shaped (S). Finally, "G" is added if the nebula is, or includes, a globule.

Dim — Greatest and least angular dimensions, in minutes of arc.

Barnard — The designation(s) of the nebula in E. E. Barnard's "Catalogue of 349 Dark Objects in the Sky," as published in Part I of *A Photographic Atlas of Selected Regions of the Milky Way* (1927).

Notes — Common names and descriptive information. The letter "B" precedes Barnard numbers. Bright nebulae and clusters adjacent to the dark clouds are listed by their Messier (M), NGC, IC, Cederblad (Ced), Gum, Sharpless (Sh2), van den Bergh (vdB), or van den Bergh-Herbst (vdBH) designation. Relative directions on the sky are north (n), south (s), preceding or west of (p), and following or east of (f), sometimes in various combinations.

Name	α 2000	δ 2000	Const	Area	Description		Dim		Barnard	Notes
	$3^h 32^m.1$:	$+31°10'$	Per		5	Ir	160'	70'	1,2,202-6	Patchy region sp NGC 1333
LDN 1470	3 44.0	+31 47	Per		5	Ir	100	70	3,4	Lies sp o Per
LDN 1471	3 48.0	+32 54	Per	0.097	5	E G	22	9	5	1° nf o Per
	4 19.0	+55 03	Cam	0.184	5	Ir	150	30	8,9,11,13	3° n of cluster NGC 1528
Be 84	4 22.1	+19 30	Tau		4	Ir	20	10		Associated with bright nebula NGC 1554-5
	4 27.4	+18 52	Tau	0.090	6	Ir	25	35		Near bright nebula Sh2-239
	4 30.0	+54 17	Cam	0.130	5	Ir	24		12	Lies sf B9 complex
	4 33:	+26 06	Tau	4.1	6	Ir	600	20	7,10,18-9,22-4,208-20	Narrow lanes roughly p-f
	4 55.2	+30 35	Aur	0.051	6	Ir	20		26-8	Several small clouds adjacent to AB Aur
	5 06.2	+31 44	Aur	0.017	6	C	10		29	1°.2 s, 2° f ι Aur
	5 06.5	-3 26	Ori	0.016	5		10	8		Near bright nebula NGC 1788
	5 29.8	+12 32	Ori	0.412	5	Ir	80	55	30-2,225	3° np λ Ori
LDN 1641	5 33:	-6:	Ori	0.018	5	Ir	300	90		Orion nebula region
	5 40.9	-2 28	Ori	0.004	4	Ir	6	4	33	Horsehead neb., part of large dark f cloud
	5 43.5	+32 39	Aur		4	C G	20		34	2° p cluster M37
	5 45.5	+9 03	Ori		5	E	20	10	35	Near FU Ori and bright nebula Ced 59
	5 49.7	+7 31	Ori	0.181	4	Ir	120		36	Narrow sp-nf lane
LDN 1622	5 54.6	+2 00	Ori	0.122	6	Ir G	15			Near reflection nebula vdB 62
	6 32.8	+10 38	Mon	1.250	5	Ir	180	40	37-9	Near bright nebulae NGC 2245,47
Be 135	7 19.0	-44 35	Pup		6	E	13	5		Contains small reflection nebula
SL 4	8 53.6	-42 13	Vel	0.071	5	Ir	60	10		In emission nebula Gum 17
Be 142	11 09.5	-77 16	Cha		6		100	40		Near reflection nebula IC 2631
	12 53:	-63:	Cru	26.2	3?	Ir	400	300		Coalsack, adjacent Southern Cross
Sa 156	12 59.1	-77 10	Cha	0.231	6					Sa 153-4,158-60 nearby; total area given
Be 146	13 57.6	-40 00	Cen		5	Ir	20	8		Adjacent to bright nebula NGC 5367
Be 145	14 48.6	-65 15	Cir		5		12	5		Near reflection nebula vdBH 63
Sa 172	15 13.2	-62 41	Cir	1.158	4					
Be 148	15 45.5:	-34 24:	Lup	0.071	6	Ir	240	20	228	Contains small, bright reflection nebula
LDN 134	15 53.6	-4 39	Lib	0.220	5	E G	22	12		
SL 11	15 57.0	-37 48	Lup	0.013	6	Ir	40	5		
SL 7	16 01.8	-41 52	Lup	0.048	6	Ir	160	10		Several small clouds
Be 149	16 09.4	-39 08	Sco	0.052	6	Ir	60	12		Contains faint reflection nebula
SL 8	16 14.2	-44 04	Nor	0.030	6	Ir	25	5		Curved
	16 14.7	-18 59	Sco	0.287	3	Ir	15		40	In bright nebula 50' nf ν Sco
	16 22:	-19 40:	Sco	1.780	6	E	200	80	41,43	Two clouds 2°.5 and 4° f ν Sco
SL 24	16 37.5	-35 12	Sco	0.004	6	Ir	50	40	231	
	16 38:	-24 06	Oph	1.560	6	Ir	600		42,44-7,51,238	Narrow lanes extending f ρ Oph
SL 25	16 44.1	-35 21	Sco	0.039	5	Ir	55	20	233	1° f B231
SL 18	16 44.8	-40 23	Sco	0.006	5	Ir	5		44a	2°.4 p bright nebula IC 4628
SL 15	16 46.6	-44 30	Sco	0.006	6	E	7	3	235	
	16 47.8	-12 05	Oph	0.167	6	Ir	150	10		Narrow n-s lane
SL 17	16 53.0	-43 35	Sco	0.013	5	Ir	15	7		Near emission nebula Gum 55
LDN 1775	16 57.2	-22 44	Oph	0.008	6	Ir G	12		46	0°.5 n of 24 Oph
SL 20	17 01.0	-40 47	Sco	0.036	5	Ir	40	15	48	1° sf bright nebula IC 4628
SL 30	17 03.0	-34 26	Sco	0.017	6	Ir	15		50	Lies 0°.5 sp CoD -33° 11706
SL 32	17 06.1	-33 15	Sco	0.016	4	Ir	30	10	53	Curved
LDN 1682	17 07.5	-32 00	Sco	0.023	5	Ir	30	10	55-6	
LDN 11	17 08.3	-22 50	Oph	0.012	6	E G	5		57	In patchy region f B44
LDN 1736	17 10.1	-28 24	Oph	0.038	5	Ir	30	20	244	Lies s of tip of Pipe nebula
SL 23	17 11.2	-40 25	Sco	0.039	6	Ir	15		58	2°.7 f bright nebula IC 4628
LDN 17	17 11.8	-22 27	Oph	0.404	3		30	20	60,246	In patchy region f B44
LDN 1749	17 12.2	-28 51	Oph	0.117	5	Ir	50	10	256	1°.5 s of stem of Pipe nebula. Curved
	17 15.2	-20 21	Oph	0.007	6	Ir	10	4	61	1° sp B63
LDN 1698	17 15.2	-32 13	Sco	0.038	5	Ir	20	5	252	
LDN 99	17 16.0	-21 23	Oph	0.090	3	Ir G	100	20	63	3° nnf θ Oph, with a globule at p end
LDN 100	17 16.2	-20 53	Oph	0.075	6	Ir	25	15	62	0°.5 sp B63
LDN 173	17 17.2	-18 32	Oph	0.020	6	Co	20		64	0°.5 p globular cluster M9
LDN 1710	17 20.7	-31 57	Sco	0.106	5	Ir	60	10		1° f B252
LDN 1773	17 21:	-27:	Oph	0.036	6	Ir	300	60	59,65-7	2° s of θ Oph. Stem of Pipe nebula
LDN 177	17 22.0	-19 19	Oph	0.034	4	Ir	30		259	0°.9 sf globular cluster M9
Be 152	17 22.0	-35 35	Sco	0.013	5		10	7	257	Faint reflection nebula at edge
LDN 102	17 22.5	-21 53	Oph	0.035	6	Ir G	16		67a	3° n of θ Oph
LDN 57	17 22.6	-23 44	Oph	0.003	6	K G	4		68	20' sp B72
LDN 55	17 22.9	-23 53	Oph	0.003	6	Ir	4		69	15' sf B68
LDN 66	17 23.5	-23 38	Oph	0.009	6	S G	4		72	The Snake, 1°.5 nnf θ Oph
LDN 54	17 23.5	-23 58	Oph	0.002	4	C?	4		70	20' sf B68
	17 25.2	-24 12	Oph		5	Ir	15	10	74	15' p 44 Oph
LDN 91	17 25.3	-22 28	Oph	0.016	5	Ir	110	40	75,261-2	Two arcs 1° nf B72
SL 22	17 26.3	-42 38	Sco	0.027	5	Ir	30		263	
LDN 69	17 28.0	-23 22	Oph	2.190	3	Ir	100	50	77,269	Faint extension of bowl of Pipe nebula
LDN 178	17 32.0	-20 32	Oph	0.672	5	Ir	120	50	268,270	
LDN 42	17 33:	-26:	Oph	1.450	6	Ir	200	140	78	2°.5 sf θ Oph. Bowl of Pipe nebula
SL 26	17 34.2	-40 25	Sco	0.019	5	E	10	5		Has fainter extensions
SL 28	17 35.3	-39 14	Sco	0.022	5	Ir	30	15		
LDN 219	17 39.5	-19 47	Oph	0.084	6	Ir	50	30	79,276	B79 is narrow, straight np extension

Name	α 2000	δ 2000	Const	Area	Description		Dim		Barnard	Notes
LDN 233	17^{h}45^m.3	−20°00′	Sgr	0.001	6	E? G	4′	′	83a	On star cloud 1°.7 n, 0°.5 f 58 Oph
LDN 235	17 46.5	−20 11	Sgr	0.038	6	Ir	30	15	84	1°.5 n, 0°.7 f 58 Oph; B83a nearby
	17 51.3	−33 53	Sco		5	Ir	90	60	283	1° nnp cluster M7
	17 54.4	−35 12	Sco		5	Ir	25	15	287	0°.5 ssf cluster M7
LDN 302	17 57.5	−17 40	Sgr	0.040	5	C	16		84a	1°.5 n of cluster M23, with faint ext. to s
	18 02.6	−23 02	Sgr						85	Dark regions in Trifid nebula
LDN 93	18 02.7	−27 50	Sgr	0.007	5	Ir G	4		86	On Sgr star cloud p cluster NGC 6520
	18 03.8	−24 23	Sgr						88−9,296	Dark regions in Lagoon nebula
LDN 1771	18 04.3	−32 30	Sgr	0.115	4	C G	12		87	Parrot's Head, 4°.5 s of cluster NGC 6520
LDN 210	18 09.2	−24 07	Sgr	0.001	5	S	1		303	In bright nebula IC 4685
LDN 227	18 10.0	−23 39	Sgr	0.050	5	K	5	2	91	Adjacent to bright nebulae IC 1274−5
LDN 108	18 10.2	−28 19	Sgr	0.003	6	Ir G	10		90	1°.5 f, 0°.4 s of cluster NGC 6520
LDN 323	18 15.5	−18 11	Sgr	0.016	6	E G	12	6	92	On star cloud 0°.7 np cluster M24
LDN 327	18 16.9	−18 04	Sgr	0.011	4	Co G	12	2	93	0°.5 f B92
LDN 406	18 25.6	−11 45	Sct	0.746	5	C G	30		95	2°.6 nf bright nebula M16
LDN 435	18 29.1	−9 56	Sct	8.4?	4	Ir	50	50	97	1° np cluster NGC 6649
LDN 379	18 30.9	−15 08	Sct	0.181	4	E	100	30	312	2°.5 f Omega nebula, with sharp n boundary
LDN 443	18 32.7	−9 08	Sct	0.045	5	Ir G	40	15	100−1	1°.4 n of cluster NGC 6649, and curved
LDN 564	18 37.6	−1 12	SerCd	0.208	5	Ir	45	15		
LDN 445	18 37.7	−9 37	Sct	0.086	5	Ir	35	25	314	1° nf cluster NGC 6649
LDN 557	18 38.6	−1 47	SerCd	0.181	6	Ir	60	10		
LDN 497	18 39.2	−6 37	Sct	0.027	6	Ir	40	40	103	On np side of Sct star cloud
LDN 532	18 47.3	−4 32	Sct	0.016	5		16	1	104	20′ n of β Sct. L-shaped
	18 49.6	−6 19	Sct		3		3		108	0°.5 p cluster M11
	18 49.7	−6 24	Sct		2		90	2	318	Narrow p−f lane just s of cluster M11
LDN 530	18 50.2	−4 46	Sct	0.124	6	Ir G	9		110	Part of B111
LDN 534	18 51:	−5:	Sct	3.185	3	Ir	120	120	111,119a	Two crescent-shaped areas n of cluster M11
	18 51.2	−6 40	Sct		4	Ir	20	20	112	Lies s of cluster M11
LDN 548	18 51.4	−4 19	Sct	1.000	5	Ir G	11		113	Part of B111
LDN 582	18 52.6	−1 56	Aql	0.283	5	Ir	50	10		
	18 53.2	−7 06	Sct	0.066	6	Ir	50	5	114−8	Chain of dark nebulae sf cluster M11
LDN 509	18 53.9	−7 27	Sct	0.003	6	C G	1		118	Member of group
LDN 617	18 57.5	+1 04	Aql	0.745	5	Ir	180	20		
	19 01.6	−5 26	Aql	0.007	5	Ir	20	5	127,129−30	Curved, lying n of 12 Aql
Be 157	19 02.9	−37 08	CrA	0.093	6	Ir	55	18		Lies sf bright nebulae NGC 6726−7
LDN 567	19 04.1	−4 28	Aql	0.024	6	Ir	16	8	132,328	40′ np λ Aql
LDN 531	19 06.1	−6 50	Aql	0.005	6	Co G	10	3	133	On Sct star cloud 2° s of λ Aql
LDN 543	19 06.9	−6 14	Aql	0.007	6	C G	6		134	1°.4 s of λ Aql
LDN 581	19 07.4	−3 55	Aql	0.072	6	Ir	50	30	135−6	1° nf λ Aql
SL 42	19 10.3	−37 08	CrA	0.033	6	E	12	8		1°.8 f bright nebulae NGC 6726−7
LDN 627	19 15.6	+0 13	Aql	2.990	3	Ir	180	10	137−8	Long, curved lane on Sct star cloud
LDN 619	19 18.1	−1 28	Aql	0.016	5	E G	10	2	139	At s tip of B137−8
LDN 673	19 20.9	+11 16	Aql	0.199	6	Ir	55	15		8° np Altair, filamentary
LDN 684	19 21.8	+12 26	Aql	0.185	5	Ir	50	10		Lies n of dark nebula LDN 673
	19 36.8	+12 27	Aql	0.086	4	Ir	40	5	334,336−7	2° np B142−3
LDN 663	19 36.9	+7 34	Aql	0.005	6	E G	4		335	3°.5 p, 1°.2 s of Altair
	19 40.7	+10 57	Aql	0.109	6	Ir	80	50	142−3	Narrow lanes 3° np Altair
LDN 857	19 59:	+35:	Cyg	0.800	1	Ir	360	180	144	Fish on the Platter, p cluster NGC 6883
LDN 865	20 02.8	+37 40	Cyg	0.075	4	G	35	6	145	Triangular, n of B144
LDN 860	20 03.5	+36 02	Cyg	0.001	6		1		146	In B144 adjacent to BD +35° 3930
LDN 880	20 13.5	+40 16	Cyg	0.015	5	Ir G	10	5	343	1°.7 p γ Cyg
LDN 889	20 24.8	+40 10	Cyg	0.800	4		90	20		Wide lane f γ Cyg
	20 26.7	+43 45	Cyg		6	K	10	4	346	In patchy area 2°.8 p, 1°.5 s of Deneb
	20 28.4	+39 55	Cyg		5		10	1	347	Narrow streak 1°.2 f, 0°.4 s of γ Cyg
LDN 906	20 40:	+42:	Cyg	15.8	3					Northern Coalsack
	20 49.1	+45 53	Cyg		6	C	3		350	14′ s of 55 Cyg
	20 49.1	+59 32	Cep	0.008	5	C	3		148−9	Two small clouds s of B150
	20 50.6	+60 18	Cep	0.111	5	Ir	60	3	150	Curved filament 1°.6 s of η Cep
LDN 935	20 56.8	+43 52	Cyg	2.250	4		150	40		Wide lane p North America nebula
	20 57.1	+45 54	Cyg		5	Ir	20	10	352	Lies n of North America nebula
LDN 970	21 12.9	+47 22	Cyg	0.123	4	C G	17		361	Has faint p extension
	21 14.5	+61 45	Cep	0.010	5	Ir	15	3	152	Double cloud 1° sp α Cep
LDN 1017	21 24.0	+50 10	Cyg	0.027	5	E G	15	8	362	Adjacent to 9th-mag. star BD +49° 3517
	21 33.6	+54 33	Cyg	0.145	5	Ir	40		364	Narrow lanes sf B157
LDN 1075	21 33.7	+54 40	Cyg	0.145	4	C G	4		157	8′ p 8th-mag. star BD +54° 2576
LDN 1090	21 34.9	+56 43	Cep	0.028	4	S	22	3	365	Lies sp bright nebula IC 1396
LDN 1088	21 38.0	+56 14	Cep	0.019	4	Ir	30	15	160	Lies s of bright nebula IC 1396
LDN 1103	21 40.3	+57 49	Cep	0.003	6	Co G	13	3	161	In n part of bright nebula IC 1396
LDN 1095	21 41.1	+56 19	Cep	0.012	4	Ir	13	2	162	Curved strip f B160
LDN 1106	21 42.2	+56 42	Cep	0.009	4	Ir G	4		163	In s part of bright nebula IC 1396
LDN 1113	21 44.4	+57 12	Cep	0.002	5	Ir G	3		367	In p part of bright nebula IC 1396
LDN 1070	21 46.5	+51 04	Cyg	0.050	5	K G	12	6	164	0°.8 f π^1 Cyg
	21 53.2	+47 12	Cyg	0.199	5	Ir	100	10	168	Narrow p−f lane, Cocoon nebula at f end
LDN 1151	21 58.9	+58 45	Cep	0.027	5	Ir	80		169−71	Narrow, curved lanes 3° nf IC 1396
LDN 1164	22 07.4	+59 10	Cep	0.019	6	Ir	40		173−4	Patchy lane nf B169

Planetary Nebulae

Planetary Nebulae

Column Headings

PK — Designation in the *Catalogue of Galactic Planetary Nebulae*, L. Perek and L. Kohoutek (1967). The number is formed by concatenating the 1950.0 galactic longitude and latitude (both truncated to the nearest degree) with an integer representing the object's position within that 1°-by-1° sky area from north to south.

NGC — Number in J. L. E. Dreyer's *New General Catalogue of Nebulae and Clusters of Stars*; if none, the *Index Catalogue* (IC) designation is given.

α 2000 and **δ 2000** — Right ascension and declination, referred to the 2000.0 equinox.

Const — Constellation in which the nebula is located.

Diam — Angular diameter in seconds of arc. If the nebula consists of a bright, compact core surrounded by a fainter, extended halo, the dimensions of both are given and are separated by a slash (/). The symbol ">" means that, while the size of the core only is listed, a larger halo exists. The symbol "<" denotes an upper limit to the planetary's diameter, as measured on a small-scale photographic plate.

m(pg) — Total photographic magnitude of the nebula. The symbol ">" means "fainter than."

RV — Radial velocity in kilometers per second, according to S. E. Schneider *et al.* (1983). Positive means receding (traveling away from the solar system). An asterisk indicates that a weighted average has been formed from discordant measurements.

EV — Expansion velocity in kilometers per second.

Type — Vorontsov-Velyaminov type, whose symbols are defined as follows.
1. *Stellar image.*
2. *Smooth disk* (*a,* brighter toward center; *b,* uniform brightness; *c,* traces of ring structure).
3. *Irregular disk* (*a,* very irregular brightness distribution; *b,* traces of ring structure).
4. *Ring structure.*
5. *Irregular form* similar to a diffuse nebula.
6. *Anomalous form.*

More complex structure is characterized by combinations of codes, such as "4 + 2" (ring and disk), or "4 + 4" (two rings).

HD — Henry Draper number of the central star. If none, the variable star designation, Smithsonian Astrophysical Observatory (SAO) number, or Cambridge Research Laboratory (CRL) infrared survey number is given.

*Mag — Photoelectric **V** magnitude of the central star. If this is not available, a photographic (p) or photoelectric **B** value is listed. The letter "v" means that the star is variable, while the symbol ">" stands for "fainter than." A colon indicates uncertainty.

*Spec — Spectral type of the central star. The abbreviation "Cont" means that the spectrum is continuous (no lines seen).

d(kpc) — Approximate distance in kiloparsecs. To convert to light-years, multiply by 3,260.

Notes — A common name, Messier number, or other information may be given here. The following words are often abbreviated: cluster (cl.); eclipsing (ecl.); photovisual (pv); the Rodgers, Campbell, and Whiteoak Hα-emission survey (RCW); Sharpless H II region (Sh2); spectroscopic (sp.); visual companion (VC).

PK	NGC	α 2000	δ 2000	Const	Diam	m(pg)	RV	EV	Type	HD	*Mag	*Spec	d(kpc)	Notes
118 +8.2	Abell 86	0ʰ01ᵐ.6	+70°43′	Cep	63″	16.7			2c					
118 +2.1		0 07.6	+64 58	Cas	125				3					
119 +6.1		0 12.6	+69 11	Cep	44	18.3			2b		19.9p			
120 +9.1	40	0 13.0	+72 32	Cep	>37	10.7	-20	28	3b+3	826	11.61	WC8	0.9	
119 -6.1	Hu 1-1	0 28.3	+55 58	Cas	5	13.3	-54	15	2		16.5	A?		Sp. binary
122 -4.1	Abell 2	0 45.6	+57 57	Cas	31	16.3	-42		2c		16			
118-74.1	246	0 47.0	-11 53	Cet	225	8.0	-46	38	3b		11.94	OVI+K	0.4	VC: V=14.28, 4″
130-10.1	650-1	1 42.4	+51 34	Per	65/290	12.2	-19	42?	3+6		17.0:		1.1	M76, Little Dumbbell. VC:
130 +1.1	IC 1747	1 57.6	+63 20	Cas	13	13.6	-67	30	3b	11758	14.54	WC4	2.1	| m(pv)=17.7, 1″.4
131 +2.1		2 12.2	+64 09	Cas	60	18.2			3b		18.3p			Sh2-189
144-15.1		2 45.4	+42 33	Per	22	16.7			3b		>17.1			
141 -7.1		2 52.3	+50 36	Per	128	>16			4		20.3p			
136 +4.1		2 58.9	+64 30	Cas	185	>15.5			2b		19.0p			
138 +2.1	IC 289	3 10.3	+61 19	Cas	>34	12.3	-13	22	4+2		>15.9	Cont	1.2	
220-53.1	1360	3 33.3	-25 51	For	390		+42	27	3		11.35	sd	0.3	Sp. binary?
159-15.1	IC 351	3 47.5	+35 03	Per	7	12.4	-9	15	2a		14.77	WR	3.9	
171-25.1	Baade 1	3 53.5	+19 28	Tau	40	13.9	-10		4		>17		2.2	
161-14.1	IC 2003	3 56.4	+33 52	Per	7	12.6	-16	18	2		16.49v?	WC7-8	4.4	Discordant magnitudes
144 +6.1	1501	4 07.0	+60 55	Cam	52	13.3	+36	39	3		14.39	WC6-7	1.2	
165-15.1	1514	4 09.2	+30 47	Tau	>114	10	+60	25	3+2	281679	9.40	A0III+sd0	0.6	Binary, period 0.41 day
147 +4.1	M 2-2	4 13.3	+56 57	Cam	12		-7		3					
206-40.1	1535	4 14.2	-12 44	Eri	18/44	9.6	-3	19	4+2c	26847	12.24	O5		
174-14.1		4 37.3	+25 03	Tau	20		-20			GL Tau	18.61v			Possible planetary
166 -6.1		4 42.9	+36 07	Aur	12		-28	18		CRL 618	17	B0-09.5	1.8	Protoplanetary, bipolar
215-30.1		5 03.2	-15 36	Lep	770	13.2	+18		3a		15.44	sdOk	0.3	
190-17.1	d. 320	5 05.6	+10 42	Ori	>7	12.9	-23	17	2+4		14.44	Cont	4.0	
167 -0.1		5 06.6	+39 08	Aur	60	16.6	+58		2b		19.6p			
173 -5.1		5 08.1	+30 48	Aur	132		+19		3		18.2p			
169 -0.1	IC 2120	5 18.2	+37 36	Aur	47		-8		2		15.05	G		
215-24.1	IC 418	5 27.5	-12 42	Lep	12	10.7	+62	22	4	35914	10.65v?	O7f	0.4	Sp. binary, period 0.2 day?
172 +0.1		5 29.0	+36 02	Aur	37	>18.9			4		20.8p			
197-14.1		5 31.8	+6 56	Ori	35	15.2			3		19.5p			
196-12.1		5 37.3	+8 16	Ori	32	17.1			2c					
196-10.1	2022	5 42.1	+9 05	Ori	>18	12.4	+14	29	4+2		15.20	Cont	2.1	
166+10.1	IC 2149	5 56.3	+46 07	Aur	>8	11.2	-31	20	3b+2	39659	11.59	O7.5p	1.0	
198 -6.1		6 02.4	+9 39	Ori	37	13.9					19.20B		2.4	
173 +3.1		6 02.6	+36 08	Aur	20		+45						4.3	Near H II region Sh2-232
204 -8.1		6 04.8	+3 57	Ori	154	>16.0			4		19.87		>0.9	
197 -3.1		6 11.2	+11 46	Ori	34	>18.2			2c		15.24	F7V?		Spectrum of foreground object?
158+17.1		6 19.6	+55 37	Lyn	20/1200	11.2					15.4p	sdO	0.1	Optical double
221-12.1	IC 2165	6 21.7	-12 59	CMa	4	12.9	+54	20	3b		>15.9	Cont	3.1	
194 +2.1	J. 900	6 25.9	+17 47	Gem	>8	12.4	+47	18	3b+2		16.26		2.3	
233-16.1		6 27.0	-25 23	CMa	34	16.3			4		15.38	sdOp	3.7	
189 +7.1	M1-7	6 37.4	+24 00	Gem	29		+2		2					
153+22.1		6 43.9	+61 47	Lyn	148	15.9			2b		17.7p			
221 -4.1		6 48.6	-9 32	Mon	44	>18.5			2c		19.9p			
204 +4.1		6 52.6	+9 58	Mon	414				3		18.2?p			
216 -0.1		6 56.2	-2 53	Mon	75	17.5			2b					
200 +8.1		6 59.9	+14 37	Gem	69	17.0			2b					
242-11.1		7 02.8	-31 35	CMa	12		+70		?+6		14.09		3.5	
223 -2.1		7 03.4	-10 34	Mon	79				2		16.8p		1.6	
212 +4.1		7 05.3	+2 47	Mon			+137		1					
234 -6.1		7 06.9	-22 02	CMa	65				3		>21p			
219 +1.1		7 07.3	-5 10	Mon	>37				4+2		18.6p			
215 +3.1	2346	7 09.4	-0 48	Mon	>55		+22	40:	3b+6		11.16v	A5V?	1.2	Ecl. binary, 16 days. Bipolar
232 -4.1		7 11.3	-19 51	CMa			+29		1		14.04v?	Cont		Possible planetary
229 -2.1		7 12.5	-16 08	CMa	66				3b		>21p			
240 -7.1	M3-2	7 14.8	-27 50	CMa			+84		3b					
241 -7.1		7 16.5	-29 19	CMa	5				2					
214 +7.1		7 23.0	+1 46	CMi	64	16.3			2c		16.56		2.4	
189+19.1	2371-2	7 25.6	+29 29	Gem	>55	13.0	+21		3a+6		>15.5	OVI?	1.2	
221 +5.1		7 26.6	-5 22	Mon	13		+95		4					
248 -8.1		7 28.9	-35 45	Pup	6				2		15.12			
205+14.1		7 29.0	+13 15	Gem			+29	32			15.99		0.2	Medusa nebula, Sh2-274
197+17.1	2392	7 29.2	+20 55	Gem	13/44	9.9	+75	54	3b+3b	59088	10.47	O7f	0.9	Eskimo nebula. Hot companion?
215+11.1		7 36.1	+2 42	CMi	87	15.4			3		19.59			
226 +5.1		7 37.3	-9 39	Mon	>3		+49		?+6		14.69		8.4	
231 +4.2	2438	7 41.8	-14 44	Pup	>66	10.1	+76		4+2		17.7:	Cont	0.9	
234 +2.1	2440	7 41.9	-18 13	Pup	14/32	10.8	+63	24	5+3		14.32		1.1	Protoplanetary?
231 +4.1		7 42.1	-14 21	Pup	32/130		+18		2b		19.1p		1.8	
243 -1.1	2452	7 47.4	-27 20	Pup	>19	12.6	+68		4+3		19.0B	WC3	2.7	RCW 17
236 +3.1		7 50.2	-19 18	Pup	37				2		>21p		3.9	
217+14.1		7 51.7	+3 00	CMi	>229	13.6	+13	14	4+3		17.18		0.4	VC: m(pv)=17.8, 3″.4
241 +2.1		7 55.2	-23 38	Pup	14		+74		4					
164+31.1		7 57.8	+53 25	Lyn	400	14	-85	35	4		16.0		0.5	This is not NGC 2474-5

PK	NGC	α 2000	δ 2000	Const	Diam	m(pg)	RV	EV	Type	HD	*Mag	*Spec	d(kpc)	Notes
245 +1.1		8h02m.5	-27°42'	Pup	7"		+64		4		15.50	O7	2.5	
224+15.1		8 06.7	-2 52	Mon	168	15.4			3b		18.94			
250 +0.1		8 09.0	-32 40	Pup	34	18.1			2					
264 -8.1		8 11.5	-48 43	Vel	46	12.4			2					
252 +4.1		8 31.8	-32 05	Pyx	49	>17.9			3b		>21p			
239+13.1	2610	8 33.4	-16 09	Hya	>37	13.6	+88		4+2		15.5v?	Cont	1.7	
244+12.1		8 40.2	-20 54	Pyx	393	14.3			4		18.35			
254 +5.1		8 40.7	-32 23	Pyx	9		+50		2a					
158+37.1		8 41.6	+58 13	UMa	284	>14.6			2b		15.9p			
208+33.1		8 46.8	+17 53	Cnc	109	15.6			2c		14.37	OVI	1.7	VC: m(pv)=17.0, 5".3
219+31.1		8 54.2	+8 55	Cnc	>980	12.2			3a		15.51	sdO6		Sh2-290
253+10.1		8 57.8	-28 57	Pyx	59		+66		2		16.8v		2.4	Ecl. binary, period 0.67 day
285-14.1	IC 2448	9 07.1	-69 57	Car	8	11.5	-24		2b		12.92B		2.8	
265 +4.1	2792	9 12.4	-42 26	Vel	11	13.5	+14		4		13.78		2.5	
261 +8.1	2818A	9 16.0	-36 38	Pyx	38	13.0	-2		3b					Open cluster member?
268 +2.1		9 16.2	-45 29	Vel	<5				1		16.09			
227+33.1		9 16.4	+3 53	Hya	120	>16.1			2b		19.4p			
278 -5.1	2867	9 21.4	-58 19	Car	11	9.7	+14		4		13.56B	WC	1.7	
277 -3.1	2899	9 27.0	-56 06	Vel	120		+4	>14						Gum 27, RCW 43
281 -5.1	IC 2501	9 38.8	-60 05	Car	<25	11.3	+33		1				1.3	
238+34.1		9 39.1	-2 48	Hya	268	13.4	+60	32	2b		14.74	sdOp	0.6	VC: m(pv)=17.0, 1".8
279 -3.1		9 43.4	-57 17	Car	<25		-7				11.3	A2III?	2.7	Sp. binary
248+29.1		9 45.6	-13 10	Hya	288	14.5			2b		16.32			
221+45.1		9 48.0	+13 17	Leo	>5		-20	15		CW Leo			0.3	Protoplan., very bright in IR
272+12.1	3132	10 07.7	-40 26	Vel	>47	8.2	-16	13	4+2	87892	10.07	A2V+sdO	0.8	VC: V=16.0, 1".6
285 -5.1	IC 2553	10 09.3	-62 37	Car	9	13.0	+37		3		12.89		2.8	
296-20.1	3195	10 09.5	-80 52	Cha	38				3					
283 -1.1		10 15.6	-58 51	Car	30				4					
286 -4.1	3211	10 17.8	-62 40	Car	12	11.8	-23		2b					
261+32.1	3242	10 24.8	-18 38	Hya	16/1250	8.6	+5	23	4+3b	90255	12.0		0.8	Ghost of Jupiter
288 -5.1		10 35.8	-64 19	Car	10	14.2		10	2a					
288 +0.1		10 54.6	-59 10	Car	33				4					
289 -0.1		10 56.2	-60 27	Car	35		-14?	95?	4	94910	6.96v	B0:Ieq	1.3	Possible planetary. AG Car
291 -4.1	IC 2621	11 00.3	-65 15	Car	5		+14		1		13.62		1.1	
290 -0.1		11 03.9	-60 36	Car	28				3					
148+57.1	3587	11 14.8	+55 01	UMa	194	12.0	+6	41	3a		15.88		0.4	M97, Owl nebula
292 +1.1	3699	11 28.0	-59 57	Cen	67		-6	>18						
290 +7.1		11 28.6	-52 56	Cen	>25		+28		3+6					
294 +4.1	3918	11 50.3	-57 11	Cen	12	8.4	-16	20	2b		10.84		0.8	
298 -4.1	4071	12 04.2	-67 18	Mus	75		+11	>11						
275+72.1		12 18.3	+11 02	Vir	690				2		17.7?p		0.2	Not planetary?
294+43.1	4361	12 24.5	-18 48	Crv	45/110	10.3	+9	38	3a+2		13.24	O6	0.8	Sp. binary?
123+34.1	IC 3568	12 32.9	+82 33	Cam	>6	11.6	-41	8	2+2a	109540	12.31	O5f	2.1	
303+40.1		12 53.6	-22 52	Hya	709	12.0	-7	4	3a	SAO 181201	9.63	G8IV+sd?	0.2	Abell 35. Parabola-shaped,
49+88.1		12 59.4	+27 38	Com		16.0	-141							ǀ star not central
304 -4.1	IC 4191	13 08.8	-67 39	Mus	5	12.0	-13	14	2				2.3	
307 -3.1	5189	13 33.5	-65 59	Mus	153	10.3	-9	31	5	117622	14.0	OVI,WC?	0.8	Gum 47, RCW 76. Binary?
307 -4.1		13 39.6	-67 23	Mus	<25	12.2	-55	10			13.65B		2.4	
318+41.1		13 40.6	-19 53	Vir	293/400	13.0	+36	28	3b+3a		11.46	sdO7	0.4	Sp. binary
312+10.1	5307	13 51.1	-51 12	Cen	13	12.1	+40	15	3					
309 -4.2	5315	13 53.9	-66 31	Cir	5	13.0	-34	>40	2		11.32	WC6	2.8	
317+19.1	5408	14 03.3	-41 23	Cen										
326+42.1	IC 972	14 04.4	-17 15	Vir	43	14.9	-27	16	2c		>16		2.0	
319+15.1	IC 4406	14 22.4	-44 09	Lup	>28	10.6	-41	6	4+3	125720	14.7B	WR	1.5	VC: 35"
331+16.1	5873	15 12.8	-38 08	Lup	3	13.3	-129	>40	2		13.55		4.8	
327+10.1	5882	15 16.8	-45 39	Lup	7	10.5	+10*	>40			12.0B		1.9	
342+27.1	me 2-1	15 22.3	-23 38	Lib	7		+46	13	2		15.4?		4.3	
322 -2.1		15 34.3	-59 09	Nor	>26	12.5	-31		4+6				2.4	RCW 93
315-13.1		15 37.2	-71 55	Aps	5		-1.2			138403	10.48B	O7(var)	1.5	Bipolar
322 -5.1	5979	15 47.7	-61 13	TrA	8		+23				13.01		3.3	
329 +2.1		15 51.7	-51 31	Nor	76	13.6	-33		4		13.87	Cont?	1.1	RCW 100
341+13.1	6026	16 01.4	-34 32	Lup	45		-112		4				2.3	
64+48.1	6058	16 04.4	+40 41	Her	>23	13.3	+3	33	3+2		13.78	O9?	2.6	
25+40.1	IC 4593	16 12.2	+12 04	Her	12/120	10.9	+22	13	2+2	145649	11.27	O7f	2.0	
342+10.1	6072	16 13.0	-36 14	Sco	40/70	14.1	+6		3a		17.5:B		1.0	
329 -2.2		16 14.5	-54 57	Nor	27	12.6	-29	19	4+3					
331 -1.1		16 17.2	-51 59	Nor	24		-21		6		14.14	O9.5-B0	2.1	Protoplanetary, bipolar
346+12.1		16 23.3	-31 45	Sco	92	>15.7	-12	17	4+2		20:p		0.9	
336 +1.1		16 23.9	-46 42	Nor	12		-76		2					
47+42.1		16 27.5	+27 54	Her	170	13.7			2c		15.69	sdO		
341 +5.1	6153	16 31.5	-40 15	Sco	25	11.5	+39		4					RCW 112
0+17.1		16 43.8	-18 57	Oph	5		-44		1		14.61		5.7	
43+37.1	6210	16 44.5	+23 49	Her	>14	9.3	-36	20	2+3b	151121	12.90	O3?	1.1	
359+15.1		16 48.6	-21 01	Oph	29	16.8			2b		19.6p		4.7	
342 +0.1		16 53.5	-42 39	Sco	16				2					

PK	NGC	α 2000 δ 2000	Const	Diam	m(pg)	RV	EV	Type	HD	*Mag	*Spec	d(kpc)	Notes
351 +7.1		16h 53.6m -31°41'	Sco	"				1					
353 +8.1		16 55.8 -29 50	Oph	<5	12.8	-44		1					
344 +0.1		16 57.4 -41 38	Sco	5				2					
0+12.1	IC 4634	17 01.6 -21 50	Oph	>9	10.7	-33	15	2a+3	153655	15p	Cont?	2.9	
351 +5.1		17 02.3 -33 10	Sco	5		-98		2					
350 +4.1		17 04.6 -33 59	Sco	5		-20		2		12.89		3.6	
353 +6.1		17 05.2 -30 32	Sco	8		-56		2					
345 +0.1	IC 4637	17 05.2 -40 53	Sco	19	13.6	+11		3	154072	12.30		1.5	
352 +5.1		17 05.5 -32 32	Sco	4		+27		2					
10+18.1		17 05.6 -10 06	Oph	4				2+3		14.7v?		3.3	
10+18.2		17 05.6 -10 08	Oph	20		+88		?+6			B1V	2.8	Bipolar. Protoplan.? Binary?
342 -2.1		17 06.4 -44 13	Sco	20				4					
344 -1.1		17 07.0 -42 41	Sco	12				4					
351 +3.1		17 07.4 -34 05	Sco					1					
345 -1.1		17 10.5 -41 53	Sco	8				2			A		
334 -9.1	IC 4642	17 11.8 -55 24	Ara	21	13.4	+44		4	154952	13.5B		3.3	
352 +3.1		17 12.2 -32 38	Sco			-14		1					
358 +7.1		17 12.6 -25 44	Oph	4		+12		2					
18+20.1		17 12.9 -3 16	Oph		13.4	+22				16.7			
349 +1.1	6302	17 13.7 -37 06	Sco	50	12.8	-39	8	6	155520		OVI	0.6	Bug nebula, Sh2-6. H II region?
9+14.1	6309	17 14.1 -12 55	Oph	14/66	10.8	-48	35	3b+6	155752	14.4	Cont	2.6	
354 +4.1		17 14.1 -31 20	Sco	4		-75		2					
352 +3.2		17 14.7 -33 25	Sco	3				2					
354 +4.2		17 15.3 -31 34	Sco					1					
359 +6.1		17 19.2 -25 17	Oph	10		-67		2a					
338 -8.1	6326	17 20.8 -51 45	Ara	14	12.2	+10		3b	156531	13.49		2.4	
358 +5.1		17 21.2 -27 12	Oph	16		+2		3+2					
2 +8.1		17 21.3 -22 19	Oph	4		+2		2					
349 -1.1	6337	17 22.3 -38 29	Sco	48		-74:		4		14.75		1.2	
357 +4.1		17 23.4 -28 59	Oph	4				2		20:p			
357 +3.1		17 24.6 -29 24	Oph	5		-191		2					
3 +7.1		17 24.8 -21 34	Oph	4				2		>21p			
358 +4.1		17 24.9 -28 06	Oph	5		+95 -		2					
359 +5.2		17 25.7 -26 12	Oph	16		-82		3		18:p		3.8	
357 +3.2		17 26.0 -29 22	Oph	4		-69		2					
352 +0.1		17 26.4 -35 02	Sco	7				2					
357 +3.4		17 27.0 -29 16	Oph	6		-250		2					
1 +5.1		17 28.1 -24 25	Oph	7		+34		2					
352 -0.1		17 28.5 -35 08	Oph	10				4					
1 +5.2		17 28.6 -24 51	Oph	5		+36		2					
358 +3.1		17 28.7 -28 27	Oph			-96		2					
357 +2.5		17 28.8 -30 08	Sco	7		+29		2		>21p		5.2	
9+10.1		17 29.0 -15 13	SerCd	12	17.2			3+2		15.95v			Binary, period 2.7 hours
6 +8.1		17 29.0 -19 16	Oph	<10		+92		1		15.32			
2 +5.1	6369	17 29.3 -23 46	Oph	30/66	12.9	-102		4+2	158269	14.66	WR(+OVI?)	1.2	
0 +4.1		17 29.4 -25 49	Oph					1		>21p			
357 +2.4		17 29.7 -29 33	Oph	<25		-204		1					
358 +3.4		17 30.0 -27 59	Oph	<25		+65		1					
358 +3.6		17 30.7 -28 04	Oph	4		+186		2					
357 +2.6		17 31.1 -30 11	Sco					1					
16+13.1		17 31.5 -8 19	Oph	58	17.8			2b		19.8p			
357 +1.1		17 32.8 -30 00	Oph	3		-72		2					
4 +6.2		17 33.6 -21 46	Oph	9		+160		2					
3 +5.1		17 34.4 -22 53	Oph	4		-59		2					
0 +3.1		17 34.9 -26 36	Oph	5		-200		2					
7 +7.1		17 35.2 -18 34	Oph	9	13.3	+13		4		>21p			
358 +1.1		17 35.2 -29 03	Oph			-292		1					
5 +6.1		17 35.4 -20 57	Oph	10		+1		2					
5 +5.1		17 36.4 -21 31	Oph	6		+18		2					
350 -3.1		17 36.5 -39 22	Sco	18		-37		4					
7 +6.1		17 37.4 -18 47	Oph	7		-65		2					
7 +6.2		17 38.2 -19 38	Oph	6		-7		2					
4 +4.1		17 38.5 -22 09	Oph	5		+26		2					
5 +5.2		17 39.9 -21 14	Oph	15		-56		3					
3 +3.1		17 40.1 -24 26	Oph	<25		+87		2		>21p		7.2	
5 +4.1		17 40.3 -22 19	Oph	5		+17		2					
1 +1.1		17 40.5 -27 01	Oph	40				3		19.7:p			Symbiotic star
3 +3.2		17 41.9 -24 11	Oph	<25		-35		1					
3 +2.1		17 41.9 -24 42	Oph	6		-63	23	3b		14.91		1.5	
45+24.1		17 42.5 +21 27	Her	48				4		13.25		3.4	Visual binary
350 -5.1		17 42.9 -39 36	Sco	8		-45		2					
6 +4.1		17 43.5 -21 10	Sgr	4		-116		2					
355 -2.1		17 44.3 -34 07	Sco	8		-76		2a					
51+25.1		17 45.0 +27 19	Her	43				3b		19.7p			
352 -4.1		17 45.1 -38 09	Sco	5		-8		2					

PK	NGC	α 2000	δ 2000	Const	Diam	m(pg)	RV	EV	Type	HD	*Mag	*Spec	d(kpc)	Notes
6 +4.2		17^{h}45^m.5	−20°58′	Sgr	4″		+100		2					
346 −8.1	IC 4663	17 45.5	−44 54	Sco	13	13.1	−49	15	4		14.00		3.0	
5 +3.1		17 45.6	−23 02	Sgr	12				2		>21p			
345 −8.1		17 45.6	−46 05	Ara	10		−84		4	161044	11.11		2.0	Possible planetary
2 +1.1		17 45.7	−25 40	Sgr	4				2		12.49			
358 −0.2		17 46.0	−30 12	Sco	4		−5		2	316248	12.11		1.6	RCW 135. Not planetary?
358 −1.1		17 46.0	−31 04	Sco	13				3+2					
11 +6.1		17 46.9	−16 17	Sgr	5		+4		2					
6 +3.1		17 47.6	−21 47	Sgr	6				2					
6 +3.2		17 47.6	−22 06	Sgr	18		+18		3+6					
359 −0.1		17 47.9	−30 00	Sgr	20	13.6	−28		2?+6	316340		Cont	1.9	
5 +2.1		17 48.1	−22 47	Sgr					1					
11 +5.1	6439	17 48.3	−16 29	Sgr	5	13.8	−94	24	2a		18.5:B		5.0	
4 +1.1		17 48.6	−24 17	Sgr	5				3					
4 +2.1		17 49.0	−23 43	Sgr	4				2					
8 +3.1	6445	17 49.2	−20 01	Sgr	>34	13.2	+16	38	3b+3		19.1:B	Cont	1.4	
356 −3.1		17 49.8	−34 01	Sco	5				2					
359 −1.1		17 50.3	−30 35	Sco	7		−62		4+2					
0 −1.1		17 50.4	−29 25	Sgr	4		+20		2		>21p		2.6	
351 −6.1		17 50.7	−39 18	Sco	9		−16		2					
353 −5.1		17 50.8	−37 24	Sco	7				2					
355 −4.1		17 51.2	−34 55	Sco	10				3					
359 −1.2		17 51.3	−30 24	Sco	4		−89		2					
358 −2.1		17 51.7	−31 36	Sco	5				2					
10 +4.1		17 52.1	−17 36	Sgr	6		−21		2					
359 −1.3		17 52.1	−30 05	Sco	6		+32		2					
357 −3.2		17 52.6	−32 46	Sco	5		+106		2					
6 +2.5		17 52.7	−22 22	Sgr	<25		+73		1					
359 −2.2		17 52.8	−30 50	Sco	7		+81		2					
357 −3.3		17 53.3	−32 41	Sco	7				2		>21p			
36+17.1		17 53.6	+10 37	Oph	156	14.7	−42		2c		14.79	O7f	2.1	
6 +2.4		17 53.6	−21 59	Sgr	4				2					
0 −1.5		17 53.8	−29 44	Sgr	6		−45		2					
53+24.1		17 54.4	+28 00	Her	5		−102	16	2		17.55		5.7	
14 +6.1		17 54.4	−12 49	SerCd	25				3					
0 −1.6		17 54.4	−29 36	Sgr	<25		+75		1					
356 −4.1		17 54.6	−34 22	Sco	<5	13.9	−271							
7 +1.1		17 55.1	−21 45	Sgr	6		+9		2		14.72v?		1.4	
359 −2.4		17 55.1	−31 12	Sco	4				3					
1 −1.3		17 56.0	−28 14	Sgr					1					
11 +4.1		17 56.3	−16 30	Sgr	7		−73		2			WN		
358 −3.2		17 56.3	−32 37	Sco	11				2?		>21p			
7 +1.2		17 57.3	−21 42	Sgr	<5		−6		1					
356 −4.2		17 57.3	−34 10	Sco	10		+76		2					
357 −4.1		17 57.4	−33 36	Sco	6	14.3	−79							
359 −3.2		17 58.2	−31 08	Sgr	6				2		>21p			
358 −3.1		17 58.2	−31 43	Sgr	4		+99		2					
357 −4.3		17 58.2	−33 48	Sco	<25		+76		1					
0 −2.6		17 58.3	−30 01	Sgr	5		+158		2					
2 −1.1		17 58.5	−27 37	Sgr	5				2		>21p			
1 −2.1		17 58.5	−28 34	Sgr					1					
357 −4.2		17 58.5	−33 29	Sgr	5		−92		2					
96+29.1	6543	17 58.6	+66 38	Dra	18/350	8.8	−66	19	3a+2	164963	11.44	O7+WR	1.1	
13 +4.1		17 59.0	−15 32	SerCd	5		−37		2					
359 −4.1		17 59.9	−31 54	Sgr	6		−11		2		>21p			
352 −7.1		18 00.2	−38 50	CrA	<25	11.4	+5		1					
2 −2.3		18 01.2	−27 38	Sgr	5				3					
357 −5.1		18 01.4	−33 18	Sgr	12		+2		2					
358 −5.1		18 01.7	−33 15	Sgr	9				2					
356 −5.2		18 02.0	−34 28	Sgr	8		+173		2					
0 −3.1		18 02.3	−30 14	Sgr	6		−54		2c					
356 −6.1		18 02.5	−35 13	Sgr	10		−47		4					
359 −4.3		18 02.8	−32 10	Sgr	13		+20		3					
3 −2.2		18 03.2	−26 58	Sgr	9		−52		2a					
3 −2.3	IC 4673	18 03.3	−27 06	Sgr	15		−16		4		14.60		2.5	
3 −2.1		18 03.7	−26 44	Sgr	5		+82		2					
358 −5.3		18 03.9	−32 42	Sgr	<10	14.5	+28							
357 −6.1		18 04.1	−34 29	Sgr	5		+24		2					
2 −3.2		18 04.5	−28 38	Sgr	5				2					
356 −6.2		18 04.5	−34 58	Sgr	13				4					
358 −5.4		18 04.9	−32 54	Sgr	10		+34		2					
354 −7.1		18 04.9	−37 38	CrA	11				2					
0 −4.1		18 05.0	−30 58	Sgr	5		−17		2					
10 +0.1	6537	18 05.2	−19 51	Sgr	>9	12.5	−17*	10	2a+6	312582			1.2	
2 −3.3		18 05.4	−28 22	Sgr	<25		+241		1					

PK	NGC	α 2000	δ 2000	Const	Diam	m(pg)	RV	EV	Type	HD	*Mag	*Spec	d(kpc)	Notes
4 −2.1		18^{h}06^m.0	−26°30′	Sgr	‹25″		+75		1					
4 −3.1		18 06.7	−26 55	Sgr	5		−114		2					
0 −4.2		18 07.1	−30 34	Sgr	10		−156		2					
342−14.1		18 07.3	−51 02	Ara	35	11.9	+45	22					2.3	
5 −2.1		18 07.9	−25 24	Sgr	10		+128		2					
1 −4.2		18 07.9	−29 45	Sgr	3		−106		2					
8 −1.1		18 08.4	−22 17	Sgr	5		−32		2		15.61		2.4	
0 −5.1		18 08.4	−31 37	Sgr	13				2		›21p			
5 −3.1		18 09.2	−26 02	Sgr					1					
6 −2.1		18 09.5	−24 12	Sgr	16		−5		5			WN?	1.5	
356 −7.2		18 09.8	−35 44	Sgr	12				2					
3 −4.2		18 10.5	−28 08	Sgr	‹10				1		12.62			
3 −4.6		18 11.0	−28 33	Sgr	12				1					
2 −4.2		18 11.1	−28 59	Sgr	8		−94		4			WC?		
3 −4.3		18 11.5	−27 46	Sgr	6				2					
11 −0.1		18 11.8	−18 46	Sgr	5				2					
3 −4.5	6565	18 11.9	−28 11	Sgr	9	13.2	−5		4	166468			3.0	
358 −7.1	6563	18 12.0	−33 52	Sgr	48	13.8	−30	11	3a	166449	18.3B		1.5	
34+11.1	6572	18 12.1	+6 51	Oph	8	9.0	−9	16	2a	166802	13.6	Of+WR	0.6	
5 −3.2		18 12.4	−26 33	Sgr	14				3b		20.0p			
4 −4.1		18 12.4	−27 29	Sgr	4				2					
3 −4.4		18 12.4	−27 52	Sgr	7				4			WC?		
3 −4.9		18 12.8	−28 20	Sgr	9		−21		1					
359 −6.1		18 13.3	−32 20	Sgr	‹25		−84		1					
11 −0.2	6567	18 13.7	−19 05	Sgr	›8	11.7	+120	18	2a+3	166935	15.0:B	Cont	1.2	
5 −4.1		18 13.7	−26 09	Sgr	8				2					
24 +5.1		18 14.3	−4 59	SerCd	42		−21		4		19.7p		1.3	
6 −3.2		18 14.5	−24 44	Sgr	5				2					
2 −6.1		18 15.1	−30 16	Sgr	4		−112		2					
1 −6.1		18 15.4	−30 32	Sgr	‹10	14.0				319167	12.5:			Symbiotic star?
4 −5.1		18 16.2	−27 15	Sgr	7		−10		4					
1 −6.2		18 16.2	−30 52	Sgr	‹5	12.0	−19	13	1	167362	11.8	Of+WR	3.2	Probable symbiotic star
10 −1.1	6578	18 16.3	−20 27	Sgr	9	13.1	+4		2a		13.97		2.9	
4 −4.2		18 16.3	−27 05	Sgr	4		−75		2		12.5:	K2III?	5.1	Possible planetary
7 −3.1		18 17.3	−23 59	Sgr	10		+70		2					
38+12.1		18 17.6	+10 09	Oph	5	12.4	+4	‹10	2		13p	WR	2.0	
0 −7.1		18 17.6	−31 57	Sgr	‹25		−21		1					
4 −5.3		18 17.7	−28 17	Sgr	10				2					
3 −6.1		18 17.7	−29 08	Sgr	7	13.0	+100		2					
8 −3.1		18 18.4	−23 25	Sgr	7		+84		2					
348−13.1	IC 4699	18 18.5	−45 59	Tel	‹10	11.9	−123	10	2	167672			5.2	
4 −5.5		18 18.6	−28 08	Sgr	7		+57		3					
0 −7.2		18 18.7	−31 55	Sgr					1					
4 −5.4		18 18.8	−28 07	Sgr	11				2					
5 −5.1		18 19.4	−26 35	Sgr	8		−72		4					
7 −4.1		18 20.1	−24 15	Sgr	‹17		+161		1					
24 +3.1		18 21.4	−6 02	SerCd	5		+87		2					
94+27.1		18 21.9	+64 22	Dra	115				3		15.09		1.6	
8 −4.1		18 22.0	−24 11	Sgr	3		+71		2					
8 −4.2		18 22.5	−24 09	Sgr	4		+157		2					
2 −7.1		18 22.6	−30 43	Sgr	14		−82		4					
5 −6.1	6620	18 22.9	−26 49	Sgr	5	15.0	+73		2b					
19 +0.1		18 24.1	−11 07	Sct	5		+35		2					
7 −6.1		18 25.0	−25 42	Sgr	8		+42		4					
9 −4.1		18 25.1	−22 35	Sgr	6		−13		2					
9 −5.1	6629	18 25.7	−23 12	Sgr	›15	11.6	+15	‹6	2a	169460	12.77	O	1.9	
27 +4.1		18 26.7	−2 43	SerCd			+95		1					
31 +5.1		18 27.1	+1 15	SerCd	9				3					
43+11.1		18 27.8	+14 29	Her			−6:		1				1.4	
16 −1.1		18 27.9	−15 33	Sct	›11				4+2				2.4	
7 −6.2		18 28.0	−26 07	Sgr	‹9	13.3	+115		1					
8 −6.1		18 28.8	−24 32	Sgr	‹7				1+2?					
11 −5.1		18 29.2	−21 47	Sgr	5		−72		2					
2 −9.1		18 29.2	−31 30	Sgr	‹5	12.6	−30	18		170124		WC		
13 −3.1		18 29.5	−19 06	Sgr	5		+140		2					
13 −4.1		18 30.0	−19 41	Sgr	5		+32		2					
18 −1.1		18 30.2	−13 54	Sct	8				6					
15 −3.1		18 30.2	−16 45	Sgr	52	17.4			2					
20 −0.1		18 30.3	−11 37	Sct	285				3b		20.1p			
32 +5.1		18 31.0	+2 25	SerCd	›12				3b+2					
55+16.1		18 31.2	+26 55	Lyr	›60	15.6			3b+2		14.87v	O9k		Ecl. binary, period 11.32 hours
34 +6.1		18 31.8	+4 05	SerCd	10			18	2					
5 −8.1		18 32.5	−28 43	Sgr	18						17.4p			
8 −7.2	6644	18 32.6	−25 08	Sgr	3	12.2	+194		2	170839	›15.9		2.8	
21 −0.1		18 32.7	−10 06	Sct	›8	›14.3	+21		2+3					

PK	NGC	α 2000	δ 2000	Const	Diam	m(pg)	RV	EV	Type	HD	*Mag	*Spec	d(kpc)	Notes
21 −0.2		18^h33^m.3	−10°15′	Sct	7″		+26		2		20.1p		3.1	
14 −4.1		18 33.3	−18 17	Sgr	5		+27		2			Cont?	3.9	
21 −1.1		18 33.5	−11 07	Sct	<10		+3		3					
10 −6.1	IC 4732	18 33.9	−22 39	Sgr	3	13.3	−145		1	171131	>16.6		4.8	
17 −2.1		18 34.0	−14 52	Sct	6		+133		2					
10 −6.2		18 34.9	−22 43	Sgr	7				2					
15 −4.1		18 35.8	−17 36	Sgr	6		+63		2					
16 −4.1		18 36.1	−17 00	Sgr	14		−41		3					
28 +1.1		18 37.6	−3 06	SerCd	8		+106		4					
27 +0.1		18 39.4	−4 20	Sct	6				2		>21p			
4−11.1		18 39.4	−30 41	Sgr	8	13.1	+50	7	2					
22 −2.1		18 40.3	−10 40	Sct	8		+92		2					
56+14.1		18 41.1	+26 55	Lyr	16				4		>21p			
17 −4.1		18 41.2	−15 34	Sct	17		+71		4		17.9p		2.6	
25 −0.1		18 42.1	−6 41	Sct	5				2					
29 +0.1		18 42.8	−3 13	Aql	42	>18.7			4					
31 +1.1		18 43.1	−0 17	Aql					1					
23 −2.1	M1-59	18 43.3	−9 05	Sct	4		+99		2					
19 −4.1		18 43.6	−13 44	Sct			+76		1					
22 −3.1		18 44.1	−11 07	Sct	6		+60		2					
9 −9.1		18 44.7	−25 20	Sgr	6		+46		2					
26 −1.2		18 45.5	−6 57	Sct	3				2					
13 −7.1		18 45.6	−20 35	Sgr	14				?+3					
2−13.1	IC 4776	18 45.8	−33 21	Sgr	8	11.7	+19		2a	173283	>14.6	WC6?	3.5	
19 −5.1		18 45.9	−14 28	Sct			+41		1					
25 −2.1		18 46.4	−7 15	Sct	5				2+3b					
24 −2.1		18 46.6	−8 28	Sct	5		+83		3					
26 −2.1		18 47.5	−6 54	Sct	7				2b					
24 −3.1		18 47.8	−9 09	Sct	6				3+2					
9−10.1		18 48.2	−25 29	Sgr	5		+174		2					
27 −2.1		18 48.8	−5 56	Sct	7				2a					
26 −2.3		18 49.7	−7 02	Sct	4				2					
51 +9.1	Hu 2-1	18 49.7	+20 51	Her	3	12.2	+14	10			13.0v?	WN	1.3	
64+15.1		18 50.0	+35 15	Lyr	17		−25		4			Cont?	2.5	
12 −9.1		18 50.4	−22 34	Sgr	4		+34		2					
21 −5.1		18 51.5	−13 11	Sct	4		+26		2					
27 −3.1		18 53.5	−6 29	Sct	35	16.7			2c		>21p			
63+13.1	6720	18 53.6	+33 02	Lyr	70/150	9.7	−19	30	4+3		14.8	Cont	0.6	M57, Ring nebula
27 −3.2		18 54.0	−6 26	Sct		13.4	+110		1					
29 −2.1		18 54.2	−4 39	Sct			+52		1					
24 −5.1		18 54.3	−10 05	Sct	21		+19		2					
25 −4.2	IC 1295	18 54.6	−8 50	Sct	>86	15.0	−36		3b+2		15:p		1.3	
38 +2.1		18 54.7	+6 02	SerCd	308				3b					RCW 181
3−14.1		18 55.6	−32 16	Sgr	4	10.9	−65		2	175194	18:p	O	1.6	
39 +2.1		18 56.3	+7 08	Aql	16				2					
43 +3.1		18 56.6	+10 52	Aql	4		+20		2		14.31		5.2	
28 −4.1		18 57.3	−6 00	Sct	8				4					
28 −3.1		18 57.8	−5 28	Sct	8				4					
32 −2.1		18 58.4	−1 04	Aql			+43		1					
78+18.1	6742	18 59.3	+48 28	Dra	30	15.0	−159		2c		19.4p		4.0	
32 −3.1		19 00.6	−2 12	Aql	5				2					
17−10.1		19 01.0	−18 13	Sgr	62	15.4			4		15.42	sdO8k	2.1	
36 −1.1	52-71	19 02.0	+2 09	Aql	>107		+25		3b+3		13.45v	B	0.8	
33 −2.1	6741	19 02.6	−0 27	Aql	6	10.8	+42	21	4		14.7		1.6	
47 +4.1		19 02.7	+14 28	Aql	7				2					
51 +6.1		19 03.6	+19 21	Sge	45				4		18.6p			
50 +5.1		19 04.5	+17 58	Aql	36	16.5			3b		17.7p			
3−17.1		19 05.6	−33 12	Sgr	5	13.4	−172*		2				5.5	
29 −5.1	6751	19 05.9	−6 00	Aql	20	12.5	−39	24	3	177656	13.89	WC6	2.0	
40 −0.1	Abell 53	19 06.7	+6 24	Aql	31	16.9			4		20.48B			
55 +6.1		19 08.7	+22 59	Vul	63	17.1			2b					
44 +1.1		19 09.8	+11 05	Aql	3				2					
33 −5.1		19 10.5	−2 21	Aql	41	15.4			3		19.9p			
62 +9.1	6765	19 11.1	+30 33	Lyr	38		−72:		5		16:		2.4	
48 +2.1		19 12.1	+15 09	Aql	6				2					
32 −6.1		19 13.0	−3 32	Aql	22				4		>21p			
37 −3.2		19 13.1	+2 53	Aql	188	>15.5			4					
49 +2.1		19 13.1	+15 47	Aql	>9				4+2					
39 −2.1		19 13.6	+4 38	Aql	8		+36		2		15.52		3.5	
48 +1.1	Hainez-249	19 13.7	+14 59	Aql	4		+14		4					
38 −3.2		19 13.9	+3 38	Aql			+7		1					
38 −3.3		19 14.3	+3 35	Aql	14				3b		19.4p			
33 −6.1	6772	19 14.6	−2 42	Aql	>62	14.2	0:		3b+2		18.4B		1.3	
27 −9.1	IC 4846	19 16.5	−9 03	Aql		12.7	+151	19	2	180324	13.7v	Of	2.8	
58 +6.1		19 17.1	+25 37	Vul	37	17.5			3b		17.66			

PK	NGC	α 2000	δ 2000	Const	Diam	m(pg)	RV	EV	Type	HD	*Mag	*Spec	d(kpc)	Notes
358−21.1	IC 1297	19ʰ17ᵐ.4	−39°37′	CrA	7″		+14	13		180206	12.91v	WR(+OVI?)	5.1	Variable RU CrA
26−11.1	Nassau 2	19 18.3	−11 06	Aql		13.3					14:p	WR		
41 −2.1	6781	19 18.4	+6 33	Aql	>109	11.8	+4	12	3b+3		14.95?		0.8	
34 −6.1	6778	19 18.4	−1 36	Aql	>16	13.3	+91	23	3+3	180871	15.0:B			
53 +3.1		19 18.7	+19 34	Sge	86	17.2			3b					
77+14.1	Abell 61	19 19.2	+46 15	Cyg	190	14.4			2b		17.39v?			
25−11.1		19 19.3	−12 16	Sgr	76	15.7			2b		18.0p			
43 −3.1		19 21.0	+7 37	Aql	>8		+48		4+3					
32 −8.1		19 21.9	−4 13	Aql					2a					
37 −6.1	6790	19 23.2	+1 31	Aql	7	10.2	+40	22	2	182083	13.53	WN	1.6	
55 +2.2		19 23.8	+21 07	Vul	>6				2+3					
45 −2.1	Vy 2-2	19 24.4	+9 54	Aql		12.7	−71		1		14.84		1.9	Discordant magnitudes
31−10.1	M 3-34	19 27.0	−6 35	Aql	5		+37:		2		14.58		5.0	
46 −3.1		19 27.8	+10 24	Aql					1					
46 −4.1	6803	19 31.3	+10 03	Aql	6	11.3	+13	14	2a	183889	15.2	WN	1.6	
45 −4.1	6804	19 31.6	+9 13	Aql	31/66	12.2	−12		4+2		14.23	O9	1.3	
9−21.1		19 32.4	−29 25	Sgr	14				3+6					
61 +3.1		19 32.9	+26 53	Vul	9				6					
47 −4.1	Abell 62	19 33.4	+10 38	Aql	156	14.8			2c		18.2p			
42 −6.1	6807	19 34.6	+5 41	Aql		13.8	−68	17	2		13.97		5.1	
64 +5.1		19 34.8	+30 31	Cyg	8	9.6	−32	26	4	184738	10.03	WC9	0.6	Campbell's star. Protoplan.?
53 −1.1		19 35.3	+17 13	Sge	4				2					
52 −2.2	Me 1-1	19 39.2	+15 57	Aql	<10	12.6	−6		4					
56 −0.1		19 39.6	+20 19	Vul	3				2					
55 −1.1		19 40.4	+18 49	Sge	3				2					
19−19.1		19 40.5	−20 27	Sgr	140				3b		18.7p			
51 −3.1	M 1-73	19 41.2	+14 57	Aql	5		+7		2					
53 −3.2		19 41.9	+16 45	Sge						HM Sge	10.9v	WC?		Protoplanetary? Slow nova?
53 −3.1		19 42.2	+17 05	Sge	34	17.1			2	UU Sge	14.67v	sdO+dK		Possible planetary. Eclipsing
52 −4.1	M 1-74	19 42.3	+15 09	Aql	<10		+10:		1			WNb	2.5	binary, period 0.465 day
25−17.1	6818	19 44.0	−14 09	Sgr	>17	9.9	−14	30	4		13.05	WNb?	1.6	
83+12.1	6826	19 44.8	+50 31	Cyg	30/140	9.8	−6	13	3a+2	186924	10.44	O6fp	1.0	Blinking planetary. Sp. bin.?
44 −9.1	Abell 64	19 45.7	+5 34	Aql	35	15.3			3		18.1p			
17−21.1		19 46.6	−23 09	Sgr	104	15.2			2a		15.90	Opk	1.8	Sh2-52
59 −1.1		19 48.5	+22 09	Vul	8				2					
82+11.1	6833	19 49.7	+48 58	Cyg	1	13.8	−109	14	2		14.77	Cont?	1.7	Discordant magnitudes
69 +3.1		19 50.0	+33 46	Cyg	>23				3b+6					
62 −0.1		19 50.5	+25 55	Vul	8				3					
65 +0.1	6842	19 55.0	+29 17	Vul	50	13.6	−5:		3b		15.98		1.3	
75 +5.1		19 57.1	+39 50	Cyg			−60:	25:		V1016 Cyg	10.5v		2.2	Protoplanetary, bipolar. Binary
19−23.1		19 57.5	−21 37	Sgr	251	14.9			3b		17.39		>1.9	
43−13.1		19 58.4	+3 03	Aql	66	16.0			2b		19.1p			
68 +1.2	Hawze 1-4	19 59.3	+31 55	Cyg	22		+12*		3b					
60 −3.1	6853	19 59.6	+22 43	Vul	350/910	7.6	−42	30	3+2		13.94	O7	0.3	M27, Dumbbell nebula. VC:
60 −4.1	Abell 68	20 00.2	+21 43	Vul	39	16.6			3		13.26			m(pv)=17.0, 6″.5
42−14.1	6852	20 00.6	+1 43	Aql	28				4					
75 +4.1		20 04.4	+39 35	Cyg	28				3b					
107+21.1		20 04.5	+74 27	Dra	216				2					
68 −0.1		20 04.8	+31 28	Cyg	>14		−9		3b+6		>21p		2.8	
79 +6.1		20 06.9	+44 15	Cyg	4				2					
79 +5.1	M 4-17	20 09.1	+43 44	Cyg	>19		−26		4+2					
82 +7.1	6884	20 10.4	+46 28	Cyg	6	12.6	−36	23	2b		>15.6	WNb	2.8	
57 −8.1	6879	20 10.5	+16 55	Sge	5	13.0	+9	23	2a		15p	Of+WR?	5.1	
74 +2.1	6881	20 10.8	+37 25	Cyg	4	14.3	−14	18	2a+3		16.71		3.2	
60 −7.1		20 11.9	+20 20	Sge	35		+39	34	3+2	FG Sge	8.9v	B4-G8	2.3	Spec. type changed over 24 years
60 −7.2	6886	20 12.7	+19 59	Sge	4	12.2	−35	19	2+3		15.7	Cont	3.1	
54−12.1	6891	20 15.2	+12 42	Del	12/74	11.7	+42	7	2a+2b	192563	12.41	O7	2.2	
69 −2.1	6894	20 16.4	+30 34	Cyg	>42	14.4	−58		4+2		15.8v?		1.4	
65 −5.1		20 17.3	+25 22	Vul	>17		−22		3+2					
58−10.1	IC 4997	20 20.2	+16 45	Sge	2	11.6	−66	15	1	193538	13v?	Of+WR	1.5	
71 −2.1		20 21.1	+32 30	Cyg	<5		−176		1		15.73		0.8	
69 −3.1		20 21.9	+30 00	Cyg	>8				?+3					
61 −9.1	6905	20 22.4	+20 07	Del	46/100	11.9	−8	27	3+3	193949	13.5B	WC6	1.3	
38−25.1	Abell 70	20 31.6	−7 06	Aql	43	14.3			4		18.6p			
85 +4.1	Abell 71	20 32.4	+47 21	Cyg	157	15.2			3b		18.95			Sh2-116. Possible planetary
59−18.1	Abell 72	20 50.1	+13 33	Del	130	14.6	−59		3b		16.12		1.5	
77 −5.1		20 51.0	+35 35	Cyg				40:		V1329 Cyg	v	M4+WN5		Ecl. binary. Symbiotic star
95 +7.1		20 56.5	+57 27	Cep	73	>17.4			4		20.5p			
93 +5.2	7008	21 00.6	+54 33	Cyg	83	13.3	−76	11	3		13.21	O7	0.8	
80 −6.1		21 02.3	+36 42	Cyg			−44	13		CRL 2688	12.25	F5Ia		Protoplanetary? Egg nebula
37−34.1	7009	21 04.2	−11 22	Aqr	25/100	8.3	−46	21	4+6	200516	11.5	Cont	0.9	Saturn nebula
89 +0.1	7026	21 06.3	+47 51	Cyg	21	12.7	−41	41	3a	201192	14.50	WC,OVI?	1.7	
84 −3.1	7027	21 07.1	+42 14	Cyg	15	10.4	+8	22	3a		11.3?B		1.1	Protoplanetary?
89 −0.1	S 1-89	21 14.0	+47 45	Cyg	>35				3+6		19.1p			
88 −1.1	7048	21 14.2	+46 16	Cyg	61	11.3	−50	11	3b		18.3B		1.2	

PK	NGC	α 2000	δ 2000	Const	Diam	m(pg)	RV	EV	Type	HD	*Mag	*Spec	d(kpc)	Notes
72-17.1	Abell 74	21h16m.8	+24°10'	Vul	831"	>12.2			2		17.11	sdOp		
89 -2.1	M1-77	21 19.1	+46 19	Cyg	7				2a					
93 +1.1		21 20.8	+51 54	Cyg	>6		-88		?+3					
101 +8.1	Abell 75	21 26.4	+62 53	Cep	57	17.0			3b		17.2p		1.9	
65-27.1		21 30.0	+12 10	Peg	3	14.9	-141	<17	1					Pease 1, in globular cl. M15
96 +2.1		21 30.0	+54 27	Cyg	6				2		14.82			
97 +3.1		21 32.2	+55 53	Cep	>37	16.4	-113		3a+3		15.90			Sh2-128. Possible planetary
89 -5.1	IC 5117	21 32.5	+44 35	Cyg	3	13.3	-26	18	2	205211	16.74	WR	2.2	Discordant magnitudes
86 -8.1	Hu 1-2	21 33.1	+39 38	Cyg	5	12.7	+10	28	2		16.1		4.0	
81-14.1	Abell 78	21 35.6	+31 41	Cyg	101	16.0	+17		4		13.32	O5fek	1.7	Binary?
66-28.1	7094	21 36.9	+12 47	Peg	95		-87		4		13.60	O	1.3	
93 -2.1	M1-79	21 37.0	+48 57	Cyg	32		-24		4		>14.4		1.2	
98 +2.1		21 39.2	+55 46	Cep	7				2					
95 -2.1		21 43.3	+50 25	Cyg			-134		1					
104 +7.1	7139	21 45.9	+63 39	Cep	78	<13	-54		3b		18.1:B		1.2	
97 -2.1	M2-50	21 57.7	+51 41	Cyg	4		-136		2					
2-52.1	IC 5148	21 59.5	-39 23	Gru	120		-28		4			Cont		IC 5148-50
103 +0.1	M2-51	22 16.1	+57 29	Cep	>41		-11		2+3		13.45		1.6	
103 +0.2	M2-52	22 20.5	+57 36	Cep	13		-92		3					
100 -5.1	IC 5217	22 23.9	+50 58	Lac	6	12.6	-99	18	2	212534	15.61	WN?	3.2	
102 -2.1	Abell 79	22 26.3	+54 49	Lac	>59	15.8			4+3		14.09			
36-57.1	7293	22 29.6	-20 48	Aqr	>769		-28	13	4+3		13.47		0.1	Helix nebula
100 -8.1		22 31.7	+47 48	Lac	1		-152		1		16.08		2.8	
104 -1.1	M2-53	22 32.3	+56 10	Lac	14		-62		3b			WNa	1.6	
102 -5.1	Abell 80	22 34.7	+52 27	Lac	132	15.2			4					
107 +2.1	7354	22 40.4	+61 17	Cep	>20	12.9	-42	26	4+3b		15.0	Cont	1.6	
117+18.1	IC 1454	22 42.6	+80 27	Cep	33	14.8			4		16.4			
107 -2.1		22 56.3	+57 09	Cep	8	14.0	-58		2			Cont?	2.5	
107-13.1		23 23.0	+46 54	And	5	13.9	-50	16	2		14.36	Cont?	7.1	
106-17.1	7662	23 25.9	+42 33	And	20/130	9.2	-13	26	4+3	220733	13.17v?	Cont	1.2	
111 -2.1	Hb 12	23 26.3	+58 11	Cas	1	14.0	-5	16			14.8	WNa	2.3	Protoplan.? Discordant mags.
116 +8.1	M2-55	23 31.9	+70 23	Cep	40		-23		3		>21		1.8	
114 +3.1		23 35.5	+64 53	Cep					3					Sh2-167
104-29.1	Jones 1	23 35.9	+30 28	Peg	332	15.1			3b		16.13			
110-12.1	K 1-20	23 39.1	+48 13	And	34				4		20.1p			
114 -4.1	Abell 82	23 45.8	+57 04	Cas	95	15.2	-31		3b		13?p			
113 -6.1		23 46.8	+54 45	Cas	40	17.6			2c		>19.7p			
112-10.1	Abell 84	23 47.7	+51 24	Cas	126	14.4			3b		18.49			
118 +8.1		23 56.6	+70 49	Cep					1		>21p			Possible planetary

Galaxies

Galaxies

Column Headings

NGC — The galaxy's number in J. L. E. Dreyer's *New General Catalogue of Nebulae and Clusters of Stars* (1888) or *Index Catalogue* (1895 and 1908). Numbers from the latter source have the prefix "IC."

UGC — The galaxy's number in P. N. Nilson's *Uppsala General Catalogue of Galaxies* (1973) or its supplement, the *Catalogue of Selected Non-UGC Galaxies* (1974); the latter are prefixed "A."

α 2000 and **δ 2000** — Right ascension and declination, referred to the 2000.0 equinox.

Const — Constellation in which the galaxy is located.

B — Total **B** (blue) magnitude from **UBV** photometry. Values followed by the letter "H" are photographic magnitudes from the Shapley-Ames catalogue (*Harvard Annals*, **88**, No. 2, 1932) that have been corrected by G. de Vaucouleurs to the **B** system. The letter "p" indicates a photographic magnitude.

V — Total visual **V** (yellow) magnitude in the **UBV** system.

B − V — Total color index in the **UBV** system.

Dim — The galaxy's major and minor diameters, expressed in minutes of arc. These dimensions extend out until the surface brightness has dropped to the equivalent of one 25th-magnitude star per square arc second, as explained in the Introduction.

Type — The Hubble type and DDO luminosity class, as described in the Introduction.

Nucl — Nuclear class according to the *Second Reference Catalogue of Bright Galaxies*, G. de Vaucouleurs *et al.* (1976). A diffuse nucleus is indicated by the prefix "D." Suffixes tell whether the nucleus is large (L), small (S), very small (VS), or extremely small (ES). The numbers 1 to 5 are used to characterize increasingly bright central regions. Seyfert galaxies are flagged by asterisks.

RV — The radial velocity, corrected for the solar system's motion around the Milky Way, in kilometers per second. Positive means receding (traveling away from the solar system). An asterisk is shown if the radial velocity is an average of discordant values. To convert to heliocentric radial velocity, subtract $300 \sin l \cos b$, where l and b are the object's galactic longitude and latitude.

Memb — Membership is indicated in these major clusters of galaxies: Cancer (Ca), Centaurus (Ce), Coma (C), Fornax I (F), Hercules (He), Hydra I (H), Perseus (P), and Virgo (V). Galaxies belonging to the Local Group are flagged with the letter "L."

Notes — Common names are placed here, along with alternate designations from the NGC, IC, and third Cambridge radio source survey (3C). Dwarf galaxies on the lists of the David Dunlap Observatory (D) or G. Reaves (GR) are identified, as well as the nearby systems of E. Holmberg (Holm.), compact blue objects catalogued by B. E. Markarian (Mrk), Vorontsov-Velyaminov interacting galaxies (VV), H. Arp's peculiar galaxies, and F. Zwicky's compact systems (1Z to 7Z).

NGC	UGC	α 2000	δ 2000	Const	B	V	B-V	Dim		Type	Nucl	RV	Memb	Notes
7805	12908	0ʰ01ᵐ4	+31°26'	Peg	14.3 p			1.4	1.0	S(B)0p		5058		VV 226, Arp 112
7806	12911	0 01.5	+31 27	Peg	14.4 p			1.3	0.9	Sb+p		4875		VV 226, Arp 112
	A444	0 02.0	−15 28	Cet	11.29	10.89	0.40	10.2	4.2	Ir+ IV-V		−38	L	WLM system, D221
7814	8	0 03.3	+16 09	Peg	11.35	10.45	0.90	6.3	2.6	Sb−		1249		
7816	16	0 03.8	+7 28	Psc	14.0 p			1.9	1.7	Sb+		5284		
7817	19	0 04.0	+20 45	Peg	12.43			3.7	1.2	Sb+:		2532		
7819	26	0 04.4	+31 29	Peg	14.3 p			2.0	1.7	SBb		5194		
		0 06.3	+20 12	Peg	14.10	13.76	0.34				*	7657		Mrk 335
2	59	0 07.3	+27 41	Peg	14.8 p			1.4	0.9	Sa		7597		
1	57	0 07.3	+27 43	Peg	13.4 p			1.9	1.5	Sb:		4779		
12	74	0 08.7	+4 37	Psc	14.5 p			1.9	1.7	SBb+:		4098		
14	75	0 08.8	+15 49	Peg	12.30			3.0	2.3	Irp+		1059		VV 80, Arp 235
13	77	0 08.8	+33 26	And	14.2 p			2.7	0.8	Sa:				
9	78	0 08.9	+23 49	Peg	14.5 p			1.5	0.9	Sbp				
16	80	0 09.1	+27 44	Peg	12.95	11.96	0.99	2.1	1.2	E3	2VS	3340		
20	84	0 09.5	+33 19	And	14.5 p			2.0	1.9	S0:				
23	89	0 09.9	+25 55	Peg	12.80	11.99	0.81	2.3	1.6	Sbp	3VS	4793		Mrk 545
24	A2	0 09.9	−24 58	Scl	12.10	11.50	0.60	5.5	1.6	Sb III		597		
26	94	0 10.4	+25 50	Peg	13.9 p			2.2	1.6	Sb−	2VS	4811		
21	98	0 10.7	+32 59	And	13.9 p			1.4	0.7	SBb+				
29	100	0 10.8	+33 21	And	13.5 p			1.8	0.9	S(B)b:				
36	106	0 11.4	+6 23	Psc	14.5 p			2.4	1.5	S(B)b		6241		
	122	0 13.3	+17 02	Peg	14.86			2.2	0.4	Ir+		1048		
48	133	0 14.0	+48 14	And	14.69			1.7	1.1	S(B)cp		2036		
45	A4	0 14.1	−23 11	Cet	11.10	10.41	0.69	8.1	5.8	S− IV-V	1	508		D223
51	138	0 14.6	+48 15	And	14.6 p			1.8	1.5	S0p				
55		0 14.9	−39 11	Scl	8.22			32.4	6.5	SBm:		98		
	156	0 16.8	+12 21	Psc	14.04			2.7	1.7	Ir+		1318		
63	167	0 17.7	+11 27	Psc	12.7 p			1.9	1.3	Sp		1351		
67A		0 18.2	+30 03	And						E5:		6585		VV 166
67		0 18.2	+30 04	And	15.5:p			0.4	0.3	E3:		6863		VV 166
68	170	0 18.3	+30 04	And	13.95	12.96	0.99	1.5	1.3	S0		5941		VV 166
71	173	0 18.4	+30 04	And	14.0	13.0	0.98	1.8	1.5	S0p		6937		VV 166
70	174	0 18.4	+30 05	And	14.5 p			1.7	1.5	Sc		7376		IC 1539, VV 166
72	176	0 18.5	+30 02	And	14.45	13.42	1.03	1.4	1.2	SBb−		7193		VV 166
	191	0 20.2	+10 53	Psc	14.06			2.5	2.1	S+ V		1320		D2
78B	194	0 20.4	+0 50	Psc	14.5 p			1.5	1.2	S0:		5389		Mrk 547
IC 10	192	0 20.4	+59 18	Cas	11.70	10.27	1.43	5.1	4.3	Ir+:		−84	L	
80	203	0 21.2	+22 21	And	13.10	12.05	1.05	2.5	2.3	S0:		5796		
83	206	0 21.4	+22 26	And	13.70	12.63	1.07	1.6	1.5	E0		6751		
91	208	0 21.8	+22 25	And	14.5 p			2.5	1.1	S(B)cp		5378		
95	214	0 22.2	+10 30	Psc	13.25	12.55	0.70	1.9	1.5	Sc	5VS	5064		Arp 65
99	230	0 24.0	+15 47	Psc	14.0 p			1.7	1.5	Scp		5343		
105	241	0 25.3	+12 53	Psc	13.9	13.2	0.71	1.2	0.8	Sb−:		5439		
	260	0 27.1	+11 35	Psc	13.53			3.0	0.8	Sc		2305		
120	267	0 27.5	−1 31	Cet	14.8 p			1.8	0.8	SB0:				
124	271	0 27.9	−1 49	Cet	13.8 p			1.6	1.0	Sc	3S			
125	286	0 28.8	+2 50	Psc	13.20	12.30	0.90	2.0	1.8	Sa:		5430		
126		0 29.1	+2 49	Psc	15.5:p			1.4	0.8	SB0:		4423		
127		0 29.20	+2 52.4	Psc	15.0	14.0	0.98	1.0	0.8	S0:		4235		
128	292	0 29.25	+2 51.9	Psc	12.6	11.6	1.00	3.4	1.0	S0p	4	4384		
131		0 29.6	−33 16	Scl				2.1	0.6	SBb:		1401		
132	301	0 30.2	+2 06	Cet	13.8 p			2.0	1.6	S(B)b+	3S			
134		0 30.4	−33 15	Scl	11.0	10.1	0.88	8.1	2.6	S(B)b+	3VS	1531		
145		0 31.7	−5 09	Cet	13.20	12.76	0.44	1.8	1.4	SBdm	1	4267		Arp 19
147	326	0 33.2	+48 30	Cas	10.24	9.30	0.94	12.9	8.1	dE4	3ES	89	L	D3
151		0 34.0	−9 42	Cet	12.28	11.54	0.74	3.7	1.9	Sb+ II:	3S	3746		NGC 153
150	A7	0 34.3	−27 48	Scl	11.75	11.13	0.62	4.2	2.3	Sc I:	4ES	1601		
148		0 34.3	−31 47	Scl	13.08	12.12	0.96	2.4	1.2	Sa		1476		
157		0 34.8	−8 24	Cet	11.03	10.42	0.61	4.3	2.9	Sc I	4VS	1749		
		0 35.4	+36 31	And	13.5					dE2			L	And III
163		0 36.0	−10 07	Cet	13.75	12.75	1.00	1.4	1.4	E0				
160	356	0 36.1	+23 57	And	13.4	12.4	0.97	3.2	1.9	S0p	2VS	5460		
165		0 36.5	−10 06	Cet	14.05	13.22	0.83	1.6	1.3	SBb+		5964		
169	365	0 36.9	+23 59	And	13.3 p			3.0	0.9	Sb−:		4682		Mrk 341, Arp 282
173	369	0 37.2	+1 56	Cet	14.5 p			3.5	3.2	Sb+		4487*		
175		0 37.4	−19 56	Cet	12.8	12.1	0.74	2.6	2.4	SBb I	4	1735		
180	380	0 38.0	+8 38	Psc	14.3 p			2.6	2.2	SBb+		5377		
182	382	0 38.2	+2 44	Psc	13.4	12.6	0.84	2.3	1.9	S(B)a:		5365		
190	397	0 38.9	+7 02	Psc	14.4 p			1.0	0.9	Sb−		12257		3Z 10
191		0 39.0	−9 00	Cet	12: p			1.5	1.3	S(B)cp		5151*		Arp 127
185	396	0 39.0	+48 20	Cas	10.07	9.17	0.90	11.5	9.8	dE0	1	4	L	
178		0 39.1	−14 10	Cet	13.1	12.6	0.50	2.0	1.2	Sc IV		1544		IC 39
192	401	0 39.3	+0 51	Cet	13.9 p			2.3	1.2	SBa:				
194	407	0 39.3	+3 02	Psc	13.00	12.08	0.92	1.9	1.9	E1		5239		

NGC	UGC	α 2000	δ 2000	Const	B	V	B–V	Dim		Type	Nucl	RV	Memb	Notes
193	408	0ʰ39ᵐ3	+3°20′	Psc	13.35	12.34	1.01	1.7	1.7	S(B)0		4356		
198	414	0 39.4	+2 48	Psc	13.5	12.8	0.67	1.3	1.3	Sc	4VS			
200	420	0 39.6	+2 53	Psc	14.0 p			2.0	1.1	SBb+	4VS	5244		
201	419	0 39.7	+0 51	Cet	14.7 p			2.2	1.9	S(B)c				
205	426	0 40.4	+41 41	And	8.85	8.01	0.84	17.4	9.8	E6:		1	L	M110
210		0 40.6	–13 52	Cet	11.65	10.85	0.80	5.4	3.7	Sb– I	5	1700		
	439	0 41.4	–1 43	Cet	13.8	13.2	0.61	1.3	1.3	Sa		5225		
216		0 41.4	–21 03	Cet	13.8	13.3	0.48	1.8	0.7	Ir–:		1597		
214	438	0 41.5	+25 30	And	12.95	12.23	0.72	2.1	1.6	Sb I	3VS	4690		
227	456	0 42.6	–1 32	Cet	13.35			2.1	1.7	E2:		5429		
221	452	0 42.7	+40 52	And	9.15	8.21	0.94	7.6	5.8	E2		21	L	M32, Arp 168
224	454	0 42.7	+41 16	And	4.38	3.47	0.91	178	63	Sb I–II		–59	L	M31, Andromeda galaxy
237	461	0 43.5	–0 07	Cet	13.7			1.8	1.2	Sc II:	3VS	4258		
		0 45.7	+38 00	And	13.9	13.2	0.68			dE0			L	And I
244	A10	0 45.8	–15 36	Cet	13.7	13.3	0.42	1.3	1.2	S0p		995		
245	476	0 46.1	–1 44	Cet	13.2 H			1.4	1.3	Ir:	4VS	4461		Mrk 555
	477	0 46.2	+19 29	Psc	14.38			3.5	0.8	Sm		2843		
247	A11	0 47.1	–20 46	Cet	9.51	8.86	0.65	20.0	7.4	S– IV	4ES	180*		
254		0 47.5	–31 25	Scl	12.70	11.82	0.88	2.1	1.1	Sa		1425		
253	A13	0 47.6	–25 17	Scl	8.04	7.07	0.97	25.1	7.4	Scp	3ES	259		Sculptor galaxy
255		0 47.8	–11 28	Cet	12.35	11.80	0.55	3.1	2.8	Sb+ II:	4S	1873		
252	491	0 48.0	+27 38	And	13.4 p			1.8	1.3	Sa:				
257	493	0 48.1	+8 19	Psc	13.7 p			2.1	1.5	Sc:		5420		
260	497	0 48.6	+27 42	And	14.3 p			1.0	1.0	Scp		5413		
		0 48.6	–12 43	Cet	13.9	13.6	0.27	1.1	0.8	SB0p		6479		
262	499	0 48.8	+31 57	And	15.0 p			1.4	1.4	Sa:	*	4399		Mrk 348
266	508	0 49.8	+32 16	Psc	12.6 p			3.2	3.2	SBb–				
268		0 50.2	–5 12	Cet	13.3 H			1.7	1.3	Sc II		5609		
271	519	0 50.8	–1 53	Cet	13.2 p			2.5	2.1	SBb–				
273		0 50.8	–6 53	Cet	13: p			2.6	0.9	S0	4VS			
274		0 51.0	–7 03	Cet	12.88			1.7	1.6	E1	5VS	1819		VV 81, Arp 140
275		0 51.1	–7 04	Cet	13.0	12.5	0.49	1.5	1.2	SBcp		1833		VV 81, Arp 140
		0 52.1	–0 28	Cet	14.78					P		1800		
278	528	0 52.1	+47 33	Cas	11.51	10.85	0.66	2.2	2.1	E0p	3L	884		
279	532	0 52.3	–2 12	Cet	14.0 p			1.8	1.5	S0		4155		Mrk 558
289		0 52.7	–31 12	Scl	11.58			3.7	2.7	Sb+ I–II:	3S	1793		
		0 52.7	–72 50	Tuc	2.79	2.29	0.50	280	160	SBmp IV		–30	L	SMC
300		0 54.9	–37 41	Scl	8.70			20.0	14.8	Sd III–IV	3ES	97*		
309		0 56.7	–9 55	Cet	12.40	11.79	0.61	3.1	2.7	Sc I–II	4S	5740		
315	597	0 57.8	+30 21	Psc	12.5 p			3.2	2.2	S0:		5218		
327		0 57.9	–5 08	Cet	13: p			1.9	0.8	Sb+:				
329		0 58.0	–5 04	Cet	13: p			1.9	0.8	Sb–:				
326	601	0 58.4	+26 52	Psc	14.35	13.19	1.16	1.5	1.5	E4p		14479		4Z 35
337		0 59.8	–7 35	Cet	12.08	11.63	0.45	2.8	2.0	Sc II–III:		1773		
		0 59.9	–33 42	Scl	10.5					dE3		197	L	Sculptor dwarf
IC 65	625	1 00.9	+47 41	And	13.10			4.3	1.3	S(B)b+		2852		
	634	1 01.4	+7 37	Psc	14.57			1.9	1.2	S(B)+ V		2347		D7
337A		1 01.6	–7 35	Cet	13.5:p			5.8	4.6	S(B)dm		489		
352		1 02.1	–4 15	Cet	12.5:p			2.7	1.2	SBb+:				
354	645	1 03.3	+22 20	Psc	14.2 p			0.9	0.5	SBp		5061		Mrk 353
357		1 03.4	–6 20	Cet	12.90	11.80	1.10	2.6	1.9	SBa	4VS	2621		
		1 03.8	+21 53	Psc	15.4			2		Ir			L	LGS 3
IC 1613		1 04.8	+2 07	Cet	9.93	9.33	0.60	12.0	11.2	Ir+ V		–125	L	D8
		1 05.1	–6 13	Cet	11.98			4.0	3.5	S(B) IV–V		1086		
375		1 07.1	+32 21	Psc	16: p			1.3	1.3	E2:		6218		Arp 331
380	682	1 07.3	+32 29	Psc	13.50	12.47	1.03	1.6	1.4	E2		4548		Arp 331
379	683	1 07.3	+32 31	Psc	13.65	12.62	1.03	1.8	1.0	S0		5581		Arp 331
384	686	1 07.4	+32 18	Psc	13.92	12.97	0.95	1.1	0.9	E3		4607		Arp 331
385	687	1 07.4	+32 19	Psc	13.80	12.87	0.93	1.5	1.3	S0:		5051		Arp 331
382	688	1 07.4	+32 24	Psc	14.2 p			0.4	0.4	E0		5363		Arp 331
383	689	1 07.4	+32 25	Psc	12.9	11.9	1.05	2.3	1.9	S0:		5095		Arp 331, 3C 31
406		1 07.4	–69 53	Tuc	12.38			3.8	1.5	Sc:		1293		
	711	1 08.7	+1 39	Cet	14.03			3.9	0.7	Sc		2088		
403	715	1 09.2	+32 45	Psc	13.3 p			2.1	0.8	Sa:		5183		
404	718	1 09.4	+35 43	And	11.05	10.11	0.94	4.4	4.2	E0	5VS	178		
407	730	1 10.6	+33 07	Psc	14.3 p			2.0	0.6	Sa:		5816		
418		1 10.6	–30 13	Scl	13: p			2.3	2.1	SBc		5659		
410	735	1 11.0	+33 09	Psc	12.6 p			2.6	1.7	SB0:		5444		
IC 1639	750	1 11.8	–0 39	Cet	14.05			0.8	0.7	P		2647		Mrk 562
434		1 12.2	–58 15	Tuc	13.0			1.9	1.2	S(B)b–		4587		
434A		1 12.5	–58 13	Tuc				1.1	0.4	SBap				
428	763	1 12.9	+0 59	Cet	11.85	11.35	0.50	4.1	3.2	Scp III–IV		1266*		
440		1 12.9	–58 17	Tuc				1.0	0.6	Sb+:		4876		
439		1 13.8	–31 45	Scl	13.0 p					E3		5610		
442	789	1 14.6	–1 01	Cet	14.5 p			1.2	0.8	Sa	3VS			

NGC	UGC	α 2000	δ 2000	Const	B	V	B-V	Dim		Type	Nucl	RV	Memb	Notes
IC 1653	796	1^{h}15^m.1	+33°23'	Psc	14.2	13.5	0.71	1'.1	1'.0	S:				4Z 42
450	806	1 15.5	−0 52	Cet	12.50	12.10	0.40	3.2	2.6	S IV–V	3VS	1851		
		1 16.4	+33 27	Psc	13.5					dE0			L	And II
467	848	1 19.2	+3 18	Psc	12.93	11.89	1.04	2.4	2.3	S0p	D3	5680		
470	858	1 19.7	+3 25	Psc	12.60	11.91	0.69	3.0	2.0	Sc	5S	2668		
473	859	1 19.9	+16 33	Psc	12.73			2.2	1.4	E4p	3VS	2406		
474	864	1 20.1	+3 25	Psc	12.0	11.1	0.93	7.9	7.2	S0	D4S	2412		Arp 227
	891	1 21.2	+12 25	Psc	14.28			3.1	1.5	S(B)+ IV–V		785		D10
485	895	1 21.4	+7 01	Psc	14.2 p			1.9	0.7	S				
491		1 21.4	−34 03	Scl	13.17			1.5	1.3	SBb:		3851		
488	907	1 21.8	+5 15	Psc	11.15	10.29	0.86	5.2	4.1	Sb− I	D4	2292		
493	914	1 22.2	+0 57	Cet	12.95	12.46	0.49	3.8	1.4	SBc:		2436		
497	915	1 22.4	−0 53	Cet	14.1 p			2.4	1.1	SBb+:		8159		Arp 8
495	920	1 22.9	+33 28	Psc	14.2	13.2	1.04	1.5	1.0	SBap		4313		
499	926	1 23.2	+33 28	Psc	13.05	12.02	1.03	2.0	1.6	S0		4574		IC 1686
509	932	1 23.4	+9 26	Psc	14.7 p			1.7	0.7	SBb:				
507	938	1 23.7	+33 15	Psc	12.15	11.16	0.99	4.3	4.3	S0		5127		VV 207, Arp 229
508	939	1 23.7	+33 17	Psc	13.90	12.84	1.06	1.6	1.6	E0:		5674		VV 207, Arp 229
516	946	1 24.1	+9 33	Psc	14.3 p			1.6	0.6	SBb:				
514	947	1 24.1	+12 55	Psc	12.5	11.9	0.58	3.5	2.9	Sc I–II:	3VS	2615		
518	952	1 24.3	+9 20	Psc	14.4 p			1.9	0.8	Sb−:				
521	962	1 24.6	+1 44	Cet	12.5			3.4	3.2	S(B)b	3VS	5099		
520	966	1 24.6	+3 48	Psc	12.05	11.20	0.85	4.8	2.1	P		2272*		VV 231, Arp 157
530	965	1 24.7	−1 35	Cet	14.0 p			1.9	0.6	SB0:		5044		IC 106
524	968	1 24.8	+9 32	Psc	11.5	10.6	0.95	3.2	3.2	E1	5	2595		
522	970	1 24.8	+10 00	Psc	14.2 p			2.8	0.5	Sb:				
532	982	1 25.3	+9 16	Psc	13.5 p			2.8	1.0	Sb:		2499		
523	979	1 25.3	+34 02	And	13.5 p			3.1	0.9	P		4931		4Z 45, Arp 158
539		1 25.3	−18 09	Cet	14: p			1.8	1.6	SBc		9672		
538	991	1 25.4	−1 33	Cet	14.7 p			1.3	0.7	Sb:		5426		
533	992	1 25.5	+1 46	Cet	12.5			3.7	2.6	E2:		5104		
535	997	1 25.5	−1 25	Cet	14.9 p			1.3	0.4	S0:		4967		
541	1004	1 25.7	−1 23	Cet	13.00	12.04	0.96	2.7	2.7	S0:		5420		Arp 133
543		1 25.8	−1 18	Cet	14.4	13.3	1.06	0.7	0.3	E6:		5267		
545	1007	1 26.0	−1 20	Cet	13.7 p			3.0	2.1	S0		5534*		3C 40, Arp 308
547	1009	1 26.0	−1 21	Cet	13.35	12.31	1.04	1.7	1.7	E1		5461		3C 40, Arp 308
536	1013	1 26.4	+34 43	And	13.2 p			3.7	1.9	SBb		5360		
550	1021	1 26.7	+2 01	Cet	13.6 p			1.7	0.9	Sb:		6058		
560	1036	1 27.4	−1 55	Cet	13.85	12.89	0.96	2.3	0.7	S0		5505		
564	1044	1 27.8	−1 53	Cet	13.45	12.48	0.97	1.8	1.6	E2:		5779*		
565	1052	1 28.2	−1 18	Cet	14.5 p			1.6	0.6	Sa:		4491		
562	1049	1 28.4	+48 23	And	14.5 p			1.5	1.3	Sc		10484		
570	1061	1 29.0	−0 57	Cet	14.2 p			2.5	1.8	SBa:		5529		
	1072	1 29.8	−1 15	Cet	13.97			2.0	1.2	S0		900		
578	A18	1 30.5	−22 40	Cet	11.50	10.93	0.57	4.8	3.2	Sc II	3VS	1689*		
584		1 31.3	−6 52	Cet	11.30	10.35	0.95	3.8	2.4	E4	D4S	1868		IC 1712
586		1 31.6	−6 54	Cet	14.1	13.2	0.94	1.6	0.8	Sa:	4VS			
	1102	1 32.6	+4 35	Psc	13.91			1.4	1.2	P		2086		VV 173-4, Arp 306
596		1 32.9	−7 02	Cet	11.78	10.87	0.91	3.5	2.2	E2		2104		
600		1 33.1	−7 19	Cet	13.0	12.4	0.59	3.5	3.0	SBd		1895		
598	1117	1 33.9	+30 39	Tri	6.26	5.71	0.55	62	39	Sc II–III		2	L	M33, Triangulum gal.
612		1 34.0	−36 30	Scl	14.15	13.15	1.00			S0p		9049		
613		1 34.3	−29 25	Scl	10.79	10.03	0.76	5.8	4.6	S(B)b+ I–II	5S	1462		
615		1 35.1	−7 20	Cet	12.3	11.5	0.85	4.0	1.7	Sb II–III	4	1910		
625		1 35.1	−41 26	Phe	12.28			3.0	1.3	SBm:		317		
622	1143	1 36.0	+0 40	Cet	14.1 p			2.0	1.7	SBb+		5483		Mrk 571
628	1149	1 36.7	+15 47	Psc	9.75	9.17	0.58	10.2	9.5	Sc I	3VS	793		M74
646		1 37.4	−64 55	Hyi						Scp		8061		
2573		1 37:	−89 21	Oct	14.0:	13.5:	0.5:	3.7	1.5	S0	4VS			Polarissima Australis
636		1 39.1	−7 31	Cet	12.30	11.33	0.97	2.3	1.9	E1		1989		
658	1192	1 42.2	+12 35	Psc	13.6 p			3.2	1.8	Sb		3078		
	1195	1 42.4	+13 58	Psc	13.16			3.0	1.1	Irp+		895		
	1197	1 42.5	+18 18	Psc	14.96			1.8	0.6	Ir+		2942		
660	1201	1 43.0	+13 38	Psc	11.62	10.78	0.84	9.1	4.1	SBap		982		
664	1210	1 43.8	+4 14	Psc	13.9 p			1.8	1.5	Sb:		5473		
661	1215	1 44.2	+28 42	Tri	13.0 p			2.2	1.8	S0:				
662	1220	1 44.5	+37 41	And	14.7	14.3	0.41	1.0	0.6	Sp		5855		5Z 98
668	1238	1 46.3	+36 27	And	13.5 p			2.3	1.7	Sb		5024		
669	1248	1 47.2	+35 33	Tri	12.9 p			3.4	0.8	Sb−		4944		
670	1250	1 47.4	+27 53	Tri	13.0	12.1	0.87	2.5	1.1	E4		3956		
IC 1727	1249	1 47.5	+27 20	Tri	12.10	11.55	0.55	6.2	2.9	SBm III–IV	1	528		VV 338
685		1 47.8	−52 47	Eri	11.79			4.1	4.0	S(B)c		1272		
672	1256	1 47.9	+27 26	Tri	11.35	10.76	0.59	6.6	2.7	SBc III	1	578		VV 338
673	1259	1 48.4	+11 32	Ari	13.3 p			2.4	1.9	S(B)c		5325		
676	1270	1 49.0	+5 54	Psc	10.99			4.3	1.5	Sa:	3VS	1601		

NGC	UGC	α 2000	δ 2000	Const	B	V	B-V	Dim		Type	Nucl	RV	Memb	Notes
681		1ʰ49ᵐ2	-10°26′	Cet	12.69	11.83	0.86	2′.8	1′.8	Sb-	4VS	1736		
	1276	1 49.3	+20 42	Ari	13.73			2.0	0.9	SBdm		2895		IC 163
678	1280	1 49.4	+22 00	Ari	13.3 p			5.0	1.1	SBb:		2985		
	1281	1 49.6	+32 35	Tri	12.63			4.7	1.1	Sdm		343		
679	1283	1 49.7	+35 47	And	13.25	12.35	0.90	2.3	2.3	S0:		5127		5Z 114
680	1286	1 49.8	+21 58	Ari	13.0 p			2.9	2.5	S0p				
684	1292	1 50.2	+27 39	Tri	13.2 p			3.9	0.9	Sb		3698		IC 165
693	1304	1 50.5	+6 09	Psc	13.5 p			2.7	1.4	Sa:				
691	1305	1 50.7	+21 46	Ari	12.39			3.5	2.7	Sb+		2807		
694	1310	1 51.0	+22 00	Ari	13.9 p			0.8	0.5	S0p		3017		Mrk 363, 5Z 122
701		1 51.1	-9 42	Cet	12.90	12.24	0.66	2.5	1.3	SBc		1869		
695	1315	1 51.2	+22 35	Ari	13.7 p			0.7	0.6	S0p				5Z 123
702		1 51.3	-4 03	Cet	14: p			1.5	1.1	SBb+p		10659		Arp 75
697	1317	1 51.3	+22 21	Ari	12.7 p			4.7	1.8	S(B)b+:				
706	1334	1 51.8	+6 18	Psc	13.2 p			2.2	1.7	Sb:		4973		
703	1346	1 52.8	+36 09	And	14.5 p			1.5	1.1	S0:		4986		
708	1348	1 52.8	+36 10	And	14.8 p			3.3	2.8	E2		5023		
720		1 53.0	-13 44	Cet	11.15	10.22	0.93	4.4	2.8	E3		1820		
718	1356	1 53.2	+4 12	Psc	12.50	11.68	0.82	2.8	2.5	Sb	5	1755		
721	1376	1 54.8	+39 23	And	14.1	13.4	0.66	2.0	1.2	SBb+				
731		1 54.9	-9 01	Cet	13: p			1.7	1.7	E0				
	1385	1 54.9	+36 55	And	14.0	13.4	0.59	0.8	0.7	SBa		5628		Mrk 2
		1 56.2	+5 36	Psc	14.70			0.1	0.1	P		900		
741	1413	1 56.4	+5 38	Psc	12.30	11.27	1.03	3.2	3.2	E1		5645		3Z 38A, VV 175
742		1 56.5	+5 38	Psc	15: p			0.2	0.2	E0:		5503		3Z 38B, VV 175
735	1411	1 56.6	+34 11	Tri	13.9 p			2.0	1.0	Sb		4927		5Z 146
736	1414	1 56.7	+33 03	Tri	13.20	12.23	0.97	2.0	1.9	S0:		4542		6Z 111
740	1421	1 56.9	+33 01	Tri	14.9 p			1.7	0.5	SBb:				
750	1430	1 57.5	+33 13	Tri	13.2	12.2	1.02	1.6	1.3	E	4	5305		VV 189, Arp 166
751	1431	1 57.6	+33 12	Tri	13.5	12.5	0.98	1.3	1.3	Ep	4	5301		VV 189, Arp 166
753	1437	1 57.7	+35 55	And	13.00	12.35	0.65	2.9	2.1	Sc II:	4VS	5020*		
761	1439	1 57.8	+33 23	Tri	14.5 p			1.6	0.6	SBb-:				
759	1440	1 57.8	+36 20	And	13.7 p			2.1	2.1	E0		4897		
782		1 57.8	-57 46	Eri	12.8			2.1	1.9	SBb	3S	5820		
746	1438	1 57.9	+44 56	And	13.11			1.9	1.3	Ir+		915		
770	1463	1 59.2	+18 57	Ari	14.05			1.3	1.0	E3:		2570		
772	1466	1 59.3	+19 01	Ari	11.10	10.33	0.77	7.1	4.5	Sb I	D4S	2562		Arp 78
779		1 59.7	-5 58	Cet	11.86	11.04	0.82	4.1	1.4	Sb- II-III	4	1466		
777	1476	2 00.2	+31 26	Tri	12.4 H			3.0	2.5	E2		5177		
788		2 01.1	-6 49	Cet	13.0	12.3	0.73	1.8	1.5	Sa	3	4170		
783	1497	2 01.1	+31 52	Tri	12.8 p			1.8	1.6	Sc		5517		IC 1765
784	1501	2 01.3	+28 50	Tri	12.26	11.76	0.50	6.2	1.7	S(B)+ IV		362		
	1546	2 03.4	+18 38	Ari	14.54			1.3	1.3	Sc		2499		
	1547	2 03.4	+22 02	Ari	14.20			2.0	1.8	Ir+ V		2788		D17
	1551	2 03.6	+24 04	Ari	13.30			3.0	1.5	SBdm		2817		
803	1554	2 03.8	+16 02	Ari	13.01	12.36	0.65	3.3	1.5	Sb II-III		2214		
	1561	2 04.1	+24 12	Ari	14.45			1.4	1.0	Irp+		740		5Z 173
812	1598	2 06.8	+44 34	And	12.8 p			3.2	1.5	Sp		5548		
821	1631	2 08.4	+11 00	Ari	11.8	10.8	0.96	3.5	2.2	E2		1874		
818	1633	2 08.7	+38 47	And	12.7 p			3.2	1.5	S(B)c:		4427		
833		2 09.3	-10 08	Cet	13.7	12.7	0.99	1.9	1.0	Sap		3914		Arp 318
835		2 09.4	-10 08	Cet	13.0	12.2	0.80	1.4	1.2	S(B)b-p		4051		Arp 318
838		2 09.6	-10 09	Cet	13.42	12.79	0.63	1.7	1.7	Irp-		3842		Arp 318
839		2 09.7	-10 11	Cet	13.85	13.02	0.83	1.8	0.9	S0p		3824		Arp 318
IC 1783		2 10.1	-32 56	For	13.1 H					Sb+ II		3222		
840	1664	2 10.2	+7 50	Cet	14.7 p			2.5	1.5	SBb		7196		
828	1655	2 10.2	+39 12	And	13.0 p			3.2	2.5	Sap		5612		6Z 177
848		2 10.3	-10 19	Cet	13.7	13.1	0.57	1.9	1.3	SBb-:		3982		
851	1680	2 11.2	+3 47	Cet	14.7 p			1.5	1.0	S0		3516		Mrk 588
841	1676	2 11.3	+37 30	And	12.8 p			2.0	1.2	S(B)b-				5Z 194
858		2 12.5	-22 27	Cet	13: p			1.6	1.4	S(B)c:		12318		
863	1727	2 14.6	-0 46	Cet	14.0 p			1.4	1.3	Sa	*	8296		Mrk 590
864	1736	2 15.5	+6 00	Cet	11.55	10.97	0.58	4.6	3.5	Sc II	4VS	1635*		
IC 1788		2 15.8	-31 12	For	13.10	12.37	0.73	2.5	1.1	Sa		3314		
IC 1784	1744	2 16.2	+32 39	Tri	14.06	13.16	0.90	1.6	1.0	Sb+p		4933		
871	1759	2 17.2	+14 33	Ari	14.1	13.5	0.61	1.3	0.5	SBc:		3833		
877	1768	2 18.0	+14 33	Ari	12.50	11.82	0.68	2.3	1.8	Sc I:	3VS	4117		
881		2 18.7	-6 35	Cet				2.9	1.7					
895		2 21.6	-5 31	Cet	12.30	11.76	0.54	3.6	2.8	Sb+ II	3VS	2319		
899		2 21.9	-20 48	Cet	13: p			1.9	1.3	Ir+		1746		
890	1823	2 22.0	+33 16	Tri	12.43			2.9	1.9	E4	3VS	4201		
891	1831	2 22.6	+42 21	And	10.90	9.98	0.92	13.5	2.8	Sb		706		
907		2 23.0	-20 43	Cet	13: p			1.9	0.7	SBd:		1546		
908	A29	2 23.1	-21 14	Cet	10.85	10.18	0.67	5.5	2.8	Sc I	3VS	1470		
898	1842	2 23.3	+41 57	And	13.8 p			2.1	0.6	Sb		5699		

NGC	UGC	α 2000	δ 2000	Const	B	V	B-V	Dim		Type	Nucl	RV	Memb	Notes
922	A30	2^{h}25^m.1	-24°47′	For	12.55	12.21	0.34	1′.9	1′.8	Scp II-III:	3	3029		
906	1868	2 25.2	+42 06	And	14.4 p			2.0	1.9	S(B)b-		4766		
910	1875	2 25.4	+41 50	And	14.5 p			2.3	2.3	S0		5315		
918	1888	2 25.9	+18 30	Ari	12.78			3.4	2.1	S(B)c		1616		
926	1901	2 26.1	-0 20	Cet	13.9 p			2.5	1.3	SBb+:		6510		
927	1908	2 26.6	+12 09	Ari	14.5 p			1.4	1.4	SBc		8343		
925	1913	2 27.3	+33 35	Tri	10.60	10.01	0.59	9.8	6.0	S(B)c II-III	1	716		
936	1929	2 27.6	-1 09	Cet	11.10	10.14	0.96	5.2	4.4	SBa	D4	1350		
941	1954	2 28.5	-1 09	Cet	12.94	12.41	0.53	2.8	2.1	Sc III-IV	3VS	1553		
947		2 28.5	-19 04	Cet	12: p			2.6	1.5					
945		2 28.6	-10 32	Cet	12: p			2.5	2.1	SBc				
948		2 28.8	-10 31	Cet	14: p			1.4	1.2	S(B)c:		4507		
942-3		2 29.2	-10 50	Cet	14: p					S0p,Irp-				VV 217, Arp 309
955	1986	2 30.6	-1 07	Cet	12.95	12.00	0.95	3.0	0.9	Sb-		1566		
958		2 30.7	-2 57	Cet	12.95	12.19	0.76	2.8	1.1	S(B)b+: I-II		5756		
949	1983	2 30.8	+37 08	Tri	12.55	11.89	0.66	2.8	1.7	S:		785		
959	2002	2 32.3	+35 30	Tri	12.91			2.2	1.4	Sm:		767		
	2023	2 33.3	+33 30	Tri	13.06			2.6	2.4	Ir+ V		763		D25
986		2 33.6	-39 02	For	11.8	11.0	0.78	3.7	2.8	SBb-	4	1943		
	2034	2 33.7	+40 32	And	13.10			3.1	2.5	Ir+ IV-V:		752		D24
976	2042	2 34.0	+20 59	Ari	13.21	12.40	0.81	1.7	1.5	Sb II:	3S	4472		
972	2045	2 34.2	+29 19	Ari	12.10	11.25	0.85	3.6	2.0	Sc II:	2S	1670		
	2053	2 34.5	+29 45	Ari	14.58			2.0	1.2	Ir- V		1173		D26
985		2 34.6	-8 47	Cet	14.10	13.45	0.65	1.3	1.3	Ring	*	12948		VV 285
988		2 35.4	-9 21	Cet	11: p			4.7	2.7	SBc:		1450		
991		2 35.5	-7 09	Cet	12.29			2.7	2.5	S IV:	3VS	1580		
	2082	2 36.3	+25 26	Ari	12.63			6.2	1.3	Sc		834		
	A34	2 36.3	+59 39	Cas						E3		223		Maffei I
IC 239	2080	2 36.5	+38 58	And	11.93	11.22	0.71	4.6	4.3	S(B)c	3VS	1064		
	2105	2 37.7	+34 26	Tri	14.2	13.5	0.72	1.5	1.0	SBa:		4951		
	2119	2 37.9	-1 51	Cet	14.2	13.5	0.74	2.0	1.1	SBb				
1015	2124	2 38.1	-1 19	Cet	12.39			3.5	3.5	SBa		2655		
1016	2128	2 38.3	+2 08	Cet	13.3 p			2.8	2.8	E0				
1019	2132	2 38.4	+1 55	Cet	14.6 p			1.2	1.1	SBb		7257		
1022		2 38.5	-6 40	Cet	12.20	11.44	0.76	2.5	2.1	Sb	4S	1505		
IC 1830	A37	2 39.1	-27 27	For	13.2	12.8	0.41	1.8	1.7	SBa:		1346		
1024	2142	2 39.2	+10 51	Ari	13.8 p			4.7	1.9	Sb-		3642		Arp 333
1012	2141	2 39.3	+30 09	Ari	12.67			2.9	1.5	P		1123		
1003	2137	2 39.3	+40 52	Per	12.02	11.46	0.56	5.4	2.1	Sc	1	790		
1032	2147	2 39.4	+1 06	Cet	13.2 p			3.6	1.4	Sa				
1035		2 39.5	-8 08	Cet	12.87			2.2	0.9	P		1319		
		2 39.9	-34 32	For	9.04			20.0	13.8	dE3		-51	L	Fornax dwarf
1042		2 40.4	-8 26	Cet	11.49	10.94	0.55	4.7	3.9	Sc	4VS	1360		
1023	2154	2 40.4	+39 04	Per	10.46	9.46	1.00	8.7	3.3	E7p	5	776*		Arp 135
1047		2 40.5	-8 09	Cet	15: p			1.5	0.6	Sa:	3S	1336		
1036	2160	2 40.5	+19 18	Ari	13.49			2.0	1.5	P:		888*		Mrk 370
1048B		2 40.6	-8 32	Cet	14: p			1.0	0.5	SBc:				
1048A		2 40.6	-8 33	Cet	14: p			1.3	0.4	SBb:				
1044		2 41.1	+8 44	Cet	14.4	13.2	1.17	0.6	0.6	S0p		6580		
1052		2 41.1	-8 15	Cet	11.55	10.59	0.96	2.9	2.0	E2		1434		
1055	2173	2 41.8	+0 26	Cet	11.40	10.55	0.85	7.6	3.0	Sb II-III		1077		
	A39	2 41.9	+59 36	Cas						Sb+ II		208		Maffei II
1068	2188	2 42.7	-0 01	Cet	9.51	8.81	0.70	6.9	5.9	Sbp	4*	1134		M77, 3C 71
1072	2208	2 43.5	+0 18	Cet	14.3 p			1.7	0.7	S(B)b:				
1058	2193	2 43.5	+37 21	Per	12.15	11.54	0.61	3.0	2.9	Sc III-IV:	4S	674		
1073	2210	2 43.7	+1 23	Cet	11.55	11.02	0.53	4.9	4.6	S(B)c II	3VS	1245		
1079		2 43.7	-29 00	For	12.30	11.37	0.93	3.1	1.9	S(B)ap	4S	2165		
1067	2204	2 43.8	+32 31	Tri	14.6 p			1.3	1.2	S(B)c		4676		
1084		2 46.0	-7 35	Eri	11.22	10.60	0.62	2.9	1.5	Sc I-II	3VS	1406		
1097A		2 46.2	-30 14	For	13.7	12.9	0.85			E5				
1097	A41	2 46.3	-30 17	For	10.25	9.25	1.00	9.3	6.6	S(B)b I-II	5S	1227		Arp 77
1085	2241	2 46.4	+3 37	Cet	13.6 p			1.7	1.4	Sb		6986		
1087	2245	2 46.4	-0 30	Cet	11.55	10.99	0.56	3.5	2.3	Sc III:		1844		
1090	2247	2 46.6	-0 15	Cet	12.60	11.93	0.67	3.8	1.8	S- IV:	3	2786		
1094	2262	2 47.5	-0 17	Cet	13.5 p			1.4	1.2	S(B)b-		6284		
	2259	2 47.9	+37 32	Per	13.42			2.8	2.3	SBdm		742		
	2275	2 48.0	+3 53	Cet	13.43			5.8	5.8	Sm		1061		D28
1104	2287	2 48.6	-0 16	Cet	14.8 p			1.3	1.0	SBa				
	2302	2 49.2	+2 08	Cet	13.24			5.8	5.8	Sm		1133		D29
		2 51.1	+4 27	Cet	14.87			0.5	0.3	Sp		1026		Mrk 600
1134	2365	2 53.6	+13 00	Ari	12.9:			2.5	0.9	S				Arp 200
1140		2 54.6	-10 02	Eri	12.85	12.46	0.39	1.4	0.9	P	4	1503		
1143	2388	2 55.16	-0 10.7	Cet	14.5	13.5	1.03	1.1	1.1	Ring		8440		VV 331, Arp 118
1144	2389	2 55.20	-0 11.1	Cet	14.2	13.3	0.86	0.9	0.6	Ring		8774		VV 331, Arp 118
		2 56.6	-54 36	Hor	11.6 H			6.9	1.7	SBd:	1			

NGC	UGC	α 2000	δ 2000	Const	B	V	B-V	Dim		Type	Nucl	RV	Memb	Notes
		2ʰ57ᵐ.7	+6°02′	Cet	14.3	13.1	1.20	0′.9	0′.4	E0:		7067*		3Z 52, 3C 75
1156	2455	2 59.7	+25 14	Ari	12.25	11.69	0.56	3.1	2.3	Ir+ IV:		485		
	2460	2 59.8	+2 46	Cet	13.43			1.7	1.7	SBb+		2721		IC 277, Mrk 602
	2463	3 00.6	+40 15	Per	13.85			3.0	2.0	S(B)m		2051		
1161	2474	3 01.2	+44 55	Per	12.6 p			3.1	2.2	S0		2081		
1160	2475	3 01.2	+44 58	Per	13.0 p			1.6	0.8	Sc:		2658		
1172		3 01.6	-14 50	Eri	12.90	12.01	0.89	2.2	2.0	E1		1621		
1167		3 01.7	+35 12	Eri	13.4	12.3	1.06	3.1	2.5	S0		4859		
1179	A48	3 02.6	-18 54	Eri	11.77			4.6	3.9	Sp	3VS	1716		
1187	A49	3 02.6	-22 52	Eri	10.90			5.0	4.1	S(B)c: I	4VS	1334		
1169	2503	3 03.6	+46 23	Per	12.5	11.7	0.82	4.4	3.0	S(B)b- II:	3VS	2564		
1199		3 03.6	-15 37	Eri	12.50	11.47	1.03	2.2	1.9	E2		2528		
1171	2510	3 04.0	+43 24	Per	12.98			3.2	1.4	Sc		2906		
1201		3 04.1	-26 04	For	11.58	10.64	0.94	4.4	2.8	Sa	D4	1630		
1175	2515	3 04.5	+42 20	Per	13.80	12.79	1.01	2.5	0.8	S	3	5613		
IC 1876		3 04.5	-27 26	For	14.7	14.4	0.29	0.7	0.7			6449		
1186	2521	3 05.5	+42 50	Per	12.5 p			3.3	1.4	SBb+:				
1209		3 06.0	-15 37	Eri	12.33	11.35	0.98	2.6	1.5	E5		2508		
IC 284	2531	3 06.2	+42 23	Per	12.64			4.9	2.5	Sm		2872		
1229		3 08.2	-22 58	Eri	14: p			1.9	1.1	SBbp		10501		VV 337
1218		3 08.4	+4 07	Cet	13.95	12.81	1.14	1.3	1.1	Sa		8669		3C 78
1222		3 08.9	-2 58	Eri	14: p			1.3	1.2			2692		Mrk 603
1232		3 09.8	-20 35	Eri	10.50	9.87	0.63	7.8	6.9	Sc I	D3VS	1644		Arp 41
1232A		3 10.0	-20 36	Eri				1.0	0.9	SBm		1702		
1249		3 10.1	-53 21	Hor	11.69			5.2	2.7	SBc	1	828		
1242		3 11.3	-8 54	Eri	14: p			1.3	0.7	SBc:	3VS	3885		VV 334, Arp 304
1241		3 11.3	-8 55	Eri	12.68			3.0	1.9	Sb+	4VS	2134		VV 334, Arp 304
1224	2578	3 11.3	+41 22	Per	15.5 p			1.7	1.5	S0:		5199	P	
1247		3 12.2	-10 29	Eri	13: p			3.6	0.7	Sb				
1233		3 12.5	+39 19	Per	14.0	13.2	0.81	2.1	0.9	Sb		4906		
1248		3 12.8	-5 13	Eri	13.5:p			1.3	1.1	S0:	4S			
1255	A60	3 13.5	-25 44	For	11.6	11.1	0.52	4.1	2.8	Sc II	4VS	1738		
1253	A62	3 14.1	-2 49	Eri	12.40			4.8	2.6	S(B)c		1695		Arp 279
1253A		3 14.4	-2 48	Eri				1.7	1.0	Ir+ V:		1825		D31
		3 14.7	-4 47	Eri	14.10			1.9	1.5	Ir+ V		2197		D32
1250	2613	3 15.4	+41 21	Per	14.2 p			2.7	1.1	S0:		6343	P	
1184	2583	3 16.6	+80 48	Cep	13.4 p			3.0	0.8	Sa				
1288		3 17.2	-32 35	For	12.80	12.11	0.69	2.3	2.2	Sb I		4371		
		3 17.3	+3 37	Cet	14.86			0.6	0.4	S0		900		
1291		3 17.3	-41 08	Eri	9.42	8.49	0.93	10.5	9.1	SBa	4	674		
1260	2634	3 17.5	+41 24	Per	14.2 p			1.7	0.8	S0:		5663	P	
1292		3 18.2	-27 37	For	12.59			3.2	1.7	Sc		1343		
1265	2651	3 18.3	+41 52	Per	14.7 p			2.1	1.9	S0		7686	P	3C 83.1
1313		3 18.3	-66 30	Ret	9.37			8.5	6.6	SBd	3VS	241		
1267	2657	3 18.7	+41 28	Per	15.4 p			1.4	1.1	S0:		5258	P	
1268	2658	3 18.8	+41 29	Per	14.5 p			1.2	0.9	Sb:		3272	P	
1270	2660	3 19.0	+41 28	Per	14.05	12.89	1.16	1.1	0.9	E2		5048	P	
1271		3 19.2	+41 21	Per	15.5:p					SB0:		5894	P	
1297		3 19.2	-19 06	Eri	12.68			2.3	2.0	E2		1471		
1272	2662	3 19.3	+41 29	Per	14.5 p			2.5	2.3	S0		4340	P	
1273		3 19.4	+41 32	Per	14.05	12.90	1.15	1.4	1.4	S0:		5497	P	
1274		3 19.7	+41 33	Per	15: p			0.6	0.4	E3		6615	P	
1300	A66	3 19.7	-19 25	Eri	11.10	10.42	0.68	6.5	4.3	SBb I	5VS	1422		
1275	2669	3 19.8	+41 31	Per	12.35	11.59	0.76	2.6	1.9	P	*	5361	P	Per A, 3C 84
1277		3 19.86	+41 34.3	Per	14.6	13.5	1.10	0.8	0.3	S0p		5117	P	
1302		3 19.9	-26 04	For	11.49			4.4	4.2	S(B)a	5	1626		
1278		3 19.90	+41 33.8	Per	13.65	12.58	1.07	1.7	1.4	E2p		6258	P	
1281		3 20.1	+41 38	Per	14.6	13.5	1.10	0.7	0.5	E5		4399	P	
1298	2683	3 20.2	-2 06	Eri	14.2 p			1.5	1.3	E1:		6619		
1282	2675	3 20.2	+41 22	Per	13.73			1.7	1.3	E1		2342	P	
1283	2676	3 20.3	+41 24	Per	15.6 p			1.3	0.8	E1:		6866	P	
	2620	3 20.4	+80 15	Cep	14.16			2.5	2.1	S(B)m		2464		
1310		3 21.0	-37 08	For				2.3	1.8	Sc:		1574	F	
1305	2697	3 21.4	-2 19	Eri	14.8 p			1.9	1.4	E2:				
1293		3 21.6	+41 24	Per	15: p			1.1	1.1	E0		4273	P	
1294	2694	3 21.7	+41 22	Per	14.2	13.1	1.09	1.8	1.6	S0:		6693	P	
1309		3 22.1	-15 24	Eri	12.00	11.55	0.45	2.3	2.2	Sc II-III	3S	2195		
1316		3 22.7	-37 12	For	9.75	8.85	0.90	7.1	5.5	S(B)0p		1632	F	For A, Arp 154
1317		3 22.8	-37 06	For	11.94	11.01	0.93	3.2	2.8	S(B)a	5	1918	F	NGC 1318
1315		3 23.1	-21 23	Eri	13: p			1.8	1.8	SB0:	3S	1640		
1319		3 23.9	-21 32	Eri	14: p			1.3	0.7	S0		4021		
1326		3 23.9	-36 28	For	11.30	10.48	0.82	4.0	3.0	SB0	4	1167*	F	
1325	A70	3 24.4	-21 33	Eri	12.32	11.55	0.77	4.6	1.8	Sb	3S	1543*		
1320		3 24.8	-3 01	Eri	14: p			1.9	0.7			2978		Mrk 607
1325A		3 24.8	-21 20	Eri	13.43	12.83	0.60	2.6	2.4	SBb:	3S			Holm. VI

NGC	UGC	α 2000	δ 2000	Const	B	V	B–V	Dim		Type	Nucl	RV	Memb	Notes
1316C		3ʰ24ᵐ9	–37°01′	For				1.7	0.7	Sa:		1923	F	
1326A		3 25.0	–36 20	For				1.4	1.3	SBm:		1700	F	
1326B		3 25.2	–36 21	For				3.8	1.1	SBd:		639	F?	
IC 1933		3 25.7	–52 47	Hor	12.77			2.5	1.5	Sc:		877		
1332	A72	3 26.3	–21 20	Eri	11.2	10.3	0.90	4.6	1.7	E7	D4	1471		
1331		3 26.5	–21 21	Eri	14: p			1.1	1.0	E2:		1233		IC 324
	2748	3 27.9	+2 34	Tau	14.75	13.64	1.11	1.5	1.1	E3:		9067		3C 88
1341		3 28.0	–37 09	For	12.98			1.6	1.5	SBa		1684	F	
1337		3 28.1	–8 23	Eri	12.28	11.72	0.56	6.8	2.0	S– IV:	2	1189		
1339		3 28.1	–32 17	For	12.58			2.3	1.9	E2		1195		
	A73	3 28.2	–17 26	Eri	14.9	14.3	0.57	0.8	0.4	E6p		1754		
1344		3 28.3	–31 04	For	11.25	10.32	0.93	3.9	2.3	E3		1088*		NGC 1340
1351A		3 28.8	–35 11	For				3.5	0.7	SBc			F?	
1345	A74	3 29.4	–17 48	Eri	14.2	13.7	0.48	1.4	1.1	Sp		1445		
1351		3 30.5	–34 52	For	12.78			1.8	1.2	S0:	4	1350	F	
1350		3 31.1	–33 38	For	11.40	10.52	0.88	4.3	2.4	SBb–		1649	F	
IC 1954		3 31.6	–51 55	Hor	11.98			3.6	1.9	SBb:	3S	929		
1353	A76	3 32.1	–20 49	Eri	12.25	11.35	0.90	3.4	1.5	Sb II-III	4VS	1605		
1357		3 33.3	–13 50	Eri	12.60	11.67	0.93	2.5	1.8	Sa	D3S	1964		
1355		3 33.4	–5 00	Eri	14.05	13.09	0.96	1.8	0.5	S0	4VS			
1365		3 33.6	–36 08	For	10.14	9.52	0.62	9.8	5.5	SBb I-II	5	1502	F	
1358		3 33.7	–5 05	Eri	13.0	12.1	0.94	2.8	2.1	S(B)b– II:	4VS	4033		
IC 1953	A78	3 33.7	–21 29	Eri	12.26			2.8	2.3	Sc II-III:	1	1902		
1359		3 33.8	–19 29	Eri	12.6	12.2	0.39	1.9	1.7	SB+p IV		1789		
1366		3 33.9	–31 12	For	12.8 H			2.7	1.2	E6		1123		
1371	A79	3 35.0	–24 56	For	11.47			5.4	4.0	S(B)a	4S	1397		NGC 1367
1375		3 35.2	–35 16	For				1.9	0.6	S0:		524	F?	
1374		3 35.3	–35 14	For	12.39			1.8	1.8	E0	3	1106	F	
1379		3 36.1	–35 27	For	12.29			2.0	1.9	E0	3	1239	F	
1380		3 36.5	–34 59	For	11.1			4.9	1.9	S0	4	1664	F	
1381		3 36.6	–35 18	For	12.48			2.9	0.8	S0	4S	1630	F	
1380A		3 36.7	–34 43	For				2.8	0.7	Sb–:	3S	1471	F	
1386		3 36.9	–36 00	Eri	12.09			3.5	1.5	Sa	3S	645	F?	
1380B		3 37.0	–35 11	For						S0:		1779	F	
1387		3 37.0	–35 31	For	11.99			2.4	2.2	S0	4	1092	F	
1376		3 37.1	–5 03	Eri	12.8 H			2.0	1.9	Sc II-III:	3S	4128*		
1389		3 37.2	–35 45	Eri	12.58			2.1	1.4	E4:	3	928	F	
1385		3 37.5	–24 30	For	11.65	11.16	0.49	3.0	2.0	Sc I-II		1389*		
1343		3 37.8	+72 34	Cas	13.3	12.3	1.02	2.8	1.9	S(B)bp		507		7Z 8
1395		3 38.5	–23 02	Eri	11.29			3.2	2.5	E3		1583		
1399		3 38.5	–35 27	For	10.85	9.90	0.95	3.2	3.1	E1p	3	1295*	F	
1393		3 38.6	–18 26	Eri	14: p			1.8	1.3	S0:				
1411		3 38.8	–44 05	Hor	11.89			2.8	2.3	S0:	4	898		
1398		3 38.9	–26 20	For	10.6	9.7	0.95	6.6	5.2	S(B)b– I	3L	1299		
1404		3 38.9	–35 35	For	11.20	10.25	0.95	2.5	2.3	E1	3	1759	F	
1394		3 39.1	–18 17	Eri	14: p			1.5	0.4	S0:				
1401		3 39.4	–22 44	Eri	13: p			2.8	0.7	SB0:		1461		
1406	A83	3 39.4	–31 19	For	12.55			3.9	1.0	Sb+		745		
1400		3 39.5	–18 41	Eri	12.08	11.11	0.97	1.9	1.7	E1	D3S	389		
1407		3 40.2	–18 35	Eri	10.8	9.8	0.97	2.5	2.5	E0	D3S	1717		
1415		3 41.0	–22 34	Eri	12.39			3.6	2.1	Sb+ III	4	1399		
1416		3 41.0	–22 43	Eri	14: p			1.5	1.5	E1:	3VS			
1409–10	2821	3 41.2	–1 17	Tau	14.7 p			1.3	1.0	S(B)0p,E1p	*	7413		3Z 55
1422		3 41.5	–21 41	Eri	13: p			2.1	0.7	SBb:				
1417		3 42.0	–4 42	Eri	12.75	12.03	0.72	2.8	1.9	Sb I-II:	4S	4057		
1433		3 42.0	–47 13	Hor	10.68	9.99	0.69	6.8	6.0	SBa	5S	802		
1425	A84	3 42.2	–29 54	For	11.70			5.4	2.7	Sb II-III	4VS	1494		
1418		3 42.3	–4 44	Eri	14.5 p			1.5	1.2	SBb:	3S	3679		
1427		3 42.3	–35 25	For	12.18			2.8	2.0	E3	3S	1416	F	
1428		3 42.4	–35 10	For				1.8	0.8	S0:	3S	1492*	F	
1421		3 42.5	–13 29	Eri	11.95	11.44	0.51	3.6	1.0	Sb+ I:	3VS	2067		
1426		3 42.8	–22 07	Eri	12.31	11.44	0.87	2.1	1.5	E2	3S	1249		
1424		3 43.2	–4 44	Eri	14.5 p			2.1	0.9	S(B)b:	4VS	5595		
1437		3 43.6	–35 52	Eri	12.48			2.9	2.1	SBa	3VS	1078	F	
1448		3 44.5	–44 39	Hor	11.30			8.1	1.8	Sc: III	3S	1005		NGC 1457
1439		3 44.8	–21 55	Eri	12.48			2.3	2.2	E1	3VS	1887		
1440		3 45.0	–18 16	Eri	12.68			2.3	1.9	S(B)a		1437		NGC 1442
IC 334	2824	3 45.2	+76 38	Cam	12.25			6.0	5.8	P		2722		
1452		3 45.4	–18 38	Eri	12.58			1.7	1.5	SBa		1805		NGC 1455
1441		3 45.7	–4 06	Eri	13.9	12.9	0.97	1.5	0.6	SBb	4S	4217		
1449		3 46.0	–4 08	Eri	14.4	13.5	0.88	1.1	0.7	S0		4131		
1451		3 46.1	–4 04	Eri	14.4	13.3	1.06	0.7	0.3	S0:		3882		
1453		3 46.4	–3 58	Eri	12.60	11.57	1.03	2.1	1.6	E2	4	3861		
IC 342	2847	3 46.8	+68 06	Cam	9.12			17.8	17.4	S(B)c I-II	4	228		
	2855	3 48.4	+70 08	Cam	12.91			6.4	3.2	S(B)c		1410		

NGC	UGC	α 2000	δ 2000	Const	B	V	B-V	Dim		Type	Nucl	RV	Memb	Notes
1461		3ʰ48ᵐ5	-16°24'	Eri	12.75	11.70	1.05	3.3	1.1	Sa	D3VS	1356		
IC 2006		3 54.1	-35 59	Eri	12.48			2.3	2.1	E1		1230	F	
1482		3 54.7	-20 30	Eri	14: p			1.5	1.1	Sa		1542		
1487		3 55.8	-42 22	Eri	12.31			2.0	1.4	P	3	537		VV 78
1493		3 57.5	-46 12	Hor	11.77			2.6	2.3	SBc	3VS	835		
1494		3 57.7	-48 54	Hor	12.18			2.6	1.8	Sd	1	906		
1511		3 59.5	-67 38	Hyi	12.05			3.3	1.2	Scp	1	1605		
1469	2909	4 00.5	+68 35	Cam	14.5 p			2.3	0.9	S0:				
1510		4 03.5	-43 25	Hor	13.48	13.00	0.48	1.0	0.9	E0p	4	782		
1512		4 03.9	-43 21	Hor	11.46	10.62	0.84	4.0	3.2	SBa	4	558		
1515		4 04.1	-54 06	Dor	11.83	11.03	0.80	5.4	1.3	S(B)b+		884		
1507	2947	4 04.5	-2 11	Eri	12.7	12.2	0.51	3.4	1.0	SBm		850		
1485	2933	4 05.1	+71 00	Cam	13.6 p			2.4	0.9	Sb:				
1518		4 06.8	-21 11	Eri	12.30	11.84	0.46	3.0	1.4	Scp		810		
IC 356	2953	4 07.8	+69 49	Cam	11.43			5.2	4.1	Sb-p		1015		Arp 213
1519		4 08.1	-17 12	Eri	13: p			2.0	0.5	Sc:				
1521		4 08.3	-21 03	Eri	12.4	11.4	0.96	2.9	1.9	E3		4097		
1527		4 08.4	-47 53	Hor	12.09			3.4	1.3	S0	4	840		
IC 2035		4 09.0	-45 31	Hor	12.18					SB0p		1264		
1533		4 09.9	-56 07	Dor	11.88	10.92	0.96	2.9	2.5	SB0	4	560		
1536		4 11.0	-56 29	Ret	13.1			2.0	1.7	SBc:	3VS	1378		
1531		4 12.0	-32 51	Eri	12.8	12.1	0.70	1.3	0.8	E6		1089		
1532		4 12.1	-32 52	Eri	11.08			5.6	1.8	Sb II-III:		1019*		
1543		4 12.8	-57 44	Ret	11.57	10.60	0.97	3.9	2.1	SB0	4	1183		
1537		4 13.7	-31 39	Eri	11.60					E4		1137		
1546		4 14.6	-56 04	Dor	12.50	11.57	0.93	3.2	1.5	S0:	3VS	935		
1549		4 15.7	-55 36	Dor	10.87	9.94	0.93	3.7	3.2	E0	4	938		
1553		4 16.2	-55 47	Dor	10.47	9.53	0.94	4.1	2.8	S0	5	1064		
IC 2056		4 16.5	-60 13	Ret	12.38			1.9	1.7	SB0p	3	868		
1559		4 17.6	-62 47	Ret	10.86	10.45	0.41	3.3	2.1	SBc	1	1025		
1566		4 20.0	-54 56	Dor	10.26	9.41	0.85	7.6	6.2	S(B)b+	4S*	1178		
1574		4 22.0	-56 58	Ret	11.3	10.5	0.82	2.0	1.9	S0	4	670		
1530	3013	4 23.4	+75 18	Cam	12.35			4.9	2.9	SBb	4VS	2661		7Z 12
1596		4 27.6	-55 02	Dor	11.90	10.97	0.93	3.9	1.2	S0:	4	1266		
IC 2082		4 29.1	-53 50	Dor	13.9	12.8	1.11			S0		11869		
1587	3063	4 30.7	+0 40	Tau	13.3 p			2.0	1.9	E1p		3826		2Z 12
1588	3064	4 30.8	+0 39	Tau	14.1 p			1.8	1.0	E1p		3314		Mrk 616, 2Z 12
1589	3065	4 30.8	+0 51	Tau	13.8 p			3.1	1.2	Sb-				
1569	3056	4 30.8	+64 51	Cam	11.95	11.18	0.77	2.9	1.5	Irp+ III-IV:		87		Arp 210, 7Z 16
1590	3071	4 31.2	+7 37	Tau	14.6 p			1.3	1.2	P		3758		2Z 13
1599		4 31.6	-4 35	Eri	14.5:p			1.1	1.0	Sb+:				
1600		4 31.7	-5 05	Eri	12.10	11.11	0.99	2.5	1.8	E2		4743		
1617		4 31.7	-54 36	Dor	11.30	10.35	0.95	4.7	2.4	SBa	4	778		
1601		4 31.8	-5 06	Eri	14.7	13.8	0.94	1.0	0.7	S0:		4910		
1560	3060	4 32.8	+71 53	Cam	12.24	11.52	0.72	9.8	2.0	Sd	1	151		IC 2062
	3087	4 33.2	+5 21	Tau	14.3	13.7	0.56	1.0	0.7	S0:	*	9944		BW Tau, 3C 120
1614		4 34.0	-8 35	Eri	13.60	12.91	0.69	1.3	1.1	SBcp	5	4643		Mrk 617, Arp 186
1573	3077	4 35.0	+73 15	Cam	13.5			2.2	1.5	E3		4440		7Z 18
1615	3096	4 36.0	+19 57	Tau	15.0 p			1.6	0.9	S0:				
1618		4 36.1	-3 09	Eri	13.35	12.55	0.80	2.8	1.2	SBb:		5267*		
1620	3103	4 36.6	-0 09	Eri	13.6 p			3.0	1.2	Sb+:		3432		
1622		4 36.6	-3 11	Eri	13.15	12.26	0.89	3.9	1.1	S(B)b-:				
1625		4 37.1	-3 18	Eri	13.2	12.4	0.80	2.7	0.8	SBb		2953		
1635	3126	4 40.1	-0 33	Eri	13.5 p			1.7	1.5	S(B)a				
	3128	4 40.6	+4 12	Tau	14.8	13.9	0.88	1.3	1.1	S(B)0		4542		
1637	A93	4 41.5	-2 51	Eri	11.52	10.87	0.65	3.3	2.9	Sc	4VS	626		
1638	3133	4 41.6	-1 49	Eri	12.95	12.06	0.89	2.5	1.9	E2	4	3223		
1640		4 42.2	-20 26	Eri	12.45	11.72	0.73	2.8	2.0	SBb- II	3S	1392		
1642	3140	4 42.9	+0 37	Tau	13.6 p			2.0	1.8	Sc:	3S			
IC 381	3130	4 44.4	+75 39	Cam	13.54			3.4	1.9	S(B)b		2683		
1672		4 45.7	-59 15	Dor	11.03			4.8	3.9	SBb I-II	4	1076		
1654	3154	4 45.8	-2 06	Eri	14.4 p			1.0	1.0	S		4453		
1653	3153	4 45.8	-2 25	Eri	12.9 p			2.0	2.0	E0		4263		
	3137	4 46.3	+76 26	Cam	14.15			4.2	1.0	Sb		1193		
1659		4 46.5	-4 47	Eri	13.19	12.49	0.70	1.7	1.3	Sc III	3VS	4444		
	3144	4 47.8	+74 55	Cam	14.56			2.1	1.1	SB+ IV-V:		1830		D33
1688		4 48.4	-59 48	Dor	12.38			2.4	1.9	SBc		995		
1666		4 48.5	-6 34	Eri	13.5:p			1.3	1.2	SB0	4			
1667		4 48.6	-6 19	Eri	12.75	12.05	0.70	1.5	1.3	Sb+ II:	3VS	4504		
		4 51.3	-17 30	Eri	14.6	13.6	0.98	0.9	0.3			9505		
1705		4 54.2	-53 22	Pic	12.80	12.30	0.50	1.8	1.4	S0p	4	398		
1690	3198	4 54.3	+1 38	Ori	14.15			1.3	1.3	E		2100		
1700		4 56.9	-4 52	Eri	11.9	11.0	0.95	2.9	2.0	E1	4	3870		
1699		4 57.0	-4 45	Eri	15: p			1.1	0.7	Sb	3VS			
IC 391	3190	4 57.4	+78 11	Cam	13.0	12.7	0.31	1.7	1.7	Sc		1687		

NGC	UGC	α 2000	δ 2000	Const	B	V	B–V	Dim		Type	Nucl	RV	Memb	Notes
IC 396	3203	4ʰ58ᵐ0	+68°19′	Cam	12.73			3ʹ0	2ʹ2	Sp		937		
1720		4 59.3	–7 52	Eri	13: p			1.8	1.2	SBb–	4VS	4078		
1723		4 59.4	–10 59	Eri	12? p			3.7	2.3	SB				
1730		4 59.6	–15 50	Lep	13: p			3.0	1.7	S				
1726		4 59.7	–7 45	Eri	13.1	12.1	0.99	1.4	1.1	E2	D4S	3953		
1744		5 00.0	–26 01	Lep	11.7	11.2	0.49	6.8	4.1	S(B)c III:		579		
1741		5 01.6	–4 16	Eri	13.80	13.38	0.42	1.7	0.9	SBm, SBmp		3949		Arp 259
IC 399		5 01.8	–4 18	Eri	14.90	14.45	0.45			Irp+		3912		
1752		5 02.1	–8 14	Eri	13: p			2.6	0.9	SBc:				
1796		5 02.7	–61 08	Dor	12.88			2.0	1.1	SBb:	1	777		
1792		5 05.2	–37 59	Col	10.85	10.18	0.67	4.0	2.1	Sb+	3VS	977		
1779		5 05.3	–9 09	Eri	13: p			2.6	1.4	SBa:				
1784		5 05.4	–11 52	Lep	12.45	11.75	0.70	4.2	2.8	S(B)c II:	3VS	2182		
1800		5 06.4	–31 57	Col	13.10	12.55	0.55	1.6	0.9	E6		522		
1808		5 07.7	–37 31	Col	10.70	9.89	0.81	7.2	4.1	S(B)a	4	769		
	A103	5 10.7	–31 36	Col	13.14			2.5	2.4	S+ IV-V:		778		D230
	A104	5 11.7	–14 47	Lep	12.7 H			2.8	2.3	S(B)c		1987		
	A106	5 12.0	–32 58	Col	13.20	12.60	0.60	3.4	2.5	Ir– V		731		D231
1832		5 12.1	–15 41	Lep	12.1	11.4	0.75	2.8	1.9	Sc II	4S	1760		
	A105	5 14.3	+62 34	Cam	13.24			6.4	4.0	Ir+		264		
	3273	5 17.7	+53 33	Aur	13.97			6.9	2.2	Sm		734		
1879	A110	5 19.8	–32 09	Col	13.15			2.2	1.9	S(B)– V		1040		D232
1875		5 21.8	+6 41	Ori	15: p					S0:		9087		VV 169, Arp 327
1888		5 22.58	–11 30.0	Lep	12.90	11.97	0.93	3.0	0.9	SBcp		2353		Arp 123
1889		5 22.59	–11 29.8	Lep	14: p			0.8	0.7	E0		2322		
		5 23.6	–69 45	Dor	0.63	0.08	0.55	650	550	SBm III-IV		13	L	Large Magellanic Cloud
1947		5 26.8	–63 46	Dor	11.83	10.80	1.03	3.0	2.6	S0p	D3	653*		
1954		5 32.8	–14 04	Lep	13: p			4.1	2.3	Sc		2914		
1964		5 33.4	–21 57	Lep	11.60	10.82	0.78	6.2	2.5	Sb II	4S	1498		
2082		5 41.8	–64 18	Dor	12.78			1.6	1.6	SBb	2S	850		
1961		5 42.1	+69 23	Cam	11.81	11.08	0.73	4.3	3.0	Sbp I:	3VS	4056		IC 2133, Arp 184
2090		5 47.0	–34 14	Col	11.75			4.5	2.3	Sc	D3S	1576		
2139		6 01.1	–23 40	Lep	12.05	11.71	0.34	2.2	1.9	Scp	3S	1591		IC 2154
2179		6 08.0	–21 45	Lep	13.40	12.45	0.95	1.5	1.0	Sa	D3	2549		
2188		6 10.1	–34 06	Col	12.3	11.8	0.48	3.7	1.1	SBm	3S	446		
2196	A121	6 12.2	–21 48	Lep	12.1	11.2	0.87	2.8	2.2	Sb I-II:	D3	2080		
2206		6 16.0	–26 46	CMa	13: p			2.3	1.4	S(B)b+		6040		
2207	A124	6 16.4	–21 22	CMa	11.35	10.65	0.70	4.3	2.9	Sc I:	4	2465		
2146	3429	6 18.7	+78 21	Cam	11.20	10.46	0.74	6.0	3.8	SBb–p		1028		
2217		6 21.7	–27 14	CMa	11.45	10.42	1.03	4.8	4.4	SBa	D4	1243		
2146A	3439	6 23.9	+78 32	Cam	13.35			3.6	1.4	Sc		1685		
2223	A129	6 24.6	–22 50	CMa	12.15	11.38	0.77	3.3	3.0	S(B)b– II:	3S	2492		
2227		6 25.9	–21 59	CMa	14: p			2.3	1.2	SBc		1999		
IC 2166	3463	6 27.0	+59 05	Lyn	12.92			2.9	2.0	S(B)b+		2811		
	3475	6 30.5	+39 29	Aur	14.15			3.9	2.2	Sm		512		
	3504	6 40.1	+60 04	Lyn	12.30			3.1	2.7	S(B)c		2222		
		6 41.6	–50 58	Car						dE			L	Carina dwarf
2272		6 42.7	–27 27	CMa	12.85	11.88	0.97	2.1	1.6	S0				
2280	A131	6 44.8	–27 38	CMa	11.78			5.6	3.2	Sb II	D3S	1662*		
2273B	3530	6 46.5	+60 21	Lyn	13.29			2.6	1.7	S(B)d		2224		
2256	3519	6 47.2	+74 14	Cam	14.0 p			2.6	2.3	S(B)0:				
2258	3523	6 47.8	+74 29	Cam	13.2 p			2.6	1.7	S0				
2273	3546	6 50.1	+60 51	Lyn	11.93			3.5	2.8	S(B)a:		2066		Mrk 620
2290	3562	6 51.0	+33 26	Gem	14.6 p			1.5	0.9	Sa:				
2291		6 51.0	+33 32	Gem	15.5:p			1.3	1.3	S0:				
2294		6 51.2	+33 32	Gem	15: p			1.1	0.4	E6:		5078*		
	3574	6 53.1	+57 11	Lyn	12.59			4.7	4.7	Sc		1545		
	3587	6 53.8	+19 18	Gem	13.43			6.3	2.3	S		1173		
2310		6 54.0	–40 52	Pup	12.39			5.0	1.2	S0	3S	917		
	3580	6 55.5	+69 35	Cam	12.11			4.6	2.2	Sa		1356		
	3522	6 55.9	+84 55	Cep	14.57			2.1	1.1	P		2346		7Z 92
	3598	6 56.5	+60 39	Lyn	14.45			2.1	1.1	Ir+		2110		
2325		7 02.7	–28 42	CMa	12.2	11.2	1.02	2.3	1.5	E4		1854		
	3647	7 04.9	+56 32	Lyn	14.60			1.3	1.1	S(B)+ V		1475		D40
	3691	7 08.0	+15 10	Gem	13.18			4.2	2.0	Sc		2102		
2339	3693	7 08.3	+18 47	Gem	12.30	11.56	0.74	2.8	2.1	Sc II-III	4VS	2334		
	3685	7 09.0	+61 36	Lyn	12.21			4.0	4.0	SBb		1913		
2329	3695	7 09.2	+48 37	Lyn	13.7 p			1.6	1.3	S0:		5820		
2341	3708	7 09.3	+20 35	Gem	13.7 p			1.1	1.0	P		5181		
2342	3709	7 09.4	+20 38	Gem	12.6 p			1.6	1.5	Sp		5221		
2337	3711	7 10.2	+44 27	Lyn	12.67			2.6	2.0	Ir+		467		
2314		7 10.5	+75 20	Cam	12.90	11.91	0.99	2.1	1.8	E1p	4	4034		
	3697	7 11.4	+71 50	Cam	13.1 p			3.5	0.3	Sbp		3323		Integral Sign galaxy
2344	3734	7 12.5	+47 10	Lyn	12.85	12.05	0.80	2.0	1.9	S(B)c:		1001		
	3730	7 14.3	+73 29	Cam	14.22	13.33	0.89			Ring		2868		VV 123, Arp 141

NGC	UGC	α 2000	δ 2000	Const	B	V	B-V	Dim		Type	Nucl	RV	Memb	Notes
2268	3653	7^{h}14^{m}3	+84°23'	Cam	12.20	11.48	0.72	3'.4	2'.2	Sb:	4VS	2466		
IC 2179	3750	7 15.5	+64 56	Cam	13.60	12.55	1.05	1.0	1.0	E1				
2347	3759	7 16.1	+64 43	Cam	13.30	12.53	0.77	2.0	1.5	Sb:	4ES	4649		
2369		7 16.6	−62 21	Car	12.7			4.6	1.7	SBa	3S	3017		
2357	3782	7 17.6	+23 22	Gem	13.91			4.5	0.8	Sc		2201		
	3808	7 21.1	+25 09	Gem	14.45			2.4	1.2	SBd		2317		
2397		7 21.3	−69 00	Vol	12.78			2.2	1.2	SBb− II-III		1027		
	3826	7 24.5	+61 42	Cam	13.09			4.0	3.6	S(B)d		1847		
2377	A132	7 24.9	−9 40	Mon	15: p			1.8	1.6	Sb+:		2142		3C 178
2336	3809	7 27.1	+80 11	Cam	11.15	10.49	0.66	6.9	4.0	Sb I	2VS	2389		
2379	3857	7 27.4	+33 49	Gem	14.45	13.49	0.96	1.1	1.1	S0:		4006		
2276	3740	7 27.4	+85 45	Cep	11.95	11.36	0.59	2.6	2.5	Sc I:	4S	2579		Arp 114, 7Z 134
2366	3851	7 28.9	+69 13	Cam	11.46	10.91	0.55	7.6	3.5	Ir+ IV-V		252		D42
2389	3872	7 29.1	+33 51	Gem	13.33	12.83	0.50	2.1	1.7	S(B)c	4VS	3792		
IC 467	3834	7 30.0	+79 52	Cam	12.44			3.5	1.4	S(B)c		2225		
2300	3798	7 32.4	+85 43	Cep	12.00	10.96	1.04	3.1	2.6	E1	3VS	2167		Arp 114
	3912	7 34.2	+4 32	CMi	13.85			3.0	1.7	Ir+		1073		
2434		7 34.9	−69 17	Vol	12.30	11.22	1.08	2.5	2.5	E0		1204		
2442		7 36.4	−69 32	Vol	11.22			6.0	5.5	SBb I-II	3VS	384		
2427		7 36.5	−47 38	Pup	12.36	11.58	0.78	5.6	2.6	S(B)dm	2VS	682		
2415	3930	7 36.9	+35 15	Lyn	12.82	12.39	0.43	1.0	1.0	Ir+:		3779		
		7 36.9	+58 46	Lyn	14.8	14.3	0.55			S0p	*	12051		Mrk 9
2403	3918	7 36.9	+65 36	Cam	8.89	8.39	0.50	17.8	11.0	Sc III	1	259		
	3946	7 38.0	+3 18	CMi	14.21			2.0	1.2	Ir+		1028		
	3964	7 40.3	−1 35	Mon	14.46			3.7	2.8	S(B)dm		1271		
2424		7 40.7	+39 14	Lyn	13.60	12.63	0.97	4.0	0.7	SBb:	3VS	3307		
	3974	7 41.9	+16 48	Gem	13.29			5.0	3.9	Ir− V:		153		D47
	3973	7 42.5	+49 49	Lyn	13.5	13.0	0.55	1.4	1.4	SBb	*	6650		Mrk 79
	3975	7 44.8	+72 48	Cam	13.92			2.4	1.9	SBdm		2639		
2445	4017	7 46.9	+39 01	Lyn	13.6	13.0	0.65	2.1	1.7	Ring		3936		VV 117, Arp 143
2444	4016	7 46.9	+39 02	Lyn	13.90	12.90	1.00	2.3	1.7	Ring		3994		VV 117, Arp 143
	4013	7 47.5	+60 56	Lyn	13.6	13.0	0.65	2.0	0.9	Sb	*	8790		Mrk 10
2441	4036	7 52.2	+73 02	Cam	13.0	12.2	0.80	2.2	2.0	Sc II:	3VS	3782		
2470	4091	7 54.3	+4 27	CMi	14.2 p			2.2	0.7	Sb−				
IC 2209	4093	7 56.2	+60 18	Cam	14.30	13.75	0.55	1.1	1.0	SBb:	3VS	1646		Mrk 13
2460	4097	7 56.9	+60 21	Cam	12.60	11.68	0.92	2.9	2.2	Sb+ III:	4	1543		
	4115	7 57.0	+14 23	Cnc	14.27			2.0	1.1	Ir+		214		
2474	4114	7 57.98	+52 51.5	Lyn	13.9 p			0.6	0.6	E0		5683		
2468	4110	7 58.0	+56 22	Lyn	14.9 p			1.3	0.7	S0:				
2475	4114	7 58.01	+52 51.7	Lyn	13.9 p			0.8	0.8	E1		5083		
2469	4111	7 58.1	+56 41	Lyn	13.2 p			1.1	0.8	Sb+p		3300		
2487	4126	7 58.3	+25 08	Gem	14.0 p			2.8	2.3	SBb		4792		
2484	4125	7 58.5	+37 47	Lyn	14.9 p			1.6	1.0	S0:		12830*		
2493	4150	8 00.4	+39 50	Lyn	13.1 p			2.5	2.5	SB0				
2507	4172	8 01.6	+15 43	Cnc	14.0 p			2.5	2.1	Sap				
2500	4165	8 01.9	+50 44	Lyn	12.20	11.60	0.60	2.9	2.7	S+ IV	3VS	552		
2514	4189	8 02.8	+15 49	Cnc	14.4 p			1.4	1.4	SBb+		4734		
2517		8 02.8	−12 19	Pup	13.25	12.30	0.95	0.8	0.6	S(B)0:				
2512	4191	8 03.1	+23 24	Cnc	14.2 p			1.7	1.2	SBb		4688		Mrk 384
2523A	4166	8 04.1	+74 03	Cam	14.5 p			1.3	0.9	SBc:				
2525	A135	8 05.6	−11 26	Pup	12.29	11.64	0.65	2.9	2.0	S(B)cp II:	3ES	1348*		
	4173	8 06.7	+80 07	Cam	14.58			3.0	0.8	Ir+		1054		
2521	4235	8 08.8	+57 46	Lyn	14.2 p			1.5	0.9	S0p				7Z 212
2532	4256	8 10.2	+33 57	Lyn	13.0	12.4	0.62	2.2	1.9	S(B)c	4VS	5211*		
2535	4264	8 11.2	+25 12	Cnc	13.1	12.6	0.52	3.0	1.7	Scp	3S	4016		VV 9, Arp 82
	4260	8 11.2	+46 28	Lyn	13.73			1.7	1.5	Ir− IV-V:		2284		D49
2536		8 11.3	+25 11	Cnc	14.7	14.1	0.58	0.9	0.6	SBcp		4061		VV 9, Arp 82
	4238	8 11.5	+76 26	Cam	12.92			2.9	1.7	SBd		1714		
2534	4268	8 12.9	+55 40	Lyn	13.8 p			1.7	1.5	E1p		3625		Mrk 85
2523B	4259	8 12.9	+73 34	Cam	14.8 p			2.3	0.4	Sb:				
2543	4273	8 13.0	+36 15	Lyn	12.7 p			2.5	1.5	SBb	4VS	2421		IC 2232
2537	4274	8 13.2	+46 00	Lyn	12.35	11.69	0.66	1.7	1.5	S:		438*		VV 138, Arp 6
2537A		8 13.7	+46 00	Lyn				0.7	0.7	SBc				VV 138, Arp 6
IC 2233	4278	8 14.0	+45 44	Lyn	13.5	13.0	0.50	4.7	0.6	SBd:		591		
2545	4287	8 14.2	+21 21	Cnc	13.20	12.40	0.80	2.2	1.3	Sb	4VS	3094	Ca	
2541	4284	8 14.7	+49 04	Lyn	12.25	11.75	0.50	6.6	3.5	S+ IV	2VS	606		
2523	4271	8 15.0	+73 35	Cam	12.65	11.95	0.70	3.0	2.0	SBb− I	D3S	3608		Arp 9
2523C	4290	8 17.8	+73 19	Cam	14.1 p			1.8	1.0	E5:				
2554	4312	8 17.9	+23 28	Cnc	13.5 p			3.4	2.6	Sap		4075	Ca	
2566		8 18.7	−25 29	Pup	15: p			3.1	2.0	SBb−:				
	4305	8 18.9	+70 43	UMa	11.27	10.75	0.52	7.6	6.2	Ir+ IV-V		305		Holm. II, D50
2549	4313	8 19.0	+57 48	Lyn	12.04	11.10	0.94	4.2	1.5	E6	5	1168		
2557	4330	8 19.2	+21 27	Cnc	14.6 p			1.4	1.3	SB0		4871	Ca	
2552	4325	8 19.3	+50 01	Lyn	12.8	12.2	0.59	3.2	2.2	Ir+ IV-V	2VS	557		
2565	4334	8 19.8	+22 02	Cnc	13.8 p			1.9	0.9	SBb+:		3514*	Ca	Mrk 386

NGC	UGC	α 2000	δ 2000	Const	B	V	B-V	Dim		Type	Nucl	RV	Memb	Notes
2560	4337	8ʰ19ᵐ9	+20°59'	Cnc	14.9 p			1.7	0.5	Sa		4827	Ca?	
2562	4345	8 20.4	+21 08	Cnc	13.85	12.85	1.00	1.4	1.0	S0:		4863	Ca?	
2563	4347	8 20.6	+21 04	Cnc	13.35	12.33	1.02	2.3	1.8	S0:		4542*	Ca	
2578		8 21.4	-13 19	Pup	13.40	12.43	0.97	2.4	1.4	SBap	3S			
2544	4312	8 21.7	+73 59	Cam	12.39			1.1	0.9	SBa:		2948		Mrk 87
2577	4367	8 22.7	+22 33	Cnc	13.43			2.1	1.3	S0:		2054	Ca?	
2575	4368	8 22.8	+24 17	Cnc	14.3 p			2.5	2.1	Sc:			Ca?	
2583		8 23.2	-5 01	Hya	14.50	13.51	0.99	0.7	0.7	E1		5697		
2550	4359	8 24.6	+74 01	Cam	13.1 p			1.2	0.8	Sb:				
2551	4362	8 24.8	+73 25	Cam	13.05	12.04	1.01	1.9	1.3	Sb III:	D3	2375		
2582	4391	8 25.3	+20 20	Cnc	14.3 p			1.3	1.3	S(B)b-		4372	Ca	
	4393	8 26.1	+45 58	Lyn	12.98			2.5	1.8	SB		2154		
2595	4422	8 27.7	+21 29	Cnc	13.9 p			3.1	2.4	S(B)c		4163*	Ca	3Z 59
	4390	8 27.8	+73 31	Cam	14.22			2.3	2.1	SBc+		2326		
2550A	4397	8 28.7	+73 45	Cam	13.5 p			1.7	1.4	Sc				
2598	4443	8 30.0	+21 29	Cnc	14.7	13.7	1.01	1.3	0.5	SBa:			Ca?	
2599	4458	8 32.2	+22 34	Cnc	13.4 p			2.4	2.4	Sa		4437	Ca?	Mrk 389
2613	A141	8 33.4	-22 58	Pyx	11.35	10.42	0.93	7.2	2.1	Sb II	4S	1444*		
2612		8 33.8	-13 09	Hya	13: p			3.2	0.7					
	4459	8 34.1	+66 10	UMa	14.38			1.7	1.3	Ir+ V		144		7Z 238, D53
2608	4484	8 35.3	+28 28	Cnc	12.80	12.09	0.71	2.5	1.6	Sc II:		2053		Arp 12
2616	4489	8 35.6	-1 51	Hya	13.6	12.6	0.99	1.3	1.3					
		8 35.7	+46 30	Lyn	14.7	14.2	0.50			S:		4361		Mrk 92
2591	4472	8 37.4	+78 02	Cam	12.83			3.2	0.8	Sc:		1509		
	4499	8 37.6	+51 39	UMa	13.23			3.0	2.4	S(B)dm		749		
	4508	8 37.9	-2 28	Hya	14.30			0.3	0.3	P		1736		
2625		8 38.4	+19 42	Cnc				0.4	0.4			4415		Mrk 625
2623	4509	8 38.4	+25 45	Cnc	14.4	13.8	0.61	0.6	0.5	P		5355		VV 79, Arp 243
	4514	8 39.6	+53 27	UMa	13.60			2.3	1.2	SBc		759		
2642		8 40.7	-4 07	Hya	12.5 H			2.3	2.2	S(B)b I	4S	4226		
2648	4541	8 42.7	+14 17	Cnc	12.6:			3.6	1.1	Sa				Arp 89
2614	4523	8 42.8	+72 59	UMa	14.0 p			2.8	2.2	Sc:				
	4543	8 43.3	+45 44	Lyn	13.91			3.6	2.0	Sdm		1983		
2639	4544	8 43.6	+50 12	UMa	12.65	11.76	0.89	2.0	1.3	Sa	4VS	3359		
2649	4555	8 44.1	+34 43	Lyn	13.1 p			1.9	1.8	S(B)b:		4040		
	4559	8 44.2	+30 07	Cnc	14.28			3.0	0.7	Sb+		2027		
2629	4569	8 47.2	+72 59	UMa	12.8 p			2.6	2.1	S0:				
2656		8 47.8	+53 52	UMa	14.45	13.30	1.15	1.5	1.5	S0p		13529		
IC 2389	4576	8 48.0	+73 32	Cam	14.0	13.4	0.64	1.7	0.4	SBb:		2743		
2633	4574	8 48.1	+74 06	Cam	12.85	11.88	0.97	2.6	1.7	SBbp	4VS	2302		Arp 80
2636	4583	8 48.4	+73 40	Cam	14.4 p			0.5	0.5	E0:	4			
2634	4581	8 48.4	+73 58	Cam	12.6 p			2.2	2.1	E1:				
2634A	4585	8 48.6	+73 56	Cam	14.3 p			1.9	0.5	SBb+:				
2654	4605	8 49.2	+60 13	UMa	12.75	11.80	0.95	4.3	0.9	Sb- II-III		1455		
2672	4619	8 49.37	+19 04.4	Cnc	12.6	11.6	0.98	2.6	2.4	E1		4109		Arp 167
2673	4620	8 49.41	+19 04.4	Cnc	13.9	12.9	0.99	1.4	1.4	E0p		3678		
2650	4603	8 50.0	+70 18	UMa	14.3 p			1.9	1.4	SBb:		3970		
2646	4604	8 50.4	+73 28	Cam	12.95	11.95	1.00	1.7	1.6	SB0:	3S	3704		
2676	4627	8 51.5	+47 33	UMa	14.3 p			1.5	1.3	S0:		5986		
2683	4641	8 52.7	+33 25	Lyn	10.61	9.72	0.89	9.3	2.5	Sb- II-III	4S	242		
2681	4645	8 53.5	+51 19	UMa	11.10	10.30	0.80	3.8	3.5	Sa	5	760		
IC 520	4630	8 53.7	+73 29	Cam	12.7 H			2.3	1.9	S(B)b-:	D3S			
2691	4664	8 54.8	+39 32	Lyn	13.9 p			1.6	1.1	Sa:	*	3969		Mrk 391
2698		8 55.6	-3 11	Hya	13: p			1.8	0.9	Sa:				
2685	4666	8 55.6	+58 44	UMa	11.90	11.05	0.85	5.2	3.0	Sbp II-III:		956		Arp 336, Helix galaxy
2655	4637	8 55.6	+78 13	Cam	10.95	10.09	0.86	5.1	4.4	S(B)a	4VL	1623*		Arp 225
2708		8 56.1	-3 22	Hya	13.5:p			3.0	1.5	Sb:		1830		
2693	4674	8 57.0	+51 21	UMa	12.70	11.71	0.99	2.2	1.7	Ep	3	5006		
2692	4675	8 57.0	+52 04	UMa	14.1 p			1.5	0.7	Sa:		3803		
2713	4691	8 57.3	+2 55	Hya	12.60	11.65	0.95	3.9	1.7	Sb II	4S	3688		
2716	4692	8 57.6	+3 05	Hya	13.3	12.4	0.86	1.8	1.4	SB0	4VS	3351		
2718	4707	8 58.9	+6 18	Hya	13.3 p			2.5	2.5	S(B)b-		3666		
2721		8 58.9	-4 46	Hya	12.5:p			2.6	1.9	S				
	4687	8 58.9	+66 28	UMa	14.2	13.5	0.70	0.8	0.7	P		3740		Mrk 100
	4704	8 59.0	+39 12	Lyn	14.45			4.0	0.6	Sdm		586		
2701	4695	8 59.1	+53 46	UMa	12.80	12.37	0.43	2.1	1.4	Sc II	3VS	2401		
2712	4708	8 59.5	+44 55	Lyn	12.70	12.02	0.68	2.9	1.7	Sb+ I	3VS	1857		
2723	4723	9 00.2	+3 11	Hya	14.5 p			1.3	1.3	S0:		3539		
2719A	4718	9 00.3	+35 43	Lyn	13.7 p			0.6	0.5	Irp+		3046		Arp 202
2719	4718	9 00.3	+35 44	Lyn	13.7 p			1.3	0.4	Irp+		3185		Arp 202
2735	4744	9 02.6	+25 56	Cnc	14.05			1.2	0.5	S(B)bp		2534		VV 40, Arp 287
2735A	4744	9 02.6	+25 56	Cnc	16.8 p			0.5	0.4	Irp+		2694		VV 40, Arp 287
IC 512	4646	9 03.9	+85 30	Cam	12.26			3.6	2.8	S(B)c		1822		
	4749	9 04.5	+51 37	UMa	13.85	13.31	0.54	0.9	0.9	S		4833		Mrk 101
2744	4757	9 04.7	+18 28	Cnc	13.8	13.4	0.44	1.8	1.2	SBb-p		3332		

NGC	UGC	α 2000	δ 2000	Const	B	V	B-V	Dim		Type	Nucl	RV	Memb	Notes
2726	4750	9ʰ04ᵐ9	+59°56′	UMa	13.17			1.5	0.5	Sa:		1526		
2749	4763	9 05.4	+18 19	Cnc	13.00	12.04	0.96	2.0	1.7	E2		4118		
2752	4772	9 05.7	+18 20	Cnc	14.8 p			2.0	0.5	SBb:		3904		
2750	4769	9 05.7	+25 26	Cnc	12.43			2.3	2.1	S(B)c		2600		
	4777	9 06.7	+34 37	Lyn	14.86			2.1	0.4	Ir+		2019		
2763		9 06.8	−15 30	Hya	12.70	12.08	0.62	2.3	2.1	S− IV:	2S	1634		
	4787	9 07.5	+33 16	Lyn	14.13			2.3	0.6	Sdm		509		
2765	4791	9 07.6	+3 23	Hya	13.3 p			2.3	1.4	S0		3642		
2742	4779	9 07.6	+60 29	UMa	12.3	11.7	0.62	3.1	1.7	Sc II	3S	1392		
	4797	9 08.1	+5 56	Hya	14.28			2.5	2.5	S− V		1140		D54
2715	4759	9 08.1	+78 05	Cam	11.95	11.38	0.57	5.0	1.9	Sc II	3VS	1304		
2764	4794	9 08.3	+21 27	Cnc	13.40	12.67	0.73	1.7	1.0	E:		2523		
2756	4796	9 09.0	+53 51	UMa	13.2 p			1.9	1.3	Sb				
2770	4806	9 09.6	+33 07	Lyn	12.06			3.7	1.3	Sc:	3VS	1908		
2742A	4803	9 10.0	+62 15	UMa	14.0 p			1.8	0.8	SBbp				
2775	4820	9 10.3	+7 02	Cnc	11.20	10.33	0.87	4.5	3.5	Sa	D3S	965		
2777	4823	9 10.7	+7 12	Cnc	13.9 p			1.0	0.7	Sb−:				
2781		9 11.5	−14 49	Hya	12.40	11.48	0.92	3.9	1.9	Sb+ II−III	3VS	1778		
2768	4821	9 11.6	+60 02	UMa	10.90	9.97	0.93	6.3	2.8	E5		1502		
	4837	9 12.1	+35 31	Lyn	14.44			2.1	1.2	Sm		1848		D55
2776	4838	9 12.2	+44 57	Lyn	12.20	11.64	0.56	2.9	2.7	Sc II	4S	2643		
2778	4840	9 12.3	+35 01	Lyn	13.01			1.5	1.1	E		2017		
2784	A152	9 12.3	−24 10	Hya	11.25	10.10	1.15	5.1	2.3	S0	D4VS	435		
2780	4843	9 12.7	+34 55	Lyn	14.05			1.0	0.8	SBp		2173		
2732	4818	9 13.4	+79 11	Cam	12.85	11.88	0.97	2.3	0.9	S:	4VS	2205		
2748	4825	9 13.7	+76 29	Cam	12.4	11.7	0.73	3.1	1.3	Sc	2S	1562		
2822		9 14.0	−69 38	Car	12.1:p			2.1	1.1	E:				
2782	4862	9 14.1	+40 07	Lyn	12.15	11.49	0.66	3.8	2.9	Sb	5	2529		Arp 215
	4841	9 14.8	+74 14	Cam	13.07			3.1	2.8	S(B)d	2VS	1289		Holm. III
2811	A155	9 16.2	−16 19	Hya	12.25	11.27	0.98	2.7	1.0	Sb+ II−III	4	2260		
2815		9 16.3	−23 38	Hya	12.7 H			3.5	1.3	Sbp III:		2278		
2793	4894	9 16.8	+34 26	Lyn	13.81			1.3	1.1	Scp		1637		
	A154	9 16.8	+53 27	UMa	14.8	14.4	0.43	0.5	0.3	P		2286		Mrk 104
2798	4905	9 17.4	+42 00	Lyn	13.0	12.3	0.74	2.8	1.1	SBap	5S	1709		VV 50, Arp 283
2799	4909	9 17.5	+42 00	Lyn	14.20			1.9	0.6	SBm:		1738		VV 50, Arp 283
2835	A157	9 17.9	−22 21	Hya	11.13			6.3	4.4	Sp	2VS	617		
		9 18.1	−12 06	Hya	14.7	13.6	1.07	0.8	0.8	S0:		15919		Hya A, 3C 218
IC 529	4888	9 18.5	+73 46	Cam	12.66			3.7	1.8	Sc:		2417		
	4922	9 18.6	+47 52	UMa	13.16			3.9	1.9	Sm		2025		
2824	4933	9 19.0	+26 16	Cnc	14.3 p			1.3	0.8	S0		9156		Mrk 394
2823	4935	9 19.3	+34 00	Lyn	15.7 p			1.0	0.6	SBa		6921		
2787	4914	9 19.3	+69 12	UMa	11.80	10.80	1.00	3.4	2.3	Sap	4S	758		
2826	4939	9 19.4	+33 37	Lyn	14.6 p			1.8	0.4	S0:		6222		
2825		9 19.4	+33 44	Lyn	15.5 p			1.1	0.5	Sa:		7806		
2830	4941	9 19.7	+33 44	Lyn	15.4 p			1.3	0.3	SBa:		6196		
2831		9 19.77	+33 44.6	Lyn	14.55	13.55	1.00	0.8	0.8	E0		5114		Arp 315
2832	4942	9 19.80	+33 44.9	Lyn	12.45	11.45	1.00	3.3	2.2	E		6905		Arp 315
2848	A160	9 20.2	−16 32	Hya	12.65	12.08	0.57	2.7	1.8	S− IV:		1789		
2805	4936	9 20.3	+64 06	UMa	11.78	11.27	0.51	6.3	5.0	S(B)d	3VS	1840		
2814	4952	9 21.2	+64 15	UMa	14.30	13.75	0.55	1.4	0.4	Ir−:		1778		
2855	A161	9 21.5	−11 55	Hya	12.45	11.55	0.90	2.7	2.4	E1	D4	1660		
IC 2458	A159	9 21.6	+64 14	UMa	14.78			0.4	0.2	P		1582		Mrk 108
2844	4971	9 21.8	+40 09	Lyn	13.65	12.85	0.80	1.9	1.0	Sa	3VS	1477		
2820	4961	9 21.8	+64 16	UMa	13.29	12.79	0.50	4.3	0.7	SBcp		1801		
2841	4966	9 22.0	+50 58	UMa	10.17	9.32	0.85	8.1	3.8	Sb− I	3L	700		
2810	4954	9 22.1	+71 50	UMa	13.4 p			2.0	2.0	E0				
2852	4986	9 23.2	+40 10	Lyn	13.55			1.2	1.2	S(B)a:		1880		
2853	4987	9 23.3	+40 12	Lyn	13.91			2.0	1.0	SB0:		1792		
2865		9 23.5	−23 10	Hya	12.35	11.43	0.92	2.0	1.5	E4	D3S	2443		
2859	5001	9 24.3	+34 31	LMi	11.65	10.72	0.93	4.8	4.2	SBa	4	1657		
2857	5000	9 24.5	+49 21	UMa	14.3 p			2.4	2.2	Sc		4919		Arp 1
2872	5018	9 25.7	+11 26	Leo	12.61			2.1	1.9	E2		2826		Arp 307
2874	5021	9 25.8	+11 26	Leo	13.5 p			2.5	0.9	SBb+		3470		Arp 307
	A164	9 26.0	+19 23	Leo	14.21			0.7	0.5	P		2381		Mrk 400
2915		9 26.2	−76 38	Cha	13.20	12.64	0.56	1.5	1.0	Ir+:	3	149		
2888		9 26.3	−28 02	Pyx	13.5	12.5	0.96	0.8	0.7	E2	3S	1952		
2884		9 26.4	−11 33	Hya	13: p			2.6	1.3	Sa:	3S			
2889		9 27.2	−11 38	Hya	12.50	11.78	0.72	2.0	1.8	Sb+ I−II:	3VS	3137		
2870	5034	9 27.8	+57 23	UMa	13.9 p			2.6	0.8	Sb+		3368		
	5028	9 27.8	+68 25	UMa	14.3	13.8	0.46	0.7	0.4	SBdmp		3796		VV 106, Arp 300
2880	5051	9 29.6	+62 30	UMa	12.50	11.62	0.88	2.6	1.6	E3	4S	1620		
2893	5060	9 30.3	+29 32	Leo	13.31			1.4	1.3	SBa		1650		Mrk 401
2902		9 30.9	−14 44	Hya	13.08			1.3	1.2	E0	D4	1816		IC 543
2907		9 31.6	−16 44	Hya	13.07			1.9	1.3	Sb−	3VS	1810		
2906	5081	9 32.2	+8 27	Leo	13.1 p			1.6	1.0	Sc:				

NGC	UGC	α 2000	δ 2000	Const	B	V	B-V	Dim		Type	Nucl	RV	Memb	Notes
2903	5079	9ʰ32ᵐ2	+21°30'	Leo	9.56	8.92	0.64	12.6	6.6	Sb+ I-II	4	467		
2892	5073	9 32.9	+67 37	UMa	14.4 p			1.7	1.7	S0p				
2911	5092	9 33.8	+10 09	Leo	12.60	11.60	1.00	4.3	3.2	E2p	4S	3026		Arp 232
2914	5096	9 34.0	+10 07	Leo	14.0	13.1	0.91	1.2	0.8	SBb-	4S	3214		Arp 137
		9 34.1	+11 01	Leo	14.87					P		2358		
2919	5102	9 34.8	+10 17	Leo	13.6 p			1.9	0.8	S(B)b:	3S			
2916	5103	9 35.0	+21 42	Leo	12.65	11.95	0.70	2.6	1.9	Sb:		3449		
2924		9 35.2	-16 24	Hya	13.28			1.6	1.5	E0		4362		
2935	A169	9 36.7	-21 08	Hya	11.93			3.5	3.0	S(B)b- I	5	1939		
2937	5131	9 37.7	+2 44	Hya	14.65	13.62	1.03	1.0	0.7	Ring		6803*		VV 316, Arp 142
2936	5130	9 37.7	+2 44	Hya	13.95	13.08	0.87	1.7	0.9	Ring		6794*		VV 316, Arp 142
2939	5134	9 38.1	+9 31	Leo	13.5 p			2.7	1.1	Sb+		3209		
2940		9 38.1	+9 36	Leo	14.54			1.3	0.8	S0		2824		
2942	5140	9 39.1	+34 00	LMi	12.8 H			2.2	1.8	Sb II:	3S	4374		
2944	5144	9 39.3	+32 19	Leo	14.7 p			1.3	0.5	SBcp		6942		VV 82, Arp 63
	5151	9 40.5	+48 20	UMa	13.48			0.7	0.6	P		450		
	5139	9 40.5	+71 11	UMa	13.41	12.94	0.47	3.5	3.0	Ir+ V		287		Holm. I, D63
2962	5167	9 40.9	+5 10	Hya	12.75	11.66	1.09	3.3	2.4	Sb+ III	D4S	1794		
2955	5166	9 41.3	+35 53	LMi	13.45	12.67	0.78	1.8	1.0	Sb II	4S	1721		
	5172	9 41.9	+48 40	UMa	14.22			2.0	2.0	Sm		2635		
2967	5180	9 42.1	+0 20	Sex	12.30	11.64	0.66	3.0	2.9	Sc III:	3VS	1963		
2974	A172	9 42.6	-3 42	Sex	11.75	10.78	0.97	3.4	2.1	Sa	4	1787		
2950	5176	9 42.6	+58 51	UMa	11.85	10.95	0.90	3.2	2.1	Sap	5	1451		
2964	5183	9 42.9	+31 51	Leo	12.05	11.34	0.71	3.0	1.7	Sc II:	3S	1261		Mrk 404
2980		9 43.2	-9 37	Sex	13: p			1.8	1.0	SBb+:				
2968	5190	9 43.2	+31 56	Leo	12.80	11.80	1.00	2.2	1.5	P	3S	1564		
2978		9 43.3	-9 45	Sex	13.5:p			1.2	1.1	Sb:				
2970		9 43.5	+31 59	Leo	14.46			1.0	0.9	E1:		1629		Mrk 405
2983	A176	9 43.7	-20 29	Hya	12.60	11.68	0.92	2.6	1.8	SBa	4VS	1752		
2977	5174	9 43.8	+74 52	Dra	12.7 p			1.7	0.8	Sb:		3236		
	A171	9 44.3	+76 21	Dra	14.87			0.8	0.3	P		2526		Mrk 118
2986		9 44.3	-21 17	Hya	11.95	10.93	1.02	2.5	2.3	E1		2132		
2959	5202	9 45.1	+68 36	UMa	13.7 p			1.5	1.4	S(B)b-p		4660		
2989		9 45.4	-18 23	Hya	13.4 H			1.4	1.1	Sb+ II		3869		
2997	A181	9 45.6	-31 11	Ant	10.6			8.1	6.5	Sc I	4VS	805*		
2992		9 45.7	-14 20	Hya	12.8	11.9	0.91	4.1	1.4	P	3S	1864		Arp 245
2993		9 45.8	-14 22	Hya	13.1	12.6	0.46	1.6	1.3	P	4	1814		Arp 245
	5224	9 45.9	+2 58	Sex	14.94			1.7	0.8	SBdm		1756		
2990	5229	9 46.3	+5 43	Sex	13.2	12.8	0.40	1.3	0.8	S:	3VS	2994		
3001	A183	9 46.3	-30 26	Ant	12.80			3.1	2.1	Sb+ I-II:		2190		
	5238	9 46.9	+0 30	Sex	14.32			2.5	0.9	S(B)d		1592		
2976	5221	9 47.3	+67 55	UMa	10.85	10.15	0.70	4.9	2.5	Scp	3VS	175		
	5249	9 47.8	+2 37	Sex	13.80			2.4	0.8	SBc		1695		
		9 47.8	+39 05	LMi	14.78					P		1617		Mrk 407
2963	5222	9 47.8	+72 58	Dra	14.3 p			1.4	0.7	SBb-		6868*		Mrk 122
		9 48.1	+32 53	Leo	14.54					P		1443		Mrk 408
3003	5251	9 48.6	+33 25	LMi	12.15	11.68	0.47	5.9	1.7	S(B)c III-IV:	1	1436		
2998	5250	9 48.7	+44 05	UMa	12.7 H			3.0	1.5	Sc II	3VS	4769		
3011	5259	9 49.7	+32 13	Leo	14.05			1.2	1.1	S0		1417		Mrk 409
3023	5269	9 49.9	+0 37	Sex	12.91			3.1	1.7	S(B)cp		1683		
3020	5271	9 50.1	+12 49	Leo	12.74			3.2	1.8	SBc:		1305		
		9 50.2	+28 01	Leo	12.3	12.0	0.33	0.8	0.7	E1:				
3059		9 50.2	-73 55	Car	11.79			3.2	3.0	SBb+		954		
	5272	9 50.3	+31 29	Leo	14.46			2.0	0.7	Ir+ IV-V:		469		D64
2985	5253	9 50.4	+72 17	UMa	11.25	10.51	0.74	4.3	3.4	Sb- I	D4S	1430		
3024	5275	9 50.5	+12 46	Leo	13.7 p			2.2	0.6	Sc:		1364		
3026	5279	9 50.9	+28 33	Leo	13.45			2.2	0.6	Ir+		1431		
3021	5280	9 51.0	+33 33	LMi	12.76			1.7	1.0	Sb	4VS	1495		
3038		9 51.3	-32 45	Ant	12.7			2.6	1.7	Sa		2404		
3032	5292	9 52.1	+29 14	Leo	12.6	11.9	0.73	2.5	2.1	Sa	4S	1507		
3039	5297	9 52.5	+2 09	Sex	14.4 p			1.3	0.7	S(B)b+:				
3041	5303	9 53.1	+16 41	Leo	12.25	11.45	0.80	3.7	2.5	Sc II-III:	3S	1294		
3044	5311	9 53.7	+1 35	Sex	12.55	12.02	0.53	4.8	0.9	Sc:	1	1146		
3052		9 54.5	-18 38	Hya	12.8 H			2.1	1.5	Sc II	3S	3329		
3054		9 54.5	-25 42	Hya	12.1 H			3.9	2.6	Sb II-III:	3S	1926		
3056		9 54.5	-28 18	Ant	12.98			2.0	1.4	E3	D3	768		
IC 2522	A189	9 55.2	-33 08	Ant	12.64			2.8	1.7	Sc IV:		2731		
3055	5328	9 55.3	+4 16	Sex	12.7	12.1	0.61	2.2	1.4	Sc II:	4VS	1616		
	A188	9 55.4	+8 24	Leo	14.95			0.8	0.4	P		1122		
3031	5318	9 55.6	+69 04	UMa	7.86	6.93	0.93	25.7	14.1	Sb I-II		95		M81
3027	5316	9 55.7	+72 12	UMa	12.45			4.7	2.3	SBd: III		1225		
3034	5322	9 55.8	+69 41	UMa	9.28	8.41	0.87	11.2	4.6	P		388		M82, 3C 231
3043	5327	9 56.2	+59 18	UMa	13.5 H			2.0	0.7	Sb-:		3016		
3061	5319	9 56.2	+75 52	Dra	13.56			2.0	1.9	SBc				
	5340	9 56.7	+28 50	Leo	14.27			2.2	1.0	Irp-		441		D68

NGC	UGC	α 2000	δ 2000	Const	B	V	B-V	Dim		Type	Nucl	RV	Memb	Notes
IC 2524		9ʰ57ᵐ5	+33°37'	LMi	14.54			0'.7	0'.4	P		1470		Mrk 411
	5349	9 58.1	+37 18	LMi	13.87			2.5	0.9	Sm		1361		
3067	5351	9 58.4	+32 22	Leo	12.70	11.72	0.98	2.5	1.0	Sb+ III		1414		
3078		9 58.4	-26 56	Hya	12.10	11.07	1.03	1.9	1.5	E3	4	2231		
		9 58.6	+13 15	Leo	14.46					P		2700		
3087		9 59.2	-34 13	Ant	12.9 H			1.9	1.3	E0		2375		
	5364	9 59.4	+30 45	Leo	13.09	12.64	0.45	4.9	3.2	Ir+ V		-26	L	Leo III, Leo A
3081		9 59.5	-22 50	Hya	12.38			2.2	1.8	Sb+ III:	4	2146		
3089		9 59.6	-28 20	Ant	12.78			1.9	1.3	Sb	4S	2375		
3074	5366	9 59.7	+35 24	LMi	14.8 p			2.5	2.3	S(B)c		5115		
	5373	10 00.0	+5 20	Sex	11.89	11.37	0.52	4.6	3.3	Ir+ IV-V		123		Sex B, D70
3095		10 00.1	-31 33	Ant	12.5			3.2	2.2	S(B):	4VS			
3091		10 00.2	-19 38	Hya	12.50			2.2	1.7	E2	3	3623		
3100		10 00.7	-31 40	Ant	12.5 p			2.8	1.7	S(B)0p	4			
3073	5374	10 00.9	+55 37	UMa	13.43			1.5	1.4	S(B)0		1131		Mrk 131
	5393	10 01.7	+33 08	LMi	14.10			2.3	1.3	SBm		1407		
	5391	10 01.7	+37 16	LMi	14.09			2.3	0.9	Sm		1550		
3065	5375	10 01.9	+72 10	UMa	12.95	12.01	0.94	2.0	1.9	Sa	4S	2116		7Z 303
3079	5387	10 02.0	+55 41	UMa	11.20	10.56	0.64	7.6	1.7	Sb II:		1212		
3066	5379	10 02.2	+72 07	UMa	13.55	12.91	0.64	1.2	1.1	S(B)b+p	3	2203		Mrk 133
3098	5397	10 02.3	+24 43	Leo	12.85	11.97	0.88	2.6	0.8	E7	3S	1257		
3109	A194	10 03.1	-26 09	Hya	10.36			14.5	3.5	Ir+ IV-V		131		D236
3077	5398	10 03.3	+68 44	UMa	10.65	9.85	0.80	4.6	3.6	E2p		148		
3104	5414	10 03.9	+40 45	LMi	12.72			3.2	2.3	Ir+		637*		VV 119, Arp 264
	5408	10 03.9	+59 26	UMa	14.6	14.1	0.54	0.6	0.5			3079		Mrk 25, 7Z 308
IC 2537	A197	10 03.9	-27 34	Ant	12.9	12.2	0.72	2.7	1.9	S- IV-V	3S	2533		
	5425	10 04.4	+13 37	Leo	13.57			0.8	0.8	Sc		2577		
3113		10 04.4	-28 27	Ant	14: p			3.2	1.5	Sd:				
3115		10 05.2	-7 43	Sex	10.10	9.15	0.95	8.3	3.2	E6	4	476		Spindle galaxy
	5423	10 05.5	+70 22	UMa	14.94			1.2	0.9	Ir+		479		M81 dwarf B
3057	5404	10 05.6	+80 17	Dra	14.2 p			2.5	1.6	S+ IV		1718		D67
3136		10 05.8	-67 23	Car	12.05	11.00	1.05	2.5	1.7	E4:	3	1410		
	A201	10 06.3	+28 57	LMi	14.6	14.3	0.35	0.6	0.5			1341		
3125		10 06.6	-29 56	Ant	13.1 H			1.5	0.8	E5				
3124		10 06.7	-19 13	Hya	12.4 H			3.2	2.7	S(B)b+ I:	3VS	3085		
IC 591	5458	10 07.5	+12 16	Leo	13.89			1.4	0.9	S		2700		
	5460	10 08.1	+51 50	UMa	13.16			2.5	2.5	SBd		1151		
	5459	10 08.1	+53 05	UMa	13.43			4.2	0.7	Sc		1184		
	5470	10 08.4	+12 18	Leo	10.81	9.84	0.97	10.7	8.3	dE3			L	Leo I
	5478	10 09.5	+30 09	LMi	14.45			1.7	1.4	Ir+ V		1326		D73
3140		10 09.5	-16 37	Hya	14: p			1.4	1.0	Sc		8209		
3143		10 10.1	-12 35	Hya	14: p			0.9	0.7	SBb				
3145		10 10.2	-12 26	Hya	12.4 H			3.3	1.7	Sb+ I	4S	3619		
	A205	10 11.1	-4 43	Sex	11.87	11.46	0.41	4.8	4.1	Ir+ V		116		Sex A, D75
3156	5503	10 12.7	+3 08	Sex	12.83			2.1	1.2	E5p		994		
3153	5505	10 12.9	+12 40	Leo	13.30			2.3	1.1	Sc:		2672		
3165	5512	10 13.5	+3 23	Sex	14.5 p			1.6	0.9	Sdm:				
3162	5510	10 13.5	+22 44	Leo	12.15	11.55	0.60	3.1	2.7	Sc II:	4S	1366		
3151		10 13.5	+38 37	LMi	15: p			1.0	0.6	S0:		7155		
3166	5516	10 13.8	+3 26	Sex	11.50	10.59	0.91	5.2	2.7	S(B)a	4VS	1203		
3158	5511	10 13.8	+38 46	LMi	12.85	11.84	1.01	2.3	2.1	E2		7015		
	5522	10 13.9	+7 01	Leo	13.48			3.0	1.8	Sc		1065		
3159		10 13.9	+38 39	LMi	14.4	13.4	0.99	1.3	1.2	E2p		6940		
3163	5517	10 14.1	+38 39	LMi	14.1	13.1	1.03	1.4	1.4	S0:		6235		
3169	5525	10 14.2	+3 28	Sex	11.25	10.45	0.80	4.8	3.2	Sb	4S	1051*		
3175	A207	10 14.7	-28 52	Ant	12.2	11.3	0.91	4.8	1.7	Sb III	3VS	849		
3144	5519	10 15.5	+74 13	Dra	14.3 p			1.4	0.9	SBap				
	5539	10 15.9	+2 41	Sex	14.32			2.0	1.0	Ir+		1097		
3177	5544	10 16.6	+21 07	Leo	13.00	12.32	0.68	1.7	1.3	Sb+ II-III	4VS	1123		
3147	5532	10 16.9	+73 24	Dra	11.45	10.65	0.80	4.0	3.5	Sb I-II:	4	2881		
3185	5554	10 17.6	+21 41	Leo	12.95	12.15	0.80	2.3	1.6	S(B)b+ III:	4S	1147		
3187	5556	10 17.8	+21 52	Leo	13.61	13.11	0.50	3.3	1.5	SBcp	1	1500		VV 307
3190	5559	10 18.1	+21 50	Leo	11.95	10.98	0.97	4.6	1.8	Sb II-III:		1216		NGC 3189, VV 307
3184	5557	10 18.3	+41 25	UMa	10.40	9.75	0.65	6.9	6.8	Sc II	4S	593*		
3193	5562	10 18.4	+21 54	Leo	11.85	10.92	0.93	2.8	2.6	E0		1278		
3200		10 18.6	-17 59	Hya	12.3 H			4.8	1.8	Sb II	3VS	3317		
3191	5565	10 19.1	+46 28	UMa	13.9 p			0.8	0.5	SBb+p		9145		
3182	5568	10 19.5	+58 12	UMa	13.0 p			2.3	1.8	Sa:				
3203		10 19.6	-26 42	Hya	12.78			3.0	0.7	S0:	4S	2153		
3188	5569	10 19.7	+57 25	UMa	14.7 p			1.1	1.0	SBb-		7863		Mrk 31
3208		10 19.7	-25 51	Hya	14: p			1.9	1.7	Sb+		2738		
3198	5572	10 19.9	+45 33	UMa	10.94	10.40	0.54	8.3	3.7	Sc II	4S	691		
3202	5581	10 20.5	+43 01	UMa	14.2 p			1.4	1.1	SBa		6728		
	5588	10 20.9	+25 22	Leo	13.89			0.4	0.3	P		1350		
3213	5590	10 21.3	+19 39	Leo	14.3 p			1.2	1.0	Sc:				

NGC	UGC	α 2000	δ 2000	Const	B	V	B-V	Dim		Type	Nucl	RV	Memb	Notes
3223		10^{h}21^{m}6	-34°16'	Ant	11.76			4.1	2.6	Sb	D3	2574		IC 2571
3206	5589	10 21.8	+56 56	UMa	12.43			3.0	2.1	SBc		1245		
3183	5582	10 21.8	+74 10	Dra	12.70			2.5	1.6	SBb+:				
3221	5601	10 22.3	+21 34	Leo	14.3 p			3.3	0.9	SBc:				
3222	5610	10 22.6	+19 53	Leo	13.7	12.8	0.93	1.3	1.1	SB0:		5475		
3214		10 23.1	+57 02	UMa	15: p			1.0	0.5	Sa:				
3226	5617	10 23.4	+19 54	Leo	12.3	11.4	0.93	2.8	2.5	E2	D3	1254		VV 209, Arp 94
3227	5620	10 23.5	+19 52	Leo	11.55	10.75	0.80	5.6	4.0	Sb	5VS*	1050*		VV 209, Arp 94
3220	5614	10 23.7	+57 02	UMa	13.7 p			1.6	0.6	Sb:				
	5612	10 24.0	+70 53	UMa	13.00			3.4	2.3	SB+ IV-V		1161		D77
3241		10 24.3	-32 29	Ant	13.6	12.7	0.92	1.5	1.1	Sb	3S	2554		
	5633	10 24.7	+14 45	Leo	14.10			2.5	1.7	SB+ IV-V		1264		D79
3239	5637	10 25.1	+17 10	Leo	12.40			5.2	3.7	Irp+		766		VV 95, Arp 263
3225	5631	10 25.1	+58 09	UMa	13.14			2.3	1.2	Sc		2227		
3250		10 26.5	-39 57	Ant	12.10	11.03	1.07	3.2	2.1	E4	3	2545		
3246	5661	10 26.7	+3 52	Sex	13.11			2.3	1.4	S(B)dm		1977		
3238	5649	10 26.7	+57 14	UMa	14.1 p			1.6	1.6	S0:				
3245A	5662	10 27.0	+28 38	LMi	15.4 p			3.5	0.4	SBb		1428		
3245	5663	10 27.3	+28 30	LMi	11.70	10.82	0.88	3.2	1.9	E5	5	1203		
3256		10 27.8	-43 54	Vel	12.00	11.33	0.67	3.5	2.0	P		2595		VV 65
IC 2574	5666	10 28.4	+68 25	UMa	11.03	10.56	0.47	12.3	5.9	S+ IV-V		185		D81, 7Z 330
3253	5674	10 28.5	+12 42	Leo	14.4 p			1.4	1.3	S(B)b+		9551		
3257		10 28.8	-35 40	Ant				1.1	1.1	S(B)0:	4	2740		
3258		10 28.9	-35 36	Ant	12.80	11.72	1.08	1.8	1.7	E1	3S	2525		
3261		10 29.0	-44 39	Vel	12.06			4.1	3.2	SBb	4	2281		
3260		10 29.1	-35 36	Ant				1.3	1.1	E2:		2130		
3254	5685	10 29.3	+29 30	LMi	12.2	11.5	0.66	5.1	1.9	Sb II	4S	1175		
3267		10 29.8	-35 19	Ant	12.9 p			2.4	1.4	S(B)0	3VS	3426		
3269		10 30.0	-35 13	Ant	12.6 p			3.1	1.5	S0	4	3472		
3268		10 30.0	-35 20	Ant	12.85	11.77	1.08	2.0	1.7	E2	3S	2479		
	5688	10 30.4	+70 04	UMa	12.91			3.6	2.2	S(B)+ V		2063		VV 294, 7Z 331
3271		10 30.5	-35 22	Ant	12.85	11.73	1.12	2.3	1.1	SB0	4	3502		IC 2585
3273		10 30.5	-35 37	Ant	13.0 p			2.3	1.1	S0	4	2136		
3275		10 30.9	-36 44	Ant	12.29			2.8	2.3	SBb-	3S	2918		
3265	5705	10 31.1	+28 48	LMi	14.1 p			1.2	1.0	E4				
	5708	10 31.2	+4 29	Sex	13.91			3.4	0.7	S(B)d		1007		
	5707	10 31.3	+43 08	UMa	13.67			2.5	1.9	S(B)c		2816		
3281		10 31.9	-34 51	Ant	12.62	11.65	0.97	3.3	1.8	Sb:		3114		
3274	5721	10 32.3	+27 40	Leo	13.15	12.75	0.40	2.2	1.1	Sc		481		
3264	5719	10 32.4	+56 05	UMa	13.42			3.4	1.5	SBm		1023		
	5720	10 32.5	+54 24	UMa	13.5	13.2	0.35	1.1	1.1	Irp+		1519*		Mrk 33, Arp 233
3259	5717	10 32.6	+65 03	UMa	12.90			2.3	1.4	Irp+ IV:	3VS	1867		
3285A		10 32.8	-27 31	Hya	14: p			1.4	1.2	S(B)d				
3277	5731	10 32.9	+28 31	LMi	12.55	11.74	0.81	2.0	1.9	Sb- II:	4VS	1403		
3266	5725	10 33.3	+64 45	UMa	13.5 p			1.7	1.4	S(B)0:	3VS			
3285		10 33.6	-27 27	Hya	13.0 H			2.5	1.7	S	5S			
3289		10 34.2	-35 20	Ant				2.2	0.5	SBc:				
3252	5732	10 34.4	+73 46	Dra	14.04			2.1	0.8	SBc:		1299		
3285B		10 34.6	-27 39	Hya	14: p			1.3	1.0	S(B)b:				
3287	5742	10 34.8	+21 39	Leo	12.90			2.2	1.1	S IV:	2VS	1058		
	5740	10 34.8	+50 45	UMa	14.57			2.3	1.3	S(B)m		707		
3290		10 35.1	-17 16	Hya	14: p			1.4	0.9	S(B)b+p		10372		Arp 53
3294	5753	10 36.3	+37 20	LMi	12.2	11.7	0.51	3.3	1.8	Sc I	D3VS	1507*		
3305		10 36.3	-27 10	Hya	14: p					E?		3819		
3307		10 36.3	-27 32	Hya	16: p			1.0	0.5	SBap			H?	
3299	5761	10 36.4	+12 42	Leo	13.29			2.1	1.7	S(B)dm	1	465		
3288	5752	10 36.4	+58 33	UMa	15.0 p			1.2	0.9	S(B)b+:		7682		NGC 3284?
3308		10 36.4	-27 26	Hya	13.25	12.21	1.04	2.0	1.4	S(B)0:	5	3406	H	
3300	5766	10 36.6	+14 10	Leo	13.19			2.1	1.1	SBa	4	2867		
3309		10 36.6	-27 31	Hya	12.9	11.9	1.02	1.9	1.7	E0		3789	H	
	5764	10 36.7	+31 33	LMi	14.51			1.8	1.1	Ir+ V		543		D83, Arp 267
3311		10 36.7	-27 32	Hya	12.7	11.6	1.09	2.5	2.2	S0:		3325*	H	
3301	5767	10 36.9	+21 53	Leo	12.24	11.38	0.86	3.6	1.2	Sa	5	1244		
3312		10 37.0	-27 34	Hya	12.49			3.6	1.5	Sb-p	5	2506*	H?	IC 629
3303	5773	10 37.1	+18 08	Leo	14.5 p			3.0	2.3	Sp,P		6252		VV 71, Arp 192
3306	5774	10 37.2	+12 39	Leo	13.7 p			1.5	0.6	SBm:		2752		
3314		10 37.2	-27 41	Hya	14: p			2.0	1.2	SBap	4ES	2763	H	
3318		10 37.3	-41 38	Vel	12.48			2.6	1.5	S(B)b	4	2623		
3304	5777	10 37.6	+37 27	LMi	14.4 p			1.7	0.7	SBa:	4S			
3316		10 37.6	-27 36	Hya	15: p			1.2	0.9	SB0	5	3703*	H	
	5776	10 38.0	+64 16	UMa	14.8	14.1	0.71	0.6	0.6			1746		Mrk 149, 7Z 339
3310	5786	10 38.7	+53 30	UMa	11.2	10.9	0.32	3.6	3.0	S(B)b+p	4VS	1063		Arp 217
3319	5789	10 39.2	+41 41	UMa	11.78	11.33	0.45	6.8	3.9	S(B)c II		759		
3320	5794	10 39.6	+47 24	UMa	12.68			2.2	1.2	Sc III-IV	3S	2367		
3338	5826	10 42.1	+13 45	Leo	11.30	10.75	0.55	5.5	3.7	Sb+ II	4S	1191		

NGC	UGC	α 2000	δ 2000	Const	B	V	B−V	Dim		Type	Nucl	RV	Memb	Notes
	A216	10^{h}42^{m}3	+47°46′	UMa	14.78			0′.7	0′.2	P		1585		Mrk 151
	5829	10 42.6	+34 27	LMi	13.15			4.6	3.8	Ir+ V		608		D84
3347		10 42.8	−36 22	Ant	12.5			4.4	2.6	SBb	4S	2642		
	5833	10 43.1	+20 25	Leo	14.78			1.9	0.6	S0		1214		Mrk 416
3354		10 43.1	−36 22	Ant	13.8									
3344	5840	10 43.5	+24 55	LMi	10.50	9.95	0.55	6.9	6.5	Sc II	3VS	513		
3358		10 43.6	−36 23	Ant	12.7			3.8	2.3	S(B)a	4	2629		
3346	5842	10 43.7	+14 52	Leo	12.49			2.8	2.5	Sc II		1137		
3351	5850	10 44.0	+11 42	Leo	10.50	9.71	0.79	7.4	5.1	S(B)b II	5	673		M95
3357		10 44.4	+14 05	Leo	14.5:p			1.9	1.9	S0:				
	5848	10 44.4	+56 25	UMa	14.33			2.1	1.4	Sm		914		
	5846	10 44.4	+60 22	UMa	14.10			1.8	1.7	Ir+ V		1122		D86
3329	5837	10 44.7	+76 49	Dra	12.54			2.1	1.3	S:	4	1865		
3353	5860	10 45.4	+55 58	UMa	13.2	12.7	0.50	1.5	1.1	S:		948		Mrk 35
3343	5863	10 46.1	+73 21	Dra	14.7 p			1.6	1.1	E3				
3365	5878	10 46.3	+1 49	Sex	12.98			4.7	0.9	Sc:		810		
3367	5880	10 46.6	+13 45	Leo	12.05	11.48	0.57	2.3	2.1	Sc I	5VS	2913*		
3359	5873	10 46.6	+63 13	UMa	11.00	10.45	0.55	6.8	4.3	S(B)c II	2VS	1124		
3368	5882	10 46.8	+11 49	Leo	10.10	9.24	0.86	7.1	5.1	Sbp	3S	773		M96
3370	5887	10 47.1	+17 16	Leo	12.24			3.1	1.9	Sc III	3S	1213		
3348	5875	10 47.2	+72 50	UMa	12.3	11.2	1.12	2.2	2.2	E1		3015		
3377A	5889	10 47.4	+14 04	Leo	14.15			2.0	1.9	S(B)m		450		D88
3377	5899	10 47.7	+13 59	Leo	11.05	10.21	0.84	4.4	2.7	E5		596		
3379	5902	10 47.8	+12 35	Leo	10.20	9.26	0.94	4.5	4.0	E1	4	756		M105
3450		10 48.1	−20 51	Hya	13.5:p			2.7	2.7	SBb				
3390		10 48.1	−31 32	Hya	13.4	12.2	1.24	4.0	0.7	Sb		2578		
3380	5906	10 48.2	+28 36	LMi	13.6 p			1.9	1.6	SBa:				
3384	5911	10 48.3	+12 38	Leo	10.87	9.96	0.91	5.9	2.6	E7	5	642		
3381	5909	10 48.4	+34 42	LMi	12.8 p			2.4	2.2	SBp		1605		
3389	5914	10 48.5	+12 32	Leo	12.35	11.80	0.55	2.7	1.5	Sc III:	3VS	1138		
3364	5890	10 48.5	+72 25	UMa	13.50			1.9	1.8	S(B)b+				
	A219	10 49.1	+52 20	UMa	14.38			0.4	0.2	P		2410		Mrk 153
3395	5931	10 49.8	+32 59	LMi	12.4	12.1	0.29	1.9	1.2	Sc	4VS	1595		VV 246, Arp 270
3396	5935	10 49.9	+32 59	LMi	12.6	12.2	0.44	2.8	1.2	P		1650*		VV 246, Arp 270
3404		10 50.3	−12 07	Hya	14: p			2.6	0.7	SBb−:	3VS			IC 2609
	5947	10 50.5	+19 39	Leo	14.70			1.2	0.7	Irp+		1158		D89
3400	5949	10 50.8	+28 28	LMi	14.3 p			1.4	0.9	SBa:				
3412	5952	10 50.9	+13 25	Leo	11.45	10.55	0.90	3.6	2.0	E5	4	737		
3423	5962	10 51.2	+5 50	Sex	11.62	11.17	0.45	3.9	3.5	Sc II	3VS	853		
3419	5964	10 51.3	+13 57	Leo	13.41			1.1	0.9	S(B)0	5	2849		
3419A	5965	10 51.3	+14 01	Leo	14.9 p			1.9	0.3	SBb:		2892		
3414	5959	10 51.3	+27 59	LMi	11.75	10.75	1.00	3.6	2.7	SBa		1395		Arp 162
3413	5960	10 51.3	+32 46	LMi	13.1 p			2.4	1.0	S0				
	5953	10 51.3	+44 34	UMa	13.6	12.9	0.69	0.5	0.3	P		1869		Mrk 155
3418	5963	10 51.4	+27 47	LMi	14.5 p			1.4	1.1	S(B)a:				
3415	5969	10 51.7	+43 43	UMa	12.84			2.4	1.5	E4		3279		
3424	5972	10 51.8	+32 54	LMi	13.2 p			3.0	0.9	SBb:		1391		
3433	5981	10 52.1	+10 09	Leo	12.45			3.5	3.2	Sb I:	3S	2582		
3430	5982	10 52.2	+32 57	LMi	12.15	11.50	0.65	3.9	2.3	Sc II:	3S	1529		IC 2613
3408	5977	10 52.2	+58 26	UMa	14.1 p			1.0	0.9	Sc:		9670		
3407	5978	10 52.3	+61 23	UMa	14.8 p			1.7	0.9	S0:		5150		
3432	5986	10 52.5	+36 37	LMi	11.73	11.25	0.48	6.2	1.5	SBm		625		VV 11, Arp 206
3437	5995	10 52.6	+22 56	Leo	12.71			2.6	0.9	Sb II-III	3S	1041		
	5979	10 52.6	+67 59	UMa	14.51			1.7	1.2	Ir+		1261		
3449		10 52.9	−32 56	Ant	13.0			2.6	1.2	Sa		2994		
3443	6000	10 53.0	+17 34	Leo	13.97			2.5	1.2	Sd		1028		
3442	6001	10 53.1	+33 55	LMi	13.24			0.7	0.6	Sa:		1714		Mrk 418
	5998	10 53.2	+50 17	UMa	14.29			0.8	0.3	P		1320		Mrk 156
3447	6006	10 53.4	+16 46	Leo	12.77			3.8	2.3	S(B)dmp		957		VV 252
3447A	6007	10 53.5	+16 47	Leo	14.3 p			1.7	1.0	Irp+		924		VV 252
3440	6009	10 53.9	+57 07	UMa	14.0 p			2.3	0.6	SBb:				
3403	5997	10 53.9	+73 41	Dra	12.79			3.1	1.3	Sb+ III		1359		
3456		10 54.1	−16 02	Crt	13: p			2.1	1.5	S(B)b+:				
3455	6028	10 54.5	+17 17	Leo	12.68			2.8	1.8	Sb	3S	987		
3454	6026	10 54.5	+17 21	Leo	13.97			2.2	0.5	SBc:		1034		
3445	6021	10 54.6	+56 59	UMa	12.8	12.4	0.38	1.6	1.5	Sc	3S	2078		VV 14, Arp 24
3448	6024	10 54.7	+54 19	UMa	12.15	11.73	0.42	5.4	1.9	P		1468		Arp 205
3464		10 54.7	−21 04	Hya	12.8 H			2.8	2.0	Sb II-III		3588		
	6029	10 55.0	+49 44	UMa	13.89			0.9	0.6	P		1442		Mrk 157
	6035	10 55.5	+17 08	Leo	14.86			1.2	1.1	Ir+		968		
3458	6037	10 56.0	+57 07	UMa	13.24			1.7	1.1	S:	4S	1891		
		10 56.1	+6 10	Leo	14.54					P		1050		
3466	6042	10 56.3	+9 45	Leo	14.6 p			1.3	0.8	SBb:				
3470	6060	10 58.7	+59 31	UMa	14.3 p			1.7	1.5	Sb:		6741		
3471	6064	10 59.1	+61 32	UMa	12.96			2.0	1.1	Sa		2188		Mrk 158

NGC	UGC	α 2000	δ 2000	Const	B	V	B−V	Dim		Type	Nucl	RV	Memb	Notes
3478	6069	10h59m.5	+46°07'	UMa	13.0 H			2.8	1.4	Sb+ I-II		6726		
3485	6077	11 00.0	+14 50	Leo	12.6 H			2.5	2.2	S(B)b+ II:	4S	1360		
3489	6082	11 00.3	+13 54	Leo	11.15	10.33	0.82	3.7	2.1	E6	4VS	577		
3486	6079	11 00.4	+28 58	LMi	10.85	10.33	0.52	6.9	5.4	Sc II	D3	674*		
3495	6098	11 01.3	+3 38	Leo	12.67			4.6	1.3	Sb II:	2VS	837		
3488	6096	11 01.4	+57 41	UMa	13.7 p			2.0	1.4	SBc:				
	6112	11 02.5	+16 44	Leo	13.84			2.3	0.8	Sd		932		
3501	6116	11 02.9	+17 59	Leo	13.43			3.7	0.6	Sc		1040		
3506	6120	11 03.2	+11 05	Leo	13.4 H			1.3	1.3	Sc	3S	6269		
3504	6118	11 03.2	+27 58	LMi	11.8	11.1	0.70	2.7	2.2	Sb	5	1479		
3499	6115	11 03.2	+56 13	UMa	14.3 p			1.0	0.9	Ir−:				
3511	A223	11 03.4	−23 05	Crt	11.57			5.4	2.2	Sc II:	2ES	976		
3507	6123	11 03.5	+18 08	Leo	11.41			3.5	3.0	SBb		879		
3510	6126	11 03.7	+28 53	LMi	13.3	12.9	0.38	3.8	0.9	SBm		664		
3513	A224	11 03.8	−23 15	Crt	11.97			2.8	2.3	S(B)c II:	5S	945		
3512	6128	11 04.0	+28 02	LMi	13.0	12.4	0.65	1.7	1.5	Sc III	4S	1352		
3509	6134	11 04.4	+4 50	Leo	14.0 p			2.3	1.1	Sb+p		7490		VV 75, Arp 335
3515	6139	11 04.6	+28 14	LMi	14.8 p			1.1	0.9	Sc:				
3517	6144	11 05.6	+56 31	UMa	13.8 p			1.2	1.1	Sb:				
3521	6150	11 05.8	−0 02	Leo	9.70	8.86	0.84	9.5	5.0	Sb+ II	4VS	640		
	6157	11 06.3	+17 30	Leo	14.15			1.9	1.7	Sdm		2859		
	6161	11 06.8	+43 44	UMa	13.47			3.0	1.7	SBdm		793		
	6162	11 06.8	+51 12	UMa	13.67			2.5	1.3	Sc		2268		
3516	6153	11 06.8	+72 34	UMa	12.45	11.63	0.82	2.3	1.8	SB0	5*	2701		
	6171	11 07.2	+18 34	Leo	14.78			2.2	0.6	Irp+		1118		
3557B		11 09.5	−37 21	Cen	13.2	12.2	0.99	1.9	1.1	E5:		2571		
3547	6209	11 09.9	+10 43	Leo	13.2	12.8	0.43	2.2	1.1	S:	2S	1414		
IC 2627	A227	11 09.9	−23 44	Crt	12.6	12.0	0.64	2.7	2.6	Sb+ II:	3S	1841		
3557		11 10.0	−37 32	Cen	11.40	10.40	1.00	4.0	2.7	E3	3	2837		
3550	6214	11 10.6	+28 46	UMa	14.2 p			1.4	1.3	P		10454		
3564		11 10.6	−37 33	Cen	13.25	12.21	1.04	2.1	0.4	S0	4S	2497		
3559	6217	11 10.7	+12 01	Leo	13.7 p			1.5	1.0	Sp				
3558		11 10.9	+28 33	UMa	15: p			1.3	1.2	E?		8891		Mrk 422
3549	6215	11 10.9	+53 23	UMa	12.47			3.2	1.3	Sb II-III	3VS	2944		
3561	6224	11 11.2	+28 43	UMa	14.7 p			1.1	0.9	Sap, S0p		8636		VV 237, Arp 105
3563	6234	11 11.4	+26 58	Leo	14.6 p			1.4	1.0	SB0:		9878		
3556	6225	11 11.5	+55 40	UMa	10.71	10.10	0.61	8.3	2.5	Sc		772		M108
3571		11 11.5	−18 17	Crt	12.8 H			3.3	1.3	Sa		3143		NGC 3544
3570	6240	11 12.1	+27 35	Leo	15.0 p			1.3	1.3	S0				
3585		11 13.3	−26 45	Hya	11.0	10.0	0.96	2.9	1.6	E5		1237		
	6251	11 13.4	+53 36	UMa	14.57			1.9	1.4	S+ IV-V		1006		D92
	6253	11 13.5	+22 10	Leo	12.38	11.48	0.90	14.5	12.9	dE0			L	Leo II, Leo B
3577	6257	11 13.8	+48 16	UMa	14.7 p			1.7	1.7	SBa	4VS	5271		
3583	6263	11 14.2	+48 19	UMa	11.74			2.8	2.0	Sc II	4VS	2213		
3592	6267	11 14.4	+17 16	Leo	14.8 p			2.0	0.6	Sb+:				
3593	6272	11 14.6	+12 49	Leo	11.7	11.0	0.75	5.8	2.5	Sb III	3	543		
3596	6277	11 15.1	+14 47	Leo	11.55			4.2	4.1	Sc II	4S	1030		
3589	6275	11 15.2	+60 42	UMa	14.5 p			1.7	1.0	Sdm:				
3599	6281	11 15.4	+18 07	Leo	12.75	11.89	0.86	2.8	2.8	S0:				
3595	6280	11 15.4	+47 27	UMa	13.0 p			1.7	0.8					
3600	6283	11 15.8	+41 36	UMa	12.35			4.3	1.0	Sa:		740		
3605	6295	11 16.8	+18 01	Leo	13.15			1.7	1.0	E3		601		
3607	6297	11 16.9	+18 03	Leo	10.95	10.01	0.94	3.7	3.2	E1	4S	841		
3608	6299	11 17.0	+18 09	Leo	11.90	10.98	0.92	3.0	2.5	E3	D3S	1118		
3611	6305	11 17.5	+4 33	Leo	12.8	12.2	0.56	2.4	2.0	Sa	4S	1603		
	6320	11 18.2	+18 51	Leo	13.57			1.0	0.8	P		1050		
3614	6318	11 18.3	+45 45	UMa	12.02			4.6	2.9	Sc II	3S	2381		
3621	A232	11 18.3	−32 49	Hya	9.90			10.0	6.5	Sc III-IV	3VS	455*		
3610	6319	11 18.4	+58 47	UMa	11.60	10.75	0.85	3.2	2.5	E2p	4VS	1868		
3613	6323	11 18.6	+58 00	UMa	11.6			3.6	2.0	E5	D3S	2154		
3623	6328	11 18.9	+13 05	Leo	10.24	9.34	0.90	10.0	3.3	Sb II:	D4S	666		M65, VV 308
3619	6330	11 19.4	+57 46	UMa	12.6			3.1	2.6	Sa	4VS	1748		
3626	6343	11 20.1	+18 21	Leo	11.7	10.9	0.83	3.1	2.2	Sb- II:	5S	1363		NGC 3622
	6345	11 20.2	+2 32	Leo	13.47			2.2	1.3	Ir+ V		1450		D94
3627	6346	11 20.2	+12 59	Leo	9.74	9.04	0.70	8.7	4.4	Sb+ II:	4S	583		M66, VV 308
3630	6349	11 20.3	+2 58	Leo	12.83			2.3	0.9	E7	4	1380		
3628	6350	11 20.3	+13 36	Leo	10.31	9.51	0.80	14.8	3.6	Sb		728		VV 308, Arp 317
3633	6351	11 20.4	+3 35	Leo	14.3 p			1.4	0.5	Sb−:				
3636		11 20.4	−10 17	Crt	13: p			1.1	1.1	E0	3S			
3629	6352	11 20.5	+26 58	Leo	12.55			2.2	1.7	Sc III:	3VS	1457		
3625	6348	11 20.5	+57 47	UMa	13.9 p			2.2	0.8	S(B)b:	3S			
3637		11 20.7	−10 16	Crt	12.68			1.7	1.6	S(B)a:	4	1649		
3631	6360	11 21.0	+53 10	UMa	11.03	10.43	0.60	4.6	4.1	Sc I	3S	1245		Arp 27
3640	6368	11 21.1	+3 14	Leo	11.25	10.31	0.94	4.1	3.4	E1	D3S	.1199		
3641	6370	11 21.2	+3 12	Leo	14.0:p			1.1	1.1	E1				

NGC	UGC	α 2000	δ 2000	Const	B	V	B–V	Dim		Type	Nucl	RV	Memb	Notes
3643		11ʰ21ᵐ4	+3°01'	Leo	15: p			1'.1	0'.5	SB0:				
3646	6376	11 21.7	+20 10	Leo	11.85	11.21	0.64	3.9	2.6	Sc I	3S	4194*		
3649	6386	11 22.2	+20 13	Leo	14.7 p			1.6	0.8	SBa	4S	4434		
3642	6385	11 22.3	+59 05	UMa	11.53	11.06	0.47	5.8	4.9	Sc I	4VS	1728		
3650	6391	11 22.6	+20 42	Leo	14.6 p			1.9	0.4	Sb				
3652	6392	11 22.6	+37 46	UMa	12.72			2.4	0.9	SBcp		2100		
3655	6396	11 22.9	+16 35	Leo	12.30	11.64	0.66	1.6	1.1	S:	3S	1384		
	6399	11 23.4	+50 55	UMa	14.10			3.3	0.8	Sm		873		
3660		11 23.6	-8 40	Crt	12.5:p			2.8	2.3	SBb+				
3656	6403	11 23.6	+53 51	UMa	12.86			1.7	1.7	Irp-		2910		VV 22, Arp 155
3662	6408	11 23.8	-1 06	Leo	13.8 p			1.6	1.0	S(B)bp				
3659	6405	11 23.8	+17 49	Leo	12.78			2.1	1.2	S:		1200		
3657	6406	11 23.9	+52 55	UMa	12.21			1.8	1.7	S(B)cp		1293		
3658	6409	11 24.0	+38 34	UMa	13.3 p			2.1	1.9	S0:				
3664A	6418	11 24.4	+3 13	Leo	15.4 p			1.1	1.1	SBmp				
3664	6419	11 24.4	+3 20	Leo	12.98			2.0	1.9	S(B)+ IV-V:		1241		VV 251, Arp 5
3666	6420	11 24.4	+11 21	Leo	12.32			4.2	1.4	Sb III		947		
3665	6426	11 24.7	+38 46	UMa	11.70	10.76	0.94	3.2	2.6	E2	D4	2012		
3672	A235	11 25.0	-9 48	Crt	11.49			4.1	2.1	Sb+ II	3S	1737		
3673	A236	11 25.2	-26 44	Hya	12.36			3.5	2.5	S(B)b II	3S	1696		
3669	6431	11 25.4	+57 43	UMa	12.99			2.3	0.7	SBc:		2041		
3675	6439	11 26.1	+43 35	UMa	10.9			5.9	3.2	Sb- II	D3S	735		
3677	6441	11 26.3	+46 58	UMa	13.5 p			2.0	1.7	S0		7525		
3674	6444	11 26.4	+57 03	UMa	13.1 p			2.0	0.7	S0	4S			
3681	6445	11 26.5	+16 52	Leo	12.40	11.70	0.70	2.5	2.4	S(B)b+	D3	1221		
	6446	11 26.6	+53 44	UMa	12.87			4.0	2.7	Sd		725		
IC 691	6447	11 26.7	+59 09	UMa	14.05			0.7	0.4	P		1425		Mrk 169
	6448	11 26.8	+64 08	UMa	14.09			1.8	0.9	P		1121		Mrk 170
3684	6453	11 27.2	+17 02	Leo	12.3	11.7	0.64	3.2	2.3	Sc III	3S	1330		
3683	6458	11 27.5	+56 53	UMa	12.84			2.0	0.9	S:		1783		
3686	6460	11 27.7	+17 13	Leo	12.0	11.4	0.58	3.3	2.6	Sc II	5VS	931		
3682	6459	11 27.7	+66 35	Dra	13.4 p			2.0	1.4	Sa:		1747		
3687	6463	11 28.0	+29 31	UMa	12.61			2.0	2.0	Sb+ II-III:		2373		Mrk 736
	6456	11 28.0	+78 59	Dra	14.45			1.4	0.8	P		96		7Z 403
3691	6464	11 28.2	+16 55	Leo	13.5 H			1.3	1.0	S		905		
3689	6467	11 28.2	+25 40	Leo	13.0	12.3	0.71	1.6	1.2	Sc II:	3S	2640		
3692	6474	11 28.4	'+9 24	Leo	12.9 p			3.3	0.9	Sb		1614		
3690	6472	11 28.5	+58 33	UMa	12.0 H			2.4	1.9	S		3104		VV 118, Arp 299
3694	6480	11 28.9	+35 24	UMa	13.48			1.0	0.7	P		2250		
3683A	6484	11 29.2	+57 08	UMa	12.6 p			2.5	1.8	SBc	3S			
	6491	11 29.4	+34 51	UMa	14.52			2.0	1.0	Sdm		2523		
3706		11 29.7	-36 25	Cen	12.28			2.9	1.9	S0		2780		
3705	6498	11 30.1	+9 17	Leo	11.46			5.0	2.3	Sb II-III	3S	891		
		11 30.3	+58 07	UMa	14.60	14.09	0.51	1.1	1.0	Sm:				
		11 30.4	+36 44	UMa	14.86					Sp		2014		Mrk 424
3717	A238	11 31.5	-30 19	Hya	12.19			5.8	1.4	Sb III	3S	1477		
3719	6521	11 32.2	+0 49	Leo	13.8 p			2.0	1.5	Sb		5723		
3720	6523	11 32.4	+0 48	Leo	13.95	13.31	0.64	1.1	1.0	E0		5084		
	6520	11 32.5	+62 30	UMa	14.2	13.3	0.90	1.6	1.0	SB:		3796		Mrk 175
3718	6524	11 32.6	+53 04	UMa	11.26	10.53	0.73	8.7	4.5	SBap	4VS	1095		Arp 214
	6531	11 32.8	+39 04	UMa	14.28			2.0	2.0	SBdm		1580		
	6534	11 33.2	+63 17	UMa	13.14			3.0	0.8	Sc		1400		
3726	6537	11 33.3	+47 02	UMa	10.95	10.44	0.51	6.0	4.5	Sc I-II	4S	818*		
	6541	11 33.5	+49 14	UMa	13.81			1.2	0.7	Ir+		314		Mrk 178
3725	6542	11 33.7	+61 53	UMa	13.6 p			1.4	1.1	SBc		3303		Mrk 179
3729	6547	11 33.8	+53 08	UMa	12.00	11.38	0.62	3.1	2.1	P	3S	1117		
3732		11 34.2	-9 51	Crt	13.17			1.3	1.2	E0	4VS	1514		NGC 3730
3733	6554	11 35.0	+54 51	UMa	12.25			4.8	2.3	S(B)c:		1278		
3737	6563	11 35.6	+54 57	UMa	13.9 p			1.3	1.3	SB0		5676		
3738	6565	11 35.8	+54 31	UMa	12.14	11.71	0.43	2.6	2.0	P	1	314		Arp 234
3735	6567	11 36.0	+70 32	Dra	12.24			4.2	1.0	Sb II:	3S	2853		
3755	6577	11 36.6	+36 25	UMa	13.18			3.2	1.5	S(B)cp	3S	1575		
3756	6579	11 36.8	+54 18	UMa	12.15	11.52	0.63	4.4	2.4	Sc II	3S	1159		
3759	6581	11 36.9	+54 49	UMa	14.3 p			1.5	1.5	S(B)0:		5620		
3759A	6582	11 37.0	+55 10	UMa	14.5 p			1.4	1.3	S(B)cp				
3769	6595	11 37.7	+47 54	UMa	11.79			3.2	1.1	Sb		772		Arp 280
3746	6597	11 37.73	+22 00.6	Leo	15.3 p			1.3	0.7	SBb		9176		
3745		11 37.74	+22 01.4	Leo				0.4	0.2	SB0:		9352		} Copeland's Septet
3748		11 37.8	+22 02	Leo				0.8	0.3	SB0:		8928		
3769A		11 37.8	+47 53	UMa	14.47			0.9	0.3	SBmp		744		Arp 280
3751	6601	11 37.9	+21 56	Leo	15.3 p			0.4	0.3	S0p		9529		
3750		11 37.9	+21 58	Leo				0.9	0.7	S(B)0:		8950		} Copeland's Septet
3753	6602	11 37.90	+21 59.0	Leo	14.6 p			2.0	0.6	Sp		8623		
3754		11 37.91	+21 59.1	Leo				0.4	0.4	SBbp		9063		
3773	6605	11 38.2	+12 07	Leo	13.01			1.6	1.4	E0p	4	904		Mrk 743

NGC	UGC	α 2000	δ 2000	Const	B	V	B-V	Dim		Type	Nucl	RV	Memb	Notes
3783		11^{h}39^{m}0	-37°45'	Cen	12.78			1.9	1.5	SBa	4VS*	2770		
3782	6618	11 39.3	+46 31	UMa	13.15			1.7	1.2	Scp		793		
3780	6615	11 39.4	+56 16	UMa	11.79			3.1	2.6	Sc II	3S	2865		
	6616	11 39.4	+58 16	UMa	13.35			3.0	2.4	Sd		1261		
3786	6621	11 39.7	+31 55	UMa	12.94			2.4	1.4	S(B)ap	D4	2745		VV 228, Arp 294
3788	6623	11 39.7	+31 56	UMa	13.25			2.2	0.8	S(B)b-p	5	2323		VV 228, Arp 294
	6628	11 40.1	+45 55	UMa	12.91			3.4	3.4	Sm		901		
3795	6629	11 40.1	+58 37	UMa	14.1 p			2.4	0.7	S				
3799	6630	11 40.2	+15 20	Leo	14.4 p			0.9	0.6	SBbp		3479		VV 350, Arp 83
3800	6634	11 40.2	+15 21	Leo	13.1 p			2.1	0.7	S(B)bp		3462		VV 350, Arp 83
IC 719	6633	11 40.3	+9 00	Vir	13.56			1.7	0.4	S0		1650		
3801	6635	11 40.3	+17 44	Leo	13.0	12.1	0.93	3.2	1.9	S0p		3174		
3802	6636	11 40.3	+17 46	Leo	14.5	13.6	0.93	1.4	0.4	S				
	6637	11 40.4	+28 23	UMa	14.29			0.8	0.3	P		2937		
3808	6643	11 40.7	+22 26	Leo	14.40	13.77	0.63	1.8	1.0	S(B)cp		6969		VV 300, Arp 87
3808A	6643	11 40.7	+22 27	Leo	12.51	11.93	0.58	0.7	0.4	Irp		7149		VV 300, Arp 87
		11 40.8	+35 12	UMa	14.63			0.5	0.3	Irp+		1510		Mrk 426
3804	6640	11 40.9	+56 12	UMa	13.11			2.4	1.7	S(B)c		1479		NGC 3794
3810	6644	11 41.0	+11 28	Leo	11.36	10.81	0.55	4.3	3.1	Sc I	3VS	862		
3813	6651	11 41.3	+36 33	UMa	12.30	11.72	0.58	2.3	1.2	Sc	4VS	1392		
3811	6650	11 41.3	+47 42	UMa	13.0 p			2.4	1.9	SBcp		3121		Mrk 185
	6655	11 41.8	+15 58	Leo	14.29			0.4	0.3	P		750		
3818	A243	11 42.0	-6 09	Vir	12.7	11.8	0.90	2.1	1.4	E2p	3S	1317		
	6670	11 42.4	+18 19	Leo	13.75			2.7	0.9	Ir+		838		
	6667	11 42.4	+51 36	UMa	14.03			3.4	0.6	Sc		1051		
3824	6676	11 42.8	+52 47	UMa	14.6 p			1.6	0.9	Sa:				
3829	6690	11 43.5	+52 43	UMa	15.0 p			1.2	0.8	SBb:				
3842	6704	11 44.0	+19 57	Leo	13.3 p			1.2	1.0	E3		6115		
3835	6703	11 44.1	+60 07	UMa	13.0 p			2.0	0.9	Sb:				
3846A	6706	11 44.2	+55 02	UMa	13.36			2.1	1.8	Smp		1525		VV 320
3838	6707	11 44.2	+57 57	UMa	12.7 p			1.7	0.8	Sa:				
	6713	11 44.5	+48 50	UMa	14.16			2.0	1.7	Sm		966		
3846	6710	11 44.5	+55 40	UMa	14.7 p			1.2	0.9	Sc:				
	6711	11 44.5	+69 44	Dra	13.73			0.7	0.4	P		2700		
3865		11 44.9	-9 14	Crt	12.9 H			2.3	1.7	Sb		5524		NGC 3854
3862	6723	11 45.1	+19 36	Leo	13.65	12.64	1.01	1.6	1.6	E0		6159		3C 264
3850	6733	11 45.6	+55 53	UMa	13.80			2.3	1.2	SBc:		1264		
3872	6738	11 45.8	+13 46	Leo	12.65	11.70	0.95	2.2	1.5	E4	3S	3011		
3870	6742	11 45.9	+50 12	UMa	13.2 p			1.3	1.0	S0p		790		Mrk 186
3877	6745	11 46.1	+47 30	UMa	11.62			5.4	1.5	Sb	1	887		
		11 46.4	+34 51	UMa	14.70					P		1346		Mrk 429
		11 46.8	-3 51	Vir	13.3 p			3.0		SBbp, SBbp, SBcp		5000		Wild's Triplet
3885		11 46.8	-27 55	Hya	12.78			1.7	0.9	Sb III:	4	1706		
3887		11 47.1	-16 51	Crt	11.6	11.0	0.64	3.3	2.7	Sc II-III:	3S	917		
3888	6765	11 47.6	+55 58	UMa	12.71			1.8	1.4	Sb II-III:	4S	2541		Mrk 188
3892		11 48.0	-10 58	Crt	12.39			2.8	2.2	S(B)a:	4VS	1532		
3893	6778	11 48.6	+48 43	UMa	11.1			4.4	2.8	Sc I:	3VS	1034		
3894	6779	11 48.8	+59 25	UMa	12.9 p			2.4	1.6	E4		3337		
	6780	11 48.9	-2 02	Vir	13.67			3.4	1.1	Sdp		1574		
3896	6781	11 48.9	+48 41	UMa	14.0 p			1.7	1.3	SBap		935		
3897	6784	11 49.0	+35 01	UMa	14.2 p			2.1	2.1	Sb+		6412		
3905		11 49.1	-9 44	Crt				1.8	1.3	SBb+		5557		
3895	6785	11 49.1	+59 26	UMa	14.0 p			1.4	1.1	SBa:	4S	3273		
3900	6786	11 49.2	+27 01	Leo	12.20	11.38	0.82	3.5	1.9	Sb+ III	3	1667		
3898	6787	11 49.2	+56 05	UMa	11.7	10.8	0.88	4.4	2.6	Sb- II	4L	1137		
3904		11 49.2	-29 17	Hya	11.95	10.96	0.99	2.2	1.7	E2	3	1369		
3902	6790	11 49.3	+26 07	Leo	14.0 p			1.8	1.5	S(B)b:				
3906	6797	11 49.7	+48 26	UMa	14.1 p			1.9	1.8	SBd				
3912	6801	11 50.1	+26 29	Leo	13.06			1.7	1.0	Sb+ III:		1724		NGC 3899
3172		11 50.2	+89 07	UMi		13.6:		0.7:	0.7:					Polarissima Borealis
3913	6813	11 50.6	+55 21	UMa	12.72			2.6	2.5	Sd:		952		IC 740
	6816	11 50.7	+56 27	UMa	14.20			1.4	1.1	Ir+		998		VV 273, D98
	6817	11 50.8	+38 52	UMa	13.15			4.1	1.8	Ir+ IV-V:		269		D99
	6818	11 50.8	+45 48	UMa	13.91			1.9	0.8	P		857		
3917	6815	11 50.8	+51 50	UMa	12.28			4.9	1.4	Sc		1041		
3916	6819	11 50.8	+55 09	UMa	14.8 p			1.8	0.5	Sb:				
3923		11 51.0	-28 48	Hya	11.1	10.1	1.00	2.9	1.9	E3	3S	1546		
3921	6823	11 51.1	+55 05	UMa	13.4 p			2.2	1.5	Sap		6070		VV 31, Arp 224
3931	6825	11 51.2	+52 00	UMa	14.6 p			1.4	1.1	S0:	3VS	952		
3930	6833	11 51.8	+38 01	UMa	12.39			3.8	3.0	S(B)c		936		
3928	6834	11 51.8	+48 41	UMa	13.01			1.8	1.8	E0		1057		Mrk 190
	6840	11 52.1	+52 06	UMa	13.39			3.0	2.6	SB- IV-V		1102		D100
3936	A248	11 52.3	-26 54	Hya	12.96			4.0	0.8	Sb III		1789		
3924	6849	11 52.6	+50 01	UMa	14.57			1.9	1.8	Sm		1077		
3938	6856	11 52.8	+44 07	UMa	10.91	10.39	0.52	5.4	4.9	Sc I	D3S	838		

NGC	UGC	α 2000	δ 2000	Const	B	V	B-V	Dim		Type	Nucl	RV	Memb	Notes
3941	6857	11ʰ52ᵐ9	+36°59′	UMa	11.36			3.8	2.5	E3	4S	969		
3945	6860	11 53.2	+60 41	UMa	11.5	10.6	0.92	5.5	3.6	SBa	3VS	1340		
3947	6863	11 53.4	+20 45	Leo	14.2 p			1.6	1.6			6180		
3952		11 53.7	−4 00	Vir	12.87			1.5	0.7	P		1458		
3949	6869	11 53.7	+47 52	UMa	11.4	11.0	0.44	3.0	1.8	Sb		745		
3953	6870	11 53.8	+52 20	UMa	10.75	10.05	0.70	6.6	3.6	Sb+ I	4S	1043		
3957		11 54.0	−19 34	Crt	12.58			3.5	0.8	S0:	3S	1620		
3956		11 54.0	−20 34	Crt	12.5 H			3.5	1.2	S− IV:	4S	506		
3955		11 54.0	−23 10	Crt	12.55	11.90	0.65	3.2	1.3	Sb III		1118*		
IC 745	6877	11 54.3	+0 08	Vir	13.64			0.7	0.6	P		1050		
3958	6880	11 54.6	+58 22	UMa	13.1 p			1.6	0.8	SBa		3491		
3962	A253	11 54.7	−13 58	Crt	11.55	10.59	0.96	2.9	2.6	E2		1622		
3963	6884	11 55.0	+58 30	UMa	12.4 H			2.8	2.6	Sc I-II	4S	3316		
3968	6895	11 55.5	+11 58	Leo	12.86			3.0	2.2	S(B)b+		6305		
	6900	11 55.6	+31 32	UMa	14.57			1.8	1.1	Ir−		541		D101
	6903	11 55.7	+1 15	Vir	13.30			2.5	2.2	S(B)c		1748		
3972	6904	11 55.8	+55 19	UMa	12.56			4.0	1.3	Sb+	3S	947		
3975		11 55.9	+60 32	UMa	15: p			0.8	0.4	Sb:		9880*		
3976	6906	11 56.0	+6 45	Vir	12.25			3.9	1.4	Sb II	D3S	2381		
3977	6909	11 56.1	+55 24	UMa	13.2 p			1.7	1.6	Sb−:	4S	5833		NGC 3980
3981	A255	11 56.1	−19 54	Crt	12.37			3.9	1.5	Sb		1499		VV 8, Arp 289
3978	6910	11 56.2	+60 31	UMa	15.2 p			1.8	1.7	S(B)b+:		10119		
	6912	11 56.3	+58 11	UMa	14.21			2.1	1.0	P		1468		VV 57, 7Z 430
	6917	11 56.5	+50 25	UMa	12.63			4.7	2.9	SBm		996		
3982	6918	11 56.5	+55 08	UMa	11.74			2.5	2.2	Sb	4S	1208		
3986	6920	11 56.7	+32 01	UMa	14.0 p			3.1	0.8	S0	3S			
3985	6921	11 56.7	+48 20	UMa	13.08			1.2	0.8	S:	1	600		
	6923	11 56.8	+53 10	UMa	13.98			2.0	0.7	Ir+		1172		
3987	6928	11 57.3	+25 12	Leo	14.4 p			2.5	0.5	Sb		4508		
	6930	11 57.3	+49 16	UMa	12.72			4.4	4.2	S(B)d		851		
3991	6933	11 57.5	+32 20	UMa	13.55	13.18	0.37	1.4	0.5	Irp+		3322		
3993	6935	11 57.6	+25 15	Leo	14.8 p			1.9	0.6	Sb:		4823		
3994	6936	11 57.6	+32 17	UMa	13.35	12.74	0.61	1.1	0.7	Scp	3S	3136		VV 249, Arp 313
3992	6937	11 57.6	+53 23	UMa	10.60	9.81	0.79	7.6	4.9	S(B)b+ I	D4	1149		M109
3990	6938	11 57.6	+55 28	UMa	13.40	12.55	0.85	1.7	1.0	S0:	4S	819		
3995	6944	11 57.7	+32 18	UMa	12.8	12.6	0.25	2.8	1.1	Sc	3S	3381		VV 249, Arp 313
3997	6942	11 57.8	+25 16	Leo	14.3 p			1.8	1.0	SBbp		4760		
3998	6946	11 57.9	+55 27	UMa	11.55	10.61	0.94	3.1	2.5	E2p	4	1237		
4004	6950	11 58.1	+27 53	Leo	14.0 p			2.2	0.8	P		3383		Mrk 432, VV 230
4008	6953	11 58.3	+28 12	Leo	12.9	12.0	0.90	2.5	1.5	Sa	3S	3552		
4016	6954	11 58.4	+27 31	Leo	14.6 p			1.5	0.8	Ir				Arp 305
	6955	11 58.4	+38 04	UMa	13.20			5.3	2.6	Ir+ V		938		D105, Zwicky 2
4013	6963	11 58.5	+43 57	UMa	12.0 H			5.2	1.3	Sb	1	754		
4024		11 58.5	−18 21	Crv	12.78			2.3	1.7	E2	3S	1482		
IC 749	6962	11 58.6	+42 44	UMa	12.80	12.22	0.58	2.5	2.1	Sc III:		827		
4010	6964	11 58.6	+47 16	UMa	12.69			4.2	1.0	SBc:		968		
4017	6967	11 58.7	+27 26	Com	13.1:			1.8	1.5	S(B)b+				Arp 305
4020	6971	11 58.9	+30 25	UMa	13.08			2.1	1.1	SBd:		742		
IC 750	6973	11 58.9	+42 43	UMa	12.80	11.82	0.98	2.9	1.4	Sb−		756		
4025	6982	11 59.2	+37 48	UMa	14.9 p			3.3	2.0	S(B)+ IV:				D107
	6983	11 59.2	+52 42	UMa	12.87			4.2	3.2	SBc		1166		
4026	6985	11 59.4	+50 58	UMa	11.7			5.1	1.4	S0	4VS	958		
4027	A260	11 59.5	−19 16	Crv	11.65	11.10	0.55	3.0	2.3	Sc		1463		VV 66, Arp 22
4030	6993	12 00.4	−1 06	Vir	11.88			4.3	3.2	Sc I	3S	1255		
4032	6995	12 00.6	+20 04	Com	12.62			2.1	2.0	Ir+ III:		1165		NGC 4042?
4033		12 00.6	−17 51	Crv	12.68			2.5	1.1	E5	4S	1312		
4037	7002	12 01.4	+13 24	Com	12.98			2.7	2.3	S(B) IV:	3VS	844		
4036	7005	12 01.4	+61 54	UMa	11.48	10.58	0.90	4.5	2.0	E6	4	1510		
	7009	12 01.8	+62 20	UMa	14.29			1.4	0.5	Ir+		1250		
4038	A264	12 01.9	−18 52	Crv	11.3	10.7	0.60	2.6	1.8	Sc		1447		VV 245 } Antennae
4039		12 01.9	−18 53	Crv	13: p			3.2	2.2	Smp		1430		VV 245 }
4041	7014	12 02.2	+62 08	UMa	11.64	11.10	0.54	2.8	2.7	Sc II		1320		
	7020	12 02.6	+64 22	UMa	14.13			1.5	0.7	S0		1586		Mrk 195
4045	7021	12 02.7	+1 59	Vir	12.65	11.84	0.81	2.8	2.0	Sp	4VS	1827		
4045A		12 02.8	+1 57	Vir	15: p			0.7	0.3	SB0:				
4047	7025	12 02.9	+48 38	UMa	13.10			1.5	1.3	Sb+ III:	4VS	3426		
4050		12 02.9	−16 22	Crv	11.99			3.1	2.2	S(B)b− II:	4VS	1700		
4051	7030	12 03.2	+44 32	UMa	10.95	10.28	0.67	5.0	4.0	Sc II	5VS*	726		
4068	7047	12 04.0	+52 35	UMa	12.79			3.2	1.8	Ir+		302		
4061	7044	12 04.1	+20 13	Com	14.11			1.5	1.1	E3:		1545		VV 179
4062	7045	12 04.1	+31 54	UMa	12.0	11.2	0.76	4.3	2.0	Sb II:	3VS	743		
4067	7048	12 04.2	+10 51	Vir	13.32			1.5	1.1	Sb:				
4064	7054	12 04.2	+18 27	Com	12.30	11.47	0.83	4.5	1.9	Sb− III		959		
4065	7050	12 04.2	+20 13	Com	13.80			1.5	1.4	E1		1122		VV 179
IC 758	7056	12 04.2	+62 30	UMa	13.48			2.0	1.9	SBc		1406		

NGC	UGC	α 2000	δ 2000	Const	B	V	B−V	Dim		Type	Nucl	RV	Memb	Notes
4073	7060	12ʰ04ᵐ.5	+1°54'	Vir	12.7 H			2.5	1.9	E1		5760		
4081	7062	12 04.6	+64 26	UMa	13.6 p			1.8	0.8	Sa:		1451		
4079	7067	12 04.8	−2 23	Vir	14.0 p			2.4	1.8	Sb+:				
4085	7075	12 05.4	+50 21	UMa	12.91	12.32	0.59	2.8	0.9	Sb III:	1	794		
4088	7081	12 05.6	+50 33	UMa	11.10	10.50	0.60	5.8	2.5	Sc I-II:	3VS	822		Arp 18
IC 2995	A268	12 05.8	−27 56	Hya	12.7 H			3.1	1.2	S IV:	1	1618		
	7089	12 05.9	+43 08	UMa	13.75			3.4	1.0	Sm		826		
4108A	7088	12 05.9	+67 15	Dra	14.7 p			1.6	0.7	S(B)b+:				
4094	A269	12 05.9	−14 32	Crv	12.48			4.2	1.8	S− IV		1232		
4096	7090	12 06.0	+47 29	UMa	11.02	10.58	0.44	6.5	2.0	Sc II:	3S	561		
4100	7095	12 06.2	+49 35	UMa	11.54			5.2	1.9	Sb I-II:	3VS	1153		
4102	7096	12 06.4	+52 43	UMa	12.3			3.2	1.9	Sc	3S	986		
4104	7099	12 06.6	+28 10	Com	13.7 p			3.0	1.7	S0		8472		
4105		12 06.7	−29 46	Hya	11.99			2.4	1.9	E2		1659		
4108	7101	12 06.8	+67 10	Dra	13.0 p			1.9	1.6	Sc:		2636		
4106		12 06.8	−29 46	Hya	12.35	11.35	1.00	1.9	1.5	E0	4	1942		
4109		12 06.9	+43 00	CVn	15: p			0.8	0.7	Sa:				
4111	7103	12 07.1	+43 04	CVn	11.65	10.77	0.88	4.8	1.1	S0:	4VS	841		
4108B	7106	12 07.2	+67 14	Dra	14.5 p			1.5	1.3	S(B)dp				7Z 439
4114		12 07.2	−14 11	Crv	13: p			2.3	1.1	S(B)a:				
4116	7111	12 07.6	+2 42	Vir	12.39	11.85	0.54	3.8	2.4	SBc III:	3VS	1190		
4117	7112	12 07.8	+43 08	CVn	14.3 p			2.8	1.1	S0:				
	A272	12 07.8	+67 23	Dra	14.46			0.5	0.5	P		2391		Mrk 197
4118		12 07.9	+43 07	CVn	15.5:p			0.9	0.5	S0:				
4121		12 07.9	+65 07	Dra	14.37			0.6	0.5	E2		1600		
4125	7118	12 08.1	+65 11	Dra	10.7	9.8	0.88	5.1	3.2	E5p		1482		
4123	7116	12 08.2	+2 53	Vir	11.85	11.23	0.62	4.5	3.5	SBb III	4VS	1207		
4124	7117	12 08.2	+10 23	Vir	12.02			4.6	1.7	Sa:	4VS	1551	V	IC 3011
4127	7122	12 08.4	+76 48	Cam	12.91			2.3	1.3	Sb+:		2006		
4128	7120	12 08.5	+68 46	Dra	12.77			2.8	1.0	Sa:	4S	2481		
4120	7121	12 08.5	+69 33	Dra	14.1 p			1.9	0.5	Sc:				
4131	7126	12 08.7	+29 18	Com	14.1 p			1.6	0.9	S		3695		
	7125	12 08.7	+36 48	CVn	13.67			4.6	0.9	Sm		1098		
4129		12 08.9	−9 02	Vir	13.2	12.6	0.59	2.6	0.8	S:		1034		
4132		12 09.0	+29 14	Com	14.5:p			1.1	0.5	S		4054		
4135		12 09.2	+44 00	CVn	15: p			1.0	0.8	S(B)b+:				
		12 09.2	+47 03	UMa	14.90	14.12	0.78	0.7	0.7	S(B)0p	*	7268*		Mrk 198
4136	7134	12 09.3	+29 56	Com	11.74			4.1	3.9	Sc III	4S	434		
4137	7135	12 09.3	+44 06	CVn	15.0 p			1.2	0.9	SBc:		9353		
4138	7139	12 09.5	+43 41	CVn	12.3			2.9	1.9	E4	3S	1090		
4142	7140	12 09.5	+53 06	UMa	13.72			2.3	1.3	SBd:		1257		
4143	7142	12 09.6	+42 32	CVn	12.1			2.9	1.8	E4	D3S	830		
4145	7154	12 10.0	+39 53	CVn	11.50	10.96	0.54	5.8	4.4	Sc: II		1035		
4144	7151	12 10.0	+46 27	UMa	11.85			5.9	1.5	Sb III	1	324		
4148	7158	12 10.1	+35 52	CVn	14.6 p			1.8	1.1	S0		4711		
IC 764	A273	12 10.2	−29 44	Hya	12.20			4.8	1.8	S IV:	2	1907		
4146	7163	12 10.3	+26 26	Com	13.8 p			1.6	1.5	SBb−:		6639		
4151	7166	12 10.5	+39 24	CVn	11.13	10.38	0.75	5.9	4.4	P	5S*	1002		
4152	7169	12 10.6	+16 02	Com	12.46	11.96	0.50	2.3	1.9	Sc II-III:	3S	2086		
	7170	12 10.6	+18 49	Com	14.86			3.0	0.4	Sc		2382		
4150	7165	12 10.6	+30 24	Com	12.45	11.66	0.79	2.5	1.8	E2	4VS	235		
4156	7173	12 10.8	+39 28	CVn	13.85	13.02	0.83	1.5	1.3	SBb	3S	6797		
	7175	12 10.9	+39 45	CVn	14.93			2.1	0.6	Sm		1202		
4157	7183	12 11.1	+50 29	UMa	11.65			6.9	1.7	Sb+ II:		907*		
4158	7182	12 11.2	+20 11	Com	13.00			2.0	1.8	Sa	4	2480		
		12 11.6	+16 29	Com	14.54					P		2400		
4162	7193	12 11.9	+24 07	Com	12.25	11.45	0.80	2.5	1.6	Sc II	4	2454		
4165	7201	12 12.2	+13 15	Vir	14.20	13.60	0.60	1.5	1.1	S(B)a:	3S	1420	V	
4169	7202	12 12.2	+29 10	Com	12.9 p			2.1	1.1	S0		3827		
4163	7199	12 12.2	+36 10	CVn	13.04			1.9	1.7	Ir+		188		
4168	7203	12 12.3	+13 12	Vir	12.25	11.26	0.99	2.8	2.6	E0		2342	V	
4173	7204	12 12.3	+29 11	Com	13.05			4.8	0.9	SBd		1113		
4174	7206	12 12.4	+29 08	Com	14.3 p			0.8	0.3			4148		
IC 769	7209	12 12.5	+12 07	Vir	13.29			2.5	1.8	Sb+		2144	V	
4175	7211	12 12.5	+29 09	Com	14.2 p			2.0	0.5	S		4048		
4178	7215	12 12.8	+10 52	Vir	11.89	11.39	0.50	5.0	2.0	SBc II:		202	V	IC 3042
4179	7214	12 12.9	+1 18	Vir	11.80	10.89	0.91	4.2	1.2	S0		1144		
	7218	12 12.9	+52 16	CVn	14.69			1.2	0.7	Irp+		882		
4180	7219	12 13.1	+7 02	Vir	13.35			1.8	0.7	Sb−:				
4183	7222	12 13.3	+43 42	CVn	12.92			5.0	0.9	S− IV:	1	981		
4190	7232	12 13.7	+36 38	CVn	13.25			1.7	1.6	Ir+		255		VV 104
4189	7235	12 13.8	+13 26	Com	12.51	11.74	0.77	2.5	2.1	Sc II:	3S	2013	V	IC 3050
4192	7231	12 13.8	+14 54	Vir	10.92	10.13	0.79	9.5	3.2	Sb I-II:	5VS	−220	V	M98
4193	7234	12 13.9	+13 10	Vir	13.23	12.36	0.87	2.3	1.2	S(B)c:	3S	2416	V	
4186	7240	12 14.1	+14 44	Com	14.42			1.4	1.1	Sb−:			V	

NGC	UGC	α 2000	δ 2000	Const	B	V	B−V	Dim		Type	Nucl	RV	Memb	Notes
	7239	12^{h}14^{m}2	+7°46′	Vir	13.85			2.2	1.9	Ir+		1126		
4194	7241	12 14.2	+54 32	UMa	12.95	12.43	0.52	2.5	1.7	Irp+	5	2629		VV 261, Arp 160
4196	7245	12 14.4	+28 25	Com	13.7 p			1.6	1.2	S0p		3966		
4197	7247	12 14.6	+5 48	Vir	13.47			3.5	0.7	Sc		1914		
	7249	12 14.6	+12 48	Vir	14.94			1.2	0.5	Ir+		541	V	D114
	7257	12 15.0	+35 57	CVn	14.05			1.7	1.0	Sm		967		
IC 3061	7255	12 15.1	+14 02	Com	14.36	13.59	0.77	2.3	0.6	SBc:		2183	V	
4203	7256	12 15.1	+33 12	Com	11.55	10.66	0.89	3.6	3.3	Ep	D4	1007		
4204	7261	12 15.2	+20 39	Com	12.77			4.2	3.6	SBdm		810		
		12 15.3	+5 46	Vir	15.0	14.6	0.39	0.6	0.4	E0p		1845		
4206	7260	12 15.3	+13 02	Vir	12.79	12.12	0.67	5.2	1.2	Sb+:	3VS	616	V	IC 3064
4210	7264	12 15.3	+65 59	Dra	13.4 p			2.2	1.7	SBb	3S	2649		
	7267	12 15.4	+51 22	CVn	14.39			2.0	0.8	Sm		561		
4207	7268	12 15.5	+9 35	Vir	13.48			1.8	1.0					
	7271	12 15.5	+43 25	CVn	14.38			2.3	0.8	SBd		599		
4211	7277	12 15.6	+28 11	Com	14.4 p			1.4	1.4	Sap,SB0p		6601		VV 199
4214	7278	12 15.6	+36 20	CVn	10.20	9.74	0.46	7.9	6.3	Ir+ III-IV		309		NGC 4228
IC 3074	7279	12 15.7	+10 41	Vir	14.41			2.3	0.3	Sm		1883	V	
4212	7275	12 15.7	+13 54	Com	11.85	11.15	0.70	3.0	2.1	Sc III	3	1947	V	NGC 4208
4217	7282	12 15.8	+47 06	CVn	11.90			5.5	1.8	Sb		1054		
4218	7283	12 15.8	+48 08	CVn	13.24			1.2	0.8	Sa:		1462		
4215	7281	12 15.9	+6 24	Vir	13.12			1.9	0.8	Sb+ III:	4ES	1980		
4216	7284	12 15.9	+13 09	Vir	10.97	9.98	0.99	8.3	2.2	Sb II	5S	55	V	
4221	7288	12 16.0	+66 14	Dra	13.6 p			2.3	1.9	SB0	4VS	2965		
4220	7290	12 16.2	+47 53	CVn	12.2			4.1	1.5	Sa:	3S	1052		
4222	7291	12 16.4	+13 19	Com	13.91			3.3	0.6	Sd:		180*	V	
4226	7297	12 16.4	+47 02	CVn	14.4 p			1.3	0.7	Sap				
4219		12 16.4	−43 20	Cen	12.09			4.5	1.7	Sb+		1739		
4227	7296	12 16.5	+33 31	CVn	13.8 p			1.8	1.1	S(B)a:		4773		
4224	7292	12 16.6	+7 28	Vir	12.95	11.84	1.11	2.4	1.0	Sa		2543		
4236	7306	12 16.7	+69 28	Dra	10.06	9.66	0.40	18.6	6.9	SB+ IV		160		
4232	7303	12 16.8	+47 26	CVn	14.6 p			1.5	0.9	S(B)b+p				
4231	7304	12 16.8	+47 27	CVn	14.5 p			1.4	1.3	S0p				
4238	7308	12 16.9	+63 25	Dra	14.2 p			1.9	0.6	Sc				
4233	7311	12 17.1	+7 37	Vir	12.97	11.88	1.09	2.3	1.1	S0		2117	V	
4234	7309	12 17.2	+3 41	Vir	13.43	12.86	0.57	1.3	1.2	SB+ IV		1952		
4235	7310	12 17.2	+7 11	Vir	12.64	11.60	1.04	4.3	1.1	Sa	D4S*	2487		IC 3098
4237	7315	12 17.2	+15 19	Com	12.50	11.65	0.85	2.3	1.6	S(B)c	3S	871	V	
4241	7319	12 17.4	+6 41	Vir	13.05	12.00	1.05	2.5	1.4	Sa:				
4250		12 17.4	+70 48	Dra	12.8 p			2.7	2.1	S(B)0	4VS	2106		7Z 447
	7321	12 17.5	+22 32	Com	13.85			5.3	0.5	Sc		362		
4244	7322	12 17.5	+37 49	CVn	10.60	10.16	0.44	16.2	2.5	S- IV:	3VS	270		
4242	7323	12 17.5	+45 37	CVn	11.54	10.99	0.55	4.8	3.8	S- IV		749		
4245	7328	12 17.6	+29 36	Com	12.25	11.36	0.89	3.3	2.6	S(B)b+ III	4VS	882		
4248	7335	12 17.8	+47 25	CVn	13.19	12.58	0.61	3.0	1.2	Ir-:		556		
	7332	12 18.0	+0 26	Vir	14.10			2.0	1.3	Ir+		805		
IC 3115	7333	12 18.0	+6 39	Vir	13.78	13.08	0.70	1.7	1.4	SBc				
4246	7334	12 18.0	+7 11	Vir	13.43	12.70	0.73	2.5	1.5	Sc	3S	3616		IC 3113
4247		12 18.0	+7 16	Vir	14.80			0.7	0.6	S(B)b-p	4	3837		
4251	7338	12 18.1	+28 10	Com	11.6			4.2	1.9	E7	D4S	999		
4253	7344	12 18.4	+29 49	Com	13.7 p			1.1	1.0	SBa:		3862		
4252	7343	12 18.5	+5 34	Vir	14.81			1.5	0.4	Sb:				
4256	7351	12 18.7	+65 54	Dra	12.43			4.6	1.0	Sb:	3S	2679		
4254	7345	12 18.8	+14 25	Com	10.42	9.84	0.58	5.4	4.8	Sc I	4S	2324	V	M99
IC 775	7350	12 18.9	+12 55	Vir	14.25	13.25	1.00	1.4	1.0	SB0p				
IC 776	7352	12 19.0	+8 51	Vir	14.01			2.1	1.2	Sdm		2363	V	
4258	7353	12 19.0	+47 18	CVn	8.99	8.31	0.68	18.2	7.9	Sb+p		537		M106
4257		12 19.1	+5 44	Vir	14.91			1.2	0.4	Sb-:				
	7354	12 19.2	+3 51	Vir	14.29			0.7	0.5	E		1461		Mrk 49
4259	7359	12 19.4	+5 23	Vir	14.55	13.56	0.99	1.1	0.5	S0		2312		
4261	7360	12 19.4	+5 49	Vir	11.32	10.33	0.99	3.9	3.2	E2		2090		3C 270
4260	7361	12 19.4	+6 06	Vir	12.70	11.77	0.93	2.6	1.4	SBb- III:	4S	1735		
4262	7365	12 19.5	+14 53	Com	12.40	11.47	0.93	2.2	2.0	E1	5	1282	V	
4264	7364	12 19.6	+5 51	Vir	13.80	12.87	0.93	1.1	0.9	SB0	3	2520		
4266	7368	12 19.7	+5 32	Vir	14.51			2.1	0.5	SBa:				
4268	7371	12 19.8	+5 17	Vir	13.73	12.69	1.04	1.6	0.6	SBa:		2203		
4270	7376	12 19.8	+5 28	Vir	13.10	12.20	0.90	2.2	1.0	Sa		2233		
4269	7372	12 19.8	+6 01	Vir	13.69			1.5	1.0	S0		2423		
4267	7373	12 19.8	+12 48	Vir	11.80	10.89	0.91	3.5	3.2	E2	D4	1177	V	
4274	7377	12 19.8	+29 37	Com	11.30	10.37	0.93	6.9	2.8	Sb+ II-III	4	715		
4272	7378	12 19.8	+30 10	Com	14.2 p			1.3	1.1	E1		8457		
4273	7380	12 19.9	+5 21	Vir	12.37	11.92	0.45	2.3	1.5	Sc III:		2188		
4275	7382	12 19.9	+27 37	Com	13.40			1.0	0.9	S:		2291		
4277		12 20.1	+5 21	Vir	14.55	13.52	1.03	0.9	0.8	S(B)a:		2384		
4278	7386	12 20.1	+29 17	Com	11.15	10.21	0.94	3.6	3.5	E1		651		

NGC	UGC	α 2000	δ 2000	Const	B	V	B-V	Dim		Type	Nucl	RV	Memb	Notes
4283	7390	12^{h}20^{m}3	+29°19'	Com	12.95	12.01	0.94	1.4	1.4	E0		1125		
4291	7397	12 20.3	+75 22	Dra	12.2			2.2	1.9	E1		1967		
4281	7389	12 20.4	+5 23	Vir	12.27	11.32	0.95	3.1	1.5	E5		2488		
4288	7399	12 20.6	+46 17	CVn	12.98			2.3	1.7	S(B)+ IV:		603		D119
4286	7398	12 20.7	+29 21	Com	14.7 p			1.9	1.2	Sa:				IC 3181
4290	7402	12 20.8	+58 06	UMa	12.49			2.5	1.9	SBb II	4VS	2885		
4289	7403	12 21.0	+3 43	Vir	14.34			3.9	0.5	Sc:				
4293	7405	12 21.2	+18 23	Com	11.19			6.0	3.0	Sap	3VS	825*	V	
4292	7404	12 21.3	+4 36	Vir	13.50			2.1	1.4	SB0:				
4294	7407	12 21.3	+11 31	Vir	12.59	12.06	0.53	3.1	1.3	S(B)c III-IV:	1	328	V	
	7408	12 21.3	+45 49	CVn	13.73			2.3	1.4	Ir+ IV-V		530		D120
4298	7412	12 21.5	+14 36	Com	12.07	11.36	0.71	3.2	1.9	Sc	3VS	1042	V	
4301	7411	12 21.6	+4 47	Vir	14.77			1.4	0.5	Sa:				
IC 783	7415	12 21.6	+15 45	Com	14.35	13.59	0.76	1.5	1.4	S(B)a:				
4299	7414	12 21.7	+11 30	Vir	12.87	12.45	0.42	1.7	1.6	S IV:	1	125	V	
4302	7418	12 21.7	+14 36	Com	12.53	11.63	0.90	5.2	1.1	Sc		1044	V	
4319	7429	12 21.7	+75 19	Dra	12.16			3.1	2.5	SBb-	4VS	1867		
4303	7420	12 21.9	+4 28	Vir	10.21	9.67	0.54	6.0	5.5	Sc I	5	1483		M61
4308	7426	12 21.9	+30 03	Com	14.13			1.0	0.9	E1:		602		
4307	7431	12 22.1	+9 02	Vir	12.79			3.7	0.9	Sb- II-III		1207	V	
4305	7432	12 22.1	+12 44	Vir	13.25	12.55	0.70	2.2	1.3	Sa		1780	V	
4306	7433	12 22.1	+12 47	Vir	13.75			1.6	1.3	SB0:		1699	V	
4309	7435	12 22.2	+7 09	Vir	13.59			2.0	1.2	S(B)0		998		
4304		12 22.2	-33 29	Hya	12.75	12.02	0.73	2.4	2.4	SBb+ I		2395		
4310	7440	12 22.4	+29 12	Com	12.92			2.6	1.4	S(B)0:		893		NGC 4311
4331	7449	12 22.4	+76 11	Dra	14.53			2.0	0.6	Irp+		1756		
4303A		12 22.5	+4 34	Vir	13.43	13.00	0.43	1.7	1.4	SBc		1136		
4312	7442	12 22.5	+15 32	Com	12.59	11.78	0.81	4.7	1.3	Sb-:		47	V	
4313	7445	12 22.6	+11 48	Vir	12.73			3.9	1.1	Sb-:	3S	1494	V	
4314	7443	12 22.6	+29 53	Com	11.35	10.51	0.84	4.8	4.3	SBa	5	879		
4318	7446	12 22.7	+8 12	Vir	14.14			0.9	0.7	E		-300	V	
4332	7453	12 22.8	+65 51	Dra	13.2 p			2.3	1.7	SBa		2993		
4321	7450	12 22.9	+15 49	Com	10.10	9.37	0.73	6.9	6.2	Sc I	4	1543	V	M100
4322		12 23.0	+15 54	Com	14.70	13.92	0.78	1.3	1.1	SB0:			V	NGC 4323
4335	7455	12 23.0	+58 27	UMa	13.7 p			2.2	1.8	E2				
4324	7451	12 23.1	+5 15	Vir	12.60			2.5	1.2	Sb+ III:	4S	1602		
4326	7454	12 23.2	+6 04	Vir	14.09			1.7	1.4	S(B)b-:		7073		
4328		12 23.3	+15 48	Com	14.25	13.45	0.80	1.5	1.4	S0:			V	
4333		12 23.4	+6 02	Vir	14.27			1.1	0.9	SBb-	3VS	6842		
4334	7458	12 23.4	+7 28	Vir	13.93			2.4	1.1	SBb-	4S	4274		
4329		12 23.4	-12 34	Crv	14: p			1.2	0.6	E5	3S			
4346	7463	12 23.5	+47 00	CVn	12.18			3.5	1.4	E6		922		
4363		12 23.5	+74 57	Dra	14.5:p			1.5	1.5	Sb:				
4339	7461	12 23.6	+6 05	Vir	12.35	11.43	0.92	2.3	2.3	E0		1169		
4340	7467	12 23.6	+16 43	Com	12.01	11.03	0.98	4.1	3.2	SBa	4S	790	V	
4344	7468	12 23.6	+17 32	Com	13.21			1.9	1.8	SB0:			V	
4343	7465	12 23.7	+6 57	Vir	13.15	12.13	1.02	2.8	0.9	Sa:	4	1000	V	
IC 3256	7466	12 23.7	+7 03	Vir	13.54	12.56	0.98	1.4	0.7	S0	5VS	609	V	NGC 4341-2?
IC 3258	7470	12 23.7	+12 28	Vir	13.75			1.9	1.7	SBmp		-517	V	
IC 3253		12 23.7	-34 38	Cen	12.38			3.3	1.6	Sc		2487		
IC 3259	7469	12 23.8	+7 11	Vir	14.35	13.59	0.76	1.8	1.1	S(B)dm:		1471	V?	
IC 3260	7472	12 23.9	+7 07	Vir	14.23	13.26	0.97	1.9	0.7	S(B)0:		998	V	
4348		12 23.9	-3 27	Vir	12.66			3.5	1.0	Sb II-III:		2055		
4353		12 24.0	+7 47	Vir	13.94			1.3	0.9	Ir+:		958		IC 3265-6
4351	7476	12 24.0	+12 12	Vir	13.04	12.65	0.39	2.0	1.4	SBb-p		2305	V	NGC 4354
4350	7473	12 24.0	+16 42	Com	11.99	11.10	0.89	3.2	1.1	E7	4VS	1121	V	
4357	7478	12 24.0	+48 46	CVn	13.5 p			3.8	1.5	Sb+		4332		
IC 3268	7477	12 24.1	+6 36	Vir	14.05			0.7	0.6	P		657		
IC 3267	7474	12 24.1	+7 03	Vir	14.23	13.45	0.78	1.1	1.1	Sc		2026	V?	
4352	7475	12 24.1	+11 13	Vir	13.55	12.65	0.90	1.9	0.9	S0:				
4359	7483	12 24.2	+31 31	Com	13.18			3.5	1.0	SBc:		1256		
	A279	12 24.3	+4 13	Vir	14.83			1.0	0.4	S0		829		Mrk 51
4360	7484	12 24.3	+9 17	Vir	13.47			1.7	1.3	E2				
	7490	12 24.3	+70 20	Dra	12.92			3.9	3.9	S+ V		636		D122
4365	7488	12 24.5	+7 19	Vir	10.51			6.2	4.6	E2		1074	V	
4386	7491	12 24.5	+75 32	Dra	12.36			3.0	1.7	Sa:		1865		
4369	7489	12 24.6	+39 23	CVn	11.83			2.5	2.4	Sa		1066		Mrk 439
4370	7492	12 24.9	+7 27	Vir	13.69			1.6	0.9	Sa		366	V	
4371	7493	12 24.9	+11 42	Vir	11.80	10.79	1.01	3.9	2.5	SBa	4	898	V	
4375	7496	12 25.0	+28 33	Com	13.9 p			1.5	1.4	SBb-p		9078		
4374	7494	12 25.1	+12 53	Vir	10.26	9.29	0.97	5.0	4.4	E1		854	V	M84, 3C 272.1
4377	7501	12 25.2	+14 46	Com	12.65	11.77	0.88	1.8	1.5	E1	4S	1281	V	3Z 65
4379	7502	12 25.2	+15 36	Com	12.62	11.65	0.97	2.1	1.8	E1		971	V	
4384	7506	12 25.2	+54 30	UMa	13.24			1.5	1.2	Sap		2528		Mrk 207
4378	7497	12 25.3	+4 55	Vir	12.16			3.3	3.1	Sa	D4S	2447		

NGC	UGC	α 2000	δ 2000	Const	B	V	B–V	Dim		Type	Nucl	RV	Memb	Notes
4391	7511	12^{h}25^{m}3	+64°56'	Dra	13.8 p			1.'4	1.'4	S0		1484		7Z 454
4373		12 25.3	−39 45	Cen	12.10	11.05	1.05	3.2	2.3	S0:	3	3172		
4380	7503	12 25.4	+10 01	Vir	12.36			3.7	2.2	Sb–	3VS	872	V	
4383	7507	12 25.4	+16 28	Com	12.68			2.2	1.2	Ep		1545	V	Mrk 769
4382	7508	12 25.4	+18 11	Com	10.10	9.22	0.88	7.1	5.2	Ep	D5	718	V	M85
4389	7514	12 25.6	+45 41	CVn	12.48			2.7	1.5	SB– IV:		786		
4385	7515	12 25.7	+0 34	Vir	13.05	12.35	0.70	2.3	1.5	SBb– II–III	4VS	2015		Mrk 52
IC 3322	7513	12 25.7	+7 13	Vir	14.14			3.4	0.5	Sc		898	V	
4387	7517	12 25.7	+12 49	Vir	12.95	12.01	0.94	1.9	1.1	E5		432	V	
4390	7519	12 25.8	+10 27	Vir	13.27			1.8	1.4	S(B)b+:				
4388	7520	12 25.8	+12 40	Vir	11.83	11.05	0.78	5.1	1.4	Sb		2535	V	
4393	7521	12 25.8	+27 33	Com	12.54			3.3	3.1	S(B)d		741		IC 3323
4395	7524	12 25.8	+33 33	CVn	10.69	10.15	0.54	12.9	11.0	S+ IV–V		307		
4394	7523	12 25.9	+18 13	Com	11.74	10.92	0.82	3.9	3.5	SBb– II	5	717	V	
4396	7526	12 26.0	+15 40	Com	13.02			3.5	1.2	Sd:				
4402	7528	12 26.1	+13 07	Vir	12.56	11.73	0.83	4.1	1.3	Sb			V	
4405	7529	12 26.1	+16 11	Com	12.99			2.0	1.4	Sa:		1700	V	
4406	7532	12 26.2	+12 57	Vir	10.11	9.18	0.93	7.4	5.5	E3		−419	V	M86
	7534	12 26.2	+58 18	UMa	14.03			3.0	1.8	Ir+		845		D123
	A281	12 26.3	+48 29	CVn	14.94	14.77	0.17	0.9	0.8	P		367*		Mrk 209, 1Z 36
4414	7539	12 26.4	+31 13	Com	10.95	10.26	0.69	3.6	2.2	Sc II:	D3	718		
4411A	7537	12 26.5	+8 52	Vir	13.55	12.83	0.72	2.2	2.0	SBdm		1168	V	
4410	7535	12 26.5	+9 01	Vir	13.6 p			1.3	0.8	Sb–p,S0p		7125		
4413	7538	12 26.5	+12 37	Vir	12.97			2.5	1.7	SBb:		14	V	
4412	7536	12 26.6	+3 58	Vir	13.14			1.5	1.4	S(B)p	3S	2185		
4415	7540	12 26.6	+8 25	Vir	13.72			1.5	1.4	Sa		399	V	
4411B	7546	12 26.8	+8 53	Vir	13.02	12.38	0.64	2.7	2.7	S(B)dm		1170	V	
4417	7542	12 26.8	+9 35	Vir	12.09	11.20	0.89	3.6	1.4	E7	2	733	V	
4419	7551	12 26.9	+15 03	Com	12.13	11.13	1.00	3.4	1.3	Ep		−342	V	
4420	7549	12 27.0	+2 30	Vir	12.67			2.2	1.2	Sc III:	1	1557		NGC 4409
4421	7554	12 27.0	+15 28	Com	12.45	11.58	0.87	2.7	2.2	SBa		1626	V	
4423		12 27.1	+5 52	Vir	14.28			2.3	0.5	Sm		984		
IC 3365	7563	12 27.1	+15 54	Com	14.17			1.9	1.1	Ir+		2271	V	
	7559	12 27.1	+37 08	CVn	13.38			3.9	2.5	Ir+ V		252		D126
	7557	12 27.2	+7 15	Vir	13.24			3.1	2.7	Sm		833	V	
4424	7561	12 27.2	+9 25	Vir	12.28	11.58	0.70	3.7	1.9	Sb III:		358	V	
4425	7562	12 27.2	+12 44	Vir	12.84	11.87	0.97	3.4	1.2	Sb– III:		1805	V	
4441	7572	12 27.3	+64 48	Dra	13.5 p			4.7	3.7	S(B)0p		1294		
4430	7566	12 27.4	+6 15	Vir	12.48			2.7	2.4	SBbp		1359		
4429	7568	12 27.4	+11 07	Vir	11.10	10.16	0.94	5.5	2.6	S0	4	1029	V	
4434	7571	12 27.5	+8 09	Vir	12.99			1.6	1.6	E0		909	V	
4428		12 27.5	−8 10	Vir	12.98			1.9	0.9	Sc III:		2874		
4431	7569	12 27.5	+12 18	Vir	13.67	12.80	0.87	2.0	1.3	S0		376	V	
4433		12 27.6	−8 17	Vir	12.78			2.3	1.1	S(B)b+		2746		
	7577	12 27.6	+43 29	CVn	12.96			3.9	2.5	Ir+ IV–V		254		D125
IC 3370		12 27.6	−39 20	Cen	12.10	11.07	1.03	2.8	2.4	E2	3	2690		
4436	7573	12 27.7	+12 19	Vir	14.03	13.12	0.91	1.9	0.9	S0			V	
4435	7575	12 27.7	+13 05	Vir	11.84	10.92	0.92	3.0	1.9	E4	4	793	V	VV 188
4438	7574	12 27.7	+13 01	Vir	10.91	10.08	0.83	9.3	3.9	Sap	4VS	182*	V	VV 188, Arp 120
4440	7581	12 27.9	+12 18	Vir	12.75	11.79	0.96	2.0	1.7	SBa	D4	627	V	
4442	7583	12 28.1	+9 48	Vir	11.40	10.48	0.92	4.6	2.4	E5p	D4	490	V	
4448	7591	12 28.2	+28 37	Com	12.0	11.1	0.92	4.0	1.6	Sb– II–III	D3	686		
4449	7592	12 28.2	+44 06	CVn	9.85	9.44	0.41	5.1	3.7	Ir+ III	2	262		
4445	7587	12 28.3	+9 26	Vir	13.66	12.83	0.83	2.8	0.6	Sb–:		162	V	
4450	7594	12 28.5	+17 05	Com	10.94	10.12	0.82	4.8	3.5	Sb	D4S	1990	V	
	7599	12 28.5	+37 13	CVn	14.52			2.0	1.0	Sm		312		D127
4451	7600	12 28.7	+9 16	Vir	13.30	12.51	0.79	1.5	1.0	S0:		600	V	
4452	7601	12 28.7	+11 45	Vir	13.30	12.41	0.89	2.4	0.6	S0		130	V	
4455	7603	12 28.7	+22 49	Com	12.95			2.8	1.0	S+ IV:		611		
	7605	12 28.7	+35 43	CVn	14.94			1.4	1.0	Ir+		329		
	7608	12 28.7	+43 13	CVn	13.10			3.5	2.9	Ir+ V		601		D129
4454	7606	12 28.8	−1 56	Vir	13.0	12.1	0.89	2.2	1.9	Sb+ II–III:	3VS	2220		
4460	7611	12 28.8	+44 52	CVn	12.30			4.4	1.4	SB0:	1	624		
4457	7609	12 29.0	+3 34	Vir	11.65	10.80	0.85	3.0	2.5	S(B)a	4	627		
4461	7613	12 29.0	+13 11	Vir	12.03	11.17	0.86	3.7	1.5	Sa	D4VS	1812	V	
4458	7610	12 29.0	+13 15	Vir	12.90	12.05	0.85	1.9	1.8	E0		308	V	
4459	7614	12 29.0	+13 59	Com	11.35	10.40	0.95	3.8	2.8	E2	4S	1039	V	
	7612	12 29.1	+2 43	Vir	14.13			2.3	1.2	SBmp		1452		D128
4462		12 29.3	−23 10	Crv	12.49			3.7	1.6	Sb III:	4	1658		
4464	7619	12 29.4	+8 10	Vir	13.70	12.74	0.96	1.1	0.9	Sa:		1103	V	
4466	7626	12 29.5	+7 42	Vir	14.62			1.4	0.5	Sb–:		913	V	
4469	7622	12 29.5	+8 45	Vir	12.22			3.9	1.5	Sp	4	404	V	
4468	7628	12 29.5	+14 03	Com	13.80	12.99	0.81	1.5	1.1	S0:			V	
4470	7627	12 29.6	+7 49	Vir	13.04			1.5	1.1	Sa:		2291	V?	
4472	7629	12 29.8	+8 00	Vir	9.31	8.37	0.94	8.9	7.4	E4		817	V	M49, Arp 134

NGC	UGC	α 2000	δ 2000	Const	B	V	B-V	Dim		Type	Nucl	RV	Memb	Notes
4473	7631	12ʰ29ᵐ8	+13°26'	Com	11.10	10.22	0.88	4.5	2.6	E4		2205	V	
4475	7632	12 29.8	+27 15	Com	14.6 p			2.1	1.2	Sb+		7351		
4474	7634	12 29.9	+14 04	Com	12.70	11.75	0.95	2.3	1.2	E6		1455	V	
	7639	12 29.9	+47 31	CVn	13.79			3.2	1.5	Ir+		462		
4476	7637	12 30.0	+12 21	Vir	13.15	12.32	0.83	1.9	1.3	E4	4	1899	V	
4477	7638	12 30.0	+13 38	Com	11.35	10.42	0.93	4.0	3.5	S(B)a		1190	V	
4478	7645	12 30.3	+12 20	Vir	12.13	11.23	0.90	2.0	1.8	E1		1404	V	
4479	7646	12 30.3	+13 35	Com	13.45	12.52	0.93	1.8	1.5	SB0:		749	V	
4480	7647	12 30.4	+4 15	Vir	13.12	12.40	0.72	2.6	1.4	S(B)c		2173		
4486B		12 30.5	+12 29	Vir	14.32	13.29	1.03	0.5	0.4	E0		1409	V	1Z 38
4485	7648	12 30.5	+41 42	CVn	12.33	11.96	0.37	2.4	1.7	Ir+ III-IV:		817		VV 30, Arp 269
4490	7651	12 30.6	+41 38	CVn	10.20	9.76	0.44	5.9	3.1	Sc III		629		VV 30, 3C 272
4483	7649	12 30.7	+9 01	Vir	13.17			1.8	1.1	S(B)a:		903	V	
4486	7654	12 30.8	+12 24	Vir	9.56	8.62	0.94	7.2	6.8	E1		1180	V	M87, Vir A, 3C 274
4489	7655	12 30.8	+16 45	Com	12.84			2.2	2.1	E1		805	V	
4488	7653	12 30.9	+8 22	Vir	12.86			3.6	1.5	SBap				
4492	7656	12 31.0	+8 05	Vir	13.17	12.53	0.64	2.0	1.9	Sa:		1639	V	IC 3438
4491	7657	12 31.0	+11 29	Vir	13.43	12.52	0.91	1.9	1.0	SBa		79	V	
4487		12 31.1	-8 03	Vir	11.49			4.1	3.0	Sc II-III	4VS	876		
4494	7662	12 31.4	+25 47	Com	10.75	9.86	0.89	4.8	3.8	E1		1289		
4500	7667	12 31.4	+57 58	UMa	13.2 p			1.9	1.3	SBa	5	3128		Mrk 213
4497	7665	12 31.5	+11 37	Vir	13.36	12.56	0.80	2.3	1.1	S(B)a:		1262	V	
IC 3453	7666	12 31.5	+14 51	Com	14.86			1.2	0.2	Ir+		2489	V	
4496A	7668	12 31.66	+3 56.3	Vir	11.7 H			3.9	3.1	S(B)c III:		1651		VV 76
4496B	7668	12 31.68	+3 55.5	Vir	12.42	11.90	0.52	1.0	0.9	Ir+:		1660		VV 76
4498	7669	12 31.7	+16 51	Com	12.62			3.2	1.9	S(B)c		1448	V	
4501	7675	12 32.0	+14 25	Com	10.27	9.52	0.75	6.9	3.9	Sb+ I	3VS	1989*	V	M88
4503	7680	12 32.1	+11 11	Vir	12.12	11.13	0.99	3.5	1.8	E2:	4S	1335	V	
4506	7682	12 32.2	+13 25	Com	13.64			1.6	1.3	Sap		608	V	
4504		12 32.3	-7 34	Vir	11.69			4.0	2.8	Sc II-III:	3VS	839		
	7690	12 32.4	+42 42	CVn	13.05			2.0	1.5	Irp+		598		
4517A	7685	12 32.5	+0 23	Vir	12.65	12.17	0.48	4.2	3.0	SBdm		1405		Reinmuth 80
		12 32.6	+45 46	CVn	14.8	14.2	0.57	0.4	0.4			5957*		Mrk 215
IC 3475	7692	12 32.7	+12 46	Vir	13.95	13.29	0.66	2.6	2.5	E:			V	D132
IC 3476	7695	12 32.7	+14 03	Com	13.17	12.70	0.47	2.2	1.9	Ir+:				
4517	7694	12 32.8	+0 07	Vir	11.20	10.47	0.73	10.2	1.9	Sc	1	1001		NGC 4437
	7699	12 32.8	+37 37	CVn	12.86			3.9	1.2	SBc		539		
IC 3481		12 32.9	+11 24	Vir	15.06	14.04	1.02	0.7	0.5	S(B)0p		7006		VV 43, Arp 175
	7698	12 32.9	+31 32	CVn	13.38			6.1	4.2	Ir+ V		343		D133
IC 3481A		12 33.0	+11 24	Vir				0.2	0.2	E1p		7224		VV 43, Arp 175
4515	7701	12 33.0	+16 15	Com	13.36			1.6	1.3	S0:				
4516	7703	12 33.1	+14 34	Com	13.67			1.9	1.1	SBb-:				
4509	7704	12 33.1	+32 06	CVn	14.1 p			1.1	0.7	Sb-p				
IC 3483		12 33.2	+11 21	Vir	15.8	15.0	0.82	0.5	0.4	S(B)bp:		28		VV 43, Arp 175
IC 3492		12 33.2	+12 51	Vir	14.94					P		300	V	
4519	7709	12 33.5	+8 39	Vir	12.35	11.70	0.65	3.1	2.2	Sc III		1078	V	
4522	7711	12 33.7	+9 10	Vir	12.73			3.7	1.1	Scp		2241	V	
IC 3499	7712	12 33.7	+10 59	Vir	14.20	13.34	0.86	1.6	0.6	Sa			V	
4520		12 33.8	-7 23	Vir						E4:				IC 799
4523	7713	12 33.8	+15 10	Com	13.62			2.6	2.5	S:+ V:		200	V	D135
4525	7714	12 33.8	+30 17	Com	12.60			2.9	1.6	Sc:		1134		
4526	7718	12 34.0	+7 42	Vir	10.58	9.64	0.94	7.2	2.3	E7	D3S	355	V	
4527	7721	12 34.1	+2 39	Vir	11.30	10.42	0.88	6.3	2.3	Sb+ II	5VS	1614		
4528	7722	12 34.1	+11 19	Vir	12.65	11.71	0.94	1.8	1.2	S0:		1257	V	
4534	7723	12 34.1	+35 31	CVn	12.25			3.0	2.5	Sd:		830		
4532	7726	12 34.3	+6 28	Vir	12.32	11.90	0.42	2.9	1.3	Ir+: III:		2059		
4535	7727	12 34.3	+8 12	Vir	10.52	9.82	0.70	6.8	5.0	S(B)c I:	5VS	1853	V	
4531	7729	12 34.3	+13 05	Vir	12.58			3.0	2.0	Sa		266	V	
4533	7725	12 34.4	+2 20	Vir	14.53			2.0	0.4	Sd:				
4536	7732	12 34.5	+2 11	Vir	10.99	10.39	0.60	7.4	3.5	Sc II:	5VS	1810*		
4539	7735	12 34.6	+18 12	Com	12.88	12.02	0.86	3.5	1.6	SBa		403	V	
4545	7747	12 34.6	+63 31	Dra	12.67			2.8	1.7	S(B)c:		2860		
	7739	12 34.7	+6 18	Vir	14.73	14.20	0.53	1.3	1.3	Ir- V		1926		Holm. VII, D137
4540	7742	12 34.8	+15 33	Com	12.49			2.0	1.6	Ir IV	1	1224	V	
4541	7749	12 35.2	-0 13	Vir	14.0 p			1.8	0.9	S(B)b+:				
4548	7753	12 35.4	+14 30	Com	10.98	10.19	0.79	5.4	4.4	SBb	D3	403	V	M91
4546	A288	12 35.5	-3 48	Vir	11.3	10.3	0.98	3.5	1.7	E6	D3S	874		
4550	7757	12 35.5	+12 13	Vir	12.50	11.59	0.91	3.5	1.1	E7		275	V	
4544	7756	12 35.6	+3 02	Vir	13.89			2.1	0.8	SBa:				
4551	7759	12 35.6	+12 16	Vir	12.85	11.90	0.95	2.0	1.6	E3:		903	V	
4562	7758	12 35.6	+25 51	Com	13.92			2.5	0.9	SBm:		1377		
4507		12 35.6	-39 55	Cen	12.8			2.3	2.0	S(B)0	4VS*	3276	Ce	
4552	7760	12 35.7	+12 33	Vir	10.81	9.81	1.00	4.2	4.2	E0		165	V	M89
4555	7762	12 35.7	+26 31	Com	13.5 p			1.7	1.5	E3		6682		
4556	7765	12 35.8	+26 54	Com	14.4 p			1.5	1.2	E4:		7392		

NGC	UGC	α 2000	δ 2000	Const	B	V	B-V	Dim		Type	Nucl	RV	Memb	Notes
4566	7769	12^{h}35^{m}9	+54°13'	UMa	13.9 p			1.5	1.1	S		5398		
4559	7766	12 36.0	+27 58	Com	10.30	9.85	0.45	10.5	4.9	Sc II-III	2	802		
4561	7768	12 36.1	+19 20	Com	13.10			1.5	1.3	Sc III-IV:	1	1430		IC 3569
4565	7772	12 36.3	+25 59	Com	10.39	9.56	0.83	16.2	2.8	Sb I:		1122		
	7774	12 36.3	+40 00	CVn	14.27			3.6	0.6	Sd		574		
4564	7773	12 36.4	+11 26	Vir	12.02	11.06	0.96	3.1	1.4	E6		942	V	
4567	7777	12 36.5	+11 15	Vir	12.08	11.32	0.76	3.0	2.1	Sc	D4	2121*	V	VV 219
IC 3576	7781	12 36.6	+6 37	Vir	13.70			3.0	3.0	S+ IV-V:		978	V	D138
4568	7776	12 36.6	+11 14	Vir	11.67	10.80	0.87	4.6	2.1	Sc	3VS	2168	V	VV 219
IC 3583	7784	12 36.7	+13 15	Vir	13.91			2.2	1.3	Ir+		1055	V	
4569	7786	12 36.8	+13 10	Vir	10.23	9.48	0.75	9.5	4.7	Sb+	5VS	-383	V	M90, Arp 76
4570	7785	12 36.9	+7 15	Vir	11.82	10.87	0.95	4.1	1.3	S0	4S	1635	V	
4571	7788	12 36.9	+14 13	Com	11.83	11.31	0.52	3.8	3.4	Sc	2VS	282	V	IC 3588
4589	7797	12 37.4	+74 12	Dra	11.8			3.0	2.7	Sa	D3	2006		
4578	7793	12 37.5	+9 33	Vir	12.27	11.35	0.92	3.6	2.8	Sa		2197	V	
4576	7792	12 37.6	+4 22	Vir	14.27	13.50	0.77	1.4	1.0	S(B)b+:				
4579	7796	12 37.7	+11 49	Vir	10.61	9.78	0.83	5.4	4.4	Sb	D3S	1730	V	M58
4580	7794	12 37.8	+5 22	Vir	12.61			2.4	1.9	Sb+ III	3VS	1187		
4575		12 37.9	-40 32	Cen	13.2 p			2.2	1.8	SBb+:		2797*	Ce	
4584	7803	12 38.3	+13 07	Vir	13.72			1.5	1.2	S(B)a:			V	
4586	7804	12 38.5	+4 19	Vir	12.60	11.59	1.01	4.4	1.6	Sb+ III	3S	722		
4592	7819	12 39.3	-0 32	Vir	11.97			4.6	1.5	Sb II-III	1	954		
IC 3617	7822	12 39.4	+7 58	Vir	14.67			1.1	0.6	Ir+ IV-V:		1995	V	D140
4593		12 39.7	-5 21	Vir	11.59			4.0	3.1	SBb- II	4	2560		
4596	7828	12 39.9	+10 11	Vir	11.47	10.48	0.99	3.9	2.8	SBa	D4	1939	V	
4595	7826	12 39.9	+15 18	Com	12.92			1.8	1.2	Sc III:		601	V	
4605	7831	12 40.0	+61 37	UMa	10.96			5.5	2.3	SBcp		286		
4594		12 40.0	-11 37	Vir	9.27	8.30	0.97	8.9	4.1	Sb-		963		M104, Sombrero galaxy
4597		12 40.2	-5 48	Vir	12.39			3.6	1.9	SBc III:		904		
4602		12 40.6	-5 08	Vir	12.18			3.6	1.4	Sc II:	3S	2417		
4603		12 40.9	-40 59	Cen	12.00			3.8	2.5	Sc:	3S	2323	Ce	
4606	7839	12 41.0	+11 55	Vir	12.74	11.90	0.84	2.8	1.5	SBa:	3	1587	V	
4617	7847	12 41.1	+50 26	CVn	14.2 p			3.1	0.7	Sb		4684		
4608	7842	12 41.2	+10 09	Vir	12.09	11.13	0.96	3.2	2.6	SBa	D4	1789	V	
4607	7843	12 41.2	+11 53	Vir	13.79	12.87	0.92	3.2	0.8	SBb:		2367	V	
4612	7850	12 41.5	+7 19	Vir	12.57			2.2	1.8	Ep	D4S	1740	V	
4618	7853	12 41.5	+41 09	CVn	11.20	10.77	0.43	4.4	3.8	Sc		613		VV 73, Arp 23
4619	7856	12 41.7	+35 04	CVn	13.5 p			1.5	1.5	SBbp				
4648	7868	12 41.8	+74 25	Dra	12.6 p			2.2	1.7	E3	4	1610		
4625	7861	12 41.9	+41 16	CVn	12.90	12.30	0.60	2.4	2.0	S(B)mp	3S	667		IC 3675
4621	7858	12 42.0	+11 39	Vir	10.75	9.79	0.96	5.1	3.4	E3	D3	341	V	M59
4620	7859	12 42.0	+12 57	Vir	13.17			2.0	1.8	S0				
4627	7860	12 42.0	+32 34	CVn	12.88	12.26	0.62	2.7	2.0	E4p	1	685		
4631	7865	12 42.1	+32 32	CVn	9.84	9.30	0.54	15.1	3.3	Sc III:		638		Arp 281
IC 3687	7866	12 42.1	+38 30	CVn	13.33			3.5	2.1	Ir+ IV-V		396		D141
4603D		12 42.1	-40 49	Cen				1.7	1.1	Sc:		2397	Ce	
4623	7862	12 42.2	+7 41	Vir	13.22			2.6	0.9	E5:		1873	V	
4616		12 42.3	-40 39	Cen				1.6	1.6	E0:		4073	Ce	
4630	7871	12 42.5	+3 58	Vir	13.14			1.7	1.3	P		587		
4632	7870	12 42.5	-0 05	Vir	12.2 H			3.2	1.3	Sc II-III	1	1572		
4633	7874	12 42.6	+14 21	Com	13.80	13.15	0.65	2.1	0.9	S(B)d:		265	V	IC 3688
4635	7876	12 42.6	+19 57	Com	13.36			2.0	1.5	S(B)d:		944		
4622		12 42.6	-40 45	Cen				2.1	2.0	Sa	3S	3986	Ce	
4634	7875	12 42.7	+14 18	Com	13.20	12.42	0.78	2.4	0.7	SBc:		150	V	
4644	7887	12 42.7	+55 09	UMa	14.75	14.00	0.75	1.8	0.7	SBbp		4953		
4636	7878	12 42.8	+2 41	Vir	10.50	9.56	0.94	6.2	5.0	E1	4	869		NGC 4624?
4638	7880	12 42.8	+11 26	Vir	12.14	11.29	0.85	2.8	1.6	E5		1006	V	
4637	7881	12 42.9	+11 26	Vir	14.82			1.5	0.8	S0:				
4639	7884	12 42.9	+13 15	Vir	12.21	11.51	0.70	2.9	2.1	S(B)b II-III:		897	V	
4646	7892	12 42.9	+54 51	UMa	13.8 p			0.7	0.4			4664		
4645A		12 43.1	-41 22	Cen				4.5	1.3	SB0:		3002	Ce	
4643	7895	12 43.3	+1 59	Vir	11.55	10.60	0.95	3.4	2.7	SBa	D4	1234		
4642	7893	12 43.3	-0 39	Vir	13.8 p			2.0	0.7	Sb				
4647	7896	12 43.5	+11 35	Vir	12.03	11.38	0.65	3.0	2.5	Sc	3VS	1286	V	VV 206, Arp 116
4645B		12 43.5	-41 22	Cen				2.1	0.7	SB0:		2372	Ce	
4649	7898	12 43.7	+11 33	Vir	9.83	8.83	1.00	7.2	6.2	E1		1128*	V	M60, VV 206
4651	7901	12 43.7	+16 24	Com	11.30	10.72	0.58	3.8	2.7	Scp II:	D4	742	V	VV 56, Arp 189
4622A		12 43.8	-40 42	Cen						S(B)0:		4539	Ce?	
4622B		12 43.8	-40 42	Cen						Sbp		4421	Ce?	
4653	7900	12 43.9	-0 34	Vir	12.82	12.29	0.53	2.6	2.4	S- IV-V		2506		
4654	7902	12 44.0	+13 08	Vir	11.10	10.46	0.64	4.7	3.0	Sc II		970	V	IC 3708
4656-7	7907	12 44.0	+32 10	CVn	10.86	10.43	0.43	13.8	3.3	Sc IV, Ir+		662*		
4645		12 44.2	-41 45	Cen	12.78			2.2	1.5	E3	3S	2373	Ce	
4650		12 44.3	-40 44	Cen						SBa		2397	Ce	
	7911	12 44.5	+0 29	Vir	13.79			2.9	2.2	SBm		1066		D144

NGC	UGC	α 2000	δ 2000	Const	B	V	B-V	Dim		Type	Nucl	RV	Memb	Notes
4660	7914	12h44m.5	+11°11'	Vir	11.94	10.99	0.95	2.8	1.9	E5		944	V	
4659	7915	12 44.5	+13 31	Com	13.08			1.8	1.3	Sa		316	V	
	A294	12 44.5	+28 29	Com	14.8	14.5	0.31	0.8	0.5	S:		957		
4662	7917	12 44.5	+37 07	CVn	14.1 p			2.5	2.1	SBb+		6962*		
4658		12 44.6	-10 05	Vir	12.58			2.2	1.0	Sc III		2250		
4650A		12 44.7	-40 42	Cen						Irp-		2239	Ce	
4669	7925	12 44.8	+54 53	UMa	15.1 p			1.7	0.5	S		5113		
4663		12 44.8	-10 12	Vir	14: p			1.4	1.1	SB0:	4			IC 811
4665	7924	12 45.1	+3 03	Vir	11.60			4.2	3.5	S(B)a	D3	678		NGC 4664
4666	7926	12 45.1	-0 28	Vir	11.55	10.76	0.79	4.5	1.5	Sc I-II:	3	1395		
4650B		12 45.2	-40 49	Cen						Sb-:		2262	Ce	
4670	7930	12 45.3	+27 08	Com	13.05	12.66	0.39	1.8	1.4	Ep	5S	1156		Arp 163
4668		12 45.5	-0 32	Vir	13.59	13.11	0.48	1.4	0.9	P		1580		
4673	7933	12 45.6	+27 04	Com	13.7 p			1.2	1.0	E1		6988		Mrk 656
4675	7935	12 45.6	+54 44	UMa	15.4 p			1.6	0.6	SBb:		4920		
	A296	12 45.6	+71 20	Dra	14.94			0.2	0.2	P		1228		Mrk 223, 7Z 48
4674		12 46.1	-8 39	Vir	15: p			1.9	0.8	SBap				
	7941	12 46.1	+64 34	Dra	13.86			4.6	0.9	Sd		2445		
4676A	7938	12 46.17	+30 44.0	Com	14.1 p			2.8	0.6	S0p		6631		IC 819 } the Mice
4676B	7939	12 46.19	+30 43.5	Com	14.1 p			2.1	1.2	SBap		6590		IC 820 }
	7943	12 46.7	+5 58	Vir	13.67			2.5	2.1	Sc		743		
4686	7946	12 46.7	+54 33	UMa	13.7 p			2.3	0.7	Sa		5128		
	A298	12 46.8	+26 34	Com	14.70			0.8	0.5	Ep		870		
4677		12 47.0	-41 35	Cen	13.7	12.7	1.03	2.9	1.0	S0	3S	2939	Ce	
4684	7951	12 47.3	-2 43	Vir	12.59			2.9	1.1	Sa	4	1461		
4682		12 47.3	-10 04	Vir	12.9 H			2.8	1.6	S- IV:		2152		
4687	7958	12 47.4	+35 21	CVn	14.13			1.2	1.0	E1		723		Mrk 442
4695	7966	12 47.5	+54 23	UMa	14.5 p			1.3	0.9	S		5041		
4679		12 47.5	-39 34	Cen	13.0 H			2.3	1.2	Sb+:				
4683		12 47.7	-41 32	Cen	14.0	12.9	1.10	2.1	0.7	S0:	3	3351	Ce	
4688	7961	12 47.8	+4 20	Vir	12.68			3.3	3.1	S IV		877		
4689	7965	12 47.8	+13 46	Com	11.59	10.93	0.66	4.0	3.5	Sb+ II:	3VS	1715	V	
4692	7967	12 47.9	+27 13	Com	13.6			1.4	1.3	E2		7910		
4691	A299	12 48.2	-3 20	Vir	11.7	11.2	0.54	3.2	2.7	SBbp		987		
4694	7969	12 48.2	+10 59	Vir	12.19			3.6	1.7	E5	4	1121	V	
4698	7970	12 48.4	+8 29	Vir	11.53	10.66	0.87	4.3	2.5	Sb- II:	D4	864*	V	
4707	7971	12 48.4	+51 10	CVn	13.97			2.3	2.2	S+ IV-V:		570		D150, 1Z 43
4697	A300	12 48.6	-5 48	Vir	10.2	9.3	0.93	6.0	3.8	E4		1170		
4704	7972	12 48.8	+41 55	CVn	14.8 p			1.2	1.1	SBb+p		8196		
4696		12 48.8	-41 19	Cen	11.75	10.71	1.04	3.5	3.2	E1p	4S	2690	Ce	
4699		12 49.0	-8 40	Vir	10.45	9.57	0.88	3.5	2.7	Sa	4VS	1359*		
4700		12 49.1	-11 25	Vir	12.37			3.0	0.7	S:	3	1248		
4701	7975	12 49.2	+3 23	Vir	12.77	12.35	0.42	3.0	2.5	Sc	3VS	624		
4703		12 49.3	-9 07	Vir	14.5:p			2.8	0.7	Sb				
4705		12 49.4	-5 12	Vir	14: p			3.0	1.2	S(B)b+:				
		12 49.4	-10 07	Vir	12.08			4.0	3.2	S- IV-V:		1164		
4710	7980	12 49.6	+15 10	Com	11.80	10.97	0.83	5.1	1.4	S0:	3S	1076	V	
4712	7977	12 49.6	+25 28	Com	13.59	13.01	0.58	2.6	1.3	S IV	3S	4453		
4708		12 49.7	-11 06	Vir	14: p			1.6	1.2	Sb-p				
4715	7986	12 49.8	+27 49	Com	15.4 p			2.0	1.7	S0		6923	C?	
4706		12 49.9	-41 17	Cen	13.8	12.7	1.06	2.2	0.9	S0	4	3444*	Ce	
4713	7985	12 50.0	+5 19	Vir	12.21	11.79	0.42	2.8	1.9	Sc III	3S	570		
4719	7987	12 50.1	+33 09	CVn	14.2 p			1.8	1.5	SBb	*	7130		Mrk 446
4750	7994	12 50.1	+72 52	Dra	11.87			2.3	2.1	Sap	4VS	1698		
4709		12 50.1	-41 23	Cen				2.7	2.5	E1	4	4390*	Ce	
4725	7989	12 50.4	+25 30	Com	9.95	9.21	0.74	11.0	7.9	S(B)b I	5VS	1131		
4718		12 50.5	-5 17	Vir	15: p			2.1	0.8	SBb:				
4736	7996	12 50.9	+41 07	CVn	8.92	8.17	0.75	11.0	9.1	Sb-p II:	5	329		M94
4724		12 50.9	-14 20	Crv	15: p			1.5	1.5	E5:				
4731		12 51.0	-6 24	Vir	11.30			6.5	3.4	SBcp II-III:		1351*		
4735		12 51.0	+28 56	Com	15: p			0.7	0.4			6684		
4727		12 51.0	-14 20	Crv	13: p			1.6	1.4	SBb+:		7455		
4733	7997	12 51.1	+10 55	Vir	12.63			2.3	2.1	S0:		963	V	
4738	7999	12 51.1	+28 48	Com	14.9 p			2.3	0.4	Sc:		4607		
4734	7998	12 51.2	+4 52	Vir	13.85			1.2	1.0	Sc:		7453		
4749	8006	12 51.2	+71 38	Dra	14.2 p			1.9	0.5	Sb:				
4739		12 51.6	-8 25	Vir	14: p			1.7	1.5	E1		3595		
4747	8005	12 51.8	+25 47	Com	13.01	12.38	0.63	3.6	1.4	P		1192		Arp 159
4742	A303	12 51.8	-10 27	Vir	12.1	11.1	1.01	2.3	1.5	E3		1168		
4746	8007	12 51.9	+12 05	Vir	13.34			2.5	0.7	Sb:		1382	V	
4754	8010	12 52.3	+11 19	Vir	11.51	10.56	0.95	4.7	2.6	SB0	D4	1393	V	
4743		12 52.3	-41 24	Cen								2781	Ce	
4753	8009	12 52.4	-1 12	Vir	10.85	9.90	0.95	5.4	2.9	P	3ES	1137		
4744		12 52.4	-41 04	Cen						SBa		3125	Ce	
4758	8014	12 52.7	+15 51	Com	13.67			3.2	0.9	SBp		1196	V?	

NGC	UGC	α 2000	δ 2000	Const	B	V	B−V	Dim		Type	Nucl	RV	Memb	Notes
4757		12^{h}52^{m}8	−10°19′	Vir	15: p			1.8	0.5	S0				
4762	8016	12 52.9	+11 14	Vir	11.12	10.22	0.90	8.7	1.6	SB0	4VS	878	V	
4756		12 52.9	−15 25	Crv	13.33			2.0	1.7	E2		3995		
4766		12 53.1	−10 23	Vir	15: p			1.6	0.4	S0				
4760		12 53.1	−10 30	Vir	13.04			1.8	1.8	E2		4246		
4765	8018	12 53.2	+4 28	Vir	13.13			1.4	1.1	S:		687		
4774		12 53.2	+36 49	CVn	14.85	14.35	0.50	0.6	0.4	Ring:		8478		1Z 45
4771	8020	12 53.4	+1 16	Vir	12.7 H			4.0	1.0	Sb III	1	1110		
4763		12 53.4	−17 00	Crv	13.3			1.7	1.3	Sa				
4772	8021	12 53.5	+2 10	Vir	12.55			3.3	1.7	Sa	D3	982		
4773		12 53.6	−8 38	Vir	15: p			2.0	1.9	E4p				
4779	8022	12 53.8	+9 44	Vir	12.83			2.3	2.0	SBb+		2753		
4775	A306	12 53.8	−6 37	Vir	11.88			2.2	2.1	Sc III	3VS	1409		
4767		12 53.9	−39 43	Cen	12.58			2.6	1.4	E5		2910	Ce	
4777		12 54.0	−8 47	Vir	14.5:p			2.3	1.3	Sa:				
4789A	8024	12 54.0	+27 09	Com	13.79			2.2	1.8	Ir+ IV-V		380		D154
4780		12 54.1	−8 37	Vir	14: p			2.1	1.4	S(B)c				
4789	8028	12 54.3	+27 04	Com	13.1	12.0	1.07	1.7	1.3	S0		8373	C?	
4781		12 54.4	−10 32	Vir	11.79			3.5	1.8	Sc II-III		744		
4786		12 54.5	−6 52	Vir	12.82			2.0	1.5	E2p		4509		
4793	8033	12 54.6	+28 56	Com	12.3	11.7	0.65	2.9	1.7	Sc III	4VS	2481		
4800	8035	12 54.6	+46 32	CVn	12.3			1.8	1.4	Sb	4	831		
4784		12 54.6	−10 37	Vir	15: p			1.8	0.4	S0				
4782		12 54.60	−12 34.4	Crv	12.75	11.73	1.02	1.5	1.5	E0		3859		VV 201, 3C 278
4783		12 54.61	−12 33.7	Crv	12.80	11.84	0.96	1.7	1.6	E0		4512		VV 201, 3C 278
4810	8034	12 54.8	+2 38	Vir	14.69			0.8	0.5	Irp+		780		VV 313
4809	8034	12 54.8	+2 39	Vir	13.93			1.9	0.8	Irp+		839		VV 313
IC 3881	8036	12 54.8	+19 11	Com	13.37			3.9	1.0	SBc		893		
4790		12 54.9	−10 15	Vir	12.58			1.8	1.3	Sc III:	3	1204		
4795	8037	12 55.0	+8 04	Vir	13.2 H			1.7	1.5	S		2788		
4798	8038	12 55.0	+27 25	Com	14.2	13.1	1.06	1.3	1.0	S0:		7676	C	
	8041	12 55.2	+0 07	Vir	12.44			3.2	2.1	SBd	3VS	1218		
4794		12 55.2	−12 37	Crv	14: p			2.3	1.1	SBa	4S			
4814	8051	12 55.4	+58 21	UMa	12.8			3.2	2.5	Sbp II:	4S	2662		
4807	8049	12 55.5	+27 31	Com	14.4 p			1.2	1.0	S(B)0p		6875	C	
4807A	8049	12 55.5	+27 32	Com				0.5	0.2	E5:		7148	C	
	8053	12 55.8	+4 01	Vir	14.11			1.7	1.3	S(B)dm		621		
4808	8054	12 55.8	+4 18	Vir	12.39			2.7	1.3	Sc III	3VS	679		
4802		12 55.8	−12 03	Crv	12: p					S0:		858		NGC 4804
4816	8057	12 56.2	+27 45	Com	14.8 p			1.6	1.3	S0:		6883	C	
	8058	12 56.2	+56 52	UMa	14.3	13.6	0.72	1.7	1.2	Scp	*	12556*		Mrk 231, 7Z 490
4821		12 56.5	+26 57	Com	15: p			0.7	0.4	E4		6976	C	
4819	8060	12 56.5	+26 59	Com	14.1 p			1.2	1.0	S(B)a:		6698	C	
IC 3896		12 56.6	−50 19	Cen	12.68			2.2	2.0	E1		2031		
4826	8062	12 56.7	+21 41	Com	9.35	8.51	0.84	9.3	5.4	Sb−	5	377		M64, Black-eye galaxy
4827	8065	12 56.7	+27 11	Com	13.9	13.0	0.95	1.7	1.5	S0		7653	C	
4818		12 56.8	−8 31	Vir	11.70			4.5	1.7	S(B)		1012		
4837	8068	12 56.9	+48 18	CVn	14.4 p			1.4	0.7	P		8833		1Z 46
4825		12 57.2	−13 40	Vir	13.1	12.1	1.04	2.0	1.5	E2p	D3	4292		
4839	8070	12 57.4	+27 30	Com	13.6 p			4.2	2.1	S0		7451	C	
4840		12 57.5	+27 37	Com	15: p			0.6	0.6	E1:		6065*	C	
4841	8072	12 57.5	+28 29	Com	12.5	11.5	1.03	1.9	1.7	S0p,E0p		6560	C	UGC 8073
4842		12 57.6	+27 29	Com	15: p			0.6	0.4	E0,E3:		7517	C	
	8074	12 57.7	+2 42	Vir	14.43			1.7	1.6	Ir+ V		811		D156
4845	8078	12 58.0	+1 35	Vir	12.10			5.0	1.6	Sb III	D3S	1124		
4848	8082	12 58.1	+28 15	Com	14.2 p			1.8	0.6	Sa:		7218	C	
4835		12 58.1	−46 15	Cen	12.49			3.4	1.0	S(B)b+: II	4	1951		
4849	8086	12 58.2	+26 24	Com	14.5 p			2.2	1.7	S0:		5932*		IC 3935
	8084	12 58.3	+2 48	Vir	14.04			1.7	1.6	SB+ IV-V		2670		D158
4850		12 58.4	+27 58	Com	15.5:p			1.6	1.6	S0		5992	C	
4853	8092	12 58.6	+27 36	Com	14.0	13.1	0.89	1.2	1.0	S0:		7556	C	2Z 67
	8091	12 58.7	+14 13	Vir	14.50			1.2	1.1	Ir+ V		171	V?	D155, GR 8
4854		12 58.8	+27 40	Com	15: p					SB0		8084	C	
IC 3946		12 58.8	+27 49	Com	15.0	13.9	1.07	1.0	0.5	S0		6108	C	
4859	8097	12 59.0	+26 49	Com	14.8 p			1.7	0.8	Sa		7058	C	
4858		12 59.0	+28 07	Com	15.5:p			0.4	0.3	SBb		9395	C	
4861	8098	12 59.0	+34 52	CVn	12.8	12.2	0.58	4.1	1.6	Ir+ IV-V		869		Mrk 59, Arp 266
4860		12 59.1	+28 07	Com	14.45	13.45	1.0	1.0	0.9	E2:		7871	C	
4868	8099	12 59.1	+37 19	CVn	13.1 H			1.7	1.6	Sb+ II:		4779		
4864		12 59.2	+27 59	Com	14.5	13.6	0.94	0.6	0.4	E2		6827	C	
4867		12 59.3	+27 58	Com	15.5:p			1.1	1.1	E3		4823	C	
4865	8100	12 59.3	+28 05	Com	14.25	13.28	0.97	1.4	0.8	E6		4652	C	
4856	A313	12 59.3	−15 02	Vir	11.4	10.4	0.97	4.6	1.6	S(B)a	D3S	1088		
4869		12 59.4	+27 55	Com	14.55	13.53	1.02	1.1	1.1	E3		6711	C	
4866	8102	12 59.5	+14 10	Vir	11.84	10.99	0.85	6.5	1.5	Sb− III:	5VS	1860	V?	

NGC	UGC	α 2000	δ 2000	Const	B	V	B−V	Dim		Type	Nucl	RV	Memb	Notes
4871		12ʰ59ᵐ.5	+27°57′	Com	15: p			0′.5	0′.4	S0		7121	C	
4875		12 59.6	+27 54	Com	15.5:p					S0		7905	C	
4872		12 59.6	+27 57	Com	14.7	13.7	1.00	1.0	1.0	SB0		7032	C	
4874	8103	12 59.6	+27 58	Com	12.9	11.9	1.01	2.7	2.7	E0		7184	C	
4876		12 59.7	+27 55	Com	15: p			0.4	0.4	E5		7000	C	
4892	8108	13 00.0	+26 54	Com	14.7 p			1.6	0.4	S		5902	C	
4881	8106	13 00.0	+28 15	Com	14.55	13.54	1.01	1.0	1.0	S0		6701	C	
4886		13 00.06	+27 59.1	Com	14.85	13.91	0.94	0.8	0.8	E0		6223	C	NGC 4882
4889	8110	13 00.13	+27 58.5	Com	12.45	11.40	1.05	3.0	2.1	E4		6476	C	NGC 4884
4880	8109	13 00.2	+12 29	Vir	12.42			3.3	2.5	Sa		1499	V?	
4895A		13 00.2	+28 10	Com	15.5:p					E5		6768	C	
4898		13 00.3	+27 57	Com	14.5:p					Ep		6814*	C	
4894		13 00.3	+27 58	Com	15.5:p					S0		4596	C	
4895	8113	13 00.3	+28 12	Com	13.8	12.8	1.01	2.3	0.9	S0p		8416	C	
4877		13 00.4	−15 17	Vir	13: p			2.7	1.3	Sb−:	D3			
4896	8117	13 00.5	+28 21	Com	14.65	13.69	0.96	1.3	0.7	S0p		5831	C	
4900	8116	13 00.6	+2 30	Vir	12.06	11.46	0.60	2.3	2.2	Sc III	4S	945		
4887		13 00.6	−14 40	Vir	14: p			1.6	0.8	S0p				
4914	8125	13 00.7	+37 19	CVn	12.3 H			3.6	2.2	E2p		4827		
IC 4045		13 00.8	+28 05	Com	14.9	13.9	1.01	0.7	0.5	E4		6537	C	
4907		13 00.8	+28 09	Com	14.2	13.4	0.83	1.4	1.3	SBb		5878	C	
4911	8128	13 00.9	+27 47	Com	13.6	12.8	0.85	1.3	1.2	S(B)b+		8014	C	
IC 4051	8129	13 00.9	+28 00	Com	14.40	13.42	0.98	1.3	1.1	E0		4941	C	
4908		13 00.9	+28 02	Com	15: p			1.1	1.0	E5		8848	C	
4891		13 00.9	−13 26	Vir	12.6			2.8	2.5	SBb+:		2475		
4899		13 00.9	−13 57	Vir	12.39			2.7	1.6	Sp	3S	2497		
4904	8121	13 01.0	−0 02	Vir	12.7	12.1	0.62	2.3	1.6	S(B)c		1182		
4902	A315	13 01.0	−14 31	Vir	11.90	11.16	0.74	3.0	2.8	SBb I	3VS	2564		
4919	8133	13 01.3	+27 48	Com	14.9 p			1.4	0.8	S0:		7119	C	
4921	8134	13 01.4	+27 53	Com	13.00	12.11	0.89	2.7	2.4	SBb−		5468	C	
4922	8135	13 01.4	+29 19	Com	14.2 p			2.1	1.3	Irp−		7372	C	
4915		13 01.5	−4 33	Vir	12.80	11.91	0.89	1.7	1.4	Sa		3028		
4923		13 01.5	+27 51	Com	14.55	13.58	0.97	1.3	1.2	S0:		5467	C	
4926	8142	13 01.9	+27 37	Com	14.00	12.91	1.09	1.4	1.3	S0		7759*	C	
4927		13 02.0	+28 00	Com	15: p					S0:		7593	C	
4926A		13 02.1	+27 39	Com	15: p			0.8	0.6	S0p		7184	C	
	8146	13 02.1	+58 42	UMa	13.91			3.7	0.6	Sc		805		
4924		13 02.2	−14 58	Vir	14: p			0.7	0.7	S(B)a:				
4929		13 02.7	+28 03	Com	15: p			1.5	1.5	E1:		6349	C	
4928		13 03.0	−8 05	Vir	13.4 H			1.3	1.0	S+ IV	4S	1544		
4931	8154	13 03.0	+28 02	Com	14.5	13.5	1.04	2.0	0.8	S0		5860	C	
	8153	13 03.1	+4 00	Vir	14.00			1.9	1.9	Sc		2779		
4944	8167	13 03.8	+28 11	Com	13.3 p			1.9	0.7	Sa:		7005	C	
4933B		13 03.90	−11 30.4	Vir	15: p			0.8	0.7	Ep	4	3158		Arp 176
4933A		13 03.94	−11 29.9	Vir	13.18			2.5	1.5	Ep	3	3040		Arp 176
4941	A321	13 04.2	−5 33	Vir	11.9	11.1	0.85	3.7	2.1	Sbp	4S	594*		
4939		13 04.2	−10 20	Vir	11.40			5.8	3.2	Sb+	4VS	2952*		
4942		13 04.3	−7 39	Vir	14: p			1.8	1.3	S(B)d:				
4936		13 04.3	−30 32	Cen	12.40	11.32	1.08	1.9	1.9	E3	D3	3065		
4948		13 04.9	−7 57	Vir	14: p			1.8	0.7	SBd:		1144		
4952	8175	13 05.0	+29 08	Com	13.6 p			1.6	1.0	E2		5882	C	
4951		13 05.1	−6 30	Vir	12.37			3.3	1.4	S(B)c	3S	1048		
4948A		13 05.1	−8 10	Vir	15: p			1.4	1.2	S(B)+ IV−V		1419		D162
4957	8178	13 05.2	+27 34	Com	14.0	12.9	1.06	1.3	1.1	E3	D3	7016	C	
4947		13 05.4	−35 20	Cen	12.38			2.8	1.7	SBb−	3S	2194		
4945		13 05.4	−49 28	Cen	9.47			20.0	4.4	SBc:		356		
4958		13 05.8	−8 01	Vir	11.40	10.53	0.87	4.1	1.4	E6	4	1381		
4961	8185	13 05.8	+27 44	Com	13.9	13.5	0.45	1.7	1.3	Sc III:	3S	2542		
IC 4182	8188	13 05.8	+37 36	CVn	12.63			6.7	5.9	Sm		380		
4966	8194	13 06.3	+29 04	Com	13.9 p			1.2	0.6	S		7119	C	
	8201	13 06.3	+67 42	Dra	12.68			3.2	2.2	Ir+ IV−V		201		7Z 499, D165
4971		13 06.9	+28 32	Com	15: p			1.3	1.3			6415	C	
4965		13 07.2	−28 14	Hya	14: p			2.6	2.3	Sdm				
4976		13 08.6	−49 30	Cen	11.17	10.20	0.97	4.3	2.6	E4p	3	1133		
4981		13 08.8	−6 47	Vir	11.98			2.8	2.2	Sc II−III	4S	1549		
4984		13 09.0	−15 31	Vir	11.79			2.8	2.2	S(B)a	5	1101		
4980		13 09.2	−28 38	Hya	15: p			2.2	1.2	S(B)a:				
4999	8236	13 09.6	+1 40	Vir	12.6			2.6	2.2	SBb	4S	3011		
4995	A329	13 09.7	−7 50	Vir	11.90	11.03	0.87	2.5	1.7	Sb	4	1577		
5000	8241	13 09.8	+28 54	Com	14.0 p			1.9	1.7	SBb+		5686	C	
	8246	13 10.0	+34 11	CVn	14.22			3.3	0.8	SBc		855		
5005	8256	13 10.9	+37 03	CVn	10.64	9.82	0.82	5.4	2.7	Sb− II	5	1069		
5004A	8259	13 11.1	+29 34	Com	15.3 p			1.6	0.9	SBb−		7211	C	
5004	8260	13 11.1	+29 38	Com	14.3 p			1.9	1.5	S0		7005	C	
5014	8271	13 11.5	+36 17	CVn	13.49			1.7	0.7	Sa:		1156		Mrk 449

NGC	UGC	α 2000	δ 2000	Const	B	V	B-V	Dim		Type			Nucl	RV	Memb	Notes
5012	8270	13^{h}11^{m}6	+22°55'	Com	12.99			2.9	1.8	Sb	II		4S	2620		
5016	8279	13 12.1	+24 06	Com	13.73			1.9	1.4	Sb	II-III		3S	2614		
IC 4213	8280	13 12.1	+35 40	CVn	13.97			2.6	0.6	Sc				865		
5023	8286	13 12.2	+44 02	CVn	12.25			6.5	1.0	Sc:				484		
		13 12.3	+26 41	Com	14.37			0.9	0.5	S				898		
5015		13 12.4	-4 19	Vir	13: p			1.9	1.5							
	8285	13 12.5	+7 11	Vir	14.85			2.0	0.6	Sm				825		
	8290	13 12.6	+22 50	Com	14.54			1.5	1.2	Smp				2599		
5017		13 12.9	-16 46	Vir	13.07			1.7	1.4	E2				2384		
5011		13 12.9	-43 06	Cen	12.58			2.0	2.0	E1			4	2838		
5018	A335	13 13.0	-19 31	Vir	11.8	10.8	1.04	2.6	2.1	Sa			4	2729		
	8303	13 13.3	+36 13	CVn	13.60	13.22	0.38	2.5	2.2	Ir+	IV-V			1009		Holm. VIII, D166
5033	8307	13 13.4	+36 36	CVn	10.60	10.06	0.54	10.5	5.6	Sb+	I-II:		D4S	961		
5022		13 13.5	-19 33	Vir	13: p			2.5	0.4	Sb			1			
	8313	13 13.8	+42 12	CVn	14.63			1.9	0.6	SBc				699		
5028		13 13.8	-13 03	Vir	14: p			1.7	1.0	E6						
5030		13 13.9	-16 30	Vir	14: p			1.9	1.2	SB0:						
5026		13 14.2	-42 58	Cen	12.7 p			2.3	1.7	SBa			3S	3607		
	8320	13 14.4	+45 55	CVn	12.63			3.8	1.5	Ir+	IV-V			291		D168
	8323	13 14.8	+34 53	CVn	14.62			1.0	0.7	Irp+				888		Mrk 450
5035		13 14.8	-16 30	Vir	14: p			1.6	1.2	S(B)0			4			
5037		13 15.0	-16 35	Vir	12.88			2.5	0.9	Sb+	III		3	1730		
5044	A341	13 15.4	-16 23	Vir	12.05	11.03	1.02	2.6	2.6	E0				2548		
	8331	13 15.5	+47 30	CVn	14.52			2.7	1.0	Ir+	IV-V:			358		D169
5042		13 15.5	-23 59	Hya	13: p			4.2	2.5	S(B)c						
5055	8334	13 15.8	+42 02	CVn	9.30	8.57	0.73	12.3	7.6	Sb+	II		4VS	587		M63
5046		13 15.8	-16 20	Vir	15: p			1.2	1.0	E2:						
5047		13 15.8	-16 31	Vir	13: p			3.0	0.8	S0			3S			
5049	A343	13 16.0	-16 24	Vir	13.6	12.9	0.69	2.1	0.7	S0				2588		
		13 16.3	+7 03	Vir	14.5	13.4	1.06	0.7	0.5	E2p						
5054	A344	13 17.0	-16 38	Vir	11.29			5.0	3.1	Sb	II:		4S	1587		
IC 875	8355	13 17.1	+57 32	UMa	13.49			1.9	1.3	S0				2834		Mrk 249
5061		13 18.1	-26 50	Hya	11.69			2.6	2.3	E2			D4	1775		
5074		13 18.4	+31 28	CVn	14.4	14.0	0.40	1.0	1.0	P:				5720		
	8365	13 18.7	+41 56	CVn	14.16			2.5	1.5	SB+	IV-V			1294		D172
5068	A345	13 18.9	-21 02	Vir	10.81			6.9	6.3	S(B)c	III-IV		2VS	402		
5064		13 19.0	-47 55	Cen	12.79			2.8	1.2	Sb:			3	2752		
5073		13 19.4	-14 52	Vir	13: p			3.5	0.7	SBc:				2567		
5077	A347	13 19.5	-12 39	Vir	12.60	11.54	1.06	2.0	1.6	E3			D4	2683		
5089	8371	13 19.6	+30 15	Com	13.79			2.1	1.0	Scp				2186		
5079		13 19.6	-12 42	Vir	12: p			1.7	1.0	SBb+p			2			
5078		13 19.8	-27 24	Hya	12: p			3.2	1.7	Sa:			D4			
5098		13 20.2	+33 09	CVn	15: p			0.9	0.9	P:				11414		
5088		13 20.3	-12 34	Vir	12.98			2.7	0.9	Sc	II:		4	1324		
5084		13 20.3	-21 50	Vir	11.98			4.8	1.3	S0			4S	1569		
5085	A349	13 20.3	-24 26	Hya	11.90			3.4	3.0	Sb	II-III		3VS	1780		
5087	A350	13 20.4	-20 37	Vir	12.0	11.0	1.03	2.3	1.5	E4			4VS	1666		
	8385	13 20.7	+9 47	Vir	13.91			2.5	1.2	S(B)+	IV-V:			1079		D173
5109	8393	13 20.9	+57 39	UMa	13.6 p			1.9	0.6	S				2270		
5090		13 21.1	-43 44	Cen	12.64			2.6	2.5	E2				2932		
5091		13 21.2	-43 44	Cen				2.2	0.4	Sp			2	3480		
5107	8396	13 21.4	+38 32	CVn	13.66			1.9	0.7	SBc:				1010		
5105		13 21.8	-13 13	Vir	13: p			2.6	2.1							
5101	A351	13 21.8	-27 26	Hya	11.70			5.5	4.9	SBa			D4	1679		
5112	8403	13 21.9	+38 44	CVn	11.93			3.9	2.9	Sc	II:			1030		
5102		13 22.0	-36 38	Cen	10.35	9.65	0.70	9.3	3.5	S0			4	247		
5116	8410	13 22.9	+26 59	Com	13.64			2.2	0.9	Sb-			3VS	2854		
5144	8420	13 22.9	+70 31	UMi	13.2 p			1.4	1.0	Scp				3187		Mrk 256, 7Z 511
	8409	13 23.0	+23 18	Com	14.39			2.5	1.2	Sdm				2817		
5117	8411	13 23.0	+28 19	CVn	13.86			2.3	1.0	SBc				2461		
5127	8419	13 23.8	+31 34	CVn	13.9 p			2.6	2.0	E2p				4869		
5121		13 24.8	-37 41	Cen	12.19			2.3	2.0	Sa			3	1324		
5141	8433	13 24.9	+36 23	CVn	13.8	12.8	0.98	1.7	1.3	S0				5283		
5142	8435	13 25.0	+36 24	CVn	14.2	13.3	0.95	1.1	0.8	S0				5245		Mrk 452
5145	8439	13 25.2	+43 15	CVn	12.98			2.3	1.8	S				1313		
5134		13 25.3	-21 08	Vir	12.28			2.8	1.7	Sb+	III		4VS	1532		
	8441	13 25.4	+57 49	UMa	13.38			3.0	2.1	Ir+				1661		D175
5128		13 25.5	-43 01	Cen	7.96	6.98	0.98	18.2	14.5	S0p				323*		Cen A, Arp 153
5135		13 25.7	-29 50	Hya	12.9 H			2.4	1.0	SBb-	II:		4	3946		
5147	8443	13 26.3	+2 06	Vir	12.29	11.79	0.50	1.8	1.5	Sc	II-III		1	1040		
5164	8458	13 27.2	+55 29	UMa	14.6 p			1.2	1.1	SBb				4907		Mrk 257
5140		13 27.2	-33 52	Cen	12.80	11.81	0.99			S(B)0:				3531		
5150		13 27.6	-29 34	Hya	13.3 H			1.4	1.4	E0				4204		
5169	8465	13 28.2	+46 40	CVn	14.47			2.4	1.0	SBb:				2585		
5173	8468	13 28.4	+46 36	CVn	13.48			1.3	1.3	E0:				2506		

NGC	UGC	α 2000	δ 2000	Const	B	V	B-V	Dim		Type	Nucl	RV	Memb	Notes
5156		13ʰ28ᵐ7	-48°55'	Cen	12.59			2.1	2.0	SBb-	3S	2722		
5161	A359	13 29.2	-33 10	Cen	12.10			5.4	2.3	S IV:		2018		
5172	8477	13 29.3	+17 03	Com	12.60	11.87	0.73	3.3	1.9	Sb+ I-II	3S	4100		
5204	8490	13 29.6	+58 25	UMa	11.75	11.26	0.49	4.8	3.0	Ir IV	1	351		
	8489	13 29.7	+45 24	CVn	14.85			2.3	0.8	S(B)+ IV-V		1401		D176
5170	A360	13 29.8	-17 58	Vir	11.79			8.1	1.3	Sb	4VS	1347		
5194	8493	13 29.9	+47 12	CVn	8.98	8.38	0.60	11.0	7.8	Sc I		565		M51, Whirlpool galaxy
5205	8501	13 29.9	+62 31	UMa	12.92			3.4	2.0	S		1940		
5195	8494	13 30.0	+47 16	CVn	10.53	9.63	0.90	5.4	4.3	P		658		Part of M51
5198	8499	13 30.2	+46 40	CVn	12.73			2.1	1.9	E2	D4VS	2592		1Z 59
	8508	13 30.8	+54 55	UMa	14.54			1.4	0.9	Ir+		202		1Z 60
	8507	13 31.0	+19 27	Com	13.89			1.4	0.8	P		1014		VV 88
5188		13 31.3	-34 47	Cen	12.7 H					SBb+p II-III		2129		
5193		13 31.9	-33 14	Cen	12.88			1.8	1.6	E2		3451		
5216	8528	13 32.1	+62 42	UMa	14.0 p			3.0	2.1	E0p				Keenan's system
5218	8529	13 32.2	+62 46	UMa	13.1	12.3	0.79	2.1	1.3	SBbp		2884		Keenan's system
5229	8550	13 34.1	+47 55	CVn	13.90			3.3	0.7	SBd:		476		
5223	8553	13 34.4	+34 42	CVn	14.4 p			1.8	1.5	E1		7224		
5238	8565	13 34.7	+51 37	CVn	13.35			2.0	1.7	S(B)dm		368		1Z 64
5227	8566	13 35.3	+1 25	Vir	14.6 p			1.9	1.7	SBb		5164		
5230	8573	13 35.5	+13 40	Vir	12.8 H			2.2	2.0	Sc I-II	3VS	6840*		
	8588	13 35.8	+45 56	CVn	14.33			1.4	1.4	S+ V		1551		D178
5243	8592	13 36.2	+38 21	CVn	14.0 p			1.7	0.6	S				
	8597	13 36.3	+46 13	CVn	14.27			1.9	1.5	SB+ V		2540		D177
IC 4296		13 36.6	-33 58	Cen	11.58	10.57	1.01			E0		3437*		
IC 4299		13 36.8	-34 04	Cen	13.70	12.67	1.03			S(B)a:		3836		
	8611	13 36.9	+44 54	CVn	14.45			1.7	1.6	S(B)d		2741		
5236		13 37.0	-29 52	Hya	8.2			11.2	10.2	Sc I-II	5	337		M83
	8614	13 37.4	+7 39	Boo	13.30			3.4	1.7	Ir+		1004		D179
5248	8616	13 37.5	+8 53	Boo	10.80	10.17	0.63	6.5	4.9	Sc I	5	1102		
5247	A368	13 38.1	-17 53	Vir	11.1	10.5	0.59	5.4	4.7	Sb+ I-II	3	1511		
5256	8632	13 38.3	+48 17	UMa	14.1 p			1.4	1.3	P		8371		Mrk 266, 1Z 67
	8639	13 39.0	+51 26	UMa	14.45			1.4	1.1	Ir+		1834		
		13 39.6	+43 03	CVn	15.0	14.5	0.48	0.6	0.4			3689		Mrk 267
5254		13 39.6	-11 30	Vir	13: p			3.2	1.8	Sc				
5257	8641	13 39.9	+0 50	Vir	13.7 p			1.9	1.1	S(B)bp	3VS	6791		VV 55, Arp 240
5253	A369	13 39.9	-31 39	Cen	10.99	10.55	0.44	4.0	1.7	E5		209		
5258	8645	13 40.0	+0 50	Vir	13.8 p			1.8	1.3	Sbp		6615		VV 55, Arp 240
	8651	13 40.0	+40 45	CVn	14.05			2.7	1.3	Ir+ IV-V:		287		D181
5260		13 40.4	-23 51	Hya	13: p			1.9	1.9	SBb+		6378		
5266A		13 40.4	-48 21	Cen	12.9 p			3.2	2.3	Sc:				
	8658	13 40.7	+54 20	UMa	13.35			2.8	1.8	SBc		2155		Holm. V
5283	8672	13 41.1	+67 40	Dra	14.13			1.3	1.2	S0:	*	2875		Mrk 270
5264		13 41.6	-29 55	Hya	13: p			1.8	1.2	Ir+ IV-V		307		D242
5278	8677	13 41.66	+55 40.2	UMa	13.6 p			1.4	1.0	Sbp	4VS	7710		VV 19, Arp 239
5279	8678	13 41.72	+55 40.4	UMa	13.6 p			1.3	0.9	SBap	4VS	7744		VV 19, Arp 239
5273	8675	13 42.1	+35 39	CVn	12.43	11.57	0.86	3.1	2.7	E1p	D3S	1090		
5276	8680	13 42.4	+35 38	CVn	14.6 p			1.1	0.6	S(B)b		5346		
	8683	13 42.6	+39 40	CVn	14.28			2.0	1.9	Ir- V		748		D182
5266		13 43.0	-48 11	Cen	12.38			3.2	2.1	S0:	3S	2865		
	8696	13 44.7	+55 53	UMa	14.8	13.9	0.88	1.3	0.3	P	*	11533		Mrk 273, 1Z 71
5289	8699	13 45.1	+41 30	CVn	13.5 p			1.9	0.7	S(B)b-:		2544		
5290	8700	13 45.3	+41 43	CVn	12.61			3.7	1.1	Sb:		2673		
5296		13 46.3	+43 51	CVn	14.70			1.2	0.7	S0:	3S	2384		
5297	8709	13 46.4	+43 52	CVn	12.17			5.6	1.4	Sb+ I	3VS	2507		
5301	8711	13 46.4	+46 06	CVn	12.62			4.4	1.1	Sb II-III:	2VS	1673		
5293	8710	13 46.9	+16 16	Boo	14.3 p			1.9	1.6	Sc	3VS	5781		
5308	8722	13 47.0	+60 58	UMa	12.20	11.31	0.89	3.5	0.8	S0	4	2132		
5291		13 47.4	-30 25	Cen				1.4	1.0	Ep		4151		
5292		13 47.7	-30 57	Cen	14: p			1.6	1.6			4266		
5303	8725	13 47.8	+38 18	CVn	12.9 p			1.1	0.6	P		1425		
5298		13 48.2	-30 27	Cen	14: p			1.4	0.8	SBb	4			
5300	8727	13 48.3	+3 57	Vir	12.49			3.9	2.7	Sc II		1123		
	8733	13 48.6	+43 24	CVn	13.60			2.5	1.5	SBc		2447		
5302		13 48.8	-30 31	Cen	13.20	12.20	1.00	1.7	1.3	SB0:	4	3115		
IC 4329		13 49.1	-30 18	Cen	12.55	11.54	1.01	3.2	1.6	E3	3	4243		
5322	8745	13 49.3	+60 12	UMa	10.85	9.97	0.88	5.5	3.9	E2	D3	2061		
IC 4329A		13 49.3	-30 19	Cen	14.15	13.10	1.05	1.3	0.4	Sa:	*	4640*		
5313	8744	13 49.7	+39 59	CVn	12.59			1.9	1.2	Sb II:		2696		
5304		13 50.0	-30 35	Cen	15: p			1.0	0.8	E4:	3	3509		
5320	8749	13 50.3	+41 22	CVn	12.68			3.5	1.9	S(B)c		2709		
5326	8764	13 50.8	+39 34	CVn	12.90			2.5	1.3	Sb- III		2653		
5324		13 52.1	-6 03	Vir	12.28			2.4	2.3	Sc II-III	D3	2954		
5334	8790	13 52.9	-1 07	Vir	12.49			4.4	3.3	S- IV-V		1311		IC 4338
5328		13 52.9	-28 29	Hya	12.6	11.8	0.85	1.7	1.4	E2		4610		

NGC	UGC	α 2000	δ 2000	Const	B	V	B−V	Dim		Type	Nucl	RV	Memb	Notes
5330		13ʰ53ᵐ0	−28°28′	Hya	15: p			′	′	E		4704		
5347	8805	13 53.3	+33 29	CVn	13.40	12.64	0.76	1.9	1.5	S(B)b+ III		2364		
5350	8810	13 53.4	+40 22	CVn	12.2	11.4	0.8	3.2	2.6	Sb I	4VS	2410		
5351	8809	13 53.5	+37 55	CVn	13.00	12.11	0.89	3.1	1.8	Sb+ II	3S	2397		
5353	8813	13 53.5	+40 17	CVn	12.05	11.06	0.99	2.8	1.5	E5	3VS	2115		
5354	8814	13 53.5	+40 18	CVn	12.45	11.46	0.99	2.3	2.0	S0		3096		
5339		13 53.8	−7 56	Vir	12: p			2.1	2.1					
5348	8821	13 54.2	+5 14	Vir	13.84			3.6	0.6	SBb+:		1411		
5368	8834	13 54.5	+54 20	UMa	13.8 p			1.1	0.9	S(B)b−:				
5452		13 54.5	+78 13	UMi	13.35			2.3	1.8	S(B)d		2272		
	8837	13 54.8	+53 54	UMa	13.10			4.0	1.4	Ir+ IV-V		284		Holm. IV, D185
5362	8835	13 54.9	+41 19	CVn	13.07			2.4	1.1	Sb		2361		
5356	8831	13 55.0	+5 20	Vir	14.1 p			3.2	1.0	S(B)b+:				
5376	8852	13 55.3	+59 30	UMa	12.89			2.1	1.4	Sa:	3S	2224		
	8839	13 55.4	+17 47	Boo	13.43			2.9	2.1	Ir−		971		D184
5379	8860	13 55.6	+59 45	UMa	14.1 p			2.2	1.1	S(B)b:	3VS	1888		
5360	8838	13 55.7	+4 59	Vir	14.09			2.0	0.9	Ir−		1132		
5371	8846	13 55.7	+40 28	CVn	11.40	10.75	0.65	4.4	3.6	Sb+ I	D3VS	2660		NGC 5390
5357		13 56.0	−30 20	Cen	14: p					E0		4806		
5363	8847	13 56.1	+5 15	Vir	11.18	10.19	0.99	4.2	2.7	Ep	3	1081		
5389	8866	13 56.1	+59 44	UMa	13.2 p			4.1	1.3	S(B)a:	4	1996		
5364	8853	13 56.2	+5 01	Vir	11.05	10.40	0.65	7.1	5.0	Sb+p I	3S	1349		
5377	8863	13 56.3	+47 14	CVn	12.0	11.2	0.82	4.6	2.7	Sap	5S	1950		
5375	8865	13 56.8	+29 10	CVn	12.25			3.5	3.0	SBb−		2444		
5378	8869	13 56.8	+37 48	CVn	13.8 p			2.7	2.3	SBa		3054		
5380	8870	13 56.9	+37 37	CVn	12.8 H			2.1	2.1	Sa		3087		
5383	8875	13 57.1	+41 51	CVn	12.05	11.41	0.64	3.5	3.1	SBb II	5	2354		Mrk 281
5374	8874	13 57.5	+6 06	Vir	13.7 p			1.9	1.7	SBb+:				
	8892	13 57.7	+57 00	UMa	14.22			2.1	1.5	Ir+		1901		
IC 4351	A376	13 57.9	−29 19	Hya	12.20			5.6	1.2	Sb III:	3	2597		
5365		13 57.9	−43 57	Cen	12.48			3.1	2.4	SB0	4S	2232		
5387	8891	13 58.4	+6 04	Vir	14.8 p			1.8	0.4	Sc:				
5395	8900	13 58.6	+37 25	CVn	12.35	11.63	0.72	3.1	1.7	Sb+ I:	3S	3587		VV 48, Arp 84
5394	8898	13 58.6	+37 27	CVn	13.65	12.99	0.66	1.9	1.1	SBbp	5	3496		VV 48, Arp 84
5403	8919	13 59.9	+38 11	CVn	14.9 p			3.2	1.0	SBb:				VV 310
5406	8925	14 00.3	+38 55	CVn	13.0 H			2.1	1.6	Sb+ I-II		5307		
5422	8935	14 00.7	+55 10	UMa	13.01			3.9	0.9	S0	4S	1986		
5430	8937	14 00.8	+59 20	UMa	12.8 H			2.4	1.5	Sb	4S	3605		Mrk 799
		14 00.8	−45 24	Cen	12.8 H			2.3	1.8	SBc:				
		14 01.4	+36 49	CVn	14.86			0.8	0.7	P		2789		Mrk 465
5398	A379	14 01.4	−33 04	Cen	12.77			2.9	1.4	SBdmp		1053		
5443	8958	14 02.2	+55 49	UMa	13.2 p			2.8	1.2	SBb:				
5433	8954	14 02.5	+32 31	CVn	14.0 p			1.7	0.5	Sm				
5448	8969	14 02.8	+49 10	UMa	12.2			4.2	2.0	Sb+ II-III	4S	2100		
5440	8963	14 03.0	+34 46	CVn	13.4 p			3.3	1.4	Sa				
5457	8981	14 03.2	+54 21	UMa	8.18	7.72	0.46	26.9	26.3	Sc I		388		M101, Pinwheel galaxy
5427	A380	14 03.4	−6 02	Vir	12.05	11.43	0.62	2.5	2.3	Sc I	4	2483*		VV 21, Arp 271
5426	A381	14 03.4	−6 04	Vir	12.75	12.16	0.59	2.9	1.6	Sc	D3	2296		VV 21, Arp 271
5444	8974	14 03.4	+35 08	CVn	12.5 H			2.7	2.3	E1		4034		
5419		14 03.7	−33 59	Cen	12.4 H					E4		4095		
5473	9011	14 04.7	+54 54	UMa	12.3	11.4	0.87	2.6	1.8	E2	4S	2156		
	8995	14 04.9	+8 49	Boo	14.09			2.3	1.2	S(B)dm		1210		
5474	9013	14 05.0	+53 40	UMa	11.35	10.85	0.50	4.5	4.2	Sc	2VS	416		VV 344
5475	9016	14 05.2	+55 45	UMa	13.4 p			2.2	0.6	Sa:				
5477	9018	14 05.6	+54 28	UMa	14.2	13.8	0.38	1.7	1.4	Ir+ IV-V		441		D186
5480	9026	14 06.4	+50 43	UMa	12.76			1.8	1.3	Sc I-II:	4S	1928		
5468	A384	14 06.6	−5 27	Vir	12.09			2.5	2.5	Sc II	4S	2764		
5481	9029	14 06.7	+50 43	Boo	12.92			1.7	1.4	S0		2239		
5484		14 06.8	+55 02	UMa	15.5:p					E2				
5472		14 06.9	−5 28	Vir	15: p			1.4	0.4	Sb−:	3S	2834		
5464		14 07.1	−30 01	Hya	13.2 H			1.0	0.7	Ir+ III:		2526		
5485	9033	14 07.2	+55 00	UMa	12.40	11.49	0.91	2.6	2.1	Sa:	3S	2136		
5486	9036	14 07.4	+55 06	UMa	13.89			1.7	1.1	Sm: III-IV:		1466		
5490	9058	14 10.0	+17 33	Boo	13.4 p			2.6	2.2	E2				
5490C		14 10.1	+17 37	Boo	15: p			1.1	0.8	SBb+				
	9057	14 10.2	−2 34	Vir	13.80			3.0	1.1	SBdm		1510		
5483		14 10.4	−43 19	Cen	12.01			3.1	2.8	Sc II	1	1628		
5492	9065	14 10.5	+19 37	Boo	13.7 p			1.8	0.5	Sbp				
	9083	14 11.4	+50 13	Boo	14.78			1.0	0.8	Sdm		2044		VV 125
5493	A386	14 11.5	−5 03	Vir	12.35	11.46	0.89	2.0	1.4	S0p	D4	2556		
5496	9079	14 11.6	−1 09	Vir	12.87			4.4	1.0	S- IV:		1471		
5494		14 12.4	−30 39	Cen	12.7			2.2	2.0	Sb II:		2480		
5506		14 13.2	−3 13	Vir	13.2 p			2.9	1.0	Sap		1690		
		14 13.2	−65 20	Cir	11.25	9.85	1.40	3.2	1.2	Sb:		176		Circinus galaxy
5507		14 13.3	−3 09	Vir	13.4 p			2.0	1.1	S(B)0:		2170		

NGC	UGC	α 2000	δ 2000	Const	B	V	B-V	Dim		Type	Nucl	RV	Memb	Notes
5523	9119	14^{h}14^{m}8	+25°19'	Boo	12.86			4.5	1.4	Sb II-III	1	1095		
5529	9127	14 15.6	+36 13	Boo	12.55			5.9	1.0	Sc:		2971		
	9128	14 15.9	+23 03	Boo	14.33			1.7	1.4	Irp+		196		D187
5533	9133	14 16.1	+35 21	Boo	12.65	11.78	0.87	3.2	2.1	S	4	3871		
5532	9137	14 16.9	+10 48	Boo	13.00	11.98	1.02	1.9	1.9	S0		7106		3C 296
5544	9142	14 17.0	+36 34	Boo	13.2 p			1.1	1.1	SBa	5	3292		VV 210, Arp 199
5545	9143	14 17.1	+36 35	Boo	13.2 p			1.3	0.5	Sb+:	4S	3302		VV 210, Arp 199
5534		14 17.7	-7 25	Vir	13.5 H			1.4	0.8	S		2590		
5548	9149	14 18.0	+25 08	Boo	13.15	12.49	0.66	1.9	1.7	Sp	5*	5034		
5557	9161	14 18.4	+36 30	Boo	12.03	11.10	0.93	2.4	2.2	E1	3	3291		
5530		14 18.5	-43 24	Lup	11.91			4.1	2.2	Sb+		174		
5607	9189	14 19.4	+71 35	UMi	13.9 p			1.1	1.0	P		7941		Mrk 286, 7Z 547
	9169	14 19.8	+9 22	Boo	13.67			3.9	0.8	Ir+		1273		
5585	9179	14 19.8	+56 44	UMa	11.37	10.87	0.50	5.5	3.7	S IV	1	462		
5560	9172	14 20.1	+4 00	Vir	13.2	12.4	0.82	3.9	0.9	SBbp	4	1682		Arp 286
5566	9175	14 20.3	+3 56	Vir	11.35	10.49	0.86	6.5	2.4	Sb+ II-III:	5	1489		Arp 286
5569	9176	14 20.5	+3 59	Vir	14.9 p			1.9	1.7	S(B)c:	2VS			Arp 286
5556	A389	14 20.6	-29 15	Hya	12.19			3.1	2.7	S(B)- IV-V:	1	1237		D243
5574	9181	14 20.9	+3 14	Vir	13.2	12.4	0.83	1.6	1.0	SB0:	4S	1685		
5576	9183	14 21.1	+3 16	Vir	11.75	10.87	0.88	3.2	2.2	E2	D3	1497		
5577	9187	14 21.2	+3 26	Vir	13.30			3.4	1.1	Sb+:	3S	1455		
5587	9202	14 22.2	+13 55	Boo	14.0 p			2.9	1.0	Sa				
5584	9201	14 22.4	-0 23	Vir	12.07			3.3	2.6	Sc II		1588		
5596	9208	14 22.5	+37 07	Boo	14.5 p			1.5	1.1	S0		4578		Mrk 470
	9211	14 22.5	+45 22	Boo	14.28			2.2	1.7	Ir+ V		819		D189
5608	9219	14 23.3	+41 46	Boo	13.41			2.7	1.4	Irp+		780		
5600	9220	14 23.8	+14 38	Boo	13.2	12.7	0.48	1.4	1.4	Sc	4	2425		
5592		14 23.9	-28 41	Hya	13.3 H			1.7	1.2	Sb I:		13367		
5614	9226	14 24.1	+34 52	Boo	12.54	11.68	0.86	2.7	2.3	S	4S	3966		VV 77, Arp 178
5613	9228	14 24.1	+34 54	Boo	16.0:p			1.1	1.0	S(B)0	4S			VV 77, Arp 178
5595		14 24.2	-16 43	Lib	12.48			2.0	1.2	Sc	3S	2568		
5597		14 24.5	-16 46	Lib	12.6	12.1	0.52	2.0	1.8	Sb	5VS	2573		
	9240	14 24.7	+44 30	Boo	13.17			1.8	1.7	Ir+ IV:		281		1Z 87, D190
5605		14 25.1	-13 10	Lib	13.2 H			1.8	1.5	Sc I		3273		
	9242	14 25.3	+39 32	Boo	14.03			5.2	0.5	Sc		1551		1Z 88
5631	9261	14 26.6	+56 35	UMa	12.50			2.2	2.1	Sa	D4	2144		
	9249	14 27.0	+8 41	Boo	14.78			2.3	0.3	Sd		1361		
5618	9250	14 27.2	-2 16	Vir	14.8 p			1.8	1.4	SBc		7098		
5619	9255	14 27.3	+4 48	Vir	14.0 p			2.5	1.4	S(B)b				
5633	9271	14 27.5	+46 09	Boo	12.85	12.30	0.55	2.3	1.4	Sb	3S	2451		1Z 89
5630	9270	14 27.6	+41 16	Boo	13.58			2.5	0.7	Sm		2777		
IC 1014	9275	14 28.3	+13 47	Boo	13.30			2.8	1.8	S(B)dm		1311		
5635	9283	14 28.5	+27 25	Boo	13.9 p			2.6	1.3	Sp				
	9291	14 28.6	+39 00	Boo	13.23			2.8	1.5	Sc		3007		
5641	9300	14 29.3	+28 49	Boo	12.9 H			2.7	1.6	S(B)b- I-II:		4538		
5638	9308	14 29.7	+3 14	Vir	12.20	11.30	0.90	2.6	2.3	E1		1654		
5636	9304	14 29.7	+3 16	Vir	14.6 p			1.9	1.4	S(B)0	3VS			
5660	9325	14 29.8	+49 37	Boo	12.3	11.8	0.46	2.8	2.6	Sc II	4S	2483		
	9324	14 29.9	+44 27	Boo	14.28			2.5	1.5	SB+ V		2877		D192
	9310	14 30.0	+3 13	Vir	14.94			2.3	0.7	SBdm		1829		
5653	9318	14 30.2	+31 13	Boo	12.90	12.20	0.7	1.8	1.5	S	4	3642		
5656	9332	14 30.4	+35 19	Boo	12.7 p			2.1	1.7	Sb-				
5667	9344	14 30.4	+59 29	Dra	13.1 p			1.8	1.0	Scp				
5645	9328	14 30.7	+7 17	Vir	12.8	12.3	0.48	2.6	1.7	Sc II-III:		1371		
5673	9347	14 31.5	+49 58	Boo	14.0 p			2.6	0.8	SBc:		2260		
IC 4444		14 31.7	-43 25	Lup	12.29			1.9	1.7	S(B)b+:		1770		
5678	9358	14 32.1	+57 55	Dra	12.06			3.2	1.7	Scp II:	3VS	2391		
5665	9352	14 32.4	+8 05	Boo	12.8 H			2.1	1.5	Sc	3S	2264		Arp 49
5672	9354	14 32.6	+31 40	Boo	14.25	13.64	0.61	1.0	0.7	Sb:		3789		
5669	9353	14 32.7	+9 53	Boo	12.25			4.1	3.2	Sc III		1372		
5643		14 32.7	-44 10	Lup	10.70			4.6	4.1	S(B)c III	4VS	962		
5676	9366	14 32.8	+49 28	Boo	11.6	10.9	0.67	3.9	2.0	Sc II:	3S	2363		
5668	9363	14 33.4	+4 27	Vir	12.15	11.49	0.66	3.3	3.1	Sc II-III	2VS	1562		
5612		14 34.1	-78 24	Aps	12.98			2.0	0.8	S0:		2538		
	9380	14 34.6	+4 16	Vir	14.33			1.9	1.0	Ir+		1692		
	9391	14 34.6	+59 21	Dra	14.44			1.9	1.2	S(B)+ V		2097		D193
5682	9388	14 34.8	+48 40	Boo	14.45	13.98	0.47	2.0	0.6	SBb		2413		
5687	9395	14 34.9	+54 29	Boo	12.6	11.8	0.85	2.6	1.9	Sa:	D3	2282		
5689	9399	14 35.5	+48 45	Boo	12.8	11.9	0.94	3.7	1.2	SBa	D4	2352		
5693	9406	14 36.2	+48 35	Boo	14.5 p			1.9	1.6	SBd		2423		
5707	9428	14 37.5	+51 34	Boo	13.3 p			2.8	0.6	Sb-:				
5690	9416	14 37.7	+2 17	Vir	12.5			3.5	1.2	Sb II-III	1	1729		
5691	9420	14 37.9	-0 24	Vir	12.54			2.0	1.7	S(B)ap		1850		
5714	9431	14 38.2	+46 38	Boo	14.05			3.0	0.5	Sc		2382		
5692	9427	14 38.3	+3 24	Vir	13.32			0.9	0.5	P		1800		

NGC	UGC	α 2000	δ 2000	Const	B	V	B-V	Dim		Type	Nucl	RV	Memb	Notes
	9432	14^h39^m.1	+2°57′	Vir	14.39			1′.4	1′.3	Ir+		1556		
5701	9436	14 39.2	+5 22	Vir	11.8			4.7	4.5	S(B): I-II	D4S	1498		
5705	9447	14 39.8	−0 43	Vir	13.55			2.8	1.9	SBd		1737		
5713	9451	14 40.2	−0 17	Vir	12.00	11.38	0.62	2.8	2.5	Sc	5	1847		
5727	9465	14 40.5	+33 59	Boo	13.91			2.3	1.2	S(B)dm		1594		
5719	9462	14 40.9	−0 19	Vir	13.8 p			3.4	1.3	S(B)b-p	4			
5716		14 41.1	−17 29	Lib	13: p			1.9	1.5	SBc:				
5729		14 42.0	−9 03	Lib	13: p			3.0	0.9					
5728		14 42.4	−17 15	Lib	12.1	11.3	0.76	2.8	1.6	S(B)b- II	4	2879		
5739	9486	14 42.5	+41 50	Boo	13.2			2.2	2.1	S		5738		
5733		14 42.8	−0 21	Vir	14.5:p			1.3	0.5	Sb:				
IC 1048	9483	14 42.9	+4 53	Vir	13.55			2.5	0.8	S		1627		
5740	9493	14 44.4	+1 41	Vir	12.60	11.91	0.69	3.1	1.7	Sb+ II-III	4	1558		
5746	9499	14 44.9	+1 57	Vir	11.55	10.60	0.95	7.9	1.7	Sb	3VS	1786		
5750	9512	14 46.2	−0 13	Vir	12.5	11.6	0.88	2.9	1.7	Sb+ III	5S	2000		
5756		14 47.6	−14 51	Lib	12.68			2.0	1.0	Scp		2106		
5757		14 47.8	−19 05	Lib	12.6 H			2.1	1.9	SBb	5	2678		
IC 1065	9553	14 49.3	+63 16	Dra	14.30	13.42	0.88	1.3	1.1	SB0		12505*		3C 305
	9560	14 50.9	+35 35	Boo	14.75	14.50	0.25	0.8	0.3	P		1262		2Z 70, VV 324
	9562	14 51.2	+35 33	Boo	14.30	13.86	0.44	1.2	1.1	P		1341		2Z 71, VV 324
5777	9568	14 51.3	+58 58	Dra	14.2 p			3.3	0.6	Sb				
5772	9566	14 51.7	+40 36	Boo	13.9 p			2.3	1.5	Sb:	4	5033		
5768		14 52.1	−2 32	Lib	12.9 H			2.0	1.6	Ir IV:	3S			
5783	9586	14 53.5	+52 05	Boo	13.23			3.0	1.9	S(B)c		2497		NGC 5785
5774	9576	14 53.7	+3 35	Vir	12.82	12.20	0.62	3.2	2.7	S(B)d	1	1588*		
5775	9579	14 54.0	+3 33	Vir	12.24	11.42	0.82	4.3	1.2	Sb:		1581		
		14 54.7	+42 01	Boo	14.54					P		2669		1Z 97
5787	9599	14 55.3	+42 30	Boo	14.1 p			1.2	0.9			5625		1Z 98
5832	9649	14 57.8	+71 41	UMi	12.30			4.0	2.6	SBb:	1	621		
	9638	14 58.2	+58 52	Dra	14.46			1.8	1.1	Ir+		2461		
5792	9631	14 58.4	−1 05	Lib	12.39			7.2	2.1	Sbp	3VS	1974		
5820		14 58.7	+53 53	Boo	12.85	11.89	0.96	2.5	2.3	E5	3VS	3443		Arp 136
5791		14 58.8	−19 16	Lib	12.86			2.4	1.4	Sa	D3	3255		
5796		14 59.4	−16 37	Lib	12.68			1.9	1.7	E0	D3	2871		
5793		14 59.4	−16 42	Lib	14.3	13.2	1.11	1.8	0.6	Sb:		3447		
5806	9645	15 00.0	+1 54	Vir	12.3	11.6	0.70	3.1	1.7	Sb+ III:	5	1298		
5811		15 00.5	+1 37	Vir	15: p			1.0	0.9	SBm:		1524		
5812	A398	15 01.0	−7 27	Lib	12.2	11.2	0.97	2.4	2.2	E1		2028		
5813	9655	15 01.2	+1 42	Vir	11.65	10.65	1.00	3.6	2.8	E1		1880		
	9663	15 01.2	+52 36	Boo	14.27			1.3	1.2	Ir+ IV-V		2594		D198
5827	9662	15 01.8	+25 58	Boo	13.7 p			1.4	1.1	Sb-p				
5829	9673	15 02.7	+23 20	Boo	14.6 p			1.9	1.7	Sc		5764		VV 7, Arp 42
5831	9678	15 04.1	+1 13	Vir	12.45	11.47	0.98	2.2	2.0	Ep		1682		
5838	9692	15 05.4	+2 06	Vir	11.80	10.83	0.97	4.2	1.6	Sa	D3	1429		
5839	9693	15 05.5	+1 38	Vir	13.6			1.4	1.4	S0:				
5845	9700	15 06.0	+1 38	Vir	13.1	12.3	0.85	0.9	0.5	E3		1786		
5846A		15 06.48	+1 35.7	Vir	14.2	13.2		0.5	0.5	E2		2268		
5846	9706	15 06.48	+1 36.3	Vir	11.25	10.23	1.02	3.4	3.2	E0	D3	1714		
5866	9723	15 06.5	+55 46	Dra	10.85	10.00	0.85	5.2	2.3	E6p	4	874*		
5860	9717	15 06.6	+42 38	Boo	14.2 p			1.0	1.0			5513		Mrk 480, 1Z 102
5850	9715	15 07.1	+1 33	Vir	11.75	10.95	0.80	4.3	3.9	SBb- I	4	2354		
5857	9724	15 07.5	+19 36	Boo	13.8	13.1	0.70	1.4	0.8	SBb	4VS	4777		
5859	9728	15 07.6	+19 35	Boo	13.25	12.42	0.83	3.0	1.0	SBb+	4VS	4735		
5854	9726	15 07.8	+2 34	Vir	12.65	11.79	0.86	2.7	0.8	Sa:	D2VS	1632		
5874	9736	15 07.9	+54 45	Boo	14.1 p			2.5	1.8	S(B)b+		3309		
	9749	15 08.8	+67 12	UMi	12.45			27.3	16.0	dE6		17	L	UMi dwarf
5858		15 08.8	−11 13	Lib	14: p			1.4	0.6	E6:				
5875	9745	15 09.2	+52 32	Boo	13.4 p			2.6	1.4	Sb:		3674		
5861		15 09.3	−11 19	Lib	12.18			3.0	1.8	Scp I:	1	1805		
5876	9747	15 09.5	+54 30	Boo	13.9 p			2.6	1.4	SBb-:				
5864	9740	15 09.6	+3 03	Vir	12.89			2.8	1.0	E7p	D3S	1632		
5879	9753	15 09.8	+57 00	Dra	12.1	11.5	0.60	4.4	1.7	Sb II-III	3VS	1019		
	9764	15 10.6	+64 54	Dra	14.29			2.5	1.7	SBdm		2448		
5894	9768	15 11.7	+59 49	Dra	13.07			3.3	0.6	SBm		2679		
	9760	15 12.0	+1 41	SerCp	14.62			2.8	0.3	Sd		2023		
5866B	9769	15 12.0	+55 48	Dra	14.28			3.0	2.3	Sdm		1029		
5888	9771	15 13.1	+41 16	Boo	14.3 p			1.5	1.0	SBb+				
5889		15 13.3	+41 20	Boo	15.5:p			0.9	0.3	SBb:				
5893	9774	15 13.6	+41 58	Boo	14.1 p			1.4	1.3	SBb		5531		
5878	A403	15 13.8	−14 16	Lib	12.3	11.5	0.80	3.5	1.7	Sb	D3	2058		
		15 13.8	−15 28	Lib	12.18			3.2	2.7	S- IV-V		2229		Fath 703
5899	9789	15 15.0	+42 03	Boo	12.6	11.8	0.80	3.0	1.3	Sb+ I:	4S	2701		
5900	9790	15 15.1	+42 13	Boo	15.0 p			1.6	0.6	Sb:				
5885		15 15.1	−10 05	Lib	12.2	11.7	0.52	3.5	3.2	S- IV-V:	3S	1968		
5905	9797	15 15.4	+55 31	Dra	12.3 H			4.2	3.3	S(B)b I:	3S	3386		

NGC	UGC	α 2000	δ 2000	Const	B	V	B–V	Dim		Type	Nucl	RV	Memb	Notes
5907	9801	15h15m9	+56°19'	Dra	11.15	10.38	0.77	12.3	1.8	Sb+ II:		780*		
	9799	15 16.7	+7 01	SerCp	14.00	12.90	1.10	2.1	1.1	E5		10537		3C 317
5908	9805	15 16.7	+55 25	Dra	12.9	11.9	0.96	3.2	1.3	Sb–	3VS	3546		
5898	A404	15 18.2	–24 06	Lib	12.60	11.52	1.08	1.7	1.7	E1p		2128		
5903	A405	15 18.6	–24 04	Lib	12.50	11.48	1.02	2.0	1.7	E1		2382		
		15 18.6	–24 07	Lib	14.6	13.5	1.06	0.6	0.4			2254		
5923	9823	15 21.2	+41 43	Boo	14.7 p			2.0	2.0	S(B)b+		5723		
5916A		15 21.2	–13 06	Lib	15: p			1.3	0.5	SBcp				
5915	A407	15 21.6	–13 06	Lib	12.38			1.6	1.2	Sb+		2230		
5916		15 21.6	–13 10	Lib	14: p			2.9	1.0	SBap	4VS			
5920		15 21.8	+7 42	SerCp	14.70	13.60	1.10	1.4	1.1	S0		13622*		3C 318.1
5921	9824	15 21.9	+5 04	SerCp	11.45	10.82	0.63	4.9	4.2	S(B)b+ I-II	5	1503		
	9837	15 23.8	+58 03	Dra	13.61			2.0	1.8	S(B)c		2857		
5929	9851	15 26.1	+41 40	Boo	13.0 p			1.1	1.0	Sb–p		2696		1Z 112, Arp 90
5930	9852	15 26.1	+41 41	Boo	13.09			2.0	1.0	S(B)bp		2875		1Z 112, Arp 90
	9858	15 26.6	+40 34	Boo	13.23			4.4	0.9	S(B)b+		2774		
5949	9866	15 28.0	+64 46	Dra	12.77			2.4	1.2	Sc	3VS	588		
5936	9867	15 30.0	+12 59	SerCp	13.00	12.41	0.59	1.5	1.4	Sc I:		4053		
	9893	15 32.9	+46 27	Boo	14.87			1.1	0.4	P		840		1Z 115
5963	9906	15 33.5	+56 35	Dra	12.16			3.5	2.8	Sp		854		
5951	9895	15 33.7	+15 00	SerCp	13.44			3.5	1.0	SBc:		1860		
5965	9914	15 34.0	+56 42	Dra	13.4 p			5.4	0.9	Sb		3615		
5953	9903	15 34.5	+15 12	SerCp	13.13			2.0	1.7	Sap		2252		VV 244, Arp 91
5954	9904	15 34.6	+15 12	SerCp	13.65			1.3	0.7	S(B)cp		2210		VV 244, Arp 91
IC 4553-4	9913	15 35.0	+23 29	SerCp	14.4 p			2.0	1.8	P		5580		Arp 220
	9912	15 35.1	+16 33	SerCp	14.16			1.7	1.6	SBdm		1117		VV 132
5961	9918	15 35.2	+30 51	CrB	13.89			0.8	0.3	P		1800		
5957	9915	15 35.4	+12 03	SerCp	13.3 p			3.0	2.9	S(B)b		1894		
		15 36.3	+30 41	CrB	14.86					P		1788		Mrk 689
5962	9926	15 36.5	+16 37	SerCp	12.05	11.39	0.66	2.8	2.1	Sc II	4S	2077		
	A410	15 37.1	+55 16	Dra	14.86			0.2	0.2	P		860		Mrk 487, 1Z 123
5964	9935	15 37.6	+5 59	SerCp	12.73			4.2	3.4	SBd	2VS	1492		IC 4551?
5981	9948	15 37.9	+59 23	Dra	13.9	13.0	0.86	2.8	0.6	Sc:	3S	1921		
	9941	15 38.4	+12 57	SerCp	14.57			1.3	1.2	Ir+		1929		
5970	9943	15 38.5	+12 11	SerCp	12.15	11.44	0.71	3.0	2.1	Sc II-III	3VS	2061		
5982	9961	15 38.7	+59 21	Dra	12.05	11.13	0.92	2.9	2.2	E3p	4	3083		
5974	9952	15 39.0	+31 45	CrB	14.14			0.5	0.3	P		1875		
5985	9969	15 39.6	+59 20	Dra	11.80	10.98	0.82	5.5	3.2	Sb I	3S	2671		
5987	9971	15 39.9	+58 05	Dra	13.3 p			4.7	1.7	Sb				
5968		15 39.9	–30 33	Lup	13: p			2.2	2.1	S(B)b–	3S	5027		
5980	9974	15 41.4	+15 47	SerCp	13.3 p			2.1	0.8	S				
	9992	15 41.8	+67 16	Dra	14.56			1.7	1.1	Ir+		646		
	9977	15 42.0	+0 42	SerCp	13.97			3.9	0.5	Sc		1944		
5984	9987	15 42.9	+14 14	SerCp	13.24			3.0	0.9	Sb II-III:		1199		
	10054	15 44.0	+81 50	UMi	13.79			3.3	1.4	SB+ IV-V		1725		D203
5992	10003	15 44.4	+41 05	Boo	14.2 p			1.1	0.9	S		9762*		Mrk 489
	10020	15 45.8	+20 35	SerCp	13.54			2.3	2.2	Sd		2195		
		15 46.5	+46 00	Boo	14.70					P		2970		Mrk 490
5967		15 48.1	–75 40	Aps	12.48			2.9	1.8	S(B)c:	3VS	2695		
	10041	15 49.0	+5 11	SerCp	13.85			3.0	1.8	S(B)dm		2224		
6015	10075	15 51.4	+62 19	Dra	11.73	11.16	0.57	5.4	2.3	Sc II	3VS	1047*		
6007	10079	15 53.4	+11 57	SerCp	14.1 p			1.9	1.4	S(B)b+:		10629		
6012	10083	15 54.2	+14 35	SerCp	13.1 p			2.3	1.7	SBb–:		2015		
	10086	15 54.6	+16 36	SerCp	14.29			0.7	0.2	P		2400		
6068A		15 54.8	+78 59	UMi	14.5:p			1.1	0.2	S0:		4255		
6068	10126	15 55.4	+79 00	UMi	13.3 p			1.2	0.8	SBb+:		4184		
6020	10100	15 57.2	+22 24	SerCp	13.55	12.58	0.97	1.7	1.2	E3				IC 1148
6018	10101	15 57.5	+15 51	SerCp	14.6 p			1.8	1.0	S0		5220		
6021	10102	15 57.5	+15 56	SerCp	14.1 p			2.1	1.3	E5		4585		
6048	10124	15 57.6	+70 42	UMi	13.6 p			2.5	1.9	E2				
6022		15 57.8	+16 16	SerCp	15: p			0.8	0.5	SBb		11325		
6023	10106	15 57.8	+16 18	SerCp	14.7 p			1.4	1.0	E3		11240		
6027	10116	15 59.2	+20 45	SerCp	13.35	12.44	0.91	2.2	1.2	Group of Sp		4400*		Seyfert's Sextet
6028	10135	16 01.4	+19 21	Her	14.0	13.2	0.85	1.7	1.5	S(B)0:		4570		1Z 133
IC 1158	10133	16 01.5	+1 43	SerCp	13.48			2.9	1.8	S(B)c		1978		
	A411	16 02.0	+18 49	Her	14.86			0.7	0.3	P		2638		Mrk 294
6032	10148	16 03.0	+20 58	Her	15.0 p			1.8	0.9	SBb:		4400		
6035	10154	16 03.4	+20 54	Her	14.46			1.2	1.1	S(B)c		2357		
6040	10165	16 04.4	+17 45	Her	14.6 p			1.6	0.8	S(B)cp,Sap		12613	He	VV 212
6041	10170	16 04.6	+17 43	Her	14.7	13.7	0.98	1.6	1.5	S(B)0,SB0		10576	He	VV 213
6051	10178	16 04.9	+23 56	SerCp	14.05	12.95	1.10	1.9	1.2	E4		9577		
6047		16 05.1	+17 43	Her	14.65	13.55	1.10	1.3	1.3	S0		9581	He	
6052	10182	16 05.2	+20 32	Her	13.45	13.00	0.45	1.0	0.7	Sc	4	4821		VV 86, Arp 209
6050	10186	16 05.4	+17 46	Her	14.9 p					S(B)c		11217	He	
6056		16 05.5	+17 58	Her	15: p					S(B)a:		11820	He	

GALAXIES

NGC	UGC	α 2000	δ 2000	Const	B	V	B–V	Dim		Type	Nucl	RV	Memb	Notes
6055	10191	16h05m.5	+18°10'	Her	15.4 p			1.5	1.0	S(B)0:		11481	He	
IC 1185		16 05.7	+17 43	Her	14.75	13.86	0.89	0.8	0.6	Sb–:		10564	He	
		16 05.7	+41 19	Her	14.70			0.7	0.3	P		2221		
	10200	16 05.7	+41 21	Her	13.56			0.7	0.5	P		2137		
6060	10196	16 05.9	+21 30	Her	14.3 p			2.2	1.3	S(B)b+		4675		
6061	10199	16 06.3	+18 15	Her	15.0 p			1.4	1.2	S0:		11332	He	
6070	10230	16 10.0	+0 43	SerCp	12.35	11.68	0.67	3.6	2.1	Sc I		2060		
6090	10267	16 11.7	+52 27	Dra	14.0 p			1.9	0.9			8962		Mrk 496, 1Z 135
6086	10270	16 12.6	+29 29	CrB	14.8 p			2.0	1.6	E2		9549		
	10288	16 14.4	−0 13	SerCp	13.66			5.3	0.8	Sc		2099		
	10290	16 14.5	+0 49	SerCp	13.85			2.0	1.9	Sm		2039		
	10310	16 16.3	+47 03	Her	13.05			2.8	2.0	Ir+		917		D204, Arp 2
6109	10316	16 17.7	+35 00	CrB	14.9 p			1.3	1.3			9054		
6106	10328	16 18.8	+7 25	Her	12.80	12.20	0.60	2.6	1.5	Sb+ II-III:	D3VS	1560		
6120	10343	16 19.8	+37 47	CrB	14.3 p			0.6	0.5	P		9400		1Z 141
6140	10359	16 20.9	+65 23	Dra	11.97			6.2	4.7	SBcp		1147		
6143	10358	16 21.7	+55 05	Dra	13.9 p			1.2	1.1	S(B)b+:	3VS	5452		
6118	10350	16 21.8	−2 17	SerCp	12.26			4.7	2.3	Sb II	2S	1618		
6137	10364	16 23.1	+37 55	CrB	14.1 p			2.2	1.4	E4		9485		
6154	10382	16 25.6	+49 50	Her	14.0 p			2.4	2.3	SBa				
6158		16 27.7	+39 23	Her	15.5:p							9101		
6160	10400	16 27.7	+40 55	Her	14.8 p			2.1	1.7	E2		9603		
6166	10409	16 28.6	+39 33	Her	13.05	11.99	1.06	2.4	1.8	E2p		9075		3C 338
6173	10421	16 29.8	+40 49	Her	14.0 p			2.2	1.7	E3		8923		
6181	10439	16 32.3	+19 50	Her	12.50	11.87	0.63	2.6	1.3	Sc I	3VS	2512		
6217	10470	16 32.6	+78 12	UMi	11.85	11.22	0.63	3.1	2.7	Sc I-II	4S	1586		Arp 185
6186	10448	16 34.4	+21 33	Her	14.2 p			1.8	1.4	SBa				
6195	10469	16 36.5	+39 02	Her	14.7 p			1.8	1.3	Sb		9196		
6196		16 37.9	+36 04	Her	13.80	12.76	1.04	1.7	1.2	S(B)0p				IC 4615
	10491	16 38.3	+41 56	Her	14.29	13.71	0.58	1.3	0.6	Ring		8500		1Z 162, Arp 125
		16 39.9	+37 10	Her	14.86					P		2100		
6211	10516	16 41.5	+57 47	Dra	13.8 p			2.0	1.5	SB0p				7Z 655
6207	10521	16 43.1	+36 50	Her	12.15	11.62	0.53	3.0	1.4	Sc II:	1	1066		
6223	10527	16 43.1	+61 34	Dra	13.1 p			3.4	2.5	P				7Z 657
6236	10546	16 44.5	+70 48	Dra	12.42			3.0	1.8	S(B)c		1525		
6248		16 46.4	+70 22	Dra	13.30			3.3	1.2	SBc		1364		
		16 49.5	+45 30	Her	14.4 p			3.5		S0p,Sap,S(B)0p		9640		Zwicky's Triplet
6246	10580	16 49.9	+55 33	Dra	14.92	14.12	0.80	1.7	0.8	SBb:		5445		
6239	10577	16 50.1	+42 44	Her	12.80	12.34	0.46	2.8	1.3	Sb		1159		
6246A	10584	16 50.2	+55 23	Dra	14.7 p			2.5	2.3	SBcp				
6215		16 51.1	−58 59	Ara	11.79			2.0	1.6	Sc III	3VS	1384		
6221		16 52.8	−59 13	Ara	11.49			3.2	2.3	SBc II	4S	1254		
6240	10592	16 53.0	+2 24	Oph	14.7 p			2.2	1.1	Irp−		7597		
6255	10606	16 54.8	+36 30	Her	13.8 p			3.5	1.5	SBc:		1026		
6275		16 55.6	+63 15	Dra	15: p							6978*		Mrk 503, 7Z 667
6285		16 58.4	+58 59	Dra	14.5:p			1.3	0.7	S0p				Arp 293
6286	10647	16 58.6	+58 58	Dra	14.2 p			1.5	1.4	Sbp				Arp 293
6306	10724	17 07.6	+60 44	Dra	14.3 p			1.2	0.4	SBb−p		3215*		
6307	10727	17 07.7	+60 45	Dra	14.0 p			1.6	1.2	SBap		3299*		
6310	10730	17 08.0	+61 00	Dra	13.8 p			2.1	0.5	Sb:		3628		
	10736	17 08.1	+69 28	Dra	13.85			3.5	1.3	S(B)dm		735		
6296	10719	17 08.7	+3 54	Oph	14.2 p			1.1	0.9	S(B)b:				
6340	10762	17 10.4	+72 18	Dra	11.90	11.03	0.87	3.4	3.0	Sap	3S	2146		
6308	10747	17 12.0	+23 23	Her	14.4 p			1.3	1.2	S(B)c:	3S	8960		
6314		17 12.6	+23 16	Her	13.85	13.06	0.79	1.8	0.9	Sa:		6928		
6315		17 12.8	+23 13	Her	14.25	13.29	0.96	0.9	0.8	SBc:		6852		
6300		17 17.0	−62 49	Ara	11.13			5.4	3.5	SBb	2	988		
6339	10790	17 17.1	+40 51	Her	13.04			3.4	1.9	SBc		2336		
6359	10804	17 17.9	+61 47	Dra	13.5	12.6	0.89	1.2	0.9	S0:	3S	3194		
6361	10851	17 18.6	+60 37	Dra	13.9 p			2.4	0.8	Sb:				Arp 124
	10806	17 18.9	+49 52	Her	13.49			2.5	1.0	SBdm		1175		
	10805	17 19.9	+14 24	Her	14.32			2.0	1.8	SB− V:		1715		D207
	10822	17 20.2	+57 55	Dra	11.87			33.5	18.9	dE3		−31	L	Draco dwarf
6395	10876	17 26.4	+71 06	Dra	12.83			2.6	0.9	Scp		1415		
6368	10856	17 27.1	+11 33	Oph	13.07			4.0	1.2	Sb		2920		
6381	10871	17 27.3	+60 01	Dra	13.6 p			1.5	1.2	Sb+:		3469		
6372	10861	17 27.5	+26 28	Her	14.1 p			1.9	1.3	Sbp		4945		
	10862	17 28.2	+7 26	Oph	13.10			3.6	3.5	SBc		1834		
6412	10897	17 29.6	+75 42	Dra	12.35	11.82	0.53	2.3	2.1	Sc II	3S	1650		Arp 38
6379	10886	17 30.6	+16 17	Her	14.6 p			1.3	1.3	Sc:		6134		
6384	10891	17 32.4	+7 04	Oph	11.30	10.57	0.73	6.0	4.3	Sb I:	3S	1801*		
6389	10893	17 32.7	+16 24	Her	13.6 p			3.1	2.1	Sb+		3276		
		17 44.1	+40 52	Her	14.70					P		1800		
6454		17 44.9	+55 42	Dra	14.5:p			1.4	1.2			9374		
IC 4662		17 47.1	−64 38	Pav	11.85	11.43	0.42	2.2	1.4	Ir+	1	167		

NGC	UGC	α 2000	δ 2000	Const	B	V	B–V	Dim		Type	Nucl	RV	Memb	Notes
6478	10998	17h48m.6	+51°09′	Dra	14.1 p			1.9	0.8	Sb+:		7110		NGC 6466
	11000	17 49.4	+36 08	Her	13.89			0.8	0.3	P		900		
6503	11012	17 49.4	+70 09	Dra	10.90	10.23	0.67	6.2	2.3	Sb III:	4VS	315		
6467	11004	17 50.7	+17 32	Her	14.5 p			2.7	1.9	S				
6482	11009	17 51.8	+23 04	Her	12.15	11.25	0.90	2.3	2.0	E3p		4127		
6500	11048	17 56.0	+18 20	Her	13.4 p			2.5	1.9	Sb−:		3174		
6504	11053	17 56.0	+33 12	Her	13.4 p			2.5	0.6	S				
6501	11049	17 56.1	+18 22	Her	13.35	12.34	1.01	2.0	1.9	SO:		3125		
	11074	17 59.1	+7 09	Oph	14.28			3.5	1.4	Sd		2059		
6509	11075	17 59.4	+6 17	Oph	13.19			2.3	1.8	Sc		1974		
6542	11092	17 59.7	+61 22	Dra	14.0 p			1.5	0.5	Sb−:				
	11093	18 01.9	+6 58	Oph	13.91			4.6	0.8	Sc		2122		
	11105	18 04.5	+21 38	Her	14.28			2.9	2.2	Sdm		2435		
6560	11117	18 05.3	+46 52	Her	14.2 p			1.3	0.9	Sp		7293		
	11113	18 05.5	+23 16	Her	14.57			1.9	1.9	S(B)c		2547		
	11124	18 07.5	+35 33	Her	13.17			2.7	2.5	SBc		1852		
6555	11121	18 07.8	+17 36	Her	12.42			2.1	1.7	S(B)c	3S	2424		
6570	11137	18 11.1	+14 06	Oph	13.24			1.8	1.1	SBm:		2478		
	11141	18 11.8	+12 05	Oph	14.10			2.8	2.8	Sdm		2426		
6574	11144	18 11.9	+14 59	Her	12.85	12.01	0.84	1.4	1.1	S:	D3S	2509		
	11152	18 12.5	+18 36	Her	14.14			2.5	1.4	SBdm		2935		
6621	11175	18 12.9	+68 22	Dra	14.1	13.1	0.99	2.5	1.1	Sbp		6533		VV 247, Arp 81
6587	11166	18 13.9	+18 50	Her	13.15	12.08	1.07	2.3	2.3	S(B)0:		3208		
	11193	18 15.0	+71 01	Dra	14.78			1.4	0.9	Ir+		1786		
6615	11196	18 18.6	+13 16	Oph	14.8 p			1.2	1.1	SBa:				
6643	11218	18 19.8	+74 34	Dra	11.75	11.07	0.68	3.9	2.1	Sc I-II	3VS	1736		
6627	11212	18 22.7	+15 42	Her	14.15	13.18	0.97	1.5	1.4	SBb	3S	5408		
6654	11238	18 24.1	+73 11	Dra	12.45	11.55	0.90	3.0	2.4	SBa	D4S	2063		7Z 793
6651	11236	18 24.3	+71 36	Dra	13.7 p			1.8	0.9	Sc		6028		
6632	11226	18 25.0	+27 32	Her	13.2 p			3.1	1.6	Sb+		4926		
6438		18 25.8	−85 25	Oct	12.5			2.6	2.0	Ep		2220		
6438A		18 26.3	−85 25	Oct						Ring		2301*		
6635	11239	18 27.6	+14 49	Her	14.5:p			1.3	1.2	SOp		5273		
IC 4710		18 28.7	−66 59	Pav	12.4 H			4.2	3.1	SBm				
6667	11269	18 30.7	+67 59	Dra	12.49			2.6	1.3	S(B)b−p		2847		
6658	11274	18 33.9	+22 53	Her	14.0	12.9	1.08	1.9	0.5	SO	4	4498		
IC 1291	11283	18 33.9	+49 16	Dra	13.60			2.4	2.2	SBdmp		2230		
IC 4721		18 34.5	−58 30	Pav	12.6			4.1	1.5	SBc:	3VS	5844		
6661	11282	18 34.6	+22 55	Her	12.95	11.94	1.01	2.0	1.3	Sa	D3	4541		
6690	11300	18 34.8	+70 32	Dra	12.6 p			3.8	1.4	Sd:	1	788		NGC 6689
6674	11308	18 38.6	+25 23	Her	12.85	12.05	0.80	4.2	2.5	SBb	4	3739		
6691	11318	18 39.3	+55 38	Dra	14.1 p			1.7	1.7	SBb+		6151		
6654A	11332	18 39.4	+73 35	Dra	13.6 p			2.7	0.9	SBdp		1679		
6701	11348	18 43.2	+60 40	Dra	12.9 p			1.8	1.6	SBa		4256		
6702	11354	18 47.0	+45 42	Lyr	13.25	12.21	1.04	2.1	1.6	E3:		4997		
6703	11356	18 47.3	+45 33	Lyr	12.45	11.44	1.01	2.6	2.5	SO	3	2590		
6711	11361	18 49.0	+47 39	Dra	14.1 p			1.3	1.2	SBb+p		4938		
6684		18 49.0	−65 11	Pav	11.35	10.45	0.90	3.7	2.7	SB0	3	688		
6710	11364	18 50.6	+26 50	Lyr	13.75	12.75	1.00	2.0	1.2	SO:		4802		
6699		18 52.1	−57 19	Pav	12.7			1.9	1.8	S(B)b	3S	3373		
IC 4797		18 56.5	−54 18	Tel	12.30	11.27	1.03	2.8	1.3	E5	3	2497		
		18 57.0	−54 33	Tel	12.80	11.83	0.97	1.6	1.2	SO:		2635		
6721		19 00.8	−57 47	Pav	13.10	12.04	1.06	1.9	1.9	E0	3S	4358		
6764	11407	19 08.3	+50 56	Cyg	12.74			2.5	1.5	SBb	*	2691		
6744		19 09.8	−63 51	Pav	9.03			15.5	10.2	S(B)b+ II	D3	519		
6754		19 11.4	−50 39	Tel	13.2 H			2.2	1.1	SBb	2S	3194		
6753		19 11.4	−57 03	Pav	11.93			2.5	2.2	Sb	4VS	3009		
6758		19 13.9	−56 19	Tel	12.61			2.1	1.9	E1		3237		
IC 4837		19 15.3	−54 40	Tel	12.48			2.8	1.4	SBcp	2	2586		
6769		19 18.4	−60 31	Pav	12.58	11.76	0.82	2.5	1.7	S(B)bp	D4	3795		VV 304
6770		19 18.7	−60 31	Pav	12.88	12.04	0.84	2.5	1.9	S(B)bp	4	3705		VV 304
6771		19 18.7	−60 33	Pav	13.54	12.48	1.06	2.6	0.6	SB0:	4	4108		
IC 4842		19 19.5	−60 39	Pav	13.30	12.29	1.01	2.4	1.3	E5:		3940		
6796	11432	19 21.5	+61 09	Dra	13.5 p			2.1	0.6	Sb+:		2371		
6780		19 22.9	−55 47	Tel	13.2 H			1.9	1.6	S(B)c	3S	3391		
6782		19 24.0	−59 55	Pav	12.71			2.7	1.6	SB0:	4	3632		
6776		19 25.4	−63 52	Pav	12.95	11.99	0.96	1.9	1.7	E2	4	5574		
		19 30.0	−17 41	Sgr	15.6					Ir−		−37	L	Sgr dwarf
IC 1302		19 30.9	+35 47	Cyg	14.0	13.3	0.73	1.0	0.7	S(B)c:		4852		
	11465	19 41.7	+50 38	Cyg	13.95	12.86	1.09	1.3	1.3	S(B)0		7446		3C 402
6814		19 42.7	−10 19	Aql	12.02	11.17	0.85	3.2	3.0	Sb+ I	4VS*	1578		
6810		19 43.6	−58 40	Pav	12.28			3.8	1.1	Sb−:	4	1713		
6824	11470	19 43.7	+56 07	Cyg	12.7	11.9	0.78	2.1	1.5	Sb:	4	3675		
6808		19 43.9	−70 39	Pav	13.39			1.7	0.7	Sa:	3S	3314		
6821		19 44.4	−6 50	Aql	14: p			1.1	1.0	SBd:	1	1680		

NGC	UGC	α 2000	δ 2000	Const	B	V	B−V	Dim		Type	Nucl	RV	Memb	Notes
6822		19^{h}44^{m}9	−14°48′	Sgr	9.35			10.2	9.5	Ir+ IV-V		65	L	Barnard's galaxy
IC 4889		19 45.3	−54 20	Tel	12.38			2.6	1.4	E5		2418		
6835		19 54.5	−12 34	Sgr	13.4	12.5	0.87	2.7	0.7	SBa	3	1715*		
6836		19 54.7	−12 41	Sgr	13.40			1.2	1.2	S(B)m	1	1762		
		19 56.2	+40 26	Cyg	14.7	13.5	1.21	0.4	0.4	E1:		5082		
6869	11506	20 00.7	+66 13	Dra	12.8 p			1.7	1.6	S0				
6851		20 03.6	−48 17	Tel	12.61			1.8	1.3	E4:	4	2994		
6854		20 05.7	−54 23	Tel	13.27			2.3	1.9	E2	3VS	5620		
6861		20 07.3	−48 22	Tel	12.10	11.10	1.00	2.7	1.4	S0:	4	2779		IC 4949
6868		20 09.9	−48 23	Tel	11.99			2.7	2.2	E2	3	2724		
6870		20 10.2	−48 17	Tel	13.15	12.24	0.91	2.5	1.0	Sb−	3S	2571		
6875		20 13.2	−46 09	Tel	12.77			2.5	1.3	E6:	4	3076		
6878		20 13.9	−44 31	Sgr	14.1	13.4	0.67	1.7	1.3	Sb I-II	D3	5826		
IC 4970		20 16.9	−70 45	Pav	14.70	13.88	0.82	0.8	0.3	S0:		4568		VV 297
6872		20 16.9	−70 46	Pav	12.45	11.59	0.86	4.7	1.4	SBbp	4	4554		VV 297
6887		20 17.2	−52 47	Tel	12.46			4.1	1.7	Sb+:	1			
6890		20 18.3	−44 48	Sgr	13.06	12.52	0.54	1.5	1.2	Sb	3VS	2399		
6876		20 18.3	−70 52	Pav	12.56			2.4	1.8	E3		3803		
6877		20 18.6	−70 51	Pav	13.85	12.73	1.12	1.4	0.7	E6		3984		
6880		20 19.5	−70 52	Pav				2.1	0.7	S(B)0:	4S	3782		
6893		20 20.8	−48 14	Tel	12.54			2.8	1.9	S(B)0	5	3098		
IC 1317	11546	20 23.3	+0 40	Aql	14.2	13.4	0.83	0.9	0.8	E2:		4172		2Z 82
6906	11548	20 23.6	+6 27	Aql	13.55	12.73	0.82	1.8	0.9	SBb+:	3	5108		
	11557	20 24.0	+60 12	Cep	13.54			3.4	3.2	S(B)dm		1684		
		20 24.0	−44 00	Sgr	12.19			2.5	1.0	S(B)a	3S	2887		
6902		20 24.5	−43 39	Sgr	12.5			2.2	1.8	SBa	4	2669		
6907		20 25.1	−24 49	Cap	12.00	11.30	0.70	3.4	3.0	S(B)b I-II:	3VS	3238		NGC 6908
6912		20 26.9	−18 38	Cap	14: p			1.6	1.3	SBc		7224		
6909		20 27.7	−47 02	Tel	12.68			2.3	1.2	E6	3	2610		
6921	11570	20 28.5	+25 43	Vul	15.0:p			1.1	0.3	Sa:		4590		
IC 5020		20 30.6	−33 29	Mic	13.1 H					Sa+		3090		
6923		20 31.7	−30 50	Mic	12.85	12.09	0.76	2.5	1.4	Sb+ II:		4651		
6927A		20 32.6	+9 53	Del				0.8	0.4	E6:		4986		
6928	11589	20 32.8	+9 56	Del	13.65	12.64	1.01	2.2	0.8	SBb−		4414		
6930	11590	20 33.0	+9 52	Del	14.0	13.1	0.87	1.4	0.6	SBb−:	3S			
6926		20 33.1	−2 01	Aql	13.10	12.35	0.75	2.1	1.4	SBcp		6229		
6929		20 33.4	−2 02	Aql	14.6	13.5	1.14	0.9	0.8	S0p				
6925		20 34.3	−31 59	Mic	12.10	11.30	0.80	4.1	1.6	Sb+ II	D3	2544*		
6946	11597	20 34.8	+60 09	Cep	9.68	8.88	0.80	11.0	9.8	Sc I	3ES	338		Arp 29
6951	11604	20 37.2	+66 06	Cep	12.2	11.1	1.10	3.8	3.3	Sbp I-II:	5ES	1627		NGC 6952
6944A		20 38.2	+6 54	Del	15: p			1.2	0.9	SBdp	3VS			
6944		20 38.4	+7 00	Del	14.25	13.25	1.00	1.7	0.8	S0:	3S	4640*		
6935		20 38.4	−52 06	Ind	13.01			2.0	1.7	S(B)a	3S	4739		
6937		20 38.8	−52 09	Ind				3.0	2.7	S(B)b−	3S	4625		
6942		20 40.8	−54 19	Ind	13.10			2.6	1.9	SB0		3898		
IC 5039		20 43.2	−29 51	Mic	13.25	12.60	0.65	2.5	0.7	Sb II-III:				
6956	11619	20 44.0	+12 31	Del	13.5 p			2.1	2.0	SBb				
6954	11618	20 44.1	+3 13	Del	14.1	13.2	0.88	1.1	0.6	Sb−:		4221		
6943		20 44.5	−68 45	Pav	11.98			4.1	2.1	S(B)c:	3S	2977		
		20 46.9	−12 51	Aqr						Ir		−26	L?	Aquarius dwarf
6962	11628	20 47.3	+0 19	Aqr	12.85	12.00	0.85	3.0	2.5	S(B)b−	D3	4382		
6963		20 47.3	+0 29	Aqr	15.0	14.1	0.92	0.7	0.7	E1:		4551		
6964	11629	20 47.4	+0 18	Aqr	13.7	12.7	1.03	1.9	1.4	E4p		4031		
6958		20 48.6	−38 00	Mic	12.30			2.4	2.2	E1	4	2760		
6972	11640	20 50.0	+9 54	Del	14.3 p			1.4	0.7	Sa		4681		
IC 5063		20 52.0	−57 05	Ind	13.05	12.01	1.04	1.9	1.5	S0:	3S	3405		
6970		20 52.1	−48 48	Ind	13.30	12.68	0.62	1.2	1.1	SBa	3S	5203*		
IC 5052		20 52.1	−69 12	Pav	12.19			5.8	0.9	SBd:	1	773		
6978		20 52.6	−5 43	Aqr	14: p			1.9	1.0			6049		
	11651	20 57.3	+25 58	Vul	14.12			3.4	1.0	Sdm		1804		
6984		20 57.9	−51 52	Ind	13.3			1.7	1.2	SBc		4469		
7013	11670	21 03.6	+29 54	Cyg	12.45			4.9	1.8	Sa	D3	883*		
7007		21 05.5	−52 33	Ind	12.87			1.9	1.2	S0:		2897		
7015	11674	21 05.7	+11 25	Equ	13.2 p			2.0	1.8	Sb+		5115		
7014		21 07.9	−47 11	Ind	13.25	12.25	1.00	1.9	1.9	E0		4722		
7020		21 11.4	−64 03	Pav	12.38			4.3	2.3	S0	4	2915		
7029		21 11.9	−49 17	Ind	12.69	11.81	0.88	1.4	0.8	E6:	5	2779		
7042	11702	21 13.8	+13 33	Peg	13.0 p			2.2	2.0	Sb		5426		
	11707	21 14.5	+26 44	Vul	14.22			3.6	2.0	Sdm		1187		
7046	11708	21 14.9	+2 50	Equ	14.2 p			2.0	1.5	SBc				
7038		21 15.2	−47 13	Ind	12.4 H			3.0	1.9	Sc:	3VS	4774		
7041		21 16.5	−48 22	Ind	12.05	11.11	0.94	3.9	1.5	S(B)0		1843		
7052	11718	21 18.6	+26 27	Vul	14.0 p			2.8	1.6	E4		5198		
7049		21 19.0	−48 34	Ind	11.80	10.74	1.06	2.8	2.2	S0	4	2122		
		21 23.3	−45 47	Ind	12.6 H			4.3	1.3	Sb+	1			

NGC	UGC	α 2000	δ 2000	Const	B	V	B−V	Dim		Type	Nucl	RV	Memb	Notes
IC 5105		21ʰ24ᵐ4	−40°37′	Mic	12.55	11.54	1.01	2.5	1.5	E4		5368		
7065		21 26.7	−7 00	Aqr				1.4	1.3	SBb−:				
7065A		21 27.0	−7 01	Aqr	15: p			1.7	1.6	S(B)c:	3VS			
7059		21 27.4	−60 01	Pav	13.16			3.2	1.8	S(B)c	D3	1702		
7064		21 29.0	−52 46	Ind	13.08			4.1	0.7	SBc:		739		
7077	11755	21 29.9	+2 25	Aqr	14.13			0.8	0.8	E		1050		
7080	11756	21 30.0	+26 43	Vul	14.1 p			2.0	1.9	SBb		5103		
7070		21 30.5	−43 06	Gru	12.84	12.29	0.55	2.1	1.9	Sc	3S	2042		
7072		21 30.6	−43 09	Gru	14.2	13.9	0.28	0.8	0.7	S(B)d:		4917		
7072A		21 30.6	−43 13	Gru				0.9	0.6	Sc:				
7070A		21 32.0	−42 48	Gru	13.4	12.4	1.01	1.7	1.0	Ir−				
7079		21 32.5	−44 05	Gru	12.5	11.6	0.89	2.8	1.7	SB0	4	2664*		
7083		21 35.7	−63 54	Ind	11.79			4.5	2.9	Sb+ II	D3S	2936		
7090		21 36.5	−54 33	Ind	11.1			7.1	1.4	SBc:		708		
	11782	21 38.2	+8 59	Peg	14.04			2.5	1.4	SBm		1343		
7102	11786	21 39.7	+6 17	Peg	13.6	13.1	0.52	1.9	1.2	SBb:	3 S			
7097		21 40.3	−42 32	Gru	12.48			2.5	1.5	E5		2398		
7096		21 41.3	−63 54	Ind	12.58			2.7	2.3	Sa		2844		
7107		21 42.4	−44 48	Gru	12.76			1.9	1.8	SBdm		2180		
7119A		21 46.3	−46 31	Gru	13.8 H			1.1	0.6	Sc:	3VS			
IC 5131		21 47.4	−34 53	PsA	13.3	12.4	0.94	1.6	1.4	SB0	5	2644		
7124		21 48.1	−50 34	Ind	13.17			2.7	1.2	SBb+	3S	4980		
7137	11815	21 48.2	+22 10	Peg	13.05	12.37	0.68	1.5	1.4	S:	4S	1670		
IC 5135		21 48.3	−34 57	PsA	13.00	12.33	0.67	1.3	1.2	P		4875		
7126		21 49.3	−60 37	Ind	13.36			2.3	1.1	Sc	3VS	2910		
7125		21 49.3	−60 43	Ind	12.84			3.2	2.1	S(B)c I		2913		
7135		21 49.8	−34 53	PsA	12.69	11.68	1.01	2.8	2.0	S0p	D3	2751		
7144		21 52.7	−48 15	Gru	11.62	10.71	0.91	3.5	3.5	E0		2062		
7145		21 53.3	−47 53	Gru	12.08	11.23	0.85	2.5	2.3	E0		1840		
7156	11843	21 54.6	+2 57	Peg	13.7	13.0	0.66	1.7	1.5	S(B)c:	3S	4178		
7155		21 56.2	−49 32	Ind	12.78			1.9	1.2	SB0		1810		
	11861	21 56.4	+73 15	Cep	13.20			4.7	3.5	S(B)dm		1760		
7163		21 59.3	−31 53	PsA	14.6	13.4	1.16	1.8	1.2	S(B)b−:		2922		
7162		21 59.8	−43 19	Gru	13.37			2.8	1.1	Sc	1			
7177	11872	22 00.7	+17 44	Peg	11.95	11.17	0.78	3.3	2.3	Sb− II	5	1441		
7166		22 00.7	−43 25	Gru	12.82	11.84	0.98	2.4	0.9	S0	5	2395		
7162A		22 00.8	−43 09	Gru	13.08			2.4	2.1	S(B)m		2264		
7171		22 01.0	−13 16	Aqr	13.00	12.28	0.72	2.8	1.7	Sb+ II	3S	2770		
7172		22 02.0	−31 52	PsA	12.82	11.90	0.92	2.2	1.3	S:		2698		
7173	A422	22 02.06	−31 58.5	PsA	13.05	12.09	0.96	1.3	1.1	E2:		2547		
7168		22 02.1	−51 45	Ind	12.58			2.0	1.6	E3	D3	2692		
7174		22 02.11	−31 59.4	PsA	13.5	12.6	0.90	1.3	0.7	S:		2824		
7176	A423	22 02.14	−31 59.3	PsA	12.9	11.9	0.99	1.3	1.3	E0p		2571		
7180		22 02.3	−20 33	Aqr	13.40	12.52	0.88	1.8	0.9	Sa:		1451		
IC 1420	11880	22 02.6	+19 45	Peg	13.85			1.6	1.6	SBp		1671		
7184	A425	22 02.7	−20 49	Aqr	12.04			5.8	1.8	Sb+ II−III	4S	2727		
IC 5152		22 02.9	−51 17	Ind	11.59			4.5	2.7	Ir+ IV:		30	L	
IC 5156		22 03.3	−33 50	PsA	13.2 H					Sb II−III		2621		
	11891	22 03.5	+43 45	Lac	13.34			3.5	2.6	Ir+		756		5Z 380
7196		22 06.0	−50 07	Ind	12.55	11.54	1.01	1.9	1.4	E3	4	2960		
	11909	22 06.3	+47 15	Lac	13.12			1.7	0.4	Sp		1404		
7201		22 06.5	−31 16	PsA	14: p			1.7	0.7	Sb−:				
7203		22 06.7	−31 10	PsA	13: p			1.9	1.5	SBa:				
7192		22 06.8	−64 19	Ind	12.49			2.5	2.5	E0		2761		
7204		22 06.9	−31 03	PsA	14: p					Irp		2680		
7217	11914	22 07.9	+31 22	Peg	11.10	10.20	0.90	3.7	3.2	Sb− II:	D4	1227		
7205		22 08.5	−57 25	Ind	11.38			4.3	2.2	Sb+	D3S	1383		
7214		22 09.1	−27 49	PsA	13.0	12.4	0.58	1.9	1.5	S(B)cp		6860		
7213		22 09.3	−47 10	Gru	11.35	10.45	0.90	1.9	1.8	Sa:	5	1737		
7218		22 10.2	−16 40	Aqr	12.55	12.06	0.49	2.5	1.3	Sc III−IV	2	1782		
7221		22 11.3	−30 37	PsA	12: p			2.2	1.9			4364		
7219		22 13.1	−64 51	Tuc	13.0 p			2.4	1.6	Sap	5			
IC 5181		22 13.4	−46 02	Gru	12.61	11.69	0.92	2.8	1.0	S0	5	2043		
7232A		22 13.7	−45 47	Gru	13.92			2.3	0.5	SBb−	2			
7229		22 14.0	−29 25	PsA	12: p			1.8	1.7			4327		
7236	11958	22 14.7	+13 51	Peg	14.55	13.51	1.04	1.0	1.0	S0		8098		Arp 169, 3C 442
7237	11958	22 14.8	+13 50	Peg	14.80	13.73	1.07	1.0	1.0	S0		8093		Arp 169, 3C 442
7240		22 15.4	+37 17	Lac	14.9	13.8	1.06	0.8	0.8	S0:		6272		
7242	11969	22 15.7	+37 18	Lac	13.15	12.03	1.12	2.6	2.0	S0:		5972		
7232		22 15.7	−45 51	Gru	12.88			3.0	0.9	SBa:		1983		
7241	11968	22 15.8	+19 14	Peg	13.8 p			3.5	1.3	SBb+p		1689		2Z 174
7233		22 15.8	−45 51	Gru				2.2	1.3	S(B)a	4S	1814		
7232B		22 16.0	−45 41	Gru				1.9	1.5	SBm				
IC 5179		22 16.1	−36 50	Gru	12.5	11.9	0.61	2.3	1.2	Sb+	3VS	3422		IC 5184
7248	11972	22 16.9	+40 30	Lac	13.6 p			2.1	1.1	S0:				

NGC	UGC	α 2000	δ 2000	Const	B	V	B-V	Dim		Type	Nucl	RV	Memb	Notes
7252		22^{h}20^{m}7	-24°41'	Aqr	12.73	12.09	0.64	2'.2	1'.8	E	5	4812		Arp 226
IC 5201		22 21.4	-46 04	Gru	11.31			8.5	4.3	SBc III		2083		
7274	12026	22 24.2	+36 08	Lac	14.2 p			1.8	1.8	E0		6111		
7280	12035	22 26.5	+16 09	Peg	12.98			2.4	1.7	S(B)0		2060		
7290	12045	22 28.4	+17 09	Peg	13.8 p			1.8	1.1	Sb+	3S	3145		
7292	12048	22 28.4	+30 18	Peg	12.95			2.1	1.7	Ir+		1234		
	12060	22 30.5	+33 49	Peg	14.16			1.8	1.7	Ir+		1168		
	12069	22 30.8	+76 30	Cep	14.57			2.4	1.7	S(B)dm		2631		
7298		22 30.8	-14 11	Aqr	14: p			1.5	1.3	Sc:	3S	5166		
7300		22 31.0	-14 00	Aqr	13.7	12.9	0.76	2.2	1.2	Sb+ I-II:	3S	4236		
	12064	22 31.3	+39 22	Lac	14.25	13.25	1.0	1.4	1.4	S0:		5417		3C 449
7302		22 32.4	-14 07	Aqr	13.2	12.1	1.07	1.9	1.3	E3	4VS	2712		IC 5228
	12074	22 33.1	+8 05	Peg	14.05			0.5	0.3	P		2100		
7307		22 33.8	-40 57	Gru	12.8 H			4.2	1.0	S(B)cp	2	1875		
	12082	22 34.3	+32 53	Peg	13.38			3.5	3.1	S+ V		1081		D213
7309		22 34.3	-10 21	Aqr	13.05	12.54	0.51	2.1	2.0	Sc II	4	4086		
7314		22 35.8	-26 03	PsA	11.60	10.91	0.69	4.6	2.3	Sc I-II	3ES	1694*		Arp 14
7316	12098	22 35.9	+20 19	Peg	13.7 p			1.3	1.1	S		5800		Mrk 307
7317		22 35.9	+33 57	Peg	14.60	13.63	0.97	1.0	0.8	E2		7015		
7318A	12099	22 35.9	+33 58	Peg	14.30	13.28	1.02	1.0	1.0	E2p		6976		
7318B	12100	22 36.0	+33 58	Peg	14.0	13.1	0.95	1.9	1.3	SBb+p		6011*		Stephan's Quintet
7320	12101	22 36.1	+33 57	Peg	13.25	12.65	0.60	2.2	1.2	Sd		1042		
7319	12102	22 36.1	+33 59	Peg	14.0	13.1	0.95	1.7	1.3	SBb+p		6878		
		22 36.5	-2 53	Aqr	13.50			2.4	2.1	S+ IV-V		1867		D214, Arp 3
7331	12113	22 37.1	+34 25	Peg	10.35	9.51	0.84	10.7	4.0	Sb I-II		1105		
7335	12116	22 37.3	+34 27	Peg	14.7 p			1.7	0.8	S0		6577		
7332	12115	22 37.4	+23 48	Peg	11.85	10.95	0.90	4.2	1.3	E7		1451		
7339	12122	22 37.8	+23 47	Peg	13.00	12.14	0.86	3.0	0.9	S(B)b+:		1536		
7343	12129	22 38.6	+34 04	Peg	14.25	13.40	0.85	1.2	1.0	SBb+:		1495		
7329		22 40.5	-66 28	Tuc	12.3			4.2	2.8	SBb	4	3058		
7348	12142	22 40.6	+11 54	Peg	14.8 p			1.3	0.8	Sc:		7537		
		22 41.9	+20 16	Peg	14.95	14.31	0.64	0.6	0.4	Ir-:		7282		Mrk 308
IC 5240		22 41.9	-44 48	Gru	12.18			3.2	2.5	SBa	3S	1476		
7361	A434	22 42.3	-30 03	PsA	12.95	12.50	0.45	3.5	1.0	S IV:		1259		
7363		22 43.3	+34 00	Peg	14.5:p			1.1	1.1	S(B)d:		1107		
	12178	22 45.1	+6 26	Peg	13.35			3.4	1.8	S(B)dm		2142		
7371		22 46.1	-11 00	Aqr	12.8	12.1	0.71	2.1	2.0	Sb III:	3S	2525		
7377		22 47.8	-22 19	Aqr	12.65	11.64	1.01	2.2	1.8	E1	D3	3499		
7383		22 49.6	+11 36	Peg	14.85	13.86	0.99	1.0	0.9	SB0		8334		
7385	12207	22 49.9	+11 36	Peg	13.35	12.29	1.06	1.6	1.4	E1p		8051		
	12221	22 49.9	+82 54	Cep	14.10			2.8	0.9	Sd		2302		
7386	12209	22 50.0	+11 42	Peg	13.60	12.54	1.06	2.1	1.7	S0:		7465		
7389		22 50.3	+11 35	Peg	14.45	13.39	1.06	1.3	0.8	SB0		8138		
7387		22 50.3	+11 38	Peg	14.95	13.98	0.97	0.7	0.6	S0:		7151		
7393		22 51.7	-5 33	Aqr	14: p			2.0	1.0	SBcp	3S	3971		VV 68, Arp 15
7392		22 51.8	-20 36	Aqr	12.65	11.85	0.80	2.0	1.3	Sb II:	4VS	3015		
7410		22 55.0	-39 40	Gru	11.3	10.4	0.92	5.5	2.0	SBa	D3	1634		
7416		22 55.7	-5 30	Aqr	13.10	12.30	0.80	3.4	1.2	SBb	3VS	2927		
7412		22 55.8	-42 39	Gru	11.9	11.4	0.52	4.0	3.1	SBb I-II	3S	1686		
7418		22 56.6	-37 02	Gru	12.0	11.4	0.61	3.3	2.8	S(B)c II	4S	1518		
7421		22 56.9	-37 21	Gru	12.70	11.98	0.72	2.2	2.1	SBb-	D3S	1838		
IC 1459		22 57.2	-36 28	Gru	10.98	10.01	0.97			E3		1634		IC 5256
IC 5267		22 57.2	-43 24	Gru	11.40	10.47	0.93	5.0	4.1	S0:	3	1691		
7428	12262	22 57.3	-1 02	Psc	13.8 p			2.6	1.5	S(B)a		3194		
7424		22 57.3	-41 04	Gru	10.98			7.6	6.8	S(B)c II	D3S	850		
IC 5269		22 57.7	-36 02	PsA	12.78			2.5	1.1	S0		2134		
IC 5271		22 58.0	-33 45	PsA	12.6 H					Sb II		1758		
7437		22 58.2	+14 18	Peg	13.47			1.9	1.9	S(B)d		2345		
7440	12276	22 58.5	+35 48	And	14.6 p			1.9	1.6	SBa		5938		
	12281	22 59.2	+13 35	Peg	14.86			3.5	0.4	Sdm		2793		
7442	12286	22 59.4	+15 33	Peg	14.2 p			1.2	1.2	Sc:				
IC 5273		22 59.5	-37 42	Gru	11.90	11.45	0.45	2.9	2.1	SBc: II	3S	1308		
7448	12294	23 00.1	+15 59	Peg	12.15	11.65	0.50	2.7	1.3	Sc II:	4S	2431		Arp 13
7443		23 00.1	-12 48	Aqr	14: p			1.8	0.8	Sa:		3173*		
7444		23 00.1	-12 50	Aqr	14: p			2.0	0.9	S0:		3043*		
7457	12306	23 01.0	+30 09	Peg	11.65	10.76	0.89	4.4	2.5	Ep	3VS	790		
7454	12305	23 01.1	+16 23	Peg	13.6 p			2.0	1.6	E4	D4S	2240		
7460	12312	23 01.8	+2 16	Psc	14.2 p			1.3	1.0	Sbp		3481		
7464	12315	23 01.9	+15 58	Peg	14.30			0.7	0.7	E1p		2108		
7463	12316	23 01.9	+15 59	Peg	13.26			3.0	0.8	S(B)bp		2676		
7465	12317	23 02.0	+15 58	Peg	13.33			1.6	1.0	SB0:		2190*		Mrk 313
7456		23 02.1	-39 35	Gru	11.87			5.9	1.8	Sc:		1205		
7462		23 02.8	-40 50	Gru	12.66			3.7	0.6	SBc:		1061		
7468	12329	23 03.0	+16 36	Peg	13.89			1.1	0.8	E3p		2322*		Mrk 314
7469	12332	23 03.3	+8 52	Peg	12.60	11.85	0.75	1.8	1.3	S	5*	5102		Arp 298

NGC	UGC	α 2000	δ 2000	Const	B	V	B-V	Dim		Type	Nucl	RV	Memb	Notes
IC 5283		23^{h}03^m.3	+8°54'	Peg	14.8	13.8	1.05	0.9	0.4	Scp		5121		
7479	12343	23 04.9	+12 19	Peg	11.7	11.0	0.70	4.1	3.2	SBb+ I	3VS	2604		
	12350	23 05.3	+16 51	Peg	14.22			3.0	1.1	Sm		2371		
7495	12391	23 09.0	+12 04	Peg	14.7 p			2.1	1.9	S(B)c		5109		
7497	12392	23 09.1	+18 11	Peg	12.80			5.0	1.5	SBc		1945		
7496		23 09.8	-43 26	Gru	11.60	11.08	0.52	3.5	2.8	SBb	4S	1443		
7499	12397	23 10.4	+7 35	Psc	14.8	13.8	1.05	1.4	0.9	S0:		12117		
7501		23 10.5	+7 35	Psc	14.9	13.8	1.14	0.8	0.6	E1:		12915		
7503		23 10.7	+7 34	Psc	14.95	13.82	1.13	0.9	0.9	E2:		13430		
7507		23 12.1	-28 32	Scl	11.3	10.4	0.93	2.6	2.6	E0		1681		
7518	12422	23 13.1	+6 19	Psc	14.30			1.5	1.4	S(B)a		3645		Mrk 527
	12423	23 13.1	+6 24	Psc	14.03			3.6	0.5	Sc		1082		
7513		23 13.2	-28 22	Scl	12.6	11.8	0.78	3.2	2.6	SBbp				
7525		23 13.7	+14 01	Peg	15: p							12448		Mrk 316
7532		23 14.4	-2 42	Psc	14.5:p			1.8	0.7			3910		Mrk 529
7537	12442	23 14.6	+4 30	Psc	13.80	13.18	0.62	2.3	0.7	Sb+:		2875		
7541	12447	23 14.7	+4 32	Psc	12.45	11.72	0.73	3.5	1.4	Sc II:		2860		
7531		23 14.8	-43 36	Gru	11.97	11.26	0.71	3.5	1.5	Sb+	D3	1550		
7547	12453	23 15.0	+18 58	Peg	14.9 p			1.3	0.6	S(B)ap		5092		
7550	12456	23 15.3	+18 57	Peg	13.9 p			1.7	1.6	S0		5221		Arp 99
7549	12457	23 15.3	+19 02	Peg	14.1 p			2.9	0.9	SBcp		5041		
7562	12464	23 16.0	+6 41	Psc	12.6	11.5	1.06	2.3	1.6	E2		4001		
7552		23 16.2	-42 35	Gru	11.40	10.71	0.69	3.5	2.5	SBb-	5	1636		IC 5294
7578A	12477	23 17.2	+18 41	Peg	15.0 p			1.6	1.6	S0p		12185		VV 181, Arp 170
7578B	12478	23 17.3	+18 42	Peg	15.0 p			1.1	1.0	E1:		12326		VV 181, Arp 170
7576		23 17.4	-4 44	Aqr	13.8	13.0	0.81	1.5	1.2	S0	5	3729		
7580	12481	23 17.6	+14 00	Peg	14.8 p			0.9	0.7	S:		5015		Mrk 318
7587	12484	23 18.0	+9 41	Peg	14.9 p			1.4	0.4	SBb-:		8978		
7585		23 18.0	-4 39	Aqr	12.6	11.7	0.90	2.3	1.9	S	D3	3504		Arp 223
7592		23 18.4	-4 25	Aqr	14: p			1.4	1.1	Sb-p		7466		
7582		23 18.4	-42 22	Gru	11.40	10.63	0.77	4.6	2.2	SBb-	3VS	1427		
7601	12487	23 18.8	+9 14	Peg	14.7 p			1.5	1.2	S(B)c:		8408		
7600		23 18.9	-7 35	Aqr	12.99			2.4	1.1	E6	4S	3529		
7590		23 18.9	-42 14	Gru	12.22	11.61	0.61	2.7	1.1	Sb+:		1411*		
7603	12493	23 19.0	+0 15	Psc	14.4 p			1.7	1.1	Sbp	*	8960		Mrk 530, Arp 92
7606		23 19.1	-8 29	Aqr	11.55	10.76	0.79	5.8	2.6	Sb+ I	D3	2361*		
7599		23 19.3	-42 15	Gru	12.01	11.41	0.60	4.4	1.5	Sc	2VS	1661*		IC 5308
7609		23 19.5	+9 29	Peg	15: p			1.5	1.3	P		12130		VV 20, Arp 150
7611	12509	23 19.6	+8 04	Psc	13.55	12.60	0.95	1.5	0.7	SBa:		3582		
7610	12511	23 19.6	+10 10	Peg	14.9 p			3.0	2.3	Scp		3752		
7615		23 19.9	+8 24	Peg	15.5:p			1.2	0.7	Sb:				
7620	12520	23 20.1	+24 13	Peg	13.5 p			1.3	1.3	Sc:		9780		Mrk 321
7617		23 20.2	+8 10	Psc	14.75	13.82	0.93	1.0	0.8	S0:		4271		
7619	12523	23 20.2	+8 12	Peg	12.1	11.1	1.03	2.9	2.6	E1		3956		
7624	12527	23 20.4	+27 19	Peg	14.1	13.4	0.66	1.1	0.9	Sc		4729		Mrk 323
7623	12526	23 20.5	+8 24	Peg	13.45	12.43	1.02	1.9	1.3	S0:		3662		
7625	12529	23 20.5	+17 14	Peg	12.80	12.10	0.70	1.8	1.7	Ep	3S	1864		VV 280, Arp 212
7626	12531	23 20.7	+8 13	Peg	12.25	11.24	1.01	2.5	2.0	E2p		3638*		
7631	12539	23 21.4	+8 13	Peg	13.8 p			1.9	1.0	Sb:		3985		
7634	12542	23 21.6	+8 52	Peg	13.7 p			1.5	1.1	SB0		3350		
7640	12554	23 22.1	+40 51	And	11.44	10.90	0.54	10.7	2.5	S(B)b+ II:	1	642		
7648	12575	23 23.8	+9 39	Peg	13.5 p			1.9	1.3	S0		4252		IC 1486, Mrk 531
7649	12579	23 24.3	+14 38	Peg	14.8	13.7	1.10	1.6	1.1	E3		12732		
	12578	23 24.4	-0 07	Psc	14.77			1.7	1.2	Sp		2863		
	12588	23 24.7	+41 20	And	13.23			1.9	1.9	Sdm		698		
7671	12602	23 27.3	+12 28	Peg	13.6	12.7	0.91	1.7	1.0	S0:		4339		
7672		23 27.5	+12 23	Peg	15: p			1.0	0.9	Sb		4603		
7673	12607	23 27.7	+23 35	Peg	13.0	12.7	0.35	1.7	1.6	Scp		3650		Mrk 325, 4Z 149
7674	12608	23 27.9	+8 47	Peg	13.6 p			1.2	1.1	Sb+p		9047		VV 343, Arp 182
7675	12608	23 28.0	+8 46	Peg	14.8 p			0.7	0.4	S(B)0:		8861		VV 343, Arp 182
7677	12610	23 28.1	+23 32	Peg	13.9 p			1.9	1.2	S(B)b+		3783		Mrk 326
7678	12614	23 28.5	+22 25	Peg	12.8	12.2	0.65	2.3	1.8	Sc I	4S	3695		Arp 28
	12613	23 28.6	+14 45	Peg	12.63	12.04	0.59	4.6	3.0	Ir+ V		38	L	Pegasus dwarf
IC 5325		23 28.7	-41 19	Phe	12.19			2.5	2.5	S(B)b+	3S	1488		
7679	12618	23 28.8	+3 31	Psc	13.25	12.68	0.57	1.9	1.3	S	5	5333		VV 329, Arp 216
7682	12622	23 29.1	+3 32	Psc	14.3	13.4	0.92	1.2	1.0	SBb-	3S	5340		VV 329, Arp 216
	12632	23 30.0	+41 00	And	13.43			4.9	3.9	S+ V:		695		D217
7685	12638	23 30.6	+3 54	Psc	15.0 p			2.1	1.7	S(B)c				
7689		23 32.7	-54 05	Phe	12.15			2.8	2.0	S(B)c	3VS	1877		
7690		23 32.9	-51 41	Phe	12.68			2.8	1.3	Sb:	4VS	1317		
IC 5328		23 33.2	-45 02	Phe	11.8			2.5	1.6	E4	D3	3006		
IC 5332		23 34.5	-36 06	Scl	11.25	10.59	0.66	6.6	5.1	Sd	3VS	706		
	12682	23 34.9	+18 14		14.00			1.4	1.1	Irp+		1621		D218
7702		23 35.4	-56 00	Phe	13.14	12.22	0.92	1.9	1.1	S0	5	6376		
7714	12699	23 36.2	+2 09	Psc	13.01			2.1	1.6	SBbp	5	2980		VV 51, Arp 284

NGC	UGC	α 2000	δ 2000	Const	B	V	B−V	Dim		Type	Nucl	RV	Memb	Notes
7715	12700	23^{h}36^m.4	+2°09'	Psc	14.62			2'.6	0'.6	Irp+		2933		VV 51, Arp 284
7716	12702	23 36.5	+0 18	Psc	12.95	12.22	0.73	2.3	1.9	Sb II:	4	2708		
IC 5338	12703	23 36.5	+21 09	Peg	15.0	13.8	1.23	1.1	0.8	E5:		16693		3C 464
7713		23 36.5	−37 56	Scl	11.58			4.3	2.0	SBd:	1	662		
	12709	23 37.5	+0 25	Psc	14.35			3.0	1.8	S+ V:		2839		D219
	12710	23 37.5	+18 00	Peg	14.45			1.3	0.8	Irp+		2743		
	12713	23 38.2	+30 43	Peg	14.75	14.30	0.45	1.3	0.7	Sa		492		
7720	12716	23 38.5	+27 02	Peg	13.65	12.58	1.07	1.9	1.5	S0p		9238*		3C 465
7721		23 38.8	−6 31	Aqr	12.30	11.76	0.54	3.4	1.5	Sc II	3VS	2177		
7723		23 38.9	−12 58	Aqr	11.85	11.14	0.71	3.6	2.6	Sb	4S	1965		
7727		23 39.9	−12 18	Aqr	11.55	10.70	0.85	4.2	3.4	S(B)ap	D3VS	1928		VV 67, Arp 222
	12732	23 40.6	+26 14	Peg	12.86			3.0	3.0	Sm		998		
7731	12737	23 41.6	+3 45	Psc	13.73			1.7	1.3	SBa		2967		
7732	12738	23 41.7	+3 44	Psc	14.5 p			2.0	0.7	Sdmp:		3092		
7741	12754	23 43.9	+26 05	Peg	11.95	11.39	0.56	4.0	2.8	SBc II	1	1018		
7738	12757	23 44.1	+0 31	Psc	14.4 p			2.2	1.7	SBb				NGC 7739
7742	12760	23 44.3	+10 46	Peg	12.25	11.54	0.71	2.0	2.0	E0p	5S	1818		
7743	12759	23 44.4	+9 56	Peg	12.2	11.2	0.98	3.1	2.6	Sa:	4	1995		
7744		23 45.0	−42 55	Phe	12.58			2.3	1.8	S(B)0:	3S	2952		IC 5348?
7750	12777	23 46.7	+3 47	Psc	13.8 p			1.8	1.0	SBcp		3053		
7752	12779	23 47.0	+29 27	Peg	14.3 p			0.7	0.4	Ir−:		5125		VV 5, Arp 86
7753	12780	23 47.1	+29 29	Peg	13.2 p			3.4	2.2	S(B)b+	3S	5328*		VV 5, Arp 86
7755	A443	23 47.9	−30 31	Scl	11.78			3.7	3.0	S(B)b II		2942		
7757	12788	23 48.8	+4 10	Psc	13.9 p			2.6	2.4	Sc		3269		Arp 68
	12791	23 48.8	+26 14	Peg	14.87			1.5	0.6	Ir+		1037		D220
7768	12806	23 50.9	+27 10	Peg	14.0 p			1.7	1.4	E2		8193		
7764		23 50.9	−40 44	Phe	12.7	12.3	0.40	1.5	1.0	Ir+		1675		
7769	12808	23 51.1	+20 09	Peg	12.8	12.1	0.70	1.8	1.8	Sc	5	4570		
7770	12813	23 51.4	+20 06	Peg	14.5 p			1.0	0.9	Sa:		4558		
7771	12815	23 51.4	+20 07	Peg	13.1	12.3	0.84	2.7	1.3	SBa	5	4510		
7779	12831	23 53.4	+7 52	Psc	13.6 p			1.6	1.3	Sa:				
7780	12833	23 53.6	+8 07	Psc	14.8 p			1.2	0.7	Sb−		5307		
7782	12834	23 53.9	+7 58	Psc	13.1			2.4	1.4	Sb	D3S	5519		
7783	12837	23 54.2	+0 23	Psc	14.1 p			1.9	0.6	S0:		7932		VV 208, Arp 323
7785	12841	23 55.3	+5 55	Psc	12.60	11.59	1.01	2.3	1.4	E5:	2	4019		
	12843	23 55.6	+17 56	Peg	13.48			3.0	1.3	S(B)dm		1989		
	12856	23 56.8	+16 50	Peg	13.86			2.2	0.8	Ir+		1990		VV 255, Arp 262
7793		23 57.8	−32 35	Scl	9.70	9.11	0.59	9.1	6.6	Sdm III−IV	4VS	214		
7796		23 59.0	−55 27	Phe	12.32			2.3	2.0	E2	4S	3396		
7798	12884	23 59.4	+20 45	Peg	12.77			1.5	1.4	S		2933		Mrk 332
7800	12885	23 59.6	+14 49	Peg	12.86			2.6	1.9	Ir−:		1949		

SUPPLEMENTARY NOTES
(*Listed by NGC number*)

1. Pair with NGC 2.

24. Pair with NGC 45 in background of Sculptor group?

45. Surface brightness among the lowest of any galaxy known; background galaxies visible through the disk.

55. Brightest in Sculptor group.

70. Forms NGC 70 group (Arp 113) with NGC 67, 67A, 69-72, and 72A.

78B. In contact with NGC 78A.

80. In group with NGC 83 and 91.

124. In group with NGC 114, 118, and 120.

128. Brightest in group with NGC 125-7 and 130. "Box-shaped" nucleus.

134. Pair with NGC 131.

147. Satellite of M31; forms a physical pair with NGC 185 about 1° away. Distance is 730 kiloparsecs.

163. Pair with NGC 165.

169. In contact with IC 1559.

178. In NGC 210 group.

185. Satellite of M31; forms a physical pair with NGC 147 about 1° away. Small dust patch silhouetted against bright nucleus.

Distance 730 kiloparsecs.

190. Brightest in a compact group.

191. Interacting with IC 1563 (type I0p).

194. In NGC 200 group.

198. In NGC 200 group.

200. Brightest in group.

205. M110, a satellite of M31. Distance 730 kiloparsecs. Observed by Charles Messier in 1773, but added to his list by K. Glyn Jones in 1967. Two small dust patches with several blue supergiants discovered near nucleus.

210. Brightest in group.

221. M32, a satellite of M31. May be causing some distortion in the latter's spiral arm structure. Distance 730 kiloparsecs.

224. M31, the Andromeda galaxy. Largest member of the Local Group; distance 730 kiloparsecs. Easily visible to the naked eye, and described as a "little cloud" by the Arabic astronomer al-Sufi in A.D. 986. Among its satellite galaxies are M32 (NGC 221), M110 (NGC 205), and Andromeda I, II, III, and IV. The last object, which is not in our tabulation, is located at 0^h 42^m.5, +40° 34' (2000.0).

247. In Sculptor group.

252. Brightest in small group.

253. The Sculptor galaxy; in Sculptor group. Its plane is inclined 12° to the line of sight; complex dust lanes. Not to be confused with the Sculptor dwarf galaxy, a satellite of the Milky Way.

260. In multiple system with NGC 252.

273. In group with NGC 274-5.

274. Interacting with NGC 275.

300. In Sculptor group.

326. Close pair of compacts in common halo.

327. Pair with NGC 329.

357. Brightest in group.

383. Brightest in a chain with NGC 375, 379-80, 382, 384-6, and 388.

404. Local Group member? Semicircular dust lane.

407. In group with NGC 410 and 414.

434. Interacting with NGC 434A?

440. Pair with NGC 434.

474. Very faint rings. In group with NGC 467 and 470.

497. Peculiar broken arm on north preceding side.

507. Brightest in group with NGC 495, 499, and 508. Extensive halo.

520. According to the *Hubble Atlas of Galaxies*, this is not a colliding pair of galaxies, but rather an irregular

system similar to M82 (NGC 3034).

530. In cluster Abell 194.

535. In cluster Abell 194.

541. In cluster Abell 194. Distorted.

543. In cluster Abell 194.

547. With NGC 545, brightest pair in cluster Abell 194.

560. In group with NGC 564, IC 119, and IC 120.

565. In cluster Abell 194.

570. In cluster Abell 194.

584. In group.

586. In group.

596. Distorted with faint tail.

598. M33, the Triangulum galaxy. In Local Group; distance 900 kiloparsecs. Can just be glimpsed without optical aid in very dark, clear skies. Brightest H II region is NGC 604, in the northeast part of the galaxy.

612. In cluster.

628. M74. Nearly face on.

646. Western arm knotty, eastern arm smooth with satellite attached.

672. Interacting with IC 1727?

678. In NGC 697 group.

680. In NGC 697 group.

691. In NGC 697 group.

694. In NGC 697 group.

695. In NGC 697 group.

697. Brightest in group.

701. Pair with IC 1738.

702. Interacting with small companion?

708. Superimposed star. In cluster Abell 262.

718. Three pairs of spiral arms, each pair of different curvature.

736. In group.

740. In group.

741. Brightest elliptical in group. In common halo with NGC 742.

750. Interacting with NGC 751.

772. Spiral arm seems to bend around NGC 770.

777. Pair with NGC 778.

788. Pair with IC 184.

828. Distorted.

833. Interacting with NGC 835. In group with NGC 838-9 and 848. Faint, diffuse streamers.

871. Pair with NGC 870.

877. Pair with NGC 876 (type Sc, photographic magnitude 16) at 2'.1. In a group with NGC 870-1.

891. In NGC 1023 group. Edge on; broad dust lane.

895. NGC 894 is part of it.

899. Interacting with IC 223 and NGC 907?

910. In cluster Abell 347.

922. Interacting with an anonymous barred spiral?

925. In NGC 1023 group.

936. Pair with NGC 941.

942. NGC 943; has distorted absorption ring.
945. Pair with NGC 948.
972. Chaotic dust lanes.
985. Ring galaxy.
991. In NGC 1052 group.
1003. In NGC 1023 group; cluster of spirals in background.
1022. In NGC 1052 group.
1023. Brightest in group.
1035. In NGC 1052 group.
1042. In NGC 1052 group.
1044. Double radio galaxy; also pair with NGC 1046 (type S0?).
1047. In NGC 1052 group.
1048A. In NGC 1052 group.
1048B. In NGC 1052 group.
1052. Brightest in group; active radio galaxy.
1055. In NGC 1068 group.
1058. In NGC 1023 group.
1068. M77, Arp 37. Brightest in group.
1073. In NGC 1068 group.
1087. In NGC 1068 group. Pair with NGC 1090.
1090. In NGC 1068 group. Pair with NGC 1087.
1094. In NGC 1068 group.
1097. Peculiar nucleus with inner spiral structure.
1097A. Small companion of NGC 1097.
1104. In NGC 1068 group.
1134. Brightest in group. Splash appearance on western side points to 15.5-magnitude Ir companion UGC 2362.
1143. Ring galaxy; interacting with NGC 1144.
1160. Pair with NGC 1161.
1169. Star superimposed near nucleus.
1175. Pair with NGC 1177.
1199. Brightest in group.
1209. In NGC 1199 group.
1229. In chain with NGC 1228, 1230, and IC 1892.
1232. NGC 1232A is satellite.
1241. Pair with NGC 1242.
1248. Pair with anonymous barred spiral.
1253. Pair with NGC 1253A.
1265. Radio galaxy with tail.
1275. Radio source Perseus A. Brightest in Perseus cluster (Abell 426). Seyfert galaxy.
1293. Pair with NGC 1294.
1315. In NGC 1332 group.
1316. Radio source Fornax A. Extended envelope with faint loops.
1316C. NGC 1316A and 1316B lie 19′ north preceding.
1319. In NGC 1332 group.
1320. Pair with NGC 1321 (Mrk 608) at 1′.7 north.
1325. Pair with NGC 1318. In NGC 1332 group.
1325A. Pair with NGC 1325.
1331. In NGC 1332 group.
1332. Pair with NGC 1331. In group.
1343. Ring structure in nucleus; faint bar.
1355. Pair with NGC 1358.
1359. Pair with anonymous barred spiral.
1365. In foreground of Fornax cluster?
1374. Pair with NGC 1375.
1376. Pair with anonymous spiral.
1393. In NGC 1407 group.
1398. Internal ring with bar.
1400. In NGC 1407 group.

1407. Brightest in group. Pair with NGC 1400.
1409. Pair with Seyfert galaxy NGC 1410. Combined magnitude and mean radial velocity are given.
1415. Pair with NGC 1416.
1417. In group with NGC 1418 and 1424.
1426. Pair with NGC 1422.
1428. A star superimposed on the nucleus may have affected some radial-velocity measurements.
1439. Pair with NGC 1426.
1440. Pair with NGC 1452.
1441. In NGC 1453 group.
1449. In NGC 1453 group.
1451. In NGC 1453 group.
1452. Pair with NGC 1440.
1453. Brightest in group.
1482. Pair with NGC 1481, of photographic magnitude 15 and 3′.3 north preceding.
1487. Peculiar colliding system?
1510. Pair with NGC 1512.
1518. Pair with NGC 1521.
1531. Pair with NGC 1532.
1549. Pair with NGC 1553.
1560. IC 2062.
1569. Much absorption.
1573. Connected pair. Brightest in group.
1587. Connected to NGC 1588.
1589. Pair with NGC 1587-8.
1590. Peculiar spiral.
1596. Pair with NGC 1602.
1599. In NGC 1600 group.
1600. Brightest in group.
1601. In NGC 1600 group.
1618. In NGC 1625 group.
1622. In NGC 1625 group.
1625. Brightest in group.
1667. Pair with NGC 1666.
1700. Pair with NGC 1699.
1726. Triple with NGC 1720 and IC 398.
1741. Peculiar interacting system with IC 399.
1779. Pair with IC 402 at 14′.5.
1800. Peculiar close pair.
1875. Extended halo. Nearby triplet of distorted galaxies.
1888. Interacting with NGC 1889.
1954. Pair with NGC 1957 (type E) at 5′.1 south following.
1961. Peculiar outer arm and streamers.
2146. Interacting with NGC 2146A?
2207. Interacting with IC 2163?
2273. Pair with NGC 2273B.
2276. Interacting with NGC 2300; arm begins straight, then bends.
2290. In group with NGC 2288-9, 2291, and 2294.
2314. Pair with IC 2174.
2329. Brightest in the cluster Abell 569.
2336. Pair with IC 467.
2341. Pair with NGC 2342.
2347. Pair with IC 2179.
2366. In M81 group.
2377. Much galactic obscuration.
2379. In NGC 2389 group.
2389. Brightest in group.
2397. Pair with NGC 2397A (type Sc?) at 10′ north preceding.
2403. In the M81 group. Similar in appearance to M33 (NGC 598).
2442. NGC 2443 is part of it.
2444. Interacting with

NGC 2445. Distorted ring.
2460. Pair with IC 2209.
2474. Pair with NGC 2475.
2493. Pair with NGC 2495 (Mrk 383), of photographic magnitude 15.5 and 1′.9 north following.
2500. In group with NGC 2541 and 2552.
2523. Pair with NGC 2523B. Bifurcated arm does not start at end of bar. Internal ring.
2536. Interacting with NGC 2535. Long distorted arm.
2537. Bear Claw or Bear Paw galaxy. Pair with NGC 2537A.
2541. In group with NGC 2500 and 2552.
2552. In group with NGC 2500 and 2541.
2566. In compact group.
2573. Polarissima Australis; so named by John Herschel in 1837 because it lies so close to the south celestial pole. His position for 1830.0 has been carefully precessed to 2000.0 here; other sources differ widely.
2583. In group with NGC 2584-5.
2608. This has a double nucleus, or a star superimposed on the nucleus.
2623. Bright knots and streamers in central region.
2629. Pair with NGC 2641.
2633. Brightest in group. Pair with NGC 2634. Absorption lane crosses arm near nucleus.
2634. Pair with NGC 2634A.
2642. Pair with anonymous spiral.
2648. Pair with 15th-magnitude spiral. Diffuse arm extends beyond companion.
2655. Very faint and diffuse outer arms; absorption on one side of nucleus.
2656. Double system.
2672. Pair with NGC 2673. Brightest in group.
2685. The Helix galaxy. Appears cigar-shaped, but has been shown to be an edge-on S0 galaxy with helical filaments along minor axis. See P. L. Schechter and J. E. Gunn, *Astronomical Journal*, **83**, 1360, 1978.
2693. Pair with NGC 2694.
2698. In group with fainter galaxies NGC 2690, 2695, 2697, 2699, 2702, and 2706.
2708. Pair with NGC 2709, of photographic magnitude 15 and 8′ north following.
2713. Pair with NGC 2716.
2719. Pair with NGC 2719A, which has a faint tail. Combined magnitude is given.
2732. Pair with an anonymous spiral having a bright nucleus.
2744. In NGC 2749 group.
2749. Brightest in group.
2752. In NGC 2749 group.
2775. Pair with NGC 2777. Almost face on.
2782. Has a peculiar nucleus and diffuse outer arms. Interacting with nearby spiral?
2798. Interacting with NGC 2799.
2805. Interacting with NGC 2814, 2820, and IC 2458.
2823. In cluster Abell 779.
2825. In cluster Abell 779.
2826. In cluster Abell 779?

2832. In cluster Abell 779 with NGC 2830-1.
2844. Brightest in a group with NGC 2852-3.
2855. Prominent dust lanes.
2859. External ring.
2865. Foreground star lies on minor axis.
2872. Pair with NGC 2874.
2889. Pair with NGC 2884.
2893. Brightest in group.
2907. Pair with an anonymous spiral.
2911. Brightest in group. A foreground star lies 25″ north preceding the nucleus. NGC 2912 is a faint galaxy 1′.3 north following.
2914. In NGC 2911 group.
2919. In NGC 2911 group.
2936. Ring. Interacting with NGC 2937.
2944. A dwarf irregular galaxy is attached.
2959. NGC 2961 lies north following (photographic magnitude 16, type Sb?).
2964. Brightest in group.
2968. In NGC 2964 group.
2970. In NGC 2964 group.
2976. In the M81 group.
2985. Pair with NGC 3027.
2986. Pair with an anonymous spiral.
2992. Interacting with NGC 2993. Faint extensions.
2998. Brightest in group.
3020. In a group with NGC 3016, 3019, and 3024.
3031. M81, brightest in group.
3032. Internal absorption ring.
3034. M82, Arp 337. In the M81 group. Explosion in nucleus?
3052. Pair with NGC 3045.
3065. Interacting with NGC 3066. Possible internal absorption ring.
3073. Pair with NGC 3079.
3077. In the M81 group. Similar to NGC 5195, the companion of the Whirlpool galaxy M51 (NGC 5194).
3091. Pair with NGC 3096.
3095. Pair with NGC 3100.
3104. Many emission knots and faint extensions.
3109. Edge on.
3115. The Spindle galaxy. Almost spherical nuclear bulge with extremely thin disk.
3145. Pair with NGC 3143. Internal ring.
3151. In NGC 3158 group.
3156. In NGC 3166 group.
3158. Brightest in group. Pair with NGC 3160.
3159. Pair with NGC 3161. In NGC 3158 group.
3163. In NGC 3158 group.
3165. In NGC 3166 group.
3166. Brightest in group. Interacting with NGC 3169.
3172. Polarissima Borealis, discovered in 1831 by John Herschel and named for its extreme proximity to the north celestial pole (only 5′ in that year). The visual magnitude estimate was made in 1978 by John Bortle, using a 12-inch reflector.
3185. In NGC 3190 group.
3187. Interacting with NGC 3190. Distorted.
3188. Pair with NGC 3188A (Mrk 30) at 0′.7 south preceding.
3190. Arp 316. Brightest in group; interacting with NGC 3187.
3193. In NGC 3190 group.
3214. Pair with NGC 3220.

3227. A large, very faint loop extends to NGC 3226.
3239. Many emission knots and faint extensions. There is a superimposed star.
3245. Pair with NGC 3245A.
3250. Five fainter spirals (?) following.
3256. Interacting with type-SBd galaxy 14′ north following?
3258. Pair with NGC 3257.
3259. Pair with NGC 3266.
3260. Pair with NGC 3258.
3267. In NGC 3267-3281 group.
3268. In NGC 3267-3281 group.
3269. In NGC 3267-3281 group.
3271. In NGC 3267-3281 group.
3273. In NGC 3267-3281 group.
3275. Pair with NGC 3275A.
3281. In NGC 3267-3281 group.
3285. Triple with NGC 3285A and 3285B.
3288. Pair with NGC 3286.
3289. In NGC 3267-3281 group.
3290. Asymmetric arms with emission knots.
3294. Pair with NGC 3304.
3299. Pair with NGC 3306.
3303. Peculiar spiral with compact companion. Faint extensions. Combined magnitude and mean radial velocity given.
3309. Brightest in Hydra I cluster (Abell 1060).
3311. Pair with NGC 3309.
3318. Pair with NGC 3318B, of type SBc and 10′.5 north following.
3329. Brightest in group of 12 galaxies.
3338. In Leo group.
3347. In group with NGC 3354 and 3358.
3351. M95. In Leo group. Internal ring with bar.
3368. M96. Brightest in the Leo group.
3377. Pair with NGC 3377A.
3379. M105. In Leo group.
3384. Pair with NGC 3379. In Leo group.
3389. In the background of M96.
3395. Interacting with NGC 3396.
3400. In NGC 3414 group.
3412. In the Leo I cloud.
3414. Brightest in group.
3418. In NGC 3414 group.
3419. Pair with NGC 3419A.
3430. In group with NGC 3413 and 3424.
3432. Dwarf companion near south-preceding end. Nearly edge on.
3440. In group with NGC 3445 and 3458.
3445. In group with NGC 3440 and 3458.
3447. In contact with NGC 3447A.
3448. Interacting with faint spiral. Diffuse extensions.
3454. Pair with NGC 3455.
3458. In group with NGC 3440 and 3445.
3489. In the Leo I cloud.
3504. Pair with NGC 3512.
3507. A star of magnitude 11 is superimposed.
3511. Pair with NGC 3513.
3512. In a group with NGC 3504 and 3515.

3550. Double system in the Leo A cluster (Abell 1185).

3556. Nearly edge on; no nucleus visible.

3557. In group with NGC 3564 and 3568.

3558. In the Leo A cluster.

3561. Multiple system in the Leo A cluster. Long, diffuse tail. Combined magnitude and mean radial velocity are given. NGC 3561C, a blue compact galaxy, is called Ambartsumian's knot.

3583. Pair with NGC 3577. Small type-E0 satellite at 0′.9 north following.

3593. In the M66 group.

3599. In NGC 3607 group.

3605. In NGC 3607 group.

3607. Brightest in group.

3608. In NGC 3607 group.

3613. Pair with NGC 3619.

3614. Pair with NGC 3614A, of type SBm and 2′.5 south preceding.

3619. In a group with NGC 3613 and 3625.

3623. M65, Arp 317. Pair with NGC 3627. Prominent absorption lane.

3625. Pair with NGC 3619.

3627. M66, Arp 16, Arp 317. Pair with NGC 3623. Large concentration at end of southern arm.

3628. Edge on.

3630. In NGC 3630-3645 group.

3631. Straight arms; absorption lane crossing southern arm.

3633. In NGC 3630-3645 group.

3637. Pair with NGC 3636. Foreground star at 1′.7.

3640. Pair with NGC 3641. In NGC 3630-3645 group.

3643. In NGC 3630-3645 group.

3646. Pair with NGC 3649.

3656. Has a peculiar dark lane and a diffuse arm.

3664. Interacting with NGC 3664A.

3665. Pair with NGC 3658.

3672. Extremely small nucleus.

3674. In a group with NGC 3683 and a faint barred spiral.

3683. Pair with NGC 3683A.

3684. In a group with NGC 3681 and 3686.

3690. Interacting with IC 694, of type SBmp(?) and 0′.6 north following.

3717. Pair with IC 2913.

3718. Interacting with NGC 3729. Unusual, bar-shaped dust lane cuts across nucleus. Arp 322, a peculiar chain of galaxies, lies 7′ south.

3720. Pair with NGC 3719.

3732. Brightest in a group with NGC 3723 and 3763.

3733. Pair with NGC 3737. In cluster Abell 1318?

3738. Pair with NGC 3756.

3745. Pair with NGC 3746. Both are part of Copeland's Septet (VV 282, Arp 320), along with NGC 3748, 3750-1, and 3753-4.

3759. Pair with IC 2943.

3769. Pair with NGC 3769A.

3780. Pair with NGC 3804.

3786. Interacting with NGC 3788. Peculiar filaments.

3799. Interacting with NGC 3800.

3800. Connected to NGC 3800A.

3801. Pair with NGC 3802.

3804. Pair with NGC 3780.

3808. An arm appears to extend around a companion.

3842. In cluster Abell 1367.

3846. In NGC 3846-3898 group.

3846A. In NGC 3846-3898 group.

3850. In NGC 3846-3898 group.

3862. Radio source 3C 264. In cluster Abell 1367.

3877. A foreground star is superimposed on the nucleus.

3888. Pair with NGC 3898. In NGC 3846-3898 group.

3893. Pair with NGC 3896.

3894. Pair with NGC 3895.

3898. Pair with NGC 3888. In NGC 3846-3898 group.

3904. Pair with NGC 3923.

3913. Interacting with NGC 3921?

3916. Pair with NGC 3921.

3917. In group with NGC 3931 and an anonymous spiral.

3921. Pair with NGC 3916. Interacting with NGC 3913? Long north-south filaments extend from a bright offset nucleus.

3923. Pair with NGC 3904.

3928. The Miniature spiral.

3931. Pair with NGC 3917.

3963. Pair with NGC 3958.

3972. Pair with NGC 3977.

3975. Pair with NGC 3978.

3977. Pair with NGC 3972.

3986. Pair with IC 2978.

3990. Pair with NGC 3998.

3991. Interacting with NGC 3994-5.

3992. M109. Observed by Charles Messier, but added to his list by Owen Gingerich in 1953.

3993. In a group with NGC 3987, 3989, and 3997.

3994. Interacting with NGC 3991 and 3995.

3998. Pair with NGC 3990.

4004. Very peculiar and asymmetric. Pair with IC 2982.

4010. Pair with NGC 4001.

4016. With distorted, patchy, figure-eight loops. Faint filaments extend to NGC 4017.

4027. Interacting with dwarf irregular NGC 4027A.

4036. Pair with NGC 4041.

4038. Colliding with NGC 4039 to form Arp 244, known as the Antennae because of the long, thin streamers.

4045. Pair with NGC 4045A.

4061. Pair with NGC 4065.

4073. In a group with NGC 4077 and other faint galaxies.

4088. Pair with NGC 4085. Faint extension to arm on preceding side.

4105. Interacting with NGC 4106.

4111. Pair with NGC 4109. Edge on, broken dust lane.

4116. Pair with NGC 4123.

4117. Pair with NGC 4118.

4125. Pair with NGC 4121.

4128. Holmberg 337a; pair with Holmberg 337b.

4131. Brightest of group.

4145. Holmberg 342a; pair with Holmberg 342b.

4151. In a group with NGC 4156 and Holmberg 345c.

4156. Double nucleus. In a group with NGC 4151 and Holmberg 345c.

4168. Pair with NGC 4165. Brightest in group.

4169. Forms the Box with

NGC 4173-5.

4186. UGC 7240; misidentified in the UGC. Pair with NGC 4192.

4189. In group with NGC 4164 and 4193.

4192. M98. In a group with NGC 4186 and Holmberg 348c.

4193. In the NGC 4189 group.

4194. Faint plumes and jets to the north.

4206. In a multiple system with NGC 4216, 4222, and IC 771.

4211. Pair connected by a diffuse tail. Combined magnitude and mean radial velocity are given.

4216. In multiple system with NGC 4206, 4222, and IC 771. Nearly edge on.

4217. Pair with NGC 4226.

4220. Pair with NGC 4218.

4222. In a multiple system with NGC 4206, 4216 and IC 771.

4231. Pair with NGC 4232.

4235. In a multiple system with NGC 4246, 4247.

4236. Holmberg 357a, pair with Holmberg 357b.

4238. In the foreground of a compact cluster of galaxies.

4244. Edge on; no nuclear bulge.

4245. Pair with NGC 4253.

4246. In multiple system with NGC 4235 and 4247.

4248. Pair with NGC 4258.

4254. M99.

4258. M106. Observed by Charles Messier, but added to his list by Helen Sawyer Hogg in 1947.

4259. In NGC 4273 group.

4261. Pair with NGC 4264.

4266. In NGC 4273 group.

4267. Pair with IC 775.

4268. In NGC 4273 group.

4269. Pair with IC 3155.

4270. In NGC 4273 group.

4273. Brightest in group.

4277. In NGC 4273 group.

4278. In group with NGC 4283 and 4286.

4281. In NGC 4273 group.

4283. In group with NGC 4278 and 4286.

4286. In group with NGC 4278 and 4283.

4288. Holmberg 371a; pair with Holmberg 371b.

4290. Pair with NGC 4284 (type Sc?).

4291. Pair with NGC 4319.

4292. Holmberg 375a; pair with Holmberg 375b.

4293. Broad absorption lanes.

4294. Pair with NGC 4299.

4298. Pair with NGC 4302.

4301. UGC 7411; misidentified in the UGC.

4303. M61. Pair with NGC 4303A.

4305. Pair with NGC 4306.

4307. Holmberg 380a; pair with Holmberg 380b.

4309. Holmberg 382a; pair with Holmberg 382b.

4312. In group with M100 (NGC 4321) and several faint galaxies.

4314. Unconnected inner and outer spiral patterns.

4319. Pair with NGC 4291. Suspected connection to the compact galaxy Mrk 205, which has a much greater redshift.

4321. M100, brightest spiral in the Virgo cluster. Pair with NGC 4312.

4340. Pair with NGC 4350.

4350. Pair with NGC 4340.

4365. Pair with NGC 4370.

4373. Pair with IC 3290.

4374. M84. Pair with NGC 4387.

4382. M85. Pair with NGC 4394. Faint, diffuse companion 3′ south.

4387. Pair with NGC 4374.

4394. Pair with NGC 4382.

4395. This galaxy of low surface brightness has four bright patches in it, each of which has an NGC designation (NGC 4395, 4399, 4400, 4401).

4406. M86. Pair with NGC 4402.

4410. Close pair in a common envelope.

4411A. Pair with NGC 4411B.

4428. Pair with NGC 4433.

4433. Chaotic dust lanes.

4438. Interacting with NGC 4435. Disrupted? Called the Eyes by Leland S. Copeland.

4440. In a group with NGC 4431 and 4436.

4450. Circular dust lanes start outside nucleus.

4457. One spiral arm brighter than the other.

4459. Internal absorption ring.

4461. Pair with NGC 4458.

4472. M49, discovered in 1771 by B. Oriani. In a group with NGC 4467, 4470, and a faint dwarf irregular galaxy. A foreground star of magnitude 13 lies just east of M49.

4478. Pair with NGC 4476.

4486. M87, the radio source Virgo A. There is a jet extending northwest from the bright nucleus.

4490. With NGC 4485, forms the interacting system Arp 269.

4496A, B. Colliding pair? Same as NGC 4505?

4501. M88.

4517. Pair with NGC 4517A.

4519. Holmberg 418a; pair with Holmberg 418b.

4526. Called the Lost galaxy by L. S. Copeland. Internal absorption ring.

4535. Holmberg 420a; pair with Holmberg 420b.

4536. Pair with NGC 4533.

4540. Pair with IC 3528.

4548. Pair with NGC 4571.

4550. Pair with NGC 4551.

4552. M89; discovered by Charles Messier in 1781.

4556. Close pair with companion in Coma cluster.

4565. Edge on.

4567. Interacting with NGC 4568 to form the Siamese Twins.

4569. M90. Interacting with dwarf companion IC 3583?

4571. Pair with NGC 4548.

4578. Holmberg 429a; pair with Holmberg 429b.

4579. M58. Forms the northern apex of an equilateral triangle with the stars ρ and 20 Vir.

4589. In a group with NGC 4572 and 4648.

4593. Broken internal ring with bar.

4594. M104, the Sombrero galaxy. Inclined only 6° to the line of sight. Prominent absorption lane; large nuclear bulge.

4596. Pair with NGC 4608.

4602. Pair with IC 804.

4603D. In Centaurus chain.

4606. Pair with NGC 4607.

4608. Pair with NGC 4596.

4616. In Centaurus chain.

4618. Interacting with NGC 4625?

4621. M59. Discovered in 1779 by J. G. Köhler while observing Comet Bode of that year.

4622. In Centaurus chain.

4622A. Close pair with NGC 4622B in Centaurus chain.

4625. Interacting with NGC 4618?

4631. Interacting with NGC 4627, which has a diffuse countertail. Edge on.

4633. Pair with NGC 4634.

4638. In a group with NGC 4635 and 4637.

4644. Pair in Haro 32 group. Combined magnitude and mean radial velocity are given.

4645A. Pair with NGC 4645B.

4646. In Haro 32 group.

4648. In group with NGC 4572 and 4589.

4649. M60, Arp 116. Discovered by J. G. Köhler while observing Comet Bode of 1779. Pair with NGC 4647, whose absorption lanes are more pronounced on the side away from M60.

4650. In Centaurus chain.

4650A. In Centaurus chain. Has the shape of a type-E6 galaxy, but with a warped gas and dust ring along the minor axis. Similar in appearance to the Helix galaxy (NGC 2685).

4650B. In Centaurus chain.

4653. In a group with NGC 4666 and 4668.

4656. Northeastern part brightest. Interacting with NGC 4657.

4658. Pair with NGC 4663.

4666. Pair with NGC 4668.

4669. In Haro 32 group.

4670. Pair with NGC 4673.

4675. In Haro 32 group.

4676A. Interacting with NGC 4676B (IC 820) to form VV 224, the Mice. Long diffuse filaments.

4686. In Haro 32 group.

4688. Holmberg 461a. Pair with small irregular galaxy Holmberg 461b.

4692. IC 823 is star at 1′.3 south preceding.

4695. In Haro 32 group.

4696. Brightest galaxy in the Centaurus cluster.

4697. Interacting with an anonymous barred spiral.

4700. Pair with an anonymous spiral.

4707. Many emission knots.

4710. Edge on; broken dust lane.

4712. Pair with NGC 4725.

4719. Weak Seyfert nucleus.

4725. Pair with NGC 4712. Internal ring.

4727. In a group with NGC 4724 and 4726 (=4740).

4728. Holmberg 469a, in a group with Holmberg 469b and 469c.

4731. Pair with a dwarf irregular galaxy.

4736. M94, discovered by P. Méchain in 1781. Has a large, bright central region without spiral arms, and also a very faint outer ring of diameter 15′.

4747. Pair with NGC 4725. Faint extensions to northeast.

4753. Chaotic dust lanes.

4754. Pair with NGC 4762.

4762. Edge on; one of the flattest galaxies known.

4760. Pair with NGC 4757.

4774. Ring galaxy.

4780. Holmberg 482a; pair with Holmberg 482b.

4781. In a group with NGC 4784 and 4790.

4782. Connected to NGC 4783.

4789. In a group with NGC 4787 and 4789A.

4790. In a group with NGC 4781 and 4784.

4794. Pair with NGC 4792, which is of type S0 and 7′ north preceding.

4807. Pair with NGC 4807A

4808. Pair with an anonymous barred spiral.

4809. Interacting with NGC 4810.

4819. Pair with NGC 4821.

4825. In a group with NGC 4820, 4823, 4829, and other galaxies.

4826. M64. Called the Black-eye galaxy because of the dark dust lane seen in silhouette against the bright nucleus.

4835. Pair with dwarf irregular galaxy.

4837. Connected pair.

4841. Pair in a common envelope. Combined magnitude and mean radial velocity are given.

4842. Pair.

4849. Pair with IC 838.

4861. IC 3961 is a bright emission patch at southwest end.

4889. Brightest member of the Coma cluster.

4898. Double system in contact.

4915. In a group with NGC 4918 and D160.

4922. Double system.

4933A. Interacting with NGC 4933B.

4936. Pair with IC 844.

4957. Pair with NGC 4961.

4958. In a group with NGC 4948, 4948A.

5004. Pair with NGC 5004A.

5005. Circular dust lanes start outside nucleus.

5012. Pair with an anonymous peculiar galaxy.

5018. Pair with NGC 5022.

5030. In NGC 5044 group.

5033. Holmberg VIII (UGC 8303) is its companion.

5037. In NGC 5044 group.

5044. Brightest in group.

5046. In NGC 5044 group.

5047. In NGC 5044 group.

5049. In NGC 5044 group.

5054. Pair with a small anonymous barred spiral.

5055. M63, the Sunflower galaxy. Discovered in 1779 by P. Méchain.

5077. Pair with NGC 5079. Brightest in group.

5078. Pair with IC 879.

5088. In NGC 5077 group.

5090. Pair with NGC 5091.

5098. Double system. Combined magnitude given.

5112. Pair with NGC 5107

5128. Radio source Centaurus A. Broad, disrupted absorption band. Colliding galaxies?

5134. Pair with IC 4237.

5135. Pair with IC 4248.

5141. Pair with NGC 5142.

5144. Distorted with condensations and jets.

5173. Pair with NGC 5169.

5193. Pair with NGC 5193A at 0′.5 south preceding.

5194. M51, the Whirlpool galaxy, whose spiral structure was first observed in 1845 with Lord Rosse's 72-inch reflector at Birr Castle, Ireland. Connected to NGC 5195. Faint plumes and extensions.

5204. In the M101 group.

5218. Connected by long streamer to NGC 5216. Known as Keenan's system (VV 33, Arp 104).

5236. M83, discovered by N. Lacaille in 1751.

5256. Double system.

5257. Connected to NGC 5258.

5273. Pair with NGC 5276.

5278. Connected to NGC 5279.

5291. Interacting with an undesignated system called the Seashell galaxy at 0′.6 away. In IC 4329 group.

5292. In IC 4329 group.

5297. Interacting with NGC 5296?

5298. Pair with an anonymous barred spiral. In IC 4329 group.

5302. In IC 4329 group.

5303. Holmberg 542a; pair with Holmberg 542b.

5304. In IC 4329 group.

5328. Pair with NGC 5330.

5350. In a group with NGC 5353-5.

5351. In a group with NGC 5341 and 5349.

5357. In IC 4329 group.

5364. In a group with NGC 5360 and 5363. Internal ring.

5376. In a group with NGC 5379 and 5389.

5380. Pair with NGC 5378.

5395. Connected to NGC 5394.

5403. Holmberg 564a; pair with Holmberg 564b.

5426. Arms linked with NGC 5427.

5457. M101, the Pinwheel galaxy. Because of an 18th-century error, the designation M102 also applies to this object.

5468. Pair with NGC 5472. NGC 5467 is either a star or a feature in the galaxy.

5474. In M101 group.

5477. In M101 group.

5480. Pair with NGC 5481.

5485. In a group with NGC 5484 and 5486.

5490. In a group with IC 982, 983, and others.

5506. Pair with NGC 5507.

5532. Often misidentified as IC 5532. Brightest in group.

5544. Colliding with NGC 5545. Spiral arms appear disturbed.

5556. Pair with an anonymous spiral.

5566. Interacting with NGC 5560 and 5569, but connection is not visible. Distorted; internal ring with bar.

5576. Pair with NGC 5574.

5585. In M101 group.

5595. Pair with NGC 5597.

5607. Peculiar spiral.

5613. External ring.

5614. Interacting with NGC 5613. NGC 5615 is a

bright knot (colliding galaxy?) in NGC 5614.

5638. Pair with NGC 5636.

5660. Pair with a faint anonymous irregular galaxy.

5665. There is an attached compact E0p galaxy.

5676. NGC 5660 at 30′.5.

5678. Pair with an anonymous E3 galaxy.

5689. Brightest in a group with NGC 5682-3, 5693, and others.

5701. A small anonymous spiral is visible between the lens and outer ring.

5713. Interacting with NGC 5719.

5728. In a group with NGC 5716 and a small anonymous barred spiral.

5740. Pair with NGC 5746.

5757. Pair with an anonymous spiral.

5774. Pair with NGC 5775.

5787. Has a faint halo.

5791. Pair with IC 1077.

5796. Pair with NGC 5793.

5806. In NGC 5846 group. Pair with NGC 5813-4.

5812. Pair with IC 1084.

5813. In NGC 5846 group. Pair with NGC 5814.

5820. Pair with NGC 5821. Faint streamers.

5827. Asymmetric, peculiar.

5829. Interacting with IC 4526.

5838. In NGC 5846 group. Pair with NGC 5848.

5839. In NGC 5846 group.

5845. In NGC 5846 group.

5846. Brightest in group. Pair with NGC 5846A.

5850. In NGC 5846 group.

5857. Interacting with NGC 5859?

5860. Close pair in common halo.

5861. Pair with NGC 5858.

5866. Thin, edge-on dust lane tilted 2° to the galactic plane. NGC 5867 is a nearby small elliptical galaxy.

5888. Pair with NGC 5889.

5893. Pair with NGC 5895-6.

5898. A triple system with NGC 5903 and an anonymous galaxy.

5899. Pair with NGC 5900.

5903. Triple system with NGC 5898 and anonymous galaxy.

5905. Pair with NGC 5908.

5907. NGC 5906 is part of it. Nearly edge on.

5915. Interacting triple system with NGC 5916 and 5916A.

5920. In group.

5930. Colliding with NGC 5929.

5953. Colliding with NGC 5954. A broad, peculiar arm connects them.

5967. Pair with NGC 5967A.

5970. Pair with IC 1131.

5981. Pair with NGC 5982.

5985. Pair with NGC 5981-2.

5992. Pair with NGC 5993.

6018. In foreground of cluster Abell 2147.

6021. In foreground of cluster Abell 2147.

6022. In Abell 2147.

6023. In cluster Abell 2147.

6027. Seyfert's Sextet. Combined magnitude given. One component (type Scp) has a radial velocity of about 20,000 km/sec, while others

are in the range 4,212 to 4,615 km/sec.

6028. Magnitude does not include the outer ring structure.

6040. Interacting pair. Combined magnitude and mean radial velocity given.

6041. Close double. IC 1170 is 0′.9 preceding.

6050. IC 1179 attached.

6051. Brightest in a cluster.

6052. NGC 6064. Pair of galaxies in contact. Chaotic, with loops.

6068. Pair with NGC 6068A.

6070. Holmberg 729a; in a group with Holmberg 729b and 729c.

6086. In cluster Abell 2162.

6090. Double system in contact.

6106. Pair with a small anonymous spiral.

6120. In cluster Abell 2199.

6158. In cluster Abell 2199.

6160. In cluster Abell 2197.

6166. Brightest in cluster Abell 2199.

6173. In cluster Abell 2197.

6196. Brightest in group. Extremely bright nucleus or superimposed star.

6215. Pair with NGC 6221.

6217. Inner arms are more curved than the outer arms.

6223. Peculiar nuclear region.

6239. Small colliding pair.

6285. Pair with NGC 6286. Faint extension following NGC 6286.

6306. Pair with NGC 6307.

6314. Pair with NGC 6315.

6359. Pair with a small barred spiral.

6361. Small compact component 1′.5 south preceding.

6438. Close pair with NGC 6438A. Incomplete irregular ring seen edge on.

6454. In a cluster.

6467. NGC 6468 is a triple star.

6501. Pair with NGC 6500.

6560. Irregular companion 4′.6 preceding.

6621. Colliding with NGC 6622, which resembles the companion of M51. Long asymmetric arm.

6661. Pair with NGC 6658.

6758. In a group with IC 4831-2, 4836-7, 4837A, and 4839-40.

6769. Interacting with NGC 6770. Brightest members in group.

6771. In a group.

6810. Pair with very faint large galaxy.

6822. Barnard's galaxy, IC 4895; in Local Group. IC 1308 is an H II region in NGC 6822. Distance, 520 kiloparsecs.

6835. Pair with NGC 6836.

6861. In group with NGC 6868, 6870, and others.

6872. Interacting with IC 4970.

6876. In a group with NGC 6877 and IC 4972.

6878. Pair with an anonymous spiral.

6880. Pair with IC 4981. In a group.

6890. In a group.

6906. Pair with NGC 6901

(= IC 5000 = IC 1316).

6921. Pair with a very faint anonymous spiral.

6926. Pair with NGC 6929.

6928. Brightest in a group with NGC 6927, 6927A, and 6930.

6935. Pair with NGC 6937.

6944. Pair with NGC 6944A.

6962. Brightest in a dense group with NGC 6963-4.

6978. In a chain of spirals with NGC 6975-7.

6984. Pair with NGC 6982 (type Sa?) at 6′ north preceding.

7013. The radial velocity is mean of discordant values.

7014. Brightest in group.

7038. In group with NGC 7014.

7049. In group with NGC 7029 and 7041.

7065. Pair with NGC 7065A.

7070. In a group with NGC 7070A, 7072, and 7072A.

7079. Radial velocity is mean of discordant values.

7097. Pair with NGC 7097A (type E5?), at 5′.9 north following.

7119A. In contact with NGC 7119B, which is of type Scp.

7125. Pair with NGC 7126.

7135. Long, thin cometlike extension, 2′.6 north following.

7144. Pair with NGC 7145.

7163. In the NGC 7163-7176 group.

7166. Pair with NGC 7162 and 7162A.

7168. Pair with an anonymous elliptical galaxy.

7171. Pair with IC 1417.

7176. Colliding with NGC 7174. In a group with NGC 7163 and 7172-4.

7196. Pair with a very small galaxy.

7201. In a small chain of galaxies with NGC 7203-4. NGC 7202 does not exist, or perhaps is a foreground star.

7205. Pair with NGC 7205A, of type S(B)d, at 8′.5 south preceding.

7217. Intermediate ring of low surface brightness.

7214. Pair with IC 5168.

7232. In a group with NGC 7232B and 7233.

7232A. Pair with IC 5181.

7236. In a common halo with NGC 7237. Faint extensions. Third component lies 0′.6 south following.

7241. Peculiar absorption lane. Pair with an anonymous Sc galaxy.

7242. Brightest in group.

7252. Faint filaments and loops around main body of galaxy.

7274. In group.

7300. In a group with NGC 7298 and 7302.

7314. Almost no nucleus. Pair with NGC 7313 (type SBb?), at 4′.3 north preceding.

7317. In Stephan's Quintet.

7318A. In Stephan's Quintet. Colliding with NGC 7318B.

7319. In Stephan's Quintet.

7320. In the NGC 7331 group and foreground of Stephan's Quintet? NGC 7320C (perhaps an outlying member of Stephan's Quintet) is 4′.1 north following.

7331. Brightest in group. In the foreground of a group of fainter galaxies. Almost edge on; dust lanes visible.

7332. Pair with NGC 7339. Nucleus somewhat "box shaped," as in NGC 128.

7335. Member of a faint group in the background of NGC 7331.

7385. Pair with NGC 7383. Brightest in cluster.

7386. In cluster.

7389. Pair with NGC 7387. In cluster.

7393. Incomplete or obscured ring? In a group.

7418. Pair with NGC 7421.

7443. Pair with NGC 7444.

7454. Pair with an anonymous barred spiral.

7457. Pair with a small anonymous spindle galaxy.

7465. Triple system with NGC 7463-4.

7469. Pair with IC 5283.

7499. Pair with NGC 7501 in the Pegasus II cluster.

7503. In the Pegasus II cluster.

7507. Pair with NGC 7513.

7518. Pair with UGC 12423.

7537. Pair with NGC 7541.

7550. Triple system with NGC 7547 and 7549. Note a difference in the arms toward and away from NGC 7549.

7552. Forms the Grus Quartet with NGC 7582, 7590, and 7599.

7562. Pair with NGC 7562A (type Sd) at 2'.3 south following.

7576. Pair with NGC 7585.

7578A. In a common envelope with NGC 7578B. In the cluster Abell 2572.

7582. Forms Grus Quartet with NGC 7552, 7590, and 7599.

7585. Faint, extended envelope. Pair with NGC 7576.

7590. Forms Grus Quartet with NGC 7552, 7582, and 7599.

7592. Colliding pair; two nuclei 0'.25 apart.

7599. Forms Grus Quartet with NGC 7552, 7582, and 7590.

7603. May be connected to a compact object, 0'.9 south following, with a very different redshift.

7609. Possibly part of a ring galaxy with an incomplete outer ring.

7611. In Pegasus I cluster.

7615. In Pegasus I cluster.

7619. Brightest member of Pegasus I cluster.

7623. In Pegasus I cluster.

7625. Narrow, chaotic absorption lanes across one end.

7626. Second brightest member of Pegasus I cluster.

7631. In Pegasus I cluster.

7640. Nearly edge on.

7648. Faint companion.

7649. In the cluster Abell 2593. Has an extended halo.

7671. Pair with NGC 7672.

7673. Pair with NGC 7677.

7674. There is a compact elliptical galaxy attached at 0'.5 north following. Pair with NGC 7675.

7678. One arm very massive and brighter than the other.

7679. Interacting with NGC 7682? Faint extensions with brighter condensations; disturbed spiral arms.

7713. Pair with NGC 7713A (type S(B)c?) at 19' north following.

7714. Interacting with NGC 7715.

7720. Brightest in cluster Abell 2634.

7727. Very faint and smooth outer arms. Pair with the barred spiral NGC 7724.

7731. Pair with NGC 7732, which is probably in the foreground.

7753. Double arm leads to NGC 7752.

7757. NGC 7756 at 4'.5 is a star. Detached, faint companion in background?

7768. In cluster Abell 2666.

7770. In a group with NGC 7769 and 7771.

7782. Brightest in group with NGC 7779-80.

7783. In a common envelope with a companion. Other companions form a chain.

7793. In Sculptor group.

7805. Interacting with NGC 7806.

The following galaxies are listed by Index Catalogue *number.*

IC 10. In Local Group; distance is 1,300 kiloparsecs.

IC 239. In NGC 1023 group.

IC 342. Brightest galaxy in the UMa-Cam cloud.

IC 356. Faint straight absorption lanes lead toward the nucleus and become triple.

IC 749. Pair with IC 750.

IC 775. Pair with NGC 4267.

IC 1065. Absorption lanes, and a very faint spiral arm.

IC 1459. Pair with IC 5264.

IC 1613. In Local Group; distance is 740 kiloparsecs. Brightest at the northeastern end.

IC 1876. Double system.

IC 1953. In NGC 1332 group?

IC 1954. In a group with IC 1920, 1928, 1933, 1942, 1946, and 1954.

IC 2082. In a cluster; may be a type-cD galaxy interacting with a faint companion.

IC 2233. Seen exactly edge on; one of the flattest galaxies known.

IC 2389. Pair with NGC 2636.

IC 2522. Pair with IC 2523.

IC 2574. Coddington's nebula; in the M81 group.

IC 3256. For a discussion of IC 3256 and NGC 4341-2 identifications, see *Publications* of the Astronomical Society of the Pacific, **79**, 627, 1967.

IC 3481. Connected to IC 3481A, but probably not to IC 3483. Sometimes called Zwicky's Triplet, but this name is more often applied to the system at $16^h 49^m.5$, $+45° 30'$ (2000.0).

IC 4296. Brightest in group.

IC 4299. In the IC 4296 group.

IC 4329. Brightest in group.

IC 4329A. In the IC 4329 group.

IC 4553. Colliding with IC 4554? Combined magnitude and mean radial velocity given. One of the brightest extragalactic infrared sources known.

IC 4721. Pair with IC 4720.

IC 4797. Pair with IC 4796.

IC 5135. Pair with IC 5131.

IC 5152. In the Local Group; distance is 1,500 kiloparsecs. Star HD 209142 lies 1'.2 north following.

IC 5179. The IC coordinates are incorrect, according to G. de Vaucouleurs.

IC 5181. Pair with NGC 7232A.

IC 5201. There is a rich cluster of faint galaxies at 12' north preceding.

IC 5269. Pair with IC 5270, whose IC coordinates are incorrect, according to G. de Vaucouleurs.

IC 5273. The coordinates given in the IC are incorrect, according to G. de Vaucouleurs.

IC 5283. Pair with NGC 7469.

IC 5328. Close pair with IC 5328A.

IC 5338. Pair with IC 5337. In the cluster Abell 2626.

The following galaxies are listed by rough 2000.0 coordinates.

0001−15. Wolf-Lundmark-Melotte (WLM) system, in the Local Group. Distance is 1,600 kiloparsecs.

0035+36. Andromeda III, a satellite of M31. Distance 730 kiloparsecs.

0045+38. Andromeda I, a satellite of M31. Distance 730 kiloparsecs.

0052−72. Small Magellanic Cloud, a satellite of the Milky Way. Distance is 60 kiloparsecs. Interacting with the Large Magellanic Cloud.

0059−33. Sculptor dwarf elliptical (spheroidal) galaxy, a satellite of the Milky Way. Distance 85 kiloparsecs. Not to be confused with the Sculptor galaxy, NGC 253, or the Sculptor dwarf irregular system.

0103+21. LGS 3, a new Local Group member confirmed by T. X. Thuan and G. E. Martin, *Astrophysical Journal Letters*, **232**, L11, 1979. At a distance of about 900 kiloparsecs, it is probably a satellite of M33. "LGS" stands for "Local Group (suspected)."

0116+33. Andromeda II, a satellite of M31. Distance is 730 kiloparsecs.

0236+59. Maffei I; highly obscured.

0239−34. Fornax dwarf elliptical (spheroidal) galaxy, a satellite of the Milky Way. Distance is 130 kiloparsecs.

0241+59. Maffei II; highly obscured. A member of the UMa-Cam Cloud?

0257+06. Two components; in the cluster Abell 400. The combined magnitude and mean radial velocity are given.

0523−69. Large Magellanic Cloud, a satellite of the Milky Way. The distance is 50 kiloparsecs. Interacting with the Small Magellanic Cloud.

0641−50. Carina dwarf elliptical (spheroidal) galaxy, a satellite of the Milky Way. Distance is 170 kiloparsecs.

0711+71. Integral Sign galaxy. Interacting with UGC 3714 at 18' south following?

0714+73. Distorted ring.

0818+70. Arp 268; in M81 group.

0918−12. The radio source Hydra A. Double nucleus; in a cluster.

0940+71. In M81 group.

0959+30. Leo III, D69; also called the Leo A dwarf irregular galaxy. Its distance is uncertain, with estimates

ranging from 1,100 to 2,300 kiloparsecs. Probably in the Local Group.

1000+05. Sextans B dwarf.

1008+12. Leo I, D74; also known as the Regulus dwarf elliptical (spheroidal) galaxy. A satellite of the Milky Way; distance 230 kiloparsecs.

1011−04. Sextans A dwarf.

1049+52. Double system.

1113+22. Leo II, D93; also called the Leo B dwarf elliptical (spheroidal) galaxy. A satellite of the Milky Way, at a distance of 230 kiloparsecs.

1146−03. Wild's Triplet. The combined magnitude and overall size are given.

1246+26. In the foreground of the Coma cluster.

1312+26. In the foreground of the Coma cluster.

1313+36. Companion of NGC 5033.

1354+53. In M101 group.

1413−65. Highly obscured.

1450+35. Pair with UGC 9562.

1508+67. Ursa Minor dwarf elliptical (spheroidal) galaxy. A satellite of the Milky Way; distance is 75 kiloparsecs.

1516+07. In cluster Abell 2052.

1605+41. Interacting pair?

1638+41. An incomplete, irregular ring. The combined magnitude and mean radial velocity of the A and B components are given.

1649+45. Zwicky's Triplet (Arp 103), although this name is sometimes also applied to the IC 3481 system. Combined magnitude and overall size given.

1720+57. Draco dwarf elliptical (spheroidal) galaxy, also designated D20. A satellite of the Milky Way, at a distance of 80 kiloparsecs.

1930−17. Sagittarius dwarf irregular galaxy, in the Local Group. Distance 1,100 kiloparsecs.

1941+50. Double system.

1956+40. Pair with an anonymous type-E0p galaxy of 16th magnitude. Much galactic obscuration.

2046−12. Aquarius dwarf galaxy; possible new Local Group member at a distance of about 1,500 kiloparsecs.

2328+14. Pegasus dwarf galaxy, D216. Distance 1,300 kiloparsecs? Probable Local Group member.

VARIABLE STARS SUPPLEMENTARY NOTES

(Continued from page 260)

SU Vir. [9.4, 13.6]

SV Vir. [9.8,]

TY Vir. Spectrum resembles those of semiregular variables in globular clusters.

TZ Vir. Variation with a 6,900-day period is probably present.

UW Vir. Period varies.

UY Vir. Period possibly varies.

AG Vir. Period varies; min. II, 8.88.

AH Vir. Period and light curve vary; min. II, 9.65.

CS Vir. Sp. binary; magnetic field varies with 9.2954-day period.

CW Vir. Maximum magnetic-field intensity occurs near light minimum.

DK Vir. Amplitude varies; long intervals of constant brightness are observed.

DL Vir. Sp. triple includes eclipsing pair; orbital period of pair and third star is probably about 2,200 days.

DM Vir. Min. II.

FF Vir. Magnetic field varies with 130.0-day period.

FM Vir. Sp. binary.

FT Vir. Light curve varies.

GG Vir. Also a 0.042-day period.

R Vol. [10.8, 13.7] May be a carbon star.

S Vol. [8.6, 13.6]

R Vul. [8.1, 12.6]

S Vul. Period varies; may be a Cepheid.

U Vul. Epoch of minimum given.

Z Vul. Period varies; min. II, 7.66.

RU Vul. [9.1, 11.2]

SV Vul. Period and light curve vary.

BW Vul. The period of this star has increased at the rate of 0.037 second per year between the years 1924 and 1970.

CK Vul. Remnant discovered in 1982.

DR Vul. Apsidal rotation with 37.8-year period; min. II. A third component is possible.

ES Vul. ADS 12287A.

FG Vul. Probable member of open cluster NGC 6940.

LT Vul. Beats possible.

MW Vul. Periods 16.148 and 16.821 days are also possible.

NT Vul. Sp. binary.

NU Vul. Shell star.

Quasi-stellar Objects (QSO's)

Quasi-stellar Objects (QSO's)

Column Headings

Designation — The coordinate designation, formed by concatenating the hours and minutes of right ascension with the declination truncated to tenths of a degree (with decimal point removed). This designation is derived from the coordinates for equinox 1950.0, since many of the objects have been described in the astronomical literature this way.

3C — The QSO's number in the revised third Cambridge radio catalogue (A. S. Bennett, 1962).

4C — The QSO's number in the fourth Cambridge radio catalogue (J. D. H. Pilkington and P. F. Scott, 1965; J. F. R. Gower *et al.,* 1967).

α 2000 and **δ 2000** — Right ascension and declination, referred to the 2000.0 equinox.

Const — Constellation in which the QSO is located.

Mag — Photoelectric visual **V** magnitude, if given to two decimal places or accompanied by an entry in the next column. Otherwise, an approximate visual brightness is given (or, if flagged by the letter "B," the photoelectric blue **B** magnitude). The letter "v" means that the brightness of the QSO varies.

B − V and **U − B** — Color indexes in the **UBV** photometric system. The bluer the QSO, the smaller its color index.

S — Radio strength (flux density) in janskys, formerly called flux units. One jansky equals 10^{-26} watt per square meter per hertz. The measurements reported in this column were carried out at a frequency of either 408 megahertz (a wavelength of 74 centimeters) or, if enclosed in parentheses, 1400 megahertz (wavelength 21 centimeters).

z — Redshift of the QSO, as determined from the emission lines in its spectrum. It is defined as the difference between the observed and rest wavelengths of a line, divided by the rest wavelength. In general, the greater the redshift, the more distant the object. BL Lacertae objects, which resemble QSO's but have exceedingly weak spectral features, are flagged by an asterisk.

Notes — Alternate names and designations from other sources, as follows: Einstein satellite X-ray survey, 1979-81 (1E); F. Zwicky's *Catalogue of Selected Compact Galaxies and of Post-Eruptive Galaxies,* 1971 (1Z to 7Z); A. Braccesi's lists of ultraviolet-excess objects, 1968, 1970 (B); second Bologna radio survey, 1970-73 (B2); B. Gaston's study of high-redshift QSO's, *Astrophysical Journal,* **277,** 411, 1983 (BG); list of blue stellar objects by A. Sandage and P. Veron, 1965 (BSO); T. Shanks *et al., Nature,* **303,** 156, 1983 (DHM); Dwingeloo-Greenbank radio survey, 1967 (DW); National Radio Astronomy Observatory 5-gigahertz radio survey, 1971 (GC); Kitt Peak QSO catalogue, 1978 (KP); W. J. Luyten's search for faint blue stars, 1962 (LB); Molonglo radio source catalogue, 1973-74 (MC); B. E. Markarian's study of compact blue objects, 1967 (Mrk).

Also, N. A. Bahcall *et al.,* "Observations of QSO's in the Direction of Clusters of Galaxies," *Astrophysical Journal,* **183,** 777, 1973 (NAB); National Radio Astronomy Observatory 1400-megahertz radio survey, 1966 (NRAO); Ohio State University radio source catalogue, 1967-74 (OA to OZ, except OO); Palomar-Green list of bright QSO's, *Astrophysical Journal,* **269,** 352, 1983 (PG); Palomar-Haro-Luyten catalogue of faint blue stars, 1962 (PHL); Parkes radio survey, 1969-75 (PKS); N. Richter and K. Sahakjan's study of blue objects near the north galactic pole, 1965 (RS); Tonantzintla catalogue of blue stellar objects, 1957-59 (Ton); Curtis Schmidt search for extragalactic emission-line objects, University of Michigan, 1977-78 (UM). If one of the abbreviations above appears without a number, the QSO is usually referred to by the catalogue label plus the coordinate designation given earlier.

Designation	3C	4C	α 2000	δ 2000	Const	Mag	B-V	U-B	S	z	Notes
0002+051			0^h05^m20^s.2	+5°24'12"	Psc	16.9				1.890	UM 18
0003+158		+15.01	0 05 59.2	+16 09 48	Peg	16.40v	+0.11	-0.70	2.25	0.450	PKS, PHL 658
0003-003	2	-00.01	0 06 22.5	-0 04 25	Psc	19.35v	+0.79	-0.96	10.5	1.037	PKS, NRAO 6
0004+024			0 07 26.8	+2 41 24	Psc	16.0				2.000?	UM 202
0014+813			0 17 08.5	+81 35 09	Cep	16.5				3.41	Very luminous
0017+257		+25.01	0 19 39.4	+26 02 52	And	15.4				0.284	B2
0017+154	9	+15.02	0 20 25.1	+15 40 55	Psc	18.21	+0.23	-0.74	7.79	2.012	PKS, PHL 2871
0024+224			0 27 15.3	+22 41 58	And	16.57	+0.33	-0.69		1.118	NAB
0026+129			0 29 13.7	+13 16 04	Psc	14.78	+0.26	-0.78		0.142	PG
0036-392			0 38 27.3	-38 59 44	Scl	16.29	+0.73	-0.53	5.80	0.592	PKS
0043+039			0 45 45.4	+4 11 04	Psc	15.88B				0.384	PG
0044+030			0 47 05.7	+3 19 55	Psc	15.97B				0.624	PKS, PG
0047-832			0 47 51.2	-82 56 49	Oct	16.9				1.112	PKS
0052+251			0 54 52.2	+25 25 39	Psc	15.42B				0.155	PG
0054-284			0 56 25.2	-28 08 32	Scl	18.25				3.61	DHM
0054+144			0 57 09.9	+14 46 11	Psc	15.71	+0.21	-0.87		0.171	PKS, PHL 909
0100+130			1 03 11.3	+13 16 16	Psc	16.57	+0.40	-0.28		2.69	PHL 957
0109+224			1 12 05.8	+22 44 39	Psc	15.5 v				*	GC
0115+027	37	+02.04	1 18 18.5	+2 58 06	Psc	17.5			6.07	0.671	PKS, NRAO 65
0117+213			1 20 17.3	+21 33 47	Psc	16.05B				1.493	PG
0118+034	39	+03.02	1 21 01.3	+3 44 13	Psc	18.09	+0.31	-0.59	3.64	0.765	PKS, NRAO 68
0119-046		-04.04	1 22 27.9	-4 21 27	Cet	16.88	+0.46	-0.72	2.93	1.955	PKS, OC-034
0122-380			1 24 17.5	-37 44 27	Scl	16.5				2.181	
0122-003		-00.10	1 25 28.8	-0 05 56	Cet	16.70v	+0.28	-0.75	1.20	1.070	PKS, UM 321
0127+233	43	+23.06	1 29 59.9	+23 38 20	Psc	20.0 v			(3.0)	1.459	PKS
0130+242		+24.02	1 33 25.3	+24 27 40	Psc	16.8				0.457	PKS, OC 250
0133+207	47	+20.07	1 36 24.4	+20 57 27	Psc	18.1 v	+0.05	-0.65	(3.6)	0.425	PKS, NRAO 78
0134+329	48	+32.08	1 37 41.3	+33 09 35	Tri	16.20v	+0.40	-0.59	34	0.367	NRAO 79
0137-010			1 40 17.0	-0 50 03	Cet	16.49v	-0.12	-0.75		0.33	UM 357
0151+045			1 54 28.0	+4 48 20	Psc	16.91	+0.10	-0.77		0.404	PHL 1226
0156+035			1 58 38.8	+3 47 46	Psc	16.9				0.66 ?	UM 153
0157-311			2 00 12.0	-30 53 28	For	19			10.3		PKS
0159-117	57		2 01 57.2	-11 32 34	Cet	16.40v	+0.14	-0.73	5.07	0.669	PKS, NRAO 88
0202-765			2 02 13.1	-76 20 06	Hyi	16.77v	+0.06	-0.77	8.46	0.389	PKS
0205+024			2 07 49.9	+2 42 56	Cet	15.41	+0.28	-0.85		0.155	Mrk 586
0219+428	66A		2 22 39.7	+43 02 08	And	15.5 v	+0.33	-0.58		0.444?*	
0226-038		-03.07	2 28 53.6	-3 37 37	Cet	16.96	+0.07	-0.82	2.23	2.064	PKS, PHL 1305
0229+341	68.1	+34.08	2 32 28.9	+34 23 46	Tri	19.0			7.8	1.238	NRAO 105
0232-042		-04.06	2 35 07.4	-4 02 06	Cet	16.46v	+0.15	-0.89	4.06	1.436	PKS, PHL 1377
0235+164			2 38 38.9	+16 37 00	Ari	15.50v	+0.96	+0.14		0.524?*	OD 160
0237-233			2 40 08.1	-23 09 18	Cet	16.63v	+0.15	-0.61	3.67	2.223	PKS, PHL 8462
0254-334			2 56 47.0	-33 15 27	For	16				1.857	
0312-770			3 11 54.8	-76 51 52	Hyi	16.10	+0.16	-0.77	1.11	0.223	PKS
0340+048	93	+04.13	3 43 30.0	+4 57 49	Tau	18.09v	+0.35	-0.50	8.65	0.357	PKS, NRAO 144
0349-146	95		3 51 28.6	-14 29 10	Eri	16.22	+0.33	-0.56	11.6	0.614	PKS, NRAO 147
0350-073	94		3 52 30.6	-7 11 02	Eri	16.49v	+0.44	-0.68	10.2	0.962	PKS, NRAO 149
0355-483			3 57 22.0	-48 12 16	Hor	16.38	+0.33	-0.88		1.005	PKS
0357+107			4 00 11.9	+10 55 13	Tau	16.78	+0.30	-0.75		0.181	1E
0405-123			4 07 48.4	-12 11 37	Eri	14.82v	+0.18	-0.60	8.17	0.574	PKS
0409+229	108	+22.08	4 12 43.9	+23 05 02	Tau	18.7	+0.9		(1.24)	1.215	PKS, NRAO 167
0414-060	110	-05.17	4 17 16.7	-5 53 45	Eri	15.94v	+0.30	-0.70	3.48	0.781	PKS, NRAO 170
0422-380			4 24 41.7	-37 56 14	Cae	16.5				0.78	PKS
0422+004			4 24 46.8	+0 36 06	Tau	16.00v	+0.53	-0.51	(1.0)	*	PKS, OF 038
0429+415	119	+41.13	4 32 36.5	+41 38 29	Per	20.0			17	0.408	
0439-433			4 41 16.2	-43 13 28	Cae	16.36	+0.28	-0.65		0.594	PKS
0448-392			4 49 42.2	-39 11 10	Cae	16.46	+0.24	-0.89		1.288	PKS
0454-220			4 56 10.0	-21 59 16	Lep	16.10	+0.06	-0.62	4.93	0.534	PKS
0454+039			4 56 47.0	+4 00 51	Ori	16.53	+0.23	-0.81	(0.37)	1.345	PKS, OF 092
0506-612			5 06 44.1	-61 09 41	Dor	16.85	+0.51	-0.52	5.03	1.093	PKS
0514-005			5 16 33.5	-0 27 16	Ori	16.18	+0.08	-0.77		0.292	1E
0518+165	138	+16.12	5 21 09.9	+16 38 21	Tau	18.84	+0.53	-0.16	17.2	0.754	PKS, NRAO 205
0521-365			5 22 57.9	-36 27 31	Col	14.62v	+0.65	-0.35	36.1	0.061 *	PKS
0530-379			5 32 30.7	-37 53 22	Col	16.7				0.29	
0537-441			5 38 49.8	-44 05 09	Pic	16.48v	+0.52	-0.48	2.56	0.894	PKS
0538+498	147	+49.14	5 42 36.1	+49 51 08	Aur	17.80v	+0.65	-0.37	62	0.545	NRAO 221
0548-322			5 50 41.9	-32 16 11	Col	15.5 v	+0.57	-0.30		0.069?*	
0610+260	154	+26.20	6 13 50.1	+26 04 37	Gem	18.0 v			9.20	0.580	PKS, NRAO 230
0622-441			6 23 31.8	-44 13 04	Pup	16.93v	+0.22	-0.73	0.97	0.688	PKS
0637-752			6 35 46.6	-75 16 17	Men	15.75	+0.33	-0.60	7.89	0.651	PKS
0642+449			6 46 32.1	+44 51 17	Aur	18.49	+1.08	+1.7		3.402	OH 471
0710+118	175	+11.26	7 13 02.5	+11 46 16	CMi	16.60v	+0.46	-0.51	8.95	0.768	PKS, NRAO 258
0725+147	181	+14.24	7 28 10.4	+14 37 37	Gem	18.92v	+0.43	-1.02	6.73	1.382	PKS, NRAO 266
0723+679	179	+67.14	7 28 10.9	+67 48 48	Cam	18.0			(2.09)	0.846	NRAO 263
0735+178			7 38 07.4	+17 42 21	Gem	14.85v	+0.47	-0.58	2.74	0.424 *	PKS, OI 158
0736-019	185	-01.18	7 38 34.0	-2 04 21	Mon	17.6			3.21	1.033	PKS

QUASI-STELLAR OBJECTS (QSO's)

Designation	3C	4C	α 2000	δ 2000	Const	Mag	B−V	U−B	S	z		Notes
0729+818			7^{h}39^{m}03^s.4	+81°46′01″	Cam	17.5			(3.38)	1.022		NRAO 271
0736+017			7 39 18.1	+1 37 04	CMi	16.47	+0.43	−0.77	2.84	0.191		PKS, OI 061
0743−673			7 43 32.4	−67 26 23	Vol	16.37	+0.24	−0.73	8.61	1.51		PKS
0740+380	186	+38.21	7 44 17.5	+37 53 17	Lyn	17.60	+0.45	−0.71	6.8	1.063		NRAO 273
0742+318		+31.30	7 45 41.7	+31 42 56	Gem	16				0.462		OI 371
0754+100			7 57 06.7	+9 56 34	Cnc	14.5 v					*	OI 090.4
0754+394			7 57 59.9	+39 20 27	Lyn	14.36	+0.38	−0.71		0.096		1E
0758+143	190	+14.25	8 01 33.6	+14 14 43	Cnc	20.32	+0.52	−0.35	7.6	1.197		PKS, NRAO 278
0802+103	191	+10.25	8 04 48.0	+10 15 22	Cnc	18.40v	+0.25	−0.84	7.32	1.956		PKS, NRAO 279
0809+483	196	+48.22	8 13 36.1	+48 13 03	Lyn	17.79v	+0.57	−0.43	33	0.871		NRAO 285
0814+227	197	+22.20	8 17 35.1	+22 37 18	Cnc	18			(1.21)	0.980		NRAO 287
0818−128			8 20 57.5	−12 58 59	Pup	15.5 v	+0.3	−0.4			*	OJ−131
0825−202			8 27 17.2	−20 26 27	Pup	18			10.3			PKS
0829+046			8 31 48.9	+4 29 39	Hya	16.5 v	+0.70	−0.37			*	PKS, OJ 049
0833+654	204	+65.09	8 37 44.9	+65 13 35	UMa	18.21	+0.55	−0.99	(1.31)	1.112		NRAO 297
0835+580	205	+58.16	8 39 06.4	+57 54 17	UMa	17.62	+0.49	−0.49	15	1.534		NRAO 298
0837−120	206		8 39 50.6	−12 14 34	Hya	15.76v	+0.02	−0.85	5.8	0.200		PKS, NRAO 299
0838+133	207	+13.38	8 40 47.6	+13 12 23	Cnc	18.15v	+0.43	−0.42	7.03	0.684		PKS, NRAO 300
0842−754			8 41 27.5	−75 40 06	Cha	18.9			13.3	0.524		PKS
0839+187			8 42 05.1	+18 35 42	Cnc	16.36	+0.27	−0.84		0.259		DW
0846+513			8 49 58.1	+51 08 29	UMa	15.72v	+0.56	−0.37		1.860	*	
0850+140	208	+14.28	8 53 08.7	+13 52 36	Cnc	17.42	+0.34	−1.00	7.75	1.109		PKS, NRAO 301
0851+202			8 54 48.9	+20 06 32	Cnc	14.0 v	+0.39	−0.64	(1.59)	0.306?	*	OJ 287
0855+143	212	+14.30	8 58 41.5	+14 09 44	Cnc	19.06	+0.90	−0.30	7.01	1.049		PKS, NRAO 310
0859−140			9 02 16.7	−14 15 30	Hya	16.59	+0.20	−0.85	3.93	1.327		PKS, OJ−199
0903+169	215	+16.26	9 06 31.9	+16 46 11	Cnc	18.27v	+0.21	−0.66	6.30	0.411		PKS, LB 9308
0906+430	216	+43.17	9 09 33.5	+42 53 46	Lyn	18.48v	+0.49	−0.60	19	0.67		NRAO 317
0906+484			9 10 10.1	+48 13 42	UMa	16.06	+0.40	−0.91		0.118		PG
0912+297			9 15 52.4	+29 33 24	Cnc	16.00v	+0.37	−0.73	(0.56)		*	OK 222
0923+201			9 25 54.8	+19 54 07	Leo	16.04B				0.190		Ton 1057
0925−203			9 27 51.9	−20 34 51	Hya	16.40v	+0.07	−0.87	1.36	0.348		PKS
0927+362	220.2	+36.15	9 30 33.5	+36 01 25	LMi	18.2			6.2	1.157		NRAO 322
0946+301			9 49 41.1	+29 55 18	Leo	16.00B				1.216		PG
0955+326	232	+32.33	9 58 21.0	+32 24 02	Leo	15.78v	+0.10	−0.68	(1.46)	0.533		Ton 469
0957+561A			10 01 20.7	+55 53 56	UMa	17.0				1.390		Comp. A of double QSO
0957+561B			10 01 20.8	+55 53 50	UMa	17.0				1.390		Comp. B of double QSO
0958+551			10 01 29.6	+54 54 38	UMa	16.00	+0.25	−0.84		1.758		Mrk 132
1001+292			10 04 04.0	+28 58 27	LMi	16.05	+0.12	−0.90		0.329		Ton 28
1001+054			10 04 20.1	+5 13 00	Sex	16.38	+0.31	−0.86		0.161		PG
1004−217			10 06 46.3	−21 59 24	Hya	16.89	+0.16	−0.83		0.330		PKS
1004+130		+13.41	10 07 26.1	+12 48 57	Leo	15.15v	+0.13	−0.82	2.74	0.240		PKS, OL 107.7
1007+417		+41.21	10 10 27.6	+41 32 27	UMa	16.5				0.611?		
1011−282			10 13 29.6	−28 31 25	Ant	16.88	−0.08	−0.83	2.60	0.253		PKS, OL−219
1011+250			10 13 53.5	+24 49 17	Leo	16.57	+0.02	−0.70		1.631		Ton 490
1012+008			10 14 54.9	+0 33 37	Sex	16.0				0.185		PG
1015+277	240	+27.21	10 17 49.5	+27 32 04	Leo	17.5			(1.3)	0.469		NRAO 351
1017+280			10 19 54.9	+27 45 55	Leo	15.69	+0.37	−0.86		1.924		Ton 34
1020−103			10 22 32.7	−10 37 42	Sex	16.11	+0.14	−0.82	1.64	0.197		PKS, OL−133
1023+067	243	+06.40	10 26 32.2	+6 28 14	Leo	16.70	+0.16	−0.84	3.43	1.699		PKS, NRAO 355
1028+313			10 30 59.1	+31 02 56	LMi	16.71v	+0.36	−0.75		0.177		B2
1038+064		+06.41	10 41 17.2	+6 10 16	Sex	16.81v	+0.16	−0.84		1.270		OL 064.5
1040+123	245	+12.37	10 42 44.7	+12 03 31	Leo	17.29	+0.46	−0.82	8.90	1.029		PKS, NRAO 358
1048−090	246		10 51 30.0	−9 18 09	Crt	16.79v	+0.06	−0.49	5.35	0.342		PKS, NRAO 359
1049+616		+61.20	10 52 32.8	+61 25 21	UMa	16.48v	+0.10	−0.76		0.422		OL 682
1050−184			10 52 34.0	−18 45 18	Crt	16.4			0.98	0.544		PKS
1100−264			11 03 25.3	−26 45 15	Hya	16.02	+0.06	−0.71	(2.37)	2.145		
1101−325			11 03 31.3	−32 51 15	Hya	16.30v	−0.01	−0.81	2.32	0.354		PKS
1100+772	249.1	+77.09	11 04 13.9	+77 01 25	Dra	15.72v	−0.02	−0.77		0.311		NRAO 363
1101+384			11 04 27.3	+38 12 32	UMa	13.5 v	+0.51	−0.55	0.98	0.031	*	Mrk 421
1103−006		−00.43	11 06 31.6	−0 52 51	Leo	16.46	−0.07	−0.77	2.44	0.426		PKS, OM−006
1104+167		+16.30	11 07 13.6	+16 27 52	Leo	15.70v	+0.21	−0.65	(1.24)	0.634		OM 109
1111+408	254	+40.28	11 14 38.7	+40 37 20	UMa	17.98	+0.15	−0.49	12	0.734		NRAO 369
1115+080C			11 18 17.0	+7 46 00	Leo	18.6				1.722		PG, comp. C of triple QSO
1115+080B						19				1.722		PG, comp. B of triple QSO
1115+080A						15.9				1.722		PG, comp. A of triple QSO
1116+215			11 19 08.7	+21 19 18	Leo	15.17B				0.177		PG
1121+422			11 24 35.5	+42 00 25	UMa	16.02B				0.234		PG
1127−145			11 30 07.1	−14 49 27	Crt	16.90	+0.27	−0.70	5.07	1.187		PKS, OM−146
1128+315			11 31 09.5	+31 14 07	UMa	16.0				0.289		Ton 580
1132+303	261	+30.22	11 34 54.5	+30 05 26	UMa	18.24	+0.24	−0.56	(1.73)	0.614		LB 10265
1133+704			11 36 26.8	+70 09 24	Dra	14.49v	+0.67	−0.22		0.046	*	Mrk 180
1136−135			11 39 10.8	−13 50 46	Crt	16.0			10.5	0.544		PKS
1137+660	263	+66.13	11 39 57.0	+65 47 49	UMa	16.32v	+0.18	−0.56	(3.1)	0.652		NRAO 381
1138+040			11 41 16.6	+3 47 00	Vir	16.05B				1.876		PG
1146−037			11 48 57.4	−4 04 11	Vir	16.90	+0.06	−0.74	0.99	0.341		PKS

Designation	3C	4C	α 2000	δ 2000	Const	Mag	B-V	U-B	S	z	Notes
1147+245			11^{h}50^{m}19^{s}2	+24°17'54"	Leo	16 v	+0.5	-0.6	(0.58)	*	OM 280
1148+387		+38.31	11 51 29.3	+38 25 53	UMa	16.2			1.70	1.303	
1150+497		+49.22	11 53 24.5	+49 31 08	UMa	16.1 v				0.334	LB 2136
1151-348			11 54 21.7	-35 05 29	Hya	17.84	+0.64	-0.52	10.9	0.258	PKS
1156+631		+63.15	11 58 39.2	+62 54 27	UMa	16.8				0.594	
1156+295		+29.45	11 59 32.2	+29 14 45	UMa	15.6 v			(1.70)	0.729	Ton 599
1202+281			12 04 42.2	+27 54 11	Com	15.51v	+0.19	-0.93		0.165	GQ Com
1206+439	268.4	+43.23	12 09 13.6	+43 39 21	CVn	18.42v	+0.58	-0.69	(2.3)	1.400	NRAO 393
1206-399			12 09 35.4	-40 16 12	Cen	16.98	+0.51	-0.85		0.966	PKS
1208+322			12 10 37.4	+31 57 08	Com	16.0	-0.2	-0.7		0.388	ON 313
1209+107			12 11 41.6	+10 30 17	Vir	16.5				2.20	KP
1215+303			12 17 52.1	+30 07 01	Com	15.5 v	+0.46	-0.61		*	ON 325
1215+113			12 18 27.2	+11 05 05	Vir	16.86	+0.09	-0.75	(0.46)	1.396	MC
1217+023			12 20 11.9	+2 03 42	Vir	16.53v	+0.02	-0.87	1.04	0.240	PKS, ON 029
1218+339	270.1	+33.29	12 20 33.9	+33 43 12	CVn	18.61	+0.19	-0.61	8.2	1.519	NRAO 396
1218+304			12 21 22.1	+30 10 36	Com	16.50	+0.65	-0.50		*	RS 4
1219+285			12 21 31.6	+28 13 58	Com	16.5 v	+0.61	-0.54	(0.44)	*	W Com
1219+755			12 21 44.1	+75 18 37	Dra	14.5				0.070	Mrk 205
1225+317			12 28 24.8	+31 28 38	CVn	15.87	+0.28	-0.68		2.230	B2
1226+023	273	+02.32	12 29 06.8	+2 03 07	Vir	12.86v	+0.21	-0.85	55.1	0.158	PKS, NRAO 400
1229-021		-02.55	12 32 00.0	-2 24 05	Vir	16.75v	+0.48	-0.66	3.87	1.038	PKS, ON-049
1241+166	275.1	+16.34	12 43 57.7	+16 22 53	Com	19.00v	+0.23	-0.43	9.35	0.557	PKS, NRAO 406
1241+176			12 44 10.8	+17 21 05	Com	15.38B				1.273	PG
1246+377			12 48 51.9	+37 30 29	CVn	16.98	+0.31	-0.78		1.242	BSO 1
1246-057			12 49 13.9	-5 59 18	Vir	16.73	+0.36	-0.30		2.212	
1248+401			12 50 48.3	+39 51 39	CVn	16.06B				1.030	PG
1250+568	277.1	+56.20	12 52 26.4	+56 34 20	UMa	17.93	-0.17	-0.78	8.5	0.321	NRAO 409
1252+119			12 54 38.3	+11 41 06	Vir	16.64v	+0.35	-0.75	0.93	0.871	PKS, ON 187
1253-055	279	-05.55	12 56 11.2	-5 47 21	Vir	17.75v	+0.26	-0.56	14.5	0.538	PKS, NRAO 413
1254+047			12 57 00.1	+4 27 34	Vir	15.84B				1.024	PG
1257+346			12 59 48.8	+34 23 22	CVn	16.79v	+0.26	-0.82		1.375	B 201
1258+404	280.1	+40.32	13 00 33.4	+40 09 07	CVn	19.44	-0.13	-0.70	5.5	1.659	NRAO 417
1302-102			13 05 33.0	-10 33 20	Vir	15.23	+0.05	-0.82		0.286	PKS, OP-106
1304+295			13 07 02.9	+29 18 42	Com	19.4				3.74	BG 57 9
1305+069	281	+06.45	13 07 53.9	+6 42 13	Vir	17.02	+0.13	-0.59	3.37	0.599	PKS, NRAO 419
1307+085			13 09 47.0	+8 19 49	Vir	15.28B				0.155	PG
1308+326			13 10 28.7	+32 20 43	CVn	15.24	+0.37	-0.57		0.996 *	B2
1317+277			13 19 56.2	+27 28 09	Com	15.98	+0.14	-0.78		1.022?	Ton 153
1318+290			13 21 15.9	+28 47 19	CVn	16.69	+0.19	-0.73		0.549?	Ton 156
1321+294			13 23 20.9	+29 10 07	CVn	16.83	+0.08	-0.74		0.960?	Ton 157
1327-214			13 30 07.1	-21 42 02	Vir	16.74v	+0.22	-0.62	5.63	0.528	PKS, OP-246
1328+254	287	+25.43	13 30 37.9	+25 09 10	Com	17.67	+0.63	-0.65	(7.0)	1.055	PKS, NRAO 424
1328+307	286	+30.26	13 31 08.3	+30 30 33	CVn	17.25	+0.26	-0.91	20	0.849	NRAO 425
1329+412			13 31 41.2	+41 01 58	CVn	16.30B				1.930	PG
1331+170			13 33 35.9	+16 49 04	Com	16.71	+0.13	-0.84		2.081	MC
1332+552		+55.27	13 34 11.6	+55 01 25	UMa	16				1.249	LB 685
1333+176			13 36 02.0	+17 25 14	Com	15.64B				0.554	PG
1340+606	288.1	+60.18	13 42 13.3	+60 21 42	UMa	18.12	+0.39	-0.82	5.5	0.961	NRAO 428
1351+640			13 53 15.8	+63 45 45	Dra	14.84	+0.34	-0.72		0.088	PG
1354+195		+19.44	13 57 04.4	+19 19 07	Boo	16.02v	+0.18	-0.55	6.0	0.720	PKS, OP 191
1356+581		+58.29	13 58 17.2	+57 52 04	UMa	16				0.321	OP 594
1355-416			13 59 00.3	-41 52 53	Cen	15.81v	-0.12	-0.86	9.52	0.313	PKS
1358+043			14 00 32.0	+4 04 58	Vir	16.31				0.427	PG
1400+162		+16.39	14 02 45.5	+15 59 56	Boo	16.74	+0.46	-0.44	(0.90)	0.245 *	OQ 100
1402+261			14 05 15.7	+25 55 40	Boo	15.57B				0.164	Ton 182
1407+265			14 09 23.8	+26 18 21	Boo	15.73B				0.944	PG
1416+067	298	+06.49	14 19 08.5	+6 28 35	Vir	16.79	+0.33	-0.70	23.4	1.436	PKS, NRAO 441
1418+546			14 19 46.6	+54 23 14	Boo	15.0 v				*	OQ 530
1421+330			14 23 26.1	+32 52 21	Boo	16.5				1.904	Mrk 679
1421-382			14 24 16.5	-38 26 47	Cen	16.87	+0.04	-0.54	6.92	0.41	PKS
1425+267			14 27 35.6	+26 32 14	Boo	15.68	+0.20	-0.73	(0.37)	0.366	Ton 202
1424-118			14 27 38.2	-12 03 51	Lib	16.52	+0.39	-0.69	4.99	0.805	PKS
1427+480			14 29 43.0	+47 47 29	Boo	16.33B				0.221	PG
1435+638			14 36 45.7	+63 36 38	Dra	15.0				2.060	
1441+522	303	+52.33	14 43 00.6	+52 01 37	Boo	19.97	+0.08	-0.59		1.57	Near N-type galaxy
1442+101			14 45 16.5	+9 58 37	Boo	17.78	+0.80	-0.37	(2.51)	3.53	OQ 172
1444+407			14 46 46.0	+40 35 09	Boo	15.95B				0.267	PG
1448-232			14 51 02.4	-23 29 30	Lib	16.96	+0.17	-0.52		2.208	PKS, OQ-279
1451-375			14 54 27.4	-37 47 33	Cen	16.69	+0.09	-0.77	2.89	0.314	PKS
1453-109			14 55 55.0	-11 08 45	Lib	17.37	+0.44	-0.76	10.3	0.940	PKS, OQ-190
1458+718	309.1	+71.15	14 59 07.6	+71 40 19	UMi	16.78v	+0.46	-0.77	(8.5)	0.905	NRAO 464
1502+602	311	+60.19	15 04 09.1	+60 00 56	Dra	18.0			(1.7)	1.022	NRAO 467
1508-055		-05.64	15 10 53.6	-5 43 07	Lib	17.18	+0.29	-0.71	7.72	1.191	PKS, OR-015
1510-089			15 12 50.6	-9 06 00	Lib	16.52v	+0.17	-0.74	3.48	0.361	PKS
1512+370		+37.43	15 14 43.0	+36 50 51	Boo	15.5			(0.99)	0.371	OR 321

Designation	3C	4C	α 2000	δ 2000	Const	Mag	B−V	U−B	S	z		Notes
1514−241			15^{h}17^{m}41^s.8	−24°22′19″	Lib	14.8 v	+0.8	−0.1		0.049	*	PKS, AP Lib
1522+101			15 24 24.6	+9 58 30	SerCp	15.74B				1.321		PG
1525+227			15 27 57.6	+22 33 04	SerCp	16.72	+0.07	−0.96	(0.25)	0.253		LB 9743
1538+477			15 39 34.5	+47 35 36	Boo	16.01B				0.770		PG
1538+149		+14.60	15 40 49.8	+14 47 49	SerCp	15.5 v	+0.52	−0.60			*	OR 165
1543+489			15 45 30.1	+48 46 12	Boo	16.05B				0.400		PG
1545+210	323.1	+21.45	15 47 43.5	+20 52 17	SerCp	16.69v	+0.11	−0.85	(2.4)	0.264		PKS, NRAO 483
1612+261			16 14 13.2	+26 04 16	CrB	15.41v	+0.65	−0.78		0.131		Ton 256
1618+177	334	+17.68	16 20 21.9	+17 36 24	Her	16.41v	+0.12	−0.79	6.80	0.555		PKS, NRAO 500
1622+238	336	+23.43	16 24 39.1	+23 45 13	Her	17.47	+0.44	−0.79	(2.5)	0.927		PKS, NRAO 501
1630+377			16 32 01.2	+37 37 51	Her	15.96B				1.471		PG
1634+706			16 34 29.0	+70 31 32	Dra	14.90B				1.334		PG
1634+628	343	+62.26	16 34 33.8	+62 45 36	Dra	20.60	+0.45	−0.65	(5.1)	0.988		NRAO 509
1634+269	342	+26.49	16 36 36.4	+26 48 09	Her	17.75	+0.26		(1.3)	0.560		PKS, NRAO 510
1634+267A			16 37 01.6	+26 36 05	Her	18.5				1.961		KP, comp. A of double QSO
1634+267B			16 37 01.6	+26 36 09	Her	20.0				1.961		KP, comp. B of double QSO
1635+119			16 37 46.7	+11 49 49	Her	16.57	+0.49	−0.74		0.146		MC
1641+399	345	+39.48	16 42 58.8	+39 48 37	Her	15.96v	+0.29	−0.50	11	0.595		NRAO 513
1652+398		+39.49	16 53 52.2	+39 45 37	Her	13.88v	+0.72	−0.28		0.034	*	Mrk 501
1656+053			16 58 33.5	+5 15 16	Oph	16.48v	+0.43	−0.63	2.34	0.879		PKS, OS 094
1700+518			17 01 24.9	+51 49 21	Dra	15.43B				0.292		PG
1704+608	351	+60.24	17 04 41.3	+60 44 30	Dra	15.28v	+0.13	−0.75	11	0.371		NRAO 522
1715+533			17 16 36.4	+53 18 13	Dra	16.30B				1.920		PG
1718+481			17 19 38.2	+48 04 13	Her	15.33B				1.084		PG
1720+246			17 22 41.2	+24 36 19	Her	16.42v	+0.40	−0.87		0.175		V396 Her
1721+343		+34.47	17 23 20.8	+34 17 58	Her	16.5 v				0.206		OT 336
1727+502			17 28 18.6	+50 13 10	Her	15.97v	+0.58	−0.46		0.055	*	1Z186
1803+676			18 03 28.8	+67 38 09	Dra	15.78	+0.26	−0.84		0.136		
1821+107			18 24 02.8	+10 44 24	Oph	16.0				1.36		PKS
1828+487	380	+48.46	18 29 31.6	+48 44 45	Dra	16.81v	+0.24	−0.59	39	0.692		NRAO 565
1912−550			19 16 39.2	−54 54 48	Tel	16.49	+0.09	−0.81	0.80	0.398		PKS
1942−571			19 46 34.6	−57 00 29	Pav	16.93	+0.25	−0.66	1.06	0.527		PKS
2000−330			20 03 24.0	−32 51 47	Sgr	19				3.78		PKS
2005−044	407	−04.76	20 08 24.1	−4 18 30	Aql	18			3.01	0.589		PKS, NRAO 623
2016+112B			20 19 18.0	+11 27 13	Del	22.5				3.273		Comp. B of double QSO
2016+112A			20 19 18.2	+11 27 15	Del	22.5				3.273		Comp. A of double QSO
2037+511	418	+51.42	20 38 36.9	+51 19 12	Cyg	20	+1.3		12	1.686		NRAO 636
2044−027	422		20 47 10.3	−2 36 23	Aqr	19.5	+1.3		5.37	0.942		PKS, NRAO 639
2112+059			21 14 52.7	+6 07 41	Equ	15.52B				0.466		PG
2115−305			21 18 10.6	−30 19 11	Mic	16.47v	+0.49	−0.54	6.89	0.98 ?		PKS, OX−325
2120+168	432	+16.72	21 22 46.3	+17 04 38	Peg	17.96	+0.22	−0.79	6.05	1.805		PKS, NRAO 656
2128−123			21 31 35.3	−12 07 04	Cap	15.98v	+0.13	−0.67	1.67	0.501		PKS
2135−147			21 37 45.2	−14 32 56	Cap	15.53v	+0.10	−0.83	8.78	0.200		PKS, PHL 1657
2141+175			21 43 35.6	+17 43 49	Peg	15.73v	+0.18	−0.73		0.213		PKS, OX 169
2145+067		+06.69	21 48 05.5	+6 57 39	Peg	16.47v	+0.41	−0.79	4.29	0.99		PKS, OX 076.1
2155−304			21 58 51.9	−30 13 30	PsA	13.09v	+0.27	−0.73		0.17	*	PKS
2200+420			22 02 43.3	+42 16 40	Lac	14.72v	+0.97	−0.20		0.07	*	BL Lac
2201+315		+31.63	22 03 14.6	+31 45 38	Peg	15.47v				0.297		B2
2204−573			22 07 53.6	−57 07 34	Ind	16.6				2.725		PKS
2216−038		−03.79	22 18 51.6	−3 35 32	Aqr	16.38v	+0.55	−0.62	1.77	0.901		PKS, OY−027
2223−052	446	−05.92	22 25 47.2	−4 57 01	Aqr	18.39v	+0.44	−0.90	11.9	1.406		PKS, NRAO 687
2227−394			22 30 32.9	−39 13 08	Gru	18.8				3.45		
2233+134			22 36 07.7	+13 43 55	Peg	16.04B				0.325		PG
2247+140		+14.82	22 50 25.3	+14 19 53	Peg	16.6			3.20	0.237		PKS, OY 181
2249+185	454	+18.67	22 51 34.7	+18 48 41	Peg	18.40v	+0.12	−0.95	5.65	1.757		PKS, NRAO 699
2251+158	454.3	+15.76	22 53 57.7	+16 08 54	Peg	16.10v	+0.47	−0.66	11.6	0.859		PKS, NRAO 701
2251+113		+11.72	22 54 10.4	+11 36 38	Peg	15.82v	+0.21	−0.84	4.16	0.323		PKS, OY 186
2252+129	455	+12.79	22 55 03.9	+13 13 34	Peg	19.7			7.25	0.543		PKS, NRAO 702
2254+074			22 57 17.3	+7 43 13	Psc	16.5 v	+0.66	−0.44	(0.39)		*	OY 091
2255−282			22 58 05.9	−27 58 21	PsA	16.7			0.98	0.926		PKS
2300−683			23 03 44.1	−68 07 37	Ind	16.4			1.23	0.512		PKS
2302+029			23 04 44.9	+3 11 46	Psc	16.03B				1.044		PG
2305+187		+18.68	23 07 45.6	+19 01 21	Peg	16.5			(0.95)	0.313		PKS, NRAO 703
2308+098		+09.72	23 11 18.4	+10 08 14	Peg	16.5			(0.47)	0.432		OZ 014
2310−322			23 13 10.0	−31 57 47	Scl	16.6			1.97	0.340		PKS
2325+269	463	+27.52	23 27 57.3	+27 15 54	Peg	17.5			(1.59)	0.875		PKS, NRAO 713
2326−477			23 29 17.6	−47 30 20	Phe	16.79	+0.25	−0.98	3.21	1.299		PKS
2344+092		+09.74	23 46 36.8	+9 30 45	Peg	15.97v	+0.23	−0.61	2.36	0.677		PKS, OZ 073.5
2345+007B			23 48 19.2	+0 57 17	Psc					2.15		Comp. B of double QSO
2345+007A			23 48 19.6	+0 57 20	Psc	19.5				2.15		Comp. A of double QSO
2350−015			23 53 21.8	−1 15 28	Psc	16				0.99 ?		UM 186
2352−342			23 55 25.4	−33 57 57	Scl	16.4				0.702		PKS

Radio Sources

Radio Sources

Column Headings

PKS — Number in the Parkes radio survey (J. A. Ekers, 1969, plus other references from 1970 to 1975).

3C — Number in the revised third Cambridge radio survey (A. S. Bennett, 1962).

4C — Number in the fourth Cambridge radio survey (J. D. H. Pilkington and P. F. Scott, 1965; J. F. R. Gower *et al.*, 1967).

α **2000** and δ **2000** — Right ascension and declination, referred to the 2000.0 equinox. These positions are derived from radio observations, and may not match the positions of optical counterparts.

Const — Constellation in which the source is located.

Type — Optical identification. With the exceptions noted below, galaxies are indicated by the letter "G" or by their Hubble type, if known. Galaxies of type D (or supergiant cD systems) are ellipticals with extensive, diffuse halos; those whose halos are less pronounced are respectively termed ED or DE types based on whether the disk or the halo is more conspicuous. Other identifications are as follows: N galaxy (N); nebula (Neb); pulsar (PSR); quasi-stellar object (QSO); Seyfert galaxy (Seyf); supernova remnant (SNR). When a positive optical identification cannot be made, a question mark and code letter are listed in this column: blank field (?B); crowded field (?C); galaxy near radio position (?G); field obscured by nebulosity (?O); starlike object near radio position (?S).

Mag — Approximate visual magnitude of the source.

z — Redshift of the source, if it is a galaxy or QSO. The redshift is determined from lines in the object's spectrum. It is defined as the difference between the observed and rest wavelengths of a line, divided by its rest wavelength. In general, the greater the redshift, the more distant the object.

S178, S408, S1400, S2700, and **S5000** — Radio strength (flux density) measured at the frequencies 178, 408, 1400, 2700, and 5000 megahertz (wavelengths of 169, 74, 21, 11, and 6 centimeters, respectively). The units are janskys, formerly called flux units, where one jansky equals 10^{-26} watt per square meter per hertz.

Notes — Alternate designations, membership in galaxy groups or clusters, and other information are listed here. In addition to Messier, NGC, and IC numbers, designations from the following sources may be given: G. O. Abell's catalogue of galaxy clusters, 1958 (A); C. S. Gum's list of southern H II regions, 1955; the 1960 catalogue of Hα-emission regions by A. W. Rodgers, C. T. Campbell, and J. B. Whiteoak (RCW); G. Westerhout's radio survey of the galaxy, 1958 (W). Some terms are abbreviated as follows: cluster (cl.); double-lobed or dumbbell source (Dbl); group (gr.); magnitude (mag.); object (obj.); supernova remnant (SNR).

PKS	3C	4C	α 2000	δ 2000	Const	Type	Mag	z	S178	S408	S1400	S2700	S5000	Notes
0000−177	465		0^{h}03^m.4	−17°27′	Cet	QSO	19.0	1.465		6.5	2.2	1.4	0.9	
0003−567			0 05.9	−56 28	Phe	G	19.5			5.2	2.0	1.1	0.6	
0003−428			0 06.0	−42 35	Phe	?B				5.2	2.1	1.0	0.5	
0003−833			0 06.1	−83 06	Oct	G	19.0			6.6	2.1	1.1	0.8	
0003−003	2	−00.01	0 06.4	−0 04	Psc	QSO	19.5	1.037	13.2	10.5	3.7	2.5	1.4	Db1
0005−062	3	−06.01	0 08.5	−5 59	Psc	N	19		8.2	4.7	1.2	0.7	0.3	
0007−446			0 10.5	−44 23	Phe	?B				6.8	1.9	0.8	0.4	
0008−421			0 10.9	−41 53	Phe	?B				6.6	4.4	2.5	1.3	
0013−634			0 16.0	−63 10	Tuc	?S				6.9	1.9	0.8	0.4	
	6.1	+78.01	0 16.5	+79 15	Cep	?G	23		17.3	8.7	3.7	1.9	1.0	Db1
0016−129			0 18.8	−12 43	Cet	?S				6.9	2.3	1.2	0.7	
0017+154	9	+15.02	0 20.4	+15 41	Psc	QSO	18	2.012	16.3	7.8	2.1	1.1	0.5	Db1
0020−253			0 23.1	−25 03	Scl	G	20			5.4	2.4	1.4	0.8	Db1
0022−297			0 24.5	−29 29	Scl	QSO?	19			7.8	2.6	1.6	1.0	
	10	+63.01	0 25.5	+64 11	Cas	SNR			145		45	21.7	11.5	Tycho's SNR
0023−263			0 25.8	−26 02	Scl	?S				17.0	8.6	5.8	3.8	
	11.1	+63.02	0 29.7	+64 00	Cas				11.6		3.0	1.6	0.8	
	13	+39.01	0 34.2	+39 24	And	?G	20		11.6	5.8	1.8	0.9	0.4	Db1, faint cluster
0032−203			0 35.1	−20 04	Cet	?S				6.9	2.4	0.9	0.5	
0033+183	14	+18.02	0 36.1	+18 38	Psc	?S	20		10.4	5.8	2.0	0.9	0.5	Db1
	14.1	+58.02	0 36.1	+58 55	Cas	?O			16.1	15	2.7	1.2	0.7	Part of 3C 14.1
0034−014	15	−01.03	0 37.1	−1 09	Cet	E	15.3	0.073	15.8	9.7	4.1	2.6	1.6	Db1
	14.1	+60.02	0 37.4	+60 30	Cas				13.2					Part of 3C 14.1
0035+130	16	+13.05	0 37.7	+13 20	Psc	G	19.5		10.5	5.8	1.8	0.9	0.5	Db1
0035−024	17	−02.03	0 38.3	−2 08	Cet	E	18.0	0.220	20.0	16.5	6.4	4.0	2.5	
0036−392			0 38.4	−39 00	Scl	QSO	16.5	0.592		5.8	1.7	0.9	0.4	
0036+030		+03.01	0 39.3	+3 20	Psc	S0	12.3	0.014	3.4	3.1	1.8	1.2	0.6	NGC 193
0038+097	18	+09.02	0 40.9	+10 03	Psc	G	18.5	0.188	19.8	11.5	4.7	2.8	1.6	Db1
	19	+32.03	0 40.9	+33 09	And	G	20.5	0.482	12.1	7.7	3.7	1.9	1.3	Db1, faint cluster?
0039−445			0 42.2	−44 14	Phe	G	19.5			10.2	3.8	2.1	1.2	
	20	+51.02	0 43.1	+52 05	Cas	?G	19		44	24	10.3	6.4	4.2	Db1
	21.1	+66.02	0 44.7	+66 19	Cas	?O			9.7					
0042−357			0 44.7	−35 31	Scl	?G				5.8	2.3	1.6	1.0	
0043−424			0 46.3	−42 08	Phe	E	18.0	0.053	21	15.6	8.1	4.9	2.9	Db1
0045−255			0 47.6	−25 17	Scl	Sc	7.1			6.1	6.0	3.5	2.1	NGC 253, triple?
	22	+50.04	0 50.9	+51 14	Cas	G	20.5		15.5	6.6	2.5	1.3	0.8	Db1
0049−433			0 52.3	−43 07	Phe	?S				8.1	2.7	1.6	0.9	
0051−038	26	−03.03	0 54.1	−3 34	Cet	E	19.0	0.210	9.8	7.0	2.1	1.2	0.6	
	28	+26.02	0 55.9	+26 26	Psc	E	17.5	0.195	16.3	6.8	1.3	0.7	0.4	Db1, in cluster A115
	27	+68.02	0 55.9	+68 21	Cas	?O			26.4		6.7	4.3	2.5	Db1
0055−016	29	−01.05	0 57.7	−1 23	Cet	E0	14.3	0.045	15.1	10.9	5.4	3.5	2.0	In cluster A119
0056−172			0 59.1	−17 01	Cet	?S				6.2	1.7	0.8	0.4	
0101−128			1 04.4	−12 35	Cet	?S				5.2	1.8	1.5	0.7	
0103−453			1 05.3	−45 06	Phe	?B				6.4	2.8	1.1	0.6	
	31	+32.05	1 07.5	+32 24	Psc	DE3	11.8	0.016	16.7	9.9	5.4	3.5	2.1	NGC 383. Db1
0105−163			1 08.3	−16 04	Cet	?S				13.2	3.8	2.3	1.1	
	33.2	+69.04	1 08.6	+69 19	Cas	?O			9.7		0.9	0.5	0.3	
0106+130	33	+13.07	1 08.9	+13 20	Psc	DE4	15.2	0.060	53	33.6	14.7	6.7	7.1	Triple source?
	33.1	+72.01	1 09.7	+73 11	Cas	G	19.5	0.181	13.0	10.0	3.2	1.7	0.9	Db1, galaxy pair
	34	+31.02	1 10.3	+31 47	Psc	G	21		11.2		2.2	0.8	0.4	Db1, faint cluster
0110−692			1 11.7	−69 00	Tuc					6.1	2.0	0.9	0.4	
	35	+49.04	1 12.0	+49 30	Cas	D3	15.0	0.068	13.5	5.0	2.3	1.3	0.6	Db1
0114−476			1 16.4	−47 22	Phe	E	16.5	0.146		10.4	2.1	0.7	0.3	Db1
0114−211			1 16.8	−20 52	Cet	E	16.5			10.6	4.1	2.4	1.2	
	36	+45.03	1 18.0	+45 35	And	QSO?	21.7		10.3	4.0	1.2	0.6	0.4	Db1
0115+027	37	+02.04	1 18.3	+2 58	Psc	QSO	17	0.671	7.5	6.1	1.5	0.9	0.6	
0116+082		+08.06	1 19.0	+8 30	Psc	G	20.5	0.594	6.6	5.2	2.2	1.5	1.1	
0117−155			1 20.5	−15 20	Cet	?S				13.5	4.7	2.7	1.6	
0119−634			1 21.7	−63 09	Tuc	QSO	18.0	0.653		5.2	1.6	0.7	0.4	
0123−016	40	−01.08	1 26.0	−1 21	Cet	cD4	12.3	0.018	25.9	16.4	5.2	2.2	1.8	Cluster A194
	41	+32.06	1 26.7	+33 12	Psc	G	21		10.6	6.7	3.7	2.2	1.5	Db1
0125−143			1 27.5	−14 03	Cet	G	20			7.4	2.5	1.5	0.8	
	42	+28.04	1 28.5	+29 02	Psc	G	20	0.395	11.2	8.5	2.8	1.6	0.8	Db1
0127+233	43	+23.06	1 30.0	+23 38	Psc	QSO	19	1.459	11.6		3.0	1.9	1.1	
0128−264			1 30.5	−26 10	Scl	?S				5.4	1.2	0.6	0.3	
0128+061	44	+06.07	1 31.4	+6 24	Psc	?B			12.0	5.2	1.5	0.8	0.3	Db1
0131−367			1 33.9	−36 29	Scl	S0p	13.2	0.030		16	7.1	5.6		NGC 612. Db1
0132+079	45		1 35.3	+8 11	Psc	G	19.5		8.2	6.0	2.4	1.4	0.7	
	46	+37.05	1 35.5	+37 54	And	G	19.5	0.437	10.2	4.1	1.1	0.7	0.3	Db1, faint cluster?
0133+207	47	+20.07	1 36.4	+20 58	Psc	QSO	18.1	0.425	26.4		3.6	2.1	1.2	Db1
	48	+32.08	1 37.7	+33 09	Tri	QSO	16.2	0.367	51.0	34	16.1	9.0	5.4	Db1?
0138+136	49	+13.10	1 41.2	+13 53	Psc	G	21		9.7	6.6	2.9	1.6	0.8	
0139−273			1 41.5	−27 07	Scl	?S				5.0		0.8	0.4	
	52	+53.02	1 48.5	+53 34	Per	G	18.5		13.7	9.0	4.0	2.3	1.5	Db1, in cluster
0148−297			1 50.6	−29 32	For	?G				7.0	2.9	1.5	0.8	

PKS	3C	4C	α 2000	δ 2000	Const	Type	Mag	z	S178	S408	S1400	S2700	S5000	Notes
	54	+43.06	1^{h}55$.^m$5	+43°45'	And	?G	23		9.6	6.0	2.8	1.0	0.6	Dbl, faint cluster?
	55	+28.05	1 57.2	+28 50	Tri	G	20.8	0.240	23	6.8	2.7	1.5	0.9	Dbl
0155−109			1 57.7	−10 44	Cet	QSO	17.1	0.616		5.4	2.0	1.2	0.7	
0157−311			2 00.2	−30 53	For	QSO?	19			10.3	3.6	2.4	1.4	
0159−117	57		2 02.0	−11 33	Cet	QSO	16	0.669		5.7	3.1	2.0	1.4	
0202−765			2 02.2	−76 20	Hyi	QSO	16.0	0.389		8.5	2.4	1.4	0.8	
0201−440			2 03.7	−43 50	Phe	?S				6.2	2.9	1.7	1.0	
0202+149		+15.05	2 04.8	+15 14	Ari	?B			5.2	5.6	3.7	3.1	2.3	
	58	+64.02	2 05.5	+64 51	Cas	SNR			26		28.0	27.9	21.4	
0208−512			2 10.8	−51 01	Eri	QSO	17.5	1.001		5.5		3.6	3.2	
0213−132	62		2 15.6	−13 00	Cet	E	18.5			11.7	4.8	2.9	1.8	
0214−480			2 16.8	−47 49	Eri	E	14.5	0.064		9.5	2.4	1.3	0.8	
0219−706			2 20.1	−70 22	Hyi	?G				6.9		1.0	0.5	
0218−021	63	−02.10	2 20.9	−1 57	Cet	G	18.5	0.175	19.1	11.8	3.5	1.8	0.9	Dbl
0219+082	64		2 22.0	+8 27	Cet	G	19.5		7.4	5.3	2.5	1.4	0.8	Dbl
	61.1		2 22.7	+86 19	Cep	G	19	0.184	31	17.6	6.3	3.7	1.9	Dbl, faint cluster
	66B	+42.07	2 23.3	+43 02	And	ED2	12.9	0.022	36	23.0	9.5	5.0	2.4	3C 66A is QSO 6' np
	65	+39.07	2 23.7	+40 01	And	G	>23		29	11	4.3	1.6	0.8	Dbl
0223−712			2 24.0	−71 00	Hyi	?S				5.3	1.5	0.6	0.2	
	67	+27.08	2 24.2	+27 50	Tri	G	18.5	0.310	10.8	7.4	3.0	1.7	0.9	
0222−234			2 25.0	−23 13	Cet	QSO?	18.5			5.4	1.8	1.3	0.8	
	68.1	+34.08	2 32.5	+34 25	Tri	QSO	19	1.238	18.4	7.8	2.7	1.4	0.8	Dbl
	68.2	+31.08	2 34.4	+31 35	Tri	?G	>23		13.0		1.4	0.4	0.2	Dbl
0235−197			2 37.7	−19 32	Cet	?B				13.3	4.4	2.4	1.4	
	69	+58.08	2 38.0	+59 12	Cas	?O			19.9	12	2.2	1.7	0.9	Dbl
0238+085		+08.11	2 41.1	+8 45	Cet	S0p	13.2	0.021	2.3	3.1	1.3	0.7	0.3	NGC 1044. Dbl
0240−002	71	−00.13	2 42.7	−0 01	Cet	Sbp	8.8	0.004	16.1	12.4	5.1	3.3	1.9	M77 (NGC 1068)
0242−514			2 43.7	−51 13	Hor	?S				8.0	2.7	0.9	0.5	
0245−558			2 46.9	−55 41	Hor	?S				8.3	2.7	1.2	0.6	
0252−712			2 52.8	−71 05	Hyi	G	19			14.1	5.9	3.1	1.5	
0254−236			2 56.3	−23 25	Eri	?S				5.9	1.5	0.6	0.3	
0255+058	75	+06.15	2 57.7	+6 03	Cet	E0:	13.1	0.024	25	16.2	6.0	3.5	2.0	Dbl, cluster A400
0300+162	76.1	+16.06	3 03.2	+16 26	Ari	DE3	15.2	0.032	10.2	5.8	3.1	1.9	1.3	Dbl
0304−122			3 07.0	−12 06	Eri	S0	16.0			5.5	1.5	0.8	0.4	
0305+039	78	+03.05	3 08.4	+4 07	Cet	DE3	12.8	0.029	19.2	13.6	7.8	5.3	3.5	NGC 1218
0307+169	79	+16.07	3 10.0	+17 06	Ari	E	18.6	0.256	24.8	14.7	4.9	2.6	1.4	N-type galaxy? Dbl
0310−150			3 12.8	−14 50	Eri	?S				6.1	2.0	1.4	0.9	
0315−685			3 16.2	−68 22	Hyi	G	19.5			5.3	1.6	0.7	0.3	
	83.1	+41.06	3 18.2	+41 55	Per	ED3	12.5	0.025	30	9.5	8.4	3.6	1.8	NGC 1265, in Per cl.
0316+162			3 19.0	+16 29	Ari	?S			7.1	8.1	8.3	4.8	2.9	Dbl?
	84	+41.07	3 19.8	+41 31	Per	Seyf	11.6	0.018	63	29.0	14	9.3	18.7	NGC 1275, Per A. Dbl
0319−453			3 21.4	−45 10	Eri	?S				9.5	3.0	1.7	0.8	
0320−374			3 22.6	−37 14	For	S0p	8.8	0.006		259		98.0		NGC 1316, For A. Dbl
0320+053		+05.14	3 23.3	+5 34	Cet	G	20		6.4	7.1	3.1	1.6	0.8	
	86	+55.06	3 27.3	+55 20	Per				22.6	21	9.5	4.5	2.9	Multiple source
0325+023	88	+02.10	3 27.9	+2 34	Tau	D	13.6	0.030	17.3	10.9	4.7	3.0	1.9	Dbl
0331−013	89	−01.12	3 34.2	−1 11	Tau	D	16.0	0.139	20.2	8.7	3.2	1.4	0.7	Dbl, in small group
0333+128	90		3 36.4	+13 03	Tau	?S			6.4	5.2	2.3	1.3	0.7	
	91	+50.10	3 37.7	+50 47	Per	?O			12.6	9	3.4	2.0	1.2	
0336−355			3 38.8	−35 23	For	E1p	9.9	0.005		7.9	2.5	1.4	0.7	NGC 1399. Dbl
0340−372			3 42.1	−37 03	Eri	?S				5.0	2.0	1.1	0.7	
0340+048	93	+04.13	3 43.5	+4 58	Tau	QSO	18.1	0.357	14.4	8.6	2.8	1.7	0.9	
0344−345			3 46.5	−34 23	For	E	16.5			9.3	3.1	2.1	0.9	Triple source
	93.1	+33.08	3 48.8	+33 56	Per	G	19	0.243	9.9	6.6	2.8	1.3	0.8	In group
0347+057		+05.16	3 49.8	+5 52	Tau	?S			11.3	7.5	3.1	2.0	1.2	
0349−146	95		3 51.5	−14 29	Eri	QSO	16.2	0.614		11.6	2.8	1.4	0.8	Dbl
0349−278			3 51.6	−27 45	Eri	E	16.8	0.066		15.8	5.3	2.9	2.2	Triple source
0350−073	94		3 52.5	−7 11	Eri	QSO	16.4	0.962		10.2	2.7	1.4	0.7	
0356+102	98	+10.12	3 58.9	+10 26	Tau	E1:	14.9	0.031	44	27.5	11.1	6.1	3.3	Dbl
0357−163			4 00.3	−16 10	Eri	?S				5.7	1.8	1.0	0.6	
0358+004	99	+00.14	4 01.1	+0 37	Tau	N	19.0	0.426	8.8	5.3	1.9	1.0	0.6	In cluster?
0403−132			4 05.6	−13 08	Eri	QSO	17.1	0.571		6.7	4.0	3.2	3.2	
0404+035	105	+03.08	4 07.4	+3 41	Tau	?G	18.5	0.089	16.2	9.4	5.1	3.3	2.0	Extended source
0405−123			4 07.8	−12 12	Eri	QSO	14.8	0.574		8.2	3.2	2.6	1.9	Dbl
	103	+42.11	4 08.1	+43 01	Per	?G			26.0	19	5.0	2.6	1.4	Dbl
0407−658			4 08.3	−65 45	Ret	?S				51.1	14.4	6.5	3.3	
0409−752			4 08.8	−75 07	Men	?S				36.0	13.4	8.1	4.3	
0406−180			4 09.1	−17 57	Eri	G?	20			5.6	1.9	1.3	0.9	
0408+070	106		4 11.2	+7 08	Tau	?S			6.1	3.7	1.6	0.9	0.5	
0409−011	107	−01.13	4 12.4	−1 00	Eri	?B			11.8	4.8	1.3	0.7	0.4	
0411−561			4 12.8	−56 01	Dor	?S				6.8	2.5	1.3	0.7	
0410+110	109	+11.18	4 13.7	+11 12	Tau	N	17.8	0.306	20.4	11.7	3.9	2.6	1.8	Dbl
0411+141		+14.11	4 14.5	+14 16	Tau	?B			6.7	5.2	2.2	1.5	0.9	
0413−210			4 16.1	−20 56	Eri	QSO	19.5	0.807		7.3	2.5	1.8	1.4	
0414−060	110	−05.17	4 17.3	−5 54	Eri	QSO	16	0.781	8.0	3.5	0.8	0.5	0.3	

PKS	3C	4C	α 2000	δ 2000	Const	Type	Mag	z	S178	S408	S1400	S2700	S5000	Notes
	111	+37.12	4^{h}18^m.4	+38°01′	Per	N	18.0	0.049	61	33.0	15	9.6	7.2	Triple source
0417+177	114	+17.24	4 20.4	+17 54	Tau	?B			10.8	4.1	0.8	0.5	0.3	Db1
0420-625			4 20.9	-62 24	Ret	G	19.5			11.5	3.0	1.7	0.8	
0427-339			4 29.1	-53 50	Dor	S0	12.8	0.039		14.6	5.6	2.9	1.6	IC 2082. Db1
0427-366			4 29.7	-36 31	Eri	?B				8.4	2.0	1.0	0.5	
0429-616			4 30.3	-61 33	Ret	?G				6.5	1.7	1.1	0.6	
	119	+41.13	4 32.6	+41 39	Per	QSO	20.0	0.408	16.4	17	8.8	5.4	3.4	Db1?
0430+052	120	+05.20	4 33.2	+5 21	Tau	S0:	13.7	0.033	6.7	6.1	3.8	4.8	5.1	BW Tau, Seyfert gal.
	123	+29.14	4 37.1	+29 40	Tau	E/cD	19.5	0.218	189	119	41	27.2	16.3	Per B, in cl. Db1
0438-436			4 40.3	-43 33	Cae	QSO	19.5	2.863		8.1	7.5	6.5	7.0	
0439+012	124	+01.11	4 42.0	+1 21	Tau	?B			9.2	4.3	1.2	0.6	0.3	
0442-282			4 44.6	-28 10	Eri	E	18.5			22	6.9	3.7	2.2	
	125	+39.15	4 46.3	+39 45	Per	?O			14.1	7.3		1.0	0.6	
	129	+44.12	4 49.0	+45 02	Per	G	19.4	0.021	23	17.0	7.6	3.0	2.0	Radio tail w of gal.
	129.1	+44.13	4 50.8	+45 03	Aur	G	19.9	0.022	11.5	2.1				
	130	+52.10	4 52.8	+52 05	Aur	DE2	16.5	0.109	13.5		5.3	1.5	0.8	Db1
	131	+31.18	4 53.4	+31 30	Aur	?O			14.8	8.0	2.5	1.5	0.9	
0453-301			4 55.2	-30 07	Cae	E	18.5			9.2	3.2	1.9	1.2	
0453-206			4 55.4	-20 34	Lep	E	14.0	0.035		11.3	4.7	2.8	1.8	
	132	+22.11	4 56.7	+22 49	Tau	E	18.5	0.214	12.4	7.9	3.7	1.9	1.0	Stellar object at 5″
0456-301			4 58.4	-30 07	Cae	E3	18			5.5	2.7	1.6	0.9	
0459+252	133	+25.16	5 03.0	+25 16	Tau	G	21		21.0		5.8	3.7	2.2	Triple source
	134	+38.15	5 04.7	+38 06	Aur	?O			77.0	45	8.0	4.2	2.0	Db1
0503-290			5 05.6	-28 56	Col	?S				5.5	1.1	0.6	0.3	
0506-612			5 06.7	-61 10	Dor	QSO	17.5	1.089		5.0	2.0	1.9	1.5	
0509-573			5 10.3	-57 20	Pic	QSO?	18.5			5.1		1.0	0.6	
0508-220			5 11.0	-22 02	Lep	N	18.5			5.1	1.9	0.9	0.7	
0511-484			5 12.8	-48 24	Pic	?S				13.2	3.6	1.8	0.8	
0511-305			5 13.6	-30 28	Col	E	16.5			8.9	2.6	1.3	0.8	Triple source?
0511+008	135	+00.18	5 14.1	+0 57	Ori	E	17.0	0.127	17.3	8.0	3.0	1.7	1.0	Galaxy pair
0512+249	136.1	+24.10	5 16.0	+24 58	Tau	D	17	0.064	14.0		3.0	1.5	0.6	
	137	+50.16	5 19.5	+50 55	Aur	?O			9.3	4.8	2.3	1.1	0.6	
0518-458			5 19.7	-45 47	Pic	E	16.0	0.035		166	66	29		Pic A. Db1
0518+165	138	+16.12	5 21.2	+16 38	Tau	QSO	18.8	0.754	18.7	17.2	9.2	6.3	4.0	Db1
0519-208			5 21.6	-20 48	Lep	?S				7.3	2.2	0.8	0.4	
	139.1		5 22.7	+34 03	Aur	Neb			22		23			IC 410 region
0521-365			5 23.0	-36 28	Col	N	16.8	0.062		36.1	16.3	12.5	9.2	
	139.2	+28.15	5 24.5	+28 13	Tau	?O			11.9		2.4	1.0	0.6	Db1
0525-696			5 25.0	-69 39	Dor	SNR				9.3		2.7	1.7	
	141	+32.18	5 26.7	+32 50	Aur	?O			14.1	6.9	1.8	1.1	0.6	IC 410 region
0528+064	142.1	+06.23	5 31.5	+6 30	Ori	?B			19.5	11.2	3.1	2.0	0.8	
	144	+21.19	5 34.5	+22 00	Tau	SNR			1530		926			M1, Crab nebula
			5 34.5	+22 01	Tau	PSR				0.8				Crab pulsar
0532-054	145	-05.21	5 35.3	-5 23	Ori	Neb				450	410			M42, Orion nebula
0534-497			5 36.2	-49 44	Pic	G	18.5			6.7	2.1	1.0	0.6	
0539-691			5 38.7	-69 05	Dor	Neb				39.1	34.0	29.8	25.1	
0539-019	147.1		5 41.7	-1 54	Ori	Neb			15.7	45	50.0	56.5	43.6	IC 434 region, Ori B
	147	+49.14	5 42.6	+49 51	Aur	QSO	17.8	0.545	59.6	62	22	13.0	8.2	Complex structure
0547-408			5 49.4	-40 51	Col	?B				8.2	2.6	1.3	0.8	
0602-319			6 04.3	-31 56	Col	QSO	18.5	0.452		6.7	3.0	1.9	1.3	
0601+203	152	+20.17	6 04.5	+20 21	Ori	?S			12.4		2.1	0.8	0.3	
0604-203			6 06.6	-20 22	Lep	G	19.5	0.164		7.4	3.2	2.0	1.0	
	153	+48.15	6 09.5	+48 04	Aur	G	18.5	0.277	19.0	10.0	3.8	2.3	1.4	Db1, in cluster?
	153.1		6 09.7	+21 11	Ori	Neb			20.5		29.1			NGC 2174-5
0610+260	154	+26.20	6 13.8	+26 05	Gem	QSO	18.0	0.580	23.1	9.2	5.5	3.3	2.1	Db1
0614-349			6 16.6	-34 56	Col	QSO?	19.9	0.329		5.3	2.7	2.0	1.3	Db1
	157		6 17.6	+22 42	Gem	SNR			230		101			IC 443
0618-371			6 20.0	-37 12	Col	G	16.6	0.033		5.3	2.6	1.8	1.3	Db1
	158	+14.17	6 21.7	+14 32	Ori	?O			16.7		2.2	1.1	0.6	
0620-526			6 21.7	-52 42	Car	D	15.5	0.051		9.3	3.5	2.4	1.2	
0625-536			6 26.3	-53 41	Car	G	15.4			20.2	6.6	3.7	1.8	Db1
0625-545			6 26.8	-54 33	Car	E	16			7.6		1.7	0.9	
0623+264	160		6 26.9	+26 23	Gem	G	19.5		4.1		1.4	1.0	0.7	
0625-354			6 27.1	-35 29	Col	E	16.5			7.8	4.1	2.9	2.1	
0624-058	161	-05.23	6 27.2	-5 53	Mon	?S			78	44.7	18.5	11.3	6.6	
0628+250	162		6 31.4	+25 01	Gem	E	16.8		5.6		1.2	0.6	0.3	
			6 32.1	+4 51	Mon	Neb					200			Rosette nebula
0637-752			6 35.8	-75 16	Men	QSO	16	0.651		7.9	4.9	5.9	5.5	
0634-205			6 36.5	-20 37	CMa	E	16.8	0.056		21	6.0	4.5	2.3	Db1
	165	+23.17	6 43.1	+23 18	Gem				13.5		2.6	1.3	0.8	Db1
	166	+21.21	6 45.4	+21 24	Gem	G	19.5	0.245	14.7	7.6	3.4	1.7	1.1	
0646-398			6 48.2	-39 57	Pup	?S				6.2	2.6	1.5	0.9	
	169.1	+45.12	6 51.3	+45 10	Aur	N?	20.9	0.633	10.3	3.8	1.2	0.5	0.4	Db1
0649+225	170		6 52.8	+22 32	Gem	G	19		6.6		1.0	0.5	0.3	
	171	+54.11	6 55.2	+54 10	Lyn	N	18.9	0.238	25	8.4	2.9	2.0	1.2	Db1

PKS	3C	4C	α 2000	δ 2000	Const	Type	Mag	z	S178	S408	S1400	S2700	S5000	Notes
0658−656			6^{h}58^m.2	−65°45'	Vol	?S				6.1	2.2	1.1	0.8	
0656−242			6 59.0	−24 17	CMa	?B				9.7	3.0	1.4	0.6	Db1
	172	+25.19	7 02.1	+25 14	Gem	G	20		15.1	6.4	2.9	1.5	0.8	Db1, galaxy pair
	173	+38.19	7 02.3	+37 57	Aur	QSO?	20		10.8	6.4	4.8	0.9	0.5	
0704−231			7 06.6	−23 12	CMa	?S				7.4	3.3	2.5	1.5	
	173.1	+74.12	7 09.3	+74 51	Cam	G	18.5	0.292	14.3	6.0	2.8	1.4	0.8	Db1
0709−206			7 11.8	−20 43	CMa	?B				7.5	2.0	1.1	0.4	
0710+118	175	+11.26	7 13.0	+11 46	CMi	QSO	16.6	0.768	17.2	8.9	2.7	1.3	0.7	Db1
0711+146	175.1	+14.21	7 14.1	+14 36	Gem	?S			12.5	4.8	2.0	1.2	0.6	Db1
0715−362			7 17.1	−36 22	Pup	E	17.8			7.1	2.2	1.1	0.6	Db1
0715−249			7 17.3	−25 05	CMa	?B				10.6	4.2	2.7	1.5	
0719−553			7 20.2	−55 25	Car	G	19.0	0.216		5.6	2.0	1.2	0.7	
0717+170	176		7 20.5	+16 59	Gem	?S			3.9	2.1	0.8	0.4	0.3	
0721+153	177	+15.19	7 24.4	+15 13	Gem	?S			10.2	3.9	1.2	0.7	0.3	Part of 3C 177. Db1
0721+161	177	+16.21	7 24.8	+16 02	Gem	D	17.5		16.2	3.7	1.5	0.8	0.5	Part of 3C 177. Db1
0722−095	178		7 24.9	−9 40	Mon	Sb+:	14.9	0.008	13		1.5	0.7	0.3	NGC 2377
0724−019	180	−02.31	7 27.1	−2 05	Mon	G	19		14.2	7.2	2.6	1.6	1.0	Db1
0725+147	181	+14.24	7 28.2	+14 38	Gem	QSO	18.9	1.382	14.0	6.7	2.1	1.2	0.7	
0725+244	182		7 28.5	+24 19	Gem	E	16.5		4.3		1.1	0.7	0.4	
0727−222			7 29.4	−22 18	Pup	?B				6.8	0.8	0.7	0.2	
	184	+70.06	7 39.4	+70 23	Cam	?G	22.5		11.8		2.6	1.2	0.6	Db1
	184.1		7 43.1	+80 27	Cam	D	17	0.118	13.0	7.6	3.4	1.9	1.1	Db1, in group
0743−673			7 43.5	−67 26	Vol	QSO	17	0.395		8.6	5.3	2.7	1.5	
	186	+38.21	7 44.3	+37 53	Lyn	QSO	17.6	1.063	14.0	6.8	1.3	0.6	0.4	
0741−063		−06.18	7 44.4	−6 30	Mon	?S			5.1	12.2	7.8	4.7	2.7	
0742+021	187	+02.21	7 45.1	+2 00	CMi	E	19.5	0.350	11.9	4.6	1.4	0.7	0.4	Db1
0745−188			7 47.4	−18 57	Pup	?S				5.2		0.6	0.3	
0745−191			7 47.5	−19 18	Pup	D	19.6	0.103		10.1		0.9	0.4	
0748−440			7 49.7	−44 12	Pup	?B				8.0	2.3	1.2	0.7	
0750−262			7 52.4	−26 26	Pup	Neb				17.2	11.3	7.4	3.8	
0758+143	190	+14.25	8 01.6	+14 15	Cnc	QSO	20	1.197	15.3	7.6	2.6	1.4	0.9	Db1
0800−098			8 02.7	−9 58	Mon	E	18.5			6.1	1.7	1.2	1.0	
0802+103	191	+10.25	8 04.8	+10 15	Cnc	QSO	18.4	1.956	11.4	7.3	2.0	0.9	0.5	
0803−008		−00.32	8 05.6	−0 58	Mon	E	15.4		7.0	5.8	1.2	0.7	0.5	
0802+243	192	+24.16	8 05.6	+24 10	Cnc	D	15.5	0.060	21.0	13.1	4.9	3.3	2.6	Triple, in cluster
0806−103	195		8 08.9	−10 28	Mon	N	18		21	13.7	4.0	2.5	1.5	Db1, faint cluster
0807−389			8 09.5	−39 05	Pup	?B				8.7	2.3	1.1	0.6	
	194	+42.25	8 10.1	+42 28	Lyn	?G	20.0		10.3	5.0	2.3	1.1	0.6	Compact, red object
	196	+48.22	8 13.6	+48 13	Lyn	QSO	17.8	0.871	73.7	33	14	7.7	4.4	Lyn A. Db1
0812−029	196.1	−02.35	8 15.5	−3 08	Hya	D	17.5	0.198	18.6	9.5	1.6	1.0	0.5	
0814−354			8 16.3	−35 35	Pup	?B				19.4	5.0	2.0	1.0	
0819−300			8 21.4	−30 11	Pup	E3	18.2			6.5	3.1	1.6	0.9	Triple source
	197.1	+47.28	8 21.6	+47 03	Lyn	DE	16.5	0.130	10.3	4.5	1.9	1.2	0.7	In group
0819+061	198	+06.30	8 22.5	+5 57	Hya	E	16.8	0.082	18.0	6.6	2.3	0.9	0.3	Db1. In small group?
0822−428			8 24.1	−43 00	Pup	SNR				198	18			Pup A
0825−202			8 27.3	−20 26	Pup	QSO?	18			10.3	3.1	2.1	1.1	
	200	+29.29	8 27.4	+29 58	Cnc	G	20	0.458	14	6.1	1.5	1.1	0.7	Triple source
0831+174	201		8 34.3	+17 19	Cnc	QSO?	19			1.9	0.6	0.3	0.2	
			8 34.5	−45 47	Vel	SNR				300				Vel X
0831+171	202		8 34.8	+17 01	Cnc	?S			8.0	5.2	2.3	1.0	0.6	
			8 35.3	−45 11	Vel	PSR				5.0	1.1			Vela pulsar
0834−196			8 37.2	−19 52	Pyx	?S				10.8	4.6	2.7	1.4	
	204	+65.09	8 37.7	+65 13	UMa	QSO	18.2	1.112	9.2		1.3	0.5	0.3	Triple source
	205	+58.16	8 39.1	+57 54	UMa	QSO	17.6	1.534	12.5	15	1.9	1.1	0.7	Db1
0837−120	206		8 39.8	−12 15	Hya	QSO	15.8	0.200		5.8	1.8	1.0	0.7	
0838+133	207	+13.38	8 40.8	+13 12	Cnc	QSO	18.2	0.684	12.0	7.0	2.8	1.7	1.3	Triple source
0842−754			8 41.5	−75 40	Cha	QSO	18	0.524		13.3	4.0	2.2	1.4	
0843−336			8 45.1	−33 48	Pyx	E3	12.3	0.008		3.2	2.2	1.4	0.9	NGC 2663
0850−206			8 53.0	−20 47	Pyx	?S				7.5	2.2	1.1	0.5	
0850+140	208	+14.28	8 53.1	+13 53	Cnc	QSO	17.4	1.109	17.0	7.8	2.2	1.1	0.5	Triple source
0851−142			8 53.8	−14 28	Hya	?S				5.2	2.4	1.0	0.4	
0851+142	208.1	+14.29	8 54.7	+14 06	Cnc	?S	20		13.5	5.5	2.2	1.2	0.8	
0852−070	209	−06.19	8 55.2	−7 15	Hya	QSO?	18		5.2	2.5	1.2	0.6	0.4	
	210	+28.21	8 58.2	+27 51	Cnc	?G	21.0		9.5	4.9	1.9	0.9	0.5	
0855+143	212	+14.30	8 58.7	+14 09	Cnc	QSO	19.5	1.049	14.1	7.0	2.8	1.4	0.8	Triple source
0857+187	213		9 00.8	+18 32	Cnc	?S			5.8	3.0	1.2	0.6	0.4	
	213.1	+29.33	9 01.1	+29 02	Cnc	G	19	0.194	11.9	5.6	2.4	1.0	0.8	In group
0859−257			9 01.8	−25 55	Pyx	?S				17.2	5.9	2.9	1.7	
0902−384			9 03.9	−38 37	Vel	?B				12.8	10.0			
0903+169	215	+16.26	9 06.5	+16 46	Cnc	QSO	18.2	0.411	9.9	6.3	1.4	0.9	0.4	Db1
0906−682			9 06.9	−68 30	Car	?S				6.1	1.7	0.9	0.4	
	217	+38.26	9 08.8	+37 49	Lyn	G	22		14.5	7.1	1.9	1.0	0.5	Db1
	216	+43.17	9 09.6	+42 51	Lyn	QSO	18.5	0.67	20.1	19	4.0	2.4	1.8	
0909−564			9 10.9	−56 37	Vel	?C			3.7	6.1	1.9	0.9	0.5	
0916−547			9 17.5	−54 56	Vel	?C				8.3	3.0	1.9	1.2	

PKS	3C	4C	α 2000	δ 2000	Const	Type	Mag	z	S178	S408	S1400	S2700	S5000	Notes
0915−118	218		9^{h}18^m.1	−12°06′	Hya	D	13.6	0.065		132	43.5	23.5	13.1	Hya A, in cl. Db1
	219	+45.19	9 21.1	+45 39	UMa	cD5	17.2	0.174	48	21.6		4.4	2.3	Lyn B. in cl. Db1
		+39.25	9 27.0	+39 02	LMi	QSO	17.9	0.698	5.4		2.9	4.5	7.6	Db1
	220.2	+36.15	9 30.6	+36 00	LMi	QSO	19.0	1.157	10.3	6.2	2.0	1.0	0.6	Db1
	220.1	+79.06	9 32.6	+79 05	Dra	G	20.5		15.8		2.3	1.1	0.5	Db1
0933+045	222	+04.32	9 36.5	+4 22	Hya	?S			11.8	5.1	0.9	0.3	0.1	
0935−289			9 38.0	−29 13	Ant	E	18.5			6.0	1.9	1.0	0.5	Db1
	220.3		9 39.4	+83 15	Cam	cD?	23		15.7		3.0	1.3	0.6	Db1, faint cluster
	223	+36.16	9 39.9	+35 53	LMi	D2	17.1	0.137	15.7	5.3	3.8	1.9	1.0	Db1, in group
	223.1	+39.28	9 41.4	+39 45	Lyn	E5	16.4	0.108	10.3	3.8	1.9	1.2	0.9	Db1, in group?
0939−115	224		9 42.0	−11 47	Hya	?S				1.8	0.6	0.4	0.1	
0939+140	225A	+14.32	9 42.2	+13 47	Leo	G	20.5		27	9.1	4.1	1.8	1.0	Comp. B: type ?B
0941+100	226	+10.27	9 44.3	+9 46	Leo	G	22		15.7	8.2	2.3	1.0	0.6	Db1
0945+076	227	+07.29	9 47.8	+7 25	Leo	N	16.3	0.086	30	22.1	7.1	4.1	2.6	Triple source
0947+145	228	+14.34	9 50.2	+14 20	Leo	G	20.5		16.1	9.3	3.5	1.9	1.1	Db1
0949+002	230	+00.32	9 52.0	−0 01	Sex	?S	15.5		19.2	12.3	3.2	1.5	0.7	
	231	+69.12	9 55.8	+69 41	UMa	Ir	8.4	0.001	14.0	11.7	8.1	5.4	3.8	M82 (NGC 3034)
	234	+29.35	10 01.8	+28 46	LMi	N	17.3	0.185	31	19.8	5.5	2.9	1.5	Triple, in group
	236	+35.22	10 06.0	+34 53	LMi	DE4	16.0	0.099	9.7	6.3	4.0	2.0	1.3	Triple source
1005+077	237	+07.30	10 08.0	+7 30	Leo	?S	20		26	15.4	6.6	4.0	1.9	Db1
1008+066	238	+06.38	10 11.0	+6 25	Leo	?B			16.3	9.3	2.8	1.4	0.7	Db1
	239	+46.20	10 11.8	+46 28	UMa	?G	23		13.0	5.7	1.5	0.6	0.3	Db1, faint cluster?
1010−647			10 12.3	−64 58	Car	?B				6.9	2.6	1.5	0.7	
1015−314			10 18.2	−31 44	Ant	?S				9.7	3.5	2.2	1.3	
1017−421			10 19.7	−42 25	Vel	?S				5.0	1.5	0.8	0.4	
1017−426			10 20.1	−42 51	Vel	?S				12.7	3.9	2.3	1.2	
	241	+22.28	10 21.9	+21 59	Leo	?S	23		11.1		1.7	0.7	0.3	
			10 23.3	−57 46	Car	Neb				310	350			Gum 29
1022+204	242		10 25.3	+20 10	Leo	?S			6.8	3.6	0.9	0.6	0.3	
1023+067	243	+06.40	10 26.5	+6 30	Leo	QSO	18.3	1.699	9.5	3.4	0.9	0.3	0.2	Db1
1030−340			10 33.2	−34 19	Ant	?S				5.6	1.4	0.6	0.3	
	244.1	+58.21	10 33.6	+58 14	UMa	G	19.0	0.428	20.2	14.6	4.0	2.0	1.1	Db1
1036−697			10 38.5	−70 03	Car	?S				6.9	2.4	1.4	0.6	
1039+029		+03.18	10 41.7	+2 43	Sex	?B			7.4	6.4	2.6	1.7	1.0	
1040+123	245	+12.37	10 42.7	+12 04	Leo	QSO	17.2	1.029	12.5	8.9	3.7	2.0	1.5	Db1
			10 44.4	−59 50	Car	Neb				610	650			NGC 3372, η Car
1046−409			10 48.7	−41 14	Vel	?S				5.0	1.9	1.4	1.0	
1048−090	246		10 51.5	−9 18	Crt	QSO	16.8	0.342		5.3	1.7	1.2	0.7	Db1
	247	+43.20	10 59.0	+43 01	UMa	G	20		16.8	8.8	4.0	1.6	0.8	Db1
1059−010	249	−00.42	11 02.1	−1 16	Leo	?B			21.6	8.0	2.7	1.4	0.7	Db1
	249.1	+77.09	11 04.2	+77 00	Dra	QSO	15.7	0.311	12.4		2.4	1.4	0.8	
1103−208			11 06.4	−21 09	Crt	?S				7.6	2.4	1.4	0.6	
	250	+25.34	11 08.9	+24 58	Leo	?G	21.5		11.9		1.6	0.5	0.3	Db1
	252	+35.25	11 11.5	+35 39	UMa	G	21.5		13.0	4.0	1.5	0.6	0.3	Db1
			11 14.0	−61 17	Car	Neb				254	370			NGC 3603, RCW 57
	254	+40.28	11 14.6	+40 37	UMa	QSO	18.0	0.734	21.7	12	3.2	1.5	0.8	Db1
1116−027	255	−02.47	11 19.4	−3 03	Leo	?B			17.8	7.7	1.6	0.6	0.2	
1118+237	256	+23.27	11 20.7	+23 28	Leo	?G	21	0.378	9.2		1.2	0.7	0.3	In group
1120+057	257	+05.49	11 23.2	+5 30	Leo	?S			9.7	5.1	1.8	0.9	0.4	
	258	+19.38	11 24.7	+19 15	Leo	G	19.5	0.165	9.7	2.7	0.8	0.5	0.4	In cluster
1123−351			11 25.9	−35 23	Hya	E	16.0	0.033		5.2	2.2	1.4	0.8	In cluster
1127−145			11 30.1	−14 49	Crt	QSO	16.9	1.187		5.1	6.4	6.0	5.5	Db1
1131−171			11 34.4	−17 28	Crt	QSO?	19.0			5.8	1.5	0.9	0.4	
1136−678			11 38.4	−68 10	Mus	?C				8.0	3.2	1.9	0.9	
1136−135			11 39.2	−13 51	Crt	QSO	16.0	0.544		10.5	4.0	2.8	2.2	
1136−321			11 39.3	−32 23	Hya	?G				6.8	2.5	1.2	0.6	
	263	+66.13	11 40.0	+65 48	UMa	QSO	16.3	0.652	13.7		4.3	1.7	1.0	Db1
1138+015		+01.32	11 41.1	+1 14	Vir	?B			9.1	5.7	2.5	1.6	1.0	
1139−285			11 41.6	−28 51	Hya	?S				6.8	2.4	1.4	0.9	
1140−114			11 42.6	−11 42	Crt	?S				5.1	1.3	0.6	0.2	
1140+223	263.1	+22.30	11 43.4	+22 07	Leo	G	20		18.3		2.9	1.7	0.8	Db1
1142+198	264	+19.40	11 45.1	+19 37	Leo	E0	12.6	0.021	26	19.3	5.6	3.1	2.4	NGC 3862, cl. A1367
	265	+31.37	11 45.5	+31 34	UMa	G	20	0.811	18.3	11.5	1.9	1.4	0.6	Db1
1143−483			11 45.5	−48 36	Cen	?S				7.5	3.1	1.8	1.2	
	266	+50.33	11 45.7	+49 46	UMa	G	22.5		10.4	4.5	1.5	0.6	0.3	Db1
1143−316			11 46.2	−31 57	Hya	?S				5.8	2.0	0.7	0.3	
1147+130	267	+13.45	11 50.0	+12 47	Leo	G	21.5		14.4	7.0	2.4	1.5	0.7	Db1
1151−348			11 54.4	−35 05	Hya	QSO	18.0	0.258		10.9	6.1	4.1	2.7	
	268.1	+73.11	12 00.4	+73 01	Dra	G	21.5		20.7		7.0	4.0	2.6	Db1
	268.2	+31.39	12 01.0	+31 32	UMa	G	19.3	0.362	9.7	4.6	1.5	0.7	0.4	Db1
	268.3	+64.14	12 06.4	+64 14	Dra	G	20.0	0.371	10.3	9.3	3.8	2.0	1.1	In cluster
	268.4	+43.23	12 09.2	+43 39	CVn	QSO	18.4	1.400	9.5		1.0	1.1	0.6	
1215+039			12 17.6	+3 38	Vir	D+E	17	0.076	5.3	7.8	1.3	1.2	0.5	Db1?
1215−457			12 18.1	−46 00	Cen	G	18.0			9.6	4.6	3.3	2.0	
1216−100			12 18.6	−10 19	Vir	D	16	0.087		7.7	2.5	1.5	0.9	Db1?

PKS	3C	4C	α 2000	δ 2000	Const	Type	Mag	z	S178	S408	S1400	S2700	S5000	Notes
1216+061	270	+06.44	12h 19m.4	+5° 50'	Vir	E2	10.3	0.007	48	41.5	17.2	9.8	4.9	NGC 4261. Virgo cl.
	270.1	+33.29	12 20.6	+33 40	CVn	QSO	18.6	1.519	12.7	8.2	2.0	1.5	0.9	Triple source
1221−423			12 23.7	−42 35	Cen	?G	17			5.1	2.7	1.6	1.0	Db1
	272	+42.35	12 24.5	+42 06	CVn	?S	>23		8.0		1.7	0.7	0.4	Db1
1222+131	272.1	+13.47	12 25.1	+12 53	Vir	E1	9.3	0.003	19.0	12.2	6.0	4.2	2.7	M84 (NGC 4374). Db1
1226+023	273	+02.32	12 29.1	+2 03	Vir	QSO	12.8	0.158	75.0	55.1	42.0	40.9	36.7	Db1
1228+126	274	+12.45	12 30.8	+12 23	Vir	E1	8.6	0.004	1050	519	220	120	67.6	M87 Vir A. Db1
1232+216	274.1	+21.36	12 35.5	+21 21	Com	G	19.5	0.422	14.3		2.8	1.6	0.8	Db1
1232−249			12 35.6	−25 12	Hya	QSO	17.2	0.355		6.4	2.3	1.1	0.6	
1233−416			12 35.7	−41 53	Cen	E	17.5			5.3	1.9	0.9	0.4	
1239−044	275	−04.43	12 42.3	−4 46	Vir	G	21	0.480	14.2	10.2	3.4	1.8	1.0	In cluster
1241+166	275.1	+16.34	12 44.0	+16 23	Com	QSO	19.0	0.557	15.6	9.4	2.9	1.5	0.9	Triple source
1245−197			12 48.4	−19 59	Crv	?B				8.6	4.9	3.9	2.5	
1246−410			12 48.8	−41 18	Cen	E1p	10.7	0.009		10.6	3.8	2.2	1.3	NGC 4696
1247−401			12 50.1	−40 26	Cen	?G				6.0	1.5	0.6	0.3	
	277	+50.35	12 51.7	+50 33	CVn	?G			13.5		1.3	0.6	0.3	Db1
	277.1	+56.20	12 52.4	+56 32	UMa	QSO	17.9	0.321	13.0	8.5	1.8	1.5	1.1	
1250−102			12 53.1	−10 29	Vir	E2	12	0.014		4.7	0.9	0.7	0.3	NGC 4760
1251+159	277.2	+15.40	12 53.6	+15 42	Com	G	21.5		9.7	5.8	1.7	1.0	0.5	Db1
	277.3	+27.22	12 54.2	+27 38	Com	G	15.9	0.086	9.0	6.3	1.7	1.9	1.2	Com A
1251−122			12 54.6	−12 33	Crv	E0	11.7	0.014		14.7	7.3	4.6	2.5	NGC 4782−3. Db1
1253−055	279	−05.55	12 56.2	−5 47	Vir	QSO	17.8	0.538	20.9	14.5	11.6	12.0	13.0	
	280	+47.36	12 57.0	+47 20	CVn	?S	>21		23.7	15	5.0	2.8	1.5	Galaxy pair nearby
	280.1	+40.32	13 00.5	+40 08	CVn	QSO	19.4	1.659	12.0	5.5	1.5	0.6	0.4	
1302−492			13 05.5	−49 28	Cen	SBc:	9.2	0.002		14	6.6	5.0	2.8	NGC 4945
1303+091		+09.45	13 05.6	+8 56	Vir	?S			9.8	5.2	1.6	0.7	0.3	
1306−095			13 08.7	−9 51	Vir	?S			7.8	4.3	2.9	1.9		
1307+000		+00.46	13 09.8	−0 13	Vir	D	19.0		9.5	5.1	1.6	0.9	0.5	
	284	+27.24	13 11.1	+27 27	Com	G	18.0	0.239	6.1	5.0	2.0	1.1	0.6	Db1, in group
1308−220	283		13 11.7	−22 17	Vir	?S				22.2	5.2	2.4	1.1	
1313+073			13 13.0	−62 40	Cen	Neb				230	260			RCW 74
		+07.32	13 16.3	+7 03	Vir	D	13.4		3.1	5.5	2.2	1.2	0.8	Db1
1318+113		+11.45	13 21.3	+11 07	Vir	QSO	19.5	2.171	8.1	5.9	2.4	1.4	0.8	
	285	+42.37	13 21.3	+42 36	CVn	E	16.0	0.079	11.4	5.0	2.9	1.2	0.7	Db1, in group
1320−446			13 23.1	−44 53	Cen	?S				6.9	3.5	1.8	1.0	
1322−428			13 25.5	−43 04	Cen	S0p	7.0	0.002		2750		128		NGC 5128, Cen A. Db1
1327−214			13 30.1	−21 42	Vir	QSO	16.7	0.526		5.6	1.9	1.3	0.9	
1328+254	287	+25.43	13 30.6	+25 09	Com	QSO	17.7	1.055	16.0		7.0	4.2	3.0	
	286	+30.26	13 31.1	+30 30	CVn	QSO	17.3	0.849	24	20	16.2	10.3	7.5	
1329−665			13 32.6	−66 47	Mus	?C				7.7		1.6	0.9	
1330+022	287.1	+02.36	13 32.9	+2 01	Vir	N	18.3	0.216	12.4	5.3	2.8	1.9	1.4	Db1, in group
1331−098			13 33.9	−10 08	Vir	?G				6.7	1.6	0.6	0.4	Triple source
1333−336			13 36.4	−33 57	Cen	E0	10.6	0.012		34		2.7	1.8	IC 4296
1334−296			13 37.0	−29 52	Hya	Sc	8.0	0.001		8.5	2.6	1.2	0.8	M83 (NGC 5236)
1335−061		−06.35	13 38.1	−6 27	Vir	QSO	17.7	0.625	18.1	9.7	3.3	1.8	1.0	
	288	+39.39	13 38.8	+38 51	CVn	G	16.5	0.246	18.7	10.6	2.8	1.8	1.0	In cluster?
	288.1	+60.18	13 42.2	+60 22	UMa	QSO	18.1	0.961	10.3	5.5	1.5	0.8	0.4	
1341−659			13 45.0	−66 10	Mus	?C				5.2		1.0	0.5	
	289	+50.37	13 45.4	+49 46	UMa	?G	>23		11.8	8.2	2.4	1.2	0.6	Db1
1344−078			13 47.0	−8 03	Vir	?S				6.0	2.0	1.1	0.6	
1345+125		+12.50	13 47.6	+12 17	Boo	S0	17.0	0.122	4.6	8.8	5.1	3.9	2.9	
1347+213	291		13 49.6	+21 08	Boo	?S			7.6		1.1	0.5	0.3	
1346−391			13 49.8	−39 23	Cen	?S				5.8	1.8	1.0	0.5	
	293	+31.43	13 52.3	+31 26	CVn	D6	14.3	0.045	12.7	10.1	5.7	2.9	1.9	
1352+164	293.1	+16.38	13 54.6	+16 15	Boo	G	19		10.2	3.9	1.3	0.4	0.2	
1354+013		+01.39	13 57.0	+1 05	Vir	?B			9.6	6.6	2.3	1.4	0.7	
1354+195		+19.44	13 57.1	+19 19	Boo	QSO	16.0	0.720	7.6	6.0	2.3	1.5	1.5	
1355−416			13 59.0	−41 53	Cen	QSO	16.5	0.313		9.5	3.9	2.5	1.4	
1358−113			14 01.7	−11 36	Vir	E	15.0	0.025		5.0	2.0	1.1	0.4	Db1, in group
1400−337			14 03.7	−33 58	Cen	E4	12.4	0.014	57	10.3	0.8	0.3	0.1	NGC 5419
	294	+34.38	14 06.7	+34 09	CVn	?B			10.3	5.4	1.4	0.5	0.3	Db1
1407+177		+17.57	14 10.0	+17 34	Boo	E2	15.5		4.4	6.5	1.6	0.7	0.6	NGC 5490
	295	+52.30	14 11.3	+52 09	Boo	cD	20.1	0.461	74.9	57	21.2	11.8	6.5	Db1, in cluster
1411−057		−05.60	14 13.8	−6 00	Vir	?S			9.3	5.5	1.3	0.6	0.3	
1413−215			14 16.7	−21 46	Vir	?S				5.6		0.6	0.3	
1414+110	296	+10.39	14 16.9	+10 48	Boo	S0	12.0	0.024	5.5	8.3	4.2	2.3	1.7	NGC 5532, in cl. Db1
1414−037	297	−03.50	14 17.4	−4 01	Vir	?B			10.3	4.2	1.7	1.0	0.6	
1416+067	298	+06.49	14 19.1	+6 29	Vir	QSO	16.8	1.436	47.1	23.4	6.3	2.8	1.5	
1417−192			14 19.8	−19 28	Vir	N	17.5	0.119		5.0	1.8	1.1	0.8	
1416−493			14 20.0	−49 37	Lup	?G				7.3	2.6	1.3	0.8	
	299	+41.27	14 21.1	+41 46	Boo	G	18.4	0.367	12.7	7.7	3.0	1.6	0.9	In faint cluster
1419−272			14 22.8	−27 28	Hya	QSO?	18.0			8.4	2.2	1.2	0.9	
1420+198	300	+19.46	14 23.0	+19 36	Boo	E	18	0.270	16.8	10.9	3.6	2.0	1.1	Db1
1421−382			14 24.3	−38 26	Cen	QSO	18	0.41		6.9	2.2	1.2	0.7	
1421−490			14 24.5	−49 14	Lup	?S				13.1	8.4	7.0	5.4	

PKS	3C	4C	α 2000	δ 2000	Const	Type	Mag	z	S178	S408	S1400	S2700	S5000	Notes
1422-297			14^{h}25^m.5	-30°00'	Hya	?S				7.2	2.5	1.3	0.8	
1424-418			14 27.9	-42 06	Cen	QSO?	18			6.4	3.2	2.6	2.1	
1425-011	300.1	-01.34	14 28.5	-1 24	Vir	?G	19	0.308	11.3	7.2	3.1	1.7	0.9	Extended source
1427+074		+07.36	14 30.0	+7 16	Vir	E	18		3.5	5.8	2.1	1.1	0.7	Db1
1434+036		+03.30	14 37.0	+3 24	Vir	?B			7.3	5.2	2.8	2.0	1.3	
1436-167			14 39.5	-16 59	Lib	D	18.7			5.6	1.9	1.0	0.6	
	303	+52.33	14 43.0	+52 00	Boo	N	17.3	0.141	13.5	6.0	3.0	1.6	0.9	Also QSO: mag. 20 with
	303.1	+77.13	14 43.0	+77 05	UMi	G	18	0.267	8.1	7.0	2.0	0.6	0.5	In group ǀ z of 1.57
	305.1	+77.14	14 47.4	+76 53	UMi	G	21	0.456	12.4		1.8	0.8	0.5	
1445-468			14 48.5	-47 02	Lup	?C				7.0	2.1	1.1	0.5	
	305	+63.21	14 49.4	+63 16	Dra	SB0	13.4	0.041	14.3	9.6	2.5	1.6	0.9	IC 1065
1451-364			14 54.5	-36 40	Cen	E	19.5			7.5	2.8	1.3	0.7	
1452+004		+00.54	14 54.7	+0 17	Vir	?S			6.2	6.7		0.2	0.1	
1452-041	306.1	-04.53	14 55.0	-4 21	Lib	E	19.0	0.442	13.5	6.7	1.9	0.9	0.5	
1453-109			14 55.9	-11 09	Lib	QSO	17.4	0.938		10.3	3.9	2.5	1.5	
	309.1	+71.15	14 59.1	+71 40	UMi	QSO	16.8	0.905	23.1		8.5	5.3	3.8	
1459-419			15 02.4	-42 06	Cen	SNR				17.4	4.2	1.8	1.2	
1502+261	310	+26.46	15 05.0	+25 57	Boo	G	15.2	0.054	55		7.2	2.9	1.3	Db1, faint cluster
	314.1	+70.16	15 10.3	+70 45	UMi	G	17		9.2		1.7	0.7	0.3	Db1? In cluster
1508-055		-05.64	15 10.9	-5 43	Lib	QSO	16	1.191	10.5	7.7	3.9	2.9	2.3	
1508+080	313	+08.44	15 11.0	+7 52	Boo	E	19.0	0.461	30	11.5	3.7	1.9	1.1	Db1
1509+015		+01.42	15 12.4	+1 21	SerCp	?B			8.4	5.6	2.2	1.3	0.7	
1511+263	315	+26.47	15 13.7	+26 07	Boo	G	16.8	0.108	18.2		4.4	2.2	1.3	Db1, faint cluster?
1514+072	317	+07.40	15 16.7	+7 01	SerCp	E	13.5	0.035	47.3	25.2	5.8	2.3	0.9	In cluster A2052
1514+186	316		15 16.9	+18 30	SerCp	G	19.0		5.9	3.4	1.3	0.7	0.4	
1517+204	318	+20.35	15 20.1	+20 16	SerCp	N	20.9	0.752	13.2		2.3	1.5	0.8	In faint group
	318.1	+07.41	15 21.8	+7 40	SerCp	S0	13.6	0.046	12.0	2.8				NGC 5920
	319	+54.34	15 24.1	+54 27	Dra	G	18.5		14.2	8.7	2.0	1.3	0.6	Db1, in group
1524-136			15 27.0	-13 51	Lib	QSO?	20			6.1	2.4	1.7	1.2	
	320	+35.36	15 31.4	+35 32	CrB	G	18	0.342	8.6	5.5	3.2	0.9	0.5	In group
1529+242	321	+24.34	15 31.7	+24 04	SerCp	G	16.0	0.096	11.2		3.5	2.0	1.1	Db1
	322	+55.31	15 35.0	+55 36	Dra	?S	22		10.2	6.8	2.5	0.8	0.5	Db1
	323	+60.21	15 41.8	+60 16	Dra	G	18.5		9.3	4.1	1.8	0.6	0.3	
1540-730			15 46.1	-73 11	Aps	?S				5.3	1.8	0.9	0.5	
1545+210	323.1	+21.45	15 47.7	+20 52	SerCp	QSO?	16.7	0.264	10.0		2.4	1.5	0.9	Db1
1547+215	324	+21.46	15 49.8	+21 26	SerCp	G	21		13.6		2.8	1.4	0.6	Db1, faint cluster?
	325	+62.25	15 50.0	+62 41	Dra	QSO?	19		14.5		4.8	1.8	0.8	Db1
1550+202	326	+20.37	15 52.4	+20 06	SerCp	G	20		10.8		2.3	1.0	0.4	Extended source
1547-795			15 55.5	-79 40	Aps	?S				10.5	4.0	2.3	1.4	
	326.1	+20.38	15 56.2	+20 05	SerCp	?B			9.0		5.1	1.4	0.8	Db1
1549-790			15 57.0	-79 13	Aps	?S				7.9	5.8	5.0	4.5	
1556-215			15 59.1	-21 39	Sco	?S				8.5	2.5	0.9	0.4	Db1
1559+021	327	+02.41	16 02.5	+1 58	SerCp	D	15.9	0.104	43	16.1	8.5	4.6	3.3	Db1, in cluster
1602+014	327.1	+01.48	16 04.8	+1 18	SerCp	?S			22.3	14.9	4.2	2.1	1.1	
1602-174			16 05.0	-17 34	Sco	?S				5.6	1.7	0.7	0.3	
1602-288			16 05.2	-28 59	Sco	?B				7.1	2.6	1.2	0.7	
1602-093			16 05.4	-9 27	Sco	?S				6.1	3.2	2.1	1.0	Triple source
1603+001		+00.58	16 06.2	0 00	SerCp	E	16.5		11.9	5.6	2.1	1.5	1.0	
1602-633			16 06.8	-63 31	TrA	G	17.5	0.059		14.6	5.2	1.9	1.8	Db1
	330	+66.17	16 09.6	+65 57	Dra	G	20.8	0.549	23.2	15.5	5.7	3.8	2.4	Db1, faint cluster
1610+224	331		16 12.3	+22 22	Her	?S				7.1		1.2	0.7	0.4	
1615+212	333		16 17.3	+21 07	Her	?S				6.9		1.8	1.0	0.6	
	332	+32.51	16 17.7	+32 21	CrB	DE3	16	0.152	9.5	6.8	2.0	1.5	0.8	Db1, in group
1610-771			16 17.8	-77 17	Aps	QSO	18.9	1.710		5.3	4.6	3.4	5.6	
1618+177	334	+17.68	16 20.4	+17 36	Her	QSO	16.4	0.555	10.9	6.8	2.2	1.0	0.6	Db1
1621-115			16 24.0	-11 41	Oph	?S				7.2	2.3	1.4	0.8	
1622+238	336	+23.43	16 24.7	+23 45	Her	QSO	17.5	0.927	12.6		2.5	1.4	0.8	
1622-310			16 25.9	-31 08	Sco	QSO?	18.5			5.2	1.6	1.1	0.7	
	341	+27.33	16 28.1	+27 42	Her	G	19.5	0.448	10.8	5.6	1.6	1.1	0.5	Db1
	338	+39.45	16 28.6	+39 33	Her	cD4	12.0	0.030	46.4	18.1	3.2	1.2	0.5	NGC 6166, cl. A2199
	337	+44.28	16 28.9	+44 18	Her	G	20		18.5	8.4	3.0	1.6	0.9	Db1, faint cluster
1627+234	340	+23.44	16 29.6	+23 20	Her	G	19	0.310	11.9		2.3	1.3	0.7	Db1
1628-268			16 31.7	-26 57	Sco	?0				5.7		1.2	0.6	
	343	+62.26	16 34.6	+62 46	Dra	QSO	20.6	0.988	12.4		5.2	2.7	1.5	Db1
1634+269	342	+26.49	16 36.6	+26 48	Her	QSO	18	0.560	6.9		1.3	0.8	0.3	
1633-681			16 38.2	-68 14	TrA	?S				6.7	1.3	0.9	0.4	
	343.1	+62.27	16 38.5	+62 35	Dra	G	20.6	0.750	11.5	13.5	3.8	2.2	1.2	
	345	+39.48	16 43.0	+39 48	Her	QSO	16.0	0.595	11.9	11	8.1	6.0	5.6	
1641+173	346	+17.70	16 43.8	+17 16	Her	E	16	0.161	11.8	8.6	3.8	2.2	1.3	
1637-771			16 44.3	-77 16	Aps	D	16	0.043	13.5	11.4	5.8	3.5	2.6	Db1
1643+134	347		16 45.5	+13 23	Her	?S				4.6	2.8	0.8	0.4	0.2	Db1
1643+022		+02.42	16 45.7	+2 12	Oph	E	17.0		8.5	5.5	1.9	1.1	0.6	
1643-223			16 46.1	-22 28	Oph	?0				5.7	2.3	1.0	0.8	
1644-106			16 47.5	-10 44	Oph	?B				6.9	2.1	1.5	0.8	
1648+050	348	+05.66	16 51.1	+5 00	Her	D	16.9	0.154	350	169	47	23	12.4	Her A. Db1

PKS	3C	4C	α 2000	δ 2000	Const	Type	Mag	z	S178	S408	S1400	S2700	S5000	Notes
1649−062			16^{h}51^m.7	−6°18′	Oph	?S				5.7	1.9	0.7	0.6	
	349	+47.45	16 59.5	+47 03	Her	G	19	0.205	12.7	8.6	4.2	1.9	1.1	Stellar obj. nearby
1655−776			17 02.8	−77 42	Aps	E	17	0.066		5.0	2.4	1.4	1.1	
	351	+60.24	17 04.7	+60 44	Dra	QSO	15.3	0.371	13.5	11	3.0	2.0	1.2	Db1
	352	+46.34	17 10.7	+46 02	Her	G	21	0.806	12.2	7.8	3.2	1.0	0.5	Db1
1716+006			17 19.4	+0 37	Oph	G	20.0			5.5	2.2	1.4	0.8	
			17 20.3	−35 50	Sco	Neb				276	323			NGC 6334, RCW 127
1717−009	353	−00.67	17 20.5	−0 59	Oph	D	15.7	0.030	220	138	54	36.3	22.9	Db1
	356	+51.36	17 24.3	+50 57	Dra	G	21		15.5	4.4	1.5	0.7	0.7	Db1
1716−800			17 25.4	−80 05	Aps	?B				5.9	2.5	1.1	0.7	
			17 25.6	−34 17	Sco	Neb				420	580			NGC 6357, W22
	357	+31.47	17 28.3	+31 46	Her	ED4	15.5	0.167	9.7	5.8	1.5	1.6	0.4	Db1, in cl. A2266
1727−214	358		17 30.7	−21 29	Oph	SNR					14.0	10.1	8.5	
1730+206	359		17 32.8	+20 38	Her	?S			5.8		0.8	0.5	0.2	
1730−130			17 33.0	−13 05	SerCd	QSO?	18.5			6.6	5.2	4.3	4.1	
1732+160		+16.49	17 34.7	+16 01	Her	QSO	18.4	1.880		5.4	1.4	0.7	0.3	
1733−565			17 37.7	−56 34	Ara	G	17.5			13.0	7.4	4.4	3.3	Db1
1737−575			17 41.5	−57 37	Pav	?C				5.1		1.2	0.6	
1737−608			17 42.0	−60 55	Pav	?S				8.3	2.9	2.0	1.1	Db1
1740−517			17 44.4	−51 45	Ara	?C				5.4		4.6	3.0	
			17 46.1	−28 51	Sgr					2500	1300			Sgr A, galactic center
	363.1		17 53.0	+5 35	Oph				97					
1756+134		+13.65	17 58.5	+13 29	Oph	?S			9.0	6.4		1.3	0.8	
1754−598			17 59.1	−59 49	Pav	?S				12.6	3.4	1.4	0.7	
			18 01.3	−23 24	Sgr	SNR				450				W28
1756−663			18 01.3	−66 23	Pav	G	19			5.5		1.0	0.6	
1759+138		+13.66	18 01.6	+13 51	Oph	?B			9.3	5.6	2.0	0.7	0.3	
			18 04.1	−24 24	Sgr	Neb				200				Lagoon nebula M8
1802+110	368	+11.54	18 05.1	+11 02	Oph	?S	>22		14.0	6.8	1.1	0.5	0.2	Db1
	371	+69.24	18 06.8	+69 59	Dra	N	14.8	0.050	10.3	3.4	2.7	1.9	1.7	Also 4C +70.20
1810+046		+04.63	18 13.3	+4 40	Oph	?C			9.4	5.5		1.2	0.7	
	372.1		18 15.3	−3 35	SerCd				44					Extended. Galactic?
1814−519			18 18.1	−51 58	Tel	QSO?	18.0			14.3	3.5	1.6	0.6	
1814−637			18 19.6	−63 46	Pav	G	18			37.0	12.3	7.4	4.2	
			18 20.5	−16 05	Sgr	Neb				250				Omega nebula M17
1817−640			18 22.3	−63 59	Pav	?S				11.5	2.6	1.4	0.7	
	379.1	+74.23	18 24.6	+74 22	Dra	E	18	0.256	10.3	4.6	1.8	1.1	0.7	Db1
1819−673			18 24.6	−67 18	Pav	?S				6.1	2.1	1.6	0.9	
1821−327			18 24.9	−32 43	Sgr	?B				11.3		3.0	1.5	
1821−583			18 25.7	−58 18	Pav	?S				7.6	1.4	0.7	0.3	
	380	+48.46	18 29.5	+48 44	Dra	QSO	16.8	0.692	63.9	39	15	9.9	7.5	
1827−360			18 31.0	−36 03	Sgr	?B				25.8	7.4	2.9	1.3	
1830−210			18 33.7	−21 03	Sgr	Neb				11.5		9.3	8.9	
	381	+47.49	18 33.8	+47 26	Lyr	E	17.2	0.161	15.2	10.0	4.6	2.3	1.3	Db1, in group
	382	+32.55	18 35.1	+32 41	Lyr	D3	14.7	0.058	22	13.7	6.2	3.5	2.2	Also 3C 384. Db1
1836+171	386	+17.81	18 38.4	+17 12	Her	D	12.1	0.018	26	14.6	7.1	4.2	2.3	Mag. includes fore-
	387	−04.71	18 41.3	−4 55	Sct				11.9		51			⏐ ground star
	390.3	+79.18	18 42.2	+79 45	Dra	N	14.5	0.057	47.5	34.3	12.3	6.6	4.0	Db1
1839−486			18 43.3	−48 37	Tel	G	17.5			9.1	3.7	2.0	1.3	
	388	+45.39	18 44.0	+45 33	Lyr	cD3	15.3	0.091	22.5	13.5	5.3	3.1	1.8	Db1, in cluster
1840−404			18 44.5	−40 22	CrA	?B				8.4	2.8	1.4	0.9	
	390.1	+06.65	18 44.6	+6 43	Oph				55					
	390	+09.62	18 45.6	+9 52	Oph	?O			20.9		3.2	2.7	1.8	
	389	−03.70	18 46.4	−3 01	Aql	?O			22.7	55				
	390.2		18 47.8	−3 09	Aql	Neb			120		80	9.1		W43
1846−009	391	−00.72	18 49.4	−0 55	Aql	?O			25.9		17.8	11.9	9.8	
	392	+01.57	18 56.1	+1 18	Aql	SNR			430	195	171			W44 (with 4C +01.58)
1853−303			18 57.2	−30 20	Sgr	?C				5.1		0.9	0.5	
1857+129	394	+12.67	18 59.4	+12 59	Aql	?S			15.9	9.4	2.6	1.5	0.9	
1859−235			19 02.8	−23 30	Sgr	?B				10.9		1.8	1.0	
1901+053	396	+05.73	19 04.1	+5 27	Aql	?O			24.5		15.3	9.1	4.3	
	396.1		19 07.1	−3 07	Aql	SNR?			32		9.2			
1904+070	397	+07.50	19 07.4	+7 06	Aql	?O			30		29	10.9		
	398	+09.63	19 11.1	+9 05	Aql	Neb			60	83	33			W49
	399.1	+30.35	19 15.9	+30 19	Lyr				13.5	3.8	2.4	1.7	1.0	
	399.2		19 18.1	+10 41	Aql				20.5					
1920−077			19 23.5	−7 42	Aql	?B				6.0		1.0	0.6	
1921+144	400		19 23.7	+14 30	Aql	?O					576	108	83	W51
	400.1	+35.46	19 24.6	+35 23	Lyr				17.8		2.3	1.3	0.7	
1921−293			19 24.9	−29 15	Sgr	QSO	17.5	0.352		5.6			10.6	
1929−397			19 33.4	−39 41	Sgr	E	16			5.3	2.6	1.5	1.0	
1932−464			19 35.9	−46 21	Tel	?S				39.6	12.6	6.5	3.4	
1933−587			19 37.5	−58 38	Pav	?S				9.1	3.1	1.7	0.9	
	400.2		19 38.7	+17 31	Sge	SNR?			24		6.8			
1934−638			19 39.4	−63 43	Pav	G	18.4	0.182		6.2	16.4	11.5	6.1	

PKS	3C	4C	α 2000	δ 2000	Const	Type	Mag	z	S178	S408	S1400	S2700	S5000	Notes
		+21.53	19^{h}39^m.6	+21°35'	Vul	PSR				2.9				Millisecond pulsar
	401	+60.29	19 40.4	+60 41	Dra	cD	18	0.201	21.6	20.9	5.0	2.8	1.4	Db1, in cluster?
1938−155			19 41.2	−15 25	Sgr	?B				16.0	7.0	4.0	2.3	
	402	+50.49	19 41.7	+50 37	Cyg	D3	12.9	0.025	14.0	6.4	5.0	1.7	0.8	Db1
1949+023	403	+02.50	19 52.3	+2 30	Aql	SO	15.4	0.059	28	13.6	5.8	3.8	2.1	Db1
1949−014	403.1	−01.51	19 52.5	−1 17	Aql	E	16	0.056	13.5	4.4	1.5	0.7	0.6	In group
	403.2		19 53.9	+32 02	Cyg	Neb			54		75			W56
1953−077	404		19 56.2	−7 37	Aql	?B			14	5.9	1.8	0.9	0.5	
1953−425			19 57.3	−42 23	Sgr	?S				10.1	3.3	1.6	0.9	
1954−552			19 58.3	−55 10	Tel	E	16.3	0.060		14.0	6.0	3.7	2.3	
	405	+40.40	19 59.5	+40 44	Cyg	cD3	15.1	0.057	8700	4400	1260			Cyg A, in cl. Db1
2001+003	406		20 03.6	+0 28	Aql	?S			5.9	2.4	0.6	0.3	0.2	
2005−044	407	−04.76	20 08.4	−4 18	Aql	QSO	18.0	0.589	6.5	3.0	1.0	0.6	0.5	
2006−565			20 10.5	−56 27	Tel	?S		0.059		12.2	1.9	0.5	0.1	
2012+234	409	+23.53	20 14.5	+23 35	Vul	?O			79		11.6	6.4	3.0	
	409.1		20 19.6	+45 22	Cyg	Neb			250		170			W63, Cyg X region
	410	+29.60	20 20.1	+29 42	Cyg	?O			33.2	18	9.1	6.0	3.8	
	410.1		20 21.4	+40 16	Cyg	Neb			440					W66, Cyg X region
2019+098	411	+09.67	20 22.1	+10 01	Del	N	19.7	0.469	16.4	10.0	3.3	1.9	0.9	
2020−575			20 24.3	−57 24	Pav	?G				9.9	3.5	1.8	1.0	
2025−155			20 28.1	−15 22	Cap	?S				5.4	1.3	0.9	0.4	
2028−078	413		20 31.0	−7 42	Aql	?S				3.8	1.4	0.8	0.4	Db1
	415.1		20 32.8	+43 28	Cyg	Neb			26					W70, Cyg X region
	415.2	+53.46	20 32.8	+53 42	Cyg				10.8	4.3	1.1	0.6	0.2	Db1
2030−230			20 33.3	−22 53	Cap	N	17.3	0.132		6.4	2.5	1.5	0.9	
	416.1		20 35.0	+47 02	Cyg	Neb			65		70			W71, Cyg X region
	416.2		20 35.8	+41 43	Cyg	Neb			170					W73, Cyg X region
2032−350			20 35.8	−34 54	Mic	?S				17.6	5.7	3.4	1.9	
	418	+51.42	20 38.6	+51 19	Cyg	QSO	20	1.686	13.5	12	5.4	4.7	3.8	
	419.1		20 40.9	+42 17	Cyg	Neb			80					W75, Cyg X region
2040−267			20 43.7	−26 33	Cap	E	13.5	0.040		6.1	2.4	1.6	0.9	Db1
2041−604			20 45.3	−60 19	Pav	?S				11.2	2.8	1.0	0.4	
2044−027		−02.80	20 47.2	−2 36	Aqr	QSO	20.0	0.942	9.0	5.4	2.3	1.5	0.8	
2045+068	424	+06.68	20 48.2	+7 01	Del	E	18.4	0.127	14.0	7.9	2.4	1.3	0.6	In group
2048−572			20 52.0	−57 04	Ind	SO	12.0	0.011		6.2	2.1	0.8	0.4	IC 5063
2049−368			20 52.3	−36 41	Mic	?S				5.8	1.4	0.6	0.3	
			20 54.6	+44 20	Cyg	Neb				600				North America nebula
2053−201			20 56.0	−19 57	Cap	E	17.8			6.2	2.6	1.6	1.0	
2058−282			21 01.6	−28 02	Mic	E	14.8	0.038		15.9	5.6	3.1	1.8	Db1, in small cl.
2058−135			21 01.7	−13 19	Aqr	E	15.2	0.030		3.7	1.3	0.6	0.5	IC 1347
	427.1	+76.13	21 04.1	+76 32	Cep	?G	21		25		4.1	1.9	0.9	Db1
2104−256			21 07.3	−25 27	Cap	E	16.8	0.037		31	11.2	6.2	4.0	Db1
	428	+49.36	21 08.4	+49 36	Cyg	?O			20.5	8.0	12	1.5	0.4	
2111−259			21 14.7	−25 42	Cap	QSO?	18.5			5.3	2.1	1.3	0.9	
2113−211			21 16.6	−20 56	Cap	?S				9.1	2.7	1.4	0.8	
2115−305			21 18.2	−30 19	Mic	QSO	16.5	0.98		6.9	2.5	1.4	0.9	
	430	+60.30	21 18.3	+60 48	Cep	ED4	15	0.055	33.9	20.0	8.1	4.5	3.0	Db1
	431	+49.38	21 18.9	+49 37	Cyg	?O			25.7	14	3.9	1.4	0.6	
2120+168	432	+16.72	21 22.8	+17 05	Peg	QSO	18.0	1.805	12.5	6.1	1.7	0.7	0.3	Db1
2120−166			21 23.0	−16 28	Cap	?S				6.1	1.3	0.8	0.4	
2120+155	434	+15.73	21 23.3	+15 48	Peg	E	20.8	0.322	11.3	3.2	1.3	0.8	0.5	Db1, in cluster
2121+248	433	+24.54	21 23.7	+25 04	Vul	G	16.2	0.102	54.4		12.4	6.9	3.6	Db1, in group
	434.1	+51.45	21 24.6	+52 06	Cyg				25		12			Db1
2126+073	435	+07.55	21 29.1	+7 33	Peg	?G	19.5	0.471	12.8	6.6	2.4	0.9	0.6	
	435.1		21 30.8	+84 37	Cep	QSO?			13.0					Triple source
2128−208			21 31.0	−20 37	Cap	?S				6.2	2.2	1.1	0.6	
2130−538			21 34.3	−53 37	Ind	?G	14	0.076		5.2	1.6	0.8	0.3	Complex structure?
2135−147			21 37.7	−14 33	Cap	QSO	15.5	0.20		8.8	3.6	2.1	1.3	Db1
2135−209			21 37.8	−20 42	Cap	?B				9.8		2.5	1.5	
2140−434			21 43.6	−43 13	Gru	?S				9.2	3.2	1.7	0.8	
	436	+27.47	21 44.2	+28 10	Peg	G	18.2	0.214	15.7	11.8	2.8	1.8	1.0	Db1
2145+151	437	+15.75	21 47.4	+15 21	Peg	?G	23.5		10.4	7.3	2.9	1.6	0.9	Db1
2140−817			21 47.4	−81 32	Oct	?S				8.7	3.0	1.6	1.0	
2146−133			21 49.5	−13 04	Cap	QSO	20	1.800		5.1	1.5	0.9	0.6	
2147+145			21 50.4	+14 50	Peg	?B				6.4	2.3	1.4	0.7	
2148+143	437.1	+14.80	21 50.7	+14 33	Peg	?B			9.6	5.4	2.0	1.3	0.7	Part of 3C 437.1
2148+135	437.1	+13.82	21 51.0	+13 48	Peg	?G			8.8	3.1	1.1	0.6	0.3	Part of 3C 437.1
2149−200			21 51.9	−19 46	Cap	QSO?	18			5.1	2.1	1.2	0.8	
2149−287			21 52.1	−28 28	PsA	?S				5.7	2.8	2.0	1.3	
2150−520			21 54.1	−51 51	Ind	?S				10.1	3.5	2.1	1.2	
	438	+37.63	21 55.9	+38 00	Cyg	G	19.3	0.292	46.3	28	8.1	3.2	1.5	Db1, in cluster
2154−184			21 57.0	−18 14	Cap	G	19.5			6.1	2.2	1.1	0.5	
2152−699			21 57.1	−69 42	Ind	D	13.8	0.027	80	30	17.9	13.4	0.5	Db1
2201−555			22 05.1	−55 18	Ind	?S				5.6	2.1	0.9	0.5	
	441	+29.65	22 06.1	+29 29	Peg	G	21		12.6	7.3	2.5	1.5	0.9	Db1, in cluster

PKS	3C	4C	α 2000	δ 2000	Const	Type	Mag	z	S178	S408	S1400	S2700	S5000	Notes
2203−188			22^{h}06^m.2	−18°36′	Aqr	QSO	19.5	0.618		9.7	6.6	5.3	4.2	
2211−172	444		22 14.4	−17 02	Aqr	D	17.8	0.153		28.7	8.6	4.6	2.1	Db1
2212+135	442		22 14.7	+13 50	Peg	S0	13.5	0.027	22	13.1	3.6	1.4	0.8	NGC 7236−7. Db1
2216−281			22 19.7	−27 57	PsA	?G				6.2	2.1	1.0	0.4	
2221−023	445	−02.83	22 23.8	−2 07	Aqr	N	15.8	0.056	23.5	17.5	5.1	3.5	2.1	
2223−052	446	−05.92	22 25.8	−4 57	Aqr	QSO	18.4	1.406	17.3	11.9	6.2	4.7	4.3	Db1
2223−528			22 27.0	−52 33	Gru	?S				8.8	2.5	1.4	0.7	
2226−411			22 29.3	−40 52	Gru	QSO?	19.5			6.8	2.7	1.9	1.0	
2226−386			22 29.8	−38 24	Gru	?S				8.2	2.0	0.9	0.4	
	449	+39.69	22 31.3	+39 19	Lac	S0p	13.3	0.018	14.6	6.4	3.8	1.8	1.0	Db1, in small group
2230+114			22 32.6	+11 44	Peg	QSO	17.3	1.037	5.5	9.2	6.9	5.4	3.5	
	452	+39.71	22 45.8	+39 41	Lac	ED	16.0	0.081	53	28.2	10.1	5.5	2.8	Triple? In cluster?
2243+174	453		22 46.0	+17 43	Peg	?S			5.3	2.4	0.6	0.2		
2247+113		+11.71	22 49.8	+11 36	Peg	E1p	12.3	0.024	5.8	7.3	2.4	1.4	1.2	NGC 7385, in gr. Db1?
	454.1	+71.20	22 50.5	+71 29	Cep	?G	21		10.2		1.7	0.6	0.2	Cluster?
2249+185	454	+18.67	22 51.6	+18 49	Peg	QSO	18.4	1.757	10.7	5.7	2.2	1.2	0.7	
	454.2	+64.24	22 52.1	+64 41	Cep	?O			10.3		2.3	1.3	0.8	
2250−412			22 53.1	−40 58	Gru	?S				13.9	4.4	2.3	1.3	
2251+158	454.3	+15.76	22 54.0	+16 09	Peg	QSO	18.0	0.859	16.2	11.6	12.2	10.2	23.3	
2252+129	455	+12.79	22 55.1	+13 13	Peg	QSO	19.7	0.543	14.0	7.3	2.8	1.5	0.8	Foreground cluster?
2252−530			22 55.8	−52 46	Gru	?S				6.4	3.5	1.7	1.0	
2253−522			22 56.8	−51 59	Gru	?G				7.2	2.5	1.3	0.8	
2259−375			23 02.4	−37 18	Gru	?S				7.5	2.5	1.5	0.9	
2308+073		+07.61	23 10.7	+7 35	Psc	E2:	13.8	0.044	4.5	3.9	1.7	1.0	0.5	NGC 7503
2309+184	457		23 12.1	+18 45	Peg	?S				8.7	4.6	2.0	0.9	0.5	Db1
2309+090	456	+09.73	23 12.5	+9 19	Peg	E	18.5	0.233	14.0	6.3	2.5	1.3	0.8	Db1
2310+050	458	+05.85	23 12.9	+5 17	Psc	E	20	0.289	10.8	7.1	2.9	1.5	1.0	Triple source
2314+038	459	+03.57	23 16.6	+4 05	Psc	N	17.5	0.220	28.8	15.8	4.3	2.5	1.3	
2317−277			23 19.9	−27 28	Scl	E	17.5			5.4	2.9	1.9	1.3	Db1
2318−166			23 21.0	−16 23	Aqr	?B				8.8	2.3	1.0	0.6	
2318+235	460	+23.58	23 21.5	+23 47	Peg	G	18.5	0.264	10.3		2.0	0.9	0.4	Db1, in cluster?
2319−550			23 22.1	−54 45	Gru	QSO?	18.0			5.7	1.7	0.9	0.5	
	461	+58.40	23 23.4	+58 50	Cas	SNR			12000	5500	2240			Cas A
2322−052		−05.96	23 25.3	−4 58	Aqr	N	18.5		8.5	5.4	1.5	0.7	0.4	
2322−123			23 25.3	−12 07	Aqr	E	15.5	0.082		7.2	2.0	0.9	0.4	In cluster A2597
2323−407			23 26.6	−40 27	Gru	?S				9.3	3.3	1.8	1.1	
2324−023			23 26.9	−2 02	Psc	E	18			5.6	2.3	1.6	1.0	
2331−417			23 34.4	−41 25	Phe	G	19.5			15.9	5.1	2.7	1.5	
2332−669			23 35.2	−66 37	Tuc	G	20			5.9	2.4	1.3	0.8	
2335+267	465	+26.64	23 38.5	+27 01	Peg	D	12.6	0.029	38		7.3	3.4	2.1	NGC 7720, cl. A2634
2337+220	466		23 40.4	+22 21	Peg	?S				6.1		2.2	1.4	0.7	
2337+132		+13.88	23 40.5	+13 33	Peg	?S				7.6	6.1	1.9	0.9	0.3	
2338+042		+04.81	23 41.0	+4 31	Psc	QSO	19.5	2.594	9.2	5.7	1.6	0.8	0.4	
2338−585			23 41.3	−58 16	Tuc	?G				8.2	2.8	1.4	0.9	
2339−353			23 41.8	−35 06	Scl	?S				5.7	1.8	1.0	0.5	
2345+184	467	+18.71	23 48.5	+18 44	Peg	N	19.5	0.632	7.6	4.7	1.8	0.8	0.4	
2347−026		−02.90	23 50.4	−2 25	Psc	?B				6.7	5.5	1.8	1.0	0.5	
	468.1	+64.25	23 50.9	+64 41	Cas	?O			37		5.2	1.9	0.9	
	469.1	+79.23	23 55.4	+79 56	Cep	?G	21		12.4		1.8	0.8	0.4	Db1, faint cluster
2354−350			23 57.0	−34 46	Scl	D	15.0	0.049		8.7	1.3	0.3	0.1	
	470	+43.59	23 58.6	+44 04	And	?S	22		9.0	6.5	4.8	1.0	0.6	Db1

X-ray Sources

X-ray Sources

Column Headings

Name — Most common name of the source. Some terms are abbreviated as follows: galactic X-ray source (GX) with the galactic longitude and latitude to the nearest degree; Large Magellanic Cloud (LMC); Small Magellanic Cloud (SMC); C. B. Stephenson and N. Sanduleak, "New Hα-Emission Stars in the Milky Way," 1977 (SS).

α 2000 and **δ 2000** — Right ascension and declination, referred to the 2000.0 equinox.

Const — Constellation in which the source is located.

S(X) — X-ray strength (flux density), in microjanskys, where a microjansky is 10^{-32} watt per square meter per hertz. The strength is averaged over the energy range of 2 to 11 kiloelectron volts (wavelength band 6.2 to 1.1 angstroms).

P(X) — Period of X-ray strength in seconds (s), minutes (m), or days (d), if the source varies regularly.

L(X) — Maximum X-ray luminosity, which is the object's rate of energy output, expressed in units of the Sun's total luminosity (about 4×10^{33} ergs per second).

R(X) — The object's maximum X-ray luminosity relative to the amount of energy it gives off at visible wavelengths.

Dist — Estimated distance in kiloparsecs, or if followed by the letter "M" in megaparsecs. To convert kiloparsecs or megaparsecs to light-years, multiply by 3,260 or 3,260,000, respectively.

Type — The source may be identified as a noneclipsing binary star (B) or eclipsing binary (E); suffixes indicate that one component is a neutron star (/NS) or white dwarf (/WD). Other types are as follows: suspected black hole (BH), X-ray burster (Bu), galaxy (G), nova (N), recurrent nova (Nr), pulsar (PSR), quasi-stellar object (QSO), supernova remnant (SNR), transient X-ray source (T).

Mag — Visual or photoelectric **V** (yellow) magnitude, unless followed by "B," which denotes a photoelectric **B** (blue) magnitude.

Spec — Spectral type, if the object is a star.

Notes — For each object, other designations and descriptive information are given on the second line. Some terms are abbreviated: eclipse (ecl.) duration; orbital period (OP); supernova remnant (SNR); G. Westerhout's radio survey of the galaxy, 1958 (W); X-ray source (XRS).

Name	α 2000	δ 2000	Const	S(X)	P(X)	L(X)	R(X)	Dist	Type	Mag	Spec
Tycho's SNR	0^{h}25^m.3	+64°09′	Cas	10		250					SNR
								3C 10, B Cas			
SMC X-3	0 52.1	-72 26	Tuc	5		13000	2	65	T	15	O9III-Ve
SMC X-2	0 54.6	-73 41	Tuc	7		2(+5)	9	65	T	16.0	B1.5Ve
								Southern component of close pair			
γ Cas	0 56.7	+60 43	Cas	11		0.8	3(-5)	0.3		2.5	B0.5IVpe
								HD 5394. Coronal emission?			
SMC X-1	1 17.1	-73 27	Tuc	57	0.71s	1.5(+5)	8	65	E/NS	13.3	B0I
								OP=3.89d, ecl. 0.61d			
V635 Cas	1 18.5	+63 44	Cas	350	3.61s	2000	1.7	3	B/NS	15.6	B
								OP=24.3d			
HD 8357?	1 22.9	+7 25	Psc	75		0.5	0.06	0.1	B	7.3	G5
								RS CVn system			
Algol	3 08.2	+40 57	Per	9		0.005	6(-5)	0.03	E	2.1	B8V+G5IV+Am
								OP=2.87d			
Per A	3 19.8	+41 31	Per	250		1(+11)		100M	G	11.6	
								Seyfert galaxy NGC 1275 and Perseus cluster			
V711 Tau	3 36.8	+0 35	Tau	100		0.08	0.05	0.03	E	5.7	G5IV+K1IV
								HR 1099, HD 22468. RS CVn system			
X Per	3 55.4	+31 03	Per	37	13.9m	2.5	0.001	0.35	B/NS	6.0	O9.5ep
								OP=580d?			
NGC 1851	5 14.1	-40 03	Col	20		1300		9.5	Bu		
								Core of globular cluster			
LMC X-2	5 20.5	-71 57	Men	44		80000	600	50		18.5B	
LMC X-4	5 32.8	-66 22	Dor	60	13.5s	1(+5)	12	50	E	14.0	O7III-V
								OP=1.4d, ecl. 0.2d			
Crab pulsar	5 34.5	+22 01	Tau	100	0.033s	250	100	2	PSR	17.7	
								Tau X-1. S(X)=1060 for nebula and pulsar			
M42	5 35.3	-5 23	Ori	5		0.8		0.46			
								Orion nebula			
HDE 245770	5 38.9	+26 19	Tau	2800	1.73m	5000	0.7	2	B/NS?	9.1	O9.7IIIe
								V725 Tau. OP>20d			
LMC X-3	5 38.9	-64 05	Dor	44		80000	140	50	BH	16.9	BIII-IV
								OP=1.7d			
LMC X-1	5 39.7	-69 45	Dor	25		50000	7	50		14.5	OB
V1055 Ori	6 17.1	+9 08	Ori	134	4.9d?	2500	1200	6	Bu?	18.5	
V616 Mon	6 22.7	-0 21	Mon	50000		50000	200	1.1	N	>20	
								Nova Mon 1975, with V=10.4			
U Gem	7 55.1	+22 00	Gem	2	0.42m	0.005	0.003	0.08	E/WD	>9	M4.5+WD
								Dwarf nova. OP=0.175d			
Pup A	8 23.3	-42 54	Pup	8		800		1.4	SNR		
Vela X	8 35.3	-45 13	Vel	49		80		0.5	SNR		
								Contains Vela pulsar			
Vela X-1	9 02.1	-40 33	Vel	1100	4.72m	1500	0.06	1.4	E/NS	6.9	B0.5Ibe
								GP Vel, HD 77581. OP=8.79d, ecl. 1.7d			
M82	9 55.8	+69 41	UMa	3		3(+7)		3M	G	8.4	
He 3-640?	11 21.2	-61 52	Cen	70	6.75m	1300	0.11	5	T	12.1	O9.5III-Ve
Cen X-3	11 21.3	-60 37	Cen	312	4.84s	20000	0.8	10	E/NS	13.4	O6-O8fp
								Krzeminski's star, V779 Cen. OP=2.087d, ecl. 0.45d			
HD 102567	11 48.0	-62 12	Cen	1000	4.87m	1500	0.7	1.5		9.0	B1Vne
								V801 Cen. Rapidly expanding shell			
BP Cru	12 26.6	-62 46	Cru	1000	11.65m	2500	0.12	2	B	10.8	B1.5Ia
								GX 301-2			

Name	α 2000	δ 2000	Const	S(X)	P(X)	L(X)	R(X)	Dist	Type	Mag	Spec	Notes
3C 273	12^h 29.1^m	+2°03'	Vir	6		3(+12)		950M	QSO	12.8		
Vir A	12 30.8	+12 24	Vir	36		8(+9)	0.003	13M	G	8.6		Galaxy M87 and Virgo cluster
EX Hya	12 52.4	−29 15	Hya	8			0.3		E	11.5		U Gem variable? OP=0.069d
GX 304−1	13 01.3	−61 36	Cen	200	4.5m	750	0.04	2.4		13.9	B2Vne	
Cen A	13 25.5	−43 01	Cen	78		2(+6)		4.4M	G	7.0		Peculiar galaxy NGC 5128
Cen X−4	14 58.4	−31 40	Cen	20000		25000	300	1.5	Nr	>13	K3−K7	V822 Cen. Bursts
Cir X−1	15 20.7	−57 10	Cir	3000		2(+5)	3	10	BH	22.5B	OI?	BR Cir. OP=16.6d? Millisecond X-ray variations
TrA X−1	15 28.3	−61 53	TrA	950			700		N	>22		KY TrA. V=17.5 in 1974
QV Nor	15 42.4	−52 23	Nor	30	8.82m	1000	0.01	7	E/NS	14.5	B0I	OP=3.7d
QX Nor	16 12.7	−52 25	Nor	1100			800		Nr			Burster
Sco X−1	16 19.9	−15 38	Sco	20000		8000	500	0.7	B	12.2		V818 Sco. First nonsolar XRS discovered. OP=0.787d
KZ TrA	16 32.3	−67 28	TrA	30	7.68s		170		B	18.5		OP=41.5m. Simultaneous X-ray, optical flares
1630−472	16 34.0	−47 23	Nor	1400					T			Four flares observed during 1970−76
V801 Ara	16 40.9	−53 45	Ara	390			1700		Bu	17.5		Simultaneous X-ray, optical flares
GX 340+0	16 45.8	−45 37	Sco	640								In a radio H II region
Her X−1	16 57.8	+35 21	Her	110	1.24s	2000	14	5	E/NS	13.2B	A9−B	HZ Her. OP=1.7d, ecl. 0.24d
V821 Ara	17 02.8	−48 47	Ara	390	0.04s		10		BH	>15		GX 339−4
HD 153919	17 03.9	−37 51	Sco	400		800	0.03	1.7	E	6.6	O6.5Iaf	V884 Sco. OP=3.41d, ecl. 1.1d
GX 349+2	17 05.7	−36 25	Sco	1140								X-ray emission varies over hours
V2107 Oph	17 08.2	−25 05	Oph	3600			800		N	21	B	Nova Oph 1977 with mag. 16.5
Terzan 2	17 27.5	−30 48	Sco	20		1300		10?	Bu			Core of globular cluster
GX 1+4	17 32.0	−24 45	Oph	220	1.95m	15000	12	10?		18.7	M6III	X-ray period decreased by 20s, 1970−78. Hot companion
GX 354−0	17 32.0	−33 50	Sco	210		13000		10?	Bu			Bursts every 4−8 hours. In globular cluster Grindlay 1?
Liller 1	17 33.4	−33 23	Sco	50		2500		9	Bu			Core of faint globular cl. Up to 4000 bursts per day
V926 Sco	17 39.0	−44 27	Sco	290		10000	1100	7	Bu	17.5		3−7s bursts every few hours
	17 45.6	−29 01	Sgr	2000		1800		12	T	18.2	K3V	
NGC 6441	17 50.2	−37 03	Sco	80		2500		8	Bu?			Core of globular cluster
GX 5−1	18 01.1	−25 05	Sgr	1400								Brightest of galactic-bulge sources
GX 17+2	18 16.0	−14 02	SerCd	1060		1300	3000	1.4		17.5	G	Bursts?
AM Her	18 16.2	+49 52	Her	7	0.129d	0.03	0.4	0.08	E/WD	12.3		OP=0.129d, ecl. 0.02d

Name	α 2000	δ 2000	Const	S(X)	P(X)	L(X)	R(X)	Dist	Type	Mag	Spec
								Notes			
NGC 6624	$18^h 23^m.7$	$-30^o 22'$	Sgr	410		20000		8	Bu		
								Core of globular cluster			
Ser X-1	18 40.0	+5 02	SerCd	310			1200		Bu	19.2B	
								MM Ser. Southern component of double star			
NGC 6712	18 53.1	-8 42	Sct	400		13000		7	Bu		
								Core of globular cluster			
1907+097	19 09.6	+9 50	Aql	275	8.4d	10000	0.6	7?	E?	16.4	OB?
								Also 0.015s X-ray variability?			
Aql X-1	19 11.3	+0 35	Aql	1300		2500	900	1.7	Nr	19.2	K0V
								V1333 Aql. V=15 during X-ray bursts			
SS 433	19 11.8	+4 59	Aql	10		180	0.09	5.1	B/NS?	14	
								In SNR W50. OP=13d. Compact companion			
Cyg X-1	19 58.4	+35 12	Cyg	1320		5000	0.15	2.5	BH	8.9	O9.7Iab
								HDE 226868. OP=5.6d. Millisecond X-ray bursts			
Cyg A	19 59.0	+40 41	Cyg	12		1(+11)			G	15.1	
								Radio galaxy			
Cyg X-3	20 32.4	+40 57	Cyg	430		25000	60	10	B		
								V1521 Cyg. Infrared star. OP=0.200d			
M15	21 30.0	+12 10	Peg	50		2000		8.0			
								Core of globular cluster			
SS Cyg	21 42.7	+43 35	Cyg	20	8.9s	0.3	0.8	0.15	Nr	>8.2	A1-dGep
								U Gem variable. OP=0.28d			
Cyg X-2	21 44.7	+38 19	Cyg	740	11.2d	25000	140	8	B	14.7	F2III:
								V1341 Cyg. OP=9.84d			
Cas A	23 23.5	+58 50	Cas	59		200		2.8	SNR